Numerical Analysis

Numerical Analysis

NINTH EDITION

Richard L. Burden
Youngstown State University

J. Douglas Faires
Youngstown State University

Australia • Brazil • Japan • Korea • Mexico • Singapore • Spain • United Kingdom • United States

Numerical Analysis, 9e, International Edition
Richard L. Burden and J. Douglas Faires

Editor in Chief: *Michelle Julet*

Publisher: *Richard Stratton*

Senior Sponsoring Editor: *Molly Taylor*

Associate Editor: *Daniel Seibert*

Senior Editorial Assistant: *Shaylin Walsh*

Associate Media Editor: *Andrew Coppola*

Senior Marketing Manager: *Jennifer Pursley Jones*

Marketing Coordinator: *Erica O'Connell*

Marketing Communications Manager: *Mary Anne Payumo*

Content Project Manager: *Jill Clark*

Senior Art Director: *Jill Ort*

Senior Manufacturing Buyer: *Diane Gibbons*

Senior Rights Acquisition Specialist: *Katie Huha*

Production Service: *Cadmus Communications*

Text Designer: *Jay Purcell*

Cover Image: *istockphoto.com © space-heater, The Pixel Palace*

Compositor: *Cadmus Communications*

Library of Congress Control Number: 2010922639

International Edition:

ISBN-13: 978-0-538-73564-3
ISBN-10: 0-538-73564-3

Cengage Learning International Offices

Asia
www.cengageasia.com
tel: (65) 6410 1200

Australia/New Zealand
www.cengage.com.au
tel: (61) 3 9685 4111

Brazil
www.cengage.com.br
tel: (55) 11 3665 9900

India
www.cengage.co.in
tel: (91) 11 4364 1111

Latin America
www.cengage.com.mx
tel: (52) 55 1500 6000

UK/Europe/Middle East/Africa
www.cengage.co.uk
tel: (44) 0 1264 332 424

Represented in Canada by Nelson Education, Ltd.
tel: (416) 752 9100 / (800) 668 0671
www.nelson.com

Cengage Learning is a leading provider of customized learning solutions with office locations around the globe, including Singapore, the United Kingdom, Australia, Mexico, Brazil, and Japan. Locate your local office at: **www.cengage.com/global**.

For product information: **www.cengage.com/international**
Visit your local office: **www.cengage.com/global**
Visit our corporate website: **www.cengage.com**

Printed in Canada
1 2 3 4 5 6 7 14 13 12 11 10

Contents

Preface

About the Text

This book was written for a sequence of courses on the theory and application of numerical approximation techniques. It is designed primarily for junior-level mathematics, science, and engineering majors who have completed at least the standard college calculus sequence. Familiarity with the fundamentals of linear algebra and differential equations is useful, but there is sufficient introductory material on these topics so that courses in these subjects are not needed as prerequisites.

Previous editions of *Numerical Analysis* have been used in a wide variety of situations. In some cases, the mathematical analysis underlying the development of approximation techniques was given more emphasis than the methods; in others, the emphasis was reversed. The book has been used as a core reference for beginning graduate level courses in engineering and computer science programs and in first-year courses in introductory analysis offered at international universities. We have adapted the book to fit these diverse users without compromising our original purpose:

> *To introduce modern approximation techniques; to explain how, why, and when they can be expected to work; and to provide a foundation for further study of numerical analysis and scientific computing.*

The book contains sufficient material for at least a full year of study, but we expect many people to use it for only a single-term course. In such a single-term course, students learn to identify the types of problems that require numerical techniques for their solution and see examples of the error propagation that can occur when numerical methods are applied. They accurately approximate the solution of problems that cannot be solved exactly and learn typical techniques for estimating error bounds for the approximations. The remainder of the text then serves as a reference for methods not considered in the course. Either the full-year or single-course treatment is consistent with the philosophy of the text.

Virtually every concept in the text is illustrated by example, and this edition contains more than 2600 class-tested exercises ranging from elementary applications of methods and algorithms to generalizations and extensions of the theory. In addition, the exercise sets include numerous applied problems from diverse areas of engineering as well as from the physical, computer, biological, economic, and social sciences. The chosen applications clearly and concisely demonstrate how numerical techniques can be, and often must be, applied in real-life situations.

A number of software packages, known as Computer Algebra Systems (CAS), have been developed to produce symbolic mathematical computations. Maple®, Mathematica®, and MATLAB® are predominant among these in the academic environment, and versions of these software packages are available for most common computer systems. In addition, Sage, a free open source system, is now available. This system was developed primarily by William Stein at the University of Washington, and was first released in February 2005. Information about Sage can be found at the site

http://www.sagemath.org .

Although there are differences among the packages, both in performance and price, all can perform standard algebra and calculus operations.

The results in most of our examples and exercises have been generated using problems for which exact solutions are known, because this permits the performance of the approximation method to be more easily monitored. For many numerical techniques the error analysis requires bounding a higher ordinary or partial derivative, which can be a tedious task and one that is not particularly instructive once the techniques of calculus have been mastered. Having a symbolic computation package available can be very useful in the study of approximation techniques, because exact values for derivatives can easily be obtained. A little insight often permits a symbolic computation to aid in the bounding process as well.

We have chosen Maple as our standard package because of its wide academic distribution and because it now has a *NumericalAnalysis* package that contains programs that parallel the methods and algorithms in our text. However, other CAS can be substituted with only minor modifications. Examples and exercises have been added whenever we felt that a CAS would be of significant benefit, and we have discussed the approximation methods that CAS employ when they are unable to solve a problem exactly.

Algorithms and Programs

In our first edition we introduced a feature that at the time was innovative and somewhat controversial. Instead of presenting our approximation techniques in a specific programming language (FORTRAN was dominant at the time), we gave algorithms in a pseudo code that would lead to a well-structured program in a variety of languages. The programs are coded and available online in most common programming languages and CAS worksheet formats. All of these are on the web site for the book:

http://www.math.ysu.edu/~faires/Numerical-Analysis/ .

For each algorithm there is a program written in FORTRAN, Pascal, C, and Java. In addition, we have coded the programs using Maple, Mathematica, and MATLAB. This should ensure that a set of programs is available for most common computing systems.

Every program is illustrated with a sample problem that is closely correlated to the text. This permits the program to be run initially in the language of your choice to see the form of the input and output. The programs can then be modified for other problems by making minor changes. The form of the input and output are, as nearly as possible, the same in each of the programming systems. This permits an instructor using the programs to discuss them generically, without regard to the particular programming system an individual student chooses to use.

The programs are designed to run on a minimally configured computer and given in ASCII format for flexibility of use. This permits them to be altered using any editor or word processor that creates standard ASCII files (commonly called "Text Only" files). Extensive README files are included with the program files so that the peculiarities of the various programming systems can be individually addressed. The README files are presented both in ASCII format and as PDF files. As new software is developed, the programs will be updated and placed on the web site for the book.

For most of the programming systems the appropriate software is needed, such as a compiler for Pascal, FORTRAN, and C, or one of the computer algebra systems (Maple,

Mathematica, and MATLAB). The Java implementations are an exception. You need the system to run the programs, but Java can be freely downloaded from various sites. The best way to obtain Java is to use a search engine to search on the name, choose a download site, and follow the instructions for that site.

New for This Edition

The first edition of this book was published more than 30 years ago, in the decade after major advances in numerical techniques were made to reflect the new widespread availability of computer equipment. In our revisions of the book we have added new techniques in order to keep our treatment current. To continue this trend, we have made a number of significant changes to the ninth edition.

- Our treatment of Numerical Linear Algebra has been extensively expanded, and constitutes one of major changes in this edition. In particular, a section on Singular Value Decomposition has been added at the end of Chapter 9. This required a complete rewrite of the early part of Chapter 9 and considerable expansion of Chapter 6 to include necessary material concerning symmetric and orthogonal matrices. Chapter 9 is approximately 40% longer than in the eighth edition, and contains a significant number of new examples and exercises. Although students would certainly benefit from a course in Linear Algebra before studying this material, sufficient background material is included in the book, and every result whose proof is not given is referenced to at least one commonly available source.

- All the Examples in the book have been rewritten to better emphasize the problem to be solved before the specific solution is presented. Additional steps have been added to many of the examples to explicitly show the computations required for the first steps of iteration processes. This gives the reader a way to test and debug programs they have written for problems similar to the examples.

- A new item designated as an Illustration has been added. This is used when discussing a specific application of a method not suitable for the problem statement-solution format of the Examples.

- The Maple code we include now follows, whenever possible, the material included in their *NumericalAnalysis* package. The statements given in the text are precisely what is needed for the Maple worksheet applications, and the output is given in the same font and color format that Maple produces.

- A number of sections have been expanded, and some divided, to make it easier for instructors to assign problems immediately after the material is presented. This is particularly true in Chapters 3, 6, 7, and 9.

- Numerous new historical notes have been added, primarily in the margins where they can be considered independent of the text material. Much of the current material used in *Numerical Analysis* was developed in middle of the 20th century, and students should be aware that mathematical discoveries are ongoing.

- The bibliographic material has been updated to reflect new editions of books that we reference. New sources have been added that were not previously available.

As always with our revisions, every sentence was examined to determine if it was phrased in a manner that best relates what is described.

Supplements

Complete solutions to all exercises in the text are available to instructors in secure, customizable online format through the Cengage Solution Builder service. Adopting instructors can sign up for access at www.cengage.com/international. Computation results in these solutions were regenerated for this edition using the programs on the web site to ensure compatibility among the various programming systems.

A set of classroom lecture slides, prepared by Professor John Carroll of Dublin City University, are available on the book's instructor companion web site. These slides, created using the Beamer package of LaTeX, are in PDF format. They present examples, hints, and step-by-step animations of important techniques in Numerical Analysis.

To access additional course materials and companion resources, please visit www.cengage.com/international. At the resulting page, search for the ISBN from the back cover of your book using the search box at the top of the page. This will take you to the product page where these companion resources can be found.

Possible Course Suggestions

Numerical Analysis is designed to give instructors flexibility in the choice of topics as well as in the level of theoretical rigor and in the emphasis on applications. In line with these aims, we provide detailed references for results not demonstrated in the text and for the applications used to indicate the practical importance of the methods. The text references cited are those most likely to be available in college libraries, and they have been updated to reflect recent editions. We also include quotations from original research papers when we feel this material is accessible to our intended audience. All referenced material has been indexed to the appropriate locations in the text, and Library of Congress information for reference material has been included to permit easy location if searching for library material.

The following flowchart indicates chapter prerequisites. Most of the possible sequences that can be generated from this chart have been taught by the authors at Youngstown State University.

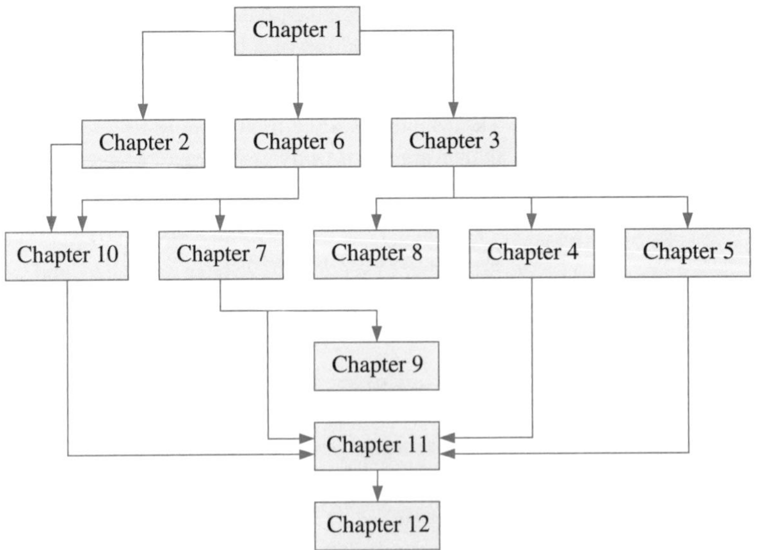

The additional material in this edition should permit instructors to prepare an undergraduate course in Numerical Linear Algebra for students who have not previously studied Numerical Analysis. This could be done by covering Chapters 1, 6, 7, and 9, and then, as time permits, including other material of the instructor's choice.

Acknowledgments

We have been fortunate to have had many of our students and colleagues give us their impressions of earlier editions of this book. We have tried to include all the suggestions that complement the philosophy of the book, and we are extremely grateful to all those who have taken the time to contact us about ways to improve subsequent versions.

We would particularly like to thank the following, whose suggestions we have used in this and previous editions.

John Carroll, Dublin City University (Ireland)

Gustav Delius, University of York (UK)

Pedro José Paúl Escolano, University of Sevilla (Spain)

Warren Hickman, Westminster College

Jozsi Jalics, Youngstown State University

Dan Kalman, American University

Robert Lantos, University of Ottawa (Canada)

Eric Rawdon, Duquesne University

Phillip Schmidt, University of Northern Kentucky

Kathleen Shannon, Salisbury University

Roy Simpson, State University of New York, Stony Brook

Dennis C. Smolarski, Santa Clara University

Richard Varga, Kent State University

James Verner, Simon Fraser University (Canada)

André Weideman, University of Stellenbosch (South Africa)

Joan Weiss, Fairfield University

Nathaniel Whitaker, University of Massachusetts at Amherst

Dick Wood, Seattle Pacific University

George Yates, Youngstown State University

As has been our practice in past editions of the book, we used undergraduate student help at Youngstown State University in preparing the ninth edition. Our assistant for this edition was Mario Sracic, who checked the new Maple code in the book and worked as our in-house copy editor. In addition, Edward Burden has been checking all the programs that accompany the text. We would like to express gratitude to our colleagues on the faculty and administration of Youngstown State University for providing us the opportunity, facilities, and encouragement to complete this project.

We would also like to thank some people who have made significant contributions to the history of numerical methods. Herman H. Goldstine has written an excellent book entitled *A History of Numerical Analysis from the 16th Through the 19th Century* [Golds]. In addition, *The words of mathematics* [Schw], by Steven Schwartzman has been a help in compiling our historical material. Another source of excellent historical mathematical knowledge is the MacTutor History of Mathematics archive at the University of St. Andrews in Scotland. It has been created by John J. O'Connor and Edmund F. Robertson and has the internet address

http://www-gap.dcs.st-and.ac.uk/~history/ .

An incredible amount of work has gone into creating the material on this site, and we have found the information to be unfailingly accurate. Finally, thanks to all the contributors to Wikipedia who have added their expertise to that site so that others can benefit from their knowledge.

In closing, thanks again to those who have spent the time and effort to contact us over the years. It has been wonderful to hear from so many students and faculty who used our book for their first exposure to the study of numerical methods. We hope this edition continues this exchange, and adds to the enjoyment of students studying numerical analysis. If you have any suggestions for improving future editions of the book, we would, as always, be grateful for your comments. We can be contacted most easily by electronic mail at the addresses listed below.

Richard L. Burden
burden@math.ysu.edu

J. Douglas Faires
faires@math.ysu.edu

CHAPTER

1 Mathematical Preliminaries and Error Analysis

Introduction

In beginning chemistry courses, we see the *ideal gas law*,

$$PV = NRT,$$

which relates the pressure P, volume V, temperature T, and number of moles N of an "ideal" gas. In this equation, R is a constant that depends on the measurement system.

Suppose two experiments are conducted to test this law, using the same gas in each case. In the first experiment,

$$P = 1.00 \text{ atm}, \qquad V = 0.100 \text{ m}^3,$$
$$N = 0.00420 \text{ mol}, \qquad R = 0.08206.$$

The ideal gas law predicts the temperature of the gas to be

$$T = \frac{PV}{NR} = \frac{(1.00)(0.100)}{(0.00420)(0.08206)} = 290.15 \text{ K} = 17°\text{C}.$$

When we measure the temperature of the gas however, we find that the true temperature is 15°C.

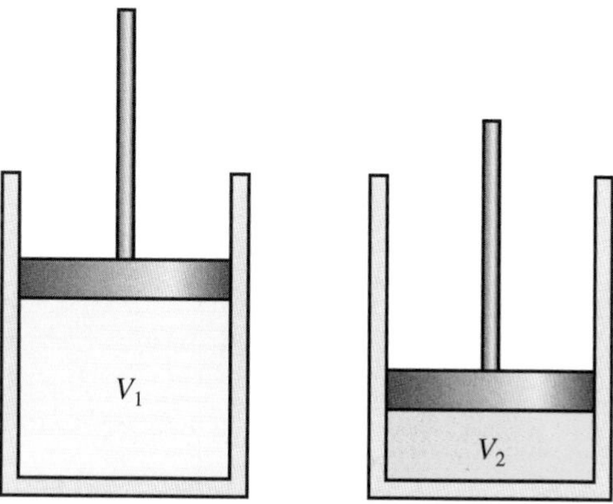

We then repeat the experiment using the same values of R and N, but increase the pressure by a factor of two and reduce the volume by the same factor. The product PV remains the same, so the predicted temperature is still 17°C. But now we find that the actual temperature of the gas is 19°C.

Clearly, the ideal gas law is suspect, but before concluding that the law is invalid in this situation, we should examine the data to see whether the error could be attributed to the experimental results. If so, we might be able to determine how much more accurate our experimental results would need to be to ensure that an error of this magnitude did not occur.

Analysis of the error involved in calculations is an important topic in numerical analysis and is introduced in Section 1.2. This particular application is considered in Exercise 28 of that section.

This chapter contains a short review of those topics from single-variable calculus that will be needed in later chapters. A solid knowledge of calculus is essential for an understanding of the analysis of numerical techniques, and more thorough review might be needed if you have been away from this subject for a while. In addition there is an introduction to convergence, error analysis, the machine representation of numbers, and some techniques for categorizing and minimizing computational error.

1.1 Review of Calculus

Limits and Continuity

The concepts of *limit* and *continuity* of a function are fundamental to the study of calculus, and form the basis for the analysis of numerical techniques.

Definition 1.1 A function f defined on a set X of real numbers has the **limit** L at x_0, written

$$\lim_{x \to x_0} f(x) = L,$$

if, given any real number $\varepsilon > 0$, there exists a real number $\delta > 0$ such that

$$|f(x) - L| < \varepsilon, \quad \text{whenever} \quad x \in X \quad \text{and} \quad 0 < |x - x_0| < \delta.$$

(See Figure 1.1.) ■

Figure 1.1

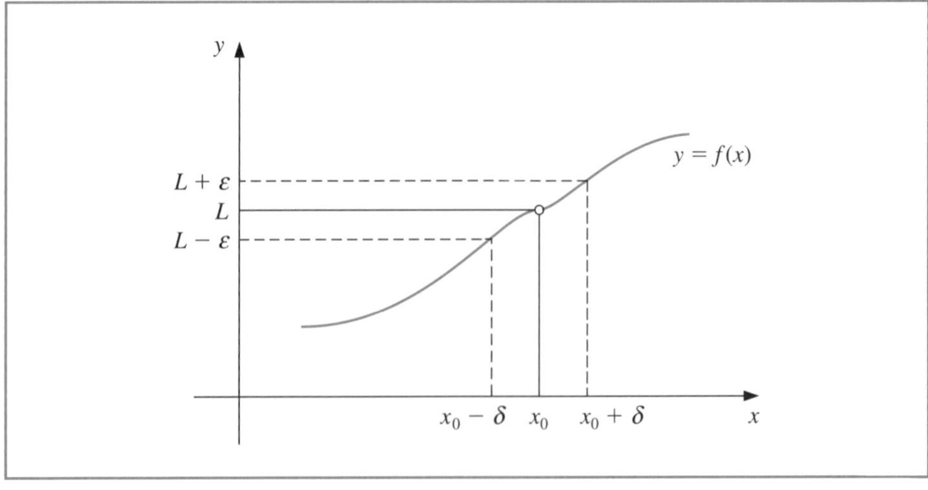

Definition 1.2 Let f be a function defined on a set X of real numbers and $x_0 \in X$. Then f is **continuous** at x_0 if

$$\lim_{x \to x_0} f(x) = f(x_0).$$

The function f is **continuous on the set** X if it is continuous at each number in X. ■

The basic concepts of calculus and its applications were developed in the late 17th and early 18th centuries, but the mathematically precise concepts of limits and continuity were not described until the time of Augustin Louis Cauchy (1789–1857), Heinrich Eduard Heine (1821–1881), and Karl Weierstrass (1815 –1897) in the latter portion of the 19th century.

The set of all functions that are continuous on the set X is denoted $C(X)$. When X is an interval of the real line, the parentheses in this notation are omitted. For example, the set of all functions continuous on the closed interval $[a, b]$ is denoted $C[a, b]$. The symbol $\mathbb{R}$ denotes the set of all real numbers, which also has the interval notation $(-\infty, \infty)$. So the set of all functions that are continuous at every real number is denoted by $C(\mathbb{R})$ or by $C(-\infty, \infty)$.

The *limit of a sequence* of real or complex numbers is defined in a similar manner.

Definition 1.3 Let $\{x_n\}_{n=1}^{\infty}$ be an infinite sequence of real numbers. This sequence has the **limit** x (**converges to** x) if, for any $\varepsilon > 0$ there exists a positive integer $N(\varepsilon)$ such that $|x_n - x| < \varepsilon$, whenever $n > N(\varepsilon)$. The notation

$$\lim_{n \to \infty} x_n = x, \quad \text{or} \quad x_n \to x \quad \text{as} \quad n \to \infty,$$

means that the sequence $\{x_n\}_{n=1}^{\infty}$ converges to x. ■

Theorem 1.4 If f is a function defined on a set X of real numbers and $x_0 \in X$, then the following statements are equivalent:

a. f is continuous at x_0;

b. If $\{x_n\}_{n=1}^{\infty}$ is any sequence in X converging to x_0, then $\lim_{n \to \infty} f(x_n) = f(x_0)$. ■

The functions we will consider when discussing numerical methods will be assumed to be continuous because this is a minimal requirement for predictable behavior. Functions that are not continuous can skip over points of interest, which can cause difficulties when attempting to approximate a solution to a problem.

Differentiability

More sophisticated assumptions about a function generally lead to better approximation results. For example, a function with a smooth graph will normally behave more predictably than one with numerous jagged features. The smoothness condition relies on the concept of the derivative.

Definition 1.5 Let f be a function defined in an open interval containing x_0. The function f is **differentiable** at x_0 if

$$f'(x_0) = \lim_{x \to x_0} \frac{f(x) - f(x_0)}{x - x_0}$$

exists. The number $f'(x_0)$ is called the **derivative** of f at x_0. A function that has a derivative at each number in a set X is **differentiable on** X. ■

The derivative of f at x_0 is the slope of the tangent line to the graph of f at $(x_0, f(x_0))$, as shown in Figure 1.2.

Figure 1.2

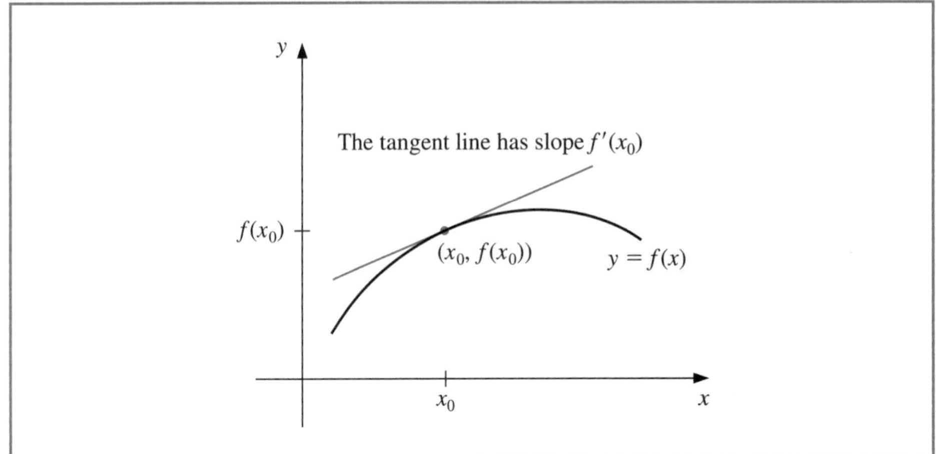

Theorem 1.6 If the function f is differentiable at x_0, then f is continuous at x_0. ■

The theorem attributed to Michel Rolle (1652–1719) appeared in 1691 in a little-known treatise entitled *Méthode pour résoundre les égalites*. Rolle originally criticized the calculus that was developed by Isaac Newton and Gottfried Leibniz, but later became one of its proponents.

The next theorems are of fundamental importance in deriving methods for error estimation. The proofs of these theorems and the other unreferenced results in this section can be found in any standard calculus text.

The set of all functions that have n continuous derivatives on X is denoted $C^n(X)$, and the set of functions that have derivatives of all orders on X is denoted $C^\infty(X)$. Polynomial, rational, trigonometric, exponential, and logarithmic functions are in $C^\infty(X)$, where X consists of all numbers for which the functions are defined. When X is an interval of the real line, we will again omit the parentheses in this notation.

Theorem 1.7 **(Rolle's Theorem)**

Suppose $f \in C[a, b]$ and f is differentiable on (a, b). If $f(a) = f(b)$, then a number c in (a, b) exists with $f'(c) = 0$. (See Figure 1.3.) ■

Figure 1.3

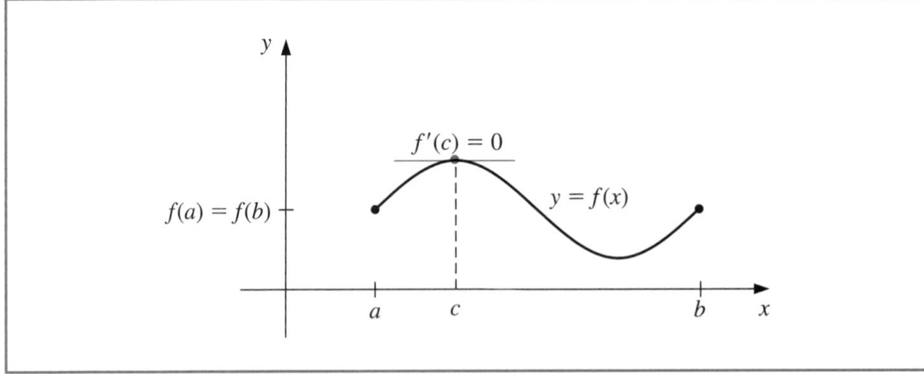

Theorem 1.8 **(Mean Value Theorem)**

If $f \in C[a, b]$ and f is differentiable on (a, b), then a number c in (a, b) exists with (See Figure 1.4.)

$$f'(c) = \frac{f(b) - f(a)}{b - a}.$$ ■

Figure 1.4

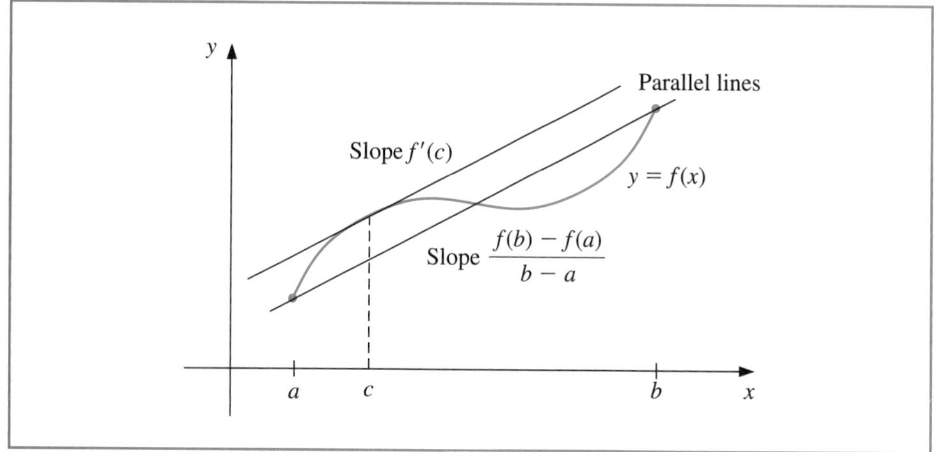

Theorem 1.9 **(Extreme Value Theorem)**

If $f \in C[a, b]$, then $c_1, c_2 \in [a, b]$ exist with $f(c_1) \leq f(x) \leq f(c_2)$, for all $x \in [a, b]$. In addition, if f is differentiable on (a, b), then the numbers c_1 and c_2 occur either at the endpoints of $[a, b]$ or where f' is zero. (See Figure 1.5.) ■

Figure 1.5

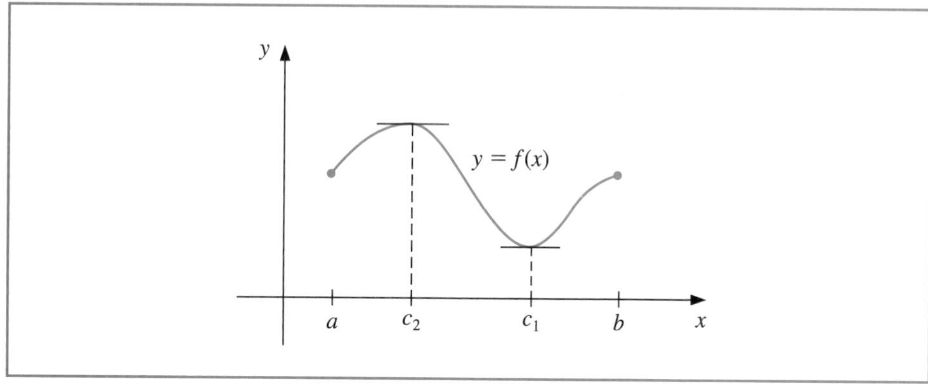

Research work on the design of algorithms and systems for performing symbolic mathematics began in the 1960s. The first system to be operational, in the 1970s, was a LISP-based system called MACSYMA.

As mentioned in the preface, we will use the computer algebra system Maple whenever appropriate. Computer algebra systems are particularly useful for symbolic differentiation and plotting graphs. Both techniques are illustrated in Example 1.

Example 1 Use Maple to find the absolute minimum and absolute maximum values of

$$f(x) = 5\cos 2x - 2x\sin 2x f(x)$$

on the intervals **(a)** $[1, 2]$, and **(b)** $[0.5, 1]$

Solution There is a choice of Text input or Math input under the Maple C 2D Math option. The Text input is used to document worksheets by adding standard text information in the document. The Math input option is used to execute Maple commands. Maple input

The Maple development project began at the University of Waterloo in late 1980. Its goal was to be accessible to researchers in mathematics, engineering, and science, but additionally to students for educational purposes. To be effective it needed to be portable, as well as space and time efficient. Demonstrations of the system were presented in 1982, and the major paper setting out the design criteria for the MAPLE system was presented in 1983 [CGGG].

can either be typed or selected from the pallets at the left of the Maple screen. We will show the input as typed because it is easier to accurately describe the commands. For pallet input instructions you should consult the Maple tutorials. In our presentation, Maple input commands appear in *italic* type, and Maple responses appear in cyan type.

To ensure that the variables we use have not been previously assigned, we first issue the command.

restart

to clear the Maple memory. We first illustrate the graphing capabilities of Maple. To access the graphing package, enter the command

with(plots)

to load the plots subpackage. Maple responds with a list of available commands in the package. This list can be suppressed by placing a colon after the *with(plots)* command.

The following command defines $f(x) = 5\cos 2x - 2x\sin 2x$ as a function of x.

$f := x \to 5\cos(2x) - 2x \cdot \sin(2x)$

and Maple responds with

$$x \to 5\cos(2x) - 2x\sin(2x)$$

We can plot the graph of f on the interval $[0.5, 2]$ with the command

plot$(f, 0.5\,..\,2)$

Figure 1.6 shows the screen that results from this command after doing a mouse click on the graph. This click tells Maple to enter its graph mode, which presents options for various views of the graph. We can determine the coordinates of a point of the graph by moving the mouse cursor to the point. The coordinates appear in the box above the left of the *plot*$(f,$ $0.5\,..\,2)$ command. This feature is useful for estimating the axis intercepts and extrema of functions.

The absolute maximum and minimum values of $f(x)$ on the interval $[a, b]$ can occur only at the endpoints, or at a critical point.

(a) When the interval is $[1, 2]$ we have

$f(1) = 5\cos 2 - 2\sin 2 = -3.899329036$ and $f(2) = 5\cos 4 - 4\sin 4 = -0.241008123.$

A critical point occurs when $f'(x) = 0$. To use Maple to find this point, we first define a function *fp* to represent f' with the command

fp $:= x \to$ *diff*$(f(x), x)$

and Maple responds with

$$x \to \frac{d}{dx} f(x)$$

To find the explicit representation of $f'(x)$ we enter the command

fp(x)

and Maple gives the derivative as

$$-12\sin(2x) - 4x\cos(2x)$$

To determine the critical point we use the command

fsolve(*fp*$(x), x, 1\,..\,2)$

Figure 1.6

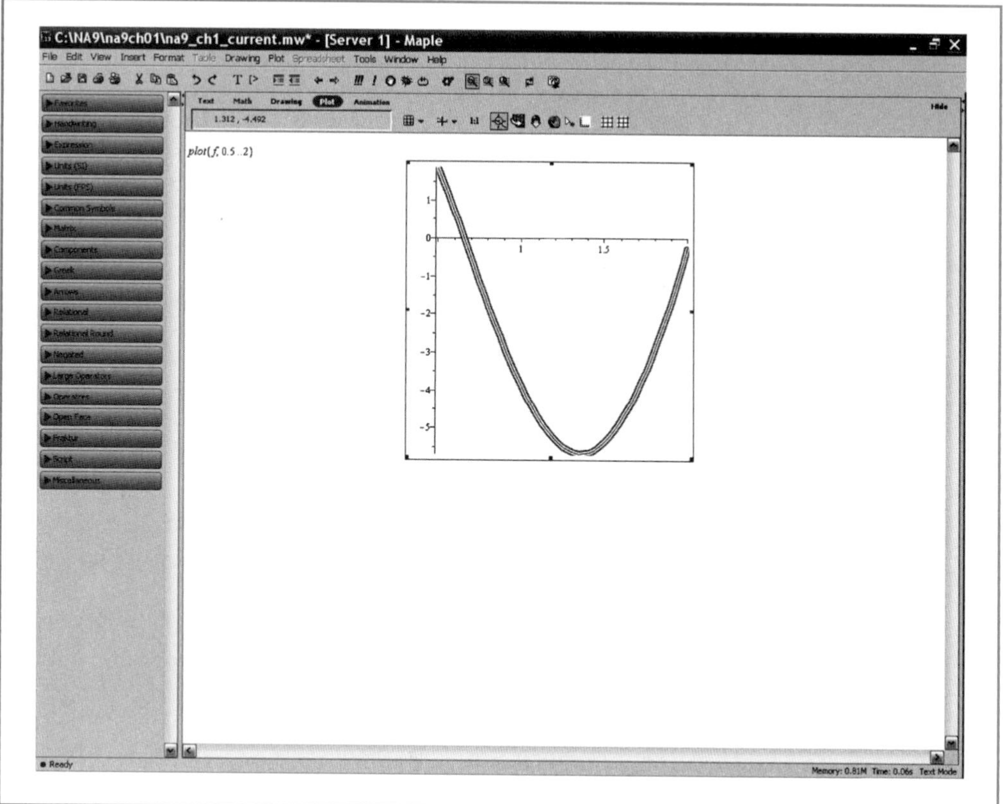

and Maple tells us that $f'(x) = fp(x) = 0$ for x in $[1, 2]$ when x is

$$1.358229874$$

We evaluate $f(x)$ at this point with the command

$f(\%)$

The % is interpreted as the last Maple response. The value of f at the critical point is

$$-5.675301338$$

As a consequence, the absolute maximum value of $f(x)$ in $[1, 2]$ is $f(2) = -0.241008123$ and the absolute minimum value is $f(1.358229874) = -5.675301338$, accurate at least to the places listed.

(b) When the interval is $[0.5, 1]$ we have the values at the endpoints given by

$$f(0.5) = 5\cos 1 - 1\sin 1 = 1.860040545 \quad \text{and} \quad f(1) = 5\cos 2 - 2\sin 2 = -3.899329036.$$

However, when we attempt to determine the critical point in the interval $[0.5, 1]$ with the command

$fsolve(fp(x), x, 0.5\,..\,1)$

Maple gives the response

$$fsolve(-12\sin(2x) - 4x\cos(2x), x, .5\,..\,1)$$

This indicates that Maple is unable to determine the solution. The reason is obvious once the graph in Figure 1.6 is considered. The function f is always decreasing on this interval, so no solution exists. Be suspicious when Maple returns the same response it is given; it is as if it was questioning your request.

In summary, on $[0.5, 1]$ the absolute maximum value is $f(0.5) = 1.86004545$ and the absolute minimum value is $f(1) = -3.899329036$, accurate at least to the places listed. ■

The following theorem is not generally presented in a basic calculus course, but is derived by applying Rolle's Theorem successively to $f, f', \ldots,$ and, finally, to $f^{(n-1)}$. This result is considered in Exercise 23.

Theorem 1.10 **(Generalized Rolle's Theorem)**

Suppose $f \in C[a, b]$ is n times differentiable on (a, b). If $f(x) = 0$ at the $n + 1$ distinct numbers $a \le x_0 < x_1 < \ldots < x_n \le b$, then a number c in (x_0, x_n), and hence in (a, b), exists with $f^{(n)}(c) = 0$. ■

We will also make frequent use of the Intermediate Value Theorem. Although its statement seems reasonable, its proof is beyond the scope of the usual calculus course. It can, however, be found in most analysis texts.

Theorem 1.11 **(Intermediate Value Theorem)**

If $f \in C[a, b]$ and K is any number between $f(a)$ and $f(b)$, then there exists a number c in (a, b) for which $f(c) = K$. ■

Figure 1.7 shows one choice for the number that is guaranteed by the Intermediate Value Theorem. In this example there are two other possibilities.

Figure 1.7

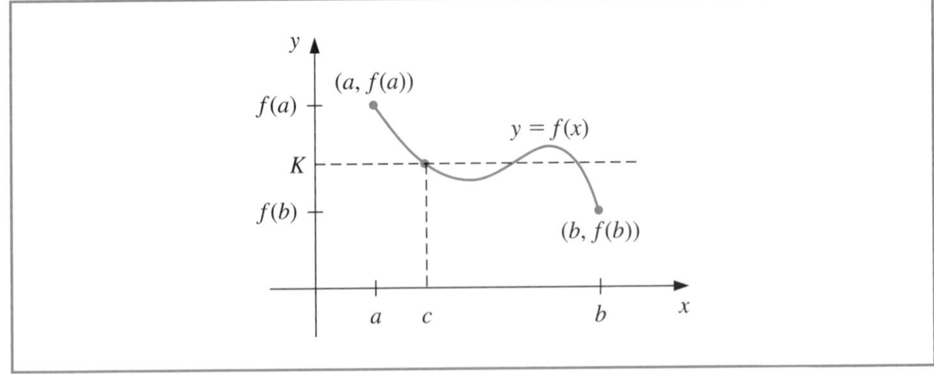

Example 2 Show that $x^5 - 2x^3 + 3x^2 - 1 = 0$ has a solution in the interval $[0, 1]$.

Solution Consider the function defined by $f(x) = x^5 - 2x^3 + 3x^2 - 1$. The function f is continuous on $[0, 1]$. In addition,

$$f(0) = -1 < 0 \quad \text{and} \quad 0 < 1 = f(1).$$

The Intermediate Value Theorem implies that a number x exists, with $0 < x < 1$, for which $x^5 - 2x^3 + 3x^2 - 1 = 0$. ■

As seen in Example 2, the Intermediate Value Theorem is used to determine when solutions to certain problems exist. It does not, however, give an efficient means for finding these solutions. This topic is considered in Chapter 2.

Integration

The other basic concept of calculus that will be used extensively is the Riemann integral.

Definition 1.12 The **Riemann integral** of the function f on the interval $[a, b]$ is the following limit, provided it exists:

$$\int_a^b f(x)\,dx = \lim_{\max \Delta x_i \to 0} \sum_{i=1}^{n} f(z_i)\,\Delta x_i,$$

where the numbers $x_0, x_1, \ldots, x_n$ satisfy $a = x_0 \le x_1 \le \cdots \le x_n = b$, where $\Delta x_i = x_i - x_{i-1}$, for each $i = 1, 2, \ldots, n$, and z_i is arbitrarily chosen in the interval $[x_{i-1}, x_i]$. ■

George Fredrich Berhard Riemann (1826–1866) made many of the important discoveries classifying the functions that have integrals. He also did fundamental work in geometry and complex function theory, and is regarded as one of the profound mathematicians of the nineteenth century.

A function f that is continuous on an interval $[a, b]$ is also Riemann integrable on $[a, b]$. This permits us to choose, for computational convenience, the points x_i to be equally spaced in $[a, b]$, and for each $i = 1, 2, \ldots, n$, to choose $z_i = x_i$. In this case,

$$\int_a^b f(x)\,dx = \lim_{n \to \infty} \frac{b-a}{n} \sum_{i=1}^{n} f(x_i),$$

where the numbers shown in Figure 1.8 as x_i are $x_i = a + i(b - a)/n$.

Figure 1.8

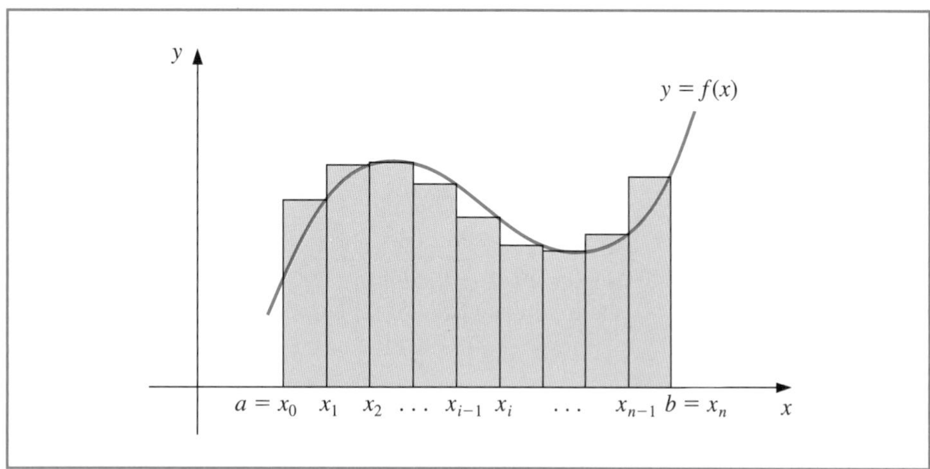

Two other results will be needed in our study of numerical analysis. The first is a generalization of the usual Mean Value Theorem for Integrals.

Theorem 1.13 **(Weighted Mean Value Theorem for Integrals)**

Suppose $f \in C[a, b]$, the Riemann integral of g exists on $[a, b]$, and $g(x)$ does not change sign on $[a, b]$. Then there exists a number c in (a, b) with

$$\int_a^b f(x)g(x)\,dx = f(c)\int_a^b g(x)\,dx. \qquad \blacksquare$$

When $g(x) \equiv 1$, Theorem 1.13 is the usual Mean Value Theorem for Integrals. It gives the **average value** of the function f over the interval $[a, b]$ as (See Figure 1.9.)

$$f(c) = \frac{1}{b-a}\int_a^b f(x)\,dx.$$

Figure 1.9

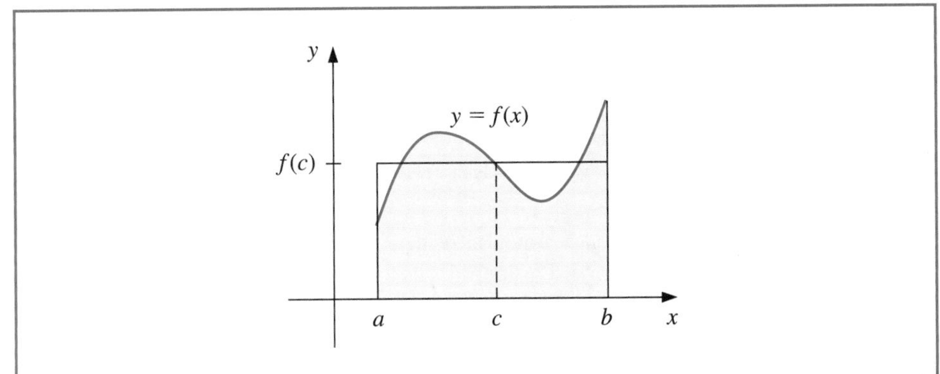

The proof of Theorem 1.13 is not generally given in a basic calculus course but can be found in most analysis texts (see, for example, [Fu], p. 162).

Taylor Polynomials and Series

The final theorem in this review from calculus describes the Taylor polynomials. These polynomials are used extensively in numerical analysis.

Theorem 1.14 **(Taylor's Theorem)**

Brook Taylor (1685–1731) described this series in 1715 in the paper *Methodus incrementorum directa et inversa*. Special cases of the result, and likely the result itself, had been previously known to Isaac Newton, James Gregory, and others.

Suppose $f \in C^n[a, b]$, that $f^{(n+1)}$ exists on $[a, b]$, and $x_0 \in [a, b]$. For every $x \in [a, b]$, there exists a number $\xi(x)$ between x_0 and x with

$$f(x) = P_n(x) + R_n(x),$$

where

$$P_n(x) = f(x_0) + f'(x_0)(x - x_0) + \frac{f''(x_0)}{2!}(x - x_0)^2 + \cdots + \frac{f^{(n)}(x_0)}{n!}(x - x_0)^n$$

$$= \sum_{k=0}^{n} \frac{f^{(k)}(x_0)}{k!}(x - x_0)^k$$

and

$$R_n(x) = \frac{f^{(n+1)}(\xi(x))}{(n+1)!}(x - x_0)^{n+1}.$$

■

Colin Maclaurin (1698–1746) is best known as the defender of the calculus of Newton when it came under bitter attack by the Irish philosopher, the Bishop George Berkeley.

Maclaurin did not discover the series that bears his name; it was known to 17th century mathematicians before he was born. However, he did devise a method for solving a system of linear equations that is known as Cramer's rule, which Cramer did not publish until 1750.

Here $P_n(x)$ is called the **nth Taylor polynomial** for f about x_0, and $R_n(x)$ is called the **remainder term** (or **truncation error**) associated with $P_n(x)$. Since the number $\xi(x)$ in the truncation error $R_n(x)$ depends on the value of x at which the polynomial $P_n(x)$ is being evaluated, it is a function of the variable x. However, we should not expect to be able to explicitly determine the function $\xi(x)$. Taylor's Theorem simply ensures that such a function exists, and that its value lies between x and x_0. In fact, one of the common problems in numerical methods is to try to determine a realistic bound for the value of $f^{(n+1)}(\xi(x))$ when x is in some specified interval.

The infinite series obtained by taking the limit of $P_n(x)$ as $n \to \infty$ is called the **Taylor series** for f about x_0. In the case $x_0 = 0$, the Taylor polynomial is often called a **Maclaurin polynomial**, and the Taylor series is often called a **Maclaurin series**.

The term **truncation error** in the Taylor polynomial refers to the error involved in using a truncated, or finite, summation to approximate the sum of an infinite series.

Example 3 Let $f(x) = \cos x$ and $x_0 = 0$. Determine

(a) the second Taylor polynomial for f about x_0; and

(b) the third Taylor polynomial for f about x_0.

Solution Since $f \in C^\infty(\mathbb{R})$, Taylor's Theorem can be applied for any $n \geq 0$. Also,

$$f'(x) = -\sin x, \quad f''(x) = -\cos x, \quad f'''(x) = \sin x, \quad \text{and} \quad f^{(4)}(x) = \cos x,$$

so

$$f(0) = 1, \quad f'(0) = 0, \quad f''(0) = -1, \quad \text{and} \quad f'''(0) = 0.$$

(a) For $n = 2$ and $x_0 = 0$, we have

$$\begin{aligned}\cos x &= f(0) + f'(0)x + \frac{f''(0)}{2!}x^2 + \frac{f'''(\xi(x))}{3!}x^3 \\ &= 1 - \frac{1}{2}x^2 + \frac{1}{6}x^3 \sin \xi(x),\end{aligned}$$

where $\xi(x)$ is some (generally unknown) number between 0 and x. (See Figure 1.10.)

Figure 1.10

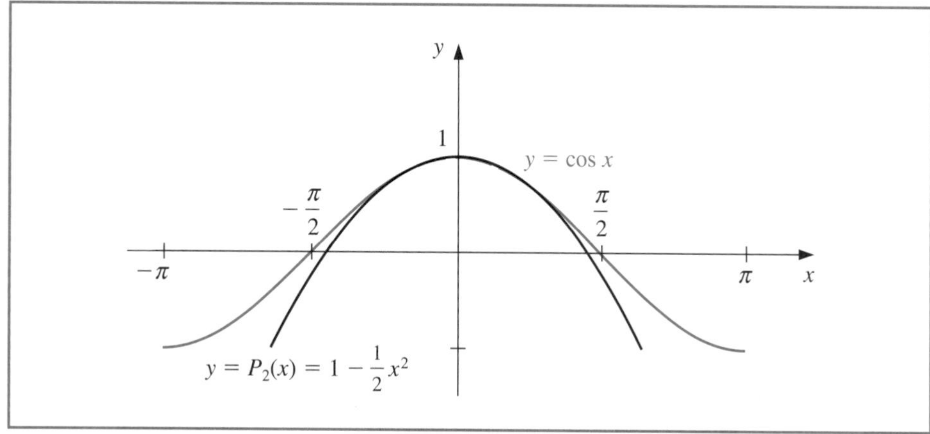

When $x = 0.01$, this becomes

$$\cos 0.01 = 1 - \frac{1}{2}(0.01)^2 + \frac{1}{6}(0.01)^3 \sin \xi(0.01) = 0.99995 + \frac{10^{-6}}{6} \sin \xi(0.01).$$

The approximation to $\cos 0.01$ given by the Taylor polynomial is therefore 0.99995. The truncation error, or remainder term, associated with this approximation is

$$\frac{10^{-6}}{6} \sin \xi(0.01) = 0.1\overline{6} \times 10^{-6} \sin \xi(0.01),$$

where the bar over the 6 in $0.1\overline{6}$ is used to indicate that this digit repeats indefinitely. Although we have no way of determining $\sin \xi(0.01)$, we know that all values of the sine lie in the interval $[-1, 1]$, so the error occurring if we use the approximation 0.99995 for the value of $\cos 0.01$ is bounded by

$$|\cos(0.01) - 0.99995| = 0.1\overline{6} \times 10^{-6} |\sin \xi(0.01)| \le 0.1\overline{6} \times 10^{-6}.$$

Hence the approximation 0.99995 matches at least the first five digits of $\cos 0.01$, and

$$\begin{aligned} 0.9999483 < 0.99995 - 1.\overline{6} \times 10^{-6} &\le \cos 0.01 \\ &\le 0.99995 + 1.\overline{6} \times 10^{-6} < 0.9999517. \end{aligned}$$

The error bound is much larger than the actual error. This is due in part to the poor bound we used for $|\sin \xi(x)|$. It is shown in Exercise 24 that for all values of x, we have $|\sin x| \le |x|$. Since $0 \le \xi < 0.01$, we could have used the fact that $|\sin \xi(x)| \le 0.01$ in the error formula, producing the bound $0.1\overline{6} \times 10^{-8}$.

(b) Since $f'''(0) = 0$, the third Taylor polynomial with remainder term about $x_0 = 0$ is

$$\cos x = 1 - \frac{1}{2}x^2 + \frac{1}{24}x^4 \cos \tilde{\xi}(x),$$

where $0 < \tilde{\xi}(x) < 0.01$. The approximating polynomial remains the same, and the approximation is still 0.99995, but we now have much better accuracy assurance. Since $|\cos \tilde{\xi}(x)| \le 1$ for all x, we have

$$\left| \frac{1}{24}x^4 \cos \tilde{\xi}(x) \right| \le \frac{1}{24}(0.01)^4(1) \approx 4.2 \times 10^{-10}.$$

So

$$|\cos 0.01 - 0.99995| \le 4.2 \times 10^{-10},$$

and

$$\begin{aligned} 0.99994999958 &= 0.99995 - 4.2 \times 10^{-10} \\ &\le \cos 0.01 \le 0.99995 + 4.2 \times 10^{-10} = 0.99995000042. \end{aligned}$$

■

Example 3 illustrates the two objectives of numerical analysis:

(i) Find an approximation to the solution of a given problem.

(ii) Determine a bound for the accuracy of the approximation.

The Taylor polynomials in both parts provide the same answer to (i), but the third Taylor polynomial gave a much better answer to (ii) than the second Taylor polynomial.

We can also use the Taylor polynomials to give us approximations to integrals.

Illustration We can use the third Taylor polynomial and its remainder term found in Example 3 to approximate $\int_0^{0.1} \cos x\, dx$. We have

$$\begin{aligned}\int_0^{0.1} \cos x\, dx &= \int_0^{0.1} \left(1 - \frac{1}{2}x^2\right) dx + \frac{1}{24}\int_0^{0.1} x^4 \cos \tilde{\xi}(x)\, dx \\ &= \left[x - \frac{1}{6}x^3\right]_0^{0.1} + \frac{1}{24}\int_0^{0.1} x^4 \cos \tilde{\xi}(x)\, dx \\ &= 0.1 - \frac{1}{6}(0.1)^3 + \frac{1}{24}\int_0^{0.1} x^4 \cos \tilde{\xi}(x)\, dx.\end{aligned}$$

Therefore

$$\int_0^{0.1} \cos x\, dx \approx 0.1 - \frac{1}{6}(0.1)^3 = 0.0998\overline{3}.$$

A bound for the error in this approximation is determined from the integral of the Taylor remainder term and the fact that $|\cos \tilde{\xi}(x)| \le 1$ for all x:

$$\begin{aligned}\frac{1}{24}\left|\int_0^{0.1} x^4 \cos \tilde{\xi}(x)\, dx\right| &\le \frac{1}{24}\int_0^{0.1} x^4 |\cos \tilde{\xi}(x)|\, dx \\ &\le \frac{1}{24}\int_0^{0.1} x^4\, dx = \frac{(0.1)^5}{120} = 8.\overline{3} \times 10^{-8}.\end{aligned}$$

The true value of this integral is

$$\int_0^{0.1} \cos x\, dx = \sin x\Big]_0^{0.1} = \sin 0.1 \approx 0.099833416647,$$

so the actual error for this approximation is 8.3314×10^{-8}, which is within the error bound. □

We can also use Maple to obtain these results. Define f by

$f := \cos(x)$

Maple allows us to place multiple statements on a line separated by either a semicolon or a colon. A semicolon will produce all the output, and a colon suppresses all but the final Maple response. For example, the third Taylor polynomial is given by

$s3 := taylor(f, x = 0, 4) : \ p3 := convert(s3, polynom)$

$$1 - \frac{1}{2}x^2$$

The first statement $s3 := taylor(f, x = 0, 4)$ determines the Taylor polynomial about $x_0 = 0$ with four terms (degree 3) and an indication of its remainder. The second $p3 := convert(s3, polynom)$ converts the series $s3$ to the polynomial $p3$ by dropping the remainder term.

Maple normally displays 10 decimal digits for approximations. To instead obtain the 11 digits we want for this illustration, enter

$Digits := 11$

and evaluate $f(0.01)$ and $P_3(0.01)$ with

$y1 := evalf(subs(x = 0.01, f)); \ y2 := evalf(subs(x = 0.01, p3)$

This produces

$$0.99995000042$$

$$0.99995000000$$

To show both the function (in black) and the polynomial (in cyan) near $x_0 = 0$, we enter

$plot\,((f, p3), x = -2\,..\,2)$

and obtain the Maple plot shown in Figure 1.11.

Figure 1.11

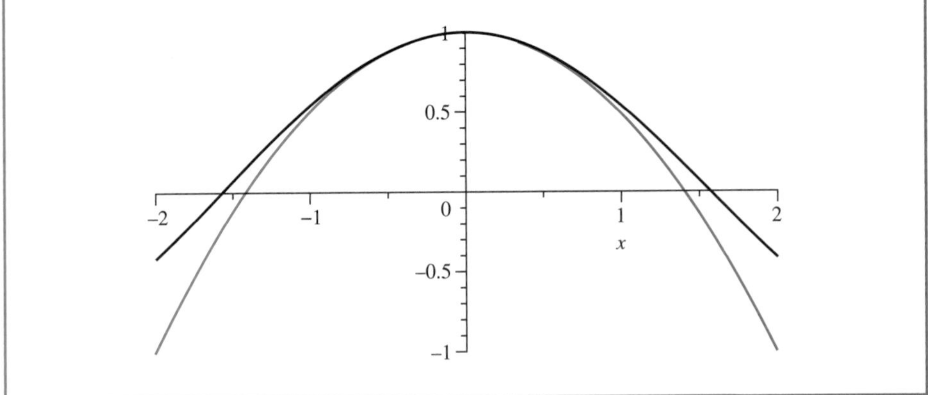

The integrals of f and the polynomial are given by

$q1 := int(f, x = 0\,..\,0.1);\ q2 := int(p3, x = 0\,..\,0.1)$

$$0.099833416647$$

$$0.099833333333$$

We assigned the names $q1$ and $q2$ to these values so that we could easily determine the error with the command

$err := |q1 - q2|$

$$8.3314\ 10^{-8}$$

There is an alternate method for generating the Taylor polynomials within the *NumericalAnalysis* subpackage of Maple's *Student* package. This subpackage will be discussed in Chapter 2.

EXERCISE SET 1.1

1. Show that the following equations have at least one solution in the given intervals.
   a. $(x-2)^2 - \ln x = 0$, $[1, 2]$ and $[e, 4]$
   b. $x\cos x - 2x^2 + 3x - 1 = 0$, $[0.2, 0.3]$ and $[1.2, 1.3]$

c. $2x\cos(2x) - (x-2)^2 = 0, \quad [2, 3]$ and $[3, 4]$

d. $x - (\ln x)^x = 0, \quad [4, 5]$

2. Find intervals containing solutions to the following equations.

a. $x - 3^{-x} = 0$

b. $4x^2 - e^x = 0$

c. $x^3 + 4.001x^2 + 4.002x + 1.101 = 0$

d. $x^3 - 2x^2 - 4x + 2 = 0$

3. Show that $f'(x)$ is 0 at least once in the given intervals.

a. $f(x) = x\sin\pi x - (x-2)\ln x, \quad [1, 2]$

b. $f(x) = (x-1)\tan x + x\sin\pi x, \quad [0, 1]$

c. $f(x) = 1 - e^x + (e-1)\sin((\pi/2)x), \quad [0, 1]$

d. $f(x) = (x-2)\sin x\ln(x+2), \quad [-1, 3]$

4. Find $\max_{a\le x\le b} |f(x)|$ for the following functions and intervals.

a. $f(x) = (2 - e^x + 2x)/3, \quad [0, 1]$

b. $f(x) = 1 + e^{-\cos(x-1)}, \quad [1, 2]$

c. $f(x) = 2x\cos(2x) - (x-2)^2, \quad [2, 4]$

d. $f(x) = (4x-3)/(x^2-2x), \quad [0.5, 1]$

5. Use the Intermediate Value Theorem 1.11 and Rolle's Theorem 1.7 to show that the graph of $f(x) = x^3 + 2x + k$ crosses the x-axis exactly once, regardless of the value of the constant k.

6. Suppose $f \in C[a, b]$ and $f'(x)$ exists on (a, b). Show that if $f'(x) \neq 0$ for all x in (a, b), then there can exist at most one number p in $[a, b]$ with $f(p) = 0$.

7. Let $f(x) = x^3$.

a. Find the second Taylor polynomial $P_2(x)$ about $x_0 = 0$.

b. Find $R_2(0.5)$ and the actual error in using $P_2(0.5)$ to approximate $f(0.5)$.

c. Repeat part (a) using $x_0 = 1$.

d. Repeat part (b) using the polynomial from part (c).

8. Find the third Taylor polynomial $P_3(x)$ for the function $f(x) = \sqrt{x+1}$ about $x_0 = 0$. Approximate $\sqrt{0.5}$, $\sqrt{0.75}$, $\sqrt{1.25}$, and $\sqrt{1.5}$ using $P_3(x)$, and find the actual errors.

9. Find the second Taylor polynomial $P_2(x)$ for the function $f(x) = e^x\cos x$ about $x_0 = 0$.

a. Use $P_2(0.5)$ to approximate $f(0.5)$. Find an upper bound for error $|f(0.5) - P_2(0.5)|$ using the error formula, and compare it to the actual error.

b. Find a bound for the error $|f(x) - P_2(x)|$ in using $P_2(x)$ to approximate $f(x)$ on the interval $[0, 1]$.

c. Approximate $\int_0^1 f(x)\,dx$ using $\int_0^1 P_2(x)\,dx$.

d. Find an upper bound for the error in (c) using $\int_0^1 |R_2(x)\,dx|$, and compare the bound to the actual error.

10. Repeat Exercise 9 using $x_0 = \pi/6$.

11. Let $f(x) = 2x\cos(2x) - (x-2)^2$ and $x_0 = 0$.

a. Find the third Taylor polynomial $P_3(x)$, and use it to approximate $f(0.4)$.

b. Use the error formula in Taylor's Theorem to find an upper bound for the error $|f(0.4) - P_3(0.4)|$. Compute the actual error.

c. Find the fourth Taylor polynomial $P_4(x)$, and use it to approximate $f(0.4)$.

d. Use the error formula in Taylor's Theorem to find an upper bound for the error $|f(0.4) - P_4(0.4)|$. Compute the actual error.

12. Find the third Taylor polynomial $P_3(x)$ for the function $f(x) = (x-1)\ln x$ about $x_0 = 1$.

a. Use $P_3(0.5)$ to approximate $f(0.5)$. Find an upper bound for error $|f(0.5) - P_3(0.5)|$ using the error formula, and compare it to the actual error.

b. Find a bound for the error $|f(x) - P_3(x)|$ in using $P_3(x)$ to approximate $f(x)$ on the interval $[0.5, 1.5]$.

c. Approximate $\int_{0.5}^{1.5} f(x)\,dx$ using $\int_{0.5}^{1.5} P_3(x)\,dx$.

d. Find an upper bound for the error in (c) using $\int_{0.5}^{1.5} |R_3(x)\,dx|$, and compare the bound to the actual error.

13. Find the fourth Taylor polynomial $P_4(x)$ for the function $f(x) = xe^{x^2}$ about $x_0 = 0$.

a. Find an upper bound for $|f(x) - P_4(x)|$, for $0 \le x \le 0.4$.

b. Approximate $\int_0^{0.4} f(x)\,dx$ using $\int_0^{0.4} P_4(x)\,dx$.

c. Find an upper bound for the error in (b) using $\int_0^{0.4} P_4(x)\,dx$.

d. Approximate $f'(0.2)$ using $P_4'(0.2)$, and find the error.

14. Use the error term of a Taylor polynomial to estimate the error involved in using $\sin x \approx x$ to approximate $\sin 1^\circ$.

15. Use a Taylor polynomial about $\pi/4$ to approximate $\cos 42^\circ$ to an accuracy of 10^{-6}.

16. Let $f(x) = e^{x/2}\sin(x/3)$. Use Maple to determine the following.

a. The third Maclaurin polynomial $P_3(x)$.

b. $f^{(4)}(x)$ and a bound for the error $|f(x) - P_3(x)|$ on $[0, 1]$.

17. Let $f(x) = \ln(x^2 + 2)$. Use Maple to determine the following.

a. The Taylor polynomial $P_3(x)$ for f expanded about $x_0 = 1$.

b. The maximum error $|f(x) - P_3(x)|$, for $0 \le x \le 1$.

c. The Maclaurin polynomial $\tilde{P}_3(x)$ for f.

d. The maximum error $|f(x) - \tilde{P}_3(x)|$, for $0 \le x \le 1$.

e. Does $P_3(0)$ approximate $f(0)$ better than $\tilde{P}_3(1)$ approximates $f(1)$?

18. Let $f(x) = (1-x)^{-1}$ and $x_0 = 0$. Find the nth Taylor polynomial $P_n(x)$ for $f(x)$ about x_0. Find a value of n necessary for $P_n(x)$ to approximate $f(x)$ to within 10^{-6} on $[0, 0.5]$.

19. Let $f(x) = e^x$ and $x_0 = 0$. Find the nth Taylor polynomial $P_n(x)$ for $f(x)$ about x_0. Find a value of n necessary for $P_n(x)$ to approximate $f(x)$ to within 10^{-6} on $[0, 0.5]$.

20. Find the nth Maclaurin polynomial $P_n(x)$ for $f(x) = \arctan x$.

21. The polynomial $P_2(x) = 1 - \frac{1}{2}x^2$ is to be used to approximate $f(x) = \cos x$ in $[-\frac{1}{2}, \frac{1}{2}]$. Find a bound for the maximum error.

22. The nth Taylor polynomial for a function f at x_0 is sometimes referred to as the polynomial of degree at most n that "best" approximates f near x_0.

a. Explain why this description is accurate.

b. Find the quadratic polynomial that best approximates a function f near $x_0 = 1$ if the tangent line at $x_0 = 1$ has equation $y = 4x - 1$, and if $f''(1) = 6$.

23. Prove the Generalized Rolle's Theorem, Theorem 1.10, by verifying the following.

a. Use Rolle's Theorem to show that $f'(z_i) = 0$ for $n-1$ numbers in $[a, b]$ with $a < z_1 < z_2 < \cdots < z_{n-1} < b$.

b. Use Rolle's Theorem to show that $f''(w_i) = 0$ for $n-2$ numbers in $[a, b]$ with $z_1 < w_1 < z_2 < w_2 \cdots w_{n-2} < z_{n-1} < b$.

c. Continue the arguments in **a.** and **b.** to show that for each $j = 1, 2, \ldots, n-1$ there are $n-j$ distinct numbers in $[a, b]$ where $f^{(j)}$ is 0.

d. Show that part **c.** implies the conclusion of the theorem.

24. In Example 3 it is stated that for all x we have $|\sin x| \le |x|$. Use the following to verify this statement.

a. Show that for all $x \ge 0$ we have $f(x) = x - \sin x$ is non-decreasing, which implies that $\sin x \le x$ with equality only when $x = 0$.

b. Use the fact that the sine function is odd to reach the conclusion.

25. A Maclaurin polynomial for e^x is used to give the approximation 2.5 to e. The error bound in this approximation is established to be $E = \frac{1}{6}$. Find a bound for the error in E.

26. The *error function* defined by

$$\text{erf}(x) = \frac{2}{\sqrt{\pi}} \int_0^x e^{-t^2}\,dt$$

gives the probability that any one of a series of trials will lie within x units of the mean, assuming that the trials have a normal distribution with mean 0 and standard deviation $\sqrt{2}/2$. This integral cannot be evaluated in terms of elementary functions, so an approximating technique must be used.

a. Integrate the Maclaurin series for e^{-x^2} to show that

$$\operatorname{erf}(x) = \frac{2}{\sqrt{\pi}} \sum_{k=0}^{\infty} \frac{(-1)^k x^{2k+1}}{(2k+1)k!}.$$

b. The error function can also be expressed in the form

$$\operatorname{erf}(x) = \frac{2}{\sqrt{\pi}} e^{-x^2} \sum_{k=0}^{\infty} \frac{2^k x^{2k+1}}{1 \cdot 3 \cdot 5 \cdots (2k+1)}.$$

Verify that the two series agree for $k = 1, 2, 3,$ and 4. [*Hint:* Use the Maclaurin series for e^{-x^2}.]

c. Use the series in part (a) to approximate erf(1) to within 10^{-7}.

d. Use the same number of terms as in part (c) to approximate erf(1) with the series in part (b).

e. Explain why difficulties occur using the series in part (b) to approximate erf(x).

27. A function $f : [a, b] \to \mathbb{R}$ is said to satisfy a *Lipschitz condition* with Lipschitz constant L on $[a, b]$ if, for every $x, y \in [a, b]$, we have $|f(x) - f(y)| \le L|x - y|$.

a. Show that if f satisfies a Lipschitz condition with Lipschitz constant L on an interval $[a, b]$, then $f \in C[a, b]$.

b. Show that if f has a derivative that is bounded on $[a, b]$ by L, then f satisfies a Lipschitz condition with Lipschitz constant L on $[a, b]$.

c. Give an example of a function that is continuous on a closed interval but does not satisfy a Lipschitz condition on the interval.

28. Suppose $f \in C[a, b]$, that x_1 and x_2 are in $[a, b]$.

a. Show that a number ξ exists between x_1 and x_2 with

$$f(\xi) = \frac{f(x_1) + f(x_2)}{2} = \frac{1}{2} f(x_1) + \frac{1}{2} f(x_2).$$

b. Suppose that c_1 and c_2 are positive constants. Show that a number ξ exists between x_1 and x_2 with

$$f(\xi) = \frac{c_1 f(x_1) + c_2 f(x_2)}{c_1 + c_2}.$$

c. Give an example to show that the result in part **b.** does not necessarily hold when c_1 and c_2 have opposite signs with $c_1 \neq -c_2$.

29. Let $f \in C[a, b]$, and let p be in the open interval (a, b).

a. Suppose $f(p) \neq 0$. Show that a $\delta > 0$ exists with $f(x) \neq 0$, for all x in $[p - \delta, p + \delta]$, with $[p - \delta, p + \delta]$ a subset of $[a, b]$.

b. Suppose $f(p) = 0$ and $k > 0$ is given. Show that a $\delta > 0$ exists with $|f(x)| \le k$, for all x in $[p - \delta, p + \delta]$, with $[p - \delta, p + \delta]$ a subset of $[a, b]$.

1.2 Round-off Errors and Computer Arithmetic

The arithmetic performed by a calculator or computer is different from the arithmetic in algebra and calculus courses. You would likely expect that we always have as true statements things such as $2+2 = 4$, $4 \cdot 8 = 32$, and $(\sqrt{3})^2 = 3$. However, with *computer* arithmetic we expect exact results for $2+2 = 4$ and $4 \cdot 8 = 32$, but we will not have precisely $(\sqrt{3})^2 = 3$. To understand why this is true we must explore the world of finite-digit arithmetic.

In our traditional mathematical world we permit numbers with an infinite number of digits. The arithmetic we use in this world *defines* $\sqrt{3}$ as that unique positive number that when multiplied by itself produces the integer 3. In the computational world, however, each representable number has only a fixed and finite number of digits. This means, for example, that only rational numbers—and not even all of these—can be represented exactly. Since $\sqrt{3}$ is not rational, it is given an approximate representation, one whose square will not be precisely 3, although it will likely be sufficiently close to 3 to be acceptable in most situations. In most cases, then, this machine arithmetic is satisfactory and passes without notice or concern, but at times problems arise because of this discrepancy.

Error due to rounding should be expected whenever computations are performed using numbers that are not powers of 2. Keeping this error under control is extremely important when the number of calculations is large.

The error that is produced when a calculator or computer is used to perform real-number calculations is called **round-off error**. It occurs because the arithmetic performed in a machine involves numbers with only a finite number of digits, with the result that calculations are performed with only approximate representations of the actual numbers. In a computer, only a relatively small subset of the real number system is used for the representation of all the real numbers. This subset contains only rational numbers, both positive and negative, and stores the fractional part, together with an exponential part.

Binary Machine Numbers

In 1985, the IEEE (Institute for Electrical and Electronic Engineers) published a report called *Binary Floating Point Arithmetic Standard 754–1985*. An updated version was published in 2008 as *IEEE 754-2008*. This provides standards for binary and decimal floating point numbers, formats for data interchange, algorithms for rounding arithmetic operations, and for the handling of exceptions. Formats are specified for single, double, and extended precisions, and these standards are generally followed by all microcomputer manufacturers using floating-point hardware.

A 64-bit (binary digit) representation is used for a real number. The first bit is a sign indicator, denoted s. This is followed by an 11-bit exponent, c, called the **characteristic**, and a 52-bit binary fraction, f, called the **mantissa**. The base for the exponent is 2.

Since 52 binary digits correspond to between 16 and 17 decimal digits, we can assume that a number represented in this system has at least 16 decimal digits of precision. The exponent of 11 binary digits gives a range of 0 to $2^{11}-1=2047$. However, using only positive integers for the exponent would not permit an adequate representation of numbers with small magnitude. To ensure that numbers with small magnitude are equally representable, 1023 is subtracted from the characteristic, so the range of the exponent is actually from -1023 to 1024.

To save storage and provide a unique representation for each floating-point number, a normalization is imposed. Using this system gives a floating-point number of the form

$$(-1)^s 2^{c-1023}(1+f).$$

Illustration Consider the machine number

$$0\ 10000000011\ 1011100100010000000000000000000000000000000000000000.$$

The leftmost bit is $s=0$, which indicates that the number is positive. The next 11 bits, 10000000011, give the characteristic and are equivalent to the decimal number

$$c = 1\cdot 2^{10} + 0\cdot 2^9 + \cdots + 0\cdot 2^2 + 1\cdot 2^1 + 1\cdot 2^0 = 1024+2+1 = 1027.$$

The exponential part of the number is, therefore, $2^{1027-1023} = 2^4$. The final 52 bits specify that the mantissa is

$$f = 1 \cdot \left(\frac{1}{2}\right)^1 + 1 \cdot \left(\frac{1}{2}\right)^3 + 1 \cdot \left(\frac{1}{2}\right)^4 + 1 \cdot \left(\frac{1}{2}\right)^5 + 1 \cdot \left(\frac{1}{2}\right)^8 + 1 \cdot \left(\frac{1}{2}\right)^{12}.$$

As a consequence, this machine number precisely represents the decimal number

$$\begin{aligned}(-1)^s 2^{c-1023}(1+f) &= (-1)^0 \cdot 2^{1027-1023}\left(1 + \left(\frac{1}{2} + \frac{1}{8} + \frac{1}{16} + \frac{1}{32} + \frac{1}{256} + \frac{1}{4096}\right)\right)\\ &= 27.56640625.\end{aligned}$$

However, the next smallest machine number is

$$0\ 10000000011\ 1011100100001111111111111111111111111111111111111111,$$

and the next largest machine number is

$$0\ 10000000011\ 1011100100010000000000000000000000000000000000000001.$$

This means that our original machine number represents not only 27.56640625, but also half of the real numbers that are between 27.56640625 and the next smallest machine number, as well as half the numbers between 27.56640625 and the next largest machine number. To be precise, it represents any real number in the interval

$$[27.566406249999998223643160599749535322189331054687 5,$$
$$27.566406250000001776356839400250464677810668945312 5). \qquad \square$$

The smallest normalized positive number that can be represented has $s = 0$, $c = 1$, and $f = 0$ and is equivalent to

$$2^{-1022} \cdot (1 + 0) \approx 0.22251 \times 10^{-307},$$

and the largest has $s = 0$, $c = 2046$, and $f = 1 - 2^{-52}$ and is equivalent to

$$2^{1023} \cdot (2 - 2^{-52}) \approx 0.17977 \times 10^{309}.$$

Numbers occurring in calculations that have a magnitude less than

$$2^{-1022} \cdot (1 + 0)$$

result in **underflow** and are generally set to zero. Numbers greater than

$$2^{1023} \cdot (2 - 2^{-52})$$

result in **overflow** and typically cause the computations to stop (unless the program has been designed to detect this occurrence). Note that there are two representations for the number zero; a positive 0 when $s = 0$, $c = 0$ and $f = 0$, and a negative 0 when $s = 1$, $c = 0$ and $f = 0$.

Decimal Machine Numbers

The use of binary digits tends to conceal the computational difficulties that occur when a finite collection of machine numbers is used to represent all the real numbers. To examine these problems, we will use more familiar decimal numbers instead of binary representation. Specifically, we assume that machine numbers are represented in the normalized *decimal* floating-point form

$$\pm 0.d_1 d_2 \ldots d_k \times 10^n, \quad 1 \le d_1 \le 9, \quad \text{and} \quad 0 \le d_i \le 9,$$

for each $i = 2, \ldots, k$. Numbers of this form are called k-digit *decimal machine numbers*.

Any positive real number within the numerical range of the machine can be normalized to the form

$$y = 0.d_1 d_2 \ldots d_k d_{k+1} d_{k+2} \ldots \times 10^n.$$

The error that results from replacing a number with its floating-point form is called **round-off error** regardless of whether the rounding or chopping method is used.

The floating-point form of y, denoted $fl(y)$, is obtained by terminating the mantissa of y at k decimal digits. There are two common ways of performing this termination. One method, called **chopping**, is to simply chop off the digits $d_{k+1}d_{k+2}\ldots$. This produces the floating-point form

$$fl(y) = 0.d_1 d_2 \ldots d_k \times 10^n.$$

The other method, called **rounding**, adds $5 \times 10^{n-(k+1)}$ to y and then chops the result to obtain a number of the form

$$fl(y) = 0.\delta_1 \delta_2 \ldots \delta_k \times 10^n.$$

For rounding, when $d_{k+1} \ge 5$, we add 1 to d_k to obtain $fl(y)$; that is, we *round up*. When $d_{k+1} < 5$, we simply chop off all but the first k digits; so we *round down*. If we round down, then $\delta_i = d_i$, for each $i = 1, 2, \ldots, k$. However, if we round up, the digits (and even the exponent) might change.

Example 1 Determine the five-digit **(a)** chopping and **(b)** rounding values of the irrational number π.

Solution The number π has an infinite decimal expansion of the form $\pi = 3.14159265\ldots$. Written in normalized decimal form, we have

$$\pi = 0.314159265\ldots \times 10^1.$$

The relative error is generally a better measure of accuracy than the absolute error because it takes into consideration the size of the number being approximated.

(a) The floating-point form of π using five-digit chopping is

$$fl(\pi) = 0.31415 \times 10^1 = 3.1415.$$

(b) The sixth digit of the decimal expansion of π is a 9, so the floating-point form of π using five-digit rounding is

$$fl(\pi) = (0.31415 + 0.00001) \times 10^1 = 3.1416.$$ ■

The following definition describes two methods for measuring approximation errors.

Definition 1.15 Suppose that p^* is an approximation to p. The **absolute error** is $|p - p^*|$, and the **relative error** is $\dfrac{|p - p^*|}{|p|}$, provided that $p \neq 0$. ■

Consider the absolute and relative errors in representing p by p^* in the following example.

Example 2 Determine the absolute and relative errors when approximating p by p^* when

(a) $p = 0.3000 \times 10^1$ and $p^* = 0.3100 \times 10^1$;

(b) $p = 0.3000 \times 10^{-3}$ and $p^* = 0.3100 \times 10^{-3}$;

(c) $p = 0.3000 \times 10^4$ and $p^* = 0.3100 \times 10^4$.

Solution

(a) For $p = 0.3000 \times 10^1$ and $p^* = 0.3100 \times 10^1$ the absolute error is 0.1, and the relative error is $0.333\overline{3} \times 10^{-1}$.

We often cannot find an accurate value for the true error in an approximation. Instead we find a bound for the error, which gives us a "worst-case" error.

(b) For $p = 0.3000 \times 10^{-3}$ and $p^* = 0.3100 \times 10^{-3}$ the absolute error is 0.1×10^{-4}, and the relative error is $0.333\overline{3} \times 10^{-1}$.

(c) For $p = 0.3000 \times 10^4$ and $p^* = 0.3100 \times 10^4$, the absolute error is 0.1×10^3, and the relative error is again $0.333\overline{3} \times 10^{-1}$.

This example shows that the same relative error, $0.333\overline{3} \times 10^{-1}$, occurs for widely varying absolute errors. As a measure of accuracy, the absolute error can be misleading and the relative error more meaningful, because the relative error takes into consideration the size of the value. ∎

The following definition uses relative error to give a measure of significant digits of accuracy for an approximation.

Definition 1.16 The number p^* is said to approximate p to t **significant digits** (or figures) if t is the largest nonnegative integer for which

$$\frac{|p - p^*|}{|p|} \leq 5 \times 10^{-t}.$$ ∎

The term significant digits is often used to loosely describe the number of decimal digits that appear to be accurate. The definition is more precise, and provides a continuous concept.

Table 1.1 illustrates the continuous nature of significant digits by listing, for the various values of p, the least upper bound of $|p - p^*|$, denoted max $|p - p^*|$, when p^* agrees with p to four significant digits.

Table 1.1

p	0.1	0.5	100	1000	5000	9990	10000
max $\lvert p - p^* \rvert$	0.00005	0.00025	0.05	0.5	2.5	4.995	5.

Returning to the machine representation of numbers, we see that the floating-point representation $fl(y)$ for the number y has the relative error

$$\left| \frac{y - fl(y)}{y} \right|.$$

If k decimal digits and chopping are used for the machine representation of

$$y = 0.d_1 d_2 \ldots d_k d_{k+1} \ldots \times 10^n,$$

then

$$\left|\frac{y - fl(y)}{y}\right| = \left|\frac{0.d_1d_2\ldots d_kd_{k+1}\ldots \times 10^n - 0.d_1d_2\ldots d_k \times 10^n}{0.d_1d_2\ldots \times 10^n}\right|$$

$$= \left|\frac{0.d_{k+1}d_{k+2}\ldots \times 10^{n-k}}{0.d_1d_2\ldots \times 10^n}\right| = \left|\frac{0.d_{k+1}d_{k+2}\ldots}{0.d_1d_2\ldots}\right| \times 10^{-k}.$$

Since $d_1 \neq 0$, the minimal value of the denominator is 0.1. The numerator is bounded above by 1. As a consequence,

$$\left|\frac{y - fl(y)}{y}\right| \leq \frac{1}{0.1} \times 10^{-k} = 10^{-k+1}.$$

In a similar manner, a bound for the relative error when using k-digit rounding arithmetic is $0.5 \times 10^{-k+1}$. (See Exercise 24.)

Note that the bounds for the relative error using k-digit arithmetic are independent of the number being represented. This result is due to the manner in which the machine numbers are distributed along the real line. Because of the exponential form of the characteristic, the same number of decimal machine numbers is used to represent each of the intervals $[0.1, 1]$, $[1, 10]$, and $[10, 100]$. In fact, within the limits of the machine, the number of decimal machine numbers in $[10^n, 10^{n+1}]$ is constant for all integers n.

Finite-Digit Arithmetic

In addition to inaccurate representation of numbers, the arithmetic performed in a computer is not exact. The arithmetic involves manipulating binary digits by various shifting, or logical, operations. Since the actual mechanics of these operations are not pertinent to this presentation, we shall devise our own approximation to computer arithmetic. Although our arithmetic will not give the exact picture, it suffices to explain the problems that occur. (For an explanation of the manipulations actually involved, the reader is urged to consult more technically oriented computer science texts, such as [Ma], *Computer System Architecture*.)

Assume that the floating-point representations $fl(x)$ and $fl(y)$ are given for the real numbers x and y and that the symbols $\oplus$, $\ominus$, $\otimes$, $\oplus$ represent machine addition, subtraction, multiplication, and division operations, respectively. We will assume a finite-digit arithmetic given by

$$x \oplus y = fl(fl(x) + fl(y)), \quad x \otimes y = fl(fl(x) \times fl(y)),$$
$$x \ominus y = fl(fl(x) - fl(y)), \quad x \oplus y = fl(fl(x) \div fl(y)).$$

This arithmetic corresponds to performing exact arithmetic on the floating-point representations of x and y and then converting the exact result to its finite-digit floating-point representation.

Rounding arithmetic is easily implemented in Maple. For example, the command

Digits := 5

causes all arithmetic to be rounded to 5 digits. To ensure that Maple uses approximate rather than exact arithmetic we use the *evalf*. For example, if $x = \pi$ and $y = \sqrt{2}$ then

evalf(*x*); *evalf*(*y*)

produces 3.1416 and 1.4142, respectively. Then $fl(fl(x) + fl(y))$ is performed using 5-digit rounding arithmetic with

evalf(*evalf*(*x*) + *evalf*(*y*))

which gives 4.5558. Implementing finite-digit chopping arithmetic is more difficult and requires a sequence of steps or a procedure. Exercise 27 explores this problem.

Example 3 Suppose that $x = \frac{5}{7}$ and $y = \frac{1}{3}$. Use five-digit chopping for calculating $x + y$, $x - y$, $x \times y$, and $x \div y$.

Solution Note that

$$x = \frac{5}{7} = 0.\overline{714285} \quad \text{and} \quad y = \frac{1}{3} = 0.\overline{3}$$

implies that the five-digit chopping values of x and y are

$$fl(x) = 0.71428 \times 10^0 \quad \text{and} \quad fl(y) = 0.33333 \times 10^0.$$

Thus

$$\begin{aligned} x \oplus y &= fl(fl(x) + fl(y)) = fl\left(0.71428 \times 10^0 + 0.33333 \times 10^0\right) \\ &= fl\left(1.04761 \times 10^0\right) = 0.10476 \times 10^1. \end{aligned}$$

The true value is $x + y = \frac{5}{7} + \frac{1}{3} = \frac{22}{21}$, so we have

$$\text{Absolute Error} = \left| \frac{22}{21} - 0.10476 \times 10^1 \right| = 0.190 \times 10^{-4}$$

and

$$\text{Relative Error} = \left| \frac{0.190 \times 10^{-4}}{22/21} \right| = 0.182 \times 10^{-4}.$$

Table 1.2 lists the values of this and the other calculations. ■

Table 1.2

Operation	Result	Actual value	Absolute error	Relative error
$x \oplus y$	0.10476×10^1	22/21	0.190×10^{-4}	0.182×10^{-4}
$x \ominus y$	0.38095×10^0	8/21	0.238×10^{-5}	0.625×10^{-5}
$x \otimes y$	0.23809×10^0	5/21	0.524×10^{-5}	0.220×10^{-4}
$x \oslash y$	0.21428×10^1	15/7	0.571×10^{-4}	0.267×10^{-4}

The maximum relative error for the operations in Example 3 is 0.267×10^{-4}, so the arithmetic produces satisfactory five-digit results. This is not the case in the following example.

Example 4 Suppose that in addition to $x = \frac{5}{7}$ and $y = \frac{1}{3}$ we have

$$u = 0.714251, \quad v = 98765.9, \quad \text{and} \quad w = 0.111111 \times 10^{-4},$$

so that

$$fl(u) = 0.71425 \times 10^0, \quad fl(v) = 0.98765 \times 10^5, \quad \text{and} \quad fl(w) = 0.11111 \times 10^{-4}.$$

Determine the five-digit chopping values of $x \ominus u$, $(x \ominus u) \oslash w$, $(x \ominus u) \otimes v$, and $u \oplus v$.

Solution These numbers were chosen to illustrate some problems that can arise with finite-digit arithmetic. Because x and u are nearly the same, their difference is small. The absolute error for $x \ominus u$ is

$$\begin{aligned}|(x-u)-(x\ominus u)| &= |(x-u)-(fl(fl(x)-fl(u)))| \\ &= \left|\left(\frac{5}{7}-0.714251\right)-\left(fl\left(0.71428\times 10^0-0.71425\times 10^0\right)\right)\right| \\ &= \left|0.347143\times 10^{-4}-fl\left(0.00003\times 10^0\right)\right| = 0.47143\times 10^{-5}.\end{aligned}$$

This approximation has a small absolute error, but a large relative error

$$\left|\frac{0.47143\times 10^{-5}}{0.347143\times 10^{-4}}\right| \le 0.136.$$

The subsequent division by the small number w or multiplication by the large number v magnifies the absolute error without modifying the relative error. The addition of the large and small numbers u and v produces large absolute error but not large relative error. These calculations are shown in Table 1.3. ■

Table 1.3

Operation	Result	Actual value	Absolute error	Relative error
$x \ominus u$	0.30000×10^{-4}	0.34714×10^{-4}	0.471×10^{-5}	0.136
$(x \ominus u) \oplus w$	0.27000×10^{1}	0.31242×10^{1}	0.424	0.136
$(x \ominus u) \otimes v$	0.29629×10^{1}	0.34285×10^{1}	0.465	0.136
$u \oplus v$	0.98765×10^{5}	0.98766×10^{5}	0.161×10^{1}	0.163×10^{-4}

One of the most common error-producing calculations involves the cancelation of significant digits due to the subtraction of nearly equal numbers. Suppose two nearly equal numbers x and y, with $x > y$, have the k-digit representations

$$fl(x) = 0.d_1d_2\ldots d_p\alpha_{p+1}\alpha_{p+2}\ldots\alpha_k \times 10^n,$$

and

$$fl(y) = 0.d_1d_2\ldots d_p\beta_{p+1}\beta_{p+2}\ldots\beta_k \times 10^n.$$

The floating-point form of $x - y$ is

$$fl(fl(x)-fl(y)) = 0.\sigma_{p+1}\sigma_{p+2}\ldots\sigma_k \times 10^{n-p},$$

where

$$0.\sigma_{p+1}\sigma_{p+2}\ldots\sigma_k = 0.\alpha_{p+1}\alpha_{p+2}\ldots\alpha_k - 0.\beta_{p+1}\beta_{p+2}\ldots\beta_k.$$

The floating-point number used to represent $x - y$ has at most $k - p$ digits of significance. However, in most calculation devices, $x - y$ will be assigned k digits, with the last p being either zero or randomly assigned. Any further calculations involving $x-y$ retain the problem of having only $k-p$ digits of significance, since a chain of calculations is no more accurate than its weakest portion.

If a finite-digit representation or calculation introduces an error, further enlargement of the error occurs when dividing by a number with small magnitude (or, equivalently, when multiplying by a number with large magnitude). Suppose, for example, that the number z has the finite-digit approximation $z + \delta$, where the error δ is introduced by representation or by previous calculation. Now divide by $\varepsilon = 10^{-n}$, where $n > 0$. Then

$$\frac{z}{\varepsilon} \approx fl\left(\frac{fl(z)}{fl(\varepsilon)}\right) = (z + \delta) \times 10^n.$$

The absolute error in this approximation, $|\delta| \times 10^n$, is the original absolute error, $|\delta|$, multiplied by the factor 10^n.

Example 5 Let $p = 0.54617$ and $q = 0.54601$. Use four-digit arithmetic to approximate $p - q$ and determine the absolute and relative errors using **(a)** rounding and **(b)** chopping.

Solution The exact value of $r = p - q$ is $r = 0.00016$.

(a) Suppose the subtraction is performed using four-digit rounding arithmetic. Rounding p and q to four digits gives $p^* = 0.5462$ and $q^* = 0.5460$, respectively, and $r^* = p^* - q^* = 0.0002$ is the four-digit approximation to r. Since

$$\frac{|r - r^*|}{|r|} = \frac{|0.00016 - 0.0002|}{|0.00016|} = 0.25,$$

the result has only one significant digit, whereas p^* and q^* were accurate to four and five significant digits, respectively.

(b) If chopping is used to obtain the four digits, the four-digit approximations to p, q, and r are $p^* = 0.5461$, $q^* = 0.5460$, and $r^* = p^* - q^* = 0.0001$. This gives

$$\frac{|r - r^*|}{|r|} = \frac{|0.00016 - 0.0001|}{|0.00016|} = 0.375,$$

which also results in only one significant digit of accuracy. ■

The loss of accuracy due to round-off error can often be avoided by a reformulation of the calculations, as illustrated in the next example.

Illustration The quadratic formula states that the roots of $ax^2 + bx + c = 0$, when $a \neq 0$, are

$$x_1 = \frac{-b + \sqrt{b^2 - 4ac}}{2a} \quad \text{and} \quad x_2 = \frac{-b - \sqrt{b^2 - 4ac}}{2a}. \tag{1.1}$$

Consider this formula applied to the equation $x^2 + 62.10x + 1 = 0$, whose roots are approximately

$$x_1 = -0.01610723 \quad \text{and} \quad x_2 = -62.08390.$$

The roots x_1 and x_2 of a general quadratic equation are related to the coefficients by the fact that

$$x_1 + x_2 = -\frac{b}{a}$$

and

$$x_1 x_2 = \frac{c}{a}.$$

This is a special case of Vièta's Formulas for the coefficients of polynomials.

We will again use four-digit rounding arithmetic in the calculations to determine the root. In this equation, b^2 is much larger than $4ac$, so the numerator in the calculation for x_1 involves the *subtraction* of nearly equal numbers. Because

$$\sqrt{b^2 - 4ac} = \sqrt{(62.10)^2 - (4.000)(1.000)(1.000)}$$
$$= \sqrt{3856. - 4.000} = \sqrt{3852.} = 62.06,$$

we have

$$fl(x_1) = \frac{-62.10 + 62.06}{2.000} = \frac{-0.04000}{2.000} = -0.02000,$$

a poor approximation to $x_1 = -0.01611$, with the large relative error

$$\frac{|-0.01611 + 0.02000|}{|-0.01611|} \approx 2.4 \times 10^{-1}.$$

On the other hand, the calculation for x_2 involves the *addition* of the nearly equal numbers $-b$ and $-\sqrt{b^2 - 4ac}$. This presents no problem since

$$fl(x_2) = \frac{-62.10 - 62.06}{2.000} = \frac{-124.2}{2.000} = -62.10$$

has the small relative error

$$\frac{|-62.08 + 62.10|}{|-62.08|} \approx 3.2 \times 10^{-4}.$$

To obtain a more accurate four-digit rounding approximation for x_1, we change the form of the quadratic formula by *rationalizing the numerator*:

$$x_1 = \frac{-b + \sqrt{b^2 - 4ac}}{2a}\left(\frac{-b - \sqrt{b^2 - 4ac}}{-b - \sqrt{b^2 - 4ac}}\right) = \frac{b^2 - (b^2 - 4ac)}{2a(-b - \sqrt{b^2 - 4ac})},$$

which simplifies to an alternate quadratic formula

$$x_1 = \frac{-2c}{b + \sqrt{b^2 - 4ac}}. \tag{1.2}$$

Using (1.2) gives

$$fl(x_1) = \frac{-2.000}{62.10 + 62.06} = \frac{-2.000}{124.2} = -0.01610,$$

which has the small relative error 6.2×10^{-4}.

The rationalization technique can also be applied to give the following alternative quadratic formula for x_2:

$$x_2 = \frac{-2c}{b - \sqrt{b^2 - 4ac}}. \tag{1.3}$$

This is the form to use if b is a negative number. In the Illustration, however, the mistaken use of this formula for x_2 would result in not only the subtraction of nearly equal numbers, but also the division by the small result of this subtraction. The inaccuracy that this combination produces,

$$fl(x_2) = \frac{-2c}{b - \sqrt{b^2 - 4ac}} = \frac{-2.000}{62.10 - 62.06} = \frac{-2.000}{0.04000} = -50.00,$$

has the large relative error 1.9×10^{-1}. □

- The lesson: Think before you compute!

Nested Arithmetic

Accuracy loss due to round-off error can also be reduced by rearranging calculations, as shown in the next example.

Example 6 Evaluate $f(x) = x^3 - 6.1x^2 + 3.2x + 1.5$ at $x = 4.71$ using three-digit arithmetic.

Solution Table 1.4 gives the intermediate results in the calculations.

Table 1.4

	x	x^2	x^3	$6.1x^2$	$3.2x$
Exact	4.71	22.1841	104.487111	135.32301	15.072
Three-digit (chopping)	4.71	22.1	104.	134.	15.0
Three-digit (rounding)	4.71	22.2	105.	135.	15.1

To illustrate the calculations, let us look at those involved with finding x^3 using three-digit rounding arithmetic. First we find

$$x^2 = 4.71^2 = 22.1841 \quad \text{which rounds to } 22.2.$$

Then we use this value of x^2 to find

$$x^3 = x^2 \cdot x = 22.2 \cdot 4.71 = 104.562 \quad \text{which rounds to } 105.$$

Also,

$$6.1x^2 = 6.1(22.2) = 135.42 \quad \text{which rounds to } 135,$$

and

$$3.2x = 3.2(4.71) = 15.072 \quad \text{which rounds to } 15.1.$$

The exact result of the evaluation is

$$\text{Exact:} \quad f(4.71) = 104.487111 - 135.32301 + 15.072 + 1.5 = -14.263899.$$

Using finite-digit arithmetic, the way in which we add the results can effect the final result. Suppose that we add left to right. Then for chopping arithmetic we have

$$\text{Three-digit (chopping):} \quad f(4.71) = ((104. - 134.) + 15.0) + 1.5 = -13.5,$$

and for rounding arithmetic we have

$$\text{Three-digit (rounding):} \quad f(4.71) = ((105. - 135.) + 15.1) + 1.5 = -13.4.$$

(You should carefully verify these results to be sure that your notion of finite-digit arithmetic is correct.) Note that the three-digit chopping values simply retain the leading three digits, with no rounding involved, and differ significantly from the three-digit rounding values.

The relative errors for the three-digit methods are

$$\text{Chopping:} \left|\frac{-14.263899 + 13.5}{-14.263899}\right| \approx 0.05, \quad \text{and Rounding:} \left|\frac{-14.263899 + 13.4}{-14.263899}\right| \approx 0.06.$$

■

Illustration As an alternative approach, the polynomial $f(x)$ in Example 6 can be written in a **nested** manner as

Remember that chopping (or rounding) is performed after each calculation.

$$f(x) = x^3 - 6.1x^2 + 3.2x + 1.5 = ((x - 6.1)x + 3.2)x + 1.5.$$

Using three-digit chopping arithmetic now produces

$$\begin{aligned} f(4.71) &= ((4.71 - 6.1)4.71 + 3.2)4.71 + 1.5 = ((-1.39)(4.71) + 3.2)4.71 + 1.5 \\ &= (-6.54 + 3.2)4.71 + 1.5 = (-3.34)4.71 + 1.5 = -15.7 + 1.5 = -14.2. \end{aligned}$$

In a similar manner, we now obtain a three-digit rounding answer of −14.3. The new relative errors are

$$\text{Three-digit (chopping):} \quad \left|\frac{-14.263899 + 14.2}{-14.263899}\right| \approx 0.0045;$$

$$\text{Three-digit (rounding):} \quad \left|\frac{-14.263899 + 14.3}{-14.263899}\right| \approx 0.0025.$$

Nesting has reduced the relative error for the chopping approximation to less than 10% of that obtained initially. For the rounding approximation the improvement has been even more dramatic; the error in this case has been reduced by more than 95%. □

Polynomials should *always* be expressed in nested form before performing an evaluation, because this form minimizes the number of arithmetic calculations. The decreased error in the Illustration is due to the reduction in computations from four multiplications and three additions to two multiplications and three additions. One way to reduce round-off error is to reduce the number of computations.

EXERCISE SET 1.2

1. Compute the absolute error and relative error in approximations of p by p^*.

a. $p = \pi, p^* = 3.1416$
b. $p = \pi, p^* = 22/7$
c. $p = \sqrt{2}, p^* = 1.414$
d. $p = e, p^* = 2.718$
e. $p = 10^\pi, p^* = 1400$
f. $p = e^{10}, p^* = 22000$
g. $p = 8!, p^* = 39900$
h. $p = 9!, p^* = \sqrt{18\pi}(9/e)^9$

2. Find the largest interval in which p^* must lie to approximate p with relative error at most 10^{-4} for each value of p.

a. $\sqrt{2}$
b. π
c. e
d. $\sqrt[3]{7}$

3. Suppose p^* must approximate p with relative error at most 10^{-3}. Find the largest interval in which p^* must lie for each value of p.

a. 150
b. 900
c. 1500
d. 90

4. Perform the following computations (i) exactly, (ii) using three-digit chopping arithmetic, and (iii) using three-digit rounding arithmetic. (iv) Compute the relative errors in parts (ii) and (iii).

a. $\frac{4}{5} + \frac{1}{3}$
b. $\frac{4}{5} \cdot \frac{1}{3}$
c. $\left(\frac{1}{3} - \frac{3}{11}\right) + \frac{3}{20}$
d. $\left(\frac{1}{3} + \frac{3}{11}\right) - \frac{3}{20}$

5. Use three-digit rounding arithmetic to perform the following calculations. Compute the absolute error and relative error with the exact value determined to at least five digits.

a. $133 + 0.921$
b. $133 - 0.499$
c. $(121 - 0.327) - 119$
d. $(121 - 119) - 0.327$
e. $\dfrac{\frac{13}{14} - \frac{6}{7}}{2e - 5.4}$
f. $-10\pi + 6e - \dfrac{3}{62}$
g. $\left(\frac{2}{9}\right) \cdot \left(\frac{9}{7}\right)$
h. $\dfrac{\pi - \frac{22}{7}}{\frac{1}{17}}$

6. Repeat Exercise 5 using four-digit rounding arithmetic.

7. Repeat Exercise 5 using three-digit chopping arithmetic.

8. Repeat Exercise 5 using four-digit chopping arithmetic.

9. The first three nonzero terms of the Maclaurin series for the arctangent function are $x - (1/3)x^3 + (1/5)x^5$. Compute the absolute error and relative error in the following approximations of π using the polynomial in place of the arctangent:

a. $4\left[\arctan\left(\frac{1}{2}\right) + \arctan\left(\frac{1}{3}\right)\right]$

b. $16\arctan\left(\frac{1}{5}\right) - 4\arctan\left(\frac{1}{239}\right)$

10. The number e can be defined by $e = \sum_{n=0}^{\infty}(1/n!)$, where $n! = n(n-1)\cdots 2\cdot 1$ for $n \neq 0$ and $0! = 1$. Compute the absolute error and relative error in the following approximations of e:

a. $\displaystyle\sum_{n=0}^{5}\frac{1}{n!}$ **b.** $\displaystyle\sum_{n=0}^{10}\frac{1}{n!}$

11. Let

$$f(x) = \frac{x\cos x - \sin x}{x - \sin x}.$$

a. Find $\lim_{x\to 0} f(x)$.

b. Use four-digit rounding arithmetic to evaluate $f(0.1)$.

c. Replace each trigonometric function with its third Maclaurin polynomial, and repeat part (b).

d. The actual value is $f(0.1) = -1.99899998$. Find the relative error for the values obtained in parts (b) and (c).

12. Let

$$f(x) = \frac{e^x - e^{-x}}{x}.$$

a. Find $\lim_{x\to 0}(e^x - e^{-x})/x$.

b. Use three-digit rounding arithmetic to evaluate $f(0.1)$.

c. Replace each exponential function with its third Maclaurin polynomial, and repeat part (b).

d. The actual value is $f(0.1) = 2.003335000$. Find the relative error for the values obtained in parts (b) and (c).

13. Use four-digit rounding arithmetic and the formulas (1.1), (1.2), and (1.3) to find the most accurate approximations to the roots of the following quadratic equations. Compute the absolute errors and relative errors.

a. $\frac{1}{3}x^2 - \frac{123}{4}x + \frac{1}{6} = 0$

b. $\frac{1}{3}x^2 + \frac{123}{4}x - \frac{1}{6} = 0$

c. $1.002x^2 - 11.01x + 0.01265 = 0$

d. $1.002x^2 + 11.01x + 0.01265 = 0$

14. Repeat Exercise 13 using four-digit chopping arithmetic.

15. Use the 64-bit long real format to find the decimal equivalent of the following floating-point machine numbers.

a. 0 10000001010 1001001100000000000000000000000000000000000000000000

b. 1 10000001010 1001001100000000000000000000000000000000000000000000

c. 0 01111111111 0101001100000000000000000000000000000000000000000000

d. 0 01111111111 0101001100000000000000000000000000000000000000000001

16. Find the next largest and smallest machine numbers in decimal form for the numbers given in Exercise 15.

17. Suppose two points (x_0, y_0) and (x_1, y_1) are on a straight line with $y_1 \neq y_0$. Two formulas are available to find the x-intercept of the line:

$$x = \frac{x_0 y_1 - x_1 y_0}{y_1 - y_0} \quad \text{and} \quad x = x_0 - \frac{(x_1 - x_0)y_0}{y_1 - y_0}.$$

a. Show that both formulas are algebraically correct.

b. Use the data $(x_0, y_0) = (1.31, 3.24)$ and $(x_1, y_1) = (1.93, 4.76)$ and three-digit rounding arithmetic to compute the x-intercept both ways. Which method is better and why?

18. The Taylor polynomial of degree n for $f(x) = e^x$ is $\sum_{i=0}^{n}(x^i/i!)$. Use the Taylor polynomial of degree nine and three-digit chopping arithmetic to find an approximation to e^{-5} by each of the following methods.

a. $e^{-5} \approx \sum_{i=0}^{9} \frac{(-5)^i}{i!} = \sum_{i=0}^{9} \frac{(-1)^i 5^i}{i!}$

b. $e^{-5} = \frac{1}{e^5} \approx \frac{1}{\sum_{i=0}^{9} \frac{5^i}{i!}}$.

c. An approximate value of e^{-5} correct to three digits is 6.74×10^{-3}. Which formula, (a) or (b), gives the most accuracy, and why?

19. The two-by-two linear system

$$ax + by = e,$$
$$cx + dy = f,$$

where a, b, c, d, e, f are given, can be solved for x and y as follows:

$$\begin{aligned} \text{set } m &= \frac{c}{a}, \quad \text{provided } a \neq 0; \\ d_1 &= d - mb; \\ f_1 &= f - me; \\ y &= \frac{f_1}{d_1}; \\ x &= \frac{(e - by)}{a}. \end{aligned}$$

Solve the following linear systems using four-digit rounding arithmetic.

a. $1.130x - 6.990y = 14.20$
$1.013x - 6.099y = 14.22$

b. $8.110x + 12.20y = -0.1370$
$-18.11x + 112.2y = -0.1376$

20. Repeat Exercise 19 using four-digit chopping arithmetic.

21. **a.** Show that the polynomial nesting technique described in Example 6 can also be applied to the evaluation of

$$f(x) = 1.01e^{4x} - 4.62e^{3x} - 3.11e^{2x} + 12.2e^x - 1.99.$$

b. Use three-digit rounding arithmetic, the assumption that $e^{1.53} = 4.62$, and the fact that $e^{nx} = (e^x)^n$ to evaluate $f(1.53)$ as given in part (a).

c. Redo the calculation in part (b) by first nesting the calculations.

d. Compare the approximations in parts (b) and (c) to the true three-digit result $f(1.53) = -7.61$.

22. A rectangular parallelepiped has sides of length 3 cm, 4 cm, and 5 cm, measured to the nearest centimeter. What are the best upper and lower bounds for the volume of this parallelepiped? What are the best upper and lower bounds for the surface area?

23. Let $P_n(x)$ be the Maclaurin polynomial of degree n for the arctangent function. Use Maple carrying 75 decimal digits to find the value of n required to approximate π to within 10^{-25} using the following formulas.

a. $4\left[P_n\left(\frac{1}{2}\right) + P_n\left(\frac{1}{3}\right)\right]$

b. $16P_n\left(\frac{1}{5}\right) - 4P_n\left(\frac{1}{239}\right)$

24. Suppose that $fl(y)$ is a k-digit rounding approximation to y. Show that

$$\left|\frac{y - fl(y)}{y}\right| \leq 0.5 \times 10^{-k+1}.$$

[*Hint:* If $d_{k+1} < 5$, then $fl(y) = 0.d_1d_2\ldots d_k \times 10^n$. If $d_{k+1} \geq 5$, then $fl(y) = 0.d_1d_2\ldots d_k \times 10^n + 10^{n-k}$.]

25. The binomial coefficient

$$\binom{m}{k} = \frac{m!}{k!\,(m-k)!}$$

describes the number of ways of choosing a subset of k objects from a set of m elements.

a. Suppose decimal machine numbers are of the form

$$\pm 0.d_1d_2d_3d_4 \times 10^n, \quad \text{with } 1 \leq d_1 \leq 9,\ 0 \leq d_i \leq 9, \text{ if } i = 2,3,4 \quad \text{and} \quad |n| \leq 15.$$

What is the largest value of m for which the binomial coefficient $\binom{m}{k}$ can be computed for all k by the definition without causing overflow?

b. Show that $\binom{m}{k}$ can also be computed by

$$\binom{m}{k} = \left(\frac{m}{k}\right)\left(\frac{m-1}{k-1}\right)\cdots\left(\frac{m-k+1}{1}\right).$$

c. What is the largest value of m for which the binomial coefficient $\binom{m}{3}$ can be computed by the formula in part (b) without causing overflow?

d. Use the equation in (b) and four-digit chopping arithmetic to compute the number of possible 5-card hands in a 52-card deck. Compute the actual and relative errors.

26. Let $f \in C[a,b]$ be a function whose derivative exists on (a,b). Suppose f is to be evaluated at x_0 in (a,b), but instead of computing the actual value $f(x_0)$, the approximate value, $\tilde{f}(x_0)$, is the actual value of f at $x_0 + \epsilon$, that is, $\tilde{f}(x_0) = f(x_0 + \epsilon)$.

a. Use the Mean Value Theorem 1.8 to estimate the absolute error $|f(x_0) - \tilde{f}(x_0)|$ and the relative error $|f(x_0) - \tilde{f}(x_0)|/|f(x_0)|$, assuming $f(x_0) \neq 0$.

b. If $\epsilon = 5 \times 10^{-6}$ and $x_0 = 1$, find bounds for the absolute and relative errors for

i. $f(x) = e^x$

ii. $f(x) = \sin x$

c. Repeat part (b) with $\epsilon = (5 \times 10^{-6})x_0$ and $x_0 = 10$.

27. The following Maple procedure chops a floating-point number x to t digits. (Use the Shift and Enter keys at the end of each line when creating the procedure.)

```
chop := proc(x, t);
    local e, x2;
    if x = 0 then 0
    else
        e := ceil (evalf (log10(abs(x))));
        x2 := evalf (trunc (x · 10^(t−e)) · 10^(e−t));
    end if
end;
```

Verify the procedure works for the following values.

a. $x = 124.031,\ t = 5$ **b.** $x = 124.036,\ t = 5$

c. $x = -124.031,\ t = 5$ **d.** $x = -124.036,\ t = 5$

e. $x = 0.00653,\ t = 2$ **f.** $x = 0.00656,\ t = 2$

g. $x = -0.00653,\ t = 2$ **h.** $x = -0.00656,\ t = 2$

28. The opening example to this chapter described a physical experiment involving the temperature of a gas under pressure. In this application, we were given $P = 1.00$ atm, $V = 0.100$ m^3, $N = 0.00420$ mol, and $R = 0.08206$. Solving for T in the ideal gas law gives

$$T = \frac{PV}{NR} = \frac{(1.00)(0.100)}{(0.00420)(0.08206)} = 290.15 \text{ K} = 17^\circ\text{C}.$$

In the laboratory, it was found that T was 15°C under these conditions, and when the pressure was doubled and the volume halved, T was 19°C. Assume that the data are rounded values accurate to the places given, and show that both laboratory figures are within the bounds of accuracy for the ideal gas law.

1.3 Algorithms and Convergence

Throughout the text we will be examining approximation procedures, called *algorithms*, involving sequences of calculations. An **algorithm** is a procedure that describes, in an unambiguous manner, a finite sequence of steps to be performed in a specified order. The object of the algorithm is to implement a procedure to solve a problem or approximate a solution to the problem.

The use of an algorithm is as old as formal mathematics, but the name derives from the Arabic mathematician Muhammad ibn-Mŝâ al-Khwarârizmî (c. 780–850). The Latin translation of his works begins with the words "Dixit Algorismi" meaning "al-Khwarârizmî says."

We use a **pseudocode** to describe the algorithms. This pseudocode specifies the form of the input to be supplied and the form of the desired output. Not all numerical procedures give satisfactory output for arbitrarily chosen input. As a consequence, a stopping technique independent of the numerical technique is incorporated into each algorithm to avoid infinite loops.

Two punctuation symbols are used in the algorithms:

- a period (.) indicates the termination of a step,
- a semicolon (;) separates tasks within a step.

Indentation is used to indicate that groups of statements are to be treated as a single entity.

Looping techniques in the algorithms are either counter-controlled, such as,

For $i = 1, 2, \ldots, n$

Set $x_i = a + i \cdot h$

or condition-controlled, such as

While $i < N$ do Steps 3–6.

To allow for conditional execution, we use the standard

If . . . then   or   If . . . then
else

constructions.

The steps in the algorithms follow the rules of structured program construction. They have been arranged so that there should be minimal difficulty translating pseudocode into any programming language suitable for scientific applications.

The algorithms are liberally laced with comments. These are written in italics and contained within parentheses to distinguish them from the algorithmic statements.

Illustration The following algorithm computes $x_1 + x_2 + \cdots + x_N = \sum_{i=1}^{N} x_i$, given N and the numbers $x_1, x_2, \ldots, x_N$.

INPUT $N, x_1, x_2, \ldots, x_n$.

OUTPUT $SUM = \sum_{i=1}^{N} x_i$.

Step 1 Set $SUM = 0$. (*Initialize accumulator.*)

Step 2 For $i = 1, 2, \ldots, N$ do
set $SUM = SUM + x_i$. (*Add the next term.*)

Step 3 OUTPUT (SUM);
STOP. □

Example 1 The Nth Taylor polynomial for $f(x) = \ln x$ expanded about $x_0 = 1$ is

$$P_N(x) = \sum_{i=1}^{N} \frac{(-1)^{i+1}}{i}(x-1)^i,$$

and the value of ln 1.5 to eight decimal places is 0.40546511. Construct an algorithm to determine the minimal value of N required for

$$|\ln 1.5 - P_N(1.5)| < 10^{-5},$$

without using the Taylor polynomial remainder term.

Solution From calculus we know that if $\sum_{n=1}^{\infty} a_n$ is an alternating series with limit A whose terms decrease in magnitude, then A and the Nth partial sum $A_N = \sum_{n=1}^{N} a_n$ differ by less than the magnitude of the $(N+1)$st term; that is,

$$|A - A_N| \leq |a_{N+1}|.$$

The following algorithm uses this bound.

INPUT value x, tolerance TOL, maximum number of iterations M.
OUTPUT degree N of the polynomial or a message of failure.
Step 1 Set $N = 1$;
$y = x - 1$;
$SUM = 0$;
$POWER = y$;
$TERM = y$;
$SIGN = -1$. (*Used to implement alternation of signs.*)

Step 2 While $N \leq M$ do Steps 3–5.

Step 3 Set $SIGN = -SIGN$; (*Alternate the signs.*)
$SUM = SUM + SIGN \cdot TERM$; (*Accumulate the terms.*)
$POWER = POWER \cdot y$;
$TERM = POWER/(N+1)$. (*Calculate the next term.*)
Step 4 If $|TERM| < TOL$ then (*Test for accuracy.*)
OUTPUT (N);
STOP. (*The procedure was successful.*)

Step 5 Set $N = N + 1$. (*Prepare for the next iteration.*)

Step 6 OUTPUT ('Method Failed'); (*The procedure was unsuccessful.*)
STOP.

The input for our problem is $x = 1.5$, $TOL = 10^{-5}$, and perhaps $M = 15$. This choice of M provides an upper bound for the number of calculations we are willing to perform, recognizing that the algorithm is likely to fail if this bound is exceeded. Whether the output is a value for N or the failure message depends on the precision of the computational device. ■

Characterizing Algorithms

We will be considering a variety of approximation problems throughout the text, and in each case we need to determine approximation methods that produce dependably accurate results for a wide class of problems. Because of the differing ways in which the approximation methods are derived, we need a variety of conditions to categorize their accuracy. Not all of these conditions will be appropriate for any particular problem.

One criterion we will impose on an algorithm whenever possible is that small changes in the initial data produce correspondingly small changes in the final results. An algorithm that satisfies this property is called **stable**; otherwise it is **unstable**. Some algorithms are stable only for certain choices of initial data, and are called **conditionally stable**. We will characterize the stability properties of algorithms whenever possible.

The word *stable* has the same root as the words stand and standard. In mathematics, the term stable applied to a problem indicates that a small change in initial data or conditions does not result in a dramatic change in the solution to the problem.

To further consider the subject of round-off error growth and its connection to algorithm stability, suppose an error with magnitude $E_0 > 0$ is introduced at some stage in the calculations and that the magnitude of the error after n subsequent operations is denoted by E_n. The two cases that arise most often in practice are defined as follows.

Definition 1.17 Suppose that $E_0 > 0$ denotes an error introduced at some stage in the calculations and E_n represents the magnitude of the error after n subsequent operations.

- If $E_n \approx CnE_0$, where C is a constant independent of n, then the growth of error is said to be **linear**.
- If $E_n \approx C^n E_0$, for some $C > 1$, then the growth of error is called **exponential**. ■

Linear growth of error is usually unavoidable, and when C and E_0 are small the results are generally acceptable. Exponential growth of error should be avoided, because the term C^n becomes large for even relatively small values of n. This leads to unacceptable inaccuracies, regardless of the size of E_0. As a consequence, an algorithm that exhibits linear growth of error is stable, whereas an algorithm exhibiting exponential error growth is unstable. (See Figure 1.12.)

Illustration For any constants c_1 and c_2,

$$p_n = c_1\left(\frac{1}{3}\right)^n + c_2 3^n,$$

is a solution to the recursive equation

$$p_n = \frac{10}{3}p_{n-1} - p_{n-2}, \quad \text{for } n = 2, 3, \ldots.$$

Figure 1.12

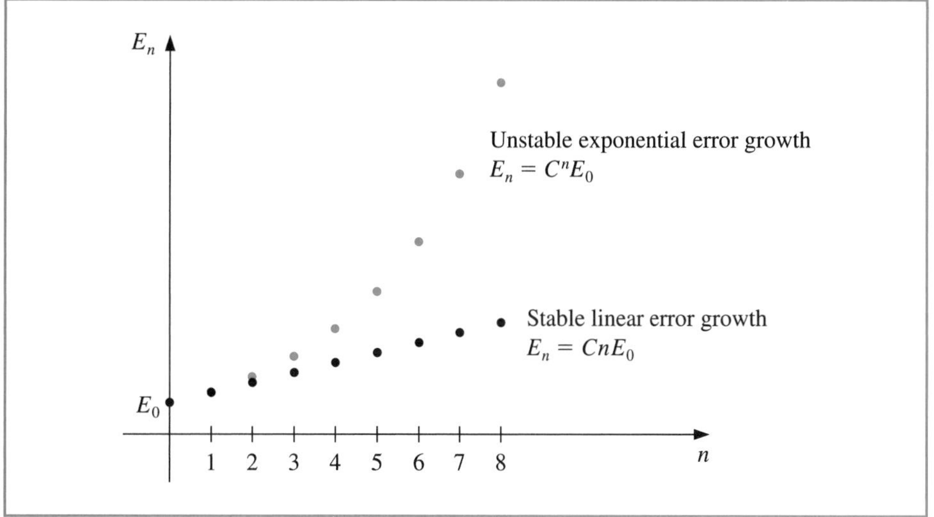

This can be seen by noting that

$$\begin{aligned}\frac{10}{3}p_{n-1} - p_{n-2} &= \frac{10}{3}\left[c_1\left(\frac{1}{3}\right)^{n-1} + c_2 3^{n-1}\right] - \left[c_1\left(\frac{1}{3}\right)^{n-2} + c_2 3^{n-2}\right] \\ &= c_1\left(\frac{1}{3}\right)^{n-2}\left[\frac{10}{3}\cdot\frac{1}{3} - 1\right] + c_2 3^{n-2}\left[\frac{10}{3}\cdot 3 - 1\right] \\ &= c_1\left(\frac{1}{3}\right)^{n-2}\left(\frac{1}{9}\right) + c_2 3^{n-2}(9) = c_1\left(\frac{1}{3}\right)^n + c_2 3^n = p_n.\end{aligned}$$

Suppose that we are given $p_0 = 1$ and $p_1 = \frac{1}{3}$. This determines unique values for the constants as $c_1 = 1$ and $c_2 = 0$. So $p_n = \left(\frac{1}{3}\right)^n$ for all n.

If five-digit rounding arithmetic is used to compute the terms of the sequence given by this equation, then $\hat{p}_0 = 1.0000$ and $\hat{p}_1 = 0.33333$, which requires modifying the constants to $\hat{c}_1 = 1.0000$ and $\hat{c}_2 = -0.12500 \times 10^{-5}$. The sequence $\{\hat{p}_n\}_{n=0}^{\infty}$ generated is then given by

$$\hat{p}_n = 1.0000\left(\frac{1}{3}\right)^n - 0.12500 \times 10^{-5}(3)^n,$$

which has round-off error,

$$p_n - \hat{p}_n = 0.12500 \times 10^{-5}(3^n),$$

This procedure is unstable because the error grows *exponentially* with n, which is reflected in the extreme inaccuracies after the first few terms, as shown in Table 1.5 on page 36.

Now consider this recursive equation:

$$p_n = 2p_{n-1} - p_{n-2}, \quad \text{for } n = 2, 3, \ldots.$$

It has the solution $p_n = c_1 + c_2 n$ for any constants c_1 and c_2, because

$$\begin{aligned}2p_{n-1} - p_{n-2} &= 2(c_1 + c_2(n-1)) - (c_1 + c_2(n-2)) \\ &= c_1(2-1) + c_2(2n - 2 - n + 2) = c_1 + c_2 n = p_n.\end{aligned}$$

Table 1.5

n	Computed $\hat{p}_n$	Correct p_n	Relative Error
0	0.10000×10^1	0.10000×10^1	
1	0.33333×10^0	0.33333×10^0	
2	0.11110×10^0	0.11111×10^0	9×10^{-5}
3	0.37000×10^{-1}	0.37037×10^{-1}	1×10^{-3}
4	0.12230×10^{-1}	0.12346×10^{-1}	9×10^{-3}
5	0.37660×10^{-2}	0.41152×10^{-2}	8×10^{-2}
6	0.32300×10^{-3}	0.13717×10^{-2}	8×10^{-1}
7	-0.26893×10^{-2}	0.45725×10^{-3}	7×10^0
8	-0.92872×10^{-2}	0.15242×10^{-3}	6×10^1

If we are given $p_0 = 1$ and $p_1 = \frac{1}{3}$, then constants in this equation are uniquely determined to be $c_1 = 1$ and $c_2 = -\frac{2}{3}$. This implies that $p_n = 1 - \frac{2}{3}n$.

If five-digit rounding arithmetic is used to compute the terms of the sequence given by this equation, then $\hat{p}_0 = 1.0000$ and $\hat{p}_1 = 0.33333$. As a consequence, the five-digit rounding constants are $\hat{c}_1 = 1.0000$ and $\hat{c}_2 = -0.66667$. Thus

$$\hat{p}_n = 1.0000 - 0.66667n,$$

which has round-off error

$$p_n - \hat{p}_n = \left(0.66667 - \frac{2}{3}\right)n.$$

This procedure is stable because the error grows grows *linearly* with n, which is reflected in the approximations shown in Table 1.6. □

Table 1.6

n	Computed $\hat{p}_n$	Correct p_n	Relative Error
0	0.10000×10^1	0.10000×10^1	
1	0.33333×10^0	0.33333×10^0	
2	-0.33330×10^0	-0.33333×10^0	9×10^{-5}
3	-0.10000×10^1	-0.10000×10^1	0
4	-0.16667×10^1	-0.16667×10^1	0
5	-0.23334×10^1	-0.23333×10^1	4×10^{-5}
6	-0.30000×10^1	-0.30000×10^1	0
7	-0.36667×10^1	-0.36667×10^1	0
8	-0.43334×10^1	-0.43333×10^1	2×10^{-5}

The effects of round-off error can be reduced by using high-order-digit arithmetic such as the double- or multiple-precision option available on most computers. Disadvantages in using double-precision arithmetic are that it takes more computation time and the growth of round-off error is not entirely eliminated.

One approach to estimating round-off error is to use interval arithmetic (that is, to retain the largest and smallest possible values at each step), so that, in the end, we obtain an interval that contains the true value. Unfortunately, a very small interval may be needed for reasonable implementation.

Rates of Convergence

Since iterative techniques involving sequences are often used, this section concludes with a brief discussion of some terminology used to describe the rate at which convergence occurs. In general, we would like the technique to converge as rapidly as possible. The following definition is used to compare the convergence rates of sequences.

Definition 1.18 Suppose $\{\beta_n\}_{n=1}^{\infty}$ is a sequence known to converge to zero, and $\{\alpha_n\}_{n=1}^{\infty}$ converges to a number α. If a positive constant K exists with

$$|\alpha_n - \alpha| \leq K|\beta_n|, \quad \text{for large } n,$$

then we say that $\{\alpha_n\}_{n=1}^{\infty}$ converges to α with **rate, or order, of convergence** $O(\beta_n)$. (This expression is read "big oh of β_n".) It is indicated by writing $\alpha_n = \alpha + O(\beta_n)$. ■

Although Definition 1.18 permits $\{\alpha_n\}_{n=1}^{\infty}$ to be compared with an arbitrary sequence $\{\beta_n\}_{n=1}^{\infty}$, in nearly every situation we use

$$\beta_n = \frac{1}{n^p},$$

for some number $p > 0$. We are generally interested in the largest value of p with $\alpha_n = \alpha + O(1/n^p)$.

Example 2 Suppose that, for $n \geq 1$,

$$\alpha_n = \frac{n+1}{n^2} \quad \text{and} \quad \hat{\alpha}_n = \frac{n+3}{n^3}.$$

Both $\lim_{n\to\infty} \alpha_n = 0$ and $\lim_{n\to\infty} \hat{\alpha}_n = 0$, but the sequence $\{\hat{\alpha}_n\}$ converges to this limit much faster than the sequence $\{\alpha_n\}$. Using five-digit rounding arithmetic we have the values shown in Table 1.7. Determine rates of convergence for these two sequences.

Table 1.7

n	1	2	3	4	5	6	7
α_n	2.00000	0.75000	0.44444	0.31250	0.24000	0.19444	0.16327
$\hat{\alpha}_n$	4.00000	0.62500	0.22222	0.10938	0.064000	0.041667	0.029155

There are numerous other ways of describing the growth of sequences and functions, some of which require bounds both above and below the sequence or function under consideration. Any good book that analyzes algorithms, for example [CLRS], will include this information.

Solution Define the sequences $\beta_n = 1/n$ and $\hat{\beta}_n = 1/n^2$. Then

$$|\alpha_n - 0| = \frac{n+1}{n^2} \leq \frac{n+n}{n^2} = 2 \cdot \frac{1}{n} = 2\beta_n$$

and

$$|\hat{\alpha}_n - 0| = \frac{n+3}{n^3} \leq \frac{n+3n}{n^3} = 4 \cdot \frac{1}{n^2} = 4\hat{\beta}_n.$$

Hence the rate of convergence of $\{\alpha_n\}$ to zero is similar to the convergence of $\{1/n\}$ to zero, whereas $\{\hat{\alpha}_n\}$ converges to zero at a rate similar to the more rapidly convergent sequence $\{1/n^2\}$. We express this by writing

$$\alpha_n = 0 + O\left(\frac{1}{n}\right) \quad \text{and} \quad \hat{\alpha}_n = 0 + O\left(\frac{1}{n^2}\right). \qquad ■$$

We also use the O (*big oh*) notation to describe the rate at which functions converge.

Definition 1.19 Suppose that $\lim_{h\to 0} G(h) = 0$ and $\lim_{h\to 0} F(h) = L$. If a positive constant K exists with

$$|F(h) - L| \le K|G(h)|, \quad \text{for sufficiently small } h,$$

then we write $F(h) = L + O(G(h))$. ■

The functions we use for comparison generally have the form $G(h) = h^p$, where $p > 0$. We are interested in the largest value of p for which $F(h) = L + O(h^p)$.

Example 3 Use the third Taylor polynomial about $h = 0$ to show that $\cos h + \frac{1}{2}h^2 = 1 + O(h^4)$.

Solution In Example 3(b) of Section 1.1 we found that this polynomial is

$$\cos h = 1 - \frac{1}{2}h^2 + \frac{1}{24}h^4 \cos \tilde{\xi}(h),$$

for some number $\tilde{\xi}(h)$ between zero and h. This implies that

$$\cos h + \frac{1}{2}h^2 = 1 + \frac{1}{24}h^4 \cos \tilde{\xi}(h).$$

Hence

$$\left|\left(\cos h + \frac{1}{2}h^2\right) - 1\right| = \left|\frac{1}{24}\cos \tilde{\xi}(h)\right| h^4 \le \frac{1}{24}h^4,$$

so as $h \to 0$, $\cos h + \frac{1}{2}h^2$ converges to its limit, 1, about as fast as h^4 converges to 0. That is,

$$\cos h + \frac{1}{2}h^2 = 1 + O(h^4). \qquad ■$$

Maple uses the O notation to indicate the form of the error in Taylor polynomials and in other situations. For example, at the end of Section 1.1 the third Taylor polynomial for $f(x) = \cos(x)$ was found by first defining

$f := \cos(x)$

and then calling the third Taylor polynomial with

$taylor(f, x = 0, 4)$

Maple responds with

$$1 - \frac{1}{2}x^2 + O(x^4)$$

to indicate that the lowest term in the truncation error is x^4.

EXERCISE SET 1.3

1. a. Use three-digit chopping arithmetic to compute the sum $\sum_{i=1}^{10}(1/i^2)$ first by $\frac{1}{1}+\frac{1}{4}+\cdots+\frac{1}{100}$ and then by $\frac{1}{100}+\frac{1}{81}+\cdots+\frac{1}{1}$. Which method is more accurate, and why?

   b. Write an algorithm to sum the finite series $\sum_{i=1}^{N} x_i$ in reverse order.

2. The number e is defined by $e=\sum_{n=0}^{\infty}(1/n!)$, where $n!=n(n-1)\cdots 2\cdot 1$ for $n\neq 0$ and $0!=1$. Use four-digit chopping arithmetic to compute the following approximations to e, and determine the absolute and relative errors.

   a. $e\approx\sum_{n=0}^{5}\frac{1}{n!}$

   b. $e\approx\sum_{j=0}^{5}\frac{1}{(5-j)!}$

   c. $e\approx\sum_{n=0}^{10}\frac{1}{n!}$

   d. $e\approx\sum_{j=0}^{10}\frac{1}{(10-j)!}$

3. The Maclaurin series for the arctangent function converges for $-1<x\leq 1$ and is given by

$$\arctan x=\lim_{n\to\infty}P_n(x)=\lim_{n\to\infty}\sum_{i=1}^{n}(-1)^{i+1}\frac{x^{2i-1}}{2i-1}.$$

   a. Use the fact that $\tan\pi/4=1$ to determine the number of n terms of the series that need to be summed to ensure that $|4P_n(1)-\pi|<10^{-3}$.

   b. The C++ programming language requires the value of π to be within 10^{-10}. How many terms of the series would we need to sum to obtain this degree of accuracy?

4. Exercise 3 details a rather inefficient means of obtaining an approximation to π. The method can be improved substantially by observing that $\pi/4=\arctan\frac{1}{2}+\arctan\frac{1}{3}$ and evaluating the series for the arctangent at $\frac{1}{2}$ and at $\frac{1}{3}$. Determine the number of terms that must be summed to ensure an approximation to π to within 10^{-3}.

5. Another formula for computing π can be deduced from the identity $\pi/4=4\arctan\frac{1}{5}-\arctan\frac{1}{239}$. Determine the number of terms that must be summed to ensure an approximation to π to within 10^{-3}.

6. Find the rates of convergence of the following sequences as $n\to\infty$.

   a. $\lim_{n\to\infty}\sin\frac{1}{n^2}=0$

   b. $\lim_{n\to\infty}\sin\frac{1}{n}=0$

   c. $\lim_{n\to\infty}\left(\sin\frac{1}{n}\right)^2=0$

   d. $\lim_{n\to\infty}[\ln(n+1)-\ln(n)]=0$

7. Find the rates of convergence of the following functions as $h\to 0$.

   a. $\lim_{h\to 0}\frac{\sin h}{h}=1$

   b. $\lim_{h\to 0}\frac{1-\cos h}{h}=0$

   c. $\lim_{h\to 0}\frac{1-e^h}{h}=-1$

   d. $\lim_{h\to 0}\frac{\sin h-h\cos h}{h}=0$

8. a. How many multiplications and additions are required to determine a sum of the form

$$\sum_{i=1}^{n}\sum_{j=1}^{i}a_ib_j?$$

   b. Modify the sum in part (a) to an equivalent form that reduces the number of computations.

9. Let $P(x)=a_nx^n+a_{n-1}x^{n-1}+\cdots+a_1x+a_0$ be a polynomial, and let x_0 be given. Construct an algorithm to evaluate $P(x_0)$ using nested multiplication.

10. Equations (1.2) and (1.3) in Section 1.2 give alternative formulas for the roots x_1 and x_2 of $ax^2+bx+c=0$. Construct an algorithm with input a,b,c and output x_1, x_2 that computes the roots x_1 and x_2 (which may be equal or be complex conjugates) using the best formula for each root.

11. Construct an algorithm that has as input an integer $n\geq 1$, numbers $x_0,x_1,\ldots,x_n$, and a number x and that produces as output the product $(x-x_0)(x-x_1)\cdots(x-x_n)$.

12. Assume that

$$\frac{1-2x}{1-x+x^2}+\frac{2x-4x^3}{1-x^2+x^4}+\frac{4x^3-8x^7}{1-x^4+x^8}+\cdots=\frac{1+2x}{1+x+x^2},$$

for $x < 1$, and let $x = 0.25$. Write and execute an algorithm that determines the number of terms needed on the left side of the equation so that the left side differs from the right side by less than 10^{-6}.

13. **a.** Suppose that $0 < q < p$ and that $\alpha_n = \alpha + O\left(n^{-p}\right)$. Show that $\alpha_n = \alpha + O\left(n^{-q}\right)$.

b. Make a table listing $1/n$, $1/n^2$, $1/n^3$, and $1/n^4$ for $n = 5, 10, 100$, and 1000, and discuss the varying rates of convergence of these sequences as n becomes large.

14. **a.** Suppose that $0 < q < p$ and that $F(h) = L + O\left(h^p\right)$. Show that $F(h) = L + O\left(h^q\right)$.

b. Make a table listing h, h^2, h^3, and h^4 for $h = 0.5, 0.1, 0.01$, and 0.001, and discuss the varying rates of convergence of these powers of h as h approaches zero.

15. Suppose that as x approaches zero,

$$F_1(x) = L_1 + O(x^\alpha) \quad \text{and} \quad F_2(x) = L_2 + O(x^\beta).$$

Let c_1 and c_2 be nonzero constants, and define

$$F(x) = c_1F_1(x) + c_2F_2(x) \quad \text{and}$$

$$G(x) = F_1(c_1x) + F_2(c_2x).$$

Show that if γ = minimum $\{\alpha, \beta\}$, then as x approaches zero,

a. $F(x) = c_1L_1 + c_2L_2 + O(x^\gamma)$

b. $G(x) = L_1 + L_2 + O(x^\gamma)$.

16. The sequence $\{F_n\}$ described by $F_0 = 1, F_1 = 1$, and $F_{n+2} = F_n + F_{n+1}$, if $n \geq 0$, is called a *Fibonacci sequence*. Its terms occur naturally in many botanical species, particularly those with petals or scales arranged in the form of a logarithmic spiral. Consider the sequence $\{x_n\}$, where $x_n = F_{n+1}/F_n$. Assuming that $\lim_{n\to\infty} x_n = x$ exists, show that $x = (1+\sqrt{5})/2$. This number is called the *golden ratio*.

17. The Fibonacci sequence also satisfies the equation

$$F_n \equiv \tilde{F}_n = \frac{1}{\sqrt{5}}\left[\left(\frac{1+\sqrt{5}}{2}\right)^n - \left(\frac{1-\sqrt{5}}{2}\right)^n\right].$$

a. Write a Maple procedure to calculate F_{100}.

b. Use Maple with the default value of *Digits* followed by *evalf* to calculate $\tilde{F}_{100}$.

c. Why is the result from part (a) more accurate than the result from part (b)?

d. Why is the result from part (b) obtained more rapidly than the result from part (a)?

e. What results when you use the command *simplify* instead of *evalf* to compute $\tilde{F}_{100}$?

18. The harmonic series $1 + \frac{1}{2} + \frac{1}{3} + \frac{1}{4} + \cdots$ diverges, but the sequence $\gamma_n = 1 + \frac{1}{2} + \cdots + \frac{1}{n} - \ln n$ converges, since $\{\gamma_n\}$ is a bounded, nonincreasing sequence. The limit $\gamma = 0.5772156649\ldots$ of the sequence $\{\gamma_n\}$ is called Euler's constant.

a. Use the default value of *Digits* in Maple to determine the value of n for γ_n to be within 10^{-2} of γ.

b. Use the default value of *Digits* in Maple to determine the value of n for γ_n to be within 10^{-3} of γ.

c. What happens if you use the default value of *Digits* in Maple to determine the value of n for γ_n to be within 10^{-4} of γ?

1.4 Numerical Software

Computer software packages for approximating the numerical solutions to problems are available in many forms. On our web site for the book

`http://www.math.ysu.edu/~faires/Numerical-Analysis/Programs.html`

we have provided programs written in C, FORTRAN, Maple, Mathematica, MATLAB, and Pascal, as well as JAVA applets. These can be used to solve the problems given in the examples and exercises, and will give satisfactory results for most problems that you may need to solve. However, they are what we call *special-purpose* programs. We use this term to distinguish these programs from those available in the standard mathematical subroutine libraries. The programs in these packages will be called *general purpose*.

The programs in general-purpose software packages differ in their intent from the algorithms and programs provided with this book. General-purpose software packages consider ways to reduce errors due to machine rounding, underflow, and overflow. They also describe the range of input that will lead to results of a certain specified accuracy. These are machine-dependent characteristics, so general-purpose software packages use parameters that describe the floating-point characteristics of the machine being used for computations.

Illustration To illustrate some differences between programs included in a general-purpose package and a program that we would provide for use in this book, let us consider an algorithm that computes the Euclidean norm of an n-dimensional vector $\mathbf{x} = (x_1, x_2, \ldots, x_n)^t$. This norm is often required within larger programs and is defined by

$$||\mathbf{x}||_2 = \left[\sum_{i=1}^{n} x_i^2\right]^{1/2}.$$

The norm gives a measure for the distance from the vector $\mathbf{x}$ to the vector $\mathbf{0}$. For example, the vector $\mathbf{x} = (2, 1, 3, -2, -1)^t$ has

$$||\mathbf{x}||_2 = [2^2 + 1^2 + 3^2 + (-2)^2 + (-1)^2]^{1/2} = \sqrt{19},$$

so its distance from $\mathbf{0} = (0, 0, 0, 0, 0)^t$ is $\sqrt{19} \approx 4.36$.

An algorithm of the type we would present for this problem is given here. It includes no machine-dependent parameters and provides no accuracy assurances, but it will give accurate results "most of the time."

INPUT $n, x_1, x_2, \ldots, x_n$.

OUTPUT $NORM$.

Step 1 Set $SUM = 0$.

Step 2 For $i = 1, 2, \ldots, n$ set $SUM = SUM + x_i^2$.

Step 3 Set $NORM = SUM^{1/2}$.

Step 4 OUTPUT ($NORM$);
STOP. □

A program based on our algorithm is easy to write and understand. However, the program could fail to give sufficient accuracy for a number of reasons. For example, the magnitude of some of the numbers might be too large or too small to be accurately represented in the floating-point system of the computer. Also, this order for performing the calculations might not produce the most accurate results, or the standard software square-root routine might not be the best available for the problem. Matters of this type are considered by algorithm designers when writing programs for general-purpose software. These programs are often used as subprograms for solving larger problems, so they must incorporate controls that we will not need.

General Purpose Algorithms

Let us now consider an algorithm for a general-purpose software program for computing the Euclidean norm. First, it is possible that although a component x_i of the vector is within the range of the machine, the square of the component is not. This can occur when some $|x_i|$ is so small that x_i^2 causes underflow or when some $|x_i|$ is so large that x_i^2 causes overflow. It is also possible for all these terms to be within the range of the machine, but overflow occurs from the addition of a square of one of the terms to the previously computed sum.

Accuracy criteria depend on the machine on which the calculations are being performed, so machine-dependent parameters are incorporated into the algorithm. Suppose we are working on a hypothetical computer with base 10, having $t \geq 4$ digits of precision, a minimum exponent *emin*, and a maximum exponent *emax*. Then the set of floating-point numbers in this machine consists of 0 and the numbers of the form

$$x = f \cdot 10^e, \quad \text{where} \quad f = \pm(f_1 10^{-1} + f_2 10^{-2} + \cdots + f_t 10^{-t}),$$

where $1 \leq f_1 \leq 9$ and $0 \leq f_i \leq 9$, for each $i = 2, \ldots, t$, and where $emin \leq e \leq emax$. These constraints imply that the smallest positive number represented in the machine is $\sigma = 10^{emin-1}$, so any computed number x with $|x| < \sigma$ causes underflow and results in x being set to 0. The largest positive number is $\lambda = (1 - 10^{-t})10^{emax}$, and any computed number x with $|x| > \lambda$ causes overflow. When underflow occurs, the program will continue, often without a significant loss of accuracy. If overflow occurs, the program will fail.

The algorithm assumes that the floating-point characteristics of the machine are described using parameters N, s, S, y, and Y. The maximum number of entries that can be summed with at least $t/2$ digits of accuracy is given by N. This implies the algorithm will proceed to find the norm of a vector $\mathbf{x} = (x_1, x_2, \ldots, x_n)^t$ only if $n \leq N$. To resolve the underflow-overflow problem, the nonzero floating-point numbers are partitioned into three groups:

- small-magnitude numbers x, those satisfying $0 < |x| < y$;
- medium-magnitude numbers x, where $y \leq |x| < Y$;
- large-magnitude numbers x, where $Y \leq |x|$.

The parameters y and Y are chosen so that there will be no underflow-overflow problem in squaring and summing the medium-magnitude numbers. Squaring small-magnitude numbers can cause underflow, so a scale factor S much greater than 1 is used with the result that $(Sx)^2$ avoids the underflow even when x^2 does not. Summing and squaring numbers having a large magnitude can cause overflow. So in this case, a positive scale factor s much smaller than 1 is used to ensure that $(sx)^2$ does not cause overflow when calculated or incorporated into a sum, even though x^2 would.

To avoid unnecessary scaling, y and Y are chosen so that the range of medium-magnitude numbers is as large as possible. The algorithm that follows is a modification of one described in [Brow, W], p. 471. It incorporates a procedure for adding scaled components of the vector that are small in magnitude until a component with medium magnitude is encountered. It then unscales the previous sum and continues by squaring and summing small and medium numbers until a component with a large magnitude is encountered. Once a component with large magnitude appears, the algorithm scales the previous sum and proceeds to scale, square, and sum the remaining numbers.

The algorithm assumes that, in transition from small to medium numbers, unscaled small numbers are negligible when compared to medium numbers. Similarly, in transition from medium to large numbers, unscaled medium numbers are negligible when compared to large numbers. Thus, the choices of the scaling parameters must be made so that numbers are equated to 0 only when they are truly negligible. Typical relationships between the machine characteristics as described by t, σ, λ, *emin*, *emax*, and the algorithm parameters N, s, S, y, and Y are given after the algorithm.

The algorithm uses three flags to indicate the various stages in the summation process. These flags are given initial values in Step 3 of the algorithm. FLAG 1 is 1 until a medium or large component is encountered; then it is changed to 0. FLAG 2 is 0 while small numbers are being summed, changes to 1 when a medium number is first encountered, and changes back to 0 when a large number is found. FLAG 3 is initially 0 and changes to 1 when a large number is first encountered. Step 3 also introduces the flag DONE, which is 0 until the calculations are complete, and then changes to 1.

INPUT $N, s, S, y, Y, \lambda, n, x_1, x_2, \ldots, x_n$.

OUTPUT *NORM* or an appropriate error message.

Step 1 If $n \le 0$ then OUTPUT ('The integer n must be positive.');
STOP.

Step 2 If $n \ge N$ then OUTPUT ('The integer n is too large.');
STOP.

Step 3 Set $SUM = 0$;
$FLAG1 = 1$; (*The small numbers are being summed.*)
$FLAG2 = 0$;
$FLAG3 = 0$;
$DONE = 0$;
$i = 1$.

Step 4 While ($i \le n$ and $FLAG1 = 1$) do Step 5.

Step 5 If $|x_i| < y$ then set $SUM = SUM + (Sx_i)^2$;
$i = i + 1$
else set $FLAG1 = 0$. (*A non-small number encountered.*)

Step 6 If $i > n$ then set $NORM = (SUM)^{1/2}/S$;
$DONE = 1$
else set $SUM = (SUM/S)/S$; (*Scale for larger numbers.*)
$FLAG2 = 1$.

Step 7 While ($i \le n$ and $FLAG2 = 1$) do Step 8. (*Sum the medium-sized numbers.*)

Step 8 If $|x_i| < Y$ then set $SUM = SUM + x_i^2$;
$i = i + 1$
else set $FLAG2 = 0$. (*A large number has been encountered.*)

Step 9 If $DONE = 0$ then
if $i > n$ then set $NORM = (SUM)^{1/2}$;
$DONE = 1$
else set $SUM = ((SUM)s)s$; (*Scale the large numbers.*)
$FLAG3 = 1$.

Step 10 While ($i \leq n$ and $FLAG3 = 1$) do Step 11.

Step 11 Set $SUM = SUM + (sx_i)^2$; (*Sum the large numbers.*)
$i = i + 1$.

Step 12 If $DONE = 0$ then
if $SUM^{1/2} < \lambda s$ then set $NORM = (SUM)^{1/2}/s$;
$DONE = 1$
else set $SUM = \lambda$. (*The norm is too large.*)

Step 13 If $DONE = 1$ then OUTPUT ('Norm is', $NORM$)
else OUTPUT ('Norm $\geq$', $NORM$, 'overflow occurred').

Step 14 STOP.

The relationships between the machine characteristics t, σ, λ, $emin$, $emax$, and the algorithm parameters N, s, S, y, and Y were chosen in [Brow, W], p. 471, as:

$N = 10^{e_N}$, where $e_N = \lfloor (t-2)/2 \rfloor$, the greatest integer less than or equal to $(t-2)/2$;

$s = 10^{e_s}$, where $e_s = \lfloor -(emax + e_N)/2 \rfloor$;

$S = 10^{e_S}$, where $e_S = \lceil (1 - emin)/2 \rceil$, the smallest integer greater than or equal to $(1 - emin)/2$;

$y = 10^{e_y}$, where $e_y = \lceil (emin + t - 2)/2 \rceil$;

$Y = 10^{e_Y}$, where $e_Y = \lfloor (emax - e_N)/2 \rfloor$.

The first portable computer was the Osborne I, produced in 1981, although it was much larger and heaver than we would currently think of as portable.

The system FORTRAN (FORmula TRANslator) was the original general-purpose scientific programming language. It is still in wide use in situations that require intensive scientific computations.

The EISPACK project was the first large-scale numerical software package to be made available in the public domain and led the way for many packages to follow.

The reliability built into this algorithm has greatly increased the complexity compared to the algorithm given earlier in the section. In the majority of cases the special-purpose and general-purpose algorithms give identical results. The advantage of the general-purpose algorithm is that it provides security for its results.

Many forms of general-purpose numerical software are available commercially and in the public domain. Most of the early software was written for mainframe computers, and a good reference for this is *Sources and Development of Mathematical Software*, edited by Wayne Cowell [Co].

Now that personal computers are sufficiently powerful, standard numerical software is available for them. Most of this numerical software is written in FORTRAN, although some packages are written in C, C++, and FORTRAN90.

ALGOL procedures were presented for matrix computations in 1971 in [WR]. A package of FORTRAN subroutines based mainly on the ALGOL procedures was then developed into the EISPACK routines. These routines are documented in the manuals published by Springer-Verlag as part of their Lecture Notes in Computer Science series [Sm,B] and [Gar]. The FORTRAN subroutines are used to compute eigenvalues and eigenvectors for a variety of different types of matrices.

LINPACK is a package of FORTRAN subroutines for analyzing and solving systems of linear equations and solving linear least squares problems. The documentation for this package is contained in [DBMS]. A step-by-step introduction to LINPACK, EISPACK, and BLAS (Basic Linear Algebra Subprograms) is given in [CV].

The LAPACK package, first available in 1992, is a library of FORTRAN subroutines that supercedes LINPACK and EISPACK by integrating these two sets of algorithms into a unified and updated package. The software has been restructured to achieve greater efficiency on vector processors and other high-performance or shared-memory multiprocessors. LAPACK is expanded in depth and breadth in version 3.0, which is available in FORTRAN, FORTRAN90, C, C++, and JAVA. C, and JAVA are only available as language interfaces or translations of the FORTRAN libraries of LAPACK. The package BLAS is not a part of LAPACK, but the code for BLAS is distributed with LAPACK.

Other packages for solving specific types of problems are available in the public domain. As an alternative to netlib, you can use Xnetlib to search the database and retrieve software. More information can be found in the article *Software Distribution using Netlib* by Dongarra, Roman, and Wade [DRW].

Software engineering was established as a laboratory discipline during the 1970s and 1980s. EISPACK was developed at Argonne Labs and LINPACK there shortly thereafter. By the early 1980s, Argonne was internationally recognized as a world leader in symbolic and numerical computation.

These software packages are highly efficient, accurate, and reliable. They are thoroughly tested, and documentation is readily available. Although the packages are portable, it is a good idea to investigate the machine dependence and read the documentation thoroughly. The programs test for almost all special contingencies that might result in error and failures. At the end of each chapter we will discuss some of the appropriate general-purpose packages.

Commercially available packages also represent the state of the art in numerical methods. Their contents are often based on the public-domain packages but include methods in libraries for almost every type of problem.

In 1970 IMSL became the first large-scale scientific library for mainframes. Since that time, the libraries have been made available for computer systems ranging from supercomputers to personal computers.

IMSL (International Mathematical and Statistical Libraries) consists of the libraries MATH, STAT, and SFUN for numerical mathematics, statistics, and special functions, respectively. These libraries contain more than 900 subroutines originally available in FORTRAN 77 and now available in C, FORTRAN90, and JAVA. These subroutines solve the most common numerical analysis problems. The libraries are available commercially from Visual Numerics.

The packages are delivered in compiled form with extensive documentation. There is an example program for each routine as well as background reference information. IMSL contains methods for linear systems, eigensystem analysis, interpolation and approximation, integration and differentiation, differential equations, transforms, nonlinear equations, optimization, and basic matrix/vector operations. The library also contains extensive statistical routines.

The Numerical Algorithms Group (NAG) was instituted in the UK in 1971 and developed the first mathematical software library. It now has over 10,000 users world-wide and contains over 1000 mathematical and statistical functions ranging from statistical, symbolic, visualisation, and numerical simulation software, to compilers and application development tools.

The Numerical Algorithms Group (NAG) has been in existence in the United Kingdom since 1970. NAG offers more than 1000 subroutines in a FORTRAN 77 library, about 400 subroutines in a C library, more than 200 subroutines in a FORTRAN 90 library, and an MPI FORTRAN numerical library for parallel machines and clusters of workstations or personal computers. A useful introduction to the NAG routines is [Ph]. The NAG library contains routines to perform most standard numerical analysis tasks in a manner similar to those in the IMSL. It also includes some statistical routines and a set of graphic routines.

The IMSL and NAG packages are designed for the mathematician, scientist, or engineer who wishes to call high-quality C, Java, or FORTRAN subroutines from within a program. The documentation available with the commercial packages illustrates the typical driver program required to use the library routines. The next three software packages are standalone environments. When activated, the user enters commands to cause the package to solve a problem. However, each package allows programming within the command language.

MATLAB was originally written to provide easy access to matrix software developed in the LINPACK and EISPACK projects. The first version was written in the late 1970s for use in courses in matrix theory, linear algebra, and numerical analysis. There are currently more than 500,000 users of MATLAB in more than 100 countries.

MATLAB is a matrix laboratory that was originally a Fortran program published by Cleve Moler [Mo] in the 1980s. The laboratory is based mainly on the EISPACK and LINPACK subroutines, although functions such as nonlinear systems, numerical integration, cubic splines, curve fitting, optimization, ordinary differential equations, and graphical tools have been incorporated. MATLAB is currently written in C and assembler, and the PC version of this package requires a numeric coprocessor. The basic structure is to perform matrix operations, such as finding the eigenvalues of a matrix entered from the command line or from an external file via function calls. This is a powerful self-contained system that is especially useful for instruction in an applied linear algebra course.

The second package is GAUSS, a mathematical and statistical system produced by Lee E. Ediefson and Samuel D. Jones in 1985. It is coded mainly in assembler and based primarily on EISPACK and LINPACK. As in the case of MATLAB, integration/differentiation, nonlinear systems, fast Fourier transforms, and graphics are available. GAUSS is oriented less toward instruction in linear algebra and more toward statistical analysis of data. This package also uses a numeric coprocessor if one is available.

The third package is Maple, a computer algebra system developed in 1980 by the Symbolic Computational Group at the University of Waterloo. The design for the original Maple system is presented in the paper by B.W. Char, K.O. Geddes, W.M. Gentlemen, and G.H. Gonnet [CGGG].

The NAG routines are compatible with Maple beginning with version 9.0.

Maple, which is written in C, has the ability to manipulate information in a symbolic manner. This symbolic manipulation allows the user to obtain exact answers instead of numerical values. Maple can give exact answers to mathematical problems such as integrals, differential equations, and linear systems. It contains a programming structure and permits text, as well as commands, to be saved in its worksheet files. These worksheets can then be loaded into Maple and the commands executed. Because of the properties of symbolic computation, numerical computation, and worksheets, Maple is the language of choice for this text. Throughout the book Maple commands, particularly from the *NumericalAnalysis* package, will be included in the text.

Although we have chosen Maple as our standard computer algebra system, the equally popular Mathematica, released in 1988, can also be used for this purpose.

Numerous packages are available that can be classified as supercalculator packages for the PC. These should not be confused, however, with the general-purpose software listed here. If you have an interest in one of these packages, you should read *Supercalculators on the PC* by B. Simon and R. M. Wilson [SW].

Additional information about software and software libraries can be found in the books by Cody and Waite [CW] and by Kockler [Ko], and in the 1995 article by Dongarra and Walker [DW]. More information about floating-point computation can be found in the book by Chaitini-Chatelin and Frayse [CF] and the article by Goldberg [Go].

Books that address the application of numerical techniques on parallel computers include those by Schendell [Sche], Phillips and Freeman [PF], Ortega [Or1], and Golub and Ortega [GO].

CHAPTER

2 Solutions of Equations in One Variable

Introduction

The growth of a population can often be modeled over short periods of time by assuming that the population grows continuously with time at a rate proportional to the number present at that time. Suppose that $N(t)$ denotes the number in the population at time t and λ denotes the constant birth rate of the population. Then the population satisfies the differential equation

$$\frac{dN(t)}{dt} = \lambda N(t),$$

whose solution is $N(t) = N_0 e^{\lambda t}$, where N_0 denotes the initial population.

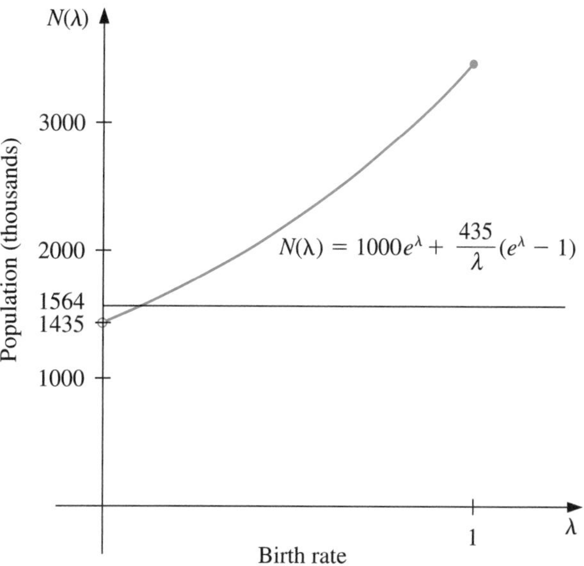

This exponential model is valid only when the population is isolated, with no immigration. If immigration is permitted at a constant rate v, then the differential equation becomes

$$\frac{dN(t)}{dt} = \lambda N(t) + v,$$

whose solution is

$$N(t) = N_0 e^{\lambda t} + \frac{v}{\lambda}(e^{\lambda t} - 1).$$

Suppose a certain population contains $N(0) = 1{,}000{,}000$ individuals initially, that 435,000 individuals immigrate into the community in the first year, and that $N(1) = 1{,}564{,}000$ individuals are present at the end of one year. To determine the birth rate of this population, we need to find λ in the equation

$$1{,}564{,}000 = 1{,}000{,}000e^{\lambda} + \frac{435{,}000}{\lambda}(e^{\lambda} - 1).$$

It is not possible to solve explicitly for λ in this equation, but numerical methods discussed in this chapter can be used to approximate solutions of equations of this type to an arbitrarily high accuracy. The solution to this particular problem is considered in Exercise 24 of Section 2.3.

2.1 The Bisection Method

In this chapter we consider one of the most basic problems of numerical approximation, the **root-finding problem**. This process involves finding a **root**, or solution, of an equation of the form $f(x) = 0$, for a given function f. A root of this equation is also called a **zero** of the function f.

The problem of finding an approximation to the root of an equation can be traced back at least to 1700 B.C.E. A cuneiform table in the Yale Babylonian Collection dating from that period gives a sexigesimal (base-60) number equivalent to 1.414222 as an approximation to $\sqrt{2}$, a result that is accurate to within 10^{-5}. This approximation can be found by applying a technique described in Exercise 19 of Section 2.2.

Bisection Technique

In computer science, the process of dividing a set continually in half to search for the solution to a problem, as the bisection method does, is known as a *binary search* procedure.

The first technique, based on the Intermediate Value Theorem, is called the **Bisection**, or **Binary-search, method**.

Suppose f is a continuous function defined on the interval $[a, b]$, with $f(a)$ and $f(b)$ of opposite sign. The Intermediate Value Theorem implies that a number p exists in (a, b) with $f(p) = 0$. Although the procedure will work when there is more than one root in the interval (a, b), we assume for simplicity that the root in this interval is unique. The method calls for a repeated halving (or bisecting) of subintervals of $[a, b]$ and, at each step, locating the half containing p.

To begin, set $a_1 = a$ and $b_1 = b$, and let p_1 be the midpoint of $[a, b]$; that is,

$$p_1 = a_1 + \frac{b_1 - a_1}{2} = \frac{a_1 + b_1}{2}.$$

- If $f(p_1) = 0$, then $p = p_1$, and we are done.
- If $f(p_1) \neq 0$, then $f(p_1)$ has the same sign as either $f(a_1)$ or $f(b_1)$.
  - If $f(p_1)$ and $f(a_1)$ have the same sign, $p \in (p_1, b_1)$. Set $a_2 = p_1$ and $b_2 = b_1$.
  - If $f(p_1)$ and $f(a_1)$ have opposite signs, $p \in (a_1, p_1)$. Set $a_2 = a_1$ and $b_2 = p_1$.

Then reapply the process to the interval $[a_2, b_2]$. This produces the method described in Algorithm 2.1. (See Figure 2.1.)

Figure 2.1

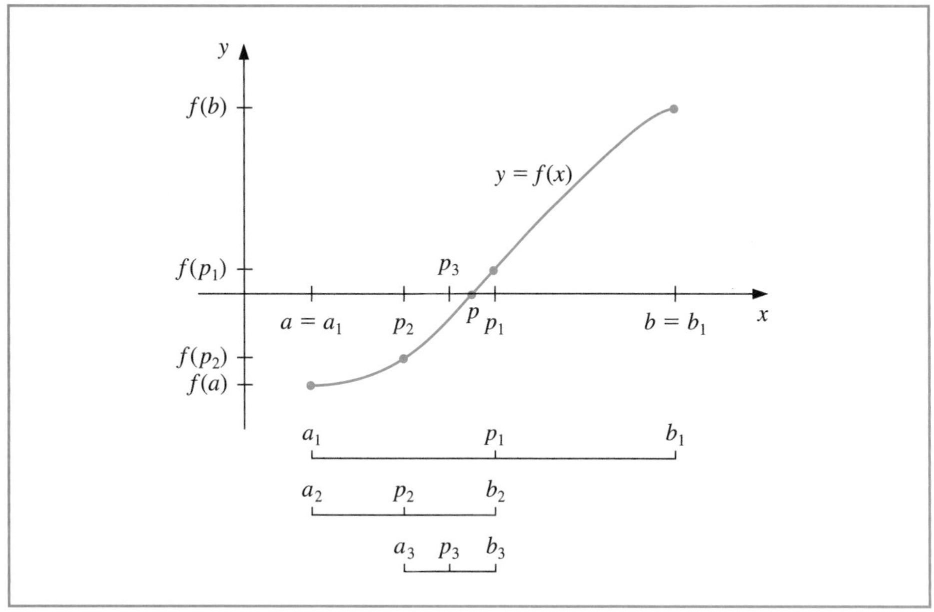

Bisection

To find a solution to $f(x) = 0$ given the continuous function f on the interval $[a, b]$, where $f(a)$ and $f(b)$ have opposite signs:

INPUT endpoints a, b; tolerance TOL; maximum number of iterations N_0.

OUTPUT approximate solution p or message of failure.

Step 1 Set $i = 1$;
$FA = f(a)$.

Step 2 While $i \le N_0$ do Steps 3–6.

Step 3 Set $p = a + (b - a)/2$; (*Compute* p_i.)
$FP = f(p)$.

Step 4 If $FP = 0$ or $(b - a)/2 < TOL$ then
OUTPUT (p); (*Procedure completed successfully.*)
STOP.

Step 5 Set $i = i + 1$.

Step 6 If $FA \cdot FP > 0$ then set $a = p$; (*Compute* a_i, b_i.)
$FA = FP$
else set $b = p$. (*FA is unchanged.*)

Step 7 OUTPUT ('Method failed after N_0 iterations, N_0 =', N_0);
(*The procedure was unsuccessful.*)
STOP.

■

Other stopping procedures can be applied in Step 4 of Algorithm 2.1 or in any of the iterative techniques in this chapter. For example, we can select a tolerance $\varepsilon > 0$ and generate $p_1, \ldots, p_N$ until one of the following conditions is met:

$$|p_N - p_{N-1}| < \varepsilon, \tag{2.1}$$

$$\frac{|p_N - p_{N-1}|}{|p_N|} < \varepsilon, \quad p_N \neq 0, \quad \text{or} \tag{2.2}$$

$$|f(p_N)| < \varepsilon. \tag{2.3}$$

Unfortunately, difficulties can arise using any of these stopping criteria. For example, there are sequences $\{p_n\}_{n=0}^{\infty}$ with the property that the differences $p_n - p_{n-1}$ converge to zero while the sequence itself diverges. (See Exercise 17.) It is also possible for $f(p_n)$ to be close to zero while p_n differs significantly from p. (See Exercise 16.) Without additional knowledge about f or p, Inequality (2.2) is the best stopping criterion to apply because it comes closest to testing relative error.

When using a computer to generate approximations, it is good practice to set an upper bound on the number of iterations. This eliminates the possibility of entering an infinite loop, a situation that can arise when the sequence diverges (and also when the program is incorrectly coded). This was done in Step 2 of Algorithm 2.1 where the bound N_0 was set and the procedure terminated if $i > N_0$.

Note that to start the Bisection Algorithm, an interval $[a, b]$ must be found with $f(a) \cdot f(b) < 0$. At each step the length of the interval known to contain a zero of f is reduced by a factor of 2; hence it is advantageous to choose the initial interval $[a, b]$ as small as possible. For example, if $f(x) = 2x^3 - x^2 + x - 1$, we have both

$$f(-4) \cdot f(4) < 0 \quad \text{and} \quad f(0) \cdot f(1) < 0,$$

so the Bisection Algorithm could be used on $[-4, 4]$ or on $[0, 1]$. Starting the Bisection Algorithm on $[0, 1]$ instead of $[-4, 4]$ will reduce by 3 the number of iterations required to achieve a specified accuracy.

The following example illustrates the Bisection Algorithm. The iteration in this example is terminated when a bound for the relative error is less than 0.0001. This is ensured by having

$$\frac{|p - p_n|}{\min\{|a_n|, |b_n|\}} < 10^{-4}.$$

Example 1 Show that $f(x) = x^3 + 4x^2 - 10 = 0$ has a root in $[1, 2]$, and use the Bisection method to determine an approximation to the root that is accurate to at least within 10^{-4}.

Solution Because $f(1) = -5$ and $f(2) = 14$ the Intermediate Value Theorem 1.11 ensures that this continuous function has a root in $[1, 2]$.

For the first iteration of the Bisection method we use the fact that at the midpoint of $[1, 2]$ we have $f(1.5) = 2.375 > 0$. This indicates that we should select the interval $[1, 1.5]$ for our second iteration. Then we find that $f(1.25) = -1.796875$ so our new interval becomes $[1.25, 1.5]$, whose midpoint is 1.375. Continuing in this manner gives the values in Table 2.1. After 13 iterations, $p_{13} = 1.365112305$ approximates the root p with an error

$$|p - p_{13}| < |b_{14} - a_{14}| = |1.365234375 - 1.365112305| = 0.000122070.$$

Since $|a_{14}| < |p|$, we have

$$\frac{|p - p_{13}|}{|p|} < \frac{|b_{14} - a_{14}|}{|a_{14}|} \leq 9.0 \times 10^{-5},$$

Table 2.1

n	a_n	b_n	p_n	$f(p_n)$
1	1.0	2.0	1.5	2.375
2	1.0	1.5	1.25	−1.79687
3	1.25	1.5	1.375	0.16211
4	1.25	1.375	1.3125	−0.84839
5	1.3125	1.375	1.34375	−0.35098
6	1.34375	1.375	1.359375	−0.09641
7	1.359375	1.375	1.3671875	0.03236
8	1.359375	1.3671875	1.36328125	−0.03215
9	1.36328125	1.3671875	1.365234375	0.000072
10	1.36328125	1.365234375	1.364257813	−0.01605
11	1.364257813	1.365234375	1.364746094	−0.00799
12	1.364746094	1.365234375	1.364990235	−0.00396
13	1.364990235	1.365234375	1.365112305	−0.00194

so the approximation is correct to at least within 10^{-4}. The correct value of p to nine decimal places is $p = 1.365230013$. Note that p_9 is closer to p than is the final approximation p_{13}. You might suspect this is true because $|f(p_9)| < |f(p_{13})|$, but we cannot be sure of this unless the true answer is known. ■

The Bisection method, though conceptually clear, has significant drawbacks. It is relatively slow to converge (that is, N may become quite large before $|p - p_N|$ is sufficiently small), and a good intermediate approximation might be inadvertently discarded. However, the method has the important property that it always converges to a solution, and for that reason it is often used as a starter for the more efficient methods we will see later in this chapter.

Theorem 2.1 Suppose that $f \in C[a, b]$ and $f(a) \cdot f(b) < 0$. The Bisection method generates a sequence $\{p_n\}_{n=1}^{\infty}$ approximating a zero p of f with

$$|p_n - p| \leq \frac{b-a}{2^n}, \quad \text{when} \quad n \geq 1.$$ ■

Proof For each $n \geq 1$, we have

$$b_n - a_n = \frac{1}{2^{n-1}}(b-a) \quad \text{and} \quad p \in (a_n, b_n).$$

Since $p_n = \frac{1}{2}(a_n + b_n)$ for all $n \geq 1$, it follows that

$$|p_n - p| \leq \frac{1}{2}(b_n - a_n) = \frac{b-a}{2^n}.$$ ■ ■ ■

Because

$$|p_n - p| \leq (b-a)\frac{1}{2^n},$$

the sequence $\{p_n\}_{n=1}^{\infty}$ converges to p with rate of convergence $O\left(\frac{1}{2^n}\right)$; that is,

$$p_n = p + O\left(\frac{1}{2^n}\right).$$

It is important to realize that Theorem 2.1 gives only a bound for approximation error and that this bound might be quite conservative. For example, this bound applied to the problem in Example 1 ensures only that

$$|p - p_9| \le \frac{2-1}{2^9} \approx 2 \times 10^{-3},$$

but the actual error is much smaller:

$$|p - p_9| = |1.365230013 - 1.365234375| \approx 4.4 \times 10^{-6}.$$

Example 2 Determine the number of iterations necessary to solve $f(x) = x^3 + 4x^2 - 10 = 0$ with accuracy 10^{-3} using $a_1 = 1$ and $b_1 = 2$.

Solution We we will use logarithms to find an integer N that satisfies

$$|p_N - p| \le 2^{-N}(b - a) = 2^{-N} < 10^{-3}.$$

Logarithms to any base would suffice, but we will use base-10 logarithms because the tolerance is given as a power of 10. Since $2^{-N} < 10^{-3}$ implies that $\log_{10} 2^{-N} < \log_{10} 10^{-3} = -3$, we have

$$-N \log_{10} 2 < -3 \quad \text{and} \quad N > \frac{3}{\log_{10} 2} \approx 9.96.$$

Hence, ten iterations will ensure an approximation accurate to within 10^{-3}.

Table 2.1 shows that the value of $p_9 = 1.365234375$ is accurate to within 10^{-4}. Again, it is important to keep in mind that the error analysis gives only a bound for the number of iterations. In many cases this bound is much larger than the actual number required. ■

Maple has a *NumericalAnalysis* package that implements many of the techniques we will discuss, and the presentation and examples in the package are closely aligned with this text. The Bisection method in this package has a number of options, some of which we will now consider. In what follows, Maple code is given in *black italic* type and Maple response in cyan.

Load the *NumericalAnalysis* package with the command

with(*Student*[*NumericalAnalysis*])

which gives access to the procedures in the package. Define the function with

$f := x^3 + 4x^2 - 10$

and use

Bisection $(f, x = [1, 2], \textit{tolerance} = 0.005)$

Maple returns

1.363281250

Note that the value that is output is the same as p_8 in Table 2.1.

The sequence of bisection intervals can be output with the command

Bisection $(f, x = [1, 2], \textit{tolerance} = 0.005, \textit{output} = \textit{sequence})$

and Maple returns the intervals containing the solution together with the solution

[1., 2.], [1., 1.500000000], [1.250000000, 1.500000000], [1.250000000, 1.375000000],

[1.312500000, 1.375000000], [1.343750000, 1.375000000], [1.359375000, 1.375000000],

[1.359375000, 1.367187500], 1.363281250

The stopping criterion can also be based on relative error by choosing the option

Bisection $(f, x = [1, 2], \textit{tolerance} = 0.005, \textit{stoppingcriterion} = \textit{relative})$

Now Maple returns

$$1.363281250$$

The option *output* = *plot* given in

$Bisection\ (f, x = [1.25, 1.5], output = plot, tolerance = 0.02)$

produces the plot shown in Figure 2.2.

Figure 2.2

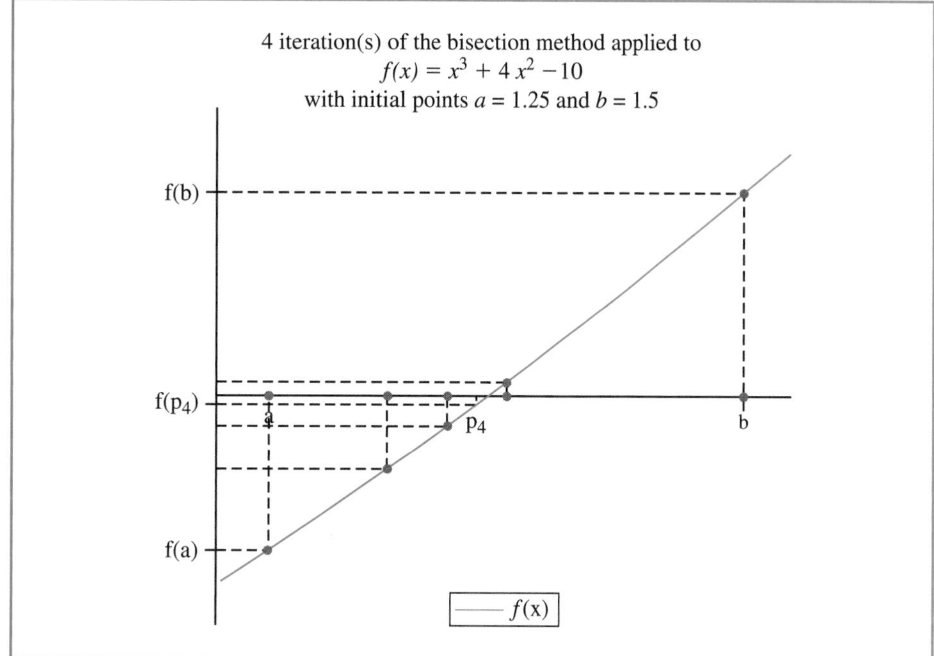

We can also set the maximum number of iterations with the option *maxiterations* = . An error message will be displayed if the stated tolerance is not met within the specified number of iterations.

The results from Bisection method can also be obtained using the command Roots. For example,

$$Roots\left(f, x = [1.0, 2.0], method = bisection, tolerance = \frac{1}{100}, output = information\right)$$

uses the Bisection method to produce the information

n	a_n	b_n	p_n	$f(p_n)$	relative error
1	1.0	2.0	1.500000000	2.37500000	0.3333333333
2	1.0	1.500000000	1.250000000	−1.796875000	0.2000000000
3	1.250000000	1.500000000	1.375000000	0.16210938	0.09090909091
4	1.250000000	1.375000000	1.312500000	−0.848388672	0.04761904762
5	1.312500000	1.375000000	1.343750000	−0.350982668	0.02325581395
6	1.343750000	1.375000000	1.359375000	−0.096408842	0.01149425287
7	1.359375000	1.375000000	1.367187500	0.03235578	0.005714285714

The bound for the number of iterations for the Bisection method assumes that the calculations are performed using infinite-digit arithmetic. When implementing the method on a computer, we need to consider the effects of round-off error. For example, the computation of the midpoint of the interval $[a_n, b_n]$ should be found from the equation

$$p_n = a_n + \frac{b_n - a_n}{2} \quad \text{instead of} \quad p_n = \frac{a_n + b_n}{2}.$$

The first equation adds a small correction, $(b_n - a_n)/2$, to the known value a_n. When $b_n - a_n$ is near the maximum precision of the machine, this correction might be in error, but the error would not significantly affect the computed value of p_n. However, when $b_n - a_n$ is near the maximum precision of the machine, it is possible for $(a_n + b_n)/2$ to return a midpoint that is not even in the interval $[a_n, b_n]$.

The Latin word *signum* means "token" or "sign". So the signum function quite naturally returns the sign of a number (unless the number is 0).

As a final remark, to determine which subinterval of $[a_n, b_n]$ contains a root of f, it is better to make use of the **signum** function, which is defined as

$$\text{sgn}(x) = \begin{cases} -1, & \text{if } x < 0, \\ 0, & \text{if } x = 0, \\ 1, & \text{if } x > 0. \end{cases}$$

The test

$$\text{sgn}\,(f(a_n))\,\text{sgn}\,(f(p_n)) < 0 \qquad \text{instead of} \qquad f(a_n)f(p_n) < 0$$

gives the same result but avoids the possibility of overflow or underflow in the multiplication of $f(a_n)$ and $f(p_n)$.

EXERCISE SET 2.1

1. Use the Bisection method to find p_3 for $f(x) = \sqrt{x} - \cos x$ on $[0, 1]$.
2. Let $f(x) = 3(x+1)(x-\frac{1}{2})(x-1)$. Use the Bisection method on the following intervals to find p_3.
   a. $[-2, 1.5]$ **b.** $[-1.25, 2.5]$
3. Use the Bisection method to find solutions accurate to within 10^{-2} for $x^4 - 2x^3 - 4x^2 + 4x + 4 = 0$ on each interval.
   a. $[-2, -1]$ **b.** $[0, 2]$ **c.** $[2, 3]$ **d.** $[-1, 0]$
4. Use the Bisection method to find solutions accurate to within 10^{-2} for $x^3 - 7x^2 + 14x - 6 = 0$ on each interval.
   a. $[0, 1]$ **b.** $[1, 3.2]$ **c.** $[3.2, 4]$
5. Use the Bisection method to find solutions, accurate to within 10^{-5} for the following problems.
   a. $3x - e^x = 0$ for $1 \le x \le 2$
   b. $2x + 3\cos x - e^x = 0$ for $0 \le x \le 1$
   c. $x^2 - 4x + 4 - \ln x = 0$ for $1 \le x \le 2$ and $2 \le x \le 4$
   d. $x + 1 - 2\sin \pi x = 0$ for $0 \le x \le 0.5$ and $0.5 \le x \le 1$
6. Use the Bisection method to find solutions accurate to within 10^{-5} for the following problems.
   a. $x - 2^{-x} = 0$ for $0 \le x \le 1$
   b. $e^x - x^2 + 3x - 2 = 0$ for $0 \le x \le 1$
   c. $2x\cos(2x) - (x+1)^2 = 0$ for $-3 \le x \le -2$ and $-1 \le x \le 0$
   d. $x\cos x - 2x^2 + 3x - 1 = 0$ for $0.2 \le x \le 0.3$ and $1.2 \le x \le 1.3$

7. a. Sketch the graphs of $y = x$ and $y = 2\sin x$.
   b. Use the Bisection method to find an approximation to within 10^{-5} to the first positive value of x with $x = 2\sin x$.
8. a. Sketch the graphs of $y = x$ and $y = \tan x$.
   b. Use the Bisection method to find an approximation to within 10^{-5} to the first positive value of x with $x = \tan x$.
9. a. Sketch the graphs of $y = e^x - 2$ and $y = \cos(e^x - 2)$.
   b. Use the Bisection method to find an approximation to within 10^{-5} to a value in $[0.5, 1.5]$ with $e^x - 2 = \cos(e^x - 2)$.
10. Let $f(x) = (x+2)(x+1)^2x(x-1)^3(x-2)$. To which zero of f does the Bisection method converge when applied on the following intervals?
    a. $[-1.5, 2.5]$ b. $[-0.5, 2.4]$ c. $[-0.5, 3]$ d. $[-3, -0.5]$
11. Let $f(x) = (x+2)(x+1)x(x-1)^3(x-2)$. To which zero of f does the Bisection method converge when applied on the following intervals?
    a. $[-3, 2.5]$ b. $[-2.5, 3]$ c. $[-1.75, 1.5]$ d. $[-1.5, 1.75]$
12. Find an approximation to $\sqrt{3}$ correct to within 10^{-4} using the Bisection Algorithm. [*Hint:* Consider $f(x) = x^2 - 3$.]
13. Find an approximation to $\sqrt[3]{25}$ correct to within 10^{-4} using the Bisection Algorithm.
14. Use Theorem 2.1 to find a bound for the number of iterations needed to achieve an approximation with accuracy 10^{-3} to the solution of $x^3 + x - 4 = 0$ lying in the interval $[1, 4]$. Find an approximation to the root with this degree of accuracy.
15. Use Theorem 2.1 to find a bound for the number of iterations needed to achieve an approximation with accuracy 10^{-4} to the solution of $x^3 - x - 1 = 0$ lying in the interval $[1, 2]$. Find an approximation to the root with this degree of accuracy.
16. Let $f(x) = (x-1)^{10}$, $p = 1$, and $p_n = 1 + 1/n$. Show that $|f(p_n)| < 10^{-3}$ whenever $n > 1$ but that $|p - p_n| < 10^{-3}$ requires that $n > 1000$.
17. Let $\{p_n\}$ be the sequence defined by $p_n = \sum_{k=1}^{n} \frac{1}{k}$. Show that $\{p_n\}$ diverges even though $\lim_{n\to\infty}(p_n - p_{n-1}) = 0$.
18. The function defined by $f(x) = \sin \pi x$ has zeros at every integer. Show that when $-1 < a < 0$ and $2 < b < 3$, the Bisection method converges to
    a. 0, if $a + b < 2$ b. 2, if $a + b > 2$ c. 1, if $a + b = 2$
19. A trough of length L has a cross section in the shape of a semicircle with radius r. (See the accompanying figure.) When filled with water to within a distance h of the top, the volume V of water is

$$V = L\left[0.5\pi r^2 - r^2 \arcsin(h/r) - h(r^2 - h^2)^{1/2}\right].$$

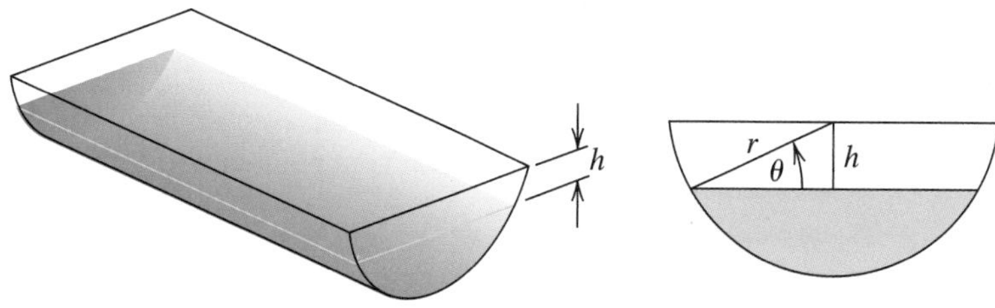

Suppose $L = 10$ ft, $r = 1$ ft, and $V = 12.4$ ft^3. Find the depth of water in the trough to within 0.01 ft.

20. A particle starts at rest on a smooth inclined plane whose angle θ is changing at a constant rate

$$\frac{d\theta}{dt} = \omega < 0.$$

At the end of t seconds, the position of the object is given by

$$x(t) = -\frac{g}{2\omega^2}\left(\frac{e^{wt} - e^{-wt}}{2} - \sin \omega t\right).$$

Suppose the particle has moved 1.7 ft in 1 s. Find, to within 10^{-5}, the rate ω at which θ changes. Assume that $g = 32.17$ ft/s^2.

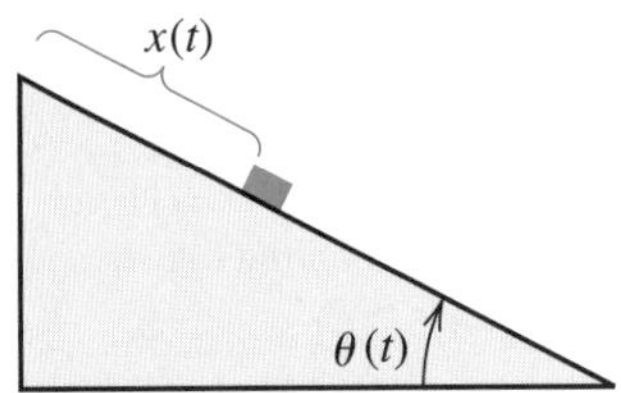

2.2 Fixed-Point Iteration

A *fixed point* for a function is a number at which the value of the function does not change when the function is applied.

Definition 2.2 The number p is a **fixed point** for a given function g if $g(p) = p$. ▪

Fixed-point results occur in many areas of mathematics, and are a major tool of economists for proving results concerning equilibria. Although the idea behind the technique is old, the terminology was first used by the Dutch mathematician L. E. J. Brouwer (1882–1962) in the early 1900s.

In this section we consider the problem of finding solutions to fixed-point problems and the connection between the fixed-point problems and the root-finding problems we wish to solve. Root-finding problems and fixed-point problems are equivalent classes in the following sense:

- Given a root-finding problem $f(p) = 0$, we can define functions g with a fixed point at p in a number of ways, for example, as

$$g(x) = x - f(x) \quad \text{or as} \quad g(x) = x + 3f(x).$$

- Conversely, if the function g has a fixed point at p, then the function defined by

$$f(x) = x - g(x)$$

  has a zero at p.

Although the problems we wish to solve are in the root-finding form, the fixed-point form is easier to analyze, and certain fixed-point choices lead to very powerful root-finding techniques.

We first need to become comfortable with this new type of problem, and to decide when a function has a fixed point and how the fixed points can be approximated to within a specified accuracy.

Example 1 Determine any fixed points of the function $g(x) = x^2 - 2$.

Solution A fixed point p for g has the property that

$$p = g(p) = p^2 - 2 \quad \text{which implies that} \quad 0 = p^2 - p - 2 = (p + 1)(p - 2).$$

A fixed point for g occurs precisely when the graph of $y = g(x)$ intersects the graph of $y = x$, so g has two fixed points, one at $p = -1$ and the other at $p = 2$. These are shown in Figure 2.3. ▪

Figure 2.3

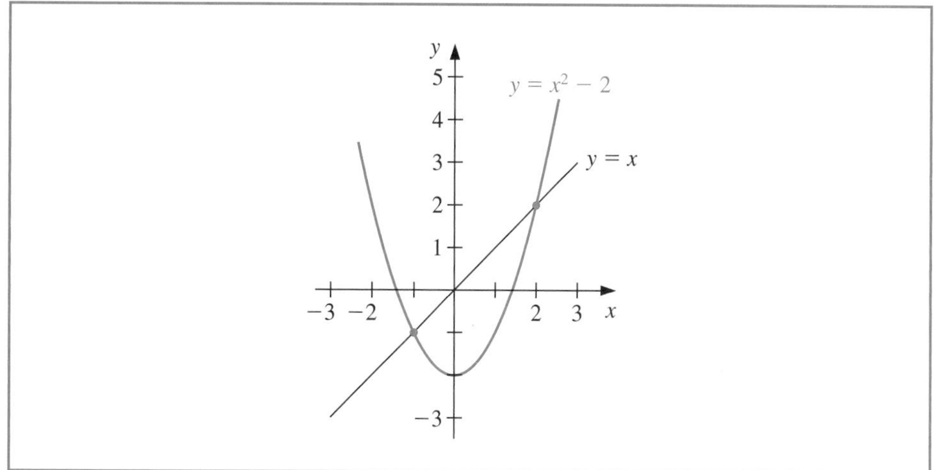

The following theorem gives sufficient conditions for the existence and uniqueness of a fixed point.

Theorem 2.3

(i) If $g \in C[a, b]$ and $g(x) \in [a, b]$ for all $x \in [a, b]$, then g has at least one fixed point in $[a, b]$.

(ii) If, in addition, $g'(x)$ exists on (a, b) and a positive constant $k < 1$ exists with

$$|g'(x)| \leq k, \quad \text{for all } x \in (a, b),$$

then there is exactly one fixed point in $[a, b]$. (See Figure 2.4.) ■

Figure 2.4

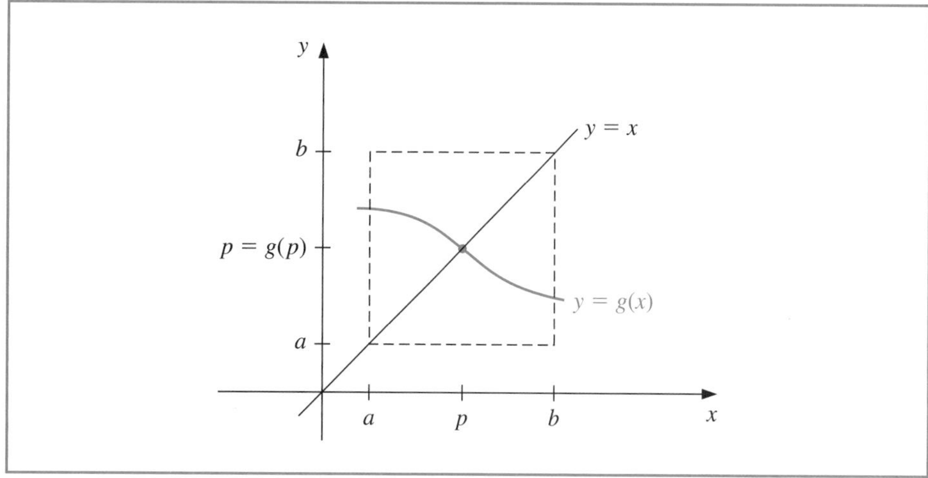

Proof

(i) If $g(a) = a$ or $g(b) = b$, then g has a fixed point at an endpoint. If not, then $g(a) > a$ and $g(b) < b$. The function $h(x) = g(x) - x$ is continuous on $[a, b]$, with

$$h(a) = g(a) - a > 0 \quad \text{and} \quad h(b) = g(b) - b < 0.$$

The Intermediate Value Theorem implies that there exists $p \in (a, b)$ for which $h(p) = 0$. This number p is a fixed point for g because

$$0 = h(p) = g(p) - p \quad \text{implies that} \quad g(p) = p.$$

(ii) Suppose, in addition, that $|g'(x)| \leq k < 1$ and that p and q are both fixed points in $[a, b]$. If $p \neq q$, then the Mean Value Theorem implies that a number ξ exists between p and q, and hence in $[a, b]$, with

$$\frac{g(p) - g(q)}{p - q} = g'(\xi).$$

Thus

$$|p - q| = |g(p) - g(q)| = |g'(\xi)||p - q| \leq k|p - q| < |p - q|,$$

which is a contradiction. This contradiction must come from the only supposition, $p \neq q$. Hence, $p = q$ and the fixed point in $[a, b]$ is unique. ■ ■ ■

Example 2 Show that $g(x) = (x^2 - 1)/3$ has a unique fixed point on the interval $[-1, 1]$.

Solution The maximum and minimum values of $g(x)$ for x in $[-1, 1]$ must occur either when x is an endpoint of the interval or when the derivative is 0. Since $g'(x) = 2x/3$, the function g is continuous and $g'(x)$ exists on $[-1, 1]$. The maximum and minimum values of $g(x)$ occur at $x = -1$, $x = 0$, or $x = 1$. But $g(-1) = 0$, $g(1) = 0$, and $g(0) = -1/3$, so an absolute maximum for $g(x)$ on $[-1, 1]$ occurs at $x = -1$ and $x = 1$, and an absolute minimum at $x = 0$.

Moreover

$$|g'(x)| = \left|\frac{2x}{3}\right| \leq \frac{2}{3}, \quad \text{for all } x \in (-1, 1).$$

So g satisfies all the hypotheses of Theorem 2.3 and has a unique fixed point in $[-1, 1]$. ■

For the function in Example 2, the unique fixed point p in the interval $[-1, 1]$ can be determined algebraically. If

$$p = g(p) = \frac{p^2 - 1}{3}, \quad \text{then} \quad p^2 - 3p - 1 = 0,$$

which, by the quadratic formula, implies, as shown on the left graph in Figure 2.4, that

$$p = \frac{1}{2}(3 - \sqrt{13}).$$

Note that g also has a unique fixed point $p = \frac{1}{2}(3 + \sqrt{13})$ for the interval $[3, 4]$. However, $g(4) = 5$ and $g'(4) = \frac{8}{3} > 1$, so g does not satisfy the hypotheses of Theorem 2.3 on $[3, 4]$. This demonstrates that the hypotheses of Theorem 2.3 are sufficient to guarantee a unique fixed point but are not necessary. (See the graph on the right in Figure 2.5.)

Figure 2.5

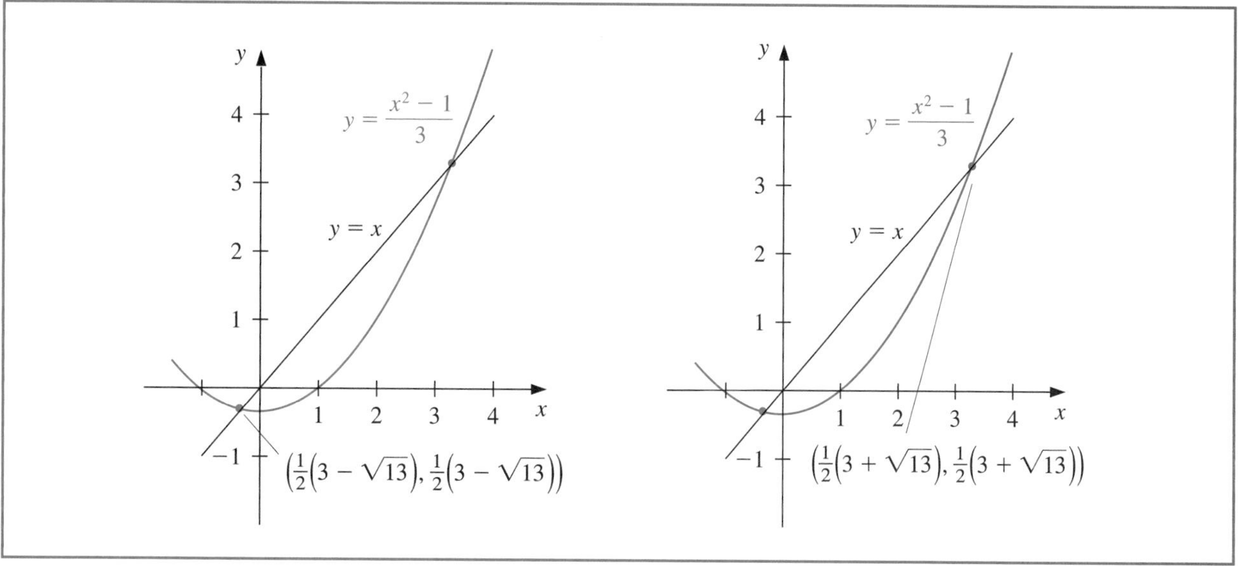

Example 3 Show that Theorem 2.3 does not ensure a unique fixed point of $g(x) = 3^{-x}$ on the interval $[0, 1]$, even though a unique fixed point on this interval does exist.

Solution $g'(x) = -3^{-x} \ln 3 < 0$ on $[0, 1]$, the function g is strictly decreasing on $[0, 1]$. So

$$g(1) = \frac{1}{3} \le g(x) \le 1 = g(0), \quad \text{for} \quad 0 \le x \le 1.$$

Thus, for $x \in [0, 1]$, we have $g(x) \in [0, 1]$. The first part of Theorem 2.3 ensures that there is at least one fixed point in $[0, 1]$.

However,

$$g'(0) = -\ln 3 = -1.098612289,$$

so $|g'(x)| \not\le 1$ on $(0, 1)$, and Theorem 2.3 cannot be used to determine uniqueness. But g is always decreasing, and it is clear from Figure 2.6 that the fixed point must be unique. ■

Figure 2.6

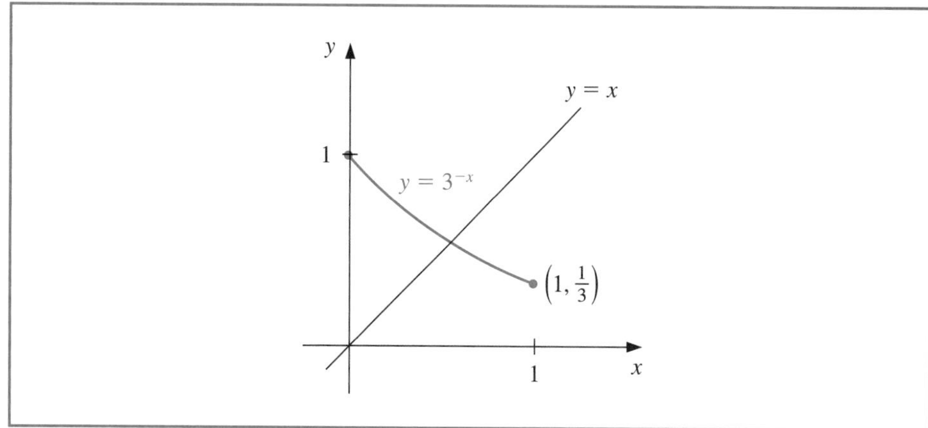

Fixed-Point Iteration

We cannot explicitly determine the fixed point in Example 3 because we have no way to solve for p in the equation $p = g(p) = 3^{-p}$. We can, however, determine approximations to this fixed point to any specified degree of accuracy. We will now consider how this can be done.

To approximate the fixed point of a function g, we choose an initial approximation p_0 and generate the sequence $\{p_n\}_{n=0}^{\infty}$ by letting $p_n = g(p_{n-1})$, for each $n \geq 1$. If the sequence converges to p and g is continuous, then

$$p = \lim_{n\to\infty} p_n = \lim_{n\to\infty} g(p_{n-1}) = g\left(\lim_{n\to\infty} p_{n-1}\right) = g(p),$$

and a solution to $x = g(x)$ is obtained. This technique is called **fixed-point**, or **functional iteration**. The procedure is illustrated in Figure 2.7 and detailed in Algorithm 2.2.

Figure 2.7

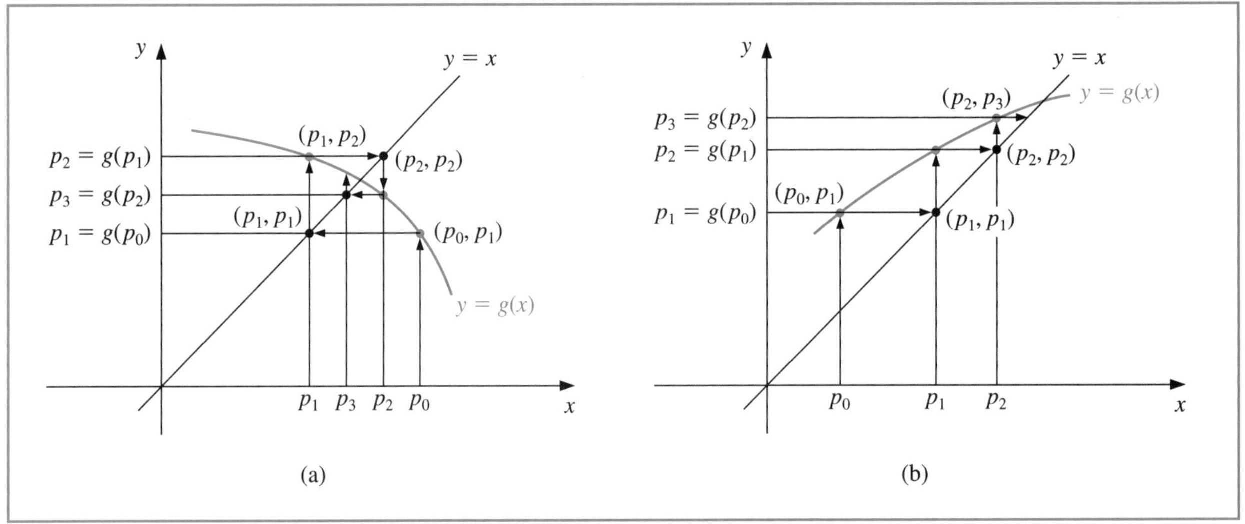

ALGORITHM 2.2

Fixed-Point Iteration

To find a solution to $p = g(p)$ given an initial approximation p_0:

INPUT initial approximation p_0; tolerance TOL; maximum number of iterations N_0.

OUTPUT approximate solution p or message of failure.

Step 1 Set $i = 1$.

Step 2 While $i \leq N_0$ do Steps 3–6.

Step 3 Set $p = g(p_0)$. *(Compute p_i.)*

Step 4 If $|p - p_0| < TOL$ then
OUTPUT (p); *(The procedure was successful.)*
STOP.

Step 5 Set $i = i + 1$.

Step 6 Set $p_0 = p$. *(Update p_0.)*

Step 7 OUTPUT ('The method failed after N_0 iterations, N_0 =', N_0);
(*The procedure was unsuccessful.*)
STOP. ■

The following illustrates some features of functional iteration.

Illustration The equation $x^3 + 4x^2 - 10 = 0$ has a unique root in $[1, 2]$. There are many ways to change the equation to the fixed-point form $x = g(x)$ using simple algebraic manipulation. For example, to obtain the function g described in part (c), we can manipulate the equation $x^3 + 4x^2 - 10 = 0$ as follows:

$$4x^2 = 10 - x^3, \quad \text{so} \quad x^2 = \frac{1}{4}(10 - x^3), \quad \text{and} \quad x = \pm\frac{1}{2}(10 - x^3)^{1/2}.$$

To obtain a positive solution, $g_3(x)$ is chosen. It is not important for you to derive the functions shown here, but you should verify that the fixed point of each is actually a solution to the original equation, $x^3 + 4x^2 - 10 = 0$.

(a) $x = g_1(x) = x - x^3 - 4x^2 + 10$

(b) $x = g_2(x) = \left(\frac{10}{x} - 4x\right)^{1/2}$

(c) $x = g_3(x) = \frac{1}{2}(10 - x^3)^{1/2}$

(d) $x = g_4(x) = \left(\frac{10}{4 + x}\right)^{1/2}$

(e) $x = g_5(x) = x - \frac{x^3 + 4x^2 - 10}{3x^2 + 8x}$

With $p_0 = 1.5$, Table 2.2 lists the results of the fixed-point iteration for all five choices of g.

Table 2.2

n	(a)	(b)	(c)	(d)	(e)
0	1.5	1.5	1.5	1.5	1.5
1	−0.875	0.8165	1.286953768	1.348399725	1.373333333
2	6.732	2.9969	1.402540804	1.367376372	1.365262015
3	−469.7	$(-8.65)^{1/2}$	1.345458374	1.364957015	1.365230014
4	1.03×10^8		1.375170253	1.365264748	1.365230013
5			1.360094193	1.365225594	
6			1.367846968	1.365230576	
7			1.363887004	1.365229942	
8			1.365916734	1.365230022	
9			1.364878217	1.365230012	
10			1.365410062	1.365230014	
15			1.365223680	1.365230013	
20			1.365230236		
25			1.365230006		
30			1.365230013		

The actual root is 1.365230013, as was noted in Example 1 of Section 2.1. Comparing the results to the Bisection Algorithm given in that example, it can be seen that excellent results have been obtained for choices (c), (d), and (e) (the Bisection method requires 27 iterations for this accuracy). It is interesting to note that choice (a) was divergent and that (b) became undefined because it involved the square root of a negative number. □

Although the various functions we have given are fixed-point problems for the same root-finding problem, they differ vastly as techniques for approximating the solution to the root-finding problem. Their purpose is to illustrate what needs to be answered:

- Question: How can we find a fixed-point problem that produces a sequence that reliably and rapidly converges to a solution to a given root-finding problem?

The following theorem and its corollary give us some clues concerning the paths we should pursue and, perhaps more importantly, some we should reject.

Theorem 2.4 **(Fixed-Point Theorem)**

Let $g \in C[a, b]$ be such that $g(x) \in [a, b]$, for all x in $[a, b]$. Suppose, in addition, that g' exists on (a, b) and that a constant $0 < k < 1$ exists with

$$|g'(x)| \leq k, \quad \text{for all } x \in (a, b).$$

Then for any number p_0 in $[a, b]$, the sequence defined by

$$p_n = g(p_{n-1}), \quad n \geq 1,$$

converges to the unique fixed point p in $[a, b]$. ■

Proof Theorem 2.3 implies that a unique point p exists in $[a, b]$ with $g(p) = p$. Since g maps $[a, b]$ into itself, the sequence $\{p_n\}_{n=0}^{\infty}$ is defined for all $n \geq 0$, and $p_n \in [a, b]$ for all n. Using the fact that $|g'(x)| \leq k$ and the Mean Value Theorem 1.8, we have, for each n,

$$|p_n - p| = |g(p_{n-1}) - g(p)| = |g'(\xi_n)||p_{n-1} - p| \leq k|p_{n-1} - p|,$$

where $\xi_n \in (a, b)$. Applying this inequality inductively gives

$$|p_n - p| \leq k|p_{n-1} - p| \leq k^2|p_{n-2} - p| \leq \cdots \leq k^n|p_0 - p|. \tag{2.4}$$

Since $0 < k < 1$, we have $\lim_{n\to\infty} k^n = 0$ and

$$\lim_{n\to\infty} |p_n - p| \leq \lim_{n\to\infty} k^n|p_0 - p| = 0.$$

Hence $\{p_n\}_{n=0}^{\infty}$ converges to p. ■ ■ ■

Corollary 2.5 If g satisfies the hypotheses of Theorem 2.4, then bounds for the error involved in using p_n to approximate p are given by

$$|p_n - p| \leq k^n \max\{p_0 - a, b - p_0\} \tag{2.5}$$

and

$$|p_n - p| \leq \frac{k^n}{1-k}|p_1 - p_0|, \quad \text{for all} \quad n \geq 1. \tag{2.6}$$

■

Proof Because $p \in [a, b]$, the first bound follows from Inequality (2.4):

$$|p_n - p| \leq k^n|p_0 - p| \leq k^n \max\{p_0 - a, b - p_0\}.$$

For $n \geq 1$, the procedure used in the proof of Theorem 2.4 implies that

$$|p_{n+1} - p_n| = |g(p_n) - g(p_{n-1})| \leq k|p_n - p_{n-1}| \leq \cdots \leq k^n|p_1 - p_0|.$$

Thus for $m > n \geq 1$,

$$\begin{aligned} |p_m - p_n| &= |p_m - p_{m-1} + p_{m-1} - \cdots + p_{n+1} - p_n| \\ &\leq |p_m - p_{m-1}| + |p_{m-1} - p_{m-2}| + \cdots + |p_{n+1} - p_n| \\ &\leq k^{m-1}|p_1 - p_0| + k^{m-2}|p_1 - p_0| + \cdots + k^n|p_1 - p_0| \\ &= k^n|p_1 - p_0| \left(1 + k + k^2 + \cdots + k^{m-n-1}\right). \end{aligned}$$

By Theorem 2.3, $\lim_{m\to\infty} p_m = p$, so

$$|p - p_n| = \lim_{m\to\infty} |p_m - p_n| \leq \lim_{m\to\infty} k^n|p_1 - p_0| \sum_{i=0}^{m-n-1} k^i \leq k^n|p_1 - p_0| \sum_{i=0}^{\infty} k^i.$$

But $\sum_{i=0}^{\infty} k^i$ is a geometric series with ratio k and $0 < k < 1$. This sequence converges to $1/(1-k)$, which gives the second bound:

$$|p - p_n| \leq \frac{k^n}{1-k}|p_1 - p_0|. \qquad \blacksquare$$

Both inequalities in the corollary relate the rate at which $\{p_n\}_{n=0}^{\infty}$ converges to the bound k on the first derivative. The rate of convergence depends on the factor k^n. The smaller the value of k, the faster the convergence, which may be very slow if k is close to 1.

Illustration Let us reconsider the various fixed-point schemes described in the preceding illustration in light of the Fixed-point Theorem 2.4 and its Corollary 2.5.

(a) For $g_1(x) = x - x^3 - 4x^2 + 10$, we have $g_1(1) = 6$ and $g_1(2) = -12$, so g_1 does not map $[1, 2]$ into itself. Moreover, $g_1'(x) = 1 - 3x^2 - 8x$, so $|g_1'(x)| > 1$ for all x in $[1, 2]$. Although Theorem 2.4 does not guarantee that the method must fail for this choice of g, there is no reason to expect convergence.

(b) With $g_2(x) = [(10/x) - 4x]^{1/2}$, we can see that g_2 does not map $[1, 2]$ into $[1, 2]$, and the sequence $\{p_n\}_{n=0}^{\infty}$ is not defined when $p_0 = 1.5$. Moreover, there is no interval containing $p \approx 1.365$ such that $|g_2'(x)| < 1$, because $|g_2'(p)| \approx 3.4$. There is no reason to expect that this method will converge.

(c) For the function $g_3(x) = \frac{1}{2}(10 - x^3)^{1/2}$, we have

$$g_3'(x) = -\frac{3}{4}x^2(10 - x^3)^{-1/2} < 0 \quad \text{on } [1, 2],$$

so g_3 is strictly decreasing on $[1, 2]$. However, $|g_3'(2)| \approx 2.12$, so the condition $|g_3'(x)| \leq k < 1$ fails on $[1, 2]$. A closer examination of the sequence $\{p_n\}_{n=0}^{\infty}$ starting with $p_0 = 1.5$ shows that it suffices to consider the interval $[1, 1.5]$ instead of $[1, 2]$. On this interval it is still true that $g_3'(x) < 0$ and g_3 is strictly decreasing, but, additionally,

$$1 < 1.28 \approx g_3(1.5) \leq g_3(x) \leq g_3(1) = 1.5,$$

for all $x \in [1, 1.5]$. This shows that g_3 maps the interval $[1, 1.5]$ into itself. It is also true that $|g_3'(x)| \leq |g_3'(1.5)| \approx 0.66$ on this interval, so Theorem 2.4 confirms the convergence of which we were already aware.

(d) For $g_4(x) = (10/(4 + x))^{1/2}$, we have

$$|g_4'(x)| = \left| \frac{-5}{\sqrt{10}(4 + x)^{3/2}} \right| \leq \frac{5}{\sqrt{10}(5)^{3/2}} < 0.15, \quad \text{for all} \quad x \in [1, 2].$$

The bound on the magnitude of $g_4'(x)$ is much smaller than the bound (found in (**c**)) on the magnitude of $g_3'(x)$, which explains the more rapid convergence using g_4.

(e) The sequence defined by

$$g_5(x) = x - \frac{x^3 + 4x^2 - 10}{3x^2 + 8x}$$

converges much more rapidly than our other choices. In the next sections we will see where this choice came from and why it is so effective. □

From what we have seen,

- Question: How can we find a fixed-point problem that produces a sequence that reliably and rapidly converges to a solution to a given root-finding problem?

might have

- Answer: Manipulate the root-finding problem into a fixed point problem that satisfies the conditions of Fixed-Point Theorem 2.4 and has a derivative that is as small as possible near the fixed point.

In the next sections we will examine this in more detail.

Maple has the fixed-point algorithm implemented in its *NumericalAnalysis* package. The options for the Bisection method are also available for fixed-point iteration. We will show only one option. After accessing the package using *with*(*Student*[*NumericalAnalysis*]): we enter the function

$$g := x - \frac{(x^3 + 4x^2 - 10)}{3x^2 + 8x}$$

and Maple returns

$$x - \frac{x^3 + 4x^2 - 10}{3x^2 + 8x}$$

Enter the command

FixedPointIteration(fixedpointiterator = *g*, *x* = 1.5, tolerance = 10^{-8}, *output* = *sequence*, *maxiterations* = 20)

and Maple returns

1.5, 1.373333333, 1.365262015, 1.365230014, 1.365230013

EXERCISE SET 2.2

1. Use algebraic manipulation to show that each of the following functions has a fixed point at p precisely when $f(p) = 0$, where $f(x) = x^4 + 2x^2 - x - 3$.

   a. $g_1(x) = \left(3 + x - 2x^2\right)^{1/4}$

   b. $g_2(x) = \left(\frac{x + 3 - x^4}{2}\right)^{1/2}$

c. $g_3(x) = \left(\dfrac{x+3}{x^2+2}\right)^{1/2}$ **d.** $g_4(x) = \dfrac{3x^4+2x^2+3}{4x^3+4x-1}$

2. **a.** Perform four iterations, if possible, on each of the functions g defined in Exercise 1. Let $p_0 = 1$ and $p_{n+1} = g(p_n)$, for $n = 0, 1, 2, 3$.

b. Which function do you think gives the best approximation to the solution?

3. The following four methods are proposed to compute $7^{1/5}$. Rank them in order, based on their apparent speed of convergence, assuming $p_0 = 1$.

a. $p_n = p_{n-1}\left(1 + \dfrac{7-p_{n-1}^5}{p_{n-1}^2}\right)^3$ **b.** $p_n = p_{n-1} - \dfrac{p_{n-1}^5 - 7}{p_{n-1}^2}$

c. $p_n = p_{n-1} - \dfrac{p_{n-1}^5 - 7}{5p_{n-1}^4}$ **d.** $p_n = p_{n-1} - \dfrac{p_{n-1}^5 - 7}{12}$

4. The following four methods are proposed to compute $21^{1/3}$. Rank them in order, based on their apparent speed of convergence, assuming $p_0 = 1$.

a. $p_n = \dfrac{20p_{n-1} + 21/p_{n-1}^2}{21}$ **b.** $p_n = p_{n-1} - \dfrac{p_{n-1}^3 - 21}{3p_{n-1}^2}$

c. $p_n = p_{n-1} - \dfrac{p_{n-1}^4 - 21p_{n-1}}{p_{n-1}^2 - 21}$ **d.** $p_n = \left(\dfrac{21}{p_{n-1}}\right)^{1/2}$

5. Use a fixed-point iteration method to determine a solution accurate to within 10^{-2} for $x^3 - x - 1 = 0$ on $[1, 2]$. Use $p_0 = 1$.

6. Use a fixed-point iteration method to determine a solution accurate to within 10^{-2} for $x^4 - 3x^2 - 3 = 0$ on $[1, 2]$. Use $p_0 = 1$.

7. Use Theorem 2.3 to show that $g(x) = \pi + 0.5\sin(x/2)$ has a unique fixed point on $[0, 2\pi]$. Use fixed-point iteration to find an approximation to the fixed point that is accurate to within 10^{-2}. Use Corollary 2.5 to estimate the number of iterations required to achieve 10^{-2} accuracy, and compare this theoretical estimate to the number actually needed.

8. Use Theorem 2.3 to show that $g(x) = 2^{-x}$ has a unique fixed point on $[\frac{1}{3}, 1]$. Use fixed-point iteration to find an approximation to the fixed point accurate to within 10^{-4}. Use Corollary 2.5 to estimate the number of iterations required to achieve 10^{-4} accuracy, and compare this theoretical estimate to the number actually needed.

9. Use a fixed-point iteration method to find an approximation to $\sqrt{3}$ that is accurate to within 10^{-4}. Compare your result and the number of iterations required with the answer obtained in Exercise 12 of Section 2.1.

10. Use a fixed-point iteration method to find an approximation to $\sqrt[3]{25}$ that is accurate to within 10^{-4}. Compare your result and the number of iterations required with the answer obtained in Exercise 13 of Section 2.1.

11. For each of the following equations, determine an interval $[a, b]$ on which fixed-point iteration will converge. Estimate the number of iterations necessary to obtain approximations accurate to within 10^{-5}, and perform the calculations.

a. $x = \dfrac{2 - e^x + x^2}{3}$ **b.** $x = \dfrac{5}{x^2} + 2$

c. $x = (e^x/3)^{1/2}$ **d.** $x = 5^{-x}$

e. $x = 6^{-x}$ **f.** $x = 0.5(\sin x + \cos x)$

12. For each of the following equations, use the given interval or determine an interval $[a, b]$ on which fixed-point iteration will converge. Estimate the number of iterations necessary to obtain approximations accurate to within 10^{-5}, and perform the calculations.

a. $2 + \sin x - x = 0$ use $[2, 3]$ **b.** $x^3 - 2x - 5 = 0$ use $[2, 3]$

c. $3x^2 - e^x = 0$ **d.** $x - \cos x = 0$

13. Find all the zeros of $f(x) = x^2 + 10\cos x$ by using the fixed-point iteration method for an appropriate iteration function g. Find the zeros accurate to within 10^{-4}.

14. Use a fixed-point iteration method to determine a solution accurate to within 10^{-4} for $x = \tan x$, for x in $[4, 5]$.

15. Use a fixed-point iteration method to determine a solution accurate to within 10^{-2} for $2 \sin \pi x + x = 0$ on $[1, 2]$. Use $p_0 = 1$.

16. Let A be a given positive constant and $g(x) = 2x - Ax^2$.

a. Show that if fixed-point iteration converges to a nonzero limit, then the limit is $p = 1/A$, so the inverse of a number can be found using only multiplications and subtractions.

b. Find an interval about $1/A$ for which fixed-point iteration converges, provided p_0 is in that interval.

17. Find a function g defined on $[0, 1]$ that satisfies none of the hypotheses of Theorem 2.3 but still has a unique fixed point on $[0, 1]$.

18. a. Show that Theorem 2.2 is true if the inequality $|g'(x)| \le k$ is replaced by $g'(x) \le k$, for all $x \in (a, b)$. [*Hint:* Only uniqueness is in question.]

b. Show that Theorem 2.3 may not hold if inequality $|g'(x)| \le k$ is replaced by $g'(x) \le k$. [*Hint:* Show that $g(x) = 1 - x^2$, for x in $[0, 1]$, provides a counterexample.]

19. a. Use Theorem 2.4 to show that the sequence defined by

$$x_n = \frac{1}{2}x_{n-1} + \frac{1}{x_{n-1}}, \quad \text{for } n \ge 1,$$

converges to $\sqrt{2}$ whenever $x_0 > \sqrt{2}$.

b. Use the fact that $0 < (x_0 - \sqrt{2})^2$ whenever $x_0 \ne \sqrt{2}$ to show that if $0 < x_0 < \sqrt{2}$, then $x_1 > \sqrt{2}$.

c. Use the results of parts (a) and (b) to show that the sequence in (a) converges to $\sqrt{2}$ whenever $x_0 > 0$.

20. a. Show that if A is any positive number, then the sequence defined by

$$x_n = \frac{1}{2}x_{n-1} + \frac{A}{2x_{n-1}}, \quad \text{for } n \ge 1,$$

converges to $\sqrt{A}$ whenever $x_0 > 0$.

b. What happens if $x_0 < 0$?

21. Replace the assumption in Theorem 2.4 that "a positive number $k < 1$ exists with $|g'(x)| \le k$" with "g satisfies a Lipschitz condition on the interval $[a, b]$ with Lipschitz constant $L < 1$." (See Exercise 27, Section 1.1.) Show that the conclusions of this theorem are still valid.

22. Suppose that g is continuously differentiable on some interval (c, d) that contains the fixed point p of g. Show that if $|g'(p)| < 1$, then there exists a $\delta > 0$ such that if $|p_0 - p| \le \delta$, then the fixed-point iteration converges.

23. An object falling vertically through the air is subjected to viscous resistance as well as to the force of gravity. Assume that an object with mass m is dropped from a height s_0 and that the height of the object after t seconds is

$$s(t) = s_0 - \frac{mg}{k}t + \frac{m^2 g}{k^2}(1 - e^{-kt/m}),$$

where $g = 32.17$ ft/s^2 and k represents the coefficient of air resistance in lb-s/ft. Suppose $s_0 = 300$ ft, $m = 0.25$ lb, and $k = 0.1$ lb-s/ft. Find, to within 0.01 s, the time it takes this quarter-pounder to hit the ground.

24. Let $g \in C^1[a, b]$ and p be in (a, b) with $g(p) = p$ and $|g'(p)| > 1$. Show that there exists a $\delta > 0$ such that if $0 < |p_0 - p| < \delta$, then $|p_0 - p| < |p_1 - p|$. Thus, no matter how close the initial approximation p_0 is to p, the next iterate p_1 is farther away, so the fixed-point iteration does not converge if $p_0 \ne p$.

2.3 Newton's Method and Its Extensions

Isaac Newton (1641–1727) was one of the most brilliant scientists of all time. The late 17th century was a vibrant period for science and mathematics and Newton's work touched nearly every aspect of mathematics. His method for solving was introduced to find a root of the equation $y^3 - 2y - 5 = 0$. Although he demonstrated the method only for polynomials, it is clear that he realized its broader applications.

Newton's (or the *Newton-Raphson*) **method** is one of the most powerful and well-known numerical methods for solving a root-finding problem. There are many ways of introducing Newton's method.

Newton's Method

If we only want an algorithm, we can consider the technique graphically, as is often done in calculus. Another possibility is to derive Newton's method as a technique to obtain faster convergence than offered by other types of functional iteration, as is done in Section 2.4. A third means of introducing Newton's method, which is discussed next, is based on Taylor polynomials. We will see there that this particular derivation produces not only the method, but also a bound for the error of the approximation.

Suppose that $f \in C^2[a, b]$. Let $p_0 \in [a, b]$ be an approximation to p such that $f'(p_0) \neq 0$ and $|p - p_0|$ is "small." Consider the first Taylor polynomial for $f(x)$ expanded about p_0 and evaluated at $x = p$.

$$f(p) = f(p_0) + (p - p_0)f'(p_0) + \frac{(p - p_0)^2}{2} f''(\xi(p)),$$

where $\xi(p)$ lies between p and p_0. Since $f(p) = 0$, this equation gives

$$0 = f(p_0) + (p - p_0)f'(p_0) + \frac{(p - p_0)^2}{2} f''(\xi(p)).$$

Newton's method is derived by assuming that since $|p - p_0|$ is small, the term involving $(p - p_0)^2$ is much smaller, so

$$0 \approx f(p_0) + (p - p_0)f'(p_0).$$

Solving for p gives

$$p \approx p_0 - \frac{f(p_0)}{f'(p_0)} \equiv p_1.$$

Joseph Raphson (1648–1715) gave a description of the method attributed to Isaac Newton in 1690, acknowledging Newton as the source of the discovery. Neither Newton nor Raphson explicitly used the derivative in their description since both considered only polynomials. Other mathematicians, particularly James Gregory (1636–1675), were aware of the underlying process at or before this time.

This sets the stage for Newton's method, which starts with an initial approximation p_0 and generates the sequence $\{p_n\}_{n=0}^{\infty}$, by

$$p_n = p_{n-1} - \frac{f(p_{n-1})}{f'(p_{n-1})}, \quad \text{for } n \geq 1. \tag{2.7}$$

Figure 2.8 on page 68 illustrates how the approximations are obtained using successive tangents. (Also see Exercise 15.) Starting with the initial approximation p_0, the approximation p_1 is the x-intercept of the tangent line to the graph of f at $(p_0, f(p_0))$. The approximation p_2 is the x-intercept of the tangent line to the graph of f at $(p_1, f(p_1))$ and so on. Algorithm 2.3 follows this procedure.

Figure 2.8

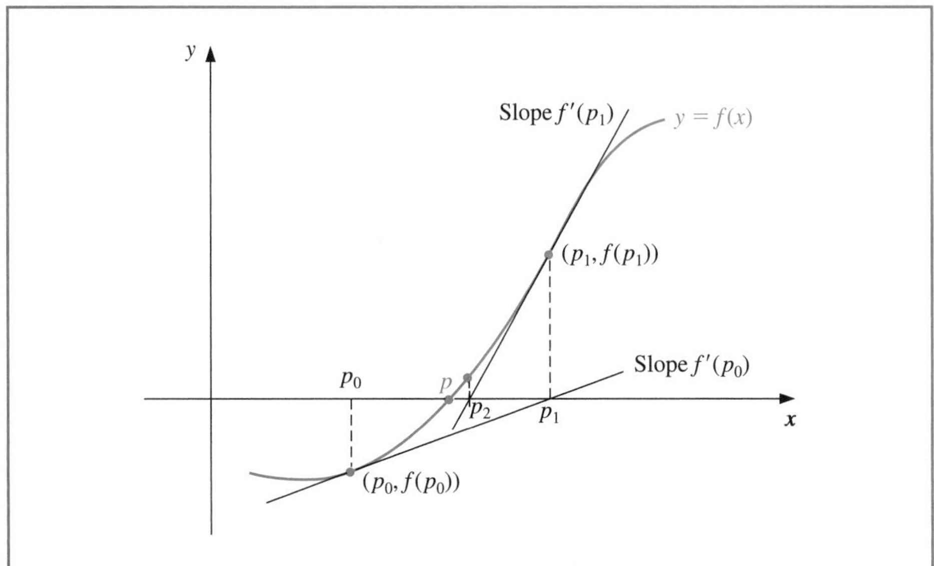

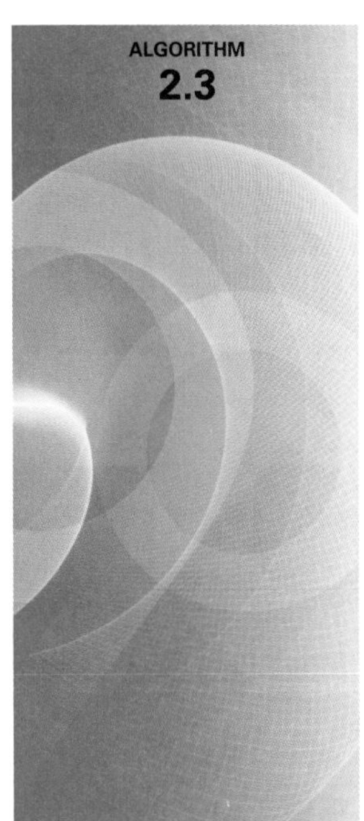

Newton's

To find a solution to $f(x) = 0$ given an initial approximation p_0:

INPUT initial approximation p_0; tolerance TOL; maximum number of iterations N_0.

OUTPUT approximate solution p or message of failure.

Step 1 Set $i = 1$.

Step 2 While $i \leq N_0$ do Steps 3–6.

Step 3 Set $p = p_0 - f(p_0)/f'(p_0)$. (*Compute* p_i.)

Step 4 If $|p - p_0| < TOL$ then
OUTPUT (p); (*The procedure was successful.*)
STOP.

Step 5 Set $i = i + 1$.

Step 6 Set $p_0 = p$. (*Update* p_0.)

Step 7 OUTPUT ('The method failed after N_0 iterations, N_0 =', N_0);
(*The procedure was unsuccessful.*)
STOP. ■

The stopping-technique inequalities given with the Bisection method are applicable to Newton's method. That is, select a tolerance $\varepsilon > 0$, and construct $p_1, \ldots p_N$ until

$$|p_N - p_{N-1}| < \varepsilon, \tag{2.8}$$

$$\frac{|p_N - p_{N-1}|}{|p_N|} < \varepsilon, \quad p_N \neq 0, \tag{2.9}$$

or

$$|f(p_N)| < \varepsilon. \tag{2.10}$$

A form of Inequality (2.8) is used in Step 4 of Algorithm 2.3. Note that none of the inequalities (2.8), (2.9), or (2.10) give precise information about the actual error $|p_N - p|$. (See Exercises 16 and 17 in Section 2.1.)

Newton's method is a functional iteration technique with $p_n = g(p_{n-1})$, for which

$$g(p_{n-1}) = p_{n-1} - \frac{f(p_{n-1})}{f'(p_{n-1})}, \quad \text{for } n \geq 1. \tag{2.11}$$

In fact, this is the functional iteration technique that was used to give the rapid convergence we saw in column (**e**) of Table 2.2 in Section 2.2.

It is clear from Equation (2.7) that Newton's method cannot be continued if $f'(p_{n-1}) = 0$ for some n. In fact, we will see that the method is most effective when f' is bounded away from zero near p.

Example 1 Consider the function $f(x) = \cos x - x = 0$. Approximate a root of f using **(a)** a fixed-point method, and **(b)** Newton's Method

Solution **(a)** A solution to this root-finding problem is also a solution to the fixed-point problem $x = \cos x$, and the graph in Figure 2.9 implies that a single fixed-point p lies in $[0, \pi/2]$.

Figure 2.9

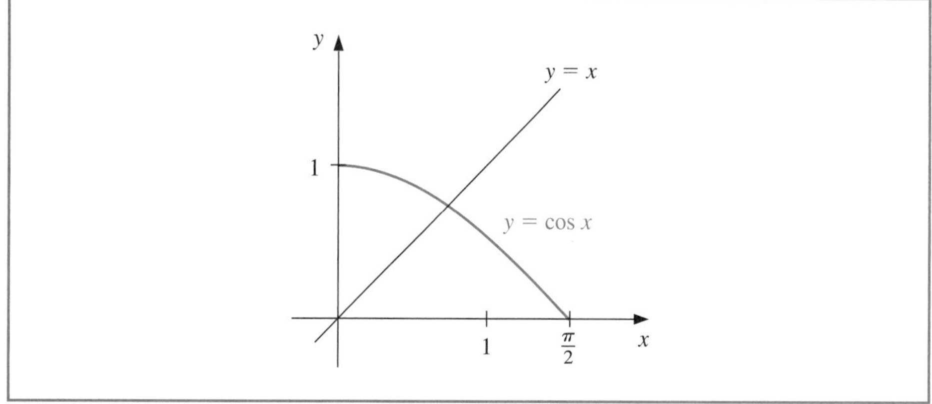

Note that the variable in the trigonometric function is in radian measure, not degrees. This will always be the case unless specified otherwise.

Table 2.3

n	p_n
0	0.7853981635
1	0.7071067810
2	0.7602445972
3	0.7246674808
4	0.7487198858
5	0.7325608446
6	0.7434642113
7	0.7361282565

Table 2.3 shows the results of fixed-point iteration with $p_0 = \pi/4$. The best we could conclude from these results is that $p \approx 0.74$.

(b) To apply Newton's method to this problem we need $f'(x) = -\sin x - 1$. Starting again with $p_0 = \pi/4$, we generate the sequence defined, for $n \geq 1$, by

$$p_n = p_{n-1} - \frac{f(p_{n-1})}{f(p'_{n-1})} = p_{n-1} - \frac{\cos p_{n-1} - p_{n-1}}{-\sin p_{n-1} - 1}.$$

This gives the approximations in Table 2.4. An excellent approximation is obtained with $n = 3$. Because of the agreement of p_3 and p_4 we could reasonably expect this result to be accurate to the places listed. ■

Table 2.4

Newton's Method

n	p_n
0	0.7853981635
1	0.7395361337
2	0.7390851781
3	0.7390851332
4	0.7390851332

Convergence using Newton's Method

Example 1 shows that Newton's method can provide extremely accurate approximations with very few iterations. For that example, only one iteration of Newton's method was needed to give better accuracy than 7 iterations of the fixed-point method. It is now time to examine Newton's method more carefully to discover why it is so effective.

The Taylor series derivation of Newton's method at the beginning of the section points out the importance of an accurate initial approximation. The crucial assumption is that the term involving $(p - p_0)^2$ is, by comparison with $|p - p_0|$, so small that it can be deleted. This will clearly be false unless p_0 is a good approximation to p. If p_0 is not sufficiently close to the actual root, there is little reason to suspect that Newton's method will converge to the root. However, in some instances, even poor initial approximations will produce convergence. (Exercises 20 and 21 illustrate some of these possibilities.)

The following convergence theorem for Newton's method illustrates the theoretical importance of the choice of p_0.

Theorem 2.6 Let $f \in C^2[a, b]$. If $p \in (a, b)$ is such that $f(p) = 0$ and $f'(p) \neq 0$, then there exists a $\delta > 0$ such that Newton's method generates a sequence $\{p_n\}_{n=1}^{\infty}$ converging to p for any initial approximation $p_0 \in [p - \delta, p + \delta]$. ■

Proof The proof is based on analyzing Newton's method as the functional iteration scheme $p_n = g(p_{n-1})$, for $n \geq 1$, with

$$g(x) = x - \frac{f(x)}{f'(x)}.$$

Let k be in $(0, 1)$. We first find an interval $[p - \delta, p + \delta]$ that g maps into itself and for which $|g'(x)| \leq k$, for all $x \in (p - \delta, p + \delta)$.

Since f' is continuous and $f'(p) \neq 0$, part (**a**) of Exercise 29 in Section 1.1 implies that there exists a $\delta_1 > 0$, such that $f'(x) \neq 0$ for $x \in [p - \delta_1, p + \delta_1] \subseteq [a, b]$. Thus g is defined and continuous on $[p - \delta_1, p + \delta_1]$. Also

$$g'(x) = 1 - \frac{f'(x)f'(x) - f(x)f''(x)}{[f'(x)]^2} = \frac{f(x)f''(x)}{[f'(x)]^2},$$

for $x \in [p - \delta_1, p + \delta_1]$, and, since $f \in C^2[a, b]$, we have $g \in C^1[p - \delta_1, p + \delta_1]$.

By assumption, $f(p) = 0$, so

$$g'(p) = \frac{f(p)f''(p)}{[f'(p)]^2} = 0.$$

Since g' is continuous and $0 < k < 1$, part (**b**) of Exercise 29 in Section 1.1 implies that there exists a δ, with $0 < \delta < \delta_1$, and

$$|g'(x)| \leq k, \quad \text{for all } x \in [p - \delta, p + \delta].$$

It remains to show that g maps $[p - \delta, p + \delta]$ into $[p - \delta, p + \delta]$. If $x \in [p - \delta, p + \delta]$, the Mean Value Theorem implies that for some number ξ between x and p, $|g(x) - g(p)| = |g'(\xi)||x - p|$. So

$$|g(x) - p| = |g(x) - g(p)| = |g'(\xi)||x - p| \leq k|x - p| < |x - p|.$$

Since $x \in [p - \delta, p + \delta]$, it follows that $|x - p| < \delta$ and that $|g(x) - p| < \delta$. Hence, g maps $[p - \delta, p + \delta]$ into $[p - \delta, p + \delta]$.

All the hypotheses of the Fixed-Point Theorem 2.4 are now satisfied, so the sequence $\{p_n\}_{n=1}^{\infty}$, defined by

$$p_n = g(p_{n-1}) = p_{n-1} - \frac{f(p_{n-1})}{f'(p_{n-1})}, \quad \text{for } n \geq 1,$$

converges to p for any $p_0 \in [p - \delta, p + \delta]$. ▪ ▪ ▪

Theorem 2.6 states that, under reasonable assumptions, Newton's method converges provided a sufficiently accurate initial approximation is chosen. It also implies that the constant k that bounds the derivative of g, and, consequently, indicates the speed of convergence of the method, decreases to 0 as the procedure continues. This result is important for the theory of Newton's method, but it is seldom applied in practice because it does not tell us how to determine δ.

In a practical application, an initial approximation is selected and successive approximations are generated by Newton's method. These will generally either converge quickly to the root, or it will be clear that convergence is unlikely.

The Secant Method

Newton's method is an extremely powerful technique, but it has a major weakness: the need to know the value of the derivative of f at each approximation. Frequently, $f'(x)$ is far more difficult and needs more arithmetic operations to calculate than $f(x)$.

To circumvent the problem of the derivative evaluation in Newton's method, we introduce a slight variation. By definition,

$$f'(p_{n-1}) = \lim_{x \to p_{n-1}} \frac{f(x) - f(p_{n-1})}{x - p_{n-1}}.$$

If p_{n-2} is close to p_{n-1}, then

$$f'(p_{n-1}) \approx \frac{f(p_{n-2}) - f(p_{n-1})}{p_{n-2} - p_{n-1}} = \frac{f(p_{n-1}) - f(p_{n-2})}{p_{n-1} - p_{n-2}}.$$

Using this approximation for $f'(p_{n-1})$ in Newton's formula gives

$$p_n = p_{n-1} - \frac{f(p_{n-1})(p_{n-1} - p_{n-2})}{f(p_{n-1}) - f(p_{n-2})}. \tag{2.12}$$

The word secant is derived from the Latin word *secan*, which means to cut. The secant method uses a secant line, a line joining two points that cut the curve, to approximate a root.

This technique is called the **Secant method** and is presented in Algorithm 2.4. (See Figure 2.10.) Starting with the two initial approximations p_0 and p_1, the approximation p_2 is the x-intercept of the line joining $(p_0, f(p_0))$ and $(p_1, f(p_1))$. The approximation p_3 is the x-intercept of the line joining $(p_1, f(p_1))$ and $(p_2, f(p_2))$, and so on. Note that only one function evaluation is needed per step for the Secant method after p_2 has been determined. In contrast, each step of Newton's method requires an evaluation of both the function and its derivative.

Figure 2.10

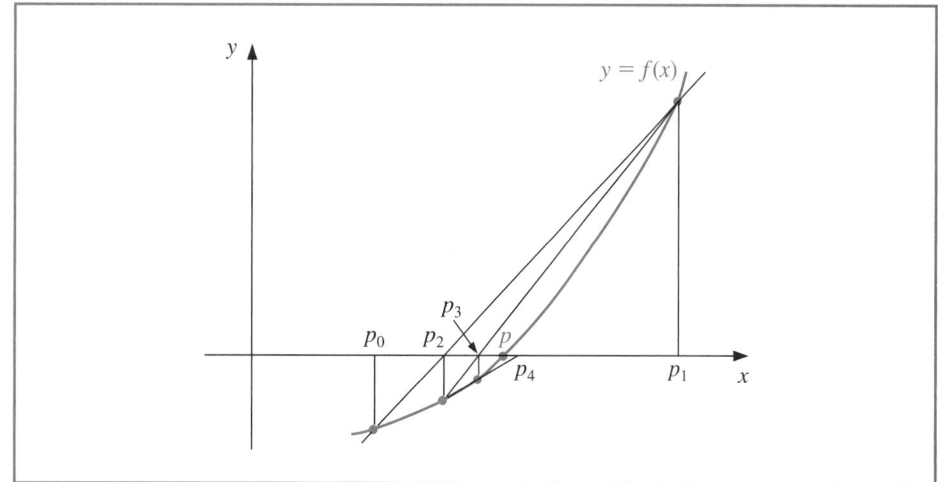

Secant

To find a solution to $f(x) = 0$ given initial approximations p_0 and p_1:

INPUT initial approximations p_0, p_1; tolerance TOL; maximum number of iterations N_0.

OUTPUT approximate solution p or message of failure.

Step 1 Set $i = 2$;
$q_0 = f(p_0)$;
$q_1 = f(p_1)$.

Step 2 While $i \le N_0$ do Steps 3–6.

Step 3 Set $p = p_1 - q_1(p_1 - p_0)/(q_1 - q_0)$. *(Compute p_i.)*

Step 4 If $|p - p_1| < TOL$ then
OUTPUT (p); *(The procedure was successful.)*
STOP.

Step 5 Set $i = i + 1$.

Step 6 Set $p_0 = p_1$; *(Update p_0, q_0, p_1, q_1.)*
$q_0 = q_1$;
$p_1 = p$;
$q_1 = f(p)$.

Step 7 OUTPUT ('The method failed after N_0 iterations, N_0 =', N_0);
(The procedure was unsuccessful.)
STOP. ■

The next example involves a problem considered in Example 1, where we used Newton's method with $p_0 = \pi/4$.

Example 2 Use the Secant method to find a solution to $x = \cos x$, and compare the approximations with those given in Example 1 which applied Newton's method.

Solution In Example 1 we compared fixed-point iteration and Newton's method starting with the initial approximation $p_0 = \pi/4$. For the Secant method we need two initial approximations. Suppose we use $p_0 = 0.5$ and $p_1 = \pi/4$. Succeeding approximations are generated by the formula

$$p_n = p_{n-1} - \frac{(p_{n-1} - p_{n-2})(\cos p_{n-1} - p_{n-1})}{(\cos p_{n-1} - p_{n-1}) - (\cos p_{n-2} - p_{n-2})}, \quad \text{for } n \ge 2.$$

These give the results in Table 2.5. ■

Table 2.5

n	Secant p_n
0	0.5
1	0.7853981635
2	0.7363841388
3	0.7390581392
4	0.7390851493
5	0.7390851332

n	Newton p_n
0	0.7853981635
1	0.7395361337
2	0.7390851781
3	0.7390851332
4	0.7390851332

Comparing the results in Table 2.5 from the Secant method and Newton's method, we see that the Secant method approximation p_5 is accurate to the tenth decimal place, whereas Newton's method obtained this accuracy by p_3. For this example, the convergence of the Secant method is much faster than functional iteration but slightly slower than Newton's method. This is generally the case. (See Exercise 14 of Section 2.4.)

Newton's method or the Secant method is often used to refine an answer obtained by another technique, such as the Bisection method, since these methods require good first approximations but generally give rapid convergence.

The Method of False Position

Each successive pair of approximations in the Bisection method brackets a root p of the equation; that is, for each positive integer n, a root lies between a_n and b_n. This implies that, for each n, the Bisection method iterations satisfy

$$|p_n - p| < \frac{1}{2}|a_n - b_n|,$$

which provides an easily calculated error bound for the approximations.

The term *Regula Falsi*, literally a false rule or false position, refers to a technique that uses results that are known to be false, but in some specific manner, to obtain convergence to a true result. False position problems can be found on the Rhind papyrus, which dates from about 1650 B.C.E.

Root bracketing is not guaranteed for either Newton's method or the Secant method. In Example 1, Newton's method was applied to $f(x) = \cos x - x$, and an approximate root was found to be 0.7390851332. Table 2.5 shows that this root is not bracketed by either p_0 and p_1 or p_1 and p_2. The Secant method approximations for this problem are also given in Table 2.5. In this case the initial approximations p_0 and p_1 bracket the root, but the pair of approximations p_3 and p_4 fail to do so.

The **method of False Position** (also called *Regula Falsi*) generates approximations in the same manner as the Secant method, but it includes a test to ensure that the root is always bracketed between successive iterations. Although it is not a method we generally recommend, it illustrates how bracketing can be incorporated.

First choose initial approximations p_0 and p_1 with $f(p_0) \cdot f(p_1) < 0$. The approximation p_2 is chosen in the same manner as in the Secant method, as the x-intercept of the line joining $(p_0, f(p_0))$ and $(p_1, f(p_1))$. To decide which secant line to use to compute p_3, consider $f(p_2) \cdot f(p_1)$, or more correctly $\text{sgn}\, f(p_2) \cdot \text{sgn}\, f(p_1)$.

- If $\text{sgn}\, f(p_2) \cdot \text{sgn}\, f(p_1) < 0$, then p_1 and p_2 bracket a root. Choose p_3 as the x-intercept of the line joining $(p_1, f(p_1))$ and $(p_2, f(p_2))$.
- If not, choose p_3 as the x-intercept of the line joining $(p_0, f(p_0))$ and $(p_2, f(p_2))$, and then interchange the indices on p_0 and p_1.

In a similar manner, once p_3 is found, the sign of $f(p_3) \cdot f(p_2)$ determines whether we use p_2 and p_3 or p_3 and p_1 to compute p_4. In the latter case a relabeling of p_2 and p_1 is performed. The relabeling ensures that the root is bracketed between successive iterations. The process is described in Algorithm 2.5, and Figure 2.11 shows how the iterations can differ from those of the Secant method. In this illustration, the first three approximations are the same, but the fourth approximations differ.

Figure 2.11

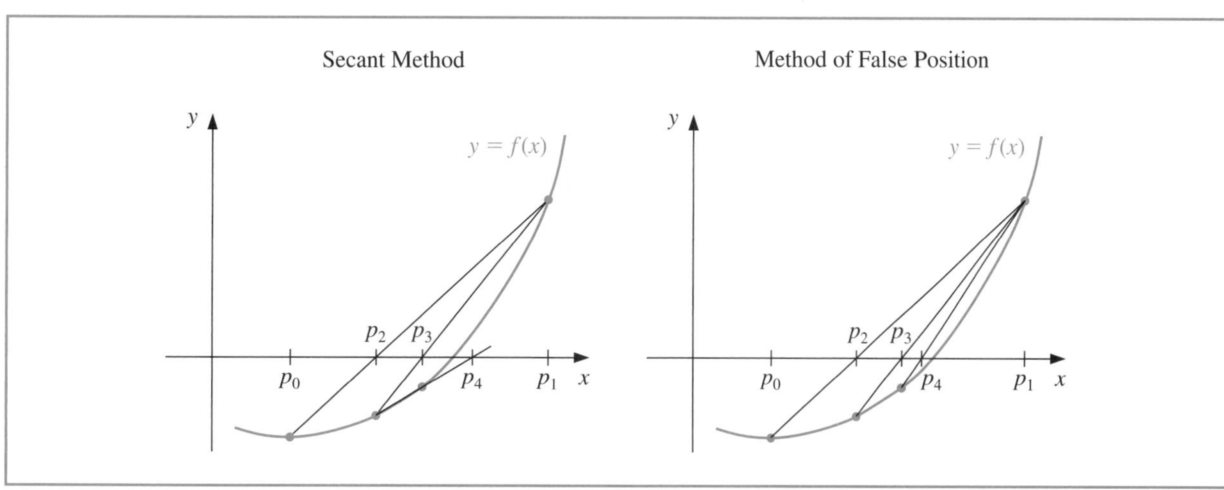

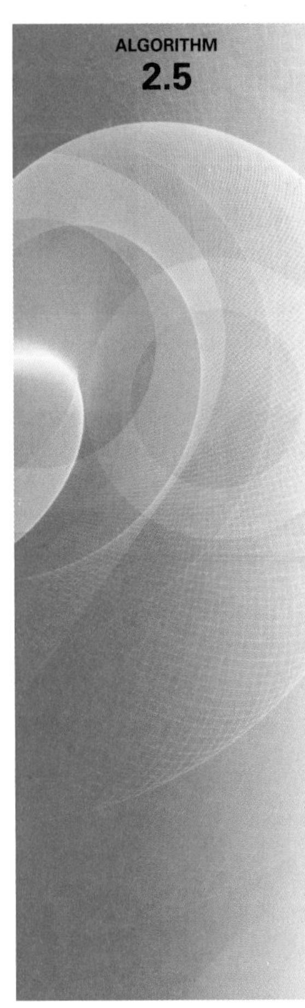

False Position

To find a solution to $f(x) = 0$ given the continuous function f on the interval $[p_0, p_1]$ where $f(p_0)$ and $f(p_1)$ have opposite signs:

INPUT initial approximations p_0, p_1; tolerance TOL; maximum number of iterations N_0.

OUTPUT approximate solution p or message of failure.

Step 1 Set $i = 2$;
$q_0 = f(p_0)$;
$q_1 = f(p_1)$.

Step 2 While $i \le N_0$ do Steps 3–7.

Step 3 Set $p = p_1 - q_1(p_1 - p_0)/(q_1 - q_0)$. *(Compute p_i.)*

Step 4 If $|p - p_1| < TOL$ then
OUTPUT (p); *(The procedure was successful.)*
STOP.

Step 5 Set $i = i + 1$;
$q = f(p)$.

Step 6 If $q \cdot q_1 < 0$ then set $p_0 = p_1$;
$q_0 = q_1$.

Step 7 Set $p_1 = p$;
$q_1 = q$.

Step 8 OUTPUT ('Method failed after N_0 iterations, N_0 =', N_0);
(The procedure unsuccessful.)
STOP. ■

Example 3 Use the method of False Position to find a solution to $x = \cos x$, and compare the approximations with those given in Example 1 which applied fixed-point iteration and Newton's method, and to those found in Example 2 which applied the Secant method.

Solution To make a reasonable comparison we will use the same initial approximations as in the Secant method, that is, $p_0 = 0.5$ and $p_1 = \pi/4$. Table 2.6 shows the results of the method of False Position applied to $f(x) = \cos x - x$ together with those we obtained using the Secant and Newton's methods. Notice that the False Position and Secant approximations agree through p_3 and that the method of False Position requires an additional iteration to obtain the same accuracy as the Secant method. ■

Table 2.6

	False Position	Secant	Newton
n	p_n	p_n	p_n
0	0.5	0.5	0.7853981635
1	0.7853981635	0.7853981635	0.7395361337
2	0.7363841388	0.7363841388	0.7390851781
3	0.7390581392	0.7390581392	0.7390851332
4	0.7390848638	0.7390851493	0.7390851332
5	0.7390851305	0.7390851332	
6	0.7390851332		

The added insurance of the method of False Position commonly requires more calculation than the Secant method, just as the simplification that the Secant method provides over Newton's method usually comes at the expense of additional iterations. Further examples of the positive and negative features of these methods can be seen by working Exercises 17 and 18.

Maple has Newton's method, the Secant method, and the method of False Position implemented in its *NumericalAnalysis* package. The options that were available for the Bisection method are also available for these techniques. For example, to generate the results in Tables 2.4, 2.5, and 2.6 we could use the commands

$with(Student[NumericalAnalysis])$

$f := \cos(x) - x$

$$Newton\left(f, x = \frac{\pi}{4.0}, tolerance = 10^{-8}, output = sequence, maxiterations = 20\right)$$

$$Secant\left(f, x = \left[0.5, \frac{\pi}{4.0}\right], tolerance = 10^{-8}, output = sequence, maxiterations = 20\right)$$

and

$$FalsePosition\left(f, x = \left[0.5, \frac{\pi}{4.0}\right], tolerance = 10^{-8}, output = sequence, maxiterations = 20\right)$$

EXERCISE SET 2.3

1. Let $f(x) = x^2 - 6$ and $p_0 = 1$. Use Newton's method to find p_2.
2. Let $f(x) = -x^3 - \cos x$ and $p_0 = -1$. Use Newton's method to find p_2. Could $p_0 = 0$ be used?
3. Let $f(x) = -x^3 - \cos x$. With $p_0 = -1$ and $p_1 = 0$, find p_3.
   a. Use the Secant method. **b.** Use the method of False Position.
4. Let $f(x) = x^2 - 6$. With $p_0 = 3$ and $p_1 = 2$, find p_3.
   a. Use the Secant method.
   b. Use the method of False Position.
   c. Which of **a.** or **b.** is closer to $\sqrt{6}$?
5. Use Newton's method to find solutions accurate to within 10^{-4} for the following problems.
   a. $x - 0.8 - 0.2\sin x = 0, \quad [0, \pi/2]$ **b.** $x^3 + 3x^2 - 1 = 0, \quad [-3, -2]$
   c. $x - \cos x = 0, \quad [0, \pi/2]$ **d.** $x^3 - 2x^2 - 5 = 0, \quad [1, 4]$
6. Use Newton's method to find solutions accurate to within 10^{-5} for the following problems.
   a. $e^x + 2^{-x} + 2\cos x - 6 = 0$ for $1 \le x \le 2$
   b. $\ln(x-1) + \cos(x-1) = 0$ for $1.3 \le x \le 2$
   c. $2x\cos 2x - (x-2)^2 = 0$ for $2 \le x \le 3$ and $3 \le x \le 4$
   d. $(x-2)^2 - \ln x = 0$ for $1 \le x \le 2$ and $e \le x \le 4$
   e. $e^x - 3x^2 = 0$ for $0 \le x \le 1$ and $3 \le x \le 5$
   f. $\sin x - e^{-x} = 0$ for $0 \le x \le 1$ $3 \le x \le 4$ and $6 \le x \le 7$
7. Repeat Exercise 5 using the Secant method.
8. Repeat Exercise 6 using the Secant method.
9. Repeat Exercise 5 using the method of False Position.
10. Repeat Exercise 6 using the method of False Position.
11. Use all three methods in this Section to find solutions to within 10^{-5} for the following problems.
    a. $3xe^x = 0$ for $1 \le x \le 2$
    b. $2x + 3\cos x - e^x = 0$ for $0 \le x \le 1$

12. Use all three methods in this Section to find solutions to within 10^{-7} for the following problems.

a. $x^2 - 4x + 4 - \ln x = 0$ for $1 \le x \le 2$ and for $2 \le x \le 4$

b. $x + 1 - 2\sin \pi x = 0$ for $0 \le x \le 1/2$ and for $1/2 \le x \le 1$

13. Use Newton's method to approximate, to within 10^{-4}, the value of x that produces the point on the graph of $y = x^2$ that is closest to $(1, 0)$. [*Hint:* Minimize $[d(x)]^2$, where $d(x)$ represents the distance from (x, x^2) to $(1, 0)$.]

14. Use Newton's method to approximate, to within 10^{-4}, the value of x that produces the point on the graph of $y = 1/x$ that is closest to $(2, 1)$.

15. The following describes Newton's method graphically: Suppose that $f'(x)$ exists on $[a, b]$ and that $f'(x) \ne 0$ on $[a, b]$. Further, suppose there exists one $p \in [a, b]$ such that $f(p) = 0$, and let $p_0 \in [a, b]$ be arbitrary. Let p_1 be the point at which the tangent line to f at $(p_0, f(p_0))$ crosses the x-axis. For each $n \ge 1$, let p_n be the x-intercept of the line tangent to f at $(p_{n-1}, f(p_{n-1}))$. Derive the formula describing this method.

16. Use Newton's method to solve the equation

$$0 = \frac{1}{2} + \frac{1}{4}x^2 - x\sin x - \frac{1}{2}\cos 2x, \quad \text{with } p_0 = \frac{\pi}{2}.$$

Iterate using Newton's method until an accuracy of 10^{-5} is obtained. Explain why the result seems unusual for Newton's method. Also, solve the equation with $p_0 = 5\pi$ and $p_0 = 10\pi$.

17. The fourth-degree polynomial

$$f(x) = 230x^4 + 18x^3 + 9x^2 - 221x - 9$$

has two real zeros, one in $[-1, 0]$ and the other in $[0, 1]$. Attempt to approximate these zeros to within 10^{-6} using the

a. Method of False Position

b. Secant method

c. Newton's method

Use the endpoints of each interval as the initial approximations in (a) and (b) and the midpoints as the initial approximation in (c).

18. The function $f(x) = \tan \pi x - 6$ has a zero at $(1/\pi)\arctan 6 \approx 0.447431543$. Let $p_0 = 0$ and $p_1 = 0.48$, and use ten iterations of each of the following methods to approximate this root. Which method is most successful and why?

a. Bisection method

b. Method of False Position

c. Secant method

19. The iteration equation for the Secant method can be written in the simpler form

$$p_n = \frac{f(p_{n-1})p_{n-2} - f(p_{n-2})p_{n-1}}{f(p_{n-1}) - f(p_{n-2})}.$$

Explain why, in general, this iteration equation is likely to be less accurate than the one given in Algorithm 2.4.

20. The equation $x^2 - 10\cos x = 0$ has two solutions, ± 1.3793646. Use Newton's method to approximate the solutions to within 10^{-5} with the following values of p_0.

a. $p_0 = -100$ **b.** $p_0 = -50$ **c.** $p_0 = -25$

d. $p_0 = 25$ **e.** $p_0 = 50$ **f.** $p_0 = 100$

21. The equation $4x^2 - e^x - e^{-x} = 0$ has two positive solutions x_1 and x_2. Use Newton's method to approximate the solution to within 10^{-5} with the following values of p_0.

a. $p_0 = -10$	**b.** $p_0 = -5$	**c.** $p_0 = -3$
d. $p_0 = -1$	**e.** $p_0 = 0$	**f.** $p_0 = 1$
g. $p_0 = 3$	**h.** $p_0 = 5$	**i.** $p_0 = 10$

22. Use Maple to determine how many iterations of Newton's method with $p_0 = \pi/4$ are needed to find a root of $f(x) = \cos x - x$ to within 10^{-100}.

23. The function described by $f(x) = \ln(x^2 + 1) - e^{0.4x} \cos \pi x$ has an infinite number of zeros.

a. Determine, within 10^{-6}, the only negative zero.

b. Determine, within 10^{-6}, the four smallest positive zeros.

c. Determine a reasonable initial approximation to find the nth smallest positive zero of f. [*Hint:* Sketch an approximate graph of f.]

d. Use part (c) to determine, within 10^{-6}, the 25th smallest positive zero of f.

24. Find an approximation for λ, accurate to within 10^{-4}, for the population equation

$$1{,}564{,}000 = 1{,}000{,}000e^{\lambda} + \frac{435{,}000}{\lambda}(e^{\lambda} - 1),$$

discussed in the introduction to this chapter. Use this value to predict the population at the end of the second year, assuming that the immigration rate during this year remains at 435,000 individuals per year.

25. The sum of two numbers is 20. If each number is added to its square root, the product of the two sums is 155.55. Determine the two numbers to within 10^{-4}.

26. The accumulated value of a savings account based on regular periodic payments can be determined from the *annuity due equation*,

$$A = \frac{P}{i}[(1 + i)^n - 1].$$

In this equation, A is the amount in the account, P is the amount regularly deposited, and i is the rate of interest per period for the n deposit periods. An engineer would like to have a savings account valued at \$750,000 upon retirement in 20 years and can afford to put \$1500 per month toward this goal. What is the minimal interest rate at which this amount can be invested, assuming that the interest is compounded monthly?

27. Problems involving the amount of money required to pay off a mortgage over a fixed period of time involve the formula

$$A = \frac{P}{i}[1 - (1 + i)^{-n}],$$

known as an *ordinary annuity equation.* In this equation, A is the amount of the mortgage, P is the amount of each payment, and i is the interest rate per period for the n payment periods. Suppose that a 30-year home mortgage in the amount of \$135,000 is needed and that the borrower can afford house payments of at most \$1000 per month. What is the maximal interest rate the borrower can afford to pay?

28. A drug administered to a patient produces a concentration in the blood stream given by $c(t) = Ate^{-t/3}$ milligrams per milliliter, t hours after A units have been injected. The maximum safe concentration is 1 mg/mL.

a. What amount should be injected to reach this maximum safe concentration, and when does this maximum occur?

b. An additional amount of this drug is to be administered to the patient after the concentration falls to 0.25 mg/mL. Determine, to the nearest minute, when this second injection should be given.

c. Assume that the concentration from consecutive injections is additive and that 75% of the amount originally injected is administered in the second injection. When is it time for the third injection?

29. Let $f(x) = 3^{3x+1} - 7 \cdot 5^{2x}$.

a. Use the Maple commands *solve* and *fsolve* to try to find all roots of f.

b. Plot $f(x)$ to find initial approximations to roots of f.

c. Use Newton's method to find roots of f to within 10^{-16}.

d. Find the exact solutions of $f(x) = 0$ without using Maple.

30. Repeat Exercise 29 using $f(x) = 2^{x^2} - 3 \cdot 7^{x+1}$.

31. The logistic population growth model is described by an equation of the form

$$P(t) = \frac{P_L}{1 - ce^{-kt}},$$

where P_L, c, and $k > 0$ are constants, and $P(t)$ is the population at time t. P_L represents the limiting value of the population since $\lim_{t\to\infty} P(t) = P_L$. Use the census data for the years 1950, 1960, and 1970 listed in the table on page 105 to determine the constants P_L, c, and k for a logistic growth model. Use the logistic model to predict the population of the United States in 1980 and in 2010, assuming $t = 0$ at 1950. Compare the 1980 prediction to the actual value.

32. The Gompertz population growth model is described by

$$P(t) = P_L e^{-ce^{-kt}},$$

where P_L, c, and $k > 0$ are constants, and $P(t)$ is the population at time t. Repeat Exercise 31 using the Gompertz growth model in place of the logistic model.

33. Player A will shut out (win by a score of 21–0) player B in a game of racquetball with probability

$$P = \frac{1+p}{2}\left(\frac{p}{1-p+p^2}\right)^{21},$$

where p denotes the probability A will win any specific rally (independent of the server). (See [Keller, J], p. 267.) Determine, to within 10^{-3}, the minimal value of p that will ensure that A will shut out B in at least half the matches they play.

34. In the design of all-terrain vehicles, it is necessary to consider the failure of the vehicle when attempting to negotiate two types of obstacles. One type of failure is called *hang-up failure* and occurs when the vehicle attempts to cross an obstacle that causes the bottom of the vehicle to touch the ground. The other type of failure is called *nose-in failure* and occurs when the vehicle descends into a ditch and its nose touches the ground.

The accompanying figure, adapted from [Bek], shows the components associated with the nose-in failure of a vehicle. In that reference it is shown that the maximum angle α that can be negotiated by a vehicle when β is the maximum angle at which hang-up failure does *not* occur satisfies the equation

$$A \sin\alpha \cos\alpha + B \sin^2\alpha - C \cos\alpha - E \sin\alpha = 0,$$

where

$$A = l \sin\beta_1, \quad B = l \cos\beta_1, \quad C = (h + 0.5D)\sin\beta_1 - 0.5D\tan\beta_1,$$

$$\text{and} \quad E = (h + 0.5D)\cos\beta_1 - 0.5D.$$

a. It is stated that when $l = 89$ in., $h = 49$ in., $D = 55$ in., and $\beta_1 = 11.5°$, angle α is approximately 33°. Verify this result.

b. Find α for the situation when l, h, and β_1 are the same as in part (a) but $D = 30$ in.

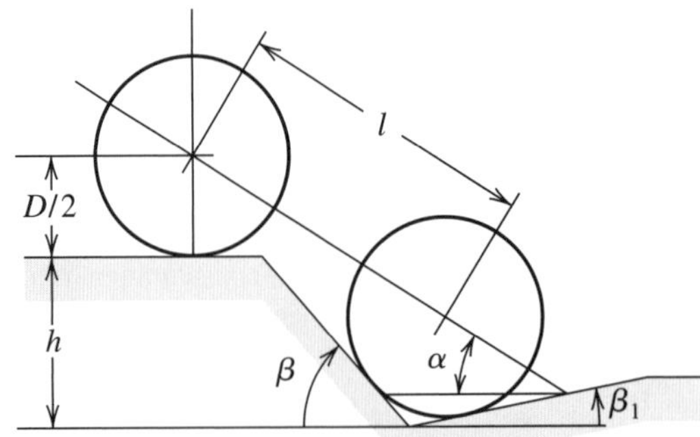

2.4 Error Analysis for Iterative Methods

In this section we investigate the order of convergence of functional iteration schemes and, as a means of obtaining rapid convergence, rediscover Newton's method. We also consider ways of accelerating the convergence of Newton's method in special circumstances. First, however, we need a new procedure for measuring how rapidly a sequence converges.

Order of Convergence

Definition 2.7 Suppose $\{p_n\}_{n=0}^{\infty}$ is a sequence that converges to p, with $p_n \neq p$ for all n. If positive constants λ and α exist with

$$\lim_{n\to\infty} \frac{|p_{n+1} - p|}{|p_n - p|^{\alpha}} = \lambda,$$

then $\{p_n\}_{n=0}^{\infty}$ **converges to p of order α, with asymptotic error constant λ.** ■

An iterative technique of the form $p_n = g(p_{n-1})$ is said to be of *order* α if the sequence $\{p_n\}_{n=0}^{\infty}$ converges to the solution $p = g(p)$ of order α.

In general, a sequence with a high order of convergence converges more rapidly than a sequence with a lower order. The asymptotic constant affects the speed of convergence but not to the extent of the order. Two cases of order are given special attention.

(i) If $\alpha = 1$ (and $\lambda < 1$), the sequence is **linearly convergent.**

(ii) If $\alpha = 2$, the sequence is **quadratically convergent.**

The next illustration compares a linearly convergent sequence to one that is quadratically convergent. It shows why we try to find methods that produce higher-order convergent sequences.

Illustration Suppose that $\{p_n\}_{n=0}^{\infty}$ is linearly convergent to 0 with

$$\lim_{n\to\infty} \frac{|p_{n+1}|}{|p_n|} = 0.5$$

and that $\{\tilde{p}_n\}_{n=0}^{\infty}$ is quadratically convergent to 0 with the same asymptotic error constant,

$$\lim_{n\to\infty} \frac{|\tilde{p}_{n+1}|}{|\tilde{p}_n|^2} = 0.5.$$

For simplicity we assume that for each n we have

$$\frac{|p_{n+1}|}{|p_n|} \approx 0.5 \quad \text{and} \quad \frac{|\tilde{p}_{n+1}|}{|\tilde{p}_n|^2} \approx 0.5.$$

For the linearly convergent scheme, this means that

$$|p_n - 0| = |p_n| \approx 0.5|p_{n-1}| \approx (0.5)^2|p_{n-2}| \approx \cdots \approx (0.5)^n|p_0|,$$

whereas the quadratically convergent procedure has

$$\begin{aligned}
|\tilde{p}_n - 0| = |\tilde{p}_n| &\approx 0.5|\tilde{p}_{n-1}|^2 \approx (0.5)[0.5|\tilde{p}_{n-2}|^2]^2 = (0.5)^3|\tilde{p}_{n-2}|^4 \\
&\approx (0.5)^3[(0.5)|\tilde{p}_{n-3}|^2]^4 = (0.5)^7|\tilde{p}_{n-3}|^8 \\
&\approx \cdots \approx (0.5)^{2^n-1}|\tilde{p}_0|^{2^n}.
\end{aligned}$$

Table 2.7 illustrates the relative speed of convergence of the sequences to 0 if $|p_0| = |\tilde{p}_0| = 1$.

Table 2.7

n	Linear Convergence Sequence $\{p_n\}_{n=0}^{\infty}$ $(0.5)^n$	Quadratic Convergence Sequence $\{\tilde{p}_n\}_{n=0}^{\infty}$ $(0.5)^{2^n-1}$
1	5.0000×10^{-1}	5.0000×10^{-1}
2	2.5000×10^{-1}	1.2500×10^{-1}
3	1.2500×10^{-1}	7.8125×10^{-3}
4	6.2500×10^{-2}	3.0518×10^{-5}
5	3.1250×10^{-2}	4.6566×10^{-10}
6	1.5625×10^{-2}	1.0842×10^{-19}
7	7.8125×10^{-3}	5.8775×10^{-39}

The quadratically convergent sequence is within 10^{-38} of 0 by the seventh term. At least 126 terms are needed to ensure this accuracy for the linearly convergent sequence. □

Quadratically convergent sequences are expected to converge much quicker than those that converge only linearly, but the next result implies that an arbitrary technique that generates a convergent sequences does so only linearly.

Theorem 2.8 Let $g \in C[a, b]$ be such that $g(x) \in [a, b]$, for all $x \in [a, b]$. Suppose, in addition, that g' is continuous on (a, b) and a positive constant $k < 1$ exists with

$$|g'(x)| \leq k, \quad \text{for all } x \in (a, b).$$

If $g'(p) \neq 0$, then for any number $p_0 \neq p$ in $[a, b]$, the sequence

$$p_n = g(p_{n-1}), \quad \text{for } n \geq 1,$$

converges only linearly to the unique fixed point p in $[a, b]$. ■

Proof We know from the Fixed-Point Theorem 2.4 in Section 2.2 that the sequence converges to p. Since g' exists on (a, b), we can apply the Mean Value Theorem to g to show that for any n,

$$p_{n+1} - p = g(p_n) - g(p) = g'(\xi_n)(p_n - p),$$

where ξ_n is between p_n and p. Since $\{p_n\}_{n=0}^{\infty}$ converges to p, we also have $\{\xi_n\}_{n=0}^{\infty}$ converging to p. Since g' is continuous on (a, b), we have

$$\lim_{n\to\infty} g'(\xi_n) = g'(p).$$

Thus

$$\lim_{n\to\infty} \frac{p_{n+1} - p}{p_n - p} = \lim_{n\to\infty} g'(\xi_n) = g'(p) \quad \text{and} \quad \lim_{n\to\infty} \frac{|p_{n+1} - p|}{|p_n - p|} = |g'(p)|.$$

Hence, if $g'(p) \neq 0$, fixed-point iteration exhibits linear convergence with asymptotic error constant $|g'(p)|$. ■ ■ ■

Theorem 2.8 implies that higher-order convergence for fixed-point methods of the form $g(p) = p$ can occur only when $g'(p) = 0$. The next result describes additional conditions that ensure the quadratic convergence we seek.

Theorem 2.9 Let p be a solution of the equation $x = g(x)$. Suppose that $g'(p) = 0$ and g'' is continuous with $|g''(x)| < M$ on an open interval I containing p. Then there exists a $\delta > 0$ such that, for $p_0 \in [p - \delta, p + \delta]$, the sequence defined by $p_n = g(p_{n-1})$, when $n \geq 1$, converges at least quadratically to p. Moreover, for sufficiently large values of n,

$$|p_{n+1} - p| < \frac{M}{2}|p_n - p|^2. \quad \blacksquare$$

Proof Choose k in $(0, 1)$ and $\delta > 0$ such that on the interval $[p-\delta, p+\delta]$, contained in I, we have $|g'(x)| \leq k$ and g'' continuous. Since $|g'(x)| \leq k < 1$, the argument used in the proof of Theorem 2.6 in Section 2.3 shows that the terms of the sequence $\{p_n\}_{n=0}^{\infty}$ are contained in $[p - \delta, p + \delta]$. Expanding $g(x)$ in a linear Taylor polynomial for $x \in [p - \delta, p + \delta]$ gives

$$g(x) = g(p) + g'(p)(x - p) + \frac{g''(\xi)}{2}(x - p)^2,$$

where ξ lies between x and p. The hypotheses $g(p) = p$ and $g'(p) = 0$ imply that

$$g(x) = p + \frac{g''(\xi)}{2}(x - p)^2.$$

In particular, when $x = p_n$,

$$p_{n+1} = g(p_n) = p + \frac{g''(\xi_n)}{2}(p_n - p)^2,$$

with ξ_n between p_n and p. Thus,

$$p_{n+1} - p = \frac{g''(\xi_n)}{2}(p_n - p)^2.$$

Since $|g'(x)| \leq k < 1$ on $[p - \delta, p + \delta]$ and g maps $[p - \delta, p + \delta]$ into itself, it follows from the Fixed-Point Theorem that $\{p_n\}_{n=0}^{\infty}$ converges to p. But ξ_n is between p and p_n for each n, so $\{\xi_n\}_{n=0}^{\infty}$ also converges to p, and

$$\lim_{n\to\infty} \frac{|p_{n+1} - p|}{|p_n - p|^2} = \frac{|g''(p)|}{2}.$$

This result implies that the sequence $\{p_n\}_{n=0}^{\infty}$ is quadratically convergent if $g''(p) \neq 0$ and of higher-order convergence if $g''(p) = 0$.

Because g'' is continuous and strictly bounded by M on the interval $[p - \delta, p + \delta]$, this also implies that, for sufficiently large values of n,

$$|p_{n+1} - p| < \frac{M}{2}|p_n - p|^2. \quad \blacksquare\ \blacksquare\ \blacksquare$$

Theorems 2.8 and 2.9 tell us that our search for quadratically convergent fixed-point methods should point in the direction of functions whose derivatives are zero at the fixed point. That is:

- For a fixed point method to converge quadratically we need to have both $g(p) = p$, and $g'(p) = 0$.

The easiest way to construct a fixed-point problem associated with a root-finding problem $f(x) = 0$ is to add or subtract a multiple of $f(x)$ from x. Consider the sequence

$$p_n = g(p_{n-1}), \quad \text{for } n \geq 1,$$

for g in the form

$$g(x) = x - \phi(x)f(x),$$

where ϕ is a differentiable function that will be chosen later.

For the iterative procedure derived from g to be quadratically convergent, we need to have $g'(p) = 0$ when $f(p) = 0$. Because

$$g'(x) = 1 - \phi'(x)f(x) - f'(x)\phi(x),$$

and $f(p) = 0$, we have

$$g'(p) = 1 - \phi'(p)f(p) - f'(p)\phi(p) = 1 - \phi'(p) \cdot 0 - f'(p)\phi(p) = 1 - f'(p)\phi(p),$$

and $g'(p) = 0$ if and only if $\phi(p) = 1/f'(p)$.

If we let $\phi(x) = 1/f'(x)$, then we will ensure that $\phi(p) = 1/f'(p)$ and produce the quadratically convergent procedure

$$p_n = g(p_{n-1}) = p_{n-1} - \frac{f(p_{n-1})}{f'(p_{n-1})}.$$

This, of course, is simply Newton's method. Hence

- If $f(p) = 0$ and $f'(p) \neq 0$, then for starting values sufficiently close to p, Newton's method will converge at least quadratically.

Multiple Roots

In the preceding discussion, the restriction was made that $f'(p) \neq 0$, where p is the solution to $f(x) = 0$. In particular, Newton's method and the Secant method will generally give problems if $f'(p) = 0$ when $f(p) = 0$. To examine these difficulties in more detail, we make the following definition.

Definition 2.10 A solution p of $f(x) = 0$ is a **zero of multiplicity** m of f if for $x \neq p$, we can write $f(x) = (x - p)^m q(x)$, where $\lim_{x\to p} q(x) \neq 0$. ■

For polynomials, p is a zero of multiplicity m of f if $f(x) = (x - p)^m q(x)$, where $q(p) \neq 0$.

In essence, $q(x)$ represents that portion of $f(x)$ that does not contribute to the zero of f. The following result gives a means to easily identify **simple** zeros of a function, those that have multiplicity one.

Theorem 2.11 The function $f \in C^1[a, b]$ has a simple zero at p in (a, b) if and only if $f(p) = 0$, but $f'(p) \neq 0$. ■

Proof If f has a simple zero at p, then $f(p) = 0$ and $f(x) = (x - p)q(x)$, where $\lim_{x\to p} q(x) \neq 0$. Since $f \in C^1[a, b]$,

$$f'(p) = \lim_{x\to p} f'(x) = \lim_{x\to p}[q(x) + (x - p)q'(x)] = \lim_{x\to p} q(x) \neq 0.$$

Conversely, if $f(p) = 0$, but $f'(p) \neq 0$, expand f in a zeroth Taylor polynomial about p. Then

$$f(x) = f(p) + f'(\xi(x))(x - p) = (x - p)f'(\xi(x)),$$

where $\xi(x)$ is between x and p. Since $f \in C^1[a, b]$,

$$\lim_{x\to p} f'(\xi(x)) = f'\left(\lim_{x\to p} \xi(x)\right) = f'(p) \neq 0.$$

Letting $q = f' \circ \xi$ gives $f(x) = (x - p)q(x)$, where $\lim_{x\to p} q(x) \neq 0$. Thus f has a simple zero at p. ■ ■ ■

The following generalization of Theorem 2.11 is considered in Exercise 12.

Theorem 2.12 The function $f \in C^m[a, b]$ has a zero of multiplicity m at p in (a, b) if and only if

$$0 = f(p) = f'(p) = f''(p) = \cdots = f^{(m-1)}(p), \quad \text{but} \quad f^{(m)}(p) \neq 0.$$ ■

The result in Theorem 2.12 implies that an interval about p exists where Newton's method converges quadratically to p for any initial approximation $p_0 = p$, provided that p is a simple zero. The following example shows that quadratic convergence might not occur if the zero is not simple.

Example 1 Let $f(x) = e^x - x - 1$. **(a)** Show that f has a zero of multiplicity 2 at $x = 0$. **(b)** Show that Newton's method with $p_0 = 1$ converges to this zero but not quadratically.

Table 2.8

n	p_n
0	1.0
1	0.58198
2	0.31906
3	0.16800
4	0.08635
5	0.04380
6	0.02206
7	0.01107
8	0.005545
9	2.7750×10^{-3}
10	1.3881×10^{-3}
11	6.9411×10^{-4}
12	3.4703×10^{-4}
13	1.7416×10^{-4}
14	8.8041×10^{-5}
15	4.2610×10^{-5}
16	1.9142×10^{-6}

Solution **(a)** We have

$$f(x) = e^x - x - 1, \qquad f'(x) = e^x - 1 \qquad \text{and} \qquad f''(x) = e^x,$$

so

$$f(0) = e^0 - 0 - 1 = 0, \qquad f'(0) = e^0 - 1 = 0 \qquad \text{and} \qquad f''(0) = e^0 = 1.$$

Theorem 2.12 implies that f has a zero of multiplicity 2 at $x = 0$.

(b) The first two terms generated by Newton's method applied to f with $p_0 = 1$ are

$$p_1 = p_0 - \frac{f(p_0)}{f'(p_0)} = 1 - \frac{e-2}{e-1} \approx 0.58198,$$

and

$$p_2 = p_1 - \frac{f(p_1)}{f'(p_1)} \approx 0.58198 - \frac{0.20760}{0.78957} \approx 0.31906.$$

The first sixteen terms of the sequence generated by Newton's method are shown in Table 2.8. The sequence is clearly converging to 0, but not quadratically. The graph of f is shown in Figure 2.12. ■

Figure 2.12

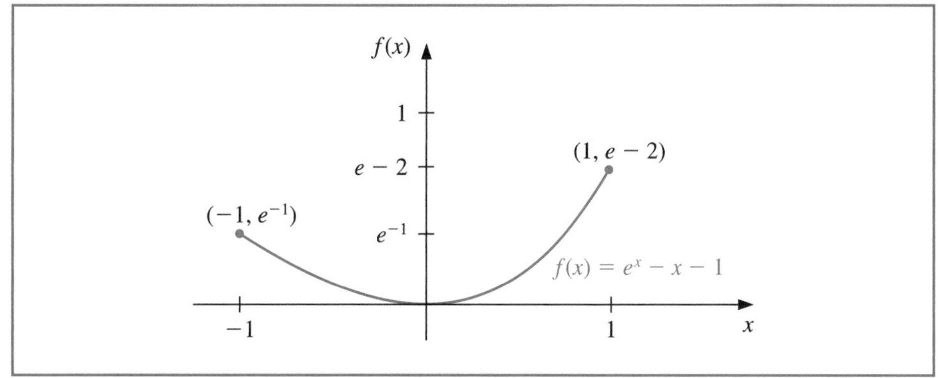

One method of handling the problem of multiple roots of a function f is to define

$$\mu(x) = \frac{f(x)}{f'(x)}.$$

If p is a zero of f of multiplicity m with $f(x) = (x-p)^m q(x)$, then

$$\begin{aligned}\mu(x) &= \frac{(x-p)^m q(x)}{m(x-p)^{m-1}q(x) + (x-p)^m q'(x)} \\ &= (x-p)\frac{q(x)}{mq(x) + (x-p)q'(x)}\end{aligned}$$

also has a zero at p. However, $q(p) \neq 0$, so

$$\frac{q(p)}{mq(p) + (p-p)q'(p)} = \frac{1}{m} \neq 0,$$

and p is a simple zero of $\mu(x)$. Newton's method can then be applied to $\mu(x)$ to give

$$g(x) = x - \frac{\mu(x)}{\mu'(x)} = x - \frac{f(x)/f'(x)}{\{[f'(x)]^2 - [f(x)][f''(x)]\}/[f'(x)]^2}$$

which simplifies to

$$g(x) = x - \frac{f(x)f'(x)}{[f'(x)]^2 - f(x)f''(x)}. \tag{2.13}$$

If g has the required continuity conditions, functional iteration applied to g will be quadratically convergent regardless of the multiplicity of the zero of f. Theoretically, the only drawback to this method is the additional calculation of $f''(x)$ and the more laborious procedure of calculating the iterates. In practice, however, multiple roots can cause serious round-off problems because the denominator of (2.13) consists of the difference of two numbers that are both close to 0.

Example 2 In Example 1 it was shown that $f(x) = e^x - x - 1$ has a zero of multiplicity 2 at $x = 0$ and that Newton's method with $p_0 = 1$ converges to this zero but not quadratically. Show that the modification of Newton's method as given in Eq. (2.13) improves the rate of convergence.

Solution Modified Newton's method gives

$$p_1 = p_0 - \frac{f(p_0)f'(p_0)}{f'(p_0)^2 - f(p_0)f''(p_0)} = 1 - \frac{(e-2)(e-1)}{(e-1)^2 - (e-2)e} \approx -2.3421061 \times 10^{-1}.$$

This is considerably closer to 0 than the first term using Newton's method, which was 0.58918. Table 2.9 lists the first five approximations to the double zero at $x = 0$. The results were obtained using a system with ten digits of precision. The relative lack of improvement in the last two entries is due to the fact that using this system both the numerator and the denominator approach 0. Consequently there is a loss of significant digits of accuracy as the approximations approach 0. ■

Table 2.9

n	p_n
1	$-2.3421061 \times 10^{-1}$
2	$-8.4582788 \times 10^{-3}$
3	$-1.1889524 \times 10^{-5}$
4	$-6.8638230 \times 10^{-6}$
5	$-2.8085217 \times 10^{-7}$

The following illustrates that the modified Newton's method converges quadratically even when in the case of a simple zero.

Illustration In Section 2.2 we found that a zero of $f(x) = x^3 + 4x^2 - 10 = 0$ is $p = 1.36523001$. Here we will compare convergence for a simple zero using both Newton's method and the modified Newton's method listed in Eq. (2.13). Let

$$\textbf{(i)} \quad p_n = p_{n-1} - \frac{p_{n-1}^3 + 4p_{n-1}^2 - 10}{3p_{n-1}^2 + 8p_{n-1}}, \quad \text{from Newton's method}$$

and, from the Modified Newton's method given by Eq. (2.13),

$$\textbf{(ii)} \quad p_n = p_{n-1} - \frac{(p_{n-1}^3 + 4p_{n-1}^2 - 10)(3p_{n-1}^2 + 8p_{n-1})}{(3p_{n-1}^2 + 8p_{n-1})^2 - (p_{n-1}^3 + 4p_{n-1}^2 - 10)(6p_{n-1} + 8)}.$$

With $p_0 = 1.5$, we have

Newton's method

$$p_1 = 1.37333333, \quad p_2 = 1.36526201, \quad \text{and} \quad p_3 = 1.36523001.$$

Modified Newton's method

$$p_1 = 1.35689898, \quad p_2 = 1.36519585, \quad \text{and} \quad p_3 = 1.36523001.$$

Both methods are rapidly convergent to the actual zero, which is given by both methods as p_3. Note, however, that in the case of a simple zero the original Newton's method requires substantially less computation. □

Maple contains Modified Newton's method as described in Eq. (2.13) in its *NumericalAnalysis* package. The options for this command are the same as those for the Bisection method. To obtain results similar to those in Table 2.9 we can use

with(Student[NumericalAnalysis])

$f := e^x - x - 1$

ModifiedNewton $(f, x = 1.0, \mathit{tolerance} = 10^{-10}, \mathit{output} = \mathit{sequence}, \mathit{maxiterations} = 20)$

Remember that there is sensitivity to round-off error in these calculations, so you might need to reset *Digits* in Maple to get the exact values in Table 2.9.

EXERCISE SET 2.4

1. Use Newton's method to find solutions accurate to within 10^{-5} to the following problems.
   a. $x^2 - 2xe^{-x} + e^{-2x} = 0, \quad \text{for } 0 \le x \le 1$
   b. $x^3 - 3x^2(2^{-x}) + 3x(4^{-x}) - 8^{-x} = 0, \quad \text{for } 0 \le x \le 1$
   c. $\cos(x + \sqrt{2}) + x(x/2 + \sqrt{2}) = 0, \quad \text{for } -2 \le x \le -1$
   d. $e^{6x} + 3(\ln 2)^2 e^{2x} - (\ln 8)e^{4x} - (\ln 2)^3 = 0, \quad \text{for } -1 \le x \le 0$
2. Use Newton's method to find solutions accurate to within 10^{-5} to the following problems.
   a. $x^2 + 6x^5 + 9x^4 - 2x^3 - 6x^2 + 1 = 0, \quad \text{for } -3 \le x \le -2$
   b. $1 - 4x\cos x + 2x^2 + \cos 2x = 0, \quad \text{for } 0 \le x \le 1$
   c. $\sin 3x + 3e^{-2x}\sin x - 3e^{-x}\sin 2x - e^{-3x} = 0, \quad \text{for } 3 \le x \le 4$
   d. $e^{3x} - 27x^6 + 27x^4e^x - 9x^2e^{2x} = 0, \quad \text{for } 3 \le x \le 5$
3. Repeat Exercise 1 using the modified Newton's method described in Eq. (2.13). Is there an improvement in speed or accuracy over Exercise 1?

4. Repeat Exercise 2 using the modified Newton's method described in Eq. (2.13). Is there an improvement in speed or accuracy over Exercise 2?

5. Use Newton's method and the modified Newton's method described in Eq. (2.13) to find a solution accurate to within 10^{-5} to the problem

$$e^{6x} + 1.441e^{2x} - 2.079e^{4x} - 0.3330 = 0, \quad \text{for } -1 \le x \le 0.$$

This is the same problem as 1(d) with the coefficients replaced by their four-digit approximations. Compare the solutions to the results in 1(d) and 2(d).

6. Show that the following sequences converge linearly to $p = 0$. How large must n be before $|p_n - p| \le 5 \times 10^{-2}$?
   a. $p_n = \dfrac{1}{n}, \quad n \ge 1$
   b. $p_n = \dfrac{1}{n^2}, \quad n \ge 1$

7. **a.** Show that for any positive integer k, the sequence defined by $p_n = 1/n^k$ converges linearly to $p = 0$.
   b. For each pair of integers k and m, determine a number N for which $1/N^k < 10^{-m}$.

8. **a.** Show that the sequence $p_n = 10^{-2^n}$ converges quadratically to 0.
   b. Show that the sequence $p_n = 10^{-n^k}$ does not converge to 0 quadratically, regardless of the size of the exponent $k > 1$.

9. **a.** Construct a sequence that converges to 0 of order 3.
   b. Suppose $\alpha > 1$. Construct a sequence that converges to 0 zero of order α.

10. Suppose p is a zero of multiplicity m of f, where $f^{(m)}$ is continuous on an open interval containing p. Show that the following fixed-point method has $g'(p) = 0$:

$$g(x) = x - \frac{mf(x)}{f'(x)}.$$

11. Show that the Bisection Algorithm 2.1 gives a sequence with an error bound that converges linearly to 0.

12. Suppose that f has m continuous derivatives. Modify the proof of Theorem 2.11 to show that f has a zero of multiplicity m at p if and only if

$$0 = f(p) = f'(p) = \cdots = f^{(m-1)}(p), \quad \text{but } f^{(m)}(p) \ne 0.$$

13. The iterative method to solve $f(x) = 0$, given by the fixed-point method $g(x) = x$, where

$$p_n = g(p_{n-1}) = p_{n-1} - \frac{f(p_{n-1})}{f'(p_{n-1})} - \frac{f''(p_{n-1})}{2f'(p_{n-1})}\left[\frac{f(p_{n-1})}{f'(p_{n-1})}\right]^2, \quad \text{for } n = 1, 2, 3, \ldots,$$

has $g'(p) = g''(p) = 0$. This will generally yield cubic ($\alpha = 3$) convergence. Expand the analysis of Example 1 to compare quadratic and cubic convergence.

14. It can be shown (see, for example, [DaB], pp. 228–229) that if $\{p_n\}_{n=0}^{\infty}$ are convergent Secant method approximations to p, the solution to $f(x) = 0$, then a constant C exists with $|p_{n+1} - p| \approx C\,|p_n - p|\,|p_{n-1} - p|$ for sufficiently large values of n. Assume $\{p_n\}$ converges to p of order α, and show that $\alpha = (1 + \sqrt{5})/2$. (*Note:* This implies that the order of convergence of the Secant method is approximately 1.62).

2.5 Accelerating Convergence

Theorem 2.8 indicates that it is rare to have the luxury of quadratic convergence. We now consider a technique called **Aitken's Δ^2 method** that can be used to accelerate the convergence of a sequence that is linearly convergent, regardless of its origin or application.

Aitken's Δ^2 Method

Alexander Aitken (1895-1967) used this technique in 1926 to accelerate the rate of convergence of a series in a paper on algebraic equations [Ai]. This process is similar to one used much earlier by the Japanese mathematician Takakazu Seki Kowa (1642-1708).

Suppose $\{p_n\}_{n=0}^{\infty}$ is a linearly convergent sequence with limit p. To motivate the construction of a sequence $\{\hat{p}_n\}_{n=0}^{\infty}$ that converges more rapidly to p than does $\{p_n\}_{n=0}^{\infty}$, let us first assume that the signs of $p_n - p$, $p_{n+1} - p$, and $p_{n+2} - p$ agree and that n is sufficiently large that

$$\frac{p_{n+1} - p}{p_n - p} \approx \frac{p_{n+2} - p}{p_{n+1} - p}.$$

Then

$$(p_{n+1} - p)^2 \approx (p_{n+2} - p)(p_n - p),$$

so

$$p_{n+1}^2 - 2p_{n+1}p + p^2 \approx p_{n+2}p_n - (p_n + p_{n+2})p + p^2$$

and

$$(p_{n+2} + p_n - 2p_{n+1})p \approx p_{n+2}p_n - p_{n+1}^2.$$

Solving for p gives

$$p \approx \frac{p_{n+2}p_n - p_{n+1}^2}{p_{n+2} - 2p_{n+1} + p_n}.$$

Adding and subtracting the terms p_n^2 and $2p_np_{n+1}$ in the numerator and grouping terms appropriately gives

$$p \approx \frac{p_np_{n+2} - 2p_np_{n+1} + p_n^2 - p_{n+1}^2 + 2p_np_{n+1} - p_n^2}{p_{n+2} - 2p_{n+1} + p_n}$$

$$= \frac{p_n(p_{n+2} - 2p_{n+1} + p_n) - (p_{n+1}^2 - 2p_np_{n+1} + p_n^2)}{p_{n+2} - 2p_{n+1} + p_n}$$

$$= p_n - \frac{(p_{n+1} - p_n)^2}{p_{n+2} - 2p_{n+1} + p_n}.$$

Aitken's Δ^2 method is based on the assumption that the sequence $\{\hat{p}_n\}_{n=0}^{\infty}$, defined by

$$\hat{p}_n = p_n - \frac{(p_{n+1} - p_n)^2}{p_{n+2} - 2p_{n+1} + p_n}, \tag{2.14}$$

converges more rapidly to p than does the original sequence $\{p_n\}_{n=0}^{\infty}$.

Table 2.10

n	p_n	$\hat{p}_n$
1	0.54030	0.96178
2	0.87758	0.98213
3	0.94496	0.98979
4	0.96891	0.99342
5	0.98007	0.99541
6	0.98614	
7	0.98981	

Example 1 The sequence $\{p_n\}_{n=1}^{\infty}$, where $p_n = \cos(1/n)$, converges linearly to $p = 1$. Determine the first five terms of the sequence given by Aitken's Δ^2 method.

Solution In order to determine a term $\hat{p}_n$ of the Aitken's Δ^2 method sequence we need to have the terms p_n, p_{n+1}, and p_{n+2} of the original sequence. So to determine $\hat{p}_5$ we need the first 7 terms of $\{p_n\}$. These are given in Table 2.10. It certainly appears that $\{\hat{p}_n\}_{n=1}^{\infty}$ converges more rapidly to $p = 1$ than does $\{p_n\}_{n=1}^{\infty}$. ■

The Δ notation associated with this technique has its origin in the following definition.

Definition 2.13 For a given sequence $\{p_n\}_{n=0}^{\infty}$, the **forward difference** Δp_n (read "delta p_n") is defined by

$$\Delta p_n = p_{n+1} - p_n, \quad \text{for } n \geq 0.$$

Higher powers of the operator Δ are defined recursively by

$$\Delta^k p_n = \Delta(\Delta^{k-1} p_n), \quad \text{for } k \geq 2.$$

■

The definition implies that

$$\Delta^2 p_n = \Delta(p_{n+1} - p_n) = \Delta p_{n+1} - \Delta p_n = (p_{n+2} - p_{n+1}) - (p_{n+1} - p_n).$$

So $\Delta^2 p_n = p_{n+2} - 2p_{n+1} + p_n$, and the formula for $\hat{p}_n$ given in Eq. (2.14) can be written as

$$\hat{p}_n = p_n - \frac{(\Delta p_n)^2}{\Delta^2 p_n}, \quad \text{for } n \geq 0. \tag{2.15}$$

To this point in our discussion of Aitken's Δ^2 method, we have stated that the sequence $\{\hat{p}_n\}_{n=0}^{\infty}$, converges to p more rapidly than does the original sequence $\{p_n\}_{n=0}^{\infty}$, but we have not said what is meant by the term "more rapid" convergence. Theorem 2.14 explains and justifies this terminology. The proof of this theorem is considered in Exercise 16.

Theorem 2.14 Suppose that $\{p_n\}_{n=0}^{\infty}$ is a sequence that converges linearly to the limit p and that

$$\lim_{n\to\infty} \frac{p_{n+1} - p}{p_n - p} < 1.$$

Then the Aitken's Δ^2 sequence $\{\hat{p}_n\}_{n=0}^{\infty}$ converges to p faster than $\{p_n\}_{n=0}^{\infty}$ in the sense that

$$\lim_{n\to\infty} \frac{\hat{p}_n - p}{p_n - p} = 0.$$

■

Steffensen's Method

Johan Frederik Steffensen (1873–1961) wrote an influential book entitled *Interpolation* in 1927.

By applying a modification of Aitken's Δ^2 method to a linearly convergent sequence obtained from fixed-point iteration, we can accelerate the convergence to quadratic. This procedure is known as Steffensen's method and differs slightly from applying Aitken's Δ^2 method directly to the linearly convergent fixed-point iteration sequence. Aitken's Δ^2 method constructs the terms in order:

$$p_0, \quad p_1 = g(p_0), \quad p_2 = g(p_1), \quad \hat{p}_0 = \{\Delta^2\}(p_0),$$
$$p_3 = g(p_2), \quad \hat{p}_1 = \{\Delta^2\}(p_1), \ldots,$$

where $\{\Delta^2\}$ indicates that Eq. (2.15) is used. Steffensen's method constructs the same first four terms, p_0, p_1, p_2, and $\hat{p}_0$. However, at this step we assume that $\hat{p}_0$ is a better approximation to p than is p_2 and apply fixed-point iteration to $\hat{p}_0$ instead of p_2. Using this notation, the sequence is

$$p_0^{(0)}, \quad p_1^{(0)} = g(p_0^{(0)}), \quad p_2^{(0)} = g(p_1^{(0)}), \quad p_0^{(1)} = \{\Delta^2\}(p_0^{(0)}), \quad p_1^{(1)} = g(p_0^{(1)}), \ldots.$$

Every third term of the Steffensen sequence is generated by Eq. (2.15); the others use fixed-point iteration on the previous term. The process is described in Algorithm 2.6.

ALGORITHM 2.6

Steffensen's

To find a solution to $p = g(p)$ given an initial approximation p_0:

INPUT initial approximation p_0; tolerance TOL; maximum number of iterations N_0.

OUTPUT approximate solution p or message of failure.

Step 1 Set $i = 1$.

Step 2 While $i \leq N_0$ do Steps 3–6.

Step 3 Set $p_1 = g(p_0)$; (*Compute* $p_1^{(i-1)}$.)
$p_2 = g(p_1)$; (*Compute* $p_2^{(i-1)}$.)
$p = p_0 - (p_1 - p_0)^2/(p_2 - 2p_1 + p_0)$. (*Compute* $p_0^{(i)}$.)

Step 4 If $|p - p_0| < TOL$ then
OUTPUT (p); (*Procedure completed successfully.*)
STOP.

Step 5 Set $i = i + 1$.

Step 6 Set $p_0 = p$. (*Update* p_0.)

Step 7 OUTPUT ('Method failed after N_0 iterations, N_0 =', N_0);
(*Procedure completed unsuccessfully.*)
STOP. ■

Note that $\Delta^2 p_n$ might be 0, which would introduce a 0 in the denominator of the next iterate. If this occurs, we terminate the sequence and select $p_2^{(n-1)}$ as the best approximation.

Illustration To solve $x^3 + 4x^2 - 10 = 0$ using Steffensen's method, let $x^3 + 4x^2 = 10$, divide by $x + 4$, and solve for x. This procedure produces the fixed-point method

$$g(x) = \left(\frac{10}{x+4}\right)^{1/2}.$$

We considered this fixed-point method in Table 2.2 column (**d**) of Section 2.2.

Applying Steffensen's procedure with $p_0 = 1.5$ gives the values in Table 2.11. The iterate $p_0^{(2)} = 1.365230013$ is accurate to the ninth decimal place. In this example, Steffensen's method gave about the same accuracy as Newton's method applied to this polynomial. These results can be seen in the Illustration at the end of Section 2.4. □

Table 2.11

k	$p_0^{(k)}$	$p_1^{(k)}$	$p_2^{(k)}$
0	1.5	1.348399725	1.367376372
1	1.365265224	1.365225534	1.365230583
2	1.365230013		

From the Illustration, it appears that Steffensen's method gives quadratic convergence without evaluating a derivative, and Theorem 2.14 states that this is the case. The proof of this theorem can be found in [He2], pp. 90–92, or [IK], pp. 103–107.

Theorem 2.15 Suppose that $x = g(x)$ has the solution p with $g'(p) \neq 1$. If there exists a $\delta > 0$ such that $g \in C^3[p - \delta, p + \delta]$, then Steffensen's method gives quadratic convergence for any $p_0 \in [p - \delta, p + \delta]$. ■

Steffensen's method can be implemented in Maple with the *NumericalAnalysis* package. For example, after entering the function

$$g := \sqrt{\frac{10}{x+4}}$$

the Maple command

Steffensen(*fixedpointiterator* = g, x = 1.5, *tolerance* = 10^{-8}, *output* = *information*, *maxiterations* = 20)

produces the results in Table 2.11, as well as an indication that the final approximation has a relative error of approximately 7.32×10^{-10}.

EXERCISE SET 2.5

1. The following sequences are linearly convergent. Generate the first five terms of the sequence $\{\hat{p}_n\}$ using Aitken's Δ^2 method.
   a. $p_0 = 0.5, \quad p_n = \cos p_{n-1}, \quad n \geq 1$
   b. $p_0 = 0.5, \quad p_n = 3^{-p_{n-1}}, \quad n \geq 1$
   c. $p_0 = 0.75, \quad p_n = (e^{p_{n-1}}/3)^{1/2}, \quad n \geq 1$
   d. $p_0 = 0.5, \quad p_n = (2 - e^{p_{n-1}} + p_{n-1}^2)/3, \quad n \geq 1$
2. Consider the function $f(x) = e^{6x} + 3(\ln 2)^2 e^{2x} - (\ln 8)e^{4x} - (\ln 2)^3$. Use Newton's method with $p_0 = 0$ to approximate a zero of f. Generate terms until $|p_{n+1} - p_n| < 0.0002$. Construct the sequence $\{\hat{p}_n\}$. Is the convergence improved?
3. Let $g(x) = \cos(x - 1)$ and $p_0^{(0)} = 2$. Use Steffensen's method to find $p_0^{(1)}$.
4. Let $g(x) = 1 + (\sin x)^2$ and $p_0^{(0)} = 1$. Use Steffensen's method to find $p_0^{(1)}$ and $p_0^{(2)}$.
5. Steffensen's method is applied to a function $g(x)$ using $p_0^{(0)} = 1$ and $p_2^{(0)} = 3$ to obtain $p_0^{(1)} = 0.75$. What is $p_1^{(0)}$?
6. Steffensen's method is applied to a function $g(x)$ using $p_0^{(0)} = 1$ and $p_1^{(0)} = \sqrt{2}$ to obtain $p_0^{(1)} = 2.7802$. What is $p_2^{(0)}$?
7. Use Steffensen's method to find, to an accuracy of 10^{-4}, the root of $x^3 - x - 1 = 0$ that lies in $[1, 2]$, and compare this to the results of Exercise 6 of Section 2.2.
8. Use Steffensen's method to find, to an accuracy of 10^{-4}, the root of $x - 2^{-x} = 0$ that lies in $[0, 1]$, and compare this to the results of Exercise 8 of Section 2.2.
9. Use Steffensen's method with $p_0 = 2$ to compute an approximation to $\sqrt{3}$ accurate to within 10^{-4}. Compare this result with those obtained in Exercise 9 of Section 2.2 and Exercise 12 of Section 2.1.
10. Use Steffensen's method with $p_0 = 3$ to compute an approximation to $\sqrt[3]{25}$ accurate to within 10^{-4}. Compare this result with those obtained in Exercise 10 of Section 2.2 and Exercise 13 of Section 2.1.
11. Use Steffensen's method to approximate the solutions of the following equations to within 10^{-5}.
    a. $x = (2 - e^x + x^2)/3$, where g is the function in Exercise 11(a) of Section 2.2.
    b. $x = 0.5(\sin x + \cos x)$, where g is the function in Exercise 11(f) of Section 2.2.
    c. $x = (e^x/3)^{1/2}$, where g is the function in Exercise 11(c) of Section 2.2.
    d. $x = 5^{-x}$, where g is the function in Exercise 11(d) of Section 2.2.
12. Use Steffensen's method to approximate the solutions of the following equations to within 10^{-5}.
    a. $2 + \sin x - x = 0$, where g is the function in Exercise 12(a) of Section 2.2.
    b. $x^3 - 2x - 5 = 0$, where g is the function in Exercise 12(b) of Section 2.2.

c. $3x^2 - e^x = 0$, where g is the function in Exercise 12(c) of Section 2.2.

d. $x - \cos x = 0$, where g is the function in Exercise 12(d) of Section 2.2.

13. The following sequences converge to 0. Use Aitken's Δ^2 method to generate $\{\hat{p}_n\}$ until $|\hat{p}_n| \leq 5\times 10^{-2}$:

a. $p_n = \dfrac{1}{n}, \quad n \geq 1$ **b.** $p_n = \dfrac{1}{n^2}, \quad n \geq 1$

14. A sequence $\{p_n\}$ is said to be **superlinearly convergent** to p if

$$\lim_{n\to\infty} \frac{|p_{n+1} - p|}{|p_n - p|} = 0.$$

a. Show that if $p_n \to p$ of order α for $\alpha > 1$, then $\{p_n\}$ is superlinearly convergent to p.

b. Show that $p_n = \frac{1}{n^n}$ is superlinearly convergent to 0 but does not converge to 0 of order α for any $\alpha > 1$.

15. Suppose that $\{p_n\}$ is superlinearly convergent to p. Show that

$$\lim_{n\to\infty} \frac{|p_{n+1} - p_n|}{|p_n - p|} = 1.$$

16. Prove Theorem 2.14. [*Hint:* Let $\delta_n = (p_{n+1} - p)/(p_n - p) - \lambda$, and show that $\lim_{n\to\infty} \delta_n = 0$. Then express $(\hat{p}_{n+1} - p)/(p_n - p)$ in terms of δ_n, δ_{n+1}, and λ.]

17. Let $P_n(x)$ be the nth Taylor polynomial for $f(x) = e^x$ expanded about $x_0 = 0$.

a. For fixed x, show that $p_n = P_n(x)$ satisfies the hypotheses of Theorem 2.14.

b. Let $x = 1$, and use Aitken's Δ^2 method to generate the sequence $\hat{p}_0, \ldots, \hat{p}_8$.

c. Does Aitken's method accelerate convergence in this situation?

2.6 Zeros of Polynomials and Müller's Method

A *polynomial of degree n* has the form

$$P(x) = a_n x^n + a_{n-1}x^{n-1} + \cdots + a_1 x + a_0,$$

where the a_i's, called the *coefficients* of P, are constants and $a_n \neq 0$. The zero function, $P(x) = 0$ for all values of x, is considered a polynomial but is assigned no degree.

Algebraic Polynomials

Theorem 2.16 **(Fundamental Theorem of Algebra)**

If $P(x)$ is a polynomial of degree $n \geq 1$ with real or complex coefficients, then $P(x) = 0$ has at least one (possibly complex) root. ■

Although the Fundamental Theorem of Algebra is basic to any study of elementary functions, the usual proof requires techniques from the study of complex function theory. The reader is referred to [SaS], p. 155, for the culmination of a systematic development of the topics needed to prove the Theorem.

Example 1 Determine all the zeros of the polynomial $P(x) = x^3 - 5x^2 + 17x - 13$.

Solution It is easily verified that $P(1) = 1 - 5 + 17 - 13 = 0$. so $x = 1$ is a zero of P and $(x - 1)$ is a factor of the polynomial. Dividing $P(x)$ by $x - 1$ gives

$$P(x) = (x - 1)(x^2 - 4x + 13).$$

Carl Friedrich Gauss (1777–1855), one of the greatest mathematicians of all time, proved the Fundamental Theorem of Algebra in his doctoral dissertation and published it in 1799. He published different proofs of this result throughout his lifetime, in 1815, 1816, and as late as 1848. The result had been stated, without proof, by Albert Girard (1595–1632), and partial proofs had been given by Jean d'Alembert (1717–1783), Euler, and Lagrange.

To determine the zeros of $x^2 - 4x + 13$ we use the quadratic formula in its standard form, which gives the complex zeros

$$\frac{-(-4) \pm \sqrt{(-4)^2 - 4(1)(13)}}{2(1)} = \frac{4 \pm \sqrt{-36}}{2} = 2 \pm 3i.$$

Hence the third-degree polynomial $P(x)$ has three zeros, $x_1 = 1$, $x_2 = 2 - 3i$, and $x_2 = 2 + 3i$. ■

In the preceding example we found that the third-degree polynomial had three distinct zeros. An important consequence of the Fundamental Theorem of Algebra is the following corollary. It states that this is always the case, provided that when the zeros are not distinct we count the number of zeros according to their multiplicities.

Corollary 2.17 If $P(x)$ is a polynomial of degree $n \geq 1$ with real or complex coefficients, then there exist unique constants $x_1, x_2, \ldots, x_k$, possibly complex, and unique positive integers $m_1, m_2, \ldots, m_k$, such that $\sum_{i=1}^{k} m_i = n$ and

$$P(x) = a_n(x - x_1)^{m_1}(x - x_2)^{m_2} \cdots (x - x_k)^{m_k}.$$ ■

By Corollary 2.17 the collection of zeros of a polynomial is unique and, if each zero x_i is counted as many times as its multiplicity m_i, a polynomial of degree n has exactly n zeros.

The following corollary of the Fundamental Theorem of Algebra is used often in this section and in later chapters.

Corollary 2.18 Let $P(x)$ and $Q(x)$ be polynomials of degree at most n. If $x_1, x_2, \ldots, x_k$, with $k > n$, are distinct numbers with $P(x_i) = Q(x_i)$ for $i = 1, 2, \ldots, k$, then $P(x) = Q(x)$ for all values of x. ■

This result implies that to show that two polynomials of degree less than or equal to n are the same, we only need to show that they agree at $n + 1$ values. This will be frequently used, particularly in Chapters 3 and 8.

Horner's Method

William Horner (1786–1837) was a child prodigy who became headmaster of a school in Bristol at age 18. Horner's method for solving algebraic equations was published in 1819 in the Philosophical Transactions of the Royal Society.

To use Newton's method to locate approximate zeros of a polynomial $P(x)$, we need to evaluate $P(x)$ and $P'(x)$ at specified values. Since $P(x)$ and $P'(x)$ are both polynomials, computational efficiency requires that the evaluation of these functions be done in the nested manner discussed in Section 1.2. Horner's method incorporates this nesting technique, and, as a consequence, requires only n multiplications and n additions to evaluate an arbitrary nth-degree polynomial.

Theorem 2.19 **(Horner's Method)**

Let

$$P(x) = a_n x^n + a_{n-1} x^{n-1} + \cdots + a_1 x + a_0.$$

Define $b_n = a_n$ and

$$b_k = a_k + b_{k+1} x_0, \quad \text{for } k = n-1, n-2, \ldots, 1, 0.$$

Then $b_0 = P(x_0)$. Moreover, if

$$Q(x) = b_n x^{n-1} + b_{n-1}x^{n-2} + \cdots + b_2 x + b_1,$$

then

$$P(x) = (x - x_0)Q(x) + b_0. \qquad \blacksquare$$

Paolo Ruffini (1765–1822) had described a similar method which won him the gold medal from the Italian Mathematical Society for Science. Neither Ruffini nor Horner was the first to discover this method; it was known in China at least 500 years earlier.

Proof By the definition of $Q(x)$,

$$\begin{aligned}(x - x_0)Q(x) + b_0 &= (x - x_0)(b_n x^{n-1} + \cdots + b_2 x + b_1) + b_0\\ &= (b_n x^n + b_{n-1}x^{n-1} + \cdots + b_2 x^2 + b_1 x)\\ &\quad - (b_n x_0 x^{n-1} + \cdots + b_2 x_0 x + b_1 x_0) + b_0\\ &= b_n x^n + (b_{n-1} - b_n x_0)x^{n-1} + \cdots + (b_1 - b_2 x_0)x + (b_0 - b_1 x_0).\end{aligned}$$

By the hypothesis, $b_n = a_n$ and $b_k - b_{k+1}x_0 = a_k$, so

$$(x - x_0)Q(x) + b_0 = P(x) \quad \text{and} \quad b_0 = P(x_0). \qquad \blacksquare\ \blacksquare\ \blacksquare$$

Example 2 Use Horner's method to evaluate $P(x) = 2x^4 - 3x^2 + 3x - 4$ at $x_0 = -2$.

Solution When we use hand calculation in Horner's method, we first construct a table, which suggests the *synthetic division* name that is often applied to the technique. For this problem, the table appears as follows:

	Coefficient of x^4	Coefficient of x^3	Coefficient of x^2	Coefficient of x	Constant term
$x_0 = -2$	$a_4 = 2$	$a_3 = 0$	$a_2 = -3$	$a_1 = 3$	$a_0 = -4$
		$b_4 x_0 = -4$	$b_3 x_0 = 8$	$b_2 x_0 = -10$	$b_1 x_0 = 14$
	$b_4 = 2$	$b_3 = -4$	$b_2 = 5$	$b_1 = -7$	$b_0 = 10$

So,

$$P(x) = (x + 2)(2x^3 - 4x^2 + 5x - 7) + 10. \qquad \blacksquare$$

The word synthetic has its roots in various languages. In standard English it generally provides the sense of something that is "false" or "substituted". But in mathematics it takes the form of something that is "grouped together". Synthetic geometry treats shapes as whole, rather than as individual objects, which is the style in analytic geometry. In synthetic division of polynomials, the various powers of the variables are not explicitly given but kept grouped together.

An additional advantage of using the Horner (or synthetic-division) procedure is that, since

$$P(x) = (x - x_0)Q(x) + b_0,$$

where

$$Q(x) = b_n x^{n-1} + b_{n-1}x^{n-2} + \cdots + b_2 x + b_1,$$

differentiating with respect to x gives

$$P'(x) = Q(x) + (x - x_0)Q'(x) \quad \text{and} \quad P'(x_0) = Q(x_0). \tag{2.16}$$

When the Newton-Raphson method is being used to find an approximate zero of a polynomial, $P(x)$ and $P'(x)$ can be evaluated in the same manner.

Example 3 Find an approximation to a zero of

$$P(x) = 2x^4 - 3x^2 + 3x - 4,$$

using Newton's method with $x_0 = -2$ and synthetic division to evaluate $P(x_n)$ and $P'(x_n)$ for each iterate x_n.

Solution With $x_0 = -2$ as an initial approximation, we obtained $P(-2)$ in Example 1 by

$$\begin{array}{r|rrrrrl} x_0 = -2 & 2 & 0 & -3 & 3 & -4 & \\ & & -4 & 8 & -10 & 14 & \\ \hline & 2 & -4 & 5 & -7 & 10 & = P(-2). \end{array}$$

Using Theorem 2.19 and Eq. (2.16),

$$Q(x) = 2x^3 - 4x^2 + 5x - 7 \quad \text{and} \quad P'(-2) = Q(-2),$$

so $P'(-2)$ can be found by evaluating $Q(-2)$ in a similar manner:

$$\begin{array}{r|rrrrl} x_0 = -2 & 2 & -4 & 5 & -7 & \\ & & -4 & 16 & -42 & \\ \hline & 2 & -8 & 21 & -49 & = Q(-2) = P'(-2) \end{array}$$

and

$$x_1 = x_0 - \frac{P(x_0)}{P'(x_0)} = x_0 - \frac{P(x_0)}{Q(x_0)} = -2 - \frac{10}{-49} \approx -1.796.$$

Repeating the procedure to find x_2 gives

$$\begin{array}{r|rrrrrl} -1.796 & 2 & 0 & -3 & 3 & -4 & \\ & & -3.592 & 6.451 & -6.197 & 5.742 & \\ \hline & 2 & -3.592 & 3.451 & -3.197 & 1.742 & = P(x_1) \\ & & -3.592 & 12.902 & -29.368 & & \\ \hline & 2 & -7.184 & 16.353 & -32.565 & = Q(x_1) & = P'(x_1). \end{array}$$

So $P(-1.796) = 1.742$, $P'(-1.796) = Q(-1.796) = -32.565$, and

$$x_2 = -1.796 - \frac{1.742}{-32.565} \approx -1.7425.$$

In a similar manner, $x_3 = -1.73897$, and an actual zero to five decimal places is -1.73896.

Note that the polynomial $Q(x)$ depends on the approximation being used and changes from iterate to iterate. ■

Algorithm 2.7 computes $P(x_0)$ and $P'(x_0)$ using Horner's method.

Horner's

To evaluate the polynomial

$$P(x) = a_n x^n + a_{n-1}x^{n-1} + \cdots + a_1 x + a_0 = (x - x_0)Q(x) + b_0$$

and its derivative at x_0:

INPUT degree n; coefficients $a_0, a_1, \ldots, a_n$; x_0.

OUTPUT $y = P(x_0)$; $z = P'(x_0)$.

Step 1 Set $y = a_n$; (*Compute b_n for P.*)
$z = a_n$. (*Compute b_{n-1} for Q.*)

Step 2 For $j = n-1, n-2, \ldots, 1$
set $y = x_0 y + a_j$; (*Compute b_j for P.*)
$z = x_0 z + y$. (*Compute b_{j-1} for Q.*)

Step 3 Set $y = x_0 y + a_0$. (*Compute b_0 for P.*)

Step 4 OUTPUT (y, z);
STOP. ■

If the Nth iterate, x_N, in Newton's method is an approximate zero for P, then

$$P(x) = (x - x_N)Q(x) + b_0 = (x - x_N)Q(x) + P(x_N) \approx (x - x_N)Q(x),$$

so $x - x_N$ is an approximate factor of $P(x)$. Letting $\hat{x}_1 = x_N$ be the approximate zero of P and $Q_1(x) \equiv Q(x)$ be the approximate factor gives

$$P(x) \approx (x - \hat{x}_1)Q_1(x).$$

We can find a second approximate zero of P by applying Newton's method to $Q_1(x)$.

If $P(x)$ is an nth-degree polynomial with n real zeros, this procedure applied repeatedly will eventually result in $(n-2)$ approximate zeros of P and an approximate quadratic factor $Q_{n-2}(x)$. At this stage, $Q_{n-2}(x) = 0$ can be solved by the quadratic formula to find the last two approximate zeros of P. Although this method can be used to find all the approximate zeros, it depends on repeated use of approximations and can lead to inaccurate results.

The procedure just described is called **deflation**. The accuracy difficulty with deflation is due to the fact that, when we obtain the approximate zeros of $P(x)$, Newton's method is used on the reduced polynomial $Q_k(x)$, that is, the polynomial having the property that

$$P(x) \approx (x - \hat{x}_1)(x - \hat{x}_2)\cdots(x - \hat{x}_k)Q_k(x).$$

An approximate zero $\hat{x}_{k+1}$ of Q_k will generally not approximate a root of $P(x) = 0$ as well as it does a root of the reduced equation $Q_k(x) = 0$, and inaccuracy increases as k increases. One way to eliminate this difficulty is to use the reduced equations to find approximations $\hat{x}_2$, $\hat{x}_3, \ldots, \hat{x}_k$ to the zeros of P, and then improve these approximations by applying Newton's method to the original polynomial $P(x)$.

Complex Zeros: Müller's Method

One problem with applying the Secant, False Position, or Newton's method to polynomials is the possibility of the polynomial having complex roots even when all the coefficients are real numbers. If the initial approximation is a real number, all subsequent approximations will also be real numbers. One way to overcome this difficulty is to begin with a complex initial approximation and do all the computations using complex arithmetic. An alternative approach has its basis in the following theorem.

Theorem 2.20 If $z = a+bi$ is a complex zero of multiplicity m of the polynomial $P(x)$ with real coefficients, then $\bar{z} = a - bi$ is also a zero of multiplicity m of the polynomial $P(x)$, and $(x^2 - 2ax + a^2 + b^2)^m$ is a factor of $P(x)$. ■

A synthetic division involving quadratic polynomials can be devised to approximately factor the polynomial so that one term will be a quadratic polynomial whose complex roots are approximations to the roots of the original polynomial. This technique was described in some detail in our second edition [BFR]. Instead of proceeding along these lines, we will now consider a method first presented by D. E. Müller [Mu]. This technique can be used for any root-finding problem, but it is particularly useful for approximating the roots of polynomials.

Müller's method is similar to the Secant method. But whereas the Secant method uses a line through two points on the curve to approximate the root, Müller's method uses a parabola through three points on the curve for the approximation.

The Secant method begins with two initial approximations p_0 and p_1 and determines the next approximation p_2 as the intersection of the x-axis with the line through $(p_0, f(p_0))$ and $(p_1, f(p_1))$. (See Figure 2.13(a).) Müller's method uses three initial approximations, p_0, p_1, and p_2, and determines the next approximation p_3 by considering the intersection of the x-axis with the parabola through $(p_0, f(p_0))$, $(p_1, f(p_1))$, and $(p_2, f(p_2))$. (See Figure 2.13(b).)

Figure 2.13

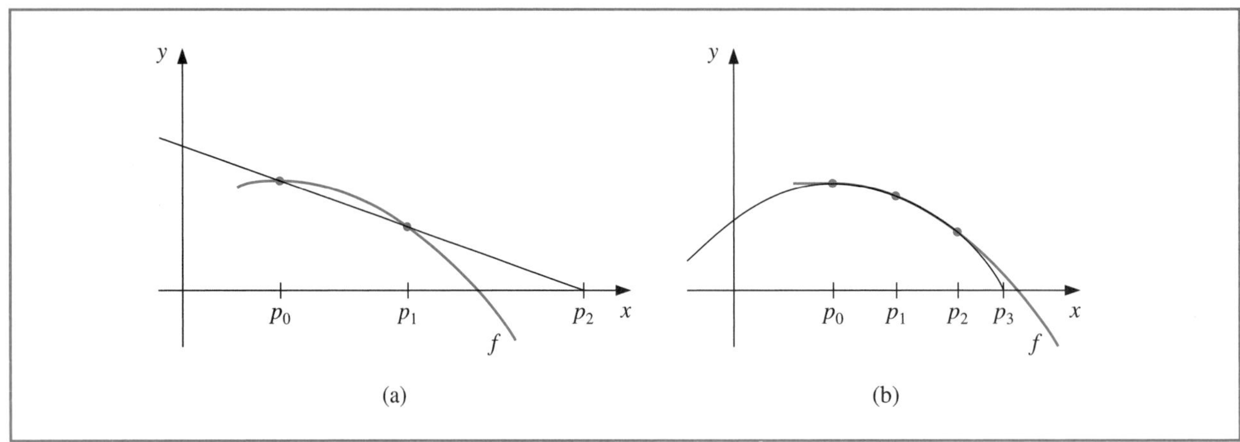

The derivation of Müller's method begins by considering the quadratic polynomial

$$P(x) = a(x - p_2)^2 + b(x - p_2) + c$$

that passes through $(p_0, f(p_0))$, $(p_1, f(p_1))$, and $(p_2, f(p_2))$. The constants a, b, and c can be determined from the conditions

$$f(p_0) = a(p_0 - p_2)^2 + b(p_0 - p_2) + c, \tag{2.17}$$

$$f(p_1) = a(p_1 - p_2)^2 + b(p_1 - p_2) + c, \tag{2.18}$$

and

$$f(p_2) = a \cdot 0^2 + b \cdot 0 + c = c \tag{2.19}$$

to be

$$c = f(p_2), \tag{2.20}$$

$$b = \frac{(p_0 - p_2)^2[f(p_1) - f(p_2)] - (p_1 - p_2)^2[f(p_0) - f(p_2)]}{(p_0 - p_2)(p_1 - p_2)(p_0 - p_1)}, \tag{2.21}$$

and

$$a = \frac{(p_1 - p_2)[f(p_0) - f(p_2)] - (p_0 - p_2)[f(p_1) - f(p_2)]}{(p_0 - p_2)(p_1 - p_2)(p_0 - p_1)}. \tag{2.22}$$

To determine p_3, a zero of P, we apply the quadratic formula to $P(x) = 0$. However, because of round-off error problems caused by the subtraction of nearly equal numbers, we apply the formula in the manner prescribed in Eq (1.2) and (1.3) of Section 1.2:

$$p_3 - p_2 = \frac{-2c}{b \pm \sqrt{b^2 - 4ac}}.$$

This formula gives two possibilities for p_3, depending on the sign preceding the radical term. In Müller's method, the sign is chosen to agree with the sign of b. Chosen in this manner, the denominator will be the largest in magnitude and will result in p_3 being selected as the closest zero of P to p_2. Thus

$$p_3 = p_2 - \frac{2c}{b + \text{sgn}(b)\sqrt{b^2 - 4ac}},$$

where a, b, and c are given in Eqs. (2.20) through (2.22).

Once p_3 is determined, the procedure is reinitialized using p_1, p_2, and p_3 in place of p_0, p_1, and p_2 to determine the next approximation, p_4. The method continues until a satisfactory conclusion is obtained. At each step, the method involves the radical $\sqrt{b^2 - 4ac}$, so the method gives approximate complex roots when $b^2 - 4ac < 0$. Algorithm 2.8 implements this procedure.

Müller's

To find a solution to $f(x) = 0$ given three approximations, p_0, p_1, and p_2:

INPUT p_0, p_1, p_2; tolerance TOL; maximum number of iterations N_0.

OUTPUT approximate solution p or message of failure.

Step 1 Set $h_1 = p_1 - p_0$;
$h_2 = p_2 - p_1$;
$\delta_1 = (f(p_1) - f(p_0))/h_1$;
$\delta_2 = (f(p_2) - f(p_1))/h_2$;
$d = (\delta_2 - \delta_1)/(h_2 + h_1)$;
$i = 3$.

Step 2 While $i \le N_0$ do Steps 3–7.

Step 3 $b = \delta_2 + h_2 d$;
$D = (b^2 - 4f(p_2)d)^{1/2}$. (*Note: May require complex arithmetic.*)

Step 4 If $|b - D| < |b + D|$ then set $E = b + D$
else set $E = b - D$.

Step 5 Set $h = -2f(p_2)/E$;
$p = p_2 + h$.

Step 6 If $|h| < TOL$ then
OUTPUT (p); (*The procedure was successful.*)
STOP.

Step 7 Set $p_0 = p_1$; (*Prepare for next iteration.*)
$p_1 = p_2$;
$p_2 = p$;
$h_1 = p_1 - p_0$;
$h_2 = p_2 - p_1$;
$\delta_1 = (f(p_1) - f(p_0))/h_1$;
$\delta_2 = (f(p_2) - f(p_1))/h_2$;
$d = (\delta_2 - \delta_1)/(h_2 + h_1)$;
$i = i + 1$.

Step 8 OUTPUT ('Method failed after N_0 iterations, N_0 =', N_0);
(*The procedure was unsuccessful.*)
STOP. ▪

Illustration Consider the polynomial $f(x) = x^4 - 3x^3 + x^2 + x + 1$, part of whose graph is shown in Figure 2.14.

Figure 2.14

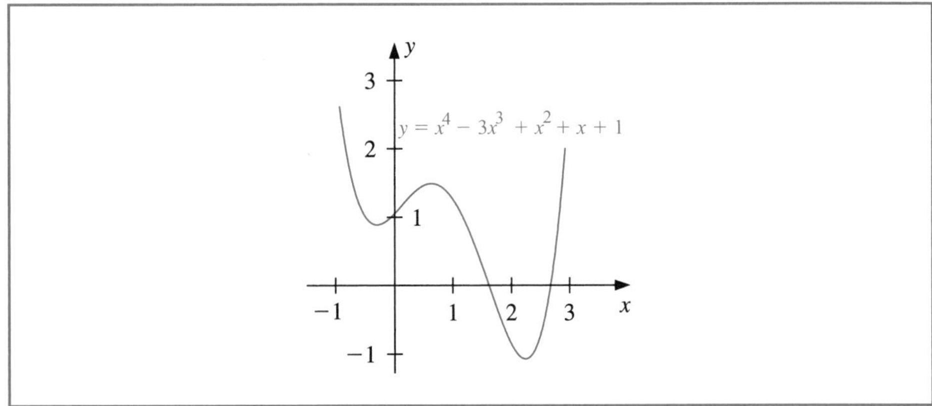

Three sets of three initial points will be used with Algorithm 2.8 and $TOL = 10^{-5}$ to approximate the zeros of f. The first set will use $p_0 = 0.5$, $p_1 = -0.5$, and $p_2 = 0$. The parabola passing through these points has complex roots because it does not intersect the x-axis. Table 2.12 gives approximations to the corresponding complex zeros of f.

Table 2.12

	$p_0 = 0.5,\ p_1 = -0.5,\ p_2 = 0$	
i	p_i	$f(p_i)$
3	$-0.100000 + 0.888819i$	$-0.01120000 + 3.014875548i$
4	$-0.492146 + 0.447031i$	$-0.1691201 - 0.7367331502i$
5	$-0.352226 + 0.484132i$	$-0.1786004 + 0.0181872213i$
6	$-0.340229 + 0.443036i$	$0.01197670 - 0.0105562185i$
7	$-0.339095 + 0.446656i$	$-0.0010550 + 0.000387261i$
8	$-0.339093 + 0.446630i$	$0.000000 + 0.000000i$
9	$-0.339093 + 0.446630i$	$0.000000 + 0.000000i$

Table 2.13 gives the approximations to the two real zeros of f. The smallest of these uses $p_0 = 0.5$, $p_1 = 1.0$, and $p_2 = 1.5$, and the largest root is approximated when $p_0 = 1.5$, $p_1 = 2.0$, and $p_2 = 2.5$.

Table 2.13

$p_0 = 0.5,$	$p_1 = 1.0,$	$p_2 = 1.5$	$p_0 = 1.5,$	$p_1 = 2.0,$	$p_2 = 2.5$
i	p_i	$f(p_i)$	i	p_i	$f(p_i)$
3	1.40637	−0.04851	3	2.24733	−0.24507
4	1.38878	0.00174	4	2.28652	−0.01446
5	1.38939	0.00000	5	2.28878	−0.00012
6	1.38939	0.00000	6	2.28880	0.00000
			7	2.28879	0.00000

The values in the tables are accurate approximations to the places listed. □

We used Maple to generate the results in Table 2.12. To find the first result in the table, define $f(x)$ with

$f := x \rightarrow x^4 - 3x^3 + x^2 + x + 1$

Then enter the initial approximations with

$p0 := 0.5; p1 := -0.5; p2 := 0.0$

and evaluate the function at these points with

$f0 := f(p0); f1 := f(p1); f2 := f(p2)$

To determine the coefficients a, b, c, and the approximate solution, enter

$c := f2;$

$$b := \frac{\left((p0 - p2)^2 \cdot (f1 - f2) - (p1 - p2)^2 \cdot (f0 - f2)\right)}{(p0 - p2) \cdot (p1 - p2) \cdot (p0 - p1)}$$

$$a := \frac{((p1 - p2) \cdot (f0 - f2) - (p0 - p2) \cdot (f1 - f2))}{(p0 - p2) \cdot (p1 - p2) \cdot (p0 - p1)}$$

$$p3 := p2 - \frac{2c}{b + \left(\frac{b}{abs(b)}\right)\sqrt{b^2 - 4a \cdot c}}$$

This produces the final Maple output

$$-0.1000000000 + 0.8888194418I$$

and evaluating at this approximation gives $f(p3)$ as

$$-0.0112000001 + 3.014875548I$$

This is our first approximation, as seen in Table 2.12.

The illustration shows that Müller's method can approximate the roots of polynomials with a variety of starting values. In fact, Müller's method generally converges to the root of a polynomial for any initial approximation choice, although problems can be constructed for which convergence will not occur. For example, suppose that for some i we have $f(p_i) = f(p_{i+1}) = f(p_{i+2}) \neq 0$. The quadratic equation then reduces to a nonzero constant function and never intersects the x-axis. This is not usually the case, however, and general-purpose software packages using Müller's method request only one initial approximation per root and will even supply this approximation as an option.

EXERCISE SET 2.6

1. Find the approximations to within 10^{-4} to all the real zeros of the following polynomials using Newton's method.
   - **a.** $f(x) = x^3 - x - 1$
   - **b.** $f(x) = x^3 - 2x^2 - 5$
   - **c.** $f(x) = x^3 + 3x^2 - 1$
   - **d.** $f(x) = x^4 + 2x^2 - x - 3$
   - **e.** $f(x) = x^3 + 4.001x^2 + 4.002x + 1.101$
   - **f.** $f(x) = x^5 - x^4 + 2x^3 - 3x^2 + x - 4$
2. Find approximations to within 10^{-5} to all the zeros of each of the following polynomials by first finding the real zeros using Newton's method and then reducing to polynomials of lower degree to determine any complex zeros.
   - **a.** $f(x) = x^4 - 2x^3 - 12x^2 + 16x - 40$
   - **b.** $f(x) = x^4 + 5x^3 - 9x^2 - 85x - 136$
   - **c.** $f(x) = x^4 + x^3 + 3x^2 + 2x + 2$
   - **d.** $f(x) = x^5 + 11x^4 - 21x^3 - 10x^2 - 21x - 5$
   - **e.** $f(x) = 16x^4 + 88x^3 + 159x^2 + 76x - 240$
   - **f.** $f(x) = x^4 - 4x^2 - 3x + 5$
   - **g.** $f(x) = x^3 - 7x^2 + 14x - 6$
   - **h.** $f(x) = x^4 - 2x^3 - 4x^2 + 4x + 4$
3. Repeat Exercise 1 using Müller's method.
4. Repeat Exercise 2 using Müller's method.
5. Use Newton's method to find, within 10^{-3}, the zeros and critical points of the following functions. Use this information to sketch the graph of f.
   - **a.** $f(x) = x^3 - 9x^2 + 12$
   - **b.** $f(x) = x^4 - 2x^3 - 5x^2 + 12x - 5$
6. $f(x) = 10x^3 - 8.3x^2 + 2.295x - 0.21141 = 0$ has a root at $x = 0.29$. Use Newton's method with an initial approximation $x_0 = 0.28$ to attempt to find this root. Explain what happens.
7. Use Maple to find a real zero of the polynomial $f(x) = x^3 + 4x - 4$.
8. Use Maple to find a real zero of the polynomial $f(x) = x^3 - 2x - 5$.
9. Use each of the following methods to find a solution in $[0.1, 1]$ accurate to within 10^{-4} for

$$600x^4 - 550x^3 + 200x^2 - 20x - 1 = 0.$$

   - **a.** Bisection method
   - **b.** Newton's method
   - **c.** Secant method
   - **d.** method of False Position
   - **e.** Müller's method

10. Two ladders crisscross an alley of width W. Each ladder reaches from the base of one wall to some point on the opposite wall. The ladders cross at a height H above the pavement. Find W given that the lengths of the ladders are $x_1 = 20$ ft and $x_2 = 30$ ft, and that $H = 8$ ft.

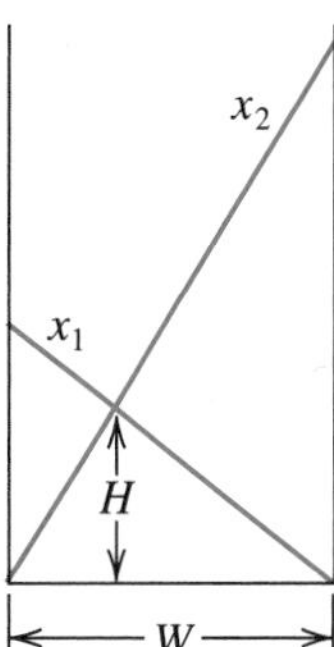

11. A can in the shape of a right circular cylinder is to be constructed to contain 1000 cm^3. The circular top and bottom of the can must have a radius of 0.25 cm more than the radius of the can so that the excess can be used to form a seal with the side. The sheet of material being formed into the side of the can must also be 0.25 cm longer than the circumference of the can so that a seal can be formed. Find, to within 10^{-4}, the minimal amount of material needed to construct the can.

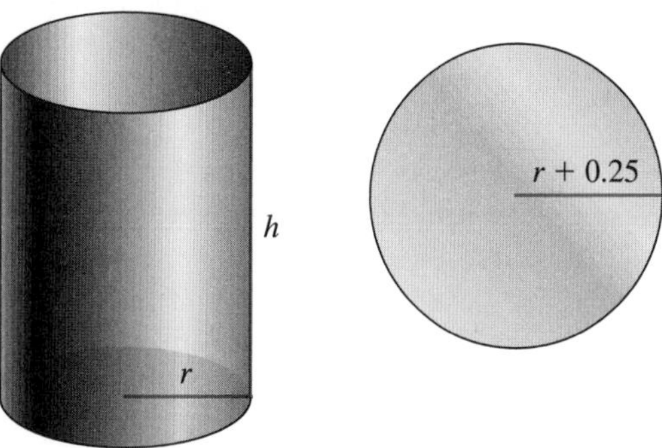

12. In 1224, Leonardo of Pisa, better known as Fibonacci, answered a mathematical challenge of John of Palermo in the presence of Emperor Frederick II: find a root of the equation $x^3 + 2x^2 + 10x = 20$. He first showed that the equation had no rational roots and no Euclidean irrational root—that is, no root in any of the forms $a \pm \sqrt{b}$, $\sqrt{a} \pm \sqrt{b}$, $\sqrt{a \pm \sqrt{b}}$, or $\sqrt{\sqrt{a} \pm \sqrt{b}}$, where a and b are rational numbers. He then approximated the only real root, probably using an algebraic technique of Omar Khayyam involving the intersection of a circle and a parabola. His answer was given in the base-60 number system as

$$1 + 22\left(\frac{1}{60}\right) + 7\left(\frac{1}{60}\right)^2 + 42\left(\frac{1}{60}\right)^3 + 33\left(\frac{1}{60}\right)^4 + 4\left(\frac{1}{60}\right)^5 + 40\left(\frac{1}{60}\right)^6.$$

How accurate was his approximation?

2.7 Survey of Methods and Software

In this chapter we have considered the problem of solving the equation $f(x) = 0$, where f is a given continuous function. All the methods begin with initial approximations and generate a sequence that converges to a root of the equation, if the method is successful. If $[a, b]$ is an interval on which $f(a)$ and $f(b)$ are of opposite sign, then the Bisection method and the method of False Position will converge. However, the convergence of these methods might be slow. Faster convergence is generally obtained using the Secant method or Newton's method. Good initial approximations are required for these methods, two for the Secant method and one for Newton's method, so the root-bracketing techniques such as Bisection or the False Position method can be used as starter methods for the Secant or Newton's method.

Müller's method will give rapid convergence without a particularly good initial approximation. It is not quite as efficient as Newton's method; its order of convergence near a root is approximately $\alpha = 1.84$, compared to the quadratic, $\alpha = 2$, order of Newton's method. However, it is better than the Secant method, whose order is approximately $\alpha = 1.62$, and it has the added advantage of being able to approximate complex roots.

Deflation is generally used with Müller's method once an approximate root of a polynomial has been determined. After an approximation to the root of the deflated equation has been determined, use either Müller's method or Newton's method in the original polynomial with this root as the initial approximation. This procedure will ensure that the root being approximated is a solution to the true equation, not to the deflated equation. We recommended Müller's method for finding all the zeros of polynomials, real or complex. Müller's method can also be used for an arbitrary continuous function.

Other high-order methods are available for determining the roots of polynomials. If this topic is of particular interest, we recommend that consideration be given to Laguerre's method, which gives cubic convergence and also approximates complex roots (see [Ho], pp. 176–179 for a complete discussion), the Jenkins-Traub method (see [JT]), and Brent's method (see [Bre]).

Another method of interest, Cauchy's method, is similar to Müller's method but avoids the failure problem of Müller's method when $f(x_i) = f(x_{i+1}) = f(x_{i+2})$, for some i. For an interesting discussion of this method, as well as more detail on Müller's method, we recommend [YG], Sections 4.10, 4.11, and 5.4.

Given a specified function f and a tolerance, an efficient program should produce an approximation to one or more solutions of $f(x) = 0$, each having an absolute or relative error within the tolerance, and the results should be generated in a reasonable amount of time. If the program cannot accomplish this task, it should at least give meaningful explanations of why success was not obtained and an indication of how to remedy the cause of failure.

IMSL has subroutines that implement Müller's method with deflation. Also included in this package is a routine due to R. P. Brent that uses a combination of linear interpolation, an inverse quadratic interpolation similar to Müller's method, and the Bisection method. Laguerre's method is also used to find zeros of a real polynomial. Another routine for finding the zeros of real polynomials uses a method of Jenkins-Traub, which is also used to find zeros of a complex polynomial.

The NAG library has a subroutine that uses a combination of the Bisection method, linear interpolation, and extrapolation to approximate a real zero of a function on a given interval. NAG also supplies subroutines to approximate all zeros of a real polynomial or complex polynomial, respectively. Both subroutines use a modified Laguerre method.

The netlib library contains a subroutine that uses a combination of the Bisection and Secant method developed by T. J. Dekker to approximate a real zero of a function in the interval. It requires specifying an interval that contains a root and returns an interval with a width that is within a specified tolerance. Another subroutine uses a combination of the bisection method, interpolation, and extrapolation to find a real zero of the function on the interval.

MATLAB has a routine to compute all the roots, both real and complex, of a polynomial, and one that computes a zero near a specified initial approximation to within a specified tolerance.

Notice that in spite of the diversity of methods, the professionally written packages are based primarily on the methods and principles discussed in this chapter. You should be able to use these packages by reading the manuals accompanying the packages to better understand the parameters and the specifications of the results that are obtained.

There are three books that we consider to be classics on the solution of nonlinear equations: those by Traub [Tr], by Ostrowski [Os], and by Householder [Ho]. In addition, the book by Brent [Bre] served as the basis for many of the currently used root-finding methods.

CHAPTER

3 Interpolation and Polynomial Approximation

Introduction

A census of the population of the United States is taken every 10 years. The following table lists the population, in thousands of people, from 1950 to 2000, and the data are also represented in the figure.

Year	1950	1960	1970	1980	1990	2000
Population (in thousands)	151,326	179,323	203,302	226,542	249,633	281,422

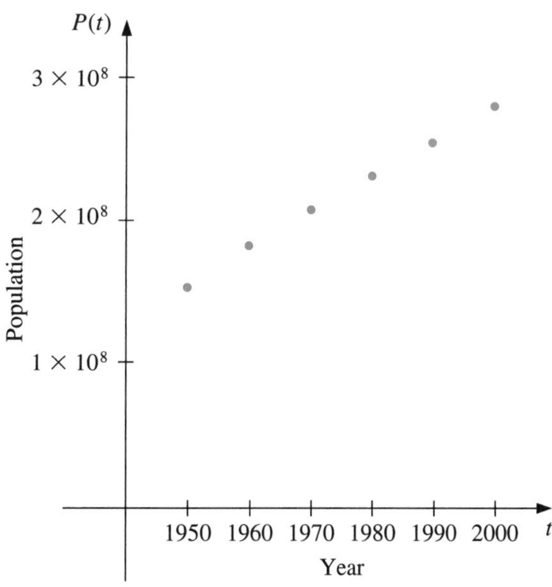

In reviewing these data, we might ask whether they could be used to provide a reasonable estimate of the population, say, in 1975 or even in the year 2020. Predictions of this type can be obtained by using a function that fits the given data. This process is called *interpolation* and is the subject of this chapter. This population problem is considered throughout the chapter and in Exercises 18 of Section 3.1, 18 of Section 3.3, and 28 of Section 3.5.

3.1 Interpolation and the Lagrange Polynomial

One of the most useful and well-known classes of functions mapping the set of real numbers into itself is the *algebraic polynomials*, the set of functions of the form

$$P_n(x) = a_n x^n + a_{n-1}x^{n-1} + \cdots + a_1 x + a_0,$$

where n is a nonnegative integer and $a_0, \ldots, a_n$ are real constants. One reason for their importance is that they uniformly approximate continuous functions. By this we mean that given any function, defined and continuous on a closed and bounded interval, there exists a polynomial that is as "close" to the given function as desired. This result is expressed precisely in the Weierstrass Approximation Theorem. (See Figure 3.1.)

Figure 3.1

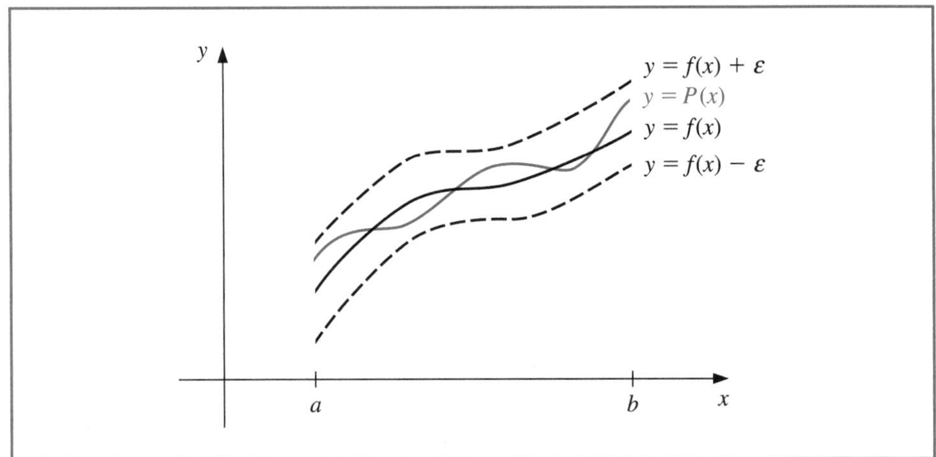

Theorem 3.1 **(Weierstrass Approximation Theorem)**

Suppose that f is defined and continuous on $[a, b]$. For each $\epsilon > 0$, there exists a polynomial $P(x)$, with the property that

$$|f(x) - P(x)| < \epsilon, \quad \text{for all } x \text{ in } [a, b].$$ ■

The proof of this theorem can be found in most elementary texts on real analysis (see, for example, [Bart], pp. 165–172).

Karl Weierstrass (1815–1897) is often referred to as the father of modern analysis because of his insistence on rigor in the demonstration of mathematical results. He was instrumental in developing tests for convergence of series, and determining ways to rigorously define irrational numbers. He was the first to demonstrate that a function could be everywhere continuous but nowhere differentiable, a result that shocked some of his contemporaries.

Another important reason for considering the class of polynomials in the approximation of functions is that the derivative and indefinite integral of a polynomial are easy to determine and are also polynomials. For these reasons, polynomials are often used for approximating continuous functions.

The Taylor polynomials were introduced in Section 1.1, where they were described as one of the fundamental building blocks of numerical analysis. Given this prominence, you might expect that polynomial interpolation would make heavy use of these functions. However this is not the case. The Taylor polynomials agree as closely as possible with a given function at a specific point, but they concentrate their accuracy near that point. A good interpolation polynomial needs to provide a relatively accurate approximation over an entire interval, and Taylor polynomials do not generally do this. For example, suppose we calculate the first six Taylor polynomials about $x_0 = 0$ for $f(x) = e^x$. Since the derivatives of $f(x)$ are all e^x, which evaluated at $x_0 = 0$ gives 1, the Taylor polynomials are

Very little of Weierstrass's work was published during his lifetime, but his lectures, particularly on the theory of functions, had significant influence on an entire generation of students.

$$P_0(x) = 1, \quad P_1(x) = 1 + x, \quad P_2(x) = 1 + x + \frac{x^2}{2}, \quad P_3(x) = 1 + x + \frac{x^2}{2} + \frac{x^3}{6},$$

$$P_4(x) = 1 + x + \frac{x^2}{2} + \frac{x^3}{6} + \frac{x^4}{24}, \quad \text{and} \quad P_5(x) = 1 + x + \frac{x^2}{2} + \frac{x^3}{6} + \frac{x^4}{24} + \frac{x^5}{120}.$$

The graphs of the polynomials are shown in Figure 3.2. (Notice that even for the higher-degree polynomials, the error becomes progressively worse as we move away from zero.)

Figure 3.2

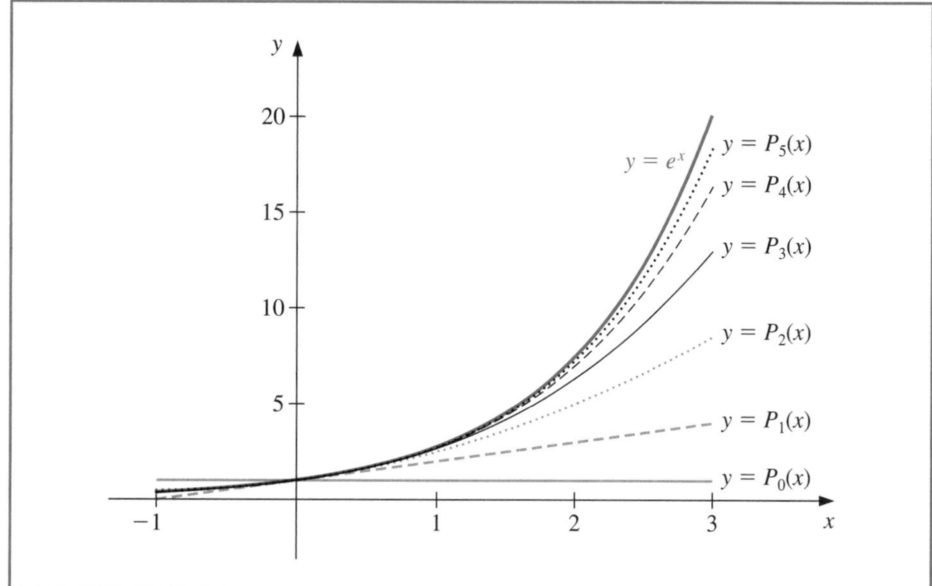

Although better approximations are obtained for $f(x) = e^x$ if higher-degree Taylor polynomials are used, this is not true for all functions. Consider, as an extreme example, using Taylor polynomials of various degrees for $f(x) = 1/x$ expanded about $x_0 = 1$ to approximate $f(3) = 1/3$. Since

$$f(x) = x^{-1}, \; f'(x) = -x^{-2}, \; f''(x) = (-1)^2 2 \cdot x^{-3},$$

and, in general,

$$f^{(k)}(x) = (-1)^k k! x^{-k-1},$$

the Taylor polynomials are

$$P_n(x) = \sum_{k=0}^{n} \frac{f^{(k)}(1)}{k!}(x-1)^k = \sum_{k=0}^{n} (-1)^k (x-1)^k.$$

To approximate $f(3) = 1/3$ by $P_n(3)$ for increasing values of n, we obtain the values in Table 3.1—rather a dramatic failure! When we approximate $f(3) = 1/3$ by $P_n(3)$ for larger values of n, the approximations become increasingly inaccurate.

Table 3.1

n	0	1	2	3	4	5	6	7
$P_n(3)$	1	−1	3	−5	11	−21	43	−85

For the Taylor polynomials all the information used in the approximation is concentrated at the single number x_0, so these polynomials will generally give inaccurate approximations as we move away from x_0. This limits Taylor polynomial approximation to the situation in which approximations are needed only at numbers close to x_0. For ordinary computational purposes it is more efficient to use methods that include information at various points. We consider this in the remainder of the chapter. The primary use of Taylor polynomials in numerical analysis is not for approximation purposes, but for the derivation of numerical techniques and error estimation.

Lagrange Interpolating Polynomials

The problem of determining a polynomial of degree one that passes through the distinct points (x_0, y_0) and (x_1, y_1) is the same as approximating a function f for which $f(x_0) = y_0$ and $f(x_1) = y_1$ by means of a first-degree polynomial **interpolating**, or agreeing with, the values of f at the given points. Using this polynomial for approximation within the interval given by the endpoints is called polynomial **interpolation**.

Define the functions

$$L_0(x) = \frac{x - x_1}{x_0 - x_1} \quad \text{and} \quad L_1(x) = \frac{x - x_0}{x_1 - x_0}.$$

The linear **Lagrange interpolating polynomial** through (x_0, y_0) and (x_1, y_1) is

$$P(x) = L_0(x)f(x_0) + L_1(x)f(x_1) = \frac{x - x_1}{x_0 - x_1}f(x_0) + \frac{x - x_0}{x_1 - x_0}f(x_1).$$

Note that

$$L_0(x_0) = 1, \quad L_0(x_1) = 0, \quad L_1(x_0) = 0, \quad \text{and} \quad L_1(x_1) = 1,$$

which implies that

$$P(x_0) = 1 \cdot f(x_0) + 0 \cdot f(x_1) = f(x_0) = y_0$$

and

$$P(x_1) = 0 \cdot f(x_0) + 1 \cdot f(x_1) = f(x_1) = y_1.$$

So P is the unique polynomial of degree at most one that passes through (x_0, y_0) and (x_1, y_1).

Example 1 Determine the linear Lagrange interpolating polynomial that passes through the points $(2, 4)$ and $(5, 1)$.

Solution In this case we have

$$L_0(x) = \frac{x - 5}{2 - 5} = -\frac{1}{3}(x - 5) \quad \text{and} \quad L_1(x) = \frac{x - 2}{5 - 2} = \frac{1}{3}(x - 2),$$

so

$$P(x) = -\frac{1}{3}(x - 5) \cdot 4 + \frac{1}{3}(x - 2) \cdot 1 = -\frac{4}{3}x + \frac{20}{3} + \frac{1}{3}x - \frac{2}{3} = -x + 6.$$

The graph of $y = P(x)$ is shown in Figure 3.3. ■

Figure 3.3

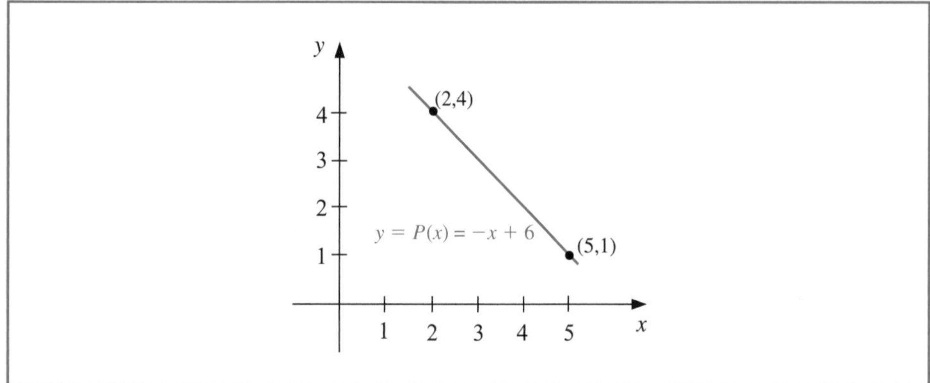

To generalize the concept of linear interpolation, consider the construction of a polynomial of degree at most n that passes through the $n+1$ points

$$(x_0, f(x_0)),\ (x_1, f(x_1)), \ldots, (x_n, f(x_n)).$$

(See Figure 3.4.)

Figure 3.4

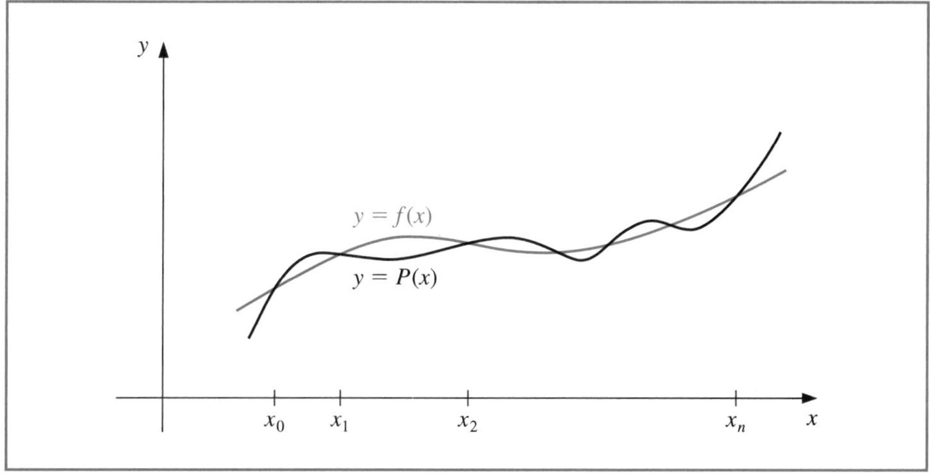

In this case we first construct, for each $k = 0, 1, \ldots, n$, a function $L_{n,k}(x)$ with the property that $L_{n,k}(x_i) = 0$ when $i \neq k$ and $L_{n,k}(x_k) = 1$. To satisfy $L_{n,k}(x_i) = 0$ for each $i \neq k$ requires that the numerator of $L_{n,k}(x)$ contain the term

$$(x - x_0)(x - x_1)\cdots(x - x_{k-1})(x - x_{k+1})\cdots(x - x_n).$$

To satisfy $L_{n,k}(x_k) = 1$, the denominator of $L_{n,k}(x)$ must be this same term but evaluated at $x = x_k$. Thus

$$L_{n,k}(x) = \frac{(x - x_0)\cdots(x - x_{k-1})(x - x_{k+1})\cdots(x - x_n)}{(x_k - x_0)\cdots(x_k - x_{k-1})(x_k - x_{k+1})\cdots(x_k - x_n)}.$$

A sketch of the graph of a typical $L_{n,k}$ (when n is even) is shown in Figure 3.5.

Figure 3.5

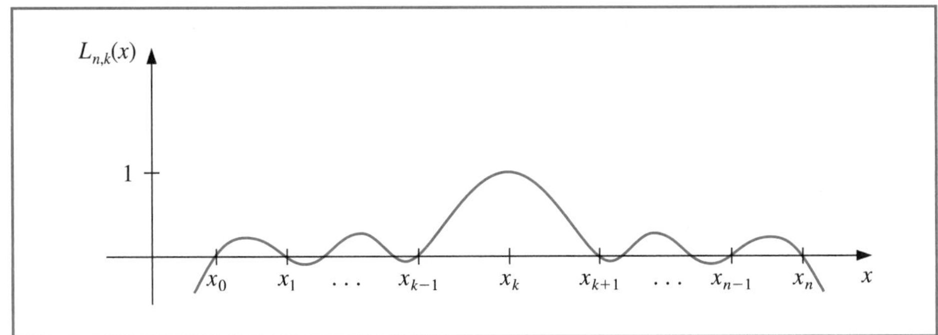

The interpolating polynomial is easily described once the form of $L_{n,k}$ is known. This polynomial, called the ***n*th Lagrange interpolating polynomial**, is defined in the following theorem.

Theorem 3.2 If $x_0, x_1, \ldots, x_n$ are $n+1$ distinct numbers and f is a function whose values are given at these numbers, then a unique polynomial $P(x)$ of degree at most n exists with

$$f(x_k) = P(x_k), \quad \text{for each } k = 0, 1, \ldots, n.$$

This polynomial is given by

$$P(x) = f(x_0)L_{n,0}(x) + \cdots + f(x_n)L_{n,n}(x) = \sum_{k=0}^{n} f(x_k)L_{n,k}(x), \tag{3.1}$$

where, for each $k = 0, 1, \ldots, n$,

$$L_{n,k}(x) = \frac{(x-x_0)(x-x_1)\cdots(x-x_{k-1})(x-x_{k+1})\cdots(x-x_n)}{(x_k-x_0)(x_k-x_1)\cdots(x_k-x_{k-1})(x_k-x_{k+1})\cdots(x_k-x_n)} \tag{3.2}$$

$$= \prod_{\substack{i=0\\ i\neq k}}^{n} \frac{(x-x_i)}{(x_k-x_i)}.$$

■

The interpolation formula named for Joseph Louis Lagrange (1736–1813) was likely known by Isaac Newton around 1675, but it appears to first have been published in 1779 by Edward Waring (1736–1798). Lagrange wrote extensively on the subject of interpolation and his work had significant influence on later mathematicians. He published this result in 1795.

The symbol $\prod$ is used to write products compactly and parallels the symbol $\sum$, which is used for writing sums.

We will write $L_{n,k}(x)$ simply as $L_k(x)$ when there is no confusion as to its degree.

Example 2 **(a)** Use the numbers (called *nodes*) $x_0 = 2$, $x_1 = 2.75$, and $x_2 = 4$ to find the second Lagrange interpolating polynomial for $f(x) = 1/x$.

(b) Use this polynomial to approximate $f(3) = 1/3$.

Solution **(a)** We first determine the coefficient polynomials $L_0(x)$, $L_1(x)$, and $L_2(x)$. In nested form they are

$$L_0(x) = \frac{(x-2.75)(x-4)}{(2-2.5)(2-4)} = \frac{2}{3}(x-2.75)(x-4),$$

$$L_1(x) = \frac{(x-2)(x-4)}{(2.75-2)(2.75-4)} = -\frac{16}{15}(x-2)(x-4),$$

and

$$L_2(x) = \frac{(x-2)(x-2.75)}{(4-2)(4-2.5)} = \frac{2}{5}(x-2)(x-2.75).$$

Also, $f(x_0) = f(2) = 1/2$, $f(x_1) = f(2.75) = 4/11$, and $f(x_2) = f(4) = 1/4$, so

$$\begin{aligned}
P(x) &= \sum_{k=0}^{2} f(x_k)L_k(x) \\
&= \frac{1}{3}(x-2.75)(x-4) - \frac{64}{165}(x-2)(x-4) + \frac{1}{10}(x-2)(x-2.75) \\
&= \frac{1}{22}x^2 - \frac{35}{88}x + \frac{49}{44}.
\end{aligned}$$

(b) An approximation to $f(3) = 1/3$ (see Figure 3.6) is

$$f(3) \approx P(3) = \frac{9}{22} - \frac{105}{88} + \frac{49}{44} = \frac{29}{88} \approx 0.32955.$$

Recall that in the opening section of this chapter (see Table 3.1) we found that no Taylor polynomial expanded about $x_0 = 1$ could be used to reasonably approximate $f(x) = 1/x$ at $x = 3$. ■

Figure 3.6

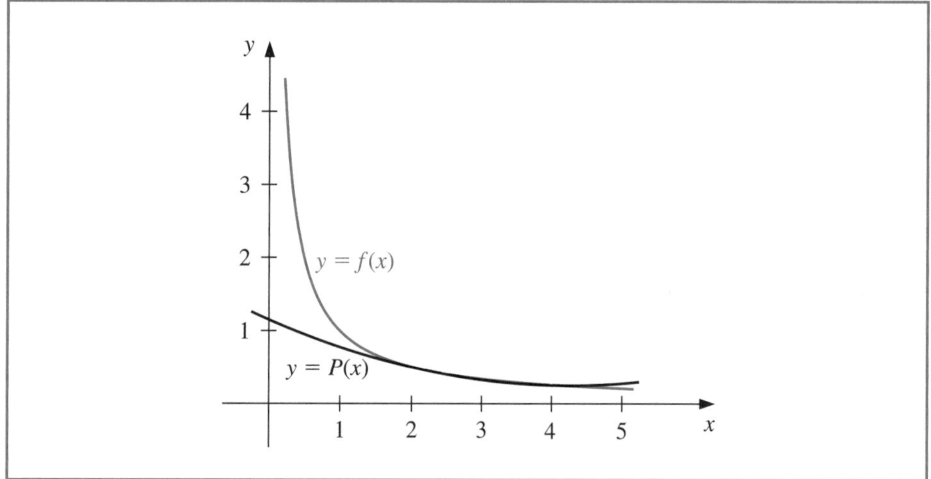

The interpolating polynomial P of degree less than or equal to 3 is defined in Maple with

$P := x \rightarrow interp([2, 11/4, 4], [1/2, 4/11, 1/4], x)$

$$x \rightarrow interp\left(\left[2, \frac{11}{4}, 4\right], \left[\frac{1}{2}, \frac{4}{11}, \frac{1}{4}\right], x\right)$$

To see the polynomial, enter

$P(x)$

$$\frac{1}{22}x^2 - \frac{35}{88}x + \frac{49}{44}$$

Evaluating $P(3)$ as an approximation to $f(3) = 1/3$, is found with

$evalf(P(3))$

$$0.3295454545$$

The interpolating polynomial can also be defined in Maple using the *CurveFitting* package and the call *PolynomialInterpolation*.

The next step is to calculate a remainder term or bound for the error involved in approximating a function by an interpolating polynomial.

Theorem 3.3 Suppose $x_0, x_1, \ldots, x_n$ are distinct numbers in the interval $[a, b]$ and $f \in C^{n+1}[a, b]$. Then, for each x in $[a, b]$, a number $\xi(x)$ (generally unknown) between $x_0, x_1, \ldots, x_n$, and hence in (a, b), exists with

$$f(x) = P(x) + \frac{f^{(n+1)}(\xi(x))}{(n+1)!}(x - x_0)(x - x_1)\cdots(x - x_n), \tag{3.3}$$

where $P(x)$ is the interpolating polynomial given in Eq. (3.1). ■

There are other ways that the error term for the Lagrange polynomial can be expressed, but this is the most useful form and the one that most closely agrees with the standard Taylor polynomial error form.

Proof Note first that if $x = x_k$, for any $k = 0, 1, \ldots, n$, then $f(x_k) = P(x_k)$, and choosing $\xi(x_k)$ arbitrarily in (a, b) yields Eq. (3.3).

If $x \neq x_k$, for all $k = 0, 1, \ldots, n$, define the function g for t in $[a, b]$ by

$$\begin{aligned} g(t) &= f(t) - P(t) - [f(x) - P(x)]\frac{(t - x_0)(t - x_1)\cdots(t - x_n)}{(x - x_0)(x - x_1)\cdots(x - x_n)} \\ &= f(t) - P(t) - [f(x) - P(x)]\prod_{i=0}^{n}\frac{(t - x_i)}{(x - x_i)}. \end{aligned}$$

Since $f \in C^{n+1}[a, b]$, and $P \in C^{\infty}[a, b]$, it follows that $g \in C^{n+1}[a, b]$. For $t = x_k$, we have

$$g(x_k) = f(x_k) - P(x_k) - [f(x) - P(x)]\prod_{i=0}^{n}\frac{(x_k - x_i)}{(x - x_i)} = 0 - [f(x) - P(x)]\cdot 0 = 0.$$

Moreover,

$$g(x) = f(x) - P(x) - [f(x) - P(x)]\prod_{i=0}^{n}\frac{(x - x_i)}{(x - x_i)} = f(x) - P(x) - [f(x) - P(x)] = 0.$$

Thus $g \in C^{n+1}[a, b]$, and g is zero at the $n + 2$ distinct numbers $x, x_0, x_1, \ldots, x_n$. By Generalized Rolle's Theorem 1.10, there exists a number ξ in (a, b) for which $g^{(n+1)}(\xi) = 0$. So

$$0 = g^{(n+1)}(\xi) = f^{(n+1)}(\xi) - P^{(n+1)}(\xi) - [f(x) - P(x)]\frac{d^{n+1}}{dt^{n+1}}\left[\prod_{i=0}^{n}\frac{(t - x_i)}{(x - x_i)}\right]_{t=\xi}. \tag{3.4}$$

However $P(x)$ is a polynomial of degree at most n, so the $(n+1)$st derivative, $P^{(n+1)}(x)$, is identically zero. Also, $\prod_{i=0}^{n}[(t - x_i)/(x - x_i)]$ is a polynomial of degree $(n + 1)$, so

$$\prod_{i=0}^{n}\frac{(t - x_i)}{(x - x_i)} = \left[\frac{1}{\prod_{i=0}^{n}(x - x_i)}\right]t^{n+1} + (\text{lower-degree terms in } t),$$

and

$$\frac{d^{n+1}}{dt^{n+1}}\prod_{i=0}^{n}\frac{(t - x_i)}{(x - x_i)} = \frac{(n+1)!}{\prod_{i=0}^{n}(x - x_i)}.$$

Equation (3.4) now becomes

$$0 = f^{(n+1)}(\xi) - 0 - [f(x) - P(x)]\frac{(n+1)!}{\prod_{i=0}^{n}(x - x_i)},$$

and, upon solving for $f(x)$, we have

$$f(x) = P(x) + \frac{f^{(n+1)}(\xi)}{(n+1)!}\prod_{i=0}^{n}(x - x_i). \qquad \blacksquare\ \blacksquare\ \blacksquare$$

The error formula in Theorem 3.3 is an important theoretical result because Lagrange polynomials are used extensively for deriving numerical differentiation and integration methods. Error bounds for these techniques are obtained from the Lagrange error formula.

Note that the error form for the Lagrange polynomial is quite similar to that for the Taylor polynomial. The nth Taylor polynomial about x_0 concentrates all the known information at x_0 and has an error term of the form

$$\frac{f^{(n+1)}(\xi(x))}{(n+1)!}(x - x_0)^{n+1}.$$

The Lagrange polynomial of degree n uses information at the distinct numbers $x_0, x_1, \ldots, x_n$ and, in place of $(x - x_0)^n$, its error formula uses a product of the $n + 1$ terms $(x - x_0)$, $(x - x_1), \ldots, (x - x_n)$:

$$\frac{f^{(n+1)}(\xi(x))}{(n+1)!}(x - x_0)(x - x_1)\cdots(x - x_n).$$

Example 3 In Example 2 we found the second Lagrange polynomial for $f(x) = 1/x$ on $[2, 4]$ using the nodes $x_0 = 2$, $x_1 = 2.75$, and $x_2 = 4$. Determine the error form for this polynomial, and the maximum error when the polynomial is used to approximate $f(x)$ for $x \in [2, 4]$.

Solution Because $f(x) = x^{-1}$, we have

$$f'(x) = -x^{-2}, \quad f''(x) = 2x^{-3}, \quad \text{and} \quad f'''(x) = -6x^{-4}.$$

As a consequence, the second Lagrange polynomial has the error form

$$\frac{f'''(\xi(x))}{3!}(x-x_0)(x-x_1)(x-x_2) = -(\xi(x))^{-4}(x-2)(x-2.75)(x-4), \quad \text{for } \xi(x) \text{ in } (2, 4).$$

The maximum value of $(\xi(x))^{-4}$ on the interval is $2^{-4} = 1/16$. We now need to determine the maximum value on this interval of the absolute value of the polynomial

$$g(x) = (x - 2)(x - 2.75)(x - 4) = x^3 - \frac{35}{4}x^2 + \frac{49}{2}x - 22.$$

Because

$$D_x\left(x^3 - \frac{35}{4}x^2 + \frac{49}{2}x - 22\right) = 3x^2 - \frac{35}{2}x + \frac{49}{2} = \frac{1}{2}(3x - 7)(2x - 7),$$

the critical points occur at

$$x = \frac{7}{3}, \text{ with } g\left(\frac{7}{3}\right) = \frac{25}{108}, \qquad \text{and} \qquad x = \frac{7}{2}, \text{ with } g\left(\frac{7}{2}\right) = -\frac{9}{16}.$$

Hence, the maximum error is

$$\frac{f'''(\xi(x))}{3!}|(x - x_0)(x - x_1)(x - x_2)| \le \frac{1}{16 \cdot 6}\left|-\frac{9}{16}\right| = \frac{3}{512} \approx 0.00586. \qquad \blacksquare$$

The next example illustrates how the error formula can be used to prepare a table of data that will ensure a specified interpolation error within a specified bound.

Example 4 Suppose a table is to be prepared for the function $f(x) = e^x$, for x in $[0, 1]$. Assume the number of decimal places to be given per entry is $d \geq 8$ and that the difference between adjacent x-values, the step size, is h. What step size h will ensure that linear interpolation gives an absolute error of at most 10^{-6} for all x in $[0, 1]$?

Solution Let $x_0, x_1, \ldots$ be the numbers at which f is evaluated, x be in [0,1], and suppose j satisfies $x_j \leq x \leq x_{j+1}$. Eq. (3.3) implies that the error in linear interpolation is

$$|f(x) - P(x)| = \left| \frac{f^{(2)}(\xi)}{2!}(x - x_j)(x - x_{j+1}) \right| = \frac{|f^{(2)}(\xi)|}{2}|(x - x_j)||(x - x_{j+1})|.$$

The step size is h, so $x_j = jh$, $x_{j+1} = (j + 1)h$, and

$$|f(x) - P(x)| \leq \frac{|f^{(2)}(\xi)|}{2!}|(x - jh)(x - (j + 1)h)|.$$

Hence

$$\begin{aligned} |f(x) - P(x)| &\leq \frac{\max_{\xi \in [0,1]} e^{\xi}}{2} \max_{x_j \leq x \leq x_{j+1}} |(x - jh)(x - (j + 1)h)| \\ &\leq \frac{e}{2} \max_{x_j \leq x \leq x_{j+1}} |(x - jh)(x - (j + 1)h)|. \end{aligned}$$

Consider the function $g(x) = (x - jh)(x - (j + 1)h)$, for $jh \leq x \leq (j + 1)h$. Because

$$g'(x) = (x - (j + 1)h) + (x - jh) = 2\left(x - jh - \frac{h}{2}\right),$$

the only critical point for g is at $x = jh + h/2$, with $g(jh + h/2) = (h/2)^2 = h^2/4$.

Since $g(jh) = 0$ and $g((j + 1)h) = 0$, the maximum value of $|g'(x)|$ in $[jh, (j + 1)h]$ must occur at the critical point which implies that

$$|f(x) - P(x)| \leq \frac{e}{2} \max_{x_j \leq x \leq x_{j+1}} |g(x)| \leq \frac{e}{2} \cdot \frac{h^2}{4} = \frac{eh^2}{8}.$$

Consequently, to ensure that the the error in linear interpolation is bounded by 10^{-6}, it is sufficient for h to be chosen so that

$$\frac{eh^2}{8} \leq 10^{-6}. \quad \text{This implies that} \quad h < 1.72 \times 10^{-3}.$$

Because $n = (1 - 0)/h$ must be an integer, a reasonable choice for the step size is $h = 0.001$. ■

EXERCISE SET 3.1

1. For the given functions $f(x)$, let $x_0 = 0$, $x_1 = 0.6$, and $x_2 = 0.9$. Construct interpolation polynomials of degree at most one and at most two to approximate $f(0.45)$, and find the absolute error.

   a. $f(x) = \sqrt{1 + x}$

   b. $f(x) = \cos x$

   c. $f(x) = \ln(x + 1)$

   d. $f(x) = \tan x$

2. For the given functions $f(x)$, let $x_0 = 1, x_1 = 1.25$, and $x_2 = 1.6$. Construct interpolation polynomials of degree at most one and at most two to approximate $f(1.4)$, and find the absolute error.
   a. $f(x) = \sin \pi x$
   c. $f(x) = e^{2x} - x$
   b. $f(x) = \sqrt[3]{x-1}$
   d. $f(x) = \log_{10}(3x - 1)$
3. Use Theorem 3.3 to find an error bound for the approximations in Exercise 1.
4. Use Theorem 3.3 to find an error bound for the approximations in Exercise 2.
5. Use appropriate Lagrange interpolating polynomials of degrees one, two, and three to approximate each of the following:
   a. $f(8.4)$ if $f(8.1) = 16.94410$, $f(8.3) = 17.56492$, $f(8.6) = 18.50515$, $f(8.7) = 18.82091$
   b. $f(0.25)$ if $f(0.1) = 0.62049958$, $f(0.2) = -0.28398668$, $f(0.3) = 0.00660095$, $f(0.4) = 0.24842440$
   c. $f\left(-\frac{1}{3}\right)$ if $f(-0.75) = -0.07181250$, $f(-0.5) = -0.02475000$, $f(-0.25) = 0.33493750$, $f(0) = 1.10100000$
   d. $f(0.9)$ if $f(0.6) = -0.17694460$, $f(0.7) = 0.01375227$, $f(0.8) = 0.22363362$, $f(1.0) = 0.65809197$
6. Use appropriate Lagrange interpolating polynomials of degrees one, two, and three to approximate each of the following:
   a. $f(0.43)$ if $f(0) = 1$, $f(0.25) = 1.64872$, $f(0.5) = 2.71828$, $f(0.75) = 4.48169$
   b. $f(0.18)$ if $f(0.1) = -0.29004986$, $f(0.2) = -0.56079734$, $f(0.3) = -0.81401972$, $f(0.4) = -1.0526302$
   c. $f(0)$ if $f(-0.5) = 1.93750$, $f(-0.25) = 1.33203$, $f(0.25) = 0.800781$, $f(0.5) = 0.687500$
   d. $f(0.25)$ if $f(-1) = 0.86199480$, $f(-0.5) = 0.95802009$, $f(0) = 1.0986123$, $f(0.5) = 1.2943767$
7. The data for Exercise 5 were generated using the following functions. Use the error formula to find a bound for the error, and compare the bound to the actual error for the cases $n = 1$ and $n = 2$.
   a. $f(x) = x \ln x$
   b. $f(x) = x \cos x - 2x^2 + 3x - 1$
   c. $f(x) = x^3 + 4.001x^2 + 4.002x + 1.101$
   d. $f(x) = \sin(e^x - 2)$
8. The data for Exercise 6 were generated using the following functions. Use the error formula to find a bound for the error, and compare the bound to the actual error for the cases $n = 1$ and $n = 2$.
   a. $f(x) = e^{2x}$
   b. $f(x) = x^2 \cos x - 3x$
   c. $f(x) = x^4 - x^3 + x^2 - x + 1$
   d. $f(x) = \ln(e^x + 2)$
9. Let $P_3(x)$ be the interpolating polynomial for the data $(0, 0)$, $(0.5, y)$, $(1, 3)$, and $(2, 2)$. The coefficient of x^3 in $P_3(x)$ is 6. Find y.
10. Let $f(x) = \sqrt{x - x^2}$ and $P_2(x)$ be the interpolation polynomial on $x_0 = 0$, x_1 and $x_2 = 1$. Find the largest value of x_1 in $(0, 1)$ for which $f(0.5) - P_2(0.5) = -0.25$.
11. Use the following values and four-digit rounding arithmetic to construct a third Lagrange polynomial approximation to $f(1.09)$. The function being approximated is $f(x) = \log_{10}(\tan x)$. Use this knowledge to find a bound for the error in the approximation.

$$f(1.00) = 0.1924 \quad f(1.05) = 0.2414 \quad f(1.10) = 0.2933 \quad f(1.15) = 0.3492$$

12. Use the Lagrange interpolating polynomial of degree three or less and four-digit chopping arithmetic to approximate $\cos 0.750$ using the following values. Find an error bound for the approximation.

$$\cos 0.698 = 0.7661 \quad \cos 0.733 = 0.7432 \quad \cos 0.768 = 0.7193 \quad \cos 0.803 = 0.6946$$

The actual value of $\cos 0.750$ is 0.7317 (to four decimal places). Explain the discrepancy between the actual error and the error bound.

13. Construct the Lagrange interpolating polynomials for the following functions, and find a bound for the absolute error on the interval $[x_0, x_n]$.

a. $f(x) = e^{2x} \cos 3x, \quad x_0 = 0, x_1 = 0.3, x_2 = 0.6, n = 2$

b. $f(x) = \sin(\ln x), \quad x_0 = 2.0, x_1 = 2.4, x_2 = 2.6, n = 2$

c. $f(x) = \ln x, \quad x_0 = 1, x_1 = 1.1, x_2 = 1.3, x_3 = 1.4, n = 3$

d. $f(x) = \cos x + \sin x, \quad x_0 = 0, x_1 = 0.25, x_2 = 0.5, x_3 = 1.0, n = 3$

14. Let $f(x) = e^x$, for $0 \leq x \leq 2$.

a. Approximate $f(0.25)$ using linear interpolation with $x_0 = 0$ and $x_1 = 0.5$.

b. Approximate $f(0.75)$ using linear interpolation with $x_0 = 0.5$ and $x_1 = 1$.

c. Approximate $f(0.25)$ and $f(0.75)$ by using the second interpolating polynomial with $x_0 = 0$, $x_1 = 1$, and $x_2 = 2$.

d. Which approximations are better and why?

15. Repeat Exercise 11 using Maple with *Digits* set to 10.

16. Repeat Exercise 12 using Maple with *Digits* set to 10.

17. Suppose you need to construct eight-decimal-place tables for the common, or base-10, logarithm function from $x = 1$ to $x = 10$ in such a way that linear interpolation is accurate to within 10^{-6}. Determine a bound for the step size for this table. What choice of step size would you make to ensure that $x = 10$ is included in the table?

18. **a.** The introduction to this chapter included a table listing the population of the United States from 1950 to 2000. Use Lagrange interpolation to approximate the population in the years 1940, 1975, and 2020.

b. The population in 1940 was approximately 132,165,000. How accurate do you think your 1975 and 2020 figures are?

19. It is suspected that the high amounts of tannin in mature oak leaves inhibit the growth of the winter moth (*Operophtera bromata L., Geometridae*) larvae that extensively damage these trees in certain years. The following table lists the average weight of two samples of larvae at times in the first 28 days after birth. The first sample was reared on young oak leaves, whereas the second sample was reared on mature leaves from the same tree.

a. Use Lagrange interpolation to approximate the average weight curve for each sample.

b. Find an approximate maximum average weight for each sample by determining the maximum of the interpolating polynomial.

Day	0	6	10	13	17	20	28
Sample 1 average weight (mg)	6.67	17.33	42.67	37.33	30.10	29.31	28.74
Sample 2 average weight (mg)	6.67	16.11	18.89	15.00	10.56	9.44	8.89

20. In Exercise 26 of Section 1.1 a Maclaurin series was integrated to approximate erf(1), where erf(x) is the normal distribution error function defined by

$$\text{erf}(x) = \frac{2}{\sqrt{\pi}} \int_0^x e^{-t^2}\, dt.$$

a. Use the Maclaurin series to construct a table for erf(x) that is accurate to within 10^{-4} for erf(x_i), where $x_i = 0.2i$, for $i = 0, 1, \ldots, 5$.

b. Use both linear interpolation and quadratic interpolation to obtain an approximation to erf($\frac{1}{3}$). Which approach seems most feasible?

21. Prove Taylor's Theorem 1.14 by following the procedure in the proof of Theorem 3.3. [*Hint:* Let

$$g(t) = f(t) - P(t) - [f(x) - P(x)] \cdot \frac{(t - x_0)^{n+1}}{(x - x_0)^{n+1}},$$

where P is the nth Taylor polynomial, and use the Generalized Rolle's Theorem 1.10.]

22. Show that $\max_{x_j \le x \le x_{j+1}} |g(x)| = h^2/4$, where $g(x) = (x - jh)(x - (j+1)h)$.

23. The Bernstein polynomial of degree n for $f \in C[0, 1]$ is given by

$$B_n(x) = \sum_{k=0}^{n} \binom{n}{k} f\left(\frac{k}{n}\right) x^k (1-x)^{n-k},$$

where $\binom{n}{k}$ denotes $n!/k!(n-k)!$. These polynomials can be used in a constructive proof of the Weierstrass Approximation Theorem 3.1 (see [Bart]) because $\lim_{n\to\infty} B_n(x) = f(x)$, for each $x \in [0, 1]$.

a. Find $B_3(x)$ for the functions

i. $f(x) = x$ **ii.** $f(x) = 1$

b. Show that for each $k \le n$,

$$\binom{n-1}{k-1} = \left(\frac{k}{n}\right)\binom{n}{k}.$$

c. Use part (b) and the fact, from (ii) in part (a), that

$$1 = \sum_{k=0}^{n} \binom{n}{k} x^k (1-x)^{n-k}, \quad \text{for each } n,$$

to show that, for $f(x) = x^2$,

$$B_n(x) = \left(\frac{n-1}{n}\right) x^2 + \frac{1}{n} x.$$

d. Use part (c) to estimate the value of n necessary for $\left|B_n(x) - x^2\right| \le 10^{-6}$ to hold for all x in $[0, 1]$.

3.2 Data Approximation and Neville's Method

In the previous section we found an explicit representation for Lagrange polynomials and their error when approximating a function on an interval. A frequent use of these polynomials involves the interpolation of tabulated data. In this case an explicit representation of the polynomial might not be needed, only the values of the polynomial at specified points. In this situation the function underlying the data might not be known so the explicit form of the error cannot be used. We will now illustrate a practical application of interpolation in such a situation.

Illustration Table 3.2 lists values of a function f at various points. The approximations to $f(1.5)$ obtained by various Lagrange polynomials that use this data will be compared to try and determine the accuracy of the approximation.

Table 3.2

x	$f(x)$
1.0	0.7651977
1.3	0.6200860
1.6	0.4554022
1.9	0.2818186
2.2	0.1103623

The most appropriate linear polynomial uses $x_0 = 1.3$ and $x_1 = 1.6$ because 1.5 is between 1.3 and 1.6. The value of the interpolating polynomial at 1.5 is

$$\begin{aligned} P_1(1.5) &= \frac{(1.5-1.6)}{(1.3-1.6)} f(1.3) + \frac{(1.5-1.3)}{(1.6-1.3)} f(1.6) \\ &= \frac{(1.5-1.6)}{(1.3-1.6)}(0.6200860) + \frac{(1.5-1.3)}{(1.6-1.3)}(0.4554022) = 0.5102968. \end{aligned}$$

Two polynomials of degree 2 can reasonably be used, one with $x_0 = 1.3$, $x_1 = 1.6$, and $x_2 = 1.9$, which gives

$$P_2(1.5) = \frac{(1.5-1.6)(1.5-1.9)}{(1.3-1.6)(1.3-1.9)}(0.6200860) + \frac{(1.5-1.3)(1.5-1.9)}{(1.6-1.3)(1.6-1.9)}(0.4554022)$$
$$+ \frac{(1.5-1.3)(1.5-1.6)}{(1.9-1.3)(1.9-1.6)}(0.2818186) = 0.5112857,$$

and one with $x_0 = 1.0$, $x_1 = 1.3$, and $x_2 = 1.6$, which gives $\hat{P}_2(1.5) = 0.5124715$.

In the third-degree case, there are also two reasonable choices for the polynomial. One with $x_0 = 1.3$, $x_1 = 1.6$, $x_2 = 1.9$, and $x_3 = 2.2$, which gives $P_3(1.5) = 0.5118302$.

The second third-degree approximation is obtained with $x_0 = 1.0$, $x_1 = 1.3$, $x_2 = 1.6$, and $x_3 = 1.9$, which gives $\hat{P}_3(1.5) = 0.5118127$. The fourth-degree Lagrange polynomial uses all the entries in the table. With $x_0 = 1.0$, $x_1 = 1.3$, $x_2 = 1.6$, $x_3 = 1.9$, and $x_4 = 2.2$, the approximation is $P_4(1.5) = 0.5118200$.

Because $P_3(1.5)$, $\hat{P}_3(1.5)$, and $P_4(1.5)$ all agree to within 2×10^{-5} units, we expect this degree of accuracy for these approximations. We also expect $P_4(1.5)$ to be the most accurate approximation, since it uses more of the given data.

The function we are approximating is actually the Bessel function of the first kind of order zero, whose value at 1.5 is known to be 0.5118277. Therefore, the true accuracies of the approximations are as follows:

$$|P_1(1.5) - f(1.5)| \approx 1.53 \times 10^{-3},$$
$$|P_2(1.5) - f(1.5)| \approx 5.42 \times 10^{-4},$$
$$|\hat{P}_2(1.5) - f(1.5)| \approx 6.44 \times 10^{-4},$$
$$|P_3(1.5) - f(1.5)| \approx 2.5 \times 10^{-6},$$
$$|\hat{P}_3(1.5) - f(1.5)| \approx 1.50 \times 10^{-5},$$
$$|P_4(1.5) - f(1.5)| \approx 7.7 \times 10^{-6}.$$

Although $P_3(1.5)$ is the most accurate approximation, if we had no knowledge of the actual value of $f(1.5)$, we would accept $P_4(1.5)$ as the best approximation since it includes the most data about the function. The Lagrange error term derived in Theorem 3.3 cannot be applied here because we have no knowledge of the fourth derivative of f. Unfortunately, this is generally the case. □

Neville's Method

A practical difficulty with Lagrange interpolation is that the error term is difficult to apply, so the degree of the polynomial needed for the desired accuracy is generally not known until computations have been performed. A common practice is to compute the results given from various polynomials until appropriate agreement is obtained, as was done in the previous Illustration. However, the work done in calculating the approximation by the second polynomial does not lessen the work needed to calculate the third approximation; nor is the fourth approximation easier to obtain once the third approximation is known, and so on. We will now derive these approximating polynomials in a manner that uses the previous calculations to greater advantage.

Definition 3.4 Let f be a function defined at $x_0, x_1, x_2, \ldots, x_n$, and suppose that $m_1, m_2, \ldots, m_k$ are k distinct integers, with $0 \le m_i \le n$ for each i. The Lagrange polynomial that agrees with $f(x)$ at the k points $x_{m_1}, x_{m_2}, \ldots, x_{m_k}$ is denoted $P_{m_1,m_2,\ldots,m_k}(x)$. ■

Example 1 Suppose that $x_0 = 1$, $x_1 = 2$, $x_2 = 3$, $x_3 = 4$, $x_4 = 6$, and $f(x) = e^x$. Determine the interpolating polynomial denoted $P_{1,2,4}(x)$, and use this polynomial to approximate $f(5)$.

Solution This is the Lagrange polynomial that agrees with $f(x)$ at $x_1 = 2$, $x_2 = 3$, and $x_4 = 6$. Hence

$$P_{1,2,4}(x) = \frac{(x-3)(x-6)}{(2-3)(2-6)}e^2 + \frac{(x-2)(x-6)}{(3-2)(3-6)}e^3 + \frac{(x-2)(x-3)}{(6-2)(6-3)}e^6.$$

So

$$\begin{aligned} f(5) \approx P(5) &= \frac{(5-3)(5-6)}{(2-3)(2-6)}e^2 + \frac{(5-2)(5-6)}{(3-2)(3-6)}e^3 + \frac{(5-2)(5-3)}{(6-2)(6-3)}e^6 \\ &= -\frac{1}{2}e^2 + e^3 + \frac{1}{2}e^6 \approx 218.105. \end{aligned}$$

■

The next result describes a method for recursively generating Lagrange polynomial approximations.

Theorem 3.5 Let f be defined at $x_0, x_1, \ldots, x_k$, and let x_j and x_i be two distinct numbers in this set. Then

$$P(x) = \frac{(x - x_j)P_{0,1,\ldots,j-1,j+1,\ldots,k}(x) - (x - x_i)P_{0,1,\ldots,i-1,i+1,\ldots,k}(x)}{(x_i - x_j)}$$

is the kth Lagrange polynomial that interpolates f at the $k+1$ points $x_0, x_1, \ldots, x_k$. ■

Proof For ease of notation, let $Q \equiv P_{0,1,\ldots,i-1,i+1,\ldots,k}$ and $\hat{Q} \equiv P_{0,1,\ldots,j-1,j+1,\ldots,k}$. Since $Q(x)$ and $\hat{Q}(x)$ are polynomials of degree $k-1$ or less, $P(x)$ is of degree at most k.

First note that $\hat{Q}(x_i) = f(x_i)$, implies that

$$P(x_i) = \frac{(x_i - x_j)\hat{Q}(x_i) - (x_i - x_i)Q(x_i)}{x_i - x_j} = \frac{(x_i - x_j)}{(x_i - x_j)}f(x_i) = f(x_i).$$

Similarly, since $Q(x_j) = f(x_j)$, we have $P(x_j) = f(x_j)$.

In addition, if $0 \le r \le k$ and r is neither i nor j, then $Q(x_r) = \hat{Q}(x_r) = f(x_r)$. So

$$P(x_r) = \frac{(x_r - x_j)\hat{Q}(x_r) - (x_r - x_i)Q(x_r)}{x_i - x_j} = \frac{(x_i - x_j)}{(x_i - x_j)}f(x_r) = f(x_r).$$

But, by definition, $P_{0,1,\ldots,k}(x)$ is the unique polynomial of degree at most k that agrees with f at $x_0, x_1, \ldots, x_k$. Thus, $P \equiv P_{0,1,\ldots,k}$. ■ ■ ■

Theorem 3.5 implies that the interpolating polynomials can be generated recursively. For example, we have

$$P_{0,1} = \frac{1}{x_1 - x_0}[(x - x_0)P_1 - (x - x_1)P_0], \qquad P_{1,2} = \frac{1}{x_2 - x_1}[(x - x_1)P_2 - (x - x_2)P_1],$$

$$P_{0,1,2} = \frac{1}{x_2 - x_0}[(x - x_0)P_{1,2} - (x - x_2)P_{0,1}],$$

and so on. They are generated in the manner shown in Table 3.3, where each row is completed before the succeeding rows are begun.

Table 3.3

x_0	P_0				
x_1	P_1	$P_{0,1}$			
x_2	P_2	$P_{1,2}$	$P_{0,1,2}$		
x_3	P_3	$P_{2,3}$	$P_{1,2,3}$	$P_{0,1,2,3}$	
x_4	P_4	$P_{3,4}$	$P_{2,3,4}$	$P_{1,2,3,4}$	$P_{0,1,2,3,4}$

The procedure that uses the result of Theorem 3.5 to recursively generate interpolating polynomial approximations is called **Neville's method**. The P notation used in Table 3.3 is cumbersome because of the number of subscripts used to represent the entries. Note, however, that as an array is being constructed, only two subscripts are needed. Proceeding down the table corresponds to using consecutive points x_i with larger i, and proceeding to the right corresponds to increasing the degree of the interpolating polynomial. Since the points appear consecutively in each entry, we need to describe only a starting point and the number of additional points used in constructing the approximation.

Eric Harold Neville (1889–1961) gave this modification of the Lagrange formula in a paper published in 1932.[N]

To avoid the multiple subscripts, we let $Q_{i,j}(x)$, for $0 \le j \le i$, denote the interpolating polynomial of degree j on the $(j+1)$ numbers $x_{i-j}, x_{i-j+1}, \ldots, x_{i-1}, x_i$; that is,

$$Q_{i,j} = P_{i-j,i-j+1,\ldots,i-1,i}.$$

Using this notation provides the Q notation array in Table 3.4.

Table 3.4

x_0	$P_0 = Q_{0,0}$				
x_1	$P_1 = Q_{1,0}$	$P_{0,1} = Q_{1,1}$			
x_2	$P_2 = Q_{2,0}$	$P_{1,2} = Q_{2,1}$	$P_{0,1,2} = Q_{2,2}$		
x_3	$P_3 = Q_{3,0}$	$P_{2,3} = Q_{3,1}$	$P_{1,2,3} = Q_{3,2}$	$P_{0,1,2,3} = Q_{3,3}$	
x_4	$P_4 = Q_{4,0}$	$P_{3,4} = Q_{4,1}$	$P_{2,3,4} = Q_{4,2}$	$P_{1,2,3,4} = Q_{4,3}$	$P_{0,1,2,3,4} = Q_{4,4}$

Example 2 Values of various interpolating polynomials at $x = 1.5$ were obtained in the Illustration at the beginning of the Section using the data shown in Table 3.5. Apply Neville's method to the data by constructing a recursive table of the form shown in Table 3.4.

Table 3.5

x	$f(x)$
1.0	0.7651977
1.3	0.6200860
1.6	0.4554022
1.9	0.2818186
2.2	0.1103623

Solution Let $x_0 = 1.0$, $x_1 = 1.3$, $x_2 = 1.6$, $x_3 = 1.9$, and $x_4 = 2.2$, then $Q_{0,0} = f(1.0)$, $Q_{1,0} = f(1.3)$, $Q_{2,0} = f(1.6)$, $Q_{3,0} = f(1.9)$, and $Q_{4,0} = f(2.2)$. These are the five polynomials of degree zero (constants) that approximate $f(1.5)$, and are the same as data given in Table 3.5.

Calculating the first-degree approximation $Q_{1,1}(1.5)$ gives

$$\begin{aligned} Q_{1,1}(1.5) &= \frac{(x - x_0)Q_{1,0} - (x - x_1)Q_{0,0}}{x_1 - x_0} \\ &= \frac{(1.5 - 1.0)Q_{1,0} - (1.5 - 1.3)Q_{0,0}}{1.3 - 1.0} \\ &= \frac{0.5(0.6200860) - 0.2(0.7651977)}{0.3} = 0.5233449. \end{aligned}$$

Similarly,

$$Q_{2,1}(1.5) = \frac{(1.5 - 1.3)(0.4554022) - (1.5 - 1.6)(0.6200860)}{1.6 - 1.3} = 0.5102968,$$

$$Q_{3,1}(1.5) = 0.5132634, \quad \text{and} \quad Q_{4,1}(1.5) = 0.5104270.$$

The best linear approximation is expected to be $Q_{2,1}$ because 1.5 is between $x_1 = 1.3$ and $x_2 = 1.6$.

In a similar manner, approximations using higher-degree polynomials are given by

$$Q_{2,2}(1.5) = \frac{(1.5 - 1.0)(0.5102968) - (1.5 - 1.6)(0.5233449)}{1.6 - 1.0} = 0.5124715,$$

$$Q_{3,2}(1.5) = 0.5112857, \quad \text{and} \quad Q_{4,2}(1.5) = 0.5137361.$$

The higher-degree approximations are generated in a similar manner and are shown in Table 3.6. ■

Table 3.6

1.0	0.7651977				
1.3	0.6200860	0.5233449			
1.6	0.4554022	0.5102968	0.5124715		
1.9	0.2818186	0.5132634	0.5112857	0.5118127	
2.2	0.1103623	0.5104270	0.5137361	0.5118302	0.5118200

If the latest approximation, $Q_{4,4}$, was not sufficiently accurate, another node, x_5, could be selected, and another row added to the table:

$$x_5 \quad Q_{5,0} \quad Q_{5,1} \quad Q_{5,2} \quad Q_{5,3} \quad Q_{5,4} \quad Q_{5,5}.$$

Then $Q_{4,4}$, $Q_{5,4}$, and $Q_{5,5}$ could be compared to determine further accuracy.

The function in Example 2 is the Bessel function of the first kind of order zero, whose value at 2.5 is -0.0483838, and the next row of approximations to $f(1.5)$ is

$$2.5 \quad -0.0483838 \quad 0.4807699 \quad 0.5301984 \quad 0.5119070 \quad 0.5118430 \quad 0.5118277.$$

The final new entry, 0.5118277, is correct to all seven decimal places.

The *NumericalAnalysis* package in Maple can be used to apply Neville's method for the values of x and $f(x) = y$ in Table 3.6. After loading the package we define the data with

$xy := [[1.0, 0.7651977], [1.3, 0.6200860], [1.6, 0.4554022], [1.9, 0.2818186]]$

Neville's method using this data gives the approximation at $x = 1.5$ with the command

$p3 := PolynomialInterpolation(xy, method = neville, extrapolate = [1.5])$

The output from Maple for this command is

$$POLYINTERP([[1.0, 0.7651977], [1.3, 0.6200860], [1.6, 0.4554022], [1.9, 0.2818186]],$$
$$method = neville, extrapolate = [1.5], INFO)$$

which isn't very informative. To display the information, we enter the command

$NevilleTable(p3, 1.5)$

and Maple returns an array with four rows and four columns. The nonzero entries corresponding to the top four rows of Table 3.6 (with the first column deleted), the zero entries are simply used to fill up the array.

To add the additional row to the table using the additional data (2.2, 0.1103623) we use the command *p3a* := *AddPoint*(*p3*, [2.2, 0.1103623])

and a new array with all the approximation entries in Table 3.6 is obtained with

NevilleTable(*p3a*, 1.5)

Example 3 Table 3.7 lists the values of $f(x) = \ln x$ accurate to the places given. Use Neville's method and four-digit rounding arithmetic to approximate $f(2.1) = \ln 2.1$ by completing the Neville table.

Table 3.7

i	x_i	$\ln x_i$
0	2.0	0.6931
1	2.2	0.7885
2	2.3	0.8329

Solution Because $x - x_0 = 0.1$, $x - x_1 = -0.1$, $x - x_2 = -0.2$, and we are given $Q_{0,0} = 0.6931$, $Q_{1,0} = 0.7885$, and $Q_{2,0} = 0.8329$, we have

$$Q_{1,1} = \frac{1}{0.2}\left[(0.1)0.7885 - (-0.1)0.6931\right] = \frac{0.1482}{0.2} = 0.7410$$

and

$$Q_{2,1} = \frac{1}{0.1}\left[(-0.1)0.8329 - (-0.2)0.7885\right] = \frac{0.07441}{0.1} = 0.7441.$$

The final approximation we can obtain from this data is

$$Q_{2,1} = \frac{1}{0.3}\left[(0.1)0.7441 - (-0.2)0.7410\right] = \frac{0.2276}{0.3} = 0.7420.$$

These values are shown in Table 3.8. ■

Table 3.8

i	x_i	$x - x_i$	Q_{i0}	Q_{i1}	Q_{i2}
0	2.0	0.1	0.6931		
1	2.2	−0.1	0.7885	0.7410	
2	2.3	−0.2	0.8329	0.7441	0.7420

In the preceding example we have $f(2.1) = \ln 2.1 = 0.7419$ to four decimal places, so the absolute error is

$$|f(2.1) - P_2(2.1)| = |0.7419 - 0.7420| = 10^{-4}.$$

However, $f'(x) = 1/x$, $f''(x) = -1/x^2$, and $f'''(x) = 2/x^3$, so the Lagrange error formula (3.3) in Theorem 3.3 gives the error bound

$$\begin{aligned}|f(2.1) - P_2(2.1)| &= \left|\frac{f'''(\xi(2.1))}{3!}(x - x_0)(x - x_1)(x - x_2)\right| \\ &= \left|\frac{1}{3\,(\xi(2.1))^3}(0.1)(-0.1)(-0.2)\right| \le \frac{0.002}{3(2)^3} = 8.\overline{3} \times 10^{-5}.\end{aligned}$$

Notice that the actual error, 10^{-4}, exceeds the error bound, $8.\overline{3} \times 10^{-5}$. This apparent contradiction is a consequence of finite-digit computations. We used four-digit rounding arithmetic, and the Lagrange error formula (3.3) assumes infinite-digit arithmetic. This caused our actual errors to exceed the theoretical error estimate.

- Remember: You cannot expect more accuracy than the arithmetic provides.

Algorithm 3.1 constructs the entries in Neville's method by rows.

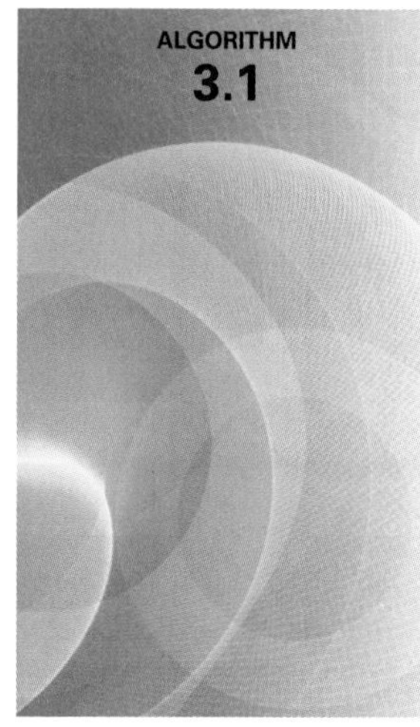

Neville's Iterated Interpolation

To evaluate the interpolating polynomial P on the $n + 1$ distinct numbers $x_0, \ldots, x_n$ at the number x for the function f:

INPUT numbers $x, x_0, x_1, \ldots, x_n$; values $f(x_0), f(x_1), \ldots, f(x_n)$ as the first column $Q_{0,0}, Q_{1,0}, \ldots, Q_{n,0}$ of Q.

OUTPUT the table Q with $P(x) = Q_{n,n}$.

Step 1 For $i = 1, 2, \ldots, n$
for $j = 1, 2, \ldots, i$

$$\text{set } Q_{i,j} = \frac{(x - x_{i-j})Q_{i,j-1} - (x - x_i)Q_{i-1,j-1}}{x_i - x_{i-j}}.$$

Step 2 OUTPUT (Q);
STOP. ■

The algorithm can be modified to allow for the addition of new interpolating nodes. For example, the inequality

$$|Q_{i,i} - Q_{i-1,i-1}| < \varepsilon$$

can be used as a stopping criterion, where ε is a prescribed error tolerance. If the inequality is true, $Q_{i,i}$ is a reasonable approximation to $f(x)$. If the inequality is false, a new interpolation point, x_{i+1}, is added.

EXERCISE SET 3.2

1. Use Neville's method to obtain the approximations for Lagrange interpolating polynomials of degrees one, two, and three to approximate each of the following:

a. $f(8.4)$ if $f(8.1) = 16.94410$, $f(8.3) = 17.56492$, $f(8.6) = 18.50515$, $f(8.7) = 18.82091$

b. $f(0.25)$ if $f(0.1) = 0.62049958$, $f(0.2) = -0.28398668$, $f(0.3) = 0.00660095$, $f(0.4) = 0.24842440$

c. $f\left(-\frac{1}{3}\right)$ if $f(-0.75) = -0.07181250$, $f(-0.5) = -0.02475000$, $f(-0.25) = 0.33493750$, $f(0) = 1.10100000$

d. $f(0.9)$ if $f(0.6) = -0.17694460$, $f(0.7) = 0.01375227$, $f(0.8) = 0.22363362$, $f(1.0) = 0.65809197$

2. Use Neville's method to obtain the approximations for Lagrange interpolating polynomials of degrees one, two, and three to approximate each of the following:

a. $f(0.43)$ if $f(0) = 1$, $f(0.25) = 1.64872$, $f(0.5) = 2.71828$, $f(0.75) = 4.48169$

b. $f(0.18)$ if $f(0.1) = -0.29004986$, $f(0.2) = -0.56079734$, $f(0.3) = -0.81401972$, $f(0.4) = -1.0526302$

c. $f(0)$ if $f(-0.5) = 1.93750$, $f(-0.25) = 1.33203$, $f(0.25) = 0.800781$, $f(0.5) = 0.687500$

d. $f(0.25)$ if $f(-1) = 0.86199480$, $f(-0.5) = 0.95802009$, $f(0) = 1.0986123$, $f(0.5) = 1.2943767$

3. Use Neville's method to approximate $\sqrt{3}$ with the following functions and values.

a. $f(x) = 3^x$ and the values $x_0 = -2$, $x_1 = -1$, $x_2 = 0$, $x_3 = 1$, and $x_4 = 2$.

b. $f(x) = \sqrt{x}$ and the values $x_0 = 0$, $x_1 = 1$, $x_2 = 2$, $x_3 = 4$, and $x_4 = 5$.

c. Compare the accuracy of the approximation in parts (a) and (b).

4. Let $P_3(x)$ be the interpolating polynomial for the data $(0, 0)$, $(0.5, y)$, $(1, 3)$, and $(2, 2)$. Use Neville's method to find y if $P_3(1.5) = 0$.

5. Neville's method is used to approximate $f(0.4)$, giving the following table.

$x_0 = 0$	$P_0 = 1$			
$x_1 = 0.25$	$P_1 = 2$	$P_{01} = 2.6$		
$x_2 = 0.5$	P_2	$P_{1,2}$	$P_{0,1,2}$	
$x_3 = 0.75$	$P_3 = 8$	$P_{2,3} = 2.4$	$P_{1,2,3} = 2.96$	$P_{0,1,2,3} = 3.016$

Determine $P_2 = f(0.5)$.

6. Neville's method is used to approximate $f(0.5)$, giving the following table.

$x_0 = 0$	$P_0 = 0$		
$x_1 = 0.4$	$P_1 = 2.8$	$P_{0,1} = 3.5$	
$x_2 = 0.7$	P_2	$P_{1,2}$	$P_{0,1,2} = \frac{27}{7}$

Determine $P_2 = f(0.7)$.

7. Suppose $x_j = j$, for $j = 0, 1, 2, 3$ and it is known that

$$P_{0,1}(x) = 2x + 1, \quad P_{0,2}(x) = x + 1, \quad \text{and} \quad P_{1,2,3}(2.5) = 3.$$

Find $P_{0,1,2,3}(2.5)$.

8. Suppose $x_j = j$, for $j = 0, 1, 2, 3$ and it is known that

$$P_{0,1}(x) = x + 1, \quad P_{1,2}(x) = 3x - 1, \quad \text{and} \quad P_{1,2,3}(1.5) = 4.$$

Find $P_{0,1,2,3}(1.5)$.

9. Neville's Algorithm is used to approximate $f(0)$ using $f(-2)$, $f(-1)$, $f(1)$, and $f(2)$. Suppose $f(-1)$ was understated by 2 and $f(1)$ was overstated by 3. Determine the error in the original calculation of the value of the interpolating polynomial to approximate $f(0)$.

10. Neville's Algorithm is used to approximate $f(0)$ using $f(-2)$, $f(-1)$, $f(1)$, and $f(2)$. Suppose $f(-1)$ was overstated by 2 and $f(1)$ was understated by 3. Determine the error in the original calculation of the value of the interpolating polynomial to approximate $f(0)$.

11. Construct a sequence of interpolating values y_n to $f(1+\sqrt{10})$, where $f(x) = (1+x^2)^{-1}$ for $-5 \le x \le 5$, as follows: For each $n = 1, 2, \ldots, 10$, let $h = 10/n$ and $y_n = P_n(1+\sqrt{10})$, where $P_n(x)$ is the interpolating polynomial for $f(x)$ at the nodes $x_0^{(n)}, x_1^{(n)}, \ldots, x_n^{(n)}$ and $x_j^{(n)} = -5 + jh$, for each $j = 0, 1, 2, \ldots, n$. Does the sequence $\{y_n\}$ appear to converge to $f(1+\sqrt{10})$?

Inverse Interpolation Suppose $f \in C^1[a,b]$, $f'(x) \neq 0$ on $[a,b]$ and f has one zero p in $[a,b]$. Let $x_0, \ldots, x_n$, be $n+1$ distinct numbers in $[a,b]$ with $f(x_k) = y_k$, for each $k = 0, 1, \ldots, n$. To approximate p construct the interpolating polynomial of degree n on the nodes $y_0, \ldots, y_n$ for f^{-1}. Since $y_k = f(x_k)$ and $0 = f(p)$, it follows that $f^{-1}(y_k) = x_k$ and $p = f^{-1}(0)$. Using iterated interpolation to approximate $f^{-1}(0)$ is called *iterated inverse interpolation.*

12. Use iterated inverse interpolation to find an approximation to the solution of $x - e^{-x} = 0$, using the data

x	0.3	0.4	0.5	0.6
e^{-x}	0.740818	0.670320	0.606531	0.548812

13. Construct an algorithm that can be used for inverse interpolation.

3.3 Divided Differences

Iterated interpolation was used in the previous section to generate successively higher-degree polynomial approximations at a specific point. Divided-difference methods introduced in this section are used to successively generate the polynomials themselves.

Suppose that $P_n(x)$ is the nth Lagrange polynomial that agrees with the function f at the distinct numbers $x_0, x_1, \dots, x_n$. Although this polynomial is unique, there are alternate algebraic representations that are useful in certain situations. The divided differences of f with respect to $x_0, x_1, \dots, x_n$ are used to express $P_n(x)$ in the form

$$P_n(x) = a_0 + a_1(x - x_0) + a_2(x - x_0)(x - x_1) + \cdots + a_n(x - x_0)\cdots(x - x_{n-1}), \quad (3.5)$$

for appropriate constants $a_0, a_1, \dots, a_n$. To determine the first of these constants, a_0, note that if $P_n(x)$ is written in the form of Eq. (3.5), then evaluating $P_n(x)$ at x_0 leaves only the constant term a_0; that is,

$$a_0 = P_n(x_0) = f(x_0).$$

As in so many areas, Isaac Newton is prominent in the study of difference equations. He developed interpolation formulas as early as 1675, using his Δ notation in tables of differences. He took a very general approach to the difference formulas, so explicit examples that he produced, including Lagrange's formulas, are often known by other names.

Similarly, when $P(x)$ is evaluated at x_1, the only nonzero terms in the evaluation of $P_n(x_1)$ are the constant and linear terms,

$$f(x_0) + a_1(x_1 - x_0) = P_n(x_1) = f(x_1);$$

so

$$a_1 = \frac{f(x_1) - f(x_0)}{x_1 - x_0}. \quad (3.6)$$

We now introduce the divided-difference notation, which is related to Aitken's Δ^2 notation used in Section 2.5. The *zeroth divided difference* of the function f with respect to x_i, denoted $f[x_i]$, is simply the value of f at x_i:

$$f[x_i] = f(x_i). \quad (3.7)$$

The remaining divided differences are defined recursively; the *first divided difference* of f with respect to x_i and x_{i+1} is denoted $f[x_i, x_{i+1}]$ and defined as

$$f[x_i, x_{i+1}] = \frac{f[x_{i+1}] - f[x_i]}{x_{i+1} - x_i}. \quad (3.8)$$

The *second divided difference*, $f[x_i, x_{i+1}, x_{i+2}]$, is defined as

$$f[x_i, x_{i+1}, x_{i+2}] = \frac{f[x_{i+1}, x_{i+2}] - f[x_i, x_{i+1}]}{x_{i+2} - x_i}.$$

Similarly, after the $(k-1)$st divided differences,

$$f[x_i, x_{i+1}, x_{i+2}, \dots, x_{i+k-1}] \quad \text{and} \quad f[x_{i+1}, x_{i+2}, \dots, x_{i+k-1}, x_{i+k}],$$

have been determined, the **kth divided difference** relative to $x_i, x_{i+1}, x_{i+2}, \dots, x_{i+k}$ is

$$f[x_i, x_{i+1}, \dots, x_{i+k-1}, x_{i+k}] = \frac{f[x_{i+1}, x_{i+2}, \dots, x_{i+k}] - f[x_i, x_{i+1}, \dots, x_{i+k-1}]}{x_{i+k} - x_i}. \quad (3.9)$$

The process ends with the single *nth divided difference*,

$$f[x_0, x_1, \dots, x_n] = \frac{f[x_1, x_2, \dots, x_n] - f[x_0, x_1, \dots, x_{n-1}]}{x_n - x_0}.$$

Because of Eq. (3.6) we can write $a_1 = f[x_0, x_1]$, just as a_0 can be expressed as $a_0 = f(x_0) = f[x_0]$. Hence the interpolating polynomial in Eq. (3.5) is

$$\begin{aligned} P_n(x) = {} & f[x_0] + f[x_0, x_1](x - x_0) + a_2(x - x_0)(x - x_1) \\ & + \cdots + a_n(x - x_0)(x - x_1)\cdots(x - x_{n-1}). \end{aligned}$$

As might be expected from the evaluation of a_0 and a_1, the required constants are

$$a_k = f[x_0, x_1, x_2, \dots, x_k],$$

for each $k = 0, 1, \dots, n$. So $P_n(x)$ can be rewritten in a form called Newton's Divided-Difference:

$$P_n(x) = f[x_0] + \sum_{k=1}^{n} f[x_0, x_1, \dots, x_k](x - x_0) \cdots (x - x_{k-1}). \tag{3.10}$$

The value of $f[x_0, x_1, \dots, x_k]$ is independent of the order of the numbers $x_0, x_1, \dots, x_k$, as shown in Exercise 21.

The generation of the divided differences is outlined in Table 3.9. Two fourth and one fifth difference can also be determined from these data.

Table 3.9

x	$f(x)$	First divided differences	Second divided differences	Third divided differences
x_0	$f[x_0]$			
		$f[x_0, x_1] = \dfrac{f[x_1] - f[x_0]}{x_1 - x_0}$		
x_1	$f[x_1]$		$f[x_0, x_1, x_2] = \dfrac{f[x_1, x_2] - f[x_0, x_1]}{x_2 - x_0}$	
		$f[x_1, x_2] = \dfrac{f[x_2] - f[x_1]}{x_2 - x_1}$		$f[x_0, x_1, x_2, x_3] = \dfrac{f[x_1, x_2, x_3] - f[x_0, x_1, x_2]}{x_3 - x_0}$
x_2	$f[x_2]$		$f[x_1, x_2, x_3] = \dfrac{f[x_2, x_3] - f[x_1, x_2]}{x_3 - x_1}$	
		$f[x_2, x_3] = \dfrac{f[x_3] - f[x_2]}{x_3 - x_2}$		$f[x_1, x_2, x_3, x_4] = \dfrac{f[x_2, x_3, x_4] - f[x_1, x_2, x_3]}{x_4 - x_1}$
x_3	$f[x_3]$		$f[x_2, x_3, x_4] = \dfrac{f[x_3, x_4] - f[x_2, x_3]}{x_4 - x_2}$	
		$f[x_3, x_4] = \dfrac{f[x_4] - f[x_3]}{x_4 - x_3}$		$f[x_2, x_3, x_4, x_5] = \dfrac{f[x_3, x_4, x_5] - f[x_2, x_3, x_4]}{x_5 - x_2}$
x_4	$f[x_4]$		$f[x_3, x_4, x_5] = \dfrac{f[x_4, x_5] - f[x_3, x_4]}{x_5 - x_3}$	
		$f[x_4, x_5] = \dfrac{f[x_5] - f[x_4]}{x_5 - x_4}$		
x_5	$f[x_5]$			

Newton's Divided-Difference Formula

To obtain the divided-difference coefficients of the interpolatory polynomial P on the $(n+1)$ distinct numbers $x_0, x_1, \dots, x_n$ for the function f:

INPUT numbers $x_0, x_1, \dots, x_n$; values $f(x_0), f(x_1), \dots, f(x_n)$ as $F_{0,0}, F_{1,0}, \dots, F_{n,0}$.

OUTPUT the numbers $F_{0,0}, F_{1,1}, \dots, F_{n,n}$ where

$$P_n(x) = F_{0,0} + \sum_{i=1}^{n} F_{i,i} \prod_{j=0}^{i-1} (x - x_j). \quad (F_{i,i} \text{ is } f[x_0, x_1, \dots, x_i].)$$

Step 1 For $i = 1, 2, \dots, n$
  For $j = 1, 2, \dots, i$
    set $F_{i,j} = \dfrac{F_{i,j-1} - F_{i-1,j-1}}{x_i - x_{i-j}}$. $\quad (F_{i,j} = f[x_{i-j}, \dots, x_i].)$

Step 2 OUTPUT $(F_{0,0}, F_{1,1}, \dots, F_{n,n})$;
STOP. ■

The form of the output in Algorithm 3.2 can be modified to produce all the divided differences, as shown in Example 1.

Example 1 Complete the divided difference table for the data used in Example 1 of Section 3.2, and reproduced in Table 3.10, and construct the interpolating polynomial that uses all this data.

Table 3.10

x	$f(x)$
1.0	0.7651977
1.3	0.6200860
1.6	0.4554022
1.9	0.2818186
2.2	0.1103623

Solution The first divided difference involving x_0 and x_1 is

$$f[x_0, x_1] = \frac{f[x_1] - f[x_0]}{x_1 - x_0} = \frac{0.6200860 - 0.7651977}{1.3 - 1.0} = -0.4837057.$$

The remaining first divided differences are found in a similar manner and are shown in the fourth column in Table 3.11.

Table 3.11

i	x_i	$f[x_i]$	$f[x_{i-1}, x_i]$	$f[x_{i-2}, x_{i-1}, x_i]$	$f[x_{i-3}, \ldots, x_i]$	$f[x_{i-4}, \ldots, x_i]$
0	1.0	0.7651977				
			−0.4837057			
1	1.3	0.6200860		−0.1087339		
			−0.5489460		0.0658784	
2	1.6	0.4554022		−0.0494433		0.0018251
			−0.5786120		0.0680685	
3	1.9	0.2818186		0.0118183		
			−0.5715210			
4	2.2	0.1103623				

The second divided difference involving x_0, x_1, and x_2 is

$$f[x_0, x_1, x_2] = \frac{f[x_1, x_2] - f[x_0, x_1]}{x_2 - x_0} = \frac{-0.5489460 - (-0.4837057)}{1.6 - 1.0} = -0.1087339.$$

The remaining second divided differences are shown in the 5th column of Table 3.11. The third divided difference involving x_0, x_1, x_2, and x_3 and the fourth divided difference involving all the data points are, respectively,

$$f[x_0, x_1, x_2, x_3] = \frac{f[x_1, x_2, x_3] - f[x_0, x_1, x_2]}{x_3 - x_0} = \frac{-0.0494433 - (-0.1087339)}{1.9 - 1.0}$$
$$= 0.0658784,$$

and

$$f[x_0, x_1, x_2, x_3, x_4] = \frac{f[x_1, x_2, x_3, x_4] - f[x_0, x_1, x_2, x_3]}{x_4 - x_0} = \frac{0.0680685 - 0.0658784}{2.2 - 1.0}$$
$$= 0.0018251.$$

All the entries are given in Table 3.11.

The coefficients of the Newton forward divided-difference form of the interpolating polynomial are along the diagonal in the table. This polynomial is

$$\begin{aligned} P_4(x) = {} & 0.7651977 - 0.4837057(x - 1.0) - 0.1087339(x - 1.0)(x - 1.3) \\ & + 0.0658784(x - 1.0)(x - 1.3)(x - 1.6) \\ & + 0.0018251(x - 1.0)(x - 1.3)(x - 1.6)(x - 1.9). \end{aligned}$$

Notice that the value $P_4(1.5) = 0.5118200$ agrees with the result in Table 3.6 for Example 2 of Section 3.2, as it must because the polynomials are the same. ∎

We can use Maple with the *NumericalAnalysis* package to create the Newton Divided-Difference table. First load the package and define the x and $f(x) = y$ values that will be used to generate the first four rows of Table 3.11.

$xy := [[1.0, 0.7651977], [1.3, 0.6200860], [1.6, 0.4554022], [1.9, 0.2818186]]$

The command to create the divided-difference table is

$p3 := PolynomialInterpolation(xy, independentvar = \text{'}x\text{'}, method = newton)$

A matrix containing the divided-difference table as its nonzero entries is created with the

$DividedDifferenceTable(p3)$

We can add another row to the table with the command

$p4 := AddPoint(p3, [2.2, 0.1103623])$

which produces the divided-difference table with entries corresponding to those in Table 3.11.

The Newton form of the interpolation polynomial is created with

$Interpolant(p4)$

which produces the polynomial in the form of $P_4(x)$ in Example 1, except that in place of the first two terms of $P_4(x)$:

$$0.7651977 - 0.4837057(x - 1.0)$$

Maple gives this as $1.248903367 - 0.4837056667x$.

The Mean Value Theorem 1.8 applied to Eq. (3.8) when $i = 0$,

$$f[x_0, x_1] = \frac{f(x_1) - f(x_0)}{x_1 - x_0},$$

implies that when f' exists, $f[x_0, x_1] = f'(\xi)$ for some number ξ between x_0 and x_1. The following theorem generalizes this result.

Theorem 3.6 Suppose that $f \in C^n[a, b]$ and $x_0, x_1, \ldots, x_n$ are distinct numbers in $[a, b]$. Then a number ξ exists in (a, b) with

$$f[x_0, x_1, \ldots, x_n] = \frac{f^{(n)}(\xi)}{n!}.$$ ■

Proof Let

$$g(x) = f(x) - P_n(x).$$

Since $f(x_i) = P_n(x_i)$ for each $i = 0, 1, \ldots, n$, the function g has $n+1$ distinct zeros in $[a, b]$. Generalized Rolle's Theorem 1.10 implies that a number ξ in (a, b) exists with $g^{(n)}(\xi) = 0$, so

$$0 = f^{(n)}(\xi) - P_n^{(n)}(\xi).$$

Since $P_n(x)$ is a polynomial of degree n whose leading coefficient is $f[x_0, x_1, \ldots, x_n]$,

$$P_n^{(n)}(x) = n! f[x_0, x_1, \ldots, x_n],$$

for all values of x. As a consequence,

$$f[x_0, x_1, \ldots, x_n] = \frac{f^{(n)}(\xi)}{n!}.$$ ■ ■ ■

Newton's divided-difference formula can be expressed in a simplified form when the nodes are arranged consecutively with equal spacing. In this case, we introduce the notation $h = x_{i+1} - x_i$, for each $i = 0, 1, \ldots, n-1$ and let $x = x_0 + sh$. Then the difference $x - x_i$ is $x - x_i = (s - i)h$. So Eq. (3.10) becomes

$$\begin{aligned} P_n(x) = P_n(x_0 + sh) &= f[x_0] + shf[x_0, x_1] + s(s-1)h^2 f[x_0, x_1, x_2] \\ &\quad + \cdots + s(s-1)\cdots(s-n+1)h^n f[x_0, x_1, \ldots, x_n] \\ &= f[x_0] + \sum_{k=1}^{n} s(s-1)\cdots(s-k+1)h^k f[x_0, x_1, \ldots, x_k]. \end{aligned}$$

Using binomial-coefficient notation,

$$\binom{s}{k} = \frac{s(s-1)\cdots(s-k+1)}{k!},$$

we can express $P_n(x)$ compactly as

$$P_n(x) = P_n(x_0 + sh) = f[x_0] + \sum_{k=1}^{n} \binom{s}{k} k!h^k f[x_0, x_i, \ldots, x_k]. \tag{3.11}$$

Forward Differences

The **Newton forward-difference formula**, is constructed by making use of the forward difference notation Δ introduced in Aitken's Δ^2 method. With this notation,

$$\begin{aligned} f[x_0, x_1] &= \frac{f(x_1) - f(x_0)}{x_1 - x_0} = \frac{1}{h}(f(x_1) - f(x_0)) = \frac{1}{h}\Delta f(x_0) \\ f[x_0, x_1, x_2] &= \frac{1}{2h}\left[\frac{\Delta f(x_1) - \Delta f(x_0)}{h}\right] = \frac{1}{2h^2}\Delta^2 f(x_0), \end{aligned}$$

and, in general,

$$f[x_0, x_1, \ldots, x_k] = \frac{1}{k!h^k}\Delta^k f(x_0).$$

Since $f[x_0] = f(x_0)$, Eq. (3.11) has the following form.

Newton Forward-Difference Formula

$$P_n(x) = f(x_0) + \sum_{k=1}^{n} \binom{s}{k} \Delta^k f(x_0) \tag{3.12}$$

Backward Differences

If the interpolating nodes are reordered from last to first as $x_n, x_{n-1}, \ldots, x_0$, we can write the interpolatory formula as

$$\begin{aligned} P_n(x) = f[x_n] &+ f[x_n, x_{n-1}](x - x_n) + f[x_n, x_{n-1}, x_{n-2}](x - x_n)(x - x_{n-1}) \\ &+ \cdots + f[x_n, \ldots, x_0](x - x_n)(x - x_{n-1})\cdots(x - x_1). \end{aligned}$$

If, in addition, the nodes are equally spaced with $x = x_n + sh$ and $x = x_i + (s+n-i)h$, then

$$\begin{aligned} P_n(x) &= P_n(x_n + sh) \\ &= f[x_n] + shf[x_n, x_{n-1}] + s(s+1)h^2 f[x_n, x_{n-1}, x_{n-2}] + \cdots \\ &\quad + s(s+1)\cdots(s+n-1)h^n f[x_n, \ldots, x_0]. \end{aligned}$$

This is used to derive a commonly applied formula known as the **Newton backward-difference formula**. To discuss this formula, we need the following definition.

Definition 3.7 Given the sequence $\{p_n\}_{n=0}^{\infty}$, define the backward difference ∇p_n (read *nabla* p_n) by

$$\nabla p_n = p_n - p_{n-1}, \quad \text{for } n \geq 1.$$

Higher powers are defined recursively by

$$\nabla^k p_n = \nabla(\nabla^{k-1} p_n), \quad \text{for } k \geq 2. \qquad \blacksquare$$

Definition 3.7 implies that

$$f[x_n, x_{n-1}] = \frac{1}{h}\nabla f(x_n), \quad f[x_n, x_{n-1}, x_{n-2}] = \frac{1}{2h^2}\nabla^2 f(x_n),$$

and, in general,

$$f[x_n, x_{n-1}, \ldots, x_{n-k}] = \frac{1}{k!h^k}\nabla^k f(x_n).$$

Consequently,

$$P_n(x) = f[x_n] + s\nabla f(x_n) + \frac{s(s+1)}{2}\nabla^2 f(x_n) + \cdots + \frac{s(s+1)\cdots(s+n-1)}{n!}\nabla^n f(x_n).$$

If we extend the binomial coefficient notation to include all real values of s by letting

$$\binom{-s}{k} = \frac{-s(-s-1)\cdots(-s-k+1)}{k!} = (-1)^k \frac{s(s+1)\cdots(s+k-1)}{k!},$$

then

$$P_n(x) = f[x_n] + (-1)^1\binom{-s}{1}\nabla f(x_n) + (-1)^2\binom{-s}{2}\nabla^2 f(x_n) + \cdots + (-1)^n\binom{-s}{n}\nabla^n f(x_n).$$

This gives the following result.

Newton Backward–Difference Formula

$$P_n(x) = f[x_n] + \sum_{k=1}^{n}(-1)^k\binom{-s}{k}\nabla^k f(x_n) \tag{3.13}$$

Illustration The divided-difference Table 3.12 corresponds to the data in Example 1.

Table 3.12

		First divided differences	Second divided differences	Third divided differences	Fourth divided differences
1.0	0.7651977				
		−0.4837057			
1.3	0.6200860		−0.1087339		
		−0.5489460		0.0658784	
1.6	0.4554022		−0.0494433		0.0018251
		−0.5786120		0.0680685	
1.9	0.2818186		0.0118183		
		−0.5715210			
2.2	0.1103623				

Only one interpolating polynomial of degree at most 4 uses these five data points, but we will organize the data points to obtain the best interpolation approximations of degrees 1, 2, and 3. This will give us a sense of accuracy of the fourth-degree approximation for the given value of x.

If an approximation to $f(1.1)$ is required, the reasonable choice for the nodes would be $x_0 = 1.0$, $x_1 = 1.3$, $x_2 = 1.6$, $x_3 = 1.9$, and $x_4 = 2.2$ since this choice makes the earliest possible use of the data points closest to $x = 1.1$, and also makes use of the fourth divided difference. This implies that $h = 0.3$ and $s = \frac{1}{3}$, so the Newton forward divided-difference formula is used with the divided differences that have a *solid* underline (__) in Table 3.12:

$$\begin{aligned}
P_4(1.1) &= P_4(1.0 + \frac{1}{3}(0.3)) \\
&= 0.7651977 + \frac{1}{3}(0.3)(-0.4837057) + \frac{1}{3}\left(-\frac{2}{3}\right)(0.3)^2(-0.1087339) \\
&\quad + \frac{1}{3}\left(-\frac{2}{3}\right)\left(-\frac{5}{3}\right)(0.3)^3(0.0658784) \\
&\quad + \frac{1}{3}\left(-\frac{2}{3}\right)\left(-\frac{5}{3}\right)\left(-\frac{8}{3}\right)(0.3)^4(0.0018251) \\
&= 0.7196460.
\end{aligned}$$

To approximate a value when x is close to the end of the tabulated values, say, $x = 2.0$, we would again like to make the earliest use of the data points closest to x. This requires using the Newton backward divided-difference formula with $s = -\frac{2}{3}$ and the divided differences in Table 3.12 that have a *wavy* underline (﹏). Notice that the fourth divided difference is used in both formulas.

$$\begin{aligned}
P_4(2.0) &= P_4\left(2.2 - \frac{2}{3}(0.3)\right) \\
&= 0.1103623 - \frac{2}{3}(0.3)(-0.5715210) - \frac{2}{3}\left(\frac{1}{3}\right)(0.3)^2(0.0118183) \\
&\quad - \frac{2}{3}\left(\frac{1}{3}\right)\left(\frac{4}{3}\right)(0.3)^3(0.0680685) - \frac{2}{3}\left(\frac{1}{3}\right)\left(\frac{4}{3}\right)\left(\frac{7}{3}\right)(0.3)^4(0.0018251) \\
&= 0.2238754.
\end{aligned}$$

□

Centered Differences

The Newton forward- and backward-difference formulas are not appropriate for approximating $f(x)$ when x lies near the center of the table because neither will permit the highest-order difference to have x_0 close to x. A number of divided-difference formulas are available for this case, each of which has situations when it can be used to maximum advantage. These methods are known as **centered-difference formulas**. We will consider only one centered-difference formula, Stirling's method.

For the centered-difference formulas, we choose x_0 near the point being approximated and label the nodes directly below x_0 as $x_1, x_2, \ldots$ and those directly above as $x_{-1}, x_{-2}, \ldots$. With this convention, **Stirling's formula** is given by

$$P_n(x) = P_{2m+1}(x) = f[x_0] + \frac{sh}{2}(f[x_{-1},x_0] + f[x_0,x_1]) + s^2h^2 f[x_{-1},x_0,x_1] \tag{3.14}$$

$$+ \frac{s(s^2-1)h^3}{2} f[x_{-2},x_{-1},x_0,x_1] + f[x_{-1},x_0,x_1,x_2])$$

$$+ \cdots + s^2(s^2-1)(s^2-4)\cdots(s^2-(m-1)^2)h^{2m} f[x_{-m},\ldots,x_m]$$

$$+ \frac{s(s^2-1)\cdots(s^2-m^2)h^{2m+1}}{2}(f[x_{-m-1},\ldots,x_m] + f[x_{-m},\ldots,x_{m+1}]),$$

James Stirling (1692–1770) published this and numerous other formulas in *Methodus Differentialis* in 1720. Techniques for accelerating the convergence of various series are included in this work.

if $n = 2m+1$ is odd. If $n = 2m$ is even, we use the same formula but delete the last line. The entries used for this formula are underlined in Table 3.13.

Table 3.13

x	$f(x)$	First divided differences	Second divided differences	Third divided differences	Fourth divided differences
x_{-2}	$f[x_{-2}]$				
		$f[x_{-2},x_{-1}]$			
x_{-1}	$f[x_{-1}]$		$f[x_{-2},x_{-1},x_0]$		
		$\underline{f[x_{-1},x_0]}$		$\underline{f[x_{-2},x_{-1},x_0,x_1]}$	
x_0	$\underline{f[x_0]}$		$\underline{f[x_{-1},x_0,x_1]}$		$\underline{f[x_{-2},x_{-1},x_0,x_1,x_2]}$
		$\underline{f[x_0,x_1]}$		$\underline{f[x_{-1},x_0,x_1,x_2]}$	
x_1	$f[x_1]$		$f[x_0,x_1,x_2]$		
		$f[x_1,x_2]$			
x_2	$f[x_2]$				

Example 2 Consider the table of data given in the previous examples. Use Stirling's formula to approximate $f(1.5)$ with $x_0 = 1.6$.

Solution To apply Stirling's formula we use the *underlined* entries in the difference Table 3.14.

Table 3.14

x	$f(x)$	First divided differences	Second divided differences	Third divided differences	Fourth divided differences
1.0	0.7651977				
		−0.4837057			
1.3	0.6200860		−0.1087339		
		<u>−0.5489460</u>		<u>0.0658784</u>	
1.6	<u>0.4554022</u>		<u>−0.0494433</u>		<u>0.0018251</u>
		<u>−0.5786120</u>		<u>0.0680685</u>	
1.9	0.2818186		0.0118183		
		−0.5715210			
2.2	0.1103623				

The formula, with $h = 0.3$, $x_0 = 1.6$, and $s = -\frac{1}{3}$, becomes

$$\begin{aligned}
f(1.5) &\approx P_4\left(1.6 + \left(-\frac{1}{3}\right)(0.3)\right) \\
&= 0.4554022 + \left(-\frac{1}{3}\right)\left(\frac{0.3}{2}\right)((-0.5489460) + (-0.5786120)) \\
&\quad + \left(-\frac{1}{3}\right)^2 (0.3)^2(-0.0494433) \\
&\quad + \frac{1}{2}\left(-\frac{1}{3}\right)\left(\left(-\frac{1}{3}\right)^2 - 1\right)(0.3)^3(0.0658784 + 0.0680685) \\
&\quad + \left(-\frac{1}{3}\right)^2\left(\left(-\frac{1}{3}\right)^2 - 1\right)(0.3)^4(0.0018251) = 0.5118200.
\end{aligned}$$

■

Most texts on numerical analysis written before the wide-spread use of computers have extensive treatments of divided-difference methods. If a more comprehensive treatment of this subject is needed, the book by Hildebrand [Hild] is a particularly good reference.

EXERCISE SET 3.3

1. Use Eq. (3.10) or Algorithm 3.2 to construct interpolating polynomials of degree one, two, and three for the following data. Approximate the specified value using each of the polynomials.
   a. $f(8.4)$ if $f(8.1) = 16.94410$, $f(8.3) = 17.56492$, $f(8.6) = 18.50515$, $f(8.7) = 18.82091$
   b. $f(0.9)$ if $f(0.6) = -0.17694460$, $f(0.7) = 0.01375227$, $f(0.8) = 0.22363362$, $f(1.0) = 0.65809197$
2. Use Eq. (3.10) or Algorithm 3.2 to construct interpolating polynomials of degree one, two, and three for the following data. Approximate the specified value using each of the polynomials.
   a. $f(0)$ if $f(-0.5) = 1.93750$, $f(-0.25) = 1.33203$, $f(0.25) = 0.800781$, $f(0.5) = 0.687500$
   b. $f(0.43)$ if $f(0) = 1$, $f(0.25) = 1.64872$, $f(0.5) = 2.71828$, $f(0.75) = 4.48169$
3. Use Newton the forward-difference formula to construct interpolating polynomials of degree one, two, and three for the following data. Approximate the specified value using each of the polynomials.
   a. $f(0.25)$ if $f(0.1) = -0.62049958$, $f(0.2) = -0.28398668$, $f(0.3) = 0.00660095$, $f(0.4) = 0.24842440$
   b. $f\left(-\frac{1}{3}\right)$ if $f(-0.75) = -0.07181250$, $f(-0.5) = -0.02475000$, $f(-0.25) = 0.33493750$, $f(0) = 1.10100000$
4. Use the Newton forward-difference formula to construct interpolating polynomials of degree one, two, and three for the following data. Approximate the specified value using each of the polynomials.
   a. $f(0.43)$ if $f(0) = 1$, $f(0.25) = 1.64872$, $f(0.5) = 2.71828$, $f(0.75) = 4.48169$
   b. $f(0.18)$ if $f(0.1) = -0.29004986$, $f(0.2) = -0.56079734$, $f(0.3) = -0.81401972$, $f(0.4) = -1.0526302$
5. Use the Newton backward-difference formula to construct interpolating polynomials of degree one, two, and three for the following data. Approximate the specified value using each of the polynomials.
   a. $f(0.25)$ if $f(0.1) = -0.62049958$, $f(0.2) = -0.28398668$, $f(0.3) = 0.00660095$, $f(0.4) = 0.24842440$
   b. $f(-1/3)$ if $f(-0.75) = -0.07181250$, $f(-0.5) = -0.02475000$, $f(-0.25) = 0.33493750$, $f(0) = 1.10100000$

6. Use the Newton backward-difference formula to construct interpolating polynomials of degree one, two, and three for the following data. Approximate the specified value using each of the polynomials.

   a. $f(0.43)$ if $f(0) = 1$, $f(0.25) = 1.64872$, $f(0.5) = 2.71828$, $f(0.75) = 4.48169$

   b. $f(0.25)$ if $f(-1) = 0.86199480$, $f(-0.5) = 0.95802009$, $f(0) = 1.0986123$, $f(0.5) = 1.2943767$

7. **a.** Use Algorithm 3.2 to construct the interpolating polynomial of degree three for the unequally spaced points given in the following table:

x	$f(x)$
−0.1	5.30000
0.0	2.00000
0.2	3.19000
0.3	1.00000

   b. Add $f(0.35) = 0.97260$ to the table, and construct the interpolating polynomial of degree four.

8. **a.** Use Algorithm 3.2 to construct the interpolating polynomial of degree four for the unequally spaced points given in the following table:

x	$f(x)$
0.0	−6.00000
0.1	−5.89483
0.3	−5.65014
0.6	−5.17788
1.0	−4.28172

   b. Add $f(1.1) = -3.99583$ to the table, and construct the interpolating polynomial of degree five.

9. **a.** Approximate $f(0.05)$ using the following data and the Newton forward-difference formula:

x	0.0	0.2	0.4	0.6	0.8
$f(x)$	1.00000	1.22140	1.49182	1.82212	2.22554

   b. Use the Newton backward-difference formula to approximate $f(0.65)$.

   c. Use Stirling's formula to approximate $f(0.43)$.

10. Show that the polynomial interpolating the following data has degree 3.

x	−2	−1	0	1	2	3
$f(x)$	1	4	11	16	13	−4

11. **a.** Show that the cubic polynomials

$$P(x) = 3 - 2(x+1) + 0(x+1)(x) + (x+1)(x)(x-1)$$

and

$$Q(x) = -1 + 4(x+2) - 3(x+2)(x+1) + (x+2)(x+1)(x)$$

both interpolate the data

x	−2	−1	0	1	2
$f(x)$	−1	3	1	−1	3

   b. Why does part (a) not violate the uniqueness property of interpolating polynomials?

12. A fourth-degree polynomial $P(x)$ satisfies $\Delta^4 P(0) = 24$, $\Delta^3 P(0) = 6$, and $\Delta^2 P(0) = 0$, where $\Delta P(x) = P(x+1) - P(x)$. Compute $\Delta^2 P(10)$.

13. The following data are given for a polynomial $P(x)$ of unknown degree.

x	0	1	2
$P(x)$	2	−1	4

Determine the coefficient of x^2 in $P(x)$ if all third-order forward differences are 1.

14. The following data are given for a polynomial $P(x)$ of unknown degree.

x	0	1	2	3
$P(x)$	4	9	15	18

Determine the coefficient of x^3 in $P(x)$ if all fourth-order forward differences are 1.

15. The Newton forward-difference formula is used to approximate $f(0.3)$ given the following data.

x	0.0	0.2	0.4	0.6
$f(x)$	15.0	21.0	30.0	51.0

Suppose it is discovered that $f(0.4)$ was understated by 10 and $f(0.6)$ was overstated by 5. By what amount should the approximation to $f(0.3)$ be changed?

16. For a function f, the Newton divided-difference formula gives the interpolating polynomial

$$P_3(x) = 1 + 4x + 4x(x - 0.25) + \frac{16}{3}x(x - 0.25)(x - 0.5),$$

on the nodes $x_0 = 0$, $x_1 = 0.25$, $x_2 = 0.5$ and $x_3 = 0.75$. Find $f(0.75)$.

17. For a function f, the forward-divided differences are given by

$x_0 = 0.0$	$f[x_0]$		
		$f[x_0, x_1]$	
$x_1 = 0.4$	$f[x_1]$		$f[x_0, x_1, x_2] = \frac{50}{7}$
		$f[x_1, x_2] = 10$	
$x_2 = 0.7$	$f[x_2] = 6$		

Determine the missing entries in the table.

18. **a.** The introduction to this chapter included a table listing the population of the United States from 1950 to 2000. Use appropriate divided differences to approximate the population in the years 1940, 1975, and 2020.

b. The population in 1940 was approximately 132,165,000. How accurate do you think your 1975 and 2020 figures are?

19. Given

$$\begin{aligned} P_n(x) = f[x_0] &+ f[x_0, x_1](x - x_0) + a_2(x - x_0)(x - x_1) \\ &+ a_3(x - x_0)(x - x_1)(x - x_2) + \cdots \\ &+ a_n(x - x_0)(x - x_1)\cdots(x - x_{n-1}), \end{aligned}$$

use $P_n(x_2)$ to show that $a_2 = f[x_0, x_1, x_2]$.

20. Show that

$$f[x_0, x_1, \ldots, x_n, x] = \frac{f^{(n+1)}(\xi(x))}{(n+1)!},$$

for some $\xi(x)$. [*Hint:* From Eq. (3.3),

$$f(x) = P_n(x) + \frac{f^{(n+1)}(\xi(x))}{(n+1)!}(x - x_0)\cdots(x - x_n).$$

Considering the interpolation polynomial of degree $n+1$ on $x_0, x_1, \ldots, x_n, x$, we have

$$f(x) = P_{n+1}(x) = P_n(x) + f[x_0, x_1, \ldots, x_n, x](x - x_0) \cdots (x - x_n).]$$

21. Let $i_0, i_1, \ldots, i_n$ be a rearrangement of the integers $0, 1, \ldots, n$. Show that $f[x_{i_0}, x_{i_1}, \ldots, x_{i_n}] = f[x_0, x_1, \ldots, x_n]$. [*Hint:* Consider the leading coefficient of the nth Lagrange polynomial on the data $\{x_0, x_1, \ldots, x_n\} = \{x_{i_0}, x_{i_1}, \ldots, x_{i_n}\}$.]

3.4 Hermite Interpolation

The Latin word *osculum*, literally a "small mouth" or "kiss", when applied to a curve indicates that it just touches and has the same shape. Hermite interpolation has this osculating property. It matches a given curve, and its derivative forces the interpolating curve to "kiss" the given curve.

Osculating polynomials generalize both the Taylor polynomials and the Lagrange polynomials. Suppose that we are given $n+1$ distinct numbers $x_0, x_1, \ldots, x_n$ in $[a, b]$ and nonnegative integers $m_0, m_1, \ldots, m_n$, and $m = \max\{m_0, m_1, \ldots, m_n\}$. The osculating polynomial approximating a function $f \in C^m[a, b]$ at x_i, for each $i = 0, \ldots, n$, is the polynomial of least degree that has the same values as the function f and all its derivatives of order less than or equal to m_i at each x_i. The degree of this osculating polynomial is at most

$$M = \sum_{i=0}^{n} m_i + n$$

because the number of conditions to be satisfied is $\sum_{i=0}^{n} m_i + (n+1)$, and a polynomial of degree M has $M+1$ coefficients that can be used to satisfy these conditions.

Definition 3.8 Let $x_0, x_1, \ldots, x_n$ be $n+1$ distinct numbers in $[a, b]$ and for $i = 0, 1, \ldots, n$ let m_i be a nonnegative integer. Suppose that $f \in C^m[a, b]$, where $m = \max_{0 \le i \le n} m_i$.

The **osculating polynomial** approximating f is the polynomial $P(x)$ of least degree such that

$$\frac{d^k P(x_i)}{dx^k} = \frac{d^k f(x_i)}{dx^k}, \quad \text{for each } i = 0, 1, \ldots, n \quad \text{and} \quad k = 0, 1, \ldots, m_i. \quad \blacksquare$$

Charles Hermite (1822–1901) made significant mathematical discoveries throughout his life in areas such as complex analysis and number theory, particularly involving the theory of equations. He is perhaps best known for proving in 1873 that e is transcendental, that is, it is not the solution to any algebraic equation having integer coefficients. This lead in 1882 to Lindemann's proof that π is also transcendental, which demonstrated that it is impossible to use the standard geometry tools of Euclid to construct a square that has the same area as a unit circle.

Note that when $n = 0$, the osculating polynomial approximating f is the m_0th Taylor polynomial for f at x_0. When $m_i = 0$ for each i, the osculating polynomial is the nth Lagrange polynomial interpolating f on $x_0, x_1, \ldots, x_n$.

Hermite Polynomials

The case when $m_i = 1$, for each $i = 0, 1, \ldots, n$, gives the **Hermite polynomials**. For a given function f, these polynomials agree with f at $x_0, x_1, \ldots, x_n$. In addition, since their first derivatives agree with those of f, they have the same "shape" as the function at $(x_i, f(x_i))$ in the sense that the *tangent lines* to the polynomial and the function agree. We will restrict our study of osculating polynomials to this situation and consider first a theorem that describes precisely the form of the Hermite polynomials.

Theorem 3.9 If $f \in C^1[a, b]$ and $x_0, \ldots, x_n \in [a, b]$ are distinct, the unique polynomial of least degree agreeing with f and f' at $x_0, \ldots, x_n$ is the Hermite polynomial of degree at most $2n+1$ given by

$$H_{2n+1}(x) = \sum_{j=0}^{n} f(x_j) H_{n,j}(x) + \sum_{j=0}^{n} f'(x_j) \hat{H}_{n,j}(x),$$

Hermite gave a description of a general osculatory polynomial in a letter to Carl W. Borchardt in 1878, to whom he regularly sent his new results. His demonstration is an interesting application of the use of complex integration techniques to solve a real-valued problem.

where, for $L_{n,j}(x)$ denoting the jth Lagrange coefficient polynomial of degree n, we have

$$H_{n,j}(x) = [1 - 2(x - x_j)L'_{n,j}(x_j)]L^2_{n,j}(x) \quad \text{and} \quad \hat{H}_{n,j}(x) = (x - x_j)L^2_{n,j}(x).$$

Moreover, if $f \in C^{2n+2}[a, b]$, then

$$f(x) = H_{2n+1}(x) + \frac{(x - x_0)^2 \ldots (x - x_n)^2}{(2n + 2)!} f^{(2n+2)}(\xi(x)),$$

for some (generally unknown) $\xi(x)$ in the interval (a, b). ■

Proof First recall that

$$L_{n,j}(x_i) = \begin{cases} 0, & \text{if } i \neq j, \\ 1, & \text{if } i = j. \end{cases}$$

Hence when $i \neq j$,

$$H_{n,j}(x_i) = 0 \qquad \text{and} \qquad \hat{H}_{n,j}(x_i) = 0,$$

whereas, for each i,

$$H_{n,i}(x_i) = [1 - 2(x_i - x_i)L'_{n,i}(x_i)] \cdot 1 = 1 \qquad \text{and} \qquad \hat{H}_{n,i}(x_i) = (x_i - x_i) \cdot 1^2 = 0.$$

As a consequence

$$H_{2n+1}(x_i) = \sum_{\substack{j=0 \\ j\neq i}}^{n} f(x_j) \cdot 0 + f(x_i) \cdot 1 + \sum_{j=0}^{n} f'(x_j) \cdot 0 = f(x_i),$$

so H_{2n+1} agrees with f at $x_0, x_1, \ldots, x_n$.

To show the agreement of H'_{2n+1} with f' at the nodes, first note that $L_{n,j}(x)$ is a factor of $H'_{n,j}(x)$, so $H'_{n,j}(x_i) = 0$ when $i \neq j$. In addition, when $i = j$ we have $L_{n,i}(x_i) = 1$, so

$$\begin{aligned} H'_{n,i}(x_i) &= -2L'_{n,i}(x_i) \cdot L^2_{n,i}(x_i) + [1 - 2(x_i - x_i)L'_{n,i}(x_i)]2L_{n,i}(x_i)L'_{n,i}(x_i) \\ &= -2L'_{n,i}(x_i) + 2L'_{n,i}(x_i) = 0. \end{aligned}$$

Hence, $H'_{n,j}(x_i) = 0$ for all i and j.

Finally,

$$\begin{aligned} \hat{H}'_{n,j}(x_i) &= L^2_{n,j}(x_i) + (x_i - x_j)2L_{n,j}(x_i)L'_{n,j}(x_i) \\ &= L_{n,j}(x_i)[L_{n,j}(x_i) + 2(x_i - x_j)L'_{n,j}(x_i)], \end{aligned}$$

so $\hat{H}'_{n,j}(x_i) = 0$ if $i \neq j$ and $\hat{H}'_{n,i}(x_i) = 1$. Combining these facts, we have

$$H'_{2n+1}(x_i) = \sum_{j=0}^{n} f(x_j) \cdot 0 + \sum_{\substack{j=0 \\ j\neq i}}^{n} f'(x_j) \cdot 0 + f'(x_i) \cdot 1 = f'(x_i).$$

Therefore, H_{2n+1} agrees with f and H'_{2n+1} with f' at $x_0, x_1, \ldots, x_n$.

The uniqueness of this polynomial and the error formula are considered in Exercise 11. ■ ■ ■

Example 1 Use the Hermite polynomial that agrees with the data listed in Table 3.15 to find an approximation of $f(1.5)$.

Table 3.15

k	x_k	$f(x_k)$	$f'(x_k)$
0	1.3	0.6200860	−0.5220232
1	1.6	0.4554022	−0.5698959
2	1.9	0.2818186	−0.5811571

Solution We first compute the Lagrange polynomials and their derivatives. This gives

$$L_{2,0}(x) = \frac{(x-x_1)(x-x_2)}{(x_0-x_1)(x_0-x_2)} = \frac{50}{9}x^2 - \frac{175}{9}x + \frac{152}{9}, \qquad L'_{2,0}(x) = \frac{100}{9}x - \frac{175}{9};$$

$$L_{2,1}(x) = \frac{(x-x_0)(x-x_2)}{(x_1-x_0)(x_1-x_2)} = \frac{-100}{9}x^2 + \frac{320}{9}x - \frac{247}{9}, \qquad L'_{2,1}(x) = \frac{-200}{9}x + \frac{320}{9};$$

and

$$L_{2,2} = \frac{(x-x_0)(x-x_1)}{(x_2-x_0)(x_2-x_1)} = \frac{50}{9}x^2 - \frac{145}{9}x + \frac{104}{9}, \qquad L'_{2,2}(x) = \frac{100}{9}x - \frac{145}{9}.$$

The polynomials $H_{2,j}(x)$ and $\hat{H}_{2,j}(x)$ are then

$$H_{2,0}(x) = [1 - 2(x-1.3)(-5)]\left(\frac{50}{9}x^2 - \frac{175}{9}x + \frac{152}{9}\right)^2$$

$$= (10x - 12)\left(\frac{50}{9}x^2 - \frac{175}{9}x + \frac{152}{9}\right)^2,$$

$$H_{2,1}(x) = 1 \cdot \left(\frac{-100}{9}x^2 + \frac{320}{9}x - \frac{247}{9}\right)^2,$$

$$H_{2,2}(x) = 10(2 - x)\left(\frac{50}{9}x^2 - \frac{145}{9}x + \frac{104}{9}\right)^2,$$

$$\hat{H}_{2,0}(x) = (x - 1.3)\left(\frac{50}{9}x^2 - \frac{175}{9}x + \frac{152}{9}\right)^2,$$

$$\hat{H}_{2,1}(x) = (x - 1.6)\left(\frac{-100}{9}x^2 + \frac{320}{9}x - \frac{247}{9}\right)^2,$$

and

$$\hat{H}_{2,2}(x) = (x - 1.9)\left(\frac{50}{9}x^2 - \frac{145}{9}x + \frac{104}{9}\right)^2.$$

Finally

$$H_5(x) = 0.6200860H_{2,0}(x) + 0.4554022H_{2,1}(x) + 0.2818186H_{2,2}(x)$$
$$- 0.5220232\hat{H}_{2,0}(x) - 0.5698959\hat{H}_{2,1}(x) - 0.5811571\hat{H}_{2,2}(x)$$

and

$$H_5(1.5) = 0.6200860\left(\frac{4}{27}\right) + 0.4554022\left(\frac{64}{81}\right) + 0.2818186\left(\frac{5}{81}\right)$$
$$- 0.5220232\left(\frac{4}{405}\right) - 0.5698959\left(\frac{-32}{405}\right) - 0.5811571\left(\frac{-2}{405}\right)$$
$$= 0.5118277,$$

a result that is accurate to the places listed. ■

Although Theorem 3.9 provides a complete description of the Hermite polynomials, it is clear from Example 1 that the need to determine and evaluate the Lagrange polynomials and their derivatives makes the procedure tedious even for small values of n.

Hermite Polynomials Using Divided Differences

There is an alternative method for generating Hermite approximations that has as its basis the Newton interpolatory divided-difference formula (3.10) at $x_0, x_1, \ldots, x_n$, that is,

$$P_n(x) = f[x_0] + \sum_{k=1}^{n} f[x_0, x_1, \ldots, x_k](x - x_0) \cdots (x - x_{k-1}).$$

The alternative method uses the connection between the nth divided difference and the nth derivative of f, as outlined in Theorem 3.6 in Section 3.3.

Suppose that the distinct numbers $x_0, x_1, \ldots, x_n$ are given together with the values of f and f' at these numbers. Define a new sequence $z_0, z_1, \ldots, z_{2n+1}$ by

$$z_{2i} = z_{2i+1} = x_i, \quad \text{for each } i = 0, 1, \ldots, n,$$

and construct the divided difference table in the form of Table 3.9 that uses $z_0, z_1, \ldots, z_{2n+1}$.

Since $z_{2i} = z_{2i+1} = x_i$ for each i, we cannot define $f[z_{2i}, z_{2i+1}]$ by the divided difference formula. However, if we assume, based on Theorem 3.6, that the reasonable substitution in this situation is $f[z_{2i}, z_{2i+1}] = f'(z_{2i}) = f'(x_i)$, we can use the entries

$$f'(x_0), f'(x_1), \ldots, f'(x_n)$$

in place of the undefined first divided differences

$$f[z_0, z_1], f[z_2, z_3], \ldots, f[z_{2n}, z_{2n+1}].$$

The remaining divided differences are produced as usual, and the appropriate divided differences are employed in Newton's interpolatory divided-difference formula. Table 3.16 shows the entries that are used for the first three divided-difference columns when determining the Hermite polynomial $H_5(x)$ for x_0, x_1, and x_2. The remaining entries are generated in the same manner as in Table 3.9. The Hermite polynomial is then given by

$$H_{2n+1}(x) = f[z_0] + \sum_{k=1}^{2n+1} f[z_0, \ldots, z_k](x - z_0)(x - z_1) \cdots (x - z_{k-1}).$$

A proof of this fact can be found in [Pow], p. 56.

Table 3.16

z	$f(z)$	First divided differences	Second divided differences
$z_0 = x_0$	$f[z_0] = f(x_0)$		
		$f[z_0, z_1] = f'(x_0)$	
$z_1 = x_0$	$f[z_1] = f(x_0)$		$f[z_0, z_1, z_2] = \dfrac{f[z_1, z_2] - f[z_0, z_1]}{z_2 - z_0}$
		$f[z_1, z_2] = \dfrac{f[z_2] - f[z_1]}{z_2 - z_1}$	
$z_2 = x_1$	$f[z_2] = f(x_1)$		$f[z_1, z_2, z_3] = \dfrac{f[z_2, z_3] - f[z_1, z_2]}{z_3 - z_1}$
		$f[z_2, z_3] = f'(x_1)$	
$z_3 = x_1$	$f[z_3] = f(x_1)$		$f[z_2, z_3, z_4] = \dfrac{f[z_3, z_4] - f[z_2, z_3]}{z_4 - z_2}$
		$f[z_3, z_4] = \dfrac{f[z_4] - f[z_3]}{z_4 - z_3}$	
$z_4 = x_2$	$f[z_4] = f(x_2)$		$f[z_3, z_4, z_5] = \dfrac{f[z_4, z_5] - f[z_3, z_4]}{z_5 - z_3}$
		$f[z_4, z_5] = f'(x_2)$	
$z_5 = x_2$	$f[z_5] = f(x_2)$		

Example 2 Use the data given in Example 1 and the divided difference method to determine the Hermite polynomial approximation at $x = 1.5$.

Solution The underlined entries in the first three columns of Table 3.17 are the data given in Example 1. The remaining entries in this table are generated by the standard divided-difference formula (3.9).

For example, for the second entry in the third column we use the second 1.3 entry in the second column and the first 1.6 entry in that column to obtain

$$\frac{0.4554022 - 0.6200860}{1.6 - 1.3} = -0.5489460.$$

For the first entry in the fourth column we use the first 1.3 entry in the third column and the first 1.6 entry in that column to obtain

$$\frac{-0.5489460 - (-0.5220232)}{1.6 - 1.3} = -0.0897427.$$

The value of the Hermite polynomial at 1.5 is

$$\begin{aligned}
H_5(1.5) &= f[1.3] + f'(1.3)(1.5 - 1.3) + f[1.3, 1.3, 1.6](1.5 - 1.3)^2 \\
&\quad + f[1.3, 1.3, 1.6, 1.6](1.5 - 1.3)^2(1.5 - 1.6) \\
&\quad + f[1.3, 1.3, 1.6, 1.6, 1.9](1.5 - 1.3)^2(1.5 - 1.6)^2 \\
&\quad + f[1.3, 1.3, 1.6, 1.6, 1.9, 1.9](1.5 - 1.3)^2(1.5 - 1.6)^2(1.5 - 1.9) \\
&= 0.6200860 + (-0.5220232)(0.2) + (-0.0897427)(0.2)^2 \\
&\quad + 0.0663657(0.2)^2(-0.1) + 0.0026663(0.2)^2(-0.1)^2 \\
&\quad + (-0.0027738)(0.2)^2(-0.1)^2(-0.4) \\
&= 0.5118277.
\end{aligned}$$

■

Table 3.17

1.3	0.6200860					
		−0.5220232				
1.3	0.6200860		−0.0897427			
		−0.5489460		0.0663657		
1.6	0.4554022		−0.0698330		0.0026663	
		−0.5698959		0.0679655		−0.0027738
1.6	0.4554022		−0.0290537		0.0010020	
		−0.5786120		0.0685667		
1.9	0.2818186		−0.0084837			
		−0.5811571				
1.9	0.2818186					

The technique used in Algorithm 3.3 can be extended for use in determining other osculating polynomials. A concise discussion of the procedures can be found in [Pow], pp. 53–57.

ALGORITHM 3.3

Hermite Interpolation

To obtain the coefficients of the Hermite interpolating polynomial $H(x)$ on the $(n+1)$ distinct numbers $x_0, \ldots, x_n$ for the function f:

INPUT numbers $x_0, x_1, \ldots, x_n$; values $f(x_0), \ldots, f(x_n)$ and $f'(x_0), \ldots, f'(x_n)$.

OUTPUT the numbers $Q_{0,0}, Q_{1,1}, \ldots, Q_{2n+1,2n+1}$ where

$$\begin{aligned} H(x) = {} & Q_{0,0} + Q_{1,1}(x-x_0) + Q_{2,2}(x-x_0)^2 + Q_{3,3}(x-x_0)^2(x-x_1) \\ & + Q_{4,4}(x-x_0)^2(x-x_1)^2 + \cdots \\ & + Q_{2n+1,2n+1}(x-x_0)^2(x-x_1)^2 \cdots (x-x_{n-1})^2(x-x_n). \end{aligned}$$

Step 1 For $i = 0, 1, \ldots, n$ do Steps 2 and 3.

Step 2 Set $z_{2i} = x_i$;
$z_{2i+1} = x_i$;
$Q_{2i,0} = f(x_i)$;
$Q_{2i+1,0} = f(x_i)$;
$Q_{2i+1,1} = f'(x_i)$.

Step 3 If $i \neq 0$ then set

$$Q_{2i,1} = \frac{Q_{2i,0} - Q_{2i-1,0}}{z_{2i} - z_{2i-1}}.$$

Step 4 For $i = 2, 3, \ldots, 2n+1$
for $j = 2, 3, \ldots, i$ set $Q_{i,j} = \dfrac{Q_{i,j-1} - Q_{i-1,j-1}}{z_i - z_{i-j}}$.

Step 5 OUTPUT $(Q_{0,0}, Q_{1,1}, \ldots, Q_{2n+1,2n+1})$;
STOP ■

The *NumericalAnalysis* package in Maple can be used to construct the Hermite coefficients. We first need to load the package and to define the data that is being used, in this case, x_i, $f(x_i)$, and $f'(x_i)$ for $i = 0, 1, \ldots, n$. This is done by presenting the data in the form $[x_i, f(x_i), f'(x_i)]$. For example, the data for Example 2 is entered as

$xy := [[1.3, 0.6200860, -0.5220232], [1.6, 0.4554022, -0.5698959],$
$[1.9, 0.2818186, -0.5811571]]$

Then the command

$h5 := PolynomialInterpolation(xy, method = hermite, independentvar = 'x')$

produces an array whose nonzero entries correspond to the values in Table 3.17. The Hermite interpolating polynomial is created with the command

$Interpolant(h5))$

This gives the polynomial in (almost) Newton forward-difference form

$$1.29871616 - 0.5220232x - 0.08974266667(x - 1.3)^2 + 0.06636555557(x - 1.3)^2(x - 1.6) + 0.002666666633(x - 1.3)^2(x - 1.6)^2 - 0.002774691277(x - 1.3)^2(x - 1.6)^2(x - 1.9)$$

If a standard representation of the polynomial is needed, it is found with

$expand(Interpolant(h5))$

giving the Maple response

$$1.001944063 - 0.0082292208x - 0.2352161732x^2 - 0.01455607812x^3 + 0.02403178946x^4 - 0.002774691277x^5$$

EXERCISE SET 3.4

1. Use Theorem 3.9 or Algorithm 3.3 to construct an approximating polynomial for the following data.

a.

x	$f(x)$	$f'(x)$
8.3	17.56492	3.116256
8.6	18.50515	3.151762

b.

x	$f(x)$	$f'(x)$
0.8	0.22363362	2.1691753
1.0	0.65809197	2.0466965

c.

x	$f(x)$	$f'(x)$
−0.5	−0.0247500	0.7510000
−0.25	0.3349375	2.1890000
0	1.1010000	4.0020000

d.

x	$f(x)$	$f'(x)$
0.1	−0.62049958	3.58502082
0.2	−0.28398668	3.14033271
0.3	0.00660095	2.66668043
0.4	0.24842440	2.16529366

2. Use Theorem 3.9 or Algorithm 3.3 to construct an approximating polynomial for the following data.

a.

x	$f(x)$	$f'(x)$
0	1.00000	2.00000
0.5	2.71828	5.43656

b.

x	$f(x)$	$f'(x)$
−0.25	1.33203	0.437500
0.25	0.800781	−0.625000

c.

x	$f(x)$	$f'(x)$
0.1	−0.29004996	−2.8019975
0.2	−0.56079734	−2.6159201
0.3	−0.81401972	−2.9734038

d.

x	$f(x)$	$f'(x)$
−1	0.86199480	0.15536240
−0.5	0.95802009	0.23269654
0	1.0986123	0.33333333
0.5	1.2943767	0.45186776

3. The data in Exercise 1 were generated using the following functions. Use the polynomials constructed in Exercise 1 for the given value of x to approximate $f(x)$, and calculate the absolute error.

   a. $f(x) = x \ln x$; approximate $f(8.4)$.

   b. $f(x) = \sin(e^x - 2)$; approximate $f(0.9)$.

   c. $f(x) = x^3 + 4.001x^2 + 4.002x + 1.101$; approximate $f(-1/3)$.

   d. $f(x) = x \cos x - 2x^2 + 3x - 1$; approximate $f(0.25)$.

4. The data in Exercise 2 were generated using the following functions. Use the polynomials constructed in Exercise 2 for the given value of x to approximate $f(x)$, and calculate the absolute error.

a. $f(x) = e^{2x}$; approximate $f(0.43)$.

b. $f(x) = x^4 - x^3 + x^2 - x + 1$; approximate $f(0)$.

c. $f(x) = x^2 \cos x - 3x$; approximate $f(0.18)$.

d. $f(x) = \ln(e^x + 2)$; approximate $f(0.25)$.

5. **a.** Use the following values and five-digit rounding arithmetic to construct the Hermite interpolating polynomial to approximate $\sin 0.34$.

x	$\sin x$	$D_x \sin x = \cos x$
0.30	0.29552	0.95534
0.32	0.31457	0.94924
0.35	0.34290	0.93937

b. Determine an error bound for the approximation in part (a), and compare it to the actual error.

c. Add $\sin 0.33 = 0.32404$ and $\cos 0.33 = 0.94604$ to the data, and redo the calculations.

6. Let $f(x) = 3xe^x - e^{2x}$.

a. Approximate $f(1.03)$ by the Hermite interpolating polynomial of degree at most three using $x_0 = 1$ and $x_1 = 1.05$. Compare the actual error to the error bound.

b. Repeat (a) with the Hermite interpolating polynomial of degree at most five, using $x_0 = 1$, $x_1 = 1.05$, and $x_2 = 1.07$.

7. Use the error formula and Maple to find a bound for the errors in the approximations of $f(x)$ in parts (a) and (c) of Exercise 3.

8. Use the error formula and Maple to find a bound for the errors in the approximations of $f(x)$ in parts (a) and (c) of Exercise 4.

9. The following table lists data for the function described by $f(x) = e^{0.1x^2}$. Approximate $f(1.25)$ by using $H_5(1.25)$ and $H_3(1.25)$, where H_5 uses the nodes $x_0 = 1$, $x_1 = 2$, and $x_2 = 3$; and H_3 uses the nodes $\bar{x}_0 = 1$ and $\bar{x}_1 = 1.5$. Find error bounds for these approximations.

x	$f(x) = e^{0.1x^2}$	$f'(x) = 0.2xe^{0.1x^2}$
$x_0 = \bar{x}_0 = 1$	1.105170918	0.2210341836
$\bar{x}_1 = 1.5$	1.252322716	0.3756968148
$x_1 = 2$	1.491824698	0.5967298792
$x_2 = 3$	2.459603111	1.475761867

10. A car traveling along a straight road is clocked at a number of points. The data from the observations are given in the following table, where the time is in seconds, the distance is in feet, and the speed is in feet per second.

Time	0	3	5	8	13
Distance	0	225	383	623	993
Speed	75	77	80	74	72

a. Use a Hermite polynomial to predict the position of the car and its speed when $t = 10$ s.

b. Use the derivative of the Hermite polynomial to determine whether the car ever exceeds a 55 mi/h speed limit on the road. If so, what is the first time the car exceeds this speed?

c. What is the predicted maximum speed for the car?

11. **a.** Show that $H_{2n+1}(x)$ is the unique polynomial of least degree agreeing with f and f' at $x_0, \ldots, x_n$. [*Hint:* Assume that $P(x)$ is another such polynomial and consider $D = H_{2n+1} - P$ and D' at $x_0, x_1, \ldots, x_n$.]

b. Derive the error term in Theorem 3.9. [*Hint:* Use the same method as in the Lagrange error derivation, Theorem 3.3, defining

$$g(t) = f(t) - H_{2n+1}(t) - \frac{(t-x_0)^2 \cdots (t-x_n)^2}{(x-x_0)^2 \cdots (x-x_n)^2}[f(x) - H_{2n+1}(x)]$$

and using the fact that $g'(t)$ has $(2n+2)$ *distinct* zeros in $[a, b]$.]

12. Let $z_0 = x_0$, $z_1 = x_0$, $z_2 = x_1$, and $z_3 = x_1$. Form the following divided-difference table.

$z_0 = x_0$	$f[z_0] = f(x_0)$			
		$f[z_0, z_1] = f'(x_0)$		
$z_1 = x_0$	$f[z_1] = f(x_0)$		$f[z_0, z_1, z_2]$	
		$f[z_1, z_2]$		$f[z_0, z_1, z_2, z_3]$
$z_2 = x_1$	$f[z_2] = f(x_1)$		$f[z_1, z_2, z_3]$	
		$f[z_2, z_3] = f'(x_1)$		
$z_3 = x_1$	$f[z_3] = f(x_1)$			

Show that the cubic Hermite polynomial $H_3(x)$ can also be written as $f[z_0] + f[z_0, z_1](x - x_0) + f[z_0, z_1, z_2](x - x_0)^2 + f[z_0, z_1, z_2, z_3](x - x_0)^2(x - x_1)$.

3.5 Cubic Spline Interpolation[1]

The previous sections concerned the approximation of arbitrary functions on closed intervals using a single polynomial. However, high-degree polynomials can oscillate erratically, that is, a minor fluctuation over a small portion of the interval can induce large fluctuations over the entire range. We will see a good example of this in Figure 3.14 at the end of this section.

An alternative approach is to divide the approximation interval into a collection of subintervals and construct a (generally) different approximating polynomial on each subinterval. This is called **piecewise-polynomial approximation.**

Piecewise-Polynomial Approximation

The simplest piecewise-polynomial approximation is **piecewise-linear** interpolation, which consists of joining a set of data points

$$\{(x_0, f(x_0)), (x_1, f(x_1)), \ldots, (x_n, f(x_n))\}$$

by a series of straight lines, as shown in Figure 3.7.

A disadvantage of linear function approximation is that there is likely no differentiability at the endpoints of the subintervals, which, in a geometrical context, means that the interpolating function is not "smooth." Often it is clear from physical conditions that smoothness is required, so the approximating function must be continuously differentiable.

An alternative procedure is to use a piecewise polynomial of Hermite type. For example, if the values of f and of f' are known at each of the points $x_0 < x_1 < \cdots < x_n$, a cubic Hermite polynomial can be used on each of the subintervals $[x_0, x_1], [x_1, x_2], \ldots, [x_{n-1}, x_n]$ to obtain a function that has a continuous derivative on the interval $[x_0, x_n]$.

[1]The proofs of the theorems in this section rely on results in Chapter 6.

Figure 3.7

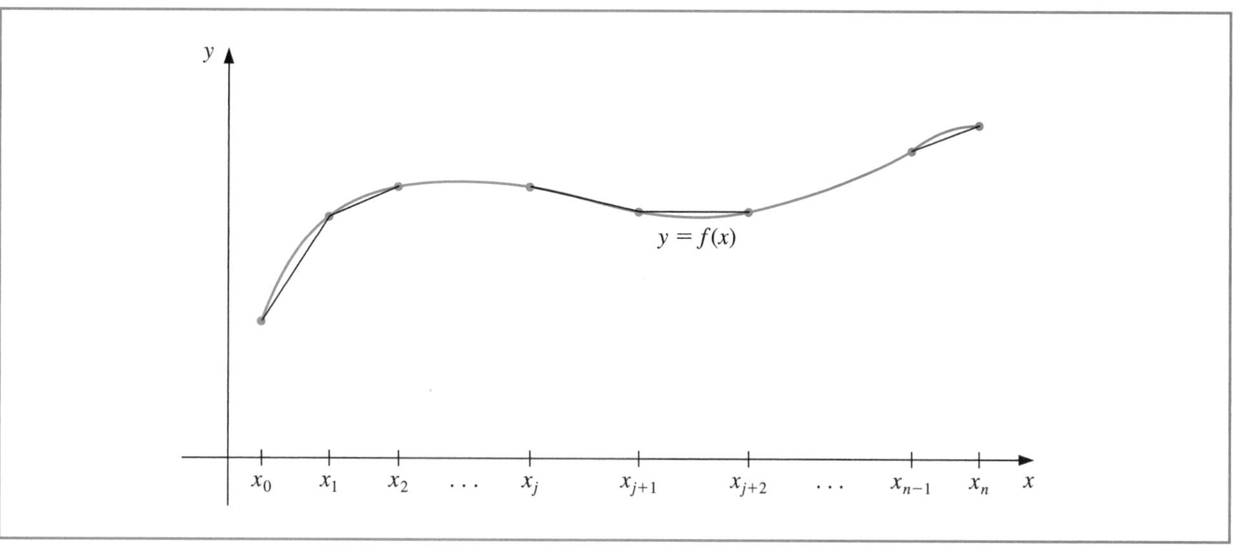

Isaac Jacob Schoenberg (1903–1990) developed his work on splines during World War II while on leave from the University of Pennsylvania to work at the Army's Ballistic Research Laboratory in Aberdeen, Maryland. His original work involved numerical procedures for solving differential equations. The much broader application of splines to the areas of data fitting and computer-aided geometric design became evident with the widespread availability of computers in the 1960s.

To determine the appropriate Hermite cubic polynomial on a given interval is simply a matter of computing $H_3(x)$ for that interval. The Lagrange interpolating polynomials needed to determine H_3 are of first degree, so this can be accomplished without great difficulty. However, to use Hermite piecewise polynomials for general interpolation, we need to know the derivative of the function being approximated, and this is frequently unavailable.

The remainder of this section considers approximation using piecewise polynomials that require no specific derivative information, except perhaps at the endpoints of the interval on which the function is being approximated.

The simplest type of differentiable piecewise-polynomial function on an entire interval $[x_0, x_n]$ is the function obtained by fitting one quadratic polynomial between each successive pair of nodes. This is done by constructing a quadratic on $[x_0, x_1]$ agreeing with the function at x_0 and x_1, another quadratic on $[x_1, x_2]$ agreeing with the function at x_1 and x_2, and so on. A general quadratic polynomial has three arbitrary constants—the constant term, the coefficient of x, and the coefficient of x^2—and only two conditions are required to fit the data at the endpoints of each subinterval. So flexibility exists that permits the quadratics to be chosen so that the interpolant has a continuous derivative on $[x_0, x_n]$. The difficulty arises because we generally need to specify conditions about the derivative of the interpolant at the endpoints x_0 and x_n. There is not a sufficient number of constants to ensure that the conditions will be satisfied. (See Exercise 26.)

The root of the word "spline" is the same as that of splint. It was originally a small strip of wood that could be used to join two boards. Later the word was used to refer to a long flexible strip, generally of metal, that could be used to draw continuous smooth curves by forcing the strip to pass through specified points and tracing along the curve.

Cubic Splines

The most common piecewise-polynomial approximation uses cubic polynomials between each successive pair of nodes and is called **cubic spline interpolation**. A general cubic polynomial involves four constants, so there is sufficient flexibility in the cubic spline procedure to ensure that the interpolant is not only continuously differentiable on the interval, but also has a continuous second derivative. The construction of the cubic spline does not, however, assume that the derivatives of the interpolant agree with those of the function it is approximating, even at the nodes. (See Figure 3.8.)

Figure 3.8

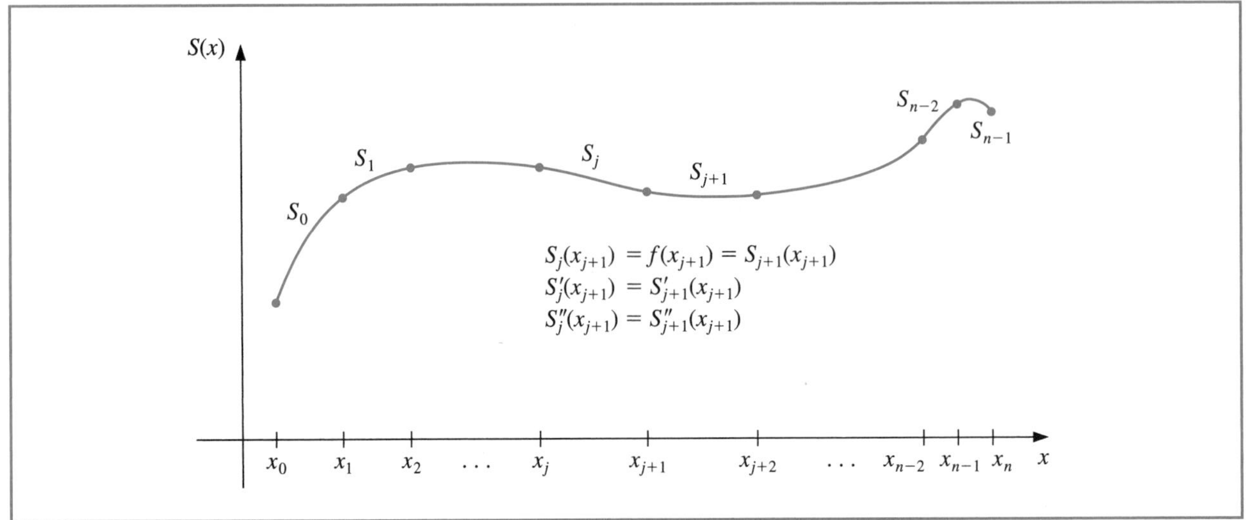

Definition 3.10 Given a function f defined on $[a, b]$ and a set of nodes $a = x_0 < x_1 < \cdots < x_n = b$, a **cubic spline interpolant** S for f is a function that satisfies the following conditions:

A natural spline has no conditions imposed for the direction at its endpoints, so the curve takes the shape of a straight line after it passes through the interpolation points nearest its endpoints. The name derives from the fact that this is the natural shape a flexible strip assumes if forced to pass through specified interpolation points with no additional constraints. (See Figure 3.9.)

(a) $S(x)$ is a cubic polynomial, denoted $S_j(x)$, on the subinterval $[x_j, x_{j+1}]$ for each $j = 0, 1, \ldots, n-1$;

(b) $S_j(x_j) = f(x_j)$ and $S_j(x_{j+1}) = f(x_{j+1})$ for each $j = 0, 1, \ldots, n-1$;

(c) $S_{j+1}(x_{j+1}) = S_j(x_{j+1})$ for each $j = 0, 1, \ldots, n-2$; (*Implied by* **(b)**.)

(d) $S'_{j+1}(x_{j+1}) = S'_j(x_{j+1})$ for each $j = 0, 1, \ldots, n-2$;

(e) $S''_{j+1}(x_{j+1}) = S''_j(x_{j+1})$ for each $j = 0, 1, \ldots, n-2$;

(f) One of the following sets of boundary conditions is satisfied:

(i) $S''(x_0) = S''(x_n) = 0$ (**natural** *(or free)* **boundary**);

(ii) $S'(x_0) = f'(x_0)$ and $S'(x_n) = f'(x_n)$ (**clamped boundary**). ■

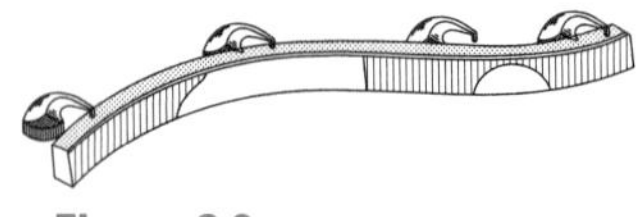

Figure 3.9

Although cubic splines are defined with other boundary conditions, the conditions given in **(f)** are sufficient for our purposes. When the free boundary conditions occur, the spline is called a **natural spline**, and its graph approximates the shape that a long flexible rod would assume if forced to go through the data points $\{(x_0, f(x_0)), (x_1, f(x_1)), \ldots, (x_n, f(x_n))\}$.

In general, clamped boundary conditions lead to more accurate approximations because they include more information about the function. However, for this type of boundary condition to hold, it is necessary to have either the values of the derivative at the endpoints or an accurate approximation to those values.

Example 1 Construct a natural cubic spline that passes through the points $(1, 2)$, $(2, 3)$, and $(3, 5)$.

Solution This spline consists of two cubics. The first for the interval $[1, 2]$, denoted

$$S_0(x) = a_0 + b_0(x-1) + c_0(x-1)^2 + d_0(x-1)^3,$$

and the other for [2, 3], denoted

$$S_1(x) = a_1 + b_1(x-2) + c_1(x-2)^2 + d_1(x-2)^3.$$

There are 8 constants to be determined, which requires 8 conditions. Four conditions come from the fact that the splines must agree with the data at the nodes. Hence

$$2 = f(1) = a_0, \quad 3 = f(2) = a_0 + b_0 + c_0 + d_0, \quad 3 = f(2) = a_1, \quad \text{and}$$
$$5 = f(3) = a_1 + b_1 + c_1 + d_1.$$

Two more come from the fact that $S_0'(2) = S_1'(2)$ and $S_0''(2) = S_1''(2)$. These are

$$S_0'(2) = S_1'(2): \quad b_0 + 2c_0 + 3d_0 = b_1 \qquad \text{and} \qquad S_0''(2) = S_1''(2): \quad 2c_0 + 6d_0 = 2c_1$$

The final two come from the natural boundary conditions:

$$S_0''(1) = 0: \quad 2c_0 = 0 \qquad \text{and} \qquad S_1''(3) = 0: \quad 2c_1 + 6d_1 = 0.$$

Solving this system of equations gives the spline

$$S(x) = \begin{cases} 2 + \frac{3}{4}(x-1) + \frac{1}{4}(x-1)^3, \text{ for } x \in [1,2] \\ 3 + \frac{3}{2}(x-2) + \frac{3}{4}(x-2)^2 - \frac{1}{4}(x-2)^3, \text{ for } x \in [2,3] \end{cases}$$

■

Construction of a Cubic Spline

As the preceding example demonstrates, a spline defined on an interval that is divided into n subintervals will require determining $4n$ constants. To construct the cubic spline interpolant for a given function f, the conditions in the definition are applied to the cubic polynomials

$$S_j(x) = a_j + b_j(x - x_j) + c_j(x - x_j)^2 + d_j(x - x_j)^3,$$

for each $j = 0, 1, \ldots, n-1$. Since $S_j(x_j) = a_j = f(x_j)$, condition **(c)** can be applied to obtain

$$a_{j+1} = S_{j+1}(x_{j+1}) = S_j(x_{j+1}) = a_j + b_j(x_{j+1} - x_j) + c_j(x_{j+1} - x_j)^2 + d_j(x_{j+1} - x_j)^3,$$

for each $j = 0, 1, \ldots, n-2$.

Clamping a spline indicates that the ends of the flexible strip are fixed so that it is forced to take a specific direction at each of its endpoints. This is important, for example, when two spline functions should match at their endpoints. This is done mathematically by specifying the values of the derivative of the curve at the endpoints of the spline.

The terms $x_{j+1} - x_j$ are used repeatedly in this development, so it is convenient to introduce the simpler notation

$$h_j = x_{j+1} - x_j,$$

for each $j = 0, 1, \ldots, n-1$. If we also define $a_n = f(x_n)$, then the equation

$$a_{j+1} = a_j + b_j h_j + c_j h_j^2 + d_j h_j^3 \tag{3.15}$$

holds for each $j = 0, 1, \ldots, n-1$.

In a similar manner, define $b_n = S'(x_n)$ and observe that

$$S_j'(x) = b_j + 2c_j(x - x_j) + 3d_j(x - x_j)^2$$

implies $S_j'(x_j) = b_j$, for each $j = 0, 1, \ldots, n-1$. Applying condition **(d)** gives

$$b_{j+1} = b_j + 2c_jh_j + 3d_jh_j^2, \tag{3.16}$$

for each $j = 0, 1, \ldots, n-1$.

Another relationship between the coefficients of S_j is obtained by defining $c_n = S''(x_n)/2$ and applying condition **(e)**. Then, for each $j = 0, 1, \ldots, n-1$,

$$c_{j+1} = c_j + 3d_jh_j. \tag{3.17}$$

Solving for d_j in Eq. (3.17) and substituting this value into Eqs. (3.15) and (3.16) gives, for each $j = 0, 1, \ldots, n-1$, the new equations

$$a_{j+1} = a_j + b_jh_j + \frac{h_j^2}{3}(2c_j + c_{j+1}) \tag{3.18}$$

and

$$b_{j+1} = b_j + h_j(c_j + c_{j+1}). \tag{3.19}$$

The final relationship involving the coefficients is obtained by solving the appropriate equation in the form of equation (3.18), first for b_j,

$$b_j = \frac{1}{h_j}(a_{j+1} - a_j) - \frac{h_j}{3}(2c_j + c_{j+1}), \tag{3.20}$$

and then, with a reduction of the index, for b_{j-1}. This gives

$$b_{j-1} = \frac{1}{h_{j-1}}(a_j - a_{j-1}) - \frac{h_{j-1}}{3}(2c_{j-1} + c_j).$$

Substituting these values into the equation derived from Eq. (3.19), with the index reduced by one, gives the linear system of equations

$$h_{j-1}c_{j-1} + 2(h_{j-1} + h_j)c_j + h_jc_{j+1} = \frac{3}{h_j}(a_{j+1} - a_j) - \frac{3}{h_{j-1}}(a_j - a_{j-1}), \tag{3.21}$$

for each $j = 1, 2, \ldots, n-1$. This system involves only the $\{c_j\}_{j=0}^n$ as unknowns. The values of $\{h_j\}_{j=0}^{n-1}$ and $\{a_j\}_{j=0}^n$ are given, respectively, by the spacing of the nodes $\{x_j\}_{j=0}^n$ and the values of f at the nodes. So once the values of $\{c_j\}_{j=0}^n$ are determined, it is a simple matter to find the remainder of the constants $\{b_j\}_{j=0}^{n-1}$ from Eq. (3.20) and $\{d_j\}_{j=0}^{n-1}$ from Eq. (3.17). Then we can construct the cubic polynomials $\{S_j(x)\}_{j=0}^{n-1}$.

The major question that arises in connection with this construction is whether the values of $\{c_j\}_{j=0}^n$ can be found using the system of equations given in (3.21) and, if so, whether these values are unique. The following theorems indicate that this is the case when either of the boundary conditions given in part **(f)** of the definition are imposed. The proofs of these theorems require material from linear algebra, which is discussed in Chapter 6.

Natural Splines

Theorem 3.11 If f is defined at $a = x_0 < x_1 < \cdots < x_n = b$, then f has a unique natural spline interpolant S on the nodes $x_0, x_1, \ldots, x_n$; that is, a spline interpolant that satisfies the natural boundary conditions $S''(a) = 0$ and $S''(b) = 0$. ■

Proof The boundary conditions in this case imply that $c_n = S''(x_n)/2 = 0$ and that

$$0 = S''(x_0) = 2c_0 + 6d_0(x_0 - x_0),$$

so $c_0 = 0$. The two equations $c_0 = 0$ and $c_n = 0$ together with the equations in (3.21) produce a linear system described by the vector equation $A\mathbf{x} = \mathbf{b}$, where A is the $(n+1) \times (n+1)$ matrix

$$A = \begin{bmatrix} 1 & 0 & 0 & \cdots & \cdots & \cdots & 0 \\ h_0 & 2(h_0+h_1) & h_1 & \ddots & & & \vdots \\ 0 & h_1 & 2(h_1+h_2) & h_2 & \ddots & & \vdots \\ \vdots & \ddots & \ddots & \ddots & \ddots & \ddots & 0 \\ \vdots & & \ddots & \ddots & h_{n-2} & 2(h_{n-2}+h_{n-1}) & h_{n-1} \\ 0 & \cdots & \cdots & \cdots & 0 & 0 & 1 \end{bmatrix},$$

and $\mathbf{b}$ and $\mathbf{x}$ are the vectors

$$\mathbf{b} = \begin{bmatrix} 0 \\ \frac{3}{h_1}(a_2 - a_1) - \frac{3}{h_0}(a_1 - a_0) \\ \vdots \\ \frac{3}{h_{n-1}}(a_n - a_{n-1}) - \frac{3}{h_{n-2}}(a_{n-1} - a_{n-2}) \\ 0 \end{bmatrix} \quad \text{and} \quad \mathbf{x} = \begin{bmatrix} c_0 \\ c_1 \\ \vdots \\ c_n \end{bmatrix}.$$

The matrix A is strictly diagonally dominant, that is, in each row the magnitude of the diagonal entry exceeds the sum of the magnitudes of all the other entries in the row. A linear system with a matrix of this form will be shown by Theorem 6.21 in Section 6.6 to have a unique solution for $c_0, c_1, \ldots, c_n$. ■ ■ ■

The solution to the cubic spline problem with the boundary conditions $S''(x_0) = S''(x_n) = 0$ can be obtained by applying Algorithm 3.4.

ALGORITHM 3.4

Natural Cubic Spline

To construct the cubic spline interpolant S for the function f, defined at the numbers $x_0 < x_1 < \cdots < x_n$, satisfying $S''(x_0) = S''(x_n) = 0$:

INPUT $n; x_0, x_1, \ldots, x_n; a_0 = f(x_0), a_1 = f(x_1), \ldots, a_n = f(x_n)$.

OUTPUT a_j, b_j, c_j, d_j for $j = 0, 1, \ldots, n-1$.

(*Note:* $S(x) = S_j(x) = a_j + b_j(x - x_j) + c_j(x - x_j)^2 + d_j(x - x_j)^3$ *for* $x_j \le x \le x_{j+1}$.)

Step 1 For $i = 0, 1, \ldots, n-1$ set $h_i = x_{i+1} - x_i$.

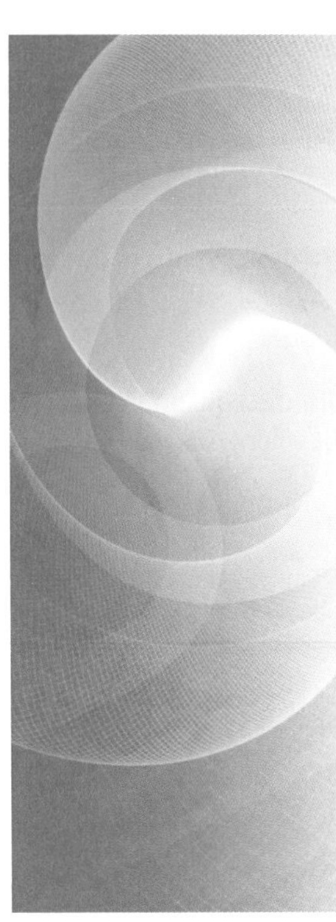

Step 2 For $i = 1, 2, \ldots, n-1$ set

$$\alpha_i = \frac{3}{h_i}(a_{i+1} - a_i) - \frac{3}{h_{i-1}}(a_i - a_{i-1}).$$

Step 3 Set $l_0 = 1$; *(Steps 3, 4, 5, and part of Step 6 solve a tridiagonal linear system using a method described in Algorithm 6.7.)*

$\mu_0 = 0$;
$z_0 = 0$.

Step 4 For $i = 1, 2, \ldots, n-1$
set $l_i = 2(x_{i+1} - x_{i-1}) - h_{i-1}\mu_{i-1}$;
$\mu_i = h_i / l_i$;
$z_i = (\alpha_i - h_{i-1} z_{i-1}) / l_i$.

Step 5 Set $l_n = 1$;
$z_n = 0$;
$c_n = 0$.

Step 6 For $j = n-1, n-2, \ldots, 0$
set $c_j = z_j - \mu_j c_{j+1}$;
$b_j = (a_{j+1} - a_j)/h_j - h_j(c_{j+1} + 2c_j)/3$;
$d_j = (c_{j+1} - c_j)/(3h_j)$.

Step 7 OUTPUT (a_j, b_j, c_j, d_j for $j = 0, 1, \ldots, n-1$);
STOP. ■

Example 2 At the beginning of Chapter 3 we gave some Taylor polynomials to approximate the exponential $f(x) = e^x$. Use the data points $(0, 1)$, $(1, e)$, $(2, e^2)$, and $(3, e^3)$ to form a natural spline $S(x)$ that approximates $f(x) = e^x$.

Solution We have $n = 3$, $h_0 = h_1 = h_2 = 1$, $a_0 = 1$, $a_1 = e$, $a_2 = e^2$, and $a_3 = e^3$. So the matrix A and the vectors $\mathbf{b}$ and $\mathbf{x}$ given in Theorem 3.11 have the forms

$$A = \begin{bmatrix} 1 & 0 & 0 & 0 \\ 1 & 4 & 1 & 0 \\ 0 & 1 & 4 & 1 \\ 0 & 0 & 0 & 1 \end{bmatrix}, \quad \mathbf{b} = \begin{bmatrix} 0 \\ 3(e^2 - 2e + 1) \\ 3(e^3 - 2e^2 + e) \\ 0 \end{bmatrix}, \quad \text{and} \quad \mathbf{x} = \begin{bmatrix} c_0 \\ c_1 \\ c_2 \\ c_3 \end{bmatrix}.$$

The vector-matrix equation $A\mathbf{x} = \mathbf{b}$ is equivalent to the system of equations

$$c_0 = 0,$$

$$c_0 + 4c_1 + c_2 = 3(e^2 - 2e + 1),$$

$$c_1 + 4c_2 + c_3 = 3(e^3 - 2e^2 + e),$$

$$c_3 = 0.$$

This system has the solution $c_0 = c_3 = 0$, and to 5 decimal places,

$$c_1 = \frac{1}{5}(-e^3 + 6e^2 - 9e + 4) \approx 0.75685, \quad \text{and} \quad c_2 = \frac{1}{5}(4e^3 - 9e^2 + 6e - 1) \approx 5.83007.$$

Solving for the remaining constants gives

$$\begin{aligned}
b_0 &= \frac{1}{h_0}(a_1 - a_0) - \frac{h_0}{3}(c_1 + 2c_0) \\
&= (e - 1) - \frac{1}{15}(-e^3 + 6e^2 - 9e + 4) \approx 1.46600, \\
b_1 &= \frac{1}{h_1}(a_2 - a_1) - \frac{h_1}{3}(c_2 + 2c_1) \\
&= (e^2 - e) - \frac{1}{15}(2e^3 + 3e^2 - 12e + 7) \approx 2.22285, \\
b_2 &= \frac{1}{h_2}(a_3 - a_2) - \frac{h_2}{3}(c_3 + 2c_2) \\
&= (e^3 - e^2) - \frac{1}{15}(8e^3 - 18e^2 + 12e - 2) \approx 8.80977, \\
d_0 &= \frac{1}{3h_0}(c_1 - c_0) = \frac{1}{15}(-e^3 + 6e^2 - 9e + 4) \approx 0.25228, \\
d_1 &= \frac{1}{3h_1}(c_2 - c_1) = \frac{1}{3}(e^3 - 3e^2 + 3e - 1) \approx 1.69107,
\end{aligned}$$

and

$$d_2 = \frac{1}{3h_2}(c_3 - c_1) = \frac{1}{15}(-4e^3 + 9e^2 - 6e + 1) \approx -1.94336.$$

The natural cubic spine is described piecewise by

$$S(x) = \begin{cases} 1 + 1.46600x + 0.25228x^3, & \text{for } x \in [0, 1], \\ 2.71828 + 2.22285(x-1) + 0.75685(x-1)^2 + 1.69107(x-1)^3, & \text{for } x \in [1, 2], \\ 7.38906 + 8.80977(x-2) + 5.83007(x-2)^2 - 1.94336(x-2)^3, & \text{for } x \in [2, 3]. \end{cases}$$

The spline and its agreement with $f(x) = e^x$ are shown in Figure 3.10. ■

Figure 3.10

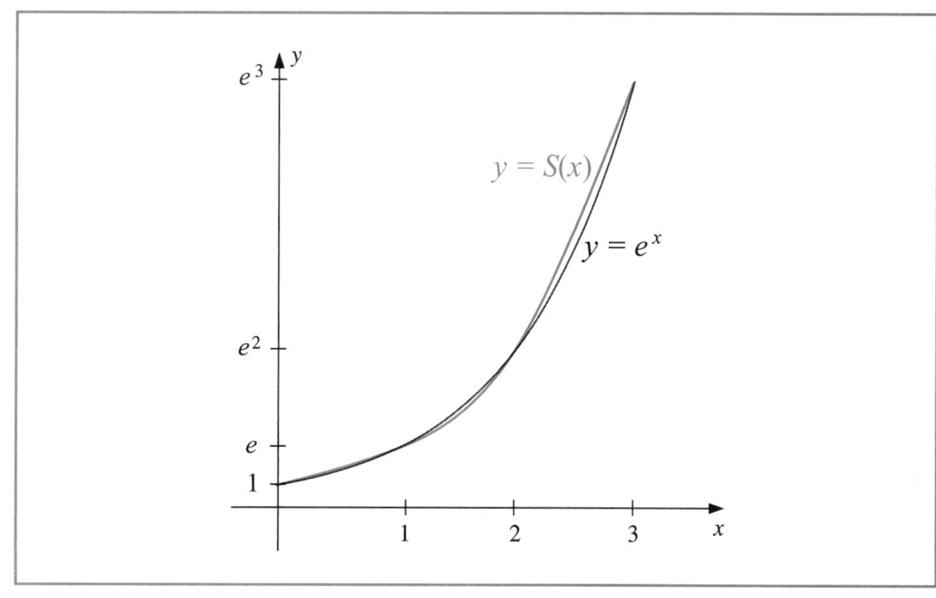

The *NumericalAnalysis* package can be used to create a cubic spline in a manner similar to other constructions in this chapter. However, the *CurveFitting* Package in Maple can also be used, and since this has not been discussed previously we will use it to create the natural spline in Example 2. First we load the package with the command

with(*CurveFitting*)

and define the function being approximated with

$f := x \rightarrow e^x$

To create a spline we need to specify the nodes, variable, the degree, and the natural endpoints. This is done with

$sn := t \rightarrow Spline([[0., 1.0], [1.0, f(1.0)], [2.0, f(2.0)], [3.0, f(3.0)]], t, degree = 3,$
$endpoints = \text{'}natural\text{'})$

Maple returns

$$t \rightarrow CurveFitting\text{:-}Spline([[0., 1.0], [1.0, f(1.0)], [2.0, f(2.0)], [3.0, f(3.0)]], t, degree = 3, endpoints = \text{'}natural\text{'})$$

The form of the natural spline is seen with the command

$sn(t)$

which produces

$$\begin{cases} 1. + 1.465998t^2 + 0.2522848t^3 & t < 1.0 \\ 0.495432 + 2.22285t + 0.756853(t-1.0)^2 + 1.691071(t-1.0)^3 & t < 2.0 \\ -10.230483 + 8.809770t + 5.830067(t-2.0)^2 - 1.943356(t-2.0)^3 & otherwise \end{cases}$$

Once we have determined a spline approximation for a function we can use it to approximate other properties of the function. The next illustration involves the integral of the spline we found in the previous example.

Illustration To approximate the integral of $f(x) = e^x$ on $[0, 3]$, which has the value

$$\int_0^3 e^x\,dx = e^3 - 1 \approx 20.08553692 - 1 = 19.08553692,$$

we can piecewise integrate the spline that approximates f on this integral. This gives

$$\int_0^3 S(x) = \int_0^1 1 + 1.46600x + 0.25228x^3\,dx$$
$$+ \int_1^2 2.71828 + 2.22285(x-1) + 0.75685(x-1)^2 + 1.69107(x-1)^3\,dx$$
$$+ \int_2^3 7.38906 + 8.80977(x-2) + 5.83007(x-2)^2 - 1.94336(x-2)^3\,dx.$$

Integrating and collecting values from like powers gives

$$\int_0^3 S(x) = \left[x + 1.46600\frac{x^2}{2} + 0.25228\frac{x^4}{4}\right]_0^1$$

$$+\left[2.71828(x-1) + 2.22285\frac{(x-1)^2}{2} + 0.75685\frac{(x-1)^3}{3} + 1.69107\frac{(x-1)^4}{4}\right]_1^2$$

$$+\left[7.38906(x-2) + 8.80977\frac{(x-2)^2}{2} + 5.83007\frac{(x-2)^3}{3} - 1.94336\frac{(x-2)^4}{4}\right]_2^3$$

$$= (1 + 2.71828 + 7.38906) + \frac{1}{2}(1.46600 + 2.22285 + 8.80977)$$

$$+\frac{1}{3}(0.75685 + 5.83007) + \frac{1}{4}(0.25228 + 1.69107 - 1.94336)$$

$$= 19.55229.$$

Because the nodes are equally spaced in this example the integral approximation is simply

$$\int_0^3 S(x)\,dx = (a_0+a_1+a_2)+\frac{1}{2}(b_0+b_1+b_2)+\frac{1}{3}(c_0+c_1+c_2)+\frac{1}{4}(d_0+d_1+d_2). \quad (3.22)$$

□

If we create the natural spline using Maple as described after Example 2, we can then use Maple's integration command to find the value in the Illustration. Simply enter

int$(sn(t), t = 0 \,..\, 3)$

19.55228648

Clamped Splines

Example 3 In Example 1 we found a natural spline S that passes through the points $(1,2)$, $(2,3)$, and $(3,5)$. Construct a clamped spline s through these points that has $s'(1) = 2$ and $s'(3) = 1$.

Solution Let

$$s_0(x) = a_0 + b_0(x-1) + c_0(x-1)^2 + d_0(x-1)^3,$$

be the cubic on $[1,2]$ and the cubic on $[2,3]$ be

$$s_1(x) = a_1 + b_1(x-2) + c_1(x-2)^2 + d_1(x-2)^3.$$

Then most of the conditions to determine the 8 constants are the same as those in Example 1. That is,

$$2 = f(1) = a_0, \quad 3 = f(2) = a_0 + b_0 + c_0 + d_0, \quad 3 = f(2) = a_1, \quad \text{and}$$
$$5 = f(3) = a_1 + b_1 + c_1 + d_1.$$

$$s_0'(2) = s_1'(2): \quad b_0 + 2c_0 + 3d_0 = b_1 \quad \text{and} \quad s_0''(2) = s_1''(2): \quad 2c_0 + 6d_0 = 2c_1$$

However, the boundary conditions are now

$$s_0'(1) = 2: \quad b_0 = 2 \quad \text{and} \quad s_1'(3) = 1: \quad b_1 + 2c_1 + 3d_1 = 1.$$

Solving this system of equations gives the spline as

$$s(x) = \begin{cases} 2 + 2(x-1) - \frac{5}{2}(x-1)^2 + \frac{3}{2}(x-1)^3, & \text{for } x \in [1,2] \\ 3 + \frac{3}{2}(x-2) + 2(x-2)^2 - \frac{3}{2}(x-2)^3, & \text{for } x \in [2,3] \end{cases}$$

■

In the case of general clamped boundary conditions we have a result that is similar to the theorem for natural boundary conditions described in Theorem 3.11.

Theorem 3.12 If f is defined at $a = x_0 < x_1 < \cdots < x_n = b$ and differentiable at a and b, then f has a unique clamped spline interpolant S on the nodes $x_0, x_1, \ldots, x_n$; that is, a spline interpolant that satisfies the clamped boundary conditions $S'(a) = f'(a)$ and $S'(b) = f'(b)$. ■

Proof Since $f'(a) = S'(a) = S'(x_0) = b_0$, Eq. (3.20) with $j = 0$ implies

$$f'(a) = \frac{1}{h_0}(a_1 - a_0) - \frac{h_0}{3}(2c_0 + c_1).$$

Consequently,

$$2h_0c_0 + h_0c_1 = \frac{3}{h_0}(a_1 - a_0) - 3f'(a).$$

Similarly,

$$f'(b) = b_n = b_{n-1} + h_{n-1}(c_{n-1} + c_n),$$

so Eq. (3.20) with $j = n - 1$ implies that

$$\begin{aligned} f'(b) &= \frac{a_n - a_{n-1}}{h_{n-1}} - \frac{h_{n-1}}{3}(2c_{n-1} + c_n) + h_{n-1}(c_{n-1} + c_n) \\ &= \frac{a_n - a_{n-1}}{h_{n-1}} + \frac{h_{n-1}}{3}(c_{n-1} + 2c_n), \end{aligned}$$

and

$$h_{n-1}c_{n-1} + 2h_{n-1}c_n = 3f'(b) - \frac{3}{h_{n-1}}(a_n - a_{n-1}).$$

Equations (3.21) together with the equations

$$2h_0c_0 + h_0c_1 = \frac{3}{h_0}(a_1 - a_0) - 3f'(a)$$

and

$$h_{n-1}c_{n-1} + 2h_{n-1}c_n = 3f'(b) - \frac{3}{h_{n-1}}(a_n - a_{n-1})$$

determine the linear system $A\mathbf{x} = \mathbf{b}$, where

$$A = \begin{bmatrix} 2h_0 & h_0 & 0 & \cdots & \cdots & 0 \\ h_0 & 2(h_0+h_1) & h_1 & \ddots & & \vdots \\ 0 & h_1 & 2(h_1+h_2) & h_2 & \ddots & \vdots \\ \vdots & \ddots & \ddots & \ddots & \ddots & 0 \\ \vdots & & \ddots & h_{n-2} & 2(h_{n-2}+h_{n-1}) & h_{n-1} \\ 0 & \cdots & \cdots & 0 & h_{n-1} & 2h_{n-1} \end{bmatrix},$$

$$\mathbf{b} = \begin{bmatrix} \frac{3}{h_0}(a_1 - a_0) - 3f'(a) \\ \frac{3}{h_1}(a_2 - a_1) - \frac{3}{h_0}(a_1 - a_0) \\ \vdots \\ \frac{3}{h_{n-1}}(a_n - a_{n-1}) - \frac{3}{h_{n-2}}(a_{n-1} - a_{n-2}) \\ 3f'(b) - \frac{3}{h_{n-1}}(a_n - a_{n-1}) \end{bmatrix}, \quad \text{and} \quad \mathbf{x} = \begin{bmatrix} c_0 \\ c_1 \\ \vdots \\ c_n \end{bmatrix}.$$

This matrix A is also strictly diagonally dominant, so it satisfies the conditions of Theorem 6.21 in Section 6.6. Therefore, the linear system has a unique solution for $c_0, c_1, \ldots, c_n$. ■ ■ ■

The solution to the cubic spline problem with the boundary conditions $S'(x_0) = f'(x_0)$ and $S'(x_n) = f'(x_n)$ can be obtained by applying Algorithm 3.5.

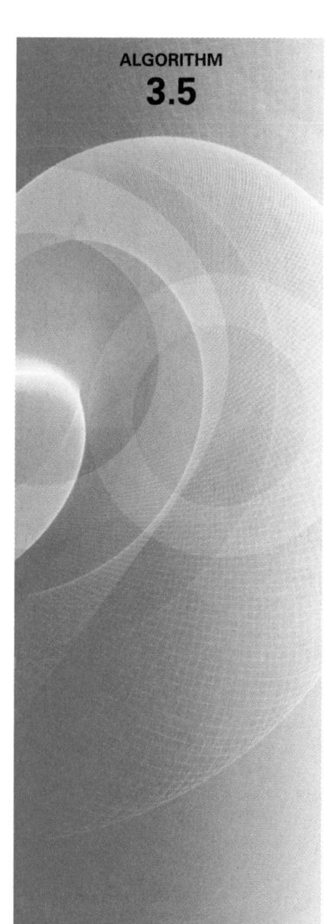

Clamped Cubic Spline

To construct the cubic spline interpolant S for the function f defined at the numbers $x_0 < x_1 < \cdots < x_n$, satisfying $S'(x_0) = f'(x_0)$ and $S'(x_n) = f'(x_n)$:

INPUT n; $x_0, x_1, \ldots, x_n$; $a_0 = f(x_0)$, $a_1 = f(x_1), \ldots, a_n = f(x_n)$; $FPO = f'(x_0)$; $FPN = f'(x_n)$.

OUTPUT a_j, b_j, c_j, d_j for $j = 0, 1, \ldots, n-1$.

(Note: $S(x) = S_j(x) = a_j + b_j(x - x_j) + c_j(x - x_j)^2 + d_j(x - x_j)^3$ for $x_j \le x \le x_{j+1}$.)

Step 1 For $i = 0, 1, \ldots, n-1$ set $h_i = x_{i+1} - x_i$.

Step 2 Set $\alpha_0 = 3(a_1 - a_0)/h_0 - 3FPO$;
$\alpha_n = 3FPN - 3(a_n - a_{n-1})/h_{n-1}$.

Step 3 For $i = 1, 2, \ldots, n-1$

$$\text{set } \alpha_i = \frac{3}{h_i}(a_{i+1} - a_i) - \frac{3}{h_{i-1}}(a_i - a_{i-1}).$$

Step 4 Set $l_0 = 2h_0$; *(Steps 4,5,6, and part of Step 7 solve a tridiagonal linear system using a method described in Algorithm 6.7.)*
$\mu_0 = 0.5$;
$z_0 = \alpha_0/l_0$.

Step 5 For $i = 1, 2, \ldots, n-1$
set $l_i = 2(x_{i+1} - x_{i-1}) - h_{i-1}\mu_{i-1}$;
$\mu_i = h_i/l_i$;
$z_i = (\alpha_i - h_{i-1}z_{i-1})/l_i$.

Step 6 Set $l_n = h_{n-1}(2 - \mu_{n-1})$;
$z_n = (\alpha_n - h_{n-1}z_{n-1})/l_n$;
$c_n = z_n$.

Step 7 For $j = n-1, n-2, \ldots, 0$
set $c_j = z_j - \mu_j c_{j+1}$;
$b_j = (a_{j+1} - a_j)/h_j - h_j(c_{j+1} + 2c_j)/3$;
$d_j = (c_{j+1} - c_j)/(3h_j)$.

Step 8 OUTPUT (a_j, b_j, c_j, d_j for $j = 0, 1, \ldots, n-1$);
STOP.

■

Example 4 Example 2 used a natural spline and the data points $(0, 1)$, $(1, e)$, $(2, e^2)$, and $(3, e^3)$ to form a new approximating function $S(x)$. Determine the clamped spline $s(x)$ that uses this data and the additional information that, since $f'(x) = e^x$, so $f'(0) = 1$ and $f'(3) = e^3$.

Solution As in Example 2, we have $n = 3$, $h_0 = h_1 = h_2 = 1$, $a_0 = 0$, $a_1 = e$, $a_2 = e^2$, and $a_3 = e^3$. This together with the information that $f'(0) = 1$ and $f'(3) = e^3$ gives the the matrix A and the vectors **b** and **x** with the forms

$$A = \begin{bmatrix} 2 & 1 & 0 & 0 \\ 1 & 4 & 1 & 0 \\ 0 & 1 & 4 & 1 \\ 0 & 0 & 1 & 2 \end{bmatrix}, \quad \mathbf{b} = \begin{bmatrix} 3(e-2) \\ 3(e^2 - 2e + 1) \\ 3(e^3 - 2e^2 + e) \\ 3e^2 \end{bmatrix}, \quad \text{and} \quad \mathbf{x} = \begin{bmatrix} c_0 \\ c_1 \\ c_2 \\ c_3 \end{bmatrix}.$$

The vector-matrix equation $A\mathbf{x} = \mathbf{b}$ is equivalent to the system of equations

$$2c_0 + c_1 = 3(e-2),$$
$$c_0 + 4c_1 + c_2 = 3(e^2 - 2e + 1),$$
$$c_1 + 4c_2 + c_3 = 3(e^3 - 2e^2 + e),$$
$$c_2 + 2c_3 = 3e^2.$$

Solving this system simultaneously for c_0, c_1, c_2 and c_3 gives, to 5 decimal places,

$$c_0 = \frac{1}{15}(2e^3 - 12e^2 + 42e - 59) = 0.44468,$$
$$c_1 = \frac{1}{15}(-4e^3 + 24e^2 - 39e + 28) = 1.26548,$$
$$c_2 = \frac{1}{15}(14e^3 - 39e^2 + 24e - 8) = 3.35087,$$
$$c_3 = \frac{1}{15}(-7e^3 + 42e^2 - 12e + 4) = 9.40815.$$

Solving for the remaining constants in the same manner as Example 2 gives

$$b_0 = 1.00000, \quad b_1 = 2.71016, \quad b_2 = 7.32652,$$

and

$$d_0 = 0.27360, \quad d_1 = 0.69513, \quad d_2 = 2.01909.$$

This gives the clamped cubic spine

$$s(x) = \begin{cases} 1 + x + 0.44468x^2 + 0.27360x^3, & \text{if } 0 \le x < 1, \\ 2.71828 + 2.71016(x-1) + 1.26548(x-1)^2 + 0.69513(x-1)^3, & \text{if } 1 \le x < 2, \\ 7.38906 + 7.32652(x-2) + 3.35087(x-2)^2 + 2.01909(x-2)^3, & \text{if } 2 \le x \le 3. \end{cases}$$

The graph of the clamped spline and $f(x) = e^x$ are so similar that no difference can be seen. ■

We can create the clamped cubic spline in Example 4 with the same commands we used for the natural spline, the only change that is needed is to specify the derivative at the endpoints. In this case we use

$sn := t \rightarrow Spline\,([[0., 1.0], [1.0, f(1.0)], [2.0, f(2.0)], [3.0, f(3.0)]], t, degree = 3,$ $endpoints = \left[1.0, e^{3.0}\right])$

giving essentially the same results as in the example.

We can also approximate the integral of f on $[0, 3]$, by integrating the clamped spline. The exact value of the integral is

$$\int_0^3 e^x\, dx = e^3 - 1 \approx 20.08554 - 1 = 19.08554.$$

Because the data is equally spaced, piecewise integrating the clamped spline results in the same formula as in (3.22), that is,

$$\int_0^3 s(x)\, dx = (a_0 + a_1 + a_2) + \frac{1}{2}(b_0 + b_1 + b_2) + \frac{1}{3}(c_0 + c_1 + c_2) + \frac{1}{4}(d_0 + d_1 + d_2).$$

Hence the integral approximation is

$$\begin{aligned} \int_0^3 s(x)\, dx &= (1 + 2.71828 + 7.38906) + \frac{1}{2}(1 + 2.71016 + 7.32652) \\ &\quad + \frac{1}{3}(0.44468 + 1.26548 + 3.35087) + \frac{1}{4}(0.27360 + 0.69513 + 2.01909) \\ &= 19.05965. \end{aligned}$$

The absolute error in the integral approximation using the clamped and natural splines are

$$\text{Natural} : |19.08554 - 19.55229| = 0.46675$$

and

$$\text{Clamped} : |19.08554 - 19.05965| = 0.02589.$$

For integration purposes the clamped spline is vastly superior. This should be no surprise since the boundary conditions for the clamped spline are exact, whereas for the natural spline we are essentially assuming that, since $f''(x) = e^x$,

$$0 = S''(0) \approx f''(0) = e^1 = 1 \quad \text{and} \quad 0 = S''(3) \approx f''(3) = e^3 \approx 20.$$

The next illustration uses a spine to approximate a curve that has no given functional representation.

Illustration Figure 3.11 shows a ruddy duck in flight. To approximate the top profile of the duck, we have chosen points along the curve through which we want the approximating curve to pass. Table 3.18 lists the coordinates of 21 data points relative to the superimposed coordinate system shown in Figure 3.12. Notice that more points are used when the curve is changing rapidly than when it is changing more slowly.

Figure 3.11

Table 3.18

x	0.9	1.3	1.9	2.1	2.6	3.0	3.9	4.4	4.7	5.0	6.0	7.0	8.0	9.2	10.5	11.3	11.6	12.0	12.6	13.0	13.3
$f(x)$	1.3	1.5	1.85	2.1	2.6	2.7	2.4	2.15	2.05	2.1	2.25	2.3	2.25	1.95	1.4	0.9	0.7	0.6	0.5	0.4	0.25

Figure 3.12

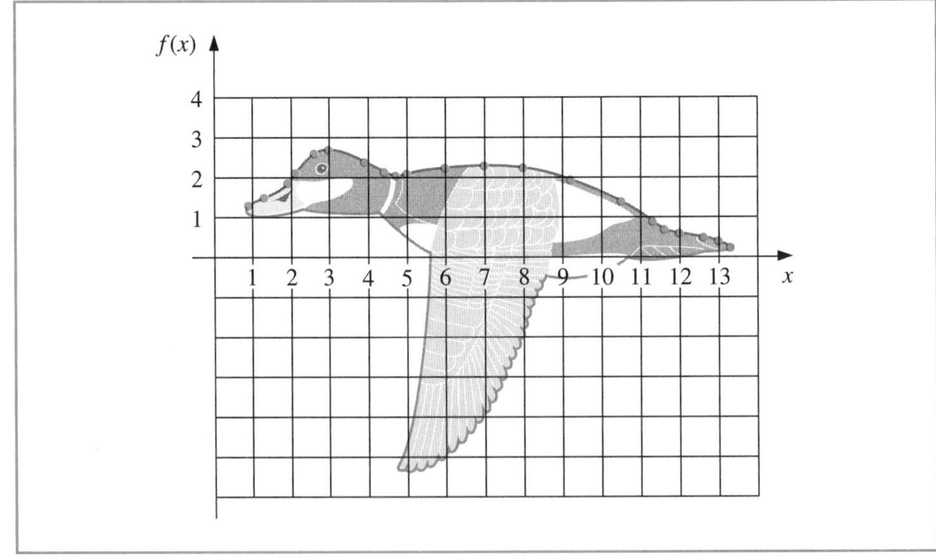

Using Algorithm 3.4 to generate the natural cubic spline for this data produces the coefficients shown in Table 3.19. This spline curve is nearly identical to the profile, as shown in Figure 3.13.

Table 3.19

j	x_j	a_j	b_j	c_j	d_j
0	0.9	1.3	5.40	0.00	−0.25
1	1.3	1.5	0.42	−0.30	0.95
2	1.9	1.85	1.09	1.41	−2.96
3	2.1	2.1	1.29	−0.37	−0.45
4	2.6	2.6	0.59	−1.04	0.45
5	3.0	2.7	−0.02	−0.50	0.17
6	3.9	2.4	−0.50	−0.03	0.08
7	4.4	2.15	−0.48	0.08	1.31
8	4.7	2.05	−0.07	1.27	−1.58
9	5.0	2.1	0.26	−0.16	0.04
10	6.0	2.25	0.08	−0.03	0.00
11	7.0	2.3	0.01	−0.04	−0.02
12	8.0	2.25	−0.14	−0.11	0.02
13	9.2	1.95	−0.34	−0.05	−0.01
14	10.5	1.4	−0.53	−0.10	−0.02
15	11.3	0.9	−0.73	−0.15	1.21
16	11.6	0.7	−0.49	0.94	−0.84
17	12.0	0.6	−0.14	−0.06	0.04
18	12.6	0.5	−0.18	0.00	−0.45
19	13.0	0.4	−0.39	−0.54	0.60
20	13.3	0.25			

Figure 3.13

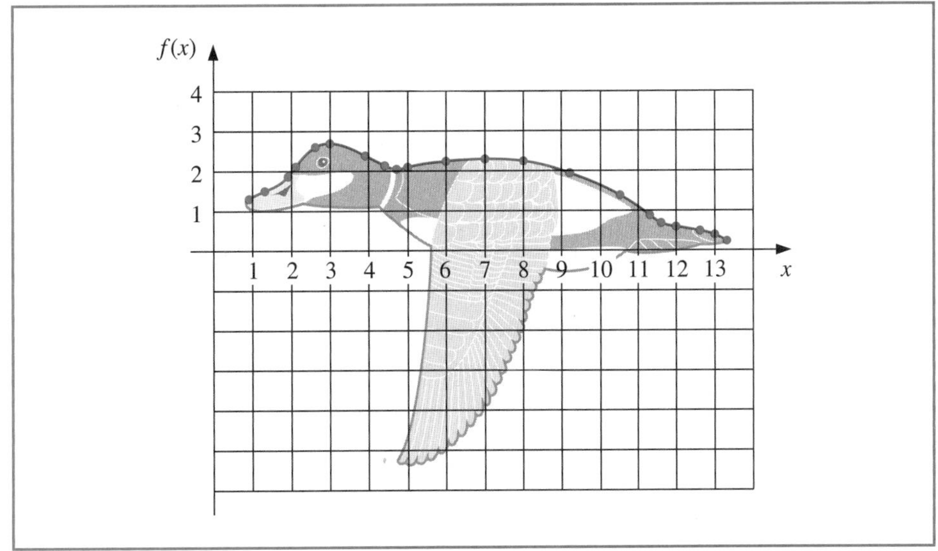

For comparison purposes, Figure 3.14 gives an illustration of the curve that is generated using a Lagrange interpolating polynomial to fit the data given in Table 3.18. The interpolating polynomial in this case is of degree 20 and oscillates wildly. It produces a very strange illustration of the back of a duck, in flight or otherwise.

Figure 3.14

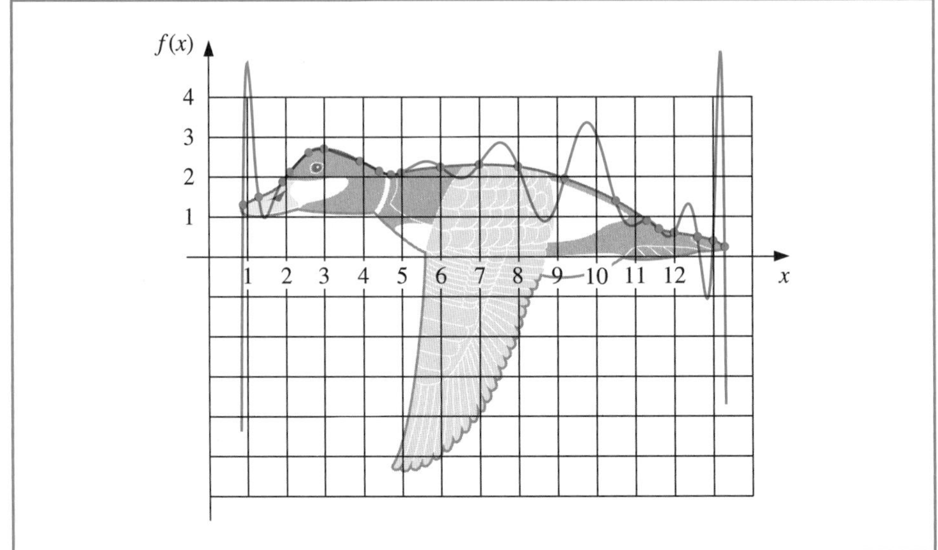

To use a clamped spline to approximate this curve we would need derivative approximations for the endpoints. Even if these approximations were available, we could expect little improvement because of the close agreement of the natural cubic spline to the curve of the top profile. □

Constructing a cubic spline to approximate the lower profile of the ruddy duck would be more difficult since the curve for this portion cannot be expressed as a function of x, and at certain points the curve does not appear to be smooth. These problems can be resolved by using separate splines to represent various portions of the curve, but a more effective approach to approximating curves of this type is considered in the next section.

The clamped boundary conditions are generally preferred when approximating functions by cubic splines, so the derivative of the function must be known or approximated at the endpoints of the interval. When the nodes are equally spaced near both endpoints, approximations can be obtained by any of the appropriate formulas given in Sections 4.1 and 4.2. When the nodes are unequally spaced, the problem is considerably more difficult.

To conclude this section, we list an error-bound formula for the cubic spline with clamped boundary conditions. The proof of this result can be found in [Schul], pp. 57–58.

Theorem 3.13 Let $f \in C^4[a,b]$ with $\max_{a\le x\le b} |f^{(4)}(x)| = M$. If S is the unique clamped cubic spline interpolant to f with respect to the nodes $a = x_0 < x_1 < \cdots < x_n = b$, then for all x in $[a,b]$,

$$|f(x) - S(x)| \le \frac{5M}{384} \max_{0\le j\le n-1} (x_{j+1} - x_j)^4.$$ ■

A fourth-order error-bound result also holds in the case of natural boundary conditions, but it is more difficult to express. (See [BD], pp. 827–835.)

The natural boundary conditions will generally give less accurate results than the clamped conditions near the ends of the interval $[x_0, x_n]$ unless the function f happens to nearly satisfy $f''(x_0) = f''(x_n) = 0$. An alternative to the natural boundary condition that does not require knowledge of the derivative of f is the *not-a-knot* condition, (see [Deb2], pp. 55–56). This condition requires that $S'''(x)$ be continuous at x_1 and at x_{n-1}.

EXERCISE SET 3.5

1. Determine the natural cubic spline S that interpolates the data $f(0) = 0$, $f(1) = 1$, and $f(2) = 2$.
2. Determine the clamped cubic spline s that interpolates the data $f(0) = 0$, $f(1) = 1$, $f(2) = 2$ and satisfies $s'(0) = s'(2) = 1$.
3. Construct the natural cubic spline for the following data.

a.

x	$f(x)$
0.8	0.22363362
1.0	0.65809197

b.

x	$f(x)$
8.3	17.56492
8.6	18.50515

c.

x	$f(x)$
−0.5	−0.0247500
−0.25	0.3349375
0	1.1010000

d.

x	$f(x)$
0.1	−0.62049958
0.2	−0.28398668
0.3	0.00660095
0.4	0.24842440

4. Construct the natural cubic spline for the following data.

a.

x	$f(x)$
0	1.00000
0.5	2.71828

b.

x	$f(x)$
−0.25	1.33203
0.25	0.800781

c.

x	$f(x)$
0.1	−0.29004996
0.2	−0.56079734
0.3	−0.81401972

d.

x	$f(x)$
−1	0.86199480
−0.5	0.95802009
0	1.0986123
0.5	1.2943767

5. The data in Exercise 3 were generated using the following functions. Use the cubic splines constructed in Exercise 3 for the given value of x to approximate $f(x)$ and $f'(x)$, and calculate the actual error.
   a. $f(x) = \sin(e^x - 2)$; approximate $f(0.9)$ and $f'(0.9)$.
   b. $f(x) = x \ln x$; approximate $f(8.4)$ and $f'(8.4)$.
   c. $f(x) = x^3 + 4.001x^2 + 4.002x + 1.101$; approximate $f(-\frac{1}{3})$ and $f'(-\frac{1}{3})$.
   d. $f(x) = x \cos x - 2x^2 + 3x - 1$; approximate $f(0.25)$ and $f'(0.25)$.
6. The data in Exercise 4 were generated using the following functions. Use the cubic splines constructed in Exercise 4 for the given value of x to approximate $f(x)$ and $f'(x)$, and calculate the actual error.
   a. $f(x) = e^{2x}$; approximate $f(0.43)$ and $f'(0.43)$.
   b. $f(x) = x^4 - x^3 + x^2 - x + 1$; approximate $f(0)$ and $f'(0)$.
   c. $f(x) = x^2 \cos x - 3x$; approximate $f(0.18)$ and $f'(0.18)$.
   d. $f(x) = \ln(e^x + 2)$; approximate $f(0.25)$ and $f'(0.25)$.
7. Construct the clamped cubic spline using the data of Exercise 3 and the fact that
   a. $f'(0.8) = 2.1691753$ and $f'(1.0) = 2.0466965$
   b. $f'(8.3) = 3.116256$ and $f'(8.6) = 3.151762$
   c. $f'(-0.5) = 0.7510000$ and $f'(0) = 4.0020000$
   d. $f'(0.1) = 3.58502082$ and $f'(0.4) = 2.16529366$
8. Construct the clamped cubic spline using the data of Exercise 4 and the fact that
   a. $f'(0) = 2$ and $f'(0.5) = 5.43656$
   b. $f'(-0.25) = 0.437500$ and $f'(0.25) = -0.625000$

c. $f'(0.1) = -2.8004996$ and $f'(0) = -2.9734038$

d. $f'(-1) = 0.15536240$ and $f'(0.5) = 0.45186276$

9. Repeat Exercise 5 using the clamped cubic splines constructed in Exercise 7.

10. Repeat Exercise 6 using the clamped cubic splines constructed in Exercise 8.

11. A natural cubic spline S on $[0, 2]$ is defined by

$$S(x) = \begin{cases} S_0(x) = 1 + 2x - x^3, & \text{if } 0 \le x < 1, \\ S_1(x) = 2 + b(x-1) + c(x-1)^2 + d(x-1)^3, & \text{if } 1 \le x \le 2. \end{cases}$$

Find b, c, and d.

12. A clamped cubic spline s for a function f is defined on $[1, 3]$ by

$$s(x) = \begin{cases} s_0(x) = 3(x-1) + 2(x-1)^2 - (x-1)^3, & \text{if } 1 \le x < 2, \\ s_1(x) = a + b(x-2) + c(x-2)^2 + d(x-2)^3, & \text{if } 2 \le x \le 3. \end{cases}$$

Given $f'(1) = f'(3)$, find a, b, c, and d.

13. A natural cubic spline S is defined by

$$S(x) = \begin{cases} S_0(x) = 1 + B(x-1) - D(x-1)^3, & \text{if } 1 \le x < 2, \\ S_1(x) = 1 + b(x-2) - \frac{3}{4}(x-2)^2 + d(x-2)^3, & \text{if } 2 \le x \le 3. \end{cases}$$

If S interpolates the data $(1, 1)$, $(2, 1)$, and $(3, 0)$, find B, D, b, and d.

14. A clamped cubic spline s for a function f is defined by

$$s(x) = \begin{cases} s_0(x) = 1 + Bx + 2x^2 - 2x^3, & \text{if } 0 \le x < 1, \\ s_1(x) = 1 + b(x-1) - 4(x-1)^2 + 7(x-1)^3, & \text{if } 1 \le x \le 2. \end{cases}$$

Find $f'(0)$ and $f'(2)$.

15. Construct a natural cubic spline to approximate $f(x) = \cos \pi x$ by using the values given by $f(x)$ at $x = 0, 0.25, 0.5, 0.75,$ and 1.0. Integrate the spline over $[0, 1]$, and compare the result to $\int_0^1 \cos \pi x \, dx = 0$. Use the derivatives of the spline to approximate $f'(0.5)$ and $f''(0.5)$. Compare these approximations to the actual values.

16. Construct a natural cubic spline to approximate $f(x) = e^{-x}$ by using the values given by $f(x)$ at $x = 0$, 0.25, 0.75, and 1.0. Integrate the spline over $[0, 1]$, and compare the result to $\int_0^1 e^{-x} \, dx = 1 - 1/e$. Use the derivatives of the spline to approximate $f'(0.5)$ and $f''(0.5)$. Compare the approximations to the actual values.

17. Repeat Exercise 15, constructing instead the clamped cubic spline with $f'(0) = f'(1) = 0$.

18. Repeat Exercise 16, constructing instead the clamped cubic spline with $f'(0) = -1$, $f'(1) = -e^{-1}$.

19. Suppose that $f(x)$ is a polynomial of degree 3. Show that $f(x)$ is its own clamped cubic spline, but that it cannot be its own natural cubic spline.

20. Suppose the data $\{x_i, f(x_i))\}_{i=1}^n$ lie on a straight line. What can be said about the natural and clamped cubic splines for the function f? [*Hint:* Take a cue from the results of Exercises 1 and 2.]

21. Given the partition $x_0 = 0$, $x_1 = 0.05$, and $x_2 = 0.1$ of $[0, 0.1]$, find the piecewise linear interpolating function F for $f(x) = e^{2x}$. Approximate $\int_0^{0.1} e^{2x} \, dx$ with $\int_0^{0.1} F(x) \, dx$, and compare the results to the actual value.

22. Let $f \in C^2[a, b]$, and let the nodes $a = x_0 < x_1 < \cdots < x_n = b$ be given. Derive an error estimate similar to that in Theorem 3.13 for the piecewise linear interpolating function F. Use this estimate to derive error bounds for Exercise 21.

23. Extend Algorithms 3.4 and 3.5 to include as output the first and second derivatives of the spline at the nodes.

24. Extend Algorithms 3.4 and 3.5 to include as output the integral of the spline over the interval $[x_0, x_n]$.

25. Given the partition $x_0 = 0$, $x_1 = 0.05$, $x_2 = 0.1$ of $[0, 0.1]$ and $f(x) = e^{2x}$:

a. Find the cubic spline s with clamped boundary conditions that interpolates f.

b. Find an approximation for $\int_0^{0.1} e^{2x} \, dx$ by evaluating $\int_0^{0.1} s(x) \, dx$.

c. Use Theorem 3.13 to estimate $\max_{0\le x\le 0.1} |f(x) - s(x)|$ and

$$\left| \int_0^{0.1} f(x)\,dx - \int_0^{0.1} s(x)\,dx \right|.$$

d. Determine the cubic spline S with natural boundary conditions, and compare $S(0.02)$, $s(0.02)$, and $e^{0.04} = 1.04081077$.

26. Let f be defined on $[a, b]$, and let the nodes $a = x_0 < x_1 < x_2 = b$ be given. A quadratic spline interpolating function S consists of the quadratic polynomial

$$S_0(x) = a_0 + b_0(x - x_0) + c_0(x - x_0)^2 \quad \text{on } [x_0, x_1]$$

and the quadratic polynomial

$$S_1(x) = a_1 + b_1(x - x_1) + c_1(x - x_1)^2 \quad \text{on } [x_1, x_2],$$

such that

i. $S(x_0) = f(x_0), S(x_1) = f(x_1)$, and $S(x_2) = f(x_2)$,

ii. $S \in C^1[x_0, x_2]$.

Show that conditions (i) and (ii) lead to five equations in the six unknowns a_0, b_0, c_0, a_1, b_1, and c_1. The problem is to decide what additional condition to impose to make the solution unique. Does the condition $S \in C^2[x_0, x_2]$ lead to a meaningful solution?

27. Determine a quadratic spline s that interpolates the data $f(0) = 0$, $f(1) = 1$, $f(2) = 2$ and satisfies $s'(0) = 2$.

28. **a.** The introduction to this chapter included a table listing the population of the United States from 1950 to 2000. Use natural cubic spline interpolation to approximate the population in the years 1940, 1975, and 2020.

b. The population in 1940 was approximately 132,165,000. How accurate do you think your 1975 and 2020 figures are?

29. A car traveling along a straight road is clocked at a number of points. The data from the observations are given in the following table, where the time is in seconds, the distance is in feet, and the speed is in feet per second.

Time	0	3	5	8	13
Distance	0	225	383	623	993
Speed	75	77	80	74	72

a. Use a clamped cubic spline to predict the position of the car and its speed when $t = 10$ s.

b. Use the derivative of the spline to determine whether the car ever exceeds a 55-mi/h speed limit on the road; if so, what is the first time the car exceeds this speed?

c. What is the predicted maximum speed for the car?

30. The 2009 Kentucky Derby was won by a horse named Mine That Bird (at more than 50:1 odds) in a time of 2:02.66 (2 minutes and 2.66 seconds) for the $1\frac{1}{4}$-mile race. Times at the quarter-mile, half-mile, and mile poles were 0:22.98, 0:47.23, and 1:37.49.

a. Use these values together with the starting time to construct a natural cubic spline for Mine That Bird's race.

b. Use the spline to predict the time at the three-quarter-mile pole, and compare this to the actual time of 1:12.09.

c. Use the spline to approximate Mine That Bird's starting speed and speed at the finish line.

31. It is suspected that the high amounts of tannin in mature oak leaves inhibit the growth of the winter moth (*Operophtera bromata L., Geometridae*) larvae that extensively damage these trees in certain years. The following table lists the average weight of two samples of larvae at times in the first 28 days after birth. The first sample was reared on young oak leaves, whereas the second sample was reared on mature leaves from the same tree.

a. Use a natural cubic spline to approximate the average weight curve for each sample.

b. Find an approximate maximum average weight for each sample by determining the maximum of the spline.

Day	0	6	10	13	17	20	28
Sample 1 average weight (mg)	6.67	17.33	42.67	37.33	30.10	29.31	28.74
Sample 2 average weight (mg)	6.67	16.11	18.89	15.00	10.56	9.44	8.89

32. The upper portion of this noble beast is to be approximated using clamped cubic spline interpolants. The curve is drawn on a grid from which the table is constructed. Use Algorithm 3.5 to construct the three clamped cubic splines.

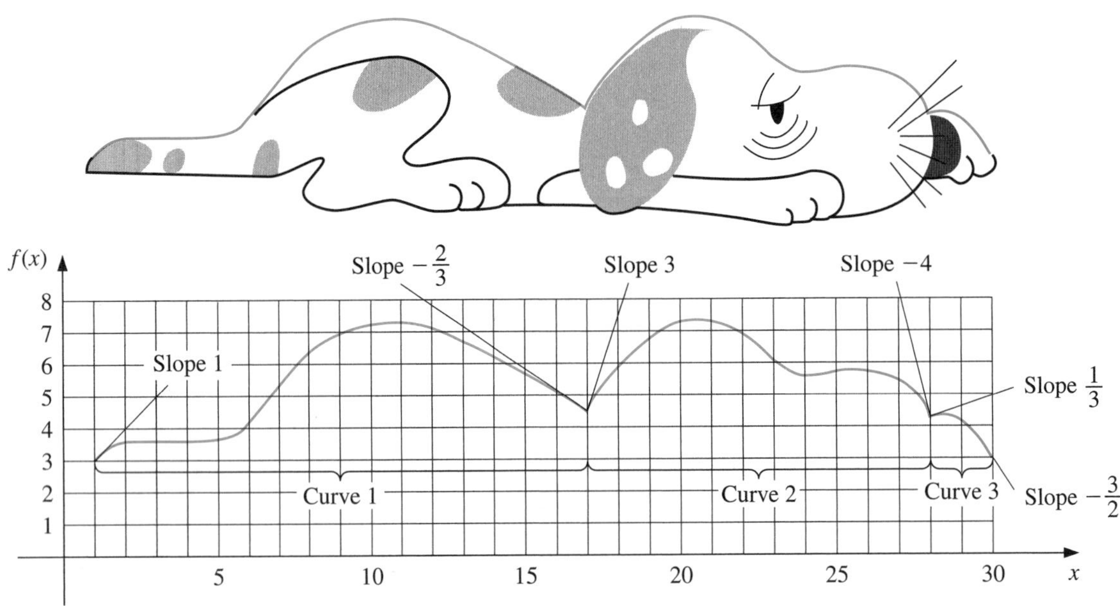

		Curve 1				Curve 2				Curve 3	
i	x_i	$f(x_i)$	$f'(x_i)$	i	x_i	$f(x_i)$	$f'(x_i)$	i	x_i	$f(x_i)$	$f'(x_i)$
0	1	3.0	1.0	0	17	4.5	3.0	0	27.7	4.1	0.33
1	2	3.7		1	20	7.0		1	28	4.3	
2	5	3.9		2	23	6.1		2	29	4.1	
3	6	4.2		3	24	5.6		3	30	3.0	−1.5
4	7	5.7		4	25	5.8					
5	8	6.6		5	27	5.2					
6	10	7.1		6	27.7	4.1	−4.0				
7	13	6.7									
8	17	4.5	−0.67								

33. Repeat Exercise 32, constructing three natural splines using Algorithm 3.4.

3.6 Parametric Curves

None of the techniques developed in this chapter can be used to generate curves of the form shown in Figure 3.15 because this curve cannot be expressed as a function of one coordinate variable in terms of the other. In this section we will see how to represent general curves by using a parameter to express both the x- and y-coordinate variables. Any good book on computer graphics will show how this technique can be extended to represent general curves and surfaces in space. (See, for example, [FVFH].)

Figure 3.15

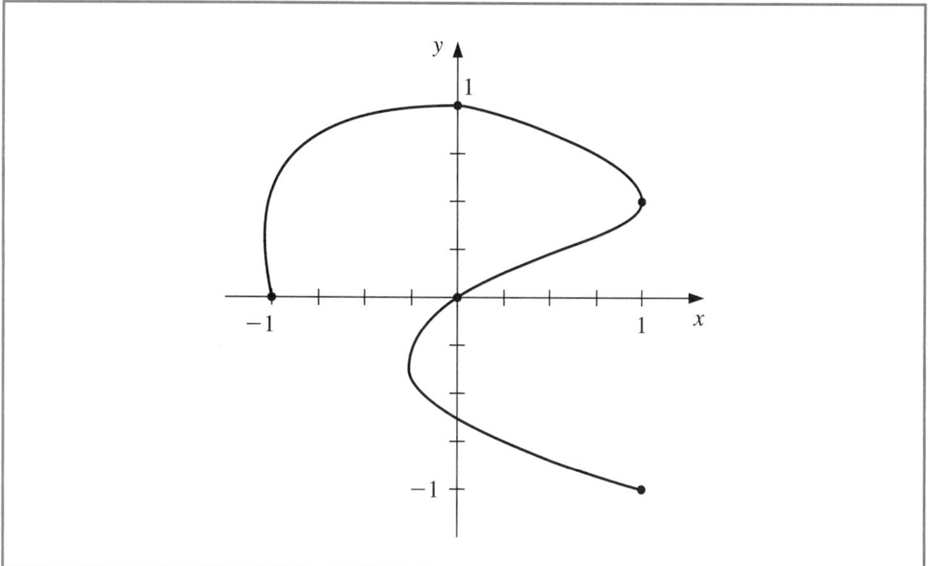

A straightforward parametric technique for determining a polynomial or piecewise polynomial to connect the points $(x_0, y_0), (x_1, y_1), \ldots, (x_n, y_n)$ in the order given is to use a parameter t on an interval $[t_0, t_n]$, with $t_0 < t_1 < \cdots < t_n$, and construct approximation functions with

$$x_i = x(t_i) \quad \text{and} \quad y_i = y(t_i), \quad \text{for each } i = 0, 1, \ldots, n.$$

The following example demonstrates the technique in the case where both approximating functions are Lagrange interpolating polynomials.

Example 1 Construct a pair of Lagrange polynomials to approximate the curve shown in Figure 3.15, using the data points shown on the curve.

Solution There is flexibility in choosing the parameter, and we will choose the points $\{t_i\}_{i=0}^4$ equally spaced in [0,1], which gives the data in Table 3.20.

Table 3.20

i	0	1	2	3	4
t_i	0	0.25	0.5	0.75	1
x_i	−1	0	1	0	1
y_i	0	1	0.5	0	−1

This produces the interpolating polynomials

$$x(t) = \left(\left(\left(64t - \tfrac{352}{3}\right)t + 60\right)t - \tfrac{14}{3}\right)t - 1 \quad \text{and} \quad y(t) = \left(\left(\left(-\tfrac{64}{3}t + 48\right)t - \tfrac{116}{3}\right)t + 11\right)t.$$

Plotting this parametric system produces the graph shown in blue in Figure 3.16. Although it passes through the required points and has the same basic shape, it is quite a crude approximation to the original curve. A more accurate approximation would require additional nodes, with the accompanying increase in computation. ■

Figure 3.16

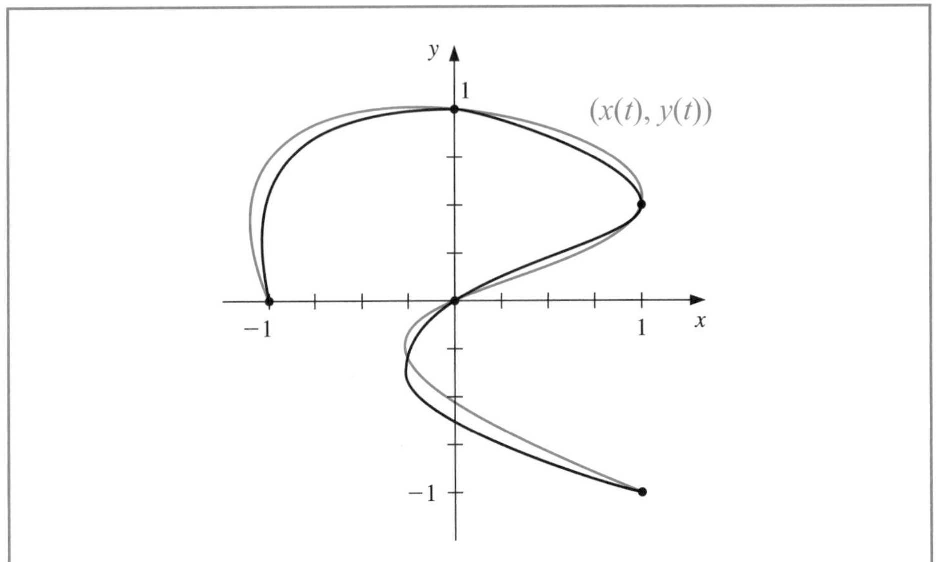

Parametric Hermite and spline curves can be generated in a similar manner, but these also require extensive computational effort.

Applications in computer graphics require the rapid generation of smooth curves that can be easily and quickly modified. For both aesthetic and computational reasons, changing one portion of these curves should have little or no effect on other portions of the curves. This eliminates the use of interpolating polynomials and splines since changing one portion of these curves affects the whole curve.

A successful computer design system needs to be based on a formal mathematical theory so that the results are predictable, but this theory should be performed in the background so that the artist can base the design on aesthetics.

The choice of curve for use in computer graphics is generally a form of the piecewise cubic Hermite polynomial. Each portion of a cubic Hermite polynomial is completely determined by specifying its endpoints and the derivatives at these endpoints. As a consequence, one portion of the curve can be changed while leaving most of the curve the same. Only the adjacent portions need to be modified to ensure smoothness at the endpoints. The computations can be performed quickly, and the curve can be modified a section at a time.

The problem with Hermite interpolation is the need to specify the derivatives at the endpoints of each section of the curve. Suppose the curve has $n + 1$ data points $(x(t_0), y(t_0)), \ldots, (x(t_n), y(t_n))$, and we wish to parameterize the cubic to allow complex features. Then we must specify $x'(t_i)$ and $y'(t_i)$, for each $i = 0, 1, \ldots, n$. This is not as difficult as it would first appear, since each portion is generated independently. We must ensure only that the derivatives at the endpoints of each portion match those in the adjacent portion. Essentially, then, we can simplify the process to one of determining a pair of cubic Hermite polynomials in the parameter t, where $t_0 = 0$ and $t_1 = 1$, given the endpoint data $(x(0), y(0))$ and $(x(1), y(1))$ and the derivatives dy/dx (at $t = 0$) and dy/dx (at $t = 1$).

Notice, however, that we are specifying only six conditions, and the cubic polynomials in $x(t)$ and $y(t)$ each have four parameters, for a total of eight. This provides flexibility in choosing the pair of cubic Hermite polynomials to satisfy the conditions, because the natural form for determining $x(t)$ and $y(t)$ requires that we specify $x'(0)$, $x'(1)$, $y'(0)$, and $y'(1)$. The explicit Hermite curve in x and y requires specifying only the quotients

$$\frac{dy}{dx}(t = 0) = \frac{y'(0)}{x'(0)} \quad \text{and} \quad \frac{dy}{dx}(t = 1) = \frac{y'(1)}{x'(1)}.$$

By multiplying $x'(0)$ and $y'(0)$ by a common scaling factor, the tangent line to the curve at $(x(0), y(0))$ remains the same, but the shape of the curve varies. The larger the scaling factor, the closer the curve comes to approximating the tangent line near $(x(0), y(0))$. A similar situation exists at the other endpoint $(x(1), y(1))$.

To further simplify the process in interactive computer graphics, the derivative at an endpoint is specified by using a second point, called a *guidepoint*, on the desired tangent line. The farther the guidepoint is from the node, the more closely the curve approximates the tangent line near the node.

In Figure 3.17, the nodes occur at (x_0, y_0) and (x_1, y_1), the guidepoint for (x_0, y_0) is $(x_0 + \alpha_0, y_0 + \beta_0)$, and the guidepoint for (x_1, y_1) is $(x_1 - \alpha_1, y_1 - \beta_1)$. The cubic Hermite polynomial $x(t)$ on $[0, 1]$ satisfies

$$x(0) = x_0, \quad x(1) = x_1, \quad x'(0) = \alpha_0, \quad \text{and} \quad x'(1) = \alpha_1.$$

Figure 3.17

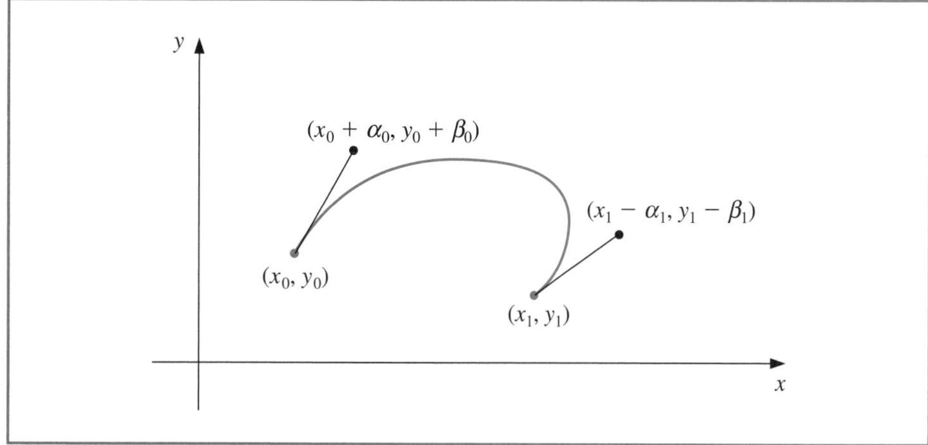

The unique cubic polynomial satisfying these conditions is

$$x(t) = [2(x_0 - x_1) + (\alpha_0 + \alpha_1)]t^3 + [3(x_1 - x_0) - (\alpha_1 + 2\alpha_0)]t^2 + \alpha_0 t + x_0. \tag{3.23}$$

In a similar manner, the unique cubic polynomial satisfying

$$y(0) = y_0, \quad y(1) = y_1, \quad y'(0) = \beta_0, \quad \text{and} \quad y'(1) = \beta_1$$

is

$$y(t) = [2(y_0 - y_1) + (\beta_0 + \beta_1)]t^3 + [3(y_1 - y_0) - (\beta_1 + 2\beta_0)]t^2 + \beta_0 t + y_0. \tag{3.24}$$

Example 2 Determine the graph of the parametric curve generated Eq. (3.23) and (3.24) when the end points are $(x_0, y_0) = (0, 0)$ and $(x_1, y_1) = (1, 0)$, and respective guide points, as shown in Figure 3.18 are $(1, 1)$ and $(0, 1)$.

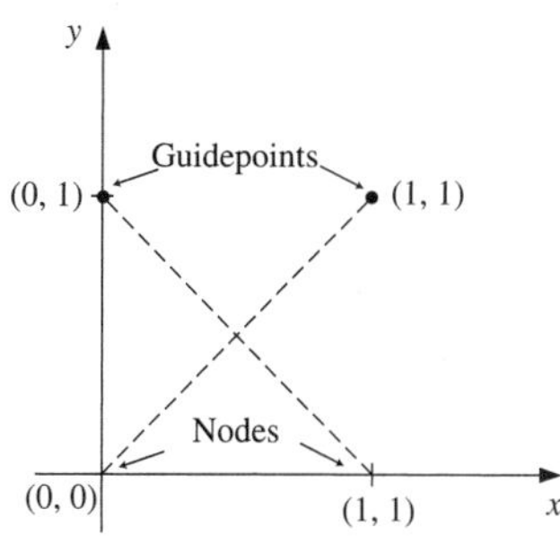

Figure 3.18

Solution The endpoint information implies that $x_0 = 0$, $x_1 = 1$, $y_0 = 0$, and $y_1 = 0$, and the guide points at $(1, 1)$ and $(0, 1)$ imply that $\alpha_0 = 1$, $\alpha_1 = 1$, $\beta_0 = 1$, and $\beta_1 = -1$. Note that the slopes of the guide lines at $(0, 0)$ and $(1, 0)$ are, respectively

$$\frac{\beta_0}{\alpha_0} = \frac{1}{1} = 1 \quad \text{and} \quad \frac{\beta_1}{\alpha_1} = \frac{-1}{1} = -1.$$

Equations (3.23) and (3.24) imply that for $t \in [0, 1]$ we have

$$x(t) = [2(0 - 1) + (1 + 1)]t^3 + [3(0 - 0) - (1 + 2 \cdot 1)]t^2 + 1 \cdot t + 0 = t$$

and

$$y(t) = [2(0 - 0) + (1 + (-1))]t^3 + [3(0 - 0) - (-1 + 2 \cdot 1)]t^2 + 1 \cdot t + 0 = -t^2 + t.$$

This graph is shown as (a) in Figure 3.19, together with some other possibilities of curves produced by Eqs. (3.23) and (3.24) when the nodes are $(0, 0)$ and $(1, 0)$ and the slopes at these nodes are 1 and -1, respectively. ■

Figure 3.19

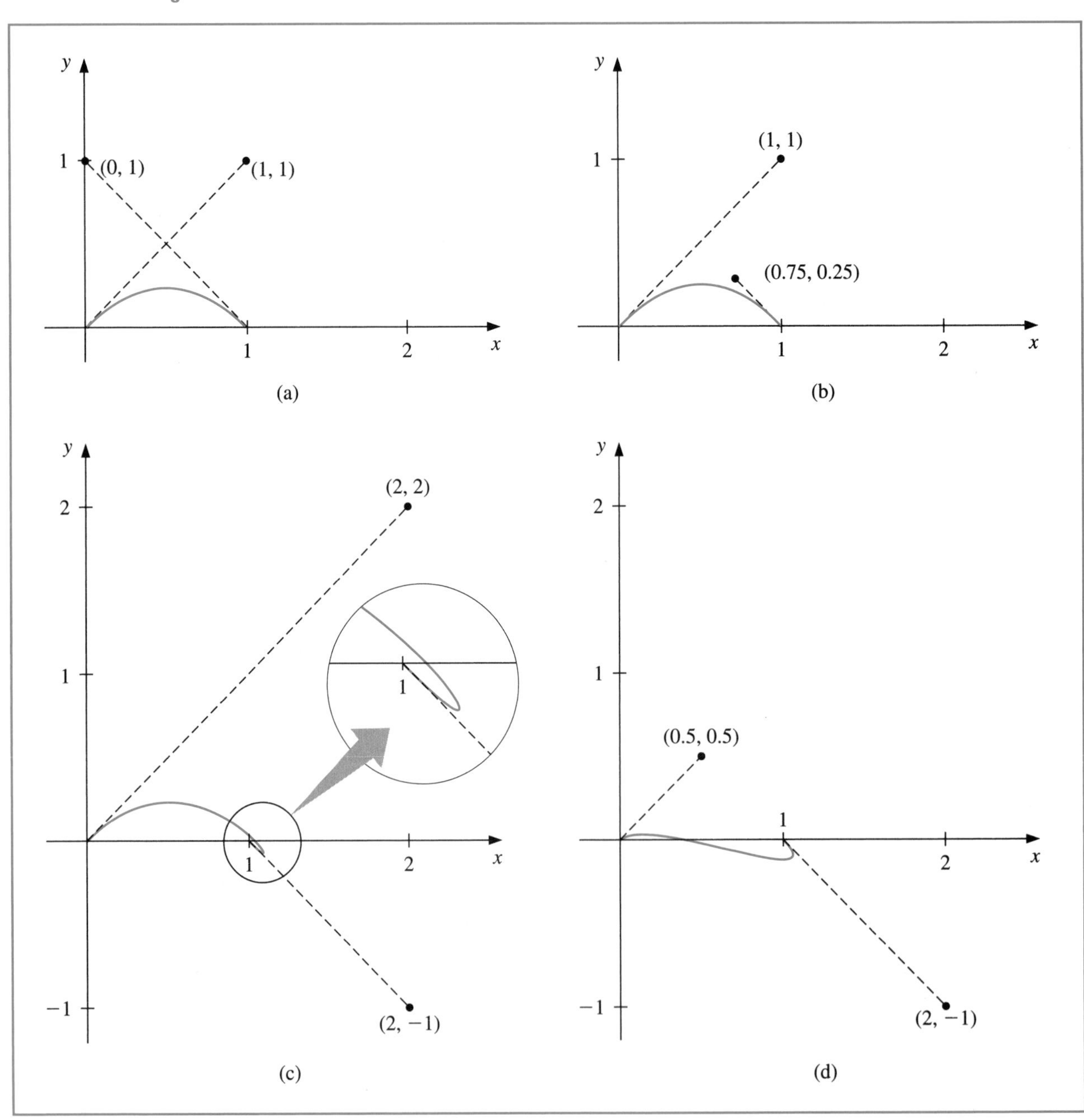

Pierre Etienne Bézier (1910–1999) was head of design and production for Renault motorcars for most of his professional life. He began his research into computer-aided design and manufacturing in 1960, developing interactive tools for curve and surface design, and initiated computer-generated milling for automobile modeling.

The Bézier curves that bear his name have the advantage of being based on a rigorous mathematical theory that does not need to be explicitly recognized by the practitioner who simply wants to make an aesthetically pleasing curve or surface. These are the curves that are the basis of the powerful Adobe Postscript system, and produce the freehand curves that are generated in most sufficiently powerful computer graphics packages.

The standard procedure for determining curves in an interactive graphics mode is to first use a mouse or touchpad to set the nodes and guidepoints to generate a first approximation to the curve. These can be set manually, but most graphics systems permit you to use your input device to draw the curve on the screen freehand and will select appropriate nodes and guidepoints for your freehand curve.

The nodes and guidepoints can then be manipulated into a position that produces an aesthetically pleasing curve. Since the computation is minimal, the curve can be determined so quickly that the resulting change is seen immediately. Moreover, all the data needed to compute the curves are imbedded in the coordinates of the nodes and guidepoints, so no analytical knowledge is required of the user.

Popular graphics programs use this type of system for their freehand graphic representations in a slightly modified form. The Hermite cubics are described as **Bézier polynomials**, which incorporate a scaling factor of 3 when computing the derivatives at the endpoints. This modifies the parametric equations to

$$x(t) = [2(x_0 - x_1) + 3(\alpha_0 + \alpha_1)]t^3 + [3(x_1 - x_0) - 3(\alpha_1 + 2\alpha_0)]t^2 + 3\alpha_0 t + x_0, \quad (3.25)$$

and

$$y(t) = [2(y_0 - y_1) + 3(\beta_0 + \beta_1)]t^3 + [3(y_1 - y_0) - 3(\beta_1 + 2\beta_0)]t^2 + 3\beta_0 t + y_0, \quad (3.26)$$

for $0 \le t \le 1$, but this change is transparent to the user of the system.

Algorithm 3.6 constructs a set of Bézier curves based on the parametric equations in Eqs. (3.25) and (3.26).

Bézier Curve

To construct the cubic Bézier curves $C_0, \ldots, C_{n-1}$ in parametric form, where C_i is represented by

$$(x_i(t), y_i(t)) = (a_0^{(i)} + a_1^{(i)}t + a_2^{(i)}t^2 + a_3^{(i)}t^3, b_0^{(i)} + b_1^{(i)}t + b_2^{(i)}t^2 + b_3^{(i)}t^3),$$

for $0 \le t \le 1$, as determined by the left endpoint (x_i, y_i), left guidepoint (x_i^+, y_i^+), right endpoint (x_{i+1}, y_{i+1}), and right guidepoint (x_{i+1}^-, y_{i+1}^-) for each $i = 0, 1, \ldots, n-1$:

INPUT n; $(x_0, y_0), \ldots, (x_n, y_n)$; $(x_0^+, y_0^+), \ldots, (x_{n-1}^+, y_{n-1}^+)$; $(x_1^-, y_1^-), \ldots, (x_n^-, y_n^-)$.

OUTPUT coefficients $\{a_0^{(i)}, a_1^{(i)}, a_2^{(i)}, a_3^{(i)}, b_0^{(i)}, b_1^{(i)}, b_2^{(i)}, b_3^{(i)}$, for $0 \le i \le n-1\}$.

Step 1 For each $i = 0, 1, \ldots, n-1$ do Steps 2 and 3.

Step 2 Set $a_0^{(i)} = x_i$;
$b_0^{(i)} = y_i$;
$a_1^{(i)} = 3(x_i^+ - x_i)$;
$b_1^{(i)} = 3(y_i^+ - y_i)$;
$a_2^{(i)} = 3(x_i + x_{i+1}^- - 2x_i^+)$;
$b_2^{(i)} = 3(y_i + y_{i+1}^- - 2y_i^+)$;
$a_3^{(i)} = x_{i+1} - x_i + 3x_i^+ - 3x_{i+1}^-$;
$b_3^{(i)} = y_{i+1} - y_i + 3y_i^+ - 3y_{i+1}^-$;

Step 3 OUTPUT $(a_0^{(i)}, a_1^{(i)}, a_2^{(i)}, a_3^{(i)}, b_0^{(i)}, b_1^{(i)}, b_2^{(i)}, b_3^{(i)})$.

Step 4 STOP. ■

Three-dimensional curves are generated in a similar manner by additionally specifying third components z_0 and z_1 for the nodes and $z_0+\gamma_0$ and $z_1-\gamma_1$ for the guidepoints. The more difficult problem involving the representation of three-dimensional curves concerns the loss of the third dimension when the curve is projected onto a two-dimensional computer screen. Various projection techniques are used, but this topic lies within the realm of computer graphics. For an introduction to this topic and ways that the technique can be modified for surface representations, see one of the many books on computer graphics methods, such as [FVFH].

EXERCISE SET 3.6

1. Let $(x_0, y_0) = (0, 0)$ and $(x_1, y_1) = (5, 2)$ be the endpoints of a curve. Use the given guidepoints to construct parametric cubic Hermite approximations $(x(t), y(t))$ to the curve, and graph the approximations.
   - **a.** $(1, 1)$ and $(6, 1)$
   - **b.** $(0.5, 0.5)$ and $(5.5, 1.5)$
   - **c.** $(1, 1)$ and $(6, 3)$
   - **d.** $(2, 2)$ and $(7, 0)$
2. Repeat Exercise 1 using cubic Bézier polynomials.
3. Construct and graph the cubic Bézier polynomials given the following points and guidepoints.
   - **a.** Point $(1, 1)$ with guidepoint $(1.5, 1.25)$ to point $(6, 2)$ with guidepoint $(7, 3)$
   - **b.** Point $(1, 1)$ with guidepoint $(1.25, 1.5)$ to point $(6, 2)$ with guidepoint $(5, 3)$
   - **c.** Point $(0, 0)$ with guidepoint $(0.5, 0.5)$ to point $(4, 6)$ with entering guidepoint $(3.5, 7)$ and exiting guidepoint $(4.5, 5)$ to point $(6, 1)$ with guidepoint $(7, 2)$
   - **d.** Point $(0, 0)$ with guidepoint $(0.5, 0.5)$ to point $(2, 1)$ with entering guidepoint $(3, 1)$ and exiting guidepoint $(3, 1)$ to point $(4, 0)$ with entering guidepoint $(5, 1)$ and exiting guidepoint $(3, -1)$ to point $(6, -1)$ with guidepoint $(6.5, -0.25)$
4. Use the data in the following table and Algorithm 3.6 to approximate the shape of the letter $\mathcal{N}$.

i	x_i	y_i	α_i	β_i	α'_i	β'_i
0	3	6	3.3	6.5		
1	2	2	2.8	3.0	2.5	2.5
2	6	6	5.8	5.0	5.0	5.8
3	5	2	5.5	2.2	4.5	2.5
4	6.5	3			6.4	2.8

5. Suppose a cubic Bézier polynomial is placed through (u_0, v_0) and (u_3, v_3) with guidepoints (u_1, v_1) and (u_2, v_2), respectively.
   - **a.** Derive the parametric equations for $u(t)$ and $v(t)$ assuming that

$$u(0) = u_0, \quad u(1) = u_3, \quad u'(0) = u_1 - u_0, \quad u'(1) = u_3 - u_2$$

and

$$v(0) = v_0, \quad v(1) = v_3, \quad v'(0) = v_1 - v_0, \quad v'(1) = v_3 - v_2.$$

   - **b.** Let $f(i/3) = u_i$, for $i = 0, 1, 2, 3$ and $g(i/3) = v_i$, for $i = 0, 1, 2, 3$. Show that the Bernstein polynomial of degree 3 in t for f is $u(t)$ and the Bernstein polynomial of degree three in t for g is $v(t)$. (See Exercise 23 of Section 3.1.)

3.7 Survey of Methods and Software

In this chapter we have considered approximating a function using polynomials and piecewise polynomials. The function can be specified by a given defining equation or by providing points in the plane through which the graph of the function passes. A set of nodes $x_0, x_1, \ldots, x_n$ is given in each case, and more information, such as the value of various derivatives, may also be required. We need to find an approximating function that satisfies the conditions specified by these data.

The interpolating polynomial $P(x)$ is the polynomial of least degree that satisfies, for a function f,

$$P(x_i) = f(x_i), \quad \text{for each } i = 0, 1, \ldots, n.$$

Although this interpolating polynomial is unique, it can take many different forms. The Lagrange form is most often used for interpolating tables when n is small and for deriving formulas for approximating derivatives and integrals. Neville's method is used for evaluating several interpolating polynomials at the same value of x. Newton's forms of the polynomial are more appropriate for computation and are also used extensively for deriving formulas for solving differential equations. However, polynomial interpolation has the inherent weaknesses of oscillation, particularly if the number of nodes is large. In this case there are other methods that can be better applied.

The Hermite polynomials interpolate a function and its derivative at the nodes. They can be very accurate but require more information about the function being approximated. When there are a large number of nodes, the Hermite polynomials also exhibit oscillation weaknesses.

The most commonly used form of interpolation is piecewise-polynomial interpolation. If function and derivative values are available, piecewise cubic Hermite interpolation is recommended. This is the preferred method for interpolating values of a function that is the solution to a differential equation. When only the function values are available, natural cubic spline interpolation can be used. This spline forces the second derivative of the spline to be zero at the endpoints. Other cubic splines require additional data. For example, the clamped cubic spline needs values of the derivative of the function at the endpoints of the interval.

Other methods of interpolation are commonly used. Trigonometric interpolation, in particular the Fast Fourier Transform discussed in Chapter 8, is used with large amounts of data when the function is assumed to have a periodic nature. Interpolation by rational functions is also used.

If the data are suspected to be inaccurate, smoothing techniques can be applied, and some form of least squares fit of data is recommended. Polynomials, trigonometric functions, rational functions, and splines can be used in least squares fitting of data. We consider these topics in Chapter 8.

Interpolation routines included in the IMSL Library are based on the book *A Practical Guide to Splines* by Carl de Boor [Deb] and use interpolation by cubic splines. There are cubic splines to minimize oscillations and to preserve concavity. Methods for two-dimensional interpolation by bicubic splines are also included.

The NAG library contains subroutines for polynomial and Hermite interpolation, for cubic spline interpolation, and for piecewise cubic Hermite interpolation. NAG also contains subroutines for interpolating functions of two variables.

The netlib library contains the subroutines to compute the cubic spline with various endpoint conditions. One package produces the Newton's divided difference coefficients for a discrete set of data points, and there are various routines for evaluating Hermite piecewise polynomials.

MATLAB can be used to interpolate a discrete set of data points, using either nearest neighbor interpolation, linear interpolation, cubic spline interpolation, or cubic interpolation. Cubic splines can also be produced.

General references to the methods in this chapter are the books by Powell [Pow] and by Davis [Da]. The seminal paper on splines is due to Schoenberg [Scho]. Important books on splines are by Schultz [Schul], De Boor [Deb2], Dierckx [Di], and Schumaker [Schum].

CHAPTER

4 Numerical Differentiation and Integration

Introduction

A sheet of corrugated roofing is constructed by pressing a flat sheet of aluminum into one whose cross section has the form of a sine wave.

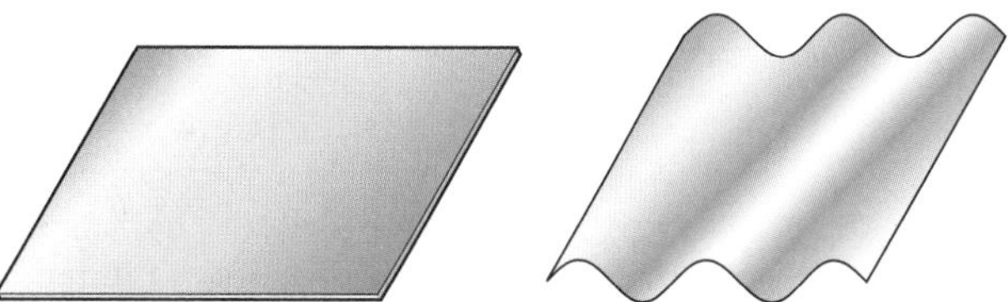

A corrugated sheet 4 ft long is needed, the height of each wave is 1 in. from the center line, and each wave has a period of approximately 2π in. The problem of finding the length of the initial flat sheet is one of determining the length of the curve given by $f(x) = \sin x$ from $x = 0$ in. to $x = 48$ in. From calculus we know that this length is

$$L = \int_0^{48} \sqrt{1 + (f'(x))^2}\, dx = \int_0^{48} \sqrt{1 + (\cos x)^2}\, dx,$$

so the problem reduces to evaluating this integral. Although the sine function is one of the most common mathematical functions, the calculation of its length involves an elliptic integral of the second kind, which cannot be evaluated explicitly. Methods are developed in this chapter to approximate the solution to problems of this type. This particular problem is considered in Exercise 25 of Section 4.4 and Exercise 12 of Section 4.5.

We mentioned in the introduction to Chapter 3 that one reason for using algebraic polynomials to approximate an arbitrary set of data is that, given any continuous function defined on a closed interval, there exists a polynomial that is arbitrarily close to the function at every point in the interval. Also, the derivatives and integrals of polynomials are easily obtained and evaluated. It should not be surprising, then, that many procedures for approximating derivatives and integrals use the polynomials that approximate the function.

4.1 Numerical Differentiation

The derivative of the function f at x_0 is

$$f'(x_0) = \lim_{h\to 0} \frac{f(x_0+h) - f(x_0)}{h}.$$

This formula gives an obvious way to generate an approximation to $f'(x_0)$; simply compute

$$\frac{f(x_0+h) - f(x_0)}{h}$$

for small values of h. Although this may be obvious, it is not very successful, due to our old nemesis round-off error. But it is certainly a place to start.

To approximate $f'(x_0)$, suppose first that $x_0 \in (a, b)$, where $f \in C^2[a, b]$, and that $x_1 = x_0 + h$ for some $h \neq 0$ that is sufficiently small to ensure that $x_1 \in [a, b]$. We construct the first Lagrange polynomial $P_{0,1}(x)$ for f determined by x_0 and x_1, with its error term:

$$\begin{aligned} f(x) &= P_{0,1}(x) + \frac{(x-x_0)(x-x_1)}{2!} f''(\xi(x)) \\ &= \frac{f(x_0)(x-x_0-h)}{-h} + \frac{f(x_0+h)(x-x_0)}{h} + \frac{(x-x_0)(x-x_0-h)}{2} f''(\xi(x)), \end{aligned}$$

for some $\xi(x)$ between x_0 and x_1. Differentiating gives

$$\begin{aligned} f'(x) &= \frac{f(x_0+h) - f(x_0)}{h} + D_x\left[\frac{(x-x_0)(x-x_0-h)}{2} f''(\xi(x))\right] \\ &= \frac{f(x_0+h) - f(x_0)}{h} + \frac{2(x-x_0) - h}{2} f''(\xi(x)) \\ &\quad + \frac{(x-x_0)(x-x_0-h)}{2} D_x(f''(\xi(x))). \end{aligned}$$

Deleting the terms involving $\xi(x)$ gives

$$f'(x) \approx \frac{f(x_0+h) - f(x_0)}{h}.$$

Difference equations were used and popularized by Isaac Newton in the last quarter of the 17th century, but many of these techniques had previously been developed by Thomas Harriot (1561–1621) and Henry Briggs (1561–1630). Harriot made significant advances in navigation techniques, and Briggs was the person most responsible for the acceptance of logarithms as an aid to computation.

One difficulty with this formula is that we have no information about $D_x f''(\xi(x))$, so the truncation error cannot be estimated. When x is x_0, however, the coefficient of $D_x f''(\xi(x))$ is 0, and the formula simplifies to

$$f'(x_0) = \frac{f(x_0+h) - f(x_0)}{h} - \frac{h}{2} f''(\xi). \tag{4.1}$$

For small values of h, the difference quotient $[f(x_0+h) - f(x_0)]/h$ can be used to approximate $f'(x_0)$ with an error bounded by $M|h|/2$, where M is a bound on $|f''(x)|$ for x between x_0 and x_0+h. This formula is known as the **forward-difference formula** if $h > 0$ (see Figure 4.1) and the **backward-difference formula** if $h < 0$.

Example 1 Use the forward-difference formula to approximate the derivative of $f(x) = \ln x$ at $x_0 = 1.8$ using $h = 0.1$, $h = 0.05$, and $h = 0.01$, and determine bounds for the approximation errors.

Solution The forward-difference formula

$$\frac{f(1.8+h) - f(1.8)}{h}$$

Figure 4.1

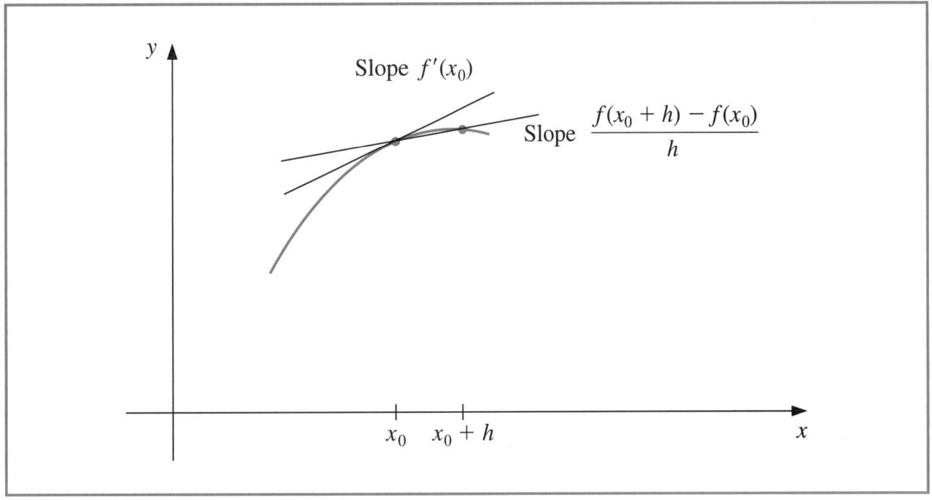

with $h = 0.1$ gives

$$\frac{\ln 1.9 - \ln 1.8}{0.1} = \frac{0.64185389 - 0.58778667}{0.1} = 0.5406722.$$

Because $f''(x) = -1/x^2$ and $1.8 < \xi < 1.9$, a bound for this approximation error is

$$\frac{|hf''(\xi)|}{2} = \frac{|h|}{2\xi^2} < \frac{0.1}{2(1.8)^2} = 0.0154321.$$

The approximation and error bounds when $h = 0.05$ and $h = 0.01$ are found in a similar manner and the results are shown in Table 4.1.

Table 4.1

h	$f(1.8+h)$	$\frac{f(1.8+h)-f(1.8)}{h}$	$\frac{\|h\|}{2(1.8)^2}$
0.1	0.64185389	0.5406722	0.0154321
0.05	0.61518564	0.5479795	0.0077160
0.01	0.59332685	0.5540180	0.0015432

Since $f'(x) = 1/x$, the exact value of $f'(1.8)$ is $0.55\overline{5}$, and in this case the error bounds are quite close to the true approximation error. ■

To obtain general derivative approximation formulas, suppose that $\{x_0, x_1, \ldots, x_n\}$ are $(n+1)$ distinct numbers in some interval I and that $f \in C^{n+1}(I)$. From Theorem 3.3 on page 112,

$$f(x) = \sum_{k=0}^{n} f(x_k)L_k(x) + \frac{(x-x_0)\cdots(x-x_n)}{(n+1)!}f^{(n+1)}(\xi(x)),$$

for some $\xi(x)$ in I, where $L_k(x)$ denotes the kth Lagrange coefficient polynomial for f at $x_0, x_1, \ldots, x_n$. Differentiating this expression gives

$$f'(x) = \sum_{k=0}^{n} f(x_k)L_k'(x) + D_x\left[\frac{(x-x_0)\cdots(x-x_n)}{(n+1!)}\right] f^{(n+1)}(\xi(x))$$
$$+ \frac{(x-x_0)\cdots(x-x_n)}{(n+1)!} D_x[f^{(n+1)}(\xi(x))].$$

We again have a problem estimating the truncation error unless x is one of the numbers x_j. In this case, the term multiplying $D_x[f^{(n+1)}(\xi(x))]$ is 0, and the formula becomes

$$f'(x_j) = \sum_{k=0}^{n} f(x_k)L_k'(x_j) + \frac{f^{(n+1)}(\xi(x_j))}{(n+1)!} \prod_{\substack{k=0 \\ k\neq j}}^{n} (x_j - x_k), \tag{4.2}$$

which is called an **$(n + 1)$-point formula** to approximate $f'(x_j)$.

In general, using more evaluation points in Eq. (4.2) produces greater accuracy, although the number of functional evaluations and growth of round-off error discourages this somewhat. The most common formulas are those involving three and five evaluation points.

We first derive some useful three-point formulas and consider aspects of their errors. Because

$$L_0(x) = \frac{(x-x_1)(x-x_2)}{(x_0-x_1)(x_0-x_2)}, \quad \text{we have} \quad L_0'(x) = \frac{2x - x_1 - x_2}{(x_0-x_1)(x_0-x_2)}.$$

Similarly,

$$L_1'(x) = \frac{2x - x_0 - x_2}{(x_1-x_0)(x_1-x_2)} \quad \text{and} \quad L_2'(x) = \frac{2x - x_0 - x_1}{(x_2-x_0)(x_2-x_1)}.$$

Hence, from Eq. (4.2),

$$f'(x_j) = f(x_0)\left[\frac{2x_j - x_1 - x_2}{(x_0-x_1)(x_0-x_2)}\right] + f(x_1)\left[\frac{2x_j - x_0 - x_2}{(x_1-x_0)(x_1-x_2)}\right]$$
$$+ f(x_2)\left[\frac{2x_j - x_0 - x_1}{(x_2-x_0)(x_2-x_1)}\right] + \frac{1}{6} f^{(3)}(\xi_j) \prod_{\substack{k=0 \\ k\neq j}}^{2} (x_j - x_k), \tag{4.3}$$

for each $j = 0, 1, 2$, where the notation ξ_j indicates that this point depends on x_j.

Three-Point Formulas

The formulas from Eq. (4.3) become especially useful if the nodes are equally spaced, that is, when

$$x_1 = x_0 + h \quad \text{and} \quad x_2 = x_0 + 2h, \quad \text{for some } h \neq 0.$$

We will assume equally-spaced nodes throughout the remainder of this section.

Using Eq. (4.3) with $x_j = x_0, x_1 = x_0 + h$, and $x_2 = x_0 + 2h$ gives

$$f'(x_0) = \frac{1}{h}\left[-\frac{3}{2}f(x_0) + 2f(x_1) - \frac{1}{2}f(x_2)\right] + \frac{h^2}{3} f^{(3)}(\xi_0).$$

Doing the same for $x_j = x_1$ gives

$$f'(x_1) = \frac{1}{h}\left[-\frac{1}{2}f(x_0) + \frac{1}{2}f(x_2)\right] - \frac{h^2}{6} f^{(3)}(\xi_1),$$

and for $x_j = x_2$,

$$f'(x_2) = \frac{1}{h}\left[\frac{1}{2}f(x_0) - 2f(x_1) + \frac{3}{2}f(x_2)\right] + \frac{h^2}{3}f^{(3)}(\xi_2).$$

Since $x_1 = x_0 + h$ and $x_2 = x_0 + 2h$, these formulas can also be expressed as

$$f'(x_0) = \frac{1}{h}\left[-\frac{3}{2}f(x_0) + 2f(x_0+h) - \frac{1}{2}f(x_0+2h)\right] + \frac{h^2}{3}f^{(3)}(\xi_0),$$

$$f'(x_0+h) = \frac{1}{h}\left[-\frac{1}{2}f(x_0) + \frac{1}{2}f(x_0+2h)\right] - \frac{h^2}{6}f^{(3)}(\xi_1),$$

and

$$f'(x_0+2h) = \frac{1}{h}\left[\frac{1}{2}f(x_0) - 2f(x_0+h) + \frac{3}{2}f(x_0+2h)\right] + \frac{h^2}{3}f^{(3)}(\xi_2).$$

As a matter of convenience, the variable substitution x_0 for $x_0 + h$ is used in the middle equation to change this formula to an approximation for $f'(x_0)$. A similar change, x_0 for $x_0 + 2h$, is used in the last equation. This gives three formulas for approximating $f'(x_0)$:

$$f'(x_0) = \frac{1}{2h}[-3f(x_0) + 4f(x_0+h) - f(x_0+2h)] + \frac{h^2}{3}f^{(3)}(\xi_0),$$

$$f'(x_0) = \frac{1}{2h}[-f(x_0-h) + f(x_0+h)] - \frac{h^2}{6}f^{(3)}(\xi_1),$$

and

$$f'(x_0) = \frac{1}{2h}[f(x_0-2h) - 4f(x_0-h) + 3f(x_0)] + \frac{h^2}{3}f^{(3)}(\xi_2).$$

Finally, note that the last of these equations can be obtained from the first by simply replacing h with $-h$, so there are actually only two formulas:

Three-Point Endpoint Formula

- $$f'(x_0) = \frac{1}{2h}[-3f(x_0) + 4f(x_0+h) - f(x_0+2h)] + \frac{h^2}{3}f^{(3)}(\xi_0), \tag{4.4}$$

where ξ_0 lies between x_0 and $x_0 + 2h$.

Three-Point Midpoint Formula

- $$f'(x_0) = \frac{1}{2h}[f(x_0+h) - f(x_0-h)] - \frac{h^2}{6}f^{(3)}(\xi_1), \tag{4.5}$$

where ξ_1 lies between $x_0 - h$ and $x_0 + h$.

Although the errors in both Eq. (4.4) and Eq. (4.5) are $O(h^2)$, the error in Eq. (4.5) is approximately half the error in Eq. (4.4). This is because Eq. (4.5) uses data on both sides of x_0 and Eq. (4.4) uses data on only one side. Note also that f needs to be evaluated at only two points in Eq. (4.5), whereas in Eq. (4.4) three evaluations are needed. Figure 4.2 on page 178 gives an illustration of the approximation produced from Eq. (4.5). The approximation in Eq. (4.4) is useful near the ends of an interval, because information about f outside the interval may not be available.

Figure 4.2

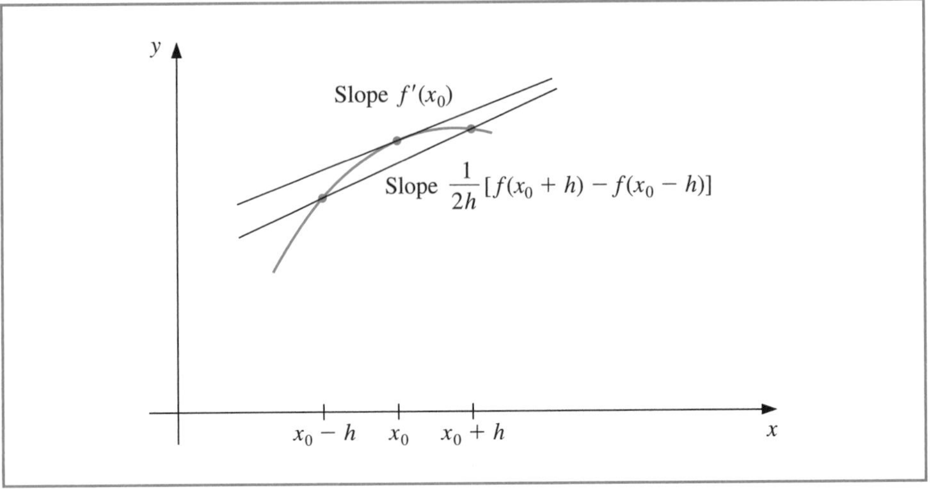

Five-Point Formulas

The methods presented in Eqs. (4.4) and (4.5) are called **three-point formulas** (even though the third point $f(x_0)$ does not appear in Eq. (4.5)). Similarly, there are **five-point formulas** that involve evaluating the function at two additional points. The error term for these formulas is $O(h^4)$. One common five-point formula is used to determine approximations for the derivative at the midpoint.

Five-Point Midpoint Formula

- $$f'(x_0) = \frac{1}{12h}[f(x_0 - 2h) - 8f(x_0 - h) + 8f(x_0 + h) - f(x_0 + 2h)] + \frac{h^4}{30}f^{(5)}(\xi), \tag{4.6}$$

  where ξ lies between $x_0 - 2h$ and $x_0 + 2h$.

The derivation of this formula is considered in Section 4.2. The other five-point formula is used for approximations at the endpoints.

Five-Point Endpoint Formula

- $$\begin{aligned} f'(x_0) = \frac{1}{12h}[&-25f(x_0) + 48f(x_0 + h) - 36f(x_0 + 2h) \\ &+ 16f(x_0 + 3h) - 3f(x_0 + 4h)] + \frac{h^4}{5}f^{(5)}(\xi), \end{aligned} \tag{4.7}$$

  where ξ lies between x_0 and $x_0 + 4h$.

Left-endpoint approximations are found using this formula with $h > 0$ and right-endpoint approximations with $h < 0$. The five-point endpoint formula is particularly useful for the clamped cubic spline interpolation of Section 3.5.

Example 2 Values for $f(x) = xe^x$ are given in Table 4.2. Use all the applicable three-point and five-point formulas to approximate $f'(2.0)$.

Table 4.2

x	$f(x)$
1.8	10.889365
1.9	12.703199
2.0	14.778112
2.1	17.148957
2.2	19.855030

Solution The data in the table permit us to find four different three-point approximations. We can use the endpoint formula (4.4) with $h = 0.1$ or with $h = -0.1$, and we can use the midpoint formula (4.5) with $h = 0.1$ or with $h = 0.2$.

Using the endpoint formula (4.4) with $h = 0.1$ gives

$$\frac{1}{0.2}[-3f(2.0) + 4f(2.1) - f(2.2] = 5[-3(14.778112) + 4(17.148957) \\ - 19.855030)] = 22.032310,$$

and with $h = -0.1$ gives 22.054525.

Using the midpoint formula (4.5) with $h = 0.1$ gives

$$\frac{1}{0.2}[f(2.1) - f(1.9)] = 5(17.148957 - 12.7703199) = 22.228790,$$

and with $h = 0.2$ gives 22.414163.

The only five-point formula for which the table gives sufficient data is the midpoint formula (4.6) with $h = 0.1$. This gives

$$\begin{aligned}\frac{1}{1.2}[f(1.8) - 8f(1.9) + 8f(2.1) - f(2.2)] &= \frac{1}{1.2}[10.889365 - 8(12.703199) \\ &\quad + 8(17.148957) - 19.855030] \\ &= 22.166999\end{aligned}$$

If we had no other information we would accept the five-point midpoint approximation using $h = 0.1$ as the most accurate, and expect the true value to be between that approximation and the three-point mid-point approximation that is in the interval [22.166, 22.229].

The true value in this case is $f'(2.0) = (2 + 1)e^2 = 22.167168$, so the approximation errors are actually:

Three-point endpoint with $h = 0.1$: 1.35×10^{-1};

Three-point endpoint with $h = -0.1$: 1.13×10^{-1};

Three-point midpoint with $h = 0.1$: -6.16×10^{-2};

Three-point midpoint with $h = 0.2$: -2.47×10^{-1};

Five-point midpoint with $h = 0.1$: 1.69×10^{-4}. ■

Methods can also be derived to find approximations to higher derivatives of a function using only tabulated values of the function at various points. The derivation is algebraically tedious, however, so only a representative procedure will be presented.

Expand a function f in a third Taylor polynomial about a point x_0 and evaluate at $x_0 + h$ and $x_0 - h$. Then

$$f(x_0 + h) = f(x_0) + f'(x_0)h + \frac{1}{2}f''(x_0)h^2 + \frac{1}{6}f'''(x_0)h^3 + \frac{1}{24}f^{(4)}(\xi_1)h^4$$

and

$$f(x_0 - h) = f(x_0) - f'(x_0)h + \frac{1}{2}f''(x_0)h^2 - \frac{1}{6}f'''(x_0)h^3 + \frac{1}{24}f^{(4)}(\xi_{-1})h^4,$$

where $x_0 - h < \xi_{-1} < x_0 < \xi_1 < x_0 + h$.

If we add these equations, the terms involving $f'(x_0)$ and $-f'(x_0)$ cancel, so

$$f(x_0 + h) + f(x_0 - h) = 2f(x_0) + f''(x_0)h^2 + \frac{1}{24}[f^{(4)}(\xi_1) + f^{(4)}(\xi_{-1})]h^4.$$

Solving this equation for $f''(x_0)$ gives

$$f''(x_0) = \frac{1}{h^2}[f(x_0 - h) - 2f(x_0) + f(x_0 + h)] - \frac{h^2}{24}[f^{(4)}(\xi_1) + f^{(4)}(\xi_{-1})]. \quad (4.8)$$

Suppose $f^{(4)}$ is continuous on $[x_0 - h, x_0 + h]$. Since $\frac{1}{2}[f^{(4)}(\xi_1) + f^{(4)}(\xi_{-1})]$ is between $f^{(4)}(\xi_1)$ and $f^{(4)}(\xi_{-1})$, the Intermediate Value Theorem implies that a number ξ exists between ξ_1 and ξ_{-1}, and hence in $(x_0 - h, x_0 + h)$, with

$$f^{(4)}(\xi) = \frac{1}{2}\left[f^{(4)}(\xi_1) + f^{(4)}(\xi_{-1})\right].$$

This permits us to rewrite Eq. (4.8) in its final form.

Second Derivative Midpoint Formula

- $$f''(x_0) = \frac{1}{h^2}[f(x_0 - h) - 2f(x_0) + f(x_0 + h)] - \frac{h^2}{12}f^{(4)}(\xi), \quad (4.9)$$

  for some ξ, where $x_0 - h < \xi < x_0 + h$.

If $f^{(4)}$ is continuous on $[x_0 - h, x_0 + h]$ it is also bounded, and the approximation is $O(h^2)$.

Example 3 In Example 2 we used the data shown in Table 4.3 to approximate the first derivative of $f(x) = xe^x$ at $x = 2.0$. Use the second derivative formula (4.9) to approximate $f''(2.0)$.

Table 4.3

x	$f(x)$
1.8	10.889365
1.9	12.703199
2.0	14.778112
2.1	17.148957
2.2	19.855030

Solution The data permits us to determine two approximations for $f''(2.0)$. Using (4.9) with $h = 0.1$ gives

$$\frac{1}{0.01}[f(1.9) - 2f(2.0) + f(2.1)] = 100[12.703199 - 2(14.778112) + 17.148957]$$
$$= 29.593200,$$

and using (4.9) with $h = 0.2$ gives

$$\frac{1}{0.04}[f(1.8) - 2f(2.0) + f(2.2)] = 25[10.889365 - 2(14.778112) + 19.855030]$$
$$= 29.704275.$$

Because $f''(x) = (x + 2)e^x$, the exact value is $f''(2.0) = 29.556224$. Hence the actual errors are -3.70×10^{-2} and -1.48×10^{-1}, respectively. ■

Round-Off Error Instability

It is particularly important to pay attention to round-off error when approximating derivatives. To illustrate the situation, let us examine the three-point midpoint formula Eq. (4.5),

$$f'(x_0) = \frac{1}{2h}[f(x_0 + h) - f(x_0 - h)] - \frac{h^2}{6}f^{(3)}(\xi_1),$$

more closely. Suppose that in evaluating $f(x_0 + h)$ and $f(x_0 - h)$ we encounter round-off errors $e(x_0 + h)$ and $e(x_0 - h)$. Then our computations actually use the values $\tilde{f}(x_0 + h)$ and $\tilde{f}(x_0 - h)$, which are related to the true values $f(x_0 + h)$ and $f(x_0 - h)$ by

$$f(x_0 + h) = \tilde{f}(x_0 + h) + e(x_0 + h) \quad \text{and} \quad f(x_0 - h) = \tilde{f}(x_0 - h) + e(x_0 - h).$$

The total error in the approximation,

$$f'(x_0) - \frac{\tilde{f}(x_0 + h) - \tilde{f}(x_0 - h)}{2h} = \frac{e(x_0 + h) - e(x_0 - h)}{2h} - \frac{h^2}{6} f^{(3)}(\xi_1),$$

is due both to round-off error, the first part, and to truncation error. If we assume that the round-off errors $e(x_0 \pm h)$ are bounded by some number $\varepsilon > 0$ and that the third derivative of f is bounded by a number $M > 0$, then

$$\left| f'(x_0) - \frac{\tilde{f}(x_0 + h) - \tilde{f}(x_0 - h)}{2h} \right| \le \frac{\varepsilon}{h} + \frac{h^2}{6} M.$$

To reduce the truncation error, $h^2M/6$, we need to reduce h. But as h is reduced, the round-off error ε/h grows. In practice, then, it is seldom advantageous to let h be too small, because in that case the round-off error will dominate the calculations.

Illustration Consider using the values in Table 4.4 to approximate $f'(0.900)$, where $f(x) = \sin x$. The true value is $\cos 0.900 = 0.62161$. The formula

$$f'(0.900) \approx \frac{f(0.900 + h) - f(0.900 - h)}{2h},$$

with different values of h, gives the approximations in Table 4.5.

Table 4.4

x	$\sin x$	x	$\sin x$
0.800	0.71736	0.901	0.78395
0.850	0.75128	0.902	0.78457
0.880	0.77074	0.905	0.78643
0.890	0.77707	0.910	0.78950
0.895	0.78021	0.920	0.79560
0.898	0.78208	0.950	0.81342
0.899	0.78270	1.000	0.84147

Table 4.5

h	Approximation to $f'(0.900)$	Error
0.001	0.62500	0.00339
0.002	0.62250	0.00089
0.005	0.62200	0.00039
0.010	0.62150	−0.00011
0.020	0.62150	−0.00011
0.050	0.62140	−0.00021
0.100	0.62055	−0.00106

The optimal choice for h appears to lie between 0.005 and 0.05. We can use calculus to verify (see Exercise 29) that a minimum for

$$e(h) = \frac{\varepsilon}{h} + \frac{h^2}{6} M,$$

occurs at $h = \sqrt[3]{3\varepsilon/M}$, where

$$M = \max_{x \in [0.800, 1.00]} |f'''(x)| = \max_{x \in [0.800, 1.00]} |\cos x| = \cos 0.8 \approx 0.69671.$$

Because values of f are given to five decimal places, we will assume that the round-off error is bounded by $\varepsilon = 5 \times 10^{-6}$. Therefore, the optimal choice of h is approximately

$$h = \sqrt[3]{\frac{3(0.000005)}{0.69671}} \approx 0.028,$$

which is consistent with the results in Table 4.6. □

In practice, we cannot compute an optimal h to use in approximating the derivative, since we have no knowledge of the third derivative of the function. But we must remain aware that reducing the step size will not always improve the approximation. □

We have considered only the round-off error problems that are presented by the three-point formula Eq. (4.5), but similar difficulties occur with all the differentiation formulas. The reason can be traced to the need to divide by a power of h. As we found in Section 1.2 (see, in particular, Example 3), division by small numbers tends to exaggerate round-off error, and this operation should be avoided if possible. In the case of numerical differentiation, we cannot avoid the problem entirely, although the higher-order methods reduce the difficulty.

Keep in mind that difference method approximations might be unstable.

As approximation methods, numerical differentiation is *unstable*, since the small values of h needed to reduce truncation error also cause the round-off error to grow. This is the first class of unstable methods we have encountered, and these techniques would be avoided if it were possible. However, in addition to being used for computational purposes, the formulas are needed for approximating the solutions of ordinary and partial-differential equations.

EXERCISE SET 4.1

1. Use the forward-difference formulas and backward-difference formulas to determine each missing entry in the following tables.

a.

x	$f(x)$	$f'(x)$
0.0	0.00000	
0.2	0.74140	
0.4	1.3718	

b.

x	$f(x)$	$f'(x)$
0.5	0.4794	
0.6	0.5646	
0.7	0.6442	

2. Use the forward-difference formulas and backward-difference formulas to determine each missing entry in the following tables.

a.

x	$f(x)$	$f'(x)$
−0.3	1.9507	
−0.2	2.0421	
−0.1	2.0601	

b.

x	$f(x)$	$f'(x)$
1.0	1.0000	
1.2	1.2625	
1.4	1.6595	

3. The data in Exercise 1 were taken from the following functions. Compute the actual errors in Exercise 1, and find error bounds using the error formulas.

a. $f(x) = e^x - 2x^2 + 3x - 1$ **b.** $f(x) = \sin x$

4. The data in Exercise 2 were taken from the following functions. Compute the actual errors in Exercise 2, and find error bounds using the error formulas.

a. $f(x) = 2\cos 2x - x$ **b.** $f(x) = x^2 \ln x + 1$

5. Use the most accurate three-point formula to determine each missing entry in the following tables.

a.

x	$f(x)$	$f'(x)$
8.1	16.94410	
8.3	17.56492	
8.5	18.19056	
8.7	18.82091	

b.

x	$f(x)$	$f'(x)$
1.1	9.025013	
1.2	11.02318	
1.3	13.46374	
1.4	16.44465	

c.

x	$f(x)$	$f'(x)$
2.0	3.6887983	
2.1	3.6905701	
2.2	3.6688192	
2.3	3.6245909	

d.

x	$f(x)$	$f'(x)$
2.9	−4.827866	
3.0	−4.240058	
3.1	−3.496909	
3.2	−2.596792	

6. Use the most accurate three-point formula to determine each missing entry in the following tables.

a.

x	$f(x)$	$f'(x)$
−0.3	−0.27652	
−0.2	−0.25074	
−0.1	−0.16134	
0	0	

b.

x	$f(x)$	$f'(x)$
7.4	−68.3193	
7.6	−71.6982	
7.8	−75.1576	
8.0	−78.6974	

c.

x	$f(x)$	$f'(x)$
1.1	1.52918	
1.2	1.64024	
1.3	1.70470	
1.4	1.71277	

d.

x	$f(x)$	$f'(x)$
−2.7	0.054797	
−2.5	0.11342	
−2.3	0.65536	
−2.1	0.98472	

7. The data in Exercise 5 were taken from the following functions. Compute the actual errors in Exercise 5, and find error bounds using the error formulas.

a. $f(x) = x \ln x$ **b.** $f(x) = e^{2x}$

c. $f(x) = 2(\ln x)^2 + 3 \sin x$ **d.** $f(x) = x \cos x - x^2 \sin x$

8. The data in Exercise 6 were taken from the following functions. Compute the actual errors in Exercise 6, and find error bounds using the error formulas.

a. $f(x) = e^{2x} - \cos 2x$ **b.** $f(x) = \ln(x+2) - (x+1)^2$

c. $f(x) = x \sin x + x^2 \cos x$ **d.** $f(x) = (\cos 3x)^2 - e^{2x}$

9. Use the formulas given in this section to determine, as accurately as possible, approximations for each missing entry in the following tables.

a.

x	$f(x)$	$f'(x)$
2.1	−1.709847	
2.2	−1.373823	
2.3	−1.119214	
2.4	−0.9160143	
2.5	−0.7470223	
2.6	−0.6015966	

b.

x	$f(x)$	$f'(x)$
−3.0	9.367879	
−2.8	8.233241	
−2.6	7.180350	
−2.4	6.209329	
−2.2	5.320305	
−2.0	4.513417	

10. Use the formulas given in this section to determine, as accurately as possible, approximations for each missing entry in the following tables.

a.

x	$f(x)$	$f'(x)$
−3.0	16.08554	
−2.8	12.64465	
−2.6	9.863738	
−2.4	7.623176	
−2.2	5.825013	
−2.0	4.389056	

b.

x	$f(x)$	$f'(x)$
1.05	−1.709847	
1.10	−1.373823	
1.15	−1.119214	
1.20	−0.9160143	
1.25	−0.7470223	
1.30	−0.6015966	

11. The data in Exercise 9 were taken from the following functions. Compute the actual errors in Exercise 9, and find error bounds using the error formulas and Maple.

a. $f(x) = e^{x/3} + x^2$ **b.** $f(x) = \tan x$

12. The data in Exercise 10 were taken from the following functions. Compute the actual errors in Exercise 10, and find error bounds using the error formulas and Maple.

a. $f(x) = e^{-x} - 1 + x$ **b.** $f(x) = \tan 2x$

13. Use the following data and the knowledge that the first five derivatives of f are bounded on $[1, 5]$ by 2, 3, 6, 12 and 23, respectively, to approximate $f'(3)$ as accurately as possible. Find a bound for the error.

x	1	2	3	4	5
$f(x)$	2.4142	2.6734	2.8974	3.0976	3.2804

14. Repeat Exercise 13, assuming instead that the third derivative of f is bounded on $[1, 5]$ by 4.

15. Repeat Exercise 1 using four-digit rounding arithmetic, and compare the errors to those in Exercise 3.

16. Repeat Exercise 5 using four-digit chopping arithmetic, and compare the errors to those in Exercise 7.

17. Repeat Exercise 9 using four-digit rounding arithmetic, and compare the errors to those in Exercise 11.

18. Consider the following table of data:

x	0.2	0.4	0.6	0.8	1.0
$f(x)$	0.9798652	0.9177710	0.808038	0.6386093	0.3843735

a. Use all the appropriate formulas given in this section to approximate $f'(0.4)$ and $f''(0.4)$.

b. Use all the appropriate formulas given in this section to approximate $f'(0.6)$ and $f''(0.6)$.

19. Let $f(x) = \cos \pi x$. Use Eq. (4.9) and the values of $f(x)$ at $x = 0.25$, 0.5, and 0.75 to approximate $f''(0.5)$. Compare this result to the exact value and to the approximation found in Exercise 15 of Section 3.5. Explain why this method is particularly accurate for this problem, and find a bound for the error.

20. Let $f(x) = 3xe^x - \cos x$. Use the following data and Eq. (4.9) to approximate $f''(1.3)$ with $h = 0.1$ and with $h = 0.01$.

x	1.20	1.29	1.30	1.31	1.40
$f(x)$	11.59006	13.78176	14.04276	14.30741	16.86187

Compare your results to $f''(1.3)$.

21. Consider the following table of data:

x	0.2	0.4	0.6	0.8	1.0
$f(x)$	0.9798652	0.9177710	0.8080348	0.6386093	0.3843735

a. Use Eq. (4.7) to approximate $f'(0.2)$.

b. Use Eq. (4.7) to approximate $f'(1.0)$.

c. Use Eq. (4.6) to approximate $f'(0.6)$.

22. Derive an $O(h^4)$ five-point formula to approximate $f'(x_0)$ that uses $f(x_0 - h)$, $f(x_0)$, $f(x_0 + h)$, $f(x_0 + 2h)$, and $f(x_0 + 3h)$. [*Hint:* Consider the expression $Af(x_0 - h) + Bf(x_0 + h) + Cf(x_0 + 2h) + Df(x_0 + 3h)$. Expand in fourth Taylor polynomials, and choose A, B, C, and D appropriately.]

23. Use the formula derived in Exercise 22 and the data of Exercise 21 to approximate $f'(0.4)$ and $f'(0.8)$.

24. **a.** Analyze the round-off errors, as in Example 4, for the formula

$$f'(x_0) = \frac{f(x_0 + h) - f(x_0)}{h} - \frac{h}{2} f''(\xi_0).$$

b. Find an optimal $h > 0$ for the function given in Example 2.

25. In Exercise 10 of Section 3.4 data were given describing a car traveling on a straight road. That problem asked to predict the position and speed of the car when $t = 10$ s. Use the following times and positions to predict the speed at each time listed.

Time	0	3	5	8	10	13
Distance	0	225	383	623	742	993

26. In a circuit with impressed voltage $\mathcal{E}(t)$ and inductance L, Kirchhoff's first law gives the relationship

$$\mathcal{E}(t) = L\frac{di}{dt} + Ri,$$

where R is the resistance in the circuit and i is the current. Suppose we measure the current for several values of t and obtain:

t	1.00	1.01	1.02	1.03	1.0
i	3.10	3.12	3.14	3.18	3.24

where t is measured in seconds, i is in amperes, the inductance L is a constant 0.98 henries, and the resistance is 0.142 ohms. Approximate the voltage $\mathcal{E}(t)$ when $t = 1.00$, 1.01, 1.02, 1.03, and 1.04.

27. All calculus students know that the derivative of a function f at x can be defined as

$$f'(x) = \lim_{h\to 0} \frac{f(x+h) - f(x)}{h}.$$

Choose your favorite function f, nonzero number x, and computer or calculator. Generate approximations $f_n'(x)$ to $f'(x)$ by

$$f_n'(x) = \frac{f(x+10^{-n}) - f(x)}{10^{-n}},$$

for $n = 1, 2, \ldots, 20$, and describe what happens.

28. Derive a method for approximating $f'''(x_0)$ whose error term is of order h^2 by expanding the function f in a fourth Taylor polynomial about x_0 and evaluating at $x_0 \pm h$ and $x_0 \pm 2h$.

29. Consider the function

$$e(h) = \frac{\varepsilon}{h} + \frac{h^2}{6}M,$$

where M is a bound for the third derivative of a function. Show that $e(h)$ has a minimum at $\sqrt[3]{3\varepsilon/M}$.

4.2 Richardson's Extrapolation

Richardson's extrapolation is used to generate high-accuracy results while using low-order formulas. Although the name attached to the method refers to a paper written by L. F. Richardson and J. A. Gaunt [RG] in 1927, the idea behind the technique is much older. An interesting article regarding the history and application of extrapolation can be found in [Joy].

Lewis Fry Richardson (1881–1953) was the first person to systematically apply mathematics to weather prediction while working in England for the Meteorological Office. As a conscientious objector during World War I, he wrote extensively about the economic futility of warfare, using systems of differential equations to model rational interactions between countries. The extrapolation technique that bears his name was the rediscovery of a technique with roots that are at least as old as Christiaan Hugyens (1629–1695), and possibly Archimedes (287–212 B.C.E.).

Extrapolation can be applied whenever it is known that an approximation technique has an error term with a predictable form, one that depends on a parameter, usually the step size h. Suppose that for each number $h \neq 0$ we have a formula $N_1(h)$ that approximates an unknown constant M, and that the truncation error involved with the approximation has the form

$$M - N_1(h) = K_1 h + K_2 h^2 + K_3 h^3 + \cdots,$$

for some collection of (unknown) constants $K_1, K_2, K_3, \ldots$.

The truncation error is $O(h)$, so unless there was a large variation in magnitude among the constants $K_1, K_2, K_3, \ldots$,

$$M - N_1(0.1) \approx 0.1K_1, \quad M - N_1(0.01) \approx 0.01K_1,$$

and, in general, $M - N_1(h) \approx K_1 h$.

The object of extrapolation is to find an easy way to combine these rather inaccurate $O(h)$ approximations in an appropriate way to produce formulas with a higher-order truncation error.

Suppose, for example, we can combine the $N_1(h)$ formulas to produce an $O(h^2)$ approximation formula, $N_2(h)$, for M with

$$M - N_2(h) = \hat{K}_2 h^2 + \hat{K}_3 h^3 + \cdots,$$

for some, again unknown, collection of constants $\hat{K}_2, \hat{K}_3, \ldots$. Then we would have

$$M - N_2(0.1) \approx 0.01\hat{K}_2, \quad M - N_2(0.01) \approx 0.0001\hat{K}_2,$$

and so on. If the constants K_1 and $\hat{K}_2$ are roughly of the same magnitude, then the $N_2(h)$ approximations would be much better than the corresponding $N_1(h)$ approximations. The extrapolation continues by combining the $N_2(h)$ approximations in a manner that produces formulas with $O(h^3)$ truncation error, and so on.

To see specifically how we can generate the extrapolation formulas, consider the $O(h)$ formula for approximating M

$$M = N_1(h) + K_1 h + K_2 h^2 + K_3 h^3 + \cdots. \tag{4.10}$$

The formula is assumed to hold for all positive h, so we replace the parameter h by half its value. Then we have a second $O(h)$ approximation formula

$$M = N_1\left(\frac{h}{2}\right) + K_1\frac{h}{2} + K_2\frac{h^2}{4} + K_3\frac{h^3}{8} + \cdots. \tag{4.11}$$

Subtracting Eq. (4.10) from twice Eq. (4.11) eliminates the term involving K_1 and gives

$$M = N_1\left(\frac{h}{2}\right) + \left[N_1\left(\frac{h}{2}\right) - N_1(h)\right] + K_2\left(\frac{h^2}{2} - h^2\right) + K_3\left(\frac{h^3}{4} - h^3\right) + \cdots. \tag{4.12}$$

Define

$$N_2(h) = N_1\left(\frac{h}{2}\right) + \left[N_1\left(\frac{h}{2}\right) - N_1(h)\right].$$

Then Eq. (4.12) is an $O(h^2)$ approximation formula for M:

$$M = N_2(h) - \frac{K_2}{2}h^2 - \frac{3K_3}{4}h^3 - \cdots. \tag{4.13}$$

Example 1 In Example 1 of Section 4.1 we use the forward-difference method with $h = 0.1$ and $h = 0.05$ to find approximations to $f'(1.8)$ for $f(x) = \ln(x)$. Assume that this formula has truncation error $O(h)$ and use extrapolation on these values to see if this results in a better approximation.

Solution In Example 1 of Section 4.1 we found that

$$\text{with } h = 0.1\text{: } f'(1.8) \approx 0.5406722, \quad \text{and} \quad \text{with } h = 0.05\text{: } f'(1.8) \approx 0.5479795.$$

This implies that

$$N_1(0.1) = 0.5406722 \quad \text{and} \quad N_1(0.05) = 0.5479795.$$

Extrapolating these results gives the new approximation

$$N_2(0.1) = N_1(0.05) + (N_1(0.05) - N_1(0.1)) = 0.5479795 + (0.5479795 - 0.5406722)$$
$$= 0.555287.$$

The $h = 0.1$ and $h = 0.05$ results were found to be accurate to within 1.5×10^{-2} and 7.7×10^{-3}, respectively. Because $f'(1.8) = 1/1.8 = 0.\overline{5}$, the extrapolated value is accurate to within 2.7×10^{-4}. ■

Extrapolation can be applied whenever the truncation error for a formula has the form

$$\sum_{j=1}^{m-1} K_j h^{\alpha_j} + O(h^{\alpha_m}),$$

for a collection of constants K_j and when $\alpha_1 < \alpha_2 < \alpha_3 < \cdots < \alpha_m$. Many formulas used for extrapolation have truncation errors that contain only even powers of h, that is, have the form

$$M = N_1(h) + K_1 h^2 + K_2 h^4 + K_3 h^6 + \cdots . \tag{4.14}$$

The extrapolation is much more effective than when all powers of h are present because the averaging process produces results with errors $O(h^2)$, $O(h^4)$, $O(h^6)$, $\ldots$, with essentially no increase in computation, over the results with errors, $O(h)$, $O(h^2)$, $O(h^3)$, $\ldots$.

Assume that approximation has the form of Eq. (4.14). Replacing h with $h/2$ gives the $O(h^2)$ approximation formula

$$M = N_1\left(\frac{h}{2}\right) + K_1\frac{h^2}{4} + K_2\frac{h^4}{16} + K_3\frac{h^6}{64} + \cdots .$$

Subtracting Eq. (4.14) from 4 times this equation eliminates the h^2 term,

$$3M = \left[4N_1\left(\frac{h}{2}\right) - N_1(h)\right] + K_2\left(\frac{h^4}{4} - h^4\right) + K_3\left(\frac{h^6}{16} - h^6\right) + \cdots .$$

Dividing this equation by 3 produces an $O(h^4)$ formula

$$M = \frac{1}{3}\left[4N_1\left(\frac{h}{2}\right) - N_1(h)\right] + \frac{K_2}{3}\left(\frac{h^4}{4} - h^4\right) + \frac{K_3}{3}\left(\frac{h^6}{16} - h^6\right) + \cdots .$$

Defining

$$N_2(h) = \frac{1}{3}\left[4N_1\left(\frac{h}{2}\right) - N_1(h)\right] = N_1\left(\frac{h}{2}\right) + \frac{1}{3}\left[N_1\left(\frac{h}{2}\right) - N_1(h)\right],$$

produces the approximation formula with truncation error $O(h^4)$:

$$M = N_2(h) - K_2\frac{h^4}{4} - K_3\frac{5h^6}{16} + \cdots . \tag{4.15}$$

Now replace h in Eq. (4.15) with $h/2$ to produce a second $O(h^4)$ formula

$$M = N_2\left(\frac{h}{2}\right) - K_2\frac{h^4}{64} - K_3\frac{5h^6}{1024} - \cdots .$$

Subtracting Eq. (4.15) from 16 times this equation eliminates the h^4 term and gives

$$15M = \left[16N_2\left(\frac{h}{2}\right) - N_2(h)\right] + K_3\frac{15h^6}{64} + \cdots .$$

Dividing this equation by 15 produces the new $O(h^6)$ formula

$$M = \frac{1}{15}\left[16N_2\left(\frac{h}{2}\right) - N_2(h)\right] + K_3\frac{h^6}{64} + \cdots .$$

We now have the $O(h^6)$ approximation formula

$$N_3(h) = \frac{1}{15}\left[16N_2\left(\frac{h}{2}\right) - N_2(h)\right] = N_2\left(\frac{h}{2}\right) + \frac{1}{15}\left[N_2\left(\frac{h}{2}\right) - N_2(h)\right].$$

Continuing this procedure gives, for each $j = 2, 3, \ldots$, the $O(h^{2j})$ approximation

$$N_j(h) = N_{j-1}\left(\frac{h}{2}\right) + \frac{N_{j-1}(h/2) - N_{j-1}(h)}{4^{j-1} - 1}.$$

Table 4.6 shows the order in which the approximations are generated when

$$M = N_1(h) + K_1 h^2 + K_2 h^4 + K_3 h^6 + \cdots. \tag{4.16}$$

It is conservatively assumed that the true result is accurate at least to within the agreement of the bottom two results in the diagonal, in this case, to within $|N_3(h) - N_4(h)|$.

Table 4.6

$O(h^2)$	$O(h^4)$	$O(h^6)$	$O(h^8)$
1: $N_1(h)$			
2: $N_1(\frac{h}{2})$	**3:** $N_2(h)$		
4: $N_1(\frac{h}{4})$	**5:** $N_2(\frac{h}{2})$	**6:** $N_3(h)$	
7: $N_1(\frac{h}{8})$	**8:** $N_2(\frac{h}{4})$	**9:** $N_3(\frac{h}{2})$	**10:** $N_4(h)$

Example 2 Taylor's theorem can be used to show that centered-difference formula in Eq. (4.5) to approximate $f'(x_0)$ can be expressed with an error formula:

$$f'(x_0) = \frac{1}{2h}[f(x_0 + h) - f(x_0 - h)] - \frac{h^2}{6} f'''(x_0) - \frac{h^4}{120} f^{(5)}(x_0) - \cdots.$$

Find approximations of order $O(h^2)$, $O(h^4)$, and $O(h^6)$ for $f'(2.0)$ when $f(x) = xe^x$ and $h = 0.2$.

Solution The constants $K_1 = -f'''(x_0)/6$, $K_2 = -f^{(5)}(x_0)/120, \cdots$, are not likely to be known, but this is not important. We only need to know that these constants exist in order to apply extrapolation.

We have the $O(h^2)$ approximation

$$f'(x_0) = N_1(h) - \frac{h^2}{6} f'''(x_0) - \frac{h^4}{120} f^{(5)}(x_0) - \cdots, \tag{4.17}$$

where

$$N_1(h) = \frac{1}{2h}[f(x_0 + h) - f(x_0 - h)].$$

This gives us the first $O(h^2)$ approximations

$$N_1(0.2) = \frac{1}{0.4}[f(2.2) - f(1.8)] = 2.5(19.855030 - 10.889365) = 22.414160,$$

and

$$N_1(0.1) = \frac{1}{0.2}[f(2.1) - f(1.9)] = 5(17.148957 - 12.703199) = 22.228786.$$

Combining these to produce the first $O(h^4)$ approximation gives

$$\begin{aligned} N_2(0.2) &= N_1(0.1) + \frac{1}{3}(N_1(0.1) - N_1(0.2)) \\ &= 22.228786 + \frac{1}{3}(22.228786 - 22.414160) = 22.166995. \end{aligned}$$

To determine an $O(h^6)$ formula we need another $O(h^4)$ result, which requires us to find the third $O(h^2)$ approximation

$$N_1(0.05) = \frac{1}{0.1}[f(2.05) - f(1.95)] = 10(15.924197 - 13.705941) = 22.182564.$$

We can now find the $O(h^4)$ approximation

$$\begin{aligned} N_2(0.1) &= N_1(0.05) + \frac{1}{3}(N_1(0.05) - N_1(0.1)) \\ &= 22.182564 + \frac{1}{3}(22.182564 - 22.228786) = 22.167157. \end{aligned}$$

and finally the $O(h^6)$ approximation

$$\begin{aligned} N_3(0.2) &= N_2(0.1) + \frac{1}{15}(N_2(0.1) - N_1(0.2)) \\ &= 22.167157 + \frac{1}{15}(22.167157 - 22.166995) = 22.167168. \end{aligned}$$

We would expect the final approximation to be accurate to at least the value 22.167 because the $N_2(0.2)$ and $N_3(0.2)$ give this same value. In fact, $N_3(0.2)$ is accurate to all the listed digits. ■

Each column beyond the first in the extrapolation table is obtained by a simple averaging process, so the technique can produce high-order approximations with minimal computational cost. However, as k increases, the round-off error in $N_1(h/2^k)$ will generally increase because the instability of numerical differentiation is related to the step size $h/2^k$. Also, the higher-order formulas depend increasingly on the entry to their immediate left in the table, which is the reason we recommend comparing the final diagonal entries to ensure accuracy.

In Section 4.1, we discussed both three- and five-point methods for approximating $f'(x_0)$ given various functional values of f. The three-point methods were derived by differentiating a Lagrange interpolating polynomial for f. The five-point methods can be obtained in a similar manner, but the derivation is tedious. Extrapolation can be used to more easily derive these formulas, as illustrated below.

Illustration Suppose we expand the function f in a fourth Taylor polynomial about x_0. Then

$$\begin{aligned} f(x) =& f(x_0) + f'(x_0)(x - x_0) + \frac{1}{2}f''(x_0)(x - x_0)^2 + \frac{1}{6}f'''(x_0)(x - x_0)^3 \\ &+ \frac{1}{24}f^{(4)}(x_0)(x - x_0)^4 + \frac{1}{120}f^{(5)}(\xi)(x - x_0)^5, \end{aligned}$$

for some number ξ between x and x_0. Evaluating f at $x_0 + h$ and $x_0 - h$ gives

$$\begin{aligned} f(x_0 + h) =& f(x_0) + f'(x_0)h + \frac{1}{2}f''(x_0)h^2 + \frac{1}{6}f'''(x_0)h^3 \\ &+ \frac{1}{24}f^{(4)}(x_0)h^4 + \frac{1}{120}f^{(5)}(\xi_1)h^5 \end{aligned} \tag{4.18}$$

and

$$\begin{aligned} f(x_0 - h) =& f(x_0) - f'(x_0)h + \frac{1}{2}f''(x_0)h^2 - \frac{1}{6}f'''(x_0)h^3 \\ &+ \frac{1}{24}f^{(4)}(x_0)h^4 - \frac{1}{120}f^{(5)}(\xi_2)h^5, \end{aligned} \tag{4.19}$$

where $x_0 - h < \xi_2 < x_0 < \xi_1 < x_0 + h$.

Subtracting Eq. (4.19) from Eq. (4.18) gives a new approximation for $f'(x)$.

$$f(x_0+h) - f(x_0-h) = 2hf'(x_0) + \frac{h^3}{3}f'''(x_0) + \frac{h^5}{120}[f^{(5)}(\xi_1) + f^{(5)}(\xi_2)], \tag{4.20}$$

which implies that

$$f'(x_0) = \frac{1}{2h}[f(x_0+h) - f(x_0-h)] - \frac{h^2}{6}f'''(x_0) - \frac{h^4}{240}[f^{(5)}(\xi_1) + f^{(5)}(\xi_2)].$$

If $f^{(5)}$ is continuous on $[x_0-h, x_0+h]$, the Intermediate Value Theorem 1.11 implies that a number $\tilde{\xi}$ in (x_0-h, x_0+h) exists with

$$f^{(5)}(\tilde{\xi}) = \frac{1}{2}\left[f^{(5)}(\xi_1) + f^{(5)}(\xi_2)\right].$$

As a consequence,we have the $O(h^2)$ approximation

$$f'(x_0) = \frac{1}{2h}[f(x_0+h) - f(x_0-h)] - \frac{h^2}{6}f'''(x_0) - \frac{h^4}{120}f^{(5)}(\tilde{\xi}). \tag{4.21}$$

Although the approximation in Eq. (4.21) is the same as that given in the three-point formula in Eq. (4.5), the unknown evaluation point occurs now in $f^{(5)}$, rather than in f'''. Extrapolation takes advantage of this by first replacing h in Eq. (4.21) with $2h$ to give the new formula

$$f'(x_0) = \frac{1}{4h}[f(x_0+2h) - f(x_0-2h)] - \frac{4h^2}{6}f'''(x_0) - \frac{16h^4}{120}f^{(5)}(\hat{\xi}), \tag{4.22}$$

where $\hat{\xi}$ is between x_0-2h and x_0+2h.

Multiplying Eq. (4.21) by 4 and subtracting Eq. (4.22) produces

$$\begin{aligned}3f'(x_0) = {} & \frac{2}{h}[f(x_0+h) - f(x_0-h)] - \frac{1}{4h}[f(x_0+2h) - f(x_0-2h)] \\ & - \frac{h^4}{30}f^{(5)}(\tilde{\xi}) + \frac{2h^4}{15}f^{(5)}(\hat{\xi}).\end{aligned}$$

Even if $f^{(5)}$ is continuous on $[x_0-2h, x_0+2h]$, the Intermediate Value Theorem 1.11 cannot be applied as we did to derive Eq. (4.21) because here we have the *difference* of terms involving $f^{(5)}$. However, an alternative method can be used to show that $f^{(5)}(\tilde{\xi})$ and $f^{(5)}(\hat{\xi})$ can still be replaced by a common value $f^{(5)}(\xi)$. Assuming this and dividing by 3 produces the five-point midpoint formula Eq. (4.6) that we saw in Section 4.1

$$f'(x_0) = \frac{1}{12h}[f(x_0-2h) - 8f(x_0-h) + 8f(x_0+h) - f(x_0+2h)] + \frac{h^4}{30}f^{(5)}(\xi). \quad \square$$

Other formulas for first and higher derivatives can be derived in a similar manner. See, for example, Exercise 8.

The technique of extrapolation is used throughout the text. The most prominent applications occur in approximating integrals in Section 4.5 and for determining approximate solutions to differential equations in Section 5.8.

EXERCISE SET 4.2

1. Apply the extrapolation process described in Example 1 to determine $N_3(h)$, an approximation to $f'(x_0)$, for the following functions and stepsizes.
   a. $f(x) = \ln x$, $x_0 = 1.0$, $h = 0.4$
   b. $f(x) = x + e^x$, $x_0 = 0.0$, $h = 0.4$
   c. $f(x) = x^3 \cos x$, $x_0 = 2.3$, $h = 0.4$
   d. $f(x) = 2^x \sin x$, $x_0 = 1.05$, $h = 0.4$
2. Add another line to the extrapolation table in Exercise 1 to obtain the approximation $N_4(h)$.
3. Repeat Exercise 1 using four-digit rounding arithmetic.
4. Repeat Exercise 2 using four-digit rounding arithmetic.
5. The following data give approximations to the integral
$$M = \int_0^{\pi} \sin x \, dx.$$
$$N_1(h) = 1.570796, \quad N_1\left(\frac{h}{2}\right) = 1.896119, \quad N_1\left(\frac{h}{4}\right) = 1.974232, \quad N_1\left(\frac{h}{8}\right) = 1.993570.$$
Assuming $M = N_1(h) + K_1h^2 + K_2h^4 + K_3h^6 + K_4h^8 + O(h^{10})$, construct an extrapolation table to determine $N_4(h)$.
6. The following data can be used to approximate the integral
$$M = \int_0^{3\pi/2} \cos x \, dx.$$
$$N_1(h) = 2.356194, \qquad N_1\left(\frac{h}{2}\right) = -0.4879837,$$
$$N_1\left(\frac{h}{4}\right) = -0.8815732, \quad N_1\left(\frac{h}{8}\right) = -0.9709157.$$
Assume a formula exists of the type given in Exercise 5 and determine $N_4(h)$.
7. Show that the five-point formula in Eq. (4.6) applied to $f(x) = xe^x$ at $x_0 = 2.0$ gives $N_2(0.2)$ in Table 4.6 when $h = 0.1$ and $N_2(0.1)$ when $h = 0.05$.
8. The forward-difference formula can be expressed as
$$f'(x_0) = \frac{1}{h}[f(x_0 + h) - f(x_0)] - \frac{h}{2}f''(x_0) - \frac{h^2}{6}f'''(x_0) + O(h^3).$$
Use extrapolation to derive an $O(h^3)$ formula for $f'(x_0)$.
9. Suppose that $N(h)$ is an approximation to M for every $h > 0$ and that
$$M = N(h) + K_1h + K_2h^2 + K_3h^3 + \cdots,$$
for some constants $K_1, K_2, K_3, \ldots$. Use the values $N(h)$, $N\left(\frac{h}{3}\right)$, and $N\left(\frac{h}{9}\right)$ to produce an $O(h^3)$ approximation to M.
10. Suppose that $N(h)$ is an approximation to M for every $h > 0$ and that
$$M = N(h) + K_1h^2 + K_2h^4 + K_3h^6 + \cdots,$$
for some constants $K_1, K_2, K_3, \ldots$. Use the values $N(h)$, $N\left(\frac{h}{3}\right)$, and $N\left(\frac{h}{9}\right)$ to produce an $O(h^6)$ approximation to M.
11. In calculus, we learn that $e = \lim_{h\to 0}(1 + h)^{1/h}$.
   a. Determine approximations to e corresponding to $h = 0.04, 0.02$, and 0.01.
   b. Use extrapolation on the approximations, assuming that constants $K_1, K_2, \ldots$ exist with $e = (1 + h)^{1/h} + K_1h + K_2h^2 + K_3h^3 + \cdots$, to produce an $O(h^3)$ approximation to e, where $h = 0.04$.
   c. Do you think that the assumption in part (b) is correct?

12. a. Show that

$$\lim_{h \to 0} \left(\frac{2+h}{2-h} \right)^{1/h} = e.$$

b. Compute approximations to e using the formula $N(h) = \left(\frac{2+h}{2-h}\right)^{1/h}$, for $h = 0.04, 0.02$, and 0.01.

c. Assume that $e = N(h) + K_1 h + K_2 h^2 + K_3 h^3 + \cdots$. Use extrapolation, with at least 16 digits of precision, to compute an $O(h^3)$ approximation to e with $h = 0.04$. Do you think the assumption is correct?

d. Show that $N(-h) = N(h)$.

e. Use part (d) to show that $K_1 = K_3 = K_5 = \cdots = 0$ in the formula

$$e = N(h) + K_1 h + K_2 h^2 + K_3 h^3 K_4 h^4 + K_5 h^5 + \cdots,$$

so that the formula reduces to

$$e = N(h) + K_2 h^2 + K_4 h^4 + K_6 h^6 + \cdots.$$

f. Use the results of part (e) and extrapolation to compute an $O(h^6)$ approximation to e with $h = 0.04$.

13. Suppose the following extrapolation table has been constructed to approximate the number M with $M = N_1(h) + K_1 h^2 + K_2 h^4 + K_3 h^6$:

$N_1(h)$		
$N_1\left(\frac{h}{2}\right)$	$N_2(h)$	
$N_1\left(\frac{h}{4}\right)$	$N_2\left(\frac{h}{2}\right)$	$N_3(h)$

a. Show that the linear interpolating polynomial $P_{0,1}(h)$ through $(h^2, N_1(h))$ and $(h^2/4, N_1(h/2))$ satisfies $P_{0,1}(0) = N_2(h)$. Similarly, show that $P_{1,2}(0) = N_2(h/2)$.

b. Show that the linear interpolating polynomial $P_{0,2}(h)$ through $(h^4, N_2(h))$ and $(h^4/16, N_2(h/2))$ satisfies $P_{0,2}(0) = N_3(h)$.

14. Suppose that $N_1(h)$ is a formula that produces $O(h)$ approximations to a number M and that

$$M = N_1(h) + K_1 h + K_2 h^2 + \cdots,$$

for a collection of positive constants $K_1, K_2, \ldots$. Then $N_1(h)$, $N_1(h/2)$, $N_1(h/4)$, $\ldots$ are all lower bounds for M. What can be said about the extrapolated approximations $N_2(h), N_3(h), \ldots$?

15. The semiperimeters of regular polygons with k sides that inscribe and circumscribe the unit circle were used by Archimedes before 200 B.C.E. to approximate π, the circumference of a semicircle. Geometry can be used to show that the sequence of inscribed and circumscribed semiperimeters $\{p_k\}$ and $\{P_k\}$, respectively, satisfy

$$p_k = k \sin\left(\frac{\pi}{k}\right) \quad \text{and} \quad P_k = k \tan\left(\frac{\pi}{k}\right),$$

with $p_k < \pi < P_k$, whenever $k \geq 4$.

a. Show that $p_4 = 2\sqrt{2}$ and $P_4 = 4$.

b. Show that for $k \geq 4$, the sequences satisfy the recurrence relations

$$P_{2k} = \frac{2p_k P_k}{p_k + P_k} \quad \text{and} \quad p_{2k} = \sqrt{p_k P_{2k}}.$$

c. Approximate π to within 10^{-4} by computing p_k and P_k until $P_k - p_k < 10^{-4}$.

d. Use Taylor Series to show that

$$\pi = p_k + \frac{\pi^3}{3!}\left(\frac{1}{k}\right)^2 - \frac{\pi^5}{5!}\left(\frac{1}{k}\right)^4 + \cdots$$

and

$$\pi = P_k - \frac{\pi^3}{3}\left(\frac{1}{k}\right)^2 + \frac{2\pi^5}{15}\left(\frac{1}{k}\right)^4 - \cdots.$$

e. Use extrapolation with $h = 1/k$ to better approximate π.

4.3 Elements of Numerical Integration

The need often arises for evaluating the definite integral of a function that has no explicit antiderivative or whose antiderivative is not easy to obtain. The basic method involved in approximating $\int_a^b f(x)\,dx$ is called **numerical quadrature**. It uses a sum $\sum_{i=0}^n a_i f(x_i)$ to approximate $\int_a^b f(x)\,dx$.

The methods of quadrature in this section are based on the interpolation polynomials given in Chapter 3. The basic idea is to select a set of distinct nodes $\{x_0, \ldots, x_n\}$ from the interval $[a, b]$. Then integrate the Lagrange interpolating polynomial

$$P_n(x) = \sum_{i=0}^n f(x_i)L_i(x)$$

and its truncation error term over $[a, b]$ to obtain

$$\begin{aligned}\int_a^b f(x)\,dx &= \int_a^b \sum_{i=0}^n f(x_i)L_i(x)\,dx + \int_a^b \prod_{i=0}^n (x - x_i)\frac{f^{(n+1)}(\xi(x))}{(n+1)!}\,dx \\ &= \sum_{i=0}^n a_i f(x_i) + \frac{1}{(n+1)!}\int_a^b \prod_{i=0}^n (x - x_i) f^{(n+1)}(\xi(x))\,dx,\end{aligned}$$

where $\xi(x)$ is in $[a, b]$ for each x and

$$a_i = \int_a^b L_i(x)\,dx, \quad \text{for each } i = 0, 1, \ldots, n.$$

The quadrature formula is, therefore,

$$\int_a^b f(x)\,dx \approx \sum_{i=0}^n a_i f(x_i),$$

with error given by

$$E(f) = \frac{1}{(n+1)!}\int_a^b \prod_{i=0}^n (x - x_i) f^{(n+1)}(\xi(x))\,dx.$$

Before discussing the general situation of quadrature formulas, let us consider formulas produced by using first and second Lagrange polynomials with equally-spaced nodes. This gives the **Trapezoidal rule** and **Simpson's rule**, which are commonly introduced in calculus courses.

The Trapezoidal Rule

To derive the Trapezoidal rule for approximating $\int_a^b f(x)\,dx$, let $x_0 = a, x_1 = b, h = b - a$ and use the linear Lagrange polynomial:

$$P_1(x) = \frac{(x - x_1)}{(x_0 - x_1)} f(x_0) + \frac{(x - x_0)}{(x_1 - x_0)} f(x_1).$$

Then

$$\int_a^b f(x)\,dx = \int_{x_0}^{x_1} \left[\frac{(x - x_1)}{(x_0 - x_1)} f(x_0) + \frac{(x - x_0)}{(x_1 - x_0)} f(x_1) \right] dx + \frac{1}{2} \int_{x_0}^{x_1} f''(\xi(x))(x - x_0)(x - x_1)\,dx. \tag{4.23}$$

The product $(x - x_0)(x - x_1)$ does not change sign on $[x_0, x_1]$, so the Weighted Mean Value Theorem for Integrals 1.13 can be applied to the error term to give, for some ξ in (x_0, x_1),

$$\begin{aligned}
\int_{x_0}^{x_1} f''(\xi(x))(x - x_0)(x - x_1)\,dx &= f''(\xi) \int_{x_0}^{x_1} (x - x_0)(x - x_1)\,dx \\
&= f''(\xi) \left[\frac{x^3}{3} - \frac{(x_1 + x_0)}{2} x^2 + x_0 x_1 x \right]_{x_0}^{x_1} \\
&= -\frac{h^3}{6} f''(\xi).
\end{aligned}$$

When we use the term *trapezoid* we mean a four-sided figure that has at least two of its sides parallel. The European term for this figure is *trapezium*. To further confuse the issue, the European word trapezoidal refers to a four-sided figure with no sides equal, and the American word for this type of figure is trapezium.

Consequently, Eq. (4.23) implies that

$$\begin{aligned}
\int_a^b f(x)\,dx &= \left[\frac{(x - x_1)^2}{2(x_0 - x_1)} f(x_0) + \frac{(x - x_0)^2}{2(x_1 - x_0)} f(x_1) \right]_{x_0}^{x_1} - \frac{h^3}{12} f''(\xi) \\
&= \frac{(x_1 - x_0)}{2} [f(x_0) + f(x_1)] - \frac{h^3}{12} f''(\xi).
\end{aligned}$$

Using the notation $h = x_1 - x_0$ gives the following rule:

Trapezoidal Rule:

$$\int_a^b f(x)\,dx = \frac{h}{2}[f(x_0) + f(x_1)] - \frac{h^3}{12} f''(\xi).$$

This is called the Trapezoidal rule because when f is a function with positive values, $\int_a^b f(x)\,dx$ is approximated by the area in a trapezoid, as shown in Figure 4.3.

Figure 4.3

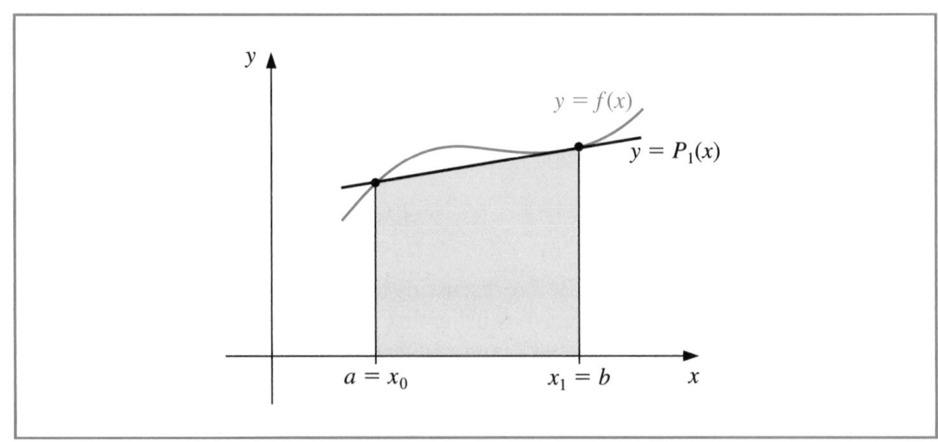

The error term for the Trapezoidal rule involves f'', so the rule gives the exact result when applied to any function whose second derivative is identically zero, that is, any polynomial of degree one or less.

Simpson's Rule

Simpson's rule results from integrating over $[a, b]$ the second Lagrange polynomial with equally-spaced nodes $x_0 = a$, $x_2 = b$, and $x_1 = a + h$, where $h = (b - a)/2$. (See Figure 4.4.)

Figure 4.4

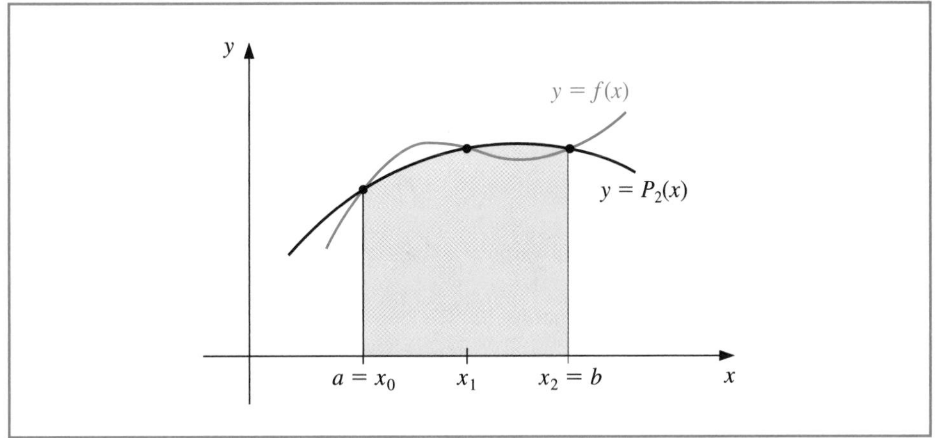

Therefore

$$\int_a^b f(x)\,dx = \int_{x_0}^{x_2} \left[\frac{(x-x_1)(x-x_2)}{(x_0-x_1)(x_0-x_2)} f(x_0) + \frac{(x-x_0)(x-x_2)}{(x_1-x_0)(x_1-x_2)} f(x_1) + \frac{(x-x_0)(x-x_1)}{(x_2-x_0)(x_2-x_1)} f(x_2) \right] dx$$
$$+ \int_{x_0}^{x_2} \frac{(x-x_0)(x-x_1)(x-x_2)}{6} f^{(3)}(\xi(x))\,dx.$$

Deriving Simpson's rule in this manner, however, provides only an $O(h^4)$ error term involving $f^{(3)}$. By approaching the problem in another way, a higher-order term involving $f^{(4)}$ can be derived.

To illustrate this alternative method, suppose that f is expanded in the third Taylor polynomial about x_1. Then for each x in $[x_0, x_2]$, a number $\xi(x)$ in (x_0, x_2) exists with

$$f(x) = f(x_1) + f'(x_1)(x-x_1) + \frac{f''(x_1)}{2}(x-x_1)^2 + \frac{f'''(x_1)}{6}(x-x_1)^3 + \frac{f^{(4)}(\xi(x))}{24}(x-x_1)^4$$

and

$$\int_{x_0}^{x_2} f(x)\,dx = \left[f(x_1)(x-x_1) + \frac{f'(x_1)}{2}(x-x_1)^2 + \frac{f''(x_1)}{6}(x-x_1)^3 + \frac{f'''(x_1)}{24}(x-x_1)^4 \right]_{x_0}^{x_2} + \frac{1}{24}\int_{x_0}^{x_2} f^{(4)}(\xi(x))(x-x_1)^4\,dx. \quad (4.24)$$

Because $(x - x_1)^4$ is never negative on $[x_0, x_2]$, the Weighted Mean Value Theorem for Integrals 1.13 implies that

$$\frac{1}{24}\int_{x_0}^{x_2} f^{(4)}(\xi(x))(x - x_1)^4\,dx = \frac{f^{(4)}(\xi_1)}{24}\int_{x_0}^{x_2}(x - x_1)^4\,dx = \frac{f^{(4)}(\xi_1)}{120}(x - x_1)^5\Bigg]_{x_0}^{x_2},$$

for some number ξ_1 in (x_0, x_2).

However, $h = x_2 - x_1 = x_1 - x_0$, so

$$(x_2 - x_1)^2 - (x_0 - x_1)^2 = (x_2 - x_1)^4 - (x_0 - x_1)^4 = 0,$$

whereas

$$(x_2 - x_1)^3 - (x_0 - x_1)^3 = 2h^3 \quad \text{and} \quad (x_2 - x_1)^5 - (x_0 - x_1)^5 = 2h^5.$$

Consequently, Eq. (4.24) can be rewritten as

$$\int_{x_0}^{x_2} f(x)\,dx = 2hf(x_1) + \frac{h^3}{3}f''(x_1) + \frac{f^{(4)}(\xi_1)}{60}h^5.$$

If we now replace $f''(x_1)$ by the approximation given in Eq. (4.9) of Section 4.1, we have

$$\int_{x_0}^{x_2} f(x)\,dx = 2hf(x_1) + \frac{h^3}{3}\left\{\frac{1}{h^2}[f(x_0) - 2f(x_1) + f(x_2)] - \frac{h^2}{12}f^{(4)}(\xi_2)\right\} + \frac{f^{(4)}(\xi_1)}{60}h^5$$

$$= \frac{h}{3}[f(x_0) + 4f(x_1) + f(x_2)] - \frac{h^5}{12}\left[\frac{1}{3}f^{(4)}(\xi_2) - \frac{1}{5}f^{(4)}(\xi_1)\right].$$

It can be shown by alternative methods (see Exercise 24) that the values ξ_1 and ξ_2 in this expression can be replaced by a common value ξ in (x_0, x_2). This gives Simpson's rule.

Thomas Simpson (1710–1761) was a self-taught mathematician who supported himself during his early years as a weaver. His primary interest was probability theory, although in 1750 he published a two-volume calculus book entitled *The Doctrine and Application of Fluxions*.

Simpson's Rule:

$$\int_{x_0}^{x_2} f(x)\,dx = \frac{h}{3}[f(x_0) + 4f(x_1) + f(x_2)] - \frac{h^5}{90}f^{(4)}(\xi).$$

The error term in Simpson's rule involves the fourth derivative of f, so it gives exact results when applied to any polynomial of degree three or less.

Example 1 Compare the Trapezoidal rule and Simpson's rule approximations to $\int_0^2 f(x)\,dx$ when $f(x)$ is

(a) x^2 **(b)** x^4 **(c)** $(x+1)^{-1}$
(d) $\sqrt{1+x^2}$ **(e)** $\sin x$ **(f)** e^x

Solution On $[0, 2]$ the Trapezoidal and Simpson's rule have the forms

$$\text{Trapezoid:} \quad \int_0^2 f(x)\,dx \approx f(0) + f(2) \quad \text{and}$$

$$\text{Simpson's:} \quad \int_0^2 f(x)\,dx \approx \frac{1}{3}[f(0) + 4f(1) + f(2)].$$

When $f(x) = x^2$ they give

$$\text{Trapezoid:} \quad \int_0^2 f(x)\,dx \approx 0^2 + 2^2 = 4 \quad \text{and}$$

$$\text{Simpson's:} \quad \int_0^2 f(x)\,dx \approx \frac{1}{3}[(0^2) + 4 \cdot 1^2 + 2^2] = \frac{8}{3}.$$

The approximation from Simpson's rule is exact because its truncation error involves $f^{(4)}$, which is identically 0 when $f(x) = x^2$.

The results to three places for the functions are summarized in Table 4.7. Notice that in each instance Simpson's Rule is significantly superior. ■

Table 4.7

	(a)	(b)	(c)	(d)	(e)	(f)
$f(x)$	x^2	x^4	$(x+1)^{-1}$	$\sqrt{1+x^2}$	$\sin x$	e^x
Exact value	2.667	6.400	1.099	2.958	1.416	6.389
Trapezoidal	4.000	16.000	1.333	3.326	0.909	8.389
Simpson's	2.667	6.667	1.111	2.964	1.425	6.421

Measuring Precision

The standard derivation of quadrature error formulas is based on determining the class of polynomials for which these formulas produce exact results. The next definition is used to facilitate the discussion of this derivation.

Definition 4.1 The **degree of accuracy**, or **precision**, of a quadrature formula is the largest positive integer n such that the formula is exact for x^k, for each $k = 0, 1, \ldots, n$. ■

The improved accuracy of Simpson's rule over the Trapezoidal rule is intuitively explained by the fact that Simpson's rule includes a midpoint evaluation that provides better balance to the approximation.

Definition 4.1 implies that the Trapezoidal and Simpson's rules have degrees of precision one and three, respectively.

Integration and summation are linear operations; that is,

$$\int_a^b (\alpha f(x) + \beta g(x))\,dx = \alpha \int_a^b f(x)\,dx + \beta \int_a^b g(x)\,dx$$

and

$$\sum_{i=0}^{n} (\alpha f(x_i) + \beta g(x_i)) = \alpha \sum_{i=0}^{n} f(x_i) + \beta \sum_{i=0}^{n} g(x_i),$$

for each pair of integrable functions f and g and each pair of real constants α and β. This implies (see Exercise 25) that:

The open and closed terminology for methods implies that the open methods use as nodes only points in the open interval, (a, b) to approximate $\int_a^b f(x)\,dx$. The closed methods include the points a and b of the closed interval $[a, b]$ as nodes.

- The degree of precision of a quadrature formula is n if and only if the error is zero for all polynomials of degree $k = 0, 1, \ldots, n$, but is not zero for some polynomial of degree $n + 1$.

The Trapezoidal and Simpson's rules are examples of a class of methods known as Newton-Cotes formulas. There are two types of Newton-Cotes formulas, open and closed.

Closed Newton-Cotes Formulas

The $(n+1)$-*point closed Newton-Cotes formula* uses nodes $x_i = x_0 + ih$, for $i = 0, 1, \ldots, n$, where $x_0 = a$, $x_n = b$ and $h = (b-a)/n$. (See Figure 4.5.) It is called closed because the endpoints of the closed interval $[a, b]$ are included as nodes.

Figure 4.5

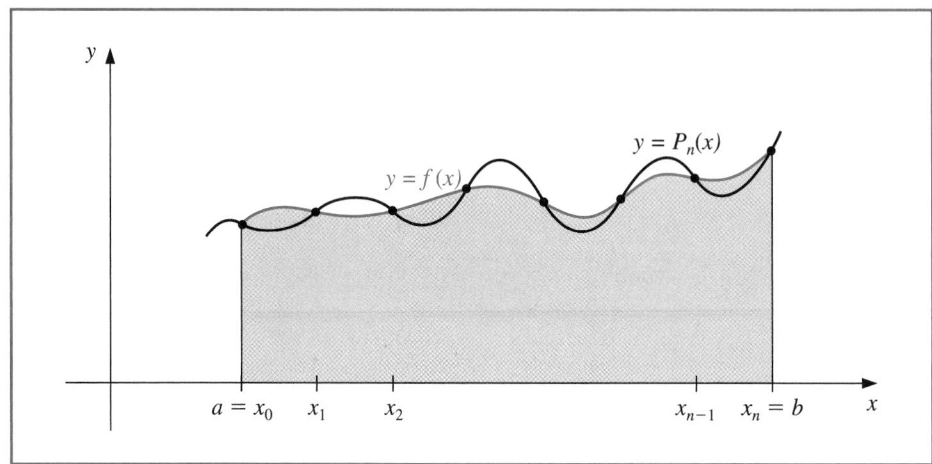

The formula assumes the form

$$\int_a^b f(x)\,dx \approx \sum_{i=0}^{n} a_i f(x_i),$$

where

$$a_i = \int_{x_0}^{x_n} L_i(x)\,dx = \int_{x_0}^{x_n} \prod_{\substack{j=0\\ j\neq i}}^{n} \frac{(x-x_j)}{(x_i-x_j)}\,dx.$$

The following theorem details the error analysis associated with the closed Newton-Cotes formulas. For a proof of this theorem, see [IK], p. 313.

Theorem 4.2 Suppose that $\sum_{i=0}^{n} a_i f(x_i)$ denotes the $(n+1)$-point closed Newton-Cotes formula with $x_0 = a$, $x_n = b$, and $h = (b-a)/n$. There exists $\xi \in (a, b)$ for which

$$\int_a^b f(x)\,dx = \sum_{i=0}^{n} a_i f(x_i) + \frac{h^{n+3} f^{(n+2)}(\xi)}{(n+2)!} \int_0^n t^2(t-1)\cdots(t-n)\,dt,$$

if n is even and $f \in C^{n+2}[a, b]$, and

$$\int_a^b f(x)\,dx = \sum_{i=0}^{n} a_i f(x_i) + \frac{h^{n+2} f^{(n+1)}(\xi)}{(n+1)!} \int_0^n t(t-1)\cdots(t-n)\,dt,$$

if n is odd and $f \in C^{n+1}[a, b]$. ■

Roger Cotes (1682–1716) rose from a modest background to become, in 1704, the first Plumian Professor at Cambridge University. He made advances in numerous mathematical areas including numerical methods for interpolation and integration. Newton is reputed to have said of Cotes ...*if he had lived we might have known something.*

Note that when n is an even integer, the degree of precision is $n + 1$, although the interpolation polynomial is of degree at most n. When n is odd, the degree of precision is only n.

Some of the common **closed Newton-Cotes formulas** with their error terms are listed. Note that in each case the unknown value ξ lies in (a, b).

$n = 1$: Trapezoidal rule

$$\int_{x_0}^{x_1} f(x)\,dx = \frac{h}{2}[f(x_0) + f(x_1)] - \frac{h^3}{12} f''(\xi), \quad \text{where} \quad x_0 < \xi < x_1. \tag{4.25}$$

$n = 2$: Simpson's rule

$$\int_{x_0}^{x_2} f(x)\,dx = \frac{h}{3}[f(x_0) + 4f(x_1) + f(x_2)] - \frac{h^5}{90} f^{(4)}(\xi), \quad \text{where} \quad x_0 < \xi < x_2. \tag{4.26}$$

$n = 3$: Simpson's Three-Eighths rule

$$\int_{x_0}^{x_3} f(x)\,dx = \frac{3h}{8}[f(x_0) + 3f(x_1) + 3f(x_2) + f(x_3)] - \frac{3h^5}{80} f^{(4)}(\xi), \tag{4.27}$$

where $x_0 < \xi < x_3$.

$n = 4$:

$$\int_{x_0}^{x_4} f(x)\,dx = \frac{2h}{45}[7f(x_0) + 32f(x_1) + 12f(x_2) + 32f(x_3) + 7f(x_4)] - \frac{8h^7}{945} f^{(6)}(\xi),$$

where $x_0 < \xi < x_4$. (4.28)

Open Newton-Cotes Formulas

The *open Newton-Cotes formulas* do not include the endpoints of $[a, b]$ as nodes. They use the nodes $x_i = x_0 + ih$, for each $i = 0, 1, \ldots, n$, where $h = (b - a)/(n + 2)$ and $x_0 = a + h$. This implies that $x_n = b - h$, so we label the endpoints by setting $x_{-1} = a$ and $x_{n+1} = b$, as shown in Figure 4.6 on page 200. Open formulas contain all the nodes used for the approximation within the open interval (a, b). The formulas become

$$\int_a^b f(x)\,dx = \int_{x_{-1}}^{x_{n+1}} f(x)\,dx \approx \sum_{i=0}^{n} a_i f(x_i),$$

where

$$a_i = \int_a^b L_i(x)\,dx.$$

Figure 4.6

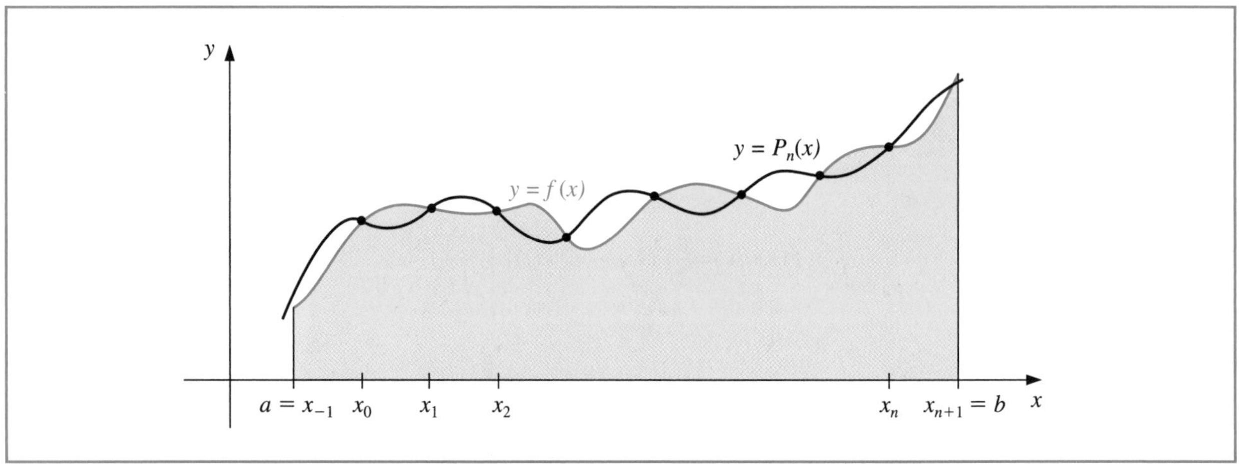

The following theorem is analogous to Theorem 4.2; its proof is contained in [IK], p. 314.

Theorem 4.3 Suppose that $\sum_{i=0}^{n} a_i f(x_i)$ denotes the $(n+1)$-point open Newton-Cotes formula with $x_{-1} = a$, $x_{n+1} = b$, and $h = (b-a)/(n+2)$. There exists $\xi \in (a, b)$ for which

$$\int_a^b f(x)\,dx = \sum_{i=0}^{n} a_i f(x_i) + \frac{h^{n+3} f^{(n+2)}(\xi)}{(n+2)!} \int_{-1}^{n+1} t^2(t-1)\cdots(t-n)\,dt,$$

if n is even and $f \in C^{n+2}[a, b]$, and

$$\int_a^b f(x)\,dx = \sum_{i=0}^{n} a_i f(x_i) + \frac{h^{n+2} f^{(n+1)}(\xi)}{(n+1)!} \int_{-1}^{n+1} t(t-1)\cdots(t-n)\,dt,$$

if n is odd and $f \in C^{n+1}[a, b]$. ■

Notice, as in the case of the closed methods, we have the degree of precision comparatively higher for the even methods than for the odd methods.

Some of the common **open Newton-Cotes** formulas with their error terms are as follows:

$n = 0$: Midpoint rule

$$\int_{x_{-1}}^{x_1} f(x)\,dx = 2hf(x_0) + \frac{h^3}{3} f''(\xi), \quad \text{where} \quad x_{-1} < \xi < x_1. \tag{4.29}$$

$n = 1$:

$$\int_{x_{-1}}^{x_2} f(x)\,dx = \frac{3h}{2}[f(x_0) + f(x_1)] + \frac{3h^3}{4} f''(\xi), \quad \text{where} \quad x_{-1} < \xi < x_2. \tag{4.30}$$

$n = 2$:

$$\int_{x_{-1}}^{x_3} f(x)\,dx = \frac{4h}{3}[2f(x_0) - f(x_1) + 2f(x_2)] + \frac{14h^5}{45} f^{(4)}(\xi), \tag{4.31}$$

where $x_{-1} < \xi < x_3$.

$n = 3$:

$$\int_{x_{-1}}^{x_4} f(x)\,dx = \frac{5h}{24}[11f(x_0) + f(x_1) + f(x_2) + 11f(x_3)] + \frac{95}{144} h^5 f^{(4)}(\xi), \tag{4.32}$$

where $x_{-1} < \xi < x_4$.

Example 2 Compare the results of the closed and open Newton-Cotes formulas listed as (4.25)–(4.28) and (4.29)–(4.32) when approximating

$$\int_0^{\pi/4} \sin x\,dx = 1 - \sqrt{2}/2 \approx 0.29289322.$$

Solution For the closed formulas we have

$$n = 1: \quad \frac{(\pi/4)}{2}\left[\sin 0 + \sin\frac{\pi}{4}\right] \approx 0.27768018$$

$$n = 2: \quad \frac{(\pi/8)}{3}\left[\sin 0 + 4\sin\frac{\pi}{8} + \sin\frac{\pi}{4}\right] \approx 0.29293264$$

$$n = 3: \quad \frac{3(\pi/12)}{8}\left[\sin 0 + 3\sin\frac{\pi}{12} + 3\sin\frac{\pi}{6} + \sin\frac{\pi}{4}\right] \approx 0.29291070$$

$$n = 4: \quad \frac{2(\pi/16)}{45}\left[7\sin 0 + 32\sin\frac{\pi}{16} + 12\sin\frac{\pi}{8} + 32\sin\frac{3\pi}{16} + 7\sin\frac{\pi}{4}\right] \approx 0.29289318$$

and for the open formulas we have

$$n = 0: \quad 2(\pi/8)\left[\sin\frac{\pi}{8}\right] \approx 0.30055887$$

$$n = 1: \quad \frac{3(\pi/12)}{2}\left[\sin\frac{\pi}{12} + \sin\frac{\pi}{6}\right] \approx 0.29798754$$

$$n = 2: \quad \frac{4(\pi/16)}{3}\left[2\sin\frac{\pi}{16} - \sin\frac{\pi}{8} + 2\sin\frac{3\pi}{16}\right] \approx 0.29285866$$

$$n = 3: \quad \frac{5(\pi/20)}{24}\left[11\sin\frac{\pi}{20} + \sin\frac{\pi}{10} + \sin\frac{3\pi}{20} + 11\sin\frac{\pi}{5}\right] \approx 0.29286923$$

Table 4.8 summarizes these results and shows the approximation errors. ■

Table 4.8

n	0	1	2	3	4
Closed formulas		0.27768018	0.29293264	0.29291070	0.29289318
Error		0.01521303	0.00003942	0.00001748	0.00000004
Open formulas	0.30055887	0.29798754	0.29285866	0.29286923	
Error	0.00766565	0.00509432	0.00003456	0.00002399	

EXERCISE SET 4.3

1. Approximate the following integrals using the Trapezoidal rule.

   a. $\int_{0.5}^{1} x^4\, dx$
   b. $\int_{0}^{0.5} \frac{2}{x-4}\, dx$
   c. $\int_{0}^{1} x^2 e^{-x}\, dx$
   d. $\int_{1}^{1.5} x^2 \ln x\, dx$
   e. $\int_{0}^{0.35} \frac{2}{x^2-4}\, dx$
   f. $\int_{1}^{1.6} \frac{2x}{x^2-4}\, dx$
   g. $\int_{0}^{\pi/4} x \sin x\, dx$
   h. $\int_{0}^{\pi/4} e^{3x} \sin 2x\, dx$

2. Approximate the following integrals using the Trapezoidal rule.

   a. $\int_{-0.25}^{0.25} (\cos x)^2\, dx$
   b. $\int_{-0.5}^{0} x \ln(x+1)\, dx$
   c. $\int_{0.75}^{1.3} \left((\sin x)^2 - 2x \sin x + 1\right)\, dx$
   d. $\int_{e}^{e+1} \frac{1}{x \ln x}\, dx$

3. Find a bound for the error in Exercise 1 using the error formula, and compare this to the actual error.
4. Find a bound for the error in Exercise 2 using the error formula, and compare this to the actual error.
5. Repeat Exercise 1 using Simpson's rule.
6. Repeat Exercise 2 using Simpson's rule.
7. Repeat Exercise 3 using Simpson's rule and the results of Exercise 5.
8. Repeat Exercise 4 using Simpson's rule and the results of Exercise 6.
9. Repeat Exercise 1 using the Midpoint rule.
10. Repeat Exercise 2 using the Midpoint rule.
11. Repeat Exercise 3 using the Midpoint rule and the results of Exercise 9.
12. Repeat Exercise 4 using the Midpoint rule and the results of Exercise 10.
13. The Trapezoidal rule applied to $\int_0^2 f(x)\, dx$ gives the value 4, and Simpson's rule gives the value 2. What is $f(1)$?
14. The Trapezoidal rule applied to $\int_0^2 f(x)\, dx$ gives the value 5, and the Midpoint rule gives the value 4. What value does Simpson's rule give?
15. Find the degree of precision of the quadrature formula

$$\int_{-1}^{1} f(x)\, dx = f\left(-\frac{\sqrt{3}}{3}\right) + f\left(\frac{\sqrt{3}}{3}\right).$$

16. Let $h = (b-a)/3$, $x_0 = a$, $x_1 = a + h$, and $x_2 = b$. Find the degree of precision of the quadrature formula

$$\int_{a}^{b} f(x)\, dx = \frac{9}{4} h f(x_1) + \frac{3}{4} h f(x_2).$$

17. The quadrature formula $\int_{-1}^{1} f(x)\, dx = c_0 f(-1) + c_1 f(0) + c_2 f(1)$ is exact for all polynomials of degree less than or equal to 2. Determine c_0, c_1, and c_2.
18. The quadrature formula $\int_{0}^{2} f(x)\, dx = c_0 f(0) + c_1 f(1) + c_2 f(2)$ is exact for all polynomials of degree less than or equal to 2. Determine c_0, c_1, and c_2.
19. Find the constants c_0, c_1, and x_1 so that the quadrature formula

$$\int_{0}^{1} f(x)\, dx = c_0 f(0) + c_1 f(x_1)$$

has the highest possible degree of precision.

20. Find the constants x_0, x_1, and c_1 so that the quadrature formula

$$\int_{0}^{1} f(x)\, dx = \frac{1}{2} f(x_0) + c_1 f(x_1)$$

has the highest possible degree of precision.

21. Approximate the following integrals using formulas (4.25) through (4.32). Are the accuracies of the approximations consistent with the error formulas? Which of parts (d) and (e) give the better approximation?

a. $\int_0^{0.1} \sqrt{1+x}\, dx$ **b.** $\int_0^{\pi/2} (\sin x)^2\, dx$

c. $\int_{1.1}^{1.5} e^x\, dx$ **d.** $\int_1^{10} \frac{1}{x}\, dx$

e. $\int_1^{5.5} \frac{1}{x}\, dx + \int_{5.5}^{10} \frac{1}{x}\, dx$ **f.** $\int_0^1 x^{1/3}\, dx$

22. Given the function f at the following values,

x	1.8	2.0	2.2	2.4	2.6
$f(x)$	3.12014	4.42569	6.04241	8.03014	10.46675

approximate $\int_{1.8}^{2.6} f(x)\, dx$ using all the appropriate quadrature formulas of this section.

23. Suppose that the data of Exercise 22 have round-off errors given by the following table.

x	1.8	2.0	2.2	2.4	2.6
Error in $f(x)$	2×10^{-6}	-2×10^{-6}	-0.9×10^{-6}	-0.9×10^{-6}	2×10^{-6}

Calculate the errors due to round-off in Exercise 22.

24. Derive Simpson's rule with error term by using

$$\int_{x_0}^{x_2} f(x)\, dx = a_0 f(x_0) + a_1 f(x_1) + a_2 f(x_2) + k f^{(4)}(\xi).$$

Find a_0, a_1, and a_2 from the fact that Simpson's rule is exact for $f(x) = x^n$ when $n = 1, 2$, and 3. Then find k by applying the integration formula with $f(x) = x^4$.

25. Prove the statement following Definition 4.1; that is, show that a quadrature formula has degree of precision n if and only if the error $E(P(x)) = 0$ for all polynomials $P(x)$ of degree $k = 0, 1, \ldots, n$, but $E(P(x)) \neq 0$ for some polynomial $P(x)$ of degree $n + 1$.

26. Derive Simpson's three-eighths rule (the closed rule with $n = 3$) with error term by using Theorem 4.2.

27. Derive the open rule with $n = 1$ with error term by using Theorem 4.3.

4.4 Composite Numerical Integration

The Newton-Cotes formulas are generally unsuitable for use over large integration intervals. High-degree formulas would be required, and the values of the coefficients in these formulas are difficult to obtain. Also, the Newton-Cotes formulas are based on interpolatory polynomials that use equally-spaced nodes, a procedure that is inaccurate over large intervals because of the oscillatory nature of high-degree polynomials.

Piecewise approximation is often effective. Recall that this was used for spline interpolation.

In this section, we discuss a *piecewise* approach to numerical integration that uses the low-order Newton-Cotes formulas. These are the techniques most often applied.

Example 1 Use Simpson's rule to approximate $\int_0^4 e^x\, dx$ and compare this to the results obtained by adding the Simpson's rule approximations for $\int_0^2 e^x\, dx$ and $\int_2^4 e^x\, dx$. Compare these approximations to the sum of Simpson's rule for $\int_0^1 e^x\, dx$, $\int_1^2 e^x\, dx$, $\int_2^3 e^x\, dx$, and $\int_3^4 e^x\, dx$.

Solution Simpson's rule on $[0, 4]$ uses $h = 2$ and gives

$$\int_0^4 e^x\,dx \approx \frac{2}{3}(e^0 + 4e^2 + e^4) = 56.76958.$$

The exact answer in this case is $e^4 - e^0 = 53.59815$, and the error -3.17143 is far larger than we would normally accept.

Applying Simpson's rule on each of the intervals $[0, 2]$ and $[2, 4]$ uses $h = 1$ and gives

$$\begin{aligned}\int_0^4 e^x\,dx &= \int_0^2 e^x\,dx + \int_2^4 e^x\,dx \\ &\approx \frac{1}{3}\left(e^0 + 4e + e^2\right) + \frac{1}{3}\left(e^2 + 4e^3 + e^4\right) \\ &= \frac{1}{3}\left(e^0 + 4e + 2e^2 + 4e^3 + e^4\right) \\ &= 53.86385.\end{aligned}$$

The error has been reduced to -0.26570.

For the integrals on $[0, 1]$,$[1, 2]$,$[3, 4]$, and $[3, 4]$ we use Simpson's rule four times with $h = \frac{1}{2}$ giving

$$\begin{aligned}\int_0^4 e^x\,dx &= \int_0^1 e^x\,dx + \int_1^2 e^x\,dx + \int_2^3 e^x\,dx + \int_3^4 e^x\,dx \\ &\approx \frac{1}{6}\left(e_0 + 4e^{1/2} + e\right) + \frac{1}{6}\left(e + 4e^{3/2} + e^2\right) \\ &\quad + \frac{1}{6}\left(e^2 + 4e^{5/2} + e^3\right) + \frac{1}{6}\left(e^3 + 4e^{7/2} + e^4\right) \\ &= \frac{1}{6}\left(e^0 + 4e^{1/2} + 2e + 4e^{3/2} + 2e^2 + 4e^{5/2} + 2e^3 + 4e^{7/2} + e^4\right) \\ &= 53.61622.\end{aligned}$$

The error for this approximation has been reduced to -0.01807. ■

To generalize this procedure for an arbitrary integral $\int_a^b f(x)\,dx$, choose an even integer n. Subdivide the interval $[a, b]$ into n subintervals, and apply Simpson's rule on each consecutive pair of subintervals. (See Figure 4.7.)

Figure 4.7

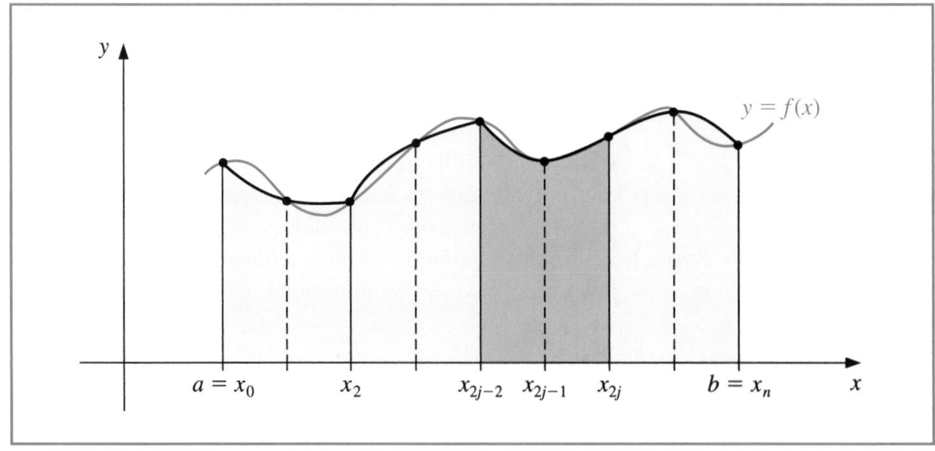

With $h = (b - a)/n$ and $x_j = a + jh$, for each $j = 0, 1, \ldots, n$, we have

$$\int_a^b f(x)\,dx = \sum_{j=1}^{n/2} \int_{x_{2j-2}}^{x_{2j}} f(x)\,dx$$

$$= \sum_{j=1}^{n/2} \left\{ \frac{h}{3}[f(x_{2j-2}) + 4f(x_{2j-1}) + f(x_{2j})] - \frac{h^5}{90} f^{(4)}(\xi_j) \right\},$$

for some ξ_j with $x_{2j-2} < \xi_j < x_{2j}$, provided that $f \in C^4[a,b]$. Using the fact that for each $j = 1, 2, \ldots, (n/2) - 1$ we have $f(x_{2j})$ appearing in the term corresponding to the interval $[x_{2j-2}, x_{2j}]$ and also in the term corresponding to the interval $[x_{2j}, x_{2j+2}]$, we can reduce this sum to

$$\int_a^b f(x)\,dx = \frac{h}{3}\left[f(x_0) + 2\sum_{j=1}^{(n/2)-1} f(x_{2j}) + 4\sum_{j=1}^{n/2} f(x_{2j-1}) + f(x_n) \right] - \frac{h^5}{90}\sum_{j=1}^{n/2} f^{(4)}(\xi_j).$$

The error associated with this approximation is

$$E(f) = -\frac{h^5}{90}\sum_{j=1}^{n/2} f^{(4)}(\xi_j),$$

where $x_{2j-2} < \xi_j < x_{2j}$, for each $j = 1, 2, \ldots, n/2$.

If $f \in C^4[a,b]$, the Extreme Value Theorem 1.9 implies that $f^{(4)}$ assumes its maximum and minimum in $[a,b]$. Since

$$\min_{x\in[a,b]} f^{(4)}(x) \le f^{(4)}(\xi_j) \le \max_{x\in[a,b]} f^{(4)}(x),$$

we have

$$\frac{n}{2}\min_{x\in[a,b]} f^{(4)}(x) \le \sum_{j=1}^{n/2} f^{(4)}(\xi_j) \le \frac{n}{2}\max_{x\in[a,b]} f^{(4)}(x)$$

and

$$\min_{x\in[a,b]} f^{(4)}(x) \le \frac{2}{n}\sum_{j=1}^{n/2} f^{(4)}(\xi_j) \le \max_{x\in[a,b]} f^{(4)}(x).$$

By the Intermediate Value Theorem 1.11, there is a $\mu \in (a,b)$ such that

$$f^{(4)}(\mu) = \frac{2}{n}\sum_{j=1}^{n/2} f^{(4)}(\xi_j).$$

Thus

$$E(f) = -\frac{h^5}{90}\sum_{j=1}^{n/2} f^{(4)}(\xi_j) = -\frac{h^5}{180} n f^{(4)}(\mu),$$

or, since $h = (b - a)/n$,

$$E(f) = -\frac{(b-a)}{180} h^4 f^{(4)}(\mu).$$

These observations produce the following result.

Theorem 4.4 Let $f \in C^4[a,b]$, n be even, $h = (b-a)/n$, and $x_j = a + jh$, for each $j = 0, 1, \ldots, n$. There exists a $\mu \in (a,b)$ for which the **Composite Simpson's rule** for n subintervals can be written with its error term as

$$\int_a^b f(x)\,dx = \frac{h}{3}\left[f(a) + 2\sum_{j=1}^{(n/2)-1} f(x_{2j}) + 4\sum_{j=1}^{n/2} f(x_{2j-1}) + f(b)\right] - \frac{b-a}{180}h^4 f^{(4)}(\mu).$$

■

Notice that the error term for the Composite Simpson's rule is $O(h^4)$, whereas it was $O(h^5)$ for the standard Simpson's rule. However, these rates are not comparable because for standard Simpson's rule we have h fixed at $h = (b-a)/2$, but for Composite Simpson's rule we have $h = (b-a)/n$, for n an even integer. This permits us to considerably reduce the value of h when the Composite Simpson's rule is used.

Algorithm 4.1 uses the Composite Simpson's rule on n subintervals. This is the most frequently used general-purpose quadrature algorithm.

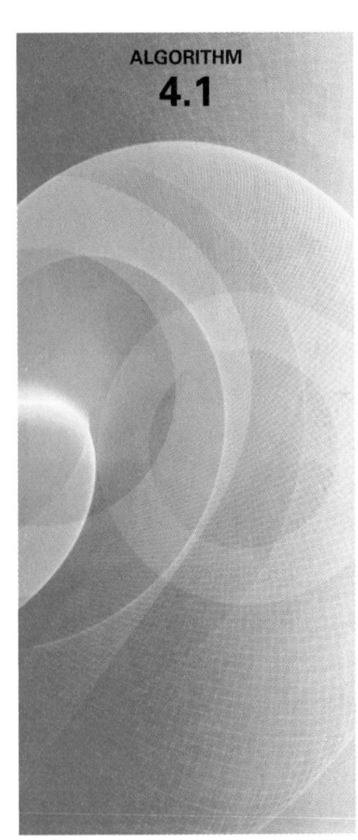

Composite Simpson's Rule

To approximate the integral $I = \int_a^b f(x)\,dx$:

INPUT endpoints a, b; even positive integer n.

OUTPUT approximation XI to I.

Step 1 Set $h = (b-a)/n$.

Step 2 Set $XI0 = f(a) + f(b)$;
$XI1 = 0$; *(Summation of $f(x_{2i-1})$.)*
$XI2 = 0$. *(Summation of $f(x_{2i})$.)*

Step 3 For $i = 1, \ldots, n-1$ do Steps 4 and 5.

Step 4 Set $X = a + ih$.

Step 5 If i is even then set $XI2 = XI2 + f(X)$
else set $XI1 = XI1 + f(X)$.

Step 6 Set $XI = h(XI0 + 2 \cdot XI2 + 4 \cdot XI1)/3$.

Step 7 OUTPUT (XI);
STOP.

■

The subdivision approach can be applied to any of the Newton-Cotes formulas. The extensions of the Trapezoidal (see Figure 4.8) and Midpoint rules are given without proof. The Trapezoidal rule requires only one interval for each application, so the integer n can be either odd or even.

Theorem 4.5 Let $f \in C^2[a,b]$, $h = (b-a)/n$, and $x_j = a + jh$, for each $j = 0, 1, \ldots, n$. There exists a $\mu \in (a,b)$ for which the **Composite Trapezoidal rule** for n subintervals can be written with its error term as

$$\int_a^b f(x)\,dx = \frac{h}{2}\left[f(a) + 2\sum_{j=1}^{n-1} f(x_j) + f(b)\right] - \frac{b-a}{12}h^2 f''(\mu).$$

■

Figure 4.8

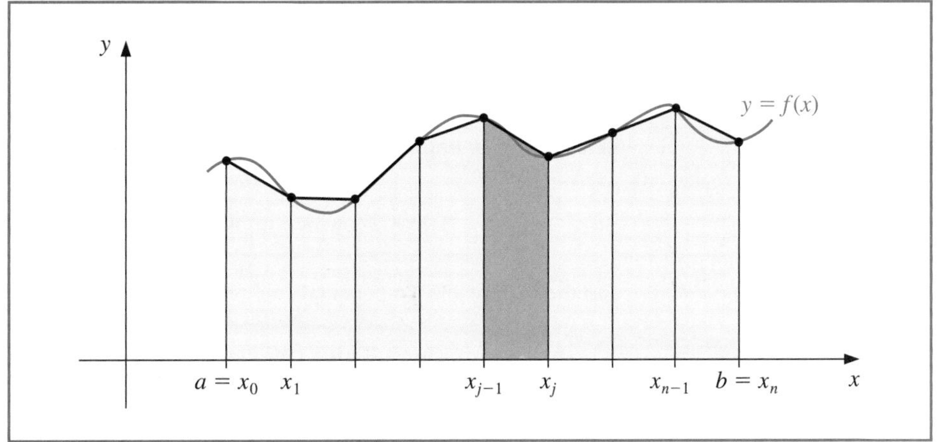

For the Composite Midpoint rule, n must again be even. (See Figure 4.9.)

Figure 4.9

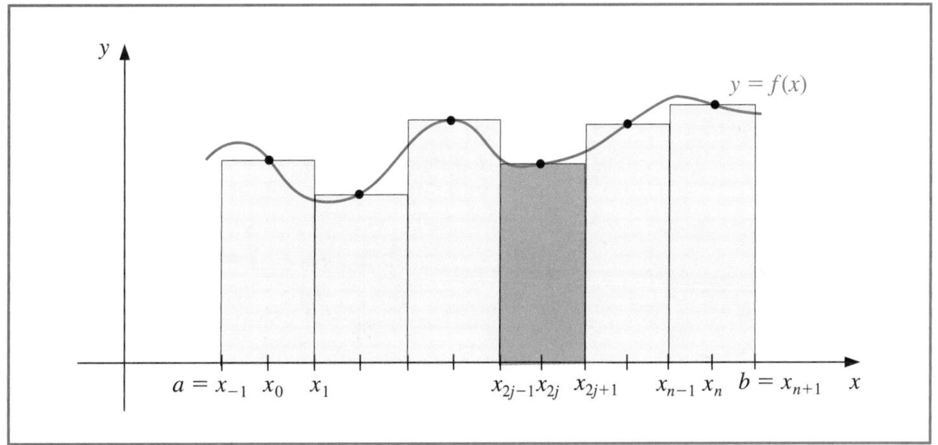

Theorem 4.6 Let $f \in C^2[a,b]$, n be even, $h = (b-a)/(n+2)$, and $x_j = a + (j+1)h$ for each $j = -1, 0, \ldots, n+1$. There exists a $\mu \in (a,b)$ for which the **Composite Midpoint rule** for $n+2$ subintervals can be written with its error term as

$$\int_a^b f(x)\,dx = 2h\sum_{j=0}^{n/2} f(x_{2j}) + \frac{b-a}{6}h^2 f''(\mu). \qquad \blacksquare$$

Example 2 Determine values of h that will ensure an approximation error of less than 0.00002 when approximating $\int_0^\pi \sin x\,dx$ and employing
(a) Composite Trapezoidal rule and **(b)** Composite Simpson's rule.

Solution **(a)** The error form for the Composite Trapezoidal rule for $f(x) = \sin x$ on $[0,\pi]$ is

$$\left|\frac{\pi h^2}{12} f''(\mu)\right| = \left|\frac{\pi h^2}{12}(-\sin\mu)\right| = \frac{\pi h^2}{12}|\sin\mu|.$$

To ensure sufficient accuracy with this technique we need to have

$$\frac{\pi h^2}{12}|\sin\mu| \le \frac{\pi h^2}{12} < 0.00002.$$

Since $h = \pi/n$ implies that $n = \pi/h$, we need

$$\frac{\pi^3}{12n^2} < 0.00002 \quad \text{which implies that} \quad n > \left(\frac{\pi^3}{12(0.00002)}\right)^{1/2} \approx 359.44.$$

and the Composite Trapezoidal rule requires $n \ge 360$.

(b) The error form for the Composite Simpson's rule for $f(x) = \sin x$ on $[0, \pi]$ is

$$\left|\frac{\pi h^4}{180} f^{(4)}(\mu)\right| = \left|\frac{\pi h^4}{180}\sin\mu\right| = \frac{\pi h^4}{180}|\sin\mu|.$$

To ensure sufficient accuracy with this technique we need to have

$$\frac{\pi h^4}{180}|\sin\mu| \le \frac{\pi h^4}{180} < 0.00002.$$

Using again the fact that $n = \pi/h$ gives

$$\frac{\pi^5}{180n^4} < 0.00002 \quad \text{which implies that} \quad n > \left(\frac{\pi^5}{180(0.00002)}\right)^{1/4} \approx 17.07.$$

So Composite Simpson's rule requires only $n \ge 18$.

Composite Simpson's rule with $n = 18$ gives

$$\int_0^\pi \sin x\,dx \approx \frac{\pi}{54}\left[2\sum_{j=1}^{8}\sin\left(\frac{j\pi}{9}\right) + 4\sum_{j=1}^{9}\sin\left(\frac{(2j-1)\pi}{18}\right)\right] = 2.0000104.$$

This is accurate to within about 10^{-5} because the true value is $-\cos(\pi) - (-\cos(0)) = 2$. ■

Composite Simpson's rule is the clear choice if you wish to minimize computation. For comparison purposes, consider the Composite Trapezoidal rule using $h = \pi/18$ for the integral in Example 2. This approximation uses the same function evaluations as Composite Simpson's rule but the approximation in this case

$$\int_0^\pi \sin x\,dx \approx \frac{\pi}{36}\left[2\sum_{j=1}^{17}\sin\left(\frac{j\pi}{18}\right) + \sin 0 + \sin\pi\right] = \frac{\pi}{36}\left[2\sum_{j=1}^{17}\sin\left(\frac{j\pi}{18}\right)\right] = 1.9949205.$$

is accurate only to about 5×10^{-3}.

Maple contains numerous procedures for numerical integration in the *NumericalAnalysis* subpackage of the *Student* package. First access the library as usual with

with(*Student*[*NumericalAnalysis*])

The command for all methods is *Quadrature* with the options in the call specifying the method to be used. We will use the Trapezoidal method to illustrate the procedure. First define the function and the interval of integration with

$f := x \to \sin(x);\ a := 0.0;\ b := \pi$

After Maple responds with the function and the interval, enter the command

$Quadrature(f(x), x = a..b, method = trapezoid, partition = 20, output = value)$

1.995885973

The value of the step size h in this instance is the width of the interval $b - a$ divided by the number specified by $partition = 20$.

Simpson's method can be called in a similar manner, except that the step size h is determined by $b - a$ divided by twice the value of *partition*. Hence, the Simpson's rule approximation using the same nodes as those in the Trapezoidal rule is called with

$Quadrature(f(x), x = a..b, method = simpson, partition = 10, output = value)$

2.000006785

Any of the Newton-Cotes methods can be called using the option

$$method = newtoncotes[open, n] \quad \text{or} \quad method = newtoncotes[closed, n]$$

Be careful to correctly specify the number in *partition* when an even number of divisions is required, and when an open method is employed.

Round-Off Error Stability

In Example 2 we saw that ensuring an accuracy of 2×10^{-5} for approximating $\int_0^{\pi} \sin x\, dx$ required 360 subdivisions of $[0, \pi]$ for the Composite Trapezoidal rule and only 18 for Composite Simpson's rule. In addition to the fact that less computation is needed for the Simpson's technique, you might suspect that because of fewer computations this method would also involve less round-off error. However, an important property shared by all the composite integration techniques is a stability with respect to round-off error. That is, the round-off error does not depend on the number of calculations performed.

Numerical integration is expected to be stable, whereas numerical differentiation is unstable.

To demonstrate this rather amazing fact, suppose we apply the Composite Simpson's rule with n subintervals to a function f on $[a, b]$ and determine the maximum bound for the round-off error. Assume that $f(x_i)$ is approximated by $\tilde{f}(x_i)$ and that

$$f(x_i) = \tilde{f}(x_i) + e_i, \quad \text{for each } i = 0, 1, \ldots, n,$$

where e_i denotes the round-off error associated with using $\tilde{f}(x_i)$ to approximate $f(x_i)$. Then the accumulated error, $e(h)$, in the Composite Simpson's rule is

$$\begin{aligned} e(h) &= \left| \frac{h}{3} \left[e_0 + 2 \sum_{j=1}^{(n/2)-1} e_{2j} + 4 \sum_{j=1}^{n/2} e_{2j-1} + e_n \right] \right| \\ &\le \frac{h}{3} \left[|e_0| + 2 \sum_{j=1}^{(n/2)-1} |e_{2j}| + 4 \sum_{j=1}^{n/2} |e_{2j-1}| + |e_n| \right]. \end{aligned}$$

If the round-off errors are uniformly bounded by ε, then

$$e(h) \le \frac{h}{3} \left[\varepsilon + 2 \left(\frac{n}{2} - 1 \right) \varepsilon + 4 \left(\frac{n}{2} \right) \varepsilon + \varepsilon \right] = \frac{h}{3} 3n\varepsilon = nh\varepsilon.$$

But $nh = b - a$, so

$$e(h) \le (b - a)\varepsilon,$$

a bound independent of h (and n). This means that, even though we may need to divide an interval into more parts to ensure accuracy, the increased computation that is required does not increase the round-off error. This result implies that the procedure is stable as h approaches zero. Recall that this was not true of the numerical differentiation procedures considered at the beginning of this chapter.

EXERCISE SET 4.4

1. Use the Composite Trapezoidal rule with the indicated values of n to approximate the following integrals.

   a. $\int_{-2}^{2} x^3 e^x \, dx, \quad n = 4$

   b. $\int_{1}^{2} x \ln x \, dx, \quad n = 4$

   c. $\int_{1}^{3} \frac{x}{x^2 + 4} \, dx, \quad n = 8$

   d. $\int_{0}^{\pi} x^2 \cos x \, dx, \quad n = 6$

   e. $\int_{0}^{2} e^{2x} \sin 3x \, dx, \quad n = 8$

   f. $\int_{0}^{2} \frac{2}{x^2 + 4} \, dx, \quad n = 6$

   g. $\int_{3}^{5} \frac{1}{\sqrt{x^2 - 4}} \, dx, \quad n = 8$

   h. $\int_{0}^{3\pi/8} \tan x \, dx, \quad n = 8$

2. Use the Composite Trapezoidal rule with the indicated values of n to approximate the following integrals.

   a. $\int_{-0.5}^{0.5} \cos^2 x \, dx, \quad n = 4$

   b. $\int_{e}^{e+2} \frac{1}{x \ln x} \, dx, \quad n = 8$

   c. $\int_{.75}^{1.75} (\sin^2 x - 2x \sin x + 1) \, dx, \quad n = 8$

   d. $\int_{-0.5}^{0.5} x \ln(x + 1) \, dx, \quad n = 6$

3. Use the Composite Simpson's rule to approximate the integrals in Exercise 1.

4. Use the Composite Simpson's rule to approximate the integrals in Exercise 2.

5. Use the Composite Midpoint rule with $n + 2$ subintervals to approximate the integrals in Exercise 1.

6. Use the Composite Midpoint rule with $n + 2$ subintervals to approximate the integrals in Exercise 2.

7. Approximate $\int_0^2 x^2 \ln(x^2 + 1) \, dx$ using $h = 0.25$. Use

   a. Composite Trapezoidal rule.

   b. Composite Simpson's rule.

   c. Composite Midpoint rule.

8. Approximate $\int_0^2 x^2 e^{-x^2} \, dx$ using $h = 0.25$. Use

   a. Composite Trapezoidal rule.

   b. Composite Simpson's rule.

   c. Composite Midpoint rule.

9. Suppose that $f(0) = 1$, $f(0.5) = 2.5$, $f(1) = 2$, and $f(0.25) = f(0.75) = \alpha$. Find α if the Composite Trapezoidal rule with $n = 4$ gives the value 1.75 for $\int_0^1 f(x) \, dx$.

10. The Midpoint rule for approximating $\int_{-1}^{1} f(x) \, dx$ gives the value 12, the Composite Midpoint rule with $n = 2$ gives 5, and Composite Simpson's rule gives 6. Use the fact that $f(-1) = f(1)$ and $f(-0.5) = f(0.5) - 1$ to determine $f(-1)$, $f(-0.5)$, $f(0)$, $f(0.5)$, and $f(1)$.

11. Determine the values of n and h required to approximate

$$\int_0^2 e^{2x} \sin 3x \, dx$$

   to within 10^{-4}. Use

   a. Composite Trapezoidal rule.

   b. Composite Simpson's rule.

   c. Composite Midpoint rule.

12. Repeat Exercise 11 for the integral $\int_0^{\pi} x^2 \cos x\, dx$.

13. Determine the values of n and h required to approximate

$$\int_0^2 \frac{1}{x+4}\, dx$$

to within 10^{-5} and compute the approximation. Use

a. Composite Trapezoidal rule.

b. Composite Simpson's rule.

c. Composite Midpoint rule.

14. Repeat Exercise 13 for the integral $\int_1^2 x \ln x\, dx$.

15. Let f be defined by

$$f(x) = \begin{cases} x^3 + 1, & 0 \le x \le 0.1, \\ 1.001 + 0.03(x - 0.1) + 0.3(x - 0.1)^2 + 2(x - 0.1)^3, & 0.1 \le x \le 0.2, \\ 1.009 + 0.15(x - 0.2) + 0.9(x - 0.2)^2 + 2(x - 0.2)^3, & 0.2 \le x \le 0.3. \end{cases}$$

a. Investigate the continuity of the derivatives of f.

b. Use the Composite Trapezoidal rule with $n = 6$ to approximate $\int_0^{0.3} f(x)\, dx$, and estimate the error using the error bound.

c. Use the Composite Simpson's rule with $n = 6$ to approximate $\int_0^{0.3} f(x)\, dx$. Are the results more accurate than in part (b)?

16. Show that the error $E(f)$ for Composite Simpson's rule can be approximated by

$$-\frac{h^4}{180}[f'''(b) - f'''(a)].$$

[*Hint:* $\sum_{j=1}^{n/2} f^{(4)}(\xi_j)(2h)$ is a Riemann Sum for $\int_a^b f^{(4)}(x)\, dx$.]

17. **a.** Derive an estimate for $E(f)$ in the Composite Trapezoidal rule using the method in Exercise 16.

b. Repeat part (a) for the Composite Midpoint rule.

18. Use the error estimates of Exercises 16 and 17 to estimate the errors in Exercise 12.

19. Use the error estimates of Exercises 16 and 17 to estimate the errors in Exercise 14.

20. In multivariable calculus and in statistics courses it is shown that

$$\int_{-\infty}^{\infty} \frac{1}{\sigma\sqrt{2\pi}} e^{-(1/2)(x/\sigma)^2}\, dx = 1,$$

for any positive σ. The function

$$f(x) = \frac{1}{\sigma\sqrt{2\pi}} e^{-(1/2)(x/\sigma)^2}$$

is the *normal density function* with *mean* $\mu = 0$ and *standard deviation* σ. The probability that a randomly chosen value described by this distribution lies in $[a, b]$ is given by $\int_a^b f(x)\, dx$. Approximate to within 10^{-5} the probability that a randomly chosen value described by this distribution will lie in

a. $[-\sigma, \sigma]$ **b.** $[-2\sigma, 2\sigma]$ **c.** $[-3\sigma, 3\sigma]$

21. Determine to within 10^{-6} the length of the graph of the ellipse with equation $4x^2 + 9y^2 = 36$.

22. A car laps a race track in 84 seconds. The speed of the car at each 6-second interval is determined by using a radar gun and is given from the beginning of the lap, in feet/second, by the entries in the following table.

Time	0	6	12	18	24	30	36	42	48	54	60	66	72	78	84
Speed	124	134	148	156	147	133	121	109	99	85	78	89	104	116	123

How long is the track?

23. A particle of mass m moving through a fluid is subjected to a viscous resistance R, which is a function of the velocity v. The relationship between the resistance R, velocity v, and time t is given by the equation

$$t = \int_{v(t_0)}^{v(t)} \frac{m}{R(u)}\, du.$$

Suppose that $R(v) = -v\sqrt{v}$ for a particular fluid, where R is in newtons and v is in meters/second. If $m = 10$ kg and $v(0) = 10$ m/s, approximate the time required for the particle to slow to $v = 5$ m/s.

24. To simulate the thermal characteristics of disk brakes (see the following figure), D. A. Secrist and R. W. Hornbeck [SH] needed to approximate numerically the "area averaged lining temperature," T, of the brake pad from the equation

$$T = \frac{\displaystyle\int_{r_e}^{r_0} T(r) r\theta_p\, dr}{\displaystyle\int_{r_e}^{r_0} r\theta_p\, dr},$$

where r_e represents the radius at which the pad-disk contact begins, r_0 represents the outside radius of the pad-disk contact, θ_p represents the angle subtended by the sector brake pads, and $T(r)$ is the temperature at each point of the pad, obtained numerically from analyzing the heat equation (see Section 12.2). Suppose $r_e = 0.308$ ft, $r_0 = 0.478$ ft, $\theta_p = 0.7051$ radians, and the temperatures given in the following table have been calculated at the various points on the disk. Approximate T.

r (ft)	$T(r)$ (°F)	r (ft)	$T(r)$ (°F)	r (ft)	$T(r)$ (°F)
0.308	640	0.376	1034	0.444	1204
0.325	794	0.393	1064	0.461	1222
0.342	885	0.410	1114	0.478	1239
0.359	943	0.427	1152		

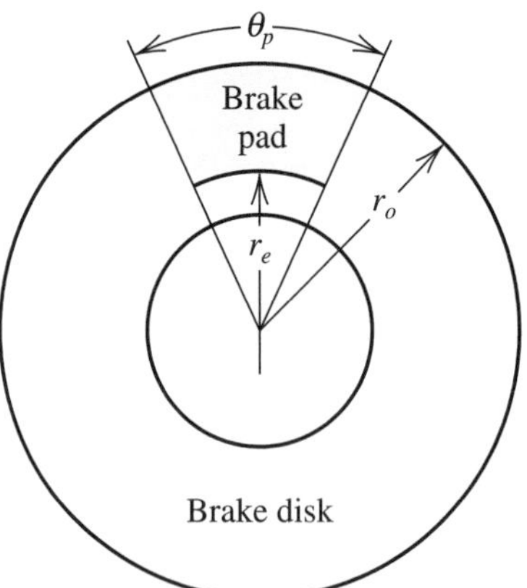

25. Find an approximation to within 10^{-4} of the value of the integral considered in the application opening this chapter:

$$\int_0^{48} \sqrt{1 + (\cos x)^2}\, dx.$$

26. The equation

$$\int_0^{x} \frac{1}{\sqrt{2\pi}} e^{-t^2/2}\, dt = 0.45$$

can be solved for x by using Newton's method with

$$f(x) = \int_0^x \frac{1}{\sqrt{2\pi}} e^{-t^2/2}\, dt - 0.45$$

and

$$f'(x) = \frac{1}{\sqrt{2\pi}} e^{-x^2/2}.$$

To evaluate f at the approximation p_k, we need a quadrature formula to approximate

$$\int_0^{p_k} \frac{1}{\sqrt{2\pi}} e^{-t^2/2}\, dt.$$

a. Find a solution to $f(x) = 0$ accurate to within 10^{-5} using Newton's method with $p_0 = 0.5$ and the Composite Simpson's rule.

b. Repeat (a) using the Composite Trapezoidal rule in place of the Composite Simpson's rule.

4.5 Romberg Integration

In this section we will illustrate how Richardson extrapolation applied to results from the Composite Trapezoidal rule can be used to obtain high accuracy approximations with little computational cost.

In Section 4.4 we found that the Composite Trapezoidal rule has a truncation error of order $O(h^2)$. Specifically, we showed that for $h = (b - a)/n$ and $x_j = a + jh$ we have

$$\int_a^b f(x)\, dx = \frac{h}{2}\left[f(a) + 2\sum_{j=1}^{n-1} f(x_j) + f(b)\right] - \frac{(b-a)f''(\mu)}{12}h^2.$$

for some number μ in (a, b).

By an alternative method it can be shown (see [RR], pp. 136–140), that if $f \in C^\infty[a, b]$, the Composite Trapezoidal rule can also be written with an error term in the form

$$\int_a^b f(x)\, dx = \frac{h}{2}\left[f(a) + 2\sum_{j=1}^{n-1} f(x_j) + f(b)\right] + K_1h^2 + K_2h^4 + K_3h^6 + \cdots, \quad (4.33)$$

where each K_i is a constant that depends only on $f^{(2i-1)}(a)$ and $f^{(2i-1)}(b)$.

Recall from Section 4.2 that Richardson extrapolation can be performed on any approximation procedure whose truncation error is of the form

$$\sum_{j=1}^{m-1} K_j h^{\alpha_j} + O(h^{\alpha_m}),$$

for a collection of constants K_j and when $\alpha_1 < \alpha_2 < \alpha_3 < \cdots < \alpha_m$. In that section we gave demonstrations to illustrate how effective this techniques is when the approximation procedure has a truncation error with only even powers of h, that is, when the truncation error has the form.

$$\sum_{j=1}^{m-1} K_j h^{2j} + O(h^{2m}).$$

Werner Romberg (1909–2003) devised this procedure for improving the accuracy of the Trapezoidal rule by eliminating the successive terms in the asymptotic expansion in 1955.

Because the Composite Trapezoidal rule has this form, it is an obvious candidate for extrapolation. This results in a technique known as **Romberg integration.**

To approximate the integral $\int_a^b f(x)\,dx$ we use the results of the Composite Trapezoidal rule with $n = 1, 2, 4, 8, 16, \ldots$, and denote the resulting approximations, respectively, by $R_{1,1}, R_{2,1}, R_{3,1}$, etc. We then apply extrapolation in the manner given in Section 4.2, that is, we obtain $O(h^4)$ approximations $R_{2,2}, R_{3,2}, R_{4,2}$, etc., by

$$R_{k,2} = R_{k,1} + \frac{1}{3}(R_{k,1} - R_{k-1,1}), \quad \text{for } k = 2, 3, \ldots$$

Then $O(h^6)$ approximations $R_{3,3}, R_{4,3}, R_{5,3}$, etc., by

$$R_{k,3} = R_{k,2} + \frac{1}{15}(R_{k,2} - R_{k-1,2}), \quad \text{for } k = 3, 4, \ldots.$$

In general, after the appropriate $R_{k,j-1}$ approximations have been obtained, we determine the $O(h^{2j})$ approximations from

$$R_{k,j} = R_{k,j-1} + \frac{1}{4^{j-1} - 1}(R_{k,j-1} - R_{k-1,j-1}), \quad \text{for } k = j, j+1, \ldots$$

Example 1 Use the Composite Trapezoidal rule to find approximations to $\int_0^\pi \sin x\,dx$ with $n = 1, 2, 4, 8$, and 16. Then perform Romberg extrapolation on the results.

The Composite Trapezoidal rule for the various values of n gives the following approximations to the true value 2.

$$R_{1,1} = \frac{\pi}{2}[\sin 0 + \sin \pi] = 0;$$

$$R_{2,1} = \frac{\pi}{4}\left[\sin 0 + 2\sin\frac{\pi}{2} + \sin \pi\right] = 1.57079633;$$

$$R_{3,1} = \frac{\pi}{8}\left[\sin 0 + 2\left(\sin\frac{\pi}{4} + \sin\frac{\pi}{2} + \sin\frac{3\pi}{4}\right) + \sin \pi\right] = 1.89611890;$$

$$R_{4,1} = \frac{\pi}{16}\left[\sin 0 + 2\left(\sin\frac{\pi}{8} + \sin\frac{\pi}{4} + \cdots + \sin\frac{3\pi}{4} + \sin\frac{7\pi}{8}\right) + \sin \pi\right] = 1.97423160;$$

$$R_{5,1} = \frac{\pi}{32}\left[\sin 0 + 2\left(\sin\frac{\pi}{16} + \sin\frac{\pi}{8} + \cdots + \sin\frac{7\pi}{8} + \sin\frac{15\pi}{16}\right) + \sin \pi\right] = 1.99357034.$$

The $O(h^4)$ approximations are

$$R_{2,2} = R_{2,1} + \frac{1}{3}(R_{2,1} - R_{1,1}) = 2.09439511; \quad R_{3,2} = R_{3,1} + \frac{1}{3}(R_{3,1} - R_{2,1}) = 2.00455976;$$

$$R_{4,2} = R_{4,1} + \frac{1}{3}(R_{4,1} - R_{3,1}) = 2.00026917; \quad R_{5,2} = R_{5,1} + \frac{1}{3}(R_{5,1} - R_{4,1}) = 2.00001659;$$

The $O(h^6)$ approximations are

$$R_{3,3} = R_{3,2} + \frac{1}{15}(R_{3,2} - R_{2,2}) = 1.99857073; \quad R_{4,3} = R_{4,2} + \frac{1}{15}(R_{4,2} - R_{3,2}) = 1.99998313;$$

$$R_{5,3} = R_{5,2} + \frac{1}{15}(R_{5,2} - R_{4,2}) = 1.99999975.$$

The two $O(h^8)$ approximations are

$$R_{4,4} = R_{4,3} + \frac{1}{63}(R_{4,3} - R_{3,3}) = 2.00000555; \quad R_{5,4} = R_{5,3} + \frac{1}{63}(R_{5,3} - R_{4,3}) = 2.00000001,$$

and the final $O(h^{10})$ approximation is

$$R_{5,5} = R_{5,4} + \frac{1}{255}(R_{5,4} - R_{4,4}) = 1.99999999.$$

These results are shown in Table 4.9. ■

Table 4.9

0				
1.57079633	2.09439511			
1.89611890	2.00455976	1.99857073		
1.97423160	2.00026917	1.99998313	2.00000555	
1.99357034	2.00001659	1.99999975	2.00000001	1.99999999

Notice that when generating the approximations for the Composite Trapezoidal rule approximations in Example 1, each consecutive approximation included all the functions evaluations from the previous approximation. That is, $R_{1,1}$ used evaluations at 0 and π, $R_{2,1}$ used these evaluations and added an evaluation at the intermediate point $\pi/2$. Then $R_{3,1}$ used the evaluations of $R_{2,1}$ and added two additional intermediate ones at $\pi/4$ and $3\pi/4$. This pattern continues with $R_{4,1}$ using the same evaluations as $R_{3,1}$ but adding evaluations at the 4 intermediate points $\pi/8$, $3\pi/8$, $5\pi/8$, and $7\pi/8$, and so on.

This evaluation procedure for Composite Trapezoidal rule approximations holds for an integral on any interval $[a, b]$. In general, the Composite Trapezoidal rule denoted $R_{k+1,1}$ uses the same evaluations as $R_{k,1}$ but adds evaluations at the 2^{k-2} intermediate points. Efficient calculation of these approximations can therefore be done in a recursive manner.

To obtain the Composite Trapezoidal rule approximations for $\int_a^b f(x)\,dx$, let $h_k = (b-a)/m_k = (b-a)/2^{k-1}$. Then

$$R_{1,1} = \frac{h_1}{2}[f(a) + f(b)] = \frac{(b-a)}{2}[f(a) + f(b)];$$

and

$$R_{2,1} = \frac{h_2}{2}[f(a) + f(b) + 2f(a + h_2)].$$

By reexpressing this result for $R_{2,1}$ we can incorporate the previously determined approximation $R_{1,1}$

$$R_{2,1} = \frac{(b-a)}{4}\left[f(a) + f(b) + 2f\left(a + \frac{(b-a)}{2}\right)\right] = \frac{1}{2}[R_{1,1} + h_1 f(a + h_2)].$$

In a similar manner we can write

$$R_{3,1} = \frac{1}{2}\{R_{2,1} + h_2[f(a + h_3) + f(a + 3h_3)]\};$$

and, in general (see Figure 4.10 on page 216), we have

$$R_{k,1} = \frac{1}{2}\left[R_{k-1,1} + h_{k-1}\sum_{i=1}^{2^{k-2}} f(a + (2i-1)h_k)\right], \tag{4.34}$$

for each $k = 2, 3, \ldots, n$. (See Exercises 14 and 15.)

Figure 4.10

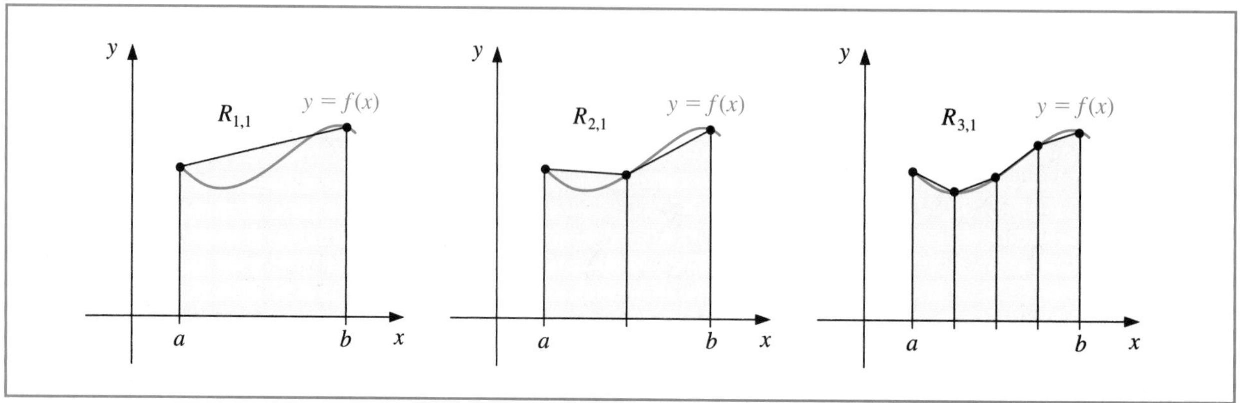

Extrapolation then is used to produce $O(h_k^{2j})$ approximations by

$$R_{k,j} = R_{k,j-1} + \frac{1}{4^{j-1}-1}(R_{k,j-1} - R_{k-1,j-1}), \quad \text{for } k = j, j+1, \ldots$$

as shown in Table 4.10.

Table 4.10

k	$O(h_k^2)$	$O(h_k^4)$	$O(h_k^6)$	$O(h_k^8)$		$O(h_k^{2n})$
1	$R_{1,1}$					
2	$R_{2,1}$	$R_{2,2}$				
3	$R_{3,1}$	$R_{3,2}$	$R_{3,3}$			
4	$R_{4,1}$	$R_{4,2}$	$R_{4,3}$	$R_{4,4}$		
$\vdots$	$\vdots$	$\vdots$	$\vdots$	$\vdots$	$\ddots$	
n	$R_{n,1}$	$R_{n,2}$	$R_{n,3}$	$R_{n,4}$	$\cdots$	$R_{n,n}$

The effective method to construct the Romberg table makes use of the highest order of approximation at each step. That is, it calculates the entries row by row, in the order $R_{1,1}$, $R_{2,1}$, $R_{2,2}$, $R_{3,1}$, $R_{3,2}$, $R_{3,3}$, etc. This also permits an entire new row in the table to be calculated by doing only one additional application of the Composite Trapezoidal rule. It then uses a simple averaging on the previously calculated values to obtain the remaining entries in the row. Remember

- Calculate the Romberg table one complete row at a time.

Example 2 Add an additional extrapolation row to Table 4.10 to approximate $\int_0^\pi \sin x\, dx$.

Solution To obtain the additional row we need the trapezoidal approximation

$$R_{6,1} = \frac{1}{2}\left[R_{5,1} + \frac{\pi}{16}\sum_{k=1}^{2^4} \sin\frac{(2k-1)\pi}{32}\right] = 1.99839336.$$

The values in Table 4.10 give

$$R_{6,2} = R_{6,1} + \frac{1}{3}(R_{6,1} - R_{5,1}) = 1.99839336 + \frac{1}{3}(1.99839336 - 1.99357035)$$
$$= 2.00000103;$$
$$R_{6,3} = R_{6,2} + \frac{1}{15}(R_{6,2} - R_{5,2}) = 2.00000103 + \frac{1}{15}(2.00000103 - 2.00001659)$$
$$= 2.00000000;$$
$$R_{6,4} = R_{6,3} + \frac{1}{63}(R_{6,3} - R_{5,3}) = 2.00000000;$$
$$R_{6,5} = R_{6,4} + \frac{1}{255}(R_{6,4} - R_{5,4}) = 2.00000000;$$

and $R_{6,6} = R_{6,5} + \frac{1}{1023}(R_{6,5} - R_{5,5}) = 2.00000000$. The new extrapolation table is shown in Table 4.11. ■

Table 4.11

0					
1.57079633	2.09439511				
1.89611890	2.00455976	1.99857073			
1.97423160	2.00026917	1.99998313	2.00000555		
1.99357034	2.00001659	1.99999975	2.00000001	1.99999999	
1.99839336	2.00000103	2.00000000	2.00000000	2.00000000	2.00000000

Notice that all the extrapolated values except for the first (in the first row of the second column) are more accurate than the best composite trapezoidal approximation (in the last row of the first column). Although there are 21 entries in Table 4.11, only the six in the left column require function evaluations since these are the only entries generated by the Composite Trapezoidal rule; the other entries are obtained by an averaging process. In fact, because of the recurrence relationship of the terms in the left column, the only function evaluations needed are those to compute the final Composite Trapezoidal rule approximation. In general, $R_{k,1}$ requires $1 + 2^{k-1}$ function evaluations, so in this case $1 + 2^5 = 33$ are needed.

Algorithm 4.2 uses the recursive procedure to find the initial Composite Trapezoidal Rule approximations and computes the results in the table row by row.

ALGORITHM 4.2

Romberg

To approximate the integral $I = \int_a^b f(x)\,dx$, select an integer $n > 0$.

INPUT endpoints a, b; integer n.

OUTPUT an array R. *(Compute R by rows; only the last 2 rows are saved in storage.)*

Step 1 Set $h = b - a$;
$R_{1,1} = \frac{h}{2}(f(a) + f(b))$.

Step 2 OUTPUT $(R_{1,1})$.

Step 3 For $i = 2, \ldots, n$ do Steps 4–8.

Step 4 Set $R_{2,1} = \frac{1}{2}\left[R_{1,1} + h\sum_{k=1}^{2^{i-2}} f(a + (k - 0.5)h)\right]$.

(*Approximation from Trapezoidal method.*)

Step 5 For $j = 2, \ldots, i$

set $R_{2,j} = R_{2,j-1} + \frac{R_{2,j-1} - R_{1,j-1}}{4^{j-1} - 1}$. (*Extrapolation.*)

Step 6 OUTPUT ($R_{2,j}$ for $j = 1, 2, \ldots, i$).

Step 7 Set $h = h/2$.

Step 8 For $j = 1, 2, \ldots, i$ set $R_{1,j} = R_{2,j}$. (*Update row 1 of R.*)

Step 9 STOP. ■

Algorithm 4.2 requires a preset integer n to determine the number of rows to be generated. We could also set an error tolerance for the approximation and generate n, within some upper bound, until consecutive diagonal entries $R_{n-1,n-1}$ and $R_{n,n}$ agree to within the tolerance. To guard against the possibility that two consecutive row elements agree with each other but not with the value of the integral being approximated, it is common to generate approximations until not only $|R_{n-1,n-1} - R_{n,n}|$ is within the tolerance, but also $|R_{n-2,n-2} - R_{n-1,n-1}|$. Although not a universal safeguard, this will ensure that two differently generated sets of approximations agree within the specified tolerance before $R_{n,n}$, is accepted as sufficiently accurate.

Romberg integration can be performed with the *Quadrature* command in the *NumericalAnalysis* subpackage of Maple's *Student* package. For example, after loading the package and defining the function and interval, the command

$Quadrature(f(x), x = a..b, method = romberg_6, output = information)$

produces the values shown in Table 4.11 together with the information that 6 applications of the Trapezoidal rule were used and 33 function evaluations were required.

The adjective *cautious* used in the description of a numerical method indicates that a check is incorporated to determine if the continuity hypotheses are likely to be true.

Romberg integration applied to a function f on the interval $[a, b]$ relies on the assumption that the Composite Trapezoidal rule has an error term that can be expressed in the form of Eq. (4.33); that is, we must have $f \in C^{2k+2}[a, b]$ for the kth row to be generated. General-purpose algorithms using Romberg integration include a check at each stage to ensure that this assumption is fulfilled. These methods are known as *cautious Romberg algorithms* and are described in [Joh]. This reference also describes methods for using the Romberg technique as an adaptive procedure, similar to the adaptive Simpson's rule that will be discussed in Section 4.6.

EXERCISE SET 4.5

1. Use Romberg integration to compute $R_{3,3}$ for the following integrals.

   a. $\int_0^1 x^2 e^{-x}\, dx$

   b. $\int_1^{1.5} x^2 \ln x\, dx$

   c. $\int_1^{1.6} \frac{2x}{x^2 - 4}\, dx$

   d. $\int_0^{\pi/4} x^2 \sin x\, dx$

e. $\int_0^{\pi/4} e^{3x} \sin 2x \, dx$

f. $\int_0^{0.35} \frac{2}{x^2-4} \, dx$

g. $\int_3^{3.5} \frac{x}{\sqrt{x^2-4}} \, dx$

h. $\int_0^{\pi/4} (\cos x)^2 \, dx$

2. Use Romberg integration to compute $R_{3,3}$ for the following integrals.

a. $\int_{-1}^{1} (\cos x)^2 \, dx$

b. $\int_e^{2e} \frac{1}{x \ln x} \, dx$

c. $\int_1^{4} \left((\sin x)^2 - 2x \sin x + 1\right) \, dx$

d. $\int_{-0.75}^{0.75} x \ln(x+1) \, dx$

3. Calculate $R_{4,4}$ for the integrals in Exercise 1.

4. Calculate $R_{4,4}$ for the integrals in Exercise 2.

5. Use Romberg integration to approximate the integrals in Exercise 1 to within 10^{-6}. Compute the Romberg table until either $|R_{n-1,n-1} - R_{n,n}| < 10^{-6}$, or $n = 10$. Compare your results to the exact values of the integrals.

6. Use Romberg integration to approximate the integrals in Exercise 2 to within 10^{-6}. Compute the Romberg table until either $|R_{n-1,n-1} - R_{n,n}| < 10^{-6}$, or $n = 10$. Compare your results to the exact values of the integrals.

7. Use the following data to approximate $\int_1^5 f(x)\, dx$ as accurately as possible.

x	1	2	3	4	5
$f(x)$	2.4142	2.6734	2.8974	3.0976	3.2804

8. Romberg integration is used to approximate

$$\int_0^1 \frac{x^2}{1+x^3} \, dx.$$

If $R_{11} = 0.250$ and $R_{22} = 0.2315$, what is R_{21}?

9. Romberg integration is used to approximate

$$\int_2^3 f(x) \, dx.$$

If $f(2) = 0.51342$, $f(3) = 0.36788$, $R_{31} = 0.43687$, and $R_{33} = 0.43662$, find $f(2.5)$.

10. Romberg integration for approximating $\int_0^1 f(x)\, dx$ gives $R_{11} = 4$ and $R_{22} = 5$. Find $f(1/2)$.

11. Romberg integration for approximating $\int_a^b f(x)\, dx$ gives $R_{11} = 8$, $R_{22} = 16/3$, and $R_{33} = 208/45$. Find R_{31}.

12. Use Romberg integration to compute the following approximations to

$$\int_0^{48} \sqrt{1 + (\cos x)^2} \, dx.$$

[*Note:* The results in this exercise are most interesting if you are using a device with between seven- and nine-digit arithmetic.]

a. Determine $R_{1,1}$, $R_{2,1}$, $R_{3,1}$, $R_{4,1}$, and $R_{5,1}$, and use these approximations to predict the value of the integral.

b. Determine $R_{2,2}$, $R_{3,3}$, $R_{4,4}$, and $R_{5,5}$, and modify your prediction.

c. Determine $R_{6,1}$, $R_{6,2}$, $R_{6,3}$, $R_{6,4}$, $R_{6,5}$, and $R_{6,6}$, and modify your prediction.

d. Determine $R_{7,7}$, $R_{8,8}$, $R_{9,9}$, and $R_{10,10}$, and make a final prediction.

e. Explain why this integral causes difficulty with Romberg integration and how it can be reformulated to more easily determine an accurate approximation.

13. Show that the approximation obtained from $R_{k,2}$ is the same as that given by the Composite Simpson's rule described in Theorem 4.4 with $h = h_k$.

14. Show that, for any k,

$$\sum_{i=1}^{2^{k-1}-1} f\left(a+\frac{i}{2}h_{k-1}\right)=\sum_{i=1}^{2^{k-2}} f\left(a+\left(i-\frac{1}{2}\right)h_{k-1}\right)+\sum_{i=1}^{2^{k-2}-1} f(a+ih_{k-1}).$$

15. Use the result of Exercise 14 to verify Eq. (4.34); that is, show that for all k,

$$R_{k,1}=\frac{1}{2}\left[R_{k-1,1}+h_{k-1}\sum_{i=1}^{2^{k-2}} f\left(a+\left(i-\frac{1}{2}\right)h_{k-1}\right)\right].$$

16. In Exercise 26 of Section 1.1, a Maclaurin series was integrated to approximate erf(1), where erf(x) is the normal distribution error function defined by

$$\operatorname{erf}(x)=\frac{2}{\sqrt{\pi}}\int_0^x e^{-t^2}\,dt.$$

Approximate erf(1) to within 10^{-7}.

4.6 Adaptive Quadrature Methods

The composite formulas are very effective in most situations, but they suffer occasionally because they require the use of equally-spaced nodes. This is inappropriate when integrating a function on an interval that contains both regions with large functional variation and regions with small functional variation.

Illustration The unique solution to the differential equation $y''+6y'+25=0$ that additionally satisfies $y(0)=0$ and $y'(0)=4$ is $y(x)=e^{-3x}\sin 4x$. Functions of this type are common in mechanical engineering because they describe certain features of spring and shock absorber systems, and in electrical engineering because they are common solutions to elementary circuit problems. The graph of $y(x)$ for x in the interval $[0,4]$ is shown in Figure 4.11.

Figure 4.11

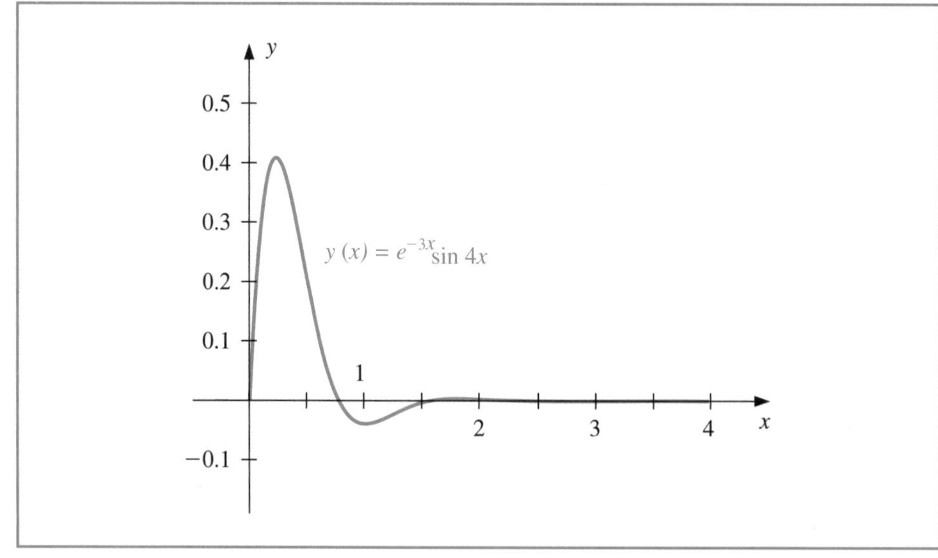

Suppose that we need the integral of $y(x)$ on $[0, 4]$. The graph indicates that the integral on $[3, 4]$ must be very close to 0, and on $[2, 3]$ would also not be expected to be large. However, on $[0, 2]$ there is significant variation of the function and it is not at all clear what the integral is on this interval. This is an example of a situation where composite integration would be inappropriate. A very low order method could be used on $[2, 4]$, but a higher-order method would be necessary on $[0, 2]$. □

The question we will consider in this section is:

- How can we determine what technique should be applied on various portions of the interval of integration, and how accurate can we expect the final approximation to be?

We will see that under quite reasonable conditions we can answer this question and also determine approximations that satisfy given accuracy requirements.

If the approximation error for an integral on a given interval is to be evenly distributed, a smaller step size is needed for the large-variation regions than for those with less variation. An efficient technique for this type of problem should predict the amount of functional variation and adapt the step size as necessary. These methods are called **Adaptive quadrature methods**. Adaptive methods are particularly popular for inclusion in professional software packages because, in addition to being efficient, they generally provide approximations that are within a given specified tolerance.

In this section we consider an Adaptive quadrature method and see how it can be used to reduce approximation error and also to predict an error estimate for the approximation that does not rely on knowledge of higher derivatives of the function. The method we discuss is based on the Composite Simpson's rule, but the technique is easily modified to use other composite procedures.

Suppose that we want to approximate $\int_a^b f(x)\,dx$ to within a specified tolerance $\varepsilon > 0$. The first step is to apply Simpson's rule with step size $h = (b - a)/2$. This produces (see Figure 4.12)

$$\int_a^b f(x)\,dx = S(a, b) - \frac{h^5}{90} f^{(4)}(\xi), \quad \text{for some } \xi \text{ in } (a, b), \tag{4.35}$$

where we denote the Simpson's rule approximation on $[a, b]$ by

$$S(a, b) = \frac{h}{3}[f(a) + 4f(a + h) + f(b)].$$

Figure 4.12

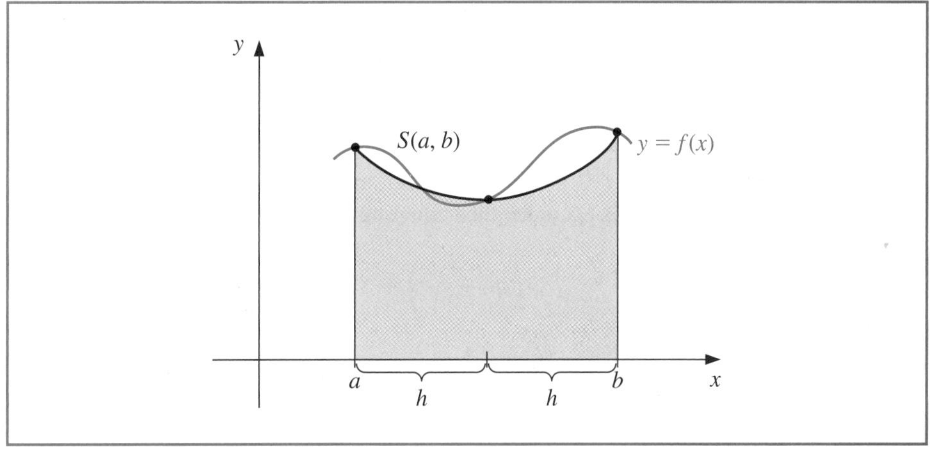

The next step is to determine an accuracy approximation that does not require $f^{(4)}(\xi)$. To do this, we apply the Composite Simpson's rule with $n = 4$ and step size $(b-a)/4 = h/2$, giving

$$\int_a^b f(x)\,dx = \frac{h}{6}\left[f(a) + 4f\left(a+\frac{h}{2}\right) + 2f(a+h) + 4f\left(a+\frac{3h}{2}\right) + f(b)\right]$$
$$-\left(\frac{h}{2}\right)^4 \frac{(b-a)}{180} f^{(4)}(\tilde{\xi}), \tag{4.36}$$

for some $\tilde{\xi}$ in (a,b). To simplify notation, let

$$S\left(a, \frac{a+b}{2}\right) = \frac{h}{6}\left[f(a) + 4f\left(a+\frac{h}{2}\right) + f(a+h)\right]$$

and

$$S\left(\frac{a+b}{2}, b\right) = \frac{h}{6}\left[f(a+h) + 4f\left(a+\frac{3h}{2}\right) + f(b)\right].$$

Then Eq. (4.36) can be rewritten (see Figure 4.13) as

$$\int_a^b f(x)\,dx = S\left(a, \frac{a+b}{2}\right) + S\left(\frac{a+b}{2}, b\right) - \frac{1}{16}\left(\frac{h^5}{90}\right) f^{(4)}(\tilde{\xi}). \tag{4.37}$$

Figure 4.13

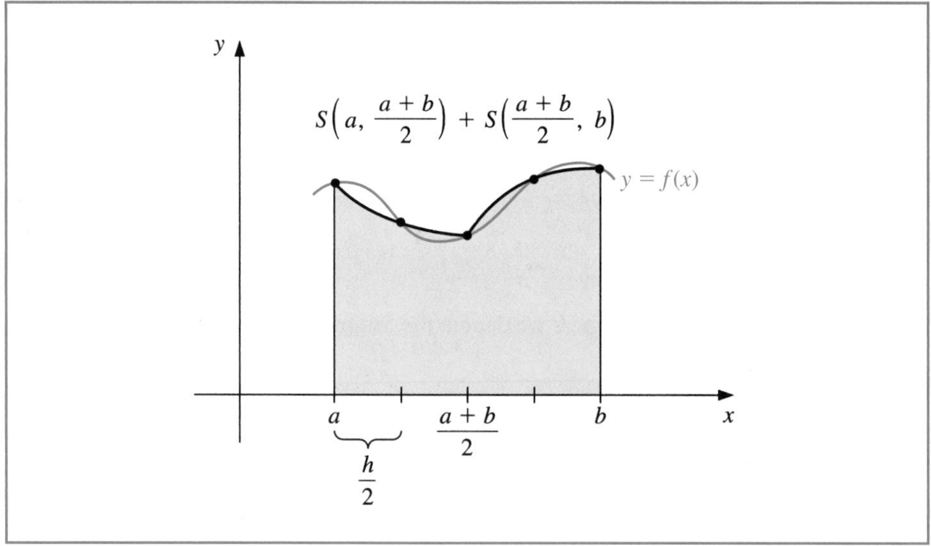

The error estimation is derived by assuming that $\xi \approx \tilde{\xi}$ or, more precisely, that $f^{(4)}(\xi) \approx f^{(4)}(\tilde{\xi})$, and the success of the technique depends on the accuracy of this assumption. If it is accurate, then equating the integrals in Eqs. (4.35) and (4.37) gives

$$S\left(a, \frac{a+b}{2}\right) + S\left(\frac{a+b}{2}, b\right) - \frac{1}{16}\left(\frac{h^5}{90}\right) f^{(4)}(\xi) \approx S(a,b) - \frac{h^5}{90} f^{(4)}(\xi),$$

so

$$\frac{h^5}{90} f^{(4)}(\xi) \approx \frac{16}{15}\left[S(a,b) - S\left(a, \frac{a+b}{2}\right) - S\left(\frac{a+b}{2}, b\right)\right].$$

Using this estimate in Eq. (4.37) produces the error estimation

$$\left|\int_a^b f(x)\,dx - S\left(a, \frac{a+b}{2}\right) - S\left(\frac{a+b}{2}, b\right)\right| \approx \frac{1}{16}\left(\frac{h^5}{90}\right) f^{(4)}(\xi)$$

$$\approx \frac{1}{15}\left|S(a,b) - S\left(a, \frac{a+b}{2}\right) - S\left(\frac{a+b}{2}, b\right)\right|.$$

This implies that $S(a,(a+b)/2) + S((a+b)/2, b)$ approximates $\int_a^b f(x)\,dx$ about 15 times better than it agrees with the computed value $S(a,b)$. Thus, if

$$\left|S(a,b) - S\left(a, \frac{a+b}{2}\right) - S\left(\frac{a+b}{2}, b\right)\right| < 15\varepsilon, \tag{4.38}$$

we expect to have

$$\left|\int_a^b f(x)\,dx - S\left(a, \frac{a+b}{2}\right) - S\left(\frac{a+b}{2}, b\right)\right| < \varepsilon, \tag{4.39}$$

and

$$S\left(a, \frac{a+b}{2}\right) + S\left(\frac{a+b}{2}, b\right)$$

is assumed to be a sufficiently accurate approximation to $\int_a^b f(x)\,dx$.

Example 1 Check the accuracy of the error estimate given in (4.38) and (4.39) when applied to the integral

$$\int_0^{\pi/2} \sin x\,dx = 1.$$

by comparing

$$\frac{1}{15}\left|S\left(0, \frac{\pi}{2}\right) - S\left(0, \frac{\pi}{4}\right) - S\left(\frac{\pi}{4}, \frac{\pi}{2}\right)\right| \quad \text{to} \quad \left|\int_0^{\pi/2} \sin x\,dx - S\left(0, \frac{\pi}{4}\right) - S\left(\frac{\pi}{4}, \frac{\pi}{2}\right)\right|.$$

Solution We have

$$S\left(0, \frac{\pi}{2}\right) = \frac{\pi/4}{3}\left[\sin 0 + 4\sin\frac{\pi}{4} + \sin\frac{\pi}{2}\right] = \frac{\pi}{12}(2\sqrt{2}+1) = 1.002279878$$

and

$$S\left(0, \frac{\pi}{4}\right) + S\left(\frac{\pi}{4}, \frac{\pi}{2}\right) = \frac{\pi/8}{3}\left[\sin 0 + 4\sin\frac{\pi}{8} + 2\sin\frac{\pi}{4} + 4\sin\frac{3\pi}{8} + \sin\frac{\pi}{2}\right]$$

$$= 1.000134585.$$

So

$$\left|S\left(0, \frac{\pi}{2}\right) - S\left(0, \frac{\pi}{4}\right) - S\left(\frac{\pi}{4}, \frac{\pi}{2}\right)\right| = |1.002279878 - 1.000134585| = 0.002145293.$$

The estimate for the error obtained when using $S(a,(a+b)) + S((a+b), b)$ to approximate $\int_a^b f(x)\,dx$ is consequently

$$\frac{1}{15}\left|S\left(0, \frac{\pi}{2}\right) - S\left(0, \frac{\pi}{4}\right) - S\left(\frac{\pi}{4}, \frac{\pi}{2}\right)\right| = 0.000143020,$$

which closely approximates the actual error

$$\left| \int_0^{\pi/2} \sin x \, dx - 1.000134585 \right| = 0.000134585,$$

even though $D_x^4 \sin x = \sin x$ varies significantly in the interval $(0, \pi/2)$. ■

When the approximations in (4.38) differ by more than 15ε, we can apply the Simpson's rule technique individually to the subintervals $[a, (a+b)/2]$ and $[(a+b)/2, b]$. Then we use the error estimation procedure to determine if the approximation to the integral on each subinterval is within a tolerance of $\varepsilon/2$. If so, we sum the approximations to produce an approximation to $\int_a^b f(x)\,dx$ within the tolerance ε.

If the approximation on one of the subintervals fails to be within the tolerance $\varepsilon/2$, then that subinterval is itself subdivided, and the procedure is reapplied to the two subintervals to determine if the approximation on each subinterval is accurate to within $\varepsilon/4$. This halving procedure is continued until each portion is within the required tolerance.

Problems can be constructed for which this tolerance will never be met, but the technique is usually successful, because each subdivision typically increases the accuracy of the approximation by a factor of 16 while requiring an increased accuracy factor of only 2.

Algorithm 4.3 details this Adaptive quadrature procedure for Simpson's rule, although some technical difficulties arise that require the implementation to differ slightly from the preceding discussion. For example, in Step 1 the tolerance has been set at 10ε rather than the 15ε figure in Inequality (4.38). This bound is chosen conservatively to compensate for error in the assumption $f^{(4)}(\xi) \approx f^{(4)}(\tilde{\xi})$. In problems where $f^{(4)}$ is known to be widely varying, this bound should be decreased even further.

It is a good idea to include a margin of safety when it is impossible to verify accuracy assumptions.

The procedure listed in the algorithm first approximates the integral on the leftmost subinterval in a subdivision. This requires the efficient storing and recalling of previously computed functional evaluations for the nodes in the right half subintervals. Steps 3, 4, and 5 contain a stacking procedure with an indicator to keep track of the data that will be required for calculating the approximation on the subinterval immediately adjacent and to the right of the subinterval on which the approximation is being generated. The method is easier to implement using a recursive programming language.

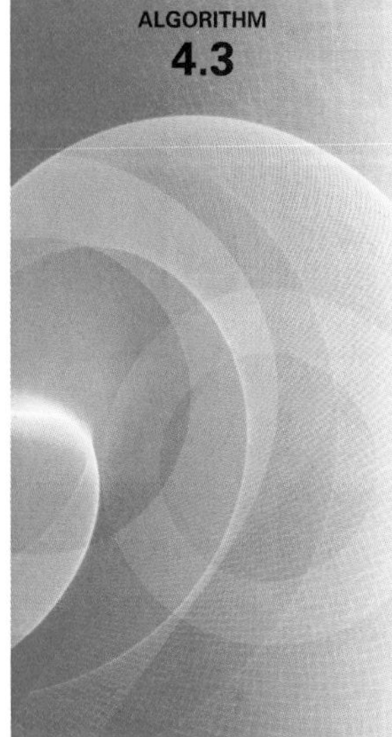

Adaptive Quadrature

To approximate the integral $I = \int_a^b f(x)\,dx$ to within a given tolerance:

INPUT endpoints a, b; tolerance TOL; limit N to number of levels.

OUTPUT approximation APP or message that N is exceeded.

Step 1 Set $APP = 0$;
$i = 1$;
$TOL_i = 10\, TOL$;
$a_i = a$;
$h_i = (b-a)/2$;
$FA_i = f(a)$;
$FC_i = f(a + h_i)$;
$FB_i = f(b)$;
$S_i = h_i(FA_i + 4FC_i + FB_i)/3$; *(Approximation from Simpson's method for entire interval.)*
$L_i = 1$.

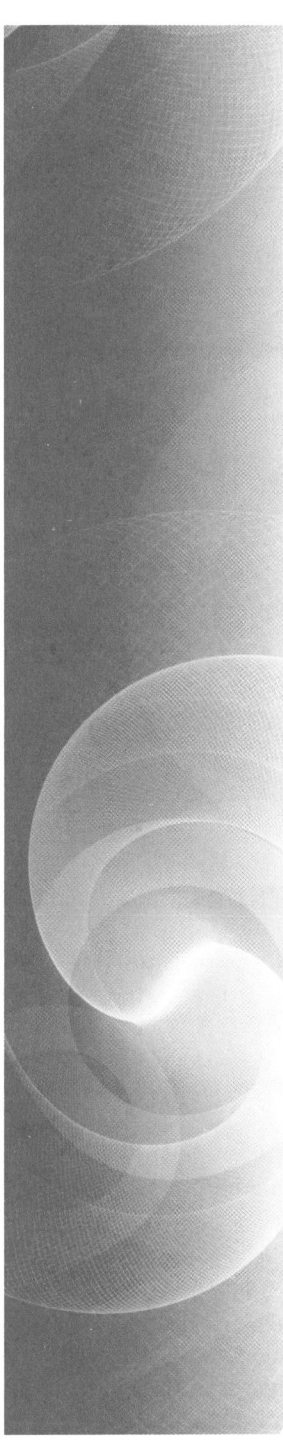

Step 2 While $i > 0$ do Steps 3–5.

Step 3 Set $FD = f(a_i + h_i/2)$;
$FE = f(a_i + 3h_i/2)$;
$S1 = h_i(FA_i + 4FD + FC_i)/6$; *(Approximations from Simpson's method for halves of subintervals.)*
$S2 = h_i(FC_i + 4FE + FB_i)/6$;
$v_1 = a_i$; *(Save data at this level.)*
$v_2 = FA_i$;
$v_3 = FC_i$;
$v_4 = FB_i$;
$v_5 = h_i$;
$v_6 = TOL_i$;
$v_7 = S_i$;
$v_8 = L_i$.

Step 4 Set $i = i - 1$. *(Delete the level.)*

Step 5 If $|S1 + S2 - v_7| < v_6$
then set $APP = APP + (S1 + S2)$
else
if $(v_8 \geq N)$
then
OUTPUT ('LEVEL EXCEEDED'); *(Procedure fails.)*
STOP.
else *(Add one level.)*
set $i = i + 1$; *(Data for right half subinterval.)*
$a_i = v_1 + v_5$;
$FA_i = v_3$;
$FC_i = FE$;
$FB_i = v_4$;
$h_i = v_5/2$;
$TOL_i = v_6/2$;
$S_i = S2$;
$L_i = v_8 + 1$;
set $i = i + 1$; *(Data for left half subinterval.)*
$a_i = v_1$;
$FA_i = v_2$;
$FC_i = FD$;
$FB_i = v_3$;
$h_i = h_{i-1}$;
$TOL_i = TOL_{i-1}$;
$S_i = S1$;
$L_i = L_{i-1}$.

Step 6 OUTPUT (APP); *(APP approximates I to within TOL.)*
STOP. ■

Illustration The graph of the function $f(x) = (100/x^2)\sin(10/x)$ for x in $[1, 3]$ is shown in Figure 4.14. Using the Adaptive Quadrature Algorithm 4.3 with tolerance 10^{-4} to approximate $\int_1^3 f(x)\,dx$ produces -1.426014, a result that is accurate to within 1.1×10^{-5}. The approximation required that Simpson's rule with $n = 4$ be performed on the 23 subintervals whose endpoints are shown on the horizontal axis in Figure 4.14. The total number of functional evaluations required for this approximation is 93.

Figure 4.14

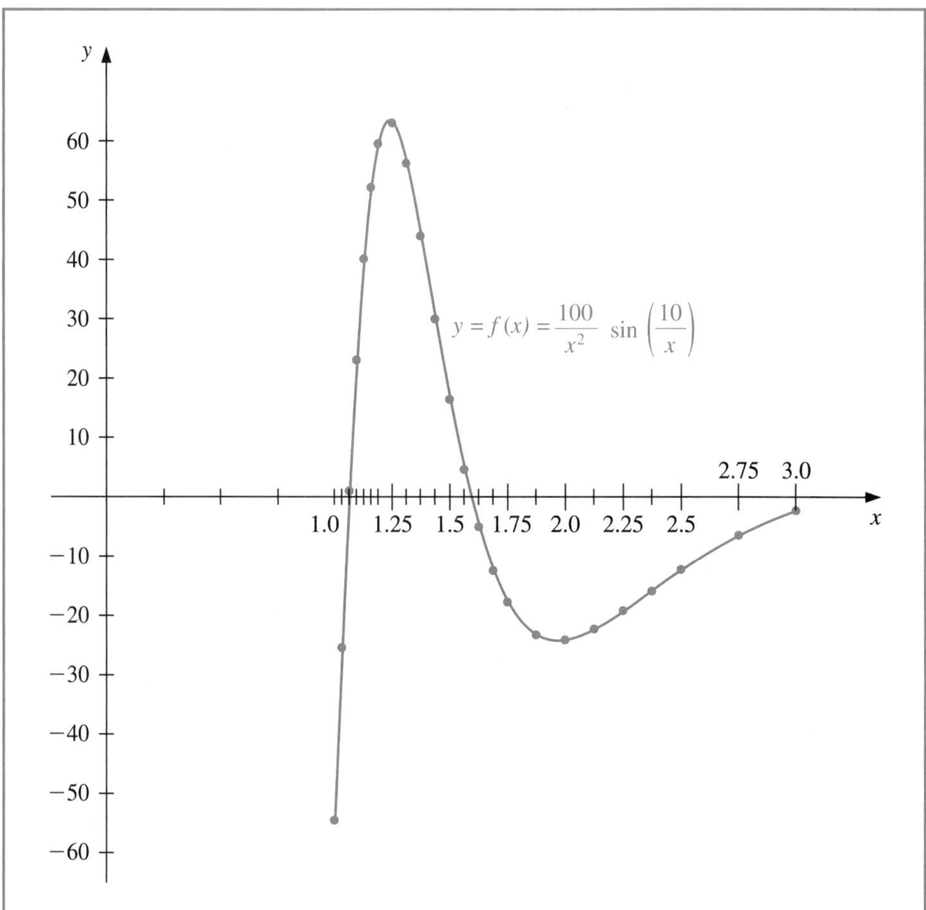

The largest value of h for which the standard Composite Simpson's rule gives 10^{-4} accuracy is $h = 1/88$. This application requires 177 function evaluations, nearly twice as many as Adaptive quadrature. □

Adaptive quadrature can be performed with the *Quadrature* command in the *Numerical-Analysis* subpackage of Maple's *Student* package. In this situation the option *adaptive* = *true* is used. For example, to produce the values in the Illustration we first load the package and define the function and interval with

$$f := x \to \frac{100}{x^2} \cdot \sin\left(\frac{10}{x}\right);\ a := 1.0;\ b := 3.0$$

Then give the *NumericalAnalysis* command

Quadrature$(f(x), x = a..b, adaptive = true, method = [simpson, 10^{-4}], output = information)$

This produces the approximation -1.42601481 and a table that lists all the intervals on which Simpson's rule was employed and whether the appropriate tolerance was satisfied (indicated by the word PASS) or was not satisfied (indicated by the word fail). It also gives what Maple thinks is the correct value of the integral to the decimal places listed, in this case -1.42602476. Then it gives the absolute and relative errors, 9.946×10^{-6} and 6.975×10^{-4}, respectively, assuming that its correct value is accurate.

EXERCISE SET 4.6

1. Compute the Simpson's rule approximations $S(a, b)$, $S(a, (a+b)/2)$, and $S((a+b)/2, b)$ for the following integrals, and verify the estimate given in the approximation formula.

   a. $\int_0^1 x^2 e^{-x}\, dx$ b. $\int_1^{1.5} x^2 \ln x\, dx$

   c. $\int_1^{1.6} \frac{2x}{x^2-4}\, dx$ d. $\int_0^{\pi/4} x^2 \sin x\, dx$

   e. $\int_0^{\pi/4} e^{3x} \sin 2x\, dx$ f. $\int_0^{0.35} \frac{2}{x^2-4}\, dx$

   g. $\int_3^{3.5} \frac{x}{\sqrt{x^2-4}}\, dx$ h. $\int_0^{\pi/4} (\cos x)^2\, dx$

2. Use Adaptive quadrature to find approximations to within 10^{-3} for the integrals in Exercise 1. Do not use a computer program to generate these results.

3. Use Adaptive quadrature to approximate the following integrals to within 10^{-5}.

   a. $\int_0^{\pi} (\sin x + \cos x)\, dx$ b. $\int_1^2 (x + \sin 4x)\, dx$

   c. $\int_{-1}^1 x \sin 4x\, dx$ d. $\int_0^{\pi/2} (6\cos 4x + 4\sin 6x)e^x\, dx$

4. Use Adaptive quadrature to approximate the following integrals to within 10^{-5}.

   a. $\int_1^3 e^{2x} \sin 3x\, dx$ b. $\int_1^3 e^{3x} \sin 2x\, dx$

   c. $\int_0^5 \left(2x\cos(2x) - (x-2)^2\right) dx$ d. $\int_0^5 \left(4x\cos(2x) - (x-2)^2\right) dx$

5. Use Simpson's Composite rule with $n = 4, 6, 8, \ldots$, until successive approximations to the following integrals agree to within 10^{-6}. Determine the number of nodes required. Use the Adaptive Quadrature Algorithm to approximate the integral to within 10^{-6}, and count the number of nodes. Did Adaptive quadrature produce any improvement?

   a. $\int_0^{\pi} x \cos x^2\, dx$ b. $\int_0^{\pi} x \sin x^2\, dx$

   c. $\int_0^{\pi} x^2 \cos x\, dx$ d. $\int_0^{\pi} x^2 \sin x\, dx$

6. Sketch the graphs of $\sin(1/x)$ and $\cos(1/x)$ on $[0.1, 2]$. Use Adaptive quadrature to approximate the following integrals to within 10^{-3}.

   a. $\int_{0.1}^2 \sin \frac{1}{x}\, dx$ b. $\int_{0.1}^2 \cos \frac{1}{x}\, dx$

7. The differential equation

$$mu''(t) + ku(t) = F_0 \cos \omega t$$

describes a spring-mass system with mass m, spring constant k, and no applied damping. The term $F_0 \cos \omega t$ describes a periodic external force applied to the system. The solution to the equation when the system is initially at rest ($u'(0) = u(0) = 0$) is

$$u(t) = \frac{F_0}{m(\omega_0^2 - \omega^2)}(\cos \omega t - \cos \omega_0 t), \quad \text{where} \quad \omega_0 = \sqrt{\frac{k}{m}} \neq \omega.$$

Sketch the graph of u when $m = 1$, $k = 9$, $F_0 = 1$, $\omega = 2$, and $t \in [0, 2\pi]$. Approximate $\int_0^{2\pi} u(t)\, dt$ to within 10^{-4}.

8. If the term $cu'(t)$ is added to the left side of the motion equation in Exercise 7, the resulting differential equation describes a spring-mass system that is damped with damping constant $c \neq 0$. The solution to this equation when the system is initially at rest is

$$u(t) = c_1 e^{r_1 t} + c_2 e^{r_2 t} + \frac{F_0}{c^2\omega^2 + m^2(\omega_0^2 - \omega^2)^2}\left(c\omega \sin \omega t + m\left(\omega_0^2 - \omega^2\right)\cos \omega t\right),$$

where

$$r_1 = \frac{-c + \sqrt{c^2 - 4\omega_0^2 m^2}}{2m} \quad \text{and} \quad r_2 = \frac{-c - \sqrt{c^2 - 4\omega_0^2 m^2}}{2m}.$$

a. Let $m = 1$, $k = 9$, $F_0 = 1$, $c = 10$, and $\omega = 2$. Find the values of c_1 and c_2 so that $u(0) = u'(0) = 0$.

b. Sketch the graph of $u(t)$ for $t \in [0, 2\pi]$ and approximate $\int_0^{2\pi} u(t)\, dt$ to within 10^{-4}.

9. Let $T(a, b)$ and $T(a, \frac{a+b}{2}) + T(\frac{a+b}{2}, b)$ be the single and double applications of the Trapezoidal rule to $\int_a^b f(x)\, dx$. Derive the relationship between

$$\left| T(a, b) - T\left(a, \frac{a+b}{2}\right) - T\left(\frac{a+b}{2}, b\right) \right|$$

and

$$\left| \int_a^b f(x)\, dx - T\left(a, \frac{a+b}{2}\right) - T\left(\frac{a+b}{2}, b\right) \right|.$$

10. The study of light diffraction at a rectangular aperture involves the Fresnel integrals

$$c(t) = \int_0^t \cos \frac{\pi}{2}\omega^2\, d\omega \quad \text{and} \quad s(t) = \int_0^t \sin \frac{\pi}{2}\omega^2\, d\omega.$$

Construct a table of values for $c(t)$ and $s(t)$ that is accurate to within 10^{-4} for values of $t = 0.1, 0.2, \ldots, 1.0$.

4.7 Gaussian Quadrature

The Newton-Cotes formulas in Section 4.3 were derived by integrating interpolating polynomials. The error term in the interpolating polynomial of degree n involves the $(n + 1)$st derivative of the function being approximated, so a Newton-Cotes formula is exact when approximating the integral of any polynomial of degree less than or equal to n.

All the Newton-Cotes formulas use values of the function at equally-spaced points. This restriction is convenient when the formulas are combined to form the composite rules we considered in Section 4.4, but it can significantly decrease the accuracy of the approximation. Consider, for example, the Trapezoidal rule applied to determine the integrals of the functions whose graphs are shown in Figure 4.15.

Figure 4.15

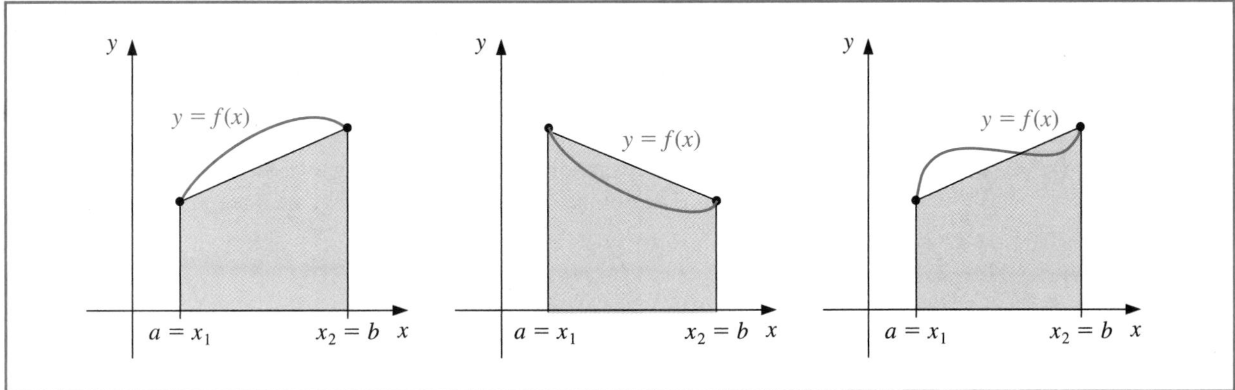

The Trapezoidal rule approximates the integral of the function by integrating the linear function that joins the endpoints of the graph of the function. But this is not likely the best line for approximating the integral. Lines such as those shown in Figure 4.16 would likely give much better approximations in most cases.

Figure 4.16

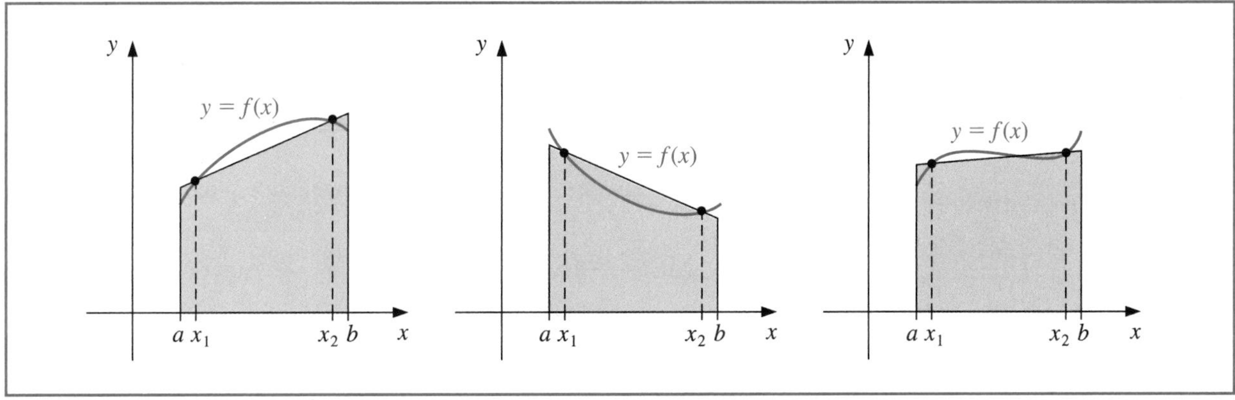

Gaussian quadrature chooses the points for evaluation in an optimal, rather than equally-spaced, way. The nodes $x_1, x_2, \ldots, x_n$ in the interval $[a, b]$ and coefficients $c_1, c_2, \ldots, c_n$, are chosen to minimize the expected error obtained in the approximation

Gauss demonstrated his method of efficient numerical integration in a paper that was presented to the Göttingen Society in 1814. He let the nodes as well as the coefficients of the function evaluations be parameters in the summation formula and found the optimal placement of the nodes. Goldstine [Golds], pp 224–232, has an interesting description of his development.

$$\int_a^b f(x)\, dx \approx \sum_{i=1}^{n} c_i f(x_i).$$

To measure this accuracy, we assume that the best choice of these values produces the exact result for the largest class of polynomials, that is, the choice that gives the greatest degree of precision.

The coefficients $c_1, c_2, \ldots, c_n$ in the approximation formula are arbitrary, and the nodes $x_1, x_2, \ldots, x_n$ are restricted only by the fact that they must lie in $[a, b]$, the interval of integration. This gives us $2n$ parameters to choose. If the coefficients of a polynomial are considered parameters, the class of polynomials of degree at most $2n - 1$ also contains $2n$ parameters. This, then, is the largest class of polynomials for which it is reasonable to expect a formula to be exact. With the proper choice of the values and constants, exactness on this set can be obtained.

To illustrate the procedure for choosing the appropriate parameters, we will show how to select the coefficients and nodes when $n = 2$ and the interval of integration is $[-1, 1]$. We will then discuss the more general situation for an arbitrary choice of nodes and coefficients and show how the technique is modified when integrating over an arbitrary interval.

Suppose we want to determine c_1, c_2, x_1, and x_2 so that the integration formula

$$\int_{-1}^{1} f(x)\, dx \approx c_1 f(x_1) + c_2 f(x_2)$$

gives the exact result whenever $f(x)$ is a polynomial of degree $2(2) - 1 = 3$ or less, that is, when

$$f(x) = a_0 + a_1 x + a_2 x^2 + a_3 x^3,$$

for some collection of constants, a_0, a_1, a_2, and a_3. Because

$$\int (a_0 + a_1 x + a_2 x^2 + a_3 x^3)\, dx = a_0 \int 1\, dx + a_1 \int x\, dx + a_2 \int x^2\, dx + a_3 \int x^3\, dx,$$

this is equivalent to showing that the formula gives exact results when $f(x)$ is 1, x, x^2, and x^3. Hence, we need c_1, c_2, x_1, and x_2, so that

$$c_1 \cdot 1 + c_2 \cdot 1 = \int_{-1}^{1} 1\, dx = 2, \qquad c_1 \cdot x_1 + c_2 \cdot x_2 = \int_{-1}^{1} x\, dx = 0,$$

$$c_1 \cdot x_1^2 + c_2 \cdot x_2^2 = \int_{-1}^{1} x^2\, dx = \frac{2}{3}, \quad \text{and} \quad c_1 \cdot x_1^3 + c_2 \cdot x_2^3 = \int_{-1}^{1} x^3\, dx = 0.$$

A little algebra shows that this system of equations has the unique solution

$$c_1 = 1, \quad c_2 = 1, \quad x_1 = -\frac{\sqrt{3}}{3}, \quad \text{and} \quad x_2 = \frac{\sqrt{3}}{3},$$

which gives the approximation formula

$$\int_{-1}^{1} f(x)\, dx \approx f\left(\frac{-\sqrt{3}}{3}\right) + f\left(\frac{\sqrt{3}}{3}\right). \tag{4.40}$$

This formula has degree of precision 3, that is, it produces the exact result for every polynomial of degree 3 or less.

Legendre Polynomials

The technique we have described could be used to determine the nodes and coefficients for formulas that give exact results for higher-degree polynomials, but an alternative method obtains them more easily. In Sections 8.2 and 8.3 we will consider various collections of orthogonal polynomials, functions that have the property that a particular definite integral of the product of any two of them is 0. The set that is relevant to our problem is the Legendre polynomials, a collection $\{P_0(x), P_1(x), \ldots, P_n(x), \ldots,\}$ with properties:

(1) For each n, $P_n(x)$ is a monic polynomial of degree n.

(2) $\int_{-1}^{1} P(x)P_n(x)\,dx = 0$ whenever $P(x)$ is a polynomial of degree less than n.

Recall that *monic* polynomials have leading coefficient 1.

The first few Legendre polynomials are

$$P_0(x) = 1, \quad P_1(x) = x, \quad P_2(x) = x^2 - \frac{1}{3},$$

$$P_3(x) = x^3 - \frac{3}{5}x, \quad \text{and} \quad P_4(x) = x^4 - \frac{6}{7}x^2 + \frac{3}{35}.$$

Adrien-Marie Legendre (1752–1833) introduced this set of polynomials in 1785. He had numerous priority disputes with Gauss, primarily due to Gauss' failure to publish many of his original results until long after he had discovered them.

The roots of these polynomials are distinct, lie in the interval $(-1, 1)$, have a symmetry with respect to the origin, and, most importantly, are the correct choice for determining the parameters that give us the nodes and coefficients for our quadrature method.

The nodes $x_1, x_2, \ldots, x_n$ needed to produce an integral approximation formula that gives exact results for any polynomial of degree less than $2n$ are the roots of the nth-degree Legendre polynomial. This is established by the following result.

Theorem 4.7 Suppose that $x_1, x_2, \ldots, x_n$ are the roots of the nth Legendre polynomial $P_n(x)$ and that for each $i = 1, 2, \ldots, n$, the numbers c_i are defined by

$$c_i = \int_{-1}^{1} \prod_{\substack{j=1 \\ j\neq i}}^{n} \frac{x - x_j}{x_i - x_j}\,dx.$$

If $P(x)$ is any polynomial of degree less than $2n$, then

$$\int_{-1}^{1} P(x)\,dx = \sum_{i=1}^{n} c_i P(x_i). \qquad \blacksquare$$

Proof Let us first consider the situation for a polynomial $P(x)$ of degree less than n. Rewrite $P(x)$ in terms of $(n-1)$st Lagrange coefficient polynomials with nodes at the roots of the nth Legendre polynomial $P_n(x)$. The error term for this representation involves the nth derivative of $P(x)$. Since $P(x)$ is of degree less than n, the nth derivative of $P(x)$ is 0, and this representation of is exact. So

$$P(x) = \sum_{i=1}^{n} P(x_i)L_i(x) = \sum_{i=1}^{n} \prod_{\substack{j=1 \\ j\neq i}}^{n} \frac{x - x_j}{x_i - x_j} P(x_i)$$

and

$$\int_{-1}^{1} P(x)\,dx = \int_{-1}^{1} \left[\sum_{i=1}^{n} \prod_{\substack{j=1 \\ j\neq i}}^{n} \frac{x - x_j}{x_i - x_j} P(x_i) \right] dx$$

$$= \sum_{i=1}^{n} \left[\int_{-1}^{1} \prod_{\substack{j=1 \\ j\neq i}}^{n} \frac{x - x_j}{x_i - x_j}\,dx \right] P(x_i) = \sum_{i=1}^{n} c_i P(x_i).$$

Hence the result is true for polynomials of degree less than n.

Now consider a polynomial $P(x)$ of degree at least n but less than $2n$. Divide $P(x)$ by the nth Legendre polynomial $P_n(x)$. This gives two polynomials $Q(x)$ and $R(x)$, each of degree less than n, with

$$P(x) = Q(x)P_n(x) + R(x).$$

Note that x_i is a root of $P_n(x)$ for each $i = 1, 2, \ldots, n$, so we have

$$P(x_i) = Q(x_i)P_n(x_i) + R(x_i) = R(x_i).$$

We now invoke the unique power of the Legendre polynomials. First, the degree of the polynomial $Q(x)$ is less than n, so (by Legendre property **(2)**),

$$\int_{-1}^{1} Q(x)P_n(x)\,dx = 0.$$

Then, since $R(x)$ is a polynomial of degree less than n, the opening argument implies that

$$\int_{-1}^{1} R(x)\,dx = \sum_{i=1}^{n} c_i R(x_i).$$

Putting these facts together verifies that the formula is exact for the polynomial $P(x)$:

$$\int_{-1}^{1} P(x)\,dx = \int_{-1}^{1} [Q(x)P_n(x) + R(x)]\,dx = \int_{-1}^{1} R(x)\,dx = \sum_{i=1}^{n} c_i R(x_i) = \sum_{i=1}^{n} c_i P(x_i).$$

▪ ▪ ▪

The constants c_i needed for the quadrature rule can be generated from the equation in Theorem 4.7, but both these constants and the roots of the Legendre polynomials are extensively tabulated. Table 4.12 lists these values for $n = 2, 3, 4,$ and 5.

Table 4.12

n	Roots $r_{n,i}$	Coefficients $c_{n,i}$
2	0.5773502692	1.0000000000
	−0.5773502692	1.0000000000
3	0.7745966692	0.5555555556
	0.0000000000	0.8888888889
	−0.7745966692	0.5555555556
4	0.8611363116	0.3478548451
	0.3399810436	0.6521451549
	−0.3399810436	0.6521451549
	−0.8611363116	0.3478548451
5	0.9061798459	0.2369268850
	0.5384693101	0.4786286705
	0.0000000000	0.5688888889
	−0.5384693101	0.4786286705
	−0.9061798459	0.2369268850

Example 1 Approximate $\int_{-1}^{1} e^x \cos x\,dx$ using Gaussian quadrature with $n = 3$.

Solution The entries in Table 4.12 give us

$$\begin{aligned}\int_{-1}^{1} e^x \cos x\,dx &\approx 0.\overline{5}e^{0.774596692}\cos 0.774596692 \\ &\quad + 0.\overline{8}\cos 0 + 0.\overline{5}e^{-0.774596692}\cos(-0.774596692) \\ &= 1.9333904.\end{aligned}$$

Integration by parts can be used to show that the true value of the integral is 1.9334214, so the absolute error is less than 3.2×10^{-5}. ▪

Gaussian Quadrature on Arbitrary Intervals

An integral $\int_a^b f(x)\,dx$ over an arbitrary $[a,b]$ can be transformed into an integral over $[-1,1]$ by using the change of variables (see Figure 4.17):

$$t = \frac{2x-a-b}{b-a} \iff x = \frac{1}{2}[(b-a)t + a + b].$$

Figure 4.17

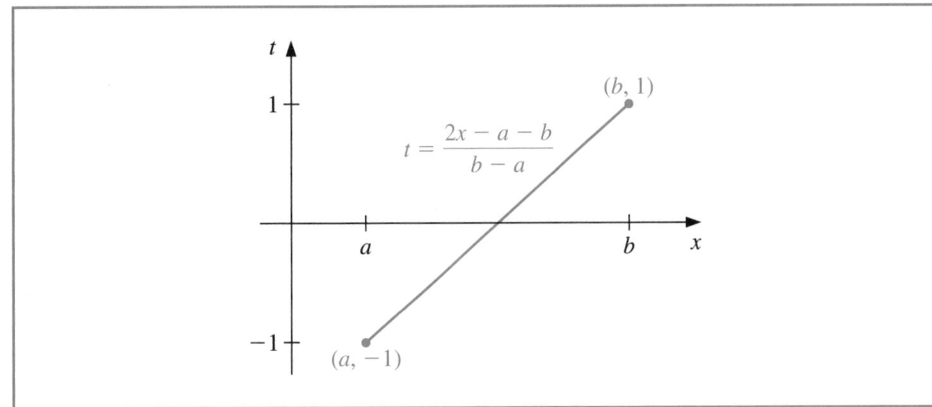

This permits Gaussian quadrature to be applied to any interval $[a,b]$, because

$$\int_a^b f(x)\,dx = \int_{-1}^1 f\left(\frac{(b-a)t+(b+a)}{2}\right)\frac{(b-a)}{2}\,dt. \tag{4.41}$$

Example 2 Consider the integral $\int_1^3 x^6 - x^2 \sin(2x)\,dx = 317.3442466.$

(a) Compare the results for the closed Newton-Cotes formula with $n=1$, the open Newton-Cotes formula with $n=1$, and Gaussian Quadrature when $n=2$.

(b) Compare the results for the closed Newton-Cotes formula with $n=2$, the open Newton-Cotes formula with $n=2$, and Gaussian Quadrature when $n=3$.

Solution **(a)** Each of the formulas in this part requires 2 evaluations of the function $f(x) = x^6 - x^2\sin(2x)$. The Newton-Cotes approximations are

$$\textbf{Closed } n = \mathbf{1}: \quad \frac{2}{2}[f(1)+f(3)] = 731.6054420;$$

$$\textbf{Open } n = \mathbf{1}: \quad \frac{3(2/3)}{2}[f(5/3)+f(7/3)] = 188.7856682.$$

Gaussian quadrature applied to this problem requires that the integral first be transformed into a problem whose interval of integration is $[-1,1]$. Using Eq. (4.41) gives

$$\int_1^3 x^6 - x^2\sin(2x)\,dx = \int_{-1}^1 (t+2)^6 - (t+2)^2\sin(2(t+2))\,dt.$$

Gaussian quadrature with $n=2$ then gives

$$\int_1^3 x^6 - x^2\sin(2x)\,dx \approx f(-0.5773502692+2) + f(0.5773502692+2) = 306.8199344;$$

(b) Each of the formulas in this part requires 3 function evaluations. The Newton-Cotes approximations are

$$\textbf{Closed } n = 2: \quad \frac{(1)}{3}[f(1) + 4f(2) + f(3)] = 333.2380940;$$

$$\textbf{Open } n = 2: \quad \frac{4(1/2)}{3}[2f(1.5) - f(2) + 2f(2.5)] = 303.5912023.$$

Gaussian quadrature with $n = 3$, once the transformation has been done, gives

$$\int_1^3 x^6 - x^2 \sin(2x)\, dx \approx 0.\bar{5}f(-0.7745966692 + 2) + 0.\bar{8}f(2)$$

$$+ 0.\bar{5}f(0.7745966692 + 2) = 317.2641516.$$

The Gaussian quadrature results are clearly superior in each instance. ■

Maple has Composite Gaussian Quadrature in the *NumericalAnalysis* subpackage of Maple's *Student* package. The default for the number of partitions in the command is 10, so the results in Example 2 would be found for $n = 2$ with

$f := x^6 - x^2 \sin(2x); a := 1; b := 3:$
Quadrature$(f(x), x = a..b, method = gaussian[2], partition = 1, output = information)$

which returns the approximation, what Maple assumes is the exact value of the integral, the absolute, and relative errors in the approximations, and the number of function evaluations.

The result when $n = 3$ is, of course, obtained by replacing the statement *method* = *gaussian*[2] with *method* = *gaussian*[3].

EXERCISE SET 4.7

1. Approximate the following integrals using Gaussian quadrature with $n = 2$, and compare your results to the exact values of the integrals.

   a. $\int_0^1 x^2 e^{-x}\, dx$ **b.** $\int_1^{1.5} x^2 \ln x\, dx$

   c. $\int_1^{1.6} \frac{2x}{x^2 - 4}\, dx$ **d.** $\int_0^{\pi/4} x^2 \sin x\, dx$

   e. $\int_0^{\pi/4} e^{3x} \sin 2x\, dx$ **f.** $\int_0^{0.35} \frac{2}{x^2 - 4}\, dx$

   g. $\int_3^{3.5} \frac{x}{\sqrt{x^2 - 4}}\, dx$ **h.** $\int_0^{\pi/4} (\cos x)^2\, dx$

2. Repeat Exercise 1 with $n = 3$.
3. Repeat Exercise 1 with $n = 4$.
4. Repeat Exercise 1 with $n = 5$.
5. Determine constants a, b, c, and d that will produce a quadrature formula

$$\int_{-1}^1 f(x)\, dx = af(-1) + bf(1) + cf'(-1) + df'(1)$$

   that has degree of precision 3.

6. Determine constants a, b, c, and d that will produce a quadrature formula

$$\int_{-1}^1 f(x)\, dx = af(-1) + bf(0) + cf(1) + df'(-1) + ef'(1)$$

   that has degree of precision 4.

7. Verify the entries for the values of $n = 2$ and 3 in Table 4.12 on page 232 by finding the roots of the respective Legendre polynomials, and use the equations preceding this table to find the coefficients associated with the values.
8. Show that the formula $Q(P) = \sum_{i=1}^{n} c_i P(x_i)$ cannot have degree of precision greater than $2n - 1$, regardless of the choice of $c_1, \ldots, c_n$ and $x_1, \ldots, x_n$. [*Hint:* Construct a polynomial that has a double root at each of the x_i's.]
9. Apply Maple's Composite Gaussian Quadrature routine to approximate $\int_{-1}^{1} x^2 e^x \, dx$ in the following manner.
   a. Use Gaussian Quadrature with $n = 8$ on the single interval $[-1, 1]$.
   b. Use Gaussian Quadrature with $n = 4$ on the intervals $[-1, 0]$ and $[0, 1]$.
   c. Use Gaussian Quadrature with $n = 2$ on the intervals $[-1, -0.5]$, $[-0.5, 0]$, $[0, 0.5]$ and $[0.5, 1]$.
   d. Give an explanation for the accuracy of the results.

4.8 Multiple Integrals

The techniques discussed in the previous sections can be modified for use in the approximation of multiple integrals. Consider the double integral

$$\iint_R f(x, y) \, dA,$$

where $R = \{(x, y) \mid a \le x \le b, c \le y \le d\}$, for some constants a, b, c, and d, is a rectangular region in the plane. (See Figure 4.18.)

Figure 4.18

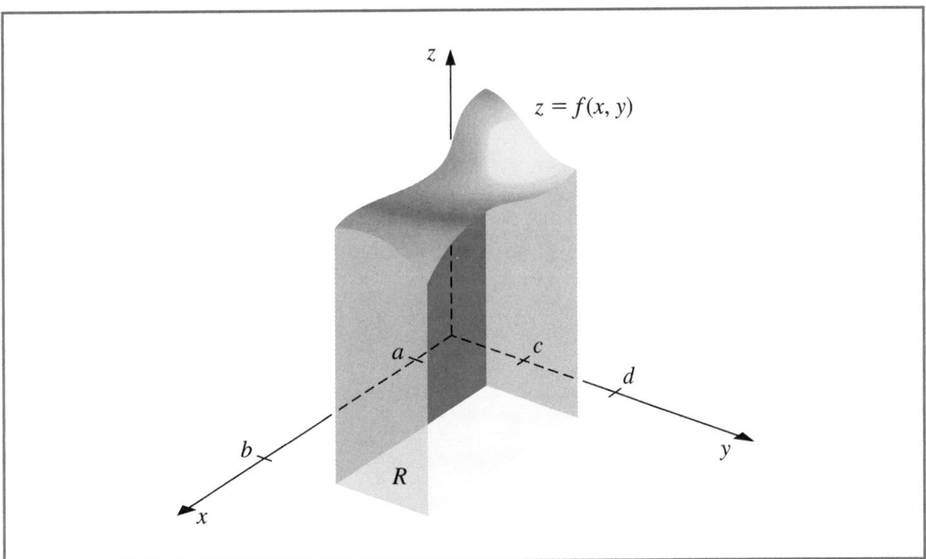

The following illustration shows how the Composite Trapezoidal rule using two subintervals in each coordinate direction would be applied to this integral.

Illustration Writing the double integral as an iterated integral gives

$$\iint_R f(x, y) \, dA = \int_a^b \left(\int_c^d f(x, y) \, dy \right) dx.$$

To simplify notation, let $k = (d-c)/2$ and $h = (b-a)/2$. Apply the Composite Trapezoidal rule to the interior integral to obtain

$$\int_c^d f(x,y)\,dy \approx \frac{k}{2}\left[f(x,c) + f(x,d) + 2f\left(x, \frac{c+d}{2}\right)\right].$$

This approximation is of order $O\left((d-c)^3\right)$. Then apply the Composite Trapezoidal rule again to approximate the integral of this function of x:

$$\begin{aligned}
\int_a^b \left(\int_c^d f(x,y)\,dy\right) dx &\approx \int_a^b \left(\frac{d-c}{4}\right)\left[f(x,c) + 2f\left(x, \frac{c+d}{2}\right) + f(d)\right] dx \\
&= \frac{b-a}{4}\left(\frac{d-c}{4}\right)\left[f(a,c) + 2f\left(a, \frac{c+d}{2}\right) + f(a,d)\right] \\
&\quad + \frac{b-a}{4}\left(2\left(\frac{d-c}{4}\right)\left[f\left(\frac{a+b}{2}, c\right)\right.\right. \\
&\quad \left.\left. + 2f\left(\frac{a+b}{2}, \frac{c+d}{2}\right) + \left(\frac{a+b}{2}, d\right)\right]\right) \\
&\quad + \frac{b-a}{4}\left(\frac{d-c}{4}\right)\left[f(b,c) + 2f\left(b, \frac{c+d}{2}\right) + f(b,d)\right] \\
&= \frac{(b-a)(d-c)}{16}\left[f(a,c) + f(a,d) + f(b,c) + f(b,d)\right. \\
&\quad + 2\left(f\left(\frac{a+b}{2}, c\right) + f\left(\frac{a+b}{2}, d\right) + f\left(a, \frac{c+d}{2}\right)\right. \\
&\quad \left.\left. + f\left(b, \frac{c+d}{2}\right)\right) + 4f\left(\frac{a+b}{2}, \frac{c+d}{2}\right)\right]
\end{aligned}$$

This approximation is of order $O\left((b-a)(d-c)\left[(b-a)^2 + (d-c)^2\right]\right)$. Figure 4.19 shows a grid with the number of functional evaluations at each of the nodes used in the approximation. □

Figure 4.19

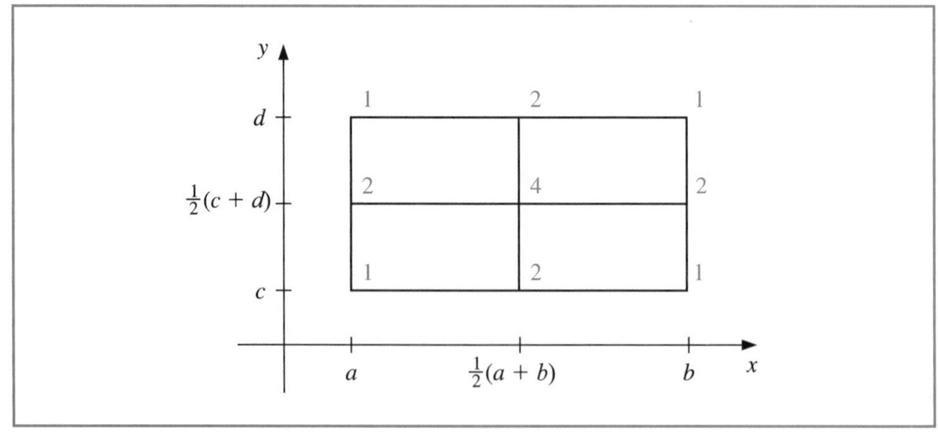

As the illustration shows, the procedure is quite straightforward. But the number of function evaluations grows with the square of the number required for a single integral. In a practical situation we would not expect to use a method as elementary as the Composite Trapezoidal rule. Instead we will employ the Composite Simpson's rule to illustrate the general approximation technique, although any other composite formula could be used in its place.

To apply the Composite Simpson's rule, we divide the region R by partitioning both $[a, b]$ and $[c, d]$ into an even number of subintervals. To simplify the notation, we choose even integers n and m and partition $[a, b]$ and $[c, d]$ with the evenly spaced mesh points $x_0, x_1, \ldots, x_n$ and $y_0, y_1, \ldots, y_m$, respectively. These subdivisions determine step sizes $h = (b - a)/n$ and $k = (d - c)/m$. Writing the double integral as the iterated integral

$$\iint_R f(x, y)\, dA = \int_a^b \left(\int_c^d f(x, y)\, dy \right) dx,$$

we first use the Composite Simpson's rule to approximate

$$\int_c^d f(x, y)\, dy,$$

treating x as a constant.

Let $y_j = c + jk$, for each $j = 0, 1, \ldots, m$. Then

$$\int_c^d f(x, y)\, dy = \frac{k}{3}\left[f(x, y_0) + 2\sum_{j=1}^{(m/2)-1} f(x, y_{2j}) + 4\sum_{j=1}^{m/2} f(x, y_{2j-1}) + f(x, y_m) \right]$$
$$- \frac{(d - c)k^4}{180} \frac{\partial^4 f}{\partial y^4}(x, \mu),$$

for some μ in (c, d). Thus

$$\int_a^b \int_c^d f(x, y)\, dy\, dx = \frac{k}{3}\Bigg[\int_a^b f(x, y_0)\, dx + 2\sum_{j=1}^{(m/2)-1} \int_a^b f(x, y_{2j})\, dx$$
$$+ 4\sum_{j=1}^{m/2} \int_a^b f(x, y_{2j-1})\, dx + \int_a^b f(x, y_m)\, dx \Bigg]$$
$$- \frac{(d - c)k^4}{180} \int_a^b \frac{\partial^4 f}{\partial y^4}(x, \mu)\, dx.$$

Composite Simpson's rule is now employed on the integrals in this equation. Let $x_i = a + ih$, for each $i = 0, 1, \ldots, n$. Then for each $j = 0, 1, \ldots, m$, we have

$$\int_a^b f(x, y_j)\, dx = \frac{h}{3}\left[f(x_0, y_j) + 2\sum_{i=1}^{(n/2)-1} f(x_{2i}, y_j) + 4\sum_{i=1}^{n/2} f(x_{2i-1}, y_j) + f(x_n, y_j) \right]$$
$$- \frac{(b - a)h^4}{180} \frac{\partial^4 f}{\partial x^4}(\xi_j, y_j),$$

for some ξ_j in (a, b). The resulting approximation has the form

$$\begin{aligned}
\int_a^b \int_c^d f(x, y)\, dy\, dx \approx \frac{hk}{9} \Bigg\{ \Bigg[& f(x_0, y_0) + 2 \sum_{i=1}^{(n/2)-1} f(x_{2i}, y_0) \\
& + 4 \sum_{i=1}^{n/2} f(x_{2i-1}, y_0) + f(x_n, y_0) \Bigg] \\
& + 2 \Bigg[\sum_{j=1}^{(m/2)-1} f(x_0, y_{2j}) + 2 \sum_{j=1}^{(m/2)-1} \sum_{i=1}^{(n/2)-1} f(x_{2i}, y_{2j}) \\
& + 4 \sum_{j=1}^{(m/2)-1} \sum_{i=1}^{n/2} f(x_{2i-1}, y_{2j}) + \sum_{j=1}^{(m/2)-1} f(x_n, y_{2j}) \Bigg] \\
& + 4 \Bigg[\sum_{j=1}^{m/2} f(x_0, y_{2j-1}) + 2 \sum_{j=1}^{m/2} \sum_{i=1}^{(n/2)-1} f(x_{2i}, y_{2j-1}) \\
& + 4 \sum_{j=1}^{m/2} \sum_{i=1}^{n/2} f(x_{2i-1}, y_{2j-1}) + \sum_{j=1}^{m/2} f(x_n, y_{2j-1}) \Bigg] \\
& + \Bigg[f(x_0, y_m) + 2 \sum_{i=1}^{(n/2)-1} f(x_{2i}, y_m) + 4 \sum_{i=1}^{n/2} f(x_{2i-1}, y_m) + f(x_n, y_m) \Bigg] \Bigg\}.
\end{aligned}$$

The error term E is given by

$$\begin{aligned}
E = \frac{-k(b-a)h^4}{540} \Bigg[& \frac{\partial^4 f}{\partial x^4}(\xi_0, y_0) + 2 \sum_{j=1}^{(m/2)-1} \frac{\partial^4 f}{\partial x^4}(\xi_{2j}, y_{2j}) + 4 \sum_{j=1}^{m/2} \frac{\partial^4 f}{\partial x^4}(\xi_{2j-1}, y_{2j-1}) \\
& + \frac{\partial^4 f}{\partial x^4}(\xi_m, y_m) \Bigg] - \frac{(d-c)k^4}{180} \int_a^b \frac{\partial^4 f}{\partial y^4}(x, \mu)\, dx.
\end{aligned}$$

If $\partial^4 f/\partial x^4$ is continuous, the Intermediate Value Theorem 1.11 can be repeatedly applied to show that the evaluation of the partial derivatives with respect to x can be replaced by a common value and that

$$E = \frac{-k(b-a)h^4}{540} \left[3m \frac{\partial^4 f}{\partial x^4}(\overline{\eta}, \overline{\mu}) \right] - \frac{(d-c)k^4}{180} \int_a^b \frac{\partial^4 f}{\partial y^4}(x, \mu)\, dx,$$

for some $(\overline{\eta}, \overline{\mu})$ in R. If $\partial^4 f/\partial y^4$ is also continuous, the Weighted Mean Value Theorem for Integrals 1.13 implies that

$$\int_a^b \frac{\partial^4 f}{\partial y^4}(x, \mu)\, dx = (b-a) \frac{\partial^4 f}{\partial y^4}(\hat{\eta}, \hat{\mu}),$$

for some $(\hat{\eta}, \hat{\mu})$ in R. Because $m = (d-c)/k$, the error term has the form

$$E = \frac{-k(b-a)h^4}{540} \left[3m \frac{\partial^4 f}{\partial x^4}(\overline{\eta}, \overline{\mu}) \right] - \frac{(d-c)(b-a)}{180} k^4 \frac{\partial^4 f}{\partial y^4}(\hat{\eta}, \hat{\mu})$$

which simplifies to

$$E = -\frac{(d-c)(b-a)}{180} \left[h^4 \frac{\partial^4 f}{\partial x^4}(\overline{\eta}, \overline{\mu}) + k^4 \frac{\partial^4 f}{\partial y^4}(\hat{\eta}, \hat{\mu}) \right],$$

for some $(\overline{\eta}, \overline{\mu})$ and $(\hat{\eta}, \hat{\mu})$ in R.

Example 1 Use Composite Simpson's rule with $n = 4$ and $m = 2$ to approximate

$$\int_{1.4}^{2.0} \int_{1.0}^{1.5} \ln(x + 2y)\, dy\, dx,$$

Solution The step sizes for this application are $h = (2.0 - 1.4)/4 = 0.15$ and $k = (1.5 - 1.0)/2 = 0.25$. The region of integration R is shown in Figure 4.20, together with the nodes (x_i, y_j), where $i = 0, 1, 2, 3, 4$ and $j = 0, 1, 2$. It also shows the coefficients $w_{i,j}$ of $f(x_i, y_i) = \ln(x_i + 2y_i)$ in the sum that gives the Composite Simpson's rule approximation to the integral.

Figure 4.20

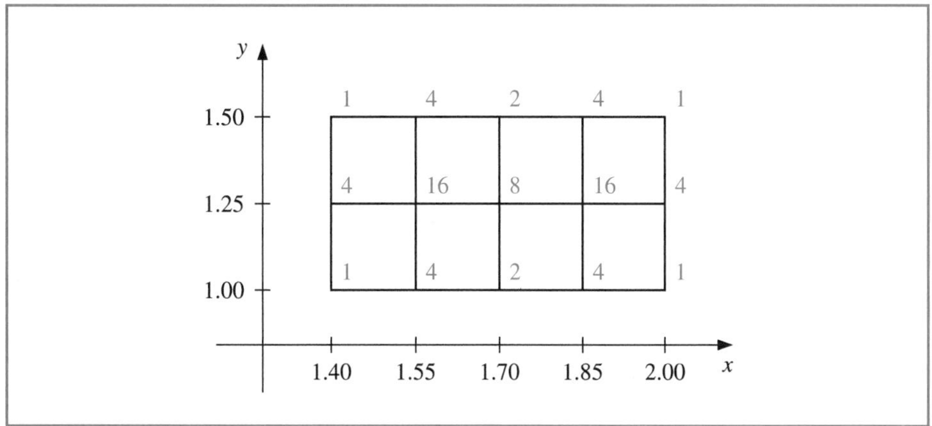

The approximation is

$$\int_{1.4}^{2.0} \int_{1.0}^{1.5} \ln(x + 2y)\, dy\, dx \approx \frac{(0.15)(0.25)}{9} \sum_{i=0}^{4} \sum_{j=0}^{2} w_{i,j} \ln(x_i + 2y_j)$$

$$= 0.4295524387.$$

We have

$$\frac{\partial^4 f}{\partial x^4}(x, y) = \frac{-6}{(x + 2y)^4} \quad \text{and} \quad \frac{\partial^4 f}{\partial y^4}(x, y) = \frac{-96}{(x + 2y)^4},$$

and the maximum values of the absolute values of these partial derivatives occur on R when $x = 1.4$ and $y = 1.0$. So the error is bounded by

$$|E| \le \frac{(0.5)(0.6)}{180} \left[(0.15)^4 \max_{(x,y)\text{in}R} \frac{6}{(x + 2y)^4} + (0.25)^4 \max_{(x,y)\text{in}R} \frac{96}{(x + 2y)^4} \right] \le 4.72 \times 10^{-6}.$$

The actual value of the integral to ten decimal places is

$$\int_{1.4}^{2.0} \int_{1.0}^{1.5} \ln(x + 2y)\, dy\, dx = 0.4295545265,$$

so the approximation is accurate to within 2.1×10^{-6}. ■

The same techniques can be applied for the approximation of triple integrals as well as higher integrals for functions of more than three variables. The number of functional evaluations required for the approximation is the product of the number of functional evaluations required when the method is applied to each variable.

Gaussian Quadrature for Double Integral Approximation

To reduce the number of functional evaluations, more efficient methods such as Gaussian quadrature, Romberg integration, or Adaptive quadrature can be incorporated in place of the Newton-Cotes formulas. The following example illustrates the use of Gaussian quadrature for the integral considered in Example 1.

Example 2 Use Gaussian quadrature with $n = 3$ in both dimensions to approximate the integral

$$\int_{1.4}^{2.0} \int_{1.0}^{1.5} \ln(x + 2y)\, dy\, dx.$$

Solution Before employing Gaussian quadrature to approximate this integral, we need to transform the region of integration

$$R = \{\, (x, y) \mid 1.4 \le x \le 2.0, 1.0 \le y \le 1.5 \,\}$$

into

$$\hat{R} = \{\, (u, v) \mid -1 \le u \le 1, -1 \le v \le 1 \,\}.$$

The linear transformations that accomplish this are

$$u = \frac{1}{2.0 - 1.4}(2x - 1.4 - 2.0) \quad \text{and} \quad v = \frac{1}{1.5 - 1.0}(2y - 1.0 - 1.5),$$

or, equivalently, $x = 0.3u + 1.7$ and $y = 0.25v + 1.25$. Employing this change of variables gives an integral on which Gaussian quadrature can be applied:

$$\int_{1.4}^{2.0} \int_{1.0}^{1.5} \ln(x + 2y)\, dy\, dx = 0.075 \int_{-1}^{1} \int_{-1}^{1} \ln(0.3u + 0.5v + 4.2)\, dv\, du.$$

The Gaussian quadrature formula for $n = 3$ in both u and v requires that we use the nodes

$$u_1 = v_1 = r_{3,2} = 0, \quad u_0 = v_0 = r_{3,1} = -0.7745966692,$$

and

$$u_2 = v_2 = r_{3,3} = 0.7745966692.$$

The associated weights are $c_{3,2} = 0.\overline{8}$ and $c_{3,1} = c_{3,3} = 0.\overline{5}$. (These are given in Table 4.12 on page 232.) The resulting approximation is

$$\begin{aligned}\int_{1.4}^{2.0} \int_{1.0}^{1.5} \ln(x + 2y)\, dy\, dx &\approx 0.075 \sum_{i=1}^{3} \sum_{j=1}^{3} c_{3,i} c_{3,j} \ln(0.3 r_{3,i} + 0.5 r_{3,j} + 4.2) \\ &= 0.4295545313.\end{aligned}$$

Although this result requires only 9 functional evaluations compared to 15 for the Composite Simpson's rule considered in Example 1, it is accurate to within 4.8×10^{-9}, compared to 2.1×10^{-6} accuracy in Example 1. ■

Non-Rectangular Regions

The use of approximation methods for double integrals is not limited to integrals with rectangular regions of integration. The techniques previously discussed can be modified to approximate double integrals of the form

$$\int_a^b \int_{c(x)}^{d(x)} f(x, y)\, dy\, dx \tag{4.42}$$

or

$$\int_c^d \int_{a(y)}^{b(y)} f(x, y)\, dx\, dy. \tag{4.43}$$

In fact, integrals on regions not of this type can also be approximated by performing appropriate partitions of the region. (See Exercise 10.)

To describe the technique involved with approximating an integral in the form

$$\int_a^b \int_{c(x)}^{d(x)} f(x, y)\, dy\, dx,$$

we will use the basic Simpson's rule to integrate with respect to both variables. The step size for the variable x is $h = (b - a)/2$, but the step size for y varies with x (see Figure 4.21) and is written

$$k(x) = \frac{d(x) - c(x)}{2}.$$

Figure 4.21

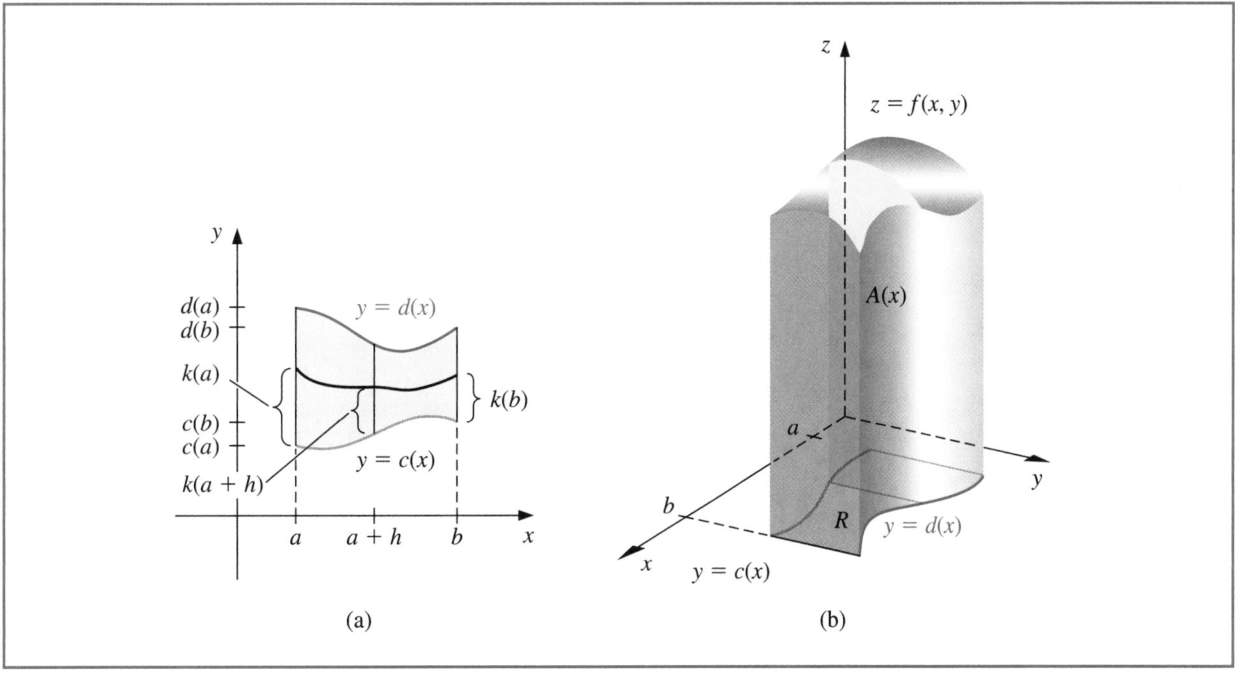

This gives

$$\int_a^b \int_{c(x)}^{d(x)} f(x, y)\,dy\,dx \approx \int_a^b \frac{k(x)}{3}[f(x, c(x)) + 4f(x, c(x) + k(x)) + f(x, d(x))]\,dx$$

$$\approx \frac{h}{3}\left\{\frac{k(a)}{3}[f(a, c(a)) + 4f(a, c(a) + k(a)) + f(a, d(a))]\right.$$

$$+ \frac{4k(a+h)}{3}[f(a+h, c(a+h)) + 4f(a+h, c(a+h)$$

$$+ k(a+h)) + f(a+h, d(a+h))]$$

$$\left. + \frac{k(b)}{3}[f(b, c(b)) + 4f(b, c(b) + k(b)) + f(b, d(b))]\right\}.$$

Algorithm 4.4 applies the Composite Simpson's rule to an integral in the form (4.42). Integrals in the form (4.43) can, of course, be handled similarly.

Simpson's Double Integral

To approximate the integral

$$I = \int_a^b \int_{c(x)}^{d(x)} f(x, y)\,dy\,dx:$$

INPUT endpoints a, b: even positive integers m, n.

OUTPUT approximation J to I.

Step 1 Set $h = (b - a)/n$;
$J_1 = 0$; (*End terms.*)
$J_2 = 0$; (*Even terms.*)
$J_3 = 0$. (*Odd terms.*)

Step 2 For $i = 0, 1, \ldots, n$ do Steps 3–8.

Step 3 Set $x = a + ih$; (*Composite Simpson's method for x.*)
$HX = (d(x) - c(x))/m$;
$K_1 = f(x, c(x)) + f(x, d(x))$; (*End terms.*)
$K_2 = 0$; (*Even terms.*)
$K_3 = 0$. (*Odd terms.*)

Step 4 For $j = 1, 2, \ldots, m - 1$ do Step 5 and 6.

Step 5 Set $y = c(x) + jHX$;
$Q = f(x, y)$.

Step 6 If j is even then set $K_2 = K_2 + Q$
else set $K_3 = K_3 + Q$.

Step 7 Set $L = (K_1 + 2K_2 + 4K_3)HX/3$.

$$\left(L \approx \int_{c(x_i)}^{d(x_i)} f(x_i, y)\,dy \quad \text{by the Composite Simpson's method.}\right)$$

Step 8 If $i = 0$ or $i = n$ then set $J_1 = J_1 + L$
else if i is even then set $J_2 = J_2 + L$
else set $J_3 = J_3 + L$.

Step 9 Set $J = h(J_1 + 2J_2 + 4J_3)/3$.

Step 10 OUTPUT (J);
STOP. ■

To apply Gaussian quadrature to the double integral

$$\int_a^b \int_{c(x)}^{d(x)} f(x,y)\,dy\,dx,$$

first requires transforming, for each x in $[a,b]$, the variable y in the interval $[c(x), d(x)]$ into the variable t in the interval $[-1,1]$. This linear transformation gives

$$f(x,y) = f\left(x, \frac{(d(x)-c(x))t + d(x) + c(x)}{2}\right) \quad \text{and} \quad dy = \frac{d(x)-c(x)}{2}\,dt.$$

The reduced calculation makes it generally worthwhile to apply Gaussian quadrature rather than a Simpson's technique when approximating double integrals.

Then, for each x in $[a,b]$, we apply Gaussian quadrature to the resulting integral

$$\int_{c(x)}^{d(x)} f(x,y)\,dy = \int_{-1}^{1} f\left(x, \frac{(d(x)-c(x))t + d(x) + c(x)}{2}\right) dt$$

to produce

$$\int_a^b \int_{c(x)}^{d(x)} f(x,y)\,dy\,dx \approx \int_a^b \frac{d(x)-c(x)}{2} \sum_{j=1}^{n} c_{n,j} f\left(x, \frac{(d(x)-c(x))r_{n,j} + d(x) + c(x)}{2}\right) dx,$$

where, as before, the roots $r_{n,j}$ and coefficients $c_{n,j}$ come from Table 4.12 on page 232. Now the interval $[a,b]$ is transformed to $[-1,1]$, and Gaussian quadrature is applied to approximate the integral on the right side of this equation. The details are given in Algorithm 4.5.

Gaussian Double Integral

To approximate the integral

$$\int_a^b \int_{c(x)}^{d(x)} f(x,y)\,dy\,dx :$$

INPUT endpoints a, b; positive integers m, n.
(The roots $r_{i,j}$ and coefficients $c_{i,j}$ need to be available for $i = \max\{m,n\}$ and for $1 \le j \le i$.)

OUTPUT approximation J to I.

Step 1 Set $h_1 = (b-a)/2$;
$h_2 = (b+a)/2$;
$J = 0$.

Step 2 For $i = 1, 2, \ldots, m$ do Steps 3–5.

Step 3 Set $JX = 0$;
$x = h_1 r_{m,i} + h_2$;
$d_1 = d(x)$;
$c_1 = c(x)$;
$k_1 = (d_1 - c_1)/2$;
$k_2 = (d_1 + c_1)/2$.

Step 4 For $j = 1, 2, \ldots, n$ do
set $y = k_1 r_{n,j} + k_2$;
$Q = f(x, y)$;
$JX = JX + c_{n,j}Q$.

Step 5 Set $J = J + c_{m,i}k_1 JX$.

Step 6 Set $J = h_1 J$.

Step 7 OUTPUT (J);
STOP. ■

Illustration The volume of the solid in Figure 4.22 is approximated by applying Simpson's Double Integral Algorithm with $n = m = 10$ to

$$\int_{0.1}^{0.5} \int_{x^3}^{x^2} e^{y/x}\, dy\, dx.$$

This requires 121 evaluations of the function $f(x, y) = e^{y/x}$ and produces the value 0.0333054, which approximates the volume of the solid shown in Figure 4.22 to nearly seven decimal places. Applying the Gaussian Quadrature Algorithm with $n = m = 5$ requires only 25 function evaluations and gives the approximation 0.03330556611, which is accurate to 11 decimal places. □

Figure 4.22

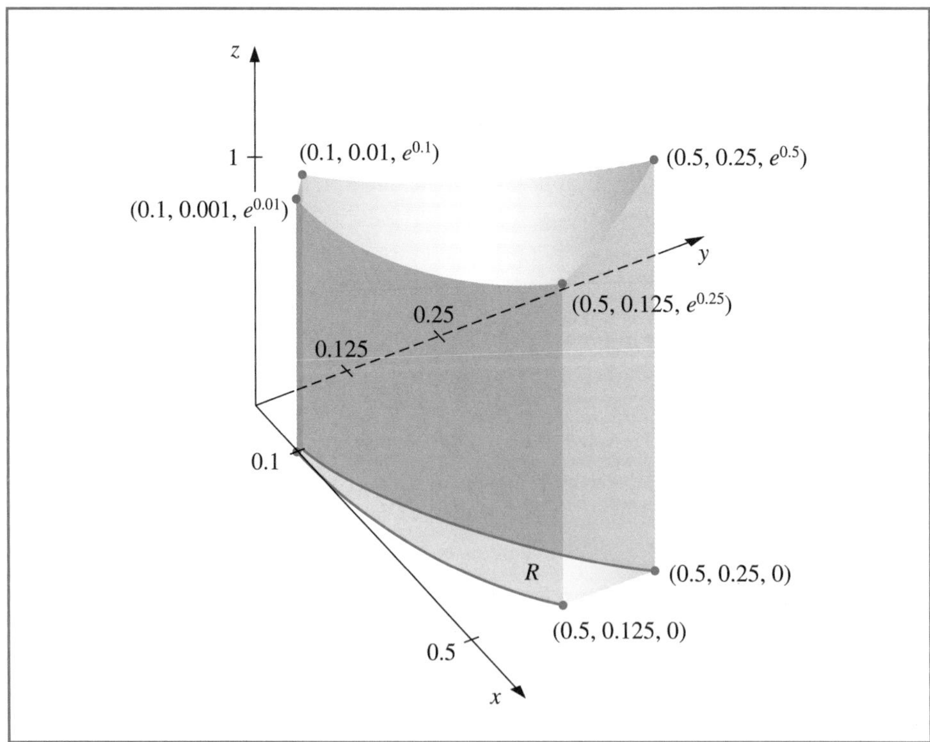

Triple Integral Approximation

Triple integrals of the form

$$\int_a^b \int_{c(x)}^{d(x)} \int_{\alpha(x,y)}^{\beta(x,y)} f(x, y, z)\, dz\, dy\, dx$$

(see Figure 4.23) are approximated in a similar manner. Because of the number of calculations involved, Gaussian quadrature is the method of choice. Algorithm 4.6 implements this procedure.

The reduced calculation makes it almost always worthwhile to apply Gaussian quadrature rather than a Simpson's technique when approximating triple or higher integrals.

Figure 4.23

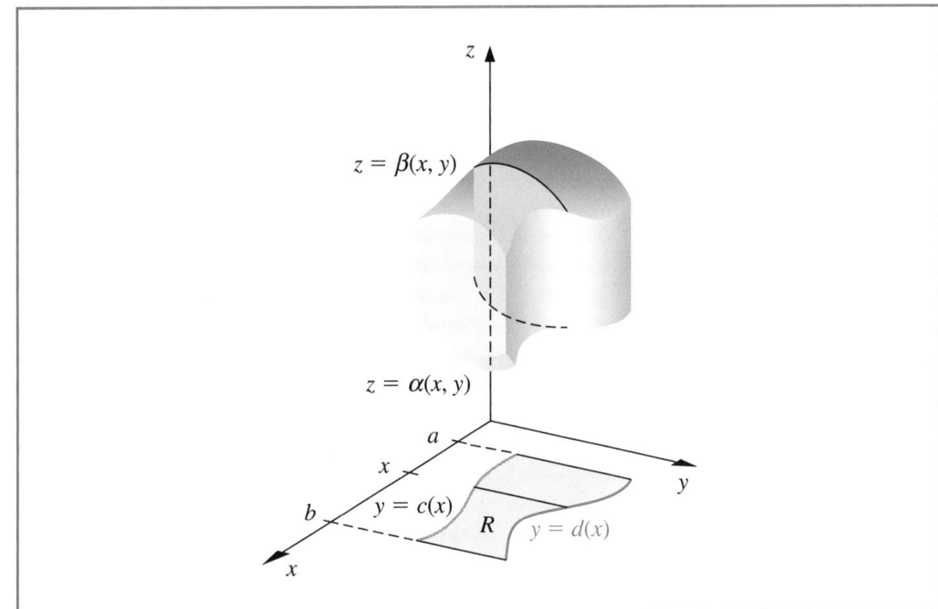

Gaussian Triple Integral

To approximate the integral

$$\int_a^b \int_{c(x)}^{d(x)} \int_{\alpha(x,y)}^{\beta(x,y)} f(x, y, z)\, dz\, dy\, dx :$$

INPUT endpoints a, b; positive integers m, n, p.
(The roots $r_{i,j}$ and coefficients $c_{i,j}$ need to be available for $i = \max\{n, m, p\}$ and for $1 \le j \le i$.)

OUTPUT approximation J to I.

Step 1 Set $h_1 = (b - a)/2$;
$h_2 = (b + a)/2$;
$J = 0$.

Step 2 For $i = 1, 2, \ldots, m$ do Steps 3–8.

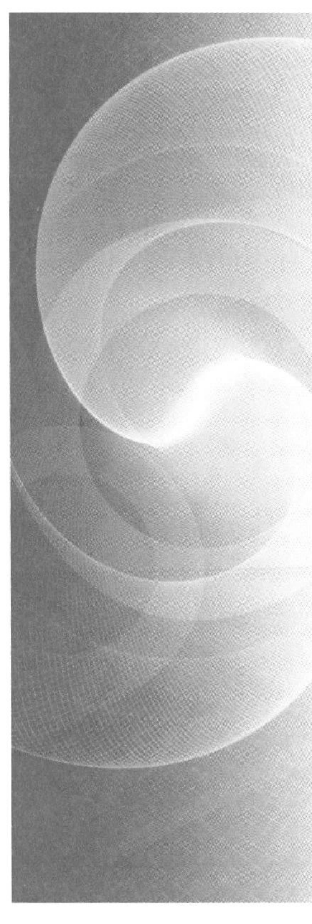

Step 3 Set $JX = 0$;
$x = h_1 r_{m,i} + h_2$;
$d_1 = d(x)$;
$c_1 = c(x)$;
$k_1 = (d_1 - c_1)/2$;
$k_2 = (d_1 + c_1)/2$.

Step 4 For $j = 1, 2, \ldots, n$ do Steps 5–7.

Step 5 Set $JY = 0$;
$y = k_1 r_{n,j} + k_2$;
$\beta_1 = \beta(x, y)$;
$\alpha_1 = \alpha(x, y)$;
$l_1 = (\beta_1 - \alpha_1)/2$;
$l_2 = (\beta_1 + \alpha_1)/2$.

Step 6 For $k = 1, 2, \ldots, p$ do
set $z = l_1 r_{p,k} + l_2$;
$Q = f(x, y, z)$;
$JY = JY + c_{p,k} Q$.

Step 7 Set $JX = JX + c_{n,j} l_1 JY$.

Step 8 Set $J = J + c_{m,i} k_1 JX$.

Step 9 Set $J = h_1 J$.

Step 10 OUTPUT (J);
STOP. ▪

The following example requires the evaluation of four triple integrals.

Illustration The center of a mass of a solid region D with density function σ occurs at

$$(\overline{x}, \overline{y}, \overline{z}) = \left(\frac{M_{yz}}{M}, \frac{M_{xz}}{M}, \frac{M_{xy}}{M} \right),$$

where

$$M_{yz} = \iiint_D x\sigma(x, y, z)\, dV, \quad M_{xz} = \iiint_D y\sigma(x, y, z)\, dV$$

and

$$M_{xy} = \iiint_D z\sigma(x, y, z)\, dV$$

are the moments about the coordinate planes and the mass of D is

$$M = \iiint_D \sigma(x, y, z)\, dV.$$

The solid shown in Figure 4.24 is bounded by the upper nappe of the cone $z^2 = x^2 + y^2$ and the plane $z = 2$. Suppose that this solid has density function given by

$$\sigma(x, y, z) = \sqrt{x^2 + y^2}.$$

Figure 4.24

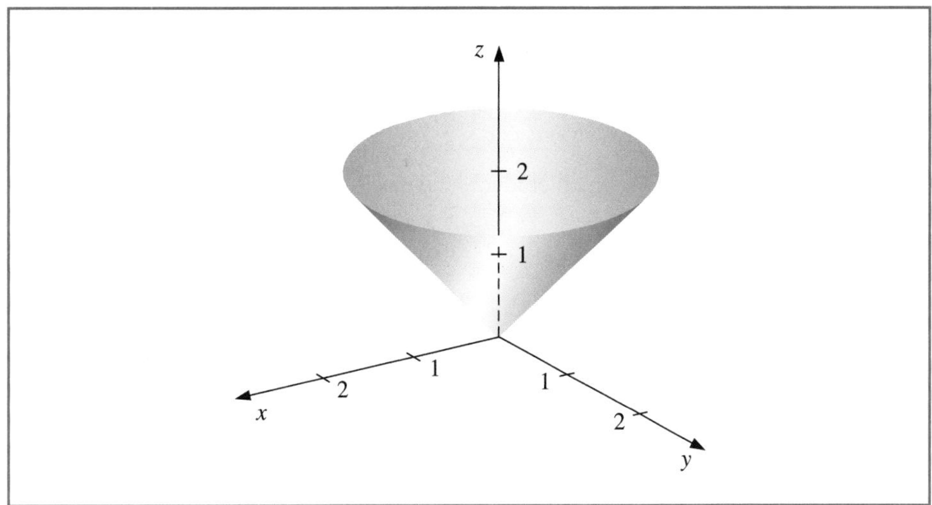

Applying the Gaussian Triple Integral Algorithm 4.6 with $n = m = p = 5$ requires 125 function evaluations per integral and gives the following approximations:

$$\begin{aligned}
M &= \int_{-2}^{2}\int_{-\sqrt{4-x^2}}^{\sqrt{4-x^2}}\int_{\sqrt{x^2+y^2}}^{2} \sqrt{x^2+y^2}\,dz\,dy\,dx \\
&= 4\int_{0}^{2}\int_{0}^{\sqrt{4-x^2}}\int_{\sqrt{x^2+y^2}}^{2} \sqrt{x^2+y^2}\,dz\,dy\,dx \approx 8.37504476, \\
M_{yz} &= \int_{-2}^{2}\int_{-\sqrt{4-x^2}}^{\sqrt{4-x^2}}\int_{\sqrt{x^2+y^2}}^{2} x\sqrt{x^2+y^2}\,dz\,dy\,dx \approx -5.55111512 \times 10^{-17}, \\
M_{xz} &= \int_{-2}^{2}\int_{-\sqrt{4-x^2}}^{\sqrt{4-x^2}}\int_{\sqrt{x^2+y^2}}^{2} y\sqrt{x^2+y^2}\,dz\,dy\,dx \approx -8.01513675 \times 10^{-17}, \\
M_{xy} &= \int_{-2}^{2}\int_{-\sqrt{4-x^2}}^{\sqrt{4-x^2}}\int_{\sqrt{x^2+y^2}}^{2} z\sqrt{x^2+y^2}\,dz\,dy\,dx \approx 13.40038156.
\end{aligned}$$

This implies that the approximate location of the center of mass is

$$(\overline{x}, \overline{y}, \overline{z}) = (0, 0, 1.60003701).$$

These integrals are quite easy to evaluate directly. If you do this, you will find that the exact center of mass occurs at $(0, 0, 1.6)$. □

Multiple integrals can be evaluated in Maple using the *MultInt* command in the *MultivariateCalculus* subpackage of the *Student* package. For example, to evaluate the multiple integral

$$\int_{2}^{4}\int_{x-1}^{x+6}\int_{-2}^{4+y^2} x^2 + y^2 + z\,dz\,dy\,dx$$

we first load the package and define the function with

$with(Student[MultivariateCalculus])\colon\ f := (x, y, z) \to x^2 + y^2 + z$

Then issue the command

$MultiInt(f(x, y, z), z = -2..4 + y^2,\ y = x - 1..x + 6,\ x = 2..4)$

which produces the result

$$1.995885970$$

EXERCISE SET 4.8

1. Use Algorithm 4.4 with $n = m = 4$ to approximate the following double integrals, and compare the results to the exact answers.

   a. $\displaystyle\int_0^{0.5}\int_0^{0.5} e^{y-x}\,dy\,dx$

   b. $\displaystyle\int_{2.1}^{2.5}\int_{1.2}^{1.4} xy^2\,dy\,dx$

   c. $\displaystyle\int_2^{2.2}\int_x^{2x} (x^2 + y^3)\,dy\,dx$

   d. $\displaystyle\int_1^{1.5}\int_0^{x} (x^2 + \sqrt{y})\,dy\,dx$

2. Find the smallest values for $n = m$ so that Algorithm 4.4 can be used to approximate the integrals in Exercise 1 to within 10^{-6} of the actual value.

3. Use Algorithm 4.4 with (i) $n = 4$, $m = 8$, (ii) $n = 8$, $m = 4$, and (iii) $n = m = 6$ to approximate the following double integrals, and compare the results to the exact answers.

   a. $\displaystyle\int_1^{e}\int_1^{x} \ln xy\,dy\,dx$

   b. $\displaystyle\int_0^{\pi/4}\int_{\sin x}^{\cos x} (2y\sin x + \cos^2 x)\,dy\,dx$

   c. $\displaystyle\int_0^{\pi}\int_0^{x} \cos y\,dy\,dx$

   d. $\displaystyle\int_0^{1}\int_x^{2x} (y^2 + x^3)\,dy\,dx$

   e. $\displaystyle\int_0^{\pi}\int_0^{x} \cos x\,dy\,dx$

   f. $\displaystyle\int_0^{1}\int_x^{2x} (x^2 + y^3)\,dy\,dx$

   g. $\displaystyle\int_0^{\pi/4}\int_0^{\sin x} \frac{1}{\sqrt{1-y^2}}\,dy\,dx$

   h. $\displaystyle\int_{-\pi}^{3\pi/2}\int_0^{2\pi} (y\sin x + x\cos y)\,dy\,dx$

4. Find the smallest values for $n = m$ so that Algorithm 4.4 can be used to approximate the integrals in Exercise 3 to within 10^{-6} of the actual value.

5. Use Algorithm 4.5 with $n = m = 2$ to approximate the integrals in Exercise 1, and compare the results to those obtained in Exercise 1.

6. Find the smallest values of $n = m$ so that Algorithm 4.5 can be used to approximate the integrals in Exercise 1 to within 10^{-6}. Do not continue beyond $n = m = 5$. Compare the number of functional evaluations required to the number required in Exercise 2.

7. Use Algorithm 4.5 with (i) $n = m = 3$, (ii) $n = 3$, $m = 4$, (iii) $n = 4$, $m = 3$, and (iv) $n = m = 4$ to approximate the integrals in Exercise 3.

8. Use Algorithm 4.5 with $n = m = 5$ to approximate the integrals in Exercise 3. Compare the number of functional evaluations required to the number required in Exercise 4.

9. Use Algorithm 4.4 with $n = m = 14$ and Algorithm 4.5 with $n = m = 4$ to approximate

$$\iint_R e^{-(x+y)}\,dA,$$

for the region R in the plane bounded by the curves $y = x^2$ and $y = \sqrt{x}$.

10. Use Algorithm 4.4 to approximate

$$\iint_R \sqrt{xy + y^2}\, dA,$$

where R is the region in the plane bounded by the lines $x + y = 6$, $3y - x = 2$, and $3x - y = 2$. First partition R into two regions R_1 and R_2 on which Algorithm 4.4 can be applied. Use $n = m = 6$ on both R_1 and R_2.

11. A plane lamina is a thin sheet of continuously distributed mass. If σ is a function describing the density of a lamina having the shape of a region R in the xy-plane, then the center of the mass of the lamina $(\overline{x}, \overline{y})$ is

$$\overline{x} = \frac{\iint_R x\sigma(x, y)\, dA}{\iint_R \sigma(x, y)\, dA}, \qquad \overline{y} = \frac{\iint_R y\sigma(x, y)\, dA}{\iint_R \sigma(x, y)\, dA}.$$

Use Algorithm 4.4 with $n = m = 14$ to find the center of mass of the lamina described by $R = \{(x, y) \mid 0 \leq x \leq 1, 0 \leq y \leq \sqrt{1 - x^2}\,\}$ with the density function $\sigma(x, y) = e^{-(x^2+y^2)}$. Compare the approximation to the exact result.

12. Repeat Exercise 11 using Algorithm 4.5 with $n = m = 5$.

13. The area of the surface described by $z = f(x, y)$ for (x, y) in R is given by

$$\iint_R \sqrt{[f_x(x, y)]^2 + [f_y(x, y)]^2 + 1}\, dA.$$

Use Algorithm 4.4 with $n = m = 8$ to find an approximation to the area of the surface on the hemisphere $x^2 + y^2 + z^2 = 9$, $z \geq 0$ that lies above the region in the plane described by $R = \{\, (x, y) \mid 0 \leq x \leq 1, 0 \leq y \leq 1\,\}$.

14. Repeat Exercise 13 using Algorithm 4.5 with $n = m = 4$.

15. Use Algorithm 4.6 with $n = m = p = 2$ to approximate the following triple integrals, and compare the results to the exact answers.

a. $\displaystyle\int_0^1 \int_1^2 \int_0^{0.5} e^{x+y+z}\, dz\, dy\, dx$

b. $\displaystyle\int_0^1 \int_x^1 \int_0^y y^2 z\, dz\, dy\, dx$

c. $\displaystyle\int_0^1 \int_{x^2}^x \int_{x-y}^{x+y} y\, dz\, dy\, dx$

d. $\displaystyle\int_0^1 \int_{x^2}^x \int_{x-y}^{x+y} z\, dz\, dy\, dx$

e. $\displaystyle\int_0^{\pi} \int_0^x \int_0^{xy} \frac{1}{y} \sin\frac{z}{y}\, dz\, dy\, dx$

f. $\displaystyle\int_0^1 \int_0^1 \int_{-xy}^{xy} e^{x^2+y^2}\, dz\, dy\, dx$

16. Repeat Exercise 15 using $n = m = p = 3$.

17. Repeat Exercise 15 using $n = m = p = 4$ and $n = m = p = 5$.

18. Use Algorithm 4.6 with $n = m = p = 4$ to approximate

$$\iiint_S xy \sin(yz)\, dV,$$

where S is the solid bounded by the coordinate planes and the planes $x = \pi$, $y = \pi/2$, $z = \pi/3$. Compare this approximation to the exact result.

19. Use Algorithm 4.6 with $n = m = p = 5$ to approximate

$$\iiint_S \sqrt{xyz}\, dV,$$

where S is the region in the first octant bounded by the cylinder $x^2 + y^2 = 4$, the sphere $x^2 + y^2 + z^2 = 4$, and the plane $x + y + z = 8$. How many functional evaluations are required for the approximation?

4.9 Improper Integrals

Improper integrals result when the notion of integration is extended either to an interval of integration on which the function is unbounded or to an interval with one or more infinite endpoints. In either circumstance, the normal rules of integral approximation must be modified.

Left Endpoint Singularity

We will first consider the situation when the integrand is unbounded at the left endpoint of the interval of integration, as shown in Figure 4.25. In this case we say that f has a **singularity** at the endpoint a. We will then show how other improper integrals can be reduced to problems of this form.

Figure 4.25

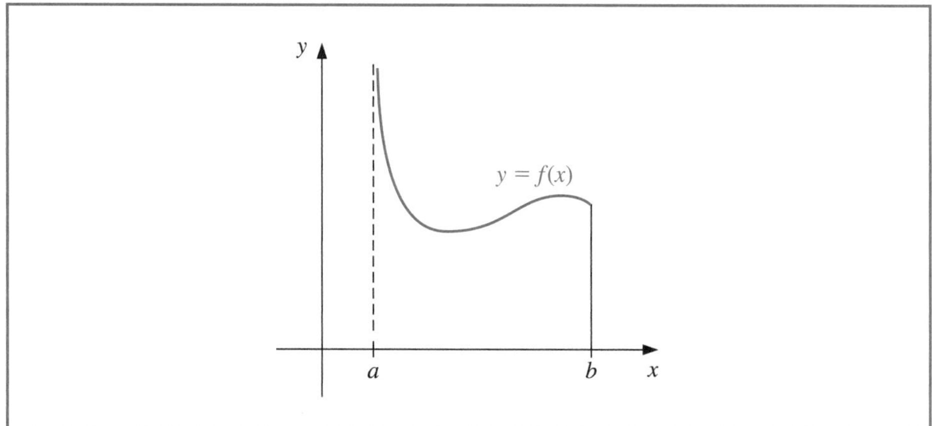

It is shown in calculus that the improper integral with a singularity at the left endpoint,

$$\int_a^b \frac{dx}{(x-a)^p},$$

converges if and only if $0 < p < 1$, and in this case, we define

$$\int_a^b \frac{1}{(x-a)^p}\,dx = \lim_{M\to a^+} \left.\frac{(x-a)^{1-p}}{1-p}\right|_{x=M}^{x=b} = \frac{(b-a)^{1-p}}{1-p}.$$

Example 1 Show that the improper integral $\int_0^1 \frac{1}{\sqrt{x}}\,dx$ converges but $\int_0^1 \frac{1}{x^2}\,dx$ diverges.

Solution For the first integral we have

$$\int_0^1 \frac{1}{\sqrt{x}}\,dx = \lim_{M\to 0^+} \int_M^1 x^{-1/2}\,dx = \lim_{M\to 0^+} 2x^{1/2}\Big|_{x=M}^{x=1} = 2 - 0 = 2,$$

but the second integral

$$\int_0^1 \frac{1}{x^2}\,dx = \lim_{M\to 0^+} \int_M^1 x^{-2}\,dx = \lim_{M\to 0^+} -x^{-1}\Big|_{x=M}^{x=1}$$

is unbounded.

If f is a function that can be written in the form

$$f(x) = \frac{g(x)}{(x-a)^p},$$

where $0 < p < 1$ and g is continuous on $[a, b]$, then the improper integral

$$\int_a^b f(x)\,dx$$

also exists. We will approximate this integral using the Composite Simpson's rule, provided that $g \in C^5[a, b]$. In that case, we can construct the fourth Taylor polynomial, $P_4(x)$, for g about a,

$$P_4(x) = g(a) + g'(a)(x-a) + \frac{g''(a)}{2!}(x-a)^2 + \frac{g'''(a)}{3!}(x-a)^3 + \frac{g^{(4)}(a)}{4!}(x-a)^4,$$

and write

$$\int_a^b f(x)\,dx = \int_a^b \frac{g(x) - P_4(x)}{(x-a)^p}\,dx + \int_a^b \frac{P_4(x)}{(x-a)^p}\,dx. \tag{4.44}$$

Because $P(x)$ is a polynomial, we can exactly determine the value of

$$\int_a^b \frac{P_4(x)}{(x-a)^p}\,dx = \sum_{k=0}^{4} \int_a^b \frac{g^{(k)}(a)}{k!}(x-a)^{k-p}\,dx = \sum_{k=0}^{4} \frac{g^{(k)}(a)}{k!(k+1-p)}(b-a)^{k+1-p}. \tag{4.45}$$

This is generally the dominant portion of the approximation, especially when the Taylor polynomial $P_4(x)$ agrees closely with $g(x)$ throughout the interval $[a, b]$.

To approximate the integral of f, we must add to this value the approximation of

$$\int_a^b \frac{g(x) - P_4(x)}{(x-a)^p}\,dx.$$

To determine this, we first define

$$G(x) = \begin{cases} \frac{g(x)-P_4(x)}{(x-a)^p}, & \text{if } a < x \le b, \\ 0, & \text{if } x = a. \end{cases}$$

■

This gives us a continuous function on $[a, b]$. In fact, $0 < p < 1$ and $P_4^{(k)}(a)$ agrees with $g^{(k)}(a)$ for each $k = 0, 1, 2, 3, 4$, so we have $G \in C^4[a, b]$. This implies that the Composite Simpson's rule can be applied to approximate the integral of G on $[a, b]$. Adding this approximation to the value in Eq. (4.45) gives an approximation to the improper integral of f on $[a, b]$, within the accuracy of the Composite Simpson's rule approximation.

Example 2 Use Composite Simpson's rule with $h = 0.25$ to approximate the value of the improper integral

$$\int_0^1 \frac{e^x}{\sqrt{x}}\,dx.$$

Solution The fourth Taylor polynomial for e^x about $x = 0$ is

$$P_4(x) = 1 + x + \frac{x^2}{2} + \frac{x^3}{6} + \frac{x^4}{24},$$

so the dominant portion of the approximation to $\int_0^1 \frac{e^x}{\sqrt{x}}\,dx$ is

$$\int_0^1 \frac{P_4(x)}{\sqrt{x}}\,dx = \int_0^1 \left(x^{-1/2} + x^{1/2} + \frac{1}{2}x^{3/2} + \frac{1}{6}x^{5/2} + \frac{1}{24}x^{7/2}\right) dx$$

$$= \lim_{M\to 0^+} \left[2x^{1/2} + \frac{2}{3}x^{3/2} + \frac{1}{5}x^{5/2} + \frac{1}{21}x^{7/2} + \frac{1}{108}x^{9/2}\right]_M^1$$

$$= 2 + \frac{2}{3} + \frac{1}{5} + \frac{1}{21} + \frac{1}{108} \approx 2.9235450.$$

For the second portion of the approximation to $\int_0^1 \frac{e^x}{\sqrt{x}}\,dx$ we need to approximate $\int_0^1 G(x)\,dx$, where

$$G(x) = \begin{cases} \frac{1}{\sqrt{x}}(e^x - P_4(x)), & \text{if } 0 < x \le 1, \\ 0, & \text{if } x = 0. \end{cases}$$

Table 4.13

x	$G(x)$
0.00	0
0.25	0.0000170
0.50	0.0004013
0.75	0.0026026
1.00	0.0099485

Table 4.13 lists the values needed for the Composite Simpson's rule for this approximation.

Using these data and the Composite Simpson's rule gives

$$\int_0^1 G(x)\,dx \approx \frac{0.25}{3}[0 + 4(0.0000170) + 2(0.0004013) + 4(0.0026026) + 0.0099485]$$

$$= 0.0017691.$$

Hence

$$\int_0^1 \frac{e^x}{\sqrt{x}}\,dx \approx 2.9235450 + 0.0017691 = 2.9253141.$$

This result is accurate to within the accuracy of the Composite Simpson's rule approximation for the function G. Because $|G^{(4)}(x)| < 1$ on $[0, 1]$, the error is bounded by

$$\frac{1-0}{180}(0.25)^4 = 0.0000217. \qquad ■$$

Right Endpoint Singularity

To approximate the improper integral with a singularity at the right endpoint, we could develop a similar technique but expand in terms of the right endpoint b instead of the left endpoint a. Alternatively, we can make the substitution

$$z = -x, \qquad dz = -\,dx$$

to change the improper integral into one of the form

$$\int_a^b f(x)\,dx = \int_{-b}^{-a} f(-z)\,dz, \tag{4.46}$$

which has its singularity at the left endpoint. Then we can apply the left endpoint singularity technique we have already developed. (See Figure 4.26.)

Figure 4.26

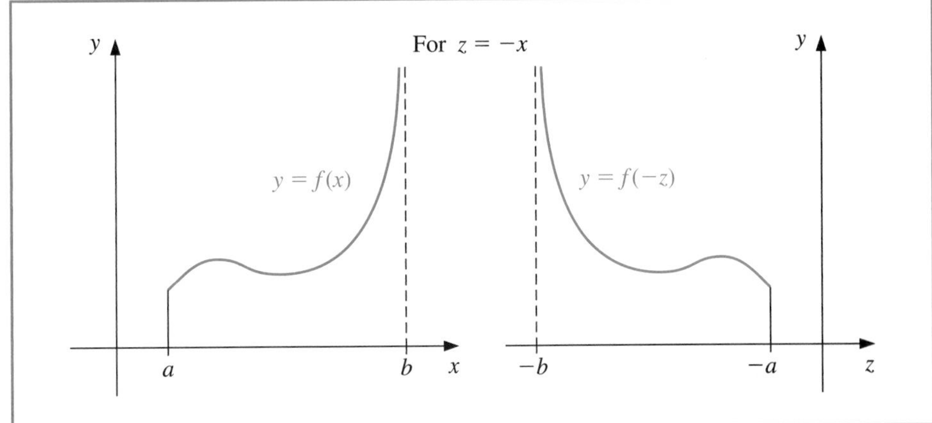

An improper integral with a singularity at c, where $a < c < b$, is treated as the sum of improper integrals with endpoint singularities since

$$\int_a^b f(x)\,dx = \int_a^c f(x)\,dx + \int_c^b f(x)\,dx.$$

Infinite Singularity

The other type of improper integral involves infinite limits of integration. The basic integral of this type has the form

$$\int_a^\infty \frac{1}{x^p}\,dx,$$

for $p > 1$. This is converted to an integral with left endpoint singularity at 0 by making the integration substitution

$$t = x^{-1}, \quad dt = -x^{-2}\,dx, \quad \text{so} \quad dx = -x^2\,dt = -t^{-2}\,dt.$$

Then

$$\int_a^\infty \frac{1}{x^p}\,dx = \int_{1/a}^0 -\frac{t^p}{t^2}\,dt = \int_0^{1/a} \frac{1}{t^{2-p}}\,dt.$$

In a similar manner, the variable change $t = x^{-1}$ converts the improper integral $\int_a^\infty f(x)\,dx$ into one that has a left endpoint singularity at zero:

$$\int_a^\infty f(x)\,dx = \int_0^{1/a} t^{-2} f\left(\frac{1}{t}\right)\,dt. \tag{4.47}$$

It can now be approximated using a quadrature formula of the type described earlier.

Example 3 Approximate the value of the improper integral

$$I = \int_1^\infty x^{-3/2} \sin\frac{1}{x}\,dx.$$

Solution We first make the variable change $t = x^{-1}$, which converts the infinite singularity into one with a left endpoint singularity. Then

$$dt = -x^{-2}\,dx, \quad \text{so} \quad dx = -x^2\,dt = -\frac{1}{t^2}\,dt,$$

and

$$I = \int_{x=1}^{x=\infty} x^{-3/2} \sin\frac{1}{x}\,dx = \int_{t=1}^{t=0} \left(\frac{1}{t}\right)^{-3/2} \sin t \left(-\frac{1}{t^2}\,dt\right) = \int_0^1 t^{-1/2} \sin t\,dt.$$

The fourth Taylor polynomial, $P_4(t)$, for $\sin t$ about 0 is

$$P_4(t) = t - \frac{1}{6}t^3,$$

so

$$G(t) = \begin{cases} \dfrac{\sin t - t + \frac{1}{6}t^3}{t^{1/2}}, & \text{if } 0 < t \le 1 \\ 0, & \text{if } t = 0 \end{cases}$$

is in $C^4[0, 1]$, and we have

$$\begin{aligned} I &= \int_0^1 t^{-1/2}\left(t - \frac{1}{6}t^3\right) dt + \int_0^1 \frac{\sin t - t + \frac{1}{6}t^3}{t^{1/2}}\,dt \\ &= \left[\frac{2}{3}t^{3/2} - \frac{1}{21}t^{7/2}\right]_0^1 + \int_0^1 \frac{\sin t - t + \frac{1}{6}t^3}{t^{1/2}}\,dt \\ &= 0.61904761 + \int_0^1 \frac{\sin t - t + \frac{1}{6}t^3}{t^{1/2}}\,dt. \end{aligned}$$

The result from the Composite Simpson's rule with $n = 16$ for the remaining integral is 0.0014890097. This gives a final approximation of

$$I = 0.0014890097 + 0.61904761 = 0.62053661,$$

which is accurate to within 4.0×10^{-8}. ■

EXERCISE SET 4.9

1. Use Simpson's Composite rule and the given values of n to approximate the following improper integrals.

   a. $\displaystyle\int_0^1 x^{-1/4} \sin x\,dx, \quad n = 4$

   b. $\displaystyle\int_0^1 \frac{e^{2x}}{\sqrt[5]{x^2}}\,dx, \quad n = 6$

   c. $\displaystyle\int_1^2 \frac{\ln x}{(x-1)^{1/5}}\,dx, \quad n = 8$

   d. $\displaystyle\int_0^1 \frac{\cos 2x}{x^{1/3}}\,dx, \quad n = 6$

2. Use the Composite Simpson's rule and the given values of n to approximate the following improper integrals.

a. $\displaystyle\int_0^1 \frac{e^{-x}}{\sqrt{1-x}}\,dx, \quad n = 6$

b. $\displaystyle\int_0^2 \frac{xe^x}{\sqrt[3]{(x-1)^2}}\,dx, \quad n = 8$

3. Use the transformation $t = x^{-1}$ and then the Composite Simpson's rule and the given values of n to approximate the following improper integrals.

a. $\displaystyle\int_1^\infty \frac{1}{x^2+9}\,dx, \quad n = 4$

b. $\displaystyle\int_1^\infty \frac{1}{1+x^4}\,dx, \quad n = 4$

c. $\displaystyle\int_1^\infty \frac{\cos x}{x^3}\,dx, \quad n = 6$

d. $\displaystyle\int_1^\infty x^{-4}\sin x\,dx, \quad n = 6$

4. The improper integral $\int_0^\infty f(x)\,dx$ cannot be converted into an integral with finite limits using the substitution $t = 1/x$ because the limit at zero becomes infinite. The problem is resolved by first writing $\int_0^\infty f(x)\,dx = \int_0^1 f(x)\,dx + \int_1^\infty f(x)\,dx$. Apply this technique to approximate the following improper integrals to within 10^{-6}.

a. $\displaystyle\int_0^\infty \frac{1}{1+x^4}\,dx$

b. $\displaystyle\int_0^\infty \frac{1}{(1+x^2)^3}\,dx$

5. Suppose a body of mass m is traveling vertically upward starting at the surface of the earth. If all resistance except gravity is neglected, the escape velocity v is given by

$$v^2 = 2gR\int_1^\infty z^{-2}\,dz, \quad \text{where } z = \frac{x}{R},$$

$R = 3960$ miles is the radius of the earth, and $g = 0.00609$ mi/s^2 is the force of gravity at the earth's surface. Approximate the escape velocity v.

6. The Laguerre polynomials $\{L_0(x), L_1(x)\ldots\}$ form an orthogonal set on $[0,\infty)$ and satisfy $\int_0^\infty e^{-x}L_i(x)L_j(x)\,dx = 0$, for $i \neq j$. (See Section 8.2.) The polynomial $L_n(x)$ has n distinct zeros $x_1, x_2, \ldots, x_n$ in $[0,\infty)$. Let

$$c_{n,i} = \int_0^\infty e^{-x}\prod_{\substack{j=1\\ j\neq i}}^{n}\frac{x-x_j}{x_i-x_j}\,dx.$$

Show that the quadrature formula

$$\int_0^\infty f(x)e^{-x}\,dx = \sum_{i=1}^{n} c_{n,i}f(x_i)$$

has degree of precision $2n-1$. (Hint: Follow the steps in the proof of Theorem 4.7.)

7. The Laguerre polynomials $L_0(x) = 1$, $L_1(x) = 1-x$, $L_2(x) = x^2-4x+2$, and $L_3(x) = -x^3 + 9x^2 - 18x + 6$ are derived in Exercise 11 of Section 8.2. As shown in Exercise 6, these polynomials are useful in approximating integrals of the form

$$\int_0^\infty e^{-x}f(x)\,dx = 0.$$

a. Derive the quadrature formula using $n = 2$ and the zeros of $L_2(x)$.

b. Derive the quadrature formula using $n = 3$ and the zeros of $L_3(x)$.

8. Use the quadrature formulas derived in Exercise 7 to approximate the integral

$$\int_0^\infty \sqrt{x}e^{-x}\,dx.$$

9. Use the quadrature formulas derived in Exercise 7 to approximate the integral

$$\int_{-\infty}^\infty \frac{1}{1+x^2}\,dx.$$

4.10 Survey of Methods and Software

In this chapter we considered approximating integrals of functions of one, two, or three variables, and approximating the derivatives of a function of a single real variable.

The Midpoint rule, Trapezoidal rule, and Simpson's rule were studied to introduce the techniques and error analysis of quadrature methods. Composite Simpson's rule is easy to use and produces accurate approximations unless the function oscillates in a subinterval of the interval of integration. Adaptive quadrature can be used if the function is suspected of oscillatory behavior. To minimize the number of nodes while maintaining accuracy, we used Gaussian quadrature. Romberg integration was introduced to take advantage of the easily applied Composite Trapezoidal rule and extrapolation.

Most software for integrating a function of a single real variable is based either on the adaptive approach or extremely accurate Gaussian formulas. Cautious Romberg integration is an adaptive technique that includes a check to make sure that the integrand is smoothly behaved over subintervals of the integral of integration. This method has been successfully used in software libraries. Multiple integrals are generally approximated by extending good adaptive methods to higher dimensions. Gaussian-type quadrature is also recommended to decrease the number of function evaluations.

The main routines in both the IMSL and NAG Libraries are based on *QUADPACK: A Subroutine Package for Automatic Integration* by R. Piessens, E. de Doncker-Kapenga, C. W. Uberhuber, and D. K. Kahaner published by Springer-Verlag in 1983 [PDUK].

The IMSL Library contains an adaptive integration scheme based on the 21-point Gaussian-Kronrod rule using the 10-point Gaussian rule for error estimation. The Gaussian rule uses the ten points $x_1, \ldots, x_{10}$ and weights $w_1, \ldots, w_{10}$ to give the quadrature formula $\sum_{i=1}^{10} w_i f(x_i)$ to approximate $\int_a^b f(x)\, dx$. The additional points $x_{11}, \ldots, x_{21}$, and the new weights $v_1, \ldots, v_{21}$, are then used in the Kronrod formula $\sum_{i=1}^{21} v_i f(x_i)$. The results of the two formulas are compared to eliminate error. The advantage in using $x_1, \ldots, x_{10}$ in each formula is that f needs to be evaluated only at 21 points. If independent 10- and 21-point Gaussian rules were used, 31 function evaluations would be needed. This procedure permits endpoint singularities in the integrand.

Other IMSL subroutines allow for endpoint singularities, user-specified singularities, and infinite intervals of integration. In addition, there are routines for applying Gauss-Kronrod rules to integrate a function of two variables, and a routine to use Gaussian quadrature to integrate a function of n variables over n intervals of the form $[a_i, b_i]$.

The NAG Library includes a routine to compute the integral of f over the interval $[a, b]$ using an adaptive method based on Gaussian Quadrature using Gauss 10-point and Kronrod 21-point rules. It also has a routine to approximate an integral using a family of Gaussian-type formulas based on $1, 3, 5, 7, 15, 31, 63, 127$, and 255 nodes. These interlacing high-precision rules are due to Patterson [Pat] and are used in an adaptive manner. NAG includes many other subroutines for approximating integrals.

MATLAB has a routine to approximate a definite integral using an adaptive Simpson's rule, and another to approximate the definite integral using an adaptive eight-panel Newton-Cotes rule.

Although numerical differentiation is unstable, derivative approximation formulas are needed for solving differential equations. The NAG Library includes a subroutine for the numerical differentiation of a function of one real variable with differentiation to the fourteenth derivative being possible. IMSL has a function that uses an adaptive change in step size for finite differences to approximate the first, second, or third, derivative of f at x to within a given tolerance. IMSL also includes a subroutine to compute the derivatives of a function defined on a set of points using quadratic interpolation. Both packages allow the differentiation and integration of interpolatory cubic splines constructed by the subroutines mentioned in Section 3.5.

For further reading on numerical integration we recommend the books by Engels [E] and by Davis and Rabinowitz [DR]. For more information on Gaussian quadrature see Stroud and Secrest [StS]. Books on multiple integrals include those by Stroud [Stro] and by Sloan and Joe [SJ].

CHAPTER

5 Initial-Value Problems for Ordinary Differential Equations

Introduction

The motion of a swinging pendulum under certain simplifying assumptions is described by the second-order differential equation

$$\frac{d^2\theta}{dt^2} + \frac{g}{L}\sin\theta = 0,$$

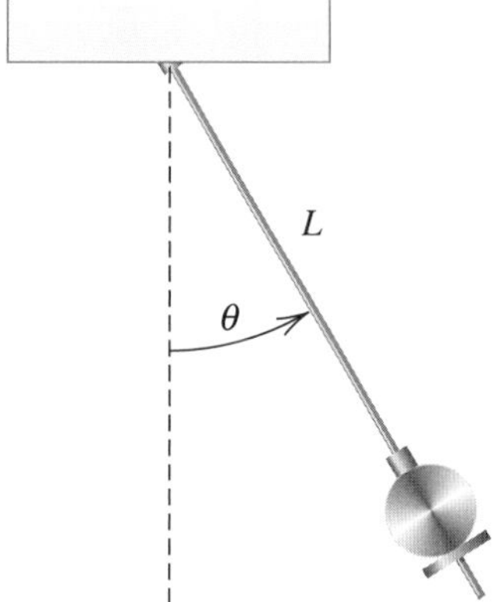

where L is the length of the pendulum, $g \approx 32.17$ ft/s^2 is the gravitational constant of the earth, and θ is the angle the pendulum makes with the vertical. If, in addition, we specify the position of the pendulum when the motion begins, $\theta(t_0) = \theta_0$, and its velocity at that point, $\theta'(t_0) = \theta_0'$, we have what is called an *initial-value problem.*

For small values of θ, the approximation $\theta \approx \sin\theta$ can be used to simplify this problem to the linear initial-value problem

$$\frac{d^2\theta}{dt^2} + \frac{g}{L}\theta = 0, \quad \theta(t_0) = \theta_0, \quad \theta'(t_0) = \theta_0'.$$

This problem can be solved by a standard differential-equation technique. For larger values of θ, the assumption that $\theta = \sin\theta$ is not reasonable so approximation methods must be used. A problem of this type is considered in Exercise 8 of Section 5.9.

Any textbook on ordinary differential equations details a number of methods for explicitly finding solutions to first-order initial-value problems. In practice, however, few of the problems originating from the study of physical phenomena can be solved exactly.

The first part of this chapter is concerned with approximating the solution $y(t)$ to a problem of the form

$$\frac{dy}{dt} = f(t, y), \quad \text{for } a \leq t \leq b,$$

subject to an initial condition $y(a) = \alpha$. Later in the chapter we deal with the extension of these methods to a system of first-order differential equations in the form

$$\begin{aligned}
\frac{dy_1}{dt} &= f_1(t, y_1, y_2, \ldots, y_n), \\
\frac{dy_2}{dt} &= f_2(t, y_1, y_2, \ldots, y_n), \\
&\vdots \\
\frac{dy_n}{dt} &= f_n(t, y_1, y_2, \ldots, y_n),
\end{aligned}$$

for $a \leq t \leq b$, subject to the initial conditions

$$y_1(a) = \alpha_1, \quad y_2(a) = \alpha_2, \quad \ldots, \quad y_n(a) = \alpha_n.$$

We also examine the relationship of a system of this type to the general nth-order initial-value problem of the form

$$y^{(n)} = f(t, y, y', y'', \ldots, y^{(n-1)}),$$

for $a \leq t \leq b$, subject to the initial conditions

$$y(a) = \alpha_1, \quad y'(a) = \alpha_2, \quad \ldots, \quad y^{n-1}(a) = \alpha_n.$$

5.1 The Elementary Theory of Initial-Value Problems

Differential equations are used to model problems in science and engineering that involve the change of some variable with respect to another. Most of these problems require the solution of an *initial-value problem*, that is, the solution to a differential equation that satisfies a given initial condition.

In common real-life situations, the differential equation that models the problem is too complicated to solve exactly, and one of two approaches is taken to approximate the solution. The first approach is to modify the problem by simplifying the differential equation to one that can be solved exactly and then use the solution of the simplified equation to approximate the solution to the original problem. The other approach, which we will examine in this chapter, uses methods for approximating the solution of the original problem. This is the approach that is most commonly taken because the approximation methods give more accurate results and realistic error information.

The methods that we consider in this chapter do not produce a continuous approximation to the solution of the initial-value problem. Rather, approximations are found at certain specified, and often equally spaced, points. Some method of interpolation, commonly Hermite, is used if intermediate values are needed.

We need some definitions and results from the theory of ordinary differential equations before considering methods for approximating the solutions to initial-value problems.

Definition 5.1 A function $f(t, y)$ is said to satisfy a **Lipschitz condition** in the variable y on a set $D \subset \mathbb{R}^2$ if a constant $L > 0$ exists with

$$|f(t, y_1) - f(t, y_2,)| \leq L|y_1 - y_2|,$$

whenever (t, y_1) and (t, y_2) are in D. The constant L is called a **Lipschitz constant** for f. ■

Example 1 Show that $f(t, y) = t|y|$ satisfies a Lipschitz condition on the interval $D = \{(t, y) \mid 1 \leq t \leq 2 \text{ and } -3 \leq y \leq 4\}$.

Solution For each pair of points (t, y_1) and (t, y_2) in D we have

$$|f(t, y_1) - f(t, y_2)| = |t|y_1| - t|y_2|| = |t|||y_1| - |y_2|| \leq 2|y_1 - y_2|.$$

Thus f satisfies a Lipschitz condition on D in the variable y with Lipschitz constant 2. The smallest value possible for the Lipschitz constant for this problem is $L = 2$, because, for example,

$$|f(2, 1) - f(2, 0)| = |2 - 0| = 2|1 - 0|.$$ ■

Definition 5.2 A set $D \subset \mathbb{R}^2$ is said to be **convex** if whenever (t_1, y_1) and (t_2, y_2) belong to D, then $((1 - \lambda)t_1 + \lambda t_2, (1 - \lambda)y_1 + \lambda y_2)$ also belongs to D for every λ in $[0, 1]$. ■

In geometric terms, Definition 5.2 states that a set is convex provided that whenever two points belong to the set, the entire straight-line segment between the points also belongs to the set. (See Figure 5.1.) The sets we consider in this chapter are generally of the form $D = \{(t, y) \mid a \leq t \leq b \text{ and } -\infty < y < \infty\}$ for some constants a and b. It is easy to verify (see Exercise 7) that these sets are convex.

Figure 5.1

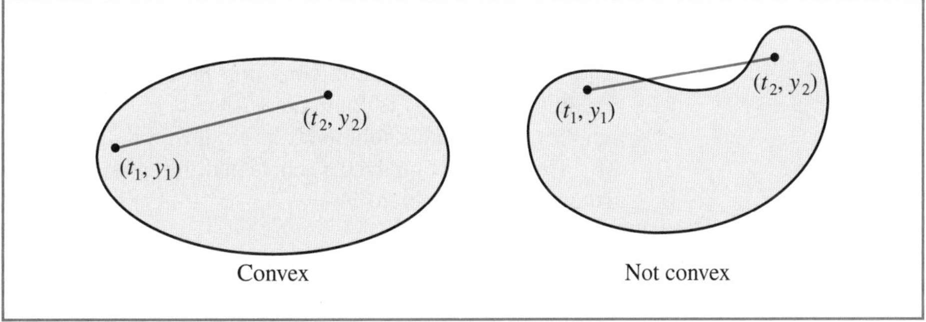

Theorem 5.3 Suppose $f(t, y)$ is defined on a convex set $D \subset \mathbb{R}^2$. If a constant $L > 0$ exists with

$$\left|\frac{\partial f}{\partial y}(t, y)\right| \leq L, \quad \text{for all } (t, y) \in D, \tag{5.1}$$

then f satisfies a Lipschitz condition on D in the variable y with Lipschitz constant L. ■

Rudolf Lipschitz (1832–1903) worked in many branches of mathematics, including number theory, Fourier series, differential equations, analytical mechanics, and potential theory. He is best known for this generalization of the work of Augustin-Louis Cauchy (1789–1857) and Guiseppe Peano (1856–1932).

The proof of Theorem 5.3 is discussed in Exercise 6; it is similar to the proof of the corresponding result for functions of one variable discussed in Exercise 27 of Section 1.1.

As the next theorem will show, it is often of significant interest to determine whether the function involved in an initial-value problem satisfies a Lipschitz condition in its second variable, and condition (5.1) is generally easier to apply than the definition. We should note, however, that Theorem 5.3 gives only sufficient conditions for a Lipschitz condition to hold. The function in Example 1, for instance, satisfies a Lipschitz condition, but the partial derivative with respect to y does not exist when $y = 0$.

The following theorem is a version of the fundamental existence and uniqueness theorem for first-order ordinary differential equations. Although the theorem can be proved with the hypothesis reduced somewhat, this form of the theorem is sufficient for our purposes. (The proof of the theorem, in approximately this form, can be found in [BiR], pp. 142–155.)

Theorem 5.4 Suppose that $D = \{(t, y) \mid a \leq t \leq b \text{ and } -\infty < y < \infty\}$ and that $f(t, y)$ is continuous on D. If f satisfies a Lipschitz condition on D in the variable y, then the initial-value problem

$$y'(t) = f(t, y), \quad a \leq t \leq b, \quad y(a) = \alpha,$$

has a unique solution $y(t)$ for $a \leq t \leq b$. ■

Example 2 Use Theorem 5.4 to show that there is a unique solution to the initial-value problem

$$y' = 1 + t\sin(ty), \quad 0 \leq t \leq 2, \quad y(0) = 0.$$

Solution Holding t constant and applying the Mean Value Theorem to the function

$$f(t, y) = 1 + t\sin(ty),$$

we find that when $y_1 < y_2$, a number ξ in (y_1, y_2) exists with

$$\frac{f(t, y_2) - f(t, y_1)}{y_2 - y_1} = \frac{\partial}{\partial y} f(t, \xi) = t^2 \cos(\xi t).$$

Thus

$$|f(t, y_2) - f(t, y_1)| = |y_2 - y_1||t^2 \cos(\xi t)| \leq 4|y_2 - y_1|,$$

and f satisfies a Lipschitz condition in the variable y with Lipschitz constant $L = 4$. Additionally, $f(t, y)$ is continuous when $0 \leq t \leq 2$ and $-\infty < y < \infty$, so Theorem 5.4 implies that a unique solution exists to this initial-value problem.

If you have completed a course in differential equations you might try to find the exact solution to this problem. ■

Well-Posed Problems

Now that we have, to some extent, taken care of the question of when initial-value problems have unique solutions, we can move to the second important consideration when approximating the solution to an initial-value problem. Initial-value problems obtained by observing physical phenomena generally only approximate the true situation, so we need to know whether small changes in the statement of the problem introduce correspondingly small changes in the solution. This is also important because of the introduction of round-off error when numerical methods are used. That is,

- Question: How do we determine whether a particular problem has the property that small changes, or perturbations, in the statement of the problem introduce correspondingly small changes in the solution?

As usual, we first need to give a workable definition to express this concept.

Definition 5.5 The initial-value problem

$$\frac{dy}{dt} = f(t, y), \quad a \le t \le b, \quad y(a) = \alpha, \tag{5.2}$$

is said to be a **well-posed problem** if:

- A unique solution, $y(t)$, to the problem exists, and
- There exist constants $\varepsilon_0 > 0$ and $k > 0$ such that for any ε, with $\varepsilon_0 > \varepsilon > 0$, whenever $\delta(t)$ is continuous with $|\delta(t)| < \varepsilon$ for all t in $[a, b]$, and when $|\delta_0| < \varepsilon$, the initial-value problem

  $$\frac{dz}{dt} = f(t, z) + \delta(t), \quad a \le t \le b, \quad z(a) = \alpha + \delta_0, \tag{5.3}$$

  has a unique solution $z(t)$ that satisfies

  $$|z(t) - y(t)| < k\varepsilon \quad \text{for all } t \text{ in } [a, b].$$ ■

The problem specified by (5.3) is called a **perturbed problem** associated with the original problem (5.2). It assumes the possibility of an error being introduced in the statement of the differential equation, as well as an error δ_0 being present in the initial condition.

Numerical methods will always be concerned with solving a perturbed problem because any round-off error introduced in the representation perturbs the original problem. Unless the original problem is well-posed, there is little reason to expect that the numerical solution to a perturbed problem will accurately approximate the solution to the original problem.

The following theorem specifies conditions that ensure that an initial-value problem is well-posed. The proof of this theorem can be found in [BiR], pp. 142–147.

Theorem 5.6 Suppose $D = \{(t, y) \mid a \le t \le b \text{ and } -\infty < y < \infty\}$. If f is continuous and satisfies a Lipschitz condition in the variable y on the set D, then the initial-value problem

$$\frac{dy}{dt} = f(t, y), \quad a \le t \le b, \quad y(a) = \alpha$$

is well-posed. ■

Example 3 Show that the initial-value problem

$$\frac{dy}{dt} = y - t^2 + 1, \quad 0 \le t \le 2, \quad y(0) = 0.5. \tag{5.4}$$

is well posed on $D = \{(t, y) \mid 0 \le t \le 2 \text{ and } -\infty < y < \infty\}$.

Solution Because

$$\left| \frac{\partial(y - t^2 + 1)}{\partial y} \right| = |1| = 1,$$

Theorem 5.3 implies that $f(t, y) = y - t^2 + 1$ satisfies a Lipschitz condition in y on D with Lipschitz constant 1. Since f is continuous on D, Theorem 5.6 implies that the problem is well-posed.

As an illustration, consider the solution to the perturbed problem

$$\frac{dz}{dt} = z - t^2 + 1 + \delta, \quad 0 \le t \le 2, \quad z(0) = 0.5 + \delta_0, \tag{5.5}$$

where δ and δ_0 are constants. The solutions to Eqs. (5.4) and (5.5) are

$$y(t) = (t+1)^2 - 0.5e^t \quad \text{and} \quad z(t) = (t+1)^2 + (\delta + \delta_0 - 0.5)e^t - \delta,$$

respectively.

Suppose that ε is a positive number. If $|\delta| < \varepsilon$ and $|\delta_0| < \varepsilon$, then

$$|y(t) - z(t)| = |(\delta + \delta_0)e^t - \delta| \le |\delta + \delta_0|e^2 + |\delta| \le (2e^2 + 1)\varepsilon,$$

for all t. This implies that problem (5.4) is well-posed with $k(\varepsilon) = 2e^2 + 1$ for all $\varepsilon > 0$. ■

Maple can be used to solve many initial-value problems. Consider the problem

$$\frac{dy}{dt} = y - t^2 + 1, \quad 0 \le t \le 2, \quad y(0) = 0.5.$$

Maple reserves the letter D to represent differentiation.

To define the differential equation and initial condition, enter

$deq := D(y)(t) = y(t) - t^2 + 1;\ init := y(0) = 0.5$

The names *deq* and *init* have been chosen by the user. The command to solve the initial-value problems is

$deqsol := dsolve\ (\{deq, init\}, y(t))$

and Maple responds with

$$y(t) = 1 + t^2 + 2t - \frac{1}{2}e^t$$

To use the solution to obtain a specific value, such as $y(1.5)$, we enter

$q := rhs(deqsol) :\ evalf(subs(t = 1.5, q))$

which gives

$$4.009155465$$

The function *rhs* (for *r*ight *h*and *s*ide) is used to assign the solution of the initial-value problem to the function q, which we then evaluate at $t = 1.5$.

The function *dsolve* can fail if an explicit solution to the initial-value problem cannot be found. For example, for the initial-value problem given in Example 2, the command

$deqsol2 := dsolve\ (\{D(y)(t) = 1 + t \cdot \sin(t \cdot y(t)), y(0) = 0\}, y(t))$

does not succeed because an explicit solution cannot be found. In this case a numerical method must be used.

EXERCISE SET 5.1

1. Use Theorem 5.4 to show that each of the following initial-value problems has a unique solution, and find the solution.
   a. $y' = y\cos t, \quad 0 \le t \le 1, \quad y(0) = 1.$
   b. $y' = -\frac{2}{t}y + t^2e^t, \quad 1 \le t \le 2, \quad y(1) = \sqrt{2}e.$
   c. $y' = \frac{2}{t}y + t^2e^t, \quad 1 \le t \le 2, \quad y(1) = 0.$
   d. $y' = \frac{4t^3y}{1+t^4}, \quad 0 \le t \le 1, \quad y(0) = 1.$

2. Show that each of the following initial-value problems has a unique solution and find the solution. Can Theorem 5.4 be applied in each case?

   a. $y' = e^{t-y}, \quad 0 \le t \le 1, \quad y(0) = 1.$

   b. $y' = t^{-2}(\sin 2t - 2ty), \quad 1 \le t \le 2, \quad y(1) = 2.$

   c. $y' = -y + ty^{1/2}, \quad 2 \le t \le 3, \quad y(2) = 2.$

   d. $y' = \dfrac{ty + y}{ty + t}, \quad 2 \le t \le 4, \quad y(2) = 4.$

3. For each choice of $f(t, y)$ given in parts (a)–(d):

   i. Does f satisfy a Lipschitz condition on $D = \{(t, y) \mid 0 \le t \le 1, -\infty < y < \infty\}$?

   ii. Can Theorem 5.6 be used to show that the initial-value problem

$$y' = f(t, y), \quad 0 \le t \le 1, \quad y(0) = 1,$$

   is well-posed?

   a. $f(t, y) = t^2 y + 1$ **b.** $f(t, y) = ty$ **c.** $f(t, y) = 1 - y$ **d.** $f(t, y) = -ty + \dfrac{4t}{y}$

4. For each choice of $f(t, y)$ given in parts (a)–(d):

   i. Does f satisfy a Lipschitz condition on $D = \{(t, y) \mid 0 \le t \le 1, -\infty < y < \infty\}$?

   ii. Can Theorem 5.6 be used to show that the initial-value problem

$$y' = f(t, y), \quad 0 \le t \le 1, \quad y(0) = 1,$$

   is well-posed?

   a. $f(t, y) = e^{t-y}$ **b.** $f(t, y) = \dfrac{1+y}{1+t}$ **c.** $f(t, y) = \cos(yt)$ **d.** $f(t, y) = \dfrac{y^2}{1+t}$

5. For the following initial-value problems, show that the given equation implicitly defines a solution. Approximate $y(2)$ using Newton's method.

   a. $y' = -\dfrac{y^3 + y}{(3y^2 + 1)t}, \quad 1 \le t \le 2, \quad y(1) = 1; \quad y^3 t + yt = 2$

   b. $y' = -\dfrac{y \cos t + 2te^y}{\sin t + t^2 e^y + 2}, \quad 1 \le t \le 2, \quad y(1) = 0; \quad y \sin t + t^2 e^y + 2y = 1$

6. Prove Theorem 5.3 by applying the Mean Value Theorem 1,8 to $f(t, y)$, holding t fixed.

7. Show that, for any constants a and b, the set $D = \{(t, y) \mid a \le t \le b, -\infty < y < \infty\}$ is convex.

8. Suppose the perturbation $\delta(t)$ is proportional to t, that is, $\delta(t) = \delta t$ for some constant δ. Show directly that the following initial-value problems are well-posed.

   a. $y' = 1 - y, \quad 0 \le t \le 2, \quad y(0) = 0$

   b. $y' = t + y, \quad 0 \le t \le 2, \quad y(0) = -1$

   c. $y' = \dfrac{2}{t}y + t^2 e^t, \quad 1 \le t \le 2, \quad y(1) = 0$

   d. $y' = -\dfrac{2}{t}y + t^2 e^t, \quad 1 \le t \le 2, \quad y(1) = \sqrt{2}e$

9. *Picard's method* for solving the initial-value problem

$$y' = f(t, y), \quad a \le t \le b, \quad y(a) = \alpha,$$

   is described as follows: Let $y_0(t) = \alpha$ for each t in $[a, b]$. Define a sequence $\{y_k(t)\}$ of functions by

$$y_k(t) = \alpha + \int_a^t f(\tau, y_{k-1}(\tau))\, d\tau, \quad k = 1, 2, \ldots.$$

   a. Integrate $y' = f(t, y(t))$, and use the initial condition to derive Picard's method.

   b. Generate $y_0(t)$, $y_1(t)$, $y_2(t)$, and $y_3(t)$ for the initial-value problem

$$y' = -y + t + 1, \quad 0 \le t \le 1, \quad y(0) = 1.$$

   c. Compare the result in part (b) to the Maclaurin series of the actual solution $y(t) = t + e^{-t}$.

5.2 Euler's Method

Euler's method is the most elementary approximation technique for solving initial-value problems. Although it is seldom used in practice, the simplicity of its derivation can be used to illustrate the techniques involved in the construction of some of the more advanced techniques, without the cumbersome algebra that accompanies these constructions.

The object of Euler's method is to obtain approximations to the well-posed initial-value problem

$$\frac{dy}{dt} = f(t, y), \quad a \le t \le b, \quad y(a) = \alpha. \tag{5.6}$$

A continuous approximation to the solution $y(t)$ will not be obtained; instead, approximations to y will be generated at various values, called **mesh points**, in the interval $[a, b]$. Once the approximate solution is obtained at the points, the approximate solution at other points in the interval can be found by interpolation.

We first make the stipulation that the mesh points are equally distributed throughout the interval $[a, b]$. This condition is ensured by choosing a positive integer N and selecting the mesh points

$$t_i = a + ih, \quad \text{for each } i = 0, 1, 2, \ldots, N.$$

The common distance between the points $h = (b - a)/N = t_{i+1} - t_i$ is called the **step size**.

The use of elementary difference methods to approximate the solution to differential equations was one of the numerous mathematical topics that was first presented to the mathematical public by the most prolific of mathematicians, Leonhard Euler (1707–1783).

We will use Taylor's Theorem to derive Euler's method. Suppose that $y(t)$, the unique solution to (5.6), has two continuous derivatives on $[a, b]$, so that for each $i = 0, 1, 2, \ldots, N - 1$,

$$y(t_{i+1}) = y(t_i) + (t_{i+1} - t_i)y'(t_i) + \frac{(t_{i+1} - t_i)^2}{2}y''(\xi_i),$$

for some number ξ_i in (t_i, t_{i+1}). Because $h = t_{i+1} - t_i$, we have

$$y(t_{i+1}) = y(t_i) + hy'(t_i) + \frac{h^2}{2}y''(\xi_i),$$

and, because $y(t)$ satisfies the differential equation (5.6),

$$y(t_{i+1}) = y(t_i) + hf(t_i, y(t_i)) + \frac{h^2}{2}y''(\xi_i). \tag{5.7}$$

Euler's method constructs $w_i \approx y(t_i)$, for each $i = 1, 2, \ldots, N$, by deleting the remainder term. Thus Euler's method is

$$\begin{aligned} w_0 &= \alpha, \\ w_{i+1} &= w_i + hf(t_i, w_i), \quad \text{for each } i = 0, 1, \ldots, N - 1. \end{aligned} \tag{5.8}$$

Illustration In Example 1 we will use an algorithm for Euler's method to approximate the solution to

$$y' = y - t^2 + 1, \quad 0 \le t \le 2, \quad y(0) = 0.5,$$

at $t = 2$. Here we will simply illustrate the steps in the technique when we have $h = 0.5$.

For this problem $f(t, y) = y - t^2 + 1$, so

$$w_0 = y(0) = 0.5;$$

$$w_1 = w_0 + 0.5\left(w_0 - (0.0)^2 + 1\right) = 0.5 + 0.5(1.5) = 1.25;$$

$$w_2 = w_1 + 0.5\left(w_1 - (0.5)^2 + 1\right) = 1.25 + 0.5(2.0) = 2.25;$$

$$w_3 = w_2 + 0.5\left(w_2 - (1.0)^2 + 1\right) = 2.25 + 0.5(2.25) = 3.375;$$

and

$$y(2) \approx w_4 = w_3 + 0.5\left(w_3 - (1.5)^2 + 1\right) = 3.375 + 0.5(2.125) = 4.4375. \qquad \square$$

Equation (5.8) is called the **difference equation** associated with Euler's method. As we will see later in this chapter, the theory and solution of difference equations parallel, in many ways, the theory and solution of differential equations. Algorithm 5.1 implements Euler's method.

Euler's

To approximate the solution of the initial-value problem

$$y' = f(t, y), \quad a \leq t \leq b, \quad y(a) = \alpha,$$

at $(N + 1)$ equally spaced numbers in the interval $[a, b]$:

INPUT endpoints a, b; integer N; initial condition α.

OUTPUT approximation w to y at the $(N + 1)$ values of t.

Step 1 Set $h = (b - a)/N$;
$t = a$;
$w = \alpha$;
OUTPUT (t, w).

Step 2 For $i = 1, 2, \ldots, N$ do Steps 3, 4.

Step 3 Set $w = w + hf(t, w)$; *(Compute w_i.)*
$t = a + ih$. *(Compute t_i.)*

Step 4 OUTPUT (t, w).

Step 5 STOP. ■

To interpret Euler's method geometrically, note that when w_i is a close approximation to $y(t_i)$, the assumption that the problem is well-posed implies that

$$f(t_i, w_i) \approx y'(t_i) = f(t_i, y(t_i)).$$

The graph of the function highlighting $y(t_i)$ is shown in Figure 5.2. One step in Euler's method appears in Figure 5.3, and a series of steps appears in Figure 5.4.

Figure 5.2

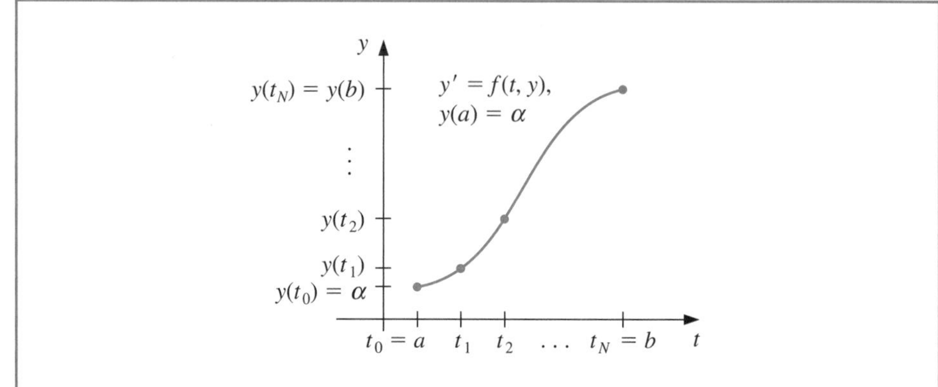

Figure 5.3

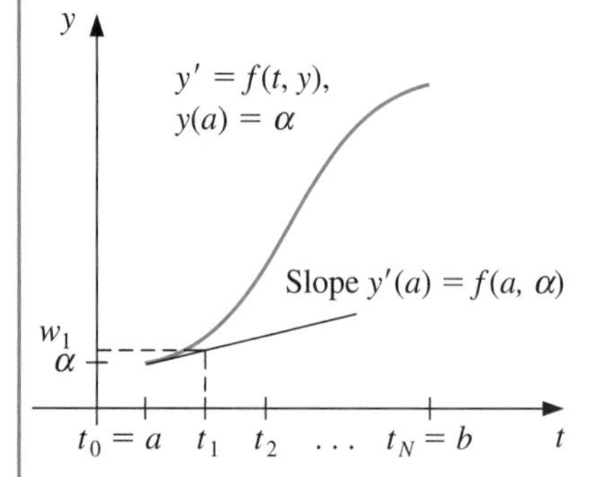

Figure 5.4

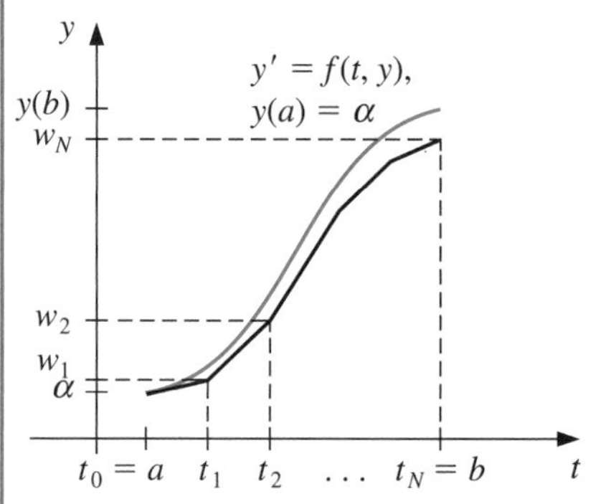

Example 1 Euler's method was used in the first illustration with $h = 0.5$ to approximate the solution to the initial-value problem

$$y' = y - t^2 + 1, \quad 0 \le t \le 2, \quad y(0) = 0.5.$$

Use Algorithm 5.1 with $N = 10$ to determine approximations, and compare these with the exact values given by $y(t) = (t + 1)^2 - 0.5e^t$.

Solution With $N = 10$ we have $h = 0.2$, $t_i = 0.2i$, $w_0 = 0.5$, and

$$w_{i+1} = w_i + h(w_i - t_i^2 + 1) = w_i + 0.2[w_i - 0.04i^2 + 1] = 1.2w_i - 0.008i^2 + 0.2,$$

for $i = 0, 1, \ldots, 9$. So

$$w_1 = 1.2(0.5) - 0.008(0)^2 + 0.2 = 0.8; \quad w_2 = 1.2(0.8) - 0.008(1)^2 + 0.2 = 1.152;$$

and so on. Table 5.1 shows the comparison between the approximate values at t_i and the actual values. ■

Table 5.1

t_i	w_i	$y_i = y(t_i)$	$\|y_i - w_i\|$
0.0	0.5000000	0.5000000	0.0000000
0.2	0.8000000	0.8292986	0.0292986
0.4	1.1520000	1.2140877	0.0620877
0.6	1.5504000	1.6489406	0.0985406
0.8	1.9884800	2.1272295	0.1387495
1.0	2.4581760	2.6408591	0.1826831
1.2	2.9498112	3.1799415	0.2301303
1.4	3.4517734	3.7324000	0.2806266
1.6	3.9501281	4.2834838	0.3333557
1.8	4.4281538	4.8151763	0.3870225
2.0	4.8657845	5.3054720	0.4396874

Note that the error grows slightly as the value of t increases. This controlled error growth is a consequence of the stability of Euler's method, which implies that the error is expected to grow in no worse than a linear manner.

Maple has implemented Euler's method as an option with the command *InitialValueProblem* within the *NumericalAnalysis* subpackage of the *Student* package. To use it for the problem in Example 1 first load the package and the differential equation.

$with(Student[NumericalAnalysis]): deq := diff(y(t), t) = y(t) - t^2 + 1$

Then issue the command

$C := InitialValueProblem(deq, y(0) = 0.5, t = 2, method = euler, numsteps = 10,$
$output = information, digits = 8)$

Maple produces

$$\left[\begin{array}{c} 1\,..\,12 \times 1\,..\,4\ Array \\ Data\ Type:\ anything \\ Storage:\ rectangular \\ Order:\ Fortran\_order \end{array}\right]$$

Double clicking on the output brings up a table that gives the values of t_i, actual solution values $y(t_i)$, the Euler approximations w_i, and the absolute errors $|\,y(t_i) - w_i|$. These agree with the values in Table 5.1.

To print the Maple table we can issue the commands

for k **from** 1 **to** 12 **do**
$print(C[k, 1], C[k, 2], C[k, 3], C[k, 4])$
end do

The options within the *InitialValueProblem* command are the specification of the first order differential equation to be solved, the initial condition, the final value of the independent variable, the choice of method, the number of steps used to determine that $h = (2 - 0)/(numsteps)$, the specification of form of the output, and the number of digits of rounding to be used in the computations. Other output options can specify a particular value of t or a plot of the solution.

Error Bounds for Euler's Method

Although Euler's method is not accurate enough to warrant its use in practice, it is sufficiently elementary to analyze the error that is produced from its application. The error analysis for the more accurate methods that we consider in subsequent sections follows the same pattern but is more complicated.

To derive an error bound for Euler's method, we need two computational lemmas.

Lemma 5.7 For all $x \geq -1$ and any positive m, we have $0 \leq (1+x)^m \leq e^{mx}$. ■

Proof Applying Taylor's Theorem with $f(x) = e^x$, $x_0 = 0$, and $n = 1$ gives

$$e^x = 1 + x + \frac{1}{2}x^2 e^{\xi},$$

where ξ is between x and zero. Thus

$$0 \leq 1 + x \leq 1 + x + \frac{1}{2}x^2 e^{\xi} = e^x,$$

and, because $1 + x \geq 0$, we have

$$0 \leq (1+x)^m \leq (e^x)^m = e^{mx}.$$

■ ■ ■

Lemma 5.8 If s and t are positive real numbers, $\{a_i\}_{i=0}^{k}$ is a sequence satisfying $a_0 \geq -t/s$, and

$$a_{i+1} \leq (1+s)a_i + t, \quad \text{for each } i = 0, 1, 2, \ldots, k-1, \tag{5.9}$$

then

$$a_{i+1} \leq e^{(i+1)s}\left(a_0 + \frac{t}{s}\right) - \frac{t}{s}.$$

■

Proof For a fixed integer i, Inequality (5.9) implies that

$$\begin{aligned}
a_{i+1} &\leq (1+s)a_i + t \\
&\leq (1+s)[(1+s)a_{i-1} + t] + t = (1+s)^2 a_{i-1} + [1 + (1+s)]t \\
&\leq (1+s)^3 a_{i-2} + \left[1 + (1+s) + (1+s)^2\right]t \\
&\ \vdots \\
&\leq (1+s)^{i+1} a_0 + \left[1 + (1+s) + (1+s)^2 + \cdots + (1+s)^i\right]t.
\end{aligned}$$

But

$$1 + (1+s) + (1+s)^2 + \cdots + (1+s)^i = \sum_{j=0}^{i}(1+s)^j$$

is a geometric series with ratio $(1+s)$ that sums to

$$\frac{1 - (1+s)^{i+1}}{1 - (1+s)} = \frac{1}{s}[(1+s)^{i+1} - 1].$$

Thus

$$a_{i+1} \leq (1+s)^{i+1}a_0 + \frac{(1+s)^{i+1} - 1}{s}t = (1+s)^{i+1}\left(a_0 + \frac{t}{s}\right) - \frac{t}{s},$$

and using Lemma 5.7 with $x = 1 + s$ gives

$$a_{i+1} \leq e^{(i+1)s}\left(a_0 + \frac{t}{s}\right) - \frac{t}{s}.$$

■ ■ ■

Theorem 5.9 Suppose f is continuous and satisfies a Lipschitz condition with constant L on

$$D = \{(t, y) \mid a \le t \le b \text{ and } -\infty < y < \infty\}$$

and that a constant M exists with

$$|y''(t)| \le M, \quad \text{for all } t \in [a, b],$$

where $y(t)$ denotes the unique solution to the initial-value problem

$$y' = f(t, y), \quad a \le t \le b, \quad y(a) = \alpha.$$

Let $w_0, w_1, \ldots, w_N$ be the approximations generated by Euler's method for some positive integer N. Then, for each $i = 0, 1, 2, \ldots, N$,

$$|y(t_i) - w_i| \le \frac{hM}{2L}\left[e^{L(t_i-a)} - 1\right]. \tag{5.10}$$

■

Proof When $i = 0$ the result is clearly true, since $y(t_0) = w_0 = \alpha$.

From Eq. (5.7), we have

$$y(t_{i+1}) = y(t_i) + hf(t_i, y(t_i)) + \frac{h^2}{2}y''(\xi_i),$$

for $i = 0, 1, \ldots, N-1$, and from the equations in (5.8),

$$w_{i+1} = w_i + hf(t_i, w_i).$$

Using the notation $y_i = y(t_i)$ and $y_{i+1} = y(t_{i+1})$, we subtract these two equations to obtain

$$y_{i+1} - w_{i+1} = y_i - w_i + h[f(t_i, y_i) - f(t_i, w_i)] + \frac{h^2}{2}y''(\xi_i)$$

Hence

$$|y_{i+1} - w_{i+1}| \le |y_i - w_i| + h|f(t_i, y_i) - f(t_i, w_i)| + \frac{h^2}{2}|y''(\xi_i)|.$$

Now f satisfies a Lipschitz condition in the second variable with constant L, and $|y''(t)| \le M$, so

$$|y_{i+1} - w_{i+1}| \le (1 + hL)|y_i - w_i| + \frac{h^2M}{2}.$$

Referring to Lemma 5.8 and letting $s = hL$, $t = h^2M/2$, and $a_j = |y_j - w_j|$, for each $j = 0, 1, \ldots, N$, we see that

$$|y_{i+1} - w_{i+1}| \le e^{(i+1)hL}\left(|y_0 - w_0| + \frac{h^2M}{2hL}\right) - \frac{h^2M}{2hL}.$$

Because $|y_0 - w_0| = 0$ and $(i+1)h = t_{i+1} - t_0 = t_{i+1} - a$, this implies that

$$|y_{i+1} - w_{i+1}| \le \frac{hM}{2L}(e^{(t_{i+1}-a)L} - 1),$$

for each $i = 0, 1, \ldots, N-1$. ■ ■ ■

The weakness of Theorem 5.9 lies in the requirement that a bound be known for the second derivative of the solution. Although this condition often prohibits us from obtaining a realistic error bound, it should be noted that if $\partial f/\partial t$ and $\partial f/\partial y$ both exist, the chain rule for partial differentiation implies that

$$y''(t) = \frac{dy'}{dt}(t) = \frac{df}{dt}(t, y(t)) = \frac{\partial f}{\partial t}(t, y(t)) + \frac{\partial f}{\partial y}(t, y(t)) \cdot f(t, y(t)).$$

So it is at times possible to obtain an error bound for $y''(t)$ without explicitly knowing $y(t)$.

Example 2 The solution to the initial-value problem

$$y' = y - t^2 + 1, \quad 0 \le t \le 2, \quad y(0) = 0.5,$$

was approximated in Example 1 using Euler's method with $h = 0.2$. Use the inequality in Theorem 5.9 to find a bounds for the approximation errors and compare these to the actual errors.

Solution Because $f(t, y) = y - t^2 + 1$, we have $\partial f(t, y)/\partial y = 1$ for all y, so $L = 1$. For this problem, the exact solution is $y(t) = (t + 1)^2 - 0.5e^t$, so $y''(t) = 2 - 0.5e^t$ and

$$|y''(t)| \le 0.5e^2 - 2, \quad \text{for all } t \in [0, 2].$$

Using the inequality in the error bound for Euler's method with $h = 0.2$, $L = 1$, and $M = 0.5e^2 - 2$ gives

$$|y_i - w_i| \le 0.1(0.5e^2 - 2)(e^{t_i} - 1).$$

Hence

$$|y(0.2) - w_1| \le 0.1(0.5e^2 - 2)(e^{0.2} - 1) = 0.03752;$$
$$|y(0.4) - w_2| \le 0.1(0.5e^2 - 2)(e^{0.4} - 1) = 0.08334;$$

and so on. Table 5.2 lists the actual error found in Example 1, together with this error bound. Note that even though the true bound for the second derivative of the solution was used, the error bound is considerably larger than the actual error, especially for increasing values of t. ■

Table 5.2

t_i	0.2	0.4	0.6	0.8	1.0	1.2	1.4	1.6	1.8	2.0
Actual Error	0.02930	0.06209	0.09854	0.13875	0.18268	0.23013	0.28063	0.33336	0.38702	0.43969
Error Bound	0.03752	0.08334	0.13931	0.20767	0.29117	0.39315	0.51771	0.66985	0.85568	1.08264

The principal importance of the error-bound formula given in Theorem 5.9 is that the bound depends linearly on the step size h. Consequently, diminishing the step size should give correspondingly greater accuracy to the approximations.

Neglected in the result of Theorem 5.9 is the effect that round-off error plays in the choice of step size. As h becomes smaller, more calculations are necessary and more round-off error is expected. In actuality then, the difference-equation form

$$w_0 = \alpha,$$
$$w_{i+1} = w_i + hf(t_i, w_i), \quad \text{for each } i = 0, 1, \ldots, N - 1,$$

is not used to calculate the approximation to the solution y_i at a mesh point t_i. We use instead an equation of the form

$$u_0 = \alpha + \delta_0,$$
$$u_{i+1} = u_i + hf(t_i, u_i) + \delta_{i+1}, \quad \text{for each } i = 0, 1, \ldots, N-1, \tag{5.11}$$

where δ_i denotes the round-off error associated with u_i. Using methods similar to those in the proof of Theorem 5.9, we can produce an error bound for the finite-digit approximations to y_i given by Euler's method.

Theorem 5.10 Let $y(t)$ denote the unique solution to the initial-value problem

$$y' = f(t, y), \quad a \le t \le b, \quad y(a) = \alpha \tag{5.12}$$

and $u_0, u_1, \ldots, u_N$ be the approximations obtained using (5.11). If $|\delta_i| < \delta$ for each $i = 0, 1, \ldots, N$ and the hypotheses of Theorem 5.9 hold for (5.12), then

$$|y(t_i) - u_i| \le \frac{1}{L}\left(\frac{hM}{2} + \frac{\delta}{h}\right)[e^{L(t_i-a)} - 1] + |\delta_0|e^{L(t_i-a)}, \tag{5.13}$$

for each $i = 0, 1, \ldots, N$. ■

The error bound (5.13) is no longer linear in h. In fact, since

$$\lim_{h\to 0}\left(\frac{hM}{2} + \frac{\delta}{h}\right) = \infty,$$

the error would be expected to become large for sufficiently small values of h. Calculus can be used to determine a lower bound for the step size h. Letting $E(h) = (hM/2) + (\delta/h)$ implies that $E'(h) = (M/2) - (\delta/h^2)$.

If $h < \sqrt{2\delta/M}$, then $E'(h) < 0$ and $E(h)$ is decreasing.

If $h > \sqrt{2\delta/M}$, then $E'(h) > 0$ and $E(h)$ is increasing.

The minimal value of $E(h)$ occurs when

$$h = \sqrt{\frac{2\delta}{M}}. \tag{5.14}$$

Decreasing h beyond this value tends to increase the total error in the approximation. Normally, however, the value of δ is sufficiently small that this lower bound for h does not affect the operation of Euler's method.

EXERCISE SET 5.2

1. Use Euler's method to approximate the solutions for each of the following initial-value problems.
   - **a.** $y' = te^{3t} - 2y, \quad 0 \le t \le 1, \quad y(0) = 0$, with $h = 0.5$
   - **b.** $y' = 1 + y/t, \quad 1 \le t \le 2, \quad y(1) = 2$, with $h = 0.25$
   - **c.** $y' = 1 + (t-y)^2, \quad 2 \le t \le 3, \quad y(2) = 1$, with $h = 0.5$
   - **d.** $y' = \cos 2t + \sin 3t, \quad 0 \le t \le 1, \quad y(0) = 1$, with $h = 0.25$

2. Use Euler's method to approximate the solutions for each of the following initial-value problems.

   a. $y' = e^{t-y}, \quad 0 \le t \le 1, \quad y(0) = 1,$ with $h = 0.5$

   b. $y' = \dfrac{1+t}{1+y}, \quad 1 \le t \le 2, \quad y(1) = 2,$ with $h = 0.5$

   c. $y' = t^{-2}(\sin 2t - 2ty), \quad 1 \le t \le 2, \quad y(1) = 2,$ with $h = 0.25$

   d. $y' = -y + ty^{1/2}, \quad 2 \le t \le 3, \quad y(2) = 2,$ with $h = 0.25$

3. The actual solutions to the initial-value problems in Exercise 1 are given here. Compare the actual error at each step to the error bound.

   a. $y(t) = \dfrac{1}{5}te^{3t} - \dfrac{1}{25}e^{3t} + \dfrac{1}{25}e^{-2t}$

   b. $y(t) = t \ln t + 2t$

   c. $y(t) = t + \dfrac{1}{1-t}$

   d. $y(t) = \dfrac{1}{2}\sin 2t - \dfrac{1}{3}\cos 3t + \dfrac{4}{3}$

4. The actual solutions to the initial-value problems in Exercise 2 are given here. Compute the actual error and compare this to the error bound if Theorem 5.9 can be applied.

   a. $y(t) = \ln(e^t + e - 1)$

   b. $y(t) = \sqrt{t^2 + 2t + 6} - 1$

   c. $y(t) = \dfrac{4 + \cos 2 - \cos 2t}{2t^2}$

   d. $y(t) = \left(t - 2 + \sqrt{2}ee^{-t/2}\right)^2$

5. Use Euler's method to approximate the solutions for each of the following initial-value problems.

   a. $y' = y/t - (y/t)^2, \quad 1 \le t \le 2, \quad y(1) = 1,$ with $h = 0.1$

   b. $y' = 1 + y/t + (y/t)^2, \quad 1 \le t \le 3, \quad y(1) = 0,$ with $h = 0.2$

   c. $y' = -(y+1)(y+3), \quad 0 \le t \le 2, \quad y(0) = -2,$ with $h = 0.2$

   d. $y' = -5y + 5t^2 + 2t, \quad 0 \le t \le 1, \quad y(0) = \frac{1}{3},$ with $h = 0.1$

6. Use Euler's method to approximate the solutions for each of the following initial-value problems.

   a. $y' = \dfrac{2 - 2ty}{t^2 + 1}, \quad 0 \le t \le 1, \quad y(0) = 1,$ with $h = 0.1$

   b. $y' = \dfrac{y^2}{1+t}, \quad 1 \le t \le 2, \quad y(1) = -(\ln 2)^{-1},$ with $h = 0.1$

   c. $y' = (y^2 + y)/t, \quad 1 \le t \le 3, \quad y(1) = -2,$ with $h = 0.2$

   d. $y' = -ty + 4ty^{-1}, \quad 0 \le t \le 1, \quad y(0) = 1,$ with $h = 0.1$

7. The actual solutions to the initial-value problems in Exercise 5 are given here. Compute the actual error in the approximations of Exercise 5.

   a. $y(t) = \dfrac{t}{1 + \ln t}$

   b. $y(t) = t \tan(\ln t)$

   c. $y(t) = -3 + \dfrac{2}{1 + e^{-2t}}$

   d. $y(t) = t^2 + \dfrac{1}{3}e^{-5t}$

8. The actual solutions to the initial-value problems in Exercise 6 are given here. Compute the actual error in the approximations of Exercise 6.

   a. $y(t) = \dfrac{2t+1}{t^2+1}$

   b. $y(t) = \dfrac{-1}{\ln(t+1)}$

   c. $y(t) = \dfrac{2t}{1-2t}$

   d. $y(t) = \sqrt{4 - 3e^{-t^2}}$

9. Given the initial-value problem

$$y' = \frac{2}{t}y + t^2e^t, \quad 1 \le t \le 2, \quad y(1) = 0,$$

with exact solution $y(t) = t^2(e^t - e)$:

   a. Use Euler's method with $h = 0.1$ to approximate the solution, and compare it with the actual values of y.

b. Use the answers generated in part (a) and linear interpolation to approximate the following values of y, and compare them to the actual values.

i. $y(1.04)$ **ii.** $y(1.55)$ **iii.** $y(1.97)$

c. Compute the value of h necessary for $|y(t_i) - w_i| \leq 0.1$, using Eq. (5.10).

10. Given the initial-value problem

$$y' = \frac{1}{t^2} - \frac{y}{t} - y^2, \quad 1 \leq t \leq 2, \quad y(1) = -1,$$

with exact solution $y(t) = -1/t$:

a. Use Euler's method with $h = 0.05$ to approximate the solution, and compare it with the actual values of y.

b. Use the answers generated in part (a) and linear interpolation to approximate the following values of y, and compare them to the actual values.

i. $y(1.052)$ **ii.** $y(1.555)$ **iii.** $y(1.978)$

c. Compute the value of h necessary for $|y(t_i) - w_i| \leq 0.05$ using Eq. (5.10).

11. Given the initial-value problem

$$y' = -y + t + 1, \quad 0 \leq t \leq 5, \quad y(0) = 1,$$

with exact solution $y(t) = e^{-t} + t$:

a. Approximate $y(5)$ using Euler's method with $h = 0.2$, $h = 0.1$, and $h = 0.05$.

b. Determine the optimal value of h to use in computing $y(5)$, assuming $\delta = 10^{-6}$ and that Eq. (5.14) is valid.

12. Consider the initial-value problem

$$y' = -10y, \quad 0 \leq t \leq 2, \quad y(0) = 1,$$

which has solution $y(t) = e^{-10t}$. What happens when Euler's method is applied to this problem with $h = 0.1$? Does this behavior violate Theorem 5.9?

13. Use the results of Exercise 5 and linear interpolation to approximate the following values of $y(t)$. Compare the approximations obtained to the actual values obtained using the functions given in Exercise 7.

a. $y(1.25)$ and $y(1.93)$ **b.** $y(2.1)$ and $y(2.75)$

c. $y(1.3)$ and $y(1.93)$ **d.** $y(0.54)$ and $y(0.94)$

14. Use the results of Exercise 6 and linear interpolation to approximate the following values of $y(t)$. Compare the approximations obtained to the actual values obtained using the functions given in Exercise 8.

a. $y(0.25)$ and $y(0.93)$ **b.** $y(1.25)$ and $y(1.93)$

c. $y(2.10)$ and $y(2.75)$ **d.** $y(0.54)$ and $y(0.94)$

15. Let $E(h) = \dfrac{hM}{2} + \dfrac{\delta}{h}$.

a. For the initial-value problem

$$y' = -y + 1, \quad 0 \leq t \leq 1, \quad y(0) = 0,$$

compute the value of h to minimize $E(h)$. Assume $\delta = 5 \times 10^{-(n+1)}$ if you will be using n-digit arithmetic in part (c).

b. For the optimal h computed in part (a), use Eq. (5.13) to compute the minimal error obtainable.

c. Compare the actual error obtained using $h = 0.1$ and $h = 0.01$ to the minimal error in part (b). Can you explain the results?

16. In a circuit with impressed voltage $\mathcal{E}$ having resistance R, inductance L, and capacitance C in parallel, the current i satisfies the differential equation

$$\frac{di}{dt} = C\frac{d^2\mathcal{E}}{dt^2} + \frac{1}{R}\frac{d\mathcal{E}}{dt} + \frac{1}{L}\mathcal{E}.$$

Suppose $C = 0.3$ farads, $R = 1.4$ ohms, $L = 1.7$ henries, and the voltage is given by

$$\mathcal{E}(t) = e^{-0.06\pi t}\sin(2t - \pi).$$

If $i(0) = 0$, find the current i for the values $t = 0.1j$, where $j = 0, 1, \ldots, 100$.

17. In a book entitled *Looking at History Through Mathematics*, Rashevsky [Ra], pp. 103–110, considers a model for a problem involving the production of nonconformists in society. Suppose that a society has a population of $x(t)$ individuals at time t, in years, and that all nonconformists who mate with other nonconformists have offspring who are also nonconformists, while a fixed proportion r of all other offspring are also nonconformist. If the birth and death rates for all individuals are assumed to be the constants b and d, respectively, and if conformists and nonconformists mate at random, the problem can be expressed by the differential equations

$$\frac{dx(t)}{dt} = (b-d)x(t) \quad \text{and} \quad \frac{dx_n(t)}{dt} = (b-d)x_n(t) + rb(x(t) - x_n(t)),$$

where $x_n(t)$ denotes the number of nonconformists in the population at time t.

a. Suppose the variable $p(t) = x_n(t)/x(t)$ is introduced to represent the proportion of nonconformists in the society at time t. Show that these equations can be combined and simplified to the single differential equation

$$\frac{dp(t)}{dt} = rb(1 - p(t)).$$

b. Assuming that $p(0) = 0.01$, $b = 0.02$, $d = 0.015$, and $r = 0.1$, approximate the solution $p(t)$ from $t = 0$ to $t = 50$ when the step size is $h = 1$ year.

c. Solve the differential equation for $p(t)$ exactly, and compare your result in part (b) when $t = 50$ with the exact value at that time.

5.3 Higher-Order Taylor Methods

Since the object of a numerical techniques is to determine accurate approximations with minimal effort, we need a means for comparing the efficiency of various approximation methods. The first device we consider is called the *local truncation error* of the method.

The local truncation error at a specified step measures the amount by which the exact solution to the differential equation fails to satisfy the difference equation being used for the approximation at that step. This might seem like an unlikely way to compare the error of various methods. We really want to know how well the approximations generated by the methods satisfy the differential equation, not the other way around. However, we don't know the exact solution so we cannot generally determine this, and the local truncation will serve quite well to determine not only the local error of a method but the actual approximation error.

Consider the initial value problem

$$y' = f(t, y), \quad a \le t \le b, \quad y(a) = \alpha.$$

Definition 5.11 The difference method

$$w_0 = \alpha$$

$$w_{i+1} = w_i + h\phi(t_i, w_i), \quad \text{for each } i = 0, 1, \ldots, N-1,$$

has **local truncation error**

$$\tau_{i+1}(h) = \frac{y_{i+1} - (y_i + h\phi(t_i, y_i))}{h} = \frac{y_{i+1} - y_i}{h} - \phi(t_i, y_i),$$

for each $i = 0, 1, \ldots, N-1$, where y_i and y_{i+1} denote the solution at t_i and t_{i+1}, respectively. ■

For example, Euler's method has local truncation error at the ith step

$$\tau_{i+1}(h) = \frac{y_{i+1} - y_i}{h} - f(t_i, y_i), \quad \text{for each } i = 0, 1, \ldots, N-1.$$

This error is a *local error* because it measures the accuracy of the method at a specific step, assuming that the method was exact at the previous step. As such, it depends on the differential equation, the step size, and the particular step in the approximation.

By considering Eq. (5.7) in the previous section, we see that Euler's method has

$$\tau_{i+1}(h) = \frac{h}{2} y''(\xi_i), \quad \text{for some } \xi_i \text{ in } (t_i, t_{i+1}).$$

When $y''(t)$ is known to be bounded by a constant M on $[a, b]$, this implies

$$|\tau_{i+1}(h)| \leq \frac{h}{2} M,$$

so the local truncation error in Euler's method is $O(h)$.

One way to select difference-equation methods for solving ordinary differential equations is in such a manner that their local truncation errors are $O(h^p)$ for as large a value of p as possible, while keeping the number and complexity of calculations of the methods within a reasonable bound.

Since Euler's method was derived by using Taylor's Theorem with $n = 1$ to approximate the solution of the differential equation, our first attempt to find methods for improving the convergence properties of difference methods is to extend this technique of derivation to larger values of n.

The methods in this section use Taylor polynomials and the knowledge of the derivative at a node to approximate the value of the function at a new node.

Suppose the solution $y(t)$ to the initial-value problem

$$y' = f(t, y), \quad a \leq t \leq b, \quad y(a) = \alpha,$$

has $(n+1)$ continuous derivatives. If we expand the solution, $y(t)$, in terms of its nth Taylor polynomial about t_i and evaluate at t_{i+1}, we obtain

$$y(t_{i+1}) = y(t_i) + hy'(t_i) + \frac{h^2}{2} y''(t_i) + \cdots + \frac{h^n}{n!} y^{(n)}(t_i) + \frac{h^{n+1}}{(n+1)!} y^{(n+1)}(\xi_i), \tag{5.15}$$

for some ξ_i in (t_i, t_{i+1}).

Successive differentiation of the solution, $y(t)$, gives

$$y'(t) = f(t, y(t)), \quad y''(t) = f'(t, y(t)), \quad \text{and, generally,} \quad y^{(k)}(t) = f^{(k-1)}(t, y(t)).$$

Substituting these results into Eq. (5.15) gives

$$\begin{aligned} y(t_{i+1}) = y(t_i) + hf(t_i, y(t_i)) + \frac{h^2}{2} f'(t_i, y(t_i)) + \cdots \\ + \frac{h^n}{n!} f^{(n-1)}(t_i, y(t_i)) + \frac{h^{n+1}}{(n+1)!} f^{(n)}(\xi_i, y(\xi_i)). \end{aligned} \tag{5.16}$$

The difference-equation method corresponding to Eq. (5.16) is obtained by deleting the remainder term involving ξ_i.

Taylor method of order n

$$w_0 = \alpha,$$
$$w_{i+1} = w_i + hT^{(n)}(t_i, w_i), \quad \text{for each } i = 0, 1, \ldots, N-1, \tag{5.17}$$

where

$$T^{(n)}(t_i, w_i) = f(t_i, w_i) + \frac{h}{2} f'(t_i, w_i) + \cdots + \frac{h^{n-1}}{n!} f^{(n-1)}(t_i, w_i).$$

Euler's method is Taylor's method of order one.

Example 1 Apply Taylor's method of orders **(a)** two and **(b)** four with $N = 10$ to the initial-value problem

$$y' = y - t^2 + 1, \quad 0 \le t \le 2, \quad y(0) = 0.5.$$

Solution **(a)** For the method of order two we need the first derivative of $f(t, y(t)) = y(t) - t^2 + 1$ with respect to the variable t. Because $y' = y - t^2 + 1$ we have

$$f'(t, y(t)) = \frac{d}{dt}(y - t^2 + 1) = y' - 2t = y - t^2 + 1 - 2t,$$

so

$$T^{(2)}(t_i, w_i) = f(t_i, w_i) + \frac{h}{2} f'(t_i, w_i) = w_i - t_i^2 + 1 + \frac{h}{2}(w_i - t_i^2 + 1 - 2t_i)$$
$$= \left(1 + \frac{h}{2}\right)(w_i - t_i^2 + 1) - ht_i$$

Because $N = 10$ we have $h = 0.2$, and $t_i = 0.2i$ for each $i = 1, 2, \ldots, 10$. Thus the second-order method becomes

$$w_0 = 0.5,$$
$$w_{i+1} = w_i + h\left[\left(1 + \frac{h}{2}\right)(w_i - t_i^2 + 1) - ht_i\right]$$
$$= w_i + 0.2\left[\left(1 + \frac{0.2}{2}\right)(w_i - 0.04i^2 + 1) - 0.04i\right]$$
$$= 1.22w_i - 0.0088i^2 - 0.008i + 0.22.$$

The first two steps give the approximations

$$y(0.2) \approx w_1 = 1.22(0.5) - 0.0088(0)^2 - 0.008(0) + 0.22 = 0.83;$$
$$y(0.4) \approx w_2 = 1.22(0.83) - 0.0088(0.2)^2 - 0.008(0.2) + 0.22 = 1.2158$$

All the approximations and their errors are shown in Table 5.3

Table 5.3

t_i	Taylor Order 2 w_i	Error $\lvert y(t_i) - w_i \rvert$
0.0	0.500000	0
0.2	0.830000	0.000701
0.4	1.215800	0.001712
0.6	1.652076	0.003135
0.8	2.132333	0.005103
1.0	2.648646	0.007787
1.2	3.191348	0.011407
1.4	3.748645	0.016245
1.6	4.306146	0.022663
1.8	4.846299	0.031122
2.0	5.347684	0.042212

(b) For Taylor's method of order four we need the first three derivatives of $f(t, y(t))$ with respect to t. Again using $y' = y - t^2 + 1$ we have

$$f'(t, y(t)) = y - t^2 + 1 - 2t,$$
$$f''(t, y(t)) = \frac{d}{dt}(y - t^2 + 1 - 2t) = y' - 2t - 2$$
$$= y - t^2 + 1 - 2t - 2 = y - t^2 - 2t - 1,$$

and

$$f'''(t, y(t)) = \frac{d}{dt}(y - t^2 - 2t - 1) = y' - 2t - 2 = y - t^2 - 2t - 1,$$

so

$$\begin{aligned} T^{(4)}(t_i, w_i) &= f(t_i, w_i) + \frac{h}{2}f'(t_i, w_i) + \frac{h^2}{6}f''(t_i, w_i) + \frac{h^3}{24}f'''(t_i, w_i) \\ &= w_i - t_i^2 + 1 + \frac{h}{2}(w_i - t_i^2 + 1 - 2t_i) + \frac{h^2}{6}(w_i - t_i^2 - 2t_i - 1) \\ &\quad + \frac{h^3}{24}(w_i - t_i^2 - 2t_i - 1) \\ &= \left(1 + \frac{h}{2} + \frac{h^2}{6} + \frac{h^3}{24}\right)(w_i - t_i^2) - \left(1 + \frac{h}{3} + \frac{h^2}{12}\right)(ht_i) \\ &\quad + 1 + \frac{h}{2} - \frac{h^2}{6} - \frac{h^3}{24}. \end{aligned}$$

Hence Taylor's method of order four is

$$\begin{aligned} w_0 &= 0.5, \\ w_{i+1} &= w_i + h\left[\left(1 + \frac{h}{2} + \frac{h^2}{6} + \frac{h^3}{24}\right)(w_i - t_i^2) - \left(1 + \frac{h}{3} + \frac{h^2}{12}\right)ht_i \right. \\ &\quad \left. + 1 + \frac{h}{2} - \frac{h^2}{6} - \frac{h^3}{24}\right], \end{aligned}$$

for $i = 0, 1, \ldots, N - 1$.

Because $N = 10$ and $h = 0.2$ the method becomes

$$\begin{aligned} w_{i+1} &= w_i + 0.2\left[\left(1 + \frac{0.2}{2} + \frac{0.04}{6} + \frac{0.008}{24}\right)(w_i - 0.04i^2) \right. \\ &\quad \left. - \left(1 + \frac{0.2}{3} + \frac{0.04}{12}\right)(0.04i) + 1 + \frac{0.2}{2} - \frac{0.04}{6} - \frac{0.008}{24}\right] \\ &= 1.2214w_i - 0.008856i^2 - 0.00856i + 0.2186, \end{aligned}$$

for each $i = 0, 1, \ldots, 9$. The first two steps give the approximations

$$y(0.2) \approx w_1 = 1.2214(0.5) - 0.008856(0)^2 - 0.00856(0) + 0.2186 = 0.8293;$$

$$y(0.4) \approx w_2 = 1.2214(0.8293) - 0.008856(0.2)^2 - 0.00856(0.2) + 0.2186 = 1.214091$$

All the approximations and their errors are shown in Table 5.4.

Table 5.4

t_i	Taylor Order 4 w_i	Error $\lvert y(t_i) - w_i \rvert$
0.0	0.500000	0
0.2	0.829300	0.000001
0.4	1.214091	0.000003
0.6	1.648947	0.000006
0.8	2.127240	0.000010
1.0	2.640874	0.000015
1.2	3.179964	0.000023
1.4	3.732432	0.000032
1.6	4.283529	0.000045
1.8	4.815238	0.000062
2.0	5.305555	0.000083

Compare these results with those of Taylor's method of order 2 in Table 5.4 and you will see that the fourth-order results are vastly superior.

The results from Table 5.4 indicate the Taylor's method of order 4 results are quite accurate at the nodes 0.2, 0.4, etc. But suppose we need to determine an approximation to an intermediate point in the table, for example, at $t = 1.25$. If we use linear interpolation on the Taylor method of order four approximations at $t = 1.2$ and $t = 1.4$, we have

$$y(1.25) \approx \left(\frac{1.25 - 1.4}{1.2 - 1.4}\right)3.1799640 + \left(\frac{1.25 - 1.2}{1.4 - 1.2}\right)3.7324321 = 3.3180810.$$

Hermite interpolation requires both the value of the function and its derivative at each node. This makes it a natural interpolation method for approximating differential equations since these data are all available.

The true value is $y(1.25) = 3.3173285$, so this approximation has an error of 0.0007525, which is nearly 30 times the average of the approximation errors at 1.2 and 1.4.

We can significantly improve the approximation by using cubic Hermite interpolation. To determine this approximation for $y(1.25)$ requires approximations to $y'(1.2)$ and $y'(1.4)$ as well as approximations to $y(1.2)$ and $y(1.4)$. However, the approximations for $y(1.2)$ and $y(1.4)$ are in the table, and the derivative approximations are available from the differential equation, because $y'(t) = f(t, y(t))$. In our example $y'(t) = y(t) - t^2 + 1$, so

$$y'(1.2) = y(1.2) - (1.2)^2 + 1 \approx 3.1799640 - 1.44 + 1 = 2.7399640$$

and

$$y'(1.4) = y(1.4) - (1.4)^2 + 1 \approx 3.7324327 - 1.96 + 1 = 2.7724321.$$

The divided-difference procedure in Section 3.4 gives the information in Table 5.5. The underlined entries come from the data, and the other entries use the divided-difference formulas.

Table 5.5

1.2	<u>3.1799640</u>			
		<u>2.7399640</u>		
1.2	<u>3.1799640</u>		0.1118825	
		2.7623405		−0.3071225
1.4	<u>3.7324321</u>		0.0504580	
		<u>2.7724321</u>		
1.4	<u>3.7324321</u>			

The cubic Hermite polynomial is

$$y(t) \approx 3.1799640 + (t - 1.2)2.7399640 + (t - 1.2)^2 0.1118825$$
$$+ (t - 1.2)^2(t - 1.4)(-0.3071225),$$

so

$$y(1.25) \approx 3.1799640 + 0.1369982 + 0.0002797 + 0.0001152 = 3.3173571,$$

a result that is accurate to within 0.0000286. This is about the average of the errors at 1.2 and at 1.4, and only 4% of the error obtained using linear interpolation. This improvement in accuracy certainly justifies the added computation required for the Hermite method. ■

Theorem 5.12 If Taylor's method of order n is used to approximate the solution to

$$y'(t) = f(t, y(t)), \quad a \le t \le b, \quad y(a) = \alpha,$$

with step size h and if $y \in C^{n+1}[a, b]$, then the local truncation error is $O(h^n)$. ■

Proof Note that Eq. (5.16) on page 277 can be rewritten

$$y_{i+1} - y_i - hf(t_i, y_i) - \frac{h^2}{2} f'(t_i, y_i) - \cdots - \frac{h^n}{n!} f^{(n-1)}(t_i, y_i) = \frac{h^{n+1}}{(n+1)!} f^{(n)}(\xi_i, y(\xi_i)),$$

for some ξ_i in (t_i, t_{i+1}). So the local truncation error is

$$\tau_{i+1}(h) = \frac{y_{i+1} - y_i}{h} - T^{(n)}(t_i, y_i) = \frac{h^n}{(n+1)!} f^{(n)}(\xi_i, y(\xi_i)),$$

for each $i = 0, 1, \ldots, N-1$. Since $y \in C^{n+1}[a, b]$, we have $y^{(n+1)}(t) = f^{(n)}(t, y(t))$ bounded on $[a, b]$ and $\tau_i(h) = O(h^n)$, for each $i = 1, 2, \ldots, N$. ■ ■ ■

Taylor's methods are options within the Maple command *InitialValueProblem*. The form and output for Taylor's methods are the same as available under Euler's method, as discussed in Section 5.1. To obtain Taylor's method of order 2 for the problem in Example 1, first load the package and the differential equation.

with(*Student*[*NumericalAnalysis*]) : $deq := diff(y(t), t) = y(t) - t^2 + 1$

Then issue

$C :=$ *InitialValueProblem*($deq, y(0) = 0.5, t = 2, method = taylor, order = 2,$
$numsteps = 10, output = information, digits = 8$)

Maple responds with an array of data similar to that produced with Euler's method. Double clicking on the output will bring up a table that gives the values of t_i, actual solution values $y(t_i)$, the Taylor approximations w_i, and the absolute errors $|\,y(t_i) - w_i|$. These agree with the values in Table 5.3.

To print the table issue the commands

for k **from** 1 **to** 12 **do**
print($C[k, 1], C[k, 2], C[k, 3], C[k, 4]$)
end do

EXERCISE SET 5.3

1. Use Taylor's method of order two to approximate the solutions for each of the following initial-value problems.
   a. $y' = te^{3t} - 2y, \quad 0 \le t \le 1, \quad y(0) = 0, \text{ with } h = 0.5$
   b. $y' = 1 + y/t, \quad 1 \le t \le 2, \quad y(1) = 2, \text{ with } h = 0.25$
   c. $y' = 1 + (t - y)^2, \quad 2 \le t \le 3, \quad y(2) = 1, \text{ with } h = 0.5$
   d. $y' = \cos 2t + \sin 3t, \quad 0 \le t \le 1, \quad y(0) = 1, \text{ with } h = 0.25$
2. Use Taylor's method of order two to approximate the solutions for each of the following initial-value problems.
   a. $y' = e^{t-y}, \quad 0 \le t \le 1, \quad y(0) = 1, \text{ with } h = 0.5$
   b. $y' = \dfrac{1+t}{1+y}, \quad 1 \le t \le 2, \quad y(1) = 2, \text{ with } h = 0.5$
   c. $y' = t^{-2}(\sin 2t - 2ty), \quad 1 \le t \le 2, \quad y(1) = 2, \text{ with } h = 0.25$
   d. $y' = -y + ty^{1/2}, \quad 2 \le t \le 3, \quad y(2) = 2, \text{ with } h = 0.25$
3. Repeat Exercise 1 using Taylor's method of order four.
4. Repeat Exercise 2 using Taylor's method of order four.
5. Use Taylor's method of order two to approximate the solution for each of the following initial-value problems.
   a. $y' = \sin t + e^{-t}, \quad 0 \le t \le 1, \quad y(0) = 0, \text{ with } h = 0.5$
   b. $y' = y/t - (y/t)^2, \quad 1 \le t \le 1.2, \quad y(1) = 1, \text{ with } h = 0.1$
   c. $y' = (y^2 + y)/t, \quad 1 \le t \le 3, \quad y(1) = -2, \text{ with } h = 0.5$
   d. $y' = -ty + 4ty^{-1}, \quad 0 \le t \le 1, \quad y(0) = 1, \text{ with } h = 0.25$
6. Use Taylor's method of order two to approximate the solution for each of the following initial-value problems.
   a. $y' = \dfrac{2 - 2ty}{t^2 + 1}, \quad 0 \le t \le 1, \quad y(0) = 1, \text{ with } h = 0.1$
   b. $y' = \dfrac{y^2}{1+t}, \quad 1 \le t \le 2, \quad y(1) = -(\ln 2)^{-1}, \text{ with } h = 0.1$

c. $y' = (y^2 + y)/t, \quad 1 \le t \le 3, \quad y(1) = -2,$ with $h = 0.2$

d. $y' = -ty + 4t/y, \quad 0 \le t \le 1, \quad y(0) = 1,$ with $h = 0.1$

7. Repeat Exercise 5 using Taylor's method of order four.

8. Repeat Exercise 6 using Taylor's method of order four.

9. Given the initial-value problem

$$y' = \frac{2}{t}y + t^2 e^t, \quad 1 \le t \le 2, \quad y(1) = 0,$$

with exact solution $y(t) = t^2(e^t - e)$:

a. Use Taylor's method of order two with $h = 0.1$ to approximate the solution, and compare it with the actual values of y.

b. Use the answers generated in part (a) and linear interpolation to approximate y at the following values, and compare them to the actual values of y.

i. $y(1.04)$ **ii.** $y(1.55)$ **iii.** $y(1.97)$

c. Use Taylor's method of order four with $h = 0.1$ to approximate the solution, and compare it with the actual values of y.

d. Use the answers generated in part (c) and piecewise cubic Hermite interpolation to approximate y at the following values, and compare them to the actual values of y.

i. $y(1.04)$ **ii.** $y(1.55)$ **iii.** $y(1.97)$

10. Given the initial-value problem

$$y' = \frac{1}{t^2} - \frac{y}{t} - y^2, \quad 1 \le t \le 2, \quad y(1) = -1,$$

with exact solution $y(t) = -1/t$:

a. Use Taylor's method of order two with $h = 0.05$ to approximate the solution, and compare it with the actual values of y.

b. Use the answers generated in part (a) and linear interpolation to approximate the following values of y, and compare them to the actual values.

i. $y(1.052)$ **ii.** $y(1.555)$ **iii.** $y(1.978)$

c. Use Taylor's method of order four with $h = 0.05$ to approximate the solution, and compare it with the actual values of y.

d. Use the answers generated in part (c) and piecewise cubic Hermite interpolation to approximate the following values of y, and compare them to the actual values.

i. $y(1.052)$ **ii.** $y(1.555)$ **iii.** $y(1.978)$

11. A projectile of mass $m = 0.11$ kg shot vertically upward with initial velocity $v(0) = 8$ m/s is slowed due to the force of gravity, $F_g = -mg$, and due to air resistance, $F_r = -kv|v|$, where $g = 9.8$ m/s^2 and $k = 0.002$ kg/m. The differential equation for the velocity v is given by

$$mv' = -mg - kv|v|.$$

a. Find the velocity after $0.1, 0.2, \ldots, 1.0$ s.

b. To the nearest tenth of a second, determine when the projectile reaches its maximum height and begins falling.

12. Use the Taylor method of order two with $h = 0.1$ to approximate the solution to

$$y' = 1 + t\sin(ty), \quad 0 \le t \le 2, \quad y(0) = 0.$$

5.4 Runge-Kutta Methods

The Taylor methods outlined in the previous section have the desirable property of high-order local truncation error, but the disadvantage of requiring the computation and evaluation of the derivatives of $f(t, y)$. This is a complicated and time-consuming procedure for most problems, so the Taylor methods are seldom used in practice.

In the later 1800s, Carl Runge (1856–1927) used methods similar to those in this section to derive numerous formulas for approximating the solution to initial-value problems.

Runge-Kutta methods have the high-order local truncation error of the Taylor methods but eliminate the need to compute and evaluate the derivatives of $f(t, y)$. Before presenting the ideas behind their derivation, we need to consider Taylor's Theorem in two variables. The proof of this result can be found in any standard book on advanced calculus (see, for example, [Fu], p. 331).

Theorem 5.13 Suppose that $f(t, y)$ and all its partial derivatives of order less than or equal to $n + 1$ are continuous on $D = \{(t, y) \mid a \le t \le b, c \le y \le d\}$, and let $(t_0, y_0) \in D$. For every $(t, y) \in D$, there exists ξ between t and t_0 and μ between y and y_0 with

$$f(t, y) = P_n(t, y) + R_n(t, y),$$

In 1901, Martin Wilhelm Kutta (1867–1944) generalized the methods that Runge developed in 1895 to incorporate systems of first-order differential equations. These techniques differ slightly from those we currently call Runge-Kutta methods.

where

$$\begin{aligned} P_n(t, y) = f(t_0, y_0) &+ \left[(t - t_0)\frac{\partial f}{\partial t}(t_0, y_0) + (y - y_0)\frac{\partial f}{\partial y}(t_0, y_0)\right] \\ &+ \left[\frac{(t - t_0)^2}{2}\frac{\partial^2 f}{\partial t^2}(t_0, y_0) + (t - t_0)(y - y_0)\frac{\partial^2 f}{\partial t \partial y}(t_0, y_0)\right. \\ &\left. + \frac{(y - y_0)^2}{2}\frac{\partial^2 f}{\partial y^2}(t_0, y_0)\right] + \cdots \\ &+ \left[\frac{1}{n!}\sum_{j=0}^{n}\binom{n}{j}(t - t_0)^{n-j}(y - y_0)^j\frac{\partial^n f}{\partial t^{n-j}\partial y^j}(t_0, y_0)\right] \end{aligned}$$

and

$$R_n(t, y) = \frac{1}{(n+1)!}\sum_{j=0}^{n+1}\binom{n+1}{j}(t - t_0)^{n+1-j}(y - y_0)^j\frac{\partial^{n+1} f}{\partial t^{n+1-j}\partial y^j}(\xi, \mu).$$

The function $P_n(t, y)$ is called the ***n*th Taylor polynomial in two variables** for the function f about (t_0, y_0), and $R_n(t, y)$ is the remainder term associated with $P_n(t, y)$. ■

Example 1 Use Maple to determine $P_2(t, y)$, the second Taylor polynomial about $(2, 3)$ for the function

$$f(t, y) = \exp\left[-\frac{(t-2)^2}{4} - \frac{(y-3)^2}{4}\right]\cos(2t + y - 7)$$

Solution To determine $P_2(t, y)$ we need the values of f and its first and second partial derivatives at $(2, 3)$. The evaluation of the function is easy

$$f(2, 3) = e^{\left(-0^2/4 - 0^2/4\right)}\cos(4 + 3 - 7) = 1,$$

but the computations involved with the partial derivatives are quite tedious. However, higher dimensional Taylor polynomials are available in the *MultivariateCalculus* subpackage of the *Student* package, which is accessed with the command

with(*Student*[*MultivariateCalculus*])

The first option of the *TaylorApproximation* command is the function, the second specifies the point (t_0, y_0) where the polynomial is centered, and the third specifies the degree of the polynomial. So we issue the command

$$TaylorApproximation\left(e^{-\frac{(t-2)^2}{4}-\frac{(y-3)^2}{4}}\cos(2t+y-7),[t,y]=[2,3],2\right)$$

The response from this Maple command is the polynomial

$$1-\frac{9}{4}(t-2)^2-2(t-2)(y-3)-\frac{3}{4}(y-3)^2$$

A plot option is also available by adding a fourth option to the *TaylorApproximation* command in the form *output* = *plot*. The plot in the default form is quite crude, however, because not many points are plotted for the function and the polynomial. A better illustration is seen in Figure 5.5.

Figure 5.5

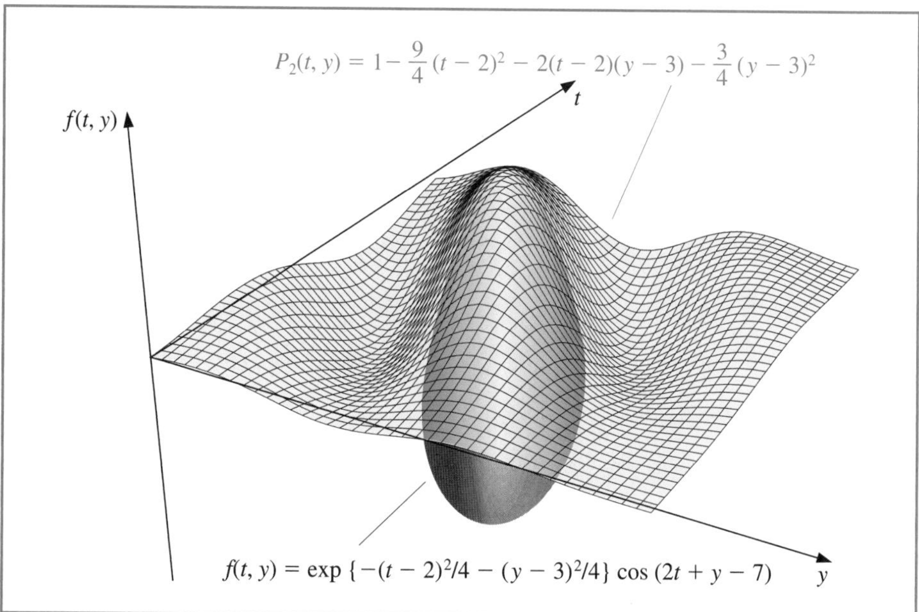

The final parameter in this command indicates that we want the second multivariate Taylor polynomial, that is, the quadratic polynomial. If this parameter is 2, we get the quadratic polynomial, and if it is 0 or 1, we get the constant polynomial 1, because there are no linear terms. When this parameter is omitted, it defaults to 6 and gives the sixth Taylor polynomial. ■

Runge-Kutta Methods of Order Two

The first step in deriving a Runge-Kutta method is to determine values for a_1, α_1, and β_1 with the property that $a_1 f(t+\alpha_1, y+\beta_1)$ approximates

$$T^{(2)}(t,y) = f(t,y) + \frac{h}{2}f'(t,y),$$

with error no greater than $O(h^2)$, which is same as the order of the local truncation error for the Taylor method of order two. Since

$$f'(t,y) = \frac{df}{dt}(t,y) = \frac{\partial f}{\partial t}(t,y) + \frac{\partial f}{\partial y}(t,y)\cdot y'(t) \quad \text{and} \quad y'(t) = f(t,y),$$

we have

$$T^{(2)}(t,y) = f(t,y) + \frac{h}{2}\frac{\partial f}{\partial t}(t,y) + \frac{h}{2}\frac{\partial f}{\partial y}(t,y) \cdot f(t,y). \tag{5.18}$$

Expanding $f(t+\alpha_1, y+\beta_1)$ in its Taylor polynomial of degree one about (t, y) gives

$$\begin{aligned} a_1 f(t+\alpha_1, y+\beta_1) = a_1 f(t,y) &+ a_1\alpha_1 \frac{\partial f}{\partial t}(t,y) \\ &+ a_1\beta_1 \frac{\partial f}{\partial y}(t,y) + a_1 \cdot R_1(t+\alpha_1, y+\beta_1), \end{aligned} \tag{5.19}$$

where

$$R_1(t+\alpha_1, y+\beta_1) = \frac{\alpha_1^2}{2}\frac{\partial^2 f}{\partial t^2}(\xi,\mu) + \alpha_1\beta_1\frac{\partial^2 f}{\partial t\partial y}(\xi,\mu) + \frac{\beta_1^2}{2}\frac{\partial^2 f}{\partial y^2}(\xi,\mu), \tag{5.20}$$

for some ξ between t and $t+\alpha_1$ and μ between y and $y+\beta_1$.

Matching the coefficients of f and its derivatives in Eqs. (5.18) and (5.19) gives the three equations

$$f(t,y):\ a_1 = 1; \quad \frac{\partial f}{\partial t}(t,y):\ a_1\alpha_1 = \frac{h}{2}; \quad \text{and} \quad \frac{\partial f}{\partial y}(t,y):\ a_1\beta_1 = \frac{h}{2}f(t,y).$$

The parameters a_1, α_1, and β_1 are therefore

$$a_1 = 1, \quad \alpha_1 = \frac{h}{2}, \quad \text{and} \quad \beta_1 = \frac{h}{2}f(t,y),$$

so

$$T^{(2)}(t,y) = f\left(t+\frac{h}{2}, y+\frac{h}{2}f(t,y)\right) - R_1\left(t+\frac{h}{2}, y+\frac{h}{2}f(t,y)\right),$$

and from Eq. (5.20),

$$\begin{aligned} R_1\left(t+\frac{h}{2}, y+\frac{h}{2}f(t,y)\right) &= \frac{h^2}{8}\frac{\partial^2 f}{\partial t^2}(\xi,\mu) + \frac{h^2}{4}f(t,y)\frac{\partial^2 f}{\partial t\partial y}(\xi,\mu) \\ &\quad + \frac{h^2}{8}(f(t,y))^2\frac{\partial^2 f}{\partial y^2}(\xi,\mu). \end{aligned}$$

If all the second-order partial derivatives of f are bounded, then

$$R_1\left(t+\frac{h}{2}, y+\frac{h}{2}f(t,y)\right)$$

is $O(h^2)$. As a consequence:

- The order of error for this new method is the same as that of the Taylor method of order two.

The difference-equation method resulting from replacing $T^{(2)}(t, y)$ in Taylor's method of order two by $f(t+(h/2), y+(h/2)f(t,y))$ is a specific Runge-Kutta method known as the *Midpoint method*.

Midpoint Method

$$w_0 = \alpha,$$

$$w_{i+1} = w_i + hf\left(t_i + \frac{h}{2}, w_i + \frac{h}{2} f(t_i, w_i)\right), \quad \text{for } i = 0, 1, \ldots, N-1.$$

Only three parameters are present in $a_1 f(t + \alpha_1, y + \beta_1)$ and all are needed in the match of $T^{(2)}$. So a more complicated form is required to satisfy the conditions for any of the higher-order Taylor methods.

The most appropriate four-parameter form for approximating

$$T^{(3)}(t, y) = f(t, y) + \frac{h}{2} f'(t, y) + \frac{h^2}{6} f''(t, y)$$

is

$$a_1 f(t, y) + a_2 f(t + \alpha_2, y + \delta_2 f(t, y)); \tag{5.21}$$

and even with this, there is insufficient flexibility to match the term

$$\frac{h^2}{6} \left[\frac{\partial f}{\partial y}(t, y)\right]^2 f(t, y),$$

resulting from the expansion of $(h^2/6) f''(t, y)$. Consequently, the best that can be obtained from using (5.21) are methods with $O(h^2)$ local truncation error.

The fact that (5.21) has four parameters, however, gives a flexibility in their choice, so a number of $O(h^2)$ methods can be derived. One of the most important is the *Modified Euler method*, which corresponds to choosing $a_1 = a_2 = \frac{1}{2}$ and $\alpha_2 = \delta_2 = h$. It has the following difference-equation form.

Modified Euler Method

$$w_0 = \alpha,$$

$$w_{i+1} = w_i + \frac{h}{2}[f(t_i, w_i) + f(t_{i+1}, w_i + hf(t_i, w_i))], \quad \text{for} \quad i = 0, 1, \ldots, N-1.$$

Example 2 Use the Midpoint method and the Modified Euler method with $N = 10$, $h = 0.2$, $t_i = 0.2i$, and $w_0 = 0.5$ to approximate the solution to our usual example,

$$y' = y - t^2 + 1, \quad 0 \le t \le 2, \quad y(0) = 0.5.$$

Solution The difference equations produced from the various formulas are

$$\text{Midpoint method:} \quad w_{i+1} = 1.22w_i - 0.0088i^2 - 0.008i + 0.218;$$

$$\text{Modified Euler method:} \quad w_{i+1} = 1.22w_i - 0.0088i^2 - 0.008i + 0.216,$$

for each $i = 0, 1, \ldots, 9$. The first two steps of these methods give

$$\text{Midpoint method:} \quad w_1 = 1.22(0.5) - 0.0088(0)^2 - 0.008(0) + 0.218 = 0.828;$$

$$\text{Modified Euler method:} \quad w_1 = 1.22(0.5) - 0.0088(0)^2 - 0.008(0) + 0.216 = 0.826,$$

and

$$\text{Midpoint method:} \quad w_2 = 1.22(0.828) - 0.0088(0.2)^2 - 0.008(0.2) + 0.218 = 1.21136;$$

$$\text{Modified Euler method:} \quad w_2 = 1.22(0.826) - 0.0088(0.2)^2 - 0.008(0.2) + 0.216 = 1.20692,$$

Table 5.6 lists all the results of the calculations. For this problem, the Midpoint method is superior to the Modified Euler method. ■

Table 5.6

t_i	$y(t_i)$	Midpoint Method	Error	Modified Euler Method	Error
0.0	0.5000000	0.5000000	0	0.5000000	0
0.2	0.8292986	0.8280000	0.0012986	0.8260000	0.0032986
0.4	1.2140877	1.2113600	0.0027277	1.2069200	0.0071677
0.6	1.6489406	1.6446592	0.0042814	1.6372424	0.0116982
0.8	2.1272295	2.1212842	0.0059453	2.1102357	0.0169938
1.0	2.6408591	2.6331668	0.0076923	2.6176876	0.0231715
1.2	3.1799415	3.1704634	0.0094781	3.1495789	0.0303627
1.4	3.7324000	3.7211654	0.0112346	3.6936862	0.0387138
1.6	4.2834838	4.2706218	0.0128620	4.2350972	0.0483866
1.8	4.8151763	4.8009586	0.0142177	4.7556185	0.0595577
2.0	5.3054720	5.2903695	0.0151025	5.2330546	0.0724173

Runge-Kutta methods are also options within the Maple command *InitialValueProblem*. The form and output for Runge-Kutta methods are the same as available under the Euler's and Taylor's methods, as discussed in Sections 5.1 and 5.2.

Higher-Order Runge-Kutta Methods

The term $T^{(3)}(t, y)$ can be approximated with error $O(h^3)$ by an expression of the form

$$f(t + \alpha_1, y + \delta_1 f(t + \alpha_2, y + \delta_2 f(t, y))),$$

involving four parameters, the algebra involved in the determination of $\alpha_1, \delta_1, \alpha_2$, and δ_2 is quite involved. The most common $O(h^3)$ is Heun's method, given by

Karl Heun (1859–1929) was a professor at the Technical University of Karlsruhe. He introduced this technique in a paper published in 1900. [Heu]

$$w_0 = \alpha$$

$$w_{i+1} = w_i + \tfrac{h}{4}\left(f(t_i, w_i) + 3f\left(t_i + \tfrac{2h}{3}, w_i + \tfrac{2h}{3} f\left(t_i + \tfrac{h}{3}, w_i + \tfrac{h}{3} f(t_i, w_i)\right)\right)\right),$$

$$\text{for} \quad i = 0, 1, \ldots, N - 1.$$

Illustration Applying Heun's method with $N = 10$, $h = 0.2$, $t_i = 0.2i$, and $w_0 = 0.5$ to approximate the solution to our usual example,

$$y' = y - t^2 + 1, \quad 0 \le t \le 2, \quad y(0) = 0.5.$$

gives the values in Table 5.7. Note the decreased error throughout the range over the Midpoint and Modified Euler approximations. □

Table 5.7

t_i	$y(t_i)$	Heun's Method	Error
0.0	0.5000000	0.5000000	0
0.2	0.8292986	0.8292444	0.0000542
0.4	1.2140877	1.2139750	0.0001127
0.6	1.6489406	1.6487659	0.0001747
0.8	2.1272295	2.1269905	0.0002390
1.0	2.6408591	2.6405555	0.0003035
1.2	3.1799415	3.1795763	0.0003653
1.4	3.7324000	3.7319803	0.0004197
1.6	4.2834838	4.2830230	0.0004608
1.8	4.8151763	4.8146966	0.0004797
2.0	5.3054720	5.3050072	0.0004648

Runge-Kutta methods of order three are not generally used. The most common Runge-Kutta method in use is of order four in difference-equation form, is given by the following.

Runge-Kutta Order Four

$$w_0 = \alpha,$$
$$k_1 = hf(t_i, w_i),$$
$$k_2 = hf\left(t_i + \frac{h}{2}, w_i + \frac{1}{2}k_1\right),$$
$$k_3 = hf\left(t_i + \frac{h}{2}, w_i + \frac{1}{2}k_2\right),$$
$$k_4 = hf(t_{i+1}, w_i + k_3),$$
$$w_{i+1} = w_i + \frac{1}{6}(k_1 + 2k_2 + 2k_3 + k_4),$$

for each $i = 0, 1, \ldots, N-1$. This method has local truncation error $O(h^4)$, provided the solution $y(t)$ has five continuous derivatives. We introduce the notation k_1, k_2, k_3, k_4 into the method is to eliminate the need for successive nesting in the second variable of $f(t, y)$. Exercise 32 shows how complicated this nesting becomes.

Algorithm 5.2 implements the Runge-Kutta method of order four.

ALGORITHM 5.2

Runge-Kutta (Order Four)

To approximate the solution of the initial-value problem

$$y' = f(t, y), \quad a \le t \le b, \quad y(a) = \alpha,$$

at $(N+1)$ equally spaced numbers in the interval $[a, b]$:

INPUT endpoints a, b; integer N; initial condition α.

OUTPUT approximation w to y at the $(N+1)$ values of t.

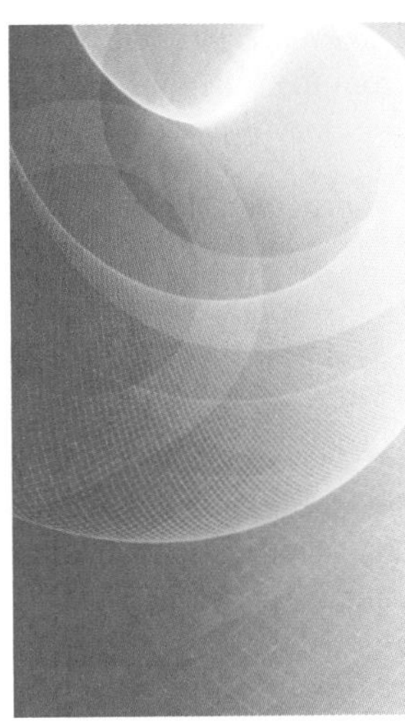

Step 1 Set $h = (b - a)/N$;
$t = a$;
$w = \alpha$;
OUTPUT (t, w).

Step 2 For $i = 1, 2, \ldots, N$ do Steps 3–5.

Step 3 Set $K_1 = hf(t, w)$;
$K_2 = hf(t + h/2, w + K_1/2)$;
$K_3 = hf(t + h/2, w + K_2/2)$;
$K_4 = hf(t + h, w + K_3)$.

Step 4 Set $w = w + (K_1 + 2K_2 + 2K_3 + K_4)/6$; (*Compute* w_i.)
$t = a + ih$. (*Compute* t_i.)

Step 5 OUTPUT (t, w).

Step 6 STOP. ∎

Example 3 Use the Runge-Kutta method of order four with $h = 0.2$, $N = 10$, and $t_i = 0.2i$ to obtain approximations to the solution of the initial-value problem

$$y' = y - t^2 + 1, \quad 0 \le t \le 2, \quad y(0) = 0.5.$$

Solution The approximation to $y(0.2)$ is obtained by

$$w_0 = 0.5$$

$$k_1 = 0.2f(0, 0.5) = 0.2(1.5) = 0.3$$

$$k_2 = 0.2f(0.1, 0.65) = 0.328$$

$$k_3 = 0.2f(0.1, 0.664) = 0.3308$$

$$k_4 = 0.2f(0.2, 0.8308) = 0.35816$$

$$w_1 = 0.5 + \frac{1}{6}(0.3 + 2(0.328) + 2(0.3308) + 0.35816) = 0.8292933.$$

The remaining results and their errors are listed in Table 5.8. ∎

Table 5.8

t_i	Exact $y_i = y(t_i)$	Runge-Kutta Order Four w_i	Error $\lvert y_i - w_i \rvert$
0.0	0.5000000	0.5000000	0
0.2	0.8292986	0.8292933	0.0000053
0.4	1.2140877	1.2140762	0.0000114
0.6	1.6489406	1.6489220	0.0000186
0.8	2.1272295	2.1272027	0.0000269
1.0	2.6408591	2.6408227	0.0000364
1.2	3.1799415	3.1798942	0.0000474
1.4	3.7324000	3.7323401	0.0000599
1.6	4.2834838	4.2834095	0.0000743
1.8	4.8151763	4.8150857	0.0000906
2.0	5.3054720	5.3053630	0.0001089

To obtain Runge-Kutta order 4 method results with *InitialValueProblem* use the option *method* = *rungekutta*, *submethod* = *rk4*. The results produced from the following call for out standard example problem agree with those in Table 5.6.

$C := InitialValueProblem(deq, y(0) = 0.5, t = 2, method = rungekutta, submethod = rk4, numsteps = 10, output = information, digits = 8)$

Computational Comparisons

The main computational effort in applying the Runge-Kutta methods is the evaluation of f. In the second-order methods, the local truncation error is $O(h^2)$, and the cost is two function evaluations per step. The Runge-Kutta method of order four requires 4 evaluations per step, and the local truncation error is $O(h^4)$. Butcher (see [But] for a summary) has established the relationship between the number of evaluations per step and the order of the local truncation error shown in Table 5.9. This table indicates why the methods of order less than five with smaller step size are used in preference to the higher-order methods using a larger step size.

Table 5.9

Evaluations per step	2	3	4	$5 \le n \le 7$	$8 \le n \le 9$	$10 \le n$
Best possible local truncation error	$O(h^2)$	$O(h^3)$	$O(h^4)$	$O(h^{n-1})$	$O(h^{n-2})$	$O(h^{n-3})$

One measure of comparing the lower-order Runge-Kutta methods is described as follows:

- The Runge-Kutta method of order four requires four evaluations per step, whereas Euler's method requires only one evaluation. Hence if the Runge-Kutta method of order four is to be superior it should give more accurate answers than Euler's method with one-fourth the step size. Similarly, if the Runge-Kutta method of order four is to be superior to the second-order Runge-Kutta methods, which require two evaluations per step, it should give more accuracy with step size h than a second-order method with step size $h/2$.

The following illustrates the superiority of the Runge-Kutta fourth-order method by this measure for the initial-value problem that we have been considering.

Illustration For the problem

$$y' = y - t^2 + 1, \quad 0 \le t \le 2, \quad y(0) = 0.5,$$

Euler's method with $h = 0.025$, the Midpoint method with $h = 0.05$, and the Runge-Kutta fourth-order method with $h = 0.1$ are compared at the common mesh points of these methods $0.1, 0.2, 0.3, 0.4,$ and 0.5. Each of these techniques requires 20 function evaluations to determine the values listed in Table 5.10 to approximate $y(0.5)$. In this example, the fourth-order method is clearly superior. □

Table 5.10

t_i	Exact	Euler $h = 0.025$	Modified Euler $h = 0.05$	Runge-Kutta Order Four $h = 0.1$
0.0	0.5000000	0.5000000	0.5000000	0.5000000
0.1	0.6574145	0.6554982	0.6573085	0.6574144
0.2	0.8292986	0.8253385	0.8290778	0.8292983
0.3	1.0150706	1.0089334	1.0147254	1.0150701
0.4	1.2140877	1.2056345	1.2136079	1.2140869
0.5	1.4256394	1.4147264	1.4250141	1.4256384

EXERCISE SET 5.4

1. Use the Modified Euler method to approximate the solutions to each of the following initial-value problems, and compare the results to the actual values.
   a. $y' = te^{3t} - 2y, \quad 0 \le t \le 1, \quad y(0) = 0$, with $h = 0.5$; actual solution $y(t) = \frac{1}{5}te^{3t} - \frac{1}{25}e^{3t} + \frac{1}{25}e^{-2t}$.
   b. $y' = 1 + y/t, \quad 1 \le t \le 2, \quad y(1) = 2$, with $h = 0.25$; actual solution $y(t) = t \ln t + 2t$.
   c. $y' = 1 + (t - y)^2, \quad 2 \le t \le 3, \quad y(2) = 1$, with $h = 0.5$; actual solution $y(t) = t + \frac{1}{1-t}$.
   d. $y' = \cos 2t + \sin 3t, \quad 0 \le t \le 1, \quad y(0) = 1$, with $h = 0.25$; actual solution $y(t) = \frac{1}{2}\sin 2t - \frac{1}{3}\cos 3t + \frac{4}{3}$.
2. Use the Modified Euler method to approximate the solutions to each of the following initial-value problems, and compare the results to the actual values.
   a. $y' = e^{t-y}, \quad 0 \le t \le 1, \quad y(0) = 1$, with $h = 0.5$; actual solution $y(t) = \ln(e^t + e - 1)$.
   b. $y' = -y + ty^{1/2}, \quad 2 \le t \le 3, \quad y(2) = 2$, with $h = 0.25$; actual solution $y(t) = \left(t - 2 + \sqrt{2}ee^{-t/2}\right)^2$.
   c. $y' = \dfrac{1+t}{1+y}, \quad 1 \le t \le 2, \quad y(1) = 2$, with $h = 0.5$; actual solution $y(t) = \sqrt{t^2 + 2t + 6} - 1$.
   d. $y' = t^{-2}(\sin 2t - 2ty), \quad 1 \le t \le 2, \quad y(1) = 2$, with $h = 0.25$; actual solution $y(t) = \frac{1}{2}t^{-2}(4 + \cos 2 - \cos 2t)$.
3. Use the Modified Euler method to approximate the solutions to each of the following initial-value problems, and compare the results to the actual values.
   a. $y' = 1 + y/t + (y/t)^2, \quad 1 \le t \le 3, \quad y(1) = 0$, with $h = 0.2$; actual solution $y(t) = t\tan(\ln t)$.
   b. $y' = y/t - (y/t)^2, \quad 1 \le t \le 2, \quad y(1) = 1$, with $h = 0.1$; actual solution $y(t) = t/(1 + \ln t)$.
   c. $y' = -(y + 1)(y + 3), \quad 0 \le t \le 2, \quad y(0) = -2$, with $h = 0.2$; actual solution $y(t) = -3 + 2(1 + e^{-2t})^{-1}$.
   d. $y' = -5y + 5t^2 + 2t, \quad 0 \le t \le 1, \quad y(0) = \frac{1}{3}$, with $h = 0.1$; actual solution $y(t) = t^2 + \frac{1}{3}e^{-5t}$.
4. Use the Modified Euler method to approximate the solutions to each of the following initial-value problems, and compare the results to the actual values.
   a. $y' = \dfrac{y^2}{1+t}, \quad 1 \le t \le 2, \quad y(1) = -(\ln 2)^{-1}$, with $h = 0.1$; actual solution $y(t) = \dfrac{-1}{\ln(t+1)}$.
   b. $y' = \dfrac{2 - 2ty}{t^2 + 1}, \quad 0 \le t \le 1, \quad y(0) = 1$, with $h = 0.1$; actual solution $y(t) = \dfrac{2t+1}{t^2+1}$.
   c. $y' = (y^2 + y)/t, \quad 1 \le t \le 3, \quad y(1) = -2$, with $h = 0.2$; actual solution $y(t) = \dfrac{2t}{1 - 2t}$.
   d. $y' = -ty + 4t/y, \quad 0 \le t \le 1, \quad y(0) = 1$, with $h = 0.1$; actual solution $y(t) = \sqrt{4 - 3e^{-t^2}}$.

5. Repeat Exercise 1 using the Midpoint method.
6. Repeat Exercise 2 using the Midpoint method.
7. Repeat Exercise 3 using the Midpoint method.
8. Repeat Exercise 4 using the Midpoint method.
9. Repeat Exercise 1 using Heun's method.
10. Repeat Exercise 2 using Heun's method.
11. Repeat Exercise 3 using Heun's method.
12. Repeat Exercise 4 using Heun's method.
13. Repeat Exercise 1 using the Runge-Kutta method of order four.
14. Repeat Exercise 2 using the Runge-Kutta method of order four.
15. Repeat Exercise 3 using the Runge-Kutta method of order four.
16. Repeat Exercise 4 using the Runge-Kutta method of order four.
17. Use the results of Exercise 3 and linear interpolation to approximate values of $y(t)$, and compare the results to the actual values.
    a. $y(2.1)$ and $y(2.75)$
    b. $y(1.25)$ and $y(1.93)$
    c. $y(1.3)$ and $y(1.93)$
    d. $y(0.54)$ and $y(0.94)$
18. Use the results of Exercise 4 and linear interpolation to approximate values of $y(t)$, and compare the results to the actual values.
    a. $y(1.25)$ and $y(1.93)$
    b. $y(0.54)$ and $y(0.94)$
    c. $y(1.3)$ and $y(2.93)$
    d. $y(0.54)$ and $y(0.94)$
19. Repeat Exercise 17 using the results of Exercise 7.
20. Repeat Exercise 18 using the results of Exercise 8.
21. Repeat Exercise 17 using the results of Exercise 11.
22. Repeat Exercise 18 using the results of Exercise 12.
23. Repeat Exercise 17 using the results of Exercise 15.
24. Repeat Exercise 18 using the results of Exercise 16.
25. Use the results of Exercise 15 and Cubic Hermite interpolation to approximate values of $y(t)$, and compare the approximations to the actual values.
    a. $y(2.1)$ and $y(2.75)$
    b. $y(1.25)$ and $y(1.93)$
    c. $y(1.3)$ and $y(1.93)$
    d. $y(0.54)$ and $y(0.94)$
26. Use the results of Exercise 16 and Cubic Hermite interpolation to approximate values of $y(t)$, and compare the approximations to the actual values.
    a. $y(1.25)$ and $y(1.93)$
    b. $y(0.54)$ and $y(0.94)$
    c. $y(1.3)$ and $y(2.93)$
    d. $y(0.54)$ and $y(0.94)$
27. Show that the Midpoint method and the Modified Euler method give the same approximations to the initial-value problem

$$y' = -y + t + 1, \quad 0 \le t \le 1, \quad y(0) = 1,$$

for any choice of h. Why is this true?

28. Water flows from an inverted conical tank with circular orifice at the rate

$$\frac{dx}{dt} = -0.6\pi r^2 \sqrt{2g}\frac{\sqrt{x}}{A(x)},$$

where r is the radius of the orifice, x is the height of the liquid level from the vertex of the cone, and $A(x)$ is the area of the cross section of the tank x units above the orifice. Suppose $r = 0.1$ ft, $g = 32.1$ ft/s^2, and the tank has an initial water level of 8 ft and initial volume of $512(\pi/3)$ ft^3. Use the Runge-Kutta method of order four to find the following.

    a. The water level after 10 min with $h = 20$ s
    b. When the tank will be empty, to within 1 min.

29. The irreversible chemical reaction in which two molecules of solid potassium dichromate ($K_2Cr_2O_7$), two molecules of water (H_2O), and three atoms of solid sulfur (S) combine to yield three molecules of the gas sulfur dioxide (SO_2), four molecules of solid potassium hydroxide (KOH), and two molecules of solid chromic oxide (Cr_2O_3) can be represented symbolically by the *stoichiometric equation*:

$$2K_2Cr_2O_7 + 2H_2O + 3S \longrightarrow 4KOH + 2Cr_2O_3 + 3SO_2.$$

If n_1 molecules of $K_2Cr_2O_7$, n_2 molecules of H_2O, and n_3 molecules of S are originally available, the following differential equation describes the amount $x(t)$ of KOH after time t:

$$\frac{dx}{dt} = k\left(n_1 - \frac{x}{2}\right)^2\left(n_2 - \frac{x}{2}\right)^2\left(n_3 - \frac{3x}{4}\right)^3,$$

where k is the velocity constant of the reaction. If $k = 6.22 \times 10^{-19}, n_1 = n_2 = 2 \times 10^3$, and $n_3 = 3 \times 10^3$, use the Runge-Kutta method of order four to determine how many units of potassium hydroxide will have been formed after 0.2 s?

30. Show that the difference method

$$w_0 = \alpha,$$
$$w_{i+1} = w_i + a_1 f(t_i, w_i) + a_2 f(t_i + \alpha_2, w_1 + \delta_2 f(t_i, w_i)),$$

for each $i = 0, 1, \ldots, N-1$, cannot have local truncation error $O(h^3)$ for any choice of constants a_1, a_2, α_2, and δ_2.

31. Show that Heun's method can be expressed in difference form, similar to that of the Runge-Kutta method of order four, as

$$w_0 = \alpha,$$
$$k_1 = hf(t_i, w_i),$$
$$k_2 = hf\left(t_i + \frac{h}{3}, w_i + \frac{1}{3}k_1\right),$$
$$k_3 = hf\left(t_i + \frac{2h}{3}, w_i + \frac{2}{3}k_2\right),$$
$$w_{i+1} = w_i + \frac{1}{4}(k_1 + 3k_3),$$

for each $i = 0, 1, \ldots, N-1$.

32. The Runge-Kutta method of order four can be written in the form

$$w_0 = \alpha,$$
$$w_{i+1} = w_i + \frac{h}{6}f(t_i, w_i) + \frac{h}{3}f(t_i + \alpha_1 h, w_i + \delta_1 h f(t_i, w_i))$$
$$+ \frac{h}{3}f(t_i + \alpha_2 h, w_i + \delta_2 h f(t_i + \gamma_2 h, w_i + \gamma_3 h f(t_i, w_i)))$$
$$+ \frac{h}{6}f(t_i + \alpha_3 h, w_i + \delta_3 h f(t_i + \gamma_4 h, w_i + \gamma_5 h f(t_i + \gamma_6 h, w_i + \gamma_7 h f(t_i, w_i)))).$$

Find the values of the constants

$$\alpha_1,\ \alpha_2,\ \alpha_3,\ \delta_1,\ \delta_2,\ \delta_3,\ \gamma_2,\ \gamma_3,\ \gamma_4,\ \gamma_5,\ \gamma_6, \text{ and } \gamma_7.$$

5.5 Error Control and the Runge-Kutta-Fehlberg Method

In Section 4.6 we saw that the appropriate use of varying step sizes for integral approximations produced efficient methods. In itself, this might not be sufficient to favor these methods due to the increased complication of applying them. However, they have another feature

You might like to review the Adaptive Quadrature material in Section 4.6 before considering this material.

that makes them worthwhile. They incorporate in the step-size procedure an estimate of the truncation error that does not require the approximation of the higher derivatives of the function. These methods are called *adaptive* because they adapt the number and position of the nodes used in the approximation to ensure that the truncation error is kept within a specified bound.

There is a close connection between the problem of approximating the value of a definite integral and that of approximating the solution to an initial-value problem. It is not surprising, then, that there are adaptive methods for approximating the solutions to initial-value problems and that these methods are not only efficient, but also incorporate the control of error.

Any one-step method for approximating the solution, $y(t)$, of the initial-value problem

$$y' = f(t, y), \quad \text{for } a \le t \le b, \quad \text{with } y(a) = \alpha$$

can be expressed in the form

$$w_{i+1} = w_i + h_i\phi(t_i, w_i, h_i), \quad \text{for } i = 0, 1, \ldots, N-1,$$

for some function ϕ.

An ideal difference-equation method

$$w_{i+1} = w_i + h_i\phi(t_i, w_i, h_i), \quad i = 0, 1, \ldots, N-1,$$

for approximating the solution, $y(t)$, to the initial-value problem

$$y' = f(t, y), \quad a \le t \le b, \quad y(a) = \alpha,$$

would have the property that, given a tolerance $\varepsilon > 0$, a minimal number of mesh points could be used to ensure that the global error, $|y(t_i) - w_i|$, did not exceed ε for any $i = 0, 1, \ldots, N$. Having a minimal number of mesh points and also controlling the global error of a difference method is, not surprisingly, inconsistent with the points being equally spaced in the interval. In this section we examine techniques used to control the error of a difference-equation method in an efficient manner by the appropriate choice of mesh points.

Although we cannot generally determine the global error of a method, we will see in Section 5.10 that there is a close connection between the local truncation error and the global error. By using methods of differing order we can predict the local truncation error and, using this prediction, choose a step size that will keep it and the global error in check.

To illustrate the technique, suppose that we have two approximation techniques. The first is obtained from an nth-order Taylor method of the form

$$y(t_{i+1}) = y(t_i) + h\phi(t_i, y(t_i), h) + O(h^{n+1}),$$

and produces approximations with local truncation error $\tau_{i+1}(h) = O(h^n)$. It is given by

$$\begin{aligned} w_0 &= \alpha \\ w_{i+1} &= w_i + h\phi(t_i, w_i, h), \quad \text{for } i > 0. \end{aligned}$$

In general, the method is generated by applying a Runge-Kutta modification to the Taylor method, but the specific derivation is unimportant.

The second method is similar but one order higher; it comes from an $(n+1)$st-order Taylor method of the form

$$y(t_{i+1}) = y(t_i) + h\tilde{\phi}(t_i, y(t_i), h) + O(h^{n+2}),$$

and produces approximations with local truncation error $\tilde{\tau}_{i+1}(h) = O(h^{n+1})$. It is given by

$$\tilde{w}_0 = \alpha$$

$$\tilde{w}_{i+1} = \tilde{w}_i + h\tilde{\phi}(t_i, \tilde{w}_i, h), \quad \text{for } i > 0.$$

We first make the assumption that $w_i \approx y(t_i) \approx \tilde{w}_i$ and choose a fixed step size h to generate the approximations w_{i+1} and $\tilde{w}_{i+1}$ to $y(t_{i+1})$. Then

$$\begin{aligned}
\tau_{i+1}(h) &= \frac{y(t_{i+1}) - y(t_i)}{h} - \phi(t_i, y(t_i), h) \\
&= \frac{y(t_{i+1}) - w_i}{h} - \phi(t_i, w_i, h) \\
&= \frac{y(t_{i+1}) - [w_i + h\phi(t_i, w_i, h)]}{h} \\
&= \frac{1}{h}(y(t_{i+1}) - w_{i+1}).
\end{aligned}$$

In a similar manner, we have

$$\tilde{\tau}_{i+1}(h) = \frac{1}{h}(y(t_{i+1}) - \tilde{w}_{i+1}).$$

As a consequence, we have

$$\begin{aligned}
\tau_{i+1}(h) &= \frac{1}{h}(y(t_{i+1}) - w_{i+1}) \\
&= \frac{1}{h}[(y(t_{i+1}) - \tilde{w}_{i+1}) + (\tilde{w}_{i+1} - w_{i+1})] \\
&= \tilde{\tau}_{i+1}(h) + \frac{1}{h}(\tilde{w}_{i+1} - w_{i+1}).
\end{aligned}$$

But $\tau_{i+1}(h)$ is $O(h^n)$ and $\tilde{\tau}_{i+1}(h)$ is $O(h^{n+1})$, so the significant portion of $\tau_{i+1}(h)$ must come from

$$\frac{1}{h}\left(\tilde{w}_{i+1} - w_{i+1}\right).$$

This gives us an easily computed approximation for the local truncation error of the $O(h^n)$ method:

$$\tau_{i+1}(h) \approx \frac{1}{h}\left(\tilde{w}_{i+1} - w_{i+1}\right).$$

The object, however, is not simply to estimate the local truncation error but to adjust the step size to keep it within a specified bound. To do this we now assume that since $\tau_{i+1}(h)$ is $O(h^n)$, a number K, independent of h, exists with

$$\tau_{i+1}(h) \approx Kh^n.$$

Then the local truncation error produced by applying the nth-order method with a new step size qh can be estimated using the original approximations w_{i+1} and $\tilde{w}_{i+1}$:

$$\tau_{i+1}(qh) \approx K(qh)^n = q^n(Kh^n) \approx q^n\tau_{i+1}(h) \approx \frac{q^n}{h}(\tilde{w}_{i+1} - w_{i+1}).$$

To bound $\tau_{i+1}(qh)$ by ε, we choose q so that

$$\frac{q^n}{h}|\tilde{w}_{i+1} - w_{i+1}| \approx |\tau_{i+1}(qh)| \le \varepsilon;$$

that is, so that

$$q \leq \left(\frac{\varepsilon h}{|\tilde{w}_{i+1} - w_{i+1}|} \right)^{1/n}. \tag{5.22}$$

Runge-Kutta-Fehlberg Method

Erwin Fehlberg developed this and other error control techniques while working for the NASA facility in Huntsville, Alabama during the 1960s. He received the Exceptional Scientific Achievement Medal from NASA in 1969.

One popular technique that uses Inequality (5.22) for error control is the **Runge-Kutta-Fehlberg method**. (See [Fe].) This technique uses a Runge-Kutta method with local truncation error of order five,

$$\tilde{w}_{i+1} = w_i + \frac{16}{135}k_1 + \frac{6656}{12825}k_3 + \frac{28561}{56430}k_4 - \frac{9}{50}k_5 + \frac{2}{55}k_6,$$

to estimate the local error in a Runge-Kutta method of order four given by

$$w_{i+1} = w_i + \frac{25}{216}k_1 + \frac{1408}{2565}k_3 + \frac{2197}{4104}k_4 - \frac{1}{5}k_5,$$

where the coefficient equations are

$$k_1 = hf(t_i, w_i),$$

$$k_2 = hf\left(t_i + \frac{h}{4}, w_i + \frac{1}{4}k_1\right),$$

$$k_3 = hf\left(t_i + \frac{3h}{8}, w_i + \frac{3}{32}k_1 + \frac{9}{32}k_2\right),$$

$$k_4 = hf\left(t_i + \frac{12h}{13}, w_i + \frac{1932}{2197}k_1 - \frac{7200}{2197}k_2 + \frac{7296}{2197}k_3\right),$$

$$k_5 = hf\left(t_i + h, w_i + \frac{439}{216}k_1 - 8k_2 + \frac{3680}{513}k_3 - \frac{845}{4104}k_4\right),$$

$$k_6 = hf\left(t_i + \frac{h}{2}, w_i - \frac{8}{27}k_1 + 2k_2 - \frac{3544}{2565}k_3 + \frac{1859}{4104}k_4 - \frac{11}{40}k_5\right).$$

An advantage to this method is that only six evaluations of f are required per step. Arbitrary Runge-Kutta methods of orders four and five used together (see Table 5.9 on page 290) require at least four evaluations of f for the fourth-order method and an additional six for the fifth-order method, for a total of at least ten function evaluations. So the Runge-Kutta-Fehlberg method has at least a 40% decrease in the number of function evaluations over the use of a pair of arbitrary fourth- and fifth-order methods.

In the error-control theory, an initial value of h at the ith step is used to find the first values of w_{i+1} and $\tilde{w}_{i+1}$, which leads to the determination of q for that step, and then the calculations were repeated. This procedure requires twice the number of function evaluations per step as without the error control. In practice, the value of q to be used is chosen somewhat differently in order to make the increased function-evaluation cost worthwhile. The value of q determined at the ith step is used for two purposes:

- When $q < 1$: to reject the initial choice of h at the ith step and repeat the calculations using qh, and
- When $q \geq 1$: to accept the computed value at the ith step using the step size h, but change the step size to qh for the $(i + 1)$st step.

Because of the penalty in terms of function evaluations that must be paid if the steps are repeated, q tends to be chosen conservatively. In fact, for the Runge-Kutta-Fehlberg method with $n = 4$, a common choice is

$$q = \left(\frac{\varepsilon h}{2|\tilde{w}_{i+1} - w_{i+1}|}\right)^{1/4} = 0.84\left(\frac{\varepsilon h}{|\tilde{w}_{i+1} - w_{i+1}|}\right)^{1/4}.$$

In Algorithm 5.3 for the Runge-Kutta-Fehlberg method, Step 9 is added to eliminate large modifications in step size. This is done to avoid spending too much time with small step sizes in regions with irregularities in the derivatives of y, and to avoid large step sizes, which can result in skipping sensitive regions between the steps. The step-size increase procedure could be omitted completely from the algorithm, and the step-size decrease procedure used only when needed to bring the error under control.

Runge-Kutta-Fehlberg

To approximate the solution of the initial-value problem

$$y' = f(t, y), \quad a \le t \le b, \quad y(a) = \alpha,$$

with local truncation error within a given tolerance:

INPUT endpoints a, b; initial condition α; tolerance TOL; maximum step size $hmax$; minimum step size $hmin$.

OUTPUT t, w, h where w approximates $y(t)$ and the step size h was used, or a message that the minimum step size was exceeded.

Step 1 Set $t = a$;
$w = \alpha$;
$h = hmax$;
$FLAG = 1$;
OUTPUT (t, w).

Step 2 While ($FLAG = 1$) do Steps 3–11.

Step 3 Set $K_1 = hf(t, w)$;

$K_2 = hf\left(t + \frac{1}{4}h, w + \frac{1}{4}K_1\right)$;

$K_3 = hf\left(t + \frac{3}{8}h, w + \frac{3}{32}K_1 + \frac{9}{32}K_2\right)$;

$K_4 = hf\left(t + \frac{12}{13}h, w + \frac{1932}{2197}K_1 - \frac{7200}{2197}K_2 + \frac{7296}{2197}K_3\right)$;

$K_5 = hf\left(t + h, w + \frac{439}{216}K_1 - 8K_2 + \frac{3680}{513}K_3 - \frac{845}{4104}K_4\right)$;

$K_6 = hf\left(t + \frac{1}{2}h, w - \frac{8}{27}K_1 + 2K_2 - \frac{3544}{2565}K_3 + \frac{1859}{4104}K_4 - \frac{11}{40}K_5\right)$.

Step 4 Set $R = \frac{1}{h}\left|\frac{1}{360}K_1 - \frac{128}{4275}K_3 - \frac{2197}{75240}K_4 + \frac{1}{50}K_5 + \frac{2}{55}K_6\right|$.

(*Note:* $R = \frac{1}{h}|\tilde{w}_{i+1} - w_{i+1}|$.)

Step 5 If $R \le TOL$ then do Steps 6 and 7.

Step 6 Set $t = t + h$; (*Approximation accepted.*)

$w = w + \frac{25}{216}K_1 + \frac{1408}{2565}K_3 + \frac{2197}{4104}K_4 - \frac{1}{5}K_5$.

Step 7 OUTPUT (t, w, h).

Step 8 Set $\delta = 0.84(TOL/R)^{1/4}$.

Step 9 If $\delta \leq 0.1$ then set $h = 0.1h$
else if $\delta \geq 4$ then set $h = 4h$
else set $h = \delta h$. *(Calculate new h.)*

Step 10 If $h > hmax$ then set $h = hmax$.

Step 11 If $t \geq b$ then set $FLAG = 0$
else if $t + h > b$ then set $h = b - t$
else if $h < hmin$ then
set $FLAG = 0$;
OUTPUT (*'minimum h exceeded'*).
(Procedure completed unsuccessfully.)

Step 12 *(The procedure is complete.)*
STOP. ■

Example 1 Use the Runge-Kutta-Fehlberg method with a tolerance $TOL = 10^{-5}$, a maximum step size $hmax = 0.25$, and a minimum step size $hmin = 0.01$ to approximate the solution to the initial-value problem

$$y' = y - t^2 + 1, \quad 0 \leq t \leq 2, \quad y(0) = 0.5,$$

and compare the results with the exact solution $y(t) = (t+1)^2 - 0.5e^t$.

Solution We will work through the first step of the calculations and then apply Algorithm 5.3 to determine the remaining results. The initial condition gives $t_0 = 0$ and $w_0 = 0.5$. To determine w_1 using $h = 0.25$, the maximum allowable stepsize, we compute

$$k_1 = hf(t_0, w_0) = 0.25\left(0.5 - 0^2 + 1\right) = 0.375;$$

$$k_2 = hf\left(t_0 + \frac{1}{4}h, w_0 + \frac{1}{4}k_1\right) = 0.25\left(\frac{1}{4}0.25, 0.5 + \frac{1}{4}0.375\right) = 0.3974609;$$

$$k_3 = hf\left(t_0 + \frac{3}{8}h, w_0 + \frac{3}{32}k_1 + \frac{9}{32}k_2\right)$$

$$= 0.25\left(0.09375, 0.5 + \frac{3}{32}0.375 + \frac{9}{32}0.3974609\right) = 0.4095383;$$

$$k_4 = hf\left(t_0 + \frac{12}{13}h, w_0 + \frac{1932}{2197}k_1 - \frac{7200}{2197}k_2 + \frac{7296}{2197}k_3\right)$$

$$= 0.25\left(0.2307692, 0.5 + \frac{1932}{2197}0.375 - \frac{7200}{2197}0.3974609 + \frac{7296}{2197}0.4095383\right)$$

$$= 0.4584971;$$

$$k_5 = hf\left(t_0 + h, w_0 + \frac{439}{216}k_1 - 8k_2 + \frac{3680}{513}k_3 - \frac{845}{4104}k_4\right)$$

$$= 0.25\left(0.25, 0.5 + \frac{439}{216}0.375 - 8(0.3974609) + \frac{3680}{513}0.4095383 - \frac{845}{4104}0.4584971\right)$$

$$= 0.4658452;$$

$$k_6 = hf\left(t_0 + \frac{1}{2}h, w_0 - \frac{8}{27}k_1 + 2k_2 - \frac{3544}{2565}k_3 + \frac{1859}{4104}k_4 - \frac{11}{40}k_5\right)$$

$$= 0.25\left(0.125, 0.5 - \frac{8}{27}0.375 + 2(0.3974609) - \frac{3544}{2565}0.4095383\right.$$

$$\left. + \frac{1859}{4104}0.4584971 - \frac{11}{40}0.4658452\right)$$

$$= 0.4204789.$$

The two approximations to $y(0.25)$ are then found to be

$$\tilde{w}_1 = w_0 + \frac{16}{135}k_1 + \frac{6656}{12825}k_3 + \frac{28561}{56430}k_4 - \frac{9}{50}k_5 + \frac{2}{55}k_6$$

$$= 0.5 + \frac{16}{135}0.375 + \frac{6656}{12825}0.4095383 + \frac{28561}{56430}0.4584971 - \frac{9}{50}0.4658452$$

$$+ \frac{2}{55}0.4204789$$

$$= 0.9204870,$$

and

$$w_1 = w_0 + \frac{25}{216}k_1 + \frac{1408}{2565}k_3 + \frac{2197}{4104}k_4 - \frac{1}{5}k_5$$

$$= 0.5 + \frac{25}{216}0.375 + \frac{1408}{2565}0.4095383 + \frac{2197}{4104}0.4584971 - \frac{1}{5}0.4658452$$

$$= 0.9204886.$$

This also implies that

$$R = \frac{1}{0.25}\left|\frac{1}{360}k_1 - \frac{128}{4275}k_3 - \frac{2197}{75240}k_4 + \frac{1}{50}k_5 + \frac{2}{55}k_6\right|$$

$$= 4\left|\frac{1}{360}0.375 - \frac{128}{4275}0.4095383 - \frac{2197}{75240}0.4584971 + \frac{1}{50}0.4658452 + \frac{2}{55}0.4204789\right|$$

$$= 0.00000621388,$$

and

$$q = 0.84\left(\frac{\varepsilon}{R}\right)^{1/4} = 0.84\left(\frac{0.00001}{0.00000621388}\right)^{1/4} = 0.9461033291.$$

Since $q < 1$ we can accept the approximation 0.9204886 for $y(0.25)$ but we should adjust the step size for the next iteration to $h = 0.9461033291(0.25) \approx 0.2365258$. However, only the leading 5 digits of this result would be expected to be accurate because R has only about 5 digits of accuracy. Because we are effectively subtracting the nearly equal numbers w_i and $\tilde{w}_i$ when we compute R, there is a good likelihood of round-off error. This is an additional reason for being conservative when computing q.

The results from the algorithm are shown in Table 5.11. Increased accuracy has been used to ensure that the calculations are accurate to all listed places. The last two columns in Table 5.11 show the results of the fifth-order method. For small values of t, the error is less than the error in the fourth-order method, but the error exceeds that of the fourth-order method when t increases. ■

Table 5.11

		RKF-4				RKF-5	
t_i	$y_i = y(t_i)$	w_i	h_i	R_i	$\lvert y_i - w_i\rvert$	$\hat{w}_i$	$\lvert y_i - \hat{w}_i\rvert$
0	0.5	0.5			0.5		
0.2500000	0.9204873	0.9204886	0.2500000	6.2×10^{-6}	1.3×10^{-6}	0.9204870	2.424×10^{-7}
0.4865522	1.3964884	1.3964910	0.2365522	4.5×10^{-6}	2.6×10^{-6}	1.3964900	1.510×10^{-6}
0.7293332	1.9537446	1.9537488	0.2427810	4.3×10^{-6}	4.2×10^{-6}	1.9537477	3.136×10^{-6}
0.9793332	2.5864198	2.5864260	0.2500000	3.8×10^{-6}	6.2×10^{-6}	2.5864251	5.242×10^{-6}
1.2293332	3.2604520	3.2604605	0.2500000	2.4×10^{-6}	8.5×10^{-6}	3.2604599	7.895×10^{-6}
1.4793332	3.9520844	3.9520955	0.2500000	7×10^{-7}	1.11×10^{-5}	3.9520954	1.096×10^{-5}
1.7293332	4.6308127	4.6308268	0.2500000	1.5×10^{-6}	1.41×10^{-5}	4.6308272	1.446×10^{-5}
1.9793332	5.2574687	5.2574861	0.2500000	4.3×10^{-6}	1.73×10^{-5}	5.2574871	1.839×10^{-5}
2.0000000	5.3054720	5.3054896	0.0206668		1.77×10^{-5}	5.3054896	1.768×10^{-5}

An implementation of the Runge-Kutta-Fehlberg method is also available in Maple using the *InitialValueProblem* command. However, it differs from our presentation because it does not require the specification of a tolerance for the solution. For our example problem it is called with

$C := InitialValueProblem(deq, y(0) = 0.5, t = 2, method = rungekutta, submethod = rkf, numsteps = 10, output = information, digits = 8)$

As usual, the information is placed in a table that is accessed by double clicking on the output. The results can be printed in the method outlined in precious sections.

EXERCISE SET 5.5

1. Use the Runge-Kutta-Fehlberg method with tolerance $TOL = 10^{-4}$, $hmax = 0.25$, and $hmin = 0.05$ to approximate the solutions to the following initial-value problems. Compare the results to the actual values.
   a. $y' = te^{3t} - 2y, \quad 0 \le t \le 1, \quad y(0) = 0$; actual solution $y(t) = \frac{1}{5}te^{3t} - \frac{1}{25}e^{3t} + \frac{1}{25}e^{-2t}$.
   b. $y' = 1 + y/t, \quad 1 \le t \le 2, \quad y(1) = 2$; actual solution $y(t) = t\ln t + 2t$.
   c. $y' = 1 + (t-y)^2, \quad 2 \le t \le 3, \quad y(2) = 1$; actual solution $y(t) = t + 1/(1-t)$.
   d. $y' = \cos 2t + \sin 3t, \quad 0 \le t \le 1, \quad y(0) = 1$; actual solution $y(t) = \frac{1}{2}\sin 2t - \frac{1}{3}\cos 3t + \frac{4}{3}$.
2. Use the Runge-Kutta Fehlberg Algorithm with tolerance $TOL = 10^{-4}$ to approximate the solution to the following initial-value problems.
   a. $y' = \sin t + e^{-t}, \quad 0 \le t \le 1, \quad y(0) = 0$, with $hmax = 0.25$ and $hmin = 0.02$.
   b. $y' = (y/t)^2 + y/t, \quad 1 \le t \le 1.2, \quad y(1) = 1$, with $hmax = 0.05$ and $hmin = 0.02$.
   c. $y' = (y^2 + y)/t, \quad 1 \le t \le 3, \quad y(1) = -2$, with $hmax = 0.5$ and $hmin = 0.02$.
   d. $y' = t^2, \quad 0 \le t \le 2, \quad y(0) = 0$, with $hmax = 0.5$ and $hmin = 0.02$.
3. Use the Runge-Kutta-Fehlberg method with tolerance $TOL = 10^{-6}$, $hmax = 0.5$, and $hmin = 0.05$ to approximate the solutions to the following initial-value problems. Compare the results to the actual values.
   a. $y' = 1 + y/t + (y/t)^2, \quad 1 \le t \le 3, \quad y(1) = 0$; actual solution $y(t) = t\tan(\ln t)$.
   b. $y' = y/t - (y/t)^2, \quad 1 \le t \le 4, \quad y(1) = 1$; actual solution $y(t) = t/(1 + \ln t)$.
   c. $y' = -(y+1)(y+3), \quad 0 \le t \le 3, \quad y(0) = -2$; actual solution $y(t) = -3 + 2(1 + e^{-2t})^{-1}$.
   d. $y' = (t + 2t^3)y^3 - ty, \quad 0 \le t \le 2, \quad y(0) = \frac{1}{3}$; actual solution $y(t) = (3 + 2t^2 + 6e^{t^2})^{-1/2}$.

4. The Runge-Kutta-Verner method (see [Ve]) is based on the formulas

$$w_{i+1} = w_i + \frac{13}{160}k_1 + \frac{2375}{5984}k_3 + \frac{5}{16}k_4 + \frac{12}{85}k_5 + \frac{3}{44}k_6 \quad \text{and}$$

$$\tilde{w}_{i+1} = w_i + \frac{3}{40}k_1 + \frac{875}{2244}k_3 + \frac{23}{72}k_4 + \frac{264}{1955}k_5 + \frac{125}{11592}k_7 + \frac{43}{616}k_8,$$

where

$$k_1 = hf(t_i, w_i),$$

$$k_2 = hf\left(t_i + \frac{h}{6}, w_i + \frac{1}{6}k_1\right),$$

$$k_3 = hf\left(t_i + \frac{4h}{15}, w_i + \frac{4}{75}k_1 + \frac{16}{75}k_2\right),$$

$$k_4 = hf\left(t_i + \frac{2h}{3}, w_i + \frac{5}{6}k_1 - \frac{8}{3}k_2 + \frac{5}{2}k_3\right),$$

$$k_5 = hf\left(t_i + \frac{5h}{6}, w_i - \frac{165}{64}k_1 + \frac{55}{6}k_2 - \frac{425}{64}k_3 + \frac{85}{96}k_4\right),$$

$$k_6 = hf\left(t_i + h, w_i + \frac{12}{5}k_1 - 8k_2 + \frac{4015}{612}k_3 - \frac{11}{36}k_4 + \frac{88}{255}k_5\right),$$

$$k_7 = hf\left(t_i + \frac{h}{15}, w_i - \frac{8263}{15000}k_1 + \frac{124}{75}k_2 - \frac{643}{680}k_3 - \frac{81}{250}k_4 + \frac{2484}{10625}k_5\right),$$

$$k_8 = hf\left(t_i + h, w_i + \frac{3501}{1720}k_1 - \frac{300}{43}k_2 + \frac{297275}{52632}k_3 - \frac{319}{2322}k_4 + \frac{24068}{84065}k_5 + \frac{3850}{26703}k_7\right).$$

The sixth-order method $\tilde{w}_{i+1}$ is used to estimate the error in the fifth-order method w_{i+1}. Construct an algorithm similar to the Runge-Kutta-Fehlberg Algorithm, and repeat Exercise 3 using this new method.

5. In the theory of the spread of contagious disease (see [Ba1] or [Ba2]), a relatively elementary differential equation can be used to predict the number of infective individuals in the population at any time, provided appropriate simplification assumptions are made. In particular, let us assume that all individuals in a fixed population have an equally likely chance of being infected and once infected remain in that state. Suppose $x(t)$ denotes the number of susceptible individuals at time t and $y(t)$ denotes the number of infectives. It is reasonable to assume that the rate at which the number of infectives changes is proportional to the product of $x(t)$ and $y(t)$ because the rate depends on both the number of infectives and the number of susceptibles present at that time. If the population is large enough to assume that $x(t)$ and $y(t)$ are continuous variables, the problem can be expressed

$$y'(t) = kx(t)y(t),$$

where k is a constant and $x(t) + y(t) = m$, the total population. This equation can be rewritten involving only $y(t)$ as

$$y'(t) = k(m - y(t))y(t).$$

a. Assuming that $m = 100{,}000$, $y(0) = 1000$, $k = 2 \times 10^{-6}$, and that time is measured in days, find an approximation to the number of infective individuals at the end of 30 days.

b. The differential equation in part (a) is called a *Bernoulli equation* and it can be transformed into a linear differential equation in $u(t) = (y(t))^{-1}$. Use this technique to find the exact solution to the equation, under the same assumptions as in part (a), and compare the true value of $y(t)$ to the approximation given there. What is $\lim_{t\to\infty} y(t)$? Does this agree with your intuition?

6. In the previous exercise, all infected individuals remained in the population to spread the disease. A more realistic proposal is to introduce a third variable $z(t)$ to represent the number of individuals

who are removed from the affected population at a given time t by isolation, recovery and consequent immunity, or death. This quite naturally complicates the problem, but it can be shown (see [Ba2]) that an approximate solution can be given in the form

$$x(t) = x(0)e^{-(k_1/k_2)z(t)} \quad \text{and} \quad y(t) = m - x(t) - z(t),$$

where k_1 is the infective rate, k_2 is the removal rate, and $z(t)$ is determined from the differential equation

$$z'(t) = k_2\left(m - z(t) - x(0)e^{-(k_1/k_2)z(t)}\right).$$

The authors are not aware of any technique for solving this problem directly, so a numerical procedure must be applied. Find an approximation to $z(30)$, $y(30)$, and $x(30)$, assuming that $m = 100{,}000$, $x(0) = 99{,}000$, $k_1 = 2 \times 10^{-6}$, and $k_2 = 10^{-4}$.

5.6 Multistep Methods

The methods discussed to this point in the chapter are called **one-step methods** because the approximation for the mesh point t_{i+1} involves information from only one of the previous mesh points, t_i. Although these methods might use function evaluation information at points between t_i and t_{i+1}, they do not retain that information for direct use in future approximations. All the information used by these methods is obtained within the subinterval over which the solution is being approximated.

The approximate solution is available at each of the mesh points $t_0, t_1, \ldots, t_i$ before the approximation at t_{i+1} is obtained, and because the error $|w_j - y(t_j)|$ tends to increase with j, so it seems reasonable to develop methods that use these more accurate previous data when approximating the solution at t_{i+1}.

Methods using the approximation at more than one previous mesh point to determine the approximation at the next point are called *multistep* methods. The precise definition of these methods follows, together with the definition of the two types of multistep methods.

Definition 5.14 An ***m*-step multistep method** for solving the initial-value problem

$$y' = f(t, y), \quad a \le t \le b, \quad y(a) = \alpha, \tag{5.23}$$

has a difference equation for finding the approximation w_{i+1} at the mesh point t_{i+1} represented by the following equation, where m is an integer greater than 1:

$$\begin{aligned} w_{i+1} &= a_{m-1}w_i + a_{m-2}w_{i-1} + \cdots + a_0 w_{i+1-m} \\ &\quad + h[b_m f(t_{i+1}, w_{i+1}) + b_{m-1} f(t_i, w_i) \\ &\quad + \cdots + b_0 f(t_{i+1-m}, w_{i+1-m})], \end{aligned} \tag{5.24}$$

for $i = m-1, m, \ldots, N-1$, where $h = (b-a)/N$, the $a_0, a_1, \ldots, a_{m-1}$ and $b_0, b_1, \ldots, b_m$ are constants, and the starting values

$$w_0 = \alpha, \quad w_1 = \alpha_1, \quad w_2 = \alpha_2, \quad \ldots, \quad w_{m-1} = \alpha_{m-1}$$

are specified. ■

When $b_m = 0$ the method is called **explicit**, or **open**, because Eq. (5.24) then gives w_{i+1} explicitly in terms of previously determined values. When $b_m \neq 0$ the method is called

implicit, or **closed**, because w_{i+1} occurs on both sides of Eq. (5.243), so w_{i+1} is specified only implicitly.

For example, the equations

$$w_0 = \alpha, \quad w_1 = \alpha_1, \quad w_2 = \alpha_2, \quad w_3 = \alpha_3,$$

$$w_{i+1} = w_i + \frac{h}{24}[55f(t_i, w_i) - 59f(t_{i-1}, w_{i-1}) + 37f(t_{i-2}, w_{i-2}) - 9f(t_{i-3}, w_{i-3})], \tag{5.25}$$

The Adams-Bashforth techniques are due to John Couch Adams (1819–1892), who did significant work in mathematics and astronomy. He developed these numerical techniques to approximate the solution of a fluid-flow problem posed by Bashforth.

for each $i = 3, 4, \ldots, N-1$, define an *explicit* four-step method known as the **fourth-order Adams-Bashforth technique**. The equations

$$w_0 = \alpha, \quad w_1 = \alpha_1, \quad w_2 = \alpha_2,$$

$$w_{i+1} = w_i + \frac{h}{24}[9f(t_{i+1}, w_{i+1}) + 19f(t_i, w_i) - 5f(t_{i-1}, w_{i-1}) + f(t_{i-2}, w_{i-2})], \tag{5.26}$$

for each $i = 2, 3, \ldots, N-1$, define an *implicit* three-step method known as the **fourth-order Adams-Moulton technique**.

Forest Ray Moulton (1872–1952) was in charge of ballistics at the Aberdeen Proving Grounds in Maryland during World War I. He was a prolific author, writing numerous books in mathematics and astronomy, and developed improved multistep methods for solving ballistic equations.

The starting values in either (5.25) or (5.26) must be specified, generally by assuming $w_0 = \alpha$ and generating the remaining values by either a Runge-Kutta or Taylor method. We will see that the implicit methods are generally more accurate then the explicit methods, but to apply an implicit method such as (5.25) directly, we must solve the implicit equation for w_{i+1}. This is not always possible,and even when it can be done the solution for w_{i+1} may not be unique.

Example 1 In Example 3 of Section 5.4 (see Table 5.8 on page 289) we used the Runge-Kutta method of order four with $h = 0.2$ to approximate the solutions to the initial value problem

$$y' = y - t^2 + 1, \quad 0 \le t \le 2, \quad y(0) = 0.5.$$

The first four approximations were found to be $y(0) = w_0 = 0.5$, $y(0.2) \approx w_1 = 0.8292933$, $y(0.4) \approx w_2 = 1.2140762$, and $y(0.6) \approx w_3 = 1.6489220$. Use these as starting values for the fourth-order Adams-Bashforth method to compute new approximations for $y(0.8)$ and $y(1.0)$, and compare these new approximations to those produced by the Runge-Kutta method of order four.

Solution For the fourth-order Adams-Bashforth we have

$$\begin{aligned} y(0.8) \approx w_4 &= w_3 + \frac{0.2}{24}(55f(0.6, w_3) - 59f(0.4, w_2) + 37f(0.2, w_1) - 9f(0, w_0)) \\ &= 1.6489220 + \frac{0.2}{24}(55f(0.6, 1.6489220) - 59f(0.4, 1.2140762) \\ &\quad + 37f(0.2, 0.8292933) - 9f(0, 0.5)) \\ &= 1.6489220 + 0.0083333(55(2.2889220) - 59(2.0540762) \\ &\quad + 37(1.7892933) - 9(1.5)) \\ &= 2.1272892, \end{aligned}$$

and

$$\begin{aligned}
y(1.0) \approx w_5 = w_4 + \frac{0.2}{24}&(55f(0.8, w_4) - 59f(0.6, w_3) + 37f(0.4, w_2) - 9f(0.2, w_1)) \\
&= 2.1272892 + \frac{0.2}{24}(55f(0.8, 2.1272892) - 59f(0.6, 1.6489220) \\
&\quad + 37f(0.4, 1.2140762) - 9f(0.2, 0.8292933)) \\
&= 2.1272892 + 0.0083333(55(2.4872892) - 59(2.2889220) \\
&\quad + 37(2.0540762) - 9(1.7892933)) \\
&= 2.6410533,
\end{aligned}$$

The error for these approximations at $t = 0.8$ and $t = 1.0$ are, respectively

$$|2.1272295 - 2.1272892| = 5.97 \times 10^{-5} \text{ and } |2.6410533 - 2.6408591| = 1.94 \times 10^{-4}.$$

The corresponding Runge-Kutta approximations had errors

$$|2.1272027 - 2.1272892| = 2.69 \times 10^{-5} \text{ and } |2.6408227 - 2.6408591| = 3.64 \times 10^{-5}.$$

■

Adams was particularly interested in the using his ability for accurate numerical calculations to investigate the orbits of the planets. He predicted the existence of Neptune by analyzing the irregularities in the planet Uranus, and developed various numerical integration techniques to assist in the approximation of the solution of differential equations.

To begin the derivation of a multistep method, note that the solution to the initial-value problem

$$y' = f(t, y), \quad a \le t \le b, \quad y(a) = \alpha,$$

if integrated over the interval $[t_i, t_{i+1}]$, has the property that

$$y(t_{i+1}) - y(t_i) = \int_{t_i}^{t_{i+1}} y'(t)\, dt = \int_{t_i}^{t_{i+1}} f(t, y(t))\, dt.$$

Consequently,

$$y(t_{i+1}) = y(t_i) + \int_{t_i}^{t_{i+1}} f(t, y(t))\, dt. \tag{5.27}$$

However we cannot integrate $f(t, y(t))$ without knowing $y(t)$, the solution to the problem, so we instead integrate an interpolating polynomial $P(t)$ to $f(t, y(t))$, one that is determined by some of the previously obtained data points $(t_0, w_0), (t_1, w_1), \ldots, (t_i, w_i)$. When we assume, in addition, that $y(t_i) \approx w_i$, Eq. (5.27) becomes

$$y(t_{i+1}) \approx w_i + \int_{t_i}^{t_{i+1}} P(t)\, dt. \tag{5.28}$$

Although any form of the interpolating polynomial can be used for the derivation, it is most convenient to use the Newton backward-difference formula, because this form more easily incorporates the most recently calculated data.

To derive an Adams-Bashforth explicit m-step technique, we form the backward-difference polynomial $P_{m-1}(t)$ through

$$(t_i, f(t_i, y(t_i))), \quad (t_{i-1}, f(t_{i-1}, y(t_{i-1}))), \ldots, \quad (t_{i+1-m}, f(t_{i+1-m}, y(t_{i+1-m}))).$$

Since $P_{m-1}(t)$ is an interpolatory polynomial of degree $m-1$, some number ξ_i in (t_{i+1-m}, t_i) exists with

$$f(t, y(t)) = P_{m-1}(t) + \frac{f^{(m)}(\xi_i, y(\xi_i))}{m!}(t - t_i)(t - t_{i-1}) \cdots (t - t_{i+1-m}).$$

Introducing the variable substitution $t = t_i + sh$, with $dt = h\,ds$, into $P_{m-1}(t)$ and the error term implies that

$$\int_{t_i}^{t_{i+1}} f(t, y(t))\,dt = \int_{t_i}^{t_{i+1}} \sum_{k=0}^{m-1} (-1)^k \binom{-s}{k} \nabla^k f(t_i, y(t_i))\,dt$$
$$+ \int_{t_i}^{t_{i+1}} \frac{f^{(m)}(\xi_i, y(\xi_i))}{m!}(t-t_i)(t-t_{i-1})\cdots(t-t_{i+1-m})\,dt$$
$$= \sum_{k=0}^{m-1} \nabla^k f(t_i, y(t_i)) h(-1)^k \int_0^1 \binom{-s}{k} ds$$
$$+ \frac{h^{m+1}}{m!} \int_0^1 s(s+1)\cdots(s+m-1) f^{(m)}(\xi_i, y(\xi_i))\,ds.$$

The integrals $(-1)^k \int_0^1 \binom{-s}{k} ds$ for various values of k are easily evaluated and are listed in Table 5.12. For example, when $k = 3$,

$$(-1)^3 \int_0^1 \binom{-s}{3} ds = -\int_0^1 \frac{(-s)(-s-1)(-s-2)}{1 \cdot 2 \cdot 3} ds$$
$$= \frac{1}{6} \int_0^1 (s^3 + 3s^2 + 2s)\,ds$$
$$= \frac{1}{6}\left[\frac{s^4}{4} + s^3 + s^2\right]_0^1 = \frac{1}{6}\left(\frac{9}{4}\right) = \frac{3}{8}.$$

Table 5.12

k	$\int_0^1 \binom{-s}{k} ds$
0	1
1	$\frac{1}{2}$
2	$\frac{5}{12}$
3	$\frac{3}{8}$
4	$\frac{251}{720}$
5	$\frac{95}{288}$

As a consequence,

$$\int_{t_i}^{t_{i+1}} f(t, y(t))\,dt = h\left[f(t_i, y(t_i)) + \frac{1}{2}\nabla f(t_i, y(t_i)) + \frac{5}{12}\nabla^2 f(t_i, y(t_i)) + \cdots\right]$$
$$+ \frac{h^{m+1}}{m!} \int_0^1 s(s+1)\cdots(s+m-1) f^{(m)}(\xi_i, y(\xi_i))\,ds. \tag{5.29}$$

Because $s(s+1)\cdots(s+m-1)$ does not change sign on $[0, 1]$, the Weighted Mean Value Theorem for Integrals can be used to deduce that for some number μ_i, where $t_{i+1-m} < \mu_i < t_{i+1}$, the error term in Eq. (5.29) becomes

$$\frac{h^{m+1}}{m!} \int_0^1 s(s+1)\cdots(s+m-1) f^{(m)}(\xi_i, y(\xi_i))\,ds$$
$$= \frac{h^{m+1} f^{(m)}(\mu_i, y(\mu_i))}{m!} \int_0^1 s(s+1)\cdots(s+m-1)\,ds.$$

Hence the error in (5.29) simplifies to

$$h^{m+1} f^{(m)}(\mu_i, y(\mu_i))(-1)^m \int_0^1 \binom{-s}{m} ds. \tag{5.30}$$

But $y(t_{i+1}) - y(t_i) = \int_{t_i}^{t_{i+1}} f(t, y(t))\,dt$, so Eq. (5.27) can be written as

$$y(t_{i+1}) = y(t_i) + h\left[f(t_i, y(t_i)) + \frac{1}{2}\nabla f(t_i, y(t_i)) + \frac{5}{12}\nabla^2 f(t_i, y(t_i)) + \cdots\right]$$
$$+ h^{m+1} f^{(m)}(\mu_i, y(\mu_i))(-1)^m \int_0^1 \binom{-s}{m} ds. \tag{5.31}$$

Example 2 Use Eq. (5.31) with $m = 3$ to derive the three-step Adams-Bashforth technique.

Solution We have

$$y(t_{i+1}) \approx y(t_i) + h\left[f(t_i, y(t_i)) + \frac{1}{2}\nabla f(t_i, y(t_i)) + \frac{5}{12}\nabla^2 f(t_i, y(t_i))\right]$$

$$= y(t_i) + h\left\{f(t_i, y(t_i)) + \frac{1}{2}[f(t_i, y(t_i)) - f(t_{i-1}, y(t_{i-1}))]\right.$$

$$\left. + \frac{5}{12}[f(t_i, y(t_i)) - 2f(t_{i-1}, y(t_{i-1})) + f(t_{i-2}, y(t_{i-2}))]\right\}$$

$$= y(t_i) + \frac{h}{12}[23f(t_i, y(t_i)) - 16f(t_{i-1}, y(t_{i-1})) + 5f(t_{i-2}, y(t_{i-2}))].$$

The three-step Adams-Bashforth method is, consequently,

$$w_0 = \alpha, \quad w_1 = \alpha_1, \quad w_2 = \alpha_2,$$

$$w_{i+1} = w_i + \frac{h}{12}[23f(t_i, w_i) - 16f(t_{i-1}, w_{i-1})] + 5f(t_{i-2}, w_{i-2})],$$

for $i = 2, 3, \ldots, N - 1$. ▪

Multistep methods can also be derived using Taylor series. An example of the procedure involved is considered in Exercise 12. A derivation using a Lagrange interpolating polynomial is discussed in Exercise 11.

The local truncation error for multistep methods is defined analogously to that of one-step methods. As in the case of one-step methods, the local truncation error provides a measure of how the solution to the differential equation fails to solve the difference equation.

Definition 5.15 If $y(t)$ is the solution to the initial-value problem

$$y' = f(t, y), \quad a \le t \le b, \quad y(a) = \alpha,$$

and

$$w_{i+1} = a_{m-1}w_i + a_{m-2}w_{i-1} + \cdots + a_0 w_{i+1-m}$$
$$+ h[b_m f(t_{i+1}, w_{i+1}) + b_{m-1} f(t_i, w_i) + \cdots + b_0 f(t_{i+1-m}, w_{i+1-m})]$$

is the $(i + 1)$st step in a multistep method, the **local truncation error** at this step is

$$\tau_{i+1}(h) = \frac{y(t_{i+1}) - a_{m-1}y(t_i) - \cdots - a_0 y(t_{i+1-m})}{h} - [b_m f(t_{i+1}, y(t_{i+1})) + \cdots + b_0 f(t_{i+1-m}, y(t_{i+1-m}))], \tag{5.32}$$

for each $i = m - 1, m, \ldots, N - 1$. ▪

Example 3 Determine the local truncation error for the three-step Adams-Bashforth method derived in Example 2.

Solution Considering the form of the error given in Eq. (5.30) and the appropriate entry in Table 5.12 gives

$$h^4 f^{(3)}(\mu_i, y(\mu_i))(-1)^3 \int_0^1 \binom{-s}{3} ds = \frac{3h^4}{8} f^{(3)}(\mu_i, y(\mu_i)).$$

Using the fact that $f^{(3)}(\mu_i, y(\mu_i)) = y^{(4)}(\mu_i)$ and the difference equation derived in Example 2, we have

$$\begin{aligned}
\tau_{i+1}(h) &= \frac{y(t_{i+1}) - y(t_i)}{h} - \frac{1}{12}[23f(t_i, y(t_i)) - 16f(t_{i-1}, y(t_{i-1})) + 5f(t_{i-2}, y(t_{i-2}))] \\
&= \frac{1}{h}\left[\frac{3h^4}{8} f^{(3)}(\mu_i, y(\mu_i))\right] = \frac{3h^3}{8} y^{(4)}(\mu_i), \quad \text{for some } \mu_i \in (t_{i-2}, t_{i+1}).
\end{aligned}$$

■

Adams-Bashforth Explicit Methods

Some of the explicit multistep methods together with their required starting values and local truncation errors are as follows. The derivation of these techniques is similar to the procedure in Examples 2 and 3.

Adams-Bashforth Two-Step Explicit Method

$$\begin{aligned}
w_0 &= \alpha, \quad w_1 = \alpha_1, \\
w_{i+1} &= w_i + \frac{h}{2}[3f(t_i, w_i) - f(t_{i-1}, w_{i-1})], \qquad (5.33)
\end{aligned}$$

where $i = 1, 2, \ldots, N-1$. The local truncation error is $\tau_{i+1}(h) = \frac{5}{12} y'''(\mu_i) h^2$, for some $\mu_i \in (t_{i-1}, t_{i+1})$.

Adams-Bashforth Three-Step Explicit Method

$$\begin{aligned}
w_0 &= \alpha, \quad w_1 = \alpha_1, \quad w_2 = \alpha_2, \\
w_{i+1} &= w_i + \frac{h}{12}[23f(t_i, w_i) - 16f(t_{i-1}, w_{i-1}) + 5f(t_{i-2}, w_{i-2})], \qquad (5.34)
\end{aligned}$$

where $i = 2, 3, \ldots, N-1$. The local truncation error is $\tau_{i+1}(h) = \frac{3}{8} y^{(4)}(\mu_i) h^3$, for some $\mu_i \in (t_{i-2}, t_{i+1})$.

Adams-Bashforth Four-Step Explicit Method

$$\begin{aligned}
w_0 &= \alpha, \quad w_1 = \alpha_1, \quad w_2 = \alpha_2, \quad w_3 = \alpha_3, \qquad (5.35) \\
w_{i+1} &= w_i + \frac{h}{24}[55f(t_i, w_i) - 59f(t_{i-1}, w_{i-1}) + 37f(t_{i-2}, w_{i-2}) - 9f(t_{i-3}, w_{i-3})],
\end{aligned}$$

where $i = 3, 4, \ldots, N-1$. The local truncation error is $\tau_{i+1}(h) = \frac{251}{720} y^{(5)}(\mu_i) h^4$, for some $\mu_i \in (t_{i-3}, t_{i+1})$.

Adams-Bashforth Five-Step Explicit Method

$$\begin{aligned}
w_0 &= \alpha, \quad w_1 = \alpha_1, \quad w_2 = \alpha_2, \quad w_3 = \alpha_3, \quad w_4 = \alpha_4, \\
w_{i+1} &= w_i + \frac{h}{720}[1901f(t_i, w_i) - 2774f(t_{i-1}, w_{i-1}) \qquad (5.36) \\
&\quad + 2616f(t_{i-2}, w_{i-2}) - 1274f(t_{i-3}, w_{i-3}) + 251f(t_{i-4}, w_{i-4})],
\end{aligned}$$

where $i = 4, 5, \ldots, N-1$. The local truncation error is $\tau_{i+1}(h) = \frac{95}{288} y^{(6)}(\mu_i) h^5$, for some $\mu_i \in (t_{i-4}, t_{i+1})$.

Adams-Moulton Implicit Methods

Implicit methods are derived by using $(t_{i+1}, f(t_{i+1}, y(t_{i+1})))$ as an additional interpolation node in the approximation of the integral

$$\int_{t_i}^{t_{i+1}} f(t, y(t))\, dt.$$

Some of the more common implicit methods are as follows.

Adams-Moulton Two-Step Implicit Method

$$w_0 = \alpha, \quad w_1 = \alpha_1,$$
$$w_{i+1} = w_i + \frac{h}{12}[5f(t_{i+1}, w_{i+1}) + 8f(t_i, w_i) - f(t_{i-1}, w_{i-1})], \tag{5.37}$$

where $i = 1, 2, \ldots, N-1$. The local truncation error is $\tau_{i+1}(h) = -\frac{1}{24}y^{(4)}(\mu_i)h^3$, for some $\mu_i \in (t_{i-1}, t_{i+1})$.

Adams-Moulton Three-Step Implicit Method

$$w_0 = \alpha, \quad w_1 = \alpha_1, \quad w_2 = \alpha_2, \tag{5.38}$$
$$w_{i+1} = w_i + \frac{h}{24}[9f(t_{i+1}, w_{i+1}) + 19f(t_i, w_i) - 5f(t_{i-1}, w_{i-1}) + f(t_{i-2}, w_{i-2})],$$

where $i = 2, 3, \ldots, N-1$. The local truncation error is $\tau_{i+1}(h) = -\frac{19}{720}y^{(5)}(\mu_i)h^4$, for some $\mu_i \in (t_{i-2}, t_{i+1})$.

Adams-Moulton Four-Step Implicit Method

$$w_0 = \alpha, \quad w_1 = \alpha_1, \quad w_2 = \alpha_2, \quad w_3 = \alpha_3,$$
$$w_{i+1} = w_i + \frac{h}{720}[251f(t_{i+1}, w_{i+1}) + 646f(t_i, w_i) \tag{5.39}$$
$$- 264f(t_{i-1}, w_{i-1}) + 106f(t_{i-2}, w_{i-2}) - 19f(t_{i-3}, w_{i-3})],$$

where $i = 3, 4, \ldots, N-1$. The local truncation error is $\tau_{i+1}(h) = -\frac{3}{160}y^{(6)}(\mu_i)h^5$, for some $\mu_i \in (t_{i-3}, t_{i+1})$.

It is interesting to compare an m-step Adams-Bashforth explicit method with an $(m-1)$-step Adams-Moulton implicit method. Both involve m evaluations of f per step, and both have the terms $y^{(m+1)}(\mu_i)h^m$ in their local truncation errors. In general, the coefficients of the terms involving f in the local truncation error are smaller for the implicit methods than for the explicit methods. This leads to greater stability and smaller round-off errors for the implicit methods.

Example 4 Consider the initial-value problem

$$y' = y - t^2 + 1, \quad 0 \le t \le 2, \quad y(0) = 0.5.$$

Use the exact values given from $y(t) = (t+1)^2 - 0.5e^t$ as starting values and $h = 0.2$ to compare the approximations from **(a)** by the explicit Adams-Bashforth four-step method and **(b)** the implicit Adams-Moulton three-step method.

Solution **(a)** The Adams-Bashforth method has the difference equation

$$w_{i+1} = w_i + \frac{h}{24}[55f(t_i, w_i) - 59f(t_{i-1}, w_{i-1}) + 37f(t_{i-2}, w_{i-2}) - 9f(t_{i-3}, w_{i-3})],$$

for $i = 3, 4, \ldots, 9$. When simplified using $f(t, y) = y - t^2 + 1$, $h = 0.2$, and $t_i = 0.2i$, it becomes

$$w_{i+1} = \frac{1}{24}[35w_i - 11.8w_{i-1} + 7.4w_{i-2} - 1.8w_{i-3} - 0.192i^2 - 0.192i + 4.736].$$

(b) The Adams-Moulton method has the difference equation

$$w_{i+1} = w_i + \frac{h}{24}[9f(t_{i+1}, w_{i+1}) + 19f(t_i, w_i) - 5f(t_{i-1}, w_{i-1}) + f(t_{i-2}, w_{i-2})],$$

for $i = 2, 3, \ldots, 9$. This reduces to

$$w_{i+1} = \frac{1}{24}[1.8w_{i+1} + 27.8w_i - w_{i-1} + 0.2w_{i-2} - 0.192i^2 - 0.192i + 4.736].$$

To use this method explicitly, we meed to solve the equation explicitly solve for w_{i+1}. This gives

$$w_{i+1} = \frac{1}{22.2}[27.8w_i - w_{i-1} + 0.2w_{i-2} - 0.192i^2 - 0.192i + 4.736],$$

for $i = 2, 3, \ldots, 9$.

The results in Table 5.13 were obtained using the exact values from $y(t) = (t + 1)^2 - 0.5e^t$ for α, α_1, α_2, and α_3 in the explicit Adams-Bashforth case and for α, α_1, and α_2 in the implicit Adams-Moulton case. Note that the implicit Adams-Moulton method gives consistently better results. ■

Table 5.13

t_i	Exact	Adams-Bashforth w_i	Error	Adams-Moulton w_i	Error
0.0	0.5000000				
0.2	0.8292986				
0.4	1.2140877				
0.6	1.6489406			1.6489341	0.0000065
0.8	2.1272295	2.1273124	0.0000828	2.1272136	0.0000160
1.0	2.6408591	2.6410810	0.0002219	2.6408298	0.0000293
1.2	3.1799415	3.1803480	0.0004065	3.1798937	0.0000478
1.4	3.7324000	3.7330601	0.0006601	3.7323270	0.0000731
1.6	4.2834838	4.2844931	0.0010093	4.2833767	0.0001071
1.8	4.8151763	4.8166575	0.0014812	4.8150236	0.0001527
2.0	5.3054720	5.3075838	0.0021119	5.3052587	0.0002132

Multistep methods are available as options of the *InitialValueProblem* command, in a manner similar to that of the one step methods. The command for the Adam Bashforth Four Step method applied to our usual example has the form

$C := InitialValueProblem(deq, y(0) = 0.5, t = 2, method = adamsbashforth,$
$submethod = step4, numsteps = 10, output = information, digits = 8)$

The output from this method is similar to the results in Table 5.13 except that the exact values were used in Table 5.13 and approximations were used as starting values for the Maple approximations.

To apply the Adams-Mouton Three Step method to this problem, the options would be changed to *method* = *adamsmoulton*, *submethod* = *step3*.

Predictor-Corrector Methods

In Example 4 the implicit Adams-Moulton method gave better results than the explicit Adams-Bashforth method of the same order. Although this is generally the case, the implicit methods have the inherent weakness of first having to convert the method algebraically to an explicit representation for w_{i+1}. This procedure is not always possible, as can be seen by considering the elementary initial-value problem

$$y' = e^y, \quad 0 \le t \le 0.25, \quad y(0) = 1.$$

Because $f(t, y) = e^y$, the three-step Adams-Moulton method has

$$w_{i+1} = w_i + \frac{h}{24}[9e^{w_{i+1}} + 19e^{w_i} - 5e^{w_{i-1}} + e^{w_{i-2}}]$$

as its difference equation, and this equation cannot be algebraically solved for w_{i+1}.

We could use Newton's method or the secant method to approximate w_{i+1}, but this complicates the procedure considerably. In practice, implicit multistep methods are not used as described above. Rather, they are used to improve approximations obtained by explicit methods. The combination of an explicit method to predict and an implicit to improve the prediction is called a **predictor-corrector method**.

Consider the following fourth-order method for solving an initial-value problem. The first step is to calculate the starting values w_0, w_1, w_2, and w_3 for the four-step explicit Adams-Bashforth method. To do this, we use a fourth-order one-step method, the Runge-Kutta method of order four. The next step is to calculate an approximation, w_{4p}, to $y(t_4)$ using the explicit Adams-Bashforth method as predictor:

$$w_{4p} = w_3 + \frac{h}{24}[55f(t_3, w_3) - 59f(t_2, w_2) + 37f(t_1, w_1) - 9f(t_0, w_0)].$$

This approximation is improved by inserting w_{4p} in the right side of the three-step implicit Adams-Moulton method and using that method as a corrector. This gives

$$w_4 = w_3 + \frac{h}{24}[9f(t_4, w_{4p}) + 19f(t_3, w_3) - 5f(t_2, w_2) + f(t_1, w_1)].$$

The only new function evaluation required in this procedure is $f(t_4, w_{4p})$ in the corrector equation; all the other values of f have been calculated for earlier approximations.

The value w_4 is then used as the approximation to $y(t_4)$, and the technique of using the Adams-Bashforth method as a predictor and the Adams-Moulton method as a corrector is repeated to find w_{5p} and w_5, the initial and final approximations to $y(t_5)$. This process is continued until we obtain an approximation w_c to $y(t_N) = y(b)$.

Improved approximations to $y(t_{i+1})$ might be obtained by iterating the Adams-Moulton formula, but these converge to the approximation given by the implicit formula rather than to the solution $y(t_{i+1})$. Hence it is usually more efficient to use a reduction in the step size if improved accuracy is needed.

Algorithm 5.4 is based on the fourth-order Adams-Bashforth method as predictor and one iteration of the Adams-Moulton method as corrector, with the starting values obtained from the fourth-order Runge-Kutta method.

ALGORITHM 5.4

Adams Fourth-Order Predictor-Corrector

To approximate the solution of the initial-value problem

$$y' = f(t, y), \quad a \le t \le b, \quad y(a) = \alpha,$$

at $(N+1)$ equally spaced numbers in the interval $[a, b]$:

INPUT endpoints a, b; integer N; initial condition α.

OUTPUT approximation w to y at the $(N+1)$ values of t.

Step 1 Set $h = (b-a)/N$;
$t_0 = a$;
$w_0 = \alpha$;
OUTPUT (t_0, w_0).

Step 2 For $i = 1, 2, 3$, do Steps 3–5.
(Compute starting values using Runge-Kutta method.)

Step 3 Set $K_1 = hf(t_{i-1}, w_{i-1})$;
$K_2 = hf(t_{i-1} + h/2, w_{i-1} + K_1/2)$;
$K_3 = hf(t_{i-1} + h/2, w_{i-1} + K_2/2)$;
$K_4 = hf(t_{i-1} + h, w_{i-1} + K_3)$.

Step 4 Set $w_i = w_{i-1} + (K_1 + 2K_2 + 2K_3 + K_4)/6$;
$t_i = a + ih$.

Step 5 OUTPUT (t_i, w_i).

Step 6 For $i = 4, \ldots, N$ do Steps 7–10.

Step 7 Set $t = a + ih$;
$w = w_3 + h[55f(t_3, w_3) - 59f(t_2, w_2) + 37f(t_1, w_1) - 9f(t_0, w_0)]/24$; *(Predict w_i.)*
$w = w_3 + h[9f(t, w) + 19f(t_3, w_3) - 5f(t_2, w_2) + f(t_1, w_1)]/24$. *(Correct w_i.)*

Step 8 OUTPUT (t, w).

Step 9 For $j = 0, 1, 2$
set $t_j = t_{j+1}$; *(Prepare for next iteration.)*
$w_j = w_{j+1}$.

Step 10 Set $t_3 = t$;
$w_3 = w$.

Step 11 STOP. ■

Example 5 Apply the Adams fourth-order predictor-corrector method with $h = 0.2$ and starting values from the Runge-Kutta fourth order method to the initial-value problem

$$y' = y - t^2 + 1, \quad 0 \le t \le 2, \quad y(0) = 0.5.$$

Solution This is continuation and modification of the problem considered in Example 1 at the beginning of the section. In that example we found that the starting approximations from Runge-Kutta are

$$y(0) = w_0 = 0.5,\ y(0.2) \approx w_1 = 0.8292933,\ y(0.4) \approx w_2 = 1.2140762, \text{ and } y(0.6) \approx w_3 = 1.6489220.$$

and the fourth-order Adams-Bashforth method gave

$$\begin{aligned} y(0.8) \approx w_{4p} &= w_3 + \frac{0.2}{24}(55f(0.6, w_3) - 59f(0.4, w_2) + 37f(0.2, w_1) - 9f(0, w_0)) \\ &= 1.6489220 + \frac{0.2}{24}(55f(0.6, 1.6489220) - 59f(0.4, 1.2140762) \\ &\quad + 37f(0.2, 0.8292933) - 9f(0, 0.5)) \\ &= 1.6489220 + 0.0083333(55(2.2889220) - 59(2.0540762) \\ &\quad + 37(1.7892933) - 9(1.5)) \\ &= 2.1272892. \end{aligned}$$

We will now use w_{4p} as the predictor of the approximation to $y(0.8)$ and determine the corrected value w_4, from the implicit Adams-Moulton method. This gives

$$\begin{aligned} y(0.8) \approx w_4 &= w_3 + \frac{0.2}{24}\big(9f(0.8, w_{4p}) + 19f(0.6, w_3) - 5f(0.4, w_2) + f(0.2, w_1)\big) \\ &= 1.6489220 + \frac{0.2}{24}(9f(0.8, 2.1272892) + 19f(0.6, 1.6489220) \\ &\quad - 5f(0.4, 1.2140762) + f(0.2, 0.8292933)) \\ &= 1.6489220 + 0.0083333(9(2.4872892) + 19(2.2889220) - 5(2.0540762) \\ &\quad + (1.7892933)) \\ &= 2.1272056. \end{aligned}$$

Now we use this approximation to determine the predictor, w_{5p}, for $y(1.0)$ as

$$\begin{aligned} y(1.0) \approx w_{5p} &= w_4 + \frac{0.2}{24}(55f(0.8, w_4) - 59f(0.6, w_3) + 37f(0.4, w_2) - 9f(0.2, w_1)) \\ &= 2.1272056 + \frac{0.2}{24}(55f(0.8, 2.1272056) - 59f(0.6, 1.6489220) \\ &\quad + 37f(0.4, 1.2140762) - 9f(0.2, 0.8292933)) \\ &= 2.1272056 + 0.0083333(55(2.4872056) - 59(2.2889220) + 37(2.0540762) \\ &\quad - 9(1.7892933)) \\ &= 2.6409314, \end{aligned}$$

and correct this with

$$\begin{aligned} y(1.0) \approx w_5 &= w_4 + \frac{0.2}{24}\big(9f(1.0, w_{5p}) + 19f(0.8, w_4) - 5f(0.6, w_3) + f(0.4, w_2)\big) \\ &= 2.1272056 + \frac{0.2}{24}(9f(1.0, 2.6409314) + 19f(0.8, 2.1272892) \\ &\quad - 5f(0.6, 1.6489220) + f(0.4, 1.2140762)) \end{aligned}$$

$$= 2.1272056 + 0.0083333(9(2.6409314) + 19(2.4872056) - 5(2.2889220) + (2.0540762))$$
$$= 2.6408286.$$

In Example 1 we found that using the explicit Adams-Bashforth method alone produced results that were inferior to those of Runge-Kutta. However, these approximations to $y(0.8)$ and $y(1.0)$ are accurate to within

$$|2.1272295 - 2.1272056| = 2.39 \times 10^{-5} \quad \text{and} \quad |2.6408286 - 2.6408591| = 3.05 \times 10^{-5}.$$

respectively, compared to those of Runge-Kutta, which were accurate, respectively, to within

$$|2.1272027 - 2.1272892| = 2.69 \times 10^{-5} \quad \text{and} \quad |2.6408227 - 2.6408591| = 3.64 \times 10^{-5}.$$

The remaining predictor-corrector approximations were generated using Algorithm 5.4 and are shown in Table 5.14. ■

Table 5.14

t_i	$y_i = y(t_i)$	w_i	Error $\lvert y_i - w_i\rvert$
0.0	0.5000000	0.5000000	0
0.2	0.8292986	0.8292933	0.0000053
0.4	1.2140877	1.2140762	0.0000114
0.6	1.6489406	1.6489220	0.0000186
0.8	2.1272295	2.1272056	0.0000239
1.0	2.6408591	2.6408286	0.0000305
1.2	3.1799415	3.1799026	0.0000389
1.4	3.7324000	3.7323505	0.0000495
1.6	4.2834838	4.2834208	0.0000630
1.8	4.8151763	4.8150964	0.0000799
2.0	5.3054720	5.3053707	0.0001013

Adams Fourth Order Predictor-Corrector method is implemented in Maple for the example problem with

$C := InitialValueProblem(deq, y(0) = 0.5, t = 2, method = adamsbashforthmoulton, submethod = step4, numsteps = 10, output = information, digits = 8)$

and generates the same values as in Table 5.14.

Other multistep methods can be derived using integration of interpolating polynomials over intervals of the form $[t_j, t_{i+1}]$, for $j \le i-1$, to obtain an approximation to $y(t_{i+1})$. When an interpolating polynomial is integrated over $[t_{i-3}, t_{i+1}]$, the result is the explicit **Milne's method**:

Edward Arthur Milne (1896–1950) worked in ballistic research during World War I, and then for the Solar Physics Observatory at Cambridge. In 1929 he was appointed the W. W. Rouse Ball chair at Wadham College in Oxford.

$$w_{i+1} = w_{i-3} + \frac{4h}{3}[2f(t_i, w_i) - f(t_{i-1}, w_{i-1}) + 2f(t_{i-2}, w_{i-2})],$$

which has local truncation error $\frac{14}{45}h^4 y^{(5)}(\xi_i)$, for some $\xi_i \in (t_{i-3}, t_{i+1})$.

Milne's method is occasionally used as a predictor for the implicit **Simpson's method**,

Simpson's name is associated with this technique because it is based on Simpson's rule for integration.

$$w_{i+1} = w_{i-1} + \frac{h}{3}[f(t_{i+1}, w_{i+1}) + 4f(t_i, w_i) + f(t_{i-1}, w_{i-1})],$$

which has local truncation error $-(h^4/90)y^{(5)}(\xi_i)$, for some $\xi_i \in (t_{i-1}, t_{i+1})$, and is obtained by integrating an interpolating polynomial over $[t_{i-1}, t_{i+1}]$.

The local truncation error involved with a predictor-corrector method of the Milne-Simpson type is generally smaller than that of the Adams-Bashforth-Moulton method. But the technique has limited use because of round-off error problems, which do not occur with the Adams procedure. Elaboration on this difficulty is given in Section 5.10.

EXERCISE SET 5.6

1. Use all the Adams-Bashforth methods to approximate the solutions to the following initial-value problems. In each case use exact starting values, and compare the results to the actual values.
   a. $y' = te^{3t} - 2y, \quad 0 \le t \le 1, \quad y(0) = 0$, with $h = 0.2$; actual solution $y(t) = \frac{1}{5}te^{3t} - \frac{1}{25}e^{3t} + \frac{1}{25}e^{-2t}$.
   b. $y' = 1 + y/t, \quad 1 \le t \le 2, \quad y(1) = 2$, with $h = 0.2$; actual solution $y(t) = t \ln t + 2t$.
   c. $y' = 1 + (t - y)^2, \quad 2 \le t \le 3, \quad y(2) = 1$, with $h = 0.2$; actual solution $y(t) = t + \frac{1}{1-t}$.
   d. $y' = \cos 2t + \sin 3t, \quad 0 \le t \le 1, \quad y(0) = 1$, with $h = 0.2$; actual solution $y(t) = \frac{1}{2}\sin 2t - \frac{1}{3}\cos 3t + \frac{4}{3}$.

2. Use each of the Adams-Bashforth methods to approximate the solutions to the following initial-value problems. In each case use starting values obtained from the Runge-Kutta method of order four. Compare the results to the actual values.
   a. $y' = \dfrac{y^2}{1+t}, \quad 1 \le t \le 2, \quad y(1) = -(\ln 2)^{-1}$, with $h = 0.1$ actual solution $y(t) = \dfrac{-1}{\ln(t+1)}$.
   b. $y' = \dfrac{2 - 2ty}{t^2 + 1}, \quad 0 \le t \le 1, \quad y(0) = 1$, with $h = 0.1$ actual solution $y(t) = \dfrac{2t+1}{t^2+2}$.
   c. $y' = (y^2 + y)/t, \quad 1 \le t \le 3, \quad y(1) = -2$, with $h = 0.2$ actual solution $y(t) = \dfrac{2t}{1-t}$.
   d. $y' = -ty + 4t/y, \quad 0 \le t \le 1, \quad y(0) = 1$, with $h = 0.1$ actual solution $y(t) = \sqrt{4 - 3e^{-t^2}}$.

3. Use each of the Adams-Bashforth methods to approximate the solutions to the following initial-value problems. In each case use starting values obtained from the Runge-Kutta method of order four. Compare the results to the actual values.
   a. $y' = 1 + y/t + (y/t)^2, \quad 1 \le t \le 3, \quad y(1) = 0$, with $h = 0.2$; actual solution $y(t) = t\tan(\ln t)$.
   b. $y' = y/t - (y/t)^2, \quad 1 \le t \le 2, \quad y(1) = 1$, with $h = 0.1$; actual solution $y(t) = \dfrac{t}{1 + \ln t}$.
   c. $y' = -(y+1)(y+3), \quad 0 \le t \le 2, \quad y(0) = -2$, with $h = 0.1$; actual solution $y(t) = -3 + 2/(1 + e^{-2t})$.
   d. $y' = -5y + 5t^2 + 2t, \quad 0 \le t \le 1, \quad y(0) = 1/3$, with $h = 0.1$; actual solution $y(t) = t^2 + \frac{1}{3}e^{-5t}$.

4. Use all the Adams-Moulton methods to approximate the solutions to the Exercises 1(a), 1(b), and 1(d). In each case use exact starting values, and explicitly solve for w_{i+1}. Compare the results to the actual values.

5. Use Algorithm 5.4 to approximate the solutions to the initial-value problems in Exercise 1.

6. Use Algorithm 5.4 to approximate the solutions to the initial-value problems in Exercise 2.

7. Use Algorithm 5.4 to approximate the solutions to the initial-value problems in Exercise 3.

8. Change Algorithm 5.4 so that the corrector can be iterated for a given number p iterations. Repeat Exercise 7 with $p = 2, 3$, and 4 iterations. Which choice of p gives the best answer for each initial-value problem?

9. The initial-value problem

$$y' = e^y, \quad 0 \le t \le 0.20, \quad y(0) = 1$$

has solution

$$y(t) = 1 - \ln(1 - et).$$

Applying the three-step Adams-Moulton method to this problem is equivalent to finding the fixed point w_{i+1} of

$$g(w) = w_i + \frac{h}{24}\left(9e^{w} + 19e^{w_i} - 5e^{w_{i-1}} + e^{w_{i-2}}\right).$$

a. With $h = 0.01$, obtain w_{i+1} by functional iteration for $i = 2, \ldots, 19$ using exact starting values w_0, w_1, and w_2. At each step use w_i to initially approximate w_{i+1}.

b. Will Newton's method speed the convergence over functional iteration?

10. Use the Milne-Simpson Predictor-Corrector method to approximate the solutions to the initial-value problems in Exercise 3.

11. **a.** Derive the Adams-Bashforth Two-Step method by using the Lagrange form of the interpolating polynomial.

b. Derive the Adams-Bashforth Four-Step method by using Newton's backward-difference form of the interpolating polynomial.

12. Derive the Adams-Bashforth Three-Step method by the following method. Set

$$y(t_{i+1}) = y(t_i) + ahf(t_i, y(t_i)) + bhf(t_{i-1}, y(t_{i-1})) + chf(t_{i-2}, y(t_{i-2})).$$

Expand $y(t_{i+1})$, $f(t_{i-2}, y(t_{i-2}))$, and $f(t_{i-1}, y(t_{i-1}))$ in Taylor series about $(t_i, y(t_i))$, and equate the coefficients of h, h^2 and h^3 to obtain a, b, and c.

13. Derive the Adams-Moulton Two-Step method and its local truncation error by using an appropriate form of an interpolating polynomial.

14. Derive Simpson's method by applying Simpson's rule to the integral

$$y(t_{i+1}) - y(t_{i-1}) = \int_{t_{i-1}}^{t_{i+1}} f(t, y(t))\, dt.$$

15. Derive Milne's method by applying the open Newton-Cotes formula (4.29) to the integral

$$y(t_{i+1}) - y(t_{i-3}) = \int_{t_{i-3}}^{t_{i+1}} f(t, y(t))\, dt.$$

16. Verify the entries in Table 5.12 on page 305.

5.7 Variable Step-Size Multistep Methods

The Runge-Kutta-Fehlberg method is used for error control because at each step it provides, at little additional cost, *two* approximations that can be compared and related to the local truncation error. Predictor-corrector techniques always generate two approximations at each step, so they are natural candidates for error-control adaptation.

To demonstrate the error-control procedure, we construct a variable step-size predictor-corrector method using the four-step explicit Adams-Bashforth method as predictor and the three-step implicit Adams-Moulton method as corrector.

The Adams-Bashforth four-step method comes from the relation

$$y(t_{i+1}) = y(t_i) + \frac{h}{24}[55f(t_i, y(t_i)) - 59f(t_{i-1}, y(t_{i-1})) + 37f(t_{i-2}, y(t_{i-2})) - 9f(t_{i-3}, y(t_{i-3}))] + \frac{251}{720}y^{(5)}(\hat{\mu}_i)h^5,$$

for some $\hat{\mu}_i \in (t_{i-3}, t_{i+1})$. The assumption that the approximations $w_0, w_1, \ldots, w_i$ are all exact implies that the Adams-Bashforth local truncation error is

$$\frac{y(t_{i+1}) - w_{i+1,p}}{h} = \frac{251}{720}y^{(5)}(\hat{\mu}_i)h^4. \tag{5.40}$$

A similar analysis of the Adams-Moulton three-step method, which comes from

$$y(t_{i+1}) = y(t_i) + \frac{h}{24}[9f(t_{i+1}, y(t_{i+1})) + 19f(t_i, y(t_i)) - 5f(t_{i-1}, y(t_{i-1})) + f(t_{i-2}, y(t_{i-2}))] - \frac{19}{720}y^{(5)}(\tilde{\mu}_i)h^5,$$

for some $\tilde{\mu}_i \in (t_{i-2}, t_{i+1})$, leads to the local truncation error

$$\frac{y(t_{i+1}) - w_{i+1}}{h} = -\frac{19}{720}y^{(5)}(\tilde{\mu}_i)h^4. \tag{5.41}$$

To proceed further, we must make the assumption that for small values of h, we have

$$y^{(5)}(\hat{\mu}_i) \approx y^{(5)}(\tilde{\mu}_i).$$

The effectiveness of the error-control technique depends directly on this assumption.

If we subtract Eq. (5.40) from Eq. (5.39), we have

$$\frac{w_{i+1} - w_{i+1,p}}{h} = \frac{h^4}{720}[251y^{(5)}(\hat{\mu}_i) + 19y^{(5)}(\tilde{\mu}_i)] \approx \frac{3}{8}h^4 y^{(5)}(\tilde{\mu}_i),$$

so

$$y^{(5)}(\tilde{\mu}_i) \approx \frac{8}{3h^5}(w_{i+1} - w_{i+1,p}). \tag{5.42}$$

Using this result to eliminate the term involving $y^{(5)}(\tilde{\mu}_i)h^4$ from Eq. (5.41) gives the approximation to the Adams-Moulton local truncation error

$$|\tau_{i+1}(h)| = \frac{|y(t_{i+1}) - w_{i+1}|}{h} \approx \frac{19h^4}{720} \cdot \frac{8}{3h^5}|w_{i+1} - w_{i+1,p}| = \frac{19|w_{i+1} - w_{i+1,p}|}{270h}.$$

Suppose we now reconsider (Eq. 5.41) with a new step size qh generating new approximations $\hat{w}_{i+1,p}$ and $\hat{w}_{i+1}$. The object is to choose q so that the local truncation error given in Eq. (5.41) is bounded by a prescribed tolerance ε. If we assume that the value $y^{(5)}(\mu)$ in Eq. (5.41) associated with qh is also approximated using Eq. (5.42), then

$$\frac{|y(t_i + qh) - \hat{w}_{i+1}|}{qh} = \frac{19q^4h^4}{720}|y^{(5)}(\mu)| \approx \frac{19q^4h^4}{720}\left[\frac{8}{3h^5}|w_{i+1} - w_{i+1,p}|\right]$$
$$= \frac{19q^4}{270}\frac{|w_{i+1} - w_{i+1,p}|}{h},$$

and we need to choose q so that

$$\frac{|y(t_i + qh) - \hat{w}_{i+1}|}{qh} \approx \frac{19q^4}{270}\frac{|w_{i+1} - w_{i+1,p}|}{h} < \varepsilon.$$

That is, choose q so that

$$q < \left(\frac{270}{19}\frac{h\varepsilon}{|w_{i+1} - w_{i+1,p}|}\right)^{1/4} \approx 2\left(\frac{h\varepsilon}{|w_{i+1} - w_{i+1,p}|}\right)^{1/4}.$$

A number of approximation assumptions have been made in this development, so in practice q is chosen conservatively, often as

$$q = 1.5\left(\frac{h\varepsilon}{|w_{i+1} - w_{i+1,p}|}\right)^{1/4}.$$

A change in step size for a multistep method is more costly in terms of function evaluations than for a one-step method, because new equally-spaced starting values must be computed. As a consequence, it is common practice to ignore the step-size change whenever the local truncation error is between $\varepsilon/10$ and ε, that is, when

$$\frac{\varepsilon}{10} < |\tau_{i+1}(h)| = \frac{|y(t_{i+1}) - w_{i+1}|}{h} \approx \frac{19|w_{i+1} - w_{i+1,p}|}{270h} < \varepsilon.$$

In addition, q is given an upper bound to ensure that a single unusually accurate approximation does not result in too large a step size. Algorithm 5.5 incorporates this safeguard with an upper bound of 4.

Remember that the multistep methods require equal step sizes for the starting values. So any change in step size necessitates recalculating new starting values at that point. In Steps 3, 16, and 19 of Algorithm 5.5 this is done by calling a Runge-Kutta subalgorithm (Algorithm 5.2), which has been set up in Step 1.

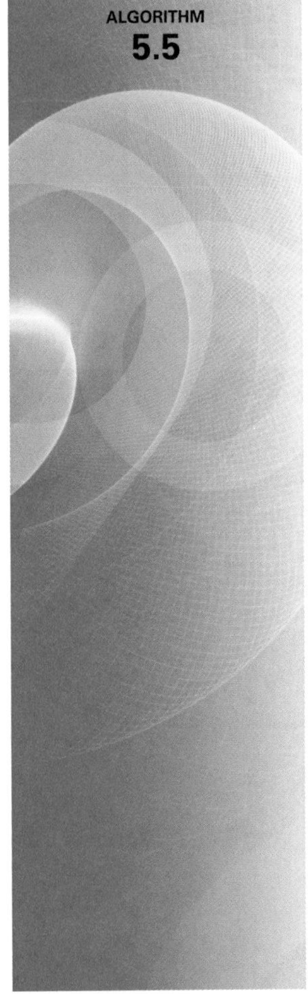

Adams Variable Step-Size Predictor-Corrector

To approximate the solution of the initial-value problem

$$y' = f(t, y), \quad a \le t \le b, \quad y(a) = \alpha$$

with local truncation error within a given tolerance:

INPUT endpoints a, b; initial condition α; tolerance TOL; maximum step size $hmax$; minimum step size $hmin$.

OUTPUT i, t_i, w_i, h where at the ith step w_i approximates $y(t_i)$ and the step size h was used, or a message that the minimum step size was exceeded.

Step 1 Set up a subalgorithm for the Runge-Kutta fourth-order method to be called $RK4(h, v_0, x_0, v_1, x_1, v_2, x_2, v_3, x_3)$ that accepts as input a step size h and starting values $v_0 \approx y(x_0)$ and returns $\{(x_j, v_j) \mid j = 1, 2, 3\}$ defined by the following:

for $j = 1, 2, 3$

set $K_1 = hf(x_{j-1}, v_{j-1})$;

$K_2 = hf(x_{j-1} + h/2, v_{j-1} + K_1/2)$

$K_3 = hf(x_{j-1} + h/2, v_{j-1} + K_2/2)$

$K_4 = hf(x_{j-1} + h, v_{j-1} + K_3)$

$v_j = v_{j-1} + (K_1 + 2K_2 + 2K_3 + K_4)/6$;

$x_j = x_0 + jh$.

Step 2 Set $t_0 = a$;

$w_0 = \alpha$;

$h = hmax$;

$FLAG = 1$; (*FLAG will be used to exit the loop in Step 4.*)

$LAST = 0$; (*LAST will indicate when the last value is calculated.*)

OUTPUT (t_0, w_0).

Step 3 Call $RK4(h, w_0, t_0, w_1, t_1, w_2, t_2, w_3, t_3)$;
Set $NFLAG = 1$; *(Indicates computation from RK4.)*
$i = 4$;
$t = t_3 + h$.

Step 4 While $(FLAG = 1)$ do Steps 5–20.

Step 5 Set $WP = w_{i-1} + \frac{h}{24}[55f(t_{i-1}, w_{i-1}) - 59f(t_{i-2}, w_{i-2})$
$+ 37f(t_{i-3}, w_{i-3}) - 9f(t_{i-4}, w_{i-4})]$; *(Predict w_i.)*
$WC = w_{i-1} + \frac{h}{24}[9f(t, WP) + 19f(t_{i-1}, w_{i-1})$
$- 5f(t_{i-2}, w_{i-2}) + f(t_{i-3}, w_{i-3})]$; *(Correct w_i.)*
$\sigma = 19|WC - WP|/(270h)$.

Step 6 If $\sigma \le TOL$ then do Steps 7–16 *(Result accepted.)*
else do Steps 17–19. *(Result rejected.)*

Step 7 Set $w_i = WC$; *(Result accepted.)*
$t_i = t$.

Step 8 If $NFLAG = 1$ then for $j = i-3, i-2, i-1, i$
OUTPUT (j, t_j, w_j, h);
(Previous results also accepted.)
else OUTPUT (i, t_i, w_i, h).
(Previous results already accepted.)

Step 9 If $LAST = 1$ then set $FLAG = 0$ *(Next step is 20.)*
else do Steps 10–16.

Step 10 Set $i = i + 1$;
$NFLAG = 0$.

Step 11 If $\sigma \le 0.1\ TOL$ or $t_{i-1} + h > b$ then do Steps 12–16.
(Increase h if it is more accurate than required or decrease h to include b as a mesh point.)

Step 12 Set $q = (TOL/(2\sigma))^{1/4}$.

Step 13 If $q > 4$ then set $h = 4h$
else set $h = qh$.

Step 14 If $h > hmax$ then set $h = hmax$.

Step 15 If $t_{i-1} + 4h > b$ then
set $h = (b - t_{i-1})/4$;
$LAST = 1$.

Step 16 Call $RK4(h, w_{i-1}, t_{i-1}, w_i, t_i, w_{i+1}, t_{i+1}, w_{i+2}, t_{i+2})$;
Set $NFLAG = 1$;
$i = i + 3$. *(True branch completed. Next step is 20.)*

Step 17 Set $q = (TOL/(2\sigma))^{1/4}$. *(False branch from Step 6: Result rejected.)*

Step 18 If $q < 0.1$ then set $h = 0.1h$
else set $h = qh$.

Step 19 If $h < hmin$ then set $FLAG = 0$;
OUTPUT ('*hmin* exceeded')
else
if $NFLAG = 1$ then set $i = i - 3$;
(*Previous results also rejected.*)
Call $RK4(h, w_{i-1}, t_{i-1}, w_i, t_i, w_{i+1}, t_{i+1}, w_{i+2}, t_{i+2})$;
set $i = i + 3$;
$NFLAG = 1$.

Step 20 Set $t = t_{i-1} + h$.

Step 21 STOP. ■

Example 1 Use Adams variable step-size predictor-corrector method with maximum step size $hmax = 0.2$, minimum step size $hmin = 0.01$, and tolerance $TOL = 10^{-5}$ to approximate the solution of the initial-value problem

$$y' = y - t^2 + 1, \quad 0 \le t \le 2, \quad y(0) = 0.5.$$

Solution We begin with $h = hmax = 0.2$, and obtain w_0, w_1, w_2 and w_3 using Runge-Kutta, then find wp_4 and wc_4 by applying the predictor-corrector method. These calculations were done in Example 5 of Section 5.6 where it was determined that the Runge-Kutta approximations are

$$y(0) = w_0 = 0.5, \; y(0.2) \approx w_1 = 0.8292933, \; y(0.4) \approx w_2 = 1.2140762, \; \text{and} \quad y(0.6) \approx w_3 = 1.6489220.$$

The predictor and corrector gave

$$y(0) = w_0 = 0.5, \; y(0.2) \approx w_1 = 0.8292933, \; y(0.4) \approx w_2 = 1.2140762, \; \text{and}$$
$$y(0.6) \approx w_3 = 1.6489220.$$
$$y(0.8) \approx w_{4p} = w_3 + \frac{0.2}{24}(55f(0.6, w_3) - 59f(0.4, w_2) + 37f(0.2, w_1) - 9f(0, w_0))$$
$$= 2.1272892,$$

and

$$y(0.8) \approx w_4 = w_3 + \frac{0.2}{24}(9f(0.8, w_{4p}) + 19f(0.6, w_3) - 5f(0.42, w_2) + f(0.2, w_1))$$
$$= 2.1272056.$$

We now need to determine if these approximations are sufficiently accurate or if there needs to be a change in the step size. First we find

$$\delta = \frac{19}{270h}|w_4 - w_{4p}| = \frac{19}{270(0.2)}|2.1272056 - 2.1272892| = 2.941 \times 10^{-5}.$$

Because this exceeds the tolerance of 10^{-5} a new step size is needed and the new step size is

$$qh = \left(\frac{10^{-5}}{2\delta}\right)^{1/4} = \left(\frac{10^{-5}}{2(2.941 \times 10^{-5})}\right)^{1/4}(0.2) = 0.642(0.2) \approx 0.128.$$

As a consequence, we need to begin the procedure again computing the Runge-Kutta values with this step size, and then use the predictor-corrector method with this same step size to compute the new values of w_{4p} and w_4. We then need to run the accuracy check on these approximations to see that we have been successful. Table 5.15 shows that this second run is successful and lists the all results obtained using Algorithm 5.5. ■

Table 5.15

t_i	$y(t_i)$	w_i	h_i	σ_i	$\lvert y(t_i) - w_i\rvert$
0	0.5	0.5			
0.1257017	0.7002323	0.7002318	0.1257017	4.051×10^{-6}	0.0000005
0.2514033	0.9230960	0.9230949	0.1257017	4.051×10^{-6}	0.0000011
0.3771050	1.1673894	1.1673877	0.1257017	4.051×10^{-6}	0.0000017
0.5028066	1.4317502	1.4317480	0.1257017	4.051×10^{-6}	0.0000022
0.6285083	1.7146334	1.7146306	0.1257017	4.610×10^{-6}	0.0000028
0.7542100	2.0142869	2.0142834	0.1257017	5.210×10^{-6}	0.0000035
0.8799116	2.3287244	2.3287200	0.1257017	5.913×10^{-6}	0.0000043
1.0056133	2.6556930	2.6556877	0.1257017	6.706×10^{-6}	0.0000054
1.1313149	2.9926385	2.9926319	0.1257017	7.604×10^{-6}	0.0000066
1.2570166	3.3366642	3.3366562	0.1257017	8.622×10^{-6}	0.0000080
1.3827183	3.6844857	3.6844761	0.1257017	9.777×10^{-6}	0.0000097
1.4857283	3.9697541	3.9697433	0.1030100	7.029×10^{-6}	0.0000108
1.5887383	4.2527830	4.2527711	0.1030100	7.029×10^{-6}	0.0000120
1.6917483	4.5310269	4.5310137	0.1030100	7.029×10^{-6}	0.0000133
1.7947583	4.8016639	4.8016488	0.1030100	7.029×10^{-6}	0.0000151
1.8977683	5.0615660	5.0615488	0.1030100	7.760×10^{-6}	0.0000172
1.9233262	5.1239941	5.1239764	0.0255579	3.918×10^{-8}	0.0000177
1.9488841	5.1854932	5.1854751	0.0255579	3.918×10^{-8}	0.0000181
1.9744421	5.2460056	5.2459870	0.0255579	3.918×10^{-8}	0.0000186
2.0000000	5.3054720	5.3054529	0.0255579	3.918×10^{-8}	0.0000191

EXERCISE SET 5.7

1. Use the Adams Variable Step-Size Predictor-Corrector Algorithm with tolerance $TOL = 10^{-4}$, $hmax = 0.25$, and $hmin = 0.025$ to approximate the solutions to the given initial-value problems. Compare the results to the actual values.
   a. $y' = te^{3t} - 2y, \quad 0 \le t \le 1, \quad y(0) = 0;$ actual solution $y(t) = \frac{1}{5}te^{3t} - \frac{1}{25}e^{3t} + \frac{1}{25}e^{-2t}$.
   b. $y' = 1 + y/t, \quad 1 \le t \le 2, \quad y(1) = 2;$ actual solution $y(t) = t \ln t + 2t$.
   c. $y' = 1 + (t - y)^2, \quad 2 \le t \le 3, \quad y(2) = 1;$ actual solution $y(t) = t + 1/(1 - t)$.
   d. $y' = \cos 2t + \sin 3t, \quad 0 \le t \le 1, \quad y(0) = 1;$ actual solution $y(t) = \frac{1}{2}\sin 2t - \frac{1}{3}\cos 3t + \frac{4}{3}$.
2. Use the Adams Variable Step-Size Predictor-Corrector Algorithm with $TOL = 10^{-4}$ to approximate the solutions to the following initial-value problems:
   a. $y' = \sin t + e^{-t}, \quad 0 \le t \le 1, \quad y(0) = 0$, with $hmax = 0.2$ and $hmin = 0.01$.
   b. $y' = (y/t)^2 + y/t, \quad 1 \le t \le 1.2, \quad y(1) = 1$, with $hmax = 0.05$ and $hmin = 0.01$.
   c. $y' = (y^2 + y)/t, \quad 1 \le t \le 3, \quad y(1) = -2$, with $hmax = 0.4$ and $hmin = 0.01$.
   d. $y' = -ty + 4t/y, \quad 0 \le t \le 1, \quad y(0) = 1$, with $hmax = 0.2$ and $hmin = 0.01$.
3. Use the Adams Variable Step-Size Predictor-Corrector Algorithm with tolerance $TOL = 10^{-6}$, $hmax = 0.5$, and $hmin = 0.02$ to approximate the solutions to the given initial-value problems. Compare the results to the actual values.
   a. $y' = 1 + y/t + (y/t)^2, \quad 1 \le t \le 3, \quad y(1) = 0;$ actual solution $y(t) = t\tan(\ln t)$.
   b. $y' = y/t - (y/t)^2, \quad 1 \le t \le 4, \quad y(1) = 1;$ actual solution $y(t) = t/(1 + \ln t)$.

c. $y' = -(y+1)(y+3), \quad 0 \le t \le 3, \quad y(0) = -2$; actual solution $y(t) = -3 + 2(1+e^{-2t})^{-1}$.

d. $y' = (t+2t^3)y^3 - ty, \quad 0 \le t \le 2, \quad y(0) = \frac{1}{3}$; actual solution $y(t) = (3 + 2t^2 + 6e^{t^2})^{-1/2}$.

4. Construct an Adams Variable Step-Size Predictor-Corrector Algorithm based on the Adams-Bashforth five-step method and the Adams-Moulton four-step method. Repeat Exercise 3 using this new method.

5. An electrical circuit consists of a capacitor of constant capacitance $C = 1.1$ farads in series with a resistor of constant resistance $R_0 = 2.1$ ohms. A voltage $\mathcal{E}(t) = 110 \sin t$ is applied at time $t = 0$. When the resistor heats up, the resistance becomes a function of the current i,

$$R(t) = R_0 + ki, \quad \text{where } k = 0.9,$$

and the differential equation for $i(t)$ becomes

$$\left(1 + \frac{2k}{R_0} i\right) \frac{di}{dt} + \frac{1}{R_0 C} i = \frac{1}{R_0 C} \frac{d\mathcal{E}}{dt}.$$

Find $i(2)$, assuming that $i(0) = 0$.

5.8 Extrapolation Methods

Extrapolation was used in Section 4.5 for the approximation of definite integrals, where we found that by correctly averaging relatively inaccurate trapezoidal approximations exceedingly accurate new approximations were produced. In this section we will apply extrapolation to increase the accuracy of approximations to the solution of initial-value problems. As we have previously seen, the original approximations must have an error expansion of a specific form for the procedure to be successful.

To apply extrapolation to solve initial-value problems, we use a technique based on the Midpoint method:

$$w_{i+1} = w_{i-1} + 2hf(t_i, w_i), \quad \text{for } i \ge 1. \tag{5.43}$$

This technique requires two starting values since both w_0 and w_1 are needed before the first midpoint approximation, w_2, can be determined. One starting value is the initial condition for $w_0 = y(a) = \alpha$. To determine the second starting value, w_1, we apply Euler's method. Subsequent approximations are obtained from (5.43). After a series of approximations of this type are generated ending at a value t, an endpoint correction is performed that involves the final two midpoint approximations. This produces an approximation $w(t, h)$ to $y(t)$ that has the form

$$y(t) = w(t, h) + \sum_{k=1}^{\infty} \delta_k h^{2k}, \tag{5.44}$$

where the δ_k are constants related to the derivatives of the solution $y(t)$. The important point is that the δ_k do not depend on the step size h. The details of this procedure can be found in the paper by Gragg [Gr].

To illustrate the extrapolation technique for solving

$$y'(t) = f(t, y), \quad a \le t \le b, \quad y(a) = \alpha,$$

assume that we have a fixed step size h. We wish to approximate $y(t_1) = y(a+h)$.

For the first extrapolation step we let $h_0 = h/2$ and use Euler's method with $w_0 = \alpha$ to approximate $y(a + h_0) = y(a + h/2)$ as

$$w_1 = w_0 + h_0 f(a, w_0).$$

We then apply the Midpoint method with $t_{i-1} = a$ and $t_i = a + h_0 = a + h/2$ to produce a first approximation to $y(a + h) = y(a + 2h_0)$,

$$w_2 = w_0 + 2h_0 f(a + h_0, w_1).$$

The endpoint correction is applied to obtain the final approximation to $y(a+h)$ for the step size h_0. This results in the $O(h_0^2)$ approximation to $y(t_1)$

$$y_{1,1} = \frac{1}{2}[w_2 + w_1 + h_0 f(a + 2h_0, w_2)].$$

We save the approximation $y_{1,1}$ and discard the intermediate results w_1 and w_2.

To obtain the next approximation, $y_{2,1}$, to $y(t_1)$, we let $h_1 = h/4$ and use Euler's method with $w_0 = \alpha$ to obtain an approximation to $y(a+h_1) = y(a+h/4)$ which we will call w_1:

$$w_1 = w_0 + h_1 f(a, w_0).$$

Next we approximate $y(a + 2h_1) = y(a + h/2)$ with w_2, $y(a + 3h_1) = y(a + 3h/4)$ with w_3, and w_4 to $y(a + 4h_1) = y(t_1)$ using the Midpoint method.

$$\begin{aligned} w_2 &= w_0 + 2h_1 f(a + h_1, w_1), \\ w_3 &= w_1 + 2h_1 f(a + 2h_1, w_2), \\ w_4 &= w_2 + 2h_1 f(a + 3h_1, w_3). \end{aligned}$$

The endpoint correction is now applied to w_3 and w_4 to produce the improved $O(h_1^2)$ approximation to $y(t_1)$,

$$y_{2,1} = \frac{1}{2}[w_4 + w_3 + h_1 f(a + 4h_1, w_4)].$$

Because of the form of the error given in (5.44), the two approximations to $y(a + h)$ have the property that

$$y(a + h) = y_{1,1} + \delta_1 \left(\frac{h}{2}\right)^2 + \delta_2 \left(\frac{h}{2}\right)^4 + \cdots = y_{1,1} + \delta_1 \frac{h^2}{4} + \delta_2 \frac{h^4}{16} + \cdots,$$

and

$$y(a + h) = y_{2,1} + \delta_1 \left(\frac{h}{4}\right)^2 + \delta_2 \left(\frac{h}{4}\right)^4 + \cdots = y_{2,1} + \delta_1 \frac{h^2}{16} + \delta_2 \frac{h^4}{256} + \cdots.$$

We can eliminate the $O(h^2)$ portion of this truncation error by averaging the two formulas appropriately. Specifically, if we subtract the first formula from 4 times the second and divide the result by 3, we have

$$y(a + h) = y_{2,1} + \frac{1}{3}(y_{2,1} - y_{1,1}) - \delta_2 \frac{h^4}{64} + \cdots.$$

So the approximation to $y(t_1)$ given by

$$y_{2,2} = y_{2,1} + \frac{1}{3}(y_{2,1} - y_{1,1})$$

has error of order $O(h^4)$.

We next let $h_2 = h/6$ and apply Euler's method once followed by the Midpoint method five times. Then we use the endpoint correction to determine the h^2 approximation, $y_{3,1}$, to $y(a+h) = y(t_1)$. This approximation can be averaged with $y_{2,1}$ to produce a second $O(h^4)$ approximation that we denote $y_{3,2}$. Then $y_{3,2}$ and $y_{2,2}$ are averaged to eliminate the $O(h^4)$ error terms and produce an approximation with error of order $O(h^6)$. Higher-order formulas are generated by continuing the process.

The only significant difference between the extrapolation performed here and that used for Romberg integration in Section 4.5 results from the way the subdivisions are chosen. In Romberg integration there is a convenient formula for representing the Composite Trapezoidal rule approximations that uses consecutive divisions of the step size by the integers $1, 2, 4, 8, 16, 32, 64, \ldots$ This procedure permits the averaging process to proceed in an easily followed manner.

We do not have a means for easily producing refined approximations for initial-value problems, so the divisions for the extrapolation technique are chosen to minimize the number of required function evaluations. The averaging procedure arising from this choice of subdivision, shown in Table 5.16, is not as elementary, but, other than that, the process is the same as that used for Romberg integration.

Table 5.16

$y_{1,1} = w(t, h_0)$		
$y_{2,1} = w(t, h_1)$	$y_{2,2} = y_{2,1} + \dfrac{h_1^2}{h_0^2 - h_1^2}(y_{2,1} - y_{1,1})$	
$y_{3,1} = w(t, h_2)$	$y_{3,2} = y_{3,1} + \dfrac{h_2^2}{h_1^2 - h_2^2}(y_{3,1} - y_{2,1})$	$y_{3,3} = y_{3,2} + \dfrac{h_2^2}{h_0^2 - h_2^2}(y_{3,2} - y_{2,2})$

Algorithm 5.6 uses nodes of the form 2^n and $2^n \cdot 3$. Other choices can be used.

Algorithm 5.6 uses the extrapolation technique with the sequence of integers

$$q_0 = 2,\ q_1 = 4,\ q_2 = 6,\ q_3 = 8,\ q_4 = 12,\ q_5 = 16,\ q_6 = 24, \quad \text{and} \quad q_7 = 32.$$

A basic step size h is selected, and the method progresses by using $h_i = h/q_i$, for each $i = 0, \ldots, 7$, to approximate $y(t+h)$. The error is controlled by requiring that the approximations $y_{1,1}, y_{2,2}, \ldots$ be computed until $|y_{i,i} - y_{i-1,i-1}|$ is less than a given tolerance. If the tolerance is not achieved by $i = 8$, then h is reduced, and the process is reapplied.

Minimum and maximum values of h, *hmin*, and *hmax*, respectively, are specified to ensure control of the method. If $y_{i,i}$ is found to be acceptable, then w_1 is set to $y_{i,i}$ and computations begin again to determine w_2, which will approximate $y(t_2) = y(a+2h)$. The process is repeated until the approximation w_N to $y(b)$ is determined.

Extrapolation

To approximate the solution of the initial-value problem

$$y' = f(t, y), \quad a \le t \le b, \quad y(a) = \alpha,$$

with local truncation error within a given tolerance:

INPUT endpoints a, b; initial condition α; tolerance *TOL*; maximum step size *hmax*; minimum step size *hmin*.

OUTPUT T, W, h where W approximates $y(t)$ and step size h was used, or a message that minimum step size was exceeded.

Step 1 Initialize the array $NK = (2, 4, 6, 8, 12, 16, 24, 32)$.

Step 2 Set $TO = a$;
$WO = \alpha$;
$h = hmax$;
$FLAG = 1$. (*FLAG is used to exit the loop in Step* 4.)

Step 3 For $i = 1, 2, \ldots, 7$
for $j = 1, \ldots, i$
set $Q_{i,j} = (NK_{i+1}/NK_j)^2$. (*Note:* $Q_{i,j} = h_j^2/h_{i+1}^2$.)

Step 4 While ($FLAG = 1$) do Steps 5–20.

Step 5 Set $k = 1$;
$NFLAG = 0$. (*When desired accuracy is achieved, NFLAG is set to* 1.)

Step 6 While ($k \le 8$ and $NFLAG = 0$) do Steps 7–14.

Step 7 Set $HK = h/NK_k$;
$T = TO$;
$W2 = WO$;
$W3 = W2 + HK \cdot f(T, W2)$; (*Euler's first step.*)
$T = TO + HK$.

Step 8 For $j = 1, \ldots, NK_k - 1$
set $W1 = W2$;
$W2 = W3$;
$W3 = W1 + 2HK \cdot f(T, W2)$; (*Midpoint method.*)
$T = TO + (j + 1) \cdot HK$.

Step 9 Set $y_k = [W3 + W2 + HK \cdot f(T, W3)]/2$.
(*Endpoint correction to compute* $y_{k,1}$.)

Step 10 If $k \ge 2$ then do Steps 11–13.
(*Note:* $y_{k-1} \equiv y_{k-1,1}, y_{k-2} \equiv y_{k-2,2}, \ldots, y_1 \equiv y_{k-1,k-1}$ *since only the previous row of the table is saved.*)

Step 11 Set $j = k$;
$v = y_1$. (*Save* $y_{k-1,k-1}$.)

Step 12 While ($j \ge 2$) do

$$\text{set } y_{j-1} = y_j + \frac{y_j - y_{j-1}}{Q_{k-1,j-1} - 1};$$

(*Extrapolation to compute* $y_{j-1} \equiv y_{k,k-j+2}$.)

$$\left(\textit{Note:} \quad y_{j-1} = \frac{h_{j-1}^2 y_j - h_k^2 y_{j-1}}{h_{j-1}^2 - h_k^2}.\right)$$

$j = j - 1$.

Step 13 If $|y_1 - v| \le TOL$ then set $NFLAG = 1$.
(y_1 *is accepted as the new* w.)

Step 14 Set $k = k + 1$.

Step 15 Set $k = k - 1$.

Step 16 If $NFLAG = 0$ then do Steps 17 and 18 *(Result rejected.)*
else do Steps 19 and 20. *(Result accepted.)*

Step 17 Set $h = h/2$. *(New value for w rejected, decrease h.)*

Step 18 If $h < hmin$ then
OUTPUT ('$hmin$ exceeded');
Set $FLAG = 0$.
(True branch completed, next step is back to Step 4.)

Step 19 Set $WO = y_1$; *(New value for w accepted.)*
$TO = TO + h$;
OUTPUT (TO, WO, h).

Step 20 If $TO \geq b$ then set $FLAG = 0$
(Procedure completed successfully.)
else if $TO + h > b$ then set $h = b - TO$
(Terminate at $t = b$.)
else if ($k \leq 3$ and $h < 0.5(hmax)$) then set $h = 2h$.
(Increase step size if possible.)

Step 21 STOP. ■

Example 1 Use the extrapolation method with maximum step size $hmax = 0.2$, minimum step size $hmin = 0.01$, and tolerance $TOL = 10^{-9}$ to approximate the solution of the initial-value problem

$$y' = y - t^2 + 1, \quad 0 \leq t \leq 2, \quad y(0) = 0.5.$$

Solution For the first step of the extrapolation method we let $w_0 = 0.5$, $t_0 = 0$ and $h = 0.2$. Then we compute

$$h_0 = h/2 = 0.1;$$

$$w_1 = w_0 + h_0 f(t_0, w_0) = 0.5 + 0.1(1.5) = 0.65;$$

$$w_2 = w_0 + 2h_0 f(t_0 + h_0, w_1) = 0.5 + 0.2(1.64) = 0.828;$$

and the first approximation to $y(0.2)$ is

$$y_{11} = \frac{1}{2}(w_2 + w_1 + h_0 f(t_0 + 2h_0, w_2)) = \frac{1}{2}(0.828 + 0.65 + 0.1 f(0.2, 0.828)) = 0.8284.$$

For the second approximation to $y(0.2)$ we compute

$$h_1 = h/4 = 0.05;$$

$$w_1 = w_0 + h_1 f(t_0, w_0) = 0.5 + 0.05(1.5) = 0.575;$$

$$w_2 = w_0 + 2h_1 f(t_0 + h_1, w_1) = 0.5 + 0.1(1.5725) = 0.65725;$$

$$w_3 = w_1 + 2h_1 f(t_0 + 2h_1, w_2) = 0.575 + 0.1(1.64725) = 0.739725;$$

$$w_4 = w_2 + 2h_1 f(t_0 + 3h_1, w_3) = 0.65725 + 0.1(1.717225) = 0.8289725.$$

Then the endpoint correction approximation is

$$y_{21} = \frac{1}{2}(w_4 + w_3 + h_1 f(t_0 + 4h_1, w_4))$$
$$= \frac{1}{2}(0.8289725 + 0.739725 + 0.05 f(0.2, 0.8289725)) = 0.8290730625.$$

This gives the first extrapolation approximation

$$y_{22} = y_{21} + \left(\frac{(1/4)^2}{(1/2)^2 - (1/4)^2}\right)(y_{21} - y_{11}) = 0.8292974167.$$

The third approximation is found by computing

$$h_2 = h/6 = 0.0\overline{3};$$
$$w_1 = w_0 + h_2 f(t_0, w_0) = 0.55;$$
$$w_2 = w_0 + 2h_2 f(t_0 + h_2, w_1) = 0.6032592593;$$
$$w_3 = w_1 + 2h_2 f(t_0 + 2h_2, w_2) = 0.6565876543;$$
$$w_4 = w_2 + 2h_2 f(t_0 + 3h_2, w_3) = 0.7130317696;$$
$$w_5 = w_3 + 2h_2 f(t_0 + 4h_2, w_4) = 0.7696045871;$$
$$w_6 = w_4 + 2h_2 f(t_0 + 5h_2, w_4) = 0.8291535569;$$

then the end-point correction approximation

$$y_{31} = \frac{1}{2}(w_6 + w_5 + h_2 f(t_0 + 6h_2, w_6) = 0.8291982979.$$

We can now find two extrapolated approximations,

$$y_{32} = y_{31} + \left(\frac{(1/6)^2}{(1/4)^2 - (1/6)^2}\right)(y_{31} - y_{21}) = 0.8292984862,$$

and

$$y_{33} = y_{32} + \left(\frac{(1/6)^2}{(1/2)^2 - (1/6)^2}\right)(y_{32} - y_{22}) = 0.8292986199.$$

Because

$$|y_{33} - y_{22}| = 1.2 \times 10^{-6}$$

does not satisfy the tolerance, we need to compute at least one more row of the extrapolation table. We use $h_3 = h/8 = 0.025$ and calculate w_1 by Euler's method, $w_2, \cdots, w_8$ by the moidpoint method and apply the endpoint correction. This will give us the new approximation y_{41} which permits us to compute the new extrapolation row

$$y_{41} = 0.8292421745 \quad y_{42} = 0.8292985873 \quad y_{43} = 0.8292986210 \quad y_{44} = 0.8292986211$$

Comparing $|y_{44} - y_{33}| = 1.2 \times 10^{-9}$ we find that the accuracy tolerance has not been reached. To obtain the entries in the next row, we use $h_4 = h/12 = 0.0\overline{6}$. First calculate w_1 by Euler's method, then w_2 through w_{12} by the Midpoint method. Finally use the endpoint correction to obtain y_{51}. The remaining entries in the fifth row are obtained using extrapolation, and are shown in Table 5.17. Because $y_{55} = 0.8292986213$ is within 10^{-9} of y_{44} it is accepted as the approximation to $y(0.2)$. The procedure begins anew to approximate $y(0.4)$. The complete set of approximations accurate to the places listed is given in Table 5.18. ■

Table 5.17

$y_{1,1} = 0.8284000000$				
$y_{2,1} = 0.8290730625$	$y_{2,2} = 0.8292974167$			
$y_{3,1} = 0.8291982979$	$y_{3,2} = 0.8292984862$	$y_{3,3} = 0.8292986199$		
$y_{4,1} = 0.8292421745$	$y_{4,2} = 0.8292985873$	$y_{4,3} = 0.8292986210$	$y_{4,4} = 0.8292986211$	
$y_{5,1} = 0.8292735291$	$y_{5,2} = 0.8292986128$	$y_{5,3} = 0.8292986213$	$y_{5,4} = 0.8292986213$	$y_{5,5} = 0.8292986213$

Table 5.18

t_i	$y_i = y(t_i)$	w_i	h_i	k
0.200	0.8292986210	0.8292986213	0.200	5
0.400	1.2140876512	1.2140876510	0.200	4
0.600	1.6489405998	1.6489406000	0.200	4
0.700	1.8831236462	1.8831236460	0.100	5
0.800	2.1272295358	2.1272295360	0.100	4
0.900	2.3801984444	2.3801984450	0.100	7
0.925	2.4446908698	2.4446908710	0.025	8
0.950	2.5096451704	2.5096451700	0.025	3
1.000	2.6408590858	2.6408590860	0.050	3
1.100	2.9079169880	2.9079169880	0.100	7
1.200	3.1799415386	3.1799415380	0.100	6
1.300	3.4553516662	3.4553516610	0.100	8
1.400	3.7324000166	3.7324000100	0.100	5
1.450	3.8709427424	3.8709427340	0.050	7
1.475	3.9401071136	3.9401071050	0.025	3
1.525	4.0780532154	4.0780532060	0.050	4
1.575	4.2152541820	4.2152541820	0.050	3
1.675	4.4862274254	4.4862274160	0.100	4
1.775	4.7504844318	4.7504844210	0.100	4
1.825	4.8792274904	4.8792274790	0.050	3
1.875	5.0052154398	5.0052154290	0.050	3
1.925	5.1280506670	5.1280506570	0.050	4
1.975	5.2473151731	5.2473151660	0.050	8
2.000	5.3054719506	5.3054719440	0.025	3

The proof that the method presented in Algorithm 5.6 converges involves results from summability theory; it can be found in the original paper of Gragg [Gr]. A number of other extrapolation procedures are available, some of which use the variable step-size techniques. For additional procedures based on the extrapolation process, see the Bulirsch and Stoer papers [BS1], [BS2], [BS3] or the text by Stetter [Stet]. The methods used by Bulirsch and Stoer involve interpolation with rational functions instead of the polynomial interpolation used in the Gragg procedure.

EXERCISE SET 5.8

1. Use the Extrapolation Algorithm with tolerance $TOL = 10^{-4}$, $hmax = 0.25$, and $hmin = 0.05$ to approximate the solutions to the following initial-value problems. Compare the results to the actual values.
   a. $y' = te^{3t} - 2y,\quad 0 \le t \le 1,\quad y(0) = 0$; actual solution $y(t) = \frac{1}{5}te^{3t} - \frac{1}{25}e^{3t} + \frac{1}{25}e^{-2t}$.
   b. $y' = 1 + y/t,\quad 1 \le t \le 2,\quad y(1) = 2$; actual solution $y(t) = t\ln t + 2t$.

c. $y' = 1 + (t - y)^2, \quad 2 \le t \le 3, \quad y(2) = 1$; actual solution $y(t) = t + 1/(1 - t)$.

d. $y' = \cos 2t + \sin 3t, \quad 0 \le t \le 1, \quad y(0) = 1$; actual solution $y(t) = \frac{1}{2}\sin 2t - \frac{1}{3}\cos 3t + \frac{4}{3}$.

2. Use the Extrapolation Algorithm with $TOL = 10^{-4}$ to approximate the solutions to the following initial-value problems:

a. $y' = \sin t + e^{-t}, \quad 0 \le t \le 1, \quad y(0) = 0$, with $hmax = 0.25$ and $hmin = 0.02$.

b. $y' = (y/t)^2 + y/t, \quad 1 \le t \le 1.2, \quad y(1) = 1$, with $hmax = 0.05$ and $hmin = 0.02$.

c. $y' = (y^2 + y)/t, \quad 1 \le t \le 3, \quad y(1) = -2$, with $hmax = 0.5$ and $hmin = 0.02$.

d. $y' = -ty + 4t/y, \quad 0 \le t \le 1, \quad y(0) = 1$, with $hmax = 0.25$ and $hmin = 0.02$.

3. Use the Extrapolation Algorithm with tolerance $TOL = 10^{-6}$, $hmax = 0.5$, and $hmin = 0.05$ to approximate the solutions to the following initial-value problems. Compare the results to the actual values.

a. $y' = 1 + y/t + (y/t)^2, \quad 1 \le t \le 3, \quad y(1) = 0$; actual solution $y(t) = t\tan(\ln t)$.

b. $y' = y/t - (y/t)^2, \quad 1 \le t \le 4, \quad y(1) = 1$; actual solution $y(t) = t/(1 + \ln t)$.

c. $y' = -(y + 1)(y + 3), \quad 0 \le t \le 3, \quad y(0) = -2$; actual solution $y(t) = -3 + 2(1 + e^{-2t})^{-1}$.

d. $y' = (t + 2t^3)y^3 - ty, \quad 0 \le t \le 2, \quad y(0) = \frac{1}{3}$; actual solution $y(t) = (3 + 2t^2 + 6e^{t^2})^{-1/2}$.

4. Let $P(t)$ be the number of individuals in a population at time t, measured in years. If the average birth rate b is constant and the average death rate d is proportional to the size of the population (due to overcrowding), then the growth rate of the population is given by the **logistic equation**

$$\frac{dP(t)}{dt} = bP(t) - k[P(t)]^2,$$

where $d = kP(t)$. Suppose $P(0) = 50,976$, $b = 2.9 \times 10^{-2}$, and $k = 1.4 \times 10^{-7}$. Find the population after 5 years.

5.9 Higher-Order Equations and Systems of Differential Equations

This section contains an introduction to the numerical solution of higher-order initial-value problems. The techniques discussed are limited to those that transform a higher-order equation into a system of first-order differential equations. Before discussing the transformation procedure, some remarks are needed concerning systems that involve first-order differential equations.

An ***m*th-order system** of first-order initial-value problems has the form

$$\begin{aligned}
\frac{du_1}{dt} &= f_1(t, u_1, u_2, \ldots, u_m), \\
\frac{du_2}{dt} &= f_2(t, u_1, u_2, \ldots, u_m), \\
&\vdots \\
\frac{du_m}{dt} &= f_m(t, u_1, u_2, \ldots, u_m),
\end{aligned} \tag{5.45}$$

for $a \le t \le b$, with the initial conditions

$$u_1(a) = \alpha_1, \quad u_2(a) = \alpha_2, \quad \ldots, \quad u_m(a) = \alpha_m. \tag{5.46}$$

The object is to find m functions $u_1(t), u_2(t), \ldots, u_m(t)$ that satisfy each of the differential equations together with all the initial conditions.

To discuss existence and uniqueness of solutions to systems of equations, we need to extend the definition of the Lipschitz condition to functions of several variables.

Definition 5.16 The function $f(t, y_1, \ldots, y_m)$, defined on the set

$$D = \{(t, u_1, \ldots, u_m) \mid a \le t \le b \text{ and } -\infty < u_i < \infty, \text{ for each } i = 1, 2, \ldots, m\}$$

is said to satisfy a **Lipschitz condition** on D in the variables $u_1, u_2, \ldots, u_m$ if a constant $L > 0$ exists with

$$|f(t, u_1, \ldots, u_m) - f(t, z_1, \ldots, z_m)| \le L \sum_{j=1}^{m} |u_j - z_j|, \tag{5.47}$$

for all $(t, u_1, \ldots, u_m)$ and $(t, z_1, \ldots, z_m)$ in D. ■

By using the Mean Value Theorem, it can be shown that if f and its first partial derivatives are continuous on D and if

$$\left| \frac{\partial f(t, u_1, \ldots, u_m)}{\partial u_i} \right| \le L,$$

for each $i = 1, 2, \ldots, m$ and all $(t, u_1, \ldots, u_m)$ in D, then f satisfies a Lipschitz condition on D with Lipschitz constant L (see [BiR], p. 141). A basic existence and uniqueness theorem follows. Its proof can be found in [BiR], pp. 152–154.

Theorem 5.17 Suppose that

$$D = \{(t, u_1, u_2, \ldots, u_m) \mid a \le t \le b \text{ and } -\infty < u_i < \infty, \text{ for each } i = 1, 2, \ldots, m\},$$

and let $f_i(t, u_1, \ldots, u_m)$, for each $i = 1, 2, \ldots, m$, be continuous and satisfy a Lipschitz condition on D. The system of first-order differential equations (5.45), subject to the initial conditions (5.46), has a unique solution $u_1(t), \ldots, u_m(t)$, for $a \le t \le b$. ■

Methods to solve systems of first-order differential equations are generalizations of the methods for a single first-order equation presented earlier in this chapter. For example, the classical Runge-Kutta method of order four given by

$$\begin{aligned}
w_0 &= \alpha, \\
k_1 &= hf(t_i, w_i), \\
k_2 &= hf\left(t_i + \frac{h}{2}, w_i + \frac{1}{2}k_1\right), \\
k_3 &= hf\left(t_i + \frac{h}{2}, w_i + \frac{1}{2}k_2\right), \\
k_4 &= hf(t_{i+1}, w_i + k_3), \\
w_{i+1} &= w_i + \frac{1}{6}(k_1 + 2k_2 + 2k_3 + k_4), \quad \text{for each } i = 0, 1, \ldots, N-1,
\end{aligned}$$

used to solve the first-order initial-value problem

$$y' = f(t, y), \quad a \le t \le b, \quad y(a) = \alpha,$$

is generalized as follows.

Let an integer $N > 0$ be chosen and set $h = (b - a)/N$. Partition the interval $[a, b]$ into N subintervals with the mesh points

$$t_j = a + jh, \quad \text{for each } j = 0, 1, \ldots, N.$$

Use the notation w_{ij}, for each $j = 0, 1, \ldots, N$ and $i = 1, 2, \ldots, m$, to denote an approximation to $u_i(t_j)$. That is, w_{ij} approximates the ith solution $u_i(t)$ of (5.45) at the jth mesh point t_j. For the initial conditions, set (see Figure 5.6)

$$w_{1,0} = \alpha_1, \; w_{2,0} = \alpha_2, \; \ldots, \; w_{m,0} = \alpha_m. \tag{5.48}$$

Figure 5.6

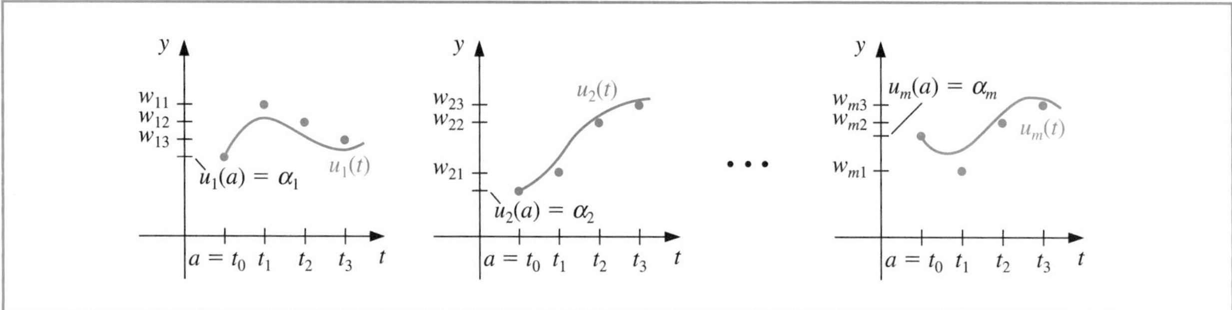

Suppose that the values $w_{1,j}, w_{2,j}, \ldots, w_{m,j}$ have been computed. We obtain $w_{1,j+1}$, $w_{2,j+1}, \ldots, w_{m,j+1}$ by first calculating

$$k_{1,i} = hf_i(t_j, w_{1,j}, w_{2,j}, \ldots, w_{m,j}), \quad \text{for each } i = 1, 2, \ldots, m; \tag{5.49}$$

$$k_{2,i} = hf_i\left(t_j + \frac{h}{2}, w_{1,j} + \frac{1}{2}k_{1,1}, w_{2,j} + \frac{1}{2}k_{1,2}, \ldots, w_{m,j} + \frac{1}{2}k_{1,m}\right), \tag{5.50}$$

for each $i = 1, 2, \ldots, m$;

$$k_{3,i} = hf_i\left(t_j + \frac{h}{2}, w_{1,j} + \frac{1}{2}k_{2,1}, w_{2,j} + \frac{1}{2}k_{2,2}, \ldots, w_{m,j} + \frac{1}{2}k_{2,m}\right), \tag{5.51}$$

for each $i = 1, 2, \ldots, m$;

$$k_{4,i} = hf_i(t_j + h, w_{1,j} + k_{3,1}, w_{2,j} + k_{3,2}, \ldots, w_{m,j} + k_{3,m}), \tag{5.52}$$

for each $i = 1, 2, \ldots, m$; and then

$$w_{i,j+1} = w_{i,j} + \frac{1}{6}(k_{1,i} + 2k_{2,i} + 2k_{3,i} + k_{4,i}), \tag{5.53}$$

for each $i = 1, 2, \ldots, m$. Note that all the values $k_{1,1}, k_{1,2}, \ldots, k_{1,m}$ must be computed before any of the terms of the form $k_{2,i}$ can be determined. In general, each $k_{l,1}, k_{l,2}, \ldots, k_{l,m}$ must be computed before any of the expressions $k_{l+1,i}$. Algorithm 5.7 implements the Runge-Kutta fourth-order method for systems of initial-value problems.

ALGORITHM 5.7

Runge-Kutta Method for Systems of Differential Equations

To approximate the solution of the mth-order system of first-order initial-value problems

$$u_j' = f_j(t, u_1, u_2, \ldots, u_m), \quad a \le t \le b, \quad \text{with} \quad u_j(a) = \alpha_j,$$

for $j = 1, 2, \ldots, m$ at $(N+1)$ equally spaced numbers in the interval $[a, b]$:

INPUT endpoints a, b; number of equations m; integer N; initial conditions $\alpha_1, \ldots, \alpha_m$.

OUTPUT approximations w_j to $u_j(t)$ at the $(N+1)$ values of t.

Step 1 Set $h = (b-a)/N$;
$t = a$.

Step 2 For $j = 1, 2, \ldots, m$ set $w_j = \alpha_j$.

Step 3 OUTPUT $(t, w_1, w_2, \ldots, w_m)$.

Step 4 For $i = 1, 2, \ldots, N$ do steps 5–11.

Step 5 For $j = 1, 2, \ldots, m$ set
$k_{1,j} = hf_j(t, w_1, w_2, \ldots, w_m)$.

Step 6 For $j = 1, 2, \ldots, m$ set
$k_{2,j} = hf_j\left(t + \frac{h}{2}, w_1 + \frac{1}{2}k_{1,1}, w_2 + \frac{1}{2}k_{1,2}, \ldots, w_m + \frac{1}{2}k_{1,m}\right)$.

Step 7 For $j = 1, 2, \ldots, m$ set
$k_{3,j} = hf_j\left(t + \frac{h}{2}, w_1 + \frac{1}{2}k_{2,1}, w_2 + \frac{1}{2}k_{2,2}, \ldots, w_m + \frac{1}{2}k_{2,m}\right)$.

Step 8 For $j = 1, 2, \ldots, m$ set
$k_{4,j} = hf_j(t + h, w_1 + k_{3,1}, w_2 + k_{3,2}, \ldots, w_m + k_{3,m})$.

Step 9 For $j = 1, 2, \ldots, m$ set
$w_j = w_j + (k_{1,j} + 2k_{2,j} + 2k_{3,j} + k_{4,j})/6$.

Step 10 Set $t = a + ih$.

Step 11 OUTPUT $(t, w_1, w_2, \ldots, w_m)$.

Step 12 STOP. ■

Illustration Kirchhoff's Law states that the sum of all instantaneous voltage changes around a closed circuit is zero. This law implies that the current $I(t)$ in a closed circuit containing a resistance of R ohms, a capacitance of C farads, an inductance of L henries, and a voltage source of $E(t)$ volts satisfies the equation

$$LI'(t) + RI(t) + \frac{1}{C}\int I(t)\,dt = E(t).$$

The currents $I_1(t)$ and $I_2(t)$ in the left and right loops, respectively, of the circuit shown in Figure 5.7 are the solutions to the system of equations

$$2I_1(t) + 6[I_1(t) - I_2(t)] + 2I_1'(t) = 12,$$

$$\frac{1}{0.5}\int I_2(t)\,dt + 4I_2(t) + 6[I_2(t) - I_1(t)] = 0.$$

Figure 5.7

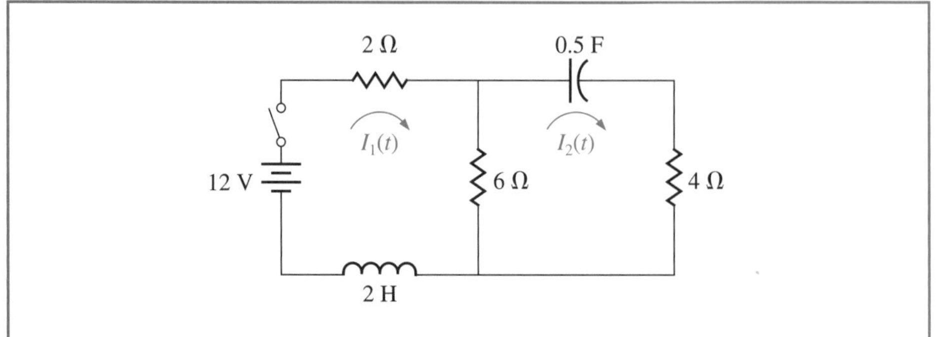

If the switch in the circuit is closed at time $t = 0$, we have the initial conditions $I_1(0) = 0$ and $I_2(0) = 0$. Solve for $I_1'(t)$ in the first equation, differentiate the second equation, and substitute for $I_1'(t)$ to get

$$I_1' = f_1(t, I_1, I_2) = -4I_1 + 3I_2 + 6, \quad I_1(0) = 0,$$
$$I_2' = f_2(t, I_1, I_2) = 0.6I_1' - 0.2I_2 = -2.4I_1 + 1.6I_2 + 3.6, \quad I_2(0) = 0.$$

The exact solution to this system is

$$I_1(t) = -3.375e^{-2t} + 1.875e^{-0.4t} + 1.5,$$
$$I_2(t) = -2.25e^{-2t} + 2.25e^{-0.4t}.$$

We will apply the Runge-Kutta method of order four to this system with $h = 0.1$. Since $w_{1,0} = I_1(0) = 0$ and $w_{2,0} = I_2(0) = 0$,

$$k_{1,1} = hf_1(t_0, w_{1,0}, w_{2,0}) = 0.1\ f_1(0, 0, 0) = 0.1\,(-4(0) + 3(0) + 6) = 0.6,$$
$$k_{1,2} = hf_2(t_0, w_{1,0}, w_{2,0}) = 0.1\ f_2(0, 0, 0) = 0.1\,(-2.4(0) + 1.6(0) + 3.6) = 0.36,$$
$$k_{2,1} = hf_1\left(t_0 + \frac{1}{2}h, w_{1,0} + \frac{1}{2}k_{1,1}, w_{2,0} + \frac{1}{2}k_{1,2}\right) = 0.1\ f_1(0.05, 0.3, 0.18)$$
$$= 0.1\,(-4(0.3) + 3(0.18) + 6) = 0.534,$$
$$k_{2,2} = hf_2\left(t_0 + \frac{1}{2}h, w_{1,0} + \frac{1}{2}k_{1,1}, w_{2,0} + \frac{1}{2}k_{1,2}\right) = 0.1\ f_2(0.05, 0.3, 0.18)$$
$$= 0.1\,(-2.4(0.3) + 1.6(0.18) + 3.6) = 0.3168.$$

Generating the remaining entries in a similar manner produces

$$k_{3,1} = (0.1)f_1(0.05, 0.267, 0.1584) = 0.54072,$$
$$k_{3,2} = (0.1)f_2(0.05, 0.267, 0.1584) = 0.321264,$$
$$k_{4,1} = (0.1)f_1(0.1, 0.54072, 0.321264) = 0.4800912,$$
$$k_{4,2} = (0.1)f_2(0.1, 0.54072, 0.321264) = 0.28162944.$$

As a consequence,

$$I_1(0.1) \approx w_{1,1} = w_{1,0} + \frac{1}{6}(k_{1,1} + 2k_{2,1} + 2k_{3,1} + k_{4,1})$$
$$= 0 + \frac{1}{6}(0.6 + 2(0.534) + 2(0.54072) + 0.4800912) = 0.5382552$$

and

$$I_2(0.1) \approx w_{2,1} = w_{2,0} + \frac{1}{6}(k_{1,2} + 2k_{2,2} + 2k_{3,2} + k_{4,2}) = 0.3196263.$$

The remaining entries in Table 5.19 are generated in a similar manner. □

Table 5.19

t_j	$w_{1,j}$	$w_{2,j}$	$\|I_1(t_j) - w_{1,j}\|$	$\|I_2(t_j) - w_{2,j}\|$
0.0	0	0	0	0
0.1	0.5382550	0.3196263	0.8285×10^{-5}	0.5803×10^{-5}
0.2	0.9684983	0.5687817	0.1514×10^{-4}	0.9596×10^{-5}
0.3	1.310717	0.7607328	0.1907×10^{-4}	0.1216×10^{-4}
0.4	1.581263	0.9063208	0.2098×10^{-4}	0.1311×10^{-4}
0.5	1.793505	1.014402	0.2193×10^{-4}	0.1240×10^{-4}

Recall that Maple reserves the letter D to represent differentiation.

Maple's *NumericalAnalysis* package does not currently approximate the solution to systems of initial value problems, but systems of first-order differential equations can by solved using *dsolve*. The system in the Illustration is defined with

$sys\,2 := D(u1)(t) = -4u1(t) + 3u2(t) + 6,\ D(u2)(t) = -2.4u1(t) + 1.6u2(t) + 3.6$

and the initial conditions with

$init\,2 := u1(0) = 0,\ u2(0) = 0$

The system is solved with the command

$sol\,2 := dsolve(\{sys\,2, init\,2\}, \{u1(t), u2(t)\})$

and Maple responds with

$$\left\{u1(t) = -\frac{27}{8}e^{-2t} + \frac{15}{8}e^{-\frac{5}{2}t} + \frac{3}{2},\ u2(t) = -\frac{9}{4}e^{-2t} + \frac{9}{4}e^{-\frac{5}{2}t}\right\}$$

To isolate the individual functions we use

$r1 := rhs(sol\,2[1]);\ r2 := rhs(sol\,2[2])$

producing

$$-\frac{27}{8}e^{-2t} + \frac{15}{8}e^{-\frac{5}{2}t} + \frac{3}{2}$$
$$-\frac{9}{4}e^{-2t} + \frac{9}{4}e^{-\frac{5}{2}t}$$

and to determine the value of the functions at $t = 0.5$ we use

$evalf(subs(t = 0.5, r1));\ evalf(subs(t = 0.5, r2))$

giving, in agreement with Table 5.19,

$$1.793527048$$

$$1.014415451$$

The command *dsolve* will fail if an explicit solution cannot be found. In that case we can use the numeric option in *dsolve*, which applies the Runge-Kutta-Fehlberg technique. This technique can also be used, of course, when the exact solution can be determined with *dsolve*. For example, with the system defined previously,

$g := dsolve(\{sys\,2, init\,2\}, \{u1(t), u2(t)\}, numeric)$

returns

$$\textbf{proc}(x\_rkf45)\ldots\textbf{end proc}$$

To approximate the solutions at $t = 0.5$, enter

$g(0.5)$

which gives approximations in the form

$$[t = 0.5,\ u2(t) = 1.014415563,\ u1(t) = 1.793527215]$$

Higher-Order Differential Equations

Many important physical problems—for example, electrical circuits and vibrating systems—involve initial-value problems whose equations have orders higher than one. New techniques are not required for solving these problems. By relabeling the variables, we can reduce a higher-order differential equation into a system of first-order differential equations and then apply one of the methods we have already discussed.

A general mth-order initial-value problem

$$y^{(m)}(t) = f(t, y, y', \ldots, y^{(m-1)}), \quad a \le t \le b,$$

with initial conditions $y(a) = \alpha_1, y'(a) = \alpha_2, \ldots, y^{(m-1)}(a) = \alpha_m$ can be converted into a system of equations in the form (5.45) and (5.46).

Let $u_1(t) = y(t), u_2(t) = y'(t), \ldots,$ and $u_m(t) = y^{(m-1)}(t)$. This produces the first-order system

$$\frac{du_1}{dt} = \frac{dy}{dt} = u_2, \quad \frac{du_2}{dt} = \frac{dy'}{dt} = u_3, \quad \ldots, \quad \frac{du_{m-1}}{dt} = \frac{dy^{(m-2)}}{dt} = u_m,$$

and

$$\frac{du_m}{dt} = \frac{dy^{(m-1)}}{dt} = y^{(m)} = f(t, y, y', \ldots, y^{(m-1)}) = f(t, u_1, u_2, \ldots, u_m),$$

with initial conditions

$$u_1(a) = y(a) = \alpha_1, \quad u_2(a) = y'(a) = \alpha_2, \quad \ldots, \quad u_m(a) = y^{(m-1)}(a) = \alpha_m.$$

Example 1 Transform the the second-order initial-value problem

$$y'' - 2y' + 2y = e^{2t}\sin t, \qquad \text{for } 0 \le t \le 1, \quad \text{with } y(0) = -0.4,\ y'(0) = -0.6$$

into a system of first order initial-value problems, and use the Runge-Kutta method with $h = 0.1$ to approximate the solution.

Solution Let $u_1(t) = y(t)$ and $u_2(t) = y'(t)$. This transforms the second-order equation into the system

$$u_1'(t) = u_2(t),$$
$$u_2'(t) = e^{2t}\sin t - 2u_1(t) + 2u_2(t),$$

with initial conditions $u_1(0) = -0.4$, $u_2(0) = -0.6$.

The initial conditions give $w_{1,0} = -0.4$ and $w_{2,0} = -0.6$. The Runge-Kutta Eqs. (5.49) through (5.52) on page 330 with $j = 0$ give

$$k_{1,1} = hf_1(t_0, w_{1,0}, w_{2,0}) = hw_{2,0} = -0.06,$$

$$k_{1,2} = hf_2(t_0, w_{1,0}, w_{2,0}) = h\left[e^{2t_0}\sin t_0 - 2w_{1,0} + 2w_{2,0}\right] = -0.04,$$

$$k_{2,1} = hf_1\left(t_0 + \frac{h}{2}, w_{1,0} + \frac{1}{2}k_{1,1}, w_{2,0} + \frac{1}{2}k_{1,2}\right) = h\left[w_{2,0} + \frac{1}{2}k_{1,2}\right] = -0.062,$$

$$k_{2,2} = hf_2\left(t_0 + \frac{h}{2}, w_{1,0} + \frac{1}{2}k_{1,1}, w_{2,0} + \frac{1}{2}k_{1,2}\right)$$
$$= h\left[e^{2(t_0+0.05)}\sin(t_0 + 0.05) - 2\left(w_{1,0} + \frac{1}{2}k_{1,1}\right) + 2\left(w_{2,0} + \frac{1}{2}k_{1,2}\right)\right]$$
$$= -0.03247644757,$$

$$k_{3,1} = h\left[w_{2,0} + \frac{1}{2}k_{2,2}\right] = -0.06162832238,$$

$$k_{3,2} = h\left[e^{2(t_0+0.05)}\sin(t_0 + 0.05) - 2\left(w_{1,0} + \frac{1}{2}k_{2,1}\right) + 2\left(w_{2,0} + \frac{1}{2}k_{2,2}\right)\right]$$
$$= -0.03152409237,$$

$$k_{4,1} = h\left[w_{2,0} + k_{3,2}\right] = -0.06315240924,$$

and

$$k_{4,2} = h\left[e^{2(t_0+0.1)}\sin(t_0 + 0.1) - 2(w_{1,0} + k_{3,1}) + 2(w_{2,0} + k_{3,2})\right] = -0.02178637298.$$

So

$$w_{1,1} = w_{1,0} + \frac{1}{6}(k_{1,1} + 2k_{2,1} + 2k_{3,1} + k_{4,1}) = -0.4617333423$$

and

$$w_{2,1} = w_{2,0} + \frac{1}{6}(k_{1,2} + 2k_{2,2} + 2k_{3,2} + k_{4,2}) = -0.6316312421.$$

The value $w_{1,1}$ approximates $u_1(0.1) = y(0.1) = 0.2e^{2(0.1)}(\sin 0.1 - 2\cos 0.1)$, and $w_{2,1}$ approximates $u_2(0.1) = y'(0.1) = 0.2e^{2(0.1)}(4\sin 0.1 - 3\cos 0.1)$.

The set of values $w_{1,j}$ and $w_{2,j}$, for $j = 0, 1, \ldots, 10$, are presented in Table 5.20 and are compared to the actual values of $u_1(t) = 0.2e^{2t}(\sin t - 2\cos t)$ and $u_2(t) = u_1'(t) = 0.2e^{2t}(4\sin t - 3\cos t)$. ■

Table 5.20

t_j	$y(t_j) = u_1(t_j)$	$w_{1,j}$	$y'(t_j) = u_2(t_j)$	$w_{2,j}$	$\|y(t_j) - w_{1,j}\|$	$\|y'(t_j) - w_{2,j}\|$
0.0	−0.40000000	−0.40000000	−0.6000000	−0.60000000	0	0
0.1	−0.46173297	−0.46173334	−0.6316304	−0.63163124	3.7×10^{-7}	7.75×10^{-7}
0.2	−0.52555905	−0.52555988	−0.6401478	−0.64014895	8.3×10^{-7}	1.01×10^{-6}
0.3	−0.58860005	−0.58860144	−0.6136630	−0.61366381	1.39×10^{-6}	8.34×10^{-7}
0.4	−0.64661028	−0.64661231	−0.5365821	−0.53658203	2.03×10^{-6}	1.79×10^{-7}
0.5	−0.69356395	−0.69356666	−0.3887395	−0.38873810	2.71×10^{-6}	5.96×10^{-7}
0.6	−0.72114849	−0.72115190	−0.1443834	−0.14438087	3.41×10^{-6}	7.75×10^{-7}
0.7	−0.71814890	−0.71815295	0.2289917	0.22899702	4.05×10^{-6}	2.03×10^{-6}
0.8	−0.66970677	−0.66971133	0.7719815	0.77199180	4.56×10^{-6}	5.30×10^{-6}
0.9	−0.55643814	−0.55644290	1.534764	1.5347815	4.76×10^{-6}	9.54×10^{-6}
1.0	−0.35339436	−0.35339886	2.578741	2.5787663	4.50×10^{-6}	1.34×10^{-5}

In Maple the nth derivative $y^{(n)}(t)$ is specified by $(D@@n)(y)(t)$.

We can also use *dsolve* from Maple on higher-order equations. To define the differential equation in Example 1, use

$def2 := (D@@2)(y)(t) - 2D(y)(t) + 2y(t) = e^{2t}\sin(t)$

and to specify the initial conditions use

$init2 := y(0) = -0.4, D(y)(0) = -0.6$

The solution is obtained with the command

$sol2 := dsolve(\{def2, init2\}, y(t))$

to obtain

$$y(t) = \frac{1}{5}e^{2t}(\sin(t) - 2\cos(t))$$

We isolate the solution in function form using

$g := rhs(sol2)$

To obtain $y(1.0) = g(1.0)$, enter

$evalf(subs(t = 1.0, g))$

which gives −0.3533943574.

Runge-Kutta-Fehlberg is also available for higher-order equations via the *dsolve* command with the numeric option. It is employed in the same manner as illustrated for systems of equations.

The other one-step methods can be extended to systems in a similar way. When error control methods like the Runge-Kutta-Fehlberg method are extended, each component of the numerical solution $(w_{1j}, w_{2j}, \ldots, w_{mj})$ must be examined for accuracy. If any of the components fail to be sufficiently accurate, the entire numerical solution $(w_{1j}, w_{2j}, \ldots, w_{mj})$ must be recomputed.

The multistep methods and predictor-corrector techniques can also be extended to systems. Again, if error control is used, each component must be accurate. The extension of the extrapolation technique to systems can also be done, but the notation becomes quite involved. If this topic is of interest, see [HNW1].

Convergence theorems and error estimates for systems are similar to those considered in Section 5.10 for the single equations, except that the bounds are given in terms of vector norms, a topic considered in Chapter 7. (A good reference for these theorems is [Ge1], pp. 45–72.)

EXERCISE SET 5.9

1. Use the Runge-Kutta method for systems to approximate the solutions of the following systems of first-order differential equations, and compare the results to the actual solutions.

   a. $u_1' = -4u_1 - 2u_2 + \cos t + 4\sin t, \quad u_1(0) = 0;$
   $u_2' = 3u_1 + u_2 - 3\sin t, \quad u_2(0) = -1; \quad 0 \le t \le 2; \quad h = 0.1;$
   actual solutions $u_1(t) = 2e^{-t} - 2e^{-2t} + \sin t$ and $u_2(t) = -3e^{-t} + 2e^{-2t}$.

   b. $u_1' = 3u_1 + 2u_2 - (2t^2 + 1)e^{2t}, \quad u_1(0) = 1;$
   $u_2' = 4u_1 + u_2 + (t^2 + 2t - 4)e^{2t}, \quad u_2(0) = 1; \quad 0 \le t \le 1; \quad h = 0.2;$
   actual solutions $u_1(t) = \frac{1}{3}e^{5t} - \frac{1}{3}e^{-t} + e^{2t}$ and $u_2(t) = \frac{1}{3}e^{5t} + \frac{2}{3}e^{-t} + t^2e^{2t}$.

   c. $u_1' = u_2, \quad u_1(0) = 1;$
   $u_2' = -u_1 - 2e^t + 1, \quad u_2(0) = 0;$
   $u_3' = -u_1 - e^t + 1, \quad u_3(0) = 1; \quad 0 \le t \le 2; \quad h = 0.5;$
   actual solutions $u_1(t) = \cos t + \sin t - e^t + 1$, $u_2(t) = -\sin t + \cos t - e^t$, and $u_3(t) = -\sin t + \cos t$.

   d. $u_1' = u_2 - u_3 + t, \quad u_1(0) = 1;$
   $u_2' = 3t^2, \quad u_2(0) = 1;$
   $u_3' = u_2 + e^{-t}, \quad u_3(0) = -1; \quad 0 \le t \le 1; \quad h = 0.1;$
   actual solutions $u_1(t) = -0.05t^5 + 0.25t^4 + t + 2 - e^{-t}$, $u_2(t) = t^3 + 1$, and $u_3(t) = 0.25t^4 + t - e^{-t}$.

2. Use the Runge-Kutta method for systems to approximate the solutions of the following systems of first-order differential equations, and compare the results to the actual solutions.

   a. $u_1' = \frac{1}{9}u_1 - \frac{2}{3}u_2 - \frac{1}{9}t^2 + \frac{2}{3}, \quad u_1(0) = -3;$
   $u_2' = u_2 + 3t - 4, \quad u_2(0) = 5; \quad 0 \le t \le 2; \quad h = 0.2;$
   actual solutions $u_1(t) = -3e^t + t^2$ and $u_2(t) = 4e^t - 3t + 1$.

   b. $u_1' = u_1 - u_2 + 2, \quad u_1(0) = -1;$
   $u_2' = -u_1 + u_2 + 4t, \quad u_2(0) = 0; \quad 0 \le t \le 1; \quad h = 0.1;$
   actual solutions $u_1(t) = -\frac{1}{2}e^{2t} + t^2 + 2t - \frac{1}{2}$ and $u_2(t) = \frac{1}{2}e^{2t} + t^2 - \frac{1}{2}$.

   c. $u_1' = u_1 + 2u_2 - 2u_3 + e^{-t}, \quad u_1(0) = 3;$
   $u_2' = u_2 + u_3 - 2e^{-t}, \quad u_2(0) = -1;$
   $u_3' = u_1 + 2u_2 + e^{-t}, \quad u_3(0) = 1; \quad 0 \le t \le 1; \quad h = 0.1;$
   actual solutions $u_1(t) = -3e^{-t} - 3\sin t + 6\cos t$, $u_2(t) = \frac{3}{2}e^{-t} + \frac{3}{10}\sin t - \frac{21}{10}\cos t - \frac{2}{5}e^{2t}$, and $u_3(t) = -e^{-t} + \frac{12}{5}\cos t + \frac{9}{5}\sin t - \frac{2}{5}e^{2t}$.

   d. $u_1' = 3u_1 + 2u_2 - u_3 - 1 - 3t - 2\sin t, \quad u_1(0) = 5;$
   $u_2' = u_1 - 2u_2 + 3u_3 + 6 - t + 2\sin t + \cos t, \quad u_2(0) = -9;$
   $u_3' = 2u_1 + 4u_3 + 8 - 2t, \quad u_3(0) = -5; \quad 0 \le t \le 2; \quad h = 0.2;$
   actual solutions $u_1(t) = 2e^{3t} + 3e^{-2t} + 1$, $u_2(t) = -8e^{-2t} + e^{4t} - 2e^{3t} + \sin t$, and $u_3(t) = 2e^{4t} - 4e^{3t} - e^{-2t} - 2$.

3. Use the Runge-Kutta for Systems Algorithm to approximate the solutions of the following higher-order differential equations, and compare the results to the actual solutions.

   a. $y'' - 2y' + y = te^t - t, \quad 0 \le t \le 1, \quad y(0) = y'(0) = 0$, with $h = 0.1$;
   actual solution $y(t) = \frac{1}{6}t^3e^t - te^t + 2e^t - t - 2$.

   b. $t^2y'' - 2ty' + 2y = t^3 \ln t, \quad 1 \le t \le 2, \quad y(1) = 1, \quad y'(1) = 0$, with $h = 0.1$;
   actual solution $y(t) = \frac{7}{4}t + \frac{1}{2}t^3 \ln t - \frac{3}{4}t^3$.

   c. $y''' + 2y'' - y' - 2y = e^t, \quad 0 \le t \le 3, \quad y(0) = 1, \quad y'(0) = 2, \quad y''(0) = 0$, with $h = 0.2$;
   actual solution $y(t) = \frac{43}{36}e^t + \frac{1}{4}e^{-t} - \frac{4}{9}e^{-2t} + \frac{1}{6}te^t$.

   d. $t^3y''' - t^2y'' + 3ty' - 4y = 5t^3 \ln t + 9t^3, \quad 1 \le t \le 2, \quad y(1) = 0, \quad y'(1) = 1, \quad y''(1) = 3$, with $h = 0.1$; actual solution $y(t) = -t^2 + t\cos(\ln t) + t\sin(\ln t) + t^3 \ln t$.

4. Use the Runge-Kutta for Systems Algorithm to approximate the solutions of the following higher-order differential equations, and compare the results to the actual solutions.

   a. $y'' - 3y' + 2y = 6e^{-t}$, $0 \le t \le 1$, $y(0) = y'(0) = 2$, with $h = 0.1$; actual solution $y(t) = 2e^{2t} - e^t + e^{-t}$.

   b. $t^2y'' + ty' - 4y = -3t$, $1 \le t \le 3$, $y(1) = 4$, $y'(1) = 3$, with $h = 0.2$; actual solution $y(t) = 2t^2 + t + t^{-2}$.

   c. $y''' + y'' - 4y' - 4y = 0$, $0 \le t \le 2$, $y(0) = 3$, $y'(0) = -1$, $y''(0) = 9$, with $h = 0.2$; actual solution $y(t) = e^{-t} + e^{2t} + e^{-2t}$.

   d. $t^3y''' + t^2y'' - 2ty' + 2y = 8t^3 - 2$, $1 \le t \le 2$, $y(1) = 2$, $y'(1) = 8$, $y''(1) = 6$, with $h = 0.1$; actual solution $y(t) = 2t - t^{-1} + t^2 + t^3 - 1$.

5. Change the Adams Fourth-Order Predictor-Corrector Algorithm to obtain approximate solutions to systems of first-order equations.

6. Repeat Exercise 2 using the algorithm developed in Exercise 5.

7. Repeat Exercise 1 using the algorithm developed in Exercise 5.

8. Suppose the swinging pendulum described in the lead example of this chapter is 2 ft long and that $g = 32.17$ ft/s^2. With $h = 0.1$ s, compare the angle θ obtained for the following two initial-value problems at $t = 0$, 1, and 2 s.

   a. $\dfrac{d^2\theta}{dt^2} + \dfrac{g}{L}\sin\theta = 0$, $\theta(0) = \dfrac{\pi}{6}$, $\theta'(0) = 0$,

   b. $\dfrac{d^2\theta}{dt^2} + \dfrac{g}{L}\theta = 0$, $\theta(0) = \dfrac{\pi}{6}$, $\theta'(0) = 0$,

9. The study of mathematical models for predicting the population dynamics of competing species has its origin in independent works published in the early part of the 20th century by A. J. Lotka and V. Volterra (see, for example, [Lo1], [Lo2], and [Vo]).

   Consider the problem of predicting the population of two species, one of which is a predator, whose population at time t is $x_2(t)$, feeding on the other, which is the prey, whose population is $x_1(t)$. We will assume that the prey always has an adequate food supply and that its birth rate at any time is proportional to the number of prey alive at that time; that is, birth rate (prey) is $k_1x_1(t)$. The death rate of the prey depends on both the number of prey and predators alive at that time. For simplicity, we assume death rate (prey) = $k_2x_1(t)x_2(t)$. The birth rate of the predator, on the other hand, depends on its food supply, $x_1(t)$, as well as on the number of predators available for reproduction purposes. For this reason, we assume that the birth rate (predator) is $k_3x_1(t)x_2(t)$. The death rate of the predator will be taken as simply proportional to the number of predators alive at the time; that is, death rate (predator) = $k_4x_2(t)$.

   Since $x_1'(t)$ and $x_2'(t)$ represent the change in the prey and predator populations, respectively, with respect to time, the problem is expressed by the system of nonlinear differential equations

$$x_1'(t) = k_1x_1(t) - k_2x_1(t)x_2(t) \quad \text{and} \quad x_2'(t) = k_3x_1(t)x_2(t) - k_4x_2(t).$$

   Solve this system for $0 \le t \le 4$, assuming that the initial population of the prey is 1000 and of the predators is 500 and that the constants are $k_1 = 3$, $k_2 = 0.002$, $k_3 = 0.0006$, and $k_4 = 0.5$. Sketch a graph of the solutions to this problem, plotting both populations with time, and describe the physical phenomena represented. Is there a stable solution to this population model? If so, for what values x_1 and x_2 is the solution stable?

10. In Exercise 9 we considered the problem of predicting the population in a predator-prey model. Another problem of this type is concerned with two species competing for the same food supply. If the numbers of species alive at time t are denoted by $x_1(t)$ and $x_2(t)$, it is often assumed that, although the birth rate of each of the species is simply proportional to the number of species alive at that time, the death rate of each species depends on the population of both species. We will assume that the population of a particular pair of species is described by the equations

$$\frac{dx_1(t)}{dt} = x_1(t)[4 - 0.0003x_1(t) - 0.0004x_2(t)] \quad \text{and} \quad \frac{dx_2(t)}{dt} = x_2(t)[2 - 0.0002x_1(t) - 0.0001x_2(t)].$$

If it is known that the initial population of each species is 10,000, find the solution to this system for $0 \leq t \leq 4$. Is there a stable solution to this population model? If so, for what values of x_1 and x_2 is the solution stable?

5.10 Stability

A number of methods have been presented in this chapter for approximating the solution to an initial-value problem. Although numerous other techniques are available, we have chosen the methods described here because they generally satisfied three criteria:

- Their development is clear enough so that you can understand how and why they work.
- One or more of the methods will give satisfactory results for most of the problems that are encountered by students in science and engineering.
- Most of the more advanced and complex techniques are based on one or a combination of the procedures described here.

One-Step Methods

In this section, we discuss why these methods are expected to give satisfactory results when some similar methods do not. Before we begin this discussion, we need to present two definitions concerned with the convergence of one-step difference-equation methods to the solution of the differential equation as the step size decreases.

Definition 5.18 A one-step difference-equation method with local truncation error $\tau_i(h)$ at the ith step is said to be **consistent** with the differential equation it approximates if

$$\lim_{h\to 0} \max_{1\leq i\leq N} |\tau_i(h)| = 0. \quad ■$$

A one-step method is consistent if the difference equation for the method approaches the differential equation as the step size goes to zero.

Note that this definition is a *local* definition since, for each of the values $\tau_i(h)$, we are assuming that the approximation w_{i-1} and the exact solution $y(t_{i-1})$ are the same. A more realistic means of analyzing the effects of making h small is to determine the *global* effect of the method. This is the maximum error of the method over the entire range of the approximation, assuming only that the method gives the exact result at the initial value.

Definition 5.19 A one-step difference-equation method is said to be **convergent** with respect to the differential equation it approximates if

$$\lim_{h\to 0} \max_{1\leq i\leq N} |w_i - y(t_i)| = 0,$$

where $y(t_i)$ denotes the exact value of the solution of the differential equation and w_i is the approximation obtained from the difference method at the ith step. ■

A method is convergent if the solution to the difference equation approaches the solution to the differential equation as the step size goes to zero.

Example 1 Show that Euler's method is convergent.

Solution Examining Inequality (5.10) on page 271, in the error-bound formula for Euler's method, we see that under the hypotheses of Theorem 5.9,

$$\max_{1\leq i\leq N} |w_i - y(t_i)| \leq \frac{Mh}{2L}|e^{L(b-a)} - 1|.$$

However, M, L, a, and b are all constants and

$$\lim_{h\to 0}\max_{1\le i\le N}|w_i - y(t_i)| \le \lim_{h\to 0}\frac{Mh}{2L}\left|e^{L(b-a)} - 1\right| = 0.$$

So Euler's method is convergent with respect to a differential equation satisfying the conditions of this definition. The rate of convergence is $O(h)$. ■

A consistent one-step method has the property that the difference equation for the method approaches the differential equation when the step size goes to zero. So the local truncation error of a consistent method approaches zero as the step size approaches zero.

The other error-bound type of problem that exists when using difference methods to approximate solutions to differential equations is a consequence of not using exact results. In practice, neither the initial conditions nor the arithmetic that is subsequently performed is represented exactly because of the round-off error associated with finite-digit arithmetic. In Section 5.2 we saw that this consideration can lead to difficulties even for the convergent Euler's method.

A method is stable when the results depend continuously on the initial data.

To analyze this situation, at least partially, we will try to determine which methods are **stable**, in the sense that small changes or perturbations in the initial conditions produce correspondingly small changes in the subsequent approximations.

The concept of stability of a one-step difference equation is somewhat analogous to the condition of a differential equation being well-posed, so it is not surprising that the Lipschitz condition appears here, as it did in the corresponding theorem for differential equations, Theorem 5.6 in Section 5.1.

Part (i) of the following theorem concerns the stability of a one-step method. The proof of this result is not difficult and is considered in Exercise 1. Part (ii) of Theorem 5.20 concerns sufficient conditions for a consistent method to be convergent. Part (iii) justifies the remark made in Section 5.5 about controlling the global error of a method by controlling its local truncation error and implies that when the local truncation error has the rate of convergence $O(h^n)$, the global error will have the same rate of convergence. The proofs of parts (ii) and (iii) are more difficult than that of part (i), and can be found within the material presented in [Ge1], pp. 57–58.

Theorem 5.20 Suppose the initial-value problem

$$y' = f(t, y), \quad a \le t \le b, \quad y(a) = \alpha,$$

is approximated by a one-step difference method in the form

$$w_0 = \alpha,$$
$$w_{i+1} = w_i + h\phi(t_i, w_i, h).$$

Suppose also that a number $h_0 > 0$ exists and that $\phi(t, w, h)$ is continuous and satisfies a Lipschitz condition in the variable w with Lipschitz constant L on the set

$$D = \{(t, w, h) \mid a \le t \le b \text{ and } -\infty < w < \infty, 0 \le h \le h_0\}.$$

Then

(i) The method is stable;

(ii) The difference method is convergent if and only if it is consistent, which is equivalent to

$$\phi(t, y, 0) = f(t, y), \quad \text{for all } a \le t \le b;$$

(iii) If a function τ exists and, for each $i = 1, 2, \ldots, N$, the local truncation error $\tau_i(h)$ satisfies $|\tau_i(h)| \le \tau(h)$ whenever $0 \le h \le h_0$, then

$$|y(t_i) - w_i| \le \frac{\tau(h)}{L} e^{L(t_i - a)}.$$

■

Example 2 The Modified Euler method is given by $w_0 = \alpha$,

$$w_{i+1} = w_i + \frac{h}{2}\left[f(t_i, w_i) + f(t_{i+1}, w_i + hf(t_i, w_i))\right], \quad \text{for } i = 0, 1, \ldots, N-1.$$

Verify that this method is stable by showing that it satisfies the hypothesis of Theorem 5.20.

Solution For this method,

$$\phi(t, w, h) = \frac{1}{2}f(t, w) + \frac{1}{2}f(t + h, w + hf(t, w)).$$

If f satisfies a Lipschitz condition on $\{(t, w) \mid a \le t \le b \text{ and } -\infty < w < \infty\}$ in the variable w with constant L, then, since

$$\phi(t, w, h) - \phi(t, \overline{w}, h) = \frac{1}{2}f(t, w) + \frac{1}{2}f(t + h, w + hf(t, w)) - \frac{1}{2}f(t, \overline{w}) - \frac{1}{2}f(t + h, \overline{w} + hf(t, \overline{w})),$$

the Lipschitz condition on f leads to

$$\begin{aligned}
|\phi(t, w, h) - \phi(t, \overline{w}, h)| &\le \frac{1}{2}L|w - \overline{w}| + \frac{1}{2}L\,|w + hf(t, w) - \overline{w} - hf(t, \overline{w})| \\
&\le L|w - \overline{w}| + \frac{1}{2}L\,|hf(t, w) - hf(t, \overline{w})| \\
&\le L|w - \overline{w}| + \frac{1}{2}hL^2|w - \overline{w}| \\
&= \left(L + \frac{1}{2}hL^2\right)|w - \overline{w}|.
\end{aligned}$$

Therefore, ϕ satisfies a Lipschitz condition in w on the set

$$\{(t, w, h) \mid a \le t \le b, -\infty < w < \infty, \text{ and } 0 \le h \le h_0\},$$

for any $h_0 > 0$ with constant

$$L' = L + \frac{1}{2}h_0L^2.$$

Finally, if f is continuous on $\{(t, w) \mid a \le t \le b, -\infty < w < \infty\}$, then ϕ is continuous on

$$\{(t, w, h) \mid a \le t \le b, -\infty < w < \infty, \text{ and } 0 \le h \le h_0\};$$

so Theorem 5.20 implies that the Modified Euler method is stable. Letting $h = 0$, we have

$$\phi(t, w, 0) = \frac{1}{2}f(t, w) + \frac{1}{2}f(t + 0, w + 0 \cdot f(t, w)) = f(t, w),$$

so the consistency condition expressed in Theorem 5.20, part (ii), holds. Thus, the method is convergent. Moreover, we have seen that for this method the local truncation error is $O(h^2)$, so the convergence of the Modified Euler method is also $O(h^2)$. ■

Multistep Methods

For multistep methods, the problems involved with consistency, convergence, and stability are compounded because of the number of approximations involved at each step. In the one-step methods, the approximation w_{i+1} depends directly only on the previous approximation w_i, whereas the multistep methods use at least two of the previous approximations, and the usual methods that are employed involve more.

The general multistep method for approximating the solution to the initial-value problem

$$y' = f(t, y), \quad a \le t \le b, \quad y(a) = \alpha, \tag{5.54}$$

has the form

$$\begin{aligned} &w_0 = \alpha, \quad w_1 = \alpha_1, \quad \ldots, \; w_{m-1} = \alpha_{m-1}, \\ &w_{i+1} = a_{m-1}w_i + a_{m-2}w_{i-1} + \cdots + a_0 w_{i+1-m} + hF(t_i, h, w_{i+1}, w_i, \ldots, w_{i+1-m}), \end{aligned} \tag{5.55}$$

for each $i = m-1, m, \ldots, N-1$, where $a_0, a_1, \ldots, a_{m+1}$ are constants and, as usual, $h = (b-a)/N$ and $t_i = a + ih$.

The local truncation error for a multistep method expressed in this form is

$$\begin{aligned} \tau_{i+1}(h) = &\frac{y(t_{i+1}) - a_{m-1}y(t_i) - \cdots - a_0 y(t_{i+1-m})}{h} \\ &- F(t_i, h, y(t_{i+1}), y(t_i), \ldots, y(t_{i+1-m})), \end{aligned}$$

for each $i = m-1, m, \ldots, N-1$. As in the one-step methods, the local truncation error measures how the solution y to the differential equation fails to satisfy the difference equation.

For the four-step Adams-Bashforth method, we have seen that

$$\tau_{i+1}(h) = \frac{251}{720} y^{(5)}(\mu_i)h^4, \quad \text{for some } \mu_i \in (t_{i-3}, t_{i+1}),$$

whereas the three-step Adams-Moulton method has

$$\tau_{i+1}(h) = -\frac{19}{720} y^{(5)}(\mu_i)h^4, \quad \text{for some } \mu_i \in (t_{i-2}, t_{i+1}),$$

provided, of course, that $y \in C^5[a, b]$.

Throughout the analysis, two assumptions will be made concerning the function F:

- If $f \equiv 0$ (that is, if the differential equation is homogeneous), then $F \equiv 0$ also.
- F satisfies a Lipschitz condition with respect to $\{w_j\}$, in the sense that a constant L exists and, for every pair of sequences $\{v_j\}_{j=0}^{N}$ and $\{\tilde{v}_j\}_{j=0}^{N}$ and for $i = m-1, m, \ldots, N-1$, we have

$$|F(t_i, h, v_{i+1}, \ldots, v_{i+1-m}) - F(t_i, h, \tilde{v}_{i+1}, \ldots, \tilde{v}_{i+1-m})| \le L \sum_{j=0}^{m} |v_{i+1-j} - \tilde{v}_{i+1-j}|.$$

The explicit Adams-Bashforth and implicit Adams-Moulton methods satisfy both of these conditions, provided f satisfies a Lipschitz condition. (See Exercise 2.)

The concept of convergence for multistep methods is the same as that for one-step methods.

- A multistep method is **convergent** if the solution to the difference equation approaches the solution to the differential equation as the step size approaches zero. This means that $\lim_{h\to 0}\max_{0\le i\le N} \mid w_i - y(t_i)| = 0$.

For consistency, however, a slightly different situation occurs. Again, we want a multistep method to be consistent provided that the difference equation approaches the differential equation as the step size approaches zero; that is, the local truncation error approaches zero at each step as the step size approaches zero. The additional condition occurs because of the number of starting values required for multistep methods. Since usually only the first starting value, $w_0 = \alpha$, is exact, we need to require that the errors in all the starting values $\{\alpha_i\}$ approach zero as the step size approaches zero. So

$$\lim_{h\to 0} |\tau_i(h)| = 0, \quad \text{for all } i = m, m+1, \ldots, N \quad \text{and} \tag{5.56}$$

$$\lim_{h\to 0} |\alpha_i - y(t_i)| = 0, \quad \text{for all } i = 1, 2, \ldots, m-1, \tag{5.57}$$

must be true for a multistep method in the form (5.55) to be **consistent**. Note that (5.57) implies that a multistep method will not be consistent unless the one-step method generating the starting values is also consistent.

The following theorem for multistep methods is similar to Theorem 5.20, part (iii), and gives a relationship between the local truncation error and global error of a multistep method. It provides the theoretical justification for attempting to control global error by controlling local truncation error. The proof of a slightly more general form of this theorem can be found in [IK], pp. 387–388.

Theorem 5.21 Suppose the initial-value problem

$$y' = f(t, y), \quad a \le t \le b, \quad y(a) = \alpha,$$

is approximated by an explicit Adams predictor-corrector method with an m-step Adams-Bashforth predictor equation

$$w_{i+1} = w_i + h[b_{m-1} f(t_i, w_i) + \cdots + b_0 f(t_{i+1-m}, w_{i+1-m})],$$

with local truncation error $\tau_{i+1}(h)$, and an $(m-1)$-step implicit Adams-Moulton corrector equation

$$w_{i+1} = w_i + h\left[\tilde{b}_{m-1} f(t_i, w_{i+1}) + \tilde{b}_{m-2} f(t_i, w_i) + \cdots + \tilde{b}_0 f(t_{i+2-m}, w_{i+2-m})\right],$$

with local truncation error $\tilde{\tau}_{i+1}(h)$. In addition, suppose that $f(t, y)$ and $f_y(t, y)$ are continuous on $D = \{(t, y) \mid a \le t \le b \text{ and } -\infty < y < \infty\}$ and that f_y is bounded. Then the local truncation error $\sigma_{i+1}(h)$ of the predictor-corrector method is

$$\sigma_{i+1}(h) = \tilde{\tau}_{i+1}(h) + \tau_{i+1}(h)\tilde{b}_{m-1}\frac{\partial f}{\partial y}(t_{i+1}, \theta_{i+1}),$$

where θ_{i+1} is a number between zero and $h\tau_{i+1}(h)$.

Moreover, there exist constants k_1 and k_2 such that

$$|w_i - y(t_i)| \le \left[\max_{0\le j\le m-1} \left|w_j - y(t_j)\right| + k_1\sigma(h)\right] e^{k_2(t_i - a)},$$

where $\sigma(h) = \max_{m\le j\le N} |\sigma_j(h)|$. ■

Before discussing connections between consistency, convergence, and stability for multistep methods, we need to consider in more detail the difference equation for a multistep method. In doing so, we will discover the reason for choosing the Adams methods as our standard multistep methods.

Associated with the difference equation (5.55) given at the beginning of this discussion,

$$w_0 = \alpha,\ w_1 = \alpha_1,\ \ldots,\ w_{m-1} = \alpha_{m-1},$$

$$w_{i+1} = a_{m-1}w_i + a_{m-2}w_{i-1} + \cdots + a_0 w_{i+1-m} + hF(t_i, h, w_{i+1}, w_i, \ldots, w_{i+1-m}),$$

is a polynomial, called the **characteristic polynomial** of the method, given by

$$P(\lambda) = \lambda^m - a_{m-1}\lambda^{m-1} - a_{m-2}\lambda^{m-2} - \cdots - a_1\lambda - a_0. \tag{5.58}$$

The stability of a multistep method with respect to round-off error is dictated the by magnitudes of the zeros of the characteristic polynomial. To see this, consider applying the standard multistep method (5.55) to the trivial initial-value problem

$$y' \equiv 0, \quad y(a) = \alpha, \quad \text{where } \alpha \neq 0. \tag{5.59}$$

This problem has exact solution $y(t) \equiv \alpha$. By examining Eqs. (5.27) and (5.28) in Section 5.6 (see page 304), we can see that any multistep method will, in theory, produce the exact solution $w_n = \alpha$ for all n. The only deviation from the exact solution is due to the round-off error of the method.

The right side of the differential equation in (5.59) has $f(t, y) \equiv 0$, so by assumption (1), we have $F(t_i, h, w_{i+1}, w_{i+2}, \ldots, w_{i+1-m}) = 0$ in the difference equation (5.55). As a consequence, the standard form of the difference equation becomes

$$w_{i+1} = a_{m-1}w_i + a_{m-2}w_{i-1} + \cdots + a_0 w_{i+1-m}. \tag{5.60}$$

Suppose λ is one of the zeros of the characteristic polynomial associated with (5.55). Then $w_n = \lambda^n$ for each n is a solution to (5.59) since

$$\lambda^{i+1} - a_{m-1}\lambda^i - a_{m-2}\lambda^{i-1} - \cdots - a_0\lambda^{i+1-m} = \lambda^{i+1-m}[\lambda^m - a_{m-1}\lambda^{m-1} - \cdots - a_0] = 0.$$

In fact, if $\lambda_1, \lambda_2, \ldots, \lambda_m$ are distinct zeros of the characteristic polynomial for (5.55), it can be shown that *every* solution to (5.60) can be expressed in the form

$$w_n = \sum_{i=1}^{m} c_i\lambda_i^n, \tag{5.61}$$

for some unique collection of constants $c_1, c_2, \ldots, c_m$.

Since the exact solution to (5.59) is $y(t) = \alpha$, the choice $w_n = \alpha$, for all n, is a solution to (5.60). Using this fact in (5.60) gives

$$0 = \alpha - \alpha a_{m-1} - \alpha a_{m-2} - \cdots - \alpha a_0 = \alpha[1 - a_{m-1} - a_{m-2} - \cdots - a_0].$$

This implies that $\lambda = 1$ is one of the zeros of the characteristic polynomial (5.58). We will assume that in the representation (5.61) this solution is described by $\lambda_1 = 1$ and $c_1 = \alpha$, so all solutions to (5.59) are expressed as

$$w_n = \alpha + \sum_{i=2}^{m} c_i\lambda_i^n. \tag{5.62}$$

If all the calculations were exact, all the constants $c_2, c_3, \ldots, c_m$ would be zero. In practice, the constants $c_2, c_3, \ldots, c_m$ are not zero due to round-off error. In fact, the round-off error grows exponentially unless $|\lambda_i| \le 1$ for each of the roots $\lambda_2, \lambda_3, \ldots, \lambda_m$. The smaller the magnitude of these roots, the more stable the method with respect to the growth of round-off error.

In deriving (5.62), we made the simplifying assumption that the zeros of the characteristic polynomial are distinct. The situation is similar when multiple zeros occur. For example, if $\lambda_k = \lambda_{k+1} = \cdots = \lambda_{k+p}$ for some k and p, it simply requires replacing the sum

$$c_k\lambda_k^n + c_{k+1}\lambda_{k+1}^n + \cdots + c_{k+p}\lambda_{k+p}^n$$

in (5.62) with

$$c_k\lambda_k^n + c_{k+1}n\lambda_k^{n-1} + c_{k+2}n(n-1)\lambda_k^{n-2} + \cdots + c_{k+p}[n(n-1)\cdots(n-p+1)]\lambda_k^{n-p}. \tag{5.63}$$

(See [He2], pp. 119–145.) Although the form of the solution is modified, the round-off error if $|\lambda_k| > 1$ still grows exponentially.

Although we have considered only the special case of approximating initial-value problems of the form (5.59), the stability characteristics for this equation determine the stability for the situation when $f(t, y)$ is not identically zero. This is because the solution to the homogeneous equation (5.59) is embedded in the solution to any equation. The following definitions are motivated by this discussion.

Definition 5.22 Let $\lambda_1, \lambda_2, \ldots, \lambda_m$ denote the (not necessarily distinct) roots of the characteristic equation

$$P(\lambda) = \lambda^m - a_{m-1}\lambda^{m-1} - \cdots - a_1\lambda - a_0 = 0$$

associated with the multistep difference method

$$w_0 = \alpha, \quad w_1 = \alpha_1, \quad \ldots, \quad w_{m-1} = \alpha_{m-1}$$

$$w_{i+1} = a_{m-1}w_i + a_{m-2}w_{i-1} + \cdots + a_0w_{i+1-m} + hF(t_i, h, w_{i+1}, w_i, \ldots, w_{i+1-m}).$$

If $|\lambda_i| \le 1$, for each $i = 1, 2, \ldots, m$, and all roots with absolute value 1 are simple roots, then the difference method is said to satisfy the **root condition**. ■

Definition 5.23

(i) Methods that satisfy the root condition and have $\lambda = 1$ as the only root of the characteristic equation with magnitude one are called **strongly stable**.

(ii) Methods that satisfy the root condition and have more than one distinct root with magnitude one are called **weakly stable**.

(iii) Methods that do not satisfy the root condition are called **unstable**. ■

Consistency and convergence of a multistep method are closely related to the round-off stability of the method. The next theorem details these connections. For the proof of this result and the theory on which it is based, see [IK], pp. 410–417.

Theorem 5.24 A multistep method of the form

$$w_0 = \alpha, \quad w_1 = \alpha_1, \quad \ldots, \quad w_{m-1} = \alpha_{m-1},$$

$$w_{i+1} = a_{m-1}w_i + a_{m-2}w_{i-1} + \cdots + a_0w_{i+1-m} + hF(t_i, h, w_{i+1}, w_i, \ldots, w_{i+1-m})$$

is stable if and only if it satisfies the root condition. Moreover, if the difference method is consistent with the differential equation, then the method is stable if and only if it is convergent. ■

Example 3 The fourth-order Adams-Bashforth method can be expressed as

$$w_{i+1} = w_i + hF(t_i, h, w_{i+1}, w_i, \ldots, w_{i-3}),$$

where

$$F(t_i, h, w_{i+1}, , \ldots, w_{i-3}) = \frac{h}{24}[55f(t_i, w_i) - 59f(t_{i-1}, w_{i-1}) + 37f(t_{i-2}, w_{i-2}) - 9f(t_{i-3}, w_{i-3})];$$

Show that this method is strongly stable.

Solution In this case we have $m = 4$, $a_0 = 0$, $a_1 = 0$, $a_2 = 0$, and $a_3 = 1$, so the characteristic equation for this Adams-Bashforth method is

$$0 = P(\lambda) = \lambda^4 - \lambda^3 = \lambda^3(\lambda - 1).$$

This polynomial has roots $\lambda_1 = 1$, $\lambda_2 = 0$, $\lambda_3 = 0$, and $\lambda_4 = 0$. Hence it satisfies the root condition and is strongly stable.

The Adams-Moulton method has a similar characteristic polynomial, $P(\lambda) = \lambda^3 - \lambda^2$, with zeros $\lambda_1 = 1$, $\lambda_2 = 0$, and $\lambda_3 = 0$, and is also strongly stable. ■

Example 4 Show that the fourth-order Milne's method, the explicit multistep method given by

$$w_{i+1} = w_{i-3} + \frac{4h}{3}\left[2f(t_i, w_i) - f(t_{i-1}, w_{i-1}) + 2f(t_{i-2}, w_{i-2})\right]$$

satisfies the root condition, but it is only weakly stable.

Solution The characteristic equation for this method, $0 = P(\lambda) = \lambda^4 - 1$, has four roots with magnitude one: $\lambda_1 = 1$, $\lambda_2 = -1$, $\lambda_3 = i$, and $\lambda_4 = -i$. Because all the roots have magnitude 1, the method satisfies the root condition. However, there are multiple roots with magnitude 1, so the method is only weakly stable. ■

Example 5 Apply the strongly stable fourth-order Adams-Bashforth method and the weakly stable Milne's method with $h = 0.1$ to the initial-value problem

$$y' = -6y + 6, \quad 0 \le t \le 1, \quad y(0) = 2,$$

which has the exact solution $y(t) = 1 + e^{-6t}$.

Solution The results in Table 5.21 show the effects of a weakly stable method versus a strongly stable method for this problem. ■

Table 5.21

t_i	Exact $y(t_i)$	Adams-Bashforth Method w_i	Error $\lvert y_i - w_i\rvert$	Milne's Method w_i	Error $\lvert y_i - w_i\rvert$
0.10000000		1.5488116		1.5488116	
0.20000000		1.3011942		1.3011942	
0.30000000		1.1652989		1.1652989	
0.40000000	1.0907180	1.0996236	8.906×10^{-3}	1.0983785	7.661×10^{-3}
0.50000000	1.0497871	1.0513350	1.548×10^{-3}	1.0417344	8.053×10^{-3}
0.60000000	1.0273237	1.0425614	1.524×10^{-2}	1.0486438	2.132×10^{-2}
0.70000000	1.0149956	1.0047990	1.020×10^{-2}	0.9634506	5.154×10^{-2}
0.80000000	1.0082297	1.0359090	2.768×10^{-2}	1.1289977	1.208×10^{-1}
0.90000000	1.0045166	0.9657936	3.872×10^{-2}	0.7282684	2.762×10^{-1}
1.00000000	1.0024788	1.0709304	6.845×10^{-2}	1.6450917	6.426×10^{-1}

The reason for choosing the Adams-Bashforth-Moulton as our standard fourth-order predictor-corrector technique in Section 5.6 over the Milne-Simpson method of the same order is that both the Adams-Bashforth and Adams-Moulton methods are strongly stable. They are more likely to give accurate approximations to a wider class of problems than is the predictor-corrector based on the Milne and Simpson techniques, both of which are weakly stable.

EXERCISE SET 5.10

1. To prove Theorem 5.20, part (i), show that the hypotheses imply that there exists a constant $K > 0$ such that

$$|u_i - v_i| \leq K|u_0 - v_0|, \quad \text{for each } 1 \leq i \leq N,$$

whenever $\{u_i\}_{i=1}^N$ and $\{v_i\}_{i=1}^N$ satisfy the difference equation $w_{i+1} = w_i + h\phi(t_i, w_i, h)$.

2. For the Adams-Bashforth and Adams-Moulton methods of order four,

a. Show that if $f = 0$, then

$$F(t_i, h, w_{i+1}, \ldots, w_{i+1-m}) = 0.$$

b. Show that if f satisfies a Lipschitz condition with constant L, then a constant C exists with

$$|F(t_i, h, w_{i+1}, \ldots, w_{i+1-m}) - F(t_i, h, v_{i+1}, \ldots, v_{i+1-m})| \leq C \sum_{j=0}^{m} |w_{i+1-j} - v_{i+1-j}|.$$

3. Use the results of Exercise 32 in Section 5.4 to show that the Runge-Kutta method of order four is consistent.

4. Consider the differential equation

$$y' = f(t, y), \quad a \leq t \leq b, \quad y(a) = \alpha.$$

a. Show that

$$y'(t_i) = \frac{-3y(t_i) + 4y(t_{i+1}) - y(t_{i+2})}{2h} + \frac{h^2}{3}y'''(\xi_1),$$

for some ξ, where $t_i < \xi_i < t_{i+2}$.

b. Part (a) suggests the difference method

$$w_{i+2} = 4w_{i+1} - 3w_i - 2hf(t_i, w_i), \quad \text{for } i = 0, 1, \ldots, N-2.$$

Use this method to solve

$$y' = 1 - y, \quad 0 \leq t \leq 1, \quad y(0) = 0,$$

with $h = 0.1$. Use the starting values $w_0 = 0$ and $w_1 = y(t_1) = 1 - e^{-0.1}$.

c. Repeat part (b) with $h = 0.01$ and $w_1 = 1 - e^{-0.01}$.

d. Analyze this method for consistency, stability, and convergence.

5. Given the multistep method

$$w_{i+1} = -\frac{3}{2}w_i + 3w_{i-1} - \frac{1}{2}w_{i-2} + 3hf(t_i, w_i), \quad \text{for } i = 2, \ldots, N-1,$$

with starting values w_0, w_1, w_2:

a. Find the local truncation error.

b. Comment on consistency, stability, and convergence.

6. Obtain an approximate solution to the differential equation

$$y' = -y, \quad 0 \leq t \leq 10, \quad y(0) = 1$$

using Milne's method with $h = 0.1$ and then $h = 0.01$, with starting values $w_0 = 1$ and $w_1 = e^{-h}$ in both cases. How does decreasing h from $h = 0.1$ to $h = 0.01$ affect the number of correct digits in the approximate solutions at $t = 1$ and $t = 10$?

7. Investigate stability for the difference method

$$w_{i+1} = -4w_i + 5w_{i-1} + 2h[f(t_i, w_i) + 2hf(t_{i-1}, w_{i-1})],$$

for $i = 1, 2, \ldots, N-1$, with starting values w_0, w_1.

8. Consider the problem $y' = 0$, $0 \leq t \leq 10$, $y(0) = 0$, which has the solution $y \equiv 0$. If the difference method of Exercise 4 is applied to the problem, then

$$w_{i+1} = 4w_i - 3w_{i-1}, \quad \text{for } i = 1, 2, \ldots, N-1,$$

$$w_0 = 0, \quad \text{and} \quad w_1 = \alpha_1.$$

Suppose $w_1 = \alpha_1 = \varepsilon$, where ε is a small rounding error. Compute w_i exactly for $i = 2, 3, \ldots, 6$ to find how the error ε is propagated.

5.11 Stiff Differential Equations

All the methods for approximating the solution to initial-value problems have error terms that involve a higher derivative of the solution of the equation. If the derivative can be reasonably bounded, then the method will have a predictable error bound that can be used to estimate the accuracy of the approximation. Even if the derivative grows as the steps increase, the error can be kept in relative control, provided that the solution also grows in magnitude. Problems frequently arise, however, when the magnitude of the derivative increases but the solution does not. In this situation, the error can grow so large that it dominates the calculations. Initial-value problems for which this is likely to occur are called **stiff equations** and are quite common, particularly in the study of vibrations, chemical reactions, and electrical circuits.

Stiff systems derive their name from the motion of spring and mass systems that have large spring constants.

Stiff differential equations are characterized as those whose exact solution has a term of the form e^{-ct}, where c is a large positive constant. This is usually only a part of the solution, called the *transient* solution. The more important portion of the solution is called the *steady-state* solution. The transient portion of a stiff equation will rapidly decay to zero as t increases, but since the nth derivative of this term has magnitude $c^n e^{-ct}$, the derivative does not decay as quickly. In fact, since the derivative in the error term is evaluated not at t, but at a number between zero and t, the derivative terms can increase as t increases– and very rapidly indeed. Fortunately, stiff equations generally can be predicted from the physical problem from which the equation is derived and, with care, the error can be kept under control. The manner in which this is done is considered in this section.

Illustration The system of initial-value problems

$$u_1' = 9u_1 + 24u_2 + 5\cos t - \frac{1}{3}\sin t, \quad u_1(0) = \frac{4}{3}$$

$$u_2' = -24u_1 - 51u_2 - 9\cos t + \frac{1}{3}\sin t, \quad u_2(0) = \frac{2}{3}$$

has the unique solution

$$u_1(t) = 2e^{-3t} - e^{-39t} + \frac{1}{3}\cos t, \quad u_2(t) = -e^{-3t} + 2e^{-39t} - \frac{1}{3}\cos t.$$

The transient term e^{-39t} in the solution causes this system to be stiff. Applying Algorithm 5.7, the Runge-Kutta Fourth-Order Method for Systems, gives results listed in Table 5.22. When $h = 0.05$, stability results and the approximations are accurate. Increasing the step size to $h = 0.1$, however, leads to the disastrous results shown in the table. □

Table 5.22

t	$u_1(t)$	$w_1(t)$ $h = 0.05$	$w_1(t)$ $h = 0.1$	$u_2(t)$	$w_2(t)$ $h = 0.05$	$w_2(t)$ $h = 0.1$
0.1	1.793061	1.712219	−2.645169	−1.032001	−0.8703152	7.844527
0.2	1.423901	1.414070	−18.45158	−0.8746809	−0.8550148	38.87631
0.3	1.131575	1.130523	−87.47221	−0.7249984	−0.7228910	176.4828
0.4	0.9094086	0.9092763	−934.0722	−0.6082141	−0.6079475	789.3540
0.5	0.7387877	9.7387506	−1760.016	−0.5156575	−0.5155810	3520.00
0.6	0.6057094	0.6056833	−7848.550	−0.4404108	−0.4403558	15697.84
0.7	0.4998603	0.4998361	−34989.63	−0.3774038	−0.3773540	69979.87
0.8	0.4136714	0.4136490	−155979.4	−0.3229535	−0.3229078	311959.5
0.9	0.3416143	0.3415939	−695332.0	−0.2744088	−0.2743673	1390664.
1.0	0.2796748	0.2796568	−3099671.	−0.2298877	−0.2298511	6199352.

Although stiffness is usually associated with systems of differential equations, the approximation characteristics of a particular numerical method applied to a stiff system can be predicted by examining the error produced when the method is applied to a simple *test equation*,

$$y' = \lambda y, \quad y(0) = \alpha, \quad \text{where } \lambda < 0. \tag{5.64}$$

The solution to this equation is $y(t) = \alpha e^{\lambda t}$, which contains the transient solution $e^{\lambda t}$. The steady-state solution is zero, so the approximation characteristics of a method are easy to determine. (A more complete discussion of the round-off error associated with stiff systems requires examining the test equation when λ is a complex number with negative real part; see [Ge1], p. 222.)

First consider Euler's method applied to the test equation. Letting $h = (b - a)/N$ and $t_j = jh$, for $j = 0, 1, 2, \ldots, N$, Eq. (5.8) on page 266 implies that

$$w_0 = \alpha, \quad \text{and} \quad w_{j+1} = w_j + h(\lambda w_j) = (1 + h\lambda)w_j,$$

so

$$w_{j+1} = (1 + h\lambda)^{j+1} w_0 = (1 + h\lambda)^{j+1}\alpha, \quad \text{for } j = 0, 1, \ldots, N - 1. \tag{5.65}$$

Since the exact solution is $y(t) = \alpha e^{\lambda t}$, the absolute error is

$$|y(t_j) - w_j| = \left|e^{jh\lambda} - (1 + h\lambda)^j\right| |\alpha| = \left|(e^{h\lambda})^j - (1 + h\lambda)^j\right| |\alpha|,$$

and the accuracy is determined by how well the term $1 + h\lambda$ approximates $e^{h\lambda}$. When $\lambda < 0$, the exact solution $(e^{h\lambda})^j$ decays to zero as j increases, but by Eq.(5.65), the approximation will have this property only if $|1 + h\lambda| < 1$, which implies that $-2 < h\lambda < 0$. This effectively restricts the step size h for Euler's method to satisfy $h < 2/|\lambda|$.

Suppose now that a round-off error δ_0 is introduced in the initial condition for Euler's method,

$$w_0 = \alpha + \delta_0.$$

At the jth step the round-off error is

$$\delta_j = (1 + h\lambda)^j \delta_0.$$

Since $\lambda < 0$, the condition for the control of the growth of round-off error is the same as the condition for controlling the absolute error, $|1+h\lambda| < 1$, which implies that $h < 2/|\lambda|$. So

- Euler's method is expected to be stable for

  $$y' = \lambda y, \quad y(0) = \alpha, \qquad \text{where } \lambda < 0,$$

  only if the step size h is less than $2/|\lambda|$.

The situation is similar for other one-step methods. In general, a function Q exists with the property that the difference method, when applied to the test equation, gives

$$w_{i+1} = Q(h\lambda)w_i. \tag{5.66}$$

The accuracy of the method depends upon how well $Q(h\lambda)$ approximates $e^{h\lambda}$, and the error will grow without bound if $|Q(h\lambda)| > 1$. An nth-order Taylor method, for example, will have stability with regard to both the growth of round-off error and absolute error, provided h is chosen to satisfy

$$\left|1 + h\lambda + \frac{1}{2}h^2\lambda^2 + \cdots + \frac{1}{n!}h^n\lambda^n\right| < 1.$$

Exercise 10 examines the specific case when the method is the classical fourth-order Runge-Kutta method,which is essentially a Taylor method of order four.

When a multistep method of the form (5.54) is applied to the test equation, the result is

$$w_{j+1} = a_{m-1}w_j + \cdots + a_0 w_{j+1-m} + h\lambda(b_m w_{j+1} + b_{m-1}w_j + \cdots + b_0 w_{j+1-m}),$$

for $j = m-1, \ldots, N-1$, or

$$(1 - h\lambda b_m)w_{j+1} - (a_{m-1} + h\lambda b_{m-1})w_j - \cdots - (a_0 + h\lambda b_0)w_{j+1-m} = 0.$$

Associated with this homogeneous difference equation is a **characteristic polynomial**

$$Q(z, h\lambda) = (1 - h\lambda b_m)z^m - (a_{m-1} + h\lambda b_{m-1})z^{m-1} - \cdots - (a_0 + h\lambda b_0).$$

This polynomial is similar to the characteristic polynomial (5.58), but it also incorporates the test equation. The theory here parallels the stability discussion in Section 5.10.

Suppose $w_0, \ldots, w_{m-1}$ are given, and, for fixed $h\lambda$, let $\beta_1, \ldots, \beta_m$ be the zeros of the polynomial $Q(z, h\lambda)$. If $\beta_1, \ldots, \beta_m$ are distinct, then $c_1, \ldots, c_m$ exist with

$$w_j = \sum_{k=1}^{m} c_k(\beta_k)^j, \quad \text{for } j = 0, \ldots, N. \tag{5.67}$$

If $Q(z, h\lambda)$ has multiple zeros, w_j is similarly defined. (See Eq. (5.63) in Section 5.10.) If w_j is to accurately approximate $y(t_j) = e^{jh\lambda} = (e^{h\lambda})^j$, then all zeros β_k must satisfy $|\beta_k| < 1$;

otherwise, certain choices of α will result in $c_k \neq 0$, and the term $c_k(\beta_k)^j$ will not decay to zero.

Illustration The test differential equation

$$y' = -30y, \quad 0 \leq t \leq 1.5, \quad y(0) = \frac{1}{3}$$

has exact solution $y = \frac{1}{3}e^{-30t}$. Using $h = 0.1$ for Euler's Algorithm 5.1, Runge-Kutta Fourth-Order Algorithm 5.2, and the Adams Predictor-Corrector Algorithm 5.4, gives the results at $t = 1.5$ in Table 5.23. □

Table 5.23

Exact solution	9.54173×10^{-21}
Euler's method	-1.09225×10^{4}
Runge-Kutta method	3.95730×10^{1}
Predictor-corrector method	8.03840×10^{5}

The inaccuracies in the Illustration are due to the fact that $|Q(h\lambda)| > 1$ for Euler's method and the Runge-Kutta method and that $Q(z, h\lambda)$ has zeros with modulus exceeding 1 for the predictor-corrector method. To apply these methods to this problem, the step size must be reduced. The following definition is used to describe the amount of step-size reduction that is required.

Definition 5.25 The **region R of absolute stability** for a one-step method is $R = \{h\lambda \in \mathcal{C} \mid |Q(h\lambda)| < 1\}$, and for a multistep method, it is $R = \{h\lambda \in \mathcal{C} \mid |\beta_k| < 1$, for all zeros β_k of $Q(z, h\lambda)\}$. ■

Equations (5.66) and (5.67) imply that a method can be applied effectively to a stiff equation only if $h\lambda$ is in the region of absolute stability of the method, which for a given problem places a restriction on the size of h. Even though the exponential term in the exact solution decays quickly to zero, λh must remain within the region of absolute stability throughout the interval of t values for the approximation to decay to zero and the growth of error to be under control. This means that, although h could normally be increased because of truncation error considerations, the absolute stability criterion forces h to remain small. Variable step-size methods are especially vulnerable to this problem because an examination of the local truncation error might indicate that the step size could increase. This could inadvertently result in λh being outside the region of absolute stability.

The region of absolute stability of a method is generally the critical factor in producing accurate approximations for stiff systems, so numerical methods are sought with as large a region of absolute stability as possible. A numerical method is said to be ***A*-stable** if its region R of absolute stability contains the entire left half-plane.

The **Implicit Trapezoidal method**, given by

This method is implicit because it involves w_{j+1} on both sides of the equation.

$$w_0 = \alpha, \tag{5.68}$$

$$w_{j+1} = w_j + \frac{h}{2}\left[f(t_{j+1}, w_{j+1}) + f(t_j, w_j)\right], \quad 0 \leq j \leq N-1,$$

is an A-stable method (see Exercise 15) and is the only A-stable multistep method. Although the Trapezoidal method does not give accurate approximations for large step sizes, its error will not grow exponentially.

The techniques commonly used for stiff systems are implicit multistep methods. Generally w_{i+1} is obtained by solving a nonlinear equation or nonlinear system iteratively, often by Newton's method. Consider, for example, the Implicit Trapezoidal method

$$w_{j+1} = w_j + \frac{h}{2}[f(t_{j+1}, w_{j+1}) + f(t_j, w_j)].$$

Having computed t_j, t_{j+1}, and w_j, we need to determine w_{j+1}, the solution to

$$F(w) = w - w_j - \frac{h}{2}[f(t_{j+1}, w) + f(t_j, w_j)] = 0. \tag{5.69}$$

To approximate this solution, select $w_{j+1}^{(0)}$, usually as w_j, and generate $w_{j+1}^{(k)}$ by applying Newton's method to (5.69),

$$\begin{aligned} w_{j+1}^{(k)} &= w_{j+1}^{(k-1)} - \frac{F(w_{j+1}^{(k-1)})}{F'(w_{j+1}^{(k-1)})} \\ &= w_{j+1}^{(k-1)} - \frac{w_{j+1}^{(k-1)} - w_j - \frac{h}{2}[f(t_j, w_j) + f(t_{j+1}, w_{j+1}^{(k-1)})]}{1 - \frac{h}{2} f_y(t_{j+1}, w_{j+1}^{(k-1)})} \end{aligned}$$

until $|w_{j+1}^{(k)} - w_{j+1}^{(k-1)}|$ is sufficiently small. This is the procedure that is used in Algorithm 5.8. Normally only three or four iterations per step are required, because of the quadratic convergence of Newton's mehod.

The Secant method can be used as an alternative to Newton's method in Eq. (5.69), but then two distinct initial approximations to w_{j+1} are required. To employ the Secant method, the usual practice is to let $w_{j+1}^{(0)} = w_j$ and obtain $w_{j+1}^{(1)}$ from some explicit multistep method. When a system of stiff equations is involved, a generalization is required for either Newton's or the Secant method. These topics are considered in Chapter 10.

ALGORITHM 5.8

Trapezoidal with Newton Iteration

To approximate the solution of the initial-value problem

$$y' = f(t, y), \quad \text{for } a \le t \le b, \quad \text{with } y(a) = \alpha$$

at $(N + 1)$ equally spaced numbers in the interval $[a, b]$:

INPUT endpoints a, b; integer N; initial condition α; tolerance TOL; maximum number of iterations M at any one step.

OUTPUT approximation w to y at the $(N + 1)$ values of t or a message of failure.

Step 1 Set $h = (b - a)/N$;
$t = a$;
$w = \alpha$;
OUTPUT (t, w).

Step 2 For $i = 1, 2, \ldots, N$ do Steps 3–7.

Step 3 Set $k_1 = w + \frac{h}{2} f(t, w)$;
$w_0 = k_1$;
$j = 1$;
$FLAG = 0$.

Step 4 While $FLAG = 0$ do Steps 5–6.

Step 5 Set $w = w_0 - \dfrac{w_0 - \frac{h}{2} f(t+h, w_0) - k_1}{1 - \frac{h}{2} f_y(t+h, w_0)}$.

Step 6 If $|w - w_0| < TOL$ then set $FLAG = 1$
else set $j = j + 1$;
$w_0 = w$;
if $j > M$ then
OUTPUT ('The maximum number of iterations exceeded');
STOP.

Step 7 Set $t = a + ih$;
OUTPUT (t, w).

Step 8 STOP. ■

Illustration The stiff initial-value problem

$$y' = 5e^{5t}(y-t)^2 + 1, \quad 0 \le t \le 1, \quad y(0) = -1$$

has solution $y(t) = t - e^{-5t}$. To show the effects of stiffness, the Implicit Trapezoidal method and the Runge-Kutta fourth-order method are applied both with $N = 4$, giving $h = 0.25$, and with $N = 5$, giving $h = 0.20$.

The Trapezoidal method performs well in both cases using $M = 10$ and $TOL = 10^{-6}$, as does Runge-Kutta with $h = 0.2$. However, $h = 0.25$ is outside the region of absolute stability of the Runge-Kutta method, which is evident from the results in Table 5.24. □

Table 5.24

	Runge–Kutta Method		Trapezoidal Method	
	$h = 0.2$		$h = 0.2$	
t_i	w_i	$\lvert y(t_i) - w_i\rvert$	w_i	$\lvert y(t_i) - w_i\rvert$
0.0	−1.0000000	0	−1.0000000	0
0.2	−0.1488521	1.9027×10^{-2}	−0.1414969	2.6383×10^{-2}
0.4	0.2684884	3.8237×10^{-3}	0.2748614	1.0197×10^{-2}
0.6	0.5519927	1.7798×10^{-3}	0.5539828	3.7700×10^{-3}
0.8	0.7822857	6.0131×10^{-4}	0.7830720	1.3876×10^{-3}
1.0	0.9934905	2.2845×10^{-4}	0.9937726	5.1050×10^{-4}
	$h = 0.25$		$h = 0.25$	
t_i	w_i	$\lvert y(t_i) - w_i\rvert$	w_i	$\lvert y(t_i) - w_i\rvert$
0.0	−1.0000000	0	−1.0000000	0
0.25	0.4014315	4.37936×10^{-1}	0.0054557	4.1961×10^{-2}
0.5	3.4374753	3.01956×10^{0}	0.4267572	8.8422×10^{-3}
0.75	1.44639×10^{23}	1.44639×10^{23}	0.7291528	2.6706×10^{-3}
1.0	Overflow		0.9940199	7.5790×10^{-4}

We have presented here only brief introduction to what the reader frequently encountering stiff differential equations should know. For further details, consult [Ge2], [Lam], or [SGe].

EXERCISE SET 5.11

1. Solve the following stiff initial-value problems using Euler's method, and compare the results with the actual solution.
   a. $y' = -9y$, $0 \le t \le 1$, $y(0) = e$, with $h = 0.1$; actual solution $y(t) = e^{1-9t}$.
   b. $y' = -20y + 20\sin t + \cos t$, $0 \le t \le 2$, $y(0) = 1$, with $h = 0.25$; actual solution $y(t) = \sin t + e^{-20t}$.
   c. $y' = -20(y-t^2)+2t$, $0 \le t \le 1$, $y(0) = \frac{1}{3}$, with $h = 0.1$; actual solution $y(t) = t^2+\frac{1}{3}e^{-20t}$.
   d. $y' = 50/y-50y$, $0 \le t \le 1$, $y(0) = \sqrt{2}$, with $h = 0.1$; actual solution $y(t) = (1+e^{-100t})^{1/2}$.
2. Solve the following stiff initial-value problems using Euler's method, and compare the results with the actual solution.
   a. $y' = -5y + 6e^t$, $0 \le t \le 1$, $y(0) = 2$, with $h = 0.1$; actual solution $y(t) = e^{-5t} + e^t$.
   b. $y' = -10y+10t+1$, $0 \le t \le 1$, $y(0) = e$, with $h = 0.1$; actual solution $y(t) = e^{-10t+1}+t$.
   c. $y' = -15(y - t^{-3}) - 3/t^4$, $1 \le t \le 3$, $y(1) = 0$, with $h = 0.25$; actual solution $y(t) = -e^{-15t} + t^{-3}$.
   d. $y' = -20y + 20\cos t - \sin t$, $0 \le t \le 2$, $y(0) = 0$, with $h = 0.25$; actual solution $y(t) = -e^{-20t} + \cos t$.
3. Repeat Exercise 1 using the Runge-Kutta fourth-order method.
4. Repeat Exercise 2 using the Runge-Kutta fourth-order method.
5. Repeat Exercise 1 using the Adams fourth-order predictor-corrector method.
6. Repeat Exercise 2 using the Adams fourth-order predictor-corrector method.
7. Repeat Exercise 1 using the Trapezoidal Algorithm with $TOL = 10^{-5}$.
8. Repeat Exercise 2 using the Trapezoidal Algorithm with $TOL = 10^{-5}$.
9. Solve the following stiff initial-value problem using the Runge-Kutta fourth-order method with (a) $h = 0.1$ and (b) $h = 0.025$.

$$u_1' = 32u_1 + 66u_2 + \frac{2}{3}t + \frac{2}{3}, \quad 0 \le t \le 0.5, \quad u_1(0) = \frac{1}{3};$$

$$u_2' = -66u_1 - 133u_2 - \frac{1}{3}t - \frac{1}{3}, \quad 0 \le t \le 0.5, \quad u_2(0) = \frac{1}{3}.$$

   Compare the results to the actual solution,

$$u_1(t) = \frac{2}{3}t + \frac{2}{3}e^{-t} - \frac{1}{3}e^{-100t} \quad \text{and} \quad u_2(t) = -\frac{1}{3}t - \frac{1}{3}e^{-t} + \frac{2}{3}e^{-100t}.$$

10. Show that the fourth-order Runge-Kutta method,

$$k_1 = hf(t_i, w_i),$$
$$k_2 = hf(t_i + h/2, w_i + k_1/2),$$
$$k_3 = hf(t_i + h/2, w_i + k_2/2),$$
$$k_4 = hf(t_i + h, w_i + k_3),$$
$$w_{i+1} = w_i + \frac{1}{6}(k_1 + 2k_2 + 2k_3 + k_4),$$

   when applied to the differential equation $y' = \lambda y$, can be written in the form

$$w_{i+1} = \left(1 + h\lambda + \frac{1}{2}(h\lambda)^2 + \frac{1}{6}(h\lambda)^3 + \frac{1}{24}(h\lambda)^4\right) w_i.$$

11. Discuss consistency, stability, and convergence for the Implicit Trapezoidal method

$$w_{i+1} = w_i + \frac{h}{2}\left(f(t_{i+1}, w_{i+1}) + f(t_i, w_i)\right), \quad \text{for } i = 0, 1, \dots, N-1,$$

with $w_0 = \alpha$ applied to the differential equation

$$y' = f(t, y), \quad a \le t \le b, \quad y(a) = \alpha.$$

12. The Backward Euler one-step method is defined by

$$w_{i+1} = w_i + hf(t_{i+1}, w_{i+1}), \quad \text{for } i = 0, \dots, N-1.$$

Show that $Q(h\lambda) = 1/(1 - h\lambda)$ for the Backward Euler method.

13. Apply the Backward Euler method to the differential equations given in Exercise 1. Use Newton's method to solve for w_{i+1}.

14. Apply the Backward Euler method to the differential equations given in Exercise 2. Use Newton's method to solve for w_{i+1}.

15. **a.** Show that the Implicit Trapezoidal method is A-stable.

b. Show that the Backward Euler method described in Exercise 12 is A-stable.

5.12 Survey of Methods and Software

In this chapter we have considered methods to approximate the solutions to initial-value problems for ordinary differential equations. We began with a discussion of the most elementary numerical technique, Euler's method. This procedure is not sufficiently accurate to be of use in applications, but it illustrates the general behavior of the more powerful techniques, without the accompanying algebraic difficulties. The Taylor methods were then considered as generalizations of Euler's method. They were found to be accurate but cumbersome because of the need to determine extensive partial derivatives of the defining function of the differential equation. The Runge-Kutta formulas simplified the Taylor methods, without increasing the order of the error. To this point we had considered only one-step methods, techniques that use only data at the most recently computed point.

Multistep methods are discussed in Section 5.6, where explicit methods of Adams-Bashforth type and implicit methods of Adams-Moulton type were considered. These culminate in predictor-corrector methods, which use an explicit method, such as an Adams-Bashforth, to predict the solution and then apply a corresponding implicit method, like an Adams-Moulton, to correct the approximation.

Section 5.9 illustrated how these techniques can be used to solve higher-order initial-value problems and systems of initial-value problems.

The more accurate adaptive methods are based on the relatively uncomplicated one-step and multistep techniques. In particular, we saw in Section 5.5 that the Runge-Kutta-Fehlberg method is a one-step procedure that seeks to select mesh spacing to keep the local error of the approximation under control. The Variable Step-Size Predictor-Corrector method presented in Section 5.7 is based on the four-step Adams-Bashforth method and three-step Adams-Moulton method. It also changes the step size to keep the local error within a given tolerance. The Extrapolation method discussed in Section 5.8 is based on a modification of the Midpoint method and incorporates extrapolation to maintain a desired accuracy of approximation.

The final topic in the chapter concerned the difficulty that is inherent in the approximation of the solution to a stiff equation, a differential equation whose exact solution contains a portion of the form $e^{-\lambda t}$, where λ is a positive constant. Special caution must be taken with problems of this type, or the results can be overwhelmed by round-off error.

Methods of the Runge-Kutta-Fehlberg type are generally sufficient for nonstiff problems when moderate accuracy is required. The extrapolation procedures are recommended for nonstiff problems where high accuracy is required. Extensions of the Implicit Trapezoidal method to variable-order and variable step-size implicit Adams-type methods are used for stiff initial-value problems.

The IMSL Library includes two subroutines for approximating the solutions of initial-value problems. Each of the methods solves a system of m first-order equations in m variables. The equations are of the form

$$\frac{du_i}{dt} = f_i(t, u_1, u_2, \ldots, u_m), \quad \text{for } i = 1, 2, \ldots, m,$$

where $u_i(t_0)$ is given for each i. A variable step-size subroutine is based on the Runge-Kutta-Verner fifth- and sixth-order methods described in Exercise 4 of Section 5.5. A subroutine of Adams type is also available to be used for stiff equations based on a method of C. William Gear. This method uses implicit multistep methods of order up to 12 and backward differentiation formulas of order up to 5.

Runge-Kutta-type procedures contained in the NAG Library are based on the Merson form of the Runge-Kutta method. A variable-order and variable step-size Adams method is also in the library, as well as a variable-order, variable step-size backward-difference method for stiff systems. Other routines incorporate the same methods but iterate until a component of the solution attains a given value or until a function of the solution is zero.

The netlib library includes several subroutines for approximating the solutions of initial-value problems in the package ODE. One subroutine is based on the Runge-Kutta-Verner fifth- and sixth-order methods, another on the Runge-Kutta-Fehlberg fourth- and fifth-order methods as described on page 297 of Section 5.5. A subroutine for stiff ordinary differential equation initial-value problems, is based on a variable coefficient backward differentiation formula.

Many books specialize in the numerical solution of initial-value problems. Two classics are by Henrici [He1] and Gear [Ge1]. Other books that survey the field are by Botha and Pinder [BP], Ortega and Poole [OP], Golub and Ortega [GO], Shampine [Sh], and Dormand [Do].

Two books by Hairer, Nörsett, and Warner provide comprehensive discussions on nonstiff [HNW1] and stiff [HNW2] problems. The book by Burrage [Bur] describes parallel and sequential methods for solving systems of initial-value problems.

CHAPTER

6 Direct Methods for Solving Linear Systems

Introduction

Kirchhoff's laws of electrical circuits state that both the net flow of current through each junction and the net voltage drop around each closed loop of a circuit are zero. Suppose that a potential of V volts is applied between the points A and G in the circuit and that i_1, i_2, i_3, i_4, and i_5 represent current flow as shown in the diagram. Using G as a reference point, Kirchhoff's laws imply that the currents satisfy the following system of linear equations:

$$\begin{aligned}
5i_1 + 5i_2 &= V,\\
i_3 - i_4 - i_5 &= 0,\\
2i_4 - 3i_5 &= 0,\\
i_1 - i_2 - i_3 &= 0,\\
5i_2 - 7i_3 - 2i_4 &= 0.
\end{aligned}$$

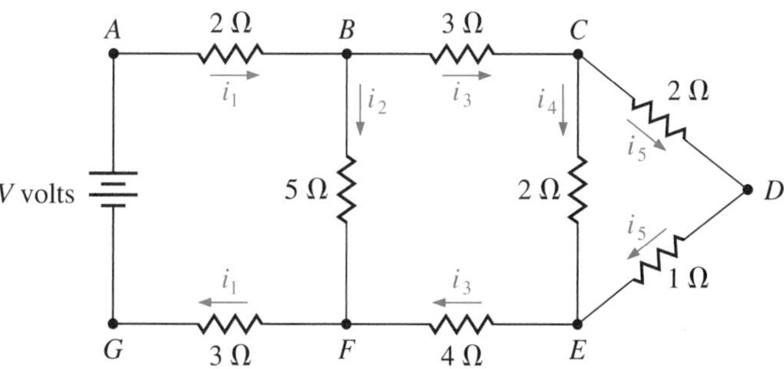

The solution of systems of this type will be considered in this chapter. This application is discussed in Exercise 29 of Section 6.6.

Linear systems of equations are associated with many problems in engineering and science, as well as with applications of mathematics to the social sciences and the quantitative study of business and economic problems.

In this chapter we consider *direct methods* for solving a linear system of n equations in n variables. Such a system has the form

$$\begin{aligned}
E_1 :&\quad a_{11}x_1 + a_{12}x_2 + \cdots + a_{1n}x_n = b_1,\\
E_2 :&\quad a_{21}x_1 + a_{22}x_2 + \cdots + a_{2n}x_n = b_2,\\
&\qquad\qquad\vdots\\
E_n :&\quad a_{n1}x_1 + a_{n2}x_2 + \cdots + a_{nn}x_n = b_n.
\end{aligned} \tag{6.1}$$

In this system we are given the constants a_{ij}, for each $i, j = 1, 2, \ldots, n$, and b_i, for each $i = 1, 2, \ldots, n$, and we need to determine the unknowns $x_1, \ldots, x_n$.

Direct techniques are methods that theoretically give the exact solution to the system in a finite number of steps. In practice, of course, the solution obtained will be contaminated by the round-off error that is involved with the arithmetic being used. Analyzing the effect of this round-off error and determining ways to keep it under control will be a major component of this chapter.

A course in linear algebra is not assumed to be prerequisite for this chapter, so we will include a number of the basic notions of the subject. These results will also be used in Chapter 7, where we consider methods of approximating the solution to linear systems using iterative methods.

6.1 Linear Systems of Equations

We use three operations to simplify the linear system given in (6.1):

1. Equation E_i can be multiplied by any nonzero constant λ with the resulting equation used in place of E_i. This operation is denoted $(\lambda E_i) \to (E_i)$.

2. Equation E_j can be multiplied by any constant λ and added to equation E_i with the resulting equation used in place of E_i. This operation is denoted $(E_i + \lambda E_j) \to (E_i)$.

3. Equations E_i and E_j can be transposed in order. This operation is denoted $(E_i) \leftrightarrow (E_j)$.

By a sequence of these operations, a linear system will be systematically transformed into to a new linear system that is more easily solved and has the same solutions. The sequence of operations is illustrated in the following.

Illustration The four equations

$$\begin{aligned}
E_1 &: \quad x_1 + x_2 \qquad\quad + 3x_4 = 4,\\
E_2 &: \quad 2x_1 + x_2 - x_3 + x_4 = 1,\\
E_3 &: \quad 3x_1 - x_2 - x_3 + 2x_4 = -3,\\
E_4 &: \quad -x_1 + 2x_2 + 3x_3 - x_4 = 4,
\end{aligned} \tag{6.2}$$

will be solved for x_1, x_2, x_3, and x_4. We first use equation E_1 to eliminate the unknown x_1 from equations E_2, E_3, and E_4 by performing $(E_2 - 2E_1) \to (E_2)$, $(E_3 - 3E_1) \to (E_3)$, and $(E_4 + E_1) \to (E_4)$. For example, in the second equation

$$(E_2 - 2E_1) \to (E_2)$$

produces

$$(2x_1 + x_2 - x_3 + x_4) - 2(x_1 + x_2 + 3x_4) = 1 - 2(4).$$

which simplifies to the result shown as E_2 in

$$\begin{aligned}
E_1 &: \quad x_1 + x_2 \qquad\quad + 3x_4 = 4,\\
E_2 &: \qquad\; - x_2 - x_3 - 5x_4 = -7,\\
E_3 &: \qquad\; - 4x_2 - x_3 - 7x_4 = -15,\\
E_4 &: \qquad\quad 3x_2 + 3x_3 + 2x_4 = 8.
\end{aligned}$$

For simplicity, the new equations are again labeled E_1, E_2, E_3, and E_4.

In the new system, E_2 is used to eliminate the unknown x_2 from E_3 and E_4 by performing $(E_3 - 4E_2) \to (E_3)$ and $(E_4 + 3E_2) \to (E_4)$. This results in

$$\begin{aligned} E_1 &: \quad x_1 + x_2 \qquad\quad + \ 3x_4 = 4, \\ E_2 &: \qquad\ - x_2 - x_3 - 5x_4 = -7, \\ E_3 &: \qquad\qquad\quad 3x_3 + 13x_4 = 13, \\ E_4 &: \qquad\qquad\qquad\ - 13x_4 = -13. \end{aligned} \tag{6.3}$$

The system of equations (6.3) is now in **triangular** (or **reduced) form** and can be solved for the unknowns by a **backward-substitution process**. Since E_4 implies $x_4 = 1$, we can solve E_3 for x_3 to give

$$x_3 = \frac{1}{3}(13 - 13x_4) = \frac{1}{3}(13 - 13) = 0.$$

Continuing, E_2 gives

$$x_2 = -(-7 + 5x_4 + x_3) = -(-7 + 5 + 0) = 2,$$

and E_1 gives

$$x_1 = 4 - 3x_4 - x_2 = 4 - 3 - 2 = -1.$$

The solution to system (6.3), and consequently to system (6.2), is therefore, $x_1 = -1$, $x_2 = 2$, $x_3 = 0$, and $x_4 = 1$. □

Matrices and Vectors

When performing the calculations in the Illustration, we would not need to write out the full equations at each step or to carry the variables x_1, x_2, x_3, and x_4 through the calculations, if they always remained in the same column. The only variation from system to system occurs in the coefficients of the unknowns and in the values on the right side of the equations. For this reason, a linear system is often replaced by a *matrix*, which contains all the information about the system that is necessary to determine its solution, but in a compact form, and one that is easily represented in a computer.

Definition 6.1 An $\boldsymbol{n \times m}$ (**$\boldsymbol{n}$ by $\boldsymbol{m}$**) **matrix** is a rectangular array of elements with n rows and m columns in which not only is the value of an element important, but also its position in the array. ■

The notation for an $n \times m$ matrix will be a capital letter such as A for the matrix and lowercase letters with double subscripts, such as a_{ij}, to refer to the entry at the intersection of the ith row and jth column; that is,

$$A = [a_{ij}] = \begin{bmatrix} a_{11} & a_{12} & \cdots & a_{1m} \\ a_{21} & a_{22} & \cdots & a_{2m} \\ \vdots & \vdots & & \vdots \\ a_{n1} & a_{n2} & \cdots & a_{nm} \end{bmatrix}.$$

Example 1 Determine the size and respective entries of the matrix

$$A = \begin{bmatrix} 2 & -1 & 7 \\ 3 & 1 & 0 \end{bmatrix}.$$

Solution The matrix has two rows and three columns so it is of size 2×3. It entries are described by $a_{11} = 2$, $a_{12} = -1$, $a_{13} = 7$, $a_{21} = 3$, $a_{22} = 1$, and $a_{23} = 0$. ■

The $1 \times n$ matrix

$$A = [a_{11} \ a_{12} \ \cdots \ a_{1n}]$$

is called an ***n*-dimensional row vector**, and an $n \times 1$ matrix

$$A = \begin{bmatrix} a_{11} \\ a_{21} \\ \vdots \\ a_{n1} \end{bmatrix}$$

is called an ***n*-dimensional column vector**. Usually the unnecessary subscripts are omitted for vectors, and a boldface lowercase letter is used for notation. Thus

$$\mathbf{x} = \begin{bmatrix} x_1 \\ x_2 \\ \vdots \\ x_n \end{bmatrix}$$

denotes a column vector, and

$$\mathbf{y} = [y_1 \ y_2 \ \ldots \ y_n]$$

a row vector. In addition, row vectors often have commas inserted between the entries to make the separation clearer. So you might see $\mathbf{y}$ written as $\mathbf{y} = [y_1, \ y_2, \ \ldots, \ y_n]$.

An $n \times (n+1)$ matrix can be used to represent the linear system

$$\begin{aligned} a_{11}x_1 + a_{12}x_2 + \cdots + a_{1n}x_n &= b_1, \\ a_{21}x_1 + a_{22}x_2 + \cdots + a_{2n}x_n &= b_2, \\ \vdots \qquad\qquad & \quad \vdots \\ a_{n1}x_1 + a_{n2}x_2 + \cdots + a_{nn}x_n &= b_n, \end{aligned}$$

by first constructing

$$A = [a_{ij}] = \begin{bmatrix} a_{11} & a_{12} & \cdots & a_{1n} \\ a_{21} & a_{22} & \cdots & a_{2n} \\ \vdots & \vdots & & \vdots \\ a_{n1} & a_{n2} & \cdots & a_{nn} \end{bmatrix} \quad \text{and} \quad \mathbf{b} = \begin{bmatrix} b_1 \\ b_2 \\ \vdots \\ b_n \end{bmatrix}$$

$$[A, \mathbf{b}] = \left[\begin{array}{cccc:c} a_{11} & a_{12} & \cdots & a_{1n} & b_1 \\ a_{21} & a_{22} & \cdots & a_{2n} & b_2 \\ \vdots & \vdots & & \vdots & \vdots \\ a_{n1} & a_{n2} & \cdots & a_{nn} & b_n \end{array}\right],$$

Augmented refers to the fact that the right-hand side of the system has been included in the matrix.

where the vertical dotted line is used to separate the coefficients of the unknowns from the values on the right-hand side of the equations. The array $[A, \mathbf{b}]$ is called an **augmented matrix**.

Repeating the operations involved in Example 1 with the matrix notation results in first considering the augmented matrix:

$$\left[\begin{array}{rrrr:r} 1 & 1 & 0 & 3 & 4 \\ 2 & 1 & -1 & 1 & 1 \\ 3 & -1 & -1 & 2 & -3 \\ -1 & 2 & 3 & -1 & 4 \end{array}\right].$$

Performing the operations as described in that example produces the augmented matrices

$$\left[\begin{array}{rrrr:r} 1 & 1 & 0 & 3 & 4 \\ 0 & -1 & -1 & -5 & -7 \\ 0 & -4 & -1 & -7 & -15 \\ 0 & 3 & 3 & 2 & 8 \end{array}\right] \quad \text{and} \quad \left[\begin{array}{rrrr:r} 1 & 1 & 0 & 3 & 4 \\ 0 & -1 & -1 & -5 & -7 \\ 0 & 0 & 3 & 13 & 13 \\ 0 & 0 & 0 & -13 & -13 \end{array}\right].$$

A technique similar to Gaussian elimination first appeared during the Han dynasty in China in the text *Nine Chapters on the Mathematical Art*, which was written about 200 B.C.E. Joseph Louis Lagrange (1736–1813) described a technique similar to this procedure in 1778 for the case when the value of each equation is 0. Gauss gave a more general description in *Theoria Motus corporum coelestium sectionibus solem ambientium*, which described the least squares technique he used in 1801 to determine the orbit of the minor planet Ceres.

The final matrix can now be transformed into its corresponding linear system, and solutions for x_1, x_2, x_3, and x_4, can be obtained. The procedure is called **Gaussian elimination with backward substitution**.

The general Gaussian elimination procedure applied to the linear system

$$\begin{aligned} E_1 &: \quad a_{11}x_1 + a_{12}x_2 + \cdots + a_{1n}x_n = b_1, \\ E_2 &: \quad a_{21}x_1 + a_{22}x_2 + \cdots + a_{2n}x_n = b_2, \\ &\qquad \vdots \qquad\qquad\qquad\qquad \vdots \\ E_n &: \quad a_{n1}x_1 + a_{n2}x_2 + \cdots + a_{nn}x_n = b_n, \end{aligned} \tag{6.4}$$

is handled in a similar manner. First form the augmented matrix $\tilde{A}$:

$$\tilde{A} = [A, \mathbf{b}] = \left[\begin{array}{cccc:c} a_{11} & a_{12} & \cdots & a_{1n} & a_{1,n+1} \\ a_{21} & a_{22} & \cdots & a_{2n} & a_{2,n+1} \\ \vdots & \vdots & & \vdots & \vdots \\ a_{n1} & a_{n2} & \cdots & a_{nn} & a_{n,n+1} \end{array}\right], \tag{6.5}$$

where A denotes the matrix formed by the coefficients. The entries in the $(n+1)$st column are the values of $\mathbf{b}$; that is, $a_{i,n+1} = b_i$ for each $i = 1, 2, \cdots, n$.

Provided $a_{11} \neq 0$, we perform the operations corresponding to

$$(E_j - (a_{j1}/a_{11})E_1) \rightarrow (E_j) \quad \text{for each } j = 2, 3, \ldots, n$$

to eliminate the coefficient of x_1 in each of these rows. Although the entries in rows $2, 3, \ldots, n$ are expected to change, for ease of notation we again denote the entry in the ith row and the jth column by a_{ij}. With this in mind, we follow a sequential procedure for $i = 2, 3, \ldots, n-1$ and perform the operation

$$(E_j - (a_{ji}/a_{ii})E_i) \rightarrow (E_j) \quad \text{for each } j = i+1, i+2, \ldots, n,$$

provided $a_{ii} \neq 0$. This eliminates (changes the coefficient to zero) x_i in each row below the ith for all values of $i = 1, 2, \ldots, n-1$. The resulting matrix has the form:

$$\tilde{\tilde{A}} = \left[\begin{array}{cccc:c} a_{11} & a_{12} & \cdots & a_{1n} & a_{1,n+1} \\ 0 & a_{22} & \cdots & a_{2n} & a_{2,n+1} \\ \vdots & \ddots & \ddots & \vdots & \vdots \\ 0 & \cdots & 0 & a_{nn} & a_{n,n+1} \end{array}\right],$$

where, except in the first row, the values of a_{ij} are not expected to agree with those in the original matrix $\tilde{A}$. The matrix $\tilde{\tilde{A}}$ represents a linear system with the same solution set as the original system.

The new linear system is triangular,

$$\begin{aligned} a_{11}x_1 + a_{12}x_2 + \cdots + a_{1n}x_n &= a_{1,n+1}, \\ a_{22}x_2 + \cdots + a_{2n}x_n &= a_{2,n+1}, \\ \ddots \qquad \vdots \quad &\qquad \vdots \\ a_{nn}x_n &= a_{n,n+1}, \end{aligned}$$

so *backward substitution* can be performed. Solving the nth equation for x_n gives

$$x_n = \frac{a_{n,n+1}}{a_{nn}}.$$

Solving the $(n-1)$st equation for x_{n-1} and using the known value for x_n yields

$$x_{n-1} = \frac{a_{n-1,n+1} - a_{n-1,n}x_n}{a_{n-1,n-1}}.$$

Continuing this process, we obtain

$$x_i = \frac{a_{i,n+1} - a_{i,n}x_n - a_{i,n-1}x_{n-1} - \cdots - a_{i,i+1}x_{i+1}}{a_{ii}} = \frac{a_{i,n+1} - \sum_{j=i+1}^{n} a_{ij}x_j}{a_{ii}},$$

for each $i = n-1, n-2, \cdots, 2, 1$.

Gaussian elimination procedure is described more precisely, although more intricately, by forming a sequence of augmented matrices $\tilde{A}^{(1)}, \tilde{A}^{(2)}, \ldots, \tilde{A}^{(n)}$, where $\tilde{A}^{(1)}$ is the matrix $\tilde{A}$ given in (6.5) and $\tilde{A}^{(k)}$, for each $k = 2, 3, \ldots, n$, has entries $a_{ij}^{(k)}$, where:

$$a_{ij}^{(k)} = \begin{cases} a_{ij}^{(k-1)}, & \text{when } i = 1, 2, \ldots, k-1 \text{ and } j = 1, 2, \ldots, n+1, \\ 0, & \text{when } i = k, k+1, \ldots, n \text{ and } j = 1, 2, \cdots, k-1, \\ a_{ij}^{(k-1)} - \dfrac{a_{i,k-1}^{(k-1)}}{a_{k-1,k-1}^{(k-1)}} a_{k-1,j}^{(k-1)}, & \text{when } i = k, k+1, \ldots, n \text{ and } j = k, k+1, \ldots, n+1. \end{cases}$$

Thus

$$\tilde{A}^{(k)} = \left[\begin{array}{cccccccccc:c} a_{11}^{(1)} & a_{12}^{(1)} & a_{13}^{(1)} & \cdots & a_{1,k-1}^{(1)} & a_{1k}^{(1)} & \cdots & a_{1n}^{(1)} & a_{1,n+1}^{(1)} \\ 0 & a_{22}^{(2)} & a_{23}^{(2)} & \cdots & a_{2,k-1}^{(2)} & a_{2k}^{(2)} & \cdots & a_{2n}^{(2)} & a_{2,n+1}^{(2)} \\ \vdots & & \ddots & & \vdots & \vdots & & \vdots & \vdots \\ \vdots & & & \ddots & a_{k-1,k-1}^{(k-1)} & a_{k-1,k}^{(k-1)} & \cdots & a_{k-1,n}^{(k-1)} & a_{k-1,n+1}^{(k-1)} \\ \vdots & & & & 0 & a_{kk}^{(k)} & \cdots & a_{kn}^{(k)} & a_{k,n+1}^{(k)} \\ \vdots & & & & \vdots & \vdots & & \vdots & \vdots \\ 0 & \cdots & \cdots & \cdots & 0 & a_{nk}^{(k)} & \cdots & a_{nn}^{(k)} & a_{n,n+1}^{(k)} \end{array}\right] \tag{6.6}$$

represents the equivalent linear system for which the variable x_{k-1} has just been eliminated from equations $E_k, E_{k+1}, \ldots, E_n$.

The procedure will fail if one of the elements $a_{11}^{(1)}, a_{22}^{(2)}, a_{33}^{(3)}, \ldots, a_{n-1,n-1}^{(n-1)}, a_{nn}^{(n)}$ is zero because the step

$$\left(E_i - \frac{a_{i,k}^{(k)}}{a_{kk}^{(k)}}(E_k)\right) \to E_i$$

either cannot be performed (this occurs if one of $a_{11}^{(1)}, \ldots, a_{n-1,n-1}^{(n-1)}$ is zero), or the backward substitution cannot be accomplished (in the case $a_{nn}^{(n)} = 0$). The system may still have a solution, but the technique for finding the solution must be altered. An illustration is given in the following example.

Example 2 Represent the linear system

$$\begin{aligned}
E_1: &\quad x_1 - x_2 + 2x_3 - x_4 = -8,\\
E_2: &\quad 2x_1 - 2x_2 + 3x_3 - 3x_4 = -20,\\
E_3: &\quad x_1 + x_2 + x_3 \qquad = -2,\\
E_4: &\quad x_1 - x_2 + 4x_3 + 3x_4 = 4,
\end{aligned}$$

as an augmented matrix and use Gaussian Elimination to find its solution.

Solution The augmented matrix is

$$\tilde{A} = \tilde{A}^{(1)} = \left[\begin{array}{cccc:c} 1 & -1 & 2 & -1 & -8 \\ 2 & -2 & 3 & -3 & -20 \\ 1 & 1 & 1 & 0 & -2 \\ 1 & -1 & 4 & 3 & 4 \end{array}\right].$$

Performing the operations

$$(E_2 - 2E_1) \to (E_2),\ (E_3 - E_1) \to (E_3), \quad \text{and} \quad (E_4 - E_1) \to (E_4),$$

gives

$$\tilde{A}^{(2)} = \left[\begin{array}{cccc:c} 1 & -1 & 2 & -1 & -8 \\ 0 & 0 & -1 & -1 & -4 \\ 0 & 2 & -1 & 1 & 6 \\ 0 & 0 & 2 & 4 & 12 \end{array}\right].$$

The pivot element for a specific column is the entry that is used to place zeros in the other entries in that column.

The diagonal entry $a_{22}^{(2)}$, called the **pivot element**, is 0, so the procedure cannot continue in its present form. But operations $(E_i) \leftrightarrow (E_j)$ are permitted, so a search is made of the elements $a_{32}^{(2)}$ and $a_{42}^{(2)}$ for the first nonzero element. Since $a_{32}^{(2)} \neq 0$, the operation $(E_2) \leftrightarrow (E_3)$ is performed to obtain a new matrix,

$$\tilde{A}^{(2)'} = \left[\begin{array}{cccc:c} 1 & -1 & 2 & -1 & -8 \\ 0 & 2 & -1 & 1 & 6 \\ 0 & 0 & -1 & -1 & -4 \\ 0 & 0 & 2 & 4 & 12 \end{array}\right].$$

Since x_2 is already eliminated from E_3 and E_4, $\tilde{A}^{(3)}$ will be $\tilde{A}^{(2)'}$, and the computations continue with the operation $(E_4 + 2E_3) \to (E_4)$, giving

$$\tilde{A}^{(4)} = \left[\begin{array}{cccc:c} 1 & -1 & 2 & -1 & -8 \\ 0 & 2 & -1 & 1 & 6 \\ 0 & 0 & -1 & -1 & -4 \\ 0 & 0 & 0 & 2 & 4 \end{array}\right].$$

Finally, the matrix is converted back into a linear system that has a solution equivalent to the solution of the original system and the backward substitution is applied:

$$x_4 = \frac{4}{2} = 2,$$

$$x_3 = \frac{[-4 - (-1)x_4]}{-1} = 2,$$

$$x_2 = \frac{[6 - x_4 - (-1)x_3]}{2} = 3,$$

$$x_1 = \frac{[-8 - (-1)x_4 - 2x_3 - (-1)x_2]}{1} = -7.$$

■

Example 2 illustrates what is done if $a_{kk}^{(k)} = 0$ for some $k = 1, 2, \ldots, n-1$. The kth column of $\tilde{A}^{(k-1)}$ from the kth row to the nth row is searched for the first nonzero entry. If $a_{pk}^{(k)} \neq 0$ for some p,with $k + 1 \leq p \leq n$, then the operation $(E_k) \leftrightarrow (E_p)$ is performed to obtain $\tilde{A}^{(k-1)'}$. The procedure can then be continued to form $\tilde{A}^{(k)}$, and so on. If $a_{pk}^{(k)} = 0$ for each p, it can be shown (see Theorem 6.17 on page 398) that the linear system does not have a unique solution and the procedure stops. Finally, if $a_{nn}^{(n)} = 0$, the linear system does not have a unique solution, and again the procedure stops.

Algorithm 6.1 summarizes Gaussian elimination with backward substitution. The algorithm incorporates pivoting when one of the pivots $a_{kk}^{(k)}$ is 0 by interchanging the kth row with the pth row, where p is the smallest integer greater than k for which $a_{pk}^{(k)} \neq 0$.

Gaussian Elimination with Backward Substitution

To solve the $n \times n$ linear system

$$\begin{aligned}
E_1 &: \ a_{11}x_1 + a_{12}x_2 + \cdots + a_{1n}x_n = a_{1,n+1}\\
E_2 &: \ a_{21}x_1 + a_{22}x_2 + \cdots + a_{2n}x_n = a_{2,n+1}\\
&\ \ \vdots \qquad \vdots \qquad\quad \vdots \qquad\qquad \vdots \qquad\quad \vdots\\
E_n &: \ a_{n1}x_1 + a_{n2}x_2 + \cdots + a_{nn}x_n = a_{n,n+1}
\end{aligned}$$

INPUT number of unknowns and equations n; augmented matrix $A = [a_{ij}]$, where $1 \leq i \leq n$ and $1 \leq j \leq n+1$.

OUTPUT solution $x_1, x_2, \ldots, x_n$ or message that the linear system has no unique solution.

Step 1 For $i = 1, \ldots, n-1$ do Steps 2–4. *(Elimination process.)*

Step 2 Let p be the smallest integer with $i \leq p \leq n$ and $a_{pi} \neq 0$.
If no integer p can be found
then OUTPUT ('no unique solution exists');
STOP.

Step 3 If $p \neq i$ then perform $(E_p) \leftrightarrow (E_i)$.

Step 4 For $j = i+1, \ldots, n$ do Steps 5 and 6.

Step 5 Set $m_{ji} = a_{ji}/a_{ii}$.

Step 6 Perform $(E_j - m_{ji}E_i) \rightarrow (E_j)$;

Step 7 If $a_{nn} = 0$ then OUTPUT ('no unique solution exists');
STOP.

Step 8 Set $x_n = a_{n,n+1}/a_{nn}$. *(Start backward substitution.)*

Step 9 For $i = n-1, \ldots, 1$ set $x_i = \left[a_{i,n+1} - \sum_{j=i+1}^{n} a_{ij}x_j\right] \Big/ a_{ii}$.

Step 10 OUTPUT $(x_1, \ldots, x_n)$; *(Procedure completed successfully.)*
STOP. ■

To define matrices and perform Gaussian elimination using Maple, first access the *LinearAlgebra* library using the command

with(LinearAlgebra)

To define the matrix $\tilde{A}^{(1)}$ of Example 2, which we will call *AA*, use the command

$AA := Matrix([[1, -1, 2, -1, -8], [2, -2, 3, -3, -20], [1, 1, 1, 0, -2], [1, -1, 4, 3, 4]])$

This lists the entries, by row, of the augmented matrix $AA \equiv \tilde{A}^{(1)}$.

The function $RowOperation(AA, [i, j], m)$ performs the operation $(E_j + mE_i) \rightarrow (E_j)$, and the same command without the last parameter, that is, $RowOperation(AA, [i, j])$ performs the operation $(E_i) \leftrightarrow (E_j)$. So the sequence of operations

$AA1 := RowOperation(AA, [2, 1], -2)$
$AA2 := RowOperation(AA1, [3, 1], -1)$
$AA3 := RowOperation(AA2, [4, 1], -1)$
$AA4 := RowOperation(AA3, [2, 3])$
$AA5 := RowOperation(AA4, [4, 3], 2)$

gives the reduction to $AA5 \equiv \tilde{A}^{(4)}$.

Gaussian Elimination is a standard routine in the *LinearAlgebra* package of Maple, and the single command

$AA5 := GaussianElimination(AA)$

returns this same reduced matrix. In either case, the final operation

$x := BackwardSubstitute(AA5)$

gives the solution $\mathbf{x}$ which has $x_1 = -7$, $x_2 = 3$, $x_3 = 2$, and $x_4 = 2$.

Illustration The purpose of this illustration is to show what can happen if Algorithm 6.1 fails. The computations will be done simultaneously on two linear systems:

$$\begin{aligned} x_1 + x_2 + x_3 &= 4, \\ 2x_1 + 2x_2 + x_3 &= 6, \\ x_1 + x_2 + 2x_3 &= 6, \end{aligned} \qquad \text{and} \qquad \begin{aligned} x_1 + x_2 + x_3 &= 4, \\ 2x_1 + 2x_2 + x_3 &= 4, \\ x_1 + x_2 + 2x_3 &= 6. \end{aligned}$$

These systems produce the augmented matrices

$$\tilde{A} = \begin{bmatrix} 1 & 1 & 1 & \vdots & 4 \\ 2 & 2 & 1 & \vdots & 6 \\ 1 & 1 & 2 & \vdots & 6 \end{bmatrix} \quad \text{and} \quad \tilde{A} = \begin{bmatrix} 1 & 1 & 1 & \vdots & 4 \\ 2 & 2 & 1 & \vdots & 4 \\ 1 & 1 & 2 & \vdots & 6 \end{bmatrix}.$$

Since $a_{11} = 1$, we perform $(E_2 - 2E_1) \rightarrow (E_2)$ and $(E_3 - E_1) \rightarrow (E_3)$ to produce

$$\tilde{A} = \begin{bmatrix} 1 & 1 & 1 & \vdots & 4 \\ 0 & 0 & -1 & \vdots & -2 \\ 0 & 0 & 1 & \vdots & 2 \end{bmatrix} \quad \text{and} \quad \tilde{A} = \begin{bmatrix} 1 & 1 & 1 & \vdots & 4 \\ 0 & 0 & -1 & \vdots & -4 \\ 0 & 0 & 1 & \vdots & 2 \end{bmatrix}.$$

At this point, $a_{22} = a_{32} = 0$. The algorithm requires that the procedure be halted, and no solution to either system is obtained. Writing the equations for each system gives

$$\begin{aligned} x_1 + x_2 + \quad x_3 &= \quad 4, \\ -x_3 &= -2, \\ x_3 &= \quad 2, \end{aligned} \qquad \text{and} \qquad \begin{aligned} x_1 + x_2 + \quad x_3 &= \quad 4, \\ -x_3 &= -4, \\ x_3 &= \quad 2. \end{aligned}$$

The first linear system has an infinite number of solutions, which can be described by $x_3 = 2$, $x_2 = 2 - x_1$, and x_1 arbitrary.

The second system leads to the contradiction $x_3 = 2$ and $x_3 = 4$, so no solution exists. In each case, however, there is no *unique* solution, as we conclude from Algorithm 6.1. □

Although Algorithm 6.1 can be viewed as the construction of the augmented matrices $\tilde{A}^{(1)}, \ldots, \tilde{A}^{(n)}$, the computations can be performed in a computer using only one $n \times (n+1)$ array for storage. At each step we simply replace the previous value of a_{ij} by the new one. In addition, we can store the multipliers m_{ji} in the locations of a_{ji} because a_{ji} has the value 0 for each $i = 1, 2, \ldots, n-1$ and $j = i+1, i+2, \ldots, n$. Thus A can be overwritten by the multipliers in the entries that are below the main diagonal (that is, the entries of the form a_{ji}, with $j > i$) and by the newly computed entries of $\tilde{A}^{(n)}$ on and above the main diagonal (the entries of the form a_{ij}, with $j \leq i$). These values can be used to solve other linear systems involving the original matrix A, as we will see in Section 6.5.

Operation Counts

Both the amount of time required to complete the calculations and the subsequent round-off error depend on the number of floating-point arithmetic operations needed to solve a routine problem. In general, the amount of time required to perform a multiplication or division on a computer is approximately the same and is considerably greater than that required to perform an addition or subtraction. The actual differences in execution time, however, depend on the particular computing system. To demonstrate the counting operations for a given method, we will count the operations required to solve a typical linear system of n equations in n unknowns using Algorithm 6.1. We will keep the count of the additions/subtractions separate from the count of the multiplications/divisions because of the time differential.

No arithmetic operations are performed until Steps 5 and 6 in the algorithm. Step 5 requires that $(n - i)$ divisions be performed. The replacement of the equation E_j by $(E_j - m_{ji}E_i)$ in Step 6 requires that m_{ji} be multiplied by each term in E_i, resulting in a total of $(n - i)(n - i + 1)$ multiplications. After this is completed, each term of the resulting equation is subtracted from the corresponding term in E_j. This requires $(n - i)(n - i + 1)$ subtractions. For each $i = 1, 2, \ldots, n - 1$, the operations required in Steps 5 and 6 are as follows.

Multiplications/divisions

$$(n - i) + (n - i)(n - i + 1) = (n - i)(n - i + 2).$$

Additions/subtractions

$$(n - i)(n - i + 1).$$

The total number of operations required by Steps 5 and 6 is obtained by summing the operation counts for each i. Recalling from calculus that

$$\sum_{j=1}^{m} 1 = m, \quad \sum_{j=1}^{m} j = \frac{m(m+1)}{2}, \quad \text{and} \quad \sum_{j=1}^{m} j^2 = \frac{m(m+1)(2m+1)}{6},$$

we have the following operation counts.

Multiplications/divisions

$$\begin{aligned}\sum_{i=1}^{n-1}(n-i)(n-i+2) &= \sum_{i=1}^{n-1}(n^2 - 2ni + i^2 + 2n - 2i)\\ &= \sum_{i=1}^{n-1}(n-i)^2 + 2\sum_{i=1}^{n-1}(n-i) = \sum_{i=1}^{n-1} i^2 + 2\sum_{i=1}^{n-1} i\\ &= \frac{(n-1)n(2n-1)}{6} + 2\frac{(n-1)n}{2} = \frac{2n^3+3n^2-5n}{6}.\end{aligned}$$

Additions/subtractions

$$\begin{aligned}\sum_{i=1}^{n-1}(n-i)(n-i+1) &= \sum_{i=1}^{n-1}(n^2 - 2ni + i^2 + n - i)\\ &= \sum_{i=1}^{n-1}(n-i)^2 + \sum_{i=1}^{n-1}(n-i) = \sum_{i=1}^{n-1} i^2 + \sum_{i=1}^{n-1} i\\ &= \frac{(n-1)n(2n-1)}{6} + \frac{(n-1)n}{2} = \frac{n^3-n}{3}.\end{aligned}$$

The only other steps in Algorithm 6.1 that involve arithmetic operations are those required for backward substitution, Steps 8 and 9. Step 8 requires one division. Step 9 requires $(n - i)$ multiplications and $(n - i - 1)$ additions for each summation term and then one subtraction and one division. The total number of operations in Steps 8 and 9 is as follows.

Multiplications/divisions

$$\begin{aligned}1 + \sum_{i=1}^{n-1}((n-i)+1) &= 1 + \left(\sum_{i=1}^{n-1}(n-i)\right) + n - 1\\ 1 &= n + \sum_{i=1}^{n-1}(n-i) = n + \sum_{i=1}^{n-1} i = \frac{n^2+n}{2}.\end{aligned}$$

Additions/subtractions

$$\sum_{i=1}^{n-1}((n-i-1)+1) = \sum_{i=1}^{n-1}(n-i) = \sum_{i=1}^{n-1} i = \frac{n^2-n}{2}$$

The total number of arithmetic operations in Algorithm 6.1 is, therefore:

Multiplications/divisions

$$\frac{2n^3 + 3n^2 - 5n}{6} + \frac{n^2 + n}{2} = \frac{n^3}{3} + n^2 - \frac{n}{3}.$$

Additions/subtractions

$$\frac{n^3 - n}{3} + \frac{n^2 - n}{2} = \frac{n^3}{3} + \frac{n^2}{2} - \frac{5n}{6}.$$

For large n, the total number of multiplications and divisions is approximately $n^3/3$, as is the total number of additions and subtractions. Thus the amount of computation and the time required increases with n in proportion to n^3, as shown in Table 6.1.

Table 6.1

n	Multiplications/Divisions	Additions/Subtractions
3	17	11
10	430	375
50	44,150	42,875
100	343,300	338,250

EXERCISE SET 6.1

1. For each of the following linear systems, obtain a solution by graphical methods, if possible. Explain the results from a geometrical standpoint.

   a. $x_1 + 2x_2 = 3,$ $2x_1 + 4x_2 = 6.$
   b. $x_1 + 2x_2 = 3,$ $x_1 - x_2 = 0.$
   c. $x_1 + 2x_2 = 0,$ $2x_1 + 4x_2 = 0.$
   d. $2x_1 + x_2 = -1,$ $4x_1 + 2x_2 = -2,$ $x_1 - 3x_2 = 5.$

2. For each of the following linear systems, obtain a solution by graphical methods, if possible. Explain the results from a geometrical standpoint.

   a. $x_1 + 2x_2 = 3,$ $-2x_1 - 4x_2 = 6.$
   b. $x_1 + 2x_2 = 0,$ $x_1 - x_2 = 0.$
   c. $2x_1 + x_2 = -1,$ $x_1 + x_2 = 2,$ $x_1 - 3x_2 = 5.$
   d. $2x_1 + x_2 + x_3 = 1,$ $2x_1 + 4x_2 - x_3 = -1.$

3. Use Gaussian elimination with backward substitution and two-digit rounding arithmetic to solve the following linear systems. Do not reorder the equations. (The exact solution to each system is $x_1 = 1, x_2 = -1, x_3 = 3$.)

   a. $4x_1 + x_2 + 2x_3 = 9,$ $2x_1 + 4x_2 - x_3 = -5,$ $x_1 + x_2 - 3x_3 = -9.$
   b. $4x_1 - x_2 + x_3 = 8,$ $2x_1 + 5x_2 + 2x_3 = 3,$ $x_1 + 2x_2 + 4x_3 = 11.$

4. Use Gaussian elimination with backward substitution and two-digit rounding arithmetic to solve the following linear systems. Do not reorder the equations. (The exact solution to each system is $x_1 = -1, x_2 = 1, x_3 = 3$.)

   a. $-x_1 + 4x_2 + x_3 = 8,$ $\frac{5}{3}x_1 + \frac{2}{3}x_2 + \frac{2}{3}x_3 = 1,$ $2x_1 + x_2 + 4x_3 = 11.$
   b. $4x_1 + 2x_2 - x_3 = -5,$ $\frac{1}{9}x_1 + \frac{1}{9}x_2 - \frac{1}{3}x_3 = -1,$ $x_1 + 4x_2 + 2x_3 = 9.$

5. Use the Gaussian Elimination Algorithm to solve the following linear systems, if possible, and determine whether row interchanges are necessary:

a. $$\begin{aligned} x_1 - x_2 + 3x_3 &= 2,\\ 3x_1 - 3x_2 + x_3 &= -1,\\ x_1 + x_2 &= 3. \end{aligned}$$

b. $$\begin{aligned} 2x_1 - 1.5x_2 + 3x_3 &= 1,\\ -x_1 + 2x_3 &= 3,\\ 4x_1 - 4.5x_2 + 5x_3 &= 1. \end{aligned}$$

c. $$\begin{aligned} 2x_1 &= 3,\\ x_1 + 1.5x_2 &= 4.5,\\ -3x_2 + 0.5x_3 &= -6.6,\\ 2x_1 - 2x_2 + x_3 + x_4 &= 0.8. \end{aligned}$$

d. $$\begin{aligned} x_1 + x_2 + x_4 &= 2,\\ 2x_1 + x_2 - x_3 + x_4 &= 1,\\ 4x_1 - x_2 - 2x_3 + 2x_4 &= 0,\\ 3x_1 - x_2 - x_3 + 2x_4 &= -3. \end{aligned}$$

6. Use the Gaussian Elimination Algorithm to solve the following linear systems, if possible, and determine whether row interchanges are necessary:

a. $$\begin{aligned} x_2 - 2x_3 &= 4,\\ x_1 - x_2 + x_3 &= 6,\\ x_1 - x_3 &= 2. \end{aligned}$$

b. $$\begin{aligned} x_1 - \tfrac{1}{2}x_2 + x_3 &= 4,\\ 2x_1 - x_2 - x_3 + x_4 &= 5,\\ x_1 + x_2 + \tfrac{1}{2}x_3 &= 2,\\ x_1 - \tfrac{1}{2}x_2 + x_3 + x_4 &= 5. \end{aligned}$$

c. $$\begin{aligned} 2x_1 - x_2 + x_3 - x_4 &= 6,\\ x_2 - x_3 + x_4 &= 5,\\ x_4 &= 5,\\ x_3 - x_4 &= 3. \end{aligned}$$

d. $$\begin{aligned} x_1 + x_2 + x_4 &= 2,\\ 2x_1 + x_2 - x_3 + x_4 &= 1,\\ -x_1 + 2x_2 + 3x_3 - x_4 &= 4,\\ 3x_1 - x_2 - x_3 + 2x_4 &= -3. \end{aligned}$$

7. Use Algorithm 6.1 and Maple with *Digits*:= 10 to solve the following linear systems.

a. $$\begin{aligned} \tfrac{1}{4}x_1 + \tfrac{1}{5}x_2 + \tfrac{1}{6}x_3 &= 9,\\ \tfrac{1}{3}x_1 + \tfrac{1}{4}x_2 + \tfrac{1}{5}x_3 &= 8,\\ \tfrac{1}{2}x_1 + x_2 + 2x_3 &= 8. \end{aligned}$$

b. $$\begin{aligned} 3.333x_1 + 15920x_2 - 10.333x_3 &= 15913,\\ 2.222x_1 + 16.71x_2 + 9.612x_3 &= 28.544,\\ 1.5611x_1 + 5.1791x_2 + 1.6852x_3 &= 8.4254. \end{aligned}$$

c. $$\begin{aligned} x_1 + \tfrac{1}{2}x_2 + \tfrac{1}{3}x_3 + \tfrac{1}{4}x_4 &= \tfrac{1}{6},\\ \tfrac{1}{2}x_1 + \tfrac{1}{3}x_2 + \tfrac{1}{4}x_3 + \tfrac{1}{5}x_4 &= \tfrac{1}{7},\\ \tfrac{1}{3}x_1 + \tfrac{1}{4}x_2 + \tfrac{1}{5}x_3 + \tfrac{1}{6}x_4 &= \tfrac{1}{8},\\ \tfrac{1}{4}x_1 + \tfrac{1}{5}x_2 + \tfrac{1}{6}x_3 + \tfrac{1}{7}x_4 &= \tfrac{1}{9}. \end{aligned}$$

d. $$\begin{aligned} 2x_1 + x_2 - x_3 + x_4 - 3x_5 &= 7,\\ x_1 + 2x_3 - x_4 + x_5 &= 2,\\ -2x_2 - x_3 + x_4 - x_5 &= -5,\\ 3x_1 + x_2 - 4x_3 + 5x_5 &= 6,\\ x_1 - x_2 - x_3 - x_4 + x_5 &= 3. \end{aligned}$$

8. Use Algorithm 6.1 and Maple with *Digits*:= 10 to solve the following linear systems.

a. $$\begin{aligned} \tfrac{1}{2}x_1 + \tfrac{1}{4}x_2 - \tfrac{1}{8}x_3 &= 0,\\ \tfrac{1}{3}x_1 - \tfrac{1}{6}x_2 + \tfrac{1}{9}x_3 &= 1,\\ \tfrac{1}{7}x_1 + \tfrac{1}{7}x_2 + \tfrac{1}{10}x_3 &= 2. \end{aligned}$$

b. $$\begin{aligned} 2.71x_1 + x_2 + 1032x_3 &= 12,\\ 4.12x_1 - x_2 + 500x_3 &= 11.49,\\ 3.33x_1 + 2x_2 - 200x_3 &= 41. \end{aligned}$$

c. $$\begin{aligned} \pi x_1 + \sqrt{2}x_2 - x_3 + x_4 &= 0,\\ ex_1 - x_2 + x_3 + 2x_4 &= 1,\\ x_1 + x_2 - \sqrt{3}x_3 + x_4 &= 2,\\ -x_1 - x_2 + x_3 - \sqrt{5}x_4 &= 3. \end{aligned}$$

d. $$\begin{aligned} x_1 + x_2 - x_3 + x_4 - x_5 &= 2,\\ 2x_1 + 2x_2 + x_3 - x_4 + x_5 &= 4,\\ 3x_1 + x_2 - 3x_3 - 2x_4 + 3x_5 &= 8,\\ 4x_1 + x_2 - x_3 + 4x_4 - 5x_5 &= 16,\\ 16x_1 - x_2 + x_3 - x_4 - x_5 &= 32. \end{aligned}$$

9. Given the linear system

$$\begin{aligned} 2x_1 - 6\alpha x_2 &= 3,\\ 3\alpha x_1 - x_2 &= \tfrac{3}{2}. \end{aligned}$$

a. Find value(s) of α for which the system has no solutions.

b. Find value(s) of α for which the system has an infinite number of solutions.

c. Assuming a unique solution exists for a given α, find the solution.

10. Given the linear system

$$\begin{aligned} x_1 - \ \ x_2 + \alpha x_3 &= -2, \\ -x_1 + 2x_2 - \alpha x_3 &= 3, \\ \alpha x_1 + \ \ x_2 + \ \ x_3 &= 2. \end{aligned}$$

a. Find value(s) of α for which the system has no solutions.

b. Find value(s) of α for which the system has an infinite number of solutions.

c. Assuming a unique solution exists for a given α, find the solution.

11. Show that the operations

a. $(\lambda E_i) \to (E_i)$ **b.** $(E_i + \lambda E_j) \to (E_i)$ **c.** $(E_i) \leftrightarrow (E_j)$

do not change the solution set of a linear system.

12. **Gauss-Jordan Method:** This method is described as follows. Use the ith equation to eliminate not only x_i from the equations $E_{i+1}, E_{i+2}, \ldots, E_n$, as was done in the Gaussian elimination method, but also from $E_1, E_2, \ldots, E_{i-1}$. Upon reducing $[A, \mathbf{b}]$ to:

$$\left[\begin{array}{cccc:c} a_{11}^{(1)} & 0 & \cdots & 0 & a_{1,n+1}^{(1)} \\ 0 & a_{22}^{(2)} & \ddots & \vdots & a_{2,n+1}^{(2)} \\ \vdots & \ddots & \ddots & 0 & \vdots \\ 0 & \cdots & 0 & a_{nn}^{(n)}) & a_{n,n+1}^{(n)} \end{array}\right],$$

the solution is obtained by setting

$$x_i = \frac{a_{i,n+1}^{(i)}}{a_{ii}^{(i)}},$$

for each $i = 1, 2, \ldots, n$. This procedure circumvents the backward substitution in the Gaussian elimination. Construct an algorithm for the Gauss-Jordan procedure patterned after that of Algorithm 6.1.

13. Use the Gauss-Jordan method and two-digit rounding arithmetic to solve the systems in Exercise 3.

14. Repeat Exercise 7 using the Gauss-Jordan method.

15. **a.** Show that the Gauss-Jordan method requires

$$\frac{n^3}{2} + n^2 - \frac{n}{2} \quad \text{multiplications/divisions}$$

and

$$\frac{n^3}{2} - \frac{n}{2} \quad \text{additions/subtractions.}$$

b. Make a table comparing the required operations for the Gauss-Jordan and Gaussian elimination methods for $n = 3, 10, 50, 100$. Which method requires less computation?

16. Consider the following Gaussian-elimination-Gauss-Jordan hybrid method for solving the system (6.4). First, apply the Gaussian-elimination technique to reduce the system to triangular form. Then use the nth equation to eliminate the coefficients of x_n in each of the first $n - 1$ rows. After this is completed use the $(n - 1)$st equation to eliminate the coefficients of x_{n-1} in the first $n - 2$ rows, etc. The system will eventually appear as the reduced system in Exercise 12.

a. Show that this method requires

$$\frac{n^3}{3} + \frac{3}{2}n^2 - \frac{5}{6}n \quad \text{multiplications/divisions}$$

and

$$\frac{n^3}{3} + \frac{n^2}{2} - \frac{5}{6}n \quad \text{additions/subtractions.}$$

b. Make a table comparing the required operations for the Gaussian elimination, Gauss-Jordan, and hybrid methods, for $n = 3, 10, 50, 100$.

17. Use the hybrid method described in Exercise 16 and two-digit rounding arithmetic to solve the systems in Exercise 3.

18. Repeat Exercise 7 using the method described in Exercise 16.

19. Suppose that in a biological system there are n species of animals and m sources of food. Let x_j represent the population of the jth species, for each $j = 1, \cdots, n$; b_i represent the available daily supply of the ith food; and a_{ij} represent the amount of the ith food consumed on the average by a member of the jth species. The linear system

$$\begin{aligned} a_{11}x_1 + a_{12}x_2 + \cdots + a_{1n}x_n &= b_1, \\ a_{21}x_1 + a_{22}x_2 + \cdots + a_{2n}x_n &= b_2, \\ \vdots \qquad \vdots \qquad\quad \vdots \quad &\quad \vdots \\ a_{m1}x_1 + a_{m2}x_2 + \cdots + a_{mn}x_n &= b_m \end{aligned}$$

represents an equilibrium where there is a daily supply of food to precisely meet the average daily consumption of each species.

a. Let

$$A = [a_{ij}] = \left[\begin{array}{ccc:c} 1 & 2 & 0 & 3 \\ 1 & 0 & 2 & 2 \\ 0 & 0 & 1 & 1 \end{array}\right],$$

$\mathbf{x} = (x_j) = [1000, 500, 350, 400]$, and $\mathbf{b} = (b_i) = [3500, 2700, 900]$. Is there sufficient food to satisfy the average daily consumption?

b. What is the maximum number of animals of each species that could be individually added to the system with the supply of food still meeting the consumption?

c. If species 1 became extinct, how much of an individual increase of each of the remaining species could be supported?

d. If species 2 became extinct, how much of an individual increase of each of the remaining species could be supported?

20. A Fredholm integral equation of the second kind is an equation of the form

$$u(x) = f(x) + \int_a^b K(x,t)u(t)\,dt,$$

where a and b and the functions f and K are given. To approximate the function u on the interval $[a, b]$, a partition $x_0 = a < x_1 < \cdots < x_{m-1} < x_m = b$ is selected and the equations

$$u(x_i) = f(x_i) + \int_a^b K(x_i, t)u(t)\,dt, \quad \text{for each } i = 0, \cdots, m,$$

are solved for $u(x_0), u(x_1), \cdots, u(x_m)$. The integrals are approximated using quadrature formulas based on the nodes $x_0, \cdots, x_m$. In our problem, $a = 0$, $b = 1$, $f(x) = x^2$, and $K(x,t) = e^{|x-t|}$.

a. Show that the linear system

$$u(0) = f(0) + \frac{1}{2}[K(0,0)u(0) + K(0,1)u(1)],$$

$$u(1) = f(1) + \frac{1}{2}[K(1,0)u(0) + K(1,1)u(1)]$$

must be solved when the Trapezoidal rule is used.

b. Set up and solve the linear system that results when the Composite Trapezoidal rule is used with $n = 4$.

c. Repeat part (b) using the Composite Simpson's rule.

6.2 Pivoting Strategies

In deriving Algorithm 6.1, we found that a row interchange was needed when one of the pivot elements $a_{kk}^{(k)}$ is 0. This row interchange has the form $(E_k) \leftrightarrow (E_p)$, where p is the smallest integer greater than k with $a_{pk}^{(k)} \neq 0$. To reduce round-off error, it is often necessary to perform row interchanges even when the pivot elements are not zero.

If $a_{kk}^{(k)}$ is small in magnitude compared to $a_{jk}^{(k)}$, then the magnitude of the multiplier

$$m_{jk} = \frac{a_{jk}^{(k)}}{a_{kk}^{(k)}}$$

will be much larger than 1. Round-off error introduced in the computation of one of the terms $a_{kl}^{(k)}$ is multiplied by m_{jk} when computing $a_{jl}^{(k+1)}$, which compounds the original error. Also, when performing the backward substitution for

$$x_k = \frac{a_{k,n+1}^{(k)} - \sum_{j=k+1}^{n} a_{kj}^{(k)}}{a_{kk}^{(k)}},$$

with a small value of $a_{kk}^{(k)}$, any error in the numerator can be dramatically increased because of the division by $a_{kk}^{(k)}$. In our next example, we will see that even for small systems, round-off error can dominate the calculations.

Example 1 Apply Gaussian elimination to the system

$$\begin{aligned} E_1: &\quad 0.003000x_1 + 59.14x_2 = 59.17 \\ E_2: &\quad 5.291x_1 - 6.130x_2 = 46.78, \end{aligned}$$

using four-digit arithmetic with rounding, and compare the results to the exact solution $x_1 = 10.00$ and $x_2 = 1.000$.

Solution The first pivot element, $a_{11}^{(1)} = 0.003000$, is small, and its associated multiplier,

$$m_{21} = \frac{5.291}{0.003000} = 1763.6\overline{6},$$

rounds to the large number 1764. Performing $(E_2 - m_{21}E_1) \rightarrow (E_2)$ and the appropriate rounding gives the system

$$\begin{aligned} 0.003000x_1 + 59.14x_2 &\approx 59.17 \\ -104300x_2 &\approx -104400, \end{aligned}$$

instead of the exact system, which is

$$\begin{aligned} 0.003000x_1 + 59.14x_2 &= 59.17 \\ -104309.37\overline{6}x_2 &= -104309.37\overline{6}. \end{aligned}$$

The disparity in the magnitudes of $m_{21}a_{13}$ and a_{23} has introduced round-off error, but the round-off error has not yet been propagated. Backward substitution yields

$$x_2 \approx 1.001,$$

which is a close approximation to the actual value, $x_2 = 1.000$. However, because of the small pivot $a_{11} = 0.003000$,

$$x_1 \approx \frac{59.17 - (59.14)(1.001)}{0.003000} = -10.00$$

contains the small error of 0.001 multiplied by

$$\frac{59.14}{0.003000} \approx 20000.$$

This ruins the approximation to the actual value $x_1 = 10.00$.

This is clearly a contrived example and the graph in Figure 6.1. shows why the error can so easily occur. For larger systems it is much more difficult to predict in advance when devastating round-off error might occur. ■

Figure 6.1

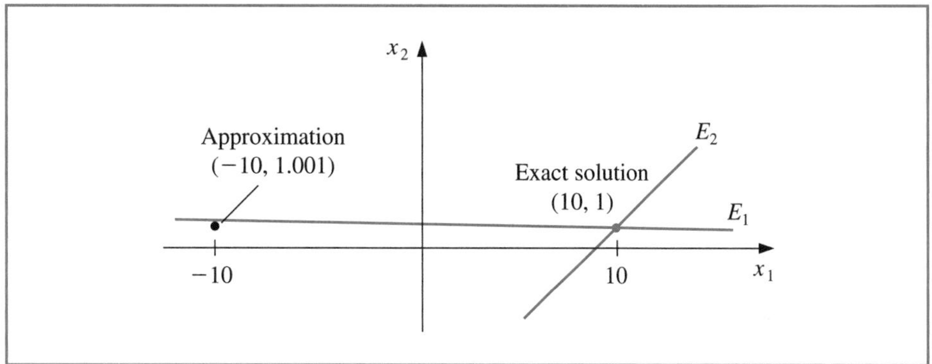

Partial Pivoting

Example 1 shows how difficulties can arise when the pivot element $a_{kk}^{(k)}$ is small relative to the entries $a_{ij}^{(k)}$, for $k \le i \le n$ and $k \le j \le n$. To avoid this problem, pivoting is performed by selecting an element $a_{pq}^{(k)}$ with a larger magnitude as the pivot, and interchanging the kth and pth rows. This can be followed by the interchange of the kth and qth columns, if necessary.

The simplest strategy is to select an element in the same column that is below the diagonal and has the largest absolute value; specifically, we determine the smallest $p \ge k$ such that

$$|a_{pk}^{(k)}| = \max_{k \le i \le n} |a_{ik}^{(k)}|$$

and perform $(E_k) \leftrightarrow (E_p)$. In this case no interchange of columns is used.

Example 2 Apply Gaussian elimination to the system

$$\begin{aligned} E_1: &\quad 0.003000x_1 + 59.14x_2 = 59.17 \\ E_2: &\quad 5.291x_1 - 6.130x_2 = 46.78, \end{aligned}$$

using partial pivoting and four-digit arithmetic with rounding, and compare the results to the exact solution $x_1 = 10.00$ and $x_2 = 1.000$.

Solution The partial-pivoting procedure first requires finding

$$\max\left\{|a_{11}^{(1)}|, |a_{21}^{(1)}|\right\} = \max\{|0.003000|, |5.291|\} = |5.291| = |a_{21}^{(1)}|.$$

This requires that the operation $(E_2) \leftrightarrow (E_1)$ be performed to produce the equivalent system

$$\begin{aligned} E_1: &\quad 5.291x_1 - 6.130x_2 = 46.78, \\ E_2: &\quad 0.003000x_1 + 59.14x_2 = 59.17. \end{aligned}$$

The multiplier for this system is

$$m_{21} = \frac{a_{21}^{(1)}}{a_{11}^{(1)}} = 0.0005670,$$

and the operation $(E_2 - m_{21}E_1) \rightarrow (E_2)$ reduces the system to

$$\begin{aligned} 5.291x_1 - 6.130x_2 &\approx 46.78, \\ 59.14x_2 &\approx 59.14. \end{aligned}$$

The four-digit answers resulting from the backward substitution are the correct values $x_1 = 10.00$ and $x_2 = 1.000$. ■

The technique just described is called **partial pivoting** (or *maximal column pivoting*) and is detailed in Algorithm 6.2. The actual row interchanging is simulated in the algorithm by interchanging the values of *NROW* in Step 5.

ALGORITHM 6.2

Gaussian Elimination with Partial Pivoting

To solve the $n \times n$ linear system

$$\begin{aligned} E_1: &\quad a_{11}x_1 + a_{12}x_2 + \cdots + a_{1n}x_n = a_{1,n+1} \\ E_2: &\quad a_{21}x_1 + a_{22}x_2 + \cdots + a_{2n}x_n = a_{2,n+1} \\ &\quad \vdots \\ E_n: &\quad a_{n1}x_1 + a_{n2}x_2 + \cdots + a_{nn}x_n = a_{n,n+1} \end{aligned}$$

INPUT number of unknowns and equations n; augmented matrix $A = [a_{ij}]$ where $1 \le i \le n$ and $1 \le j \le n+1$.

OUTPUT solution $x_1, \ldots, x_n$ or message that the linear system has no unique solution.

Step 1 For $i = 1, \ldots, n$ set $NROW(i) = i$. (*Initialize row pointer.*)

Step 2 For $i = 1, \ldots, n-1$ do Steps 3–6. (*Elimination process.*)

Step 3 Let p be the smallest integer with $i \le p \le n$ and $|a(NROW(p), i)| = \max_{i \le j \le n} |a(NROW(j), i)|$. (*Notation:* $a(NROW(i), j) \equiv a_{NROW_i, j}$.)

Step 4 If $a(NROW(p), i) = 0$ then OUTPUT ('no unique solution exists'); STOP.

Step 5 If $NROW(i) \ne NROW(p)$ then set $NCOPY = NROW(i)$; $NROW(i) = NROW(p)$; $NROW(p) = NCOPY$.
(*Simulated row interchange.*)

Step 6 For $j = i+1, \ldots, n$ do Steps 7 and 8.

Step 7 Set $m(NROW(j), i) = a(NROW(j), i)/a(NROW(i), i)$.

Step 8 Perform $(E_{NROW(j)} - m(NROW(j), i) \cdot E_{NROW(i)}) \rightarrow (E_{NROW(j)})$.

Step 9 If $a(NROW(n), n) = 0$ then OUTPUT ('no unique solution exists'); STOP.

Step 10 Set $x_n = a(NROW(n), n+1)/a(NROW(n), n)$.
(*Start backward substitution.*)

Step 11 For $i = n-1, \ldots, 1$

$$\text{set } x_i = \frac{a(NROW(i), n+1) - \sum_{j=i+1}^{n} a(NROW(i), j) \cdot x_j}{a(NROW(i), i)}.$$

Step 12 OUTPUT $(x_1, \ldots, x_n)$; (*Procedure completed successfully.*)
STOP. ■

Each multiplier m_{ji} in the partial pivoting algorithm has magnitude less than or equal to 1. Although this strategy is sufficient for many linear systems, situations do arise when it is inadequate.

Illustration The linear system

$$\begin{aligned} E_1: &\quad 30.00x_1 + 591400x_2 = 591700, \\ E_2: &\quad 5.291x_1 - 6.130x_2 = 46.78, \end{aligned}$$

is the same as that in Examples 1 and 2 except that all the entries in the first equation have been multiplied by 10^4. The partial pivoting procedure described in Algorithm 6.2 with four-digit arithmetic leads to the same results as obtained in Example 1. The maximal value in the first column is 30.00, and the multiplier

$$m_{21} = \frac{5.291}{30.00} = 0.1764$$

leads to the system

$$\begin{aligned} 30.00x_1 + 591400x_2 &\approx 591700, \\ -104300x_2 &\approx -104400, \end{aligned}$$

which has the same inaccurate solutions as in Example 1: $x_2 \approx 1.001$ and $x_1 \approx -10.00$. □

Scaled Partial Pivoting

Scaled partial pivoting (or *scaled-column pivoting*) is needed for the system in the Illustration. It places the element in the pivot position that is largest relative to the entries in its row. The first step in this procedure is to define a scale factor s_i for each row as

$$s_i = \max_{1 \le j \le n} |a_{ij}|.$$

If we have $s_i = 0$ for some i, then the system has no unique solution since all entries in the ith row are 0. Assuming that this is not the case, the appropriate row interchange to place zeros in the first column is determined by choosing the least integer p with

$$\frac{|a_{p1}|}{s_p} = \max_{1 \le k \le n} \frac{|a_{k1}|}{s_k}$$

and performing $(E_1) \leftrightarrow (E_p)$. The effect of scaling is to ensure that the largest element in each row has a *relative* magnitude of 1 before the comparison for row interchange is performed.

In a similar manner, before eliminating the variable x_i using the operations

$$E_k - m_{ki}E_i, \quad \text{for } k = i+1, \dots, n,$$

we select the smallest integer $p \geq i$ with

$$\frac{|a_{pi}|}{s_p} = \max_{i \leq k \leq n} \frac{|a_{ki}|}{s_k}$$

and perform the row interchange $(E_i) \leftrightarrow (E_p)$ if $i \neq p$. The scale factors $s_1, \dots, s_n$ are computed only once, at the start of the procedure. They are row dependent, so they must also be interchanged when row interchanges are performed.

Illustration Applying scaled partial pivoting to the previous Illustration gives

$$s_1 = \max\{|30.00|, |591400|\} = 591400$$

and

$$s_2 = \max\{|5.291|, |-6.130|\} = 6.130.$$

Consequently

$$\frac{|a_{11}|}{s_1} = \frac{30.00}{591400} = 0.5073 \times 10^{-4}, \qquad \frac{|a_{21}|}{s_2} = \frac{5.291}{6.130} = 0.8631,$$

and the interchange $(E_1) \leftrightarrow (E_2)$ is made.

Applying Gaussian elimination to the new system

$$5.291x_1 - 6.130x_2 = 46.78$$
$$30.00x_1 + 591400x_2 = 591700$$

produces the correct results: $x_1 = 10.00$ and $x_2 = 1.000$. □

Algorithm 6.3 implements scaled partial pivoting.

ALGORITHM 6.3

Gaussian Elimination with Scaled Partial Pivoting

The only steps in this algorithm that differ from those of Algorithm 6.2 are:

Step 1 For $i = 1, \dots, n$ set $s_i = \max_{1 \leq j \leq n} |a_{ij}|$;
if $s_i = 0$ then OUTPUT ('no unique solution exists');
STOP.
set $NROW(i) = i$.

Step 2 For $i = 1, \dots, n-1$ do Steps 3–6. *(Elimination process.)*
Step 3 Let p be the smallest integer with $i \leq p \leq n$ and

$$\frac{|a(NROW(p), i)|}{s(NROW(p))} = \max_{i \leq j \leq n} \frac{|a(NROW(j), i)|}{s(NROW(j))}.$$ ■

The next example demonstrates using Maple and the *LinearAlgebra* library to perform scaled partial pivoting with finite-digit rounding arithmetic.

Example 3 Solve the linear system using three-digit rounding arithmetic in Maple with the *LinearAlgebra* library.

$$\begin{aligned}
2.11x_1 - 4.21x_2 + 0.921x_3 &= 2.01,\\
4.01x_1 + 10.2x_2 - 1.12x_3 &= -3.09,\\
1.09x_1 + 0.987x_2 + 0.832x_3 &= 4.21.
\end{aligned}$$

Solution To obtain three-digit rounding arithmetic, enter

$Digits := 3$

We have $s_1 = 4.21$, $s_2 = 10.2$, and $s_3 = 1.09$. So

$$\frac{|a_{11}|}{s_1} = \frac{2.11}{4.21} = 0.501, \quad \frac{|a_{21}|}{s_1} = \frac{4.01}{10.2} = 0.393, \quad \text{and} \quad \frac{|a_{31}|}{s_3} = \frac{1.09}{1.09} = 1.$$

Next we load the *LinearAlgebra* library.

with(*LinearAlgebra*)

The augmented matrix *AA* is defined by

$AA := Matrix([[2.11, -4.21, 0.921, 2.01], [4.01, 10.2, -1.12, -3.09],$
$[1.09, 0.987, 0.832, 4.21]])$

which gives

$$\begin{bmatrix} 2.11 & -4.21 & .921 & 2.01 \\ 4.01 & 10.2 & -1.12 & -3.09 \\ 1.09 & .987 & .832 & 4.21 \end{bmatrix}.$$

Since $|a_{31}|/s_3$ is largest, we perform $(E_1) \leftrightarrow (E_3)$ using

$AA1 := RowOperation(AA, [1, 3])$

to obtain

$$\begin{bmatrix} 1.09 & .987 & .832 & 4.21 \\ 4.01 & 10.2 & -1.12 & -3.09 \\ 2.11 & -4.21 & .921 & 2.01 \end{bmatrix}.$$

Compute the multipliers

$$m21 := \frac{AA1[2, 1]}{AA1[1, 1]}; m31 := \frac{AA1[3, 1]}{AA1[1, 1]}$$

giving

$$3.68$$

$$1.94$$

Perform the first two eliminations using

$AA2 := RowOperation(AA1, [2, 1], -m21): AA3 := RowOperation(AA2, [3, 1], -m31)$

to produce

$$\begin{bmatrix} 1.09 & .987 & .832 & 4.21 \\ 0 & 6.57 & -4.18 & -18.6 \\ 0 & -6.12 & -.689 & -6.16 \end{bmatrix}.$$

Since

$$\frac{|a_{22}|}{s_2} = \frac{6.57}{10.2} = 0.644 \quad \text{and} \quad \frac{|a_{32}|}{s_3} = \frac{6.12}{4.21} = 1.45,$$

we perform

$AA4 := RowOperation(AA3, [2, 3])$

giving

$$\begin{bmatrix} 1.09 & .987 & .832 & 4.21 \\ 0 & -6.12 & -.689 & -6.16 \\ 0 & 6.57 & -4.18 & -18.6 \end{bmatrix}.$$

The multiplier m_{32} is computed by

$$m32 := \frac{AA4[3,2]}{AA4[2,2]}$$

$$-1.07$$

and the elimination step

$AA5 := RowOperation(AA4, [3, 2], -m32)$

results in the matrix

$$\begin{bmatrix} 1.09 & .987 & .832 & 4.21 \\ 0 & -6.12 & -.689 & -6.16 \\ 0 & .02 & -4.92 & -25.2 \end{bmatrix}.$$

We cannot use *BackwardSubstitute* on this matrix because of the entry .02 in the last row of the second column, that is, which Maple knows as the (3, 2) position. This entry is nonzero due to rounding, but we can remedy this minor problem setting it to 0 with the command

$AA5[3, 2] := 0$

You can verify this is correct with the command *evalm*(*AA*5)

Finally, backward substitution gives the solution $\mathbf{x}$, which to 3 decimal digits is $x_1 = -0.436$, $x_2 = 0.430$, and $x_3 = 5.12$. ■

The first additional computations required for scaled partial pivoting result from the determination of the scale factors; there are $(n-1)$ comparisons for each of the n rows, for a total of

$$n(n-1) \text{ comparisons.}$$

To determine the correct first interchange, n divisions are performed, followed by $n-1$ comparisons. So the first interchange determination adds

$$n \text{ divisions and } (n-1) \text{ comparisons.}$$

The scaling factors are computed only once, so the second step requires

$$(n-1) \text{ divisions and } (n-2) \text{ comparisons.}$$

We proceed in a similar manner until there are zeros below the main diagonal in all but the nth row. The final step requires that we perform

$$2 \text{ divisions and } 1 \text{ comparison.}$$

As a consequence, scaled partial pivoting adds a total of

$$n(n-1) + \sum_{k=1}^{n-1} k = n(n-1) + \frac{(n-1)n}{2} = \frac{3}{2}n(n-1) \quad \text{comparisons} \tag{6.7}$$

and

$$\sum_{k=2}^{n} k = \left(\sum_{k=1}^{n} k\right) - 1 = \frac{n(n+1)}{2} - 1 = \frac{1}{2}(n-1)(n+2) \quad \text{divisions}$$

to the Gaussian elimination procedure. The time required to perform a comparison is about the same as an addition/subtraction. Since the total time to perform the basic Gaussian elimination procedure is $O(n^3/3)$ multiplications/divisions and $O(n^3/3)$ additions/subtractions, scaled partial pivoting does not add significantly to the computational time required to solve a system for large values of n.

To emphasize the importance of choosing the scale factors only once, consider the amount of additional computation that would be required if the procedure were modified so that new scale factors were determined each time a row interchange decision was to be made. In this case, the term $n(n-1)$ in Eq. (6.7) would be replaced by

$$\sum_{k=2}^{n} k(k-1) = \frac{1}{3}n(n^2-1).$$

As a consequence, this pivoting technique would add $O(n^3/3)$ comparisons, in addition to the $[n(n+1)/2] - 1$ divisions.

Complete Pivoting

Pivoting can incorporate the interchange of both rows and columns. **Complete** (or *maximal*) **pivoting** at the kth step searches all the entries a_{ij}, for $i = k, k+1, \ldots, n$ and $j = k, k+1, \ldots, n$, to find the entry with the largest magnitude. Both row and column interchanges are performed to bring this entry to the pivot position. The first step of total pivoting requires that $n^2 - 1$ comparisons be performed, the second step requires $(n-1)^2 - 1$ comparisons, and so on. The total additional time required to incorporate complete pivoting into Gaussian elimination is

$$\sum_{k=2}^{n} (k^2 - 1) = \frac{n(n-1)(2n+5)}{6}$$

comparisons. Complete pivoting is, consequently, the strategy recommended only for systems where accuracy is essential and the amount of execution time needed for this method can be justified.

EXERCISE SET 6.2

1. Find the row interchanges that are required to solve the following linear systems using Algorithm 6.1.

   a. $\begin{aligned} x_1 - 5x_2 + x_3 &= 7, \\ 10x_1 + 20x_3 &= 6, \\ 5x_1 - x_3 &= 4. \end{aligned}$

   b. $\begin{aligned} 2x_1 - 3x_2 + 2x_3 &= 5, \\ -4x_1 + 2x_2 - 6x_3 &= 14, \\ 2x_1 + 2x_2 + 4x_3 &= 8. \end{aligned}$

   c. $\begin{aligned} x_1 + x_2 - x_3 &= 1, \\ x_1 + x_2 + 4x_3 &= 2, \\ 2x_1 - x_2 + 2x_3 &= 3. \end{aligned}$

   d. $\begin{aligned} x_2 + x_3 &= 6, \\ x_1 - 2x_2 - x_3 &= 4, \\ x_1 - x_2 + x_3 &= 5. \end{aligned}$

2. Find the row interchanges that are required to solve the following linear systems using Algorithm 6.1.

a. $13x_1 + 17x_2 + x_3 = 5,$
$x_2 + 19x_3 = 1,$
$12x_2 - x_3 = 0.$

b. $x_1 + x_2 - x_3 = 0,$
$12x_2 - x_3 = 4,$
$2x_1 + x_2 + x_3 = 5.$

c. $5x_1 + x_2 - 6x_3 = 7,$
$2x_1 + x_2 - x_3 = 8,$
$6x_1 + 12x_2 + x_3 = 9.$

d. $x_1 - x_2 + x_3 = 5,$
$7x_1 + 5x_2 - x_3 = 8,$
$2x_1 + x_2 + x_3 = 7.$

3. Repeat Exercise 1 using Algorithm 6.2.

4. Repeat Exercise 2 using Algorithm 6.2.

5. Repeat Exercise 1 using Algorithm 6.3.

6. Repeat Exercise 2 using Algorithm 6.3.

7. Repeat Exercise 1 using complete pivoting.

8. Repeat Exercise 2 using complete pivoting.

9. Use Gaussian elimination and three-digit chopping arithmetic to solve the following linear systems, and compare the approximations to the actual solution.

a. $0.03x_1 + 58.9x_2 = 59.2,$
$5.31x_1 - 6.10x_2 = 47.0.$
Actual solution $[10, 1]$.

b. $3.03x_1 - 12.1x_2 + 14x_3 = -119,$
$-3.03x_1 + 12.1x_2 - 7x_3 = 120,$
$6.11x_1 - 14.2x_2 + 21x_3 = -139.$
Actual solution $[0, 10, \frac{1}{7}]$.

c. $1.19x_1 + 2.11x_2 - 100x_3 + x_4 = 1.12,$
$14.2x_1 - 0.122x_2 + 12.2x_3 - x_4 = 3.44,$
$100x_2 - 99.9x_3 + x_4 = 2.15,$
$15.3x_1 + 0.110x_2 - 13.1x_3 - x_4 = 4.16.$
Actual solution $[0.176, 0.0126, -0.0206, -1.18]$.

d. $\pi x_1 - ex_2 + \sqrt{2}x_3 - \sqrt{3}x_4 = \sqrt{11},$
$\pi^2 x_1 + ex_2 - e^2 x_3 + \frac{3}{7}x_4 = 0,$
$\sqrt{5}x_1 - \sqrt{6}x_2 + x_3 - \sqrt{2}x_4 = \pi,$
$\pi^3 x_1 + e^2 x_2 - \sqrt{7}x_3 + \frac{1}{9}x_4 = \sqrt{2}.$
Actual solution $[0.788, -3.12, 0.167, 4.55]$.

10. Use Gaussian elimination and three-digit chopping arithmetic to solve the following linear systems, and compare the approximations to the actual solution.

a. $58.9x_1 + 0.03x_2 = 59.2,$
$-6.10x_1 + 5.31x_2 = 47.0.$
Actual solution $[1, 10]$.

b. $3.3330x_1 + 15920x_2 + 10.333x_3 = 7953,$
$2.2220x_1 + 16.710x_2 + 9.6120x_3 = 0.965,$
$-1.5611x_1 + 5.1792x_2 - 1.6855x_3 = 2.714.$
Actual solution $[1, 0.5, -1]$.

c. $2.12x_1 - 2.12x_2 + 51.3x_3 + 100x_4 = \pi,$
$0.333x_1 - 0.333x_2 - 12.2x_3 + 19.7x_4 = \sqrt{2},$
$6.19x_1 + 8.20x_2 - 1.00x_3 - 2.01x_4 = 0,$
$-5.73x_1 + 6.12x_2 + x_3 - x_4 = -1.$
Actual solution $[0.0998, -0.0683, -0.0363, 0.0465]$.

d. $\pi x_1 + \sqrt{2}x_2 - x_3 + x_4 = 0,$
$ex_1 - x_2 + x_3 + 2x_4 = 1,$
$x_1 + x_2 - \sqrt{3}x_3 + x_4 = 2,$
$-x_1 - x_2 + x_3 - \sqrt{5}x_4 = 3.$
Actual solution $[1.35, -4.68, -4.03, -1.66]$.

11. Repeat Exercise 9 using three-digit rounding arithmetic.
12. Repeat Exercise 10 using three-digit rounding arithmetic.
13. Repeat Exercise 9 using Gaussian elimination with partial pivoting.
14. Repeat Exercise 10 using Gaussian elimination with partial pivoting.
15. Repeat Exercise 9 using Gaussian elimination with partial pivoting and three-digit rounding arithmetic.
16. Repeat Exercise 10 using Gaussian elimination with partial pivoting and three-digit rounding arithmetic.
17. Repeat Exercise 9 using Gaussian elimination with scaled partial pivoting.
18. Repeat Exercise 10 using Gaussian elimination with scaled partial pivoting.
19. Repeat Exercise 9 using Gaussian elimination with scaled partial pivoting and three-digit rounding arithmetic.
20. Repeat Exercise 10 using Gaussian elimination with scaled partial pivoting and three-digit rounding arithmetic.
21. Repeat Exercise 9 using Algorithm 6.1 in Maple with *Digits*:= 10.
22. Repeat Exercise 10 using Algorithm 6.1 in Maple with *Digits*:= 10.
23. Repeat Exercise 9 using Algorithm 6.2 in Maple with *Digits*:= 10.
24. Repeat Exercise 10 using Algorithm 6.2 in Maple with *Digits*:= 10.
25. Repeat Exercise 9 using Algorithm 6.3 in Maple with *Digits*:= 10.
26. Repeat Exercise 10 using Algorithm 6.3 in Maple with *Digits*:= 10.
27. Repeat Exercise 9 using Gaussian elimination with complete pivoting.
28. Repeat Exercise 10 using Gaussian elimination with complete pivoting.
29. Repeat Exercise 9 using Gaussian elimination with complete pivoting and three-digit rounding arithmetic.
30. Repeat Exercise 10 using Gaussian elimination with complete pivoting and three-digit rounding arithmetic.
31. Suppose that

$$2x_1 + x_2 + 3x_3 = 1,$$
$$4x_1 + 6x_2 + 8x_3 = 5,$$
$$6x_1 + \alpha x_2 + 10x_3 = 5,$$

with $|\alpha| < 10$. For which of the following values of α will there be no row interchange required when solving this system using scaled partial pivoting?

a. $\alpha = 6$ **b.** $\alpha = 9$ **c.** $\alpha = -3$

32. Construct an algorithm for the complete pivoting procedure discussed in the text.
33. Use the complete pivoting algorithm to repeat Exercise 9 Maple with *Digits*:= 10.
34. Use the complete pivoting algorithm to repeat Exercise 10 Maple with *Digits*:= 10.

6.3 Linear Algebra and Matrix Inversion

Matrices were introduced in Section 6.1 as a convenient method for expressing and manipulating linear systems. In this section we consider some algebra associated with matrices and show how it can be used to solve problems involving linear systems.

Definition 6.2 Two matrices A and B are **equal** if they have the same number of rows and columns, say $n \times m$, and if $a_{ij} = b_{ij}$, for each $i = 1, 2, \ldots, n$ and $j = 1, 2, \ldots, m$. ■

This definition means, for example, that

$$\begin{bmatrix} 2 & -1 & 7 \\ 3 & 1 & 0 \end{bmatrix} \neq \begin{bmatrix} 2 & 3 \\ -1 & 1 \\ 7 & 0 \end{bmatrix},$$

because they differ in dimension.

Matrix Arithmetic

Two important operations performed on matrices are the sum of two matrices and the multiplication of a matrix by a real number.

Definition 6.3 If A and B are both $n \times m$ matrices, then the **sum** of A and B, denoted $A + B$, is the $n \times m$ matrix whose entries are $a_{ij} + b_{ij}$, for each $i = 1, 2, \ldots, n$ and $j = 1, 2, \ldots, m$. ■

Definition 6.4 If A is an $n \times m$ matrix and λ is a real number, then the **scalar multiplication** of λ and A, denoted λA, is the $n \times m$ matrix whose entries are λa_{ij}, for each $i = 1, 2, \ldots, n$ and $j = 1, 2, \ldots, m$. ■

Example 1 Determine $A + B$ and λA when

$$A = \begin{bmatrix} 2 & -1 & 7 \\ 3 & 1 & 0 \end{bmatrix}, \quad B = \begin{bmatrix} 4 & 2 & -8 \\ 0 & 1 & 6 \end{bmatrix}, \quad \text{and } \lambda = -2.$$

Solution We have

$$A + B = \begin{bmatrix} 2+4 & -1+2 & 7-8 \\ 3+0 & 1+1 & 0+6 \end{bmatrix} = \begin{bmatrix} 6 & 1 & -1 \\ 3 & 2 & 6 \end{bmatrix},$$

and

$$\lambda A = \begin{bmatrix} -2(2) & -2(-1) & -2(7) \\ -2(3) & -2(1) & -2(0) \end{bmatrix} = \begin{bmatrix} -4 & 2 & -14 \\ -6 & -2 & 0 \end{bmatrix}.$$ ■

We have the following general properties for matrix addition and scalar multiplication. These properties are sufficient to classify the set of all $n \times m$ matrices with real entries as a **vector space** over the field of real numbers.

- We let O denote a matrix all of whose entries are 0 and $-A$ denote the matrix whose entries are $-a_{ij}$.

Theorem 6.5 Let A, B, and C be $n \times m$ matrices and λ and μ be real numbers. The following properties of addition and scalar multiplication hold:

(i) $A + B = B + A$, **(ii)** $(A + B) + C = A + (B + C)$,

(iii) $A + O = O + A = A$, **(iv)** $A + (-A) = -A + A = 0$,

(v) $\lambda(A + B) = \lambda A + \lambda B$, **(vi)** $(\lambda + \mu)A = \lambda A + \mu A$,

(vii) $\lambda(\mu A) = (\lambda\mu)A$, **(viii)** $1A = A$.

All these properties follow from similar results concerning the real numbers. ■

Matrix-Vector Products

The product of matrices can also be defined in certain instances. We will first consider the product of an $n \times m$ matrix and a $m \times 1$ column vector.

Definition 6.6 Let A be an $n \times m$ matrix and $\mathbf{b}$ an m-dimensional column vector. The **matrix-vector product** of A and $\mathbf{b}$, denoted $A\mathbf{b}$, is an n-dimensional column vector given by

$$A\mathbf{b} = \begin{bmatrix} a_{11} & a_{12} & \cdots & a_{1m} \\ a_{21} & a_{22} & \cdots & a_{2m} \\ \vdots & \vdots & & \vdots \\ a_{n1} & a_{n2} & \cdots & a_{nm} \end{bmatrix} \begin{bmatrix} b_1 \\ b_2 \\ \vdots \\ b_m \end{bmatrix} = \begin{bmatrix} \sum_{i=1}^m a_{1i}b_i \\ \sum_{i=1}^m a_{2i}b_i \\ \vdots \\ \sum_{i=1}^m a_{ni}b_i \end{bmatrix}.$$ ■

For this product to be defined the number of columns of the matrix A must match the number of rows of the vector $\mathbf{b}$, and the result is another column vector with the number of rows matching the number of rows in the matrix.

Example 2 Determine the product $A\mathbf{b}$ if $A = \begin{bmatrix} 3 & 2 \\ -1 & 1 \\ 6 & 4 \end{bmatrix}$ and $\mathbf{b} = \begin{bmatrix} 3 \\ -1 \end{bmatrix}$.

Solution Because A has dimension 3×2 and $\mathbf{b}$ has dimension 2×1, the product is defined and is a vector with three rows. These are

$$3(3) + 2(-1) = 7, \quad (-1)(3) + 1(-1) = -4, \quad \text{and} \quad 6(3) + 4(-1) = 14.$$

That is,

$$A\mathbf{b} = \begin{bmatrix} 3 & 2 \\ -1 & 1 \\ 6 & 4 \end{bmatrix} \begin{bmatrix} 3 \\ -1 \end{bmatrix} = \begin{bmatrix} 7 \\ -4 \\ 14 \end{bmatrix}$$ ■

The introduction of the matrix-vector product permits us to view the linear system

$$\begin{aligned} a_{11}x_1 + a_{12}x_2 + \cdots + a_{1n}x_n &= b_1, \\ a_{21}x_1 + a_{22}x_2 + \cdots + a_{2n}x_n &= b_2, \\ &\vdots \\ a_{n1}x_1 + a_{n2}x_2 + \cdots + a_{nn}x_n &= b_n, \end{aligned}$$

as the matrix equation

$$A\mathbf{x} = \mathbf{b},$$

where

$$A = \begin{bmatrix} a_{11} & a_{12} & \cdots & a_{1n} \\ a_{21} & a_{22} & \cdots & a_{2n} \\ \vdots & \vdots & & \vdots \\ a_{n1} & a_{n2} & \cdots & a_{nn} \end{bmatrix}, \quad \mathbf{x} = \begin{bmatrix} x_1 \\ x_2 \\ \vdots \\ x_n \end{bmatrix}, \quad \text{and} \quad \mathbf{b} = \begin{bmatrix} b_1 \\ b_2 \\ \vdots \\ b_n \end{bmatrix},$$

because all the entries in the product $A\mathbf{x}$ must match the corresponding entries in the vector $\mathbf{b}$. In essence, then, an $n \times m$ matrix is a function with domain the set of m-dimensional column vectors and range a subset of the n-dimensional column vectors.

Matrix-Matrix Products

We can use this matrix-vector multiplication to define general matrix-matrix multiplication.

Definition 6.7 Let A be an $n \times m$ matrix and B an $m \times p$ matrix. The **matrix product** of A and B, denoted AB, is an $n \times p$ matrix C whose entries c_{ij} are

$$c_{ij} = \sum_{k=1}^{m} a_{ik}b_{kj} = a_{i1}b_{1j} + a_{i2}b_{2j} + \cdots + a_{im}b_{mj},$$

for each $i = 1, 2, \cdots n$, and $j = 1, 2, \cdots, p$. ■

The computation of c_{ij} can be viewed as the multiplication of the entries of the ith row of A with corresponding entries in the jth column of B, followed by a summation; that is,

$$[a_{i1}, a_{i2}, \cdots, a_{im}]\begin{bmatrix} b_{1j} \\ b_{2j} \\ \vdots \\ b_{mj} \end{bmatrix} = c_{ij},$$

where

$$c_{ij} = a_{i1}b_{1j} + a_{i2}b_{2j} + \cdots + a_{im}b_{mj} = \sum_{k=1}^{m} a_{ik}b_{kj}.$$

This explains why the number of columns of A must equal the number of rows of B for the product AB to be defined.

The following example should serve to clarify the matrix multiplication process.

Example 3 Determine all possible products of the matrices

$$A = \begin{bmatrix} 3 & 2 \\ -1 & 1 \\ 1 & 4 \end{bmatrix}, \quad B = \begin{bmatrix} 2 & 1 & -1 \\ 3 & 1 & 2 \end{bmatrix},$$

$$C = \begin{bmatrix} 2 & 1 & 0 & 1 \\ -1 & 3 & 2 & 1 \\ 1 & 1 & 2 & 0 \end{bmatrix}, \quad \text{and} \quad D = \begin{bmatrix} 1 & -1 \\ 2 & -1 \end{bmatrix}.$$

Solution The size of the matrices are

$$A : 3 \times 2, \quad B : 2 \times 3, \quad C : 3 \times 4, \quad \text{and} \quad D : 2 \times 2.$$

The products that can be defined, and their dimensions, are:

$$AB : 3 \times 3, \quad BA : 2 \times 2, \quad AD : 3 \times 2, \quad BC : 2 \times 4, \quad DB : 2 \times 3, \quad \text{and} \quad DD : 2 \times 2.$$

These products are

$$AB = \begin{bmatrix} 12 & 5 & 1 \\ 1 & 0 & 3 \\ 14 & 5 & 7 \end{bmatrix}, \quad BA = \begin{bmatrix} 4 & 1 \\ 10 & 15 \end{bmatrix}, \quad AD = \begin{bmatrix} 7 & -5 \\ 1 & 0 \\ 9 & -5 \end{bmatrix},$$

$$BC = \begin{bmatrix} 2 & 4 & 0 & 3 \\ 7 & 8 & 6 & 4 \end{bmatrix}, \quad DB = \begin{bmatrix} -1 & 0 & -3 \\ 1 & 1 & -4 \end{bmatrix}, \quad \text{and} \quad DD = \begin{bmatrix} -1 & 0 \\ 0 & -1 \end{bmatrix}.$$

■

Notice that although the matrix products AB and BA are both defined, their results are very different; they do not even have the same dimension. In mathematical language, we say that the matrix product operation is *not commutative*, that is, products in reverse order can differ. This is the case even when both products are defined and are of the same dimension. Almost any example will show this, for example,

$$\begin{bmatrix} 1 & 1 \\ 1 & 0 \end{bmatrix}\begin{bmatrix} 0 & 1 \\ 1 & 1 \end{bmatrix} = \begin{bmatrix} 1 & 2 \\ 0 & 1 \end{bmatrix} \quad \text{whereas} \quad \begin{bmatrix} 0 & 1 \\ 1 & 1 \end{bmatrix}\begin{bmatrix} 1 & 1 \\ 1 & 0 \end{bmatrix} = \begin{bmatrix} 1 & 0 \\ 2 & 1 \end{bmatrix}$$

Certain important operations involving matrix product do hold, however, as indicated in the following result.

Theorem 6.8 Let A be an $n \times m$ matrix, B be an $m \times k$ matrix, C be a $k \times p$ matrix, D be an $m \times k$ matrix, and λ be a real number. The following properties hold:

(a) $A(BC) = (AB)C$; **(b)** $A(B + D) = AB + AD$; **(c)** $\lambda(AB) = (\lambda A)B = A(\lambda B)$. ■

Proof The verification of the property in part **(a)** is presented to show the method involved. The other parts can be shown in a similar manner.

To show that $A(BC) = (AB)C$, compute the sj-entry of each side of the equation. BC is an $m \times p$ matrix with sj-entry

$$(BC)_{sj} = \sum_{l=1}^{k} b_{sl}c_{lj}.$$

Thus, $A(BC)$ is an $n \times p$ matrix with entries

$$[A(BC)]_{ij} = \sum_{s=1}^{m} a_{is}(BC)_{sj} = \sum_{s=1}^{m} a_{is}\left(\sum_{l=1}^{k} b_{sl}c_{lj}\right) = \sum_{s=1}^{m}\sum_{l=1}^{k} a_{is}b_{sl}c_{lj}.$$

Similarly, AB is an $n \times k$ matrix with entries

$$(AB)_{il} = \sum_{s=1}^{m} a_{is}b_{sl},$$

so $(AB)C$ is an $n \times p$ matrix with entries

$$[(AB)C]_{ij} = \sum_{l=1}^{k} (AB)_{il}c_{lj} = \sum_{l=1}^{k}\left(\sum_{s=1}^{m} a_{is}b_{sl}\right)c_{lj} = \sum_{l=1}^{k}\sum_{s=1}^{m} a_{is}b_{sl}c_{lj}.$$

Interchanging the order of summation on the right side gives

$$[(AB)C]_{ij} = \sum_{s=1}^{m}\sum_{l=1}^{k} a_{is}b_{sl}c_{lj} = [A(BC)]_{ij},$$

for each $i = 1, 2, \ldots, n$ and $j = 1, 2, \ldots, p$. So $A(BC) = (AB)C$. ■ ■ ■

Square Matrices

Matrices that have the same number of rows as columns are important in applications.

Definition 6.9

The term diagonal applied to a matrix refers to the entries in the diagonal that runs from the top left entry to the bottom right entry.

(i) A **square** matrix has the same number of rows as columns.

(ii) A **diagonal** matrix $D = [d_{ij}]$ is a square matrix with $d_{ij} = 0$ whenever $i \neq j$.

(iii) The **identity matrix of order** n, $I_n = [\delta_{ij}]$, is a diagonal matrix whose diagonal entries are all 1s. When the size of I_n is clear, this matrix is generally written simply as I. ▪

For example, the identity matrix of order three is

$$I = \begin{bmatrix} 1 & 0 & 0 \\ 0 & 1 & 0 \\ 0 & 0 & 1 \end{bmatrix}.$$

Definition 6.10 An **upper-triangular** $n \times n$ matrix $U = [u_{ij}]$ has, for each $j = 1, 2, \cdots, n$, the entries

$$u_{ij} = 0, \quad \text{for each } i = j+1, j+2, \cdots, n;$$

A triangular matrix is one that has all zero entries except either on and above (upper) or on and below (lower) the main diagonal.

and a **lower-triangular** matrix $L = [l_{ij}]$ has, for each $j = 1, 2, \cdots, n$, the entries

$$l_{ij} = 0, \quad \text{for each } i = 1, 2, \cdots, j-1.$$

▪

A diagonal matrix, then, is both both upper triangular and lower triangular because its only nonzero entries must lie on the main diagonal.

Illustration Consider the identity matrix of order three,

$$I_3 = \begin{bmatrix} 1 & 0 & 0 \\ 0 & 1 & 0 \\ 0 & 0 & 1 \end{bmatrix}.$$

If A is any 3×3 matrix, then

$$AI_3 = \begin{bmatrix} a_{11} & a_{12} & a_{13} \\ a_{21} & a_{22} & a_{23} \\ a_{31} & a_{32} & a_{33} \end{bmatrix} \begin{bmatrix} 1 & 0 & 0 \\ 0 & 1 & 0 \\ 0 & 0 & 1 \end{bmatrix} = \begin{bmatrix} a_{11} & a_{12} & a_{13} \\ a_{21} & a_{22} & a_{23} \\ a_{31} & a_{32} & a_{33} \end{bmatrix} = A.$$

□

The identity matrix I_n commutes with any $n \times n$ matrix A; that is, the order of multiplication does not matter,

$$I_n A = A = AI_n.$$

Keep in mind that this property is not true in general, even for square matrices.

Inverse Matrices

Related to the linear systems is the **inverse of a matrix**.

Definition 6.11 An $n \times n$ matrix A is said to be **nonsingular** (or *invertible*) if an $n \times n$ matrix A^{-1} exists with $AA^{-1} = A^{-1}A = I$. The matrix A^{-1} is called the **inverse** of A. A matrix without an inverse is called **singular** (or *noninvertible*). ▪

The word singular means something that deviates from the ordinary. Hence a singular matrix does *not* have an inverse.

The following properties regarding matrix inverses follow from Definition 6.11. The proofs of these results are considered in Exercise 5.

Theorem 6.12 For any nonsingular $n \times n$ matrix A:

(i) A^{-1} is unique.

(ii) A^{-1} is nonsingular and $(A^{-1})^{-1} = A$.

(iii) If B is also a nonsingular $n \times n$ matrix, then $(AB)^{-1} = B^{-1}A^{-1}$. ■

Example 4 Let

$$A = \begin{bmatrix} 1 & 2 & -1 \\ 2 & 1 & 0 \\ -1 & 1 & 2 \end{bmatrix} \quad \text{and} \quad B = \begin{bmatrix} -\frac{2}{9} & \frac{5}{9} & -\frac{1}{9} \\ \frac{4}{9} & -\frac{1}{9} & \frac{2}{9} \\ -\frac{1}{3} & \frac{1}{3} & \frac{1}{3} \end{bmatrix}.$$

Show that $B = A^{-1}$, and that the solution to the linear system described by

$$\begin{aligned} x_1 + 2x_2 - x_3 &= 2, \\ 2x_1 + x_2 \phantom{{}+2x_3} &= 3, \\ -x_1 + x_2 + 2x_3 &= 4. \end{aligned}$$

is given by the entries in $B\mathbf{b}$, where $\mathbf{b}$ is the column vector with entries 2, 3, and 4.

Solution First note that

$$AB = \begin{bmatrix} 1 & 2 & -1 \\ 2 & 1 & 0 \\ -1 & 1 & 2 \end{bmatrix} \cdot \begin{bmatrix} -\frac{2}{9} & \frac{5}{9} & -\frac{1}{9} \\ \frac{4}{9} & -\frac{1}{9} & \frac{2}{9} \\ -\frac{1}{3} & \frac{1}{3} & \frac{1}{3} \end{bmatrix} = \begin{bmatrix} 1 & 0 & 0 \\ 0 & 1 & 0 \\ 0 & 0 & 1 \end{bmatrix} = I_3.$$

In a similar manner, $BA = I_3$, so A and B are both nonsingular with $B = A^{-1}$ and $A = B^{-1}$.

Now convert the given linear system to the matrix equation

$$\begin{bmatrix} 1 & 2 & -1 \\ 2 & 1 & 0 \\ -1 & 1 & 2 \end{bmatrix} \begin{bmatrix} x_1 \\ x_2 \\ x_3 \end{bmatrix} = \begin{bmatrix} 2 \\ 3 \\ 4 \end{bmatrix},$$

and multiply both sides by B, the inverse of A. Because we have both

$$B(A\mathbf{x}) = (BA)\mathbf{x} = I_3\mathbf{x} = \mathbf{x} \quad \text{and} \quad B(A\mathbf{x}) = \mathbf{b},$$

we have

$$BA\mathbf{x} = \left(\begin{bmatrix} -\frac{2}{9} & \frac{5}{9} & -\frac{1}{9} \\ \frac{4}{9} & -\frac{1}{9} & \frac{2}{9} \\ -\frac{3}{9} & \frac{3}{9} & \frac{3}{9} \end{bmatrix} \begin{bmatrix} 1 & 2 & -1 \\ 2 & 1 & 0 \\ -1 & 1 & 2 \end{bmatrix} \right) \mathbf{x} = \mathbf{x}$$

and

$$BA\mathbf{x} = B(\mathbf{b}) = \begin{bmatrix} -\frac{2}{9} & \frac{5}{9} & -\frac{1}{9} \\ \frac{4}{9} & -\frac{1}{9} & \frac{2}{9} \\ -\frac{1}{3} & \frac{1}{3} & \frac{1}{3} \end{bmatrix} \begin{bmatrix} 2 \\ 3 \\ 4 \end{bmatrix} = \begin{bmatrix} \frac{7}{9} \\ \frac{13}{9} \\ \frac{5}{3} \end{bmatrix}$$

This implies that $\mathbf{x} = B\mathbf{b}$ and gives the solution $x_1 = 7/9$, $x_2 = 13/9$, and $x_3 = 5/3$. ■

Although it is easy to solve a linear system of the form $A\mathbf{x} = \mathbf{b}$ if A^{-1} is known, it is not computationally efficient to determine A^{-1} in order to solve the system. (See

Exercise 8.) Even so, it is useful from a conceptual standpoint to describe a method for determining the inverse of a matrix.

To find a method of computing A^{-1} assuming A is nonsingular, let us look again at matrix multiplication. Let B_j be the jth column of the $n \times n$ matrix B,

$$B_j = \begin{bmatrix} b_{1j} \\ b_{2j} \\ \vdots \\ b_{nj} \end{bmatrix}.$$

If $AB = C$, then the jth column of C is given by the product

$$\begin{bmatrix} c_{1j} \\ c_{2j} \\ \vdots \\ c_{nj} \end{bmatrix} = C_j = AB_j = \begin{bmatrix} a_{11} & a_{12} & \cdots & a_{1n} \\ a_{21} & a_{22} & \cdots & a_{2n} \\ \vdots & \vdots & & \vdots \\ a_{n1} & a_{n2} & \cdots & a_{nn} \end{bmatrix} \begin{bmatrix} b_{1j} \\ b_{2j} \\ \vdots \\ b_{nj} \end{bmatrix} = \begin{bmatrix} \sum_{k=1}^{n} a_{1k}b_{kj} \\ \sum_{k=1}^{n} a_{2k}b_{kj} \\ \vdots \\ \sum_{k=1}^{n} a_{nk}b_{kj} \end{bmatrix}.$$

Suppose that A^{-1} exists and that $A^{-1} = B = (b_{ij})$. Then $AB = I$ and

$$AB_j = \begin{bmatrix} 0 \\ \vdots \\ 0 \\ 1 \\ 0 \\ \vdots \\ 0 \end{bmatrix}, \quad \text{where the value 1 appears in the } j\text{th row.}$$

To find B we need to solve n linear systems in which the jth column of the inverse is the solution of the linear system with right-hand side the jth column of I. The next illustration demonstrates this method.

Illustration To determine the inverse of the matrix

$$A = \begin{bmatrix} 1 & 2 & -1 \\ 2 & 1 & 0 \\ -1 & 1 & 2 \end{bmatrix},$$

let us first consider the product AB, where B is an arbitrary 3×3 matrix.

$$AB = \begin{bmatrix} 1 & 2 & -1 \\ 2 & 1 & 0 \\ -1 & 1 & 2 \end{bmatrix} \begin{bmatrix} b_{11} & b_{12} & b_{13} \\ b_{21} & b_{22} & b_{23} \\ b_{31} & b_{32} & b_{33} \end{bmatrix}$$

$$= \begin{bmatrix} b_{11} + 2b_{21} - b_{31} & b_{12} + 2b_{22} - b_{32} & b_{13} + 2b_{23} - b_{33} \\ 2b_{11} + b_{21} & 2b_{12} + b_{22} & 2b_{13} + b_{23} \\ -b_{11} + b_{21} + 2b_{31} & -b_{12} + b_{22} + 2b_{32} & -b_{13} + b_{23} + 2b_{33} \end{bmatrix}.$$

If $B = A^{-1}$, then $AB = I$, so

$$\begin{aligned} b_{11} + 2b_{21} - b_{31} &= 1, \\ 2b_{11} + b_{21} &= 0, \\ -b_{11} + b_{21} + 2b_{31} &= 0, \end{aligned} \qquad \begin{aligned} b_{12} + 2b_{22} - b_{32} &= 0, \\ 2b_{12} + b_{22} &= 1, \\ -b_{12} + b_{22} + 2b_{32} &= 0, \end{aligned} \qquad \text{and} \qquad \begin{aligned} b_{13} + 2b_{23} - b_{33} &= 0, \\ 2b_{13} + b_{23} &= 0, \\ -b_{13} + b_{23} + 2b_{33} &= 1. \end{aligned}$$

Notice that the coefficients in each of the systems of equations are the same, the only change in the systems occurs on the right side of the equations. As a consequence, Gaussian elimination can be performed on a larger augmented matrix formed by combining the matrices for each of the systems:

$$\left[\begin{array}{rrr|rrr} 1 & 2 & -1 & 1 & 0 & 0 \\ 2 & 1 & 0 & 0 & 1 & 0 \\ -1 & 1 & 2 & 0 & 0 & 1 \end{array}\right].$$

First, performing $(E_2 - 2E_1) \to (E_2)$ and $(E_3 + E_1) \to (E_3)$, followed by $(E_3 + E_2) \to (E_3)$ produces

$$\left[\begin{array}{rrr|rrr} 1 & 2 & -1 & 1 & 0 & 0 \\ 0 & -3 & 2 & -2 & 1 & 0 \\ 0 & 3 & 1 & 1 & 0 & 1 \end{array}\right] \quad \text{and} \quad \left[\begin{array}{rrr|rrr} 1 & 2 & -1 & 1 & 0 & 0 \\ 0 & -3 & 2 & -2 & 1 & 0 \\ 0 & 0 & 3 & -1 & 1 & 1 \end{array}\right].$$

Backward substitution is performed on each of the three augmented matrices,

$$\left[\begin{array}{rrr|r} 1 & 2 & -1 & 1 \\ 0 & -3 & 2 & -2 \\ 0 & 0 & 3 & -1 \end{array}\right], \left[\begin{array}{rrr|r} 1 & 2 & -1 & 0 \\ 0 & -3 & 2 & 1 \\ 0 & 0 & 3 & 1 \end{array}\right], \left[\begin{array}{rrr|r} 1 & 2 & -1 & 0 \\ 0 & -3 & 2 & 0 \\ 0 & 0 & 3 & 1 \end{array}\right],$$

to eventually give

$$\begin{array}{llll} b_{11} = -\frac{2}{9}, & b_{12} = \frac{5}{9}, & & b_{13} = -\frac{1}{9}, \\ b_{21} = \frac{4}{9}, & b_{22} = -\frac{1}{9}, & \text{and} & b_{23} = \frac{2}{9}, \\ b_{31} = -\frac{1}{3}, & b_{32} = \frac{1}{3}, & & b_{32} = \frac{1}{3}. \end{array}$$

As shown in Example 4, these are the entries of A^{-1}:

$$B = A^{-1} = \begin{bmatrix} -\frac{2}{9} & \frac{5}{9} & -\frac{1}{9} \\ \frac{4}{9} & -\frac{1}{9} & \frac{2}{9} \\ -\frac{1}{3} & \frac{1}{3} & \frac{1}{3} \end{bmatrix} \qquad \square$$

As we saw in the illustration, in order to compute A^{-1} it is convenient to set up a larger augmented matrix,

$$[\, A \;\vdots\; I \,].$$

Upon performing the elimination in accordance with Algorithm 6.1, we obtain an augmented matrix of the form

$$[\, U \;\vdots\; Y \,],$$

where U is an upper-triangular matrix and Y is the matrix obtained by performing the same operations on the identity I that were performed to take A into U.

Gaussian elimination with backward substitution requires

$$\frac{4}{3}n^3 - \frac{1}{3}n \text{ multiplications/divisions} \quad \text{and} \quad \frac{4}{3}n^3 - \frac{3}{2}n^2 + \frac{n}{6} \text{ additions/subtractions.}$$

to solve the n linear systems (see Exercise 8(a)). Special care can be taken in the implementation to note the operations that need *not* be performed, as, for example, a multiplication when one of the multipliers is known to be unity or a subtraction when the subtrahend is known to be 0. The number of multiplications/divisions required can then be reduced to n^3 and the number of additions/subtractions reduced to $n^3 - 2n^2 + n$ (see Exercise 8(d)).

Transpose of a Matrix

Another important matrix associated with a given matrix A is its *transpose*, denoted A^t.

Definition 6.13 The **transpose** of an $n \times m$ matrix $A = [a_{ij}]$ is the $m \times n$ matrix $A^t = [a_{ji}]$, where for each i, the ith column of A^t is the same as the ith row of A. A square matrix A is called **symmetric** if $A = A^t$. ■

Illustration The matrices

$$A = \begin{bmatrix} 7 & 2 & 0 \\ 3 & 5 & -1 \\ 0 & 5 & -6 \end{bmatrix}, \quad B = \begin{bmatrix} 2 & 4 & 7 \\ 3 & -5 & -1 \end{bmatrix}, \quad C = \begin{bmatrix} 6 & 4 & -3 \\ 4 & -2 & 0 \\ -3 & 0 & 1 \end{bmatrix}$$

have transposes

$$A^t = \begin{bmatrix} 7 & 3 & 0 \\ 2 & 5 & 5 \\ 0 & -1 & -6 \end{bmatrix}, \quad B^t = \begin{bmatrix} 2 & 3 \\ 4 & -5 \\ 7 & -1 \end{bmatrix}, \quad C^t = \begin{bmatrix} 6 & 4 & -3 \\ 4 & -2 & 0 \\ -3 & 0 & 1 \end{bmatrix}.$$

The matrix C is symmetric because $C^t = C$. The matrices A and B are not symmetric. □

The proof of the next result follows directly from the definition of the transpose.

Theorem 6.14 The following operations involving the transpose of a matrix hold whenever the operation is possible:

(i) $(A^t)^t = A$, **(iii)** $(AB)^t = B^tA^t$,

(ii) $(A + B)^t = A^t + B^t$, **(iv)** if A^{-1} exists, then $(A^{-1})^t = (A^t)^{-1}$. ■

Matrix arithmetic is performed in Maple using the *LinearAlgebra* package whenever the operations are defined. For example, the addition of two $n \times m$ matrices A and B is done in Maple with the command $A + B$, and scalar multiplication by a number c is defined by $c\,A$.

If A is $n \times m$ and B is $m \times p$, then the $n \times p$ matrix AB is produced with the command $A.B$. Matrix transposition is achieved with *Transpose*(A) and matrix inversion, with *MatrixInverse*(A).

EXERCISE SET 6.3

1. Perform the following matrix-vector multiplications:

a. $\begin{bmatrix} 3 & 0 \\ 2 & 1 \end{bmatrix}\begin{bmatrix} 1 \\ -2 \end{bmatrix}$

b. $\begin{bmatrix} 3 & 2 \\ 6 & 4 \end{bmatrix}\begin{bmatrix} 1 \\ -1 \end{bmatrix}$

c. $\begin{bmatrix} 2 & 1 & 0 \\ 1 & -1 & 2 \\ 0 & 2 & 4 \end{bmatrix}\begin{bmatrix} 2 \\ 5 \\ 1 \end{bmatrix}$

d. $[2 \quad -2 \quad 1]\begin{bmatrix} 3 & -2 & 0 \\ -2 & 3 & 1 \\ 0 & 1 & -2 \end{bmatrix}$

2. Perform the following matrix-vector multiplications:

a. $\begin{bmatrix} 2 & 1 \\ -4 & 3 \end{bmatrix}\begin{bmatrix} 3 \\ -2 \end{bmatrix}$ **b.** $\begin{bmatrix} 2 & -2 \\ -4 & 4 \end{bmatrix}\begin{bmatrix} 1 \\ 1 \end{bmatrix}$

c. $\begin{bmatrix} 2 & 0 & 0 \\ 3 & -1 & 2 \\ 0 & 2 & -3 \end{bmatrix}\begin{bmatrix} 2 \\ 5 \\ 1 \end{bmatrix}$ **d.** $[-4 \ \ 0 \ \ 0]\begin{bmatrix} 1 & -2 & 4 \\ -2 & 3 & 1 \\ 4 & 1 & 0 \end{bmatrix}$

3. Perform the following matrix-matrix multiplications:

a. $\begin{bmatrix} 2 & -3 \\ 3 & -1 \end{bmatrix}\begin{bmatrix} 1 & 5 & -4 \\ -3 & 2 & 0 \end{bmatrix}$ **b.** $\begin{bmatrix} 2 & -3 \\ 3 & -1 \end{bmatrix}\begin{bmatrix} 1 & 5 \\ 2 & 0 \end{bmatrix}$

c. $\begin{bmatrix} 2 & -3 & 1 \\ 4 & 3 & 0 \\ 5 & 2 & -4 \end{bmatrix}\begin{bmatrix} 0 & 1 & -2 \\ 1 & 0 & -1 \\ 2 & 3 & -2 \end{bmatrix}$ **d.** $\begin{bmatrix} 2 & 1 & 2 \\ -2 & 3 & 0 \\ 2 & -1 & 3 \end{bmatrix}\begin{bmatrix} 1 & -2 \\ -4 & 1 \\ 0 & 2 \end{bmatrix}$

4. Perform the following matrix-matrix multiplications:

a. $\begin{bmatrix} -1 & 3 \\ -2 & 4 \end{bmatrix}\begin{bmatrix} 2 & -2 & 3 \\ -3 & 2 & 2 \end{bmatrix}$ **b.** $\begin{bmatrix} -2 & 3 \\ 0 & 3 \end{bmatrix}\begin{bmatrix} 2 & -5 \\ -5 & 2 \end{bmatrix}$

c. $\begin{bmatrix} 2 & -3 & -2 \\ -3 & 4 & 1 \\ -2 & 1 & -4 \end{bmatrix}\begin{bmatrix} 2 & -3 & 4 \\ -3 & 4 & -1 \\ 4 & -1 & -2 \end{bmatrix}$ **d.** $\begin{bmatrix} 3 & -1 & 0 \\ 2 & -2 & 3 \\ -2 & 1 & 4 \end{bmatrix}\begin{bmatrix} -1 & 2 \\ 4 & -1 \\ 3 & -5 \end{bmatrix}$

5. Determine which of the following matrices are nonsingular, and compute the inverse of these matrices:

a. $\begin{bmatrix} 4 & 2 & 6 \\ 3 & 0 & 7 \\ -2 & -1 & -3 \end{bmatrix}$ **b.** $\begin{bmatrix} 1 & 2 & 0 \\ 2 & 1 & -1 \\ 3 & 1 & 1 \end{bmatrix}$

c. $\begin{bmatrix} 1 & 1 & -1 & 1 \\ 1 & 2 & -4 & -2 \\ 2 & 1 & 1 & 5 \\ -1 & 0 & -2 & -4 \end{bmatrix}$ **d.** $\begin{bmatrix} 4 & 0 & 0 & 0 \\ 6 & 7 & 0 & 0 \\ 9 & 11 & 1 & 0 \\ 5 & 4 & 1 & 1 \end{bmatrix}$

6. Determine which of the following matrices are nonsingular, and compute the inverse of these matrices:

a. $\begin{bmatrix} 1 & 2 & -1 \\ 0 & 1 & 2 \\ -1 & 4 & 3 \end{bmatrix}$ **b.** $\begin{bmatrix} 4 & 0 & 0 \\ 0 & 0 & 0 \\ 0 & 0 & 3 \end{bmatrix}$

c. $\begin{bmatrix} 1 & 2 & 3 & 4 \\ 2 & 1 & -1 & 1 \\ -3 & 2 & 0 & 1 \\ 0 & 5 & 2 & 6 \end{bmatrix}$ **d.** $\begin{bmatrix} 2 & 0 & 1 & 2 \\ 1 & 1 & 0 & 2 \\ 2 & -1 & 3 & 1 \\ 3 & -1 & 4 & 3 \end{bmatrix}$

7. Given the two 4×4 linear systems having the same coefficient matrix:

$$\begin{aligned} x_1 - x_2 + 2x_3 - x_4 &= 6, \\ x_1 \qquad - x_3 + x_4 &= 4, \\ 2x_1 + x_2 + 3x_3 - 4x_4 &= -2, \\ -x_2 + x_3 - x_4 &= 5; \end{aligned} \qquad \begin{aligned} x_1 - x_2 + 2x_3 - x_4 &= 1, \\ x_1 - x_3 + x_4 &= 1, \\ 2x_1 + x_2 + 3x_3 - 4x_4 &= 2, \\ -x_2 + x_3 - x_4 &= -1. \end{aligned}$$

a. Solve the linear systems by applying Gaussian elimination to the augmented matrix

$$\left[\begin{array}{rrrr:rr} 1 & -1 & 2 & -1 & 6 & 1 \\ 1 & 0 & -1 & 1 & 4 & 1 \\ 2 & 1 & 3 & -4 & -2 & 2 \\ 0 & -1 & 1 & -1 & 5 & -1 \end{array}\right].$$

b. Solve the linear systems by finding and multiplying by the inverse of

$$\begin{bmatrix} 1 & -1 & 2 & -1 \\ 1 & 0 & -1 & 1 \\ 2 & 1 & 3 & -4 \\ 0 & -1 & 1 & -1 \end{bmatrix}.$$

c. Which method requires more operations?

8. Consider the four 3×3 linear systems having the same coefficient matrix:

$$\begin{aligned} 2x_1 - 3x_2 + x_3 &= 2, \\ x_1 + x_2 - x_3 &= -1, \\ -x_1 + x_2 - 3x_3 &= 0; \end{aligned} \qquad \begin{aligned} 2x_1 - 3x_2 + x_3 &= 6, \\ x_1 + x_2 - x_3 &= 4, \\ -x_1 + x_2 - 3x_3 &= 5; \end{aligned}$$

$$\begin{aligned} 2x_1 - 3x_2 + x_3 &= 0, \\ x_1 + x_2 - x_3 &= 1, \\ -x_1 + x_2 - 3x_3 &= -3; \end{aligned} \qquad \begin{aligned} 2x_1 - 3x_2 + x_3 &= -1, \\ x_1 + x_2 - x_3 &= 0, \\ -x_1 + x_2 - 3x_3 &= 0. \end{aligned}$$

a. Solve the linear systems by applying Gaussian elimination to the augmented matrix

$$\left[\begin{array}{rrr:rrrr} 2 & -3 & 1 & 2 & 6 & 0 & -1 \\ 1 & 1 & -1 & -1 & 4 & 1 & 0 \\ -1 & 1 & -3 & 0 & 5 & -3 & 0 \end{array}\right].$$

b. Solve the linear systems by finding and multiplying by the inverse of

$$A = \begin{bmatrix} 2 & -3 & 1 \\ 1 & 1 & -1 \\ -1 & 1 & -3 \end{bmatrix}.$$

c. Which method requires more operations?

9. The following statements are needed to prove Theorem 6.12.

a. Show that if A^{-1} exists, it is unique.

b. Show that if A is nonsingular, then $(A^{-1})^{-1} = A$.

c. Show that if A and B are nonsingular nxn matrices, then $(AB)^{-1} = B^{-1}A^{-1}$.

10. Prove the following statements or provide counterexamples to show they are not true.

a. The product of two symmetric matrices is symmetric.

b. The inverse of a nonsingular symmetric matrix is a nonsingular symmetric matrix.

c. If A and B are $n \times n$ matrices, then $(AB)^t = A^tB^t$.

11. **a.** Show that the product of two $n \times n$ lower triangular matrices is lower triangular.

b. Show that the product of two $n \times n$ upper triangular matrices is upper triangular.

c. Show that the inverse of a nonsingular $n \times n$ lower triangular matrix is lower triangular.

12. Suppose m linear systems

$$A\mathbf{x}^{(p)} = \mathbf{b}^{(p)}, \quad p = 1, 2, \ldots, m,$$

are to be solved, each with the $n \times n$ coefficient matrix A.

a. Show that Gaussian elimination with backward substitution applied to the aug- mented matrix

$$[A : \quad \mathbf{b}^{(1)}\mathbf{b}^{(2)} \cdots \mathbf{b}^{(m)}]$$

requires

$$\frac{1}{3}n^3 + mn^2 - \frac{1}{3}n \quad \text{multiplications/ divisions}$$

and

$$\frac{1}{3}n^3 + mn^2 - \frac{1}{2}n^2 - mn + \frac{1}{6}n \quad \text{additions/subtractions.}$$

b. Show that the Gauss-Jordan method (see Exercise 12, Section 6.1) applied to the augmented matrix

$$[A: \quad \mathbf{b}^{(1)}\mathbf{b}^{(2)} \cdots \mathbf{b}^{(m)}]$$

requires

$$\frac{1}{2}n^3 + mn^2 - \frac{1}{2}n \quad \text{multiplications/divisions}$$

and

$$\frac{1}{2}n^3 + (m-1)n^2 + \left(\frac{1}{2} - m\right)n \quad \text{additions/subtractions.}$$

c. For the special case

$$\mathbf{b}^{(p)} = \begin{bmatrix} 0 \\ \vdots \\ 0 \\ 1 \\ \vdots \\ 0 \end{bmatrix} \leftarrow p\text{th row},$$

for each $p = 1, \ldots, m$, with $m = n$, the solution $x^{(p)}$ is the pth column of A^{-1}. Show that Gaussian elimination with backward substitution requires

$$\frac{4}{3}n^3 - \frac{1}{3}n \quad \text{multiplications/divisions}$$

and

$$\frac{4}{3}n^3 - \frac{3}{2}n^2 + \frac{1}{6}n \quad \text{additions/subtractions}$$

for this application, and that the Gauss-Jordan method requires

$$\frac{3}{2}n^3 - \frac{1}{2}n \quad \text{multiplications/divisions}$$

and

$$\frac{3}{2}n^3 - 2n^2 + \frac{1}{2}n \quad \text{additions/subtractions.}$$

d. Construct an algorithm using Gaussian elimination to find A^{-1}, but do not per- form multiplications when one of the multipliers is known to be 1, and do not per- form additions/subtractions when one of the elements involved is known to be 0. Show that the required computations are reduced to n^3 multiplications/divisions and $n^3 - 2n^2 + n$ additions/subtractions.

e. Show that solving the linear system $Ax = \mathbf{b}$, when A^{-1} is known, still requires n^2 multiplications/divisions and $n^2 - n$ additions/subtractions.

f. Show that solving m linear systems $Ax^{(p)} = \mathbf{b}^{(p)}$, for $p = 1, 2, \ldots, m$, by the method $x^{(p)} = A^{-1}\mathbf{b}(p)$ requires mn^2 multiplications and $m(n^2 - n)$ additions, if A^{-1} is known.

g. Let A be an $n \times n$ matrix. Compare the number of operations required to solve n linear systems involving A by Gaussian elimination with backward substitution and by first inverting A and then multiplying $Ax = \mathbf{b}$ by A^{-1}, for $n = 3, 10, 50, 100$. Is it ever advantageous to compute A^{-1} for the purpose of solving linear systems?

13. Use the algorithm developed in Exercise 8(d) to find the inverses of the nonsingular matrices in Exercise 5.

14. It is often useful to partition matrices into a collection of submatrices. For example, the matrices

$$A = \begin{bmatrix} 1 & 2 & -1 \\ 3 & -4 & -3 \\ 6 & 5 & 0 \end{bmatrix} \quad \text{and} \quad B = \begin{bmatrix} 2 & -1 & 7 & 0 \\ 3 & 0 & 4 & 5 \\ -2 & 1 & -3 & 1 \end{bmatrix}$$

can be partitioned into

$$\left[\begin{array}{cc:c} 1 & 2 & -1 \\ 3 & -4 & -3 \\ \hdashline 6 & 5 & 0 \end{array}\right] = \left[\begin{array}{c:c} A_{11} & A_{12} \\ \hdashline A_{21} & A_{22} \end{array}\right] \quad \text{and} \quad \left[\begin{array}{ccc:c} 2 & -1 & 7 & 0 \\ 3 & 0 & 4 & 5 \\ \hdashline -2 & 1 & -3 & 1 \end{array}\right] = \left[\begin{array}{c:c} B_{11} & B_{12} \\ \hdashline B_{21} & B_{22} \end{array}\right]$$

a. Show that the product of A and B in this case is

$$AB = \left[\begin{array}{c:c} A_{11}B_{11} + A_{12}B_{21} & A_{11}B_{12} + A_{12}B_{22} \\ \hdashline A_{21}B_{11} + A_{22}B_{21} & A_{21}B_{12} + A_{22}B_{22} \end{array}\right]$$

b. If B were instead partitioned into

$$B = \left[\begin{array}{ccc:c} 2 & -1 & 7 & 0 \\ \hdashline 3 & 0 & 4 & 5 \\ -2 & 1 & -3 & 1 \end{array}\right] = \left[\begin{array}{c:c} B_{11} & B_{12} \\ \hdashline B_{21} & B_{22} \end{array}\right],$$

would the result in part (a) hold?

c. Make a conjecture concerning the conditions necessary for the result in part (a) to hold in the general case.

15. In a paper entitled "Population Waves," Bernadelli [Ber] (see also [Se]) hypothesizes a type of simplified beetle that has a natural life span of 3 years. The female of this species has a survival rate of $\frac{1}{2}$ in the first year of life, has a survival rate of $\frac{1}{3}$ from the second to third years, and gives birth to an average of six new females before expiring at the end of the third year. A matrix can be used to show the contribution an individual female beetle makes, in a probabilistic sense, to the female population of the species by letting a_{ij} in the matrix $A = [a_{ij}]$ denote the contribution that a single female beetle of age j will make to the next year's female population of age i; that is,

$$A = \begin{bmatrix} 0 & 0 & 6 \\ \frac{1}{2} & 0 & 0 \\ 0 & \frac{1}{3} & 0 \end{bmatrix}.$$

a. The contribution that a female beetle makes to the population 2 years hence is determined from the entries of A^2, of 3 years hence from A^3, and so on. Construct A^2 and A^3, and try to make a general statement about the contribution of a female beetle to the population in n years' time for any positive integral value of n.

b. Use your conclusions from part (a) to describe what will occur in future years to a population of these beetles that initially consists of 6000 female beetles in each of the three age groups.

c. Construct A^{-1}, and describe its significance regarding the population of this species.

16. The study of food chains is an important topic in the determination of the spread and accumulation of environmental pollutants in living matter. Suppose that a food chain has three links. The first link consists of vegetation of types $v_1, v_2, \ldots, v_n$, which provide all the food requirements for herbivores of species $h_1, h_2, \ldots, h_m$ in the second link. The third link consists of carnivorous animals $c_1, c_2, \ldots, c_k$, which depend entirely on the herbivores in the second link for their food supply. The coordinate a_{ij} of the matrix

$$A = \begin{bmatrix} a_{11} & a_{12} & \cdots & a_{1m} \\ a_{21} & a_{22} & \cdots & a_{2m} \\ \vdots & \vdots & & \vdots \\ a_{n1} & a_{n2} & \cdots & a_{nm} \end{bmatrix}$$

represents the total number of plants of type v_i eaten by the herbivores in the species h_j, whereas b_{ij} in

$$B = \begin{bmatrix} b_{11} & b_{12} & \cdots & b_{1k} \\ b_{21} & b_{22} & \cdots & b_{2k} \\ \vdots & \vdots & & \vdots \\ b_{m1} & b_{m2} & \cdots & b_{mk} \end{bmatrix}$$

describes the number of herbivores in species h_i that are devoured by the animals of type c_j.

a. Show that the number of plants of type v_i that eventually end up in the animals of species c_j is given by the entry in the ith row and jth column of the matrix AB.

b. What physical significance is associated with the matrices $A-1, B-1$, and $(AB)-1 = B-1A-1$?

17. In Section 3.6 we found that the parametric form $(x(t), y(t))$ of the cubic Hermite polynomials through $(x(0), y(0)) = (x_0, y_0)$ and $(x(1), y(1)) = (x_1, y_1)$ with guide points $(x_0+\alpha_0, y_0+\beta_0)$ and $(x_1-\alpha_1, y_1-\beta_1)$, respectively, are given by

$$x(t) = (2(x_0 - x_1) + (\alpha_0 + \alpha_1))t^3 + (3(x_1 - x_0) - \alpha_1 - 2\alpha_0)t^2 + \alpha_0 t + x_0,$$

and

$$y(t) = (2(y_0 - y_1) + (\beta_0 + \beta_1))t^3 + (3(y_1 - y_0) - \beta_1 - 2\beta_0)t^2 + \beta_0 t + y_0.$$

The Bézier cubic polynomials have the form

$$\hat{x}(t) = (2(x_0 - x_1) + 3(\alpha_0 + \alpha_1))t^3 + (3(x_1 - x_0) - 3(\alpha_1 + 2\alpha_0))t^2, +3\alpha_0 t + x_0$$

and

$$\hat{y}(t) = (2(y_0 - y_1) + 3(\beta_0 + \beta_1))t^3 + (3(y_1 - y_0) - 3(\beta_1 + 2\beta_0))t^2 + 3\beta_0 t + y_0.$$

a. Show that the matrix

$$A = \begin{bmatrix} 7 & 4 & 4 & 0 \\ -6 & -3 & -6 & 0 \\ 0 & 0 & 3 & 0 \\ 0 & 0 & 0 & 1 \end{bmatrix}$$

transforms the Hermite polynomial coefficients into the Bézier polynomial coefficients.

b. Determine a matrix B that transforms the Bézier polynomial coefficients into the Hermite polynomial coefficients.

18. Consider the 2×2 linear system $(A + iB)(x + iy) = c + i\mathbf{d}$ with complex entries in component form:

$$(a_{11} + ib_{11})(x_1 + iy_1) + (a_{12} + ib_{12})(x_2 + iy_2) = c_1 + id_1,$$
$$(a_{11} + ib_{21})(x_1 + iy_1) + (a_{22} + ib_{22})(x_2 + iy_2) = c_2 + id_2.$$

a. Use the properties of complex numbers to convert this system to the equivalent 4×4 real linear system

$$A\mathbf{x} - B\mathbf{y} = \mathbf{c},$$
$$B\mathbf{x} + A\mathbf{y} = \mathbf{d}.$$

b. Solve the linear system

$$(1 - 2i)(x_1 + iy_1) + (3 + 2i)(x_2 + iy_2) = 5 + 2i,$$
$$(2 + i)(x_1 + iy_1) + (4 + 3i)(x_2 + iy_2) = 4 - i.$$

6.4 The Determinant of a Matrix

The *determinant* of a matrix provides existence and uniqueness results for linear systems having the same number of equations and unknowns. We will denote the determinant of a square matrix A by $\det A$, but it is also common to use the notation $|A|$.

Definition 6.15 Suppose that A is a square matrix.

(i) If $A = [a]$ is a 1×1 matrix, then $\det A = a$.

(ii) If A is an $n \times n$ matrix, with $n > 1$ the **minor** M_{ij} is the determinant of the $(n-1) \times (n-1)$ submatrix of A obtained by deleting the ith row and jth column of the matrix A.

(iii) The **cofactor** A_{ij} associated with M_{ij} is defined by $A_{ij} = (-1)^{i+j} M_{ij}$.

(iv) The **determinant** of the $n \times n$ matrix A, when $n > 1$, is given either by

$$\det A = \sum_{j=1}^{n} a_{ij} A_{ij} = \sum_{j=1}^{n} (-1)^{i+j} a_{ij} M_{ij}, \quad \text{for any } i = 1, 2, \cdots, n,$$

or by

$$\det A = \sum_{i=1}^{n} a_{ij} A_{ij} = \sum_{i=1}^{n} (-1)^{i+j} a_{ij} M_{ij}, \quad \text{for any } j = 1, 2, \cdots, n.$$ ■

The notion of a determinant appeared independently in 1683 both in Japan and Europe, although neither Takakazu Seki Kowa (1642–1708) nor Gottfried Leibniz (1646–1716) appear to have used the term determinant.

It can be shown (see Exercise 9) that to calculate the determinant of a general $n \times n$ matrix by this definition requires $O(n!)$ multiplications/divisions and additions/subtractions. Even for relatively small values of n, the number of calculations becomes unwieldy.

Although it appears that there are $2n$ different definitions of $\det A$, depending on which row or column is chosen, all definitions give the same numerical result. The flexibility in the definition is used in the following example. It is most convenient to compute $\det A$ across the row or down the column with the most zeros.

Example 1 Find the determinant of the matrix

$$A = \begin{bmatrix} 2 & -1 & 3 & 0 \\ 4 & -2 & 7 & 0 \\ -3 & -4 & 1 & 5 \\ 6 & -6 & 8 & 0 \end{bmatrix}.$$

using the row or column with the most zero entries.

Solution To compute $\det A$, it is easiest to use the fourth column:

$$\det A = a_{14}A_{14} + a_{24}A_{24} + a_{34}A_{34} + a_{44}A_{44} = 5A_{34} = -5M_{34}.$$

Eliminating the third row and the fourth column gives

$$\det A = -5 \det \begin{bmatrix} 2 & -1 & 3 \\ 4 & -2 & 7 \\ 6 & -6 & 8 \end{bmatrix}$$

$$= -5 \left\{ 2 \det \begin{bmatrix} -2 & 7 \\ -6 & 8 \end{bmatrix} - (-1) \det \begin{bmatrix} 4 & 7 \\ 6 & 8 \end{bmatrix} + 3 \det \begin{bmatrix} 4 & -2 \\ 6 & -6 \end{bmatrix} \right\} = -30.$$

■

The determinant of an $n \times n$ matrix of can be computed in Maple with the *LinearAlgebra* package using the command *Determinant*(A).

The following properties are useful in relating linear systems and Gaussian elimination to determinants. These are proved in any standard linear algebra text.

Theorem 6.16 Suppose A is an $n \times n$ matrix:

(i) If any row or column of A has only zero entries, then $\det A = 0$.

(ii) If A has two rows or two columns the same, then $\det A = 0$.

(iii) If $\tilde{A}$ is obtained from A by the operation $(E_i) \leftrightarrow (E_j)$, with $i \neq j$, then $\det \tilde{A} = -\det A$.

(iv) If $\tilde{A}$ is obtained from A by the operation $(\lambda E_i) \to (E_i)$, then $\det \tilde{A} = \lambda \det A$.

(v) If $\tilde{A}$ is obtained from A by the operation $(E_i + \lambda E_j) \to (E_i)$ with $i \neq j$, then $\det \tilde{A} = \det A$.

(vi) If B is also an $n \times n$ matrix, then $\det AB = \det A \det B$.

(vii) $\det A^t = \det A$.

(viii) When A^{-1} exists, $\det A^{-1} = (\det A)^{-1}$.

(ix) If A is an upper triangular, lower triangular, or diagonal matrix, then $\det A = \prod_{i=1}^{n} a_{ii}$. ■

As part (ix) of Theorem 6.16 indicates, the determinant of a triangular matrix is simply the product of its diagonal elements. By employing the row operations given in parts (iii), f(iv), and (v) we can reduce a given square matrix to triangular form to find its determinant.

Example 2 Compute the determinant of the matrix

$$A = \begin{bmatrix} 2 & 1 & -1 & 1 \\ 1 & 1 & 0 & 3 \\ -1 & 2 & 3 & -1 \\ 3 & -1 & -1 & 2 \end{bmatrix}$$

using parts (iii), (iv), and (v) of Theorem 6.16, doing the computations in Maple with the *LinearAlgebra* package.

Solution Matrix A is defined in Maple by

$A := \mathit{Matrix}([[2, 1, -1, 1], [1, 1, 0, 3], [-1, 2, 3, -1], [3, -1, -1, 2]])$

The sequence of operations in Table 6.2 produces the matrix

$$A8 = \begin{bmatrix} 1 & \frac{1}{2} & -\frac{1}{2} & \frac{1}{2} \\ 0 & 1 & 1 & 5 \\ 0 & 0 & 3 & 13 \\ 0 & 0 & 0 & -13 \end{bmatrix}.$$

By part (ix), $\det A8 = -39$, so $\det A = 39$. ■

Table 6.2

Operation	Maple	Effect
$\frac{1}{2}E_1 \to E_1$	$A1 := RowOperation(A, 1, \frac{1}{2})$	$\det A1 = \frac{1}{2}\det A$
$E_2 - E_1 \to E_2$	$A2 := RowOperation(A1, [2, 1], -1)$	$\det A2 = \det A1 = \frac{1}{2}\det A$
$E_3 + E_1 \to E_3$	$A3 := RowOperation(A2, [3, 1], 1)$	$\det A3 = \det A2 = \frac{1}{2}\det A$
$E_4 - 3E_1 \to E_4$	$A4 := RowOperation(A3, [4, 1], -3)$	$\det A4 = \det A3 = \frac{1}{2}\det A$
$2E_2 \to E_2$	$A5 := RowOperation(A4, 2, 2)$	$\det A5 = 2\det A4 = \det A$
$E_3 - \frac{5}{2}E_2 \to E_3$	$A6 := RowOperation(A5, [3, 2], -\frac{5}{2})$	$\det A6 = \det A5 = \det A$
$E_4 + \frac{5}{2}E_2 \to E_4$	$A7 := RowOperation(A6, [4, 2], \frac{5}{2})$	$\det A7 = \det A6 = \det A$
$E_3 \leftrightarrow E_4$	$A8 := RowOperation(A7, [3, 4])$	$\det A8 = -\det A7 = -\det A$

The key result relating nonsingularity, Gaussian elimination, linear systems, and determinants is that the following statements are equivalent.

Theorem 6.17 The following statements are equivalent for any $n \times n$ matrix A:

(i) The equation $A\mathbf{x} = \mathbf{0}$ has the unique solution $\mathbf{x} = \mathbf{0}$.

(ii) The system $A\mathbf{x} = \mathbf{b}$ has a unique solution for any n-dimensional column vector $\mathbf{b}$.

(iii) The matrix A is nonsingular; that is, A^{-1} exists.

(iv) $\det A \neq 0$.

(v) Gaussian elimination with row interchanges can be performed on the system $A\mathbf{x} = \mathbf{b}$ for any n-dimensional column vector $\mathbf{b}$. ■

The following Corollary to Theorem 6.17 illustrates how the determinant can be used to show important properties about square matrices.

Corollary 6.18 Suppose that A and B are both $n \times n$ matrices with either $AB = I$ or $BA = I$. Then $B = A^{-1}$ (and $A = B^{-1}$). ■

Proof Suppose that $AB = I$. Then by part (vi) of Theorem 6.16,

$$1 = \det(I) = \det(AB) = \det(A) \cdot \det(B), \quad \text{so} \quad \det(A) \neq 0 \text{ and } \det(B) \neq 0.$$

The equivalence of parts (iii) and (iv) of Theorem 6.17 imply that both A^{-1} and B^{-1} exist. Hence

$$A^{-1} = A^{-1} \cdot I = A^{-1} \cdot (AB) = \left(A^{-1}A\right) \cdot B = I \cdot B = B.$$

The roles of A and B are similar, so this also establishes that $BA = I$. Hence $B = A^{-1}$. ■ ■ ■

EXERCISE SET 6.4

1. Use Definition 6.15 to compute the determinants of the following matrices:

a. $\begin{bmatrix} 4 & 2 & 6 \\ -1 & 0 & 4 \\ 2 & 1 & 7 \end{bmatrix}$ b. $\begin{bmatrix} 2 & 2 & 1 \\ 3 & 4 & -1 \\ 3 & 0 & 5 \end{bmatrix}$

c. $\begin{bmatrix} 1 & 1 & 2 & 1 \\ 2 & -1 & 2 & 0 \\ 3 & 4 & 1 & 1 \\ -1 & 5 & 2 & 3 \end{bmatrix}$ d. $\begin{bmatrix} 1 & 2 & 3 & 4 \\ 2 & 1 & -1 & 1 \\ -3 & 2 & 0 & 1 \\ 0 & 5 & 2 & 6 \end{bmatrix}$

2. Use Definition 6.15 to compute the determinants of the following matrices:

a. $\begin{bmatrix} 1 & 2 & 0 \\ 2 & 1 & -1 \\ 3 & 1 & 1 \end{bmatrix}$ b. $\begin{bmatrix} 4 & 0 & 1 \\ 2 & 1 & 0 \\ 2 & 2 & 3 \end{bmatrix}$

c. $\begin{bmatrix} 1 & 1 & -1 & 1 \\ 1 & 2 & -4 & -2 \\ 2 & 1 & 1 & 5 \\ -1 & 0 & -2 & -4 \end{bmatrix}$ d. $\begin{bmatrix} 2 & 0 & 1 & 2 \\ 1 & 1 & 0 & 2 \\ 2 & -1 & 3 & 1 \\ 3 & -1 & 4 & 3 \end{bmatrix}$

3. Repeat Exercise 2 using the method of Example 2.
4. Repeat Exercise 1 using the method of Example 2.
5. Find all values of α that make the following matrix singular.

$$A = \begin{bmatrix} 1 & -1 & \alpha \\ 2 & 2 & 1 \\ 0 & \alpha & -\frac{3}{2} \end{bmatrix}.$$

6. Find all values of α that make the following matrix singular.

$$A = \begin{bmatrix} 1 & 2 & -1 \\ 1 & \alpha & 1 \\ 2 & \alpha & -1 \end{bmatrix}.$$

7. Find all values of α so that the following linear system has no solutions.

$$\begin{aligned} 2x_1 - x_2 + 3x_3 &= 5, \\ 4x_1 + 2x_2 + 2x_3 &= 6, \\ -2x_1 + \alpha x_2 + 3x_3 &= 4. \end{aligned}$$

8. Find all values of α so that the following linear system has an infinite number of solutions.

$$\begin{aligned} 2x_1 - x_2 + 3x_3 &= 5, \\ 4x_1 + 2x_2 + 2x_3 &= 6, \\ -2x_1 + \alpha x_2 + 3x_3 &= 1. \end{aligned}$$

9. Use mathematical induction to show that when $n > 1$, the evaluation of the determinant of an $n \times n$ matrix using the definition requires

$$n! \sum_{k=1}^{n-1} \frac{1}{k!} \text{ multiplications/divisions} \quad \text{and} \quad n! - 1 \text{ additions/subtractions.}$$

10. Let A be a 3×3 matrix. Show that if $\tilde{A}$ is the matrix obtained from A using any of the operations

$$(E_1) \leftrightarrow (E_2), \quad (E_1) \leftrightarrow (E_3), \quad \text{or} \quad (E_2) \leftrightarrow (E_3),$$

then $\det \tilde{A} = -\det A$.

11. Prove that AB is nonsingular if and only if both A and B are nonsingular.

12. The solution by **Cramer's rule** to the linear system

$$\begin{aligned} a_{11}x_1 + a_{12}x_2 + a_{13}x_3 &= b_1, \\ a_{21}x_1 + a_{22}x_2 + a_{23}x_3 &= b_2, \\ a_{31}x_1 + a_{32}x_2 + a_{33}x_3 &= b_3, \end{aligned}$$

has

$$x_1 = \frac{1}{D}\det\begin{bmatrix} b_1 & a_{12} & a_{13} \\ b_2 & a_{22} & a_{23} \\ b_3 & a_{32} & a_{33} \end{bmatrix} \equiv \frac{D_1}{D}, \quad x_2 = \frac{1}{D}\det\begin{bmatrix} a_{11} & b_1 & a_{13} \\ a_{21} & b_2 & a_{23} \\ a_{31} & b_3 & a_{33} \end{bmatrix} \equiv \frac{D_2}{D},$$

and

$$x_3 = \frac{1}{D}\det\begin{bmatrix} a_{11} & a_{12} & b_1 \\ a_{21} & a_{22} & b_2 \\ a_{31} & a_{32} & b_3 \end{bmatrix} \equiv \frac{D_3}{D}, \quad \text{where} \quad D = \det\begin{bmatrix} a_{11} & a_{12} & a_{13} \\ a_{21} & a_{22} & a_{23} \\ a_{31} & a_{32} & a_{33} \end{bmatrix}.$$

a. Find the solution to the linear system

$$\begin{aligned} 2x_1 + 3x_2 - x_3 &= 4, \\ x_1 - 2x_2 + x_3 &= 6, \\ x_1 - 12x_2 + 5x_3 &= 10, \end{aligned}$$

by Cramer's rule.

b. Show that the linear system

$$\begin{aligned} 2x_1 + 3x_2 - x_3 &= 4, \\ x_1 - 2x_2 + x_3 &= 6, \\ -x_1 - 12x_2 + 5x_3 &= 9 \end{aligned}$$

does not have a solution. Compute D_1, D_2, and D_3.

c. Show that the linear system

$$\begin{aligned} 2x_1 + 3x_2 - x_3 &= 4, \\ x_1 - 2x_2 + x_3 &= 6, \\ -x_1 - 12x_2 + 5x_3 &= 10 \end{aligned}$$

has an infinite number of solutions. Compute D_1, D_2, and D_3.

d. Prove that if a 3×3 linear system with $D = 0$ has solutions, then $D_1 = D_2 = D_3 = 0$.

e. Determine the number of multiplications/divisions and additions/subtractions required for Cramer's rule on a 3×3 system.

13. **a.** Generalize Cramer's rule to an $n \times n$ linear system.

b. Use the result in Exercise 9 to determine the number of multiplications/divisions and additions/subtractions required for Cramer's rule on an $n \times n$ system.

6.5 Matrix Factorization

Gaussian elimination is the principal tool in the direct solution of linear systems of equations, so it should be no surprise that it appears in other guises. In this section we will see that the steps used to solve a system of the form $A\mathbf{x} = \mathbf{b}$ can be used to factor a matrix. The factorization is particularly useful when it has the form $A = LU$, where L is lower triangular and U is upper triangular. Although not all matrices have this type of representation, many do that occur frequently in the application of numerical techniques.

In Section 6.1 we found that Gaussian elimination applied to an arbitrary linear system $A\mathbf{x} = \mathbf{b}$ requires $O(n^3/3)$ arithmetic operations to determine $\mathbf{x}$. However, to solve a linear system that involves an upper-triangular system requires only backward substitution, which takes $O(n^2)$ operations. The number of operations required to solve a lower-triangular systems is similar.

Suppose that A has been factored into the triangular form $A = LU$, where L is lower triangular and U is upper triangular. Then we can solve for $\mathbf{x}$ more easily by using a two-step process.

- First we let $\mathbf{y} = U\mathbf{x}$ and solve the lower triangular system $L\mathbf{y} = \mathbf{b}$ for $\mathbf{y}$. Since L is triangular, determining $\mathbf{y}$ from this equation requires only $O(n^2)$ operations.

- Once $\mathbf{y}$ is known, the upper triangular system $U\mathbf{x} = \mathbf{y}$ requires only an additional $O(n^2)$ operations to determine the solution $\mathbf{x}$.

Solving a linear system $A\mathbf{x} = \mathbf{b}$ in factored form means that the number of operations needed to solve the system $A\mathbf{x} = \mathbf{b}$ is reduced from $O(n^3/3)$ to $O(2n^2)$.

Example 1 Compare the approximate number of operations required to determine the solution to a linear system using a technique requiring $O(n^3/3)$ operations and one requiring $O(2n^2)$ when $n = 20$, $n = 100$, and $n = 1000$.

Solution Table 6.3 gives the results of these calculations. ■

Table 6.3

n	$n^3/3$	$2n^2$	% Reduction
10	$3.\overline{3} \times 10^2$	2×10^2	40
100	$3.\overline{3} \times 10^5$	2×10^4	94
1000	$3.\overline{3} \times 10^8$	2×10^6	99.4

As the example illustrates, the reduction factor increases dramatically with the size of the matrix. Not surprisingly, the reductions from the factorization come at a cost; determining the specific matrices L and U requires $O(n^3/3)$ operations. But once the factorization is determined, systems involving the matrix A can be solved in this simplified manner for any number of vectors $\mathbf{b}$.

To see which matrices have an LU factorization and to find how it is determined, first suppose that Gaussian elimination can be performed on the system $A\mathbf{x} = \mathbf{b}$ without row interchanges. With the notation in Section 6.1, this is equivalent to having nonzero pivot elements $a_{ii}^{(i)}$, for each $i = 1, 2, \ldots, n$.

The first step in the Gaussian elimination process consists of performing, for each $j = 2, 3, \ldots, n$, the operations

$$(E_j - m_{j,1}E_1) \to (E_j), \quad \text{where} \quad m_{j,1} = \frac{a_{j1}^{(1)}}{a_{11}^{(1)}}. \tag{6.8}$$

These operations transform the system into one in which all the entries in the first column below the diagonal are zero.

The system of operations in (6.8) can be viewed in another way. It is simultaneously accomplished by multiplying the original matrix A on the left by the matrix

$$M^{(1)} = \begin{bmatrix} 1 & 0 & \cdots & \cdots & 0 \\ -m_{21} & 1 & \ddots & & \vdots \\ \vdots & 0 & \ddots & \ddots & \vdots \\ \vdots & \vdots & \ddots & \ddots & 0 \\ -m_{n1} & 0 & \cdots & 0 & 1 \end{bmatrix}.$$

Matrix factorization is another of the important techniques that Gauss seems to be the first to have discovered. It is included in his two-volume treatise on celestial mechanics *Theoria motus corporum coelestium in sectionibus conicis Solem ambientium*, which was published in 1809.

This is called the **first Gaussian transformation matrix**. We denote the product of this matrix with $A^{(1)} \equiv A$ by $A^{(2)}$ and with $\mathbf{b}$ by $\mathbf{b}^{(2)}$, so

$$A^{(2)}\mathbf{x} = M^{(1)}A\mathbf{x} = M^{(1)}\mathbf{b} = \mathbf{b}^{(2)}.$$

In a similar manner we construct $M^{(2)}$, the identity matrix with the entries below the diagonal in the second column replaced by the negatives of the multipliers

$$m_{j,2} = \frac{a_{j2}^{(2)}}{a_{22}^{(2)}}.$$

The product of this matrix with $A^{(2)}$ has zeros below the diagonal in the first two columns, and we let

$$A^{(3)}\mathbf{x} = M^{(2)}A^{(2)}\mathbf{x} = M^{(2)}M^{(1)}A\mathbf{x} = M^{(2)}M^{(1)}\mathbf{b} = \mathbf{b}^{(3)}.$$

In general, with $A^{(k)}\mathbf{x} = \mathbf{b}^{(k)}$ already formed, multiply by the ***k*th Gaussian transformation matrix**

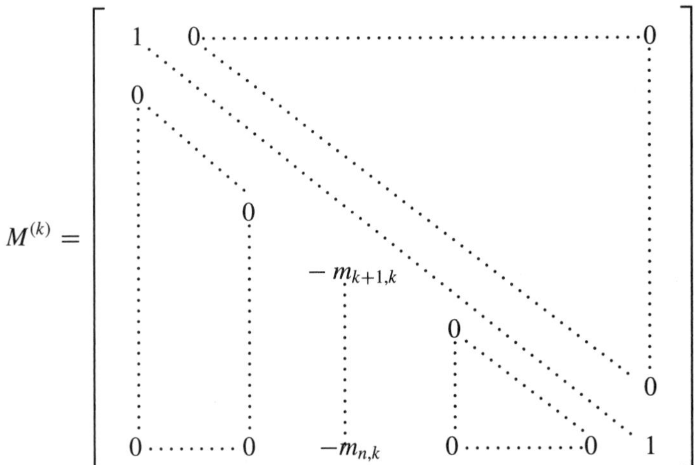

to obtain

$$A^{(k+1)}\mathbf{x} = M^{(k)}A^{(k)}\mathbf{x} = M^{(k)}\cdots M^{(1)}A\mathbf{x} = M^{(k)}\mathbf{b}^{(k)} = \mathbf{b}^{(k+1)} = M^{(k)}\cdots M^{(1)}\mathbf{b}. \quad (6.9)$$

The process ends with the formation of $A^{(n)}\mathbf{x} = \mathbf{b}^{(n)}$, where $A^{(n)}$ is the upper triangular matrix

$$A^{(n)} = \begin{bmatrix} a_{11}^{(1)} & a_{12}^{(1)} & \cdots & \cdots & a_{1n}^{(1)} \\ 0 & a_{22}^{(2)} & \ddots & & \vdots \\ \vdots & \ddots & \ddots & \ddots & \vdots \\ \vdots & & \ddots & \ddots & a_{n-1,n}^{(n-1)} \\ 0 & \cdots & \cdots & 0 & a_{nn}^{(n)} \end{bmatrix},$$

given by

$$A^{(n)} = M^{(n-1)}M^{(n-2)}\cdots M^{(1)}A.$$

This process forms the $U = A^{(n)}$ portion of the matrix factorization $A = LU$. To determine the complementary lower triangular matrix L, first recall the multiplication of $A^{(k)}\mathbf{x} = \mathbf{b}^{(k)}$ by the Gaussian transformation of $M^{(k)}$ used to obtain (6.9):

$$A^{(k+1)}\mathbf{x} = M^{(k)}A^{(k)}\mathbf{x} = M^{(k)}\mathbf{b}^{(k)} = \mathbf{b}^{(k+1)},$$

where $M^{(k)}$ generates the row operations

$$(E_j - m_{j,k}E_k) \to (E_j), \quad \text{for } j = k+1, \ldots, n.$$

To reverse the effects of this transformation and return to $A^{(k)}$ requires that the operations $(E_j + m_{j,k}E_k) \to (E_j)$ be performed for each $j = k+1, \ldots, n$. This is equivalent to multiplying by the inverse of the matrix $M^{(k)}$, the matrix

$$L^{(k)} = [M^{(k)}]^{-1} = \begin{bmatrix} 1 & 0 & \cdots & & & & \cdots & 0 \\ 0 & \ddots & & & & & & \vdots \\ \vdots & & 0 & & & & & \\ \vdots & & \vdots & m_{k+1,k} & \ddots & & & \\ \vdots & & \vdots & \vdots & 0 & \ddots & & \\ \vdots & & \vdots & \vdots & \vdots & & \ddots & 0 \\ 0 & \cdots & 0 & m_{n,k} & 0 & \cdots & 0 & 1 \end{bmatrix}.$$

The lower-triangular matrix L in the factorization of A, then, is the product of the matrices $L^{(k)}$:

$$L = L^{(1)}L^{(2)}\cdots L^{(n-1)} = \begin{bmatrix} 1 & 0 & \cdots & 0 \\ m_{21} & 1 & \ddots & \vdots \\ \vdots & \ddots & \ddots & 0 \\ m_{n1} & \cdots & m_{n,n-1} & 1 \end{bmatrix},$$

since the product of L with the upper-triangular matrix $U = M^{(n-1)}\cdots M^{(2)}M^{(1)}A$ gives

$$\begin{aligned} LU &= L^{(1)}L^{(2)}\cdots L^{(n-3)}L^{(n-2)}L^{(n-1)} \cdot M^{(n-1)}M^{(n-2)}M^{(n-3)}\cdots M^{(2)}M^{(1)}A \\ &= [M^{(1)}]^{-1}[M^{(2)}]^{-1}\cdots[M^{(n-2)}]^{-1}[M^{(n-1)}]^{-1}\cdot M^{(n-1)}M^{(n-2)}\cdots M^{(2)}M^{(1)}A = A. \end{aligned}$$

Theorem 6.19 follows from these observations.

Theorem 6.19 If Gaussian elimination can be performed on the linear system $A\mathbf{x} = \mathbf{b}$ without row interchanges, then the matrix A can be factored into the product of a lower-triangular matrix L and an upper-triangular matrix U, that is, $A = LU$, where $m_{ji} = a_{ji}^{(i)}/a_{ii}^{(i)}$,

$$U = \begin{bmatrix} a_{11}^{(1)} & a_{12}^{(1)} & \cdots & a_{1n}^{(1)} \\ 0 & a_{22}^{(2)} & \ddots & \vdots \\ \vdots & \ddots & \ddots & a_{n-1,n}^{(n-1)} \\ 0 & \cdots & 0 & a_{nn}^{(n)} \end{bmatrix}, \quad \text{and} \quad L = \begin{bmatrix} 1 & 0 & \cdots & 0 \\ m_{21} & 1 & \ddots & \vdots \\ \vdots & \ddots & \ddots & 0 \\ m_{n1} & \cdots & m_{n,n-1} & 1 \end{bmatrix}.$$

■

Example 2 **(a)** Determine the LU factorization for matrix A in the linear system $A\mathbf{x} = \mathbf{b}$, where

$$A = \begin{bmatrix} 1 & 1 & 0 & 3 \\ 2 & 1 & -1 & 1 \\ 3 & -1 & -1 & 2 \\ -1 & 2 & 3 & -1 \end{bmatrix} \quad \text{and} \quad \mathbf{b} = \begin{bmatrix} 1 \\ 1 \\ -3 \\ 4 \end{bmatrix}.$$

(b) Then use the factorization to solve the system

$$\begin{aligned} x_1 + x_2 \phantom{{}- x_3} + 3x_4 &= 8, \\ 2x_1 + x_2 - x_3 + x_4 &= 7, \\ 3x_1 - x_2 - x_3 + 2x_4 &= 14, \\ -x_1 + 2x_2 + 3x_3 - x_4 &= -7. \end{aligned}$$

Solution **(a)** The original system was considered in Section 6.1, where we saw that the sequence of operations $(E_2 - 2E_1) \to (E_2)$, $(E_3 - 3E_1) \to (E_3)$, $(E_4 - (-1)E_1) \to (E_4)$, $(E_3 - 4E_2) \to (E_3)$, $(E_4 - (-3)E_2) \to (E_4)$ converts the system to the triangular system

$$\begin{aligned} x_1 + x_2 \phantom{{}- x_3} + 3x_4 &= 4, \\ -x_2 - x_3 - 5x_4 &= -7, \\ 3x_3 + 13x_4 &= 13, \\ -13x_4 &= -13. \end{aligned}$$

The multipliers m_{ij} and the upper triangular matrix produce the factorization

$$A = \begin{bmatrix} 1 & 1 & 0 & 3 \\ 2 & 1 & -1 & 1 \\ 3 & -1 & -1 & 2 \\ -1 & 2 & 3 & -1 \end{bmatrix} = \begin{bmatrix} 1 & 0 & 0 & 0 \\ 2 & 1 & 0 & 0 \\ 3 & 4 & 1 & 0 \\ -1 & -3 & 0 & 1 \end{bmatrix} \begin{bmatrix} 1 & 1 & 0 & 3 \\ 0 & -1 & -1 & -5 \\ 0 & 0 & 3 & 13 \\ 0 & 0 & 0 & -13 \end{bmatrix} = LU.$$

(b) To solve

$$A\mathbf{x} = LU\mathbf{x} = \begin{bmatrix} 1 & 0 & 0 & 0 \\ 2 & 1 & 0 & 0 \\ 3 & 4 & 1 & 0 \\ -1 & -3 & 0 & 1 \end{bmatrix} \begin{bmatrix} 1 & 1 & 0 & 3 \\ 0 & -1 & -1 & -5 \\ 0 & 0 & 3 & 13 \\ 0 & 0 & 0 & -13 \end{bmatrix} \begin{bmatrix} x_1 \\ x_2 \\ x_3 \\ x_4 \end{bmatrix} = \begin{bmatrix} 8 \\ 7 \\ 14 \\ -7 \end{bmatrix},$$

we first introduce the substitution $\mathbf{y} = U\mathbf{x}$. Then $\mathbf{b} = L(U\mathbf{x}) = L\mathbf{y}$. That is,

$$L\mathbf{y} = \begin{bmatrix} 1 & 0 & 0 & 0 \\ 2 & 1 & 0 & 0 \\ 3 & 4 & 1 & 0 \\ -1 & -3 & 0 & 1 \end{bmatrix} \begin{bmatrix} y_1 \\ y_2 \\ y_3 \\ y_4 \end{bmatrix} = \begin{bmatrix} 8 \\ 7 \\ 14 \\ -7 \end{bmatrix}.$$

This system is solved for $\mathbf{y}$ by a simple forward-substitution process:

$$\begin{aligned} y_1 &= 8; \\ 2y_1 + y_2 &= 7, \quad \text{so } y_2 = 7 - 2y_1 = -9; \\ 3y_1 + 4y_2 + y_3 &= 14, \quad \text{so } y_3 = 14 - 3y_1 - 4y_2 = 26; \\ -y_1 - 3y_2 + y_4 &= -7, \quad \text{so } y_4 = -7 + y_1 + 3y_2 = -26. \end{aligned}$$

We then solve $U\mathbf{x} = \mathbf{y}$ for $\mathbf{x}$, the solution of the original system; that is,

$$\begin{bmatrix} 1 & 1 & 0 & 3 \\ 0 & -1 & -1 & -5 \\ 0 & 0 & 3 & 13 \\ 0 & 0 & 0 & -13 \end{bmatrix} \begin{bmatrix} x_1 \\ x_2 \\ x_3 \\ x_4 \end{bmatrix} = \begin{bmatrix} 8 \\ -9 \\ 26 \\ -26 \end{bmatrix}.$$

Using backward substitution we obtain $x_4 = 2, x_3 = 0, x_2 = -1, x_1 = 3$. ■

The *NumericalAnalysis* subpackage of Maple can be used to perform the matrix factorization in Example 2. First load the package

with(*Student*[*NumericalAnalysis*])

and the matrix A

$A := Matrix([[1, 1, 0, 3], [2, 1, -1, 1], [3, -1, -1, 2], [-1, 2, 3, -1]])$

The factorization is performed with the command

Lower, *Upper* := *MatrixDecomposition*(*A*, *method* = *LU*, *output* = ['*L*', '*U*'])

giving

$$\begin{bmatrix} 1 & 0 & 0 & 0 \\ 2 & 1 & 0 & 0 \\ 3 & 4 & 1 & 0 \\ -1 & -3 & 0 & 1 \end{bmatrix}, \begin{bmatrix} 1 & 1 & 0 & 3 \\ 0 & -1 & -1 & -5 \\ 0 & 0 & 3 & 13 \\ 0 & 0 & 0 & -13 \end{bmatrix}$$

To use the factorization to solve the system $A\mathbf{x} = \mathbf{b}$, define $\mathbf{b}$ by

$b := Vector([8, 7, 14, -7])$

Then perform the forward substitution to determine $\mathbf{y}$ with $U\mathbf{x} = y$, followed by backward substitution to determine $\mathbf{x}$ with $U\mathbf{x} = \mathbf{y}$.

$y :=$ *ForwardSubstitution*(*Lower*, *b*): $x :=$ *BackSubstitution*(*Upper*, *y*)

The solution agrees with that in Example 2.

The factorization used in Example 2 is called *Doolittle's method* and requires that 1s be on the diagonal of L, which results in the factorization described in Theorem 6.19. In Section 6.6, we consider *Crout's method*, a factorization which requires that 1s be on the diagonal elements of U, and *Cholesky's method*, which requires that $l_{ii} = u_{ii}$, for each i.

A general procedure for factoring matrices into a product of triangular matrices is contained in Algorithm 6.4. Although new matrices L and U are constructed, the generated values can replace the corresponding entries of A that are no longer needed.

Algorithm 6.4 permits either the diagonal of L or the diagonal of U to be specified.

LU Factorization

To factor the $n \times n$ matrix $A = [a_{ij}]$ into the product of the lower-triangular matrix $L = [l_{ij}]$ and the upper-triangular matrix $U = [u_{ij}]$; that is, $A = LU$, where the main diagonal of either L or U consists of all ones:

INPUT dimension n; the entries a_{ij}, $1 \leq i, j \leq n$ of A; the diagonal $l_{11} = \cdots = l_{nn} = 1$ of L or the diagonal $u_{11} = \cdots = u_{nn} = 1$ of U.

OUTPUT the entries l_{ij}, $1 \leq j \leq i$, $1 \leq i \leq n$ of L and the entries, u_{ij}, $i \leq j \leq n$, $1 \leq i \leq n$ of U.

Step 1 Select l_{11} and u_{11} satisfying $l_{11}u_{11} = a_{11}$.
If $l_{11}u_{11} = 0$ then OUTPUT ('Factorization impossible');
STOP.

Step 2 For $j = 2, \ldots, n$ set $u_{1j} = a_{1j}/l_{11}$; (*First row of* U.)
$l_{j1} = a_{j1}/u_{11}$. (*First column of* L.)

Step 3 For $i = 2, \ldots, n-1$ do Steps 4 and 5.

Step 4 Select l_{ii} and u_{ii} satisfying $l_{ii}u_{ii} = a_{ii} - \sum_{k=1}^{i-1} l_{ik}u_{ki}$.
If $l_{ii}u_{ii} = 0$ then OUTPUT ('Factorization impossible');
STOP.

Step 5 For $j = i+1, \ldots, n$

set $u_{ij} = \frac{1}{l_{ii}}\left[a_{ij} - \sum_{k=1}^{i-1} l_{ik}u_{kj}\right]$; (*ith row of* U.)

$l_{ji} = \frac{1}{u_{ii}}\left[a_{ji} - \sum_{k=1}^{i-1} l_{jk}u_{ki}\right]$. (*ith column of* L.)

Step 6 Select l_{nn} and u_{nn} satisfying $l_{nn}u_{nn} = a_{nn} - \sum_{k=1}^{n-1} l_{nk}u_{kn}$.
(*Note: If* $l_{nn}u_{nn} = 0$, *then* $A = LU$ *but* A *is singular.*)

Step 7 OUTPUT (l_{ij} for $j = 1, \ldots, i$ and $i = 1, \ldots, n$);
OUTPUT (u_{ij} for $j = i, \ldots, n$ and $i = 1, \ldots, n$);
STOP. ■

Once the matrix factorization is complete, the solution to a linear system of the form $A\mathbf{x} = LU\mathbf{x} = \mathbf{b}$ is found by first letting $\mathbf{y} = U\mathbf{x}$ and solving $L\mathbf{y} = \mathbf{b}$ for $\mathbf{y}$. Since L is lower triangular, we have

$$y_1 = \frac{b_1}{l_{11}},$$

and, for each $i = 2, 3, \ldots, n$,

$$y_i = \frac{1}{l_{ii}}\left[b_i - \sum_{j=1}^{i-1} l_{ij}y_j\right].$$

After $\mathbf{y}$ is found by this forward-substitution process, the upper-triangular system $U\mathbf{x} = \mathbf{y}$ is solved for $\mathbf{x}$ by backward substitution using the equations

$$x_n = \frac{y_n}{u_{nn}} \quad \text{and} \quad x_i = \frac{1}{u_{ii}}\left[y_i - \sum_{j=i+1}^{n} u_{ij}x_j\right].$$

Permutation Matrices

In the previous discussion we assumed that $A\mathbf{x} = \mathbf{b}$ can be solved using Gaussian elimination without row interchanges. From a practical standpoint, this factorization is useful only when row interchanges are not required to control the round-off error resulting from the use of finite-digit arithmetic. Fortunately, many systems we encounter when using approximation methods are of this type, but we will now consider the modifications that must be made when row interchanges are required. We begin the discussion with the introduction of a class of matrices that are used to rearrange, or permute, rows of a given matrix.

An $n \times n$ **permutation matrix** $P = [p_{ij}]$ is a matrix obtained by rearranging the rows of I_n, the identity matrix. This gives a matrix with precisely one nonzero entry in each row and in each column, and each nonzero entry is a 1.

Illustration The matrix

$$P = \begin{bmatrix} 1 & 0 & 0 \\ 0 & 0 & 1 \\ 0 & 1 & 0 \end{bmatrix}$$

is a 3×3 permutation matrix. For any 3×3 matrix A, multiplying on the left by P has the effect of interchanging the second and third rows of A:

$$PA = \begin{bmatrix} 1 & 0 & 0 \\ 0 & 0 & 1 \\ 0 & 1 & 0 \end{bmatrix} \begin{bmatrix} a_{11} & a_{12} & a_{13} \\ a_{21} & a_{22} & a_{23} \\ a_{31} & a_{32} & a_{33} \end{bmatrix} = \begin{bmatrix} a_{11} & a_{12} & a_{13} \\ a_{31} & a_{32} & a_{33} \\ a_{21} & a_{22} & a_{23} \end{bmatrix}.$$

Similarly, multiplying A on the right by P interchanges the second and third columns of A. □

Two useful properties of permutation matrices relate to Gaussian elimination, the first of which is illustrated in the previous example. Suppose $k_1, \cdots, k_n$ is a permutation of the integers $1, \cdots, n$ and the permutation matrix $P = (p_{ij})$ is defined by

$$p_{ij} = \begin{cases} 1, & \text{if } j = k_i, \\ 0, & \text{otherwise.} \end{cases}$$

Then

The matrix multiplication AP permutes the columns of A.

- PA permutes the rows of A; that is,

$$PA = \begin{bmatrix} a_{k_1 1} & a_{k_1 2} & \cdots & a_{k_1 n} \\ a_{k_2 1} & a_{k_2 2} & \cdots & a_{k_2 n} \\ \vdots & \vdots & \ddots & \vdots \\ a_{k_n 1} & a_{k_n 2} & \cdots & a_{k_n n} \end{bmatrix}.$$

- P^{-1} exists and $P^{-1} = P^t$.

At the end of Section 6.4 we saw that for any nonsingular matrix A, the linear system $A\mathbf{x} = \mathbf{b}$ can be solved by Gaussian elimination, with the possibility of row interchanges. If we knew the row interchanges that were required to solve the system by Gaussian elimination, we could arrange the original equations in an order that would ensure that no row interchanges are needed. Hence there *is* a rearrangement of the equations in the system that permits Gaussian elimination to proceed *without* row interchanges. This implies that for any nonsingular matrix A, a permutation matrix P exists for which the system

$$PA\mathbf{x} = P\mathbf{b}$$

can be solved without row interchanges. As a consequence, this matrix PA can be factored into

$$PA = LU,$$

where L is lower triangular and U is upper triangular. Because $P^{-1} = P^t$, this produces the factorization

$$A = P^{-1}LU = (P^tL)U.$$

The matrix U is still upper triangular, but P^tL is not lower triangular unless $P = I$.

Example 3 Determine a factorization in the form $A = (P^tL)U$ for the matrix

$$A = \begin{bmatrix} 0 & 0 & -1 & 1 \\ 1 & 1 & -1 & 2 \\ -1 & -1 & 2 & 0 \\ 1 & 2 & 0 & 2 \end{bmatrix}.$$

Solution The matrix A cannot have an LU factorization because $a_{11} = 0$. However, using the row interchange $(E_1) \leftrightarrow (E_2)$, followed by $(E_3 + E_1) \to (E_3)$ and $(E_4 - E_1) \to (E_4)$, produces

$$\begin{bmatrix} 1 & 1 & -1 & 2 \\ 0 & 0 & -1 & 1 \\ 0 & 0 & 1 & 2 \\ 0 & 1 & 1 & 0 \end{bmatrix}.$$

Then the row interchange $(E_2) \leftrightarrow (E_4)$, followed by $(E_4 + E_3) \to (E_4)$, gives the matrix

$$U = \begin{bmatrix} 1 & 1 & -1 & 2 \\ 0 & 1 & 1 & 0 \\ 0 & 0 & 1 & 2 \\ 0 & 0 & 0 & 3 \end{bmatrix}.$$

The permutation matrix associated with the row interchanges $(E_1) \leftrightarrow (E_2)$ and $(E_2) \leftrightarrow (E_4)$ is

$$P = \begin{bmatrix} 0 & 1 & 0 & 0 \\ 0 & 0 & 0 & 1 \\ 0 & 0 & 1 & 0 \\ 1 & 0 & 0 & 0 \end{bmatrix},$$

and

$$PA = \begin{bmatrix} 1 & 1 & -1 & 2 \\ 1 & 2 & 0 & 2 \\ -1 & -1 & 2 & 0 \\ 0 & 0 & -1 & 1 \end{bmatrix}.$$

Gaussian elimination is performed on PA using the same operations as on A, except without the row interchanges. That is, $(E_2 - E_1) \rightarrow (E_2)$, $(E_3 + E_1) \rightarrow (E_3)$, followed by $(E_4 + E_3) \rightarrow (E_4)$. The nonzero multipliers for PA are consequently,

$$m_{21} = 1, \quad m_{31} = -1, \quad \text{and} \quad m_{43} = -1,$$

and the LU factorization of PA is

$$PA = \begin{bmatrix} 1 & 0 & 0 & 0 \\ 1 & 1 & 0 & 0 \\ -1 & 0 & 1 & 0 \\ 0 & 0 & -1 & 1 \end{bmatrix} \begin{bmatrix} 1 & 1 & -1 & 2 \\ 0 & 1 & 1 & 0 \\ 0 & 0 & 1 & 2 \\ 0 & 0 & 0 & 3 \end{bmatrix} = LU.$$

Multiplying by $P^{-1} = P^t$ produces the factorization

$$A = P^{-1}(LU) = P^t(LU) = (P^tL)U = \begin{bmatrix} 0 & 0 & -1 & 1 \\ 1 & 0 & 0 & 0 \\ -1 & 0 & 1 & 0 \\ 1 & 1 & 0 & 0 \end{bmatrix} \begin{bmatrix} 1 & 1 & -1 & 2 \\ 0 & 1 & 1 & 0 \\ 0 & 0 & 1 & 2 \\ 0 & 0 & 0 & 3 \end{bmatrix}. \blacksquare$$

A matrix factorization of the form $A = PLU$ for a matrix A can be obtained using the *LinearAlgebra* package of Maple with the command

LUDecomposition(A)

The function call

$(P, L, U) :=$ *LUDecomposition*(A)

gives the factorization, and stores the permutation matrix as P, the lower triangular matrix as L, and the upper triangular matrix as U.

EXERCISE SET 6.5

1. Solve the following linear systems:

 a. $\begin{bmatrix} 1 & 0 & 0 \\ -2 & 1 & 0 \\ 3 & 0 & 1 \end{bmatrix} \begin{bmatrix} 2 & 1 & -1 \\ 0 & 4 & 2 \\ 0 & 0 & 5 \end{bmatrix} \begin{bmatrix} x_1 \\ x_2 \\ x_3 \end{bmatrix} = \begin{bmatrix} 1 \\ 0 \\ -5 \end{bmatrix}$

 b. $\begin{bmatrix} 1 & 0 & 0 \\ 2 & 1 & 0 \\ -3 & 2 & 1 \end{bmatrix} \begin{bmatrix} 1 & 2 & -3 \\ 0 & 1 & 2 \\ 0 & 0 & 1 \end{bmatrix} \begin{bmatrix} x_1 \\ x_2 \\ x_3 \end{bmatrix} = \begin{bmatrix} 4 \\ 6 \\ 8 \end{bmatrix}$

2. Solve the following linear systems:

 a. $\begin{bmatrix} 1 & 0 & 0 \\ 2 & 1 & 0 \\ -1 & 0 & 1 \end{bmatrix} \begin{bmatrix} 2 & 3 & -1 \\ 0 & -2 & 1 \\ 0 & 0 & 3 \end{bmatrix} \begin{bmatrix} x_1 \\ x_2 \\ x_3 \end{bmatrix} = \begin{bmatrix} 2 \\ -1 \\ 1 \end{bmatrix}$

 b. $\begin{bmatrix} 2 & 0 & 0 \\ -1 & 1 & 0 \\ 3 & 2 & -1 \end{bmatrix} \begin{bmatrix} 1 & 1 & 1 \\ 0 & 1 & 2 \\ 0 & 0 & 1 \end{bmatrix} \begin{bmatrix} x_1 \\ x_2 \\ x_3 \end{bmatrix} = \begin{bmatrix} -1 \\ 3 \\ 0 \end{bmatrix}$

3. Consider the following matrices. Find the permutation matrix P so that PA can be factored into the product LU, where L is lower triangular with 1s on its diagonal and U is upper triangular for these matrices.

a. $A = \begin{bmatrix} 0 & 2 & -1 \\ 1 & -1 & 2 \\ 1 & -1 & 4 \end{bmatrix}$ **b.** $A = \begin{bmatrix} 1 & 2 & -1 \\ 2 & 4 & 7 \\ -1 & 2 & 5 \end{bmatrix}$

c. $A = \begin{bmatrix} 1 & 1 & -1 & 2 \\ -1 & -1 & 1 & 5 \\ 2 & 2 & 3 & 7 \\ 2 & 3 & 4 & 5 \end{bmatrix}$ **d.** $A = \begin{bmatrix} 1 & 1 & -1 & 2 \\ 2 & 2 & 4 & 5 \\ 1 & -1 & 1 & 7 \\ 2 & 3 & 4 & 6 \end{bmatrix}$

4. Consider the following matrices. Find the permutation matrix P so that PA can be factored into the product LU, where L is lower triangular with 1s on its diagonal and U is upper triangular for these matrices.

a. $A = \begin{bmatrix} 1 & 2 & -1 \\ 2 & 4 & 0 \\ 0 & 1 & -1 \end{bmatrix}$ **b.** $A = \begin{bmatrix} 0 & 1 & 1 \\ 1 & -2 & -1 \\ 1 & -1 & 1 \end{bmatrix}$

c. $A = \begin{bmatrix} 1 & 1 & -1 & 0 \\ 1 & 1 & 4 & 3 \\ 2 & -1 & 2 & 4 \\ 2 & -1 & 2 & 3 \end{bmatrix}$ **d.** $A = \begin{bmatrix} 0 & 1 & 1 & 2 \\ 0 & 1 & 1 & -1 \\ 1 & 2 & -1 & 3 \\ 1 & 1 & 2 & 0 \end{bmatrix}$

5. Factor the following matrices into the LU decomposition using the LU Factorization Algorithm with $l_{ii} = 1$ for all i.

a. $\begin{bmatrix} 2 & -1 & 1 \\ 3 & 3 & 9 \\ 3 & 3 & 5 \end{bmatrix}$ **b.** $\begin{bmatrix} 1.012 & -2.132 & 3.104 \\ -2.132 & 4.096 & -7.013 \\ 3.104 & -7.013 & 0.014 \end{bmatrix}$

c. $\begin{bmatrix} 2 & 0 & 0 & 0 \\ 1 & 1.5 & 0 & 0 \\ 0 & -3 & 0.5 & 0 \\ 2 & -2 & 1 & 1 \end{bmatrix}$ **d.** $\begin{bmatrix} 2.1756 & 4.0231 & -2.1732 & 5.1967 \\ -4.0231 & 6.0000 & 0 & 1.1973 \\ -1.0000 & -5.2107 & 1.1111 & 0 \\ 6.0235 & 7.0000 & 0 & -4.1561 \end{bmatrix}$

6. Factor the following matrices into the LU decomposition using the LU Factorization Algorithm with $l_{ii} = 1$ for all i.

a. $\begin{bmatrix} 1 & -1 & 0 \\ 2 & 2 & 3 \\ -1 & 3 & 2 \end{bmatrix}$ **b.** $\begin{bmatrix} \frac{1}{3} & \frac{1}{2} & -\frac{1}{4} \\ \frac{1}{5} & \frac{2}{3} & \frac{3}{8} \\ \frac{2}{5} & -\frac{2}{3} & \frac{5}{8} \end{bmatrix}$

c. $\begin{bmatrix} 2 & 1 & 0 & 0 \\ -1 & 3 & 3 & 0 \\ 2 & -2 & 1 & 4 \\ -2 & 2 & 2 & 5 \end{bmatrix}$ **d.** $\begin{bmatrix} 2.121 & -3.460 & 0 & 5.217 \\ 0 & 5.193 & -2.197 & 4.206 \\ 5.132 & 1.414 & 3.141 & 0 \\ -3.111 & -1.732 & 2.718 & 5.212 \end{bmatrix}$

7. Modify the LU Factorization Algorithm so that it can be used to solve a linear system, and then solve the following linear systems.

a.
$$\begin{aligned} 2x_1 - x_2 + x_3 &= -1, \\ 3x_1 + 3x_2 + 9x_3 &= 0, \\ 3x_1 + 3x_2 + 5x_3 &= 4. \end{aligned}$$

b.
$$\begin{aligned} 1.012x_1 - 2.132x_2 + 3.104x_3 &= 1.984, \\ -2.132x_1 + 4.096x_2 - 7.013x_3 &= -5.049, \\ 3.104x_1 - 7.013x_2 + 0.014x_3 &= -3.895. \end{aligned}$$

c.
$$\begin{aligned} 2x_1 \qquad\qquad\qquad\qquad &= 3, \\ x_1 + 1.5x_2 \qquad\qquad\quad &= 4.5, \\ -\ 3x_2 + 0.5x_3 \qquad &= -6.6, \\ 2x_1 - 2x_2 + x_3 + x_4 &= 0.8. \end{aligned}$$

d.
$$\begin{aligned} 2.1756x_1 + 4.0231x_2 - 2.1732x_3 + 5.1967x_4 &= 17.102, \\ -4.0231x_1 + 6.0000x_2 \qquad\qquad + 1.1973x_4 &= -6.1593, \\ -1.0000x_1 - 5.2107x_2 + 1.1111x_3 \qquad\qquad &= 3.0004, \\ 6.0235x_1 + 7.0000x_2 \qquad\qquad - 4.1561x_4 &= 0.0000. \end{aligned}$$

8. Modify the *LU* Factorization Algorithm so that it can be used to solve a linear system, and then solve the following linear systems.

a. $x_1 - x_2 = 2,$
$2x_1 + 2x_2 + 3x_3 = -1,$
$-x_1 + 3x_2 + 2x_3 = 4.$

b. $\frac{1}{3}x_1 + \frac{1}{2}x_2 - \frac{1}{4}x_3 = 1,$
$\frac{1}{5}x_1 + \frac{2}{3}x_2 + \frac{3}{8}x_3 = 2,$
$\frac{2}{5}x_1 - \frac{2}{3}x_2 + \frac{5}{8}x_3 = -3.$

b. $2x_1 + x_2 = 0,$
$-x_1 + 3x_2 + 3x_3 = 5,$
$2x_1 - 2x_2 + x_3 + 4x_4 = -2,$
$-2x_1 + 2x_2 + 2x_3 + 5x_4 = 6.$

d. $2.121x_1 - 3.460x_2 + 5.217x_4 = 1.909,$
$5.193x_2 - 2.197x_3 + 4.206x_4 = 0,$
$5.132x_1 + 1.414x_2 + 3.141x_3 = -2.101,$
$-3.111x_1 - 1.732x_2 + 2.718x_3 + 5.212x_4 = 6.824.$

9. Obtain factorizations of the form $A = P^tLU$ for the following matrices.

a. $A = \begin{bmatrix} 0 & 2 & 3 \\ 1 & 1 & -1 \\ 0 & -1 & 1 \end{bmatrix}$

b. $A = \begin{bmatrix} 1 & 2 & -1 \\ 1 & 2 & 3 \\ 2 & -1 & 4 \end{bmatrix}$

c. $A = \begin{bmatrix} 1 & -2 & 3 & 0 \\ 3 & -6 & 9 & 3 \\ 2 & 1 & 4 & 1 \\ 1 & -2 & 2 & -2 \end{bmatrix}$

d. $A = \begin{bmatrix} 1 & -2 & 3 & 0 \\ 1 & -2 & 3 & 1 \\ 1 & -2 & 2 & -2 \\ 2 & 1 & 3 & -1 \end{bmatrix}$

10. Suppose $A = P^tLU$, where P is a permutation matrix, L is a lower-triangular matrix with ones on the diagonal, and U is an upper-triangular matrix.

a. Count the number of operations needed to compute P^tLU for a given matrix A.

b. Show that if P contains k row interchanges, then

$$\det P = \det P^t = (-1)^k.$$

c. Use $\det A = \det P^t \det L \det U = (-1)^k \det U$ to count the number of operations for determining $\det A$ by factoring.

d. Compute $\det A$ and count the number of operations when

$$A = \begin{bmatrix} 0 & 2 & 1 & 4 & -1 & 3 \\ 1 & 2 & -1 & 3 & 4 & 0 \\ 0 & 1 & 1 & -1 & 2 & -1 \\ 2 & 3 & -4 & 2 & 0 & 5 \\ 1 & 1 & 1 & 3 & 0 & 2 \\ -1 & -1 & 2 & -1 & 2 & 0 \end{bmatrix}.$$

11. a. Show that the *LU* Factorization Algorithm requires

$$\frac{1}{3}n^3 - \frac{1}{3}n \text{ multiplications/divisions} \quad \text{and} \quad \frac{1}{3}n^3 - \frac{1}{2}n^2 + \frac{1}{6}n \text{ additions/subtractions.}$$

b. Show that solving $L\mathbf{y} = \mathbf{b}$, where L is a lower-triangular matrix with $l_{ii} = 1$ for all i, requires

$$\frac{1}{2}n^2 - \frac{1}{2}n \text{ multiplications/divisions} \quad \text{and} \quad \frac{1}{2}n^2 - \frac{1}{2}n \text{ additions/subtractions.}$$

c. Show that solving $A\mathbf{x} = \mathbf{b}$ by first factoring A into $A = LU$ and then solving $L\mathbf{y} = \mathbf{b}$ and $U\mathbf{x} = \mathbf{y}$ requires the same number of operations as the Gaussian Elimination Algorithm 6.1.

d. Count the number of operations required to solve m linear systems $A\mathbf{x}^{(k)} = \mathbf{b}^{(k)}$ for $k = 1, \ldots, m$ by first factoring A and then using the method of part (c) m times.

6.6 Special Types of Matrices

We now turn attention to two classes of matrices for which Gaussian elimination can be performed effectively without row interchanges.

Diagonally Dominant Matrices

The first class is described in the following definition.

Definition 6.20 The $n \times n$ matrix A is said to be **diagonally dominant** when

$$|a_{ii}| \geq \sum_{\substack{j=1,\\ j\neq i}}^{n} |a_{ij}| \quad \text{holds for each } i = 1, 2, \cdots, n. \tag{6.10}$$

Each main diagonal entry in a strictly diagonally dominant matrix has a magnitude that is strictly greater that the sum of the magnitudes of all the other entries in that row.

A diagonally dominant matrix is said to be **strictly diagonally dominant** when the inequality in (6.10) is strict for each n, that is, when

$$|a_{ii}| > \sum_{\substack{j=1,\\ j\neq i}}^{n} |a_{ij}| \quad \text{holds for each } i = 1, 2, \cdots, n.$$ ■

Illustration Consider the matrices

$$A = \begin{bmatrix} 7 & 2 & 0 \\ 3 & 5 & -1 \\ 0 & 5 & -6 \end{bmatrix} \quad \text{and} \quad B = \begin{bmatrix} 6 & 4 & -3 \\ 4 & -2 & 0 \\ -3 & 0 & 1 \end{bmatrix}.$$

The nonsymmetric matrix A is strictly diagonally dominant because

$$|7| > |2| + |0|, \quad |5| > |3| + |-1|, \quad \text{and} \quad |-6| > |0| + |5|.$$

The symmetric matrix B is not strictly diagonally dominant because, for example, in the first row the absolute value of the diagonal element is $|6| < |4| + |-3| = 7$. It is interesting to note that A^t is not strictly diagonally dominant, because the middle row of A^t is [2 5 5], nor, of course, is B^t because $B^t = B$. □

The following theorem was used in Section 3.5 to ensure that there are unique solutions to the linear systems needed to determine cubic spline interpolants.

Theorem 6.21 A strictly diagonally dominant matrix A is nonsingular. Moreover, in this case, Gaussian elimination can be performed on any linear system of the form $A\mathbf{x} = \mathbf{b}$ to obtain its unique solution without row or column interchanges, and the computations will be stable with respect to the growth of round-off errors. ■

Proof We first use proof by contradiction to show that A is nonsingular. Consider the linear system described by $A\mathbf{x} = \mathbf{0}$, and suppose that a nonzero solution $\mathbf{x} = (x_i)$ to this system exists. Let k be an index for which

$$0 < |x_k| = \max_{1\leq j\leq n} |x_j|.$$

Because $\sum_{j=1}^{n} a_{ij}x_j = 0$ for each $i = 1, 2, \ldots, n$, we have, when $i = k$,

$$a_{kk}x_k = -\sum_{\substack{j=1,\\ j\neq k}}^{n} a_{kj}x_j.$$

From the triangle inequality we have

$$|a_{kk}||x_k| \le \sum_{\substack{j=1,\\ j\ne k}}^{n} |a_{kj}||x_j|, \quad \text{so} \quad |a_{kk}| \le \sum_{\substack{j=1,\\ j\ne k}}^{n} |a_{kj}|\frac{|x_j|}{|x_k|} \le \sum_{\substack{j=1,\\ j\ne k}}^{n} |a_{kj}|.$$

This inequality contradicts the strict diagonal dominance of A. Consequently, the only solution to $A\mathbf{x} = \mathbf{0}$ is $\mathbf{x} = \mathbf{0}$. This is shown in Theorem 6.17 on page 398 to be equivalent to the nonsingularity of A.

To prove that Gaussian elimination can be performed without row interchanges, we show that each of the matrices $A^{(2)}, A^{(3)}, \ldots, A^{(n)}$ generated by the Gaussian elimination process (and described in Section 6.5) is strictly diagonally dominant. This will ensure that at each stage of the Gaussian elimination process the pivot element is nonzero.

Since A is strictly diagonally dominant, $a_{11} \ne 0$ and $A^{(2)}$ can be formed. Thus for each $i = 2, 3, \ldots, n$,

$$a_{ij}^{(2)} = a_{ij}^{(1)} - \frac{a_{1j}^{(1)} a_{i1}^{(1)}}{a_{11}^{(1)}}, \quad \text{for} \quad 2 \le j \le n.$$

First, $a_{i1}^{(2)} = 0$. The triangle inequality implies that

$$\sum_{\substack{j=2\\ j\ne i}}^{n} |a_{ij}^{(2)}| = \sum_{\substack{j=2\\ j\ne i}}^{n} \left| a_{ij}^{(1)} - \frac{a_{1j}^{(1)} a_{i1}^{(1)}}{a_{11}^{(1)}} \right| \le \sum_{\substack{j=2\\ j\ne i}}^{n} |a_{ij}^{(1)}| + \sum_{\substack{j=2\\ j\ne i}}^{n} \left| \frac{a_{1j}^{(1)} a_{i1}^{(1)}}{a_{11}^{(1)}} \right|.$$

But since A is strictly diagonally dominant,

$$\sum_{\substack{j=2\\ j\ne i}}^{n} |a_{ij}^{(1)}| < |a_{ii}^{(1)}| - |a_{i1}^{(1)}| \quad \text{and} \quad \sum_{\substack{j=2\\ j\ne i}}^{n} |a_{1j}^{(1)}| < |a_{11}^{(1)}| - |a_{1i}^{(1)}|,$$

so

$$\sum_{\substack{j=2\\ j\ne i}}^{n} |a_{ij}^{(2)}| < |a_{ii}^{(1)}| - |a_{i1}^{(1)}| + \frac{|a_{i1}^{(1)}|}{|a_{11}^{(1)}|}(|a_{11}^{(1)}| - |a_{1i}^{(1)}|) = |a_{ii}^{(1)}| - \frac{|a_{i1}^{(1)}||a_{1i}^{(1)}|}{|a_{11}^{(1)}|}.$$

The triangle inequality also implies that

$$|a_{ii}^{(1)}| - \frac{|a_{i1}^{(1)}||a_{1i}^{(1)}|}{|a_{11}^{(1)}|} \le \left| a_{ii}^{(1)} - \frac{|a_{i1}^{(1)}||a_{1i}^{(1)}|}{|a_{11}^{(1)}|} \right| = |a_{ii}^{(2)}|.$$

which gives

$$\sum_{\substack{j=2\\ j\ne i}}^{n} |a_{ij}^{(2)}| < |a_{ii}^{(2)}|.$$

This establishes the strict diagonal dominance for rows $2, \ldots, n$. But the first row of $A^{(2)}$ and A are the same, so $A^{(2)}$ is strictly diagonally dominant.

This process is continued inductively until the upper-triangular and strictly diagonally dominant $A^{(n)}$ is obtained. This implies that all the diagonal elements are nonzero, so Gaussian elimination can be performed without row interchanges.

The demonstration of stability for this procedure can be found in [We]. ■ ■ ■

Positive Definite Matrices

The next special class of matrices is called *positive definite*.

Definition 6.22 A matrix A is **positive definite** if it is symmetric and if $\mathbf{x}^t A\mathbf{x} > 0$ for every n-dimensional vector $\mathbf{x} \neq \mathbf{0}$. ■

The name positive definite refers to the fact that the number $\mathbf{x}^t A\mathbf{x}$ must be positive whenever $\mathbf{x} \neq \mathbf{0}$.

Not all authors require symmetry of a positive definite matrix. For example, Golub and Van Loan [GV], a standard reference in matrix methods, requires only that $\mathbf{x}^t A\mathbf{x} > 0$ for each $\mathbf{x} \neq 0$. Matrices we call positive definite are called symmetric positive definite in [GV]. Keep this discrepancy in mind if you are using material from other sources.

To be precise, Definition 6.22 should specify that the 1×1 matrix generated by the operation $\mathbf{x}^t A\mathbf{x}$ has a positive value for its only entry since the operation is performed as follows:

$$\mathbf{x}^t A\mathbf{x} = [x_1, x_2, \cdots, x_n] \begin{bmatrix} a_{11} & a_{12} & \cdots & a_{1n} \\ a_{21} & a_{22} & \cdots & a_{2n} \\ \vdots & \vdots & & \vdots \\ a_{n1} & a_{n2} & \cdots & a_{nn} \end{bmatrix} \begin{bmatrix} x_1 \\ x_2 \\ \vdots \\ x_n \end{bmatrix}$$

$$= [x_1, x_2, \cdots, x_n] \begin{bmatrix} \sum_{j=1}^n a_{1j}x_j \\ \sum_{j=1}^n a_{2j}x_j \\ \vdots \\ \sum_{j=1}^n a_{nj}x_j \end{bmatrix} = \left[\sum_{i=1}^n \sum_{j=1}^n a_{ij}x_i x_j \right].$$

Example 1 Show that the matrix

$$A = \begin{bmatrix} 2 & -1 & 0 \\ -1 & 2 & -1 \\ 0 & -1 & 2 \end{bmatrix}$$

is positive definite

Solution Suppose $\mathbf{x}$ is any three-dimensional column vector. Then

$$\mathbf{x}^t A\mathbf{x} = [x_1, x_2, x_3] \begin{bmatrix} 2 & -1 & 0 \\ -1 & 2 & -1 \\ 0 & -1 & 2 \end{bmatrix} \begin{bmatrix} x_1 \\ x_2 \\ x_3 \end{bmatrix}$$

$$= [x_1, x_2, x_3] \begin{bmatrix} 2x_1 - x_2 \\ -x_1 + 2x_2 - x_3 \\ -x_2 + 2x_3 \end{bmatrix}$$

$$= 2x_1^2 - 2x_1x_2 + 2x_2^2 - 2x_2x_3 + 2x_3^2.$$

Rearranging the terms gives

$$\mathbf{x}^t A\mathbf{x} = x_1^2 + (x_1^2 - 2x_1x_2 + x_2^2) + (x_2^2 - 2x_2x_3 + x_3^2) + x_3^2$$

$$= x_1^2 + (x_1 - x_2)^2 + (x_2 - x_3)^2 + x_3^2,$$

which implies that

$$x_1^2 + (x_1 - x_2)^2 + (x_2 - x_3)^2 + x_3^2 > 0$$

unless $x_1 = x_2 = x_3 = 0$. ■

It should be clear from the example that using the definition to determine if a matrix is positive definite can be difficult. Fortunately, there are more easily verified criteria, which are presented in Chapter 9, for identifying members of this important class. The next result provides some necessary conditions that can be used to eliminate certain matrices from consideration.

Theorem 6.23 If A is an $n \times n$ positive definite matrix, then

(i) A has an inverse; **(ii)** $a_{ii} > 0$, for each $i = 1, 2, \ldots, n$;

(iii) $\max_{1 \le k,j \le n} |a_{kj}| \le \max_{1 \le i \le n} |a_{ii}|$; **(iv)** $(a_{ij})^2 < a_{ii}a_{jj}$, for each $i \ne j$. ■

Proof

(i) If $\mathbf{x}$ satisfies $A\mathbf{x} = \mathbf{0}$, then $\mathbf{x}^t A\mathbf{x} = 0$. Since A is positive definite, this implies $\mathbf{x} = \mathbf{0}$. Consequently, $A\mathbf{x} = \mathbf{0}$ has only the zero solution. By Theorem 6.17 on page 398, this is equivalent to A being nonsingular.

(ii) For a given i, let $\mathbf{x} = (x_j)$ be defined by $x_i = 1$ and $x_j = 0$, if $j \ne i$. Since $\mathbf{x} \ne \mathbf{0}$,

$$0 < \mathbf{x}^t A\mathbf{x} = a_{ii}.$$

(iii) For $k \ne j$, define $\mathbf{x} = (x_i)$ by

$$x_i = \begin{cases} 0, & \text{if } i \ne j \text{ and } i \ne k, \\ 1, & \text{if } i = j, \\ -1, & \text{if } i = k. \end{cases}$$

Since $\mathbf{x} \ne \mathbf{0}$,

$$0 < \mathbf{x}^t A\mathbf{x} = a_{jj} + a_{kk} - a_{jk} - a_{kj}.$$

But $A^t = A$, so $a_{jk} = a_{kj}$, which implies that

$$2a_{kj} < a_{jj} + a_{kk}. \tag{6.11}$$

Now define $\mathbf{z} = (z_i)$ by

$$z_i = \begin{cases} 0, & \text{if } i \ne j \text{ and } i \ne k, \\ 1, & \text{if } i = j \text{ or } i = k. \end{cases}$$

Then $\mathbf{z}^t A\mathbf{z} > 0$, so

$$-2a_{kj} < a_{kk} + a_{jj}. \tag{6.12}$$

Equations (6.11) and (6.12) imply that for each $k \ne j$,

$$|a_{kj}| < \frac{a_{kk} + a_{jj}}{2} \le \max_{1 \le i \le n} |a_{ii}|, \quad \text{so} \quad \max_{1 \le k,j \le n} |a_{kj}| \le \max_{1 \le i \le n} |a_{ii}|.$$

(iv) For $i \ne j$, define $\mathbf{x} = (x_k)$ by

$$x_k = \begin{cases} 0, & \text{if } k \ne j \text{ and } k \ne i, \\ \alpha, & \text{if } k = i, \\ 1, & \text{if } k = j, \end{cases}$$

where α represents an arbitrary real number. Because $\mathbf{x} \ne \mathbf{0}$,

$$0 < \mathbf{x}^t A\mathbf{x} = a_{ii}\alpha^2 + 2a_{ij}\alpha + a_{jj}.$$

As a quadratic polynomial in α with no real roots, the discriminant of $P(\alpha) = a_{ii}\alpha^2 + 2a_{ij}\alpha + a_{jj}$ must be negative. Thus

$$4a_{ij}^2 - 4a_{ii}a_{jj} < 0 \quad \text{and} \quad a_{ij}^2 < a_{ii}a_{jj}.$$ ■ ■ ■

Although Theorem 6.23 provides some important conditions that must be true of positive definite matrices, it does not ensure that a matrix satisfying these conditions is positive definite.

The following notion will be used to provide a necessary and sufficient condition.

Definition 6.24 A **leading principal submatrix** of a matrix A is a matrix of the form

$$A_k = \begin{bmatrix} a_{11} & a_{12} & \cdots & a_{1k} \\ a_{21} & a_{22} & \cdots & a_{2k} \\ \vdots & \vdots & & \vdots \\ a_{k1} & a_{k2} & \cdots & a_{kk} \end{bmatrix},$$

for some $1 \le k \le n$. ■

A proof of the following result can be found in [Stew2], p. 250.

Theorem 6.25 A symmetric matrix A is positive definite if and only if each of its leading principal submatrices has a positive determinant. ■

Example 2 In Example 1 we used the definition to show that the symmetric matrix

$$A = \begin{bmatrix} 2 & -1 & 0 \\ -1 & 2 & -1 \\ 0 & -1 & 2 \end{bmatrix}$$

is positive definite. Confirm this using Theorem 6.25.

Solution Note that

$$\det A_1 = \det[2] = 2 > 0,$$

$$\det A_2 = \det\begin{bmatrix} 2 & -1 \\ -1 & 2 \end{bmatrix} = 4 - 1 = 3 > 0,$$

and

$$\det A_3 = \det\begin{bmatrix} 2 & -1 & 0 \\ -1 & 2 & -1 \\ 0 & -1 & 2 \end{bmatrix} = 2\det\begin{bmatrix} 2 & -1 \\ -1 & 2 \end{bmatrix} - (-1)\det\begin{bmatrix} -1 & -1 \\ 0 & 2 \end{bmatrix}$$

$$= 2(4-1) + (-2+0) = 4 > 0.$$

in agreement with Theorem 6.25. ■

The next result extends part **(i)** of Theorem 6.23 and parallels the strictly diagonally dominant results presented in Theorem 6.21 on page 412. We will not give a proof of this theorem because it requires introducing terminology and results that are not needed for any other purpose. The development and proof can be found in [We], pp. 120 ff.

Theorem 6.26 The symmetric matrix A is positive definite if and only if Gaussian elimination without row interchanges can be performed on the linear system $A\mathbf{x} = \mathbf{b}$ with all pivot elements positive. Moreover, in this case, the computations are stable with respect to the growth of round-off errors. ■

Some interesting facts that are uncovered in constructing the proof of Theorem 6.26 are presented in the following corollaries.

Corollary 6.27 The matrix A is positive definite if and only if A can be factored in the form LDL^t, where L is lower triangular with 1s on its diagonal and D is a diagonal matrix with positive diagonal entries. ■

Corollary 6.28 The matrix A is positive definite if and only if A can be factored in the form LL^t, where L is lower triangular with nonzero diagonal entries. ■

The matrix L in this Corollary is not the same as the matrix L in Corollary 6.27. A relationship between them is presented in Exercise 32.

Algorithm 6.5 is based on the LU Factorization Algorithm 6.4 and obtains the LDL^t factorization described in Corollary 6.27.

LDL^t Factorization

To factor the positive definite $n \times n$ matrix A into the form LDL^t, where L is a lower triangular matrix with 1s along the diagonal and D is a diagonal matrix with positive entries on the diagonal:

INPUT the dimension n; entries a_{ij}, for $1 \le i, j \le n$ of A.

OUTPUT the entries l_{ij}, for $1 \le j < i$ and $1 \le i \le n$ of L, and d_i, for $1 \le i \le n$ of D.

Step 1 For $i = 1, \ldots, n$ do Steps 2–4.

Step 2 For $j = 1, \ldots, i-1$, set $v_j = l_{ij} d_j$.

Step 3 Set $d_i = a_{ii} - \sum_{j=1}^{i-1} l_{ij} v_j$.

Step 4 For $j = i+1, \ldots, n$ set $l_{ji} = (a_{ji} - \sum_{k=1}^{i-1} l_{jk} v_k)/d_i$.

Step 5 OUTPUT (l_{ij} for $j = 1, \ldots, i-1$ and $i = 1, \ldots, n$);
OUTPUT (d_i for $i = 1, \ldots, n$);
STOP. ■

The *NumericalAnalysis* subpackage factors a positive definite matrix A as LDL^t with the command

$L, DD, Lt := MatrixDecomposition(A, method = LDLt)$

Corollary 6.27 has a counterpart when A is symmetric but not necessarily positive definite. This result is widely applied because symmetric matrices are common and easily recognized.

Corollary 6.29 Let A be a symmetric $n \times n$ matrix for which Gaussian elimination can be applied without row interchanges. Then A can be factored into LDL^t, where L is lower triangular with 1s on its diagonal and D is the diagonal matrix with $a_{11}^{(1)}, \ldots, a_{nn}^{(n)}$ on its diagonal. ■

Example 3 Determine the LDL^t factorization of the positive definite matrix

$$A = \begin{bmatrix} 4 & -1 & 1 \\ -1 & 4.25 & 2.75 \\ 1 & 2.75 & 3.5 \end{bmatrix}.$$

Solution The LDL^t factorization has 1s on the diagonal of the lower triangular matrix L so we need to have

$$A = \begin{bmatrix} a_{11} & a_{21} & a_{31} \\ a_{21} & a_{22} & a_{32} \\ a_{31} & a_{32} & a_{33} \end{bmatrix} = \begin{bmatrix} 1 & 0 & 0 \\ l_{21} & 1 & 0 \\ l_{31} & l_{32} & 1 \end{bmatrix} \begin{bmatrix} d_1 & 0 & 0 \\ 0 & d_2 & 0 \\ 0 & 0 & d_3 \end{bmatrix} \begin{bmatrix} 1 & l_{21} & l_{31} \\ 0 & 1 & l_{32} \\ 0 & 0 & 1 \end{bmatrix}$$

$$= \begin{bmatrix} d_1 & d_1 l_{21} & d_1 l_{31} \\ d_1 l_{21} & d_2 + d_1 l_{21}^2 & d_2 l_{32} + d_1 l_{21} l_{31} \\ d_1 l_{31} & d_1 l_{21} l_{31} + d_2 l_{32} & d_1 l_{31}^2 + d_2 l_{32}^2 + d_3 \end{bmatrix}$$

Thus

$a_{11} : 4 = d_1 \Longrightarrow d_1 = 4,$ $\quad a_{21} : -1 = d_1 l_{21} \Longrightarrow l_{21} = -0.25$

$a_{31} : 1 = d_1 l_{31} \Longrightarrow l_{31} = 0.25,$ $\quad a_{22} : 4.25 = d_2 + d_1 l_{21}^2 \Longrightarrow d_2 = 4$

$a_{32} : 2.75 = d_1 l_{21} l_{31} + d_2 l_{32} \Longrightarrow l_{32} = 0.75,$ $\quad a_{33} : 3.5 = d_1 l_{31}^2 + d_2 l_{32}^2 + d_3 \Longrightarrow d_3 = 1,$

and we have

$$A = LDL^t = \begin{bmatrix} 1 & 0 & 0 \\ -0.25 & 1 & 0 \\ 0.25 & 0.75 & 1 \end{bmatrix} \begin{bmatrix} 4 & 0 & 0 \\ 0 & 4 & 0 \\ 0 & 0 & 1 \end{bmatrix} \begin{bmatrix} 1 & -0.25 & 0.25 \\ 0 & 1 & 0.75 \\ 0 & 0 & 1 \end{bmatrix}.$$ ■

Algorithm 6.5 is easily modified to factor the symmetric matrices described in Corollary 6.29. It simply requires adding a check to ensure that the diagonal elements are nonzero. The Cholesky Algorithm 6.6 produces the LL^t factorization described in Corollary 6.28.

ALGORITHM **6.6**

Cholesky

To factor the positive definite $n \times n$ matrix A into LL^t, where L is lower triangular:

INPUT the dimension n; entries a_{ij}, for $1 \le i, j \le n$ of A.

OUTPUT the entries l_{ij}, for $1 \le j \le i$ and $1 \le i \le n$ of L. (*The entries of* $U = L^t$ *are* $u_{ij} = l_{ji}$, *for* $i \le j \le n$ *and* $1 \le i \le n$.)

Step 1 Set $l_{11} = \sqrt{a_{11}}$.

Step 2 For $j = 2, \ldots, n$, set $l_{j1} = a_{j1}/l_{11}$.

Step 3 For $i = 2, \ldots, n-1$ do Steps 4 and 5.

Step 4 Set $l_{ii} = \left(a_{ii} - \sum_{k=1}^{i-1} l_{ik}^2\right)^{1/2}$.

Step 5 For $j = i+1, \ldots, n$

set $l_{ji} = \left(a_{ji} - \sum_{k=1}^{i-1} l_{jk} l_{ik}\right) / l_{ii}$.

Step 6 Set $l_{nn} = \left(a_{nn} - \sum_{k=1}^{n-1} l_{nk}^2\right)^{1/2}$.

Step 7 OUTPUT (l_{ij} for $j = 1, \ldots, i$ and $i = 1, \ldots, n$);
STOP. ■

Andre-Louis Cholesky (1875-1918) was a French military officer involved in geodesy and surveying in the early 1900s. He developed this factorization method to compute solutions to least squares problems.

The Cholesky factorization of A is computed in the *LinearAlgebra* library of Maple using the statement

$L := LUDecomposition(A, method = \text{'}Cholesky\text{'})$

and gives the lower triangular matrix L as its output.

Example 4 Determine the Cholesky LL^t factorization of the positive definite matrix

$$A = \begin{bmatrix} 4 & -1 & 1 \\ -1 & 4.25 & 2.75 \\ 1 & 2.75 & 3.5 \end{bmatrix}.$$

Solution The LL^t factorization does not necessarily has 1s on the diagonal of the lower triangular matrix L so we need to have

$$A = \begin{bmatrix} a_{11} & a_{21} & a_{31} \\ a_{21} & a_{22} & a_{32} \\ a_{31} & a_{32} & a_{33} \end{bmatrix} = \begin{bmatrix} l_{11} & 0 & 0 \\ l_{21} & l_{22} & 0 \\ l_{31} & l_{32} & l_{33} \end{bmatrix} \begin{bmatrix} l_{11} & l_{21} & l_{31} \\ 0 & l_{22} & l_{32} \\ 0 & 0 & l_{33} \end{bmatrix}$$

$$= \begin{bmatrix} l_{11}^2 & l_{11}l_{21} & l_{11}l_{31} \\ l_{11}l_{21} & l_{21}^2 + l_{22}^2 & l_{21}l_{31} + l_{22}l_{32} \\ l_{11}l_{31} & l_{21}l_{31} + l_{22}l_{32} & l_{31}^2 + l_{32}^2 + l_{33}^2 \end{bmatrix}$$

Thus

$$a_{11}: \quad 4 = l_{11}^2 \implies l_{11} = 2, \qquad a_{21}: \quad -1 = l_{11}l_{21} \implies l_{21} = -0.5$$

$$a_{31}: \quad 1 = l_{11}l_{31} \implies l_{31} = 0.5, \qquad a_{22}: \quad 4.25 = l_{21}^2 + l_{22}^2 \implies l_{22} = 2$$

$$a_{32}: \quad 2.75 = l_{21}l_{31} + l_{22}l_{32} \implies l_{32} = 1.5, \quad a_{33}: \quad 3.5 = l_{31}^2 + l_{32}^2 + l_{33}^2 \implies l_{33} = 1,$$

and we have

$$A = LL^t = \begin{bmatrix} 2 & 0 & 0 \\ -0.5 & 2 & 0 \\ 0.5 & 1.5 & 1 \end{bmatrix} \begin{bmatrix} 2 & -0.5 & 0.5 \\ 0 & 2 & 1.5 \\ 0 & 0 & 1 \end{bmatrix}.$$ ■

The LDL^t factorization described in Algorithm 6.5 requires

$$\frac{1}{6}n^3 + n^2 - \frac{7}{6}n \text{ multiplications/divisions} \quad \text{and} \quad \frac{1}{6}n^3 - \frac{1}{6}n \text{ additions/subtractions.}$$

The LL^t Cholesky factorization of a positive definite matrix requires only

$$\frac{1}{6}n^3 + \frac{1}{2}n^2 - \frac{2}{3}n \text{ multiplications/divisions} \quad \text{and} \quad \frac{1}{6}n^3 - \frac{1}{6}n \text{ additions/subtractions.}$$

This computational advantage of Cholesky's factorization is misleading, because it requires extracting n square roots. However, the number of operations required for computing the n square roots is a linear factor of n and will decrease in significance as n increases.

Algorithm 6.5 provides a stable method for factoring a positive definite matrix into the form $A = LDL^t$, but it must be modified to solve the linear system $A\mathbf{x} = \mathbf{b}$. To do this, we delete the STOP statement from Step 5 in the algorithm and add the following steps to solve the lower triangular system $L\mathbf{y} = \mathbf{b}$:

Step 6 Set $y_1 = b_1$.

Step 7 For $i = 2, \ldots, n$ set $y_i = b_i - \sum_{j=1}^{i-1} l_{ij} y_j$.

The linear system $D\mathbf{z} = \mathbf{y}$ can then be solved by

Step 8 For $i = 1, \ldots, n$ set $z_i = y_i / d_i$.

Finally, the upper-triangular system $L^t\mathbf{x} = \mathbf{z}$ is solved with the steps given by

Step 9 Set $x_n = z_n$.

Step 10 For $i = n-1, \ldots, 1$ set $x_i = z_i - \sum_{j=i+1}^{n} l_{ji} x_j$.

Step 11 OUTPUT (x_i for $i = 1, \ldots, n$);
STOP.

Table 6.4 shows the additional operations required to solve the linear system.

Table 6.4

Step	Multiplications/Divisions	Additions/Subtractions
6	0	0
7	$n(n-1)/2$	$n(n-1)/2$
8	n	0
9	0	0
10	$n(n-1)/2$	$n(n-1)/2$
Total	n^2	$n^2 - n$

If the Cholesky factorization given in Algorithm 6.6 is preferred, the additional steps for solving the system $A\mathbf{x} = \mathbf{b}$ are as follows. First delete the STOP statement from Step 7. Then add

Step 8 Set $y_1 = b_1 / l_{11}$.

Step 9 For $i = 2, \ldots, n$ set $y_i = \left(b_i - \sum_{j=1}^{i-1} l_{ij} y_j\right) \Big/ l_{ii}$.

Step 10 Set $x_n = y_n / l_{nn}$.

Step 11 For $i = n-1, \ldots, 1$ set $x_i = \left(y_i - \sum_{j=i+1}^{n} l_{ji} x_j\right) \Big/ l_{ii}$.

Step 12 OUTPUT (x_i for $i = 1, \ldots, n$);
STOP.

Steps 8–12 require $n^2 + n$ multiplications/divisions and $n^2 - n$ additions/ subtractions.

Band Matrices

The last class of matrices considered are *band matrices*. In many applications, the band matrices are also strictly diagonally dominant or positive definite.

Definition 6.30 An $n \times n$ matrix is called a **band matrix** if integers p and q, with $1 < p, q < n$, exist with the property that $a_{ij} = 0$ whenever $p \le j - i$ or $q \le i - j$. The **band width** of a band matrix is defined as $w = p + q - 1$. ■

The name for a band matrix comes from the fact that all the nonzero entries lie in a band which is centered on the main diagonal.

The number p describes the number of diagonals above, and including, the main diagonal on which nonzero entries may lie. The number q describes the number of diagonals below, and including, the main diagonal on which nonzero entries may lie. For example, the matrix

$$A = \begin{bmatrix} 7 & 2 & 0 \\ 3 & 5 & -1 \\ 0 & -5 & -6 \end{bmatrix}$$

is a band matrix with $p = q = 2$ and bandwidth $2 + 2 - 1 = 3$.

The definition of band matrix forces those matrices to concentrate all their nonzero entries about the diagonal. Two special cases of band matrices that occur frequently have $p = q = 2$ and $p = q = 4$.

Tridiagonal Matrices

Matrices of bandwidth 3 occurring when $p = q = 2$ are called **tridiagonal** because they have the form

$$A = \begin{bmatrix} a_{11} & a_{12} & 0 & \cdots & \cdots & 0 \\ a_{21} & a_{22} & a_{23} & \ddots & & \vdots \\ 0 & a_{32} & a_{33} & a_{34} & \ddots & \vdots \\ \vdots & \ddots & \ddots & \ddots & \ddots & 0 \\ \vdots & & \ddots & \ddots & \ddots & a_{n-1,n} \\ 0 & \cdots & \cdots & 0 & a_{n,n-1} & a_{nn} \end{bmatrix}.$$

Tridiagonal matrices are also considered in Chapter 11 in connection with the study of piecewise linear approximations to boundary-value problems. The case of $p = q = 4$ will be used for the solution of boundary-value problems when the approximating functions assume the form of cubic splines.

The factorization algorithms can be simplified considerably in the case of band matrices because a large number of zeros appear in these matrices in regular patterns. It is particularly interesting to observe the form the Crout or Doolittle method assumes in this case.

To illustrate the situation, suppose a tridiagonal matrix A can be factored into the triangular matrices L and U. Then A has at most $(3n - 2)$ nonzero entries. Then there are only $(3n - 2)$ conditions to be applied to determine the entries of L and U, provided, of course, that the zero entries of A are also obtained.

Suppose that the matrices L and U also have tridiagonal form, that is,

$$L = \begin{bmatrix} l_{11} & 0 & \cdots & \cdots & 0 \\ l_{21} & l_{22} & \ddots & & \vdots \\ 0 & \ddots & \ddots & \ddots & \vdots \\ \vdots & \ddots & \ddots & \ddots & 0 \\ 0 & \cdots & 0 & l_{n,n-1} & l_{nn} \end{bmatrix} \quad \text{and} \quad U = \begin{bmatrix} 1 & u_{12} & 0 & \cdots & 0 \\ 0 & 1 & \ddots & \ddots & \vdots \\ \vdots & \ddots & \ddots & \ddots & 0 \\ \vdots & & \ddots & \ddots & u_{n-1,n} \\ 0 & \cdots & \cdots & 0 & 1 \end{bmatrix}.$$

There are $(2n-1)$ undetermined entries of L and $(n-1)$ undetermined entries of U, which totals $(3n-2)$, the number of possible nonzero entries of A. The 0 entries of A are obtained automatically.

The multiplication involved with $A = LU$ gives, in addition to the 0 entries,

$$a_{11} = l_{11};$$

$$a_{i,i-1} = l_{i,i-1}, \quad \text{for each } i = 2, 3, \ldots, n; \tag{6.13}$$

$$a_{ii} = l_{i,i-1}u_{i-1,i} + l_{ii}, \quad \text{for each } i = 2, 3, \ldots, n; \tag{6.14}$$

and

$$a_{i,i+1} = l_{ii}u_{i,i+1}, \quad \text{for each } i = 1, 2, \ldots, n-1. \tag{6.15}$$

A solution to this system is found by first using Eq. (6.13) to obtain all the nonzero off-diagonal terms in L and then using Eqs. (6.14) and (6.15) to alternately obtain the remainder of the entries in U and L. Once an entry L or U is computed, the corresponding entry in A is not needed. So the entries in A can be overwritten by the entries in L and U with the result that no new storage is required.

Algorithm 6.7 solves an $n \times n$ system of linear equations whose coefficient matrix is tridiagonal. This algorithm requires only $(5n-4)$ multiplications/divisions and $(3n-3)$ additions/subtractions. Consequently, it has considerable computational advantage over the methods that do not consider the tridiagonality of the matrix.

ALGORITHM 6.7

Crout Factorization for Tridiagonal Linear Systems

To solve the $n \times n$ linear system

$$\begin{aligned}
E_1 &: \quad a_{11}x_1 + a_{12}x_2 &&= a_{1,n+1},\\
E_2 &: \quad a_{21}x_1 + a_{22}x_2 + a_{23}x_3 &&= a_{2,n+1},\\
&\;\;\vdots \qquad\qquad\qquad \vdots && \;\;\vdots\\
E_{n-1} &: \quad a_{n-1,n-2}x_{n-2} + a_{n-1,n-1}x_{n-1} + a_{n-1,n}x_n &&= a_{n-1,n+1},\\
E_n &: \quad a_{n,n-1}x_{n-1} + a_{nn}x_n &&= a_{n,n+1},
\end{aligned}$$

which is assumed to have a unique solution:

INPUT the dimension n; the entries of A.

OUTPUT the solution $x_1, \ldots, x_n$.

(Steps 1–3 set up and solve $L\mathbf{z} = \mathbf{b}$.)

Step 1 Set $l_{11} = a_{11}$;
$u_{12} = a_{12}/l_{11}$;
$z_1 = a_{1,n+1}/l_{11}$.

Step 2 For $i = 2, \ldots, n-1$ set $l_{i,i-1} = a_{i,i-1}$; *(ith row of L.)*
$l_{ii} = a_{ii} - l_{i,i-1}u_{i-1,i}$;
$u_{i,i+1} = a_{i,i+1}/l_{ii}$; *((i + 1)th column of U.)*
$z_i = (a_{i,n+1} - l_{i,i-1}z_{i-1})/l_{ii}$.

Step 3 Set $l_{n,n-1} = a_{n,n-1}$; *(nth row of L.)*
$l_{nn} = a_{nn} - l_{n,n-1}u_{n-1,n}$.
$z_n = (a_{n,n+1} - l_{n,n-1}z_{n-1})/l_{nn}$.

(Steps 4 and 5 solve $U\mathbf{x} = \mathbf{z}$.)

Step 4 Set $x_n = z_n$.

Step 5 For $i = n-1, \ldots, 1$ set $x_i = z_i - u_{i,i+1}x_{i+1}$.

Step 6 OUTPUT $(x_1, \ldots, x_n)$;
STOP. ■

Example 5 Determine the Crout factorization of the symmetric tridiagonal matrix

$$\begin{bmatrix} 2 & -1 & 0 & 0 \\ -1 & 2 & -1 & 0 \\ 0 & -1 & 2 & -1 \\ 0 & 0 & -1 & 2 \end{bmatrix},$$

and use this factorization to solve the linear system

$$\begin{aligned} 2x_1 - x_2 \qquad\qquad\quad &= 1, \\ -x_1 + 2x_2 - x_3 \qquad &= 0, \\ - x_2 + 2x_3 - x_4 &= 0, \\ - x_3 + 2x_4 &= 1. \end{aligned}$$

Solution The LU factorization of A has the form

$$A = \begin{bmatrix} a_{11} & a_{12} & 0 & 0 \\ a_{21} & a_{22} & a_{23} & 0 \\ 0 & a_{32} & a_{33} & a_{34} \\ 0 & 0 & a_{43} & a_{44} \end{bmatrix} = \begin{bmatrix} l_{11} & 0 & 0 & 0 \\ l_{21} & l_{22} & 0 & 0 \\ 0 & l_{32} & l_{33} & 0 \\ 0 & 0 & l_{43} & l_{44} \end{bmatrix} \begin{bmatrix} 1 & u_{12} & 0 & 0 \\ 0 & 1 & u_{23} & 0 \\ 0 & 0 & 1 & u_{34} \\ 0 & 0 & 0 & 1 \end{bmatrix}$$

$$= \begin{bmatrix} l_{11} & l_{11}u_{12} & 0 & 0 \\ l_{21} & l_{22} + l_{21}u_{12} & l_{22}u_{23} & 0 \\ 0 & l_{32} & l_{33} + l_{32}u_{23} & l_{33}u_{34} \\ 0 & 0 & l_{43} & l_{44} + l_{43}u_{34} \end{bmatrix}.$$

Thus

$$\begin{aligned} &a_{11}: \; 2 = l_{11} \implies l_{11} = 2, && a_{12}: \; -1 = l_{11}u_{12} \implies u_{12} = -\tfrac{1}{2}, \\ &a_{21}: \; -1 = l_{21} \implies l_{21} = -1, && a_{22}: \; 2 = l_{22} + l_{21}u_{12} \implies l_{22} = -\tfrac{3}{2}, \\ &a_{23}: \; -1 = l_{22}u_{23} \implies u_{23} = -\tfrac{2}{3}, && a_{32}: \; -1 = l_{32} \implies l_{32} = -1, \\ &a_{33}: \; 2 = l_{33} + l_{32}u_{23} \implies l_{33} = \tfrac{4}{3}, && a_{34}: \; -1 = l_{33}u_{34} \implies u_{34} = -\tfrac{3}{4}, \\ &a_{43}: \; -1 = l_{43} \implies l_{43} = -1, && a_{44}: \; 2 = l_{44} + l_{43}u_{34} \implies l_{44} = \tfrac{5}{4}. \end{aligned}$$

This gives the Crout factorization

$$A = \begin{bmatrix} 2 & -1 & 0 & 0 \\ -1 & 2 & -1 & 0 \\ 0 & -1 & 2 & -1 \\ 0 & 0 & -1 & 2 \end{bmatrix} = \begin{bmatrix} 2 & 0 & 0 & 0 \\ -1 & \frac{3}{2} & 0 & 0 \\ 0 & -1 & \frac{4}{3} & 0 \\ 0 & 0 & -1 & \frac{5}{4} \end{bmatrix} \begin{bmatrix} 1 & -\frac{1}{2} & 0 & 0 \\ 0 & 1 & -\frac{2}{3} & 0 \\ 0 & 0 & 1 & -\frac{3}{4} \\ 0 & 0 & 0 & 1 \end{bmatrix} = LU.$$

Solving the system

$$L\mathbf{z} = \begin{bmatrix} 2 & 0 & 0 & 0 \\ -1 & \frac{3}{2} & 0 & 0 \\ 0 & -1 & \frac{4}{3} & 0 \\ 0 & 0 & -1 & \frac{5}{4} \end{bmatrix} \begin{bmatrix} z_1 \\ z_2 \\ z_3 \\ z_4 \end{bmatrix} = \begin{bmatrix} 1 \\ 0 \\ 0 \\ 1 \end{bmatrix} \quad \text{gives} \quad \begin{bmatrix} z_1 \\ z_2 \\ z_3 \\ z_4 \end{bmatrix} = \begin{bmatrix} \frac{1}{2} \\ \frac{1}{3} \\ \frac{1}{4} \\ 1 \end{bmatrix},$$

and then solving

$$U\mathbf{x} = \begin{bmatrix} 1 & -\frac{1}{2} & 0 & 0 \\ 0 & 1 & -\frac{2}{3} & 0 \\ 0 & 0 & 1 & -\frac{3}{4} \\ 0 & 0 & 0 & 1 \end{bmatrix} \begin{bmatrix} x_1 \\ x_2 \\ x_3 \\ x_4 \end{bmatrix} = \begin{bmatrix} \frac{1}{2} \\ \frac{1}{3} \\ \frac{1}{4} \\ 1 \end{bmatrix} \quad \text{gives} \quad \begin{bmatrix} x_1 \\ x_2 \\ x_3 \\ x_4 \end{bmatrix} = \begin{bmatrix} 1 \\ 1 \\ 1 \\ 1 \end{bmatrix}. \quad ■$$

The Crout Factorization Algorithm can be applied whenever $l_{ii} \neq 0$ for each $i = 1, 2, \ldots, n$. Two conditions, either of which ensure that this is true, are that the coefficient matrix of the system is positive definite or that it is strictly diagonally dominant. An additional condition that ensures this algorithm can be applied is given in the next theorem, whose proof is considered in Exercise 28.

Theorem 6.31 Suppose that $A = [a_{ij}]$ is tridiagonal with $a_{i,i-1}a_{i,i+1} \neq 0$, for each $i = 2, 3, \ldots, n-1$. If $|a_{11}| > |a_{12}|$, $|a_{ii}| \geq |a_{i,i-1}| + |a_{i,i+1}|$, for each $i = 2, 3, \ldots, n-1$, and $|a_{nn}| > |a_{n,n-1}|$, then A is nonsingular and the values of l_{ii} described in the Crout Factorization Algorithm are nonzero for each $i = 1, 2, \ldots, n$. ■

The *LinearAlgebra* package of Maple supports a number of commands that test properties for matrices. The return in each case is *true* if the property holds for the matrix and is *false* if it does not hold. For example,

IsDefinite(*A*, *query* = '*positive_definite*')

would return *true* for the positive matrix

$$A = \begin{bmatrix} 2 & -1 & 0 \\ -1 & 2 & -1 \\ 0 & -1 & 2 \end{bmatrix}$$

but would return *false* for the matrix

$$A = \begin{bmatrix} -1 & 2 \\ 2 & -1 \end{bmatrix}.$$

Consistent with our definition, symmetry is required for a *true* result.

The *NumericalAnalysis* subpackage also has query commands for matrices. Some of these are

IsMatrixShape(*A*, '*diagonal*')
IsMatrixShape(*A*, '*symmetric*')
IsMatrixShape(*A*, '*positivedefinite*')
IsMatrixShape(*A*, '*diagonallydominant*')
IsMatrixShape(*A*, '*strictlydiagonallydominant*')
IsMatrixShape(*A*, '*triangular*'$_{'upper'}$)
IsMatrixShape(*A*, '*triangular*'$_{'lower'}$)

EXERCISE SET 6.6

1. Determine which of the following matrices are (i) symmetric, (ii) singular, (iii) strictly diagonally dominant, (iv) positive definite.

a. $\begin{bmatrix} -2 & 1 \\ 1 & -3 \end{bmatrix}$ **b.** $\begin{bmatrix} 2 & 1 & 0 \\ 0 & 3 & 2 \\ 1 & 2 & 4 \end{bmatrix}$

c. $\begin{bmatrix} 2 & -1 & 0 \\ -1 & 4 & 2 \\ 0 & 2 & 2 \end{bmatrix}$ **d.** $\begin{bmatrix} 2 & 3 & 1 & 2 \\ -2 & 4 & -1 & 5 \\ 3 & 7 & 1.5 & 1 \\ 6 & -9 & 3 & 7 \end{bmatrix}$

2. Determine which of the following matrices are (i) symmetric, (ii) singular, (iii) strictly diagonally dominant, (iv) positive definite.

a. $\begin{bmatrix} 2 & 1 \\ 1 & 3 \end{bmatrix}$ **b.** $\begin{bmatrix} 2 & 1 & 0 \\ 0 & 3 & 0 \\ 1 & 0 & 4 \end{bmatrix}$

c. $\begin{bmatrix} 4 & 2 & 6 \\ 3 & 0 & 7 \\ -2 & -1 & -3 \end{bmatrix}$ **d.** $\begin{bmatrix} 4 & 0 & 0 & 0 \\ 6 & 7 & 0 & 0 \\ 9 & 11 & 1 & 0 \\ 5 & 4 & 1 & 1 \end{bmatrix}$

3. Use the LDL^t Factorization Algorithm to find a factorization of the form $A = LDL^t$ for the following matrices:

a. $A = \begin{bmatrix} 4 & -1 & 1 \\ -1 & 3 & 0 \\ 1 & 0 & 2 \end{bmatrix}$ **b.** $A = \begin{bmatrix} 4 & 2 & 2 \\ 2 & 6 & 2 \\ 2 & 2 & 5 \end{bmatrix}$

c. $A = \begin{bmatrix} 4 & 0 & 2 & 1 \\ 0 & 3 & -1 & 1 \\ 2 & -1 & 6 & 3 \\ 1 & 1 & 3 & 8 \end{bmatrix}$ **d.** $A = \begin{bmatrix} 4 & 1 & 1 & 1 \\ 1 & 3 & 0 & -1 \\ 1 & 0 & 2 & 1 \\ 1 & -1 & 1 & 4 \end{bmatrix}$

4. Use the LDL^t Factorization Algorithm to find a factorizaton of the form $A = LDL^t$ for the following matrices:

a. $A = \begin{bmatrix} 2 & -1 & 0 \\ -1 & 2 & -1 \\ 0 & -1 & 2 \end{bmatrix}$ **b.** $A = \begin{bmatrix} 4 & 1 & 1 & 1 \\ 1 & 3 & -1 & 1 \\ 1 & -1 & 2 & 0 \\ 1 & 1 & 0 & 2 \end{bmatrix}$

c. $A = \begin{bmatrix} 4 & 1 & -1 & 0 \\ 1 & 3 & -1 & 0 \\ -1 & -1 & 5 & 2 \\ 0 & 0 & 2 & 4 \end{bmatrix}$ **d.** $A = \begin{bmatrix} 6 & 2 & 1 & -1 \\ 2 & 4 & 1 & 0 \\ 1 & 1 & 4 & -1 \\ -1 & 0 & -1 & 3 \end{bmatrix}$

5. Use the Cholesky Algorithm to find a factorization of the form $A = LL^t$ for the matrices in Exercise 3.

6. Use the Cholesky Algorithm to find a factorization of the form $A = LL^t$ for the matrices in Exercise 4.

7. Modify the LDL^t Factorization Algorithm as suggested in the text so that it can be used to solve linear systems. Use the modified algorithm to solve the following linear systems.

a.
$$\begin{aligned} 2x_1 - x_2 \quad\quad &= 3, \\ -x_1 + 2x_2 - x_3 &= -3, \\ - x_2 + 2x_3 &= 1. \end{aligned}$$

b.
$$\begin{aligned} 4x_1 + x_2 + x_3 + x_4 &= 0.65, \\ x_1 + 3x_2 - x_3 + x_4 &= 0.05, \\ x_1 - x_2 + 2x_3 \quad\quad &= 0, \\ x_1 + x_2 \quad\quad + 2x_4 &= 0.5. \end{aligned}$$

c.
$$\begin{aligned} 4x_1 + x_2 - x_3 &= 7, \\ x_1 + 3x_2 - x_3 &= 8, \\ -x_1 - x_2 + 5x_3 + 2x_4 &= -4, \\ 2x_3 + 4x_4 &= 6. \end{aligned}$$

d.
$$\begin{aligned} 6x_1 + 2x_2 + x_3 - x_4 &= 0, \\ 2x_1 + 4x_2 + x_3 &= 7, \\ x_1 + x_2 + 4x_3 - x_4 &= -1, \\ -x_1 - x_3 + 3x_4 &= -2. \end{aligned}$$

8. Use the modified algorithm from Exercise 7 to solve the following linear systems.

a.
$$\begin{aligned} 4x_1 - x_2 + x_3 &= -1, \\ -x_1 + 3x_2 &= 4, \\ x_1 + 2x_3 &= 5. \end{aligned}$$

b.
$$\begin{aligned} 4x_1 + 2x_2 + 2x_3 &= 0, \\ 2x_1 + 6x_2 + 2x_3 &= 1, \\ 2x_1 + 2x_2 + 5x_3 &= 0. \end{aligned}$$

c.
$$\begin{aligned} 4x_1 + 2x_3 + x_4 &= -2, \\ 3x_2 - x_3 + x_4 &= 0, \\ 2x_1 - x_2 + 6x_3 + 3x_4 &= 7, \\ x_1 + x_2 + 3x_3 + 8x_4 &= -2. \end{aligned}$$

d.
$$\begin{aligned} 4x_1 + x_2 + x_3 + x_4 &= 2, \\ x_1 + 3x_2 - x_4 &= 2, \\ x_1 + 2x_3 + x_4 &= 1, \\ x_1 - x_2 + x_3 + 4x_4 &= 1. \end{aligned}$$

9. Modify the Cholesky Algorithm as suggested in the text so that it can be used to solve linear systems, and use the modified algorithm to solve the linear systems in Exercise 7.

10. Use the modified algorithm developed in Exercise 9 to solve the linear systems in Exercise 8.

11. Use Crout factorization for tridiagonal systems to solve the following linear systems.

a.
$$\begin{aligned} x_1 - x_2 &= 0, \\ -2x_1 + 4x_2 - 2x_3 &= -1, \\ -x_2 + 2x_3 &= 1.5. \end{aligned}$$

b.
$$\begin{aligned} 3x_1 + x_2 &= -1, \\ 2x_1 + 4x_2 + x_3 &= 7, \\ 2x_2 + 5x_3 &= 9. \end{aligned}$$

c.
$$\begin{aligned} 2x_1 - x_2 &= 3, \\ -x_1 + 2x_2 - x_3 &= -3, \\ -x_2 + 2x_3 &= 1. \end{aligned}$$

d.
$$\begin{aligned} 0.5x_1 + 0.25x_2 &= 0.35, \\ 0.35x_1 + 0.8x_2 + 0.4x_3 &= 0.77, \\ 0.25x_2 + x_3 + 0.5x_4 &= -0.5, \\ x_3 - 2x_4 &= -2.25. \end{aligned}$$

12. Use Crout factorization for tridiagonal systems to solve the following linear systems.

a.
$$\begin{aligned} 2x_1 + x_2 &= 3, \\ x_1 + 2x_2 + x_3 &= -2, \\ 2x_2 + 3x_3 &= 0. \end{aligned}$$

b.
$$\begin{aligned} 2x_1 - x_2 &= 5, \\ -x_1 + 3x_2 + x_3 &= 4, \\ x_2 + 4x_3 &= 0. \end{aligned}$$

c.
$$\begin{aligned} 2x_1 - x_2 &= 3, \\ x_1 + 2x_2 - x_3 &= 4, \\ x_2 - 2x_3 + x_4 &= 0, \\ x_3 + 2x_4 &= 6. \end{aligned}$$

d.
$$\begin{aligned} 2x_1 - x_2 &= 1, \\ x_1 + 2x_2 - x_3 &= 2, \\ 2x_2 + 4x_3 - x_4 &= -1, \\ 2x_4 - x_5 &= -2, \\ x_4 + 2x_5 &= -1. \end{aligned}$$

13. Let A be the 10×10 tridiagonal matrix given by $a_{ii} = 2, a_{i,i+1} = a_{i,i-1} = -1$, for each $i = 2, \cdots, 9$, and $a_{11} = a_{10,10} = 2, a_{12} = a_{10,9} = -1$. Let $\mathbf{b}$ be the ten-dimensional column vector given by $b_1 = b_{10} = 1$ and $b_i = 0$, for each $i = 2, 3, \cdots, 9$. Solve $A\mathbf{x} = \mathbf{b}$ using the Crout factorization for tridiagonal systems.

14. Modify the LDL^t factorization to factor a symmetric matrix A. [*Note:* The factorization may not always be possible.] Apply the new algorithm to the following matrices:

a. $A = \begin{bmatrix} 3 & -3 & 6 \\ -3 & 2 & -7 \\ 6 & -7 & 13 \end{bmatrix}$

b. $A = \begin{bmatrix} 3 & -6 & 9 \\ -6 & 14 & -20 \\ 9 & -20 & 29 \end{bmatrix}$

c. $A = \begin{bmatrix} -1 & 2 & 0 & 1 \\ 2 & -3 & 2 & -1 \\ 0 & 2 & 5 & 6 \\ 1 & -1 & 6 & 12 \end{bmatrix}$

d. $A = \begin{bmatrix} 2 & -2 & 4 & -4 \\ -2 & 3 & -4 & 5 \\ 4 & -4 & 10 & -10 \\ -4 & 5 & -10 & 14 \end{bmatrix}$

15. Which of the symmetric matrices in Exercise 14 are positive definite?

16. Find all α so that $A = \begin{bmatrix} \alpha & 1 & -1 \\ 1 & 2 & 1 \\ -1 & 1 & 4 \end{bmatrix}$ is positive definite.

17. Find all α so that $A = \begin{bmatrix} 2 & \alpha & -1 \\ \alpha & 2 & 1 \\ -1 & 1 & 4 \end{bmatrix}$ is positive definite.

18. Find all α and $\beta > 0$ so that the matrix

$$A = \begin{bmatrix} 4 & \alpha & 1 \\ 2\beta & 5 & 4 \\ \beta & 2 & \alpha \end{bmatrix}$$

is strictly diagonally dominant.

19. Find all $\alpha > 0$ and $\beta > 0$ so that the matrix

$$A = \begin{bmatrix} 3 & 2 & \beta \\ \alpha & 5 & \beta \\ 2 & 1 & \alpha \end{bmatrix}$$

is strictly diagonally dominant.

20. Suppose that A and B are strictly diagonally dominant $n \times n$ matrices. Which of the following must be strictly diagonally dominant?

a. $-A$ **b.** A^t **c.** $A + B$ **d.** A^2 **e.** $A - B$

21. Suppose that A and B are positive definite $n \times n$ matrices. Which of the following must be positive definite?

a. $-A$ **b.** A^t **c.** $A + B$ **d.** A^2 **e.** $A - B$

22. Let

$$A = \begin{bmatrix} 1 & 0 & -1 \\ 0 & 1 & 1 \\ -1 & 1 & \alpha \end{bmatrix}.$$

Find all values of α for which

a. A is singular.
b. A is strictly diagonally dominant.
c. A is symmetric.
d. A is positive definite.

23. Let

$$A = \begin{bmatrix} \alpha & 1 & 0 \\ \beta & 2 & 1 \\ 0 & 1 & 2 \end{bmatrix}.$$

Find all values of α and β for which

a. A is singular.
b. A is strictly diagonally dominant.
c. A is symmetric.
d. A is positive definite.

24. Suppose A and B commute, that is, $AB = BA$. Must A^t and B^t also commute?

25. Construct a matrix A that is nonsymmetric but for which $\mathbf{x}^t A\mathbf{x} > 0$ for all $\mathbf{x} \neq 0$.

26. Show that Gaussian elimination can be performed on A without row interchanges if and only if all leading principal submatrices of A are nonsingular. [*Hint:* Partition each matrix in the equation

$$A^{(k)} = M^{(k-1)}M^{(k-2)} \cdots M^{(1)}A$$

vertically between the kth and $(k+1)$st columns and horizontally between the kth and $(k+1)$st rows (see Exercise 14 of Section 6.3). Show that the nonsingularity of the leading principal submatrix of A is equivalent to $a_{k,k}^{(k)} \neq 0$.]

27. Tridiagonal matrices are usually labeled by using the notation

$$A = \begin{bmatrix} a_1 & c_1 & 0 & \cdots & 0 \\ b_2 & a_2 & c_2 & \ddots & \vdots \\ 0 & b_3 & \ddots & \ddots & 0 \\ \vdots & \ddots & \ddots & \ddots & c_{n-1} \\ 0 & \cdots & 0 & b_n & a_n \end{bmatrix}$$

to emphasize that it is not necessary to consider all the matrix entries. Rewrite the Crout Factorization Algorithm using this notation, and change the notation of the l_{ij} and u_{ij} in a similar manner.

28. Prove Theorem 6.31. [*Hint:* Show that $|u_{i,i+1}| < 1$, for each $i = 1, 2, \ldots, n-1$, and that $|l_{ii}| > 0$, for each $i = 1, 2, \ldots, n$. Deduce that $\det A = \det L \cdot \det U \neq 0$.]

29. Suppose $V = 5.5$ volts in the lead example of this chapter. By reordering the equations, a tridiagonal linear system can be formed. Use the Crout Factorization Algorithm to find the solution of the modified system.

30. Construct the operation count for solving an $n \times n$ linear system using the Crout Factorization Algorithm.

31. In a paper by Dorn and Burdick [DoB], it is reported that the average wing length that resulted from mating three mutant varieties of fruit flies (*Drosophila melanogaster*) can be expressed in the symmetric matrix form

$$A = \begin{bmatrix} 1.59 & 1.69 & 2.13 \\ 1.69 & 1.31 & 1.72 \\ 2.13 & 1.72 & 1.85 \end{bmatrix},$$

where a_{ij} denotes the average wing length of an offspring resulting from the mating of a male of type i with a female of type j.

a. What physical significance is associated with the symmetry of this matrix?

b. Is this matrix positive definite? If so, prove it; if not, find a nonzero vector $\mathbf{x}$ for which $\mathbf{x}^t A\mathbf{x} \leq 0$.

32. Suppose that the positive definite matrix A has the Cholesky factorization $A = LL^t$ and also the factorization $A = \hat{L}D\hat{L}^t$, where D is the diagonal matrix with positive diagonal entries $d_{11}, d_{22}, \ldots, d_{nn}$. Let $D^{1/2}$ be the diagonal matrix with diagonal entries $\sqrt{d_{11}}, \sqrt{d_{22}}, \ldots, \sqrt{d_{nn}}$.

a. Show that $D = D^{1/2}D^{1/2}$. **b.** Show that $L = \hat{L}D^{1/2}$.

6.7 Survey of Methods and Software

In this chapter we have looked at direct methods for solving linear systems. A linear system consists of n equations in n unknowns expressed in matrix notation as $A\mathbf{x} = \mathbf{b}$. These techniques use a finite sequence of arithmetic operations to determine the exact solution of the system subject only to round-off error. We found that the linear system $A\mathbf{x} = \mathbf{b}$ has a unique solution if and only if A^{-1} exists, which is equivalent to $\det A \neq 0$. When A^{-1} is known, the solution of the linear system is the vector $\mathbf{x} = A^{-1}\mathbf{b}$.

Pivoting techniques were introduced to minimize the effects of round-off error, which can dominate the solution when using direct methods. We studied partial pivoting, scaled partial pivoting, and briefly discussed complete pivoting. We recommend the partial or scaled partial pivoting methods for most problems because these decrease the effects of round-off error without adding much extra computation. Complete pivoting should be used if round-off error is suspected to be large. In Section 5 of Chapter 7 we will see some procedures for estimating this round-off error.

Gaussian elimination with minor modifications was shown to yield a factorization of the matrix A into LU, where L is lower triangular with 1s on the diagonal and U is upper triangular. This process is called Doolittle factorization. Not all nonsingular matrices can be factored this way, but a permutation of the rows will always give a factorization of the form $PA = LU$, where P is the permutation matrix used to rearrange the rows of A. The advantage of the factorization is that the work is significantly reduced when solving linear systems $A\mathbf{x} = \mathbf{b}$ with the same coefficient matrix A and different vectors $\mathbf{b}$.

Factorizations take a simpler form when the matrix A is positive definite. For example, the Choleski factorization has the form $A = LL^t$, where L is lower triangular. A symmetric matrix that has an LU factorization can also be factored in the form $A = LDL^t$, where D is diagonal and L is lower triangular with 1s on the diagonal. With these factorizations, manipulations involving A can be simplified. If A is tridiagonal, the LU factorization takes a particularly simple form, with U having 1s on the main diagonal and 0s elsewhere, except on the diagonal immediately above the main diagonal. In addition, L has its only nonzero entries on the main diagonal and one diagonal below. Another important method of matrix factorization is considered in Section 6 of Chapter 9.

The direct methods are the methods of choice for most linear systems. For tridiagonal, banded, and positive definite matrices, the special methods are recommended. For the general case, Gaussian elimination or LU factorization methods, which allow pivoting, are recommended. In these cases, the effects of round-off error should be monitored. In Section 7.5 we discuss estimating errors in direct methods.

Large linear systems with primarily 0 entries occurring in regular patterns can be solved efficiently using an iterative procedure such as those discussed in Chapter 7. Systems of this type arise naturally, for example, when finite-difference techniques are used to solve boundary-value problems, a common application in the numerical solution of partial-differential equations.

It can be very difficult to solve a large linear system that has primarily nonzero entries or one where the 0 entries are not in a predictable pattern. The matrix associated with the system can be placed in secondary storage in partitioned form and portions read into main memory only as needed for calculation. Methods that require secondary storage can be either iterative or direct, but they generally require techniques from the fields of data structures and graph theory. The reader is referred to [BuR] and [RW] for a discussion of the current techniques.

The software for matrix operations and the direct solution of linear systems implemented in IMSL and NAG is based on LAPACK, a subroutine package in the public domain. There is excellent documentation available with it and from the books written about it. We will focus on several of the subroutines that are available in all three sources.

Accompanying LAPACK is a set of lower-level operations called Basic Linear Algebra Subprograms (BLAS). Level 1 of BLAS generally consists of vector-vector operations such as vector additions with input data and operation counts of $O(n)$. Level 2 consists of the matrix-vector operations such as the product of a matrix and a vector with input data and operation counts of $O(n^2)$. Level 3 consists of the matrix-matrix operations such as matrix products with input data and operation counts of $O(n^3)$.

The subroutines in LAPACK for solving linear systems first factor the matrix A. The factorization depends on the type of matrix in the following way:

1. General matrix $PA = LU$;
2. Positive definite matrix $A = LL^t$;
3. Symmetric matrix $A = LDL^t$;
4. Tridiagonal matrix $A = LU$ (in banded form).

In addition, inverses and determinants can be computed.

Many of the subroutines in LINPACK, and its successor LAPACK, can be implemented using MATLAB. A nonsingular matrix A can be factored into the form $PA = LU$, where P is the permutation matrix defined by performing partial pivoting to solve a linear system involving A. A system of the form $A\mathbf{x} = \mathbf{b}$ is found by solving a lower triangular system followed by the solution to an upper triangular system.

Other MATLAB commands include computing the inverse, transpose, and determinant of matrix A by issuing the commands $\text{inv}(A)$, A', and $\det(A)$, respectively.

The IMSL Library includes counterparts to almost all the LAPACK subroutines and some extensions as well. The NAG Library has numerous subroutines for direct methods of solving linear systems similar to those in LAPACK and IMSL.

Further information on the numerical solution of linear systems and matrices can be found in Golub and Van Loan [GV], Forsythe and Moler [FM], and Stewart [Stew1]. The use of direct techniques for solving large sparse systems is discussed in detail in George and Liu [GL] and in Pissanetzky [Pi]. Coleman and Van Loan [CV] consider the use of BLAS, LINPACK, and MATLAB.

CHAPTER

7 Iterative Techniques in Matrix Algebra

Introduction

Trusses are lightweight structures capable of carrying heavy loads. In bridge design, the individual members of the truss are connected with rotatable pin joints that permit forces to be transferred from one member of the truss to another. The accompanying figure shows a truss that is held stationary at the lower left endpoint ①, is permitted to move horizontally at the lower right endpoint ④, and has pin joints at ①, ②, ③, and ④. A load of 10,000 newtons (N) is placed at joint ③, and the resulting forces on the joints are given by f_1, f_2, f_3, f_4, and f_5, as shown. When positive, these forces indicate tension on the truss elements, and when negative, compression. The stationary support member could have both a horizontal force component F_1 and a vertical force component F_2, but the movable support member has only a vertical force component F_3.

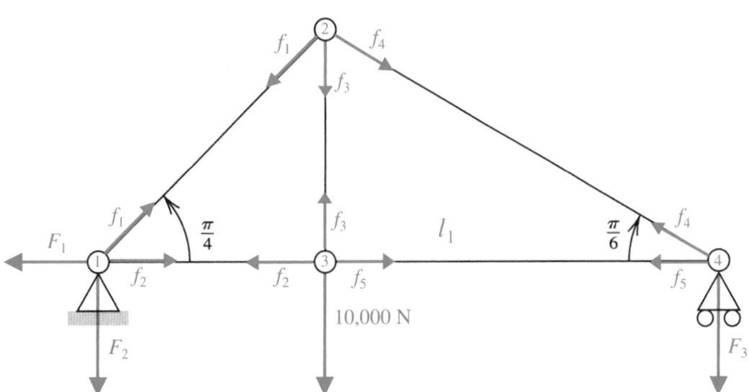

If the truss is in static equilibrium, the forces at each joint must add to the zero vector, so the sum of the horizontal and vertical components at each joint must be 0. This produces the system of linear equations shown in the accompanying table. An 8×8 matrix describing this system has 47 zero entries and only 17 nonzero entries. Matrices with a high percentage of zero entries are called *sparse* and are often solved using iterative, rather than direct, techniques. The iterative solution to this system is considered in Exercise 18 of Section 7.3 and Exercise 10 in Section 7.4.

Joint	Horizontal Component	Vertical Component
①	$-F_1 + \frac{\sqrt{2}}{2} f_1 + f_2 = 0$	$\frac{\sqrt{2}}{2} f_1 - F_2 = 0$
②	$-\frac{\sqrt{2}}{2} f_1 + \frac{\sqrt{3}}{2} f_4 = 0$	$-\frac{\sqrt{2}}{2} f_1 - f_3 - \frac{1}{2} f_4 = 0$
③	$-f_2 + f_5 = 0$	$f_3 - 10{,}000 = 0$
④	$-\frac{\sqrt{3}}{2} f_4 - f_5 = 0$	$\frac{1}{2} f_4 - F_3 = 0$

The methods presented in Chapter 6 used direct techniques to solve a system of $n \times n$ linear equations of the form $A\mathbf{x} = \mathbf{b}$. In this chapter, we present iterative methods to solve a system of this type.

7.1 Norms of Vectors and Matrices

In Chapter 2 we described iterative techniques for finding roots of equations of the form $f(x) = 0$. An initial approximation (or approximations) was found, and new approximations are then determined based on how well the previous approximations satisfied the equation. The objective is to find a way to minimize the difference between the approximations and the exact solution.

To discuss iterative methods for solving linear systems, we first need to determine a way to measure the distance between n-dimensional column vectors. This will permit us to determine whether a sequence of vectors converges to a solution of the system.

A scalar is a real (or complex) number generally denoted using italic or Greek letters. Vectors are denoted using boldface letters.

In actuality, this measure is also needed when the solution is obtained by the direct methods presented in Chapter 6. Those methods required a large number of arithmetic operations, and using finite-digit arithmetic leads only to an approximation to an actual solution of the system.

Vector Norms

Let $\mathbb{R}^n$ denote the set of all n-dimensional column vectors with real-number components. To define a distance in $\mathbb{R}^n$ we use the notion of a norm, which is the generalization of the absolute value on $\mathbb{R}$, the set of real numbers.

Definition 7.1 A **vector norm** on $\mathbb{R}^n$ is a function, $\|\cdot\|$, from $\mathbb{R}^n$ into $\mathbb{R}$ with the following properties:

(i) $\|\mathbf{x}\| \geq 0$ for all $\mathbf{x} \in \mathbb{R}^n$,

(ii) $\|\mathbf{x}\| = 0$ if and only if $\mathbf{x} = \mathbf{0}$,

(iii) $\|\alpha\mathbf{x}\| = |\alpha|\|\mathbf{x}\|$ for all $\alpha \in \mathbb{R}$ and $\mathbf{x} \in \mathbb{R}^n$,

(iv) $\|\mathbf{x} + \mathbf{y}\| \leq \|\mathbf{x}\| + \|\mathbf{y}\|$ for all $\mathbf{x}, \mathbf{y} \in \mathbb{R}^n$. ■

Vectors in $\mathbb{R}^n$ are column vectors, and it is convenient to use the transpose notation presented in Section 6.3 when a vector is represented in terms of its components. For example, the vector

$$\mathbf{x} = \begin{bmatrix} x_1 \\ x_2 \\ \vdots \\ x_n \end{bmatrix}$$

will be written $\mathbf{x} = (x_1, x_2, \ldots, x_n)^t$.

We will need only two specific norms on $\mathbb{R}^n$, although a third norm on $\mathbb{R}^n$ is presented in Exercise 2.

Definition 7.2 The l_2 and l_∞ norms for the vector $\mathbf{x} = (x_1, x_2, \ldots, x_n)^t$ are defined by

$$\|\mathbf{x}\|_2 = \left\{\sum_{i=1}^{n} x_i^2\right\}^{1/2} \quad \text{and} \quad \|\mathbf{x}\|_\infty = \max_{1\leq i\leq n} |x_i|. \qquad ■$$

Note that each of these norms reduces to the absolute value in the case $n = 1$.

The l_2 norm is called the **Euclidean norm** of the vector $\mathbf{x}$ because it represents the usual notion of distance from the origin in case $\mathbf{x}$ is in $\mathbb{R}^1 \equiv \mathbb{R}$, $\mathbb{R}^2$, or $\mathbb{R}^3$. For example, the l_2 norm of the vector $\mathbf{x} = (x_1, x_2, x_3)^t$ gives the length of the straight line joining the points $(0, 0, 0)$ and (x_1, x_2, x_3). Figure 7.1 shows the boundary of those vectors in $\mathbb{R}^2$ and $\mathbb{R}^3$ that have l_2 norm less than 1. Figure 7.2 is a similar illustration for the l_∞ norm.

Figure 7.1

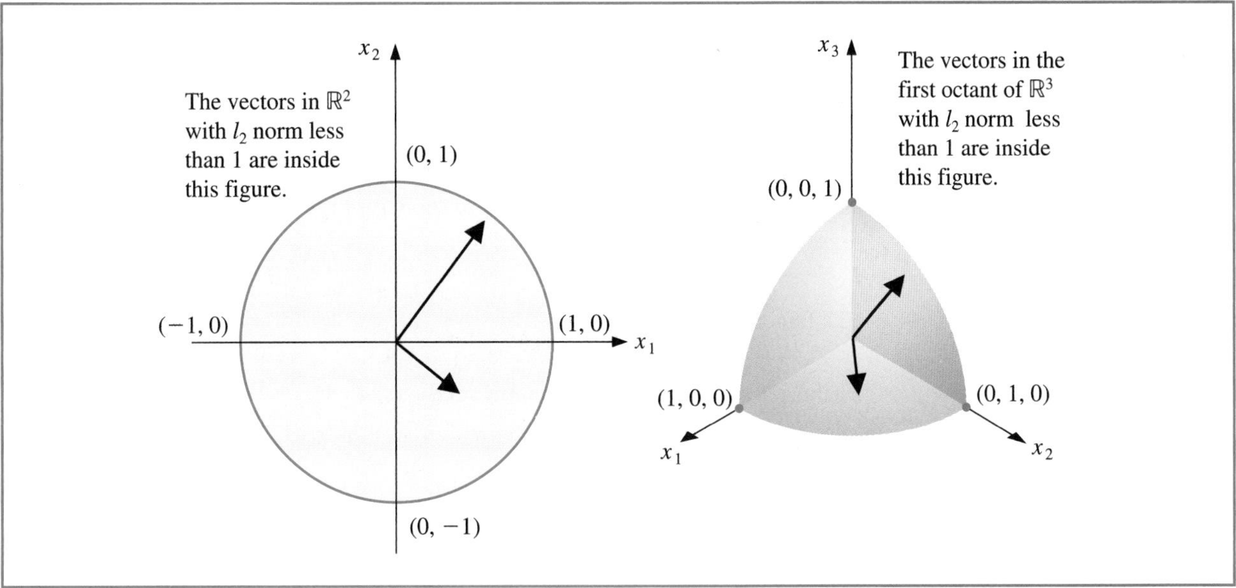

Figure 7.2

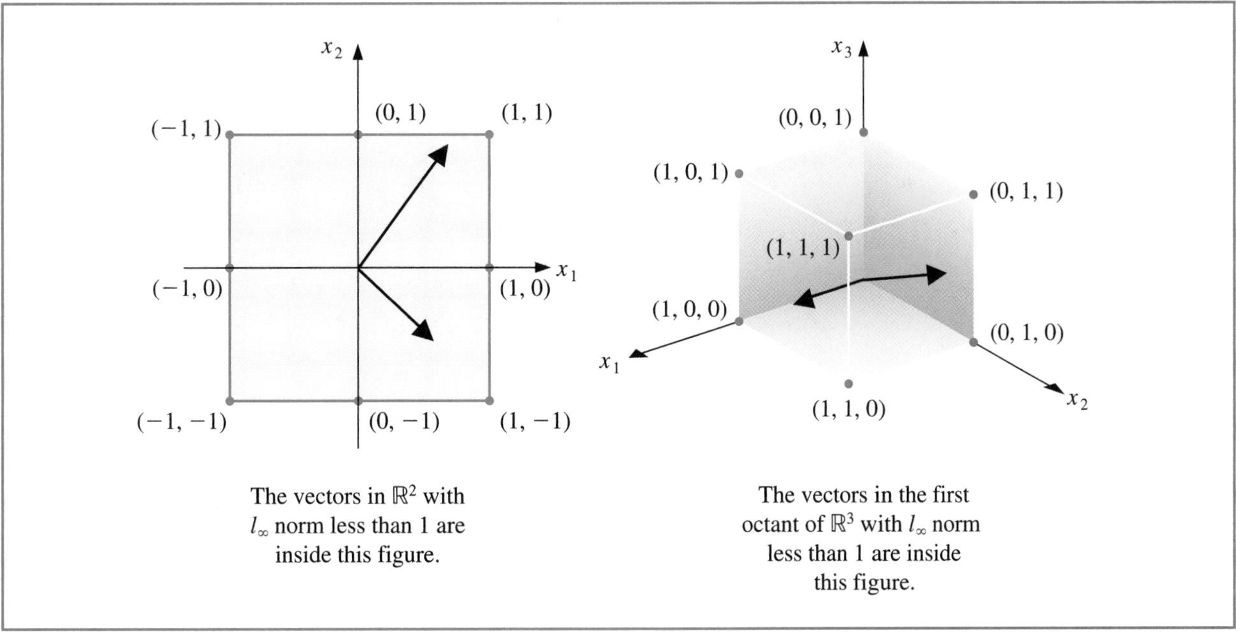

Example 1 Determine the l_2 norm and the l_∞ norm of the vector $\mathbf{x} = (-1, 1, -2)^t$.

Solution The vector $\mathbf{x} = (-1, 1, -2)^t$ in $\mathbb{R}^3$ has norms

$$\|\mathbf{x}\|_2 = \sqrt{(-1)^2 + (1)^2 + (-2)^2} = \sqrt{6}$$

and

$$\|\mathbf{x}\|_\infty = \max\{|-1|, |1|, |-2|\} = 2. \qquad \blacksquare$$

It is easy to show that the properties in Definition 7.1 hold for the l_∞ norm because they follow from similar results for absolute values. The only property that requires much demonstration is (iv), and in this case if $\mathbf{x} = (x_1, x_2, \ldots, x_n)^t$ and $\mathbf{y} = (y_1, y_2, \ldots, y_n)^t$, then

$$\|\mathbf{x}+\mathbf{y}\|_\infty = \max_{1\le i\le n} |x_i + y_i| \le \max_{1\le i\le n} (|x_i| + |y_i|) \le \max_{1\le i\le n} |x_i| + \max_{1\le i\le n} |y_i| = \|\mathbf{x}\|_\infty + \|\mathbf{y}\|_\infty.$$

The first three conditions also are easy to show for the l_2 norm. But to show that

$$\|\mathbf{x}+\mathbf{y}\|_2 \le \|\mathbf{x}\|_2 + \|\mathbf{y}\|_2, \quad \text{for each } \mathbf{x}, \mathbf{y} \in \mathbb{R}_n,$$

we need a famous inequality.

Theorem 7.3 **(Cauchy-Bunyakovsky-Schwarz Inequality for Sums)**
For each $\mathbf{x} = (x_1, x_2, \ldots, x_n)^t$ and $\mathbf{y} = (y_1, y_2, \ldots, y_n)^t$ in $\mathbb{R}^n$,

$$\mathbf{x}^t\mathbf{y} = \sum_{i=1}^n x_i y_i \le \left\{\sum_{i=1}^n x_i^2\right\}^{1/2} \left\{\sum_{i=1}^n y_i^2\right\}^{1/2} = \|\mathbf{x}\|_2 \cdot \|\mathbf{y}\|_2. \tag{7.1}$$

$\blacksquare$

There are many forms of this inequality, hence many discoverers. Augustin Louis Cauchy (1789–1857) describes this inequality in 1821 in *Cours d'Analyse Algébrique*, the first rigorous calculus book. An integral form of the equality appears in the work of Viktor Yakovlevich Bunyakovsky (1804–1889) in 1859, and Hermann Amandus Schwarz (1843–1921) used a double integral form of this inequality in 1885. More details on the history can be found in [Stee].

Proof If $\mathbf{y} = \mathbf{0}$ or $\mathbf{x} = \mathbf{0}$, the result is immediate because both sides of the inequality are zero.

Suppose $\mathbf{y} \ne \mathbf{0}$ and $\mathbf{x} \ne \mathbf{0}$. Note that for each $\lambda \in \mathbb{R}$ we have

$$0 \le ||\mathbf{x} - \lambda\mathbf{y}||_2^2 = \sum_{i=1}^n (x_i - \lambda y_i)^2 = \sum_{i=1}^n x_i^2 - 2\lambda \sum_{i=1}^n x_i y_i + \lambda^2 \sum_{i=1}^n y_i^2,$$

so that

$$2\lambda \sum_{i=1}^n x_i y_i \le \sum_{i=1}^n x_i^2 + \lambda^2 \sum_{i=1}^n y_i^2 = \|\mathbf{x}\|_2^2 + \lambda^2\|\mathbf{y}\|_2^2.$$

However $\|\mathbf{x}\|_2 > 0$ and $\|\mathbf{y}\|_2 > 0$, so we can let $\lambda = \|\mathbf{x}\|_2/\|\mathbf{y}\|_2$ to give

$$\left(2\frac{\|\mathbf{x}\|_2}{\|\mathbf{y}\|_2}\right)\left(\sum_{i=1}^n x_i y_i\right) \le \|\mathbf{x}\|_2^2 + \frac{\|\mathbf{x}\|_2^2}{\|\mathbf{y}\|_2^2}\|\mathbf{y}\|_2^2 = 2\|\mathbf{x}\|_2^2.$$

Hence

$$2\sum_{i=1}^n x_i y_i \le 2\|\mathbf{x}\|_2^2 \frac{\|\mathbf{y}\|_2}{\|\mathbf{x}\|_2} = 2\|\mathbf{x}\|_2\|\mathbf{y}\|_2,$$

and

$$\mathbf{x}^t\mathbf{y} = \sum_{i=1}^{n} x_i y_i \leq \|\mathbf{x}\|_2 \|\mathbf{y}\|_2 = \left\{\sum_{i=1}^{n} x_i^2\right\}^{1/2} \left\{\sum_{i=1}^{n} y_i^2\right\}^{1/2}. \quad \blacksquare \blacksquare \blacksquare$$

With this result we see that for each $\mathbf{x}, \mathbf{y} \in \mathbb{R}^n$,

$$\|\mathbf{x}+\mathbf{y}\|_2^2 = \sum_{i=1}^{n}(x_i+y_i)^2 = \sum_{i=1}^{n} x_i^2 + 2\sum_{i=1}^{n} x_i y_i + \sum_{i=1}^{n} y_i^2 \leq \|\mathbf{x}\|_2^2 + 2\|\mathbf{x}\|_2\|\mathbf{y}\|_2 + \|\mathbf{y}\|_2^2,$$

which gives norm property **(iv)**:

$$\|\mathbf{x}+\mathbf{y}\|_2 \leq \left(\|\mathbf{x}\|_2^2 + 2\|\mathbf{x}\|_2\|\mathbf{y}\|_2 + \|\mathbf{y}\|_2^2\right)^{1/2} = \|\mathbf{x}\|_2 + \|\mathbf{y}\|_2.$$

Distance between Vectors in $\mathbb{R}^n$

The norm of a vector gives a measure for the distance between an arbitrary vector and the zero vector, just as the absolute value of a real number describes its distance from 0. Similarly, the **distance between two vectors** is defined as the norm of the difference of the vectors just as distance between two real numbers is the absolute value of their difference.

Definition 7.4 If $\mathbf{x} = (x_1, x_2, \ldots, x_n)^t$ and $\mathbf{y} = (y_1, y_2, \ldots, y_n)^t$ are vectors in $\mathbb{R}^n$, the l_2 and l_∞ distances between $\mathbf{x}$ and $\mathbf{y}$ are defined by

$$\|\mathbf{x}-\mathbf{y}\|_2 = \left\{\sum_{i=1}^{n}(x_i-y_i)^2\right\}^{1/2} \quad \text{and} \quad \|\mathbf{x}-\mathbf{y}\|_\infty = \max_{1\leq i\leq n}|x_i - y_i|. \quad \blacksquare$$

Example 2 The linear system

$$\begin{aligned}
3.3330x_1 + 15920x_2 - 10.333x_3 &= 15913,\\
2.2220x_1 + 16.710x_2 + 9.6120x_3 &= 28.544,\\
1.5611x_1 + 5.1791x_2 + 1.6852x_3 &= 8.4254
\end{aligned}$$

has the exact solution $\mathbf{x} = (x_1, x_2, x_3)^t = (1, 1, 1)^t$, and Gaussian elimination performed using five-digit rounding arithmetic and partial pivoting (Algorithm 6.2), produces the approximate solution

$$\tilde{\mathbf{x}} = (\tilde{x}_1, \tilde{x}_2, \tilde{x}_3)^t = (1.2001, 0.99991, 0.92538)^t.$$

Determine the l_2 and l_∞ distances between the exact and approximate solutions.

Solution Measurements of $\mathbf{x} - \tilde{\mathbf{x}}$ are given by

$$\begin{aligned}
\|\mathbf{x}-\tilde{\mathbf{x}}\|_\infty &= \max\{|1-1.2001|, |1-0.99991|, |1-0.92538|\}\\
&= \max\{0.2001, 0.00009, 0.07462\} = 0.2001
\end{aligned}$$

and

$$\begin{aligned}
\|\mathbf{x}-\tilde{\mathbf{x}}\|_2 &= \left[(1-1.2001)^2 + (1-0.99991)^2 + (1-0.92538)^2\right]^{1/2}\\
&= [(0.2001)^2 + (0.00009)^2 + (0.07462)^2]^{1/2} = 0.21356.
\end{aligned}$$

Although the components $\tilde{x}_2$ and $\tilde{x}_3$ are good approximations to x_2 and x_3, the component $\tilde{x}_1$ is a poor approximation to x_1, and $|x_1 - \tilde{x}_1|$ dominates both norms. $\blacksquare$

The concept of distance in $\mathbb{R}^n$ is also used to define a limit of a sequence of vectors in this space.

Definition 7.5 A sequence $\{\mathbf{x}^{(k)}\}_{k=1}^{\infty}$ of vectors in $\mathbb{R}^n$ is said to **converge** to $\mathbf{x}$ with respect to the norm $\|\cdot\|$ if, given any $\varepsilon > 0$, there exists an integer $N(\varepsilon)$ such that

$$\|\mathbf{x}^{(k)} - \mathbf{x}\| < \varepsilon, \quad \text{for all } k \geq N(\varepsilon). \qquad \blacksquare$$

Theorem 7.6 The sequence of vectors $\{\mathbf{x}^{(k)}\}$ converges to $\mathbf{x}$ in $\mathbb{R}^n$ with respect to the l_∞ norm if and only if $\lim_{k\to\infty} x_i^{(k)} = x_i$, for each $i = 1, 2, \ldots, n$. ■

Proof Suppose $\{\mathbf{x}^{(k)}\}$ converges to $\mathbf{x}$ with respect to the l_∞ norm. Given any $\varepsilon > 0$, there exists an integer $N(\varepsilon)$ such that for all $k \geq N(\varepsilon)$,

$$\max_{i=1,2,\ldots,n} |x_i^{(k)} - x_i| = \|\mathbf{x}^{(k)} - \mathbf{x}\|_\infty < \varepsilon.$$

This result implies that $|x_i^{(k)} - x_i| < \varepsilon$, for each $i = 1, 2, \ldots, n$, so $\lim_{k\to\infty} x_i^{(k)} = x_i$ for each i.

Conversely, suppose that $\lim_{k\to\infty} x_i^{(k)} = x_i$, for every $i = 1, 2, \ldots, n$. For a given $\varepsilon > 0$, let $N_i(\varepsilon)$ for each i represent an integer with the property that

$$|x_i^{(k)} - x_i| < \varepsilon,$$

whenever $k \geq N_i(\varepsilon)$.

Define $N(\varepsilon) = \max_{i=1,2,\ldots,n} N_i(\varepsilon)$. If $k \geq N(\varepsilon)$, then

$$\max_{i=1,2,\ldots,n} |x_i^{(k)} - x_i| = \|\mathbf{x}^{(k)} - \mathbf{x}\|_\infty < \varepsilon.$$

This implies that $\{\mathbf{x}^{(k)}\}$ converges to $\mathbf{x}$ with respect to the l_∞ norm. ■ ■ ■

Example 3 Show that

$$\mathbf{x}^{(k)} = (x_1^{(k)}, x_2^{(k)}, x_3^{(k)}, x_4^{(k)})^t = \left(1, 2 + \frac{1}{k}, \frac{3}{k^2}, e^{-k}\sin k\right)^t.$$

converges to $\mathbf{x} = (1, 2, 0, 0)^t$ with respect to the l_∞ norm.

Solution Because

$$\lim_{k\to\infty} 1 = 1, \quad \lim_{k\to\infty} (2 + 1/k) = 2, \quad \lim_{k\to\infty} 3/k^2 = 0 \quad \text{and} \quad \lim_{k\to\infty} e^{-k}\sin k = 0,$$

Theorem 7.6 implies that the sequence $\{\mathbf{x}^{(k)}\}$ converges to $(1, 2, 0, 0)^t$ with respect to the l_∞ norm. ■

To show directly that the sequence in Example 3 converges to $(1, 2, 0, 0)^t$ with respect to the l_2 norm is quite complicated. It is better to prove the next result and apply it to this special case.

Theorem 7.7 For each $\mathbf{x} \in \mathbb{R}^n$,

$$\|\mathbf{x}\|_\infty \leq \|\mathbf{x}\|_2 \leq \sqrt{n}\|\mathbf{x}\|_\infty. \qquad \blacksquare$$

Proof Let x_j be a coordinate of $\mathbf{x}$ such that $\|\mathbf{x}\|_\infty = \max_{1\leq i\leq n} |x_i| = |x_j|$. Then

$$\|\mathbf{x}\|_\infty^2 = |x_j|^2 = x_j^2 \leq \sum_{i=1}^{n} x_i^2 = \|\mathbf{x}\|_2^2,$$

and

$$\|\mathbf{x}\|_\infty \leq \|\mathbf{x}\|_2.$$

So

$$\|\mathbf{x}\|_2^2 = \sum_{i=1}^{n} x_i^2 \leq \sum_{i=1}^{n} x_j^2 = nx_j^2 = n\|\mathbf{x}\|_\infty^2,$$

and $\|\mathbf{x}\|_2 \leq \sqrt{n}\|\mathbf{x}\|_\infty$. ■ ■ ■

Figure 7.3 illustrates this result when $n = 2$.

Figure 7.3

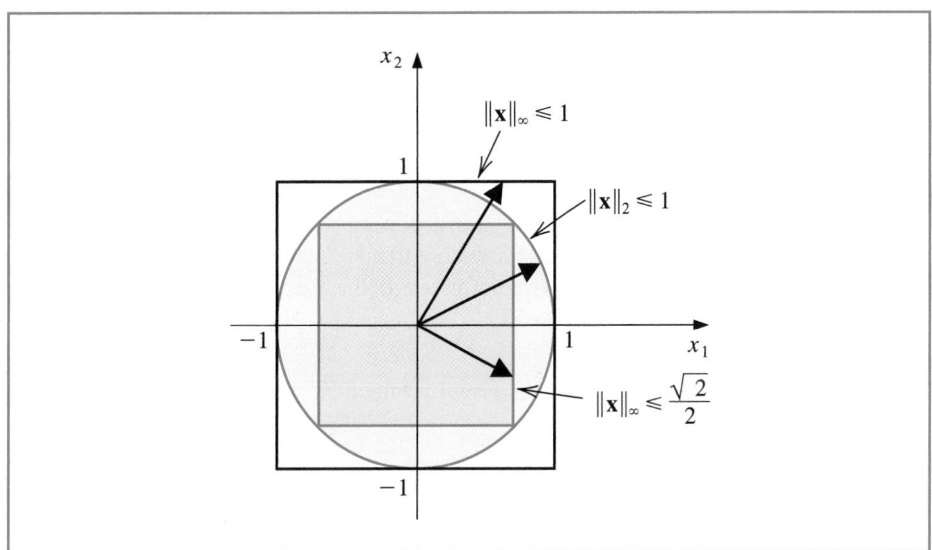

Example 4 In Example 3, we found that the sequence $\{\mathbf{x}^{(k)}\}$, defined by

$$\mathbf{x}^{(k)} = \left(1, 2 + \frac{1}{k}, \frac{3}{k^2}, e^{-k}\sin k\right)^t,$$

converges to $\mathbf{x} = (1, 2, 0, 0)^t$ with respect to the l_∞ norm. Show that this sequence also converges to $\mathbf{x}$ with respect to the l_2 norm.

Solution Given any $\varepsilon > 0$, there exists an integer $N(\varepsilon/2)$ with the property that

$$\|\mathbf{x}^{(k)} - \mathbf{x}\|_\infty < \frac{\varepsilon}{2},$$

whenever $k \geq N(\varepsilon/2)$. By Theorem 7.7, this implies that

$$\|\mathbf{x}^{(k)} - \mathbf{x}\|_2 \leq \sqrt{4}\|\mathbf{x}^{(k)} - \mathbf{x}\|_\infty \leq 2(\varepsilon/2) = \varepsilon,$$

when $k \geq N(\varepsilon/2)$. So $\{\mathbf{x}^{(k)}\}$ also converges to $\mathbf{x}$ with respect to the l_2 norm. ■

It can be shown that all norms on $\mathbb{R}^n$ are equivalent with respect to convergence; that is, if $\|\cdot\|$ and $\|\cdot\|'$ are any two norms on $\mathbb{R}^n$ and $\{\mathbf{x}^{(k)}\}_{k=1}^{\infty}$ has the limit $\mathbf{x}$ with respect to $\|\cdot\|$, then $\{\mathbf{x}^{(k)}\}_{k=1}^{\infty}$ also has the limit $\mathbf{x}$ with respect to $\|\cdot\|'$. The proof of this fact for the general case can be found in [Or2], p. 8. The case for the l_2 and l_∞ norms follows from Theorem 7.7.

Matrix Norms and Distances

In the subsequent sections of this and later chapters, we will need methods for determining the distance between $n \times n$ matrices. This again requires the use of a norm.

Definition 7.8 A **matrix norm** on the set of all $n \times n$ matrices is a real-valued function, $\|\cdot\|$, defined on this set, satisfying for all $n \times n$ matrices A and B and all real numbers α:

(i) $\|A\| \geq 0$;

(ii) $\|A\| = 0$, if and only if A is O, the matrix with all 0 entries;

(iii) $\|\alpha A\| = |\alpha|\,\|A\|$;

(iv) $\|A + B\| \leq \|A\| + \|B\|$;

(v) $\|AB\| \leq \|A\|\,\|B\|$. ■

The **distance between $n \times n$ matrices** A and B with respect to this matrix norm is $\|A - B\|$.

Although matrix norms can be obtained in various ways, the norms considered most frequently are those that are natural consequences of the vector norms l_2 and l_∞.

These norms are defined using the following theorem, whose proof is considered in Exercise 13.

Theorem 7.9 If $\|\cdot\|$ is a vector norm on $\mathbb{R}^n$, then

$$\|A\| = \max_{\|\mathbf{x}\|=1} \|A\mathbf{x}\| \tag{7.2}$$

is a matrix norm. ■

Every vector norm produces an associated natural matrix norm.

Matrix norms defined by vector norms are called the **natural**, or *induced*, **matrix norm** associated with the vector norm. In this text, all matrix norms will be assumed to be natural matrix norms unless specified otherwise.

For any $\mathbf{z} \neq \mathbf{0}$, the vector $\mathbf{x} = \mathbf{z}/\|\mathbf{z}\|$ is a unit vector. Hence

$$\max_{\|\mathbf{x}\|=1} \|A\mathbf{x}\| = \max_{\mathbf{z}\neq\mathbf{0}} \left\| A\left(\frac{\mathbf{z}}{\|\mathbf{z}\|}\right) \right\| = \max_{\mathbf{z}\neq\mathbf{0}} \frac{\|A\mathbf{z}\|}{\|\mathbf{z}\|},$$

and we can alternatively write

$$\|A\| = \max_{\mathbf{z}\neq\mathbf{0}} \frac{\|A\mathbf{z}\|}{\|\mathbf{z}\|}. \tag{7.3}$$

The following corollary to Theorem 7.9 follows from this representation of $\|A\|$.

Corollary 7.10 For any vector $\mathbf{z} \neq \mathbf{0}$, matrix A, and any natural norm $\|\cdot\|$, we have

$$\|A\mathbf{z}\| \leq \|A\| \cdot \|\mathbf{z}\|.$$ ■

The measure given to a matrix under a natural norm describes how the matrix stretches unit vectors relative to that norm. The maximum stretch is the norm of the matrix. The matrix norms we will consider have the forms

$$\|A\|_\infty = \max_{\|\mathbf{x}\|_\infty = 1} \|A\mathbf{x}\|_\infty, \quad \text{the } l_\infty \text{ norm,}$$

and

$$\|A\|_2 = \max_{\|\mathbf{x}\|_2 = 1} \|A\mathbf{x}\|_2, \quad \text{the } l_2 \text{ norm.}$$

An illustration of these norms when $n = 2$ is shown in Figures 7.4 and 7.5 for the matrix

$$A = \begin{bmatrix} 0 & -2 \\ 2 & 0 \end{bmatrix}$$

Figure 7.4

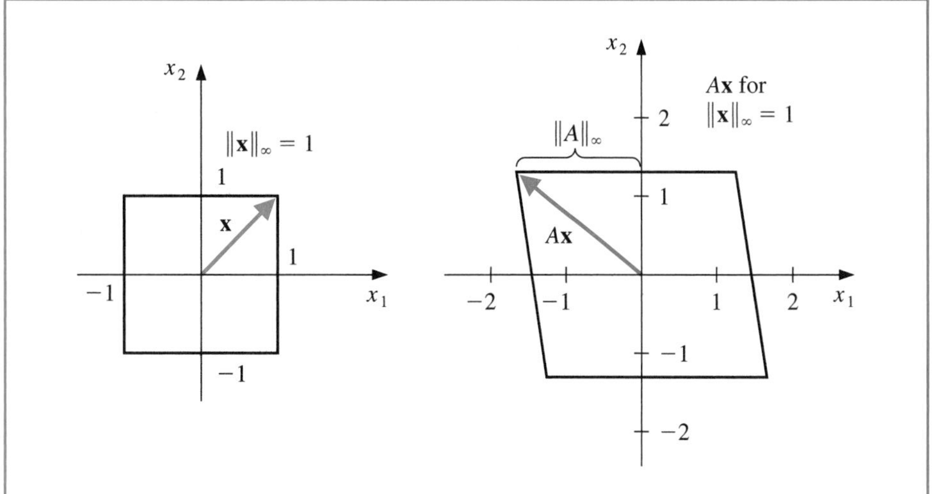

Figure 7.5

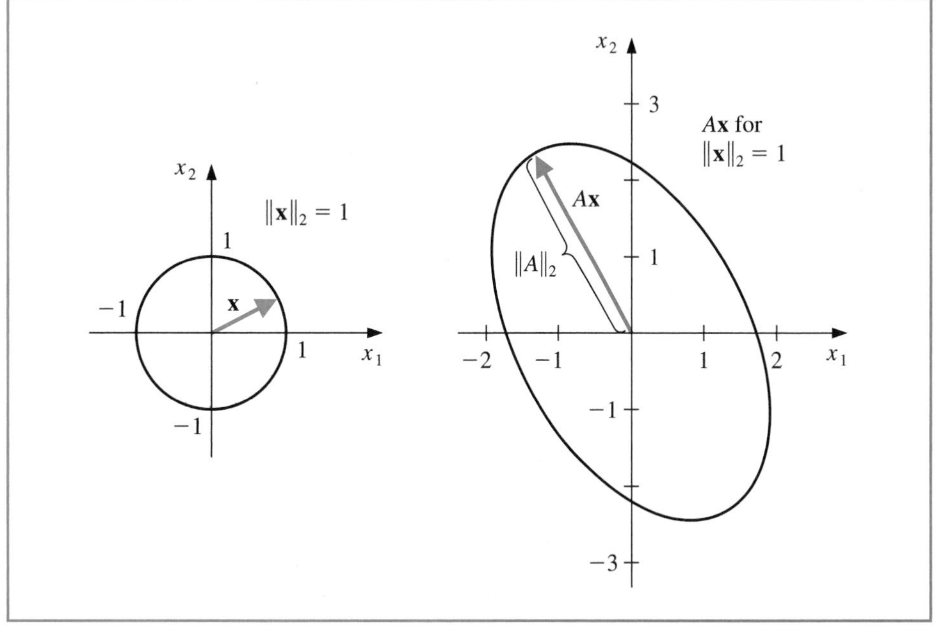

The l_∞ norm of a matrix can be easily computed from the entries of the matrix.

Theorem 7.11 If $A = (a_{ij})$ is an $n \times n$ matrix, then

$$\|A\|_\infty = \max_{1\le i\le n} \sum_{j=1}^{n} |a_{ij}|.$$ ■

Proof First we show that $\|A\|_\infty \le \max_{1\le i\le n} \sum_{j=1}^{n} |a_{ij}|$.

Let $\mathbf{x}$ be an n-dimensional vector with $1 = \|\mathbf{x}\|_\infty = \max_{1\le i\le n} |x_i|$. Since $A\mathbf{x}$ is also an n-dimensional vector,

$$\|A\mathbf{x}\|_\infty = \max_{1\le i\le n} |(A\mathbf{x})_i| = \max_{1\le i\le n} \left| \sum_{j=1}^{n} a_{ij}x_j \right| \le \max_{1\le i\le n} \sum_{j=1}^{n} |a_{ij}| \max_{1\le j\le n} |x_j|.$$

But $\max_{1\le j\le n} |x_j| = \|\mathbf{x}\|_\infty = 1$, so

$$\|A\mathbf{x}\|_\infty \le \max_{1\le i\le n} \sum_{j=1}^{n} |a_{ij}|,$$

and consequently,

$$\|A\|_\infty = \max_{\|\mathbf{x}\|_\infty=1} \|A\mathbf{x}\|_\infty \le \max_{1\le i\le n} \sum_{j=1}^{n} |a_{ij}|. \tag{7.4}$$

Now we will show the opposite inequality. Let p be an integer with

$$\sum_{j=1}^{n} |a_{pj}| = \max_{1\le i\le n} \sum_{j=1}^{n} |a_{ij}|,$$

and $\mathbf{x}$ be the vector with components

$$x_j = \begin{cases} 1, & \text{if } a_{pj} \ge 0, \\ -1, & \text{if } a_{pj} < 0. \end{cases}$$

Then $\|\mathbf{x}\|_\infty = 1$ and $a_{pj}x_j = |a_{pj}|$, for all $j = 1, 2, \ldots, n$, so

$$\|A\mathbf{x}\|_\infty = \max_{1\le i\le n} \left| \sum_{j=1}^{n} a_{ij}x_j \right| \ge \left| \sum_{j=1}^{n} a_{pj}x_j \right| = \left| \sum_{j=1}^{n} |a_{pj}| \right| = \max_{1\le i\le n} \sum_{j=1}^{n} |a_{ij}|.$$

This result implies that

$$\|A\|_\infty = \max_{\|\mathbf{x}\|_\infty=1} \|A\mathbf{x}\|_\infty \ge \max_{1\le i\le n} \sum_{j=1}^{n} |a_{ij}|.$$

Putting this together with Inequality (7.4) gives $\|A\|_\infty = \max_{1\le i\le n} \sum_{j=1}^{n} |a_{ij}|$. ■ ■ ■

Example 5 Determine $\|A\|_\infty$ for the matrix

$$A = \begin{bmatrix} 1 & 2 & -1 \\ 0 & 3 & -1 \\ 5 & -1 & 1 \end{bmatrix}.$$

Solution We have

$$\sum_{j=1}^{3} |a_{1j}| = |1| + |2| + |-1| = 4, \quad \sum_{j=1}^{3} |a_{2j}| = |0| + |3| + |-1| = 4,$$

and

$$\sum_{j=1}^{3} |a_{3j}| = |5| + |-1| + |1| = 7.$$

So Theorem 7.11 implies that $\|A\|_\infty = \max\{4, 4, 7\} = 7$. ■

In the next section, we will discover an alternative method for finding the l_2 norm of a matrix.

EXERCISE SET 7.1

1. Find l_∞ and l_2 norms of the vectors.
   a. $\mathbf{x} = (2, 1, -3, 4)^t$
   b. $\mathbf{x} = (3, -4, 0, \frac{3}{2})^t$
   c. $\mathbf{x} = (\sin k, \cos k, 2^k)^t$ for a fixed positive integer k
   d. $\mathbf{x} = (4/(k+1), 2/k^2, k^2 e^{-k})^t$ for a fixed positive integer k
2. a. Find $\|\mathbf{x}\|_1$ for the vectors given in Exercise 1.
   b. Verify that the function $\|\cdot\|_1$, defined on $\mathbb{R}^n$ by

   $$\|\mathbf{x}\|_1 = \sum_{i=1}^{n} |x_i|,$$

   is a norm on $\mathbb{R}^n$.
   c. Prove that for all $\mathbf{x} \in \mathbb{R}^n$, $\|\mathbf{x}\|_1 \geq \|\mathbf{x}\|_2$.
3. Prove that the following sequences are convergent, and find their limits.
   a. $\mathbf{x}^{(k)} = (1/k, e^{1-k}, -2/k^2)^t$
   b. $\mathbf{x}^{(k)} = \left(e^{-k}\cos k, k\sin(1/k), 3 + k^{-2}\right)^t$
   c. $\mathbf{x}^{(k)} = (ke^{-k^2}, (\cos k)/k, \sqrt{k^2+k} - k)^t$
   d. $\mathbf{x}^{(k)} = (e^{1/k}, (k^2+1)/(1-k^2), (1/k^2)(1+3+5+\cdots+(2k-1)))^t$
4. Find the l_∞ norm of the matrices.
   a. $\begin{bmatrix} 10 & 15 \\ 0 & 1 \end{bmatrix}$
   b. $\begin{bmatrix} 10 & 0 \\ 15 & 1 \end{bmatrix}$
   c. $\begin{bmatrix} 2 & -1 & 0 \\ -1 & 2 & -1 \\ 0 & -1 & 2 \end{bmatrix}$
   d. $\begin{bmatrix} 4 & -1 & 7 \\ -1 & 4 & 0 \\ -7 & 0 & 4 \end{bmatrix}$

5. The following linear systems $A\mathbf{x} = \mathbf{b}$ have $\mathbf{x}$ as the actual solution and $\tilde{\mathbf{x}}$ as an approximate solution. Compute $\|\mathbf{x} - \tilde{\mathbf{x}}\|_\infty$ and $\|A\tilde{\mathbf{x}} - \mathbf{b}\|_\infty$.

a. $\frac{1}{2}x_1 + \frac{1}{3}x_2 = \frac{1}{63}$,
$\frac{1}{3}x_1 + \frac{1}{4}x_2 = \frac{1}{168}$,
$\mathbf{x} = \left(\frac{1}{7}, -\frac{1}{6}\right)^t$,
$\tilde{\mathbf{x}} = (0.142, -0.166)^t$.

b. $x_1 + 2x_2 + 3x_3 = 1$,
$2x_1 + 3x_2 + 4x_3 = -1$,
$3x_1 + 4x_2 + 6x_3 = 2$,
$\mathbf{x} = (0, -7, 5)^t$,
$\tilde{\mathbf{x}} = (-0.2, -7.5, 5.4)^t$.

c. $x_1 + 2x_2 + 3x_3 = 1$,
$2x_1 + 3x_2 + 4x_3 = -1$,
$3x_1 + 4x_2 + 6x_3 = 2$,
$\mathbf{x} = (0, -7, 5)^t$,
$\tilde{\mathbf{x}} = (-0.33, -7.9, 5.8)^t$.

d. $0.04x_1 + 0.01x_2 - 0.01x_3 = 0.06$,
$0.2x_1 + 0.5x_2 - 0.2x_3 = 0.3$,
$x_1 + 2x_2 + 4x_3 = 11$,
$\mathbf{x} = (1.827586, 0.6551724, 1.965517)^t$,
$\tilde{\mathbf{x}} = (1.8, 0.64, 1.9)^t$.

6. The matrix norm $\|\cdot\|_1$, defined by $\|A\|_1 = \max_{\|\mathbf{x}\|_1=1} \|A\mathbf{x}\|_1$, can be computed using the formula

$$\|A\|_1 = \max_{1\le j\le n} \sum_{i=1}^{n} |a_{ij}|,$$

where the vector norm $\|\cdot\|_1$ is defined in Exercise 2. Find $\|\cdot\|_1$ for the matrices in Exercise 4.

7. Show by example that $\|\cdot\|_\circledast$, defined by $\|A\|_\circledast = \max_{1\le i,j\le n} |a_{ij}|$, does not define a matrix norm.

8. Show that $\|\cdot\|_①$, defined by

$$\|A\|_① = \sum_{i=1}^{n}\sum_{j=1}^{n} |a_{ij}|,$$

is a matrix norm. Find $\|\cdot\|_①$ for the matrices in Exercise 4.

9. **a.** The Frobenius norm (which is not a natural norm) is defined for an $n \times n$ matrix A by

$$\|A\|_F = \left(\sum_{i=1}^{n}\sum_{j=1}^{n} |a_{ij}|^2\right)^{1/2}.$$

Show that $\|\cdot\|_F$ is a matrix norm.

b. Find $\|\cdot\|_F$ for the matrices in Exercise 4.

c. For any matrix A, show that $\|A\|_2 \le \|A\|_F \le n^{1/2}\|A\|_2$.

10. In Exercise 9 the Frobenius norm of a matrix was defined. Show that for any $n \times n$ matrix A and vector $\mathbf{x}$ in $\mathbb{R}^n$, $\|A\mathbf{x}\|_2 \le \|A\|_F\|\mathbf{x}\|_2$.

11. Let S be a positive definite $n \times n$ matrix. For any $\mathbf{x}$ in $\mathbb{R}^n$ define $\|\mathbf{x}\| = (\mathbf{x}^t S\mathbf{x})^{1/2}$. Show that this defines a norm on $\mathbb{R}^n$. [*Hint:* Use the Cholesky factorization of S to show that $\mathbf{x}^t S\mathbf{y} = \mathbf{y}^t S\mathbf{x} \le (\mathbf{x}^t S\mathbf{x})^{1/2}(\mathbf{y}^t S\mathbf{y})^{1/2}$.]

12. Let S be a real and nonsingular matrix, and let $\|\cdot\|$ be any norm on $\mathbb{R}^n$. Define $\|\cdot\|'$ by $\|\mathbf{x}\|' = \|S\mathbf{x}\|$. Show that $\|\cdot\|'$ is also a norm on $\mathbb{R}^n$.

13. Prove that if $\|\cdot\|$ is a vector norm on $\mathbb{R}^n$, then $\|A\| = \max_{\|\mathbf{x}\|=1} \|A\mathbf{x}\|$ is a matrix norm.

14. The following excerpt from the *Mathematics Magazine* [Sz] gives an alternative way to prove the Cauchy-Buniakowsky-Schwarz Inequality.

a. Show that when $\mathbf{x} \ne \mathbf{0}$ and $\mathbf{y} \ne \mathbf{0}$, we have

$$\frac{\sum_{i=1}^{n} x_i y_i}{\left(\sum_{i=1}^{n} x_i^2\right)^{1/2}\left(\sum_{i=1}^{n} y_i^2\right)^{1/2}} = 1 - \frac{1}{2}\sum_{i=1}^{n}\left(\frac{x_i}{\left(\sum_{j=1}^{n} x_j^2\right)^{1/2}} - \frac{y_i}{\left(\sum_{j=1}^{n} y_j^2\right)^{1/2}}\right)^2.$$

b. Use the result in part (a) to show that

$$\sum_{i=1}^{n} x_i y_i \leq \left(\sum_{i=1}^{n} x_i^2\right)^{1/2} \left(\sum_{i=1}^{n} y_i^2\right)^{1/2}.$$

15. Show that the Cauchy-Buniakowsky-Schwarz Inequality can be strengthened to

$$\sum_{i=1}^{n} x_i y_i \leq \sum_{i=1}^{n} |x_i y_i| \leq \left(\sum_{i=1}^{n} x_i^2\right)^{1/2} \left(\sum_{i=1}^{n} y_i^2\right)^{1/2}.$$

7.2 Eigenvalues and Eigenvectors

An $n \times m$ matrix can be considered as a function that uses matrix multiplication to take m-dimensional column vectors into n-dimensional column vectors. So an $n \times m$ matrix is actually a linear function from $\mathbb{R}^m$ to $\mathbb{R}^n$. A square matrix A takes the set of n-dimensional vectors into itself, which gives a linear function from $\mathbb{R}^n$ to $\mathbb{R}^n$. In this case, certain nonzero vectors $\mathbf{x}$ might be parallel to $A\mathbf{x}$, which means that a constant λ exists with $A\mathbf{x} = \lambda\mathbf{x}$. For these vectors, we have $(A - \lambda I)\mathbf{x} = \mathbf{0}$. There is a close connection between these numbers λ and the likelihood that an iterative method will converge. We will consider this connection in this section.

Definition 7.12 If A is a square matrix, the **characteristic polynomial** of A is defined by

$$p(\lambda) = \det(A - \lambda I). \quad \blacksquare$$

It is not difficult to show (see Exercise 13) that p is an nth-degree polynomial and, consequently, has at most n distinct zeros, some of which might be complex. If λ is a zero of p, then, since $\det(A - \lambda I) = 0$, Theorem 6.17 on page 398 implies that the linear system defined by $(A - \lambda I)\mathbf{x} = \mathbf{0}$ has a solution with $\mathbf{x} \neq \mathbf{0}$. We wish to study the zeros of p and the nonzero solutions corresponding to these systems.

Definition 7.13 If p is the characteristic polynomial of the matrix A, the zeros of p are **eigenvalues**, or characteristic values, of the matrix A. If λ is an eigenvalue of A and $\mathbf{x} \neq \mathbf{0}$ satisfies $(A - \lambda I)\mathbf{x} = \mathbf{0}$, then $\mathbf{x}$ is an **eigenvector**, or characteristic vector, of A corresponding to the eigenvalue λ. ■

The prefix *eigen* comes from the German adjective meaning "to own", and is synonymous in English with the word *characteristic*. Each matrix has its own eigen- or characteristic equation, with corresponding eigen- or characteristic values and functions.

To determine the eigenvalues of a matrix, we can use the fact that

- λ is an eigenvalue of A if and only if $\det(A - \lambda I) = 0$.

Once an eigenvalue λ has been found a corresponding eigenvector $\mathbf{x} \neq \mathbf{0}$ is determined by solving the system

- $(A - \lambda I)\mathbf{x} = \mathbf{0}$.

Example 1 Show that there are no nonzero vectors $\mathbf{x}$ in $\mathbb{R}^2$ with $A\mathbf{x}$ parallel to $\mathbf{x}$ if

$$A = \begin{bmatrix} 0 & 1 \\ -1 & 0 \end{bmatrix}.$$

Solution The eigenvalues of A are the solutions to the characteristic polynomial

$$0 = \det(A - \lambda I) = \det \begin{bmatrix} -\lambda & 1 \\ -1 & -\lambda \end{bmatrix} = \lambda^2 + 1,$$

so the eigenvalues of A are the complex numbers $\lambda_1 = i$ and $\lambda_2 = -i$. A corresponding eigenvector $\mathbf{x}$ for λ_1 needs to satisfy

$$\begin{bmatrix} 0 \\ 0 \end{bmatrix} = \begin{bmatrix} -i & 1 \\ -1 & -i \end{bmatrix} \begin{bmatrix} x_1 \\ x_2 \end{bmatrix} = \begin{bmatrix} -ix_1 + x_2 \\ -x_1 - ix_2 \end{bmatrix},$$

that is, $0 = -ix_1 + x_2$, so $x_2 = ix_1$, and $0 = -x_1 - ix_2$. Hence if $\mathbf{x}$ is an eigenvector of A, then exactly one of its components is real and the other is complex. As a consequence, there are no nonzero vectors $\mathbf{x}$ in $\mathbb{R}^2$ with $A\mathbf{x}$ parallel to $\mathbf{x}$. ■

If $\mathbf{x}$ is an eigenvector associated with the real eigenvalue λ, then $A\mathbf{x} = \lambda\mathbf{x}$, so the matrix A takes the vector $\mathbf{x}$ into a scalar multiple of itself.

- If λ is real and $\lambda > 1$, then A has the effect of stretching $\mathbf{x}$ by a factor of λ, as illustrated in Figure 7.6(a).
- If $0 < \lambda < 1$, then A shrinks $\mathbf{x}$ by a factor of λ (see Figure 7.6(b)).
- If $\lambda < 0$, the effects are similar (see Figure 7.6(c) and (d)), although the direction of $A\mathbf{x}$ is reversed.

Figure 7.6

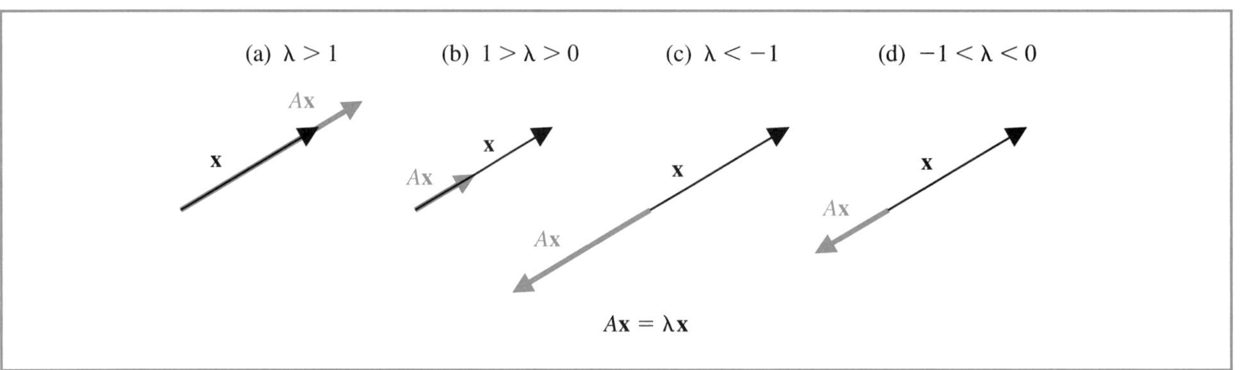

Notice also that if $\mathbf{x}$ is an eigenvector of A associated with the eigenvalue λ and α is any nonzero constant, then $\alpha\mathbf{x}$ is also an eigenvector since

$$A(\alpha\mathbf{x}) = \alpha(A\mathbf{x}) = \alpha(\lambda\mathbf{x}) = \lambda(\alpha\mathbf{x}).$$

An important consequence of this is that for any vector norm $||\cdot||$ we could choose the constant $\alpha = \pm||\mathbf{x}||^{-1}$, which would result in $\alpha\mathbf{x}$ being an eigenvector with norm 1. So

- For every eigenvalue and any vector norm there are eigenvectors with norm 1.

Example 2 Determine the eigenvalues and eigenvectors for the matrix

$$A = \begin{bmatrix} 2 & 0 & 0 \\ 1 & 1 & 2 \\ 1 & -1 & 4 \end{bmatrix}.$$

Solution The characteristic polynomial of A is

$$p(\lambda) = \det(A - \lambda I) = \det \begin{bmatrix} 2-\lambda & 0 & 0 \\ 1 & 1-\lambda & 2 \\ 1 & -1 & 4-\lambda \end{bmatrix}$$

$$= -(\lambda^3 - 7\lambda^2 + 16\lambda - 12) = -(\lambda - 3)(\lambda - 2)^2,$$

so there are two eigenvalues of A: $\lambda_1 = 3$ and $\lambda_2 = 2$.

An eigenvector $\mathbf{x}_1$ corresponding to the eigenvalue $\lambda_1 = 3$ is a solution to the vector-matrix equation $(A - 3 \cdot I)\mathbf{x}_1 = \mathbf{0}$, so

$$\begin{bmatrix} 0 \\ 0 \\ 0 \end{bmatrix} = \begin{bmatrix} -1 & 0 & 0 \\ 1 & -2 & 2 \\ 1 & -1 & 1 \end{bmatrix} \cdot \begin{bmatrix} x_1 \\ x_2 \\ x_3 \end{bmatrix},$$

which implies that $x_1 = 0$ and $x_2 = x_3$.

Any nonzero value of x_3 produces an eigenvector for the eigenvalue $\lambda_1 = 3$. For example, when $x_3 = 1$ we have the eigenvector $\mathbf{x}_1 = (0, 1, 1)^t$, and any eigenvector of A corresponding to $\lambda = 3$ is a nonzero multiple of $\mathbf{x}_1$.

An eigenvector $\mathbf{x} \neq \mathbf{0}$ of A associated with $\lambda_2 = 2$ is a solution of the system $(A - 2 \cdot I)\mathbf{x} = \mathbf{0}$, so

$$\begin{bmatrix} 0 \\ 0 \\ 0 \end{bmatrix} = \begin{bmatrix} 0 & 0 & 0 \\ 1 & -1 & 2 \\ 1 & -1 & 2 \end{bmatrix} \cdot \begin{bmatrix} x_1 \\ x_2 \\ x_3 \end{bmatrix}.$$

In this case the eigenvector has only to satisfy the equation

$$x_1 - x_2 + 2x_3 = 0,$$

which can be done in various ways. For example, when $x_1 = 0$ we have $x_2 = 2x_3$, so one choice would be $\mathbf{x}_2 = (0, 2, 1)^t$. We could also choose $x_2 = 0$, which requires that $x_1 = -2x_3$. Hence $\mathbf{x}_3 = (-2, 0, 1)^t$ gives a second eigenvector for the eigenvalue $\lambda_2 = 2$ that is not a multiple of $\mathbf{x}_2$. The eigenvectors of A corresponding to the eigenvalue $\lambda_2 = 2$ generate an entire plane. This plane is described by all vectors of the form

$$\alpha\mathbf{x}_2 + \beta\mathbf{x}_3 = (-2\beta, 2\alpha, \alpha + \beta)^t,$$

for arbitrary constants α and β, provided that at least one of the constants is nonzero. ■

The package *LinearAlgebra* in Maple provides the function *Eigenvalues* to compute eigenvalues. The function *Eigenvectors* gives both the eigenvalues and the corresponding eigenvectors of a matrix. To produce results for the matrix in Example 2, we first load the package with

with(*LinearAlgebra*)

Then we enter the matrix

$A := ([[2, 0, 0], [1, 1, 2], [1, -1, 4]])$

giving

$$\begin{bmatrix} 2 & 0 & 0 \\ 1 & 1 & 2 \\ 1 & -1 & 4 \end{bmatrix}$$

To determine the eigenvalues and eigenvectors we use

evalf(Eigenvectors(A))

which returns

$$\begin{bmatrix} 3 \\ 2 \\ 2 \end{bmatrix}, \begin{bmatrix} 0 & -2 & 1 \\ 1 & 0 & 1 \\ 1 & 1 & 0 \end{bmatrix}$$

implying that the eigenvalues are 3, 2, and 2 with corresponding eigenvectors given by the respective columns as $(0, 1, 1)^t$, $(-2, 0, 1)^t$, and $(1, 1, 0)^t$.

The *LinearAlgebra* package also contains the command *CharacteristicPolynomial*, so the eigenvalues could also be obtained with

$p :=$ *CharacteristicPolynomial*(A, λ); *factor*(p)

This gives

$$-12 + \lambda^3 - 7\lambda^2 + 16\lambda$$

$$(\lambda - 3)(\lambda - 2)^2$$

The notions of eigenvalues and eigenvectors are introduced here for a specific computational convenience, but these concepts arise frequently in the study of physical systems. In fact, they are of sufficient interest that Chapter 9 is devoted to their numerical approximation.

Spectral Radius

Definition 7.14 The **spectral radius** $\rho(A)$ of a matrix A is defined by

$$\rho(A) = \max |\lambda|, \qquad \text{where } \lambda \text{ is an eigenvalue of } A.$$

(For complex $\lambda = \alpha + \beta i$, we define $|\lambda| = (\alpha^2 + \beta^2)^{1/2}$.) ■

For the matrix considered in Example 2, $\rho(A) = \max\{2, 3\} = 3$.

The spectral radius is closely related to the norm of a matrix, as shown in the following theorem.

Theorem 7.15 If A is an $n \times n$ matrix, then

(i) $\|A\|_2 = [\rho(A^tA)]^{1/2}$,

(ii) $\rho(A) \leq \|A\|$, for any natural norm $\|\cdot\|$. ■

Proof The proof of part (i) requires more information concerning eigenvalues than we presently have available. For the details involved in the proof, see [Or2], p. 21.

To prove part (ii), suppose λ is an eigenvalue of A with eigenvector $\mathbf{x}$ and $\|\mathbf{x}\| = 1$. Then $A\mathbf{x} = \lambda\mathbf{x}$ and

$$|\lambda| = |\lambda| \cdot \|\mathbf{x}\| = \|\lambda\mathbf{x}\| = \|A\mathbf{x}\| \leq \|A\|\|\mathbf{x}\| = \|A\|.$$

Thus

$$\rho(A) = \max |\lambda| \leq \|A\|.$$

■ ■ ■

Part (i) of Theorem 7.15 implies that if A is symmetric, then $\|A\|_2 = \rho(A)$ (see Exercise 14).

An interesting and useful result, which is similar to part (ii) of Theorem 7.15, is that for any matrix A and any $\varepsilon > 0$, there exists a natural norm $\|\cdot\|$ with the property that $\rho(A) < \|A\| < \rho(A) + \varepsilon$. Consequently, $\rho(A)$ is the greatest lower bound for the natural norms on A. The proof of this result can be found in [Or2], p. 23.

Example 3 Determine the l_2 norm of

$$A = \begin{bmatrix} 1 & 1 & 0 \\ 1 & 2 & 1 \\ -1 & 1 & 2 \end{bmatrix}.$$

Solution To apply Theorem 7.15 we need to calculate $\rho(A^tA)$, so we first need the eigenvalues of A^tA.

$$A^tA = \begin{bmatrix} 1 & 1 & -1 \\ 1 & 2 & 1 \\ 0 & 1 & 2 \end{bmatrix} \begin{bmatrix} 1 & 1 & 0 \\ 1 & 2 & 1 \\ -1 & 1 & 2 \end{bmatrix} = \begin{bmatrix} 3 & 2 & -1 \\ 2 & 6 & 4 \\ -1 & 4 & 5 \end{bmatrix}.$$

If

$$0 = \det(A^tA - \lambda I) = \det \begin{bmatrix} 3-\lambda & 2 & -1 \\ 2 & 6-\lambda & 4 \\ -1 & 4 & 5-\lambda \end{bmatrix}$$

$$= -\lambda^3 + 14\lambda^2 - 42\lambda = -\lambda(\lambda^2 - 14\lambda + 42),$$

then $\lambda = 0$ or $\lambda = 7 \pm \sqrt{7}$. By Theorem 7.15 we have

$$||A||_2 = \sqrt{\rho(A^tA)} = \sqrt{\max\{0, 7-\sqrt{7}, 7+\sqrt{7}\}} = \sqrt{7+\sqrt{7}} \approx 3.106.$$ ■

The operations in Example 3 can also be performed using the *LinearAlgebra* package in Maple by first loading the package and then entering the matrix.

with(*LinearAlgebra*): $A :=$ *Matrix*([[1, 1, 0], [1, 2, 1], [−1, 1, 2]])

Maple will respond by showing the matrix that was entered. To determine the transpose of A we use

$B :=$ *Transpose*(A)

which gives

$$\begin{bmatrix} 1 & 1 & -1 \\ 1 & 2 & 1 \\ 0 & 1 & 2 \end{bmatrix}$$

Then we can compute the product AB with

$C := A.B$

which produces

$$\begin{bmatrix} 3 & 2 & -1 \\ 2 & 6 & 4 \\ -1 & 4 & 5 \end{bmatrix}$$

The command

evalf(Eigenvalues(C))

gives the vector

$$\begin{bmatrix} 0. \\ 9.645751311 \\ 4.354248689 \end{bmatrix}$$

Since $||A||_2 = \sqrt{\rho(A^tA)} = \sqrt{\rho(C)}$, we have

$$||A||_2 = \sqrt{9.645751311} = 3.105760987,$$

which we could also find with *evalf(Norm(A, 2))*.

To determine the l_∞ norm of A, replace the last command with *evalf(Norm(A, infinity))* which Maple gives as 4. This is seen to be correct because it is the sum of the magnitude of the entries in the second row.

Convergent Matrices

In studying iterative matrix techniques, it is of particular importance to know when powers of a matrix become small (that is, when all the entries approach zero). Matrices of this type are called *convergent*.

Definition 7.16 We call an $n \times n$ matrix A **convergent** if

$$\lim_{k\to\infty} (A^k)_{ij} = 0, \quad \text{for each } i = 1, 2, \ldots, n \text{ and } j = 1, 2, \ldots, n.$$ ■

Example 4 Show that

$$A = \begin{bmatrix} \frac{1}{2} & 0 \\ \frac{1}{4} & \frac{1}{2} \end{bmatrix}$$

is a convergent matrix.

Solution Computing powers of A, we obtain:

$$A^2 = \begin{bmatrix} \frac{1}{4} & 0 \\ \frac{1}{4} & \frac{1}{4} \end{bmatrix}, \quad A^3 = \begin{bmatrix} \frac{1}{8} & 0 \\ \frac{3}{16} & \frac{1}{8} \end{bmatrix}, \quad A^4 = \begin{bmatrix} \frac{1}{16} & 0 \\ \frac{1}{8} & \frac{1}{16} \end{bmatrix},$$

and, in general,

$$A^k = \begin{bmatrix} (\frac{1}{2})^k & 0 \\ \frac{k}{2^{k+1}} & (\frac{1}{2})^k \end{bmatrix}.$$

So A is a convergent matrix because

$$\lim_{k\to\infty} \left(\frac{1}{2}\right)^k = 0 \quad \text{and} \quad \lim_{k\to\infty} \frac{k}{2^{k+1}} = 0.$$ ■

Notice that the convergent matrix A in Example 4 has $\rho(A) = \frac{1}{2}$, because $\frac{1}{2}$ is the only eigenvalue of A. This illustrates an important connection that exists between the spectral radius of a matrix and the convergence of the matrix, as detailed in the following result.

Theorem 7.17 The following statements are equivalent.

(i) A is a convergent matrix.

(ii) $\lim_{n\to\infty} \|A^n\| = 0$, for some natural norm.

(iii) $\lim_{n\to\infty} \|A^n\| = 0$, for all natural norms.

(iv) $\rho(A) < 1$.

(v) $\lim_{n\to\infty} A^n\mathbf{x} = \mathbf{0}$, for every $\mathbf{x}$. ■

The proof of this theorem can be found in [IK], p. 14.

EXERCISE SET 7.2

1. Compute the eigenvalues and associated eigenvectors of the following matrices.

a. $\begin{bmatrix} 1 & 1 \\ -2 & -2 \end{bmatrix}$ **b.** $\begin{bmatrix} -1 & -1 \\ \frac{1}{3} & \frac{1}{6} \end{bmatrix}$ **c.** $\begin{bmatrix} 3 & 4 \\ 1 & 0 \end{bmatrix}$

d. $\begin{bmatrix} 3 & 2 & -1 \\ 1 & -2 & 3 \\ 2 & 0 & 4 \end{bmatrix}$ **e.** $\begin{bmatrix} \frac{1}{2} & 0 & 0 \\ -1 & \frac{1}{2} & 0 \\ 2 & 2 & -\frac{1}{3} \end{bmatrix}$ **f.** $\begin{bmatrix} 2 & -1 & 0 \\ 0 & 2 & 4 \\ 0 & 0 & 2 \end{bmatrix}$

2. Compute the eigenvalues and associated eigenvectors of the following matrices.

a. $\begin{bmatrix} 2 & -1 \\ -1 & 2 \end{bmatrix}$ **b.** $\begin{bmatrix} 0 & 1 \\ 1 & 1 \end{bmatrix}$ **c.** $\begin{bmatrix} 0 & \frac{1}{2} \\ \frac{1}{2} & 0 \end{bmatrix}$

d. $\begin{bmatrix} 2 & 1 & 0 \\ 1 & 2 & 0 \\ 0 & 0 & 3 \end{bmatrix}$ **e.** $\begin{bmatrix} -1 & 2 & 0 \\ 0 & 3 & 4 \\ 0 & 0 & 7 \end{bmatrix}$ **f.** $\begin{bmatrix} 2 & 1 & 1 \\ 2 & 3 & 2 \\ 1 & 1 & 2 \end{bmatrix}$

3. Find the complex eigenvalues and associated eigenvectors for the following matrices.

a. $\begin{bmatrix} 2 & 2 \\ -1 & 2 \end{bmatrix}$ **b.** $\begin{bmatrix} 1 & 2 \\ -1 & 2 \end{bmatrix}$

4. Find the complex eigenvalues and associated eigenvectors for the following matrices.

a. $\begin{bmatrix} 1 & 0 & 2 \\ 0 & 1 & -1 \\ -1 & 1 & 1 \end{bmatrix}$ **b.** $\begin{bmatrix} 0 & 1 & -2 \\ 1 & 0 & 0 \\ 1 & 1 & 1 \end{bmatrix}$

5. Find the spectral radius for each matrix in Exercise 1.

6. Find the spectral radius for each matrix in Exercise 2.

7. Which of the matrices in Exercise 1 are convergent?

8. Which of the matrices in Exercise 2 are convergent?

9. Find the l_2 norm for the matrices in Exercise 1.

10. Find the l_2 norm for the matrices in Exercise 2.

11. Let $A_1 = \begin{bmatrix} 1 & 0 \\ \frac{1}{4} & \frac{1}{2} \end{bmatrix}$ and $A_2 = \begin{bmatrix} \frac{1}{2} & 0 \\ 16 & \frac{1}{2} \end{bmatrix}$. Show that A_1 is not convergent, but A_2 is convergent.

12. An $n \times n$ matrix A is called *nilpotent* if an integer m exists with $A^m = O_n$. Show that if λ is an eigenvalue of a nilpotent matrix, then $\lambda = 0$.

13. Show that the characteristic polynomial $p(\lambda) = \det(A - \lambda I)$ for the $n \times n$ matrix A is an nth-degree polynomial. [*Hint:* Expand $\det(A - \lambda I)$ along the first row, and use mathematical induction on n.]

14. **a.** Show that if A is an $n \times n$ matrix, then

$$\det A = \prod_{i=1}^{n} \lambda_i,$$

where $\lambda_i, \ldots, \lambda_n$ are the eigenvalues of A. [*Hint:* Consider $p(0)$.]

b. Show that A is singular if and only if $\lambda = 0$ is an eigenvalue of A.

15. Let λ be an eigenvalue of the $n \times n$ matrix A and $\mathbf{x} \neq \mathbf{0}$ be an associated eigenvector.

a. Show that λ is also an eigenvalue of A^t.

b. Show that for any integer $k \geq 1$, λ^k is an eigenvalue of A^k with eigenvector $\mathbf{x}$.

c. Show that if A^{-1} exists, then $1/\lambda$ is an eigenvalue of A^{-1} with eigenvector $\mathbf{x}$.

d. Generalize parts (b) and (c) to $(A^{-1})^k$ for integers $k \geq 2$.

e. Given the polynomial $q(x) = q_0 + q_1 x + \cdots + q_k x^k$, define $q(A)$ to be the matrix $q(A) = q_0 I + q_1 A + \cdots + q_k A^k$. Show that $q(\lambda)$ is an eigenvalue of $q(A)$ with eigenvector $\mathbf{x}$.

f. Let $\alpha \neq \lambda$ be given. Show that if $A - \alpha I$ is nonsingular, then $1/(\lambda - \alpha)$ is an eigenvalue of $(A - \alpha I)^{-1}$ with eigenvector $\mathbf{x}$.

16. Show that if A is symmetric, then $||A||_2 = \rho(A)$.

17. In Exercise 15 of Section 6.3, we assumed that the contribution a female beetle of a certain type made to the future years' beetle population could be expressed in terms of the matrix

$$A = \begin{bmatrix} 0 & 0 & 6 \\ \frac{1}{2} & 0 & 0 \\ 0 & \frac{1}{3} & 0 \end{bmatrix},$$

where the entry in the ith row and jth column represents the probabilistic contribution of a beetle of age j onto the next year's female population of age i.

a. Does the matrix A have any real eigenvalues? If so, determine them and any associated eigenvectors.

b. If a sample of this species was needed for laboratory test purposes that would have a constant proportion in each age group from year to year, what criteria could be imposed on the initial population to ensure that this requirement would be satisfied?

18. Find matrices A and B for which $\rho(A+B) > \rho(A) + \rho(B)$. (This shows that $\rho(A)$ cannot be a matrix norm.)

19. Show that if $||\cdot||$ is any natural norm, then $(||A^{-1}||)^{-1} \leq |\lambda| \leq ||A||$ for any eigenvalue λ of the nonsingular matrix A.

7.3 The Jacobi and Gauss-Siedel Iterative Techniques

In this section we describe the Jacobi and the Gauss-Seidel iterative methods, classic methods that date to the late eighteenth century. Iterative techniques are seldom used for solving linear systems of small dimension since the time required for sufficient accuracy exceeds that required for direct techniques such as Gaussian elimination. For large systems with a high percentage of 0 entries, however, these techniques are efficient in terms of both computer storage and computation. Systems of this type arise frequently in circuit analysis and in the numerical solution of boundary-value problems and partial-differential equations.

An iterative technique to solve the $n \times n$ linear system $A\mathbf{x} = \mathbf{b}$ starts with an initial approximation $\mathbf{x}^{(0)}$ to the solution $\mathbf{x}$ and generates a sequence of vectors $\{\mathbf{x}^{(k)}\}_{k=0}^{\infty}$ that converges to $\mathbf{x}$.

Jacobi's Method

The **Jacobi iterative method** is obtained by solving the ith equation in $A\mathbf{x} = \mathbf{b}$ for x_i to obtain (provided $a_{ii} \neq 0$)

$$x_i = \sum_{\substack{j=1 \\ j \neq i}}^{n} \left(-\frac{a_{ij} x_j}{a_{ii}} \right) + \frac{b_i}{a_{ii}}, \qquad \text{for } i = 1, 2, \ldots, n.$$

For each $k \geq 1$, generate the components $x_i^{(k)}$ of $\mathbf{x}^{(k)}$ from the components of $\mathbf{x}^{(k-1)}$ by

$$x_i^{(k)} = \frac{1}{a_{ii}}\left[\sum_{\substack{j=1\\ j\neq i}}^{n}\left(-a_{ij}x_j^{(k-1)}\right) + b_i\right], \qquad \text{for } i = 1, 2, \ldots, n. \tag{7.5}$$

Example 1 The linear system $A\mathbf{x} = \mathbf{b}$ given by

Carl Gustav Jacob Jacobi (1804–1851) was initially recognized for his work in the area of number theory and elliptic functions, but his mathematical interests and abilities were very broad. He had a strong personality that was influential in establishing a research-oriented attitude that became the nucleus of a revival of mathematics at German universities in the 19th century.

$$\begin{aligned}
E_1 :&\quad 10x_1 - x_2 + 2x_3 \phantom{{}+3x_4} = 6,\\
E_2 :&\quad -x_1 + 11x_2 - x_3 + 3x_4 = 25,\\
E_3 :&\quad 2x_1 - x_2 + 10x_3 - x_4 = -11,\\
E_4 :&\quad 3x_2 - x_3 + 8x_4 = 15
\end{aligned}$$

has the unique solution $\mathbf{x} = (1, 2, -1, 1)^t$. Use Jacobi's iterative technique to find approximations $\mathbf{x}^{(k)}$ to $\mathbf{x}$ starting with $\mathbf{x}^{(0)} = (0, 0, 0, 0)^t$ until

$$\frac{\|\mathbf{x}^{(k)} - \mathbf{x}^{(k-1)}\|_\infty}{\|\mathbf{x}^{(k)}\|_\infty} < 10^{-3}.$$

Solution We first solve equation E_i for x_i, for each $i = 1, 2, 3, 4$, to obtain

$$\begin{aligned}
x_1 &= \frac{1}{10}x_2 - \frac{1}{5}x_3 + \frac{3}{5},\\
x_2 &= \frac{1}{11}x_1 + \frac{1}{11}x_3 - \frac{3}{11}x_4 + \frac{25}{11},\\
x_3 &= -\frac{1}{5}x_1 + \frac{1}{10}x_2 + \frac{1}{10}x_4 - \frac{11}{10},\\
x_4 &= -\frac{3}{8}x_2 + \frac{1}{8}x_3 + \frac{15}{8}.
\end{aligned}$$

From the initial approximation $\mathbf{x}^{(0)} = (0, 0, 0, 0)^t$ we have $\mathbf{x}^{(1)}$ given by

$$\begin{aligned}
x_1^{(1)} &= \frac{1}{10}x_2^{(0)} - \frac{1}{5}x_3^{(0)} + \frac{3}{5} = 0.6000,\\
x_2^{(1)} &= \frac{1}{11}x_1^{(0)} + \frac{1}{11}x_3^{(0)} - \frac{3}{11}x_4^{(0)} + \frac{25}{11} = 2.2727,\\
x_3^{(1)} &= -\frac{1}{5}x_1^{(0)} + \frac{1}{10}x_2^{(0)} + \frac{1}{10}x_4^{(0)} - \frac{11}{10} = -1.1000,\\
x_4^{(1)} &= -\frac{3}{8}x_2^{(0)} + \frac{1}{8}x_3^{(0)} + \frac{15}{8} = 1.8750.
\end{aligned}$$

Additional iterates, $\mathbf{x}^{(k)} = (x_1^{(k)}, x_2^{(k)}, x_3^{(k)}, x_4^{(k)})^t$, are generated in a similar manner and are presented in Table 7.1.

Table 7.1

k	0	1	2	3	4	5	6	7	8	9	10
$x_1^{(k)}$	0.0000	0.6000	1.0473	0.9326	1.0152	0.9890	1.0032	0.9981	1.0006	0.9997	1.0001
$x_2^{(k)}$	0.0000	2.2727	1.7159	2.053	1.9537	2.0114	1.9922	2.0023	1.9987	2.0004	1.9998
$x_3^{(k)}$	0.0000	−1.1000	−0.8052	−1.0493	−0.9681	−1.0103	−0.9945	−1.0020	−0.9990	−1.0004	−0.9998
$x_4^{(k)}$	0.0000	1.8750	0.8852	1.1309	0.9739	1.0214	0.9944	1.0036	0.9989	1.0006	0.9998

We stopped after ten iterations because

$$\frac{\|\mathbf{x}^{(10)} - \mathbf{x}^{(9)}\|_\infty}{\|\mathbf{x}^{(10)}\|_\infty} = \frac{8.0 \times 10^{-4}}{1.9998} < 10^{-3}.$$

In fact, $\|\mathbf{x}^{(10)} - \mathbf{x}\|_\infty = 0.0002$. ■

In general, iterative techniques for solving linear systems involve a process that converts the system $A\mathbf{x} = \mathbf{b}$ into an equivalent system of the form $\mathbf{x} = T\mathbf{x} + \mathbf{c}$ for some fixed matrix T and vector $\mathbf{c}$. After the initial vector $\mathbf{x}^{(0)}$ is selected, the sequence of approximate solution vectors is generated by computing

$$\mathbf{x}^{(k)} = T\mathbf{x}^{(k-1)} + \mathbf{c},$$

for each $k = 1, 2, 3, \ldots$. This should be reminiscent of the fixed-point iteration studied in Chapter 2.

The Jacobi method can be written in the form $\mathbf{x}^{(k)} = T\mathbf{x}^{(k-1)} + \mathbf{c}$ by splitting A into its diagonal and off-diagonal parts. To see this, let D be the diagonal matrix whose diagonal entries are those of A, $-L$ be the strictly lower-triangular part of A, and $-U$ be the strictly upper-triangular part of A. With this notation,

$$A = \begin{bmatrix} a_{11} & a_{12} & \cdots & a_{1n} \\ a_{21} & a_{22} & \cdots & a_{2n} \\ \vdots & \vdots & & \vdots \\ a_{n1} & a_{n2} & \cdots & a_{nn} \end{bmatrix}$$

is split into

$$A = \begin{bmatrix} a_{11} & 0 & \cdots & 0 \\ 0 & a_{22} & \ddots & \vdots \\ \vdots & \ddots & \ddots & 0 \\ 0 & \cdots & 0 & a_{nn} \end{bmatrix} - \begin{bmatrix} 0 & \cdots & \cdots & 0 \\ -a_{21} & \ddots & & \vdots \\ \vdots & \ddots & \ddots & \vdots \\ -a_{n1} & \cdots & -a_{n,n-1} & 0 \end{bmatrix} - \begin{bmatrix} 0 & -a_{12} & \cdots & -a_{1n} \\ \vdots & \ddots & \ddots & \vdots \\ \vdots & & \ddots & -a_{n-1,n} \\ 0 & \cdots & \cdots & 0 \end{bmatrix}$$

$$= D - L - U.$$

The equation $A\mathbf{x} = \mathbf{b}$, or $(D - L - U)\mathbf{x} = \mathbf{b}$, is then transformed into

$$D\mathbf{x} = (L + U)\mathbf{x} + \mathbf{b},$$

and, if D^{-1} exists, that is, if $a_{ii} \neq 0$ for each i, then

$$\mathbf{x} = D^{-1}(L + U)\mathbf{x} + D^{-1}\mathbf{b}.$$

This results in the matrix form of the Jacobi iterative technique:

$$\mathbf{x}^{(k)} = D^{-1}(L + U)\mathbf{x}^{(k-1)} + D^{-1}\mathbf{b}, \quad k = 1, 2, \ldots. \tag{7.6}$$

Introducing the notation $T_j = D^{-1}(L + U)$ and $\mathbf{c}_j = D^{-1}\mathbf{b}$ gives the Jacobi technique the form

$$\mathbf{x}^{(k)} = T_j\mathbf{x}^{(k-1)} + \mathbf{c}_j. \tag{7.7}$$

In practice, Eq. (7.5) is used in computation and Eq. (7.7) for theoretical purposes.

Example 2 Express the Jacobi iteration method for the linear system $A\mathbf{x} = \mathbf{b}$ given by

$$\begin{aligned}
E_1 &: \quad 10x_1 - x_2 + 2x_3 = 6, \\
E_2 &: \quad -x_1 + 11x_2 - x_3 + 3x_4 = 25, \\
E_3 &: \quad 2x_1 - x_2 + 10x_3 - x_4 = -11, \\
E_4 &: \quad 3x_2 - x_3 + 8x_4 = 15
\end{aligned}$$

in the form $\mathbf{x}^{(k)} = T\mathbf{x}^{(k-1)} + \mathbf{c}$.

Solution We saw in Example 1 that the Jacobi method for this system has the form

$$\begin{aligned}
x_1 &= \frac{1}{10}x_2 - \frac{1}{5}x_3 + \frac{3}{5}, \\
x_2 &= \frac{1}{11}x_1 + \frac{1}{11}x_3 - \frac{3}{11}x_4 + \frac{25}{11}, \\
x_3 &= -\frac{1}{5}x_1 + \frac{1}{10}x_2 + \frac{1}{10}x_4 - \frac{11}{10}, \\
x_4 &= -\frac{3}{8}x_2 + \frac{1}{8}x_3 + \frac{15}{8}.
\end{aligned}$$

Hence we have

$$T = \begin{bmatrix} 0 & \frac{1}{10} & -\frac{1}{5} & 0 \\ \frac{1}{11} & 0 & \frac{1}{11} & -\frac{3}{11} \\ -\frac{1}{5} & \frac{1}{10} & 0 & \frac{1}{10} \\ 0 & -\frac{3}{8} & \frac{1}{8} & 0 \end{bmatrix} \quad \text{and} \quad \mathbf{c} = \begin{bmatrix} \frac{3}{5} \\ \frac{25}{11} \\ -\frac{11}{10} \\ \frac{15}{8} \end{bmatrix}.$$ ■

Algorithm 7.1 implements the Jacobi iterative technique.

Jacobi Iterative

To solve $A\mathbf{x} = \mathbf{b}$ given an initial approximation $\mathbf{x}^{(0)}$:

INPUT the number of equations and unknowns n; the entries a_{ij}, $1 \le i, j \le n$ of the matrix A; the entries b_i, $1 \le i \le n$ of $\mathbf{b}$; the entries XO_i, $1 \le i \le n$ of $\mathbf{XO} = \mathbf{x}^{(0)}$; tolerance TOL; maximum number of iterations N.

OUTPUT the approximate solution $x_1, \ldots, x_n$ or a message that the number of iterations was exceeded.

Step 1 Set $k = 1$.

Step 2 While ($k \le N$) do Steps 3–6.

Step 3 For $i = 1, \ldots, n$

$$\text{set } x_i = \frac{1}{a_{ii}}\left[-\sum_{\substack{j=1 \\ j \ne i}}^{n} (a_{ij}XO_j) + b_i\right].$$

Step 4 If $||\mathbf{x} - \mathbf{XO}|| < TOL$ then OUTPUT $(x_1, \ldots, x_n)$;
(*The procedure was successful.*)
STOP.

Step 5 Set $k = k + 1$.

Step 6 For $i = 1, \ldots, n$ set $XO_i = x_i$.

Step 7 OUTPUT ('Maximum number of iterations exceeded');
(*The procedure was successful.*)
STOP. ■

Step 3 of the algorithm requires that $a_{ii} \neq 0$, for each $i = 1, 2, \ldots, n$. If one of the a_{ii} entries is 0 and the system is nonsingular, a reordering of the equations can be performed so that no $a_{ii} = 0$. To speed convergence, the equations should be arranged so that a_{ii} is as large as possible. This subject is discussed in more detail later in this chapter.

Another possible stopping criterion in Step 4 is to iterate until

$$\frac{\|\mathbf{x}^{(k)} - \mathbf{x}^{(k-1)}\|}{\|\mathbf{x}^{(k)}\|}$$

is smaller than some prescribed tolerance. For this purpose, any convenient norm can be used, the usual being the l_∞ norm.

The *NumericalAnalysis* subpackage of the Maple *Student* package implements the Jacobi iterative method. To illustrate this with our example we first enter both *NumericalAnalysis* and *LinearAlgebra.*

with(*Student*[*NumericalAnalysis*]): *with*(*LinearAlgebra*):

Colons are used at the end of the commands to suppress output for both packages. Enter the matrix with

$A := Matrix([[10, -1, 2, 0, 6], [-1, 11, -1, 3, 25], [2, -1, 10, -1, -11], [0, 3, -1, 8, 15]])$

The following command gives a collection of output that is in agreement with the results in Table 7.1.

IterativeApproximate(*A*, *initialapprox* = *Vector*([0., 0., 0., 0.]), *tolerance* = 10^{-3},
maxiterations = 20, *stoppingcriterion* = *relative*(*infinity*), *method* = *jacobi*,
output = *approximates*)

If the option *output* = *approximates* is omitted, then only the final approximation result is output. Notice that the initial approximations was specified by [0., 0., 0., 0.], with decimal points placed after the entries. This was done so that Maple will give the results as 10-digit decimals. If the specification had simply been [0, 0, 0, 0], the output would have been given in fractional form.

Phillip Ludwig Seidel (1821–1896) worked as an assistant to Jacobi solving problems on systems of linear equations that resulted from Gauss's work on least squares. These equations generally had off-diagonal elements that were much smaller than those on the diagonal, so the iterative methods were particularly effective. The iterative techniques now known as Jacobi and Gauss-Seidel were both known to Gauss before being applied in this situation, but Gauss's results were not often widely communicated.

The Gauss-Seidel Method

A possible improvement in Algorithm 7.1 can be seen by reconsidering Eq. (7.5). The components of $\mathbf{x}^{(k-1)}$ are used to compute all the components $x_i^{(k)}$ of $\mathbf{x}^{(k)}$. But, for $i > 1$, the components $x_1^{(k)}, \ldots, x_{i-1}^{(k)}$ of $\mathbf{x}^{(k)}$ have already been computed and are expected to be better approximations to the actual solutions $x_1, \ldots, x_{i-1}$ than are $x_1^{(k-1)}, \ldots, x_{i-1}^{(k-1)}$. It seems reasonable, then, to compute $x_i^{(k)}$ using these most recently calculated values. That is, to use

$$x_i^{(k)} = \frac{1}{a_{ii}}\left[-\sum_{j=1}^{i-1}(a_{ij}x_j^{(k)}) - \sum_{j=i+1}^{n}(a_{ij}x_j^{(k-1)}) + b_i\right], \qquad (7.8)$$

for each $i = 1, 2, \ldots, n$, instead of Eq. (7.5). This modification is called the **Gauss-Seidel iterative technique** and is illustrated in the following example.

Example 3 Use the Gauss-Seidel iterative technique to find approximate solutions to

$$\begin{aligned}
10x_1 - x_2 + 2x_3 \quad &= 6,\\
-x_1 + 11x_2 - x_3 + 3x_4 &= 25,\\
2x_1 - x_2 + 10x_3 - x_4 &= -11,\\
3x_2 - x_3 + 8x_4 &= 15
\end{aligned}$$

starting with $\mathbf{x} = (0, 0, 0, 0)^t$ and iterating until

$$\frac{\|\mathbf{x}^{(k)} - \mathbf{x}^{(k-1)}\|_\infty}{\|\mathbf{x}^{(k)}\|_\infty} < 10^{-3}.$$

Solution The solution $\mathbf{x} = (1, 2, -1, 1)^t$ was approximated by Jacobi's method in Example 1. For the Gauss-Seidel method we write the system, for each $k = 1, 2, \ldots$ as

$$\begin{aligned}
x_1^{(k)} &= \frac{1}{10}x_2^{(k-1)} - \frac{1}{5}x_3^{(k-1)} + \frac{3}{5},\\
x_2^{(k)} &= \frac{1}{11}x_1^{(k)} + \frac{1}{11}x_3^{(k-1)} - \frac{3}{11}x_4^{(k-1)} + \frac{25}{11},\\
x_3^{(k)} &= -\frac{1}{5}x_1^{(k)} + \frac{1}{10}x_2^{(k)} + \frac{1}{10}x_4^{(k-1)} - \frac{11}{10},\\
x_4^{(k)} &= -\frac{3}{8}x_2^{(k)} + \frac{1}{8}x_3^{(k)} + \frac{15}{8}.
\end{aligned}$$

When $\mathbf{x}^{(0)} = (0, 0, 0, 0)^t$, we have $\mathbf{x}^{(1)} = (0.6000, 2.3272, -0.9873, 0.8789)^t$. Subsequent iterations give the values in Table 7.2.

Table 7.2

k	0	1	2	3	4	5
$x_1^{(k)}$	0.0000	0.6000	1.030	1.0065	1.0009	1.0001
$x_2^{(k)}$	0.0000	2.3272	2.037	2.0036	2.0003	2.0000
$x_3^{(k)}$	0.0000	−0.9873	−1.014	−1.0025	−1.0003	−1.0000
$x_4^{(k)}$	0.0000	0.8789	0.9844	0.9983	0.9999	1.0000

Because

$$\frac{\|\mathbf{x}^{(5)} - \mathbf{x}^{(4)}\|_\infty}{\|\mathbf{x}^{(5)}\|_\infty} = \frac{0.0008}{2.000} = 4 \times 10^{-4},$$

$\mathbf{x}^{(5)}$ is accepted as a reasonable approximation to the solution. Note that Jacobi's method in Example 1 required twice as many iterations for the same accuracy. ■

To write the Gauss-Seidel method in matrix form, multiply both sides of Eq. (7.8) by a_{ii} and collect all kth iterate terms, to give

$$a_{i1}x_1^{(k)} + a_{i2}x_2^{(k)} + \cdots + a_{ii}x_i^{(k)} = -a_{i,i+1}x_{i+1}^{(k-1)} - \cdots - a_{in}x_n^{(k-1)} + b_i,$$

for each $i = 1, 2, \ldots, n$. Writing all n equations gives

$$\begin{aligned}
a_{11}x_1^{(k)} &= -a_{12}x_2^{(k-1)} - a_{13}x_3^{(k-1)} - \cdots - a_{1n}x_n^{(k-1)} + b_1,\\
a_{21}x_1^{(k)} + a_{22}x_2^{(k)} &= -a_{23}x_3^{(k-1)} - \cdots - a_{2n}x_n^{(k-1)} + b_2,\\
&\vdots\\
a_{n1}x_1^{(k)} + a_{n2}x_2^{(k)} + \cdots + a_{nn}x_n^{(k)} &= b_n;
\end{aligned}$$

with the definitions of D, L, and U given previously, we have the Gauss-Seidel method represented by

$$(D - L)\mathbf{x}^{(k)} = U\mathbf{x}^{(k-1)} + \mathbf{b}$$

and

$$\mathbf{x}^{(k)} = (D - L)^{-1}U\mathbf{x}^{(k-1)} + (D - L)^{-1}\mathbf{b}, \quad \text{for each } k = 1, 2, \ldots. \tag{7.9}$$

Letting $T_g = (D - L)^{-1}U$ and $\mathbf{c}_g = (D - L)^{-1}\mathbf{b}$, gives the Gauss-Seidel technique the form

$$\mathbf{x}^{(k)} = T_g\mathbf{x}^{(k-1)} + \mathbf{c}_g. \tag{7.10}$$

For the lower-triangular matrix $D - L$ to be nonsingular, it is necessary and sufficient that $a_{ii} \neq 0$, for each $i = 1, 2, \ldots, n$.

Algorithm 7.2 implements the Gauss-Seidel method.

Gauss-Seidel Iterative

To solve $A\mathbf{x} = \mathbf{b}$ given an initial approximation $\mathbf{x}^{(0)}$:

INPUT the number of equations and unknowns n; the entries a_{ij}, $1 \leq i, j \leq n$ of the matrix A; the entries b_i, $1 \leq i \leq n$ of $\mathbf{b}$; the entries XO_i, $1 \leq i \leq n$ of $\mathbf{XO} = \mathbf{x}^{(0)}$; tolerance TOL; maximum number of iterations N.

OUTPUT the approximate solution $x_1, \ldots, x_n$ or a message that the number of iterations was exceeded.

Step 1 Set $k = 1$.

Step 2 While $(k \leq N)$ do Steps 3–6.

Step 3 For $i = 1, \ldots, n$

$$\text{set } x_i = \frac{1}{a_{ii}}\left[-\sum_{j=1}^{i-1} a_{ij}x_j - \sum_{j=i+1}^{n} a_{ij}XO_j + b_i\right].$$

Step 4 If $||\mathbf{x} - \mathbf{XO}|| < TOL$ then OUTPUT $(x_1, \ldots, x_n)$;
(*The procedure was successful.*)
STOP.

Step 5 Set $k = k + 1$.

Step 6 For $i = 1, \ldots, n$ set $XO_i = x_i$.

Step 7 OUTPUT ('Maximum number of iterations exceeded');
(*The procedure was successful.*)
STOP. ■

The comments following Algorithm 7.1 regarding reordering and stopping criteria also apply to the Gauss-Seidel Algorithm 7.2.

The results of Examples 1 and 2 appear to imply that the Gauss-Seidel method is superior to the Jacobi method. This is almost always true, but there are linear systems for which the Jacobi method converges and the Gauss-Seidel method does not (see Exercises 9 and 10).

The *NumericalAnalysis* subpackage of the Maple *Student* package implements the Gauss-Siedel method in a manner similar to that of the Jacobi iterative method. The results in Table 7.2 are obtained by loading both *NumericalAnalysis* and *LinearAlgebra*, the matrix A, and then using the command

IterativeApproximate(A, initialapprox = *Vector*([0., 0., 0., 0.]), *tolerance* = 10^{-3}, *maxiterations* = 20, *stoppingcriterion* = *relative*(*infinity*), *method* = *gaussseidel*, *output* = *approximates*)

If we change the final option to *output* = [*approximates, distances*], the output also includes the l_∞ distances between the approximations and the actual solution.

General Iteration Methods

To study the convergence of general iteration techniques, we need to analyze the formula

$$\mathbf{x}^{(k)} = T\mathbf{x}^{(k-1)} + \mathbf{c}, \quad \text{for each } k = 1, 2, \ldots,$$

where $\mathbf{x}^{(0)}$ is arbitrary. The next lemma and Theorem 7.17 on page 449 provide the key for this study.

Lemma 7.18 If the spectral radius satisfies $\rho(T) < 1$, then $(I - T)^{-1}$ exists, and

$$(I - T)^{-1} = I + T + T^2 + \cdots = \sum_{j=0}^{\infty} T^j.$$ ■

Proof Because $T\mathbf{x} = \lambda\mathbf{x}$ is true precisely when $(I - T)\mathbf{x} = (1 - \lambda)\mathbf{x}$, we have λ as an eigenvalue of T precisely when $1 - \lambda$ is an eigenvalue of $I - T$. But $|\lambda| \le \rho(T) < 1$, so $\lambda = 1$ is not an eigenvalue of T, and 0 cannot be an eigenvalue of $I - T$. Hence, $(I - T)^{-1}$ exists.

Let $S_m = I + T + T^2 + \cdots + T^m$. Then

$$(I - T)S_m = (1 + T + T^2 + \cdots + T^m) - (T + T^2 + \cdots + T^{m+1}) = I - T^{m+1},$$

and, since T is convergent, Theorem 7.17 implies that

$$\lim_{m\to\infty} (I - T)S_m = \lim_{m\to\infty} (I - T^{m+1}) = I.$$

Thus, $(I - T)^{-1} = \lim_{m\to\infty} S_m = I + T + T^2 + \cdots = \sum_{j=0}^{\infty} T^j$. ■ ■ ■

Theorem 7.19 For any $\mathbf{x}^{(0)} \in \mathbb{R}^n$, the sequence $\{\mathbf{x}^{(k)}\}_{k=0}^{\infty}$ defined by

$$\mathbf{x}^{(k)} = T\mathbf{x}^{(k-1)} + \mathbf{c}, \quad \text{for each } k \ge 1, \tag{7.11}$$

converges to the unique solution of $\mathbf{x} = T\mathbf{x} + \mathbf{c}$ if and only if $\rho(T) < 1$. ■

Proof First assume that $\rho(T) < 1$. Then,

$$\begin{aligned}
\mathbf{x}^{(k)} &= T\mathbf{x}^{(k-1)} + \mathbf{c} \\
&= T(T\mathbf{x}^{(k-2)} + \mathbf{c}) + \mathbf{c} \\
&= T^2\mathbf{x}^{(k-2)} + (T + I)\mathbf{c} \\
&\ \ \vdots \\
&= T^k\mathbf{x}^{(0)} + (T^{k-1} + \cdots + T + I)\mathbf{c}.
\end{aligned}$$

Because $\rho(T) < 1$, Theorem 7.17 implies that T is convergent, and

$$\lim_{k\to\infty} T^k\mathbf{x}^{(0)} = \mathbf{0}.$$

Lemma 7.18 implies that

$$\lim_{k\to\infty} \mathbf{x}^{(k)} = \lim_{k\to\infty} T^k\mathbf{x}^{(0)} + \left(\sum_{j=0}^{\infty} T^j\right)\mathbf{c} = \mathbf{0} + (I-T)^{-1}\mathbf{c} = (I-T)^{-1}\mathbf{c}.$$

Hence, the sequence $\{\mathbf{x}^{(k)}\}$ converges to the vector $\mathbf{x} \equiv (I-T)^{-1}\mathbf{c}$ and $\mathbf{x} = T\mathbf{x} + \mathbf{c}$.

To prove the converse, we will show that for any $\mathbf{z} \in \mathbb{R}^n$, we have $\lim_{k\to\infty} T^k\mathbf{z} = \mathbf{0}$. By Theorem 7.17, this is equivalent to $\rho(T) < 1$.

Let $\mathbf{z}$ be an arbitrary vector, and $\mathbf{x}$ be the unique solution to $\mathbf{x} = T\mathbf{x} + \mathbf{c}$. Define $\mathbf{x}^{(0)} = \mathbf{x} - \mathbf{z}$, and, for $k \geq 1$, $\mathbf{x}^{(k)} = T\mathbf{x}^{(k-1)} + \mathbf{c}$. Then $\{\mathbf{x}^{(k)}\}$ converges to $\mathbf{x}$. Also,

$$\mathbf{x} - \mathbf{x}^{(k)} = (T\mathbf{x} + \mathbf{c}) - \left(T\mathbf{x}^{(k-1)} + \mathbf{c}\right) = T\left(\mathbf{x} - \mathbf{x}^{(k-1)}\right),$$

so

$$\mathbf{x} - \mathbf{x}^{(k)} = T\left(\mathbf{x} - \mathbf{x}^{(k-1)}\right) = T^2\left(\mathbf{x} - \mathbf{x}^{(k-2)}\right) = \cdots = T^k\left(\mathbf{x} - \mathbf{x}^{(0)}\right) = T^k\mathbf{z}.$$

Hence $\lim_{k\to\infty} T^k\mathbf{z} = \lim_{k\to\infty} T^k\left(\mathbf{x} - \mathbf{x}^{(0)}\right) = \lim_{k\to\infty}\left(\mathbf{x} - \mathbf{x}^{(k)}\right) = \mathbf{0}$.

But $\mathbf{z} \in \mathbb{R}^n$ was arbitrary, so by Theorem 7.17, T is convergent and $\rho(T) < 1$. ▪ ▪ ▪

The proof of the following corollary is similar to the proofs in Corollary 2.5 on page 62. It is considered in Exercise 13.

Corollary 7.20 If $\|T\| < 1$ for any natural matrix norm and $\mathbf{c}$ is a given vector, then the sequence $\{\mathbf{x}^{(k)}\}_{k=0}^{\infty}$ defined by $\mathbf{x}^{(k)} = T\mathbf{x}^{(k-1)} + \mathbf{c}$ converges, for any $\mathbf{x}^{(0)} \in \mathbb{R}^n$, to a vector $\mathbf{x} \in \mathbb{R}^n$, with $\mathbf{x} = T\mathbf{x} + \mathbf{c}$, and the following error bounds hold:

(i) $\|\mathbf{x} - \mathbf{x}^{(k)}\| \leq \|T\|^k\|\mathbf{x}^{(0)} - \mathbf{x}\|$; **(ii)** $\|\mathbf{x} - \mathbf{x}^{(k)}\| \leq \frac{\|T\|^k}{1-\|T\|}\|\mathbf{x}^{(1)} - \mathbf{x}^{(0)}\|$. ▪

We have seen that the Jacobi and Gauss-Seidel iterative techniques can be written

$$\mathbf{x}^{(k)} = T_j\mathbf{x}^{(k-1)} + \mathbf{c}_j \quad \text{and} \quad \mathbf{x}^{(k)} = T_g\mathbf{x}^{(k-1)} + \mathbf{c}_g,$$

using the matrices

$$T_j = D^{-1}(L+U) \quad \text{and} \quad T_g = (D-L)^{-1}U.$$

If $\rho(T_j)$ or $\rho(T_g)$ is less than 1, then the corresponding sequence $\{\mathbf{x}^{(k)}\}_{k=0}^{\infty}$ will converge to the solution $\mathbf{x}$ of $A\mathbf{x} = \mathbf{b}$. For example, the Jacobi scheme has

$$\mathbf{x}^{(k)} = D^{-1}(L+U)\mathbf{x}^{(k-1)} + D^{-1}\mathbf{b},$$

and, if $\{\mathbf{x}^{(k)}\}_{k=0}^{\infty}$ converges to $\mathbf{x}$, then

$$\mathbf{x} = D^{-1}(L+U)\mathbf{x} + D^{-1}\mathbf{b}.$$

This implies that

$$D\mathbf{x} = (L+U)\mathbf{x} + \mathbf{b} \quad \text{and} \quad (D-L-U)\mathbf{x} = \mathbf{b}.$$

Since $D - L - U = A$, the solution $\mathbf{x}$ satisfies $A\mathbf{x} = \mathbf{b}$.

We can now give easily verified sufficiency conditions for convergence of the Jacobi and Gauss-Seidel methods. (To prove convergence for the Jacobi scheme see Exercise 14, and for the Gauss-Seidel scheme see [Or2], p. 120.)

Theorem 7.21 If A is strictly diagonally dominant, then for any choice of $\mathbf{x}^{(0)}$, both the Jacobi and Gauss-Seidel methods give sequences $\{\mathbf{x}^{(k)}\}_{k=0}^{\infty}$ that converge to the unique solution of $A\mathbf{x} = \mathbf{b}$. ■

The relationship of the rapidity of convergence to the spectral radius of the iteration matrix T can be seen from Corollary 7.20. The inequalities hold for any natural matrix norm, so it follows from the statement after Theorem 7.15 on page 446 that

$$\|\mathbf{x}^{(k)} - \mathbf{x}\| \approx \rho(T)^k \|\mathbf{x}^{(0)} - \mathbf{x}\|. \tag{7.12}$$

Thus we would like to select the iterative technique with minimal $\rho(T) < 1$ for a particular system $A\mathbf{x} = \mathbf{b}$. No general results exist to tell which of the two techniques, Jacobi or Gauss-Seidel, will be most successful for an arbitrary linear system. In special cases, however, the answer is known, as is demonstrated in the following theorem. The proof of this result can be found in [Y], pp. 120–127.

Theorem 7.22 **(Stein-Rosenberg)**

If $a_{ij} \le 0$, for each $i \ne j$ and $a_{ii} > 0$, for each $i = 1, 2, \ldots, n$, then one and only one of the following statements holds:

(i) $0 \le \rho(T_g) < \rho(T_j) < 1$; **(ii)** $1 < \rho(T_j) < \rho(T_g)$;
(iii) $\rho(T_j) = \rho(T_g) = 0$; **(iv)** $\rho(T_j) = \rho(T_g) = 1$. ■

For the special case described in Theorem 7.22, we see from part (i) that when one method gives convergence, then both give convergence, and the Gauss-Seidel method converges faster than the Jacobi method. Part (ii) indicates that when one method diverges then both diverge, and the divergence is more pronounced for the Gauss-Seidel method.

EXERCISE SET 7.3

1. Find the first two iterations of the Jacobi method for the following linear systems, using $\mathbf{x}^{(0)} = \mathbf{0}$:

a.
$$\begin{aligned} 4x_1 + x_2 - x_3 &= 5,\\ -x_1 + 3x_2 + x_3 &= -4,\\ 2x_1 + 2x_2 + 5x_3 &= 1. \end{aligned}$$

b.
$$\begin{aligned} -2x_1 + x_2 + \tfrac{1}{2}x_3 &= 4,\\ x_1 - 2x_2 - \tfrac{1}{2}x_3 &= -4,\\ x_2 + 2x_3 &= 0. \end{aligned}$$

c.
$$\begin{aligned} 4x_1 + x_2 - x_3 + x_4 &= -2,\\ x_1 + 4x_2 - x_3 - x_4 &= -1,\\ -x_1 - x_2 + 5x_3 + x_4 &= 0,\\ x_1 - x_2 + x_3 + 3x_4 &= 1. \end{aligned}$$

d.
$$\begin{aligned} 4x_1 - x_2 - x_4 &= 0,\\ -x_1 + 4x_2 - x_3 - x_5 &= 5,\\ -x_2 + 4x_3 - x_6 &= 0,\\ -x_1 + 4x_4 - x_5 &= 6,\\ -x_2 - x_4 + 4x_5 - x_6 &= -2,\\ -x_3 - x_5 + 4x_6 &= 6. \end{aligned}$$

2. Find the first two iterations of the Jacobi method for the following linear systems, using $\mathbf{x}^{(0)} = \mathbf{0}$:

a.
$$\begin{aligned} 3x_1 - x_2 + x_3 &= 1,\\ 3x_1 + 6x_2 + 2x_3 &= 0,\\ 3x_1 + 3x_2 + 7x_3 &= 4. \end{aligned}$$

b.
$$\begin{aligned} 10x_1 - x_2 &= 9,\\ -x_1 + 10x_2 - 2x_3 &= 7,\\ -2x_2 + 10x_3 &= 6. \end{aligned}$$

c.
$$\begin{aligned} 10x_1 + 5x_2 &= 6,\\ 5x_1 + 10x_2 - 4x_3 &= 25,\\ -4x_2 + 8x_3 - x_4 &= -11,\\ -x_3 + 5x_4 &= -11. \end{aligned}$$

d.
$$\begin{aligned} 4x_1 + x_2 + x_3 + x_5 &= 6,\\ -x_1 - 3x_2 + x_3 + x_4 &= 6,\\ 2x_1 + x_2 + 5x_3 - x_4 - x_5 &= 6,\\ -x_1 - x_2 - x_3 + 4x_4 &= 6,\\ 2x_2 - x_3 + x_4 + 4x_5 &= 6. \end{aligned}$$

3. Repeat Exercise 1 using the Gauss-Seidel method.

4. Repeat Exercise 2 using the Gauss-Seidel method.

5. Use the Jacobi method to solve the linear systems in Exercise 1, with $TOL = 10^{-3}$ in the l_∞ norm.

6. Use the Jacobi method to solve the linear systems in Exercise 2, with $TOL = 10^{-3}$ in the l_∞ norm.

7. Use the Gauss-Seidel method to solve the linear systems in Exercise 1, with $TOL = 10^{-3}$ in the l_∞ norm.

8. Use the Gauss-Seidel method to solve the linear systems in Exercise 2, with $TOL = 10^{-3}$ in the l_∞ norm.

9. The linear system

$$\begin{aligned} 2x_1 - x_2 + x_3 &= -1, \\ 2x_1 + 2x_2 + 2x_3 &= 4, \\ -x_1 - x_2 + 2x_3 &= -5 \end{aligned}$$

has the solution $(1, 2, -1)^t$.

a. Show that $\rho(T_j) = \frac{\sqrt{5}}{2} > 1$.

b. Show that the Jacobi method with $\mathbf{x}^{(0)} = \mathbf{0}$ fails to give a good approximation after 25 iterations.

c. Show that $\rho(T_g) = \frac{1}{2}$.

d. Use the Gauss-Seidel method with $\mathbf{x}^{(0)} = \mathbf{0}$ to approximate the solution to the linear system to within 10^{-5} in the l_∞ norm.

10. The linear system

$$\begin{aligned} x_1 + 2x_2 - 2x_3 &= 7, \\ x_1 + x_2 + x_3 &= 2, \\ 2x_1 + 2x_2 + x_3 &= 5 \end{aligned}$$

has the solution $(1, 2, -1)^t$.

a. Show that $\rho(T_j) = 0$.

b. Use the Jacobi method with $\mathbf{x}^{(0)} = \mathbf{0}$ to approximate the solution to the linear system to within 10^{-5} in the l_∞ norm.

c. Show that $\rho(T_g) = 2$.

d. Show that the Gauss-Seidel method applied as in part (b) fails to give a good approximation in 25 iterations.

11. The linear system

$$\begin{aligned} x_1 \qquad - x_3 &= 0.2, \\ -\frac{1}{2}x_1 + x_2 - \frac{1}{4}x_3 &= -1.425, \\ x_1 - \frac{1}{2}x_2 + x_3 &= 2. \end{aligned}$$

has the solution $(0.9, -0.8, 0.7)^t$.

a. Is the coefficient matrix

$$A = \begin{bmatrix} 1 & 0 & -1 \\ -\frac{1}{2} & 1 & -\frac{1}{4} \\ 1 & -\frac{1}{2} & 1 \end{bmatrix}$$

strictly diagonally dominant?

b. Compute the spectral radius of the Gauss-Seidel matrix T_g.

c. Use the Gauss-Seidel iterative method to approximate the solution to the linear system with a tolerance of 10^{-2} and a maximum of 300 iterations.

d. What happens in part (c) when the system is changed to

$$\begin{aligned} x_1 \qquad - 2x_3 &= 0.2, \\ -\frac{1}{2}x_1 + x_2 - \frac{1}{4}x_3 &= -1.425, \\ x_1 - \frac{1}{2}x_2 + x_3 &= 2. \end{aligned}$$

12. Repeat Exercise 11 using the Jacobi method.

13. **a.** Prove that

$$\|\mathbf{x}^{(k)} - \mathbf{x}\| \leq \|T\|^k \|\mathbf{x}^{(0)} - \mathbf{x}\| \quad \text{and} \quad \|\mathbf{x}^{(k)} - \mathbf{x}\| \leq \frac{\|T\|^k}{1 - \|T\|}\|\mathbf{x}^{(1)} - \mathbf{x}^{(0)}\|,$$

where T is an $n \times n$ matrix with $\|T\| < 1$ and

$$\mathbf{x}^{(k)} = T\mathbf{x}^{(k-1)} + \mathbf{c}, \quad k = 1, 2, \ldots,$$

with $\mathbf{x}^{(0)}$ arbitrary, $\mathbf{c} \in \mathbb{R}^n$, and $\mathbf{x} = T\mathbf{x} + \mathbf{c}$.

b. Apply the bounds to Exercise 1, when possible, using the l_∞ norm.

14. Show that if A is strictly diagonally dominant, then $||T_j||_\infty < 1$.

15. Use **(a)** the Jacobi and **(b)** the Gauss-Seidel methods to solve the linear system $A\mathbf{x} = \mathbf{b}$ to within 10^{-5} in the l_∞ norm, where the entries of A are

$$a_{i,j} = \begin{cases} 2i, & \text{when } j = i \text{ and } i = 1, 2, \ldots, 80, \\ 0.5i, & \text{when } \begin{cases} j = i + 2 \text{ and } i = 1, 2, \ldots, 78, \\ j = i - 2 \text{ and } i = 3, 4, \ldots, 80, \end{cases} \\ 0.25i, & \text{when } \begin{cases} j = i + 4 \text{ and } i = 1, 2, \ldots, 76, \\ j = i - 4 \text{ and } i = 5, 6, \ldots, 80, \end{cases} \\ 0, & \text{otherwise,} \end{cases}$$

and those of $\mathbf{b}$ are $b_i = \pi$, for each $i = 1, 2, \ldots, 80$.

16. Suppose that an object can be at any one of $n+1$ equally spaced points $x_0, x_1, \ldots, x_n$. When an object is at location x_i, it is equally likely to move to either x_{i-1} or x_{i+1} and cannot directly move to any other location. Consider the probabilities $\{P_i\}_{i=0}^n$ that an object starting at location x_i will reach the left endpoint x_0 before reaching the right endpoint x_n. Clearly, $P_0 = 1$ and $P_n = 0$. Since the object can move to x_i only from x_{i-1} or x_{i+1} and does so with probability $\frac{1}{2}$ for each of these locations,

$$P_i = \frac{1}{2}P_{i-1} + \frac{1}{2}P_{i+1}, \quad \text{for each } i = 1, 2, \ldots, n-1.$$

a. Show that

$$\begin{bmatrix} 1 & -\frac{1}{2} & 0 & \cdots & \cdots & 0 \\ -\frac{1}{2} & 1 & -\frac{1}{2} & \ddots & & \vdots \\ 0 & -\frac{1}{2} & 1 & \ddots & \ddots & \vdots \\ \vdots & \ddots & \ddots & \ddots & \ddots & 0 \\ \vdots & & \ddots & -\frac{1}{2} & 1 & -\frac{1}{2} \\ 0 & \cdots & \cdots & 0 & -\frac{1}{2} & 1 \end{bmatrix} \begin{bmatrix} P_1 \\ P_2 \\ \vdots \\ P_{n-1} \end{bmatrix} = \begin{bmatrix} \frac{1}{2} \\ 0 \\ \vdots \\ 0 \end{bmatrix}.$$

b. Solve this system using $n = 10$, 50, and 100.

c. Change the probabilities to α and $1 - \alpha$ for movement to the left and right, respectively, and derive the linear system similar to the one in part (a).

d. Repeat part (b) with $\alpha = \frac{1}{3}$.

17. Suppose that A is a positive definite.

a. Show that we can write $A = D - L - L^t$, where D is diagonal with $d_{ii} > 0$ for each $1 \leq i \leq n$ and L is lower triangular. Further, show that $D - L$ is nonsingular.

b. Let $T_g = (D - L)^{-1}L^t$ and $P = A - T_g^t A T_g$. Show that P is symmetric.

c. Show that T_g can also be written as $T_g = I - (D - L)^{-1}A$.

d. Let $Q = (D - L)^{-1}A$. Show that $T_g = I - Q$ and $P = Q^t[AQ^{-1} - A + (Q^t)^{-1}A]Q$.

e. Show that $P = Q^tDQ$ and P is positive definite.

f. Let λ be an eigenvalue of T_g with eigenvector $\mathbf{x} \neq \mathbf{0}$. Use part (b) to show that $\mathbf{x}^tP\mathbf{x} > 0$ implies that $|\lambda| < 1$.

g. Show that T_g is convergent and prove that the Gauss-Seidel method converges.

18. The forces on the bridge truss described in the opening to this chapter satisfy the equations in the following table:

Joint	Horizontal Component	Vertical Component
①	$-F_1 + \frac{\sqrt{2}}{2}f_1 + f_2 = 0$	$\frac{\sqrt{2}}{2}f_1 - F_2 = 0$
②	$-\frac{\sqrt{2}}{2}f_1 + \frac{\sqrt{3}}{2}f_4 = 0$	$-\frac{\sqrt{2}}{2}f_1 - f_3 - \frac{1}{2}f_4 = 0$
③	$-f_2 + f_5 = 0$	$f_3 - 10{,}000 = 0$
④	$-\frac{\sqrt{3}}{2}f_4 - f_5 = 0$	$\frac{1}{2}f_4 - F_3 = 0$

This linear system can be placed in the matrix form

$$\begin{bmatrix} -1 & 0 & 0 & \frac{\sqrt{2}}{2} & 1 & 0 & 0 & 0 \\ 0 & -1 & 0 & \frac{\sqrt{2}}{2} & 0 & 0 & 0 & 0 \\ 0 & 0 & -1 & 0 & 0 & 0 & \frac{1}{2} & 0 \\ 0 & 0 & 0 & -\frac{\sqrt{2}}{2} & 0 & -1 & -\frac{1}{2} & 0 \\ 0 & 0 & 0 & 0 & -1 & 0 & 0 & 1 \\ 0 & 0 & 0 & 0 & 0 & 1 & 0 & 0 \\ 0 & 0 & 0 & -\frac{\sqrt{2}}{2} & 0 & 0 & \frac{\sqrt{3}}{2} & 0 \\ 0 & 0 & 0 & 0 & 0 & 0 & -\frac{\sqrt{3}}{2} & -1 \end{bmatrix} \begin{bmatrix} F_1 \\ F_2 \\ F_3 \\ f_1 \\ f_2 \\ f_3 \\ f_4 \\ f_5 \end{bmatrix} = \begin{bmatrix} 0 \\ 0 \\ 0 \\ 0 \\ 0 \\ 10{,}000 \\ 0 \\ 0 \end{bmatrix}.$$

a. Explain why the system of equations was reordered.

b. Approximate the solution of the resulting linear system to within 10^{-2} in the l_∞ norm using as initial approximation the vector all of whose entries are 1s with **(i)** the Jacobi method and **(ii)** the Gauss-Seidel method.

7.4 Relaxation Techniques for Solving Linear Systems

We saw in Section 7.3 that the rate of convergence of an iterative technique depends on the spectral radius of the matrix associated with the method. One way to select a procedure to accelerate convergence is to choose a method whose associated matrix has minimal spectral radius. Before describing a procedure for selecting such a method, we need to introduce a new means of measuring the amount by which an approximation to the solution to a linear system differs from the true solution to the system. The method makes use of the vector described in the following definition.

Definition 7.23 Suppose $\tilde{\mathbf{x}} \in \mathbb{R}^n$ is an approximation to the solution of the linear system defined by $A\mathbf{x} = \mathbf{b}$. The **residual vector** for $\tilde{\mathbf{x}}$ with respect to this system is $\mathbf{r} = \mathbf{b} - A\tilde{\mathbf{x}}$. ■

The word residual means what is left over, which is an appropriate name for this vector.

In procedures such as the Jacobi or Gauss-Seidel methods, a residual vector is associated with each calculation of an approximate component to the solution vector. The true objective is to generate a sequence of approximations that will cause the residual vectors to converge rapidly to zero. Suppose we let

$$\mathbf{r}_i^{(k)} = (r_{1i}^{(k)}, r_{2i}^{(k)}, \ldots, r_{ni}^{(k)})^t$$

denote the residual vector for the Gauss-Seidel method corresponding to the approximate solution vector $\mathbf{x}_i^{(k)}$ defined by

$$\mathbf{x}_i^{(k)} = (x_1^{(k)}, x_2^{(k)}, \ldots, x_{i-1}^{(k)}, x_i^{(k-1)}, \ldots, x_n^{(k-1)})^t.$$

The mth component of $\mathbf{r}_i^{(k)}$ is

$$r_{mi}^{(k)} = b_m - \sum_{j=1}^{i-1} a_{mj} x_j^{(k)} - \sum_{j=i}^{n} a_{mj} x_j^{(k-1)}, \tag{7.13}$$

or, equivalently,

$$r_{mi}^{(k)} = b_m - \sum_{j=1}^{i-1} a_{mj} x_j^{(k)} - \sum_{j=i+1}^{n} a_{mj} x_j^{(k-1)} - a_{mi} x_i^{(k-1)},$$

for each $m = 1, 2, \ldots, n$.

In particular, the ith component of $\mathbf{r}_i^{(k)}$ is

$$r_{ii}^{(k)} = b_i - \sum_{j=1}^{i-1} a_{ij} x_j^{(k)} - \sum_{j=i+1}^{n} a_{ij} x_j^{(k-1)} - a_{ii} x_i^{(k-1)},$$

so

$$a_{ii} x_i^{(k-1)} + r_{ii}^{(k)} = b_i - \sum_{j=1}^{i-1} a_{ij} x_j^{(k)} - \sum_{j=i+1}^{n} a_{ij} x_j^{(k-1)}. \tag{7.14}$$

Recall, however, that in the Gauss-Seidel method, $x_i^{(k)}$ is chosen to be

$$x_i^{(k)} = \frac{1}{a_{ii}} \left[b_i - \sum_{j=1}^{i-1} a_{ij} x_j^{(k)} - \sum_{j=i+1}^{n} a_{ij} x_j^{(k-1)} \right], \tag{7.15}$$

so Eq. (7.14) can be rewritten as

$$a_{ii} x_i^{(k-1)} + r_{ii}^{(k)} = a_{ii} x_i^{(k)}.$$

Consequently, the Gauss-Seidel method can be characterized as choosing $x_i^{(k)}$ to satisfy

$$x_i^{(k)} = x_i^{(k-1)} + \frac{r_{ii}^{(k)}}{a_{ii}}. \tag{7.16}$$

We can derive another connection between the residual vectors and the Gauss-Seidel technique. Consider the residual vector $\mathbf{r}_{i+1}^{(k)}$, associated with the vector $\mathbf{x}_{i+1}^{(k)} = (x_1^{(k)}, \ldots, x_i^{(k)}, x_{i+1}^{(k-1)}, \ldots, x_n^{(k-1)})^t$. By Eq. (7.13) the ith component of $\mathbf{r}_{i+1}^{(k)}$ is

$$\begin{aligned} r_{i,i+1}^{(k)} &= b_i - \sum_{j=1}^{i} a_{ij} x_j^{(k)} - \sum_{j=i+1}^{n} a_{ij} x_j^{(k-1)} \\ &= b_i - \sum_{j=1}^{i-1} a_{ij} x_j^{(k)} - \sum_{j=i+1}^{n} a_{ij} x_j^{(k-1)} - a_{ii} x_i^{(k)}. \end{aligned}$$

By the manner in which $x_i^{(k)}$ is defined in Eq. (7.15) we see that $r_{i,i+1}^{(k)} = 0$. In a sense, then, the Gauss-Seidel technique is characterized by choosing each $x_{i+1}^{(k)}$ in such a way that the ith component of $\mathbf{r}_{i+1}^{(k)}$ is zero.

Choosing $x_{i+1}^{(k)}$ so that one coordinate of the residual vector is zero, however, is not necessarily the most efficient way to reduce the norm of the vector $\mathbf{r}_{i+1}^{(k)}$. If we modify the Gauss-Seidel procedure, as given by Eq. (7.16), to

$$x_i^{(k)} = x_i^{(k-1)} + \omega \frac{r_{ii}^{(k)}}{a_{ii}}, \tag{7.17}$$

then for certain choices of positive ω we can reduce the norm of the residual vector and obtain significantly faster convergence.

Methods involving Eq. (7.17) are called **relaxation methods**. For choices of ω with $0 < \omega < 1$, the procedures are called **under-relaxation methods**. We will be interested in choices of ω with $1 < \omega$, and these are called **over-relaxation methods**. They are used to accelerate the convergence for systems that are convergent by the Gauss-Seidel technique. The methods are abbreviated **SOR**, for **Successive Over-Relaxation**, and are particularly useful for solving the linear systems that occur in the numerical solution of certain partial-differential equations.

Before illustrating the advantages of the SOR method, we note that by using Eq. (7.14), we can reformulate Eq. (7.17) for calculation purposes as

$$x_i^{(k)} = (1-\omega)x_i^{(k-1)} + \frac{\omega}{a_{ii}}\left[b_i - \sum_{j=1}^{i-1} a_{ij}x_j^{(k)} - \sum_{j=i+1}^{n} a_{ij}x_j^{(k-1)}\right].$$

To determine the matrix form of the SOR method, we rewrite this as

$$a_{ii}x_i^{(k)} + \omega\sum_{j=1}^{i-1} a_{ij}x_j^{(k)} = (1-\omega)a_{ii}x_i^{(k-1)} - \omega\sum_{j=i+1}^{n} a_{ij}x_j^{(k-1)} + \omega b_i,$$

so that in vector form, we have

$$(D-\omega L)\mathbf{x}^{(k)} = [(1-\omega)D + \omega U]\mathbf{x}^{(k-1)} + \omega\mathbf{b}.$$

That is,

$$\mathbf{x}^{(k)} = (D-\omega L)^{-1}[(1-\omega)D + \omega U]\mathbf{x}^{(k-1)} + \omega(D-\omega L)^{-1}\mathbf{b}. \tag{7.18}$$

Letting $T_\omega = (D-\omega L)^{-1}[(1-\omega)D + \omega U]$ and $\mathbf{c}_\omega = \omega(D-\omega L)^{-1}\mathbf{b}$, gives the SOR technique the form

$$\mathbf{x}^{(k)} = T_\omega\mathbf{x}^{(k-1)} + \mathbf{c}_\omega. \tag{7.19}$$

Example 1 The linear system $A\mathbf{x} = \mathbf{b}$ given by

$$\begin{aligned} 4x_1 + 3x_2 \quad\quad &= 24,\\ 3x_1 + 4x_2 - \ x_3 &= 30,\\ -\ x_2 + 4x_3 &= -24, \end{aligned}$$

has the solution $(3, 4, -5)^t$. Compare the iterations from the Gauss-Seidel method and the SOR method with $\omega = 1.25$ using $\mathbf{x}^{(0)} = (1, 1, 1)^t$ for both methods.

Solution For each $k = 1, 2, \ldots$, the equations for the Gauss-Seidel method are

$$x_1^{(k)} = -0.75x_2^{(k-1)} + 6,$$
$$x_2^{(k)} = -0.75x_1^{(k)} + 0.25x_3^{(k-1)} + 7.5,$$
$$x_3^{(k)} = 0.25x_2^{(k)} - 6,$$

and the equations for the SOR method with $\omega = 1.25$ are

$$x_1^{(k)} = -0.25x_1^{(k-1)} - 0.9375x_2^{(k-1)} + 7.5,$$
$$x_2^{(k)} = -0.9375x_1^{(k)} - 0.25x_2^{(k-1)} + 0.3125x_3^{(k-1)} + 9.375,$$
$$x_3^{(k)} = 0.3125x_2^{(k)} - 0.25x_3^{(k-1)} - 7.5.$$

The first seven iterates for each method are listed in Tables 7.3 and 7.4. For the iterates to be accurate to seven decimal places, the Gauss-Seidel method requires 34 iterations, as opposed to 14 iterations for the SOR method with $\omega = 1.25$. ■

Table 7.3

k	0	1	2	3	4	5	6	7
$x_1^{(k)}$	1	5.250000	3.1406250	3.0878906	3.0549316	3.0343323	3.0214577	3.0134110
$x_2^{(k)}$	1	3.812500	3.8828125	3.9267578	3.9542236	3.9713898	3.9821186	3.9888241
$x_3^{(k)}$	1	−5.046875	−5.0292969	−5.0183105	−5.0114441	−5.0071526	−5.0044703	−5.0027940

Table 7.4

k	0	1	2	3	4	5	6	7
$x_1^{(k)}$	1	6.312500	2.6223145	3.1333027	2.9570512	3.0037211	2.9963276	3.0000498
$x_2^{(k)}$	1	3.5195313	3.9585266	4.0102646	4.0074838	4.0029250	4.0009262	4.0002586
$x_3^{(k)}$	1	−6.6501465	−4.6004238	−5.0966863	−4.9734897	−5.0057135	−4.9982822	−5.0003486

An obvious question to ask is how the appropriate value of ω is chosen when the SOR method is used. Although no complete answer to this question is known for the general $n \times n$ linear system, the following results can be used in certain important situations.

Theorem 7.24 **(Kahan)**

If $a_{ii} \neq 0$, for each $i = 1, 2, \ldots, n$, then $\rho(T_\omega) \geq |\omega - 1|$. This implies that the SOR method can converge only if $0 < \omega < 2$. ■

The proof of this theorem is considered in Exercise 9. The proof of the next two results can be found in [Or2], pp. 123–133. These results will be used in Chapter 12.

Theorem 7.25 **(Ostrowski-Reich)**

If A is a positive definite matrix and $0 < \omega < 2$, then the SOR method converges for any choice of initial approximate vector $\mathbf{x}^{(0)}$. ■

Theorem 7.26 If A is positive definite and tridiagonal, then $\rho(T_g) = [\rho(T_j)]^2 < 1$, and the optimal choice of ω for the SOR method is

$$\omega = \frac{2}{1 + \sqrt{1 - [\rho(T_j)]^2}}.$$

With this choice of ω, we have $\rho(T_\omega) = \omega - 1$. ▪

Example 2 Find the optimal choice of ω for the SOR method for the matrix

$$A = \begin{bmatrix} 4 & 3 & 0 \\ 3 & 4 & -1 \\ 0 & -1 & 4 \end{bmatrix}.$$

Solution This matrix is clearly tridiagonal, so we can apply the result in Theorem 7.26 if we can also who that it is positive definite. Because the matrix is symmetric, Theorem 6.24 on page 416 states that it is positive definite if and only if all its leading principle submatrices has a positive determinant. This is easily seen to be the case because

$$\det(A) = 24, \quad \det\left(\begin{bmatrix} 4 & 3 \\ 3 & 4 \end{bmatrix}\right) = 7, \quad \text{and} \quad \det([4]) = 4.$$

Because

$$T_j = D^{-1}(L+U) = \begin{bmatrix} \frac{1}{4} & 0 & 0 \\ 0 & \frac{1}{4} & 0 \\ 0 & 0 & \frac{1}{4} \end{bmatrix} \begin{bmatrix} 0 & -3 & 0 \\ -3 & 0 & 1 \\ 0 & 1 & 0 \end{bmatrix} = \begin{bmatrix} 0 & -0.75 & 0 \\ -0.75 & 0 & 0.25 \\ 0 & 0.25 & 0 \end{bmatrix},$$

we have

$$T_j - \lambda I = \begin{bmatrix} -\lambda & -0.75 & 0 \\ -0.75 & -\lambda & 0.25 \\ 0 & 0.25 & -\lambda \end{bmatrix},$$

so

$$\det(T_j - \lambda I) = -\lambda(\lambda^2 - 0.625).$$

Thus

$$\rho(T_j) = \sqrt{0.625}$$

and

$$\omega = \frac{2}{1 + \sqrt{1 - [\rho(T_j)]^2}} = \frac{2}{1 + \sqrt{1 - 0.625}} \approx 1.24.$$

This explains the rapid convergence obtained in Example 1 when using $\omega = 1.25$. ▪

We close this section with Algorithm 7.3 for the SOR method.

SOR

To solve $A\mathbf{x} = \mathbf{b}$ given the parameter ω and an initial approximation $\mathbf{x}^{(0)}$:

INPUT the number of equations and unknowns n; the entries a_{ij}, $1 \le i, j \le n$, of the matrix A; the entries b_i, $1 \le i \le n$, of $\mathbf{b}$; the entries XO_i, $1 \le i \le n$, of $\mathbf{XO} = \mathbf{x}^{(0)}$; the parameter ω; tolerance TOL; maximum number of iterations N.

OUTPUT the approximate solution $x_1, \ldots, x_n$ or a message that the number of iterations was exceeded.

Step 1 Set $k = 1$.

Step 2 While $(k \le N)$ do Steps 3–6.

Step 3 For $i = 1, \ldots, n$

$$\text{set } x_i = (1-\omega)XO_i + \frac{1}{a_{ii}}\left[\omega\left(-\sum_{j=1}^{i-1} a_{ij}x_j - \sum_{j=i+1}^{n} a_{ij}XO_j + b_i\right)\right].$$

Step 4 If $||\mathbf{x} - \mathbf{XO}|| < TOL$ then OUTPUT $(x_1, \ldots, x_n)$;
(*The procedure was successful.*)
STOP.

Step 5 Set $k = k + 1$.

Step 6 For $i = 1, \ldots, n$ set $XO_i = x_i$.

Step 7 OUTPUT ('Maximum number of iterations exceeded');
(*The procedure was successful.*)
STOP. ■

The *NumericalAnalysis* subpackage of the Maple *Student* package implements the SOR method in a manner similar to that of the Jacobi and Gauss-Seidel methods. The SOR results in Table 7.4 are obtained by loading both *NumericalAnalysis* and *LinearAlgebra*, the matrix A, the vector $\mathbf{b} = [24, 30, -24]^t$, and then using the command

IterativeApproximate(A, $\mathbf{b}$, *initialapprox* = *Vector*([1., 1., 1., 1.]), *tolerance* = 10^{-3}, *maxiterations* = 20, *stoppingcriterion* = *relative*(*infinity*), *method* = *SOR*(1.25), *output* = *approximates*)

The input *method* = *SOR*(1.25) indicates that the SOR method should use the value $\omega = 1.25$.

EXERCISE SET 7.4

1. Find the first two iterations of the SOR method with $\omega = 1.1$ for the following linear systems, using $\mathbf{x}^{(0)} = \mathbf{0}$:

a.
$$\begin{aligned} 4x_1 + x_2 - x_3 &= 5,\\ -x_1 + 3x_2 + x_3 &= -4,\\ 2x_1 + 2x_2 + 5x_3 &= 1. \end{aligned}$$

b.
$$\begin{aligned} -2x_1 + x_2 + \tfrac{1}{2}x_3 &= 4,\\ x_1 - 2x_2 - \tfrac{1}{2}x_3 &= -4,\\ x_2 + 2x_3 &= 0. \end{aligned}$$

c.
$$\begin{aligned} 4x_1 + x_2 - x_3 + x_4 &= -2,\\ x_1 + 4x_2 - x_3 - x_4 &= -1,\\ -x_1 - x_2 + 5x_3 + x_4 &= 0,\\ x_1 - x_2 + x_3 + 3x_4 &= 1. \end{aligned}$$

d.
$$\begin{aligned} 4x_1 - x_2 &= 0,\\ -x_1 + 4x_2 - x_3 &= 5,\\ - x_2 + 4x_3 &= 0,\\ + 4x_4 - x_5 &= 6,\\ - x_4 + 4x_5 - x_6 &= -2,\\ - x_5 + 4x_6 &= 6. \end{aligned}$$

2. Find the first two iterations of the SOR method with $\omega = 1.1$ for the following linear systems, using $\mathbf{x}^{(0)} = \mathbf{0}$:

a.
$$\begin{aligned} 3x_1 - x_2 + x_3 &= 1, \\ 3x_1 + 6x_2 + 2x_3 &= 0, \\ 3x_1 + 3x_2 + 7x_3 &= 4. \end{aligned}$$

b.
$$\begin{aligned} 10x_1 - x_2 &= 9, \\ -x_1 + 10x_2 - 2x_3 &= 7, \\ - 2x_2 + 10x_3 &= 6. \end{aligned}$$

c.
$$\begin{aligned} 10x_1 + 5x_2 &= 6, \\ 5x_1 + 10x_2 - 4x_3 &= 25, \\ - 4x_2 + 8x_3 - x_4 &= -11, \\ - x_3 + 5x_4 &= -11. \end{aligned}$$

d.
$$\begin{aligned} 4x_1 + x_2 + x_3 + x_5 &= 6, \\ -x_1 - 3x_2 + x_3 + x_4 &= 6, \\ 2x_1 + x_2 + 5x_3 - x_4 - x_5 &= 6, \\ -x_1 - x_2 - x_3 + 4x_4 &= 6, \\ 2x_2 - x_3 + x_4 + 4x_5 &= 6. \end{aligned}$$

3. Repeat Exercise 1 using $\omega = 1.3$.

4. Repeat Exercise 2 using $\omega = 1.3$.

5. Use the SOR method with $\omega = 1.2$ to solve the linear systems in Exercise 1 with a tolerance $TOL = 10^{-3}$ in the l_∞ norm.

6. Use the SOR method with $\omega = 1.2$ to solve the linear systems in Exercise 2 with a tolerance $TOL = 10^{-3}$ in the l_∞ norm.

7. Determine which matrices in Exercise 1 are tridiagonal and positive definite. Repeat Exercise 1 for these matrices using the optimal choice of ω.

8. Determine which matrices in Exercise 2 are tridiagonal and positive definite. Repeat Exercise 2 for these matrices using the optimal choice of ω.

9. Prove Kahan's Theorem 7.24. [*Hint:* If $\lambda_1, \ldots, \lambda_n$ are eigenvalues of T_ω, then $\det T_\omega = \prod_{i=1}^{n} \lambda_i$. Since $\det D^{-1} = \det(D - \omega L)^{-1}$ and the determinant of a product of matrices is the product of the determinants of the factors, the result follows from Eq. (7.18).]

10. The forces on the bridge truss described in the opening to this chapter satisfy the equations in the following table:

Joint	Horizontal Component	Vertical Component
①	$-F_1 + \frac{\sqrt{2}}{2} f_1 + f_2 = 0$	$\frac{\sqrt{2}}{2} f_1 - F_2 = 0$
②	$-\frac{\sqrt{2}}{2} f_1 + \frac{\sqrt{3}}{2} f_4 = 0$	$-\frac{\sqrt{2}}{2} f_1 - f_3 - \frac{1}{2} f_4 = 0$
③	$-f_2 + f_5 = 0$	$f_3 - 10{,}000 = 0$
④	$-\frac{\sqrt{3}}{2} f_4 - f_5 = 0$	$\frac{1}{2} f_4 - F_3 = 0$

This linear system can be placed in the matrix form

$$\begin{bmatrix} -1 & 0 & 0 & \frac{\sqrt{2}}{2} & 1 & 0 & 0 & 0 \\ 0 & -1 & 0 & \frac{\sqrt{2}}{2} & 0 & 0 & 0 & 0 \\ 0 & 0 & -1 & 0 & 0 & 0 & \frac{1}{2} & 0 \\ 0 & 0 & 0 & -\frac{\sqrt{2}}{2} & 0 & -1 & -\frac{1}{2} & 0 \\ 0 & 0 & 0 & 0 & -1 & 0 & 0 & 1 \\ 0 & 0 & 0 & 0 & 0 & 1 & 0 & 0 \\ 0 & 0 & 0 & -\frac{\sqrt{2}}{2} & 0 & 0 & \frac{\sqrt{3}}{2} & 0 \\ 0 & 0 & 0 & 0 & 0 & 0 & -\frac{\sqrt{3}}{2} & -1 \end{bmatrix} \begin{bmatrix} F_1 \\ F_2 \\ F_3 \\ f_1 \\ f_2 \\ f_3 \\ f_4 \\ f_5 \end{bmatrix} = \begin{bmatrix} 0 \\ 0 \\ 0 \\ 0 \\ 0 \\ 10{,}000 \\ 0 \\ 0 \end{bmatrix}.$$

a. Explain why the system of equations was reordered.

b. Approximate the solution of the resulting linear system to within 10^{-2} in the l_∞ norm using as initial approximation the vector all of whose entries are 1s and the SOR method with $\omega = 1.25$.

11. Use the SOR method to solve the linear system $A\mathbf{x} = \mathbf{b}$ to within 10^{-5} in the l_∞ norm, where the entries of A are

$$a_{i,j} = \begin{cases} 2i, & \text{when } j = i \text{ and } i = 1, 2, \ldots, 80, \\ 0.5i, & \text{when } \begin{cases} j = i + 2 \text{ and } i = 1, 2, \ldots, 78, \\ j = i - 2 \text{ and } i = 3, 4, \ldots, 80, \end{cases} \\ 0.25i, & \text{when } \begin{cases} j = i + 4 \text{ and } i = 1, 2, \ldots, 76, \\ j = i - 4 \text{ and } i = 5, 6, \ldots, 80, \end{cases} \\ 0, & \text{otherwise,} \end{cases}$$

and those of $\mathbf{b}$ are $b_i = \pi$, for each $i = 1, 2, \ldots, 80$.

12. In Exercise 17 of Section 7.3 a technique was outlined to prove that the Gauss-Seidel method converges when A is a positive definite matrix. Extend this method of proof to show that in this case there is also convergence for the SOR method with $0 < \omega < 2$.

7.5 Error Bounds and Iterative Refinement

It seems intuitively reasonable that if $\tilde{\mathbf{x}}$ is an approximation to the solution $\mathbf{x}$ of $A\mathbf{x} = \mathbf{b}$ and the residual vector $\mathbf{r} = \mathbf{b} - A\tilde{\mathbf{x}}$ has the property that $\|\mathbf{r}\|$ is small, then $\|\mathbf{x} - \tilde{\mathbf{x}}\|$ would be small as well. This is often the case, but certain systems, which occur frequently in practice, fail to have this property.

Example 1 The linear system $A\mathbf{x} = \mathbf{b}$ given by

$$\begin{bmatrix} 1 & 2 \\ 1.0001 & 2 \end{bmatrix} \begin{bmatrix} x_1 \\ x_2 \end{bmatrix} = \begin{bmatrix} 3 \\ 3.0001 \end{bmatrix}$$

has the unique solution $\mathbf{x} = (1, 1)^t$. Determine the residual vector for the poor approximation $\tilde{\mathbf{x}} = (3, -0.0001)^t$.

Solution We have

$$\mathbf{r} = \mathbf{b} - A\tilde{\mathbf{x}} = \begin{bmatrix} 3 \\ 3.0001 \end{bmatrix} - \begin{bmatrix} 1 & 2 \\ 1.0001 & 2 \end{bmatrix} \begin{bmatrix} 3 \\ -0.0001 \end{bmatrix} = \begin{bmatrix} 0.0002 \\ 0 \end{bmatrix},$$

so $\|\mathbf{r}\|_\infty = 0.0002$. Although the norm of the residual vector is small, the approximation $\tilde{\mathbf{x}} = (3, -0.0001)^t$ is obviously quite poor; in fact, $\|\mathbf{x} - \tilde{\mathbf{x}}\|_\infty = 2$. ■

The difficulty in Example 1 is explained quite simply by noting that the solution to the system represents the intersection of the lines

$$l_1: \quad x_1 + 2x_2 = 3 \quad \text{and} \quad l_2: \quad 1.0001x_1 + 2x_2 = 3.0001.$$

The point $(3, -0.0001)$ lies on l_2, and the lines are nearly parallel. This implies that $(3, -0.0001)$ also lies close to l_1, even though it differs significantly from the solution of the system, given by the intersection point $(1, 1)$. (See Figure 7.7.)

Figure 7.7

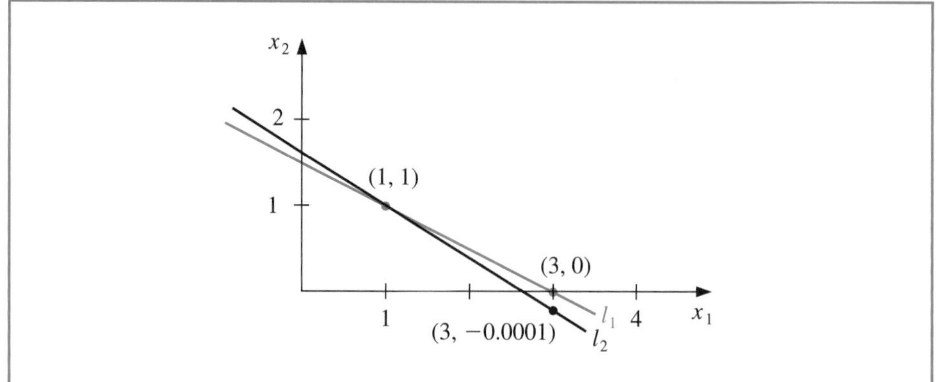

Example 1 was clearly constructed to show the difficulties that can—and, in fact, do—arise. Had the lines not been nearly coincident, we would expect a small residual vector to imply an accurate approximation.

In the general situation, we cannot rely on the geometry of the system to give an indication of when problems might occur. We can, however, obtain this information by considering the norms of the matrix A and its inverse.

Theorem 7.27 Suppose that $\tilde{\mathbf{x}}$ is an approximation to the solution of $A\mathbf{x} = \mathbf{b}$, A is a nonsingular matrix, and $\mathbf{r}$ is the residual vector for $\tilde{\mathbf{x}}$. Then for any natural norm,

$$\|\mathbf{x} - \tilde{\mathbf{x}}\| \le \|\mathbf{r}\| \cdot \|A^{-1}\|$$

and if $\mathbf{x} \neq \mathbf{0}$ and $\mathbf{b} \neq \mathbf{0}$,

$$\frac{\|\mathbf{x} - \tilde{\mathbf{x}}\|}{\|\mathbf{x}\|} \le \|A\| \cdot \|A^{-1}\| \frac{\|\mathbf{r}\|}{\|\mathbf{b}\|}. \tag{7.20}$$

■

Proof Since $\mathbf{r} = \mathbf{b} - A\tilde{\mathbf{x}} = A\mathbf{x} - A\tilde{\mathbf{x}}$ and A is nonsingular, we have $\mathbf{x} - \tilde{\mathbf{x}} = A^{-1}\mathbf{r}$. Theorem 7.11 on page 440 implies that

$$\|\mathbf{x} - \tilde{\mathbf{x}}\| = \|A^{-1}\mathbf{r}\| \le \|A^{-1}\| \cdot \|\mathbf{r}\|.$$

Moreover, since $\mathbf{b} = A\mathbf{x}$, we have $\|\mathbf{b}\| \le \|A\| \cdot \|\mathbf{x}\|$. So $1/\|\mathbf{x}\| \le \|A\|/\|\mathbf{b}\|$ and

$$\frac{\|\mathbf{x} - \tilde{\mathbf{x}}\|}{\|\mathbf{x}\|} \le \frac{\|A\| \cdot \|A^{-1}\|}{\|\mathbf{b}\|} \|\mathbf{r}\|.$$

■ ■ ■

Condition Numbers

The inequalities in Theorem 7.27 imply that $\|A^{-1}\|$ and $\|A\| \cdot \|A^{-1}\|$ provide an indication of the connection between the residual vector and the accuracy of the approximation. In general, the relative error $\|\mathbf{x} - \tilde{\mathbf{x}}\|/\|\mathbf{x}\|$ is of most interest, and, by Inequality (7.20), this error is bounded by the product of $\|A\| \cdot \|A^{-1}\|$ with the relative residual for this approximation, $\|\mathbf{r}\|/\|\mathbf{b}\|$. Any convenient norm can be used for this approximation; the only requirement is that it be used consistently throughout.

Definition 7.28 The **condition number** of the nonsingular matrix A relative to a norm $\|\cdot\|$ is

$$K(A) = \|A\| \cdot \|A^{-1}\|.$$

■

With this notation, the inequalities in Theorem 7.27 become

$$\|\mathbf{x}-\tilde{\mathbf{x}}\| \le K(A)\frac{\|\mathbf{r}\|}{\|A\|}$$

and

$$\frac{\|\mathbf{x}-\tilde{\mathbf{x}}\|}{\|\mathbf{x}\|} \le K(A)\frac{\|\mathbf{r}\|}{\|\mathbf{b}\|}.$$

For any nonsingular matrix A and natural norm $\|\cdot\|$,

$$1 = \|I\| = \|A\cdot A^{-1}\| \le \|A\|\cdot\|A^{-1}\| = K(A).$$

A matrix A is **well-conditioned** if $K(A)$ is close to 1, and is **ill-conditioned** when $K(A)$ is significantly greater than 1. Conditioning in this context refers to the relative security that a small residual vector implies a correspondingly accurate approximate solution.

Example 2 Determine the condition number for the matrix

$$A = \begin{bmatrix} 1 & 2 \\ 1.0001 & 2 \end{bmatrix}.$$

Solution We saw in Example 1 that the very poor approximation $(3, -0.0001)^t$ to the exact solution $(1, 1)^t$ had a residual vector with small norm, so we should expect the condition number of A to be large. We have $\|A\|_\infty = \max\{|1|+|2|, |1.001|+|2|\} = 3.0001$, which would not be considered large. However,

$$A^{-1} = \begin{bmatrix} -10000 & 10000 \\ 5000.5 & -5000 \end{bmatrix}, \quad \text{so} \quad \|A^{-1}\|_\infty = 20000,$$

and for the infinity norm, $K(A) = (20000)(3.0001) = 60002$. The size of the condition number for this example should certainly keep us from making hasty accuracy decisions based on the residual of an approximation. ■

The condition number K_∞ can be computed in Maple by first loading the *LinearAlgebra* package and the matrix. Then the command *ConditionNumber*(A) gives the condition number in the l_∞ norm. For example, we can obtain the condition number of the matrix A in Example 2 with

$A :=$ *Matrix*([[1, 2], [1.0001, 2]]): *ConditionNumber*(A)

60002.00000

Although the condition number of a matrix depends totally on the norms of the matrix and its inverse, the calculation of the inverse is subject to roundoff error and is dependent on the accuracy with which the calculations are performed. If the operations involve arithmetic with t digits of accuracy, the approximate condition number for the matrix A is the norm of the matrix times the norm of the approximation to the inverse of A, which is obtained using t-digit arithmetic. In fact, this condition number also depends on the method used to calculate the inverse of A. In addition, because of the number of calculations needed to compute the inverse, we need to be able to estimate the condition number without directly determining the inverse.

If we assume that the approximate solution to the linear system $A\mathbf{x} = \mathbf{b}$ is being determined using t-digit arithmetic and Gaussian elimination, it can be shown (see [FM], pp. 45–47) that the residual vector $\mathbf{r}$ for the approximation $\tilde{\mathbf{x}}$ has

$$\|\mathbf{r}\| \approx 10^{-t}\|A\|\cdot\|\tilde{\mathbf{x}}\|. \tag{7.21}$$

From this approximation, an estimate for the effective condition number in t-digit arithmetic can be obtained without the need to invert the matrix A. In actuality, this approximation assumes that all the arithmetic operations in the Gaussian elimination technique are performed using t-digit arithmetic but that the operations needed to determine the residual are done in double-precision (that is, $2t$-digit) arithmetic. This technique does not add significantly to the computational effort and eliminates much of the loss of accuracy involved with the subtraction of the nearly equal numbers that occur in the calculation of the residual.

The approximation for the t-digit condition number $K(A)$ comes from consideration of the linear system

$$A\mathbf{y} = \mathbf{r}.$$

The solution to this system can be readily approximated because the multipliers for the Gaussian elimination method have already been calculated. So A can be factored in the form P^tLU as described in Section 5 of Chapter 6. In fact $\tilde{\mathbf{y}}$, the approximate solution of $A\mathbf{y} = \mathbf{r}$, satisfies

$$\tilde{\mathbf{y}} \approx A^{-1}\mathbf{r} = A^{-1}(\mathbf{b} - A\tilde{\mathbf{x}}) = A^{-1}\mathbf{b} - A^{-1}A\tilde{\mathbf{x}} = \mathbf{x} - \tilde{\mathbf{x}}; \tag{7.22}$$

and

$$\mathbf{x} \approx \tilde{\mathbf{x}} + \tilde{\mathbf{y}}.$$

So $\tilde{\mathbf{y}}$ is an estimate of the error produced when $\tilde{\mathbf{x}}$ approximates the solution $\mathbf{x}$ to the original system. Equations (7.21) and (7.22) imply that

$$\|\tilde{\mathbf{y}}\| \approx \|\mathbf{x} - \tilde{\mathbf{x}}\| = \|A^{-1}\mathbf{r}\| \le \|A^{-1}\| \cdot \|\mathbf{r}\| \approx \|A^{-1}\| \left(10^{-t}\|A\| \cdot \|\tilde{\mathbf{x}}\|\right) = 10^{-t}\|\tilde{\mathbf{x}}\|K(A).$$

This gives an approximation for the condition number involved with solving the system $A\mathbf{x} = \mathbf{b}$ using Gaussian elimination and the t-digit type of arithmetic just described:

$$K(A) \approx \frac{\|\tilde{\mathbf{y}}\|}{\|\tilde{\mathbf{x}}\|}10^t. \tag{7.23}$$

Illustration The linear system given by

$$\begin{bmatrix} 3.3330 & 15920 & -10.333 \\ 2.2220 & 16.710 & 9.6120 \\ 1.5611 & 5.1791 & 1.6852 \end{bmatrix} \begin{bmatrix} x_1 \\ x_2 \\ x_3 \end{bmatrix} = \begin{bmatrix} 15913 \\ 28.544 \\ 8.4254 \end{bmatrix}$$

has the exact solution $\mathbf{x} = (1, 1, 1)^t$.

Using Gaussian elimination and five-digit rounding arithmetic leads successively to the augmented matrices

$$\begin{bmatrix} 3.3330 & 15920 & -10.333 & 15913 \\ 0 & -10596 & 16.501 & 10580 \\ 0 & -7451.4 & 6.5250 & -7444.9 \end{bmatrix}$$

and

$$\begin{bmatrix} 3.3330 & 15920 & -10.333 & 15913 \\ 0 & -10596 & 16.501 & -10580 \\ 0 & 0 & -5.0790 & -4.7000 \end{bmatrix}.$$

The approximate solution to this system is

$$\tilde{\mathbf{x}} = (1.2001, 0.99991, 0.92538)^t.$$

The residual vector corresponding to $\tilde{\mathbf{x}}$ is computed in double precision to be

$$\mathbf{r} = \mathbf{b} - A\tilde{\mathbf{x}}$$

$$= \begin{bmatrix} 15913 \\ 28.544 \\ 8.4254 \end{bmatrix} - \begin{bmatrix} 3.3330 & 15920 & -10.333 \\ 2.2220 & 16.710 & 9.6120 \\ 1.5611 & 5.1791 & 1.6852 \end{bmatrix} \begin{bmatrix} 1.2001 \\ 0.99991 \\ 0.92538 \end{bmatrix}$$

$$= \begin{bmatrix} 15913 \\ 28.544 \\ 8.4254 \end{bmatrix} - \begin{bmatrix} 15913.00518 \\ 28.26987086 \\ 8.611560367 \end{bmatrix} = \begin{bmatrix} -0.00518 \\ 0.27412914 \\ -0.186160367 \end{bmatrix},$$

so

$$\|\mathbf{r}\|_\infty = 0.27413.$$

The estimate for the condition number given in the preceding discussion is obtained by first solving the system $A\mathbf{y} = \mathbf{r}$ for $\tilde{\mathbf{y}}$:

$$\begin{bmatrix} 3.3330 & 15920 & -10.333 \\ 2.2220 & 16.710 & 9.6120 \\ 1.5611 & 5.1791 & 1.6852 \end{bmatrix} \begin{bmatrix} y_1 \\ y_2 \\ y_3 \end{bmatrix} = \begin{bmatrix} -0.00518 \\ 0.27413 \\ -0.18616 \end{bmatrix}.$$

This implies that $\tilde{\mathbf{y}} = (-0.20008, 8.9987 \times 10^{-5}, 0.074607)^t$. Using the estimate in Eq. (7.23) gives

$$K(A) \approx \frac{\|\tilde{\mathbf{y}}\|_\infty}{\|\tilde{\mathbf{x}}\|_\infty} 10^5 = \frac{0.20008}{1.2001} 10^5 = 16672. \tag{7.24}$$

To determine the *exact* condition number of A, we first must find A^{-1}. Using five-digit rounding arithmetic for the calculations gives the approximation:

$$A^{-1} \approx \begin{bmatrix} -1.1701 \times 10^{-4} & -1.4983 \times 10^{-1} & 8.5416 \times 10^{-1} \\ 6.2782 \times 10^{-5} & 1.2124 \times 10^{-4} & -3.0662 \times 10^{-4} \\ -8.6631 \times 10^{-5} & 1.3846 \times 10^{-1} & -1.9689 \times 10^{-1} \end{bmatrix}.$$

Theorem 7.11 on page 440 implies that $\|A^{-1}\|_\infty = 1.0041$ and $\|A\|_\infty = 15934$.

As a consequence, the ill-conditioned matrix A has

$$K(A) = (1.0041)(15934) = 15999.$$

The estimate in (7.24) is quite close to $K(A)$ and requires considerably less computational effort.

Since the actual solution $\mathbf{x} = (1, 1, 1)^t$ is known for this system, we can calculate both

$$\|\mathbf{x} - \tilde{\mathbf{x}}\|_\infty = 0.2001 \quad \text{and} \quad \frac{\|\mathbf{x} - \tilde{\mathbf{x}}\|_\infty}{\|\mathbf{x}\|_\infty} = \frac{0.2001}{1} = 0.2001.$$

The error bounds given in Theorem 7.27 for these values are

$$\|\mathbf{x} - \tilde{\mathbf{x}}\|_\infty \le K(A)\frac{\|\mathbf{r}\|_\infty}{\|A\|_\infty} = \frac{(15999)(0.27413)}{15934} = 0.27525$$

and

$$\frac{\|\mathbf{x} - \tilde{\mathbf{x}}\|_\infty}{\|\mathbf{x}\|_\infty} \le K(A)\frac{\|\mathbf{r}\|_\infty}{\|\mathbf{b}\|_\infty} = \frac{(15999)(0.27413)}{15913} = 0.27561.$$

□

Iterative Refinement

In Eq. (7.22), we used the estimate $\tilde{\mathbf{y}} \approx \mathbf{x} - \tilde{\mathbf{x}}$, where $\tilde{\mathbf{y}}$ is the approximate solution to the system $A\mathbf{y} = \mathbf{r}$. In general, $\tilde{\mathbf{x}} + \tilde{\mathbf{y}}$ is a more accurate approximation to the solution of the linear system $A\mathbf{x} = \mathbf{b}$ than the original approximation $\tilde{\mathbf{x}}$. The method using this assumption is called **iterative refinement**, or *iterative improvement*, and consists of performing iterations on the system whose right-hand side is the residual vector for successive approximations until satisfactory accuracy results.

If the process is applied using t-digit arithmetic and if $K_\infty(A) \approx 10^q$, then after k iterations of iterative refinement the solution has approximately the smaller of t and $k(t-q)$ correct digits. If the system is well-conditioned, one or two iterations will indicate that the solution is accurate. There is the possibility of significant improvement on ill-conditioned systems unless the matrix A is so ill-conditioned that $K_\infty(A) > 10^t$. In that situation, increased precision should be used for the calculations.

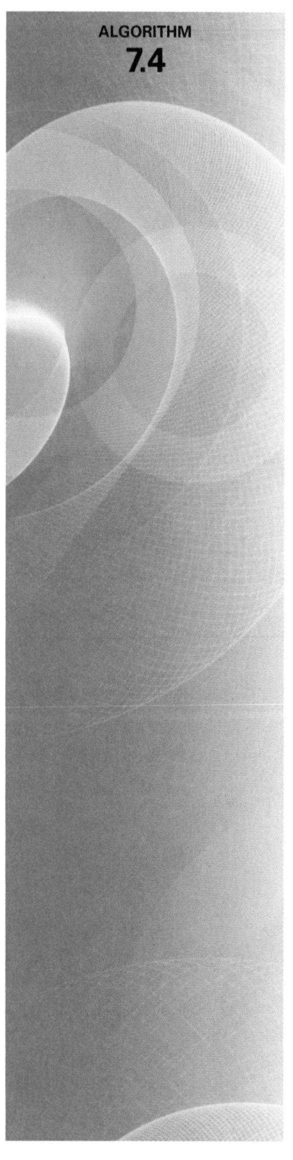

Iterative Refinement

To approximate the solution to the linear system $A\mathbf{x} = \mathbf{b}$:

INPUT the number of equations and unknowns n; the entries a_{ij}, $1 \le i, j \le n$ of the matrix A; the entries b_i, $1 \le i \le n$ of $\mathbf{b}$; the maximum number of iterations N; tolerance TOL; number of digits of precision t.

OUTPUT the approximation $\mathbf{xx} = (xx_i, \ldots, xx_n)^t$ or a message that the number of iterations was exceeded, and an approximation $COND$ to $K_\infty(A)$.

Step 0 Solve the system $A\mathbf{x} = \mathbf{b}$ for $x_1, \ldots, x_n$ by Gaussian elimination saving the multipliers $m_{ji}, j = i+1, i+2, \ldots, n$, $i = 1, 2, \ldots, n-1$ and noting row interchanges.

Step 1 Set $k = 1$.

Step 2 While $(k \le N)$ do Steps 3–9.

Step 3 For $i = 1, 2, \ldots, n$ (*Calculate* $\mathbf{r}$.)

$$\text{set } r_i = b_i - \sum_{j=1}^{n} a_{ij}x_j.$$

(*Perform the computations in double-precision arithmetic.*)

Step 4 Solve the linear system $A\mathbf{y} = \mathbf{r}$ by using Gaussian elimination in the same order as in Step 0.

Step 5 For $i = 1, \ldots, n$ set $xx_i = x_i + y_i$.

Step 6 If $k = 1$ then set $COND = \dfrac{\|\mathbf{y}\|_\infty}{\|\mathbf{xx}\|_\infty} 10^t$.

Step 7 If $\|\mathbf{x} - \mathbf{xx}\|_\infty < TOL$ then OUTPUT ($\mathbf{xx}$);
OUTPUT ($COND$);
(*The procedure was successful.*)
STOP.

Step 8 Set $k = k + 1$.

Step 9 For $i = 1, \ldots, n$ set $x_i = xx_i$.

Step 10 OUTPUT ('Maximum number of iterations exceeded');
OUTPUT (*COND*);
(*The procedure was unsuccessful.*)
STOP. ■

If t-digit arithmetic is used, a recommended stopping procedure in Step 7 is to iterate until $|y_i^{(k)}| \leq 10^{-t}$, for each $i = 1, 2, \ldots, n$.

Illustration In our earlier illustration we found the approximation to the linear system

$$\begin{bmatrix} 3.3330 & 15920 & -10.333 \\ 2.2220 & 16.710 & 9.6120 \\ 1.5611 & 5.1791 & 1.6852 \end{bmatrix} \begin{bmatrix} x_1 \\ x_2 \\ x_3 \end{bmatrix} = \begin{bmatrix} 15913 \\ 28.544 \\ 8.4254 \end{bmatrix}$$

using five-digit arithmetic and Gaussian elimination, to be

$$\tilde{\mathbf{x}}^{(1)} = (1.2001, 0.99991, 0.92538)^t$$

and the solution to $A\mathbf{y} = \mathbf{r}^{(1)}$ to be

$$\tilde{\mathbf{y}}^{(1)} = (-0.20008, 8.9987 \times 10^{-5}, 0.074607)^t.$$

By Step 5 in this algorithm,

$$\tilde{\mathbf{x}}^{(2)} = \tilde{\mathbf{x}}^{(1)} + \tilde{\mathbf{y}}^{(1)} = (1.0000, 1.0000, 0.99999)^t,$$

and the actual error in this approximation is

$$\|\mathbf{x} - \tilde{\mathbf{x}}^{(2)}\|_\infty = 1 \times 10^{-5}.$$

Using the suggested stopping technique for the algorithm, we compute $\mathbf{r}^{(2)} = \mathbf{b} - A\tilde{\mathbf{x}}^{(2)}$ and solve the system $A\mathbf{y}^{(2)} = \mathbf{r}^{(2)}$, which gives

$$\tilde{\mathbf{y}}^{(2)} = (1.5002 \times 10^{-9}, 2.0951 \times 10^{-10}, 1.0000 \times 10^{-5})^t.$$

Since $\|\tilde{\mathbf{y}}^{(2)}\|_\infty \leq 10^{-5}$, we conclude that

$$\tilde{\mathbf{x}}^{(3)} = \tilde{\mathbf{x}}^{(2)} + \tilde{\mathbf{y}}^{(2)} = (1.0000, 1.0000, 1.0000)^t$$

is sufficiently accurate, which is certainly correct. ■

Throughout this section it has been assumed that in the linear system $A\mathbf{x} = \mathbf{b}$, A and $\mathbf{b}$ can be represented exactly. Realistically, the entries a_{ij} and b_j will be altered or perturbed by an amount δa_{ij} and δb_j, causing the linear system

$$(A + \delta A)\mathbf{x} = \mathbf{b} + \delta\mathbf{b}$$

to be solved in place of $A\mathbf{x} = \mathbf{b}$. Normally, if $\|\delta A\|$ and $\|\delta\mathbf{b}\|$ are small (on the order of 10^{-t}), the t-digit arithmetic should yield a solution $\tilde{\mathbf{x}}$ for which $\|\mathbf{x} - \tilde{\mathbf{x}}\|$ is correspondingly small. However, in the case of ill-conditioned systems, we have seen that even if A and $\mathbf{b}$ are represented exactly, rounding errors can cause $\|\mathbf{x} - \tilde{\mathbf{x}}\|$ to be large. The following theorem relates the perturbations of linear systems to the condition number of a matrix. The proof of this result can be found in [Or2], p. 33.

Theorem 7.29 Suppose A is nonsingular and

$$\|\delta A\| < \frac{1}{\|A^{-1}\|}.$$

The solution $\tilde{\mathbf{x}}$ to $(A + \delta A)\tilde{\mathbf{x}} = \mathbf{b} + \delta\mathbf{b}$ approximates the solution $\mathbf{x}$ of $A\mathbf{x} = \mathbf{b}$ with the error estimate

$$\frac{\|\mathbf{x} - \tilde{\mathbf{x}}\|}{\|\mathbf{x}\|} \leq \frac{K(A)\|A\|}{\|A\| - K(A)\|\delta A\|}\left(\frac{\|\delta\mathbf{b}\|}{\|\mathbf{b}\|} + \frac{\|\delta A\|}{\|A\|}\right). \tag{7.25}$$

■

The estimate in inequality (7.25) states that if the matrix A is well-conditioned (that is, $K(A)$ is not too large), then small changes in A and $\mathbf{b}$ produce correspondingly small changes in the solution $\mathbf{x}$. If, on the other hand, A is ill-conditioned, then small changes in A and $\mathbf{b}$ may produce large changes in $\mathbf{x}$.

James Hardy Wilkinson (1919–1986) is best known for his extensive work in numerical methods for solving linear equations and eigenvalues problems. He also developed the technique of backward error analysis.

The theorem is independent of the particular numerical procedure used to solve $A\mathbf{x} = \mathbf{b}$. It can be shown, by means of a backward error analysis (see [Wil1] or [Wil2]), that if Gaussian elimination with pivoting is used to solve $A\mathbf{x} = \mathbf{b}$ in t-digit arithmetic, the numerical solution $\tilde{\mathbf{x}}$ is the actual solution of a linear system:

$$(A + \delta A)\tilde{\mathbf{x}} = \mathbf{b}, \quad \text{where } \|\delta A\|_\infty \leq f(n)10^{1-t}\max_{i,j,k}|a_{ij}^{(k)}|.$$

for some function $f(n)$. Wilkinson found that in practice $f(n) \approx n$ and, at worst, $f(n) \leq 1.01(n^3 + 3n^2)$.

EXERCISE SET 7.5

1. Compute the condition numbers of the following matrices relative to $\|\cdot\|_\infty$.

   a. $\begin{bmatrix} \frac{1}{2} & \frac{1}{3} \\ \frac{1}{3} & \frac{1}{4} \end{bmatrix}$ **b.** $\begin{bmatrix} 1 & 2 \\ 1.00001 & 2 \end{bmatrix}$

   c. $\begin{bmatrix} 3.9 & 1.6 \\ 6.8 & 2.9 \end{bmatrix}$ **d.** $\begin{bmatrix} 1.003 & 58.09 \\ 5.550 & 321.8 \end{bmatrix}$

2. Compute the condition numbers of the following matrices relative to $\|\cdot\|_\infty$.

   a. $\begin{bmatrix} 0.03 & 58.9 \\ 5.31 & -6.10 \end{bmatrix}$ **b.** $\begin{bmatrix} 58.9 & 0.03 \\ -6.10 & 5.31 \end{bmatrix}$

   c. $\begin{bmatrix} 1 & -1 & -1 \\ 0 & 1 & -1 \\ 0 & 0 & -1 \end{bmatrix}$ **d.** $\begin{bmatrix} 0.04 & 0.01 & -0.01 \\ 0.2 & 0.5 & -0.2 \\ 1 & 2 & 4 \end{bmatrix}$

3. The following linear systems $A\mathbf{x} = \mathbf{b}$ have $\mathbf{x}$ as the actual solution and $\tilde{\mathbf{x}}$ as an approximate solution. Using the results of Exercise 1, compute

$$\|\mathbf{x} - \tilde{\mathbf{x}}\|_\infty \quad \text{and} \quad K_\infty(A)\frac{\|\mathbf{b} - A\tilde{\mathbf{x}}\|_\infty}{\|A\|_\infty}.$$

a. $\frac{1}{2}x_1 + \frac{1}{3}x_2 = \frac{1}{63}$,
$\frac{1}{3}x_1 + \frac{1}{4}x_2 = \frac{1}{168}$,
$\mathbf{x} = \left(\frac{1}{7}, -\frac{1}{6}\right)^t$,
$\tilde{\mathbf{x}} = (0.142, -0.166)^t$.

b. $x_1 + 2x_2 = 3$,
$1.0001x_1 + 2x_2 = 3.0001$,
$\mathbf{x} = (1, 1)^t$,
$\tilde{\mathbf{x}} = (0.96, 1.02)^t$.

c. $3.9x_1 + 1.6x_2 = 5.5$,
$6.8x_1 + 2.9x_2 = 9.7$,
$\mathbf{x} = (1, 1)^t$,
$\tilde{\mathbf{x}} = (0.98, 1.1)^t$.

d. $1.003x_1 + 58.09x_2 = 68.12$,
$5.550x_1 + 321.8x_2 = 377.3$,
$\mathbf{x} = (10, 1)^t$,
$\tilde{\mathbf{x}} = (-10, 1)^t$.

4. The following linear systems $A\mathbf{x} = \mathbf{b}$ have $\mathbf{x}$ as the actual solution and $\tilde{\mathbf{x}}$ as an approximate solution. Using the results of Exercise 2, compute

$$\|\mathbf{x} - \tilde{\mathbf{x}}\|_\infty \quad \text{and} \quad K_\infty(A)\frac{\|\mathbf{b} - A\tilde{\mathbf{x}}\|_\infty}{\|A\|_\infty}.$$

a. $0.03x_1 + 58.9x_2 = 59.2$,
$5.31x_1 - 6.10x_2 = 47.0$,
$\mathbf{x} = (10, 1)^t$,
$\tilde{\mathbf{x}} = (30.0, 0.990)^t$.

b. $58.9x_1 + 0.03x_2 = 59.2$,
$-6.10x_1 + 5.31x_2 = 47.0$,
$\mathbf{x} = (1, 10)^t$,
$\tilde{\mathbf{x}} = (1.02, 9.98)^t$.

c. $x_1 - x_2 - x_3 = 2\pi$,
$x_2 - x_3 = 0$,
$-x_3 = \pi$.
$\mathbf{x} = (0, -\pi, -\pi)^t$,
$\tilde{\mathbf{x}} = (-0.1, -3.15, -3.14)^t$.

d. $0.04x_1 + 0.01x_2 - 0.01x_3 = 0.06$,
$0.2x_1 + 0.5x_2 - 0.2x_3 = 0.3$,
$x_1 + 2x_2 + 4x_3 = 11$,
$\mathbf{x} = (1.827586, 0.6551724, 1.965517)^t$,
$\tilde{\mathbf{x}} = (1.8, 0.64, 1.9)^t$.

5. **(i)** Use Gaussian elimination and three-digit rounding arithmetic to approximate the solutions to the following linear systems. **(ii)** Then use one iteration of iterative refinement to improve the approximation, and compare the approximations to the actual solutions.

a. $0.03x_1 + 58.9x_2 = 59.2$,
$5.31x_1 - 6.10x_2 = 47.0$.
Actual solution $(10, 1)^t$.

b. $3.3330x_1 + 15920x_2 + 10.333x_3 = 7953$,
$2.2220x_1 + 16.710x_2 + 9.6120x_3 = 0.965$,
$-1.5611x_1 + 5.1792x_2 - 1.6855x_3 = 2.714$.
Actual solution $(1, 0.5, -1)^t$.

c. $1.19x_1 + 2.11x_2 - 100x_3 + x_4 = 1.12$,
$14.2x_1 - 0.122x_2 + 12.2x_3 - x_4 = 3.44$,
$100x_2 - 99.9x_3 + x_4 = 2.15$,
$15.3x_1 + 0.110x_2 - 13.1x_3 - x_4 = 4.16$.
Actual solution $(0.17682530, 0.01269269, -0.02065405, -1.18260870)^t$.

d. $\pi x_1 - ex_2 + \sqrt{2}x_3 - \sqrt{3}x_4 = \sqrt{11}$,
$\pi^2 x_1 + ex_2 - e^2 x_3 + \frac{3}{7}x_4 = 0$,
$\sqrt{5}x_1 - \sqrt{6}x_2 + x_3 - \sqrt{2}x_4 = \pi$,
$\pi^3 x_1 + e^2 x_2 - \sqrt{7}x_3 + \frac{1}{9}x_4 = \sqrt{2}$.
Actual solution $(0.78839378, -3.12541367, 0.16759660, 4.55700252)^t$.

6. Repeat Exercise 5 using four-digit rounding arithmetic.

7. The linear system

$$\begin{bmatrix} 1 & 2 \\ 1.0001 & 2 \end{bmatrix}\begin{bmatrix} x_1 \\ x_2 \end{bmatrix} = \begin{bmatrix} 3 \\ 3.0001 \end{bmatrix}$$

has solution $(1, 1)^t$. Change A slightly to

$$\begin{bmatrix} 1 & 2 \\ 0.9999 & 2 \end{bmatrix},$$

and consider the linear system

$$\begin{bmatrix} 1 & 2 \\ 0.9999 & 2 \end{bmatrix}\begin{bmatrix} x_1 \\ x_2 \end{bmatrix} = \begin{bmatrix} 3 \\ 3.0001 \end{bmatrix}.$$

Compute the new solution using five-digit rounding arithmetic, and compare the actual error to the estimate (7.25). Is A ill-conditioned?

8. The linear system $A\mathbf{x} = \mathbf{b}$ given by

$$\begin{bmatrix} 1 & 2 \\ 1.00001 & 2 \end{bmatrix}\begin{bmatrix} x_1 \\ x_2 \end{bmatrix} = \begin{bmatrix} 3 \\ 3.00001 \end{bmatrix}$$

has solution $(1, 1)^t$. Use seven-digit rounding arithmetic to find the solution of the perturbed system

$$\begin{bmatrix} 1 & 2 \\ 1.000011 & 2 \end{bmatrix}\begin{bmatrix} x_1 \\ x_2 \end{bmatrix} = \begin{bmatrix} 3.00001 \\ 3.00003 \end{bmatrix},$$

and compare the actual error to the estimate (7.25). Is A ill-conditioned?

9. Show that if B is singular, then

$$\frac{1}{K(A)} \le \frac{||A - B||}{||A||}.$$

[*Hint:* There exists a vector with $||\mathbf{x}|| = 1$, such that $B\mathbf{x} = \mathbf{0}$. Derive the estimate using $||A\mathbf{x}|| \ge ||\mathbf{x}|| \,/\, ||A^{-1}||$.]

10. Using Exercise 9, estimate the condition numbers for the following matrices:

a. $\begin{bmatrix} 1 & 2 \\ 1.0001 & 2 \end{bmatrix}$ **b.** $\begin{bmatrix} 3.9 & 1.6 \\ 6.8 & 2.9 \end{bmatrix}$

11. The $n \times n$ *Hilbert matrix* $H^{(n)}$ (see page 512) defined by

$$H_{ij}^{(n)} = \frac{1}{i + j - 1}, \quad 1 \le i, j \le n,$$

is an ill-conditioned matrix that arises in solving the normal equations for the coefficients of the least-squares polynomial (see Example 1 of Section 8.2).

a. Show that

$$[H^{(4)}]^{-1} = \begin{bmatrix} 16 & -120 & 240 & -140 \\ -120 & 1200 & -2700 & 1680 \\ 240 & -2700 & 6480 & -4200 \\ -140 & 1680 & -4200 & 2800 \end{bmatrix},$$

and compute $K_\infty(H^{(4)})$.

b. Show that

$$[H^{(5)}]^{-1} = \begin{bmatrix} 25 & -300 & 1050 & -1400 & 630 \\ -300 & 4800 & -18900 & 26880 & -12600 \\ 1050 & -18900 & 79380 & -117600 & 56700 \\ -1400 & 26880 & -117600 & 179200 & -88200 \\ 630 & -12600 & 56700 & -88200 & 44100 \end{bmatrix},$$

and compute $K_\infty(H^{(5)})$.

c. Solve the linear system

$$H^{(4)} \begin{bmatrix} x_1 \\ x_2 \\ x_3 \\ x_4 \end{bmatrix} = \begin{bmatrix} 1 \\ 0 \\ 0 \\ 1 \end{bmatrix}$$

using five-digit rounding arithmetic, and compare the actual error to that estimated in (7.25).

12. Use four-digit rounding arithmetic to compute the inverse H^{-1} of the 3×3 Hilbert matrix H, and then compute $\hat{H} = (H^{-1})^{-1}$. Determine $||H - \hat{H}||_\infty$.

7.6 The Conjugate Gradient Method

The conjugate gradient method of Hestenes and Stiefel [HS] was originally developed as a direct method designed to solve an $n \times n$ positive definite linear system. As a direct method it is generally inferior to Gaussian elimination with pivoting. Both methods require n steps to determine a solution, and the steps of the conjugate gradient method are more computationally expensive than those of Gaussian elimination.

Magnus Hestenes (1906–1991) and Eduard Steifel (1907–1998) published the original paper on the conjugate gradient method in 1952 while working at the Institute for Numerical Analysis on the campus of UCLA.

However, the conjugate gradient method is useful when employed as an iterative approximation method for solving large sparse systems with nonzero entries occurring in predictable patterns. These problems frequently arise in the solution of boundary-value problems. When the matrix has been preconditioned to make the calculations more effective, good results are obtained in only about $\sqrt{n}$ iterations. Employed in this way, the method is preferred over Gaussian elimination and the previously-discussed iterative methods.

Throughout this section we assume that the matrix A is positive definite. We will use the *inner product* notation

$$\langle \mathbf{x}, \mathbf{y} \rangle = \mathbf{x}^t \mathbf{y}, \tag{7.26}$$

where $\mathbf{x}$ and $\mathbf{y}$ are n-dimensional vectors. We will also need some additional standard results from linear algebra. A review of this material is found in Section 9.1.

The next result follows easily from the properties of transposes (see Exercise 12).

Theorem 7.30 For any vectors $\mathbf{x}$, $\mathbf{y}$, and $\mathbf{z}$ and any real number α, we have

(a) $\langle \mathbf{x}, \mathbf{y} \rangle = \langle \mathbf{y}, \mathbf{x} \rangle$;

(b) $\langle \alpha\mathbf{x}, \mathbf{y} \rangle = \langle \mathbf{x}, \alpha\mathbf{y} \rangle = \alpha\langle \mathbf{x}, \mathbf{y} \rangle$;

(c) $\langle \mathbf{x} + \mathbf{z}, \mathbf{y} \rangle = \langle \mathbf{x}, \mathbf{y} \rangle + \langle \mathbf{z}, \mathbf{y} \rangle$;

(d) $\langle \mathbf{x}, \mathbf{x} \rangle \geq 0$;

(e) $\langle \mathbf{x}, \mathbf{x} \rangle = 0$ if and only if $\mathbf{x} = \mathbf{0}$. ■

When A is positive definite, $\langle \mathbf{x}, A\mathbf{x} \rangle = \mathbf{x}^t A\mathbf{x} > 0$ unless $\mathbf{x} = \mathbf{0}$. Also, since A is symmetric, we have $\mathbf{x}^t A\mathbf{y} = \mathbf{x}^t A^t \mathbf{y} = (A\mathbf{x})^t \mathbf{y}$, so in addition to the results in Theorem 7.30, we have for each $\mathbf{x}$ and $\mathbf{y}$,

$$\langle \mathbf{x}, A\mathbf{y} \rangle = (A\mathbf{x})^t \mathbf{y} = \mathbf{x}^t A^t \mathbf{y} = \mathbf{x}^t A\mathbf{y} = \langle A\mathbf{x}, \mathbf{y} \rangle. \tag{7.27}$$

The following result is a basic tool in the development of the conjugate gradient method.

Theorem 7.31 The vector $\mathbf{x}^*$ is a solution to the positive definite linear system $A\mathbf{x} = \mathbf{b}$ if and only if $\mathbf{x}^*$ produces the minimal value of

$$g(\mathbf{x}) = \langle \mathbf{x}, A\mathbf{x} \rangle - 2\langle \mathbf{x}, \mathbf{b} \rangle.$$ ■

Proof Let $\mathbf{x}$ and $\mathbf{v} \neq \mathbf{0}$ be fixed vectors and t a real number variable. We have

$$\begin{aligned} g(\mathbf{x}+t\mathbf{v}) &= \langle \mathbf{x}+t\mathbf{v}, A\mathbf{x}+tA\mathbf{v}\rangle - 2\langle \mathbf{x}+t\mathbf{v}, \mathbf{b}\rangle \\ &= \langle \mathbf{x}, A\mathbf{x}\rangle + t\langle \mathbf{v}, A\mathbf{x}\rangle + t\langle \mathbf{x}, A\mathbf{v}\rangle + t^2\langle \mathbf{v}, A\mathbf{v}\rangle - 2\langle \mathbf{x}, \mathbf{b}\rangle - 2t\langle \mathbf{v}, \mathbf{b}\rangle \\ &= \langle \mathbf{x}, A\mathbf{x}\rangle - 2\langle \mathbf{x}, \mathbf{b}\rangle + 2t\langle \mathbf{v}, A\mathbf{x}\rangle - 2t\langle \mathbf{v}, \mathbf{b}\rangle + t^2\langle \mathbf{v}, A\mathbf{v}\rangle, \end{aligned}$$

so

$$g(\mathbf{x}+t\mathbf{v}) = g(\mathbf{x}) - 2t\langle \mathbf{v}, \mathbf{b}-A\mathbf{x}\rangle + t^2\langle \mathbf{v}, A\mathbf{v}\rangle. \tag{7.28}$$

With $\mathbf{x}$ and $\mathbf{v}$ fixed we can define the quadratic function h in t by

$$h(t) = g(\mathbf{x}+t\mathbf{v}).$$

Then h assumes a minimal value when $h'(t) = 0$, because its t^2 coefficient, $\langle \mathbf{v}, A\mathbf{v}\rangle$, is positive. Because

$$h'(t) = -2\langle \mathbf{v}, \mathbf{b}-A\mathbf{x}\rangle + 2t\langle \mathbf{v}, A\mathbf{v}\rangle,$$

the minimum occurs when

$$\hat{t} = \frac{\langle \mathbf{v}, \mathbf{b}-A\mathbf{x}\rangle}{\langle \mathbf{v}, A\mathbf{v}\rangle},$$

and, from Equation (7.28),

$$\begin{aligned} h(\hat{t}) &= g(\mathbf{x}+\hat{t}\mathbf{v}) \\ &= g(\mathbf{x}) - 2\hat{t}\langle \mathbf{v}, \mathbf{b}-A\mathbf{x}\rangle + \hat{t}^2\langle \mathbf{v}, A\mathbf{v}\rangle \\ &= g(\mathbf{x}) - 2\frac{\langle \mathbf{v}, \mathbf{b}-A\mathbf{x}\rangle}{\langle \mathbf{v}, A\mathbf{v}\rangle}\langle \mathbf{v}, \mathbf{b}-A\mathbf{x}\rangle + \left(\frac{\langle \mathbf{v}, \mathbf{b}-A\mathbf{x}\rangle}{\langle \mathbf{v}, A\mathbf{v}\rangle}\right)^2 \langle \mathbf{v}, A\mathbf{v}\rangle \\ &= g(\mathbf{x}) - \frac{\langle \mathbf{v}, \mathbf{b}-A\mathbf{x}\rangle^2}{\langle \mathbf{v}, A\mathbf{v}\rangle}. \end{aligned}$$

So for any vector $\mathbf{v} \neq \mathbf{0}$, we have $g(\mathbf{x}+\hat{t}\mathbf{v}) < g(\mathbf{x})$ unless $\langle \mathbf{v}, \mathbf{b}-A\mathbf{x}\rangle = 0$, in which case $g(\mathbf{x}) = g(\mathbf{x}+\hat{t}\mathbf{v})$. This is the basic result we need to prove Theorem 7.31.

Suppose $\mathbf{x}^*$ satisfies $A\mathbf{x}^* = \mathbf{b}$. Then $\langle \mathbf{v}, \mathbf{b}-A\mathbf{x}^*\rangle = 0$ for any vector $\mathbf{v}$, and $g(\mathbf{x})$ cannot be made any smaller than $g(\mathbf{x}^*)$. Thus, $\mathbf{x}^*$ minimizes g.

On the other hand, suppose that $\mathbf{x}^*$ is a vector that minimizes g. Then for any vector $\mathbf{v}$, we have $g(\mathbf{x}^*+\hat{t}\mathbf{v}) \geq g(\mathbf{x}^*)$. Thus, $\langle \mathbf{v}, \mathbf{b}-A\mathbf{x}^*\rangle = 0$. This implies that $\mathbf{b}-A\mathbf{x}^* = \mathbf{0}$ and, consequently, that $A\mathbf{x}^* = \mathbf{b}$. ▪ ▪ ▪

To begin the conjugate gradient method, we choose $\mathbf{x}$, an approximate solution to $A\mathbf{x}^* = \mathbf{b}$, and $\mathbf{v} \neq \mathbf{0}$, which gives a *search direction* in which to move away from $\mathbf{x}$ to improve the approximation. Let $\mathbf{r} = \mathbf{b} - A\mathbf{x}$ be the residual vector associated with $\mathbf{x}$ and

$$t = \frac{\langle \mathbf{v}, \mathbf{b}-A\mathbf{x}\rangle}{\langle \mathbf{v}, A\mathbf{v}\rangle} = \frac{\langle \mathbf{v}, \mathbf{r}\rangle}{\langle \mathbf{v}, A\mathbf{v}\rangle}.$$

If $\mathbf{r} \neq \mathbf{0}$ and if $\mathbf{v}$ and $\mathbf{r}$ are not orthogonal, then $\mathbf{x}+t\mathbf{v}$ gives a smaller value for g than $g(\mathbf{x})$ and is presumably closer to $\mathbf{x}^*$ than is $\mathbf{x}$. This suggests the following method.

Let $\mathbf{x}^{(0)}$ be an initial approximation to $\mathbf{x}^*$, and let $\mathbf{v}^{(1)} \neq \mathbf{0}$ be an initial search direction. For $k = 1, 2, 3, \ldots,$ we compute

$$t_k = \frac{\langle \mathbf{v}^{(k)}, \mathbf{b}-A\mathbf{x}^{(k-1)}\rangle}{\langle \mathbf{v}^{(k)}, A\mathbf{v}^{(k)}\rangle},$$

$$\mathbf{x}^{(k)} = \mathbf{x}^{(k-1)} + t_k\mathbf{v}^{(k)}$$

and choose a new search direction $\mathbf{v}^{(k+1)}$. The object is to make this selection so that the sequence of approximations $\{\mathbf{x}^{(k)}\}$ converges rapidly to $\mathbf{x}^*$.

To choose the search directions, we view g as a function of the components of $\mathbf{x} = (x_1, x_2, \ldots, x_n)^t$. Thus,

$$g(x_1, x_2, \ldots, x_n) = \langle \mathbf{x}, A\mathbf{x}\rangle - 2\langle \mathbf{x}, \mathbf{b}\rangle = \sum_{i=1}^{n}\sum_{j=1}^{n} a_{ij}x_i x_j - 2\sum_{i=1}^{n} x_i b_i.$$

Taking partial derivatives with respect to the component variables x_k gives

$$\frac{\partial g}{\partial x_k}(\mathbf{x}) = 2\sum_{i=1}^{n} a_{ki}x_i - 2b_k,$$

which is the kth component of the vector $2(A\mathbf{x} - \mathbf{b})$. Therefore, the gradient of g is

$$\nabla g(\mathbf{x}) = \left(\frac{\partial g}{\partial x_1}(\mathbf{x}), \frac{\partial g}{\partial x_2}(\mathbf{x}), \ldots, \frac{\partial g}{\partial x_n}(\mathbf{x})\right)^t = 2(A\mathbf{x} - \mathbf{b}) = -2\mathbf{r},$$

where the vector $\mathbf{r}$ is the residual vector for $\mathbf{x}$.

From multivariable calculus, we know that the direction of greatest decrease in the value of $g(\mathbf{x})$ is the direction given by $-\nabla g(\mathbf{x})$; that is, in the direction of the residual $\mathbf{r}$. The method that chooses

$$\mathbf{v}^{(k+1)} = \mathbf{r}^{(k)} = \mathbf{b} - A\mathbf{x}^{(k)}$$

is called the *method of steepest descent.* Although we will see in Section 10.4 that this method has merit for nonlinear systems and optimization problems, it is not used for linear systems because of slow convergence.

An alternative approach uses a set of nonzero direction vectors $\{\mathbf{v}^{(1)}, \ldots, \mathbf{v}^{(n)}\}$ that satisfy

$$\langle \mathbf{v}^{(i)}, A\mathbf{v}^{(j)}\rangle = 0, \quad \text{if} \quad i \neq j.$$

This is called an **A-orthogonality condition**, and the set of vectors $\{\mathbf{v}^{(1)}, \ldots, \mathbf{v}^{(n)}\}$ is said to be **A-orthogonal**. It is not difficult to show that a set of A-orthogonal vectors associated with the positive definite matrix A is linearly independent. (See Exercise 13(a).) This set of search directions gives

$$t_k = \frac{\langle \mathbf{v}^{(k)}, \mathbf{b} - A\mathbf{x}^{(k-1)}\rangle}{\langle \mathbf{v}^{(k)}, A\mathbf{v}^{(k)}\rangle} = \frac{\langle \mathbf{v}^{(k)}, \mathbf{r}^{(k-1)}\rangle}{\langle \mathbf{v}^{(k)}, A\mathbf{v}^{(k)}\rangle}$$

and $\mathbf{x}^{(k)} = \mathbf{x}^{(k-1)} + t_k\mathbf{v}^{(k)}$.

The following theorem shows that this choice of search directions gives convergence in at most n-steps, so as a direct method it produces the exact solution, assuming that the arithmetic is exact.

Theorem 7.32 Let $\{\mathbf{v}^{(1)}, \ldots, \mathbf{v}^{(n)}\}$ be an A-orthogonal set of nonzero vectors associated with the positive definite matrix A, and let $\mathbf{x}^{(0)}$ be arbitrary. Define

$$t_k = \frac{\langle \mathbf{v}^{(k)}, \mathbf{b} - A\mathbf{x}^{(k-1)}\rangle}{\langle \mathbf{v}^{(k)}, A\mathbf{v}^{(k)}\rangle} \quad \text{and} \quad \mathbf{x}^{(k)} = \mathbf{x}^{(k-1)} + t_k\mathbf{v}^{(k)},$$

for $k = 1, 2, \ldots, n$. Then, assuming exact arithmetic, $A\mathbf{x}^{(n)} = \mathbf{b}$. ■

Proof Since, for each $k = 1, 2, \ldots, n$, $\mathbf{x}^{(k)} = \mathbf{x}^{(k-1)} + t_k\mathbf{v}^{(k)}$, we have

$$\begin{aligned}
A\mathbf{x}^{(n)} &= A\mathbf{x}^{(n-1)} + t_nA\mathbf{v}^{(n)} \\
&= (A\mathbf{x}^{(n-2)} + t_{n-1}A\mathbf{v}^{(n-1)}) + t_nA\mathbf{v}^{(n)} \\
&\ \ \vdots \\
&= A\mathbf{x}^{(0)} + t_1A\mathbf{v}^{(1)} + t_2A\mathbf{v}^{(2)} + \cdots + t_nA\mathbf{v}^{(n)}.
\end{aligned}$$

Subtracting $\mathbf{b}$ from this result yields

$$A\mathbf{x}^{(n)} - \mathbf{b} = A\mathbf{x}^{(0)} - \mathbf{b} + t_1A\mathbf{v}^{(1)} + t_2A\mathbf{v}^{(2)} + \cdots + t_nA\mathbf{v}^{(n)}.$$

We now take the inner product of both sides with the vector $\mathbf{v}^{(k)}$ and use the properties of inner products and the fact that A is symmetric to obtain

$$\begin{aligned}
\langle A\mathbf{x}^{(n)} - \mathbf{b}, \mathbf{v}^{(k)}\rangle &= \langle A\mathbf{x}^{(0)} - \mathbf{b}, \mathbf{v}^{(k)}\rangle + t_1\langle A\mathbf{v}^{(1)}, \mathbf{v}^{(k)}\rangle + \cdots + t_n\langle A\mathbf{v}^{(n)}, \mathbf{v}^{(k)}\rangle \\
&= \langle A\mathbf{x}^{(0)} - \mathbf{b}, \mathbf{v}^{(k)}\rangle + t_1\langle \mathbf{v}^{(1)}, A\mathbf{v}^{(k)}\rangle + \cdots + t_n\langle \mathbf{v}^{(n)}, A\mathbf{v}^{(k)}\rangle.
\end{aligned}$$

The A-orthogonality property gives, for each k,

$$\langle A\mathbf{x}^{(n)} - \mathbf{b}, \mathbf{v}^{(k)}\rangle = \langle A\mathbf{x}^{(0)} - \mathbf{b}, \mathbf{v}^{(k)}\rangle + t_k\langle \mathbf{v}^{(k)}, A\mathbf{v}^{(k)}\rangle. \tag{7.29}$$

However $t_k\langle \mathbf{v}^{(k)}, A\mathbf{v}^{(k)}\rangle = \langle \mathbf{v}^{(k)}, \mathbf{b} - A\mathbf{x}^{(k-1)}\rangle$ so

$$\begin{aligned}
t_k\langle \mathbf{v}^{(k)}, A\mathbf{v}^{(k)}\rangle &= \langle \mathbf{v}^{(k)}, \mathbf{b} - A\mathbf{x}^{(0)} + A\mathbf{x}^{(0)} - A\mathbf{x}^{(1)} + \cdots - A\mathbf{x}^{(k-2)} + A\mathbf{x}^{(k-2)} - A\mathbf{x}^{(k-1)}\rangle \\
&= \langle \mathbf{v}^{(k)}, \mathbf{b} - A\mathbf{x}^{(0)}\rangle + \langle \mathbf{v}^{(k)}, A\mathbf{x}^{(0)} - A\mathbf{x}^{(1)}\rangle + \cdots + \langle \mathbf{v}^{(k)}, A\mathbf{x}^{(k-2)} - A\mathbf{x}^{(k-1)}\rangle.
\end{aligned}$$

But for any i,

$$\mathbf{x}^{(i)} = \mathbf{x}^{(i-1)} + t_i\mathbf{v}^{(i)} \quad \text{and} \quad A\mathbf{x}^{(i)} = A\mathbf{x}^{(i-1)} + t_iA\mathbf{v}^{(i)},$$

so

$$A\mathbf{x}^{(i-1)} - A\mathbf{x}^{(i)} = -t_iA\mathbf{v}^{(i)}.$$

Thus

$$t_k\langle \mathbf{v}^{(k)}, A\mathbf{v}^{(k)}\rangle = \langle \mathbf{v}^{(k)}, \mathbf{b} - A\mathbf{x}^{(0)}\rangle - t_1\langle \mathbf{v}^{(k)}, A\mathbf{v}^{(1)}\rangle - \cdots - t_{k-1}\langle \mathbf{v}^{(k)}, A\mathbf{v}^{(k-1)}\rangle.$$

Because of the A-orthogonality, $\langle \mathbf{v}^{(k)}, A\mathbf{v}^{(i)}\rangle = 0$, for $i \neq k$, so

$$\langle \mathbf{v}^{(k)}, A\mathbf{v}^{(k)}\rangle t_k = \langle \mathbf{v}^{(k)}, \mathbf{b} - A\mathbf{x}^{(0)}\rangle.$$

From Eq.(7.29),

$$\begin{aligned}
\langle A\mathbf{x}^{(n)} - \mathbf{b}, \mathbf{v}^{(k)}\rangle &= \langle A\mathbf{x}^{(0)} - \mathbf{b}, \mathbf{v}^{(k)}\rangle + \langle \mathbf{v}^{(k)}, \mathbf{b} - A\mathbf{x}^{(0)}\rangle \\
&= \langle A\mathbf{x}^{(0)} - \mathbf{b}, \mathbf{v}^{(k)}\rangle + \langle \mathbf{b} - A\mathbf{x}^{(0)}, \mathbf{v}^{(k)}\rangle \\
&= \langle A\mathbf{x}^{(0)} - \mathbf{b}, \mathbf{v}^{(k)}\rangle - \langle A\mathbf{x}^{(0)} - \mathbf{b}, \mathbf{v}^{(k)}\rangle = 0.
\end{aligned}$$

Hence the vector $A\mathbf{x}^{(n)} - \mathbf{b}$ is orthogonal to the A-orthogonal set of vectors $\{\mathbf{v}^{(1)}, \ldots, \mathbf{v}^{(n)}\}$. From this, it follows (see Exercise 13(b)) that $A\mathbf{x}^{(n)} - \mathbf{b} = \mathbf{0}$, so $A\mathbf{x}^{(n)} = \mathbf{b}$. ▪ ▪ ▪

Example 1 The linear system

$$\begin{aligned} 4x_1 + 3x_2 \qquad &= 24, \\ 3x_1 + 4x_2 - \ x_3 &= 30, \\ - \ x_2 + 4x_3 &= -24 \end{aligned}$$

has the exact solution $\mathbf{x}^* = (3, 4, -5)^t$. Show that the procedure described in Theorem 7.32 with $\mathbf{x}^{(0)} = (0, 0, 0)^t$ produces this exact solution after three iterations.

Solution We established in Example 2 of Section 7.4 that the coefficient matrix

$$A = \begin{bmatrix} 4 & 3 & 0 \\ 3 & 4 & -1 \\ 0 & -1 & 4 \end{bmatrix}.$$

of this system is positive definite. Let $\mathbf{v}^{(1)} = (1, 0, 0)^t$, $\mathbf{v}^{(2)} = (-3/4, 1, 0)^t$, and $\mathbf{v}^{(3)} = (-3/7, 4/7, 1)^t$. Then

$$\langle \mathbf{v}^{(1)}, A\mathbf{v}^{(2)} \rangle = \mathbf{v}^{(1)t} A\mathbf{v}^{(2)} = (1, 0, 0) \begin{bmatrix} 4 & 3 & 0 \\ 3 & 4 & -1 \\ 0 & -1 & 4 \end{bmatrix} \begin{bmatrix} -\frac{3}{4} \\ 1 \\ 0 \end{bmatrix} = 0,$$

$$\langle \mathbf{v}^{(1)}, A\mathbf{v}^{(3)} \rangle = (1, 0, 0) \begin{bmatrix} 4 & 3 & 0 \\ 3 & 4 & -1 \\ 0 & -1 & 4 \end{bmatrix} \begin{bmatrix} -\frac{3}{7} \\ \frac{4}{7} \\ 1 \end{bmatrix} = 0,$$

and

$$\langle \mathbf{v}^{(2)}, A\mathbf{v}^{(3)} \rangle = \left(-\frac{3}{4}, 1, 0 \right) \begin{bmatrix} 4 & 3 & 0 \\ 3 & 4 & -1 \\ 0 & -1 & 4 \end{bmatrix} \begin{bmatrix} -\frac{3}{7} \\ \frac{4}{7} \\ 1 \end{bmatrix} = 0.$$

Hence $\{\mathbf{v}^{(1)}, \mathbf{v}^{(2)}, \mathbf{v}^{(3)}\}$ is an A-orthogonal set.

Applying the iterations described in Theorem 7.22 for A with $\mathbf{x}^{(0)} = (0, 0, 0)^t$ and $\mathbf{b} = (24, 30, -24)^t$ gives

$$\mathbf{r}^{(0)} = \mathbf{b} - A\mathbf{x}^{(0)} = \mathbf{b} = (24, 30, -24)^t,$$

so

$$\langle \mathbf{v}^{(1)}, \mathbf{r}^{(0)} \rangle = \mathbf{v}^{(1)t} \mathbf{r}^{(0)} = 24, \quad \langle \mathbf{v}^{(1)}, A\mathbf{v}^{(1)} \rangle = 4, \quad \text{and} \quad t_0 = \frac{24}{4} = 6.$$

Hence

$$\mathbf{x}^{(1)} = \mathbf{x}^{(0)} + t_0 \mathbf{v}^{(1)} = (0, 0, 0)^t + 6(1, 0, 0)^t = (6, 0, 0)^t.$$

Continuing, we have

$$\mathbf{r}^{(1)} = \mathbf{b} - A\mathbf{x}^{(1)} = (0, 12, -24)^t; \quad t_1 = \frac{\langle \mathbf{v}^{(2)}, \mathbf{r}^{(1)} \rangle}{\langle \mathbf{v}^{(2)}, A\mathbf{v}^{(2)} \rangle} = \frac{12}{7/4} = \frac{48}{7};$$

$$\mathbf{x}^{(2)} = \mathbf{x}^{(1)} + t_1 \mathbf{v}^{(2)} = (6, 0, 0)^t + \frac{48}{7} \left(-\frac{3}{4}, 1, 0 \right)^t = \left(\frac{6}{7}, \frac{48}{7}, 0 \right)^t;$$

$$\mathbf{r}^{(2)} = \mathbf{b} - A\mathbf{x}^{(2)} = \left(0, 0, -\frac{120}{7} \right); \quad t_2 = \frac{\langle \mathbf{v}^{(3)}, \mathbf{r}^{(2)} \rangle}{\langle \mathbf{v}^{(3)}, A\mathbf{v}^{(3)} \rangle} = \frac{-120/7}{24/7} = -5;$$

and

$$\mathbf{x}^{(3)} = \mathbf{x}^{(2)} + t_2\mathbf{v}^{(3)} = \left(\frac{6}{7}, \frac{48}{7}, 0\right)^t + (-5)\left(-\frac{3}{7}, \frac{4}{7}, 1\right)^t = (3, 4, -5)^t.$$

Since we applied the technique $n = 3$ times, this must be the actual solution. ■

Before discussing how to determine the A-orthogonal set, we will continue the development. The use of an A-orthogonal set $\{\mathbf{v}^{(1)}, \ldots, \mathbf{v}^{(n)}\}$ of direction vectors gives what is called a *conjugate direction* method. The following theorem shows the orthogonality of the residual vectors $\mathbf{r}^{(k)}$ and the direction vectors $\mathbf{v}^{(j)}$. A proof of this result using mathematical induction is considered in Exercise 14.

Theorem 7.33 The residual vectors $\mathbf{r}^{(k)}$, where $k = 1, 2, \ldots, n$, for a conjugate direction method, satisfy the equations

$$\langle \mathbf{r}^{(k)}, \mathbf{v}^{(j)} \rangle = 0, \quad \text{for each} \quad j = 1, 2, \ldots, k.$$

■

The conjugate gradient method of Hestenes and Stiefel chooses the search directions $\{\mathbf{v}^{(k)}\}$ during the iterative process so that the residual vectors $\{\mathbf{r}^{(k)}\}$ are mutually orthogonal. To construct the direction vectors $\{\mathbf{v}^{(1)}, \mathbf{v}^{(2)}, \ldots\}$ and the approximations $\{\mathbf{x}^{(1)}, \mathbf{x}^{(2)}, \ldots\}$, we start with an initial approximation $\mathbf{x}^{(0)}$ and use the steepest descent direction $\mathbf{r}^{(0)} = \mathbf{b} - A\mathbf{x}^{(0)}$ as the first search direction $\mathbf{v}^{(1)}$.

Assume that the conjugate directions $\mathbf{v}^{(1)}, \ldots, \mathbf{v}^{(k-1)}$ and the approximations $\mathbf{x}^{(1)}, \ldots, \mathbf{x}^{(k-1)}$ have been computed with

$$\mathbf{x}^{(k-1)} = \mathbf{x}^{(k-2)} + t_{k-1}\mathbf{v}^{(k-1)},$$

where

$$\langle \mathbf{v}^{(i)}, A\mathbf{v}^{(j)} \rangle = 0 \quad \text{and} \quad \langle \mathbf{r}^{(i)}, \mathbf{r}^{(j)} \rangle = 0, \quad \text{for} \quad i \neq j.$$

If $\mathbf{x}^{(k-1)}$ is the solution to $A\mathbf{x} = \mathbf{b}$, we are done. Otherwise, $\mathbf{r}^{(k-1)} = \mathbf{b} - A\mathbf{x}^{(k-1)} \neq \mathbf{0}$ and Theorem 7.33 implies that $\langle \mathbf{r}^{(k-1)}, \mathbf{v}^{(i)} \rangle = 0$, for each $i = 1, 2, \ldots, k-1$.

We use $\mathbf{r}^{(k-1)}$ to generate $\mathbf{v}^{(k)}$ by setting

$$\mathbf{v}^{(k)} = \mathbf{r}^{(k-1)} + s_{k-1}\mathbf{v}^{(k-1)}.$$

We want to choose s_{k-1} so that

$$\langle \mathbf{v}^{(k-1)}, A\mathbf{v}^{(k)} \rangle = 0.$$

Since

$$A\mathbf{v}^{(k)} = A\mathbf{r}^{(k-1)} + s_{k-1}A\mathbf{v}^{(k-1)}$$

and

$$\langle \mathbf{v}^{(k-1)}, A\mathbf{v}^{(k)} \rangle = \langle \mathbf{v}^{(k-1)}, A\mathbf{r}^{(k-1)} \rangle + s_{k-1}\langle \mathbf{v}^{(k-1)}, A\mathbf{v}^{(k-1)} \rangle,$$

we will have $\langle \mathbf{v}^{(k-1)}, A\mathbf{v}^{(k)} \rangle = 0$ when

$$s_{k-1} = -\frac{\langle \mathbf{v}^{(k-1)}, A\mathbf{r}^{(k-1)} \rangle}{\langle \mathbf{v}^{(k-1)}, A\mathbf{v}^{(k-1)} \rangle}.$$

It can also be shown that with this choice of s_{k-1} we have $\langle \mathbf{v}^{(k)}, A\mathbf{v}^{(i)} \rangle = 0$, for each $i = 1, 2, \ldots, k-2$ (see [Lu], p. 245). Thus $\{\mathbf{v}^{(1)}, \ldots \mathbf{v}^{(k)}\}$ is an A-orthogonal set.

Having chosen $\mathbf{v}^{(k)}$, we compute

$$t_k = \frac{\langle \mathbf{v}^{(k)}, \mathbf{r}^{(k-1)} \rangle}{\langle \mathbf{v}^{(k)}, A\mathbf{v}^{(k)} \rangle} = \frac{\langle \mathbf{r}^{(k-1)} + s_{k-1}\mathbf{v}^{(k-1)}, \mathbf{r}^{(k-1)} \rangle}{\langle \mathbf{v}^{(k)}, A\mathbf{v}^{(k)} \rangle}$$

$$= \frac{\langle \mathbf{r}^{(k-1)}, \mathbf{r}^{(k-1)} \rangle}{\langle \mathbf{v}^{(k)}, A\mathbf{v}^{(k)} \rangle} + s_{k-1}\frac{\langle \mathbf{v}^{(k-1)}, \mathbf{r}^{(k-1)} \rangle}{\langle \mathbf{v}^{(k)}, A\mathbf{v}^{(k)} \rangle}.$$

By Theorem 7.33, $\langle \mathbf{v}^{(k-1)}, \mathbf{r}^{(k-1)} \rangle = 0$, so

$$t_k = \frac{\langle \mathbf{r}^{(k-1)}, \mathbf{r}^{(k-1)} \rangle}{\langle \mathbf{v}^{(k)}, A\mathbf{v}^{(k)} \rangle}. \tag{7.30}$$

Thus

$$\mathbf{x}^{(k)} = \mathbf{x}^{(k-1)} + t_k\mathbf{v}^{(k)}.$$

To compute $\mathbf{r}^{(k)}$, we multiply by A and subtract $\mathbf{b}$ to obtain

$$A\mathbf{x}^{(k)} - \mathbf{b} = A\mathbf{x}^{(k-1)} - \mathbf{b} + t_kA\mathbf{v}^{(k)}$$

or

$$\mathbf{r}^{(k)} = \mathbf{r}^{(k-1)} - t_kA\mathbf{v}^{(k)}.$$

This gives

$$\langle \mathbf{r}^{(k)}, \mathbf{r}^{(k)} \rangle = \langle \mathbf{r}^{(k-1)}, \mathbf{r}^{(k)} \rangle - t_k\langle A\mathbf{v}^{(k)}, \mathbf{r}^{(k)} \rangle = -t_k\langle \mathbf{r}^{(k)}, A\mathbf{v}^{(k)} \rangle.$$

Further, from Eq. (7.30),

$$\langle \mathbf{r}^{(k-1)}, \mathbf{r}^{(k-1)} \rangle = t_k\langle \mathbf{v}^{(k)}, A\mathbf{v}^{(k)} \rangle,$$

so

$$s_k = -\frac{\langle \mathbf{v}^{(k)}, A\mathbf{r}^{(k)} \rangle}{\langle \mathbf{v}^{(k)}, A\mathbf{v}^{(k)} \rangle} = -\frac{\langle \mathbf{r}^{(k)}, A\mathbf{v}^{(k)} \rangle}{\langle \mathbf{v}^{(k)}, A\mathbf{v}^{(k)} \rangle} = \frac{(1/t_k)\langle \mathbf{r}^{(k)}, \mathbf{r}^{(k)} \rangle}{(1/t_k)\langle \mathbf{r}^{(k-1)}, \mathbf{r}^{(k-1)} \rangle} = \frac{\langle \mathbf{r}^{(k)}, \mathbf{r}^{(k)} \rangle}{\langle \mathbf{r}^{(k-1)}, \mathbf{r}^{(k-1)} \rangle}.$$

In summary, we have

$$\mathbf{r}^{(0)} = \mathbf{b} - A\mathbf{x}^{(0)}; \quad \mathbf{v}^{(1)} = \mathbf{r}^{(0)};$$

and, for $k = 1, 2, \ldots, n$,

$$t_k = \frac{\langle \mathbf{r}^{(k-1)}, \mathbf{r}^{(k-1)} \rangle}{\langle \mathbf{v}^{(k)}, A\mathbf{v}^{(k)} \rangle}, \quad \mathbf{x}^{(k)} = \mathbf{x}^{(k-1)} + t_k\mathbf{v}^{(k)}, \quad \mathbf{r}^{(k)} = \mathbf{r}^{(k-1)} - t_kA\mathbf{v}^{(k)}, \quad s_k = \frac{\langle \mathbf{r}^{(k)}, \mathbf{r}^{(k)} \rangle}{\langle \mathbf{r}^{(k-1)}, \mathbf{r}^{(k-1)} \rangle},$$

and

$$\mathbf{v}^{(k+1)} = \mathbf{r}^{(k)} + s_k\mathbf{v}^{(k)}. \tag{7.31}$$

Preconditioning

Rather than presenting an algorithm for the conjugate gradient method using these formulas, we extend the method to include *preconditioning*. If the matrix A is ill-conditioned, the conjugate gradient method is highly susceptible to rounding errors. So, although the exact answer should be obtained in n steps, this is not usually the case. As a direct method the conjugate gradient method is not as good as Gaussian elimination with pivoting. The main use of the conjugate gradient method is as an iterative method applied to a better-conditioned system. In this case an acceptable approximate solution is often obtained in about $\sqrt{n}$ steps.

Preconditioning replaces a given system with one having the same solutions but with better convergence characteristics.

When preconditioning is used, the conjugate gradient method is not applied directly to the matrix A but to another positive definite matrix that a smaller condition number. We need to do this in such a way that once the solution to this new system is found it will be easy to obtain the solution to the original system. The expectation is that this will reduce the rounding error when the method is applied. To maintain the positive definiteness of the resulting matrix, we need to multiply on each side by a nonsingular matrix. We will denote this matrix by C^{-1}, and consider

$$\tilde{A} = C^{-1}A(C^{-1})^t,$$

with the hope that $\tilde{A}$ has a lower condition number than A. To simplify the notation, we use the matrix notation $C^{-t} \equiv \left(C^{-1}\right)^t$. Later in the section we will see a reasonable way to select C, but first we will consider the conjugate applied to $\tilde{A}$.

Consider the linear system

$$\tilde{A}\tilde{\mathbf{x}} = \tilde{\mathbf{b}},$$

where $\tilde{\mathbf{x}} = C^t\mathbf{x}$ and $\tilde{\mathbf{b}} = C^{-1}\mathbf{b}$. Then

$$\tilde{A}\tilde{\mathbf{x}} = (C^{-1}AC^{-t})(C^t\mathbf{x}) = C^{-1}A\mathbf{x}.$$

Thus, we could solve $\tilde{A}\tilde{\mathbf{x}} = \tilde{\mathbf{b}}$ for $\tilde{\mathbf{x}}$ and then obtain $\mathbf{x}$ by multiplying by C^{-t}. However, instead of rewriting equations (7.31) using $\tilde{\mathbf{r}}^{(k)}$, $\tilde{\mathbf{v}}^{(k)}$, $\tilde{t}_k$, $\tilde{\mathbf{x}}^{(k)}$, and $\tilde{s}_k$, we incorporate the preconditioning implicitly.

Since

$$\tilde{\mathbf{x}}^{(k)} = C^t\mathbf{x}^{(k)},$$

we have

$$\tilde{\mathbf{r}}^{(k)} = \tilde{\mathbf{b}} - \tilde{A}\tilde{\mathbf{x}}^{(k)} = C^{-1}\mathbf{b} - (C^{-1}AC^{-t})C^t\mathbf{x}^{(k)} = C^{-1}(\mathbf{b} - A\mathbf{x}^{(k)}) = C^{-1}\mathbf{r}^{(k)}.$$

Let $\tilde{\mathbf{v}}^{(k)} = C^t\mathbf{v}^{(k)}$ and $\mathbf{w}^{(k)} = C^{-1}\mathbf{r}^{(k)}$. Then

$$\tilde{s}_k = \frac{\langle \tilde{\mathbf{r}}^{(k)}, \tilde{\mathbf{r}}^{(k)} \rangle}{\langle \tilde{\mathbf{r}}^{(k-1)}, \tilde{\mathbf{r}}^{(k-1)} \rangle} = \frac{\langle C^{-1}\mathbf{r}^{(k)}, C^{-1}\mathbf{r}^{(k)} \rangle}{\langle C^{-1}\mathbf{r}^{(k-1)}, C^{-1}\mathbf{r}^{(k-1)} \rangle},$$

so

$$\tilde{s}_k = \frac{\langle \mathbf{w}^{(k)}, \mathbf{w}^{(k)} \rangle}{\langle \mathbf{w}^{(k-1)}, \mathbf{w}^{(k-1)} \rangle}. \tag{7.32}$$

Thus

$$\tilde{t}_k = \frac{\langle \tilde{\mathbf{r}}^{(k-1)}, \tilde{\mathbf{r}}^{(k-1)} \rangle}{\langle \tilde{\mathbf{v}}^{(k)}, \tilde{A}\tilde{\mathbf{v}}^{(k)} \rangle} = \frac{\langle C^{-1}\mathbf{r}^{(k-1)}, C^{-1}\mathbf{r}^{(k-1)} \rangle}{\langle C^t\mathbf{v}^{(k)}, C^{-1}AC^{-t}C^t\mathbf{v}^{(k)} \rangle} = \frac{\langle \mathbf{w}^{(k-1)}, \mathbf{w}^{(k-1)} \rangle}{\langle C^t\mathbf{v}^{(k)}, C^{-1}A\mathbf{v}^{(k)} \rangle}$$

and, since

$$\begin{aligned}\langle C^t\mathbf{v}^{(k)}, C^{-1}A\mathbf{v}^{(k)}\rangle &= [C^t\mathbf{v}^{(k)}]^tC^{-1}A\mathbf{v}^{(k)}\\ &= [\mathbf{v}^{(k)}]^tCC^{-1}A\mathbf{v}^{(k)} = [\mathbf{v}^{(k)}]^tA\mathbf{v}^{(k)} = \langle \mathbf{v}^{(k)}, A\mathbf{v}^{(k)}\rangle,\end{aligned}$$

we have

$$\tilde{t}_k = \frac{\langle \mathbf{w}^{(k-1)}, \mathbf{w}^{(k-1)}\rangle}{\langle \mathbf{v}^{(k)}, A\mathbf{v}^{(k)}\rangle}. \tag{7.33}$$

Further,

$$\tilde{\mathbf{x}}^{(k)} = \tilde{\mathbf{x}}^{(k-1)} + \tilde{t}_k\tilde{\mathbf{v}}^{(k)}, \quad \text{so} \quad C^t\mathbf{x}^{(k)} = C^t\mathbf{x}^{(k-1)} + \tilde{t}_kC^t\mathbf{v}^{(k)}$$

and

$$\mathbf{x}^{(k)} = \mathbf{x}^{(k-1)} + \tilde{t}_k\mathbf{v}^{(k)}. \tag{7.34}$$

Continuing,

$$\tilde{\mathbf{r}}^{(k)} = \tilde{\mathbf{r}}^{(k-1)} - \tilde{t}_k\tilde{A}\tilde{\mathbf{v}}^{(k)},$$

so

$$C^{-1}\mathbf{r}^{(k)} = C^{-1}\mathbf{r}^{(k-1)} - \tilde{t}_kC^{-1}AC^{-t}\tilde{v}^{(k)}, \qquad \mathbf{r}^{(k)} = \mathbf{r}^{(k-1)} - \tilde{t}_kAC^{-t}C^t\mathbf{v}^{(k)},$$

and

$$\mathbf{r}^{(k)} = \mathbf{r}^{(k-1)} - \tilde{t}_kA\mathbf{v}^{(k)}. \tag{7.35}$$

Finally,

$$\tilde{\mathbf{v}}^{(k+1)} = \tilde{\mathbf{r}}^{(k)} + \tilde{s}_k\tilde{\mathbf{v}}^{(k)} \quad \text{and} \quad C^t\mathbf{v}^{(k+1)} = C^{-1}\mathbf{r}^{(k)} + \tilde{s}_kC^t\mathbf{v}^{(k)},$$

so

$$\mathbf{v}^{(k+1)} = C^{-t}C^{-1}\mathbf{r}^{(k)} + \tilde{s}_k\mathbf{v}^{(k)} = C^{-t}\mathbf{w}^{(k)} + \tilde{s}_k\mathbf{v}^{(k)}. \tag{7.36}$$

The preconditioned conjugate gradient method is based on using equations (7.32)–(7.36) in the order (7.33), (7.34), (7.35), (7.32), and (7.36). Algorithm 7.5 implements this procedure.

ALGORITHM 7.5

Preconditioned Conjugate Gradient Method

To solve $A\mathbf{x} = \mathbf{b}$ given the preconditioning matrix C^{-1} and the initial approximation $\mathbf{x}^{(0)}$:

INPUT the number of equations and unknowns n; the entries a_{ij}, $1 \le i, j \le n$ of the matrix A; the entries b_j, $1 \le j \le n$ of the vector $\mathbf{b}$; the entries γ_{ij}, $1 \le i, j \le n$ of the preconditioning matrix C^{-1}, the entries x_i, $1 \le i \le n$ of the initial approximation $\mathbf{x} = \mathbf{x}^{(0)}$, the maximum number of iterations N; tolerance TOL.

OUTPUT the approximate solution $x_1, \ldots x_n$ and the residual $r_1, \ldots r_n$ or a message that the number of iterations was exceeded.

Step 1 Set $\mathbf{r} = \mathbf{b} - A\mathbf{x}$; (*Compute* $\mathbf{r}^{(0)}$.)
$\mathbf{w} = C^{-1}\mathbf{r}$; (*Note:* $\mathbf{w} = \mathbf{w}^{(0)}$)
$\mathbf{v} = C^{-t}\mathbf{w}$; (*Note:* $\mathbf{v} = \mathbf{v}^{(1)}$)
$\alpha = \sum_{j=1}^{n} w_j^2$.

Step 2 Set $k = 1$.

Step 3 While ($k \leq N$) do Steps 4–7.

Step 4 If $\|\mathbf{v}\| < TOL$, then
OUTPUT ('Solution vector'; $x_1, \ldots, x_n$);
OUTPUT ('with residual'; $r_1, \ldots, r_n$);
(*The procedure was successful.*)
STOP

Step 5 Set $\mathbf{u} = A\mathbf{v}$; (*Note:* $\mathbf{u} = A\mathbf{v}^{(k)}$)
$t = \dfrac{\alpha}{\sum_{j=1}^{n} v_j u_j}$; (*Note:* $t = t_k$)
$\mathbf{x} = \mathbf{x} + t\mathbf{v}$; (*Note:* $\mathbf{x} = \mathbf{x}^{(k)}$)
$\mathbf{r} = \mathbf{r} - t\mathbf{u}$; (*Note:* $\mathbf{r} = \mathbf{r}^{(k)}$)
$\mathbf{w} = C^{-1}\mathbf{r}$; (*Note:* $\mathbf{w} = \mathbf{w}^{(k)}$)
$\beta = \sum_{j=1}^{n} w_j^2$. (*Note:* $\beta = \langle \mathbf{w}^{(k)}, \mathbf{w}^{(k)} \rangle$)

Step 6 If $|\beta| < TOL$ then
if $\|\mathbf{r}\| < TOL$ then
OUTPUT('Solution vector'; $x_1, \ldots, x_n$);
OUTPUT('with residual'; $r_1, \ldots, r_n$);
(*The procedure was successful.*)
STOP

Step 7 Set $s = \beta/\alpha$; ($s = s_k$)
$\mathbf{v} = C^{-t}\mathbf{w} + s\mathbf{v}$; (*Note:* $\mathbf{v} = \mathbf{v}^{(k+1)}$)
$\alpha = \beta$; (*Update* α.)
$k = k + 1$.

Step 8 If ($k > n$) then
OUTPUT ('The maximum number of iterations was exceeded.');
(*The procedure was unsuccessful.*)
STOP. ■

The next example illustrates the calculations for an elementary problem.

Example 2 The linear system $A\mathbf{x} = \mathbf{b}$ given by

$$\begin{aligned} 4x_1 + 3x_2 &= 24, \\ 3x_1 + 4x_2 - x_3 &= 30, \\ - x_2 + 4x_3 &= -24 \end{aligned}$$

has solution $(3, 4, -5)^t$. Use the conjugate gradient method with $\mathbf{x}^{(0)} = (0, 0, 0)^t$ and no preconditioning, that is, with $C = C^{-1} = I$, to approximate the solution.

Solution The solution was considered in Example 2 of Section 7.4 where the SOR method were used with a nearly optimal value of $\omega = 1.25$.

For the conjugate gradient method we start with

$$\mathbf{r}^{(0)} = \mathbf{b} - A\mathbf{x}^{(0)} = \mathbf{b} = (24, 30, -24)^t;$$
$$\mathbf{w} = C^{-1}\mathbf{r}^{(0)} = (24, 30, -24)^t;$$
$$\mathbf{v}^{(1)} = C^{-t}\mathbf{w} = (24, 30, -24)^t;$$
$$\alpha = \langle \mathbf{w}, \mathbf{w} \rangle = 2052.$$

We start the first iteration with $k = 1$. Then

$$\mathbf{u} = A\mathbf{v}^{(1)} = (186.0, 216.0, -126.0)^t;$$
$$t_1 = \frac{\alpha}{\langle \mathbf{v}^{(1)}, \mathbf{u} \rangle} = 0.1469072165;$$
$$\mathbf{x}^{(1)} = \mathbf{x}^{(0)} + t_1\mathbf{v}^{(1)} = (3.525773196, 4.407216495, -3.525773196)^t;$$
$$\mathbf{r}^{(1)} = \mathbf{r}^{(0)} - t_1\mathbf{u} = (-3.32474227, -1.73195876, -5.48969072)^t;$$
$$\mathbf{w} = C^{-1}\mathbf{r}^{(1)} = \mathbf{r}^{(1)};$$
$$\beta = \langle \mathbf{w}, \mathbf{w} \rangle = 44.19029651;$$
$$s_1 = \frac{\beta}{\alpha} = 0.02153523222;$$
$$\mathbf{v}^{(2)} = C^{-t}\mathbf{w} + s_1\mathbf{v}^{(1)} = (-2.807896697, -1.085901793, -6.006536293)^t.$$

Set

$$\alpha = \beta = 44.19029651.$$

For the second iteration we have

$$\mathbf{u} = A\mathbf{v}^{(2)} = (-14.48929217, -6.760760967, -22.94024338)^t;$$
$$t_2 = 0.2378157558;$$
$$\mathbf{x}^{(2)} = (2.858011121, 4.148971939, -4.954222164)^t;$$
$$\mathbf{r}^{(2)} = (0.121039698, -0.124143281, -0.034139402)^t;$$
$$\mathbf{w} = C^{-1}\mathbf{r}^{(2)} = \mathbf{r}^{(2)};$$
$$\beta = 0.03122766148;$$
$$s_2 = 0.0007066633163;$$
$$\mathbf{v}^{(3)} = (0.1190554504, -0.1249106480, -0.03838400086)^t.$$

Set $\alpha = \beta = 0.03122766148$.

The third iteration gives

$$\mathbf{u} = A\mathbf{v}^{(3)} = (0.1014898976, -0.1040922099, -0.0286253554)^t;$$
$$t_3 = 1.192628008;$$
$$\mathbf{x}^{(3)} = (2.999999998, 4.000000002, -4.999999998)^t;$$
$$\mathbf{r}^{(3)} = (0.36 \times 10^{-8}, 0.39 \times 10^{-8}, -0.141 \times 10^{-8})^t.$$

Since $\mathbf{x}^{(3)}$ is nearly the exact solution, rounding error did not significantly effect the result. In Example 2 of Section 7.4, the SOR method with $\omega = 1.25$ required 14 iterations for an accuracy of 10^{-7}. It should be noted, however, that in this example, we are really comparing a direct method to iterative methods. ■

The next example illustrates the effect of preconditioning on a poorly conditioned matrix. In this example, we use $D^{-1/2}$ to represent the diagonal matrix whose entries are the reciprocals of the square roots of the diagonal entries of the coefficient matrix A. This is used as the preconditioner. Because the matrix A is positive definite we expect the eigenvalues of $D^{-1/2}AD^{-1/2}$ to be close to 1, with the result that the condition number of this matrix will be small relative to the condition number of A.

Example 3 Use Maple to find the eigenvalues and condition number of the matrix

$$A = \begin{bmatrix} 0.2 & 0.1 & 1 & 1 & 0 \\ 0.1 & 4 & -1 & 1 & -1 \\ 1 & -1 & 60 & 0 & -2 \\ 1 & 1 & 0 & 8 & 4 \\ 0 & -1 & -2 & 4 & 700 \end{bmatrix}$$

and compare these with the eigenvalues and condition number of the preconditioned matrix $D^{-1/2}AD^{-t/2}$.

Solution We first need to load the *LinearAlgebra* package and then enter the matrix.

with(*LinearAlgebra*):
$A := Matrix([[0.2, 0.1, 1, 1, 0], [0.1, 4, -1, 1, -1], [1, -1, 60, 0, -2],$
$[1, 1, 0, 8, 4], [0, -1, -2, 4, 700]])$

To determine the preconditioned matrix we first need the diagonal matrix, which being symmetric is also its transpose. its diagonal entries are specified by

$$a1 := \frac{1}{\sqrt{0.2}};\ a2 := \frac{1}{\sqrt{4.0}};\ a3 := \frac{1}{\sqrt{60.0}};\ a4 := \frac{1}{\sqrt{8.0}};\ a5 := \frac{1}{\sqrt{700.0}}$$

and the preconditioning matrix is

$CI := Matrix([[a1, 0, 0, 0, 0], [0, a2, 0, 0, 0], [0, 0, a3, 0, 0], [0, 0, 0, a4, 0], [0, 0, 0, 0, a5]])$

which Maple returns as

$$\begin{bmatrix} 2.23607 & 0 & 0 & 0 & 0 \\ 0 & .500000 & 0 & 0 & 0 \\ 0 & 0 & .129099 & 0 & 0 \\ 0 & 0 & 0 & .353553 & 0 \\ 0 & 0 & 0 & 0 & 0.0377965 \end{bmatrix}$$

The preconditioned matrix is

$AH := CI.A.Transpose(CI)$

$$\begin{bmatrix} 1.000002 & 0.1118035 & 0.2886744 & 0.7905693 & 0 \\ 0.1118035 & 1 & -0.0645495 & 0.1767765 & -0.0188983 \\ 0.2886744 & -0.0645495 & 0.9999931 & 0 & -0.00975898 \\ 0.7905693 & 0.1767765 & 0 & 0.9999964 & 0.05345219 \\ 0 & -0.0188983 & -0.00975898 & 0.05345219 & 1.000005 \end{bmatrix}$$

The eigenvalues of A and AH are found with

Eigenvalues(*A*); *Eigenvalues*(*AH*)

Maple gives these as

$$\textit{Eigenvalues of } A : 700.031, 60.0284, 0.0570747, 8.33845, 3.74533$$

$$\textit{Eigenvalues of } AH : 1.88052, 0.156370, 0.852686, 1.10159, 1.00884$$

The condition numbers of A and AH in the l_∞ norm are found with

ConditionNumber(A); *ConditionNumber*(AH)

which Maple gives as 13961.7 for A and 16.1155 for AH. It is certainly true in this case that AH is better conditioned that the original matrix A. ■

Illustration The linear system $A\mathbf{x} = \mathbf{b}$ with

$$A = \begin{bmatrix} 0.2 & 0.1 & 1 & 1 & 0 \\ 0.1 & 4 & -1 & 1 & -1 \\ 1 & -1 & 60 & 0 & -2 \\ 1 & 1 & 0 & 8 & 4 \\ 0 & -1 & -2 & 4 & 700 \end{bmatrix} \quad \text{and} \quad \mathbf{b} = \begin{bmatrix} 1 \\ 2 \\ 3 \\ 4 \\ 5 \end{bmatrix}$$

has the solution

$$\mathbf{x}^* = (7.859713071, 0.4229264082, -0.07359223906, -0.5406430164, 0.01062616286)^t.$$

Table 7.5 lists the results obtained by using the Jacobi, Gauss-Seidel, and SOR (with $\omega = 1.25$) iterative methods applied to the system with A with a tolerance of 0.01, as well as those when the Conjugate Gradient method is applied both in its unpreconditioned form and using the preconditioning matrix described in Example 3. The preconditioned conjugate gradient method not only gives the most accurate approximations, it also uses the smallest number of iterations. □

Table 7.5

Method	Number of Iterations	$\mathbf{x}^{(k)}$	$\|\mathbf{x}^* - \mathbf{x}^{(k)}\|_\infty$
Jacobi	49	$(7.86277141, 0.42320802, -0.07348669, -0.53975964, 0.01062847)^t$	0.00305834
Gauss-Seidel	15	$(7.83525748, 0.42257868, -0.07319124, -0.53753055, 0.01060903)^t$	0.02445559
SOR ($\omega = 1.25$)	7	$(7.85152706, 0.42277371, -0.07348303, -0.53978369, 0.01062286)^t$	0.00818607
Conjugate Gradient	5	$(7.85341523, 0.42298677, -0.07347963, -0.53987920, 0.008628916)^t$	0.00629785
Conjugate Gradient (Preconditioned)	4	$(7.85968827, 0.42288329, -0.07359878, -0.54063200, 0.01064344)^t$	0.00009312

The preconditioned conjugate gradient method is often used in the solution of large linear systems in which the matrix is sparse and positive definite. These systems must be solved to approximate solutions to boundary-value problems in ordinary-differential equations (Sections 11.3, 11.4, 11.5). The larger the system, the more impressive the conjugate gradient method becomes because it significantly reduces the number of iterations required. In these systems, the preconditioning matrix C is approximately equal to L in the Cholesky factorization LL^t of A. Generally, small entries in A are ignored and Cholesky's method is applied to obtain what is called an incomplete LL^t factorization of A. Thus, $C^{-t}C^{-1} \approx A^{-1}$ and a good approximation is obtained. More information about the conjugate gradient method can be found in [Kelley].

EXERCISE SET 7.6

1. The linear system

$$x_1 + \frac{1}{2}x_2 = \frac{5}{21},$$
$$\frac{1}{2}x_1 + \frac{1}{3}x_2 = \frac{11}{84}$$

has solution $(x_1, x_2)^t = (1/6, 1/7)^t$.

   a. Solve the linear system using Gaussian elimination with two-digit rounding arithmetic.
   b. Solve the linear system using the conjugate gradient method ($C = C^{-1} = I$) with two-digit rounding arithmetic.
   c. Which method gives the better answer?
   d. Choose $C^{-1} = D^{-1/2}$. Does this choice improve the conjugate gradient method?

2. The linear system

$$0.1x_1 + 0.2x_2 = 0.3,$$
$$0.2x_1 + 113x_2 = 113.2$$

has solution $(x_1, x_2)^t = (1, 1)^t$. Repeat the directions for Exercise 1 on this linear system.

3. The linear system

$$x_1 + \frac{1}{2}x_2 + \frac{1}{3}x_3 = \frac{5}{6},$$
$$\frac{1}{2}x_1 + \frac{1}{3}x_2 + \frac{1}{4}x_3 = \frac{5}{12},$$
$$\frac{1}{3}x_1 + \frac{1}{4}x_2 + \frac{1}{5}x_3 = \frac{17}{60}$$

has solution $(1, -1, 1)^t$.

   a. Solve the linear system using Gaussian elimination with three-digit rounding arithmetic.
   b. Solve the linear system using the conjugate gradient method with three-digit rounding arithmetic.
   c. Does pivoting improve the answer in (a)?
   d. Repeat part (b) using $C^{-1} = D^{-1/2}$. Does this improve the answer in (b)?

4. Repeat Exercise 3 using single-precision arithmetic on a computer.

5. Perform only two steps of the conjugate gradient method with $C = C^{-1} = I$ on each of the following linear systems. Compare the results in parts (b) and (c) to the results obtained in parts (b) and (c) of Exercise 2 of Section 7.3 and Exercise 2 of Section 7.4.

   a.
$$3x_1 - x_2 + x_3 = 1,$$
$$-x_1 + 6x_2 + 2x_3 = 0,$$
$$x_1 + 2x_2 + 7x_3 = 4.$$

   b.
$$10x_1 - x_2 = 9,$$
$$-x_1 + 10x_2 - 2x_3 = 7,$$
$$-2x_2 + 10x_3 = 6.$$

c.
$$\begin{aligned} 10x_1 + 5x_2 &= 6, \\ 5x_1 + 10x_2 - 4x_3 &= 25, \\ -\ 4x_2 + 8x_3 - x_4 &= -11, \\ -\ x_3 + 5x_4 &= -11. \end{aligned}$$

d.
$$\begin{aligned} 4x_1 + x_2 - x_3 + x_4 &= -2, \\ x_1 + 4x_2 - x_3 - x_4 &= -1, \\ -x_1 - x_2 + 5x_3 + x_4 &= 0, \\ x_1 - x_2 + x_3 + 3x_4 &= 1. \end{aligned}$$

e.
$$\begin{aligned} 4x_1 + x_2 + x_3 + x_5 &= 6, \\ x_1 + 3x_2 + x_3 + x_4 &= 6, \\ x_1 + x_2 + 5x_3 - x_4 - x_5 &= 6, \\ x_2 - x_3 + 4x_4 &= 6, \\ x_1 - x_3 + + 4x_5 &= 6. \end{aligned}$$

f.
$$\begin{aligned} 4x_1 - x_2 - x_4 &= 0, \\ -x_1 + 4x_2 - x_3 - x_5 &= 5, \\ -\ x_2 + 4x_3 - x_6 &= 0, \\ -x_1 + 4x_4 - x_5 &= 6, \\ -\ x_2 - x_4 + 4x_5 - x_6 &= -2, \\ -\ x_3 - x_5 + 4x_6 &= 6. \end{aligned}$$

6. Repeat Exercise 5 using $C^{-1} = D^{-1/2}$.
7. Repeat Exercise 5 with $TOL = 10^{-3}$ in the l_∞ norm. Compare the results in parts (b) and (c) to those obtained in Exercises 6 and 8 of Section 7.3 and Exercise 6 of Section 7.4.
8. Repeat Exercise 7 using $C^{-1} = D^{-1/2}$.
9. Approximate solutions to the following linear systems $A\mathbf{x} = \mathbf{b}$ to within 10^{-5} in the l_∞ norm.

(i)

$$a_{i,j} = \begin{cases} 4, & \text{when } j = i \text{ and } i = 1, 2, \ldots, 16, \\ -1, & \text{when } \begin{cases} j = i + 1 \text{ and } i = 1, 2, 3, 5, 6, 7, 9, 10, 11, 13, 14, 15, \\ j = i - 1 \text{ and } i = 2, 3, 4, 6, 7, 8, 10, 11, 12, 14, 15, 16, \\ j = i + 4 \text{ and } i = 1, 2, \ldots, 12, \\ j = i - 4 \text{ and } i = 5, 6, \ldots, 16, \end{cases} \\ 0, & \text{otherwise} \end{cases}$$

and

$$\begin{aligned} \mathbf{b} = (&1.902207, 1.051143, 1.175689, 3.480083, 0.819600, -0.264419, \\ &-0.412789, 1.175689, 0.913337, -0.150209, -0.264419, 1.051143, \\ &1.966694, 0.913337, 0.819600, 1.902207)^t \end{aligned}$$

(ii)

$$a_{i,j} = \begin{cases} 4, & \text{when } j = i \text{ and } i = 1, 2, \ldots, 25, \\ -1, & \text{when } \begin{cases} j = i + 1 \text{ and } i = \begin{cases} 1, 2, 3, 4, 6, 7, 8, 9, 11, 12, 13, 14, \\ 16, 17, 18, 19, 21, 22, 23, 24, \end{cases} \\ j = i - 1 \text{ and } i = \begin{cases} 2, 3, 4, 5, 7, 8, 9, 10, 12, 13, 14, 15, \\ 17, 18, 19, 20, 22, 23, 24, 25, \end{cases} \\ j = i + 5 \text{ and } i = 1, 2, \ldots, 20, \\ j = i - 5 \text{ and } i = 6, 7, \ldots, 25, \end{cases} \\ 0, & \text{otherwise} \end{cases}$$

and

$$\mathbf{b} = (1, 0, -1, 0, 2, 1, 0, -1, 0, 2, 1, 0, -1, 0, 2, 1, 0, -1, 0, 2, 1, 0, -1, 0, 2)^t$$

(iii)

$$a_{i,j} = \begin{cases} 2i, & \text{when } j = i \text{ and } i = 1, 2, \ldots, 40, \\ -1, & \text{when } \begin{cases} j = i+1 \text{ and } i = 1, 2, \ldots, 39, \\ j = i-1 \text{ and } i = 2, 3, \ldots, 40, \end{cases} \\ 0, & \text{otherwise} \end{cases}$$

and $b_i = 1.5i - 6$, for each $i = 1, 2, \ldots, 40$

a. Use the Jacobi method, **b.** Use the Gauss-Seidel method,

c. Use the SOR method with $\omega = 1.3$ in (i), $\omega = 1.2$ in (ii), and $\omega = 1.1$ in (iii).

d. Use the conjugate gradient method and preconditioning with $C^{-1} = D^{-1/2}$.

10. Solve the linear system in Exercise 16(b) of Exercise Set 7.3 using the conjugate gradient method with $C^{-1} = I$.

11. Let

$$A_1 = \begin{bmatrix} 4 & -1 & 0 & 0 \\ -1 & 4 & -1 & 0 \\ 0 & -1 & 4 & -1 \\ 0 & 0 & -1 & 4 \end{bmatrix}, \quad -I = \begin{bmatrix} -1 & 0 & 0 & 0 \\ 0 & -1 & 0 & 0 \\ 0 & 0 & -1 & 0 \\ 0 & 0 & 0 & -1 \end{bmatrix}, \quad \text{and}$$

$$O = \begin{bmatrix} 0 & 0 & 0 & 0 \\ 0 & 0 & 0 & 0 \\ 0 & 0 & 0 & 0 \\ 0 & 0 & 0 & 0 \end{bmatrix}.$$

Form the 16×16 matrix A in partitioned form,

$$A = \begin{bmatrix} A_1 & -I & O & O \\ -I & A_1 & -I & O \\ O & -I & A_1 & -I \\ O & O & -I & A_1 \end{bmatrix}.$$

Let $\mathbf{b} = (1, 2, 3, 4, 5, 6, 7, 8, 9, 0, 1, 2, 3, 4, 5, 6)^t$.

a. Solve $A\mathbf{x} = \mathbf{b}$ using the conjugate gradient method with tolerance 0.05.

b. Solve $A\mathbf{x} = \mathbf{b}$ using the preconditioned conjugate gradient method with $C^{-1} = D^{-1/2}$ and tolerance 0.05.

c. Is there any tolerance for which the methods of part (a) and part (b) require a different number of iterations?

12. Use the transpose properties given in Theorem 6.14 on page 390 to prove Theorem 7.30.

13. a. Show that an A-orthogonal set of nonzero vectors associated with a positive definite matrix is linearly independent.

b. Show that if $\{\mathbf{v}^{(1)}, \mathbf{v}^{(2)}, \ldots, \mathbf{v}^{(n)}\}$ is a set of A-orthogonal nonzero vectors in $\mathbb{R}$ and $\mathbf{z}^t\mathbf{v}^{(i)} = \mathbf{0}$, for each $i = 1, 2, \ldots, n$, then $\mathbf{z} = \mathbf{0}$.

14. Prove Theorem 7.33 using mathematical induction as follows:

a. Show that $\langle \mathbf{r}^{(1)}, \mathbf{v}^{(1)} \rangle = 0$.

b. Assume that $\langle \mathbf{r}^{(k)}, \mathbf{v}^{(j)} \rangle = 0$, for each $k \leq l$ and $j = 1, 2, \ldots, k$, and show that this implies that $\langle \mathbf{r}^{(l+1)}, \mathbf{v}^{(j)} \rangle = 0$, for each $j = 1, 2, \ldots, l$.

c. Show that $\langle \mathbf{r}^{(l+1)}, \mathbf{v}^{(l+1)} \rangle = 0$.

15. In Example 3 the eigenvalues were found for the matrix A and the conditioned matrix AH. Use these to determine the condition numbers of A and AH in the l_2 norm, and compare your results to those given with the Maple commands *ConditionNumber*(A,2) and *ConditionNumber*(AH,2).

7.7 Survey of Methods and Software

In this chapter we studied iterative techniques to approximate the solution of linear systems. We began with the Jacobi method and the Gauss-Seidel method to introduce the iterative methods. Both methods require an arbitrary initial approximation $\mathbf{x}^{(0)}$ and generate a sequence of vectors $\mathbf{x}^{(i+1)}$ using an equation of the form

$$\mathbf{x}^{(i+1)} = T\mathbf{x}^{(i)} + \mathbf{c}.$$

It was noted that the method will converge if and only if the spectral radius of the iteration matrix $\rho(T) < 1$, and the smaller the spectral radius, the faster the convergence. Analysis of the residual vectors of the Gauss-Seidel technique led to the SOR iterative method, which involves a parameter ω to speed convergence.

Aleksei Nikolaevich Krylov (1863–1945) worked in applied mathematics, primarily in the areas of boundary value problems, the acceleration of convergence of Fourier series, and various classical problems involving mechanical systems. During the early 1930s he was the Director of the Physics-Mathematics Institute of the Soviet Academy of Sciences.

These iterative methods and modifications are used extensively in the solution of linear systems that arise in the numerical solution of boundary value problems and partial differential equations (see Chapters 11 and 12). These systems are often very large, on the order of 10,000 equations in 10,000 unknowns, and are sparse with their nonzero entries in predictable positions. The iterative methods are also useful for other large sparse systems and are easily adapted for efficient use on parallel computers.

Almost all commercial and public domain packages that contain iterative methods for the solution of a linear system of equations require a preconditioner to be used with the method. Faster convergence of iterative solvers is often achieved by using a preconditioner. A preconditioner produces an equivalent system of equations that hopefully exhibits better convergence characteristics than the original system. The IMSL Library has a preconditioned conjugate gradient method, and the NAG Library has several subroutines for the iterative solution of linear systems.

All of the subroutines are based on Krylov subspaces. Saad [Sa2] has a detailed description of Krylov subspace methods. The packages LINPACK and LAPACK contain only direct methods for the solution of linear systems; however, the packages do contain many subroutines that are used by the iterative solvers. The public domain packages IML++, ITPACK, SLAP, and Templates, contain iterative methods. MATLAB contains several iterative methods that are also based on Krylov subspaces.

The concepts of condition number and poorly conditioned matrices were introduced in Section 7.5. Many of the subroutines for solving a linear system or for factoring a matrix into an LU factorization include checks for ill-conditioned matrices and also give an estimate of the condition number. LAPACK has numerous routines that include the estimate of a condition number, as do the ISML and NAG libraries.

LAPACK, LINPACK, the IMSL Library, and the NAG Library have subroutines that improve on a solution to a linear system that is poorly conditioned. The subroutines test the condition number and then use iterative refinement to obtain the most accurate solution possible given the precision of the computer.

More information on the use of iterative methods for solving linear systems can be found in Varga [Var1], Young [Y], Hageman and Young [HY], and Axelsson [Ax]. Iterative methods for large sparse systems are discussed in Barrett et al [Barr], Hackbusch [Hac], Kelley [Kelley], and Saad [Sa2].

CHAPTER

8 Approximation Theory

Introduction

Hooke's law states that when a force is applied to a spring constructed of uniform material, the length of the spring is a linear function of that force. We can write the linear function as $F(l) = k(l - E)$, where $F(l)$ represents the force required to stretch the spring l units, the constant E represents the length of the spring with no force applied, and the constant k is the spring constant.

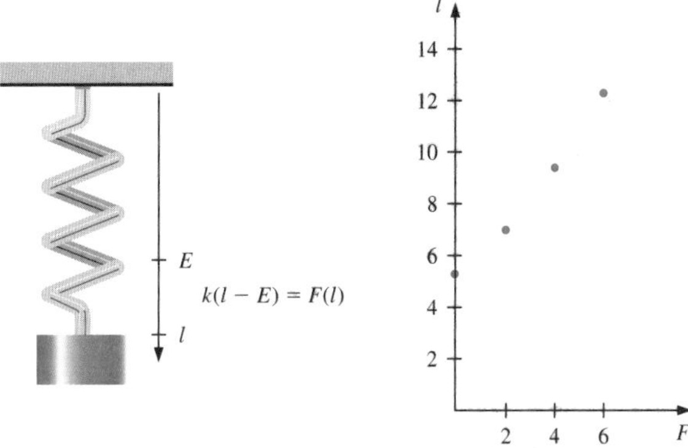

Suppose we want to determine the spring constant for a spring that has initial length 5.3 in. We apply forces of 2, 4, and 6 lb to the spring and find that its length increases to 7.0, 9.4, and 12.3 in., respectively. A quick examination shows that the points $(0, 5.3)$, $(2, 7.0)$, $(4, 9.4)$, and $(6, 12.3)$ do not quite lie in a straight line. Although we could use a random pair of these data points to approximate the spring constant, it would seem more reasonable to find the line that *best* approximates all the data points to determine the constant. This type of approximation will be considered in this chapter, and this spring application can be found in Exercise 7 of Section 8.1.

Approximation theory involves two general types of problems. One problem arises when a function is given explicitly, but we wish to find a "simpler" type of function, such as a polynomial, to approximate values of the given function. The other problem in approximation theory is concerned with fitting functions to given data and finding the "best" function in a certain class to represent the data.

Both problems have been touched upon in Chapter 3. The nth Taylor polynomial about the number x_0 is an excellent approximation to an $(n + 1)$-times differentiable function f in a small neighborhood of x_0. The Lagrange interpolating polynomials, or, more generally, osculatory polynomials, were discussed both as approximating polynomials and as polynomials to fit certain data. Cubic splines were also discussed in that chapter. In this chapter, limitations to these techniques are considered, and other avenues of approach are discussed.

8.1 Discrete Least Squares Approximation

Consider the problem of estimating the values of a function at nontabulated points, given the experimental data in Table 8.1.

Table 8.1

x_i	y_i	x_i	y_i
1	1.3	6	8.8
2	3.5	7	10.1
3	4.2	8	12.5
4	5.0	9	13.0
5	7.0	10	15.6

Figure 8.1 shows a graph of the values in Table 8.1. From this graph, it appears that the actual relationship between x and y is linear. The likely reason that no line precisely fits the data is because of errors in the data. So it is unreasonable to require that the approximating function agree exactly with the data. In fact, such a function would introduce oscillations that were not originally present. For example, the graph of the ninth-degree interpolating polynomial shown in unconstrained mode for the data in Table 8.1 is obtained in Maple using the commands

$p := interp([1, 2, 3, 4, 5, 6, 7, 8, 9, 10], [1.3, 3.5, 4.2, 5.0, 7.0, 8.8, 10.1, 12.5, 13.0, 15.6], x):$
$plot(p, x = 1..10)$

Figure 8.1

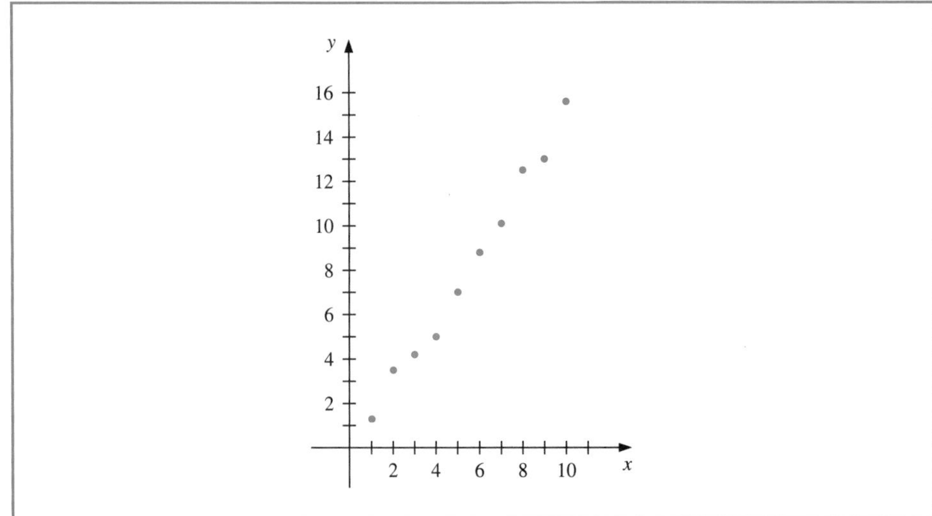

The plot obtained (with the data points added) is shown in Figure 8.2.

Figure 8.2

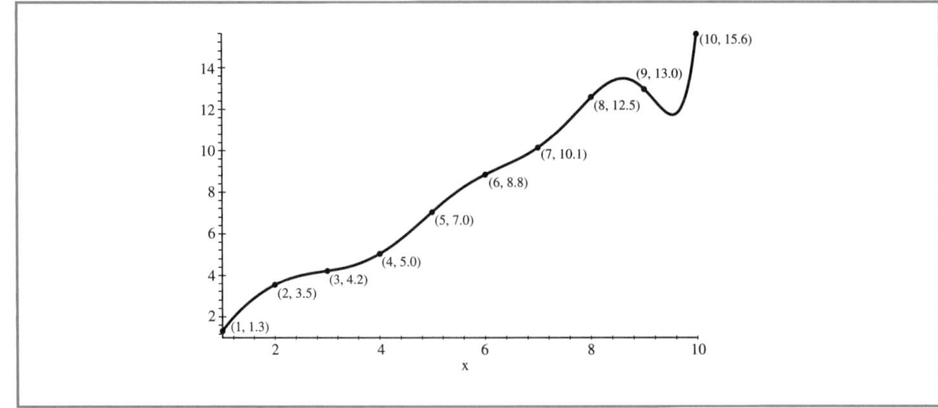

This polynomial is clearly a poor predictor of information between a number of the data points. A better approach would be to find the "best" (in some sense) approximating line, even if it does not agree precisely with the data at any point.

Let $a_1 x_i + a_0$ denote the ith value on the approximating line and y_i be the ith given y-value. We assume throughout that the independent variables, the x_i, are exact, it is the dependent variables, the y_i, that are suspect. This is a reasonable assumption in most experimental situations.

The problem of finding the equation of the best linear approximation in the absolute sense requires that values of a_0 and a_1 be found to minimize

$$E_\infty(a_0, a_1) = \max_{1\le i\le 10} \{|y_i - (a_1 x_i + a_0)|\}.$$

This is commonly called a **minimax** problem and cannot be handled by elementary techniques.

Another approach to determining the best linear approximation involves finding values of a_0 and a_1 to minimize

$$E_1(a_0, a_1) = \sum_{i=1}^{10} |y_i - (a_1 x_i + a_0)|.$$

This quantity is called the **absolute deviation**. To minimize a function of two variables, we need to set its partial derivatives to zero and simultaneously solve the resulting equations. In the case of the absolute deviation, we need to find a_0 and a_1 with

$$0 = \frac{\partial}{\partial a_0} \sum_{i=1}^{10} |y_i - (a_1 x_i + a_0)| \quad \text{and} \quad 0 = \frac{\partial}{\partial a_1} \sum_{i=1}^{10} |y_i - (a_1 x_i + a_0)|.$$

The problem is that the absolute-value function is not differentiable at zero, and we might not be able to find solutions to this pair of equations.

Linear Least Squares

The **least squares** approach to this problem involves determining the best approximating line when the error involved is the sum of the squares of the differences between the y-values on the approximating line and the given y-values. Hence, constants a_0 and a_1 must be found that minimize the least squares error:

$$E_2(a_0, a_1) = \sum_{i=1}^{10} \left[y_i - (a_1 x_i + a_0)\right]^2.$$

The least squares method is the most convenient procedure for determining best linear approximations, but there are also important theoretical considerations that favor it. The minimax approach generally assigns too much weight to a bit of data that is badly in error, whereas the absolute deviation method does not give sufficient weight to a point that is considerably out of line with the approximation. The least squares approach puts substantially more weight on a point that is out of line with the rest of the data, but will not permit that point to completely dominate the approximation. An additional reason for considering the least squares approach involves the study of the statistical distribution of error. (See [Lar], pp. 463–481.)

The general problem of fitting the best least squares line to a collection of data $\{(x_i, y_i)\}_{i=1}^m$ involves minimizing the total error,

$$E \equiv E_2(a_0, a_1) = \sum_{i=1}^{m} \left[y_i - (a_1 x_i + a_0)\right]^2,$$

with respect to the parameters a_0 and a_1. For a minimum to occur, we need both

$$\frac{\partial E}{\partial a_0} = 0 \quad \text{and} \quad \frac{\partial E}{\partial a_1} = 0,$$

that is,

$$0 = \frac{\partial}{\partial a_0} \sum_{i=1}^{m} \left[(y_i - (a_1 x_i - a_0)\right]^2 = 2 \sum_{i=1}^{m} (y_i - a_1 x_i - a_0)(-1)$$

and

$$0 = \frac{\partial}{\partial a_1} \sum_{i=1}^{m} \left[y_i - (a_1 x_i + a_0)\right]^2 = 2 \sum_{i=1}^{m} (y_i - a_1 x_i - a_0)(-x_i).$$

The word normal as used here implies perpendicular. The normal equations are obtained by finding perpendicular directions to a multidimensional surface.

These equations simplify to the **normal equations**:

$$a_0 \cdot m + a_1 \sum_{i=1}^{m} x_i = \sum_{i=1}^{m} y_i \quad \text{and} \quad a_0 \sum_{i=1}^{m} x_i + a_1 \sum_{i=1}^{m} x_i^2 = \sum_{i=1}^{m} x_i y_i.$$

The solution to this system of equations is

$$a_0 = \frac{\displaystyle\sum_{i=1}^{m} x_i^2 \sum_{i=1}^{m} y_i - \sum_{i=1}^{m} x_i y_i \sum_{i=1}^{m} x_i}{m\left(\displaystyle\sum_{i=1}^{m} x_i^2\right) - \left(\displaystyle\sum_{i=1}^{m} x_i\right)^2} \tag{8.1}$$

and

$$a_1 = \frac{m \displaystyle\sum_{i=1}^{m} x_i y_i - \sum_{i=1}^{m} x_i \sum_{i=1}^{m} y_i}{m\left(\displaystyle\sum_{i=1}^{m} x_i^2\right) - \left(\displaystyle\sum_{i=1}^{m} x_i\right)^2}. \tag{8.2}$$

Example 1 Find the least squares line approximating the data in Table 8.1.

Solution We first extend the table to include x_i^2 and $x_i y_i$ and sum the columns. This is shown in Table 8.2.

Table 8.2

x_i	y_i	x_i^2	$x_i y_i$	$P(x_i) = 1.538x_i - 0.360$
1	1.3	1	1.3	1.18
2	3.5	4	7.0	2.72
3	4.2	9	12.6	4.25
4	5.0	16	20.0	5.79
5	7.0	25	35.0	7.33
6	8.8	36	52.8	8.87
7	10.1	49	70.7	10.41
8	12.5	64	100.0	11.94
9	13.0	81	117.0	13.48
10	15.6	100	156.0	15.02
55	81.0	385	572.4	$E = \sum_{i=1}^{10}(y_i - P(x_i))^2 \approx 2.34$

The normal equations (8.1) and (8.2) imply that

$$a_0 = \frac{385(81) - 55(572.4)}{10(385) - (55)^2} = -0.360$$

and

$$a_1 = \frac{10(572.4) - 55(81)}{10(385) - (55)^2} = 1.538,$$

so $P(x) = 1.538x - 0.360$. The graph of this line and the data points are shown in Figure 8.3. The approximate values given by the least squares technique at the data points are in Table 8.2. ■

Figure 8.3

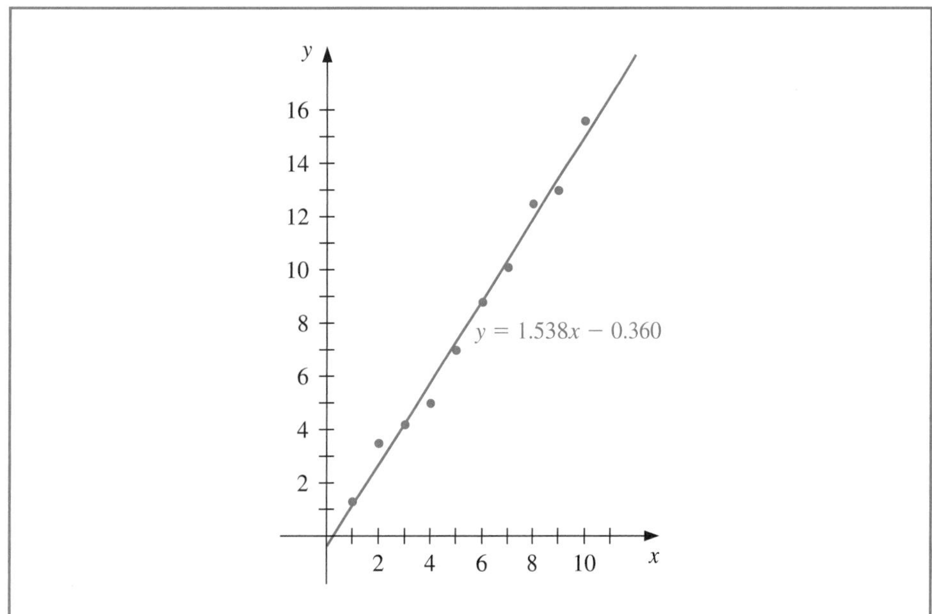

Polynomial Least Squares

The general problem of approximating a set of data, $\{(x_i, y_i) \mid i = 1, 2, \ldots, m\}$, with an algebraic polynomial

$$P_n(x) = a_n x^n + a_{n-1}x^{n-1} + \cdots + a_1 x + a_0,$$

of degree $n < m - 1$, using the least squares procedure is handled similarly. We choose the constants $a_0, a_1, \ldots, a_n$ to minimize the least squares error $E = E_2(a_0, a_1, \ldots, a_n)$, where

$$\begin{aligned} E &= \sum_{i=1}^{m} (y_i - P_n(x_i))^2 \\ &= \sum_{i=1}^{m} y_i^2 - 2\sum_{i=1}^{m} P_n(x_i)y_i + \sum_{i=1}^{m} (P_n(x_i))^2 \end{aligned}$$

$$= \sum_{i=1}^{m} y_i^2 - 2\sum_{i=1}^{m}\left(\sum_{j=0}^{n} a_j x_i^j\right) y_i + \sum_{i=1}^{m}\left(\sum_{j=0}^{n} a_j x_i^j\right)^2$$

$$= \sum_{i=1}^{m} y_i^2 - 2\sum_{j=0}^{n} a_j\left(\sum_{i=1}^{m} y_i x_i^j\right) + \sum_{j=0}^{n}\sum_{k=0}^{n} a_j a_k\left(\sum_{i=1}^{m} x_i^{j+k}\right).$$

As in the linear case, for E to be minimized it is necessary that $\partial E/\partial a_j = 0$, for each $j = 0, 1, \ldots, n$. Thus, for each j, we must have

$$0 = \frac{\partial E}{\partial a_j} = -2\sum_{i=1}^{m} y_i x_i^j + 2\sum_{k=0}^{n} a_k \sum_{i=1}^{m} x_i^{j+k}.$$

This gives $n + 1$ **normal equations** in the $n + 1$ unknowns a_j. These are

$$\sum_{k=0}^{n} a_k \sum_{i=1}^{m} x_i^{j+k} = \sum_{i=1}^{m} y_i x_i^j, \quad \text{for each } j = 0, 1, \ldots, n. \tag{8.3}$$

It is helpful to write the equations as follows:

$$a_0\sum_{i=1}^{m} x_i^0 + a_1\sum_{i=1}^{m} x_i^1 + a_2\sum_{i=1}^{m} x_i^2 + \cdots + a_n\sum_{i=1}^{m} x_i^n = \sum_{i=1}^{m} y_i x_i^0,$$

$$a_0\sum_{i=1}^{m} x_i^1 + a_1\sum_{i=1}^{m} x_i^2 + a_2\sum_{i=1}^{m} x_i^3 + \cdots + a_n\sum_{i=1}^{m} x_i^{n+1} = \sum_{i=1}^{m} y_i x_i^1,$$

$$\vdots$$

$$a_0\sum_{i=1}^{m} x_i^n + a_1\sum_{i=1}^{m} x_i^{n+1} + a_2\sum_{i=1}^{m} x_i^{n+2} + \cdots + a_n\sum_{i=1}^{m} x_i^{2n} = \sum_{i=1}^{m} y_i x_i^n.$$

These *normal equations* have a unique solution provided that the x_i are distinct (see Exercise 14).

Example 2 Fit the data in Table 8.3 with the discrete least squares polynomial of degree at most 2.

Table 8.3

i	x_i	y_i
1	0	1.0000
2	0.25	1.2840
3	0.50	1.6487
4	0.75	2.1170
5	1.00	2.7183

Solution For this problem, $n = 2, m = 5$, and the three normal equations are

$$\begin{aligned} 5a_0 + \quad 2.5a_1 + \quad 1.875a_2 &= 8.7680,\\ 2.5a_0 + \quad 1.875a_1 + 1.5625a_2 &= 5.4514,\\ 1.875a_0 + 1.5625a_1 + 1.3828a_2 &= 4.4015. \end{aligned}$$

To solve this system using Maple, we first define the equations

$eq1 := 5a0 + 2.5a1 + 1.875a2 = 8.7680:$
$eq2 := 2.5a0 + 1.875a1 + 1.5625a2 = 5.4514:$
$eq3 := 1.875a0 + 1.5625a1 + 1.3828a2 = 4.4015$

and then solve the system with

$solve(\{eq1, eq2, eq3\}, \{a0, a1, a2\})$

This gives

$$\{a_0 = 1.005075519, \quad a_1 = 0.8646758482, \quad a_2 = .8431641518\}$$

Thus the least squares polynomial of degree 2 fitting the data in Table 8.3 is

$$P_2(x) = 1.0051 + 0.86468x + 0.84316x^2,$$

whose graph is shown in Figure 8.4. At the given values of x_i we have the approximations shown in Table 8.4.

Figure 8.4

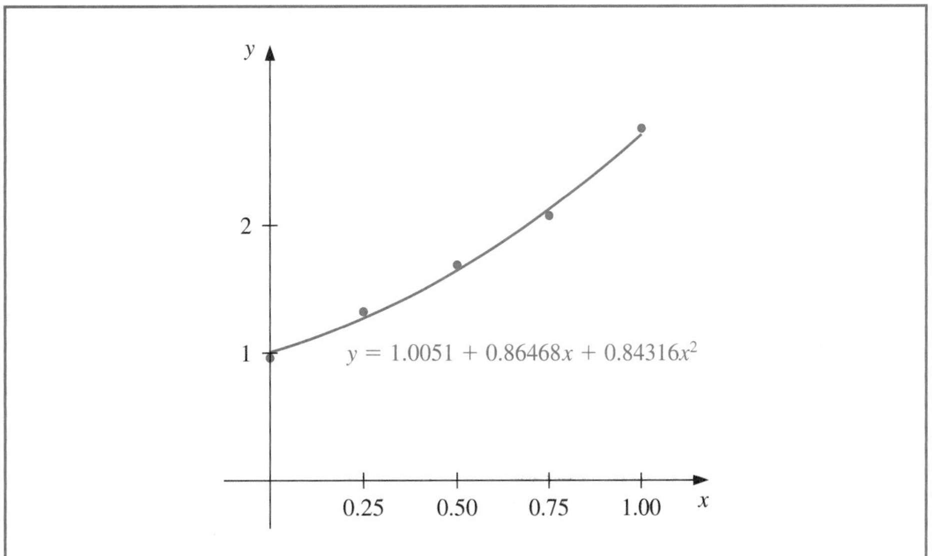

Table 8.4

i	1	2	3	4	5
x_i	0	0.25	0.50	0.75	1.00
y_i	1.0000	1.2840	1.6487	2.1170	2.7183
$P(x_i)$	1.0051	1.2740	1.6482	2.1279	2.7129
$y_i - P(x_i)$	−0.0051	0.0100	0.0004	−0.0109	0.0054

The total error,

$$E = \sum_{i=1}^{5} (y_i - P(x_i))^2 = 2.74 \times 10^{-4},$$

is the least that can be obtained by using a polynomial of degree at most 2. ■

Maple has a function called *LinearFit* within the *Statistics* package which can be used to compute the discrete least squares approximations. To compute the approximation in Example 2 we first load the package and define the data

with(*Statistics*): *xvals* := *Vector*([0, 0.25, 0.5, 0.75, 1]): *yvals* := *Vector*([1, 1.284, 1.6487, 2.117, 2.7183]):

To define the least squares polynomial for this data we enter the command

$P := x \rightarrow LinearFit([1, x, x^2], xvals, yvals, x)$: $P(x)$

Maple returns a result which rounded to 5 decimal places is

$$1.00514 + 0.86418x + 0.84366x^2$$

The approximation at a specific value, for example at $x = 1.7$, is found with $P(1.7)$

$$4.91242$$

At times it is appropriate to assume that the data are exponentially related. This requires the approximating function to be of the form

$$y = be^{ax} \tag{8.4}$$

or

$$y = bx^a, \tag{8.5}$$

for some constants a and b. The difficulty with applying the least squares procedure in a situation of this type comes from attempting to minimize

$$E = \sum_{i=1}^{m}(y_i - be^{ax_i})^2, \quad \text{in the case of Eq. (8.4),}$$

or

$$E = \sum_{i=1}^{m}(y_i - bx_i^a)^2, \quad \text{in the case of Eq. (8.5).}$$

The normal equations associated with these procedures are obtained from either

$$0 = \frac{\partial E}{\partial b} = 2\sum_{i=1}^{m}(y_i - be^{ax_i})(-e^{ax_i})$$

and

$$0 = \frac{\partial E}{\partial a} = 2\sum_{i=1}^{m}(y_i - be^{ax_i})(-bx_ie^{ax_i}), \quad \text{in the case of Eq. (8.4);}$$

or

$$0 = \frac{\partial E}{\partial b} = 2\sum_{i=1}^{m}(y_i - bx_i^a)(-x_i^a)$$

and

$$0 = \frac{\partial E}{\partial a} = 2\sum_{i=1}^{m}(y_i - bx_i^a)(-b(\ln x_i)x_i^a), \quad \text{in the case of Eq. (8.5).}$$

No exact solution to either of these systems in a and b can generally be found.

The method that is commonly used when the data are suspected to be exponentially related is to consider the logarithm of the approximating equation:

$$\ln y = \ln b + ax, \quad \text{in the case of Eq. (8.4),}$$

and

$$\ln y = \ln b + a\ln x, \quad \text{in the case of Eq. (8.5).}$$

In either case, a linear problem now appears, and solutions for $\ln b$ and a can be obtained by appropriately modifying the normal equations (8.1) and (8.2).

However, the approximation obtained in this manner is *not* the least squares approximation for the original problem, and this approximation can in some cases differ significantly from the least squares approximation to the original problem. The application in Exercise 13 describes such a problem. This application will be reconsidered as Exercise 11 in Section 10.3, where the exact solution to the exponential least squares problem is approximated by using methods suitable for solving nonlinear systems of equations.

Illustration Consider the collection of data in the first three columns of Table 8.5.

Table 8.5

i	x_i	y_i	$\ln y_i$	x_i^2	$x_i \ln y_i$
1	1.00	5.10	1.629	1.0000	1.629
2	1.25	5.79	1.756	1.5625	2.195
3	1.50	6.53	1.876	2.2500	2.814
4	1.75	7.45	2.008	3.0625	3.514
5	2.00	8.46	2.135	4.0000	4.270
	7.50		9.404	11.875	14.422

If x_i is graphed with $\ln y_i$, the data appear to have a linear relation, so it is reasonable to assume an approximation of the form

$$y = be^{ax}, \quad \text{which implies that} \quad \ln y = \ln b + ax.$$

Extending the table and summing the appropriate columns gives the remaining data in Table 8.5.

Using the normal equations (8.1) and (8.2),

$$a = \frac{(5)(14.422) - (7.5)(9.404)}{(5)(11.875) - (7.5)^2} = 0.5056$$

and

$$\ln b = \frac{(11.875)(9.404) - (14.422)(7.5)}{(5)(11.875) - (7.5)^2} = 1.122.$$

With $\ln b = 1.122$ we have $b = e^{1.122} = 3.071$, and the approximation assumes the form

$$y = 3.071e^{0.5056x}.$$

At the data points this gives the values in Table 8.6. (See Figure 8.5.) □

Table 8.6

i	x_i	y_i	$3.071e^{0.5056x_i}$	$\lvert y_i - 3.071e^{0.5056x_i}\rvert$
1	1.00	5.10	5.09	0.01
2	1.25	5.79	5.78	0.01
3	1.50	6.53	6.56	0.03
4	1.75	7.45	7.44	0.01
5	2.00	8.46	8.44	0.02

Figure 8.5

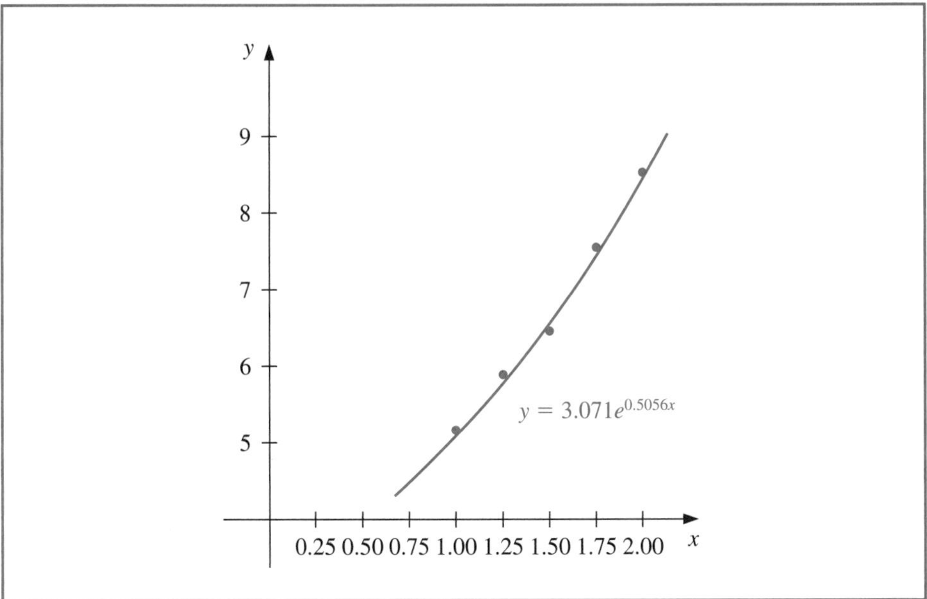

Exponential and other nonlinear discrete least squares approximations can be obtain in the *Statistics* package by using the commands *ExponentialFit* and *NonlinearFit*.

For example, the approximation in the Illustration can be obtained by first defining the data with

$X := Vector([1, 1.25, 1.5, 1.75, 2])$: $Y := Vector([5.1, 5.79, 6.53, 7.45, 8.46])$:

and then issuing the command

ExponentialFit(X, Y, x)

gives the result, rounded to 5 decimal places,

$$3.07249e^{0.50572x}$$

If instead the *NonlinearFit* command is issued, the approximation produced uses methods of Chapter 10 for solving a system of nonlinear equations. The approximation that Maple gives in this case is

$$3.06658(1.66023)^x \approx 3.06658e^{0.50695}.$$

EXERCISE SET 8.1

1. Compute the linear least squares polynomial for the data of Example 2.
2. Compute the least squares polynomial of degree 2 for the data of Example 1, and compare the total error E for the two polynomials.
3. Find the least squares polynomials of degrees 1, 2, and 3 for the data in the following table. Compute the error E in each case. Graph the data and the polynomials.

x_i	0	0.15	0.31	0.5	0.6	0.75
y_i	1.0	1.004	1.031	1.117	1.223	1.422

4. Find the least squares polynomials of degrees 1, 2, and 3 for the data in the following table. Compute the error E in each case. Graph the data and the polynomials.

x_i	1.0	1.1	1.3	1.5	1.9	2.1
y_i	1.84	1.96	2.21	2.45	2.94	3.18

5. Given the data:

x_i	4.0	4.2	4.5	4.7	5.1	5.5	5.9	6.3	6.8	7.1
y_i	102.56	113.18	130.11	142.05	167.53	195.14	224.87	256.73	299.50	326.72

a. Construct the least squares polynomial of degree 1, and compute the error.

b. Construct the least squares polynomial of degree 2, and compute the error.

c. Construct the least squares polynomial of degree 3, and compute the error.

d. Construct the least squares approximation of the form be^{ax}, and compute the error.

e. Construct the least squares approximation of the form bx^a, and compute the error.

6. Repeat Exercise 5 for the following data.

x_i	0.2	0.3	0.6	0.9	1.1	1.3	1.4	1.6
y_i	0.050446	0.098426	0.33277	0.72660	1.0972	1.5697	1.8487	2.5015

7. In the lead example of this chapter, an experiment was described to determine the spring constant k in Hooke's law:

$$F(l) = k(l - E).$$

The function F is the force required to stretch the spring l units, where the constant $E = 5.3$ in. is the length of the unstretched spring.

a. Suppose measurements are made of the length l, in inches, for applied weights $F(l)$, in pounds, as given in the following table.

$F(l)$	l
2	7.0
4	9.4
6	12.3

Find the least squares approximation for k.

b. Additional measurements are made, giving more data:

$F(l)$	l
3	8.3
5	11.3
8	14.4
10	15.9

Compute the new least squares approximation for k. Which of (a) or (b) best fits the total experimental data?

8. The following list contains homework grades and the final-examination grades for 30 numerical analysis students. Find the equation of the least squares line for this data, and use this line to determine the homework grade required to predict minimal A (90%) and D (60%) grades on the final.

Homework	Final	Homework	Final
302	45	323	83
325	72	337	99
285	54	337	70
339	54	304	62
334	79	319	66
322	65	234	51
331	99	337	53
279	63	351	100
316	65	339	67
347	99	343	83
343	83	314	42
290	74	344	79
326	76	185	59
233	57	340	75
254	45	316	45

9. The following table lists the college grade-point averages of 20 mathematics and computer science majors, together with the scores that these students received on the mathematics portion of the ACT (American College Testing Program) test while in high school. Plot these data, and find the equation of the least squares line for this data.

ACT score	Grade-point average	ACT score	Grade-point average
28	3.84	29	3.75
25	3.21	28	3.65
28	3.23	27	3.87
27	3.63	29	3.75
28	3.75	21	1.66
33	3.20	28	3.12
28	3.41	28	2.96
29	3.38	26	2.92
23	3.53	30	3.10
27	2.03	24	2.81

10. The following set of data, presented to the Senate Antitrust Subcommittee, shows the comparative crash-survivability characteristics of cars in various classes. Find the least squares line that approximates these data. (The table shows the percent of accident-involved vehicles in which the most severe injury was fatal or serious.)

Type	Average Weight	Percent Occurrence
1. Domestic luxury regular	4800 lb	3.1
2. Domestic intermediate regular	3700 lb	4.0
3. Domestic economy regular	3400 lb	5.2
4. Domestic compact	2800 lb	6.4
5. Foreign compact	1900 lb	9.6

11. To determine a relationship between the number of fish and the number of species of fish in samples taken for a portion of the Great Barrier Reef, P. Sale and R. Dybdahl [SD] fit a linear least squares polynomial to the following collection of data, which were collected in samples over a 2-year period. Let x be the number of fish in the sample and y be the number of species in the sample.

x	y	x	y	x	y
13	11	29	12	60	14
15	10	30	14	62	21
16	11	31	16	64	21
21	12	36	17	70	24
22	12	40	13	72	17
23	13	42	14	100	23
25	13	55	22	130	34

Determine the linear least squares polynomial for these data.

12. To determine a functional relationship between the attenuation coefficient and the thickness of a sample of taconite, V. P. Singh [Si] fits a collection of data by using a linear least squares polynomial. The following collection of data is taken from a graph in that paper. Find the linear least squares polynomial fitting these data.

Thickness (cm)	Attenuation coefficient (dB/cm)
0.040	26.5
0.041	28.1
0.055	25.2
0.056	26.0
0.062	24.0
0.071	25.0
0.071	26.4
0.078	27.2
0.082	25.6
0.090	25.0
0.092	26.8
0.100	24.8
0.105	27.0
0.120	25.0
0.123	27.3
0.130	26.9
0.140	26.2

13. In a paper dealing with the efficiency of energy utilization of the larvae of the modest sphinx moth (*Pachysphinx modesta*), L. Schroeder [Schr1] used the following data to determine a relation between W, the live weight of the larvae in grams, and R, the oxygen consumption of the larvae in milliliters/hour. For biological reasons, it is assumed that a relationship in the form of $R = bW^a$ exists between W and R.

a. Find the logarithmic linear least squares polynomial by using

$$\ln R = \ln b + a \ln W.$$

b. Compute the error associated with the approximation in part (a):

$$E = \sum_{i=1}^{37} (R_i - bW_i^a)^2.$$

c. Modify the logarithmic least squares equation in part (a) by adding the quadratic term $c(\ln W_i)^2$, and determine the logarithmic quadratic least squares polynomial.

d. Determine the formula for and compute the error associated with the approximation in part (c).

W	R	W	R	W	R	W	R	W	R
0.017	0.154	0.025	0.23	0.020	0.181	0.020	0.180	0.025	0.234
0.087	0.296	0.111	0.357	0.085	0.260	0.119	0.299	0.233	0.537
0.174	0.363	0.211	0.366	0.171	0.334	0.210	0.428	0.783	1.47
1.11	0.531	0.999	0.771	1.29	0.87	1.32	1.15	1.35	2.48
1.74	2.23	3.02	2.01	3.04	3.59	3.34	2.83	1.69	1.44
4.09	3.58	4.28	3.28	4.29	3.40	5.48	4.15	2.75	1.84
5.45	3.52	4.58	2.96	5.30	3.88			4.83	4.66
5.96	2.40	4.68	5.10					5.53	6.94

14. Show that the normal equations (8.3) resulting from discrete least squares approximation yield a symmetric and nonsingular matrix and hence have a unique solution. [*Hint:* Let $A = (a_{ij})$, where

$$a_{ij} = \sum_{k=1}^{m} x_k^{i+j-2}$$

and $x_1, x_2, \ldots, x_m$ are distinct with $n < m - 1$. Suppose A is singular and that $\mathbf{c} \neq \mathbf{0}$ is such that $\mathbf{c}^t A\mathbf{c} = 0$. Show that the nth-degree polynomial whose coefficients are the coordinates of $\mathbf{c}$ has more than n roots, and use this to establish a contradiction.]

8.2 Orthogonal Polynomials and Least Squares Approximation

The previous section considered the problem of least squares approximation to fit a collection of data. The other approximation problem mentioned in the introduction concerns the approximation of functions.

Suppose $f \in C[a, b]$ and that a polynomial $P_n(x)$ of degree at most n is required that will minimize the error

$$\int_a^b [f(x) - P_n(x)]^2\, dx.$$

To determine a least squares approximating polynomial; that is, a polynomial to minimize this expression, let

$$P_n(x) = a_n x^n + a_{n-1}x^{n-1} + \cdots + a_1 x + a_0 = \sum_{k=0}^{n} a_k x^k,$$

and define, as shown in Figure 8.6,

$$E \equiv E_2(a_0, a_1, \ldots, a_n) = \int_a^b \left(f(x) - \sum_{k=0}^{n} a_k x^k\right)^2 dx.$$

The problem is to find real coefficients $a_0, a_1, \ldots, a_n$ that will minimize E. A necessary condition for the numbers $a_0, a_1, \ldots, a_n$ to minimize E is that

$$\frac{\partial E}{\partial a_j} = 0, \quad \text{for each } j = 0, 1, \ldots, n.$$

Figure 8.6

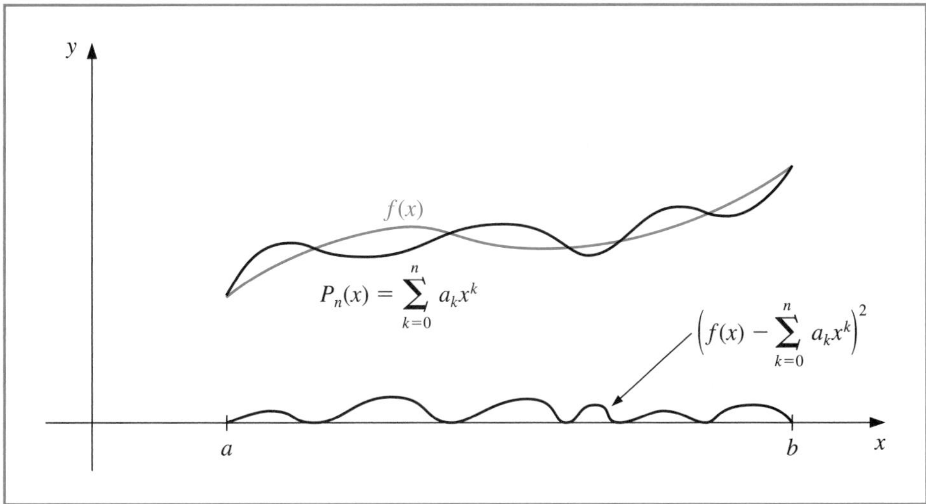

Since

$$E = \int_a^b [f(x)]^2\,dx - 2\sum_{k=0}^{n} a_k \int_a^b x^k f(x)\,dx + \int_a^b \left(\sum_{k=0}^{n} a_k x^k\right)^2 dx,$$

we have

$$\frac{\partial E}{\partial a_j} = -2\int_a^b x^j f(x)\,dx + 2\sum_{k=0}^{n} a_k \int_a^b x^{j+k}\,dx.$$

Hence, to find $P_n(x)$, the $(n+1)$ linear **normal equations**

$$\sum_{k=0}^{n} a_k \int_a^b x^{j+k}\,dx = \int_a^b x^j f(x)\,dx, \quad \text{for each } j = 0, 1, \ldots, n, \tag{8.6}$$

must be solved for the $(n+1)$ unknowns a_j. The normal equations always have a unique solution provided that $f \in C[a,b]$. (See Exercise 15.)

Example 1 Find the least squares approximating polynomial of degree 2 for the function $f(x) = \sin \pi x$ on the interval [0, 1].

Solution The normal equations for $P_2(x) = a_2x^2 + a_1x + a_0$ are

$$a_0 \int_0^1 1\,dx + a_1 \int_0^1 x\,dx + a_2 \int_0^1 x^2\,dx = \int_0^1 \sin \pi x\,dx,$$

$$a_0 \int_0^1 x\,dx + a_1 \int_0^1 x^2\,dx + a_2 \int_0^1 x^3\,dx = \int_0^1 x \sin \pi x\,dx,$$

$$a_0 \int_0^1 x^2\,dx + a_1 \int_0^1 x^3\,dx + a_2 \int_0^1 x^4\,dx = \int_0^1 x^2 \sin \pi x\,dx.$$

Performing the integration yields

$$a_0 + \frac{1}{2}a_1 + \frac{1}{3}a_2 = \frac{2}{\pi}, \quad \frac{1}{2}a_0 + \frac{1}{3}a_1 + \frac{1}{4}a_2 = \frac{1}{\pi}, \quad \frac{1}{3}a_0 + \frac{1}{4}a_1 + \frac{1}{5}a_2 = \frac{\pi^2 - 4}{\pi^3}.$$

These three equations in three unknowns can be solved to obtain

$$a_0 = \frac{12\pi^2 - 120}{\pi^3} \approx -0.050465 \quad \text{and} \quad a_1 = -a_2 = \frac{720 - 60\pi^2}{\pi^3} \approx 4.12251.$$

Consequently, the least squares polynomial approximation of degree 2 for $f(x) = \sin \pi x$ on $[0, 1]$ is $P_2(x) = -4.12251x^2 + 4.12251x - 0.050465$. (See Figure 8.7.) ■

Figure 8.7

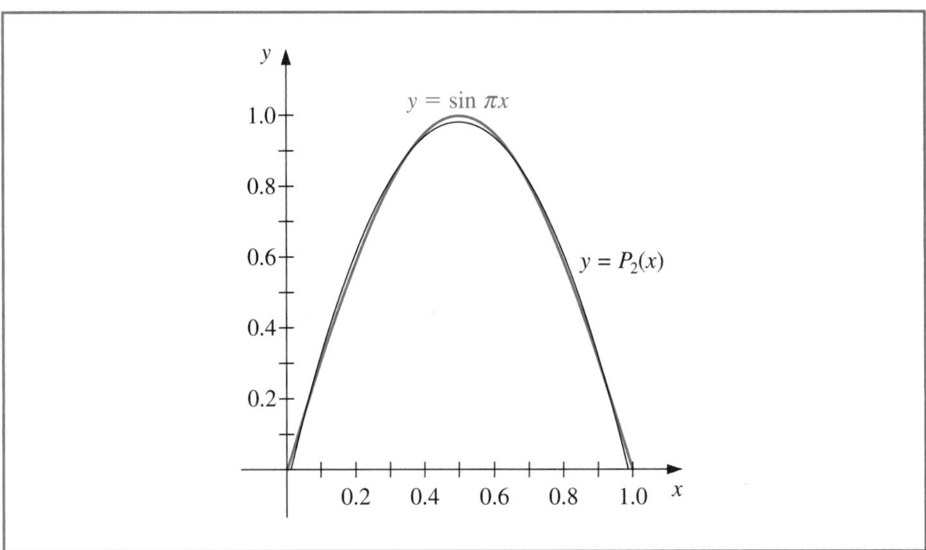

Example 1 illustrates a difficulty in obtaining a least squares polynomial approximation. An $(n + 1) \times (n + 1)$ linear system for the unknowns $a_0, \ldots, a_n$ must be solved, and the coefficients in the linear system are of the form

$$\int_a^b x^{j+k}\, dx = \frac{b^{j+k+1} - a^{j+k+1}}{j + k + 1},$$

a linear system that does not have an easily computed numerical solution. The matrix in the linear system is known as a **Hilbert matrix**, which is a classic example for demonstrating round-off error difficulties. (See Exercise 11 of Section 7.5.)

David Hilbert (1862–1943) was the dominant mathematician at the turn of the 20th century. He is best remembered for giving a talk at the International Congress of Mathematicians in Paris in 1900 in which he posed 23 problems that he thought would be important for mathematicians in the next century.

Another disadvantage is similar to the situation that occurred when the Lagrange polynomials were first introduced in Section 3.1. The calculations that were performed in obtaining the best nth-degree polynomial, $P_n(x)$, do not lessen the amount of work required to obtain $P_{n+1}(x)$, the polynomial of next higher degree.

Linearly Independent Functions

A different technique to obtain least squares approximations will now be considered. This turns out to be computationally efficient, and once $P_n(x)$ is known, it is easy to determine $P_{n+1}(x)$. To facilitate the discussion, we need some new concepts.

Definition 8.1 The set of functions $\{\phi_0, \ldots, \phi_n\}$ is said to be **linearly independent** on $[a, b]$ if, whenever

$$c_0\phi_0(x) + c_1\phi_1(x) + \cdots + c_n\phi_n(x) = 0, \quad \text{for all } x \in [a, b],$$

we have $c_0 = c_1 = \cdots = c_n = 0$. Otherwise the set of functions is said to be **linearly dependent**. ■

Theorem 8.2 Suppose that, for each $j = 0, 1, \ldots, n$, $\phi_j(x)$ is a polynomial of degree j. Then $\{\phi_0, \ldots, \phi_n\}$ is linearly independent on any interval $[a, b]$. ■

Proof Let $c_0, \ldots, c_n$ be real numbers for which

$$P(x) = c_0\phi_0(x) + c_1\phi_1(x) + \cdots + c_n\phi_n(x) = 0, \quad \text{for all } x \in [a, b].$$

The polynomial $P(x)$ vanishes on $[a, b]$, so it must be the zero polynomial, and the coefficients of all the powers of x are zero. In particular, the coefficient of x^n is zero. But $c_n\phi_n(x)$ is the only term in $P(x)$ that contains x^n, so we must have $c_n = 0$. Hence

$$P(x) = \sum_{j=0}^{n-1} c_j\phi_j(x).$$

In this representation of $P(x)$, the only term that contains a power of x^{n-1} is $c_{n-1}\phi_{n-1}(x)$, so this term must also be zero and

$$P(x) = \sum_{j=0}^{n-2} c_j\phi_j(x).$$

In like manner, the remaining constants $c_{n-2}, c_{n-3}, \ldots, c_1, c_0$ are all zero, which implies that $\{\phi_0, \phi_1, \ldots, \phi_n\}$ is linearly independent on $[a, b]$. ■ ■ ■

Example 2 Let $\phi_0(x) = 2$, $\phi_1(x) = x - 3$, and $\phi_2(x) = x^2 + 2x + 7$, and $Q(x) = a_0 + a_1x + a_2x^2$. Show that there exist constants c_0, c_1, and c_2 such that $Q(x) = c_0\phi_0(x) + c_1\phi_1(x) + c_2\phi_2(x)$.

Solution By Theorem 8.2, $\{\phi_0, \phi_1, \phi_2\}$ is linearly independent on any interval $[a, b]$. First note that

$$1 = \frac{1}{2}\phi_0(x), \quad x = \phi_1(x) + 3 = \phi_1(x) + \frac{3}{2}\phi_0(x),$$

and

$$\begin{aligned} x^2 &= \phi_2(x) - 2x - 7 = \phi_2(x) - 2\left[\phi_1(x) + \frac{3}{2}\phi_0(x)\right] - 7\left[\frac{1}{2}\phi_0(x)\right] \\ &= \phi_2(x) - 2\phi_1(x) - \frac{13}{2}\phi_0(x). \end{aligned}$$

Hence

$$\begin{aligned} Q(x) &= a_0\left[\frac{1}{2}\phi_0(x)\right] + a_1\left[\phi_1(x) + \frac{3}{2}\phi_0(x)\right] + a_2\left[\phi_2(x) - 2\phi_1(x) - \frac{13}{2}\phi_0(x)\right] \\ &= \left(\frac{1}{2}a_0 + \frac{3}{2}a_1 - \frac{13}{2}a_2\right)\phi_0(x) + [a_1 - 2a_2]\phi_1(x) + a_2\phi_2(x). \end{aligned}$$ ■

The situation illustrated in Example 2 holds in a much more general setting. Let $\prod_n$ denote the **set of all polynomials of degree at most n**. The following result is used extensively in many applications of linear algebra. Its proof is considered in Exercise 13.

Theorem 8.3 Suppose that $\{\phi_0(x), \phi_1(x), \ldots, \phi_n(x)\}$ is a collection of linearly independent polynomials in $\prod_n$. Then any polynomial in $\prod_n$ can be written uniquely as a linear combination of $\phi_0(x)$, $\phi_1(x), \ldots, \phi_n(x)$. ■

Orthogonal Functions

To discuss general function approximation requires the introduction of the notions of weight functions and orthogonality.

Definition 8.4 An integrable function w is called a **weight function** on the interval I if $w(x) \geq 0$, for all x in I, but $w(x) \not\equiv 0$ on any subinterval of I. ■

The purpose of a weight function is to assign varying degrees of importance to approximations on certain portions of the interval. For example, the weight function

$$w(x) = \frac{1}{\sqrt{1-x^2}}$$

places less emphasis near the center of the interval $(-1, 1)$ and more emphasis when $|x|$ is near 1 (see Figure 8.8). This weight function is used in the next section.

Figure 8.8

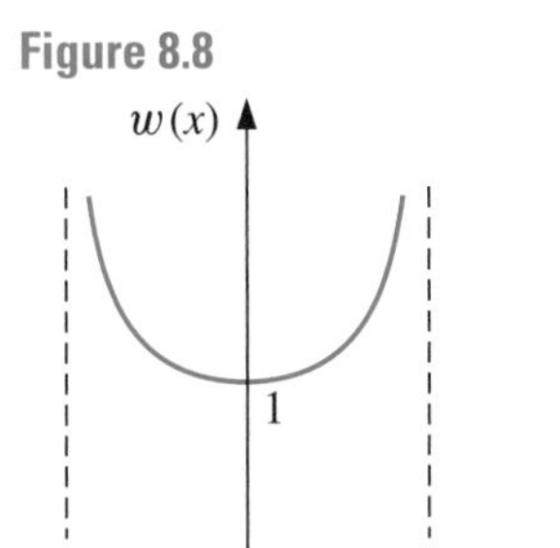

Suppose $\{\phi_0, \phi_1, \ldots, \phi_n\}$ is a set of linearly independent functions on $[a, b]$ and w is a weight function for $[a, b]$. Given $f \in C[a, b]$, we seek a linear combination

$$P(x) = \sum_{k=0}^{n} a_k \phi_k(x)$$

to minimize the error

$$E = E(a_0, \ldots, a_n) = \int_a^b w(x)\left[f(x) - \sum_{k=0}^{n} a_k\phi_k(x)\right]^2 dx.$$

This problem reduces to the situation considered at the beginning of this section in the special case when $w(x) \equiv 1$ and $\phi_k(x) = x^k$, for each $k = 0, 1, \ldots, n$.

The normal equations associated with this problem are derived from the fact that for each $j = 0, 1, \ldots, n$,

$$0 = \frac{\partial E}{\partial a_j} = 2\int_a^b w(x)\left[f(x) - \sum_{k=0}^{n} a_k\phi_k(x)\right]\phi_j(x)\, dx.$$

The system of normal equations can be written

$$\int_a^b w(x) f(x)\phi_j(x)\, dx = \sum_{k=0}^{n} a_k \int_a^b w(x)\phi_k(x)\phi_j(x)\, dx, \quad \text{for } j = 0, 1, \ldots, n.$$

If the functions $\phi_0, \phi_1, \ldots, \phi_n$ can be chosen so that

$$\int_a^b w(x)\phi_k(x)\phi_j(x)\, dx = \begin{cases} 0, & \text{when } j \neq k, \\ \alpha_j > 0, & \text{when } j = k, \end{cases} \tag{8.7}$$

then the normal equations will reduce to

$$\int_a^b w(x) f(x)\phi_j(x)\, dx = a_j \int_a^b w(x)[\phi_j(x)]^2\, dx = a_j\alpha_j,$$

for each $j = 0, 1, \ldots, n$. These are easily solved to give

$$a_j = \frac{1}{\alpha_j}\int_a^b w(x) f(x)\phi_j(x)\, dx.$$

The word orthogonal means right-angled. So in a sense, orthogonal functions are perpendicular to one another.

Hence the least squares approximation problem is greatly simplified when the functions $\phi_0, \phi_1, \ldots, \phi_n$ are chosen to satisfy the *orthogonality* condition in Eq. (8.7). The remainder of this section is devoted to studying collections of this type.

Definition 8.5 $\{\phi_0, \phi_1, \dots, \phi_n\}$ is said to be an **orthogonal set of functions** for the interval $[a, b]$ with respect to the weight function w if

$$\int_a^b w(x)\phi_k(x)\phi_j(x)\,dx = \begin{cases} 0, & \text{when } j \neq k, \\ \alpha_j > 0, & \text{when } j = k. \end{cases}$$

If, in addition, $\alpha_j = 1$ for each $j = 0, 1, \dots, n$, the set is said to be **orthonormal.** ■

This definition, together with the remarks preceding it, produces the following theorem.

Theorem 8.6 If $\{\phi_0, \dots, \phi_n\}$ is an orthogonal set of functions on an interval $[a, b]$ with respect to the weight function w, then the least squares approximation to f on $[a, b]$ with respect to w is

$$P(x) = \sum_{j=0}^{n} a_j\phi_j(x),$$

where, for each $j = 0, 1, \dots, n$,

$$a_j = \frac{\int_a^b w(x)\phi_j(x)f(x)\,dx}{\int_a^b w(x)[\phi_j(x)]^2\,dx} = \frac{1}{\alpha_j}\int_a^b w(x)\phi_j(x)f(x)\,dx.$$ ■

Although Definition 8.5 and Theorem 8.6 allow for broad classes of orthogonal functions, we will consider only orthogonal sets of polynomials. The next theorem, which is based on the **Gram-Schmidt process**, describes how to construct orthogonal polynomials on $[a, b]$ with respect to a weight function w.

Theorem 8.7 The set of polynomial functions $\{\phi_0, \phi_1, \dots, \phi_n\}$ defined in the following way is orthogonal on $[a, b]$ with respect to the weight function w.

Erhard Schmidt (1876–1959) received his doctorate under the supervision of David Hilbert in 1905 for a problem involving integral equations. Schmidt published a paper in 1907 in which he gave what is now called the Gram-Schmidt process for constructing an orthonormal basis for a set of functions. This generalized results of Jorgen Pedersen Gram (1850–1916) who considered this problem when studying least squares. Laplace, however, presented a similar process much earlier than either Gram or Schmidt.

$$\phi_0(x) \equiv 1, \quad \phi_1(x) = x - B_1, \quad \text{for each } x \text{ in } [a, b],$$

where

$$B_1 = \frac{\int_a^b xw(x)[\phi_0(x)]^2\,dx}{\int_a^b w(x)[\phi_0(x)]^2\,dx},$$

and when $k \geq 2$,

$$\phi_k(x) = (x - B_k)\phi_{k-1}(x) - C_k\phi_{k-2}(x), \quad \text{for each } x \text{ in } [a, b],$$

where

$$B_k = \frac{\int_a^b xw(x)[\phi_{k-1}(x)]^2\,dx}{\int_a^b w(x)[\phi_{k-1}(x)]^2\,dx}$$

and

$$C_k = \frac{\int_a^b xw(x)\phi_{k-1}(x)\phi_{k-2}(x)\,dx}{\int_a^b w(x)[\phi_{k-2}(x)]^2\,dx}.$$ ■

Theorem 8.7 provides a recursive procedure for constructing a set of orthogonal polynomials. The proof of this theorem follows by applying mathematical induction to the degree of the polynomial $\phi_n(x)$.

Corollary 8.8 For any $n > 0$, the set of polynomial functions $\{\phi_0, \ldots, \phi_n\}$ given in Theorem 8.7 is linearly independent on $[a, b]$ and

$$\int_a^b w(x)\phi_n(x)Q_k(x)\, dx = 0,$$

for any polynomial $Q_k(x)$ of degree $k < n$. ▪

Proof For each $k = 0, 1, \ldots, n$, $\phi_k(x)$ is a polynomial of degree k. So Theorem 8.2 implies that $\{\phi_0, \ldots, \phi_n\}$ is a linearly independent set.

Let $Q_k(x)$ be a polynomial of degree $k < n$. By Theorem 8.3 there exist numbers $c_0, \ldots, c_k$ such that

$$Q_k(x) = \sum_{j=0}^{k} c_j\phi_j(x).$$

Because ϕ_n is orthogonal to ϕ_j for each $j = 0, 1, \ldots, k$ we have

$$\int_a^b w(x)Q_k(x)\phi_n(x)\, dx = \sum_{j=0}^{k} c_j \int_a^b w(x)\phi_j(x)\phi_n(x)\, dx = \sum_{j=0}^{k} c_j \cdot 0 = 0.$$ ▪ ▪ ▪

Illustration The set of **Legendre polynomials**, $\{P_n(x)\}$, is orthogonal on $[-1, 1]$ with respect to the weight function $w(x) \equiv 1$. The classical definition of the Legendre polynomials requires that $P_n(1) = 1$ for each n, and a recursive relation is used to generate the polynomials when $n \geq 2$. This normalization will not be needed in our discussion, and the least squares approximating polynomials generated in either case are essentially the same.

Using the Gram-Schmidt process with $P_0(x) \equiv 1$ gives

$$B_1 = \frac{\int_{-1}^{1} x\, dx}{\int_{-1}^{1} dx} = 0 \quad \text{and} \quad P_1(x) = (x - B_1)P_0(x) = x.$$

Also,

$$B_2 = \frac{\int_{-1}^{1} x^3\, dx}{\int_{-1}^{1} x^2\, dx} = 0 \quad \text{and} \quad C_2 = \frac{\int_{-1}^{1} x^2\, dx}{\int_{-1}^{1} 1\, dx} = \frac{1}{3},$$

so

$$P_2(x) = (x - B_2)P_1(x) - C_2P_0(x) = (x - 0)x - \frac{1}{3} \cdot 1 = x^2 - \frac{1}{3}.$$

The higher-degree Legendre polynomials shown in Figure 8.9 are derived in the same manner. Although the integration can be tedious, it is not difficult with a Computer Algebra System.

Figure 8.9

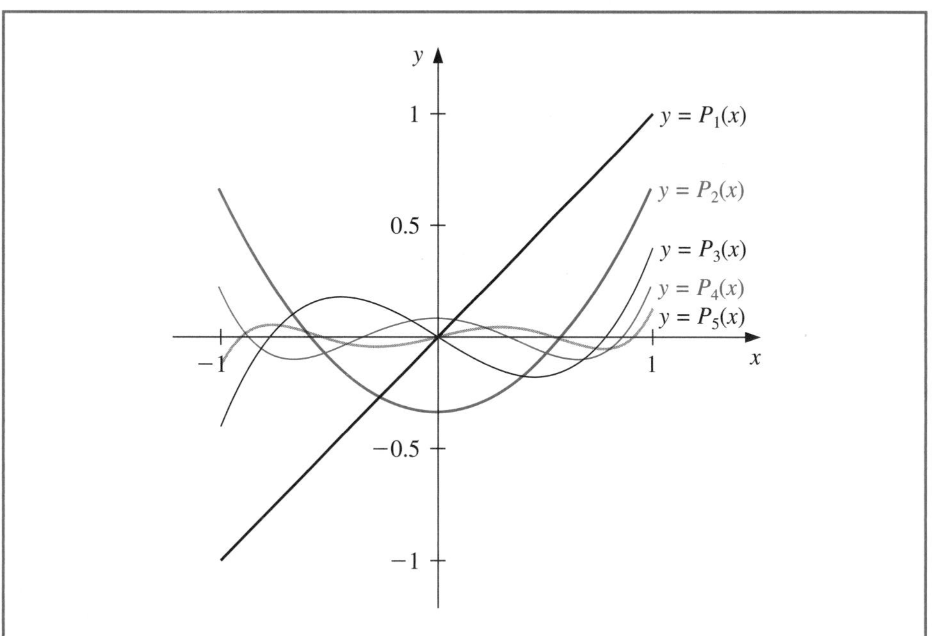

For example, the Maple command *int* is used to compute the integrals B_3 and C_3:

$$B3 := \frac{int\left(x\left(x^2-\frac{1}{3}\right)^2, x=-1..1\right)}{int\left(\left(x^2-\frac{1}{3}\right)^2, x=-1..1\right)}; \quad C3 := \frac{int\left(x\left(x^2-\frac{1}{3}\right), x=-1..1\right)}{int(x^2, x=-1..1)}$$

$$0$$

$$\frac{4}{15}$$

Thus

$$P_3(x) = xP_2(x) - \frac{4}{15}P_1(x) = x^3 - \frac{1}{3}x - \frac{4}{15}x = x^3 - \frac{3}{5}x.$$

The next two Legendre polynomials are

$$P_4(x) = x^4 - \frac{6}{7}x^2 + \frac{3}{35} \quad \text{and} \quad P_5(x) = x^5 - \frac{10}{9}x^3 + \frac{5}{21}x. \qquad \square$$

The Legendre polynomials were introduced in Section 4.7, where their roots, given on page 232, were used as the nodes in Gaussian quadrature.

EXERCISE SET 8.2

1. Find the linear least squares polynomial approximation to $f(x)$ on the indicated interval if
   a. $f(x) = x^2 + 3x + 2$, $[0, 1]$;
   b. $f(x) = x^3$, $[0, 2]$;
   c. $f(x) = e^x$, $[0, 2]$;
   d. $f(x) = \frac{1}{x}$, $[1, 3]$;
   e. $f(x) = \frac{1}{2}\cos x + \frac{1}{3}\sin 2x$, $[0, 1]$;
   f. $f(x) = x \ln x$, $[1, 3]$.
2. Find the linear least squares polynomial approximation on the interval $[-1, 1]$ for the following functions.
   a. $f(x) = x^2 - 2x + 3$
   b. $f(x) = x^3$
   c. $f(x) = e^x$
   d. $f(x) = \frac{1}{x+2}$
   e. $f(x) = \frac{1}{2}\cos x + \frac{1}{3}\sin 2x$
   f. $f(x) = \ln(x + 2)$
3. Find the least squares polynomial approximation of degree two to the functions and intervals in Exercise 1.
4. Find the least squares polynomial approximation of degree 2 on the interval $[-1, 1]$ for the functions in Exercise 3.
5. Compute the error E for the approximations in Exercise 3.
6. Compute the error E for the approximations in Exercise 4.
7. Use the Gram-Schmidt process to construct $\phi_0(x)$, $\phi_1(x)$, $\phi_2(x)$, and $\phi_3(x)$ for the following intervals.
   a. $[0, 1]$
   b. $[0, 2]$
   c. $[1, 3]$
8. Repeat Exercise 1 using the results of Exercise 7.
9. Obtain the least squares approximation polynomial of degree 3 for the functions in Exercise 1 using the results of Exercise 7.
10. Repeat Exercise 3 using the results of Exercise 7.
11. Use the Gram-Schmidt procedure to calculate L_1, L_2, and L_3, where $\{L_0(x), L_1(x), L_2(x), L_3(x)\}$ is an orthogonal set of polynomials on $(0, \infty)$ with respect to the weight functions $w(x) = e^{-x}$ and $L_0(x) \equiv 1$. The polynomials obtained from this procedure are called the **Laguerre polynomials**.
12. Use the Laguerre polynomials calculated in Exercise 11 to compute the least squares polynomials of degree one, two, and three on the interval $(0, \infty)$ with respect to the weight function $w(x) = e^{-x}$ for the following functions:
    a. $f(x) = x^2$
    b. $f(x) = e^{-x}$
    c. $f(x) = x^3$
    d. $f(x) = e^{-2x}$
13. Suppose $\{\phi_0, \phi_1, \ldots, \phi_n\}$ is any linearly independent set in $\prod_n$. Show that for any element $Q \in \prod_n$, there exist unique constants $c_0, c_1, \ldots, c_n$, such that

$$Q(x) = \sum_{k=0}^{n} c_k \phi_k(x).$$

14. Show that if $\{\phi_0, \phi_1, \ldots, \phi_n\}$ is an orthogonal set of functions on $[a, b]$ with respect to the weight function w, then $\{\phi_0, \phi_1, \ldots, \phi_n\}$ is a linearly independent set.
15. Show that the normal equations (8.6) have a unique solution. [*Hint:* Show that the only solution for the function $f(x) \equiv 0$ is $a_j = 0, j = 0, 1, \ldots, n$. Multiply Eq. (8.6) by a_j, and sum over all j. Interchange the integral sign and the summation sign to obtain $\int_a^b [P(x)]^2 dx = 0$. Thus, $P(x) \equiv 0$, so $a_j = 0$, for $j = 0, \ldots, n$. Hence, the coefficient matrix is nonsingular, and there is a unique solution to Eq. (8.6).]

8.3 Chebyshev Polynomials and Economization of Power Series

The Chebyshev polynomials $\{T_n(x)\}$ are orthogonal on $(-1, 1)$ with respect to the weight function $w(x) = (1 - x^2)^{-1/2}$. Although they can be derived by the method in the previous

Pafnuty Lvovich Chebyshev (1821–1894) did exceptional mathematical work in many areas, including applied mathematics, number theory, approximation theory, and probability. In 1852 he traveled from St. Petersburg to visit mathematicians in France, England, and Germany. Lagrange and Legendre had studied individual sets of orthogonal polynomials, but Chebyshev was the first to see the important consequences of studying the theory in general. He developed the Chebyshev polynomials to study least squares approximation and probability, then applied his results to interpolation, approximate quadrature, and other areas.

section, it is easier to give their definition and then show that they satisfy the required orthogonality properties.

For $x \in [-1, 1]$, define

$$T_n(x) = \cos[n \arccos x], \quad \text{for each } n \geq 0. \tag{8.8}$$

It might not be obvious from this definition that for each n, $T_n(x)$ is a polynomial in x, but we will now show this. First note that

$$T_0(x) = \cos 0 = 1 \quad \text{and} \quad T_1(x) = \cos(\arccos x) = x.$$

For $n \geq 1$, we introduce the substitution $\theta = \arccos x$ to change this equation to

$$T_n(\theta(x)) \equiv T_n(\theta) = \cos(n\theta), \quad \text{where } \theta \in [0, \pi].$$

A recurrence relation is derived by noting that

$$T_{n+1}(\theta) = \cos(n+1)\theta = \cos\theta \, \cos(n\theta) - \sin\theta \, \sin(n\theta)$$

and

$$T_{n-1}(\theta) = \cos(n-1)\theta = \cos\theta \, \cos(n\theta) + \sin\theta \, \sin(n\theta)$$

Adding these equations gives

$$T_{n+1}(\theta) = 2\cos\theta \, \cos(n\theta) - T_{n-1}(\theta).$$

Returning to the variable $x = \cos\theta$, we have, for $n \geq 1$,

$$T_{n+1}(x) = 2x\cos(n \arccos x) - T_{n-1}(x),$$

that is,

$$T_{n+1}(x) = 2xT_n(x) - T_{n-1}(x). \tag{8.9}$$

Because $T_0(x) = 1$ and $T_1(x) = x$, the recurrence relation implies that the next three Chebyshev polynomials are

$$T_2(x) = 2xT_1(x) - T_0(x) = 2x^2 - 1,$$

$$T_3(x) = 2xT_2(x) - T_1(x) = 4x^3 - 3x,$$

and

$$T_4(x) = 2xT_3(x) - T_2(x) = 8x^4 - 8x^2 + 1.$$

The recurrence relation also implies that when $n \geq 1$, $T_n(x)$ is a polynomial of degree n with leading coefficient 2^{n-1}. The graphs of T_1, T_2, T_3, and T_4 are shown in Figure 8.10.

Figure 8.10

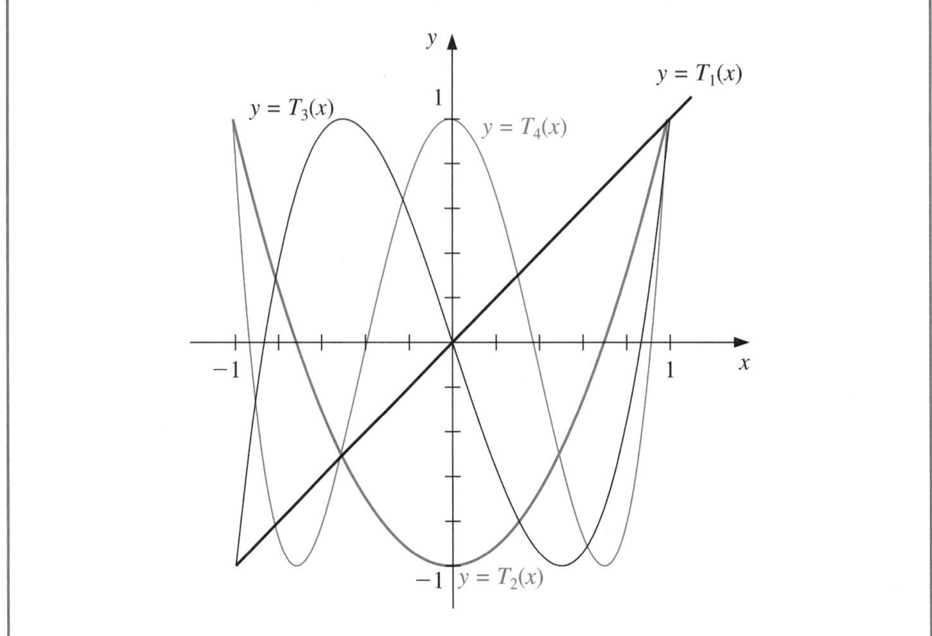

To show the orthogonality of the Chebyshev polynomials with respect to the weight function $w(x) = (1 - x^2)^{-1/2}$, consider

$$\int_{-1}^{1} \frac{T_n(x)T_m(x)}{\sqrt{1-x^2}}\, dx = \int_{-1}^{1} \frac{\cos(n \arccos x)\cos(m \arccos x)}{\sqrt{1-x^2}}\, dx.$$

Reintroducing the substitution $\theta = \arccos x$ gives

$$d\theta = -\frac{1}{\sqrt{1-x^2}}\, dx$$

and

$$\int_{-1}^{1} \frac{T_n(x)T_m(x)}{\sqrt{1-x^2}}\, dx = -\int_{\pi}^{0} \cos(n\theta)\cos(m\theta)\, d\theta = \int_{0}^{\pi} \cos(n\theta)\cos(m\theta)\, d\theta.$$

Suppose $n \neq m$. Since

$$\cos(n\theta)\cos(m\theta) = \frac{1}{2}[\cos(n+m)\theta + \cos(n-m)\theta],$$

we have

$$\begin{aligned}\int_{-1}^{1} \frac{T_n(x)T_m(x)}{\sqrt{1-x^2}}\, dx &= \frac{1}{2}\int_{0}^{\pi} \cos((n+m)\theta)\, d\theta + \frac{1}{2}\int_{0}^{\pi} \cos((n-m)\theta)\, d\theta \\ &= \left[\frac{1}{2(n+m)}\sin((n+m)\theta) + \frac{1}{2(n-m)}\sin((n-m)\theta)\right]_0^{\pi} = 0.\end{aligned}$$

By a similar technique (see Exercise 9), we also have

$$\int_{-1}^{1} \frac{[T_n(x)]^2}{\sqrt{1-x^2}}\, dx = \frac{\pi}{2}, \quad \text{for each } n \geq 1. \tag{8.10}$$

The Chebyshev polynomials are used to minimize approximation error. We will see how they are used to solve two problems of this type:

- an optimal placing of interpolating points to minimize the error in Lagrange interpolation;
- a means of reducing the degree of an approximating polynomial with minimal loss of accuracy.

The next result concerns the zeros and extreme points of $T_n(x)$.

Theorem 8.9 The Chebyshev polynomial $T_n(x)$ of degree $n \geq 1$ has n simple zeros in $[-1, 1]$ at

$$\bar{x}_k = \cos\left(\frac{2k-1}{2n}\pi\right), \quad \text{for each } k = 1, 2, \ldots, n.$$

Moreover, $T_n(x)$ assumes its absolute extrema at

$$\bar{x}'_k = \cos\left(\frac{k\pi}{n}\right) \quad \text{with} \quad T_n(\bar{x}'_k) = (-1)^k, \quad \text{for each} \quad k = 0, 1, \ldots, n.$$ ■

Proof Let

$$\bar{x}_k = \cos\left(\frac{2k-1}{2n}\pi\right), \quad \text{for } k = 1, 2, \ldots, n.$$

Then

$$T_n(\bar{x}_k) = \cos(n \arccos \bar{x}_k) = \cos\left(n \arccos\left(\cos\left(\frac{2k-1}{2n}\pi\right)\right)\right) = \cos\left(\frac{2k-1}{2}\pi\right) = 0.$$

But the $\bar{x}_k$ are distinct (see Exercise 10) and $T_n(x)$ is a polynomial of degree n, so all the zeros of $T_n(x)$ must have this form.

To show the second statement, first note that

$$T'_n(x) = \frac{d}{dx}[\cos(n \arccos x)] = \frac{n \sin(n \arccos x)}{\sqrt{1-x^2}},$$

and that, when $k = 1, 2, \ldots, n-1$,

$$T'_n(\bar{x}'_k) = \frac{n \sin\left(n \arccos\left(\cos\left(\frac{k\pi}{n}\right)\right)\right)}{\sqrt{1-\left[\cos\left(\frac{k\pi}{n}\right)\right]^2}} = \frac{n \sin(k\pi)}{\sin\left(\frac{k\pi}{n}\right)} = 0.$$

Since $T_n(x)$ is a polynomial of degree n, its derivative $T'_n(x)$ is a polynomial of degree $(n-1)$, and all the zeros of $T'_n(x)$ occur at these $n-1$ distinct points (that they are distinct is considered in Exercise 11). The only other possibilities for extrema of $T_n(x)$ occur at the endpoints of the interval $[-1, 1]$; that is, at $\bar{x}'_0 = 1$ and at $\bar{x}'_n = -1$.

For any $k = 0, 1, \ldots, n$ we have

$$T_n(\bar{x}'_k) = \cos\left(n \arccos\left(\cos\left(\frac{k\pi}{n}\right)\right)\right) = \cos(k\pi) = (-1)^k.$$

So a maximum occurs at each even value of k and a minimum at each odd value. ■ ■ ■

The monic (polynomials with leading coefficient 1) Chebyshev polynomials $\tilde{T}_n(x)$ are derived from the Chebyshev polynomials $T_n(x)$ by dividing by the leading coefficient 2^{n-1}. Hence

$$\tilde{T}_0(x) = 1 \quad \text{and} \quad \tilde{T}_n(x) = \frac{1}{2^{n-1}} T_n(x), \quad \text{for each } n \geq 1. \tag{8.11}$$

The recurrence relationship satisfied by the Chebyshev polynomials implies that

$$\tilde{T}_2(x) = x\tilde{T}_1(x) - \frac{1}{2}\tilde{T}_0(x) \quad \text{and} \tag{8.12}$$

$$\tilde{T}_{n+1}(x) = x\tilde{T}_n(x) - \frac{1}{4}\tilde{T}_{n-1}(x), \quad \text{for each } n \geq 2.$$

The graphs of $\tilde{T}_1$, $\tilde{T}_2$, $\tilde{T}_3$, $\tilde{T}_4$, and $\tilde{T}_5$ are shown in Figure 8.11.

Figure 8.11

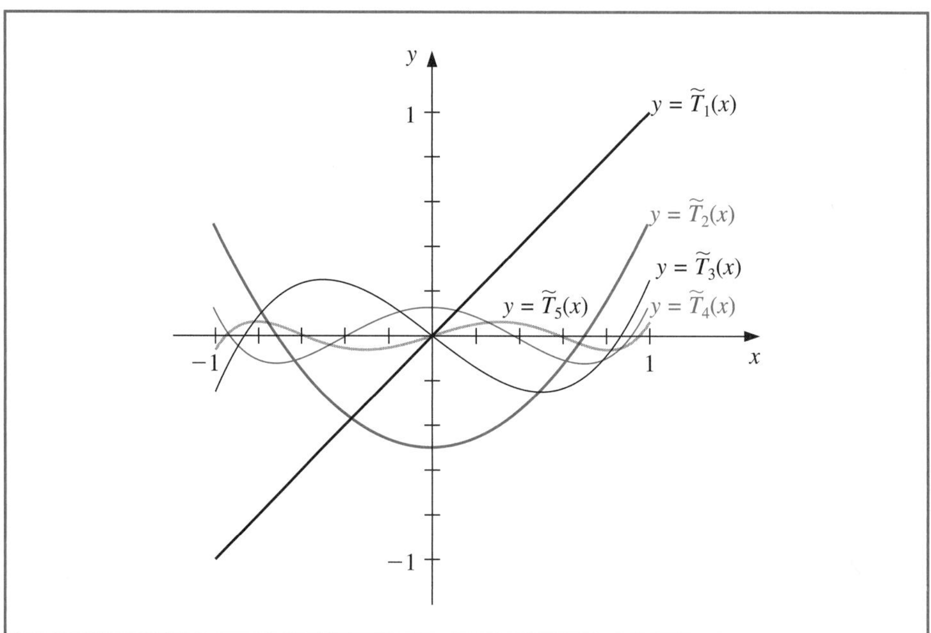

Because $\tilde{T}_n(x)$ is just a multiple of $T_n(x)$, Theorem 8.9 implies that the zeros of $\tilde{T}_n(x)$ also occur at

$$\bar{x}_k = \cos\left(\frac{2k-1}{2n}\pi\right), \quad \text{for each } k = 1, 2, \ldots, n,$$

and the extreme values of $\tilde{T}_n(x)$, for $n \geq 1$, occur at

$$\bar{x}'_k = \cos\left(\frac{k\pi}{n}\right), \quad \text{with} \quad \tilde{T}_n(\bar{x}'_k) = \frac{(-1)^k}{2^{n-1}}, \quad \text{for each } k = 0, 1, 2, \ldots, n. \tag{8.13}$$

Let $\widetilde{\prod}_n$ denote **the set of all monic polynomials of degree n**. The relation expressed in Eq. (8.13) leads to an important minimization property that distinguishes $\tilde{T}_n(x)$ from the other members of $\widetilde{\prod}_n$.

Theorem 8.10 The polynomials of the form $\tilde{T}_n(x)$, when $n \geq 1$, have the property that

$$\frac{1}{2^{n-1}} = \max_{x\in[-1,1]} |\tilde{T}_n(x)| \leq \max_{x\in[-1,1]} |P_n(x)|, \quad \text{for all } P_n(x) \in \widetilde{\prod}_n.$$

Moreover, equality occurs only if $P_n \equiv \tilde{T}_n$. ■

Proof Suppose that $P_n(x) \in \tilde{\prod}_n$ and that

$$\max_{x\in[-1,1]} |P_n(x)| \le \frac{1}{2^{n-1}} = \max_{x\in[-1,1]} |\tilde{T}_n(x)|.$$

Let $Q = \tilde{T}_n - P_n$. Then $\tilde{T}_n(x)$ and $P_n(x)$ are both monic polynomials of degree n, so $Q(x)$ is a polynomial of degree at most $(n-1)$. Moreover, at the $n+1$ extreme points $\bar{x}'_k$ of $\tilde{T}_n(x)$, we have

$$Q(\bar{x}'_k) = \tilde{T}_n(\bar{x}'_k) - P_n(\bar{x}'_k) = \frac{(-1)^k}{2^{n-1}} - P_n(\bar{x}'_k).$$

However

$$|P_n(\bar{x}'_k)| \le \frac{1}{2^{n-1}}, \quad \text{for each } k = 0, 1, \ldots, n,$$

so we have

$$Q(\bar{x}'_k) \le 0, \quad \text{when } k \text{ is odd} \quad \text{and} \quad Q(\bar{x}'_k) \ge 0, \quad \text{when } k \text{ is even.}$$

Since Q is continuous, the Intermediate Value Theorem implies that for each $j = 0, 1, \ldots, n-1$ the polynomial $Q(x)$ has at least one zero between $\bar{x}'_j$ and $\bar{x}'_{j+1}$. Thus, Q has at least n zeros in the interval $[-1, 1]$. But the degree of $Q(x)$ is less than n, so $Q \equiv 0$. This implies that $P_n \equiv \tilde{T}_n$. ■ ■ ■

Minimizing Lagrange Interpolation Error

Theorem 8.10 can be used to answer the question of where to place interpolating nodes to minimize the error in Lagrange interpolation. Theorem 3.3 on page 112 applied to the interval $[-1, 1]$ states that, if $x_0, \ldots, x_n$ are distinct numbers in the interval $[-1, 1]$ and if $f \in C^{n+1}[-1, 1]$, then, for each $x \in [-1, 1]$, a number $\xi(x)$ exists in $(-1, 1)$ with

$$f(x) - P(x) = \frac{f^{(n+1)}(\xi(x))}{(n+1)!}(x-x_0)(x-x_1)\cdots(x-x_n),$$

where $P(x)$ is the Lagrange interpolating polynomial. Generally, there is no control over $\xi(x)$, so to minimize the error by shrewd placement of the nodes $x_0, \ldots, x_n$, we choose $x_0, \ldots, x_n$ to minimize the quantity

$$|(x-x_0)(x-x_1)\cdots(x-x_n)|$$

throughout the interval $[-1, 1]$.

Since $(x-x_0)(x-x_1)\cdots(x-x_n)$ is a monic polynomial of degree $(n+1)$, we have just seen that the minimum is obtained when

$$(x-x_0)(x-x_1)\cdots(x-x_n) = \tilde{T}_{n+1}(x).$$

The maximum value of $|(x-x_0)(x-x_1)\cdots(x-x_n)|$ is smallest when x_k is chosen for each $k = 0, 1, \ldots, n$ to be the $(k+1)$st zero of $\tilde{T}_{n+1}$. Hence we choose x_k to be

$$\bar{x}_{k+1} = \cos\left(\frac{2k+1}{2(n+1)}\pi\right).$$

Because $\max_{x\in[-1,1]} |\tilde{T}_{n+1}(x)| = 2^{-n}$, this also implies that

$$\frac{1}{2^n} = \max_{x\in[-1,1]} |(x-\bar{x}_1)\cdots(x-\bar{x}_{n+1})| \le \max_{x\in[-1,1]} |(x-x_0)\cdots(x-x_n)|,$$

for any choice of $x_0, x_1, \ldots, x_n$ in the interval $[-1, 1]$. The next corollary follows from these observations.

Corollary 8.11 Suppose that $P(x)$ is the interpolating polynomial of degree at most n with nodes at the zeros of $T_{n+1}(x)$. Then

$$\max_{x\in[-1,1]} |f(x) - P(x)| \leq \frac{1}{2^n(n+1)!} \max_{x\in[-1,1]} |f^{(n+1)}(x)|, \quad \text{for each } f \in C^{n+1}[-1,1]. \quad \blacksquare$$

Minimizing Approximation Error on Arbitrary Intervals

The technique for choosing points to minimize the interpolating error is extended to a general closed interval $[a, b]$ by using the change of variables

$$\tilde{x} = \frac{1}{2}[(b-a)x + a + b]$$

to transform the numbers $\bar{x}_k$ in the interval $[-1, 1]$ into the corresponding number $\tilde{x}_k$ in the interval $[a, b]$, as shown in the next example.

Example 1 Let $f(x) = xe^x$ on [0, 1.5]. Compare the values given by the Lagrange polynomial with four equally-spaced nodes with those given by the Lagrange polynomial with nodes given by zeros of the fourth Chebyshev polynomial.

Solution The equally-spaced nodes $x_0 = 0, x_1 = 0.5, x_2 = 1$, and $x_3 = 1.5$ give

$$L_0(x) = -1.3333x^3 + 4.0000x^2 - 3.6667x + 1,$$

$$L_1(x) = 4.0000x^3 - 10.000x^2 + 6.0000x,$$

$$L_2(x) = -4.0000x^3 + 8.0000x^2 - 3.0000x,$$

$$L_3(x) = 1.3333x^3 - 2.000x^2 + 0.66667x,$$

which produces the polynomial

$$P_3(x) = L_0(x)(0) + L_1(x)(0.5e^{0.5}) + L_2(x)e^1 + L_3(x)(1.5e^{1.5}) = 1.3875x^3 + 0.057570x^2 + 1.2730x.$$

For the second interpolating polynomial, we shift the zeros $\bar{x}_k = \cos((2k+1)/8)\pi$, for $k = 0, 1, 2, 3$, of $\tilde{T}_4$ from $[-1, 1]$ to $[0, 1.5]$, using the linear transformation

$$\tilde{x}_k = \frac{1}{2}[(1.5-0)\bar{x}_k + (1.5+0)] = 0.75 + 0.75\bar{x}_k.$$

Because

$$\bar{x}_0 = \cos\frac{\pi}{8} = 0.92388, \quad \bar{x}_1 = \cos\frac{3\pi}{8} = 0.38268,$$

$$\bar{x}_2 = \cos\frac{5\pi}{8} = -0.38268, \quad \text{and} \bar{x}_4 = \cos\frac{7\pi}{8} = -0.92388,$$

we have

$$\tilde{x}_0 = 1.44291, \quad \tilde{x}_1 = 1.03701, \quad \tilde{x}_2 = 0.46299, \quad \text{and} \quad \tilde{x}_3 = 0.05709.$$

The Lagrange coefficient polynomials for this set of nodes are

$$\tilde{L}_0(x) = 1.8142x^3 - 2.8249x^2 + 1.0264x - 0.049728,$$

$$\tilde{L}_1(x) = -4.3799x^3 + 8.5977x^2 - 3.4026x + 0.16705,$$

$$\tilde{L}_2(x) = 4.3799x^3 - 11.112x^2 + 7.1738x - 0.37415,$$

$$\tilde{L}_3(x) = -1.8142x^3 + 5.3390x^2 - 4.7976x + 1.2568.$$

The functional values required for these polynomials are given in the last two columns of Table 8.7. The interpolation polynomial of degree at most 3 is

$$\tilde{P}_3(x) = 1.3811x^3 + 0.044652x^2 + 1.3031x - 0.014352.$$

Table 8.7

x	$f(x) = xe^x$	$\tilde{x}$	$f(\tilde{x}) = xe^x$
$x_0 = 0.0$	0.00000	$\tilde{x}_0 = 1.44291$	6.10783
$x_1 = 0.5$	0.824361	$\tilde{x}_1 = 1.03701$	2.92517
$x_2 = 1.0$	2.71828	$\tilde{x}_2 = 0.46299$	0.73560
$x_3 = 1.5$	6.72253	$\tilde{x}_3 = 0.05709$	0.060444

For comparison, Table 8.8 lists various values of x, together with the values of $f(x)$, $P_3(x)$, and $\tilde{P}_3(x)$. It can be seen from this table that, although the error using $P_3(x)$ is less than using $\tilde{P}_3(x)$ near the middle of the table, the maximum error involved with using $\tilde{P}_3(x)$, 0.0180, is considerably less than when using $P_3(x)$, which gives the error 0.0290. (See Figure 8.12.) ■

Table 8.8

x	$f(x) = xe^x$	$P_3(x)$	$\lvert xe^x - P_3(x)\rvert$	$\tilde{P}_3(x)$	$\lvert xe^x - \tilde{P}_3(x)\rvert$
0.15	0.1743	0.1969	0.0226	0.1868	0.0125
0.25	0.3210	0.3435	0.0225	0.3358	0.0148
0.35	0.4967	0.5121	0.0154	0.5064	0.0097
0.65	1.245	1.233	0.012	1.231	0.014
0.75	1.588	1.572	0.016	1.571	0.017
0.85	1.989	1.976	0.013	1.974	0.015
1.15	3.632	3.650	0.018	3.644	0.012
1.25	4.363	4.391	0.028	4.382	0.019
1.35	5.208	5.237	0.029	5.224	0.016

Figure 8.12

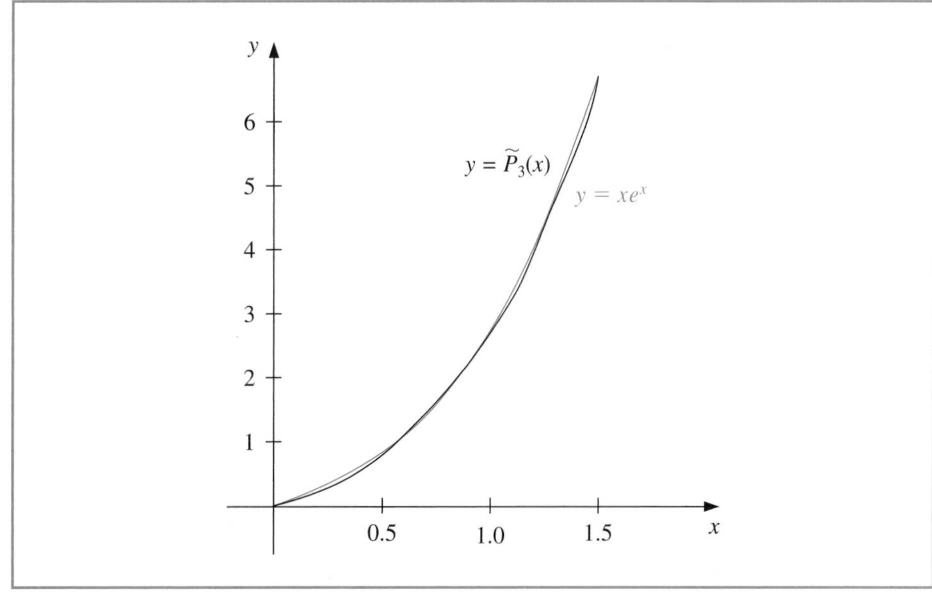

Reducing the Degree of Approximating Polynomials

Chebyshev polynomials can also be used to reduce the degree of an approximating polynomial with a minimal loss of accuracy. Because the Chebyshev polynomials have a minimum maximum-absolute value that is spread uniformly on an interval, they can be used to reduce the degree of an approximation polynomial without exceeding the error tolerance.

Consider approximating an arbitrary nth-degree polynomial

$$P_n(x) = a_n x^n + a_{n-1}x^{n-1} + \cdots + a_1 x + a_0$$

on $[-1, 1]$ with a polynomial of degree at most $n - 1$. The object is to choose $P_{n-1}(x)$ in $\prod_{n-1}$ so that

$$\max_{x\in[-1,1]} |P_n(x) - P_{n-1}(x)|$$

is as small as possible.

We first note that $(P_n(x) - P_{n-1}(x))/a_n$ is a monic polynomial of degree n, so applying Theorem 8.10 gives

$$\max_{x\in[-1,1]} \left| \frac{1}{a_n}(P_n(x) - P_{n-1}(x)) \right| \geq \frac{1}{2^{n-1}}.$$

Equality occurs precisely when

$$\frac{1}{a_n}(P_n(x) - P_{n-1}(x)) = \tilde{T}_n(x).$$

This means that we should choose

$$P_{n-1}(x) = P_n(x) - a_n \tilde{T}_n(x),$$

and with this choice we have the minimum value of

$$\max_{x\in[-1,1]} |P_n(x) - P_{n-1}(x)| = |a_n| \max_{x\in[-1,1]} \left| \frac{1}{a_n}(P_n(x) - P_{n-1}(x)) \right| = \frac{|a_n|}{2^{n-1}}.$$

Illustration The function $f(x) = e^x$ is approximated on the interval $[-1, 1]$ by the fourth Maclaurin polynomial

$$P_4(x) = 1 + x + \frac{x^2}{2} + \frac{x^3}{6} + \frac{x^4}{24},$$

which has truncation error

$$|R_4(x)| = \frac{|f^{(5)}(\xi(x))||x^5|}{120} \leq \frac{e}{120} \approx 0.023, \quad \text{for } -1 \leq x \leq 1.$$

Suppose that an error of 0.05 is tolerable and that we would like to reduce the degree of the approximating polynomial while staying within this bound.

The polynomial of degree 3 or less that best uniformly approximates $P_4(x)$ on $[-1, 1]$ is

$$P_3(x) = P_4(x) - a_4\tilde{T}_4(x) = 1 + x + \frac{x^2}{2} + \frac{x^3}{6} + \frac{x^4}{24} - \frac{1}{24}\left(x^4 - x^2 + \frac{1}{8}\right)$$

$$= \frac{191}{192} + x + \frac{13}{24}x^2 + \frac{1}{6}x^3.$$

With this choice, we have

$$|P_4(x) - P_3(x)| = |a_4\tilde{T}_4(x)| \le \frac{1}{24} \cdot \frac{1}{2^3} = \frac{1}{192} \le 0.0053.$$

Adding this error bound to the bound for the Maclaurin truncation error gives

$$0.023 + 0.0053 = 0.0283,$$

which is within the permissible error of 0.05.

The polynomial of degree 2 or less that best uniformly approximates $P_3(x)$ on $[-1, 1]$ is

$$\begin{aligned} P_2(x) &= P_3(x) - \frac{1}{6}\tilde{T}_3(x) \\ &= \frac{191}{192} + x + \frac{13}{24}x^2 + \frac{1}{6}x^3 - \frac{1}{6}(x^3 - \frac{3}{4}x) = \frac{191}{192} + \frac{9}{8}x + \frac{13}{24}x^2. \end{aligned}$$

However,

$$|P_3(x) - P_2(x)| = \left|\frac{1}{6}\tilde{T}_3(x)\right| = \frac{1}{6}\left(\frac{1}{2}\right)^2 = \frac{1}{24} \approx 0.042,$$

which—when added to the already accumulated error bound of 0.0283—exceeds the tolerance of 0.05. Consequently, the polynomial of least degree that best approximates e^x on $[-1, 1]$ with an error bound of less than 0.05 is

$$P_3(x) = \frac{191}{192} + x + \frac{13}{24}x^2 + \frac{1}{6}x^3.$$

Table 8.9 lists the function and the approximating polynomials at various points in $[-1, 1]$. Note that the tabulated entries for P_2 are well within the tolerance of 0.05, even though the error bound for $P_2(x)$ exceeded the tolerance. □

Table 8.9

x	e^x	$P_4(x)$	$P_3(x)$	$P_2(x)$	$\lvert e^x - P_2(x)\rvert$
−0.75	0.47237	0.47412	0.47917	0.45573	0.01664
−0.25	0.77880	0.77881	0.77604	0.74740	0.03140
0.00	1.00000	1.00000	0.99479	0.99479	0.00521
0.25	1.28403	1.28402	1.28125	1.30990	0.02587
0.75	2.11700	2.11475	2.11979	2.14323	0.02623

EXERCISE SET 8.3

1. Use the zeros of $\tilde{T}_3$ to construct an interpolating polynomial of degree 2 for the following functions on the interval $[-1, 1]$.

   a. $f(x) = e^x$ **b.** $f(x) = \sin x$ **c.** $f(x) = \ln(x + 2)$ **d.** $f(x) = x^4$

2. Use the zeros of $\tilde{T}_4$ to construct an interpolating polynomial of degree 3 for the functions in Exercise 1.
3. Find a bound for the maximum error of the approximation in Exercise 1 on the interval $[-1, 1]$.
4. Repeat Exercise 3 for the approximations computed in Exercise 3.

5. Use the zeros of $\tilde{T}_3$ and transformations of the given interval to construct an interpolating polynomial of degree 2 for the following functions.

   a. $f(x) = \dfrac{1}{x}, \quad [1, 3]$
   b. $f(x) = e^{-x}, \quad [0, 2]$
   c. $f(x) = \dfrac{1}{2}\cos x + \dfrac{1}{3}\sin 2x, \quad [0, 1]$
   d. $f(x) = x \ln x, \quad [1, 3]$

6. Find the sixth Maclaurin polynomial for xe^x, and use Chebyshev economization to obtain a lesser-degree polynomial approximation while keeping the error less than 0.01 on $[-1, 1]$.

7. Find the sixth Maclaurin polynomial for $\sin x$, and use Chebyshev economization to obtain a lesser-degree polynomial approximation while keeping the error less than 0.01 on $[-1, 1]$.

8. Show that for any positive integers i and j with $i > j$, we have $T_i(x)T_j(x) = \frac{1}{2}[T_{i+j}(x) + T_{i-j}(x)]$.

9. Show that for each Chebyshev polynomial $T_n(x)$, we have

$$\int_{-1}^{1} \frac{[T_n(x)]^2}{\sqrt{1-x^2}}\, dx = \frac{\pi}{2}.$$

10. Show that for each n, the Chebyshev polynomial $T_n(x)$ has n distinct zeros in $(-1, 1)$.

11. Show that for each n, the derivative of the Chebyshev polynomial $T_n(x)$ has $n - 1$ distinct zeros in $(-1, 1)$.

8.4 Rational Function Approximation

The class of algebraic polynomials has some distinct advantages for use in approximation:

- There are a sufficient number of polynomials to approximate any continuous function on a closed interval to within an arbitrary tolerance;
- Polynomials are easily evaluated at arbitrary values; and
- The derivatives and integrals of polynomials exist and are easily determined.

The disadvantage of using polynomials for approximation is their tendency for oscillation. This often causes error bounds in polynomial approximation to significantly exceed the average approximation error, because error bounds are determined by the maximum approximation error. We now consider methods that spread the approximation error more evenly over the approximation interval. These techniques involve rational functions.

A **rational function** r of degree N has the form

$$r(x) = \frac{p(x)}{q(x)},$$

where $p(x)$ and $q(x)$ are polynomials whose degrees sum to N.

Every polynomial is a rational function (simply let $q(x) \equiv 1$), so approximation by rational functions gives results that are no worse than approximation by polynomials. However, rational functions whose numerator and denominator have the same or nearly the same degree often produce approximation results superior to polynomial methods for the same amount of computation effort. (This statement is based on the assumption that the amount of computation effort required for division is approximately the same as for multiplication.)

Rational functions have the added advantage of permitting efficient approximation of functions with infinite discontinuities near, but outside, the interval of approximation. Polynomial approximation is generally unacceptable in this situation.

Padé Approximation

Henri Padé (1863–1953) gave a systematic study of what we call today Padé approximations in his doctoral thesis in 1892. He proved results on their general structure and also clearly set out the connection between Padé approximations and continued fractions. These ideas, however, had been studied by Daniel Bernoulli (1700–1782) and others as early as 1730. James Stirling (1692–1770) gave a similar method in *Methodus differentialis* published in the same year, and Euler used Padé-type approximation to find the sum of a series.

Suppose r is a rational function of degree $N = n + m$ of the form

$$r(x) = \frac{p(x)}{q(x)} = \frac{p_0 + p_1x + \cdots + p_nx^n}{q_0 + q_1x + \cdots + q_mx^m},$$

that is used to approximate a function f on a closed interval I containing zero. For r to be defined at zero requires that $q_0 \neq 0$. In fact, we can assume that $q_0 = 1$, for if this is not the case we simply replace $p(x)$ by $p(x)/q_0$ and $q(x)$ by $q(x)/q_0$. Consequently, there are $N + 1$ parameters $q_1, q_2, \ldots, q_m, p_0, p_1, \ldots, p_n$ available for the approximation of f by r.

The **Padé approximation technique**, is the extension of Taylor polynomial approximation to rational functions. It chooses the $N + 1$ parameters so that $f^{(k)}(0) = r^{(k)}(0)$, for each $k = 0, 1, \ldots, N$. When $n = N$ and $m = 0$, the Padé approximation is simply the Nth Maclaurin polynomial.

Consider the difference

$$f(x) - r(x) = f(x) - \frac{p(x)}{q(x)} = \frac{f(x)q(x) - p(x)}{q(x)} = \frac{f(x)\sum_{i=0}^{m} q_ix^i - \sum_{i=0}^{n} p_ix^i}{q(x)},$$

and suppose f has the Maclaurin series expansion $f(x) = \sum_{i=0}^{\infty} a_ix^i$. Then

$$f(x) - r(x) = \frac{\sum_{i=0}^{\infty} a_ix^i \sum_{i=0}^{m} q_ix^i - \sum_{i=0}^{n} p_ix^i}{q(x)}. \tag{8.14}$$

The object is to choose the constants $q_1, q_2, \ldots, q_m$ and $p_0, p_1, \ldots, p_n$ so that

$$f^{(k)}(0) - r^{(k)}(0) = 0, \quad \text{for each } k = 0, 1, \ldots, N.$$

In Section 2.4 (see, in particular, Exercise 10 on page 86) we found that this is equivalent to $f - r$ having a zero of multiplicity $N + 1$ at $x = 0$. As a consequence, we choose $q_1, q_2, \ldots, q_m$ and $p_0, p_1, \ldots, p_n$ so that the numerator on the right side of Eq. (8.14),

$$(a_0 + a_1x + \cdots)(1 + q_1x + \cdots + q_mx^m) - (p_0 + p_1x + \cdots + p_nx^n), \tag{8.15}$$

has no terms of degree less than or equal to N.

To simplify notation, we define $p_{n+1} = p_{n+2} = \cdots = p_N = 0$ and $q_{m+1} = q_{m+2} = \cdots = q_N = 0$. We can then express the coefficient of x^k in expression (8.15) more compactly as

$$\left(\sum_{i=0}^{k} a_iq_{k-i}\right) - p_k.$$

The rational function for Padé approximation results from the solution of the $N + 1$ linear equations

$$\sum_{i=0}^{k} a_iq_{k-i} = p_k, \quad k = 0, 1, \ldots, N$$

in the $N + 1$ unknowns $q_1, q_2, \ldots, q_m, p_0, p_1, \ldots, p_n$.

Example 1 The Maclaurin series expansion for e^{-x} is

$$\sum_{i=0}^{\infty} \frac{(-1)^i}{i!}x^i.$$

Find the Padé approximation to e^{-x} of degree 5 with $n = 3$ and $m = 2$.

Solution To find the Padé approximation we need to choose p_0, p_1, p_2, p_3, q_1, and q_2 so that the coefficients of x^k for $k = 0, 1, \ldots, 5$ are 0 in the expression

$$\left(1 - x + \frac{x^2}{2} - \frac{x^3}{6} + \cdots\right)(1 + q_1x + q_2x^2) - (p_0 + p_1x + p_2x^2 + p_3x^3).$$

Expanding and collecting terms produces

$$\begin{aligned}
x^5:&\quad -\frac{1}{120} + \frac{1}{24}q_1 - \frac{1}{6}q_2 = 0; &\qquad x^2:&\quad \frac{1}{2} - q_1 + q_2 = p_2;\\
x^4:&\quad \frac{1}{24} - \frac{1}{6}q_1 + \frac{1}{2}q_2 = 0; & x^1:&\quad -1 + q_1 = p_1;\\
x^3:&\quad -\frac{1}{6} + \frac{1}{2}q_1 - q_2 = p_3; & x^0:&\quad 1 = p_0.
\end{aligned}$$

To solve the system in Maple, we use the following commands:

$eq\,1 := -1 + q1 = p1$:
$eq\,2 := \frac{1}{2} - q1 + q2 = p2$:
$eq\,3 := -\frac{1}{6} + \frac{1}{2}q1 - q2 = p3$:
$eq\,4 := \frac{1}{24} - \frac{1}{6}q1 + \frac{1}{2}q2 = 0$:
$eq\,5 := -\frac{1}{120} + \frac{1}{24}q1 - \frac{1}{6}q2 = 0$:
$solve(\{eq1, eq2, eq3, eq4, eq5\}, \{q1, q2, p1, p2, p3\})$

This gives

$$\left\{p_1 = -\frac{3}{5},\ p_2 = \frac{3}{20},\ p_3 = -\frac{1}{60},\ q_1 = \frac{2}{5},\ q_2 = \frac{1}{20}\right\}$$

So the Padé approximation is

$$r(x) = \frac{1 - \frac{3}{5}x + \frac{3}{20}x^2 - \frac{1}{60}x^3}{1 + \frac{2}{5}x + \frac{1}{20}x^2}.$$

Table 8.10 lists values of $r(x)$ and $P_5(x)$, the fifth Maclaurin polynomial. The Padé approximation is clearly superior in this example. ■

Table 8.10

x	e^{-x}	$P_5(x)$	$\lvert e^{-x} - P_5(x)\rvert$	$r(x)$	$\lvert e^{-x} - r(x)\rvert$
0.2	0.81873075	0.81873067	8.64×10^{-8}	0.81873075	7.55×10^{-9}
0.4	0.67032005	0.67031467	5.38×10^{-6}	0.67031963	4.11×10^{-7}
0.6	0.54881164	0.54875200	5.96×10^{-5}	0.54880763	4.00×10^{-6}
0.8	0.44932896	0.44900267	3.26×10^{-4}	0.44930966	1.93×10^{-5}
1.0	0.36787944	0.36666667	1.21×10^{-3}	0.36781609	6.33×10^{-5}

Maple can also be used directly to compute a Padé approximation. We first compute the Maclaurin series with the call

$series(exp(-x), x)$

to obtain

$$1 - x + \frac{1}{2}x^2 - \frac{1}{6}x^3 + \frac{1}{24}x^4 - \frac{1}{120}x^5 + O(x^6)$$

The Padé approximation $r(x)$ with $n = 3$ and $m = 2$ is found using the command

$r := x \rightarrow convert(\%, ratpoly, 3, 2);$

where the % refers to the result of the preceding calculation, namely, the series. The Maple result is

$$x \rightarrow \frac{1 - \frac{3}{5}x + \frac{3}{20}x^2 - \frac{1}{60}x^3}{1 + \frac{2}{5}x + \frac{1}{20}x^2}$$

We can then compute, for example, $r(0.8)$ by entering

$r(0.8)$

which produces the approximation 0.4493096647 to $e^{-0.8} = 0.449328964$.

Algorithm 8.1 implements the Padé approximation technique.

Padé Rational Approximation

To obtain the rational approximation

$$r(x) = \frac{p(x)}{q(x)} = \frac{\sum_{i=0}^{n} p_i x^i}{\sum_{j=0}^{m} q_j x^j}$$

for a given function $f(x)$:

INPUT nonnegative integers m and n.

OUTPUT coefficients $q_0, q_1, \ldots, q_m$ and $p_0, p_1, \ldots, p_n$.

Step 1 Set $N = m + n$.

Step 2 For $i = 0, 1, \ldots, N$ set $a_i = \dfrac{f^{(i)}(0)}{i!}$.
(*The coefficients of the Maclaurin polynomial are $a_0, \ldots, a_N$, which could be input instead of calculated.*)

Step 3 Set $q_0 = 1$;
$p_0 = a_0$.

Step 4 For $i = 1, 2, \ldots, N$ do Steps 5–10. (*Set up a linear system with matrix B.*)

Step 5 For $j = 1, 2, \ldots, i - 1$
if $j \leq n$ then set $b_{i,j} = 0$.

Step 6 If $i \leq n$ then set $b_{i,i} = 1$.

Step 7 For $j = i + 1, i + 2, \ldots, N$ set $b_{i,j} = 0$.

Step 8 For $j = 1, 2, \ldots, i$
if $j \leq m$ then set $b_{i,n+j} = -a_{i-j}$.

Step 9 For $j = n + i + 1, n + i + 2, \ldots, N$ set $b_{i,j} = 0$.

Step 10 Set $b_{i,N+1} = a_i$.

(*Steps 11–22 solve the linear system using partial pivoting.*)

Step 11 For $i = n + 1, n + 2, \ldots, N - 1$ do Steps 12–18.

Step 12 Let k be the smallest integer with $i \leq k \leq N$ and $|b_{k,i}| = \max_{i \leq j \leq N} |b_{j,i}|$.
(*Find pivot element.*)

Step 13 If $b_{k,i} = 0$ then OUTPUT ("The system is singular ");
STOP.

Step 14 If $k \neq i$ then (*Interchange row i and row k.*)
for $j = i, i+1, \ldots, N+1$ set

$$b_{COPY} = b_{i,j};$$
$$b_{i,j} = b_{k,j};$$
$$b_{k,j} = b_{COPY}.$$

Step 15 For $j = i+1, i+2, \ldots, N$ do Steps 16–18. (*Perform elimination.*)

Step 16 Set $xm = \dfrac{b_{j,i}}{b_{i,i}}$.

Step 17 For $k = i+1, i+2, \ldots, N+1$
set $b_{j,k} = b_{j,k} - xm \cdot b_{i,k}$.

Step 18 Set $b_{j,i} = 0$.

Step 19 If $b_{N,N} = 0$ then OUTPUT ("The system is singular");
STOP.

Step 20 If $m > 0$ then set $q_m = \dfrac{b_{N,N+1}}{b_{N,N}}$. (*Start backward substitution.*)

Step 21 For $i = N-1, N-2, \ldots, n+1$ set $q_{i-n} = \dfrac{b_{i,N+1} - \sum_{j=i+1}^{N} b_{i,j}q_{j-n}}{b_{i,i}}$.

Step 22 For $i = n, n-1, \ldots, 1$ set $p_i = b_{i,N+1} - \sum_{j=n+1}^{N} b_{i,j}q_{j-n}$.

Step 23 OUTPUT $(q_0, q_1, \ldots, q_m, p_0, p_1, \ldots, p_n)$;
STOP. (*The procedure was successful.*) ■

Continued Fraction Approximation

It is interesting to compare the number of arithmetic operations required for calculations of $P_5(x)$ and $r(x)$ in Example 1. Using nested multiplication, $P_5(x)$ can be expressed as

$$P_5(x) = \left(\left(\left(\left(-\frac{1}{120}x + \frac{1}{24}\right)x - \frac{1}{6}\right)x + \frac{1}{2}\right)x - 1\right)x + 1.$$

Assuming that the coefficients of $1, x, x^2, x^3, x^4$, and x^5 are represented as decimals, a single calculation of $P_5(x)$ in nested form requires five multiplications and five additions/subtractions.

Using nested multiplication, $r(x)$ is expressed as

$$r(x) = \frac{\left(\left(-\frac{1}{60}x + \frac{3}{20}\right)x - \frac{3}{5}\right)x + 1}{\left(\frac{1}{20}x + \frac{2}{5}\right)x + 1},$$

so a single calculation of $r(x)$ requires five multiplications, five additions/subtractions, and one division. Hence, computational effort appears to favor the polynomial approximation. However, by reexpressing $r(x)$ by continued division, we can write

$$r(x) = \frac{1 - \frac{3}{5}x + \frac{3}{20}x^2 - \frac{1}{60}x^3}{1 + \frac{2}{5}x + \frac{1}{20}x^2}$$
$$= \frac{-\frac{1}{3}x^3 + 3x^2 - 12x + 20}{x^2 + 8x + 20}$$

$$= -\frac{1}{3}x + \frac{17}{3} + \frac{(-\frac{152}{3}x - \frac{280}{3})}{x^2 + 8x + 20}$$

$$= -\frac{1}{3}x + \frac{17}{3} + \frac{-\frac{152}{3}}{\left(\frac{x^2+8x+20}{x+(35/19)}\right)}$$

or

$$r(x) = -\frac{1}{3}x + \frac{17}{3} + \frac{-\frac{152}{3}}{\left(x + \frac{117}{19} + \frac{3125/361}{(x+(35/19))}\right)}. \tag{8.16}$$

Written in this form, a single calculation of $r(x)$ requires one multiplication, five additions/subtractions, and two divisions. If the amount of computation required for division is approximately the same as for multiplication, the computational effort required for an evaluation of the polynomial $P_5(x)$ significantly exceeds that required for an evaluation of the rational function $r(x)$.

Using continued fractions for rational approximation is a subject that has its roots in the works of Christopher Clavius (1537–1612). It was employed in the 18th and 19th centuries by, for example, Euler, Lagrange, and Hermite.

Expressing a rational function approximation in a form such as Eq. (8.16) is called **continued-fraction** approximation. This is a classical approximation technique of current interest because of the computational efficiency of this representation. It is, however, a specialized technique that we will not discuss further. A rather extensive treatment of this subject and of rational approximation in general can be found in [RR], pp. 285–322.

Although the rational-function approximation in Example 1 gave results superior to the polynomial approximation of the same degree, note that the approximation has a wide variation in accuracy. The approximation at 0.2 is accurate to within 8×10^{-9}, but at 1.0 the approximation and the function agree only to within 7×10^{-5}. This accuracy variation is expected because the Padé approximation is based on a Taylor polynomial representation of e^{-x}, and the Taylor representation has a wide variation of accuracy in [0.2, 1.0].

Chebyshev Rational Function Approximation

To obtain more uniformly accurate rational-function approximations we use Chebyshev polynomials, a class that exhibits more uniform behavior. The general Chebyshev rational-function approximation method proceeds in the same manner as Padé approximation, except that each x^k term in the Padé approximation is replaced by the kth-degree Chebyshev polynomial $T_k(x)$.

Suppose we want to approximate the function f by an Nth-degree rational function r written in the form

$$r(x) = \frac{\sum_{k=0}^{n} p_k T_k(x)}{\sum_{k=0}^{m} q_k T_k(x)}, \quad \text{where } N = n + m \text{ and } q_0 = 1.$$

Writing $f(x)$ in a series involving Chebyshev polynomials as

$$f(x) = \sum_{k=0}^{\infty} a_k T_k(x),$$

gives

$$f(x) - r(x) = \sum_{k=0}^{\infty} a_k T_k(x) - \frac{\sum_{k=0}^{n} p_k T_k(x)}{\sum_{k=0}^{m} q_k T_k(x)}$$

or

$$f(x) - r(x) = \frac{\sum_{k=0}^{\infty} a_k T_k(x) \sum_{k=0}^{m} q_k T_k(x) - \sum_{k=0}^{n} p_k T_k(x)}{\sum_{k=0}^{m} q_k T_k(x)}. \tag{8.17}$$

The coefficients $q_1, q_2, \ldots, q_m$ and $p_0, p_1, \ldots, p_n$ are chosen so that the numerator on the right-hand side of this equation has zero coefficients for $T_k(x)$ when $k = 0, 1, \ldots, N$. This implies that the series

$$\begin{aligned}&(a_0 T_0(x) + a_1 T_1(x) + \cdots)(T_0(x) + q_1 T_1(x) + \cdots + q_m T_m(x)) \\ &\quad - (p_0 T_0(x) + p_1 T_1(x) + \cdots + p_n T_n(x))\end{aligned}$$

has no terms of degree less than or equal to N.

Two problems arise with the Chebyshev procedure that make it more difficult to implement than the Padé method. One occurs because the product of the polynomial $q(x)$ and the series for $f(x)$ involves products of Chebyshev polynomials. This problem is resolved by making use of the relationship

$$T_i(x)T_j(x) = \frac{1}{2}\left[T_{i+j}(x) + T_{|i-j|}(x)\right]. \tag{8.18}$$

(See Exercise 8 of Section 8.3.) The other problem is more difficult to resolve and involves the computation of the Chebyshev series for $f(x)$. In theory, this is not difficult for if

$$f(x) = \sum_{k=0}^{\infty} a_k T_k(x),$$

then the orthogonality of the Chebyshev polynomials implies that

$$a_0 = \frac{1}{\pi}\int_{-1}^{1} \frac{f(x)}{\sqrt{1-x^2}}\,dx \quad \text{and} \quad a_k = \frac{2}{\pi}\int_{-1}^{1} \frac{f(x)T_k(x)}{\sqrt{1-x^2}}\,dx, \quad \text{where } k \geq 1.$$

Practically, however, these integrals can seldom be evaluated in closed form, and a numerical integration technique is required for each evaluation.

Example 2 The first five terms of the Chebyshev expansion for e^{-x} are

$$\begin{aligned}\tilde{P}_5(x) = {} & 1.266066T_0(x) - 1.130318T_1(x) + 0.271495T_2(x) - 0.044337T_3(x) \\ & + 0.005474T_4(x) - 0.000543T_5(x).\end{aligned}$$

Determine the Chebyshev rational approximation of degree 5 with $n = 3$ and $m = 2$.

Solution Finding this approximation requires choosing p_0, p_1, p_2, p_3, q_1, and q_2 so that for $k = 0, 1, 2, 3, 4,$ and 5, the coefficients of $T_k(x)$ are 0 in the expansion

$$\tilde{P}_5(x)[T_0(x) + q_1 T_1(x) + q_2 T_2(x)] - [p_0 T_0(x) + p_1 T_1(x) + p_2 T_2(x) + p_3 T_3(x)].$$

Using the relation (8.18) and collecting terms gives the equations

$$\begin{aligned}
T_0 &: \quad 1.266066 - 0.565159q_1 + 0.1357485q_2 = p_0, \\
T_1 &: \quad -1.130318 + 1.401814q_1 - 0.587328q_2 = p_1, \\
T_2 &: \quad 0.271495 - 0.587328q_1 + 1.268803q_2 = p_2, \\
T_3 &: \quad -0.044337 + 0.138485q_1 - 0.565431q_2 = p_3, \\
T_4 &: \quad 0.005474 - 0.022440q_1 + 0.135748q_2 = 0, \\
T_5 &: \quad -0.000543 + 0.002737q_1 - 0.022169q_2 = 0.
\end{aligned}$$

The solution to this system produces the rational function

$$r_T(x) = \frac{1.055265T_0(x) - 0.613016T_1(x) + 0.077478T_2(x) - 0.004506T_3(x)}{T_0(x) + 0.378331T_1(x) + 0.022216T_2(x)}.$$

We found at the beginning of Section 8.3 that

$$T_0(x) = 1,\ T_1(x) = x,\ T_2(x) = 2x^2 - 1,\ T_3(x) = 4x^3 - 3x.$$

Using these to convert to an expression involving powers of x gives

$$r_T(x) = \frac{0.977787 - 0.599499x + 0.154956x^2 - 0.018022x^3}{0.977784 + 0.378331x + 0.044432x^2}.$$

Table 8.11 lists values of $r_T(x)$ and, for comparison purposes, the values of $r(x)$ obtained in Example 1. Note that the approximation given by $r(x)$ is superior to that of $r_T(x)$ for $x = 0.2$ and 0.4, but that the maximum error for $r(x)$ is 6.33×10^{-5} compared to 9.13×10^{-6} for $r_T(x)$. ■

Table 8.11

x	e^{-x}	$r(x)$	$\lvert e^{-x} - r(x)\rvert$	$r_T(x)$	$\lvert e^{-x} - r_T(x)\rvert$
0.2	0.81873075	0.81873075	7.55×10^{-9}	0.81872510	5.66×10^{-6}
0.4	0.67032005	0.67031963	4.11×10^{-7}	0.67031310	6.95×10^{-6}
0.6	0.54881164	0.54880763	4.00×10^{-6}	0.54881292	1.28×10^{-6}
0.8	0.44932896	0.44930966	1.93×10^{-5}	0.44933809	9.13×10^{-6}
1.0	0.36787944	0.36781609	6.33×10^{-5}	0.36787155	7.89×10^{-6}

The Chebyshev approximation can be generated using Algorithm 8.2.

Chebyshev Rational Approximation

To obtain the rational approximation

$$r_T(x) = \frac{\sum_{k=0}^{n} p_k T_k(x)}{\sum_{k=0}^{m} q_k T_k(x)}$$

for a given function $f(x)$:

INPUT nonnegative integers m and n.

OUTPUT coefficients $q_0, q_1, \ldots, q_m$ and $p_0, p_1, \ldots, p_n$.

Step 1 Set $N = m + n$.

Step 2 Set $a_0 = \frac{2}{\pi}\int_0^{\pi} f(\cos\theta)\, d\theta$; *(The coefficient a_0 is doubled for computational efficiency.)*

For $k = 1, 2, \ldots, N + m$ set

$$a_k = \frac{2}{\pi}\int_0^{\pi} f(\cos\theta)\cos k\theta\, d\theta.$$

(The integrals can be evaluated using a numerical integration procedure or the coefficients can be input directly.)

Step 3 Set $q_0 = 1$.

Step 4 For $i = 0, 1, \ldots, N$ do Steps 5–9. *(Set up a linear system with matrix B.)*

Step 5 For $j = 0, 1, \ldots, i$
if $j \le n$ then set $b_{i,j} = 0$.

Step 6 If $i \le n$ then set $b_{i,i} = 1$.

Step 7 For $j = i + 1, i + 2, \ldots, n$ set $b_{i,j} = 0$.

Step 8 For $j = n + 1, n + 2, \ldots, N$
if $i \ne 0$ then set $b_{i,j} = -\frac{1}{2}(a_{i+j-n} + a_{|i-j+n|})$
else set $b_{i,j} = -\frac{1}{2}a_{j-n}$.

Step 9 If $i \ne 0$ then set $b_{i,N+1} = a_i$
else set $b_{i,N+1} = \frac{1}{2}a_i$.

(Steps 10–21 solve the linear system using partial pivoting.)

Step 10 For $i = n + 1, n + 2, \ldots, N - 1$ do Steps 11–17.

Step 11 Let k be the smallest integer with $i \le k \le N$ and
$|b_{k,i}| = \max_{i \le j \le N} |b_{j,i}|$. *(Find pivot element.)*

Step 12 If $b_{k,i} = 0$ then OUTPUT ("The system is singular");
STOP.

Step 13 If $k \ne i$ then *(Interchange row i and row k.)*
for $j = i, i + 1, \ldots, N + 1$ set

$$b_{COPY} = b_{i,j};$$
$$b_{i,j} = b_{k,j};$$
$$b_{k,j} = b_{COPY}.$$

Step 14 For $j = i + 1, i + 2, \ldots, N$ do Steps 15–17. *(Perform elimination.)*

Step 15 Set $xm = \dfrac{b_{j,i}}{b_{i,i}}$.

Step 16 For $k = i + 1, i + 2, \ldots, N + 1$
set $b_{j,k} = b_{j,k} - xm \cdot b_{i,k}$.

Step 17 Set $b_{j,i} = 0$.

Step 18 If $b_{N,N} = 0$ then OUTPUT ("The system is singular");
STOP.

Step 19 If $m > 0$ then set $q_m = \dfrac{b_{N,N+1}}{b_{N,N}}$. *(Start backward substitution.)*

Step 20 For $i = N - 1, N - 2, \ldots, n + 1$ set $q_{i-n} = \dfrac{b_{i,N+1} - \sum_{j=i+1}^{N} b_{i,j}q_{j-n}}{b_{i,i}}$.

Step 21 For $i = n, n - 1, \ldots, 0$ set $p_i = b_{i,N+1} - \sum_{j=n+1}^{N} b_{i,j}q_{j-n}$.

Step 22 OUTPUT $(q_0, q_1, \ldots, q_m, p_0, p_1, \ldots, p_n)$;
STOP. *(The procedure was successful.)* ■

We can obtain both the Chebyshev series expansion and the Chebyshev rational approximation using Maple using the *orthopoly* and *numapprox* packages. Load the packages and then enter the command

$g := chebyshev(e^{-x}, x, 0.00001)$

The parameter 0.000001 tells Maple to truncate the series when the remaining coefficients divided by the largest coefficient is smaller that 0.000001. Maple returns

$$1.266065878T(0,x) - 1.130318208T(1,x) + .2714953396T(2,x) - 0.04433684985T(3,x)$$
$$+ 0.005474240442T(4,x) - 0.0005429263119T(5,x) + 0.00004497732296T(6,x)$$
$$- 0.000003198436462T(7,x)$$

The approximation to $e^{-0.8} = 0.449328964$ is found with

evalf(subs(x = .8, g))

$$0.4493288893$$

To obtain the Chebyshev rational approximation enter

gg := *convert(chebyshev(e*$^{-x}$, *x*, 0.00001), *ratpoly*, 3, 2)

resulting in

$$gg := \frac{0.9763521942 - 0.5893075371x + 0.1483579430x^2 - 0.01643823341x^3}{0.9763483269 + 0.3870509565x + 0.04730334625x^2}$$

In 1930, Evgeny Remez (1896–1975) developed general computational methods of Chebyshev approximation for polynomials. He later developed a similar algorithm for the rational approximation of continuous functions defined on an interval with a prescribed degree of accuracy. His work encompassed various areas of approximation theory as well as the methods for approximating the solutions of differential equations.

We can evaluate $g(0.8)$ by

evalf(subs(x = 0.8, g))

which gives 0.4493317577 as an approximation to $e^{-0.8} = 0.449328964$.

The Chebyshev method does not produce the best rational function approximation in the sense of the approximation whose maximum approximation error is minimal. The method can, however, be used as a starting point for an iterative method known as the second Remez' algorithm that converges to the best approximation. A discussion of the techniques involved with this procedure and an improvement on this algorithm can be found in [RR], pp. 292–305, or in [Pow], pp. 90–92.

EXERCISE SET 8.4

1. Determine all degree 2 Padé approximations for $f(x) = e^{2x}$. Compare the results at $x_i = 0.2i$, for $i = 1, 2, 3, 4, 5$, with the actual values $f(x_i)$.
2. Determine all degree 3 Padé approximations for $f(x) = x\ln(x+1)$. Compare the results at $x_i = 0.2i$, for $i = 1, 2, 3, 4, 5$, with the actual values $f(x_i)$.
3. Determine the Padé approximation of degree 5 with $n = 2$ and $m = 3$ for $f(x) = e^x$. Compare the results at $x_i = 0.2i$, for $i = 1, 2, 3, 4, 5$, with those from the fifth Maclaurin polynomial.
4. Repeat Exercise 3 using instead the Padé approximation of degree 5 with $n = 3$ and $m = 2$. Compare the results at each x_i with those computed in Exercise 3.
5. Determine the Padé approximation of degree 6 with $n = m = 3$ for $f(x) = \sin x$. Compare the results at $x_i = 0.1i$, for $i = 0, 1, \ldots, 5$, with the exact results and with the results of the sixth Maclaurin polynomial.
6. Determine the Padé approximations of degree 6 with (a) $n = 2, m = 4$ and (b) $n = 4$, $m = 2$ for $f(x) = \sin x$. Compare the results at each x_i to those obtained in Exercise 5.
7. Table 8.10 lists results of the Padé approximation of degree 5 with $n = 3$ and $m = 2$, the fifth Maclaurin polynomial, and the exact values of $f(x) = e^{-x}$ when $x_i = 0.2i$, for $i = 1, 2, 3, 4,$

and 5. Compare these results with those produced from the other Padé approximations of degree five.

a. $n = 0, m = 5$ **b.** $n = 1, m = 4$ **c.** $n = 3, m = 2$ **d.** $n = 4, m = 1$

8. Express the following rational functions in continued-fraction form:

a. $\dfrac{x^2 + 3x + 2}{x^2 - x + 1}$ **b.** $\dfrac{4x^2 + 3x - 7}{2x^3 + x^2 - x + 5}$

c. $\dfrac{2x^3 - 3x^2 + 4x - 5}{x^2 + 2x + 4}$ **d.** $\dfrac{2x^3 + x^2 - x + 3}{3x^3 + 2x^2 - x + 1}$

9. Find all the Chebyshev rational approximations of degree 2 for $f(x) = e^{-x}$. Which give the best approximations to $f(x) = e^{-x}$ at $x = 0.25, 0.5$, and 1?

10. Find all the Chebyshev rational approximations of degree 3 for $f(x) = \cos x$. Which give the best approximations to $f(x) = \cos x$ at $x = \pi/4$ and $\pi/3$?

11. Find the Chebyshev rational approximation of degree 4 with $n = m = 2$ for $f(x) = \sin x$. Compare the results at $x_i = 0.1i$, for $i = 0, 1, 2, 3, 4, 5$, from this approximation with those obtained in Exercise 5 using a sixth-degree Padé approximation.

12. Find all Chebyshev rational approximations of degree 5 for $f(x) = e^x$. Compare the results at $x_i = 0.2i$, for $i = 1, 2, 3, 4, 5$, with those obtained in Exercises 3 and 4.

13. To accurately approximate $f(x) = e^x$ for inclusion in a mathematical library, we first restrict the domain of f. Given a real number x, divide by $\ln\sqrt{10}$ to obtain the relation

$$x = M \cdot \ln\sqrt{10} + s,$$

where M is an integer and s is a real number satisfying $|s| \leq \frac{1}{2}\ln\sqrt{10}$.

a. Show that $e^x = e^s \cdot 10^{M/2}$.

b. Construct a rational function approximation for e^s using $n = m = 3$. Estimate the error when $0 \leq |s| \leq \frac{1}{2}\ln\sqrt{10}$.

c. Design an implementation of e^x using the results of part (a) and (b) and the approximations

$$\frac{1}{\ln\sqrt{10}} = 0.8685889638 \quad \text{and} \quad \sqrt{10} = 3.162277660.$$

14. To accurately approximate $\sin x$ and $\cos x$ for inclusion in a mathematical library, we first restrict their domains. Given a real number x, divide by π to obtain the relation

$$|x| = M\pi + s, \quad \text{where } M \text{ is an integer and } |s| \leq \frac{\pi}{2}.$$

a. Show that $\sin x = \text{sgn}(x) \cdot (-1)^M \cdot \sin s$.

b. Construct a rational approximation to $\sin s$ using $n = m = 4$. Estimate the error when $0 \leq |s| \leq \pi/2$.

c. Design an implementation of $\sin x$ using parts (a) and (b).

d. Repeat part (c) for $\cos x$ using the fact that $\cos x = \sin(x + \pi/2)$.

8.5 Trigonometric Polynomial Approximation

The use of series of sine and cosine functions to represent arbitrary functions had its beginnings in the 1750s with the study of the motion of a vibrating string. This problem was considered by Jean d'Alembert and then taken up by the foremost mathematician of the time, Leonhard Euler. But it was Daniel Bernoulli who first advocated the use of the infinite sums of sine and cosines as a solution to the problem, sums that we now know as Fourier series. In the early part of the 19th century, Jean Baptiste Joseph Fourier used these series to study the flow of heat and developed quite a complete theory of the subject.

During the late 17th and early 18th centuries, the Bernoulli family produced no less than 8 important mathematicians and physicists. Daniel Bernoulli's most important work involved the pressure, density, and velocity of fluid flow, which produced what is known as the *Bernoulli principle*.

The first observation in the development of Fourier series is that, for each positive integer n, the set of functions $\{\phi_0, \phi_1, \ldots, \phi_{2n-1}\}$, where

$$\phi_0(x) = \frac{1}{2},$$

$$\phi_k(x) = \cos kx, \quad \text{for each } k = 1, 2, \ldots, n,$$

and

$$\phi_{n+k}(x) = \sin kx, \quad \text{for each } k = 1, 2, \ldots, n-1,$$

is an orthogonal set on $[-\pi, \pi]$ with respect to $w(x) \equiv 1$. This orthogonality follows from the fact that for every integer j, the integrals of $\sin jx$ and $\cos jx$ over $[-\pi, \pi]$ are 0, and we can rewrite products of sine and cosine functions as sums by using the three trigonometric identities

$$\begin{aligned} \sin t_1 \sin t_2 &= \frac{1}{2}[\cos(t_1 - t_2) - \cos(t_1 + t_2)], \\ \cos t_1 \cos t_2 &= \frac{1}{2}[\cos(t_1 - t_2) + \cos(t_1 + t_2)], \\ \sin t_1 \cos t_2 &= \frac{1}{2}[\sin(t_1 - t_2) + \sin(t_1 + t_2)]. \end{aligned} \tag{8.19}$$

Orthogonal Trigonometric Polynomials

Let $\mathcal{T}_n$ denote the set of all linear combinations of the functions $\phi_0, \phi_1, \ldots, \phi_{2n-1}$. This set is called the set of **trigonometric polynomials** of degree less than or equal to n. (Some sources also include an additional function in the set, $\phi_{2n}(x) = \sin nx$.)

For a function $f \in C[-\pi, \pi]$, we want to find the *continuous least squares* approximation by functions in $\mathcal{T}_n$ in the form

$$S_n(x) = \frac{a_0}{2} + a_n \cos nx + \sum_{k=1}^{n-1} (a_k \cos kx + b_k \sin kx).$$

Since the set of functions $\{\phi_0, \phi_1, \ldots, \phi_{2n-1}\}$ is orthogonal on $[-\pi, \pi]$ with respect to $w(x) \equiv 1$, it follows from Theorem 8.6 on page 515 and the equations in (8.19) that the appropriate selection of coefficients is

$$a_k = \frac{\int_{-\pi}^{\pi} f(x) \cos kx \, dx}{\int_{-\pi}^{\pi} (\cos kx)^2 \, dx} = \frac{1}{\pi} \int_{-\pi}^{\pi} f(x) \cos kx \, dx, \quad \text{for each } k = 0, 1, 2, \ldots, n, \tag{8.20}$$

and

$$b_k = \frac{\int_{-\pi}^{\pi} f(x) \sin kx \, dx}{\int_{-\pi}^{\pi} (\sin kx)^2 \, dx} = \frac{1}{\pi} \int_{-\pi}^{\pi} f(x) \sin kx \, dx, \quad \text{for each } k = 1, 2, \ldots, n-1. \tag{8.21}$$

Joseph Fourier (1768–1830) published his theory of trigonometric series in *Théorie analytique de la chaleur* to solve the problem of steady state heat distribution in a solid.

The limit of $S_n(x)$ when $n \to \infty$ is called the **Fourier series** of f. Fourier series are used to describe the solution of various ordinary and partial-differential equations that occur in physical situations.

Example 1 Determine the trigonometric polynomial from $\mathcal{T}_n$ that approximates

$$f(x) = |x|, \quad \text{for } -\pi < x < \pi.$$

Solution We first need to find the coefficients

$$a_0 = \frac{1}{\pi}\int_{-\pi}^{\pi} |x|\,dx = -\frac{1}{\pi}\int_{-\pi}^{0} x\,dx + \frac{1}{\pi}\int_{0}^{\pi} x\,dx = \frac{2}{\pi}\int_{0}^{\pi} x\,dx = \pi,$$

$$a_k = \frac{1}{\pi}\int_{-\pi}^{\pi} |x|\cos kx\,dx = \frac{2}{\pi}\int_{0}^{\pi} x\cos kx\,dx = \frac{2}{\pi k^2}\left[(-1)^k - 1\right],$$

for each $k = 1, 2, \ldots, n$, and

$$b_k = \frac{1}{\pi}\int_{-\pi}^{\pi} |x|\sin kx\,dx = 0, \quad \text{for each } k = 1, 2, \ldots, n-1.$$

That the b_k's are all 0 follows from the fact that $g(x) = |x|\sin kx$ is an odd function for each k, and the integral of a continuous odd function over an interval of the form $[-a, a]$ is 0. (See Exercises 13 and 14.) The trigonometric polynomial from $\mathcal{T}_n$ approximating f is therefore,

$$S_n(x) = \frac{\pi}{2} + \frac{2}{\pi}\sum_{k=1}^{n} \frac{(-1)^k - 1}{k^2}\cos kx.$$

The first few trigonometric polynomials for $f(x) = |x|$ are shown in Figure 8.13. ▪

Figure 8.13

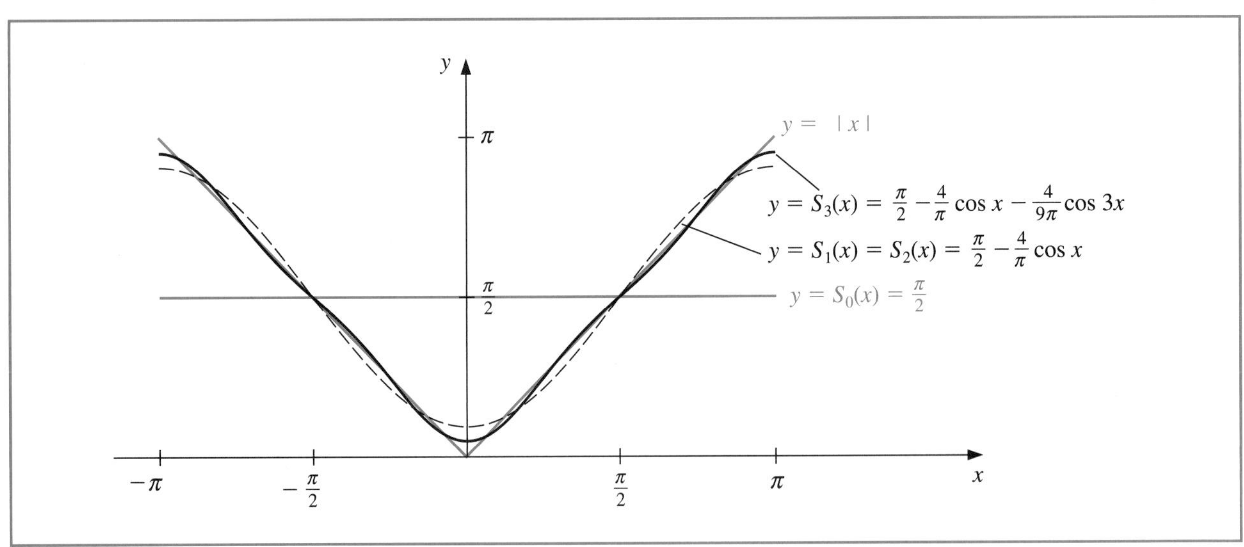

The Fourier series for f is

$$S(x) = \lim_{n\to\infty} S_n(x) = \frac{\pi}{2} + \frac{2}{\pi}\sum_{k=1}^{\infty} \frac{(-1)^k - 1}{k^2}\cos kx.$$

Since $|\cos kx| \le 1$ for every k and x, the series converges, and $S(x)$ exists for all real numbers x.

Discrete Trigonometric Approximation

There is a discrete analog that is useful for the *discrete least squares* approximation and the interpolation of large amounts of data.

Suppose that a collection of $2m$ paired data points $\{(x_j, y_j)\}_{j=0}^{2m-1}$ is given, with the first elements in the pairs equally partitioning a closed interval. For convenience, we assume that the interval is $[-\pi, \pi]$, so, as shown in Figure 8.14,

$$x_j = -\pi + \left(\frac{j}{m}\right)\pi, \quad \text{for each } j = 0, 1, \ldots, 2m-1. \tag{8.22}$$

If it is not $[-\pi, \pi]$, a simple linear transformation could be used to transform the data into this form.

Figure 8.14

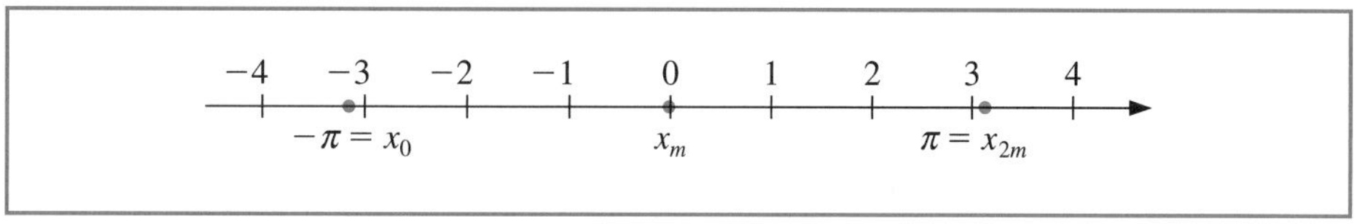

The goal in the discrete case is to determine the trigonometric polynomial $S_n(x)$ in $\mathcal{T}_n$ that will minimize

$$E(S_n) = \sum_{j=0}^{2m-1} [y_j - S_n(x_j)]^2.$$

To do this we need to choose the constants $a_0, a_1, \ldots, a_n, b_1, b_2, \ldots, b_{n-1}$ to minimize

$$E(S_n) = \sum_{j=0}^{2m-1} \left\{ y_j - \left[\frac{a_0}{2} + a_n \cos nx_j + \sum_{k=1}^{n-1} (a_k \cos kx_j + b_k \sin kx_j) \right] \right\}^2. \tag{8.23}$$

The determination of the constants is simplified by the fact that the set $\{\phi_0, \phi_1, \ldots, \phi_{2n-1}\}$ is orthogonal with respect to summation over the equally spaced points $\{x_j\}_{j=0}^{2m-1}$ in $[-\pi, \pi]$. By this we mean that for each $k \neq l$,

$$\sum_{j=0}^{2m-1} \phi_k(x_j)\phi_l(x_j) = 0. \tag{8.24}$$

To show this orthogonality, we use the following lemma.

Lemma 8.12 Suppose that the integer r is not a multiple of $2m$. Then

- $\displaystyle\sum_{j=0}^{2m-1} \cos rx_j = 0 \quad \text{and} \quad \sum_{j=0}^{2m-1} \sin rx_j = 0.$

Moreover, if r is not a multiple of m, then

- $\displaystyle\sum_{j=0}^{2m-1} (\cos rx_j)^2 = m \quad \text{and} \quad \sum_{j=0}^{2m-1} (\sin rx_j)^2 = m.$ ■

Euler first used the symbol i in 1794 to represent $\sqrt{-1}$ in his memoir *De Formulis Differentialibus Angularibus.*

Proof *Euler's Formula* states that with $i^2 = -1$, we have, for every real number z,

$$e^{iz} = \cos z + i \sin z. \tag{8.25}$$

Applying this result gives

$$\sum_{j=0}^{2m-1} \cos rx_j + i \sum_{j=0}^{2m-1} \sin rx_j = \sum_{j=0}^{2m-1} \left(\cos rx_j + i \sin rx_j\right) = \sum_{j=0}^{2m-1} e^{irx_j}.$$

But

$$e^{irx_j} = e^{ir(-\pi + j\pi/m)} = e^{-ir\pi} \cdot e^{irj\pi/m},$$

so

$$\sum_{j=0}^{2m-1} \cos rx_j + i \sum_{j=0}^{2m-1} \sin rx_j = e^{-ir\pi} \sum_{j=0}^{2m-1} e^{irj\pi/m}.$$

Since $\sum_{j=0}^{2m-1} e^{irj\pi/m}$ is a geometric series with first term 1 and ratio $e^{ir\pi/m} \neq 1$, we have

$$\sum_{j=0}^{2m-1} e^{irj\pi/m} = \frac{1 - (e^{ir\pi/m})^{2m}}{1 - e^{ir\pi/m}} = \frac{1 - e^{2ir\pi}}{1 - e^{ir\pi/m}}.$$

But $e^{2ir\pi} = \cos 2r\pi + i \sin 2r\pi = 1$, so $1 - e^{2ir\pi} = 0$ and

$$\sum_{j=0}^{2m-1} \cos rx_j + i \sum_{j=0}^{2m-1} \sin rx_j = e^{-ir\pi} \sum_{j=0}^{2m-1} e^{irj\pi/m} = 0.$$

This implies that both the real and imaginary parts are zero, so

$$\sum_{j=0}^{2m-1} \cos rx_j = 0 \quad \text{and} \quad \sum_{j=0}^{2m-1} \sin rx_j = 0.$$

In addition, if r is not a multiple of m, these sums imply that

$$\sum_{j=0}^{2m-1} (\cos rx_j)^2 = \sum_{j=0}^{2m-1} \frac{1}{2}\left(1 + \cos 2rx_j\right) = \frac{1}{2}\left[2m + \sum_{j=0}^{2m-1} \cos 2rx_j\right] = \frac{1}{2}(2m + 0) = m$$

and, similarly, that

$$\sum_{j=0}^{2m-1} (\sin rx_j)^2 = \sum_{j=0}^{2m-1} \frac{1}{2}\left(1 - \cos 2rx_j\right) = m.$$

▪ ▪ ▪

We can now show the orthogonality stated in (8.24). Consider, for example, the case

$$\sum_{j=0}^{2m-1} \phi_k(x_j)\phi_{n+l}(x_j) = \sum_{j=0}^{2m-1} (\cos kx_j)(\sin lx_j).$$

Since

$$\cos kx_j \sin lx_j = \frac{1}{2}[\sin(l+k)x_j + \sin(l-k)x_j]$$

and $(l+k)$ and $(l-k)$ are both integers that are not multiples of $2m$, Lemma 8.12 implies that

$$\sum_{j=0}^{2m-1} (\cos kx_j)(\sin lx_j) = \frac{1}{2}\left[\sum_{j=0}^{2m-1} \sin(l+k)x_j + \sum_{j=0}^{2m-1} \sin(l-k)x_j\right] = \frac{1}{2}(0+0) = 0.$$

This technique is used to show that the orthogonality condition is satisfied for any pair of the functions and to produce the following result.

Theorem 8.13 The constants in the summation

$$S_n(x) = \frac{a_0}{2} + a_n \cos nx + \sum_{k=1}^{n-1}(a_k \cos kx + b_k \sin kx)$$

that minimize the least squares sum

$$E(a_0, \ldots, a_n, b_1, \ldots, b_{n-1}) = \sum_{j=0}^{2m-1} (y_j - S_n(x_j))^2$$

are

- $$a_k = \frac{1}{m}\sum_{j=0}^{2m-1} y_j \cos kx_j, \quad \text{for each } k = 0, 1, \ldots, n,$$

and

- $$b_k = \frac{1}{m}\sum_{j=0}^{2m-1} y_j \sin kx_j, \quad \text{for each } k = 1, 2, \ldots, n-1.$$ ■

The theorem is proved by setting the partial derivatives of E with respect to the a_k's and the b_k's to zero, as was done in Sections 8.1 and 8.2, and applying the orthogonality to simplify the equations. For example,

$$0 = \frac{\partial E}{\partial b_k} = 2\sum_{j=0}^{2m-1}[y_j - S_n(x_j)](-\sin kx_j),$$

so

$$\begin{aligned}
0 &= \sum_{j=0}^{2m-1} y_j \sin kx_j - \sum_{j=0}^{2m-1} S_n(x_j)\sin kx_j \\
&= \sum_{j=0}^{2m-1} y_j \sin kx_j - \frac{a_0}{2}\sum_{j=0}^{2m-1} \sin kx_j - a_n \sum_{j=0}^{2m-1} \sin kx_j \cos nx_j \\
&\quad - \sum_{l=1}^{n-1} a_l \sum_{j=0}^{2m-1} \sin kx_j \cos lx_j - \sum_{\substack{l=1,\\ l\neq k}}^{n-1} b_l \sum_{j=0}^{2m-1} \sin kx_j \sin lx_j - b_k \sum_{j=0}^{2m-1} (\sin kx_j)^2.
\end{aligned}$$

The orthogonality implies that all but the first and last sums on the right side are zero, and Lemma 8.12 states the final sum is m. Hence

$$0 = \sum_{j=0}^{2m-1} y_j \sin kx_j - mb_k,$$

which implies that

$$b_k = \frac{1}{m}\sum_{j=0}^{2m-1} y_j \sin kx_j.$$

The result for the a_k's is similar but need an additional step to determine a_0 (See Exercise 17.)

Example 2 Find $S_2(x)$, the discrete least squares trigonometric polynomial of degree 2 for $f(x) = 2x^2 - 9$ when x is in $[-\pi, \pi]$.

Solution We have $m = 2(2) - 1 = 3$, so the nodes are

$$x_j = \pi + \frac{j}{m}\pi \quad \text{and} \quad y_j = f(x_j) = 2x_j^2 - 9, \quad \text{for } j = 0, 1, 2, 3, 4, 5.$$

The trigonometric polynomial is

$$S_2(x) = \frac{1}{2}a_0 + a_2 \cos 2x + (a_1 \cos x + b_1 \sin x),$$

where

$$a_k = \frac{1}{3}\sum_{j=0}^{5} y_j \cos kx_j, \text{ for } k = 0, 1, 2, \quad \text{and} \quad b_1 = \frac{1}{3}\sum_{j=0}^{5} y_j \sin x_j.$$

The coefficients are

$$a_0 = \frac{1}{3}\left(f(-\pi) + f\left(-\frac{2\pi}{3}\right) + f\left(-\frac{\pi}{3}\right) f(0) + f\left(\frac{\pi}{3}\right) + f\left(\frac{2\pi}{3}\right)\right) = -4.10944566,$$

$$a_1 = \frac{1}{3}\left(f(-\pi)\cos(-\pi) + f\left(-\frac{2\pi}{3}\right)\cos\left(-\frac{2\pi}{3}\right) + f\left(-\frac{\pi}{3}\right)\cos\left(-\frac{\pi}{3}\right) f(0)\cos 0\right.$$
$$\left. + f\left(\frac{\pi}{3}\right)\cos\left(\frac{\pi}{3}\right) + f\left(\frac{2\pi}{3}\right)\cos\left(\frac{2\pi}{3}\right)\right) = -8.77298169,$$

$$a_2 = \frac{1}{3}\left(f(-\pi)\cos(-2\pi) + f\left(-\frac{2\pi}{3}\right)\cos\left(-\frac{4\pi}{3}\right) + f\left(-\frac{\pi}{3}\right)\cos\left(-\frac{2\pi}{3}\right) f(0)\cos 0\right.$$
$$\left. + f\left(\frac{\pi}{3}\right)\cos\left(\frac{2\pi}{3}\right) + f\left(\frac{2\pi}{3}\right)\cos\left(\frac{4\pi}{3}\right)\right) = 2.92432723,$$

and

$$b_1 = \frac{1}{3}\left(f(-\pi)\sin(-\pi) + f\left(-\frac{2\pi}{3}\right)\sin\left(-\frac{\pi}{3}\right) + f\left(-\frac{\pi}{3}\right)\left(-\frac{\pi}{3}\right) f(0)\sin 0\right.$$
$$\left. + f\left(\frac{\pi}{3}\right)\left(\frac{\pi}{3}\right) + f\left(\frac{2\pi}{3}\right)\left(\frac{2\pi}{3}\right)\right) = 0.$$

Thus

$$S_2(x) = \frac{1}{2}(-4.10944562) - 8.77298169\cos x + 2.92432723\cos 2x.$$

Figure 8.15 shows $f(x)$ and the discrete least squares trigonometric polynomial $S_2(x)$. ■

Figure 8.15

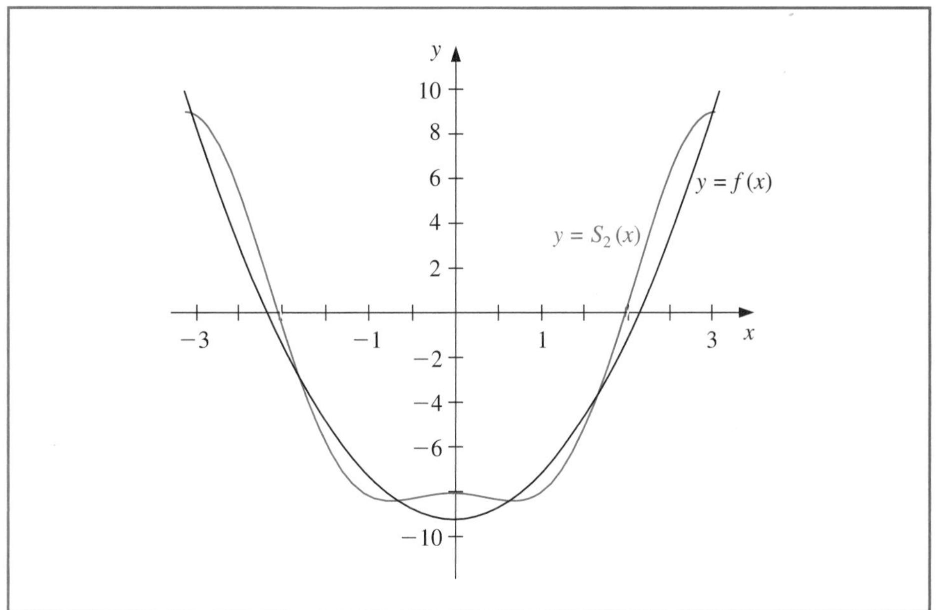

The next example gives an illustration of finding a least-squares approximation for a function that is defined on a closed interval other than $[-\pi, \pi]$.

Example 3 Find the discrete least squares approximation $S_3(x)$ for

$$f(x) = x^4 - 3x^3 + 2x^2 - \tan x(x-2)$$

using the data $\{(x_j, y_j)\}_{j=0}^{9}$, where $x_j = j/5$ and $y_j = f(x_j)$.

Solution We first need the linear transformation from $[0, 2]$ to $[-\pi, \pi]$ given by

$$z_j = \pi(x_j - 1).$$

Then the transformed data have the form

$$\left\{\left(z_j, f\left(1 + \frac{z_j}{\pi}\right)\right)\right\}_{j=0}^{9}.$$

The least squares trigonometric polynomial is consequently,

$$S_3(z) = \left[\frac{a_0}{2} + a_3 \cos 3z + \sum_{k=1}^{2}(a_k \cos kz + b_k \sin kz)\right],$$

where

$$a_k = \frac{1}{5}\sum_{j=0}^{9} f\left(1 + \frac{z_j}{\pi}\right)\cos kz_j, \quad \text{for } k = 0, 1, 2, 3,$$

and

$$b_k = \frac{1}{5}\sum_{j=0}^{9} f\left(1 + \frac{z_j}{\pi}\right)\sin kz_j, \quad \text{for } k = 1, 2.$$

Evaluating these sums produces the approximation

$$S_3(z) = 0.76201 + 0.77177 \cos z + 0.017423 \cos 2z + 0.0065673 \cos 3z$$
$$- 0.38676 \sin z + 0.047806 \sin 2z,$$

and converting back to the variable x gives

$$S_3(x) = 0.76201 + 0.77177 \cos \pi(x-1) + 0.017423 \cos 2\pi(x-1)$$
$$+ 0.0065673 \cos 3\pi(x-1) - 0.38676 \sin \pi(x-1) + 0.047806 \sin 2\pi(x-1).$$

Table 8.12 lists values of $f(x)$ and $S_3(x)$. ■

Table 8.12

x	$f(x)$	$S_3(x)$	$\lvert f(x) - S_3(x)\rvert$
0.125	0.26440	0.24060	2.38×10^{-2}
0.375	0.84081	0.85154	1.07×10^{-2}
0.625	1.36150	1.36248	9.74×10^{-4}
0.875	1.61282	1.60406	8.75×10^{-3}
1.125	1.36672	1.37566	8.94×10^{-3}
1.375	0.71697	0.71545	1.52×10^{-3}
1.625	0.07909	0.06929	9.80×10^{-3}
1.875	−0.14576	−0.12302	2.27×10^{-2}

EXERCISE SET 8.5

1. Find the continuous least squares trigonometric polynomial $S_n(x)$ for $f(x) = x$ on $[-\pi, \pi]$.
2. Find the continuous least squares trigonometric polynomial $S_2(x)$ for $f(x) = x^2$ on $[-\pi, \pi]$.
3. Find the continuous least squares trigonometric polynomial $S_3(x)$ for $f(x) = e^x$ on $[-\pi, \pi]$.
4. Find the general continuous least squares trigonometric polynomial $S_n(x)$ for $f(x) = e^x$ on $[-\pi, \pi]$.
5. Find the general continuous least squares trigonometric polynomial $S_n(x)$ in for

$$f(x) = \begin{cases} -1, & \text{if } -\pi < x < 0. \\ 1, & \text{if } 0 \le x \le \pi. \end{cases}$$

6. Find the general continuous least squares trigonometric polynomial $S_n(x)$ for

$$f(x) = \begin{cases} 0, & \text{if } -\pi < x \le 0, \\ 1, & \text{if } 0 < x < \pi. \end{cases}$$

7. Determine the discrete least squares trigonometric polynomial $S_n(x)$ on the interval $[-\pi, \pi]$ for the following functions, using the given values of m and n:
   a. $f(x) = \cos 2x,\ m = 4, n = 2$
   b. $f(x) = \cos 3x,\ m = 4, n = 2$
   c. $f(x) = x^2 \cos x,\ m = 6, n = 3$
   d. $f(x) = \sin \frac{x}{2} + 2\cos \frac{x}{3},\ m = 6, n = 3$
8. Compute the error $E(S_n)$ for each of the functions in Exercise 7.
9. Determine the discrete least squares trigonometric polynomial $S_3(x)$, using $m = 4$ for $f(x) = e^x \cos 2x$ on the interval $[-\pi, \pi]$. Compute the error $E(S_3)$.
10. Repeat Exercise 9 using $m = 8$. Compare the values of the approximating polynomials with the values of f at the points $\xi_j = -\pi + 0.2j\pi$, for $0 \le j \le 10$. Which approximation is better?

11. Let $f(x) = 2\tan x - \sec 2x$, for $2 \leq x \leq 4$. Determine the discrete least squares trigonometric polynomials $S_n(x)$, using the values of n and m as follows, and compute the error in each case.
   a. $n = 3, \quad m = 6$ **b.** $n = 4, \quad m = 6$

12. **a.** Determine the discrete least squares trigonometric polynomial $S_4(x)$, using $m = 16$, for $f(x) = x^2 \sin x$ on the interval $[0, 1]$.
   b. Compute $\int_0^1 S_4(x)\,dx$.
   c. Compare the integral in part (b) to $\int_0^1 x^2 \sin x\,dx$.

13. Show that for any continuous odd function f defined on the interval $[-a, a]$, we have $\int_{-a}^{a} f(x)\,dx = 0$.

14. Show that for any continuous even function f defined on the interval $[-a, a]$, we have $\int_{-a}^{a} f(x)\,dx = 2\int_0^a f(x)\,dx$.

15. Show that the functions $\phi_0(x) = 1/2, \phi_1(x) = \cos x, \ldots, \phi_n(x) = \cos nx, \phi_{n+1}(x) = \sin x, \ldots, \phi_{2n-1}(x) = \sin(n-1)x$ are orthogonal on $[-\pi, \pi]$ with respect to $w(x) \equiv 1$.

16. In Example 1 the Fourier series was determined for $f(x) = |x|$. Use this series and the assumption that it represents f at zero to find the value of the convergent infinite series $\sum_{k=0}^{\infty}(1/(2k+1)^2)$.

17. Show that the form of the constants a_k for $k = 0, \ldots, n$ in Theorem 8.13 is correct as stated.

8.6 Fast Fourier Transforms

In the latter part of Section 8.5, we determined the form of the discrete least squares polynomial of degree n on the $2m$ data points $\{(x_j, y_j)\}_{j=0}^{2m-1}$, where $x_j = -\pi + (j/m)\pi$, for each $j = 0, 1, \ldots, 2m - 1$.

The *interpolatory* trigonometric polynomial in $\mathcal{T}_m$ on these $2m$ data points is nearly the same as the least squares polynomial. This is because the least squares trigonometric polynomial minimizes the error term

$$E(S_m) = \sum_{j=0}^{2m-1} \left(y_j - S_m(x_j)\right)^2,$$

and for the interpolatory trigonometric polynomial, this error is 0, hence minimized, when the $S_m(x_j) = y_j$, for each $j = 0, 1, \ldots, 2m - 1$.

A modification is needed to the form of the polynomial, however, if we want the coefficients to assume the same form as in the least squares case. In Lemma 8.12 we found that if r is not a multiple of m, then

$$\sum_{j=0}^{2m-1} (\cos rx_j)^2 = m.$$

Interpolation requires computing instead

$$\sum_{j=0}^{2m-1} (\cos mx_j)^2,$$

which (see Exercise 8) has the value $2m$. This requires the interpolatory polynomial to be written as

$$S_m(x) = \frac{a_0 + a_m \cos mx}{2} + \sum_{k=1}^{m-1} (a_k \cos kx + b_k \sin kx), \tag{8.26}$$

if we want the form of the constants a_k and b_k to agree with those of the discrete least squares polynomial; that is,

- $$a_k = \frac{1}{m} \sum_{j=0}^{2m-1} y_j \cos kx_j, \quad \text{for each } k = 0, 1, \ldots, m, \text{ and}$$

- $$b_k = \frac{1}{m} \sum_{j=0}^{2m-1} y_j \sin kx_j \quad \text{for each } k = 1, 2, \ldots, m-1.$$

The interpolation of large amounts of equally-spaced data by trigonometric polynomials can produce very accurate results. It is the appropriate approximation technique in areas involving digital filters, antenna field patterns, quantum mechanics, optics, and in numerous simulation problems. Until the middle of the 1960s, however, the method had not been extensively applied due to the number of arithmetic calculations required for the determination of the constants in the approximation.

The interpolation of $2m$ data points by the direct-calculation technique requires approximately $(2m)^2$ multiplications and $(2m)^2$ additions. The approximation of many thousands of data points is not unusual in areas requiring trigonometric interpolation, so the direct methods for evaluating the constants require multiplication and addition operations numbering in the millions. The roundoff error associated with this number of calculations generally dominates the approximation.

In 1965, a paper by J. W. Cooley and J. W. Tukey in the journal *Mathematics of Computation* [CT] described a different method of calculating the constants in the interpolating trigonometric polynomial. This method requires only $O(m \log_2 m)$ multiplications and $O(m \log_2 m)$ additions, provided m is chosen in an appropriate manner. For a problem with thousands of data points, this reduces the number of calculations from millions to thousands. The method had actually been discovered a number of years before the Cooley-Tukey paper appeared but had gone largely unnoticed. ([Brigh], pp. 8–9, contains a short, but interesting, historical summary of the method.)

The method described by Cooley and Tukey is known either as the **Cooley-Tukey algorithm** or the **fast Fourier transform (FFT) algorithm** and has led to a revolution in the use of interpolatory trigonometric polynomials. The method consists of organizing the problem so that the number of data points being used can be easily factored, particularly into powers of two.

Instead of directly evaluating the constants a_k and b_k, the fast Fourier transform procedure computes the complex coefficients c_k in

$$\frac{1}{m} \sum_{k=0}^{2m-1} c_k e^{ikx}, \tag{8.27}$$

where

$$c_k = \sum_{j=0}^{2m-1} y_j e^{ik\pi j/m}, \quad \text{for each } k = 0, 1, \ldots, 2m-1. \tag{8.28}$$

Leonhard Euler first gave this formula in 1748 in *Introductio in analysin infinitorum*, which made the ideas of Johann Bernoulli more precise. This work bases the calculus on the theory of elementary functions rather than curves.

Once the constants c_k have been determined, a_k and b_k can be recovered by using *Euler's Formula*,

$$e^{iz} = \cos z + i \sin z.$$

For each $k = 0, 1, \ldots, m$ we have

$$\frac{1}{m}c_k(-1)^k = \frac{1}{m}c_k e^{-i\pi k} = \frac{1}{m}\sum_{j=0}^{2m-1} y_j e^{ik\pi j/m} e^{-i\pi k} = \frac{1}{m}\sum_{j=0}^{2m-1} y_j e^{ik(-\pi+(\pi j/m))}$$

$$= \frac{1}{m}\sum_{j=0}^{2m-1} y_j \left(\cos k\left(-\pi + \frac{\pi j}{m}\right) + i \sin k\left(-\pi + \frac{\pi j}{m}\right)\right)$$

$$= \frac{1}{m}\sum_{j=0}^{2m-1} y_j(\cos kx_j + i \sin kx_j).$$

So, given c_k we have

$$a_k + ib_k = \frac{(-1)^k}{m}c_k. \tag{8.29}$$

For notational convenience, b_0 and b_m are added to the collection, but both are 0 and do not contribute to the resulting sum.

The operation-reduction feature of the fast Fourier transform results from calculating the coefficients c_k in clusters, and uses as a basic relation the fact that for any integer n,

$$e^{n\pi i} = \cos n\pi + i \sin n\pi = (-1)^n.$$

Suppose $m = 2^p$ for some positive integer p. For each $k = 0, 1, \ldots, m-1$ we have

$$c_k + c_{m+k} = \sum_{j=0}^{2m-1} y_j e^{ik\pi j/m} + \sum_{j=0}^{2m-1} y_j e^{i(m+k)\pi j/m} = \sum_{j=0}^{2m-1} y_j e^{ik\pi j/m}(1 + e^{\pi ij}).$$

But

$$1 + e^{i\pi j} = \begin{cases} 2, & \text{if } j \text{ is even}, \\ 0, & \text{if } j \text{ is odd}, \end{cases}$$

so there are only m nonzero terms to be summed.

If j is replaced by $2j$ in the index of the sum, we can write the sum as

$$c_k + c_{m+k} = 2\sum_{j=0}^{m-1} y_{2j} e^{ik\pi(2j)/m};$$

that is,

$$c_k + c_{m+k} = 2\sum_{j=0}^{m-1} y_{2j} e^{ik\pi j/(m/2)}. \tag{8.30}$$

In a similar manner,

$$c_k - c_{m+k} = 2e^{ik\pi/m}\sum_{j=0}^{m-1} y_{2j+1} e^{ik\pi j/(m/2)}. \tag{8.31}$$

Since c_k and c_{m+k} can both be recovered from Eqs. (8.30) and (8.31), these relations determine all the coefficients c_k. Note also that the sums in Eqs. (8.30) and (8.31) are of the same form as the sum in Eq. (8.28), except that the index m has been replaced by $m/2$.

There are $2m$ coefficients $c_0, c_1, \ldots, c_{2m-1}$ to be calculated. Using the basic formula (8.28) requires $2m$ complex multiplications per coefficient, for a total of $(2m)^2$ operations. Equation (8.30) requires m complex multiplications for each $k = 0, 1, \ldots, m-1$, and (8.31) requires $m+1$ complex multiplications for each $k = 0, 1, \ldots, m-1$. Using these equations to compute $c_0, c_1, \ldots, c_{2m-1}$ reduces the number of complex multiplications from $(2m)^2 = 4m^2$ to

$$m \cdot m + m(m+1) = 2m^2 + m.$$

The sums in (8.30) and (8.31) have the same form as the original and m is a power of 2, so the reduction technique can be reapplied to the sums in (8.30) and (8.31). Each of these is replaced by two sums from $j = 0$ to $j = (m/2) - 1$. This reduces the $2m^2$ portion of the sum to

$$2\left[\frac{m}{2} \cdot \frac{m}{2} + \frac{m}{2} \cdot \left(\frac{m}{2} + 1\right)\right] = m^2 + m.$$

So a total of

$$(m^2 + m) + m = m^2 + 2m$$

complex multiplications are now needed, instead of $(2m)^2$.

Applying the technique one more time gives us 4 sums each with $m/4$ terms and reduces the m^2 portion of this total to

$$4\left[\left(\frac{m}{4}\right)^2 + \frac{m}{4}\left(\frac{m}{4} + 1\right)\right] = \frac{m^2}{2} + m,$$

for a new total of $(m^2/2) + 3m$ complex multiplications. Repeating the process r times reduces the total number of required complex multiplications to

$$\frac{m^2}{2^{r-2}} + mr.$$

The process is complete when $r = p + 1$, because we then have $m = 2^p$ and $2m = 2^{p+1}$. As a consequence, after $r = p + 1$ reductions of this type, the number of complex multiplications is reduced from $(2m)^2$ to

$$\frac{(2^p)^2}{2^{p-1}} + m(p+1) = 2m + pm + m = 3m + m\log_2 m = O(m \log_2 m).$$

Because of the way the calculations are arranged, the number of required complex additions is comparable.

To illustrate the significance of this reduction, suppose we have $m = 2^{10} = 1024$. The direct calculation of the c_k, for $k = 0, 1, \ldots, 2m-1$, would require

$$(2m)^2 = (2048)^2 \approx 4{,}200{,}000$$

calculations. The fast Fourier transform procedure reduces the number of calculations to

$$3(1024) + 1024 \log_2 1024 \approx 13{,}300.$$

Illustration Consider the fast Fourier transform technique applied to $8 = 2^3$ data points $\{(x_j, y_j)\}_{j=0}^{7}$, where $x_j = -\pi + j\pi/4$, for each $j = 0, 1, \ldots, 7$. In this case $2m = 8$, so $m = 4 = 2^2$ and $p = 2$.

From Eq. (8.26) we have

$$S_4(x) = \frac{a_0 + a_4 \cos 4x}{2} + \sum_{k=1}^{3} (a_k \cos kx + b_k \sin kx),$$

where

$$a_k = \frac{1}{4} \sum_{j=0}^{7} y_j \cos kx_j \quad \text{and} \quad b_k = \frac{1}{4} \sum_{j=0}^{7} y_j \sin kx_j, \quad k = 0, 1, 2, 3, 4.$$

Define the Fourier transform as

$$\frac{1}{4} \sum_{j=0}^{7} c_k e^{ikx},$$

where

$$c_k = \sum_{j=0}^{7} y_j e^{ik\pi j/4}, \quad \text{for } k = 0, 1, \ldots, 7.$$

Then by Eq. (8.31), for $k = 0, 1, 2, 3, 4$, we have

$$\frac{1}{4} c_k e^{-ik\pi} = a_k + ib_k.$$

By direct calculation, the complex constants c_k are given by

$$\begin{aligned}
c_0 =& y_0 + y_1 + y_2 + y_3 + y_4 + y_5 + y_6 + y_7;\\
c_1 =& y_0 + \left(\frac{i+1}{\sqrt{2}}\right) y_1 + iy_2 + \left(\frac{i-1}{\sqrt{2}}\right) y_3 - y_4 - \left(\frac{i+1}{\sqrt{2}}\right) y_5 - iy_6 - \left(\frac{i-1}{\sqrt{2}}\right) y_7;\\
c_2 =& y_0 + iy_1 - y_2 - iy_3 + y_4 + iy_5 - y_6 - iy_7;\\
c_3 =& y_0 + \left(\frac{i-1}{\sqrt{2}}\right) y_1 - iy_2 + \left(\frac{i+1}{\sqrt{2}}\right) y_3 - y_4 - \left(\frac{i-1}{\sqrt{2}}\right) y_5 + iy_6 - \left(\frac{i+1}{\sqrt{2}}\right) y_7;\\
c_4 =& y_0 - y_1 + y_2 - y_3 + y_4 - y_5 + y_6 - y_7;\\
c_5 =& y_0 - \left(\frac{i+1}{\sqrt{2}}\right) y_1 + iy_2 - \left(\frac{i-1}{\sqrt{2}}\right) y_3 - y_4 + \left(\frac{i+1}{\sqrt{2}}\right) y_5 - iy_6 + \left(\frac{i-1}{\sqrt{2}}\right) y_7;\\
c_6 =& y_0 - iy_1 - y_2 + iy_3 + y_4 - iy_5 - y_6 + iy_7;\\
c_7 =& y_0 - \left(\frac{i-1}{\sqrt{2}}\right) y_1 - iy_2 - \left(\frac{i+1}{\sqrt{2}}\right) y_3 - y_4 + \left(\frac{i-1}{\sqrt{2}}\right) y_5 + iy_6 + \left(\frac{i+1}{\sqrt{2}}\right) y_7.
\end{aligned}$$

Because of the small size of the collection of data points, many of the coefficients of the y_j in these equations are 1 or -1. This frequency will decrease in a larger application, so to count the computational operations accurately, multiplication by 1 or -1 will be included, even though it would not be necessary in this example. With this understanding, 64 multiplications/divisions and 56 additions/subtractions are required for the direct computation of $c_0, c_1, \ldots, c_7$.

To apply the fast Fourier transform procedure with $r = 1$, we first define

$$d_0 = \frac{c_0 + c_4}{2} = y_0 + y_2 + y_4 + y_6; \qquad d_4 = \frac{c_2 + c_6}{2} = y_0 - y_2 + y_4 - y_6;$$

$$d_1 = \frac{c_0 - c_4}{2} = y_1 + y_3 + y_5 + y_7; \qquad d_5 = \frac{c_2 - c_6}{2} = i(y_1 - y_3 + y_5 - y_7);$$

$$d_2 = \frac{c_1 + c_5}{2} = y_0 + iy_2 - y_4 - iy_6; \qquad d_6 = \frac{c_3 + c_7}{2} = y_0 - iy_2 - y_4 + iy_6;$$

$$d_3 = \frac{c_1 - c_5}{2} = \left(\frac{i+1}{\sqrt{2}}\right)(y_1 + iy_3 - y_5 - iy_7); \qquad d_7 = \frac{c_3 - c_7}{2} = \left(\frac{i-1}{\sqrt{2}}\right)(y_1 - iy_3 - y_5 + iy_7).$$

We then define, for $r = 2$,

$$e_0 = \frac{d_0 + d_4}{2} = y_0 + y_4; \qquad e_4 = \frac{d_2 + d_6}{2} = y_0 - y_4;$$

$$e_1 = \frac{d_0 - d_4}{2} = y_2 + y_6; \qquad e_5 = \frac{d_2 - d_6}{2} = i(y_2 - y_6);$$

$$e_2 = \frac{id_1 + d_5}{2} = i(y_1 + y_5); \qquad e_6 = \frac{id_3 + d_7}{2} = \left(\frac{i-1}{\sqrt{2}}\right)(y_1 - y_5);$$

$$e_3 = \frac{id_1 - d_5}{2} = i(y_3 + y_7); \qquad e_7 = \frac{id_3 - d_7}{2} = i\left(\frac{i-1}{\sqrt{2}}\right)(y_3 - y_7).$$

Finally, for $r = p + 1 = 3$, we define

$$f_0 = \frac{e_0 + e_4}{2} = y_0; \qquad f_4 = \frac{((i+1)/\sqrt{2})e_2 + e_6}{2} = \left(\frac{i-1}{\sqrt{2}}\right)y_1;$$

$$f_1 = \frac{e_0 - e_4}{2} = y_4; \qquad f_5 = \frac{((i+1)/\sqrt{2})e_2 - e_6}{2} = \left(\frac{i-1}{\sqrt{2}}\right)y_5;$$

$$f_2 = \frac{ie_1 + e_5}{2} = iy_2; \qquad f_6 = \frac{((i-1)/\sqrt{2})e_3 + e_7}{2} = \left(\frac{-i-1}{\sqrt{2}}\right)y_3;$$

$$f_3 = \frac{ie_1 - e_5}{2} = iy_6; \qquad f_7 = \frac{((i-1)/\sqrt{2})e_3 - e_7}{2} = \left(\frac{-i-1}{\sqrt{2}}\right)y_7.$$

The $c_0, \ldots, c_7$, $d_0, \ldots, d_7$, $e_0, \ldots, e_7$, and $f_0, \ldots, f_7$ are independent of the particular data points; they depend only on the fact that $m = 4$. For each m there is a unique set of constants $\{c_k\}_{k=0}^{2m-1}$, $\{d_k\}_{k=0}^{2m-1}$, $\{e_k\}_{k=0}^{2m-1}$, and $\{f_k\}_{k=0}^{2m-1}$. This portion of the work is not needed for a particular application, only the following calculations are required:

The f_k:

$$f_0 = y_0; \quad f_1 = y_4; \quad f_2 = iy_2; \quad f_3 = iy_6;$$

$$f_4 = \left(\frac{i-1}{\sqrt{2}}\right)y_1; \quad f_5 = \left(\frac{i-1}{\sqrt{2}}\right)y_5; \quad f_6 = -\left(\frac{i+1}{\sqrt{2}}\right)y_3; \quad f_7 = -\left(\frac{i+1}{\sqrt{2}}\right)y_7.$$

The e_k:

$$e_0 = f_0 + f_1; \; e_1 = -i(f_2 + f_3); \; e_2 = -\left(\frac{i-1}{\sqrt{2}}\right)(f_4 + f_5);$$

$$e_3 = -\left(\frac{i+1}{\sqrt{2}}\right)(f_6 + f_7); \; e_4 = f_0 - f_1; \; e_5 = f_2 - f_3; \; e_6 = f_4 - f_5; \; e_7 = f_6 - f_7.$$

The d_k:

$$d_0 = e_0 + e_1; \quad d_1 = -i(e_2 + e_3); \quad d_2 = e_4 + e_5; \quad d_3 = -i(e_6 + e_7);$$

$$d_4 = e_0 - e_1; \quad d_5 = e_2 - e_3; \quad d_6 = e_4 - e_5; \quad d_7 = e_6 - e_7.$$

The c_k:

$$c_0 = d_0 + d_1; \quad c_1 = d_2 + d_3; \quad c_2 = d_4 + d_5; \quad c_3 = d_6 + d_7;$$

$$c_4 = d_0 - d_1; \quad c_5 = d_2 - d_3; \quad c_6 = d_4 - d_5; \quad c_7 = d_6 - d_7.$$

Computing the constants $c_0, c_1, \ldots, c_7$ in this manner requires the number of operations shown in Table 8.13. Note again that multiplication by 1 or -1 has been included in the count, even though this does not require computational effort.

Table 8.13

Step	Multiplications/divisions	Additions/subtractions
(The f_k:)	8	0
(The e_k:)	8	8
(The d_k:)	8	8
(The c_k:)	0	8
Total	24	24

The lack of multiplications/divisions when finding the c_k reflects the fact that for any m, the coefficients $\{c_k\}_{k=0}^{2m-1}$ are computed from $\{d_k\}_{k=0}^{2m-1}$ in the same manner:

$$c_k = d_{2k} + d_{2k+1} \quad \text{and} \quad c_{k+m} = d_{2k} - d_{2k+1}, \quad \text{for } k = 0, 1, \ldots, m-1,$$

so no complex multiplication is involved.

In summary, the direct computation of the coefficients $c_0, c_1, \ldots, c_7$ requires 64 multiplications/divisions and 56 additions/subtractions. The fast Fourier transform technique reduces the computations to 24 multiplications/divisions and 24 additions/subtractions. □

Algorithm 8.3 performs the fast Fourier transform when $m = 2^p$ for some positive integer p. Modifications of the technique can be made when m takes other forms.

Fast Fourier Transform

To compute the coefficients in the summation

$$\frac{1}{m}\sum_{k=0}^{2m-1} c_k e^{ikx} = \frac{1}{m}\sum_{k=0}^{2m-1} c_k(\cos kx + i \sin kx), \quad \text{where } i = \sqrt{-1},$$

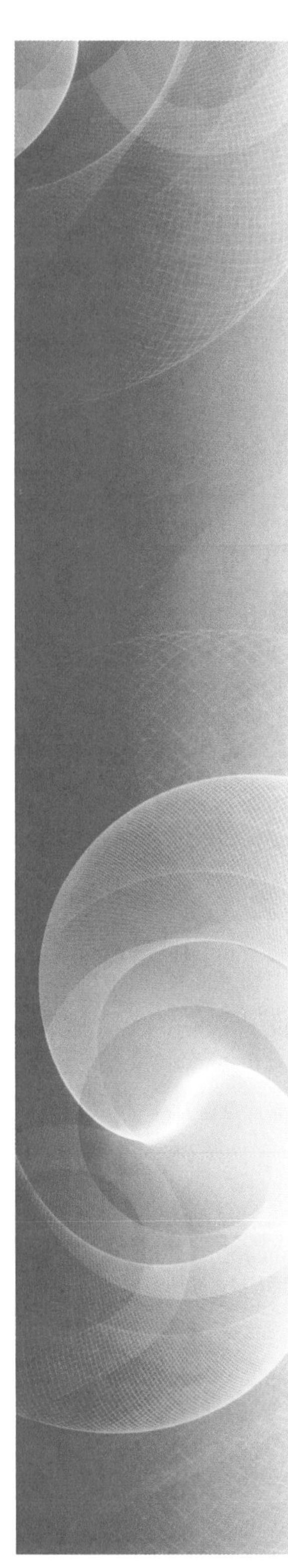

for the data $\{(x_j, y_j)\}_{j=0}^{2m-1}$ where $m = 2^p$ and $x_j = -\pi + j\pi/m$ for $j = 0, 1, \ldots, 2m-1$:

INPUT $m, p; y_0, y_1, \ldots, y_{2m-1}$.

OUTPUT complex numbers $c_0, \ldots, c_{2m-1}$; real numbers $a_0, \ldots, a_m; b_1, \ldots, b_{m-1}$.

Step 1 Set $M = m$;
$q = p$;
$\zeta = e^{\pi i/m}$.

Step 2 For $j = 0, 1, \ldots, 2m-1$ set $c_j = y_j$.

Step 3 For $j = 1, 2, \ldots, M$ set $\xi_j = \zeta^j$;
$\xi_{j+M} = -\xi_j$.

Step 4 Set $K = 0$;
$\xi_0 = 1$.

Step 5 For $L = 1, 2, \ldots, p+1$ do Steps 6–12.

Step 6 While $K < 2m-1$ do Steps 7–11.

Step 7 For $j = 1, 2, \ldots, M$ do Steps 8–10.

Step 8 Let $K = k_p \cdot 2^p + k_{p-1} \cdot 2^{p-1} + \cdots + k_1 \cdot 2 + k_0$;
(*Decompose k.*)
set $K_1 = K/2^q = k_p \cdot 2^{p-q} + \cdots + k_{q+1} \cdot 2 + k_q$;
$K_2 = k_q \cdot 2^p + k_{q+1} \cdot 2^{p-1} + \cdots + k_p \cdot 2^q$.

Step 9 Set $\eta = c_{K+M}\xi_{K_2}$;
$c_{K+M} = c_K - \eta$;
$c_K = c_K + \eta$.

Step 10 Set $K = K + 1$.

Step 11 Set $K = K + M$.

Step 12 Set $K = 0$;
$M = M/2$;
$q = q - 1$.

Step 13 While $K < 2m-1$ do Steps 14–16.

Step 14 Let $K = k_p \cdot 2^p + k_{p-1} \cdot 2^{p-1} + \cdots + k_1 \cdot 2 + k_0$; (*Decompose k.*)
set $j = k_0 \cdot 2^p + k_1 \cdot 2^{p-1} + \cdots + k_{p-1} \cdot 2 + k_p$.

Step 15 If $j > K$ then interchange c_j and c_k.

Step 16 Set $K = K + 1$.

Step 17 Set $a_0 = c_0/m$;
$a_m = \text{Re}(e^{-i\pi m} c_m/m)$.

Step 18 For $j = 1, \ldots, m-1$ set $a_j = \text{Re}(e^{-i\pi j} c_j/m)$;
$b_j = \text{Im}(e^{-i\pi j} c_j/m)$.

Step 19 OUTPUT $(c_0, \ldots, c_{2m-1}; a_0, \ldots, a_m; b_1, \ldots, b_{m-1})$;
STOP. ■

Example 1 Find the interpolating trigonometric polynomial of degree 2 on $[-\pi, \pi]$ for the data $\left\{(x_j, f(x_j))\right\}_{j=0}^{3}$, where

$$a_k = \frac{1}{2}\sum_{j=0}^{3} f(x_j)\cos(kx_j) \quad \text{for } k = 0, 1, 2 \quad \text{and} \quad b_1 = \frac{1}{2}\sum_{j=0}^{3} f(x_j)\sin(x_j).$$

Solution We have

$$a_0 = \frac{1}{2}\left(f(-\pi) + f\left(-\frac{\pi}{2}\right) + f(0) + f\left(\frac{\pi}{2}\right)\right) = -3.19559339,$$

$$a_1 = \frac{1}{2}\left(f(-\pi)\cos(-\pi) + f\left(-\frac{\pi}{2}\right)\cos\left(-\frac{\pi}{2}\right) + f(0)\cos 0 + f\left(\frac{\pi}{2}\right)\right)\cos\left(\frac{\pi}{2}\right)$$

$$= -9.86960441,$$

$$a_2 = \frac{1}{2}\left(f(-\pi)\cos(-2\pi) + f\left(-\frac{\pi}{2}\right)\cos(-\pi) + f(0)\cos 0 + f\left(\frac{\pi}{2}\right)\right)\cos(\pi)$$

$$= 4.93480220,$$

and

$$b_1 = \frac{1}{2}\left(f(-\pi)\sin(-\pi) + f\left(-\frac{\pi}{2}\right)\sin\left(-\frac{\pi}{2}\right) + f(0)\sin 0 + f\left(\frac{\pi}{2}\right)\sin\left(\frac{\pi}{2}\right)\right) = 0.$$

So

$$S_2(x) = \frac{1}{2}(-3.19559339 + 4.93480220\cos 2x) - 9.86960441\cos x.$$

Figure 8.16 shows $f(x)$ and the interpolating trigonometric polynomial $S_2(x)$. ∎

Figure 8.16

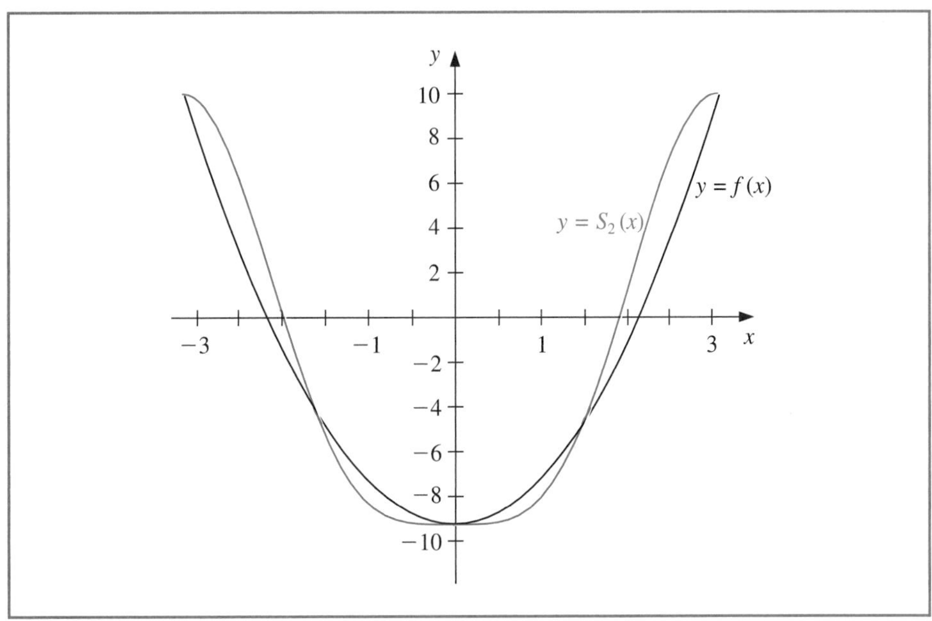

The next example gives an illustration of finding an interpolating trigonometric polynomial for a function that is defined on a closed interval other than $[-\pi, \pi]$.

Example 2 Determine the trigonometric interpolating polynomial of degree 4 on $[0, 2]$ for the data $\{(j/4, f(j/4))\}_{j=0}^{7}$, where $f(x) = x^4 - 3x^3 + 2x^2 - \tan x(x-2)$.

Solution We first need to transform the interval $[0, 2]$ to $[-\pi, \pi]$. This is given by

$$z_j = \pi(x_j - 1),$$

so that the input data to Algorithm 8.3 are

$$\left\{z_j, f\left(1 + \frac{z_j}{\pi}\right)\right\}_{j=0}^{7}.$$

The interpolating polynomial in z is

$$S_4(z) = 0.761979 + 0.771841 \cos z + 0.0173037 \cos 2z + 0.00686304 \cos 3z$$
$$- 0.000578545 \cos 4z - 0.386374 \sin z + 0.0468750 \sin 2z - 0.0113738 \sin 3z.$$

The trigonometric polynomial $S_4(x)$ on $[0, 2]$ is obtained by substituting $z = \pi(x - 1)$ into $S_4(z)$. The graphs of $y = f(x)$ and $y = S_4(x)$ are shown in Figure 8.17. Values of $f(x)$ and $S_4(x)$ are given in Table 8.14. ■

Figure 8.17

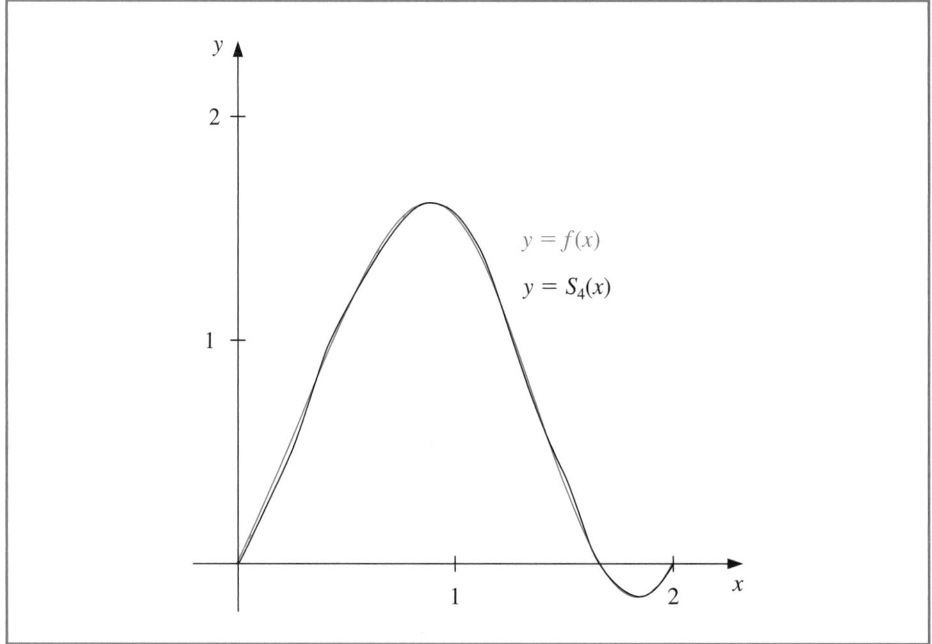

Table 8.14

x	$f(x)$	$S_4(x)$	$\lvert f(x) - S_4(x)\rvert$
0.125	0.26440	0.25001	1.44×10^{-2}
0.375	0.84081	0.84647	5.66×10^{-3}
0.625	1.36150	1.35824	3.27×10^{-3}
0.875	1.61282	1.61515	2.33×10^{-3}
1.125	1.36672	1.36471	2.02×10^{-3}
1.375	0.71697	0.71931	2.33×10^{-3}
1.625	0.07909	0.07496	4.14×10^{-3}
1.875	−0.14576	−0.13301	1.27×10^{-2}

More details on the verification of the validity of the fast Fourier transform procedure can be found in [Ham], which presents the method from a mathematical approach, or in [Brac], where the presentation is based on methods more likely to be familiar to engineers. [AHU], pp. 252–269, is a good reference for a discussion of the computational aspects of the method. Modification of the procedure for the case when m is not a power of 2 can be found in [Win]. A presentation of the techniques and related material from the point of view of applied abstract algebra is given in [Lau, pp. 438–465].

EXERCISE SET 8.6

1. Determine the trigonometric interpolating polynomial $S_2(x)$ of degree 2 on $[-\pi, \pi]$ for the following functions, and graph $f(x) - S_2(x)$:

   a. $f(x) = x(\pi - x)$ **b.** $f(x) = \pi(x - \pi)$

   c. $f(x) = |x|$ **d.** $f(x) = \begin{cases} -1, & -\pi \le x \le 0 \\ 1, & 0 < x \le \pi \end{cases}$

2. Determine the trigonometric interpolating polynomial of degree 4 for $f(x) = x(\pi - x)$ on the interval $[-\pi, \pi]$ using:

   a. Direct calculation; **b.** The Fast Fourier Transform Algorithm.

3. Use the Fast Fourier Transform Algorithm to compute the trigonometric interpolating polynomial of degree 4 on $[-\pi, \pi]$ for the following functions.

   a. $f(x) = \pi(x - \pi)$ **b.** $f(x) = |x|$

   c. $f(x) = \cos \pi x - 2 \sin \pi x$ **d.** $f(x) = x \cos x^2 + e^x \cos e^x$

4. **a.** Determine the trigonometric interpolating polynomial $S_4(x)$ of degree 4 for $f(x) = x^2 \sin x$ on the interval $[0, 1]$.

   b. Compute $\int_0^1 S_4(x)\, dx$.

   c. Compare the integral in part (b) to $\int_0^1 x^2 \sin x\, dx$.

5. Use the approximations obtained in Exercise 3 to approximate the following integrals, and compare your results to the actual values.

   a. $\displaystyle\int_{-\pi}^{\pi} \pi(x - \pi)\, dx$ **b.** $\displaystyle\int_{-\pi}^{\pi} |x|\, dx$

   c. $\displaystyle\int_{-\pi}^{\pi} (\cos \pi x - 2 \sin \pi x)\, dx$ **d.** $\displaystyle\int_{-\pi}^{\pi} (x \cos x^2 + e^x \cos e^x)\, dx$

6. Use the Fast Fourier Transform Algorithm to determine the trigonometric interpolating polynomial of degree 16 for $f(x) = x^2 \cos x$ on $[-\pi, \pi]$.

7. Use the Fast Fourier Transform Algorithm to determine the trigonometric interpolating polynomial of degree 64 for $f(x) = x^2 \cos x$ on $[-\pi, \pi]$.

8. Use a trigonometric identity to show that $\sum_{j=0}^{2m-1} (\cos mx_j)^2 = 2m$.

9. Show that $c_0, \ldots, c_{2m-1}$ in Algorithm 8.3 are given by

$$\begin{bmatrix} c_0 \\ c_1 \\ c_2 \\ \vdots \\ c_{2m-1} \end{bmatrix} = \begin{bmatrix} 1 & 1 & 1 & \cdots & 1 \\ 1 & \zeta & \zeta^2 & \cdots & \zeta^{2m-1} \\ 1 & \zeta^2 & \zeta^4 & \cdots & \zeta^{4m-2} \\ \vdots & \vdots & \vdots & & \vdots \\ 1 & \zeta^{2m-1} & \zeta^{4m-2} & \cdots & \zeta^{(2m-1)^2} \end{bmatrix} \begin{bmatrix} y_0 \\ y_1 \\ y_2 \\ \vdots \\ y_{2m-1} \end{bmatrix},$$

where $\zeta = e^{\pi i/m}$.

10. In the discussion preceding Algorithm 8.3, an example for $m = 4$ was explained. Define vectors **c**, **d**, **e**, f, and **y** as

$$\mathbf{c} = (c_0, \ldots, c_7)^t, \ \mathbf{d} = (d_0, \ldots, d_7)^t, \ \mathbf{e} = (e_0, \ldots, e_7)^t, \ \mathbf{f} = (f_0, \ldots, f_7)^t, \ \mathbf{y} = (y_0, \ldots, y_7)^t.$$

Find matrices A, B, C, and D so that $\mathbf{c} = A\mathbf{d}$, $\mathbf{d} = B\mathbf{e}$, $\mathbf{e} = C\,\mathbf{f}$, and $\mathbf{f} = D\mathbf{y}$.

8.7 Survey of Methods and Software

In this chapter we have considered approximating data and functions with elementary functions. The elementary functions used were polynomials, rational functions, and trigonometric polynomials. We considered two types of approximations, discrete and continuous. Discrete approximations arise when approximating a finite set of data with an elementary function. Continuous approximations are used when the function to be approximated is known.

Discrete least squares techniques are recommended when the function is specified by giving a set of data that may not exactly represent the function. Least squares fit of data can take the form of a linear or other polynomial approximation or even an exponential form. These approximations are computed by solving sets of normal equations, as given in Section 8.1.

If the data are periodic, a trigonometric least squares fit may be appropriate. Because of the orthonormality of the trigonometric basis functions, the least squares trigonometric approximation does not require the solution of a linear system. For large amounts of periodic data, interpolation by trigonometric polynomials is also recommended. An efficient method of computing the trigonometric interpolating polynomial is given by the fast Fourier transform.

When the function to be approximated can be evaluated at any required argument, the approximations seek to minimize an integral instead of a sum. The continuous least squares polynomial approximations were considered in Section 8.2. Efficient computation of least squares polynomials lead to orthonormal sets of polynomials, such as the Legendre and Chebyshev polynomials. Approximation by rational functions was studied in Section 8.4, where Padé approximation as a generalization of the Maclaurin polynomial and its extension to Chebyshev rational approximation were presented. Both methods allow a more uniform method of approximation than polynomials. Continuous least squares approximation by trigonometric functions was discussed in Section 8.5, especially as it relates to Fourier series.

The IMSL Library provides a number of routines for approximation including

1. Linear least squares fit of data with statistics;
2. Discrete least squares fit of data with the user's choice of basis functions;
3. Cubic spline least squares approximation;
4. Rational weighted Chebyshev approximation;
5. Fast Fourier transform fit of data.

The NAG Library provides routines that include computing

1. Least square polynomial approximation using a technique to minimize round-off error;
2. Cubic spline least squares approximation;

3. Best fit in the l_1 sense;

4. Best fit in the l_∞ sense;

5. Fast Fourier transform fit of data.

The netlib library contains a routine to compute the polynomial least squares approximation to a discrete set of points, and a routine to evaluate this polynomial and any of its derivatives at a given point.

For further information on the general theory of approximation theory see Powell [Pow], Davis [Da], or Cheney [Ch]. A good reference for methods of least squares is Lawson and Hanson [LH], and information about Fourier transforms can be found in Van Loan [Van] and in Briggs and Hanson [BH].

CHAPTER

9 Approximating Eigenvalues

Introduction

The longitudinal vibrations of an elastic bar of local stiffness $p(x)$ and density $\rho(x)$ are described by the partial differential equation

$$\rho(x)\frac{\partial^2 v}{\partial t^2}(x,t) = \frac{\partial}{\partial x}\left[p(x)\frac{\partial v}{\partial x}(x,t)\right],$$

where $v(x,t)$ is the mean longitudinal displacement of a section of the bar from its equilibrium position x at time t. The vibrations can be written as a sum of simple harmonic vibrations:

$$v(x,t) = \sum_{k=0}^{\infty} c_k u_k(x) \cos\sqrt{\lambda_k}(t - t_0),$$

where

$$\frac{d}{dx}\left[p(x)\frac{du_k}{dx}(x)\right] + \lambda_k \rho(x) u_k(x) = 0.$$

If the bar has length l and is fixed at its ends, then this differential equation holds for $0 < x < l$ and $v(0) = v(l) = 0$.

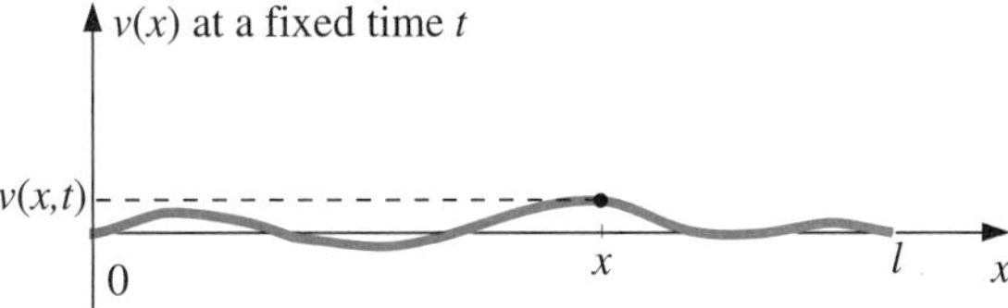

A system of these differential equations is called a Sturm-Liouville system, and the numbers λ_k are eigenvalues with corresponding eigenfunctions $u_k(x)$.

Suppose the bar is 1 m long with uniform stiffness $p(x) = p$ and uniform density $\rho(x) = \rho$. To approximate u and λ, let $h = 0.2$. Then $x_j = 0.2j$, for $0 \le j \le 5$, and we can use the midpoint formula (4.5) in Section 4.1 to approximate the first derivatives. This gives the linear system

$$A\mathbf{w} = \begin{bmatrix} 2 & -1 & 0 & 0 \\ -1 & 2 & -1 & 0 \\ 0 & -1 & 2 & -1 \\ 0 & 0 & -1 & 2 \end{bmatrix}\begin{bmatrix} w_1 \\ w_2 \\ w_3 \\ w_4 \end{bmatrix} = -0.04\frac{\rho}{p}\lambda\begin{bmatrix} w_1 \\ w_2 \\ w_3 \\ w_4 \end{bmatrix} = -0.04\frac{\rho}{p}\lambda\mathbf{w}.$$

In this system, $w_j \approx u(x_j)$, for $1 \le j \le 4$, and $w_0 = w_5 = 0$. The four eigenvalues of A approximate the eigenvalues of the *Sturm-Liouville system*. It is the approximation of eigenvalues that we will consider in this chapter. A Sturm-Liouville application is discussed in Exercise 13 of Section 9.5.

9.1 Linear Algebra and Eigenvalues

Eigenvalues and eigenvectors were introduced in Chapter 7 in connection with the convergence of iterative methods for approximating the solution to a linear system. To determine the eigenvalues of an $n \times n$ matrix A, we construct the characteristic polynomial

$$p(\lambda) = \det(A - \lambda I)$$

and then determine its zeros. Finding the determinant of an $n \times n$ matrix is computationally expensive, and finding good approximations to the roots of $p(\lambda)$ is also difficult. In this chapter we will explore other means for approximating the eigenvalues of a matrix. In Section 9.6 we give an introduction to a technique for factoring a general $m \times n$ matrix into a form that has valuable applications in a number of areas.

In Chapter 7 we found that an iterative technique for solving a linear system will converge if all the eigenvalues associated with the problem have magnitude less than 1. The exact values of the eigenvalues in this case are not of primary importance—only the region of the complex plane in which they lie. An important result in this regard was first discovered by S. A. Geršgorin. It is the subject of a very interesting book by Richard Varga. [Var2]

Theorem 9.1 **(Geršgorin Circle)**

Semyon Aranovich Geršgorin (1901–1933) worked at the Petrograd Technological Institute until 1930, when he moved to the Leningrad Mechanical Engineering Institute. His 1931 paper *Über die Abgrenzung der Eigenwerte einer Matrix* ([Ger]) included what is now known as his Circle Theorem.

Let A be an $n \times n$ matrix and R_i denote the circle in the complex plane with center a_{ii} and radius $\sum_{j=1, j\neq i}^{n} |a_{ij}|$; that is,

$$R_i = \left\{ z \in \mathcal{C} \,\middle|\, |z - a_{ii}| \le \sum_{j=1, j\neq i}^{n} |a_{ij}| \right\},$$

where $\mathcal{C}$ denotes the complex plane. The eigenvalues of A are contained within the union of these circles, $R = \cup_{i=1}^{n} R_i$. Moreover, the union of any k of the circles that do not intersect the remaining $(n-k)$ contains precisely k (counting multiplicities) of the eigenvalues. ■

Proof Suppose that λ is an eigenvalue of A with associated eigenvector $\mathbf{x}$, where $\|\mathbf{x}\|_\infty = 1$. Since $A\mathbf{x} = \lambda\mathbf{x}$, the equivalent component representation is

$$\sum_{j=1}^{n} a_{ij} x_j = \lambda x_i, \quad \text{for each } i = 1, 2, \ldots, n. \tag{9.1}$$

Let k be an integer with $|x_k| = \|\mathbf{x}\|_\infty = 1$. When $i = k$, Eq. (9.1) implies that

$$\sum_{j=1}^{n} a_{kj} x_j = \lambda x_k.$$

Thus

$$\sum_{\substack{j=1,\\ j\neq k}}^{n} a_{kj} x_j = \lambda x_k - a_{kk} x_k = (\lambda - a_{kk}) x_k,$$

and

$$|\lambda - a_{kk}| \cdot |x_k| = \left| \sum_{\substack{j=1,\\ j\neq k}}^{n} a_{kj}x_j \right| \leq \sum_{\substack{j=1,\\ j\neq k}}^{n} |a_{kj}||x_j|.$$

But $|x_k| = \|\mathbf{x}\|_\infty = 1$, so $|x_j| \leq |x_k| = 1$ for all $j = 1, 2, \ldots, n$. Hence

$$|\lambda - a_{kk}| \leq \sum_{\substack{j=1,\\ j\neq k}}^{n} |a_{kj}|.$$

This proves the first assertion in the theorem, that $\lambda \in R_k$. A proof of the second statement is contained in [Var2], p. 8, or in [Or2], p. 48. ■ ■ ■

Example 1 Determine the Geršgorin circles for the matrix

$$A = \begin{bmatrix} 4 & 1 & 1 \\ 0 & 2 & 1 \\ -2 & 0 & 9 \end{bmatrix},$$

and use these to find bounds for the spectral radius of A.

Solution The circles in the Geršgorin Theorem are (see Figure 9.1)

$R_1 = \{z \in \mathcal{C} \mid |z-4| \leq 2\}$, $R_2 = \{z \in \mathcal{C} \mid |z-2| \leq 1\}$, and $R_3 = \{z \in \mathcal{C} \mid |z-9| \leq 2\}$.

Because R_1 and R_2 are disjoint from R_3, there are precisely two eigenvalues within $R_1 \cup R_2$ and one within R_3. Moreover, $\rho(A) = \max_{1\leq i\leq 3} |\lambda_i|$, so $7 \leq \rho(A) \leq 11$. ■

Figure 9.1

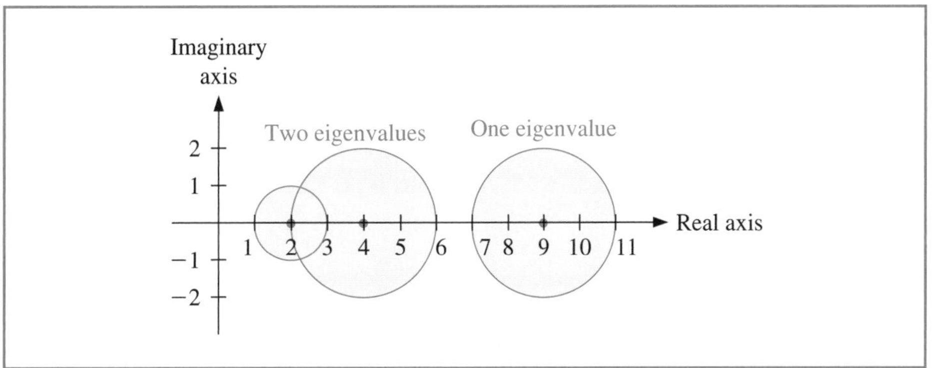

Even when we need to find the eigenvalues, many techniques for their approximation are iterative. Determining regions in which they lie is the first step for finding the approximation, because it provides us with an initial approximations.

Before considering further results concerning eigenvalues and eigenvectors, we need some definitions and results from linear algebra. All the general results that will be needed in the remainder of this chapter are listed here for ease of reference. The proofs of many of the results that are not given are considered in the exercises, and all can be be found in most standard texts on linear algebra (see, for example, [ND], [Poo], or [DG]).

The first definition parallels the definition for the linear independence of functions described in Section 8.2. In fact, much of what we will see in this section parallels material in Chapter 8.

Definition 9.2 Let $\{\mathbf{v}^{(1)}, \mathbf{v}^{(2)}, \mathbf{v}^{(3)}, \ldots, \mathbf{v}^{(k)}\}$ be a set of vectors. The set is **linearly independent** if whenever

$$\mathbf{0} = \alpha_1\mathbf{v}^{(1)} + \alpha_2\mathbf{v}^{(2)} + \alpha_3\mathbf{v}^{(3)} + \cdots + \alpha_k\mathbf{v}^{(k)},$$

then $\alpha_i = 0$, for each $i = 0, 1, \ldots, k$. Otherwise the set of vectors is **linearly dependent.** ■

Note that any set of vectors containing the zero vector is linearly dependent.

Theorem 9.3 Suppose that $\{\mathbf{v}^{(1)}, \mathbf{v}^{(2)}, \mathbf{v}^{(3)}, \ldots, \mathbf{v}^{(n)}\}$ is a set of n linearly independent vectors in $\mathbb{R}^n$. Then for any vector $\mathbf{x} \in \mathbb{R}^n$ a unique collection of constants $\beta_1, \beta_2, \ldots, \beta_n$ exists with

$$\mathbf{x} = \beta_1\mathbf{v}^{(1)} + \beta_2\mathbf{v}^{(2)} + \beta_3\mathbf{v}^{(3)} + \cdots + \beta_n\mathbf{v}^{(n)}.$$ ■

Proof Let A be the matrix whose columns are the vectors $\mathbf{v}^{(1)}, \mathbf{v}^{(2)}, \ldots, \mathbf{v}^{(n)}$. Then the set $\{\mathbf{v}^{(1)}, \mathbf{v}^{(2)}, \ldots, \mathbf{v}^{(n)}\}$ is linearly independent if and only if the matrix equation

$$A(\alpha_1, \alpha_2, \ldots, \alpha_n)^t = \mathbf{0} \quad \text{has the unique solution} \quad (\alpha_1, \alpha_2, \ldots, \alpha_n)^t = \mathbf{0}.$$

But by Theorem 6.16 on page 397, this is equivalent to the matrix equation $A(\beta_1, \beta_2, \ldots, \beta_n)^t = \mathbf{x}$, having a unique solution for any vector $\mathbf{x} \in \mathbb{R}^n$. This, in turn, is equivalent to the statement that for any vector $\mathbf{x} \in \mathbb{R}^n$ a unique collection of constants $\beta_1, \beta_2, \ldots, \beta_n$ exists with

$$\mathbf{x} = \beta_1\mathbf{v}^{(1)} + \beta_2\mathbf{v}^{(2)} + \beta_3\mathbf{v}^{(3)} + \cdots + \beta_n\mathbf{v}^{(n)}.$$ ■ ■ ■

Definition 9.4 Any collection of n linearly independent vectors in $\mathbb{R}^n$ is called a **basis** for $\mathbb{R}^n$. ■

Example 2 **(a)** Show that $\mathbf{v}^{(1)} = (1, 0, 0)^t, \mathbf{v}^{(2)} = (-1, 1, 1)^t$, and $\mathbf{v}^{(3)} = (0, 4, 2)^t$ is a basis for $\mathbb{R}^3$, and **(b)** given an arbitrary vector $\mathbf{x} \in \mathbb{R}^3$ find β_1, β_2, and β_3 with

$$\mathbf{x} = \beta_1\mathbf{v}^{(1)} + \beta_2\mathbf{v}^{(2)} + \beta_3\mathbf{v}^{(3)}.$$

Solution **(a)** Let α_1, α_2, and α_3 be numbers with $\mathbf{0} = \alpha_1\mathbf{v}^{(1)} + \alpha_2\mathbf{v}^{(2)} + \alpha_3\mathbf{v}^{(3)}$. Then

$$\begin{aligned}(0, 0, 0)^t &= \alpha_1(1, 0, 0)^t + \alpha_2(-1, 1, 1)^t + \alpha_3(0, 4, 2)^t \\ &= (\alpha_1 - \alpha_2, \alpha_2 + 4\alpha_3, \alpha_2 + 2\alpha_3)^t,\end{aligned}$$

so $\alpha_1 - \alpha_2 = 0, \quad \alpha_2 + 4\alpha_3 = 0, \quad$ and $\quad \alpha_2 + 2\alpha_3 = 0.$

The only solution to this system is $\alpha_1 = \alpha_2 = \alpha_3 = 0$, so this set $\{\mathbf{v}^{(1)}, \mathbf{v}^{(2)}, \mathbf{v}^{(3)}\}$ of 3 linearly independent vectors in $\mathbb{R}^3$ is a basis for $\mathbb{R}^3$.

(b) Let $\mathbf{x} = (x_1, x_2, x_3)^t$ be a vector in $\mathbb{R}^3$. Solving

$$\begin{aligned}\mathbf{x} &= \beta_1\mathbf{v}^{(1)} + \beta_2\mathbf{v}^{(2)} + \beta_3\mathbf{v}^{(3)} \\ &= \beta_1(1, 0, 0)^t + \beta_2(-1, 1, 1)^t + \beta_3((0, 4, 2)^t \\ &= (\beta_1 - \beta_2, \beta_2 + 4\beta_3, \beta_2 + 2\beta_3)^t\end{aligned}$$

is equivalent to solving for β_1, β_2, and β_3 in the system

$$\beta_1 - \beta_2 = x_1, \quad \beta_2 + 4\beta_3 = x_2, \beta_2 + 2\beta_3 = x_3.$$

This system has the unique solution

$$\beta_1 = x_1 - x_2 + 2x_3, \quad \beta_2 = 2x_3 - x_2, \quad \text{and} \quad \beta_3 = \frac{1}{2}(x_2 - x_3).$$ ■

The next result will be used in Section 9.3 to develop the Power method for approximating eigenvalues. A proof of this result is considered in Exercise 10.

Theorem 9.5 If A is a matrix and $\lambda_1, \ldots, \lambda_k$ are distinct eigenvalues of A with associated eigenvectors $\mathbf{x}^{(1)}, \mathbf{x}^{(2)}, \ldots, \mathbf{x}^{(k)}$, then $\{\mathbf{x}^{(1)}, \mathbf{x}^{(2)}, \ldots, \mathbf{x}^{(k)}\}$ is a linearly independent set. ■

Example 3 Show that a basis can be formed for $\mathbb{R}^3$ using the eigenvectors of the 3×3 matrix

$$A = \begin{bmatrix} 2 & 0 & 0 \\ 1 & 1 & 2 \\ 1 & -1 & 4 \end{bmatrix}.$$

Solution In Example 2 of Section 7.2 we found that A has the characteristic polynomial

$$p(\lambda) = \det(A - \lambda I) = (\lambda - 3)(\lambda - 2)^2.$$

Hence there are two distinct eigenvalues of A: $\lambda_1 = 3$ and $\lambda_2 = 2$. In that example we also found that $\lambda_1 = 3$ has the eigenvector $\mathbf{x}_1 = (0, 1, 1)^t$, and that there are two linearly independent eigenvectors $\mathbf{x}_2 = (0, 2, 1)^t$ and $\mathbf{x}_3 = (-2, 0, 1)^t$ corresponding to $\lambda_2 = 2$.

It is not difficult to show (see Exercise 8) that this set of three eigenvectors

$$\{\mathbf{x}_1, \mathbf{x}_2, \mathbf{x}_3\} = \{(0, 1, 1)^t, (0, 2, 1)^t, (-2, 0, 1)^t\}$$

is linearly independent and hence forms a basis for $\mathbb{R}^3$. ■

In the next example we will see a matrix whose eigenvalues are the same as those in Example 3 but whose eigenvectors have a different character.

Example 4 Show that no collection of eigenvectors of the 3×3 matrix

$$B = \begin{bmatrix} 2 & 1 & 0 \\ 0 & 2 & 0 \\ 0 & 0 & 3 \end{bmatrix}$$

can form a basis for $\mathbb{R}^3$.

Solution This matrix also has the same characteristic polynomial as the matrix A in Example 3:

$$p(\lambda) = \det \begin{bmatrix} 2-\lambda & 1 & 0 \\ 0 & 2-\lambda & 0 \\ 0 & 0 & 3-\lambda \end{bmatrix} = (\lambda - 3)(\lambda - 2)^2,$$

so its eigenvalues are the same as those of A in Example 3, that is, $\lambda_1 = 3$ and $\lambda_2 = 2$.

To determine eigenvectors for B corresponding to the eigenvalue $\lambda_1 = 3$, we need to solve the system $(B - 3I)\mathbf{x} = \mathbf{0}$, so

$$\begin{bmatrix} 0 \\ 0 \\ 0 \end{bmatrix} = (B - 3I) \begin{bmatrix} x_1 \\ x_2 \\ x_3 \end{bmatrix} = \begin{bmatrix} -1 & 1 & 0 \\ 0 & -1 & 0 \\ 0 & 0 & 0 \end{bmatrix} \begin{bmatrix} x_1 \\ x_2 \\ x_3 \end{bmatrix} = \begin{bmatrix} -x_1 + x_2 \\ -x_2 \\ 0, \end{bmatrix}.$$

Hence $x_2 = 0$, $x_1 = x_2 = 0$, and x_3 is arbitrary. Setting $x_3 = 1$ gives the only linearly independent eigenvector $(0, 0, 1)^t$ corresponding to $\lambda_1 = 3$.

Consider $\lambda_2 = 2$. If

$$\begin{bmatrix} 0 \\ 0 \\ 0 \end{bmatrix} = (B - 2\lambda) \begin{bmatrix} x_1 \\ x_2 \\ x_3 \end{bmatrix} = \begin{bmatrix} 0 & 1 & 0 \\ 0 & 0 & 0 \\ 0 & 0 & 1 \end{bmatrix} \cdot \begin{bmatrix} x_1 \\ x_2 \\ x_3 \end{bmatrix} = \begin{bmatrix} x_2 \\ 0 \\ x_3, \end{bmatrix},$$

then $x_2 = 0, x_3 = 0$, and x_1 is arbitrary. There is only one linearly independent eigenvector corresponding to $\lambda_2 = 2$, which can be expressed as $(1, 0, 0)^t$, even though $\lambda_2 = 2$ was a zero of multiplicity 2 of the characteristic polynomial of B.

These two eigenvectors are clearly not sufficient to form a basis for $\mathbb{R}^3$. In particular, $(0, 1, 0)^t$ is not a linear combination of $\{(0, 0, 1)^t, (1, 0, 0)^t\}$. ■

We will see that when the number of linearly independent eigenvectors does not match the size of the matrix, as is the case in Example 4, there can be difficulties with the approximation methods for finding eigenvalues.

Orthogonal Vectors

In Section 8.2 we considered orthogonal and orthonormal sets of functions. Vectors with these properties are defined in a similar manner.

Definition 9.6 A set of vectors $\{\mathbf{v}^{(1)}, \mathbf{v}^{(2)}, \ldots, \mathbf{v}^{(n)}\}$ is called **orthogonal** if $(\mathbf{v}^{(i)})^t\mathbf{v}^{(j)} = 0$, for all $i \neq j$. If, in addition, $(\mathbf{v}^{(i)})^t\mathbf{v}^{(i)} = 1$, for all $i = 1, 2, \ldots, n$, then the set is called **orthonormal**. ■

Because $\mathbf{x}^t\mathbf{x} = \|\mathbf{x}\|_2^2$ for any $\mathbf{x}$ in $\mathbb{R}^n$, a set of orthogonal vectors $\{\mathbf{v}^{(1)}, \mathbf{v}^{(2)}, \ldots, \mathbf{v}^{(n)}\}$ is orthonormal if and only if

$$\|\mathbf{v}^{(i)}\|_2 = 1, \quad \text{for each } i = 1, 2, \ldots, n.$$

Example 5 **(a)** Show that the vectors $\mathbf{v}^{(1)} = (0, 4, 2)^t$, $\mathbf{v}^{(2)} = (-5, -1, 2)^t$, and $\mathbf{v}^{(3)} = (1, -1, 2)^t$ form an orthogonal set, and **(b)** use these to determine a set of orthonormal vectors.

Solution **(a)** We have $(\mathbf{v}^{(1)})^t\mathbf{v}^{(2)} = 0(-5) + 4(-1) + 2(2) = 0$,

$$(\mathbf{v}^{(1)})^t\mathbf{v}^{(3)} = 0(1) + 4(-1) + 2(2) = 0, \quad \text{and} \quad (\mathbf{v}^{(2)})^t\mathbf{v}^{(3)} = -5(1) - 1(-1) + 2(2) = 0,$$

so the vectors are orthogonal, and form a basis for $\mathbb{R}^n$. The l_2 norms of these vectors are

$$\|\mathbf{v}^{(1)}\|_2 = 2\sqrt{5}, \quad \|\mathbf{v}^{(2)}\|_2 = \sqrt{30}, \quad \text{and} \quad \|\mathbf{v}^{(3)}\|_2 = \sqrt{6}.$$

(b) The vectors

$$\mathbf{u}^{(1)} = \frac{\mathbf{v}^{(1)}}{\|\mathbf{v}^{(1)}\|_2} = \left(\frac{0}{2\sqrt{5}}, \frac{4}{2\sqrt{5}}, \frac{2}{2\sqrt{5}}\right)^t = \left(0, \frac{2\sqrt{5}}{5}, \frac{\sqrt{5}}{5}\right)^t,$$

$$\mathbf{u}^{(2)} = \frac{\mathbf{v}^{(2)}}{\|\mathbf{v}^{(2)}\|_2} = \left(\frac{-5}{\sqrt{30}}, \frac{-1}{\sqrt{30}}, \frac{2}{\sqrt{30}}\right)^t = \left(-\frac{\sqrt{30}}{6}, -\frac{\sqrt{30}}{30}, \frac{\sqrt{30}}{15}\right)^t,$$

$$\mathbf{u}^{(3)} = \frac{\mathbf{v}^{(3)}}{\|\mathbf{v}^{(3)}\|_2} = \left(\frac{1}{\sqrt{6}}, \frac{-1}{\sqrt{6}}, \frac{2}{\sqrt{6}}\right)^t = \left(\frac{\sqrt{6}}{6}, -\frac{\sqrt{6}}{6}, \frac{\sqrt{6}}{3}\right)^t$$

form an orthonormal set, since they inherit orthogonality from $\mathbf{v}^{(1)}, \mathbf{v}^{(2)}$, and $\mathbf{v}^{(3)}$, and additionally,

$$\|\mathbf{u}^{(1)}\|_2 = \|\mathbf{u}^{(2)}\|_2 = \|\mathbf{u}^{(3)}\|_2 = 1.$$

■

The proof of the next result is considered in Exercise 9.

Theorem 9.7 An orthogonal set of nonzero vectors is linearly independent. ■

The **Gram-Schmidt** process for constructing a set of polynomials that are orthogonal with respect to a given weight function was described in Theorem 8.7 of Section 8.2 (see page 515). There is a parallel process, also known as Gram-Schmidt, that permits us to construct an orthogonal basis for $\mathbb{R}^n$ given a set of n linearly independent vectors in $\mathbb{R}^n$.

Theorem 9.8 Let $\{\mathbf{x}_1, \mathbf{x}_2, \ldots, \mathbf{x}_k\}$ be a set of k linearly independent vectors in $\mathbb{R}^n$. Then $\{\mathbf{v}_1, \mathbf{v}_2, \ldots, \mathbf{v}_k\}$ defined by

$$\mathbf{v}_1 = \mathbf{x}_1,$$

$$\mathbf{v}_2 = \mathbf{x}_2 - \left(\frac{\mathbf{v}_1^t\mathbf{x}_2}{\mathbf{v}_1^t\mathbf{v}_1}\right)\mathbf{v}_1,$$

$$\mathbf{v}_3 = \mathbf{x}_3 - \left(\frac{\mathbf{v}_1^t\mathbf{x}_3}{\mathbf{v}_1^t\mathbf{v}_1}\right)\mathbf{v}_1 - \left(\frac{\mathbf{v}_2^t\mathbf{x}_3}{\mathbf{v}_2^t\mathbf{v}_2}\right)\mathbf{v}_2,$$

$$\vdots$$

$$\mathbf{v}_k = \mathbf{x}_k - \sum_{i=1}^{k-1}\left(\frac{\mathbf{v}_i^t\mathbf{x}_k}{\mathbf{v}_i^t\mathbf{v}_i}\right)\mathbf{v}_i.$$

is set of k orthogonal vectors in $\mathbb{R}^n$. ■

The proof of this theorem, discussed in Exercise 16, is a direct verification of the fact that for each $1 \le i \le k$ and $1 \le j \le k$, with $i \ne j$, we have $\mathbf{v}_i^t\mathbf{v}_j = 0$.

Note that when the original set of vectors forms a basis for $\mathbb{R}^n$, that is, when $k = n$, then the constructed vectors form an orthogonal basis for $\mathbb{R}^n$. From this we can form an orthonormal basis $\{\mathbf{u}_1, \mathbf{u}_2, \ldots, \mathbf{u}_n\}$ simply by defining for each $i = 1, 2, \ldots, n$

$$\mathbf{u}_i = \frac{\mathbf{v}_i}{||\mathbf{v}_i||_2}.$$

The following example illustrates how an orthonormal basis for $\mathbb{R}^3$ can be constructed from three linearly independent vectors in $\mathbb{R}^3$.

Example 6 Use the Gram-Schmidt process to determine a set of orthogonal vectors from the linearly independent vectors

$$\mathbf{x}^{(1)} = (1, 0, 0)^t, \quad \mathbf{x}^{(2)} = (1, 1, 0)^t, \quad \text{and} \quad \mathbf{x}^{(3)} = (1, 1, 1)^t.$$

Solution We have the orthogonal vectors $\mathbf{v}^{(1)}$, $\mathbf{v}^{(2)}$, and $\mathbf{v}^{(3)}$, given by

$$\mathbf{v}^{(1)} = \mathbf{x}^{(1)} = (1, 0, 0)^t$$

$$\mathbf{v}^{(2)} = (1, 1, 0)^t - \left(\frac{((1,0,0)^t)^t(1,1,0)^t}{((1,0,0)^t)^t(1,0,0)^t}\right)(1,0,0)^t = (1,1,0)^t - (1,0,0)^t = (0,1,0)^t$$

$$\mathbf{v}^{(3)} = (1, 1, 1)^t - \left(\frac{((1,0,0)^t)^t(1,1,1)^t}{((1,0,0)^t)^t(1,0,0)^t}\right)(1,0,0)^t - \left(\frac{((0,1,0)^t)^t(1,1,1)^t}{((0,1,0)^t)^t(0,1,0)^t}\right)(0,1,0)^t$$

$$= (1,1,1)^t - (1,0,0)^t - (0,1,0)^t = (0,0,1)^t.$$

The set $\{\mathbf{v}^{(1)}, \mathbf{v}^{(2)}, \mathbf{v}^{(3)}\}$ happens to be orthonormal as well as orthogonal, but this is not commonly the situation. ■

The LinearAlgebra package in Maple has a Gram-Schmidt command that returns an orthogonal set of vectors, or even an orthonormal set. These commands

GramSchmidt$(\{x1, x2, x3\})$

gives an orthogonal set of vectors, and the command

GramSchmidt$(\{x1, x2, x3\}, normalized)$

produces the orthonormal set.

EXERCISE SET 9.1

1. Find the eigenvalues and associated eigenvectors of the following 3×3 matrices. Is there a set of linearly independent eigenvectors?

   a. $A = \begin{bmatrix} 2 & -3 & 6 \\ 0 & 3 & -4 \\ 0 & 2 & -3 \end{bmatrix}$ **b.** $A = \begin{bmatrix} 2 & 0 & 1 \\ 0 & 2 & 0 \\ 1 & 0 & 2 \end{bmatrix}$

   c. $A = \begin{bmatrix} 1 & 1 & 1 \\ 1 & 1 & 0 \\ 1 & 0 & 1 \end{bmatrix}$ **d.** $A = \begin{bmatrix} 2 & 1 & -1 \\ 0 & 2 & 1 \\ 0 & 0 & 3 \end{bmatrix}$

2. Find the eigenvalues and associated eigenvectors of the following 3×3 matrices. Is there a set of linearly independent eigenvectors?

   a. $A = \begin{bmatrix} 1 & 0 & 0 \\ -1 & 0 & 1 \\ -1 & -1 & 2 \end{bmatrix}$ **b.** $A = \begin{bmatrix} 2 & -1 & -1 \\ -1 & 2 & -1 \\ -1 & -1 & 2 \end{bmatrix}$

   c. $A = \begin{bmatrix} 2 & 1 & 1 \\ 1 & 2 & 1 \\ 1 & 1 & 2 \end{bmatrix}$ **d.** $A = \begin{bmatrix} 2 & 1 & 1 \\ 0 & 3 & 1 \\ 0 & 0 & 2 \end{bmatrix}$

3. Use the Geršgorin Circle Theorem to determine bounds for the eigenvalues, and the spectral radius of the following matrices.

   a. $\begin{bmatrix} 1 & 0 & 0 \\ -1 & 0 & 1 \\ -1 & -1 & 2 \end{bmatrix}$ **b.** $\begin{bmatrix} 4 & -1 & 0 \\ -1 & 4 & -1 \\ -1 & -1 & 4 \end{bmatrix}$

   c. $\begin{bmatrix} 3 & 2 & 1 \\ 2 & 3 & 0 \\ 1 & 0 & 3 \end{bmatrix}$ **d.** $\begin{bmatrix} 4.75 & 2.25 & -0.25 \\ 2.25 & 4.75 & 1.25 \\ -0.25 & 1.25 & 4.75 \end{bmatrix}$

4. Use the Geršgorin Circle Theorem to determine bounds for the eigenvalues, and the spectral radius of the following matrices.

   a. $\begin{bmatrix} -4 & 0 & 1 & 3 \\ 0 & -4 & 2 & 1 \\ 1 & 2 & -2 & 0 \\ 3 & 1 & 0 & -4 \end{bmatrix}$ **b.** $\begin{bmatrix} 1 & 0 & -1 & 1 \\ 2 & 2 & -1 & 1 \\ 0 & 1 & 3 & -2 \\ 1 & 0 & 1 & 4 \end{bmatrix}$

   c. $\begin{bmatrix} 1 & 1 & 0 & 0 \\ 1 & 2 & 0 & 1 \\ 0 & 0 & 3 & 3 \\ 0 & 1 & 3 & 2 \end{bmatrix}$ **d.** $\begin{bmatrix} 3 & -1 & 0 & 1 \\ -1 & 3 & 1 & 0 \\ 0 & 1 & 9 & 2 \\ 1 & 0 & 2 & 9 \end{bmatrix}$

5. For the matrices in Exercise 1 that have 3 linearly independent eigenvectors form the factorization $A = PDP^{-1}$.

6. For the matrices in Exercise 2 that have 3 linearly independent eigenvectors form the factorization $A = PDP^{-1}$.

7. Show that $\mathbf{v}_1 = (2, -1)^t$, $\mathbf{v}_2 = (1, 1)^t$, and $\mathbf{v}_3 = (1, 3)^t$ are linearly dependent.

8. Show that the three eigenvectors in Example 3 are linearly independent.

9. Show that a set $\{\mathbf{v}_1, \ldots, \mathbf{v}_k\}$ of k nonzero orthogonal vectors is linearly independent.

10. Show that if A is a matrix and $\lambda_1, \lambda_2, \ldots, \lambda_k$ are distinct eigenvalues with associated eigenvectors $\mathbf{x}_1$, $\mathbf{x}_2, \ldots, \mathbf{x}_k$, then $\{\mathbf{x}_1, \mathbf{x}_2, \ldots, \mathbf{x}_k\}$ is a linearly independent set.

11. Let $\{\mathbf{v}_1, \ldots, \mathbf{v}_n\}$ be a set of orthonormal nonzero vectors in $\mathbb{R}^n$ and $\mathbf{x} \in \mathbb{R}^n$. Determine the values of c_k, for $k = 1, 2, \ldots, n$, if

$$\mathbf{x} = \sum_{k=1}^{n} c_k \mathbf{v}_k.$$

12. Assume that $\{\mathbf{x}_1, \mathbf{x}_2\}$, $\{\mathbf{x}_1, \mathbf{x}_3\}$, and $\{\mathbf{x}_2, \mathbf{x}_3\}$, are all linearly independent. Must $\{\mathbf{x}_1, \mathbf{x}_2, \mathbf{x}_3\}$ be linearly independent?

13. Consider the follow sets of vectors. **(i)** Show that the set is linearly independent; **(ii)** use the Gram-Schmidt process to find a set of orthogonal vectors; **(iii)** determine a set of orthonormal vectors from the vectors in (ii).

  a. $\mathbf{v}_1 = (1, 1)^t$, $\mathbf{v}_2 = (-2, 1)^t$

  b. $\mathbf{v}_1 = (1, 1, 0)^t$, $\mathbf{v}_2 = (1, 0, 1)^t$, $\mathbf{v}_3 = (0, 1, 1)^t$

  c. $\mathbf{v}_1 = (1, 1, 1, 1)^t$, $\mathbf{v}_2 = (0, 2, 2, 2)^t$, $\mathbf{v}_3 = (1, 0, 0, 1)^t$

  d. $\mathbf{v}_1 = (2, 2, 3, 2, 3)^t$, $\mathbf{v}_2 = (2, -1, 0, -1, 0)^t$, $\mathbf{v}_3 = (0, 0, 1, 0, -1)^t$, $\mathbf{v}_4 = (1, 2, -1, 0, -1)^t$, $\mathbf{v}_5 = (0, 1, 0, -1, 0)^t$

14. Consider the follow sets of vectors. **(i)** Show that the set is linearly independent; **(ii)** use the Gram-Schmidt process to find a set of orthogonal vectors; **(iii)** determine a set of orthonormal vectors from the vectors in (ii).

  a. $\mathbf{v}_1 = (2, -1)^t$, $\mathbf{v}_2 = (1, 3)^t$

  b. $\mathbf{v}_1 = (2, -1, 1)^t$, $\mathbf{v}_2 = (1, 0, 1)^t$, $\mathbf{v}_3 = (0, 2, 0)^t$

  c. $\mathbf{v}_1 = (1, 1, 1, 1)^t$, $\mathbf{v}_2 = (0, 1, 1, 1)^t$, $\mathbf{v}_3 = (0, 0, 1, 0)^t$

  d. $\mathbf{v}_1 = (2, 2, 0, 2, 1)^t$, $\mathbf{v}_2 = (-1, 2, 0, -1, 1)^t$, $\mathbf{v}_3 = (0, 1, 0, 1, 0)^t$, $\mathbf{v}_4 = (-1, 0, 0, 1, 1)^t$

15. Use the Geršgorin Circle Theorem to show that a strictly diagonally dominant matrix must be nonsingular.

16. Prove that the set of vectors $\{\mathbf{v}_1, \mathbf{v}_2, \ldots, \mathbf{v}_k\}$ described in the Gram-Schmidt Theorem is orthogonal.

17. A **persymmetric matrix** is a matrix that is symmetric about both diagonals; that is, an $N \times N$ matrix $A = (a_{ij})$ is persymmetric if $a_{ij} = a_{ji} = a_{N+1-i,N+1-j}$, for all $i = 1, 2, \ldots, N$ and $j = 1, 2, \ldots, N$. A number of problems in communication theory have solutions that involve the eigenvalues and eigenvectors of matrices that are in persymmetric form. For example, the eigenvector corresponding to the minimal eigenvalue of the 4×4 persymmetric matrix

$$A = \begin{bmatrix} 2 & -1 & 0 & 0 \\ -1 & 2 & -1 & 0 \\ 0 & -1 & 2 & -1 \\ 0 & 0 & -1 & 2 \end{bmatrix}$$

gives the unit energy-channel impulse response for a given error sequence of length 2, and subsequently the minimum weight of any possible error sequence.

  a. Use the Geršgorin Circle Theorem to show that if A is the matrix given above and λ is its minimal eigenvalue, then $|\lambda - 4| = \rho(A - 4I)$, where ρ denotes the spectral radius.

  b. Find the minimal eigenvalue of the matrix A by finding all the eigenvalues $A - 4I$ and computing its spectral radius. Then find the corresponding eigenvector.

  c. Use the Geršgorin Circle Theorem to show that if λ is the minimal eigenvalue of the matrix

$$B = \begin{bmatrix} 3 & -1 & -1 & 1 \\ -1 & 3 & -1 & -1 \\ -1 & -1 & 3 & -1 \\ 1 & -1 & -1 & 3 \end{bmatrix},$$

  then $|\lambda - 6| = \rho(B - 6I)$.

  d. Repeat part (b) using the matrix B and the result in part (c).

9.2 Orthogonal Matrices and Similarity Transformations

In this section we will consider the connection between sets of vectors and matrices formed using these vectors as their columns. We first consider some results about a class of special matrices. The terminology in the next definition follows from the fact that the columns of an orthogonal matrix will form an orthogonal set of vectors.

Definition 9.9 A matrix Q is said to be **orthogonal** if its columns $\{\mathbf{q}_1^t, \mathbf{q}_2^t, \ldots, \mathbf{q}_n^t\}$ form an orthonormal set in $\mathbb{R}^n$. ■

It would probably be better to call orthogonal matrices *orthonormal* because the columns form not just an orthogonal but an orthonormal set of vectors.

The Maple command *IsOrthogonal*(A) in the *LinearAlgebra* package returns *true* if A is orthogonal and *false* otherwise.

The following important properties of orthogonal matrices are considered in Exercise 16.

Theorem 9.10 Suppose that Q is an orthogonal $n \times n$ matrix. Then

(i) Q is invertible with $Q^{-1} = Q^t$;

(ii) For any $\mathbf{x}$ and $\mathbf{y}$ in $\mathbb{R}^n$, $(Q\mathbf{x})^t Q\mathbf{y} = \mathbf{x}^t\mathbf{y}$;

(iii) For any $\mathbf{x}$ in $\mathbb{R}^n$, $||Q\mathbf{x}||_2 = ||\mathbf{x}||_2$. ■

In addition, the converse of part **(i)** holds. (See Exercise 18.) That is,

- any invertible matrix Q with $Q^{-1} = Q^t$ is orthogonal.

As an example, the permutation matrices discussed in Section 6.5 have this property, so they are orthogonal.

Property (iii) of Theorem 9.10 is often expressed by stating that orthogonal matrices are l_2-norm preserving. As an immediate consequence of this property, every orthogonal matrix Q has $||Q||_2 = 1$.

Example 1 Show that the matrix

$$Q = [\mathbf{u}^{(1)}, \mathbf{u}^{(2)}, \mathbf{u}^{(3)}] = \begin{bmatrix} 0 & -\frac{\sqrt{30}}{6} & \frac{\sqrt{6}}{6} \\ \frac{2\sqrt{5}}{5} & -\frac{\sqrt{30}}{30} & -\frac{\sqrt{6}}{6} \\ \frac{\sqrt{5}}{5} & \frac{\sqrt{30}}{15} & \frac{\sqrt{6}}{3} \end{bmatrix}$$

formed from the orthonormal set of vectors found in Example 5 of Section 9.1 is an orthogonal matrix.

Solution Note that

$$QQ^t = \begin{bmatrix} 0 & \frac{-\sqrt{30}}{6} & \frac{\sqrt{6}}{6} \\ \frac{2\sqrt{5}}{5} & -\frac{\sqrt{30}}{30} & -\frac{\sqrt{6}}{6} \\ \frac{\sqrt{5}}{5} & \frac{\sqrt{30}}{15} & \frac{\sqrt{6}}{3} \end{bmatrix} \cdot \begin{bmatrix} 0 & \frac{2\sqrt{5}}{5} & \frac{\sqrt{5}}{5} \\ -\frac{\sqrt{30}}{6} & -\frac{\sqrt{30}}{30} & \frac{\sqrt{30}}{15} \\ \frac{\sqrt{6}}{6} & -\frac{\sqrt{6}}{6} & \frac{\sqrt{6}}{3} \end{bmatrix} = \begin{bmatrix} 1 & 0 & 0 \\ 0 & 1 & 0 \\ 0 & 0 & 1 \end{bmatrix} = I.$$

By Corollary 6.18 in Section 6.4 (see page 398) this is sufficient to ensure that $Q^t = Q^{-1}$. So Q is an orthogonal matrix. ■

The next definition provides the basis for many of the techniques for determining the eigenvalues of a matrix.

Definition 9.11 Two matrices A and B are said to be **similar** if a nonsingular matrix S exists with $A = S^{-1}BS$. ■

An important feature of similar matrices is that they have the same eigenvalues.

Theorem 9.12 Suppose A and B are similar matrices with $A = S^{-1}BS$ and λ is an eigenvalue of A with associated eigenvector $\mathbf{x}$. Then λ is an eigenvalue of B with associated eigenvector $S\mathbf{x}$. ■

Proof Let $\mathbf{x} \neq \mathbf{0}$ be such that

$$S^{-1}BS\mathbf{x} = A\mathbf{x} = \lambda\mathbf{x}.$$

Multiplying on the left by the matrix S gives

$$BS\mathbf{x} = \lambda S\mathbf{x}.$$

Since $\mathbf{x} \neq \mathbf{0}$ and S is nonsingular, $S\mathbf{x} \neq \mathbf{0}$. Hence, $S\mathbf{x}$ is an eigenvector of B corresponding to its eigenvalue λ. ■ ■ ■

The Maple command *IsSimilar*(A, B) in the *LinearAlgebra* package returns *true* if A and B are similar and *false* otherwise.

A particularly important use of similarity occurs when an $n \times n$ matrix A is similar to diagonal matrix. That is, when a diagonal matrix D and an invertible matrix S exists with

$$A = S^{-1}DS \quad \text{or equivalently} \quad D = SAS^{-1}.$$

In this case the matrix A is said to be *diagonalizable*. The following result is considered in Exercise 19.

Theorem 9.13 An $n \times n$ matrix A is similar to a diagonal matrix D if and only if A has n linearly independent eigenvectors. In this case, $D = S^{-1}AS$, where the columns of S consist of the eigenvectors, and the ith diagonal element of D is the eigenvalue of A that corresponds to the ith column of S. ■

The pair of matrices S and D is not unique. For example, any reordering of the columns of S and corresponding reordering of the diagonal elements of D will give a distinct pair. See Exercise 15 for an illustration.

We saw in Theorem 9.3 that the eigenvectors of a matrix that correspond to distinct eigenvalues form a linearly independent set. As a consequence we have the follow Corollary to Theorem 9.13.

Corollary 9.14 An $n \times n$ matrix A that has n distinct eigenvalues is similar to a diagonal matrix. ■

In fact, we do not need the similarity matrix to be diagonal for this concept to be useful. Suppose that A is similar to a triangular matrix B. The determination of eigenvalues is easy for a triangular matrix B, for in this case λ is a solution to the equation

$$0 = \det(B - \lambda I) = \prod_{i=1}^{n}(b_{ii} - \lambda)$$

if and only if $\lambda = b_{ii}$ for some i. The next result describes a relationship, called a **similarity transformation**, between arbitrary matrices and triangular matrices.

Theorem 9.15 **(Schur)**

Let A be an arbitrary matrix. A nonsingular matrix U exists with the property that

$$T = U^{-1}AU,$$

where T is an upper-triangular matrix whose diagonal entries consist of the eigenvalues of A. ▪

Issai Schur (1875–1941) is mainly known for his work in group theory but he also worked in number theory, analysis, and other areas. He published what is now known as Schur's Theorem in 1909.

The l_2 norm of a unitary matrix is 1.

The matrix U whose existence is ensured in Theorem 9.15 satisfies the condition $\|U\mathbf{x}\|_2 = \|\mathbf{x}\|_2$ for any vector $\mathbf{x}$. Matrices with this property are called **unitary**. Although we will not make use of this norm-preserving property, it does significantly increase the application of Schur's Theorem.

Theorem 9.15 is an existence theorem that ensures that the triangular matrix T exists, but it does not provide a constructive means for finding T, since it requires a knowledge of the eigenvalues of A. In most instances, the similarity transformation U is too difficult to determine.

The following result for symmetric matrices reduces the complication, because in this case the transformation matrix is orthogonal.

Theorem 9.16 The $n \times n$ matrix A is symmetric if and only if there exists a diagonal matrix D and an orthogonal matrix Q with $A = QDQ^t$. ▪

Proof First suppose that $A = QDQ^t$, where Q is orthogonal and D is diagonal. Then

$$A^t = \left(QDQ^t\right)^t = \left(Q^t\right)^t DQ^t = QDQ^t = A,$$

and A is symmetric.

To prove that every symmetric matrix A can be written in the form $A = QDQ^t$, first consider the distinct eigenvalues of A. If $A\mathbf{v}_1 = \lambda_1\mathbf{v}_1$ and $A\mathbf{v}_2 = \lambda_2\mathbf{v}_2$, with $\lambda_1 \neq \lambda_2$, then since $A^t = A$ we have

$$(\lambda_1 - \lambda_2)\mathbf{v}_1^t\mathbf{v}_2 = (\lambda_1\mathbf{v}_1)^t\mathbf{v}_2 - \mathbf{v}_1^t(\lambda_2\mathbf{v}_2) = (A\mathbf{v}_1)^t\mathbf{v}_2 - \mathbf{v}_1^t(A\mathbf{v}_2) = \mathbf{v}_1^tA^t\mathbf{v}_2 - \mathbf{v}_1^tA\mathbf{v}_2 = 0,$$

so $\mathbf{v}_1^t\mathbf{v}_2 = 0$. Hence we can choose orthonormal vectors for distinct eigenvalues by simply normalizing all these orthogonal eigenvectors. When the eigenvalues are not distinct, there will be subspaces of eigenvectors for each of the multiple eigenvalues, and with the help of the Gram-Schmidt orthogonalization process we can find a full set of n orthonormal eigenvectors. ▪ ▪ ▪

The following corollaries to Theorem 9.16 demonstrate some of the interesting properties of symmetric matrices.

Corollary 9.17 Suppose that A is a symmetric $n \times n$ matrix. There exist n eigenvectors of A that form an orthonormal set, and the eigenvalues of A are real numbers. ▪

Proof If $Q = (q_{ij})$ and $D = (d_{ij})$ are the matrices specified in Theorem 9.16, then

$$D = Q^tAQ = Q^{-1}AQ \quad \text{implies that} \quad AQ = QD.$$

Let $1 \leq i \leq n$ and $\mathbf{v}_i = (q_{1i}, q_{2i}, \ldots, q_{ni})^t$ be the ith column of Q. Then

$$A\mathbf{v}_i = d_{ii}\mathbf{v}_i,$$

and d_{ii} is an eigenvalue of A with eigenvector, $\mathbf{v}_i$, the ith column of Q. The columns of Q are orthonormal, so the eigenvectors of A are orthonormal.

Multiplying this equation on the left by $\mathbf{v}_i^t$ gives

$$\mathbf{v}_i^t A\mathbf{v}_i = d_{ii}\mathbf{v}_i^t\mathbf{v}_i.$$

Since $\mathbf{v}_i^t A\mathbf{v}_i$ and $\mathbf{v}_i^t\mathbf{v}_i$ are real numbers and $\mathbf{v}_i^t\mathbf{v}_i = 1$, the eigenvalue $d_{ii} = \mathbf{v}_i^t A\mathbf{v}_i$ is a real number, for each $i = 1, 2, \ldots, n$. ■ ■ ■

A symmetric matrix whose eigenvalues are all nonnegative real numbers is sometimes called *nonnegative definite* (or *positive semidefinite*).

Recall from Section 6.6 that a symmetric matrix A is called positive definite if for all nonzero vectors $\mathbf{x}$ we have $\mathbf{x}^t A\mathbf{x} > 0$. The following theorem characterizes positive definite matrices in terms of eigenvalues. This eigenvalue property makes positive definite matrices important in applications.

Theorem 9.18 A symmetric matrix A is positive definite if and only if all the eigenvalues of A are positive. ■

Proof First suppose that A is positive definite and that λ is an eigenvalue of A with associated eigenvector $\mathbf{x}$, with $||\mathbf{x}||_2 = 1$. Then

$$0 < \mathbf{x}^t A\mathbf{x} = \lambda\mathbf{x}^t\mathbf{x} = \lambda\|\mathbf{x}\|_2^2 = \lambda.$$

To show the converse, suppose that A is symmetric with positive eigenvalues. By Corollary 9.17, A has n eigenvectors, $\mathbf{v}^{(1)}, \mathbf{v}^{(2)}, \ldots, \mathbf{v}^{(n)}$, that form an orthonormal and, by Theorem 9.7, linearly independent set. Hence, for any $\mathbf{x} \neq \mathbf{0}$ there exists a unique set of nonzero constants $\beta_1, \beta_2, \ldots, \beta_n$ for which

$$\mathbf{x} = \sum_{i=1}^{n} \beta_i \mathbf{v}^{(i)}.$$

Multiplying by $\mathbf{x}^t A$ gives

$$\mathbf{x}^t A\mathbf{x} = \mathbf{x}^t\left(\sum_{i=1}^{n}\beta_i A\mathbf{v}^{(i)}\right) = \mathbf{x}^t\left(\sum_{i=1}^{n}\beta_i\lambda_i\mathbf{v}^{(i)}\right) = \sum_{j=1}^{n}\sum_{i=1}^{n}\beta_j\beta_i\lambda_i(\mathbf{v}^{(j)})^t\mathbf{v}^{(i)}.$$

But the vectors $\mathbf{v}^{(1)}, \mathbf{v}^{(2)}, \ldots, \mathbf{v}^{(n)}$ form an orthonormal set, so

$$(\mathbf{v}^{(j)})^t\mathbf{v}^{(i)} = \begin{cases} 0, & \text{if } i \neq j, \\ 1, & \text{if } i = j. \end{cases}$$

This, together with the fact that the λ_i are all positive, implies that

$$\mathbf{x}^t A\mathbf{x} = \sum_{j=1}^{n}\sum_{i=1}^{n}\beta_j\beta_i\lambda_i(\mathbf{v}^{(j)})^t\mathbf{v}^{(i)} = \sum_{i=1}^{n}\lambda_i\beta_i^2 > 0.$$

Hence, A is positive definite. ■ ■ ■

EXERCISE SET 9.2

1. Show that the following pairs of matrices are not similar.

   a. $A = \begin{bmatrix} 2 & 1 \\ 1 & 2 \end{bmatrix}$ and $B = \begin{bmatrix} 1 & 2 \\ 2 & 1 \end{bmatrix}$

   b. $A = \begin{bmatrix} 2 & 0 \\ 1 & 3 \end{bmatrix}$ and $B = \begin{bmatrix} 4 & -1 \\ -2 & 2 \end{bmatrix}$

c. $A = \begin{bmatrix} 1 & 2 & 1 \\ 0 & 1 & 2 \\ 0 & 0 & 2 \end{bmatrix}$ and $B = \begin{bmatrix} 1 & 2 & 0 \\ 0 & 1 & 2 \\ 1 & 0 & 2 \end{bmatrix}$

d. $A = \begin{bmatrix} 1 & 2 & 1 \\ -3 & 2 & 2 \\ 0 & 1 & 2 \end{bmatrix}$ and $B = \begin{bmatrix} 1 & 2 & 1 \\ 0 & 1 & 2 \\ -3 & 2 & 2 \end{bmatrix}$

2. Show that the following pairs of matrices are not similar.

a. $A = \begin{bmatrix} 1 & 1 \\ 0 & 3 \end{bmatrix}$ and $B = \begin{bmatrix} 2 & 2 \\ 1 & 2 \end{bmatrix}$

b. $A = \begin{bmatrix} 1 & 1 \\ 2 & -2 \end{bmatrix}$ and $B = \begin{bmatrix} -1 & 2 \\ 1 & 2 \end{bmatrix}$

c. $A = \begin{bmatrix} 1 & 1 & -1 \\ -1 & 0 & 1 \\ 0 & 1 & 1 \end{bmatrix}$ and $B = \begin{bmatrix} 2 & -2 & 0 \\ -2 & 0 & 2 \\ 2 & 2 & -2 \end{bmatrix}$

d. $A = \begin{bmatrix} 1 & 1 & -1 \\ 2 & -2 & 2 \\ -3 & 3 & 3 \end{bmatrix}$ and $B = \begin{bmatrix} 1 & 2 & 1 \\ 2 & 3 & 2 \\ 0 & 1 & 0 \end{bmatrix}$

3. Define $A = PDP^{-1}$ for the following matrices D and P. Determine A^3.

a. $P = \begin{bmatrix} 2 & -1 \\ 3 & 1 \end{bmatrix}$ and $D = \begin{bmatrix} 1 & 0 \\ 0 & 2 \end{bmatrix}$

b. $P = \begin{bmatrix} -1 & 2 \\ 1 & 0 \end{bmatrix}$ and $D = \begin{bmatrix} -2 & 0 \\ 0 & 1 \end{bmatrix}$

c. $P = \begin{bmatrix} 1 & 2 & -1 \\ 2 & 1 & 0 \\ 1 & 0 & 2 \end{bmatrix}$ and $D = \begin{bmatrix} 0 & 0 & 0 \\ 0 & 1 & 0 \\ 0 & 0 & -1 \end{bmatrix}$

d. $P = \begin{bmatrix} 2 & -1 & 0 \\ -1 & 2 & -1 \\ 0 & -1 & 2 \end{bmatrix}$ and $D = \begin{bmatrix} 2 & 0 & 0 \\ 0 & 2 & 0 \\ 0 & 0 & 2 \end{bmatrix}$

4. Determine A^4 for the matrices in Exercise 3.

5. For each of the following matrices determine if it diagonalizable and, if so, find P and D with $A = PDP^{-1}$.

a. $A = \begin{bmatrix} 4 & -1 \\ -4 & 1 \end{bmatrix}$

b. $A = \begin{bmatrix} 2 & -1 \\ -1 & 2 \end{bmatrix}$

c. $A = \begin{bmatrix} 2 & 0 & 1 \\ 0 & 1 & 0 \\ 1 & 0 & 2 \end{bmatrix}$

d. $A = \begin{bmatrix} 1 & 1 & 1 \\ 1 & 1 & 0 \\ 1 & 0 & 1 \end{bmatrix}$

6. For each of the following matrices determine if it diagonalizable and, if so, find P and D with $A = PDP^{-1}$.

a. $A = \begin{bmatrix} 2 & 1 \\ 0 & 1 \end{bmatrix}$

b. $A = \begin{bmatrix} 2 & 1 \\ 1 & 2 \end{bmatrix}$

c. $A = \begin{bmatrix} 2 & 1 & 1 \\ 1 & 2 & 1 \\ 1 & 1 & 2 \end{bmatrix}$

d. $A = \begin{bmatrix} 2 & 1 & 1 \\ 0 & 3 & 1 \\ 0 & 0 & 2 \end{bmatrix}$

7. **(i)** Determine if the following matrices are positive definite, and if so, **(ii)** construct an orthogonal matrix Q for which $Q^tAQ = D$, where D is a diagonal matrix.

a. $A = \begin{bmatrix} 2 & 1 \\ 1 & 2 \end{bmatrix}$

b. $A = \begin{bmatrix} 1 & 2 \\ 2 & 1 \end{bmatrix}$

c. $A = \begin{bmatrix} 2 & 0 & 1 \\ 0 & 2 & 0 \\ 1 & 0 & 2 \end{bmatrix}$

d. $A = \begin{bmatrix} 1 & 1 & 1 \\ 1 & 1 & 0 \\ 1 & 0 & 1 \end{bmatrix}$

8. **(i)** Determine if the following matrices are positive definite, and if so, **(ii)** construct an orthogonal matrix Q for which $Q^tAQ = D$, where D is a diagonal matrix.

a. $A = \begin{bmatrix} 4 & 2 & 1 \\ 2 & 4 & 0 \\ 1 & 0 & 4 \end{bmatrix}$ **b.** $A = \begin{bmatrix} 3 & 2 & 1 \\ 2 & 2 & 0 \\ 1 & 0 & 1 \end{bmatrix}$

c. $A = \begin{bmatrix} 1 & -1 & -1 & 1 \\ -1 & 2 & -1 & -2 \\ -1 & -1 & 3 & 0 \\ 1 & -2 & 0 & 4 \end{bmatrix}$ **d.** $A = \begin{bmatrix} 8 & 4 & 2 & 1 \\ 4 & 8 & 2 & 1 \\ 2 & 2 & 8 & 1 \\ 1 & 1 & 1 & 8 \end{bmatrix}$

9. Show that each of the following matrices is nonsingular but not diagonalizable.

a. $A = \begin{bmatrix} 2 & 1 & 0 \\ 0 & 2 & 0 \\ 0 & 0 & 3 \end{bmatrix}$ **b.** $A = \begin{bmatrix} 2 & -3 & 6 \\ 0 & 3 & -4 \\ 0 & 2 & -3 \end{bmatrix}$

c. $A = \begin{bmatrix} 2 & 1 & -1 \\ 0 & 2 & 1 \\ 0 & 0 & 3 \end{bmatrix}$ **d.** $A = \begin{bmatrix} 1 & 0 & 0 \\ -1 & 0 & 1 \\ -1 & -1 & 2 \end{bmatrix}$

10. Show that the following matrices are singular but are diagonalizable.

a. $A = \begin{bmatrix} 2 & -1 & 0 \\ -1 & 2 & 0 \\ 0 & 0 & 0 \end{bmatrix}$ **b.** $A = \begin{bmatrix} 2 & -1 & -1 \\ -1 & 2 & -1 \\ -1 & -1 & 2 \end{bmatrix}$

11. In Exercise 31 of Section 6.6, a symmetric matrix

$$A = \begin{bmatrix} 1.59 & 1.69 & 2.13 \\ 1.69 & 1.31 & 1.72 \\ 2.13 & 1.72 & 1.85 \end{bmatrix}$$

was used to describe the average wing lengths of fruit flies that were offspring resulting from the mating of three mutants of the flies. The entry a_{ij} represents the average wing length of a fly that is the offspring of a male fly of type i and a female fly of type j.

a. Find the eigenvalues and associated eigenvectors of this matrix.

b. Is this matrix positive definite?

12. Suppose that A and B are nonsingular $n \times n$ matrices. Prove the AB is similar to BA.

13. Show that if A is similar to B and B is similar to C, then A is similar to C.

14. Show that if A is similar to B, then

a. $\det(A) = \det(B)$.

b. The characteristic polynomial of A is the same as the characteristic polynomial of B.

c. A is nonsingular if and only if B is nonsingular.

d. If A is nonsingular, show that A^{-1} is similar to B^{-1}.

e. A^t is similar to B^t.

15. Show that the matrix given in Example 3 of Section 9.1,

$$A = \begin{bmatrix} 2 & 0 & 0 \\ 1 & 1 & 2 \\ 1 & -1 & 4 \end{bmatrix}$$

is similar to the diagonal matrices

$$D_1 = \begin{bmatrix} 3 & 0 & 0 \\ 0 & 2 & 0 \\ 0 & 0 & 2 \end{bmatrix}, \quad D_2 = \begin{bmatrix} 2 & 0 & 0 \\ 0 & 3 & 0 \\ 0 & 0 & 2 \end{bmatrix}, \quad \text{and} \quad D_3 = \begin{bmatrix} 2 & 0 & 0 \\ 0 & 2 & 0 \\ 0 & 0 & 3 \end{bmatrix}.$$

16. Prove Theorem 9.10.

17. Show that there is no diagonal matrix similar to the matrix given in Example 4 of Section 9.1,

$$B = \begin{bmatrix} 2 & 1 & 0 \\ 0 & 2 & 0 \\ 0 & 0 & 3 \end{bmatrix}.$$

18. Prove that if Q is nonsingular matrix with $Q^t = Q^{-1}$, then Q is orthogonal.

19. Prove Theorem 9.13.

9.3 The Power Method

The **Power method** is an iterative technique used to determine the dominant eigenvalue of a matrix—that is, the eigenvalue with the largest magnitude. By modifying the method slightly, it can also used to determine other eigenvalues. One useful feature of the Power method is that it produces not only an eigenvalue, but also an associated eigenvector. In fact, the Power method is often applied to find an eigenvector for an eigenvalue that is determined by some other means.

The name for the Power method is derived from the fact that the iterations exaggerate the relative size of the magnitudes of the eigenvalues.

To apply the Power method, we assume that the $n \times n$ matrix A has n eigenvalues $\lambda_1, \lambda_2, \ldots, \lambda_n$ with an associated collection of linearly independent eigenvectors $\{\mathbf{v}^{(1)}, \mathbf{v}^{(2)}, \mathbf{v}^{(3)}, \ldots, \mathbf{v}^{(n)}\}$. Moreover, we assume that A has precisely one eigenvalue, λ_1, that is largest in magnitude, so that

$$|\lambda_1| > |\lambda_2| \geq |\lambda_3| \geq \cdots \geq |\lambda_n| \geq 0.$$

Example 4 of Section 9.1 illustrates that an $n \times n$ matrix need not have n linearly independent eigenvectors. When it does not the Power method may still be successful, but it is not guaranteed to be.

If $\mathbf{x}$ is any vector in $\mathbb{R}^n$, the fact that $\{\mathbf{v}^{(1)}, \mathbf{v}^{(2)}, \mathbf{v}^{(3)}, \ldots, \mathbf{v}^{(n)}\}$ is linearly independent implies that constants $\beta_1, \beta_2, \ldots, \beta_n$ exist with

$$\mathbf{x} = \sum_{j=1}^{n} \beta_j \mathbf{v}^{(j)}.$$

Multiplying both sides of this equation by $A, A^2, \ldots, A^k, \ldots$ gives

$$A\mathbf{x} = \sum_{j=1}^{n} \beta_j A\mathbf{v}^{(j)} = \sum_{j=1}^{n} \beta_j \lambda_j \mathbf{v}^{(j)}, \quad A^2\mathbf{x} = \sum_{j=1}^{n} \beta_j \lambda_j A\mathbf{v}^{(j)} = \sum_{j=1}^{n} \beta_j \lambda_j^2 \mathbf{v}^{(j)},$$

and generally, $A^k\mathbf{x} = \sum_{j=1}^{n} \beta_j \lambda_j^k \mathbf{v}^{(j)}$.

If λ_1^k is factored from each term on the right side of the last equation, then

$$A^k\mathbf{x} = \lambda_1^k \sum_{j=1}^{n} \beta_j \left(\frac{\lambda_j}{\lambda_1}\right)^k \mathbf{v}^{(j)}.$$

Since $|\lambda_1| > |\lambda_j|$, for all $j = 2, 3, \ldots, n$, we have $\lim_{k\to\infty}(\lambda_j/\lambda_1)^k = 0$, and

$$\lim_{k\to\infty} A^k\mathbf{x} = \lim_{k\to\infty} \lambda_1^k \beta_1 \mathbf{v}^{(1)}. \tag{9.2}$$

The sequence in Eq. (9.2) converges to $\mathbf{0}$ if $|\lambda_1| < 1$ and diverges if $|\lambda_1| > 1$, provided, of course, that $\beta_1 \neq 0$. As a consequence, the entries in the $A^k\mathbf{x}$ will grow with k if $|\lambda_1| > 1$ and will go to 0 if $|\lambda_1| < 1$, perhaps resulting in overflow or underflow. To take care of that possibility, we scale the powers of $A^k\mathbf{x}$ in an appropriate manner to ensure that the limit in Eq. (9.2) is finite and nonzero. The scaling begins by choosing $\mathbf{x}$ to be a unit vector $\mathbf{x}^{(0)}$ relative to $\|\cdot\|_\infty$ and choosing a component $x_{p_0}^{(0)}$ of $\mathbf{x}^{(0)}$ with

$$x_{p_0}^{(0)} = 1 = \|\mathbf{x}^{(0)}\|_\infty.$$

Let $\mathbf{y}^{(1)} = A\mathbf{x}^{(0)}$, and define $\mu^{(1)} = y_{p_0}^{(1)}$. Then

$$\mu^{(1)} = y_{p_0}^{(1)} = \frac{y_{p_0}^{(1)}}{x_{p_0}^{(0)}} = \frac{\beta_1\lambda_1 v_{p_0}^{(1)} + \sum_{j=2}^n \beta_j\lambda_j v_{p_0}^{(j)}}{\beta_1 v_{p_0}^{(1)} + \sum_{j=2}^n \beta_j v_{p_0}^{(j)}} = \lambda_1\left[\frac{\beta_1 v_{p_0}^{(1)} + \sum_{j=2}^n \beta_j(\lambda_j/\lambda_1)v_{p_0}^{(j)}}{\beta_1 v_{p_0}^{(1)} + \sum_{j=2}^n \beta_j v_{p_0}^{(j)}}\right].$$

Let p_1 be the least integer such that

$$|y_{p_1}^{(1)}| = \|\mathbf{y}^{(1)}\|_\infty,$$

and define $\mathbf{x}^{(1)}$ by

$$\mathbf{x}^{(1)} = \frac{1}{y_{p_1}^{(1)}}\mathbf{y}^{(1)} = \frac{1}{y_{p_1}^{(1)}}A\mathbf{x}^{(0)}.$$

Then

$$x_{p_1}^{(1)} = 1 = \|\mathbf{x}^{(1)}\|_\infty.$$

Now define

$$\mathbf{y}^{(2)} = A\mathbf{x}^{(1)} = \frac{1}{y_{p_1}^{(1)}}A^2\mathbf{x}^{(0)}$$

and

$$\mu^{(2)} = y_{p_1}^{(2)} = \frac{y_{p_1}^{(2)}}{x_{p_1}^{(1)}} = \frac{\left[\beta_1\lambda_1^2 v_{p_1}^{(1)} + \sum_{j=2}^n \beta_j\lambda_j^2 v_{p_1}^{(j)}\right]\Big/ y_{p_1}^{(1)}}{\left[\beta_1\lambda_1 v_{p_1}^{(1)} + \sum_{j=2}^n \beta_j\lambda_j v_{p_1}^{(j)}\right]\Big/ y_{p_1}^{(1)}}$$

$$= \lambda_1\left[\frac{\beta_1 v_{p_1}^{(1)} + \sum_{j=2}^n \beta_j(\lambda_j/\lambda_1)^2 v_{p_1}^{(j)}}{\beta_1 v_{p_1}^{(1)} + \sum_{j=2}^n \beta_j(\lambda_j/\lambda_1)v_{p_1}^{(j)}}\right].$$

Let p_2 be the smallest integer with

$$|y_{p_2}^{(2)}| = \|\mathbf{y}^{(2)}\|_\infty,$$

and define

$$\mathbf{x}^{(2)} = \frac{1}{y_{p_2}^{(2)}}\mathbf{y}^{(2)} = \frac{1}{y_{p_2}^{(2)}}A\mathbf{x}^{(1)} = \frac{1}{y_{p_2}^{(2)}y_{p_1}^{(1)}}A^2\mathbf{x}^{(0)}.$$

In a similar manner, define sequences of vectors $\{\mathbf{x}^{(m)}\}_{m=0}^\infty$ and $\{\mathbf{y}^{(m)}\}_{m=1}^\infty$, and a sequence of scalars $\{\mu^{(m)}\}_{m=1}^\infty$ inductively by

$$\mathbf{y}^{(m)} = A\mathbf{x}^{(m-1)},$$

$$\mu^{(m)} = y_{p_{m-1}}^{(m)} = \lambda_1\left[\frac{\beta_1 v_{p_{m-1}}^{(1)} + \sum_{j=2}^n (\lambda_j/\lambda_1)^m \beta_j v_{p_{m-1}}^{(j)}}{\beta_1 v_{p_{m-1}}^{(1)} + \sum_{j=2}^n (\lambda_j/\lambda_1)^{m-1}\beta_j v_{p_{m-1}}^{(j)}}\right], \tag{9.3}$$

and

$$\mathbf{x}^{(m)} = \frac{\mathbf{y}^{(m)}}{y_{p_m}^{(m)}} = \frac{A^m\mathbf{x}^{(0)}}{\prod_{k=1}^m y_{p_k}^{(k)}},$$

where at each step, p_m is used to represent the smallest integer for which

$$|y_{p_m}^{(m)}| = \|\mathbf{y}^{(m)}\|_\infty.$$

By examining Eq. (9.3), we see that since $|\lambda_j/\lambda_1| < 1$, for each $j = 2, 3, \ldots, n$, $\lim_{m\to\infty} \mu^{(m)} = \lambda_1$, provided that $\mathbf{x}^{(0)}$ is chosen so that $\beta_1 \neq 0$. Moreover, the sequence of vectors $\{\mathbf{x}^{(m)}\}_{m=0}^{\infty}$ converges to an eigenvector associated with λ_1 that has l_∞ norm equal to one.

Illustration The matrix

$$A = \begin{bmatrix} -2 & -3 \\ 6 & 7 \end{bmatrix}$$

Has eigenvalues $\lambda_1 = 4$ and $\lambda_2 = 1$ with corresponding eigenvectors $\mathbf{v}_1 = (1, -2)^t$ and $\mathbf{v}_2 = (1, -1)^t$. If we start with the arbitrary vector $\mathbf{x}_0 = (1, 1)^t$ and multiply by the matrix A we obtain

$$\mathbf{x}_1 = A\mathbf{x}_0 = \begin{bmatrix} -5 \\ 13 \end{bmatrix}, \quad \mathbf{x}_2 = A\mathbf{x}_1 = \begin{bmatrix} -29 \\ 61 \end{bmatrix}, \quad \mathbf{x}_3 = A\mathbf{x}_2 = \begin{bmatrix} -125 \\ 253 \end{bmatrix},$$

$$\mathbf{x}_4 = A\mathbf{x}_3 = \begin{bmatrix} -509 \\ 1021 \end{bmatrix}, \quad \mathbf{x}_5 = A\mathbf{x}_4 = \begin{bmatrix} -2045 \\ 4093 \end{bmatrix}, \quad \mathbf{x}_6 = A\mathbf{x}_5 = \begin{bmatrix} -8189 \\ 16381 \end{bmatrix}.$$

As a consequence, approximations to the dominant eigenvalue $\lambda_1 = 4$ are

$$\lambda_1^{(1)} = \frac{61}{13} = 4.6923, \quad \lambda_1^{(2)} = \frac{253}{61} = 4.14754, \quad \lambda_1^{(3)} = \frac{1021}{253} = 4.03557,$$

$$\lambda_1^{(4)} = \frac{4093}{1021} = 4.00881, \quad \lambda_1^{(5)} = \frac{16381}{4093} = 4.00200.$$

An approximate eigenvector corresponding to $\lambda_1^{(5)} = \dfrac{16381}{4093} = 4.00200$ is

$$\mathbf{x}_6 = \begin{bmatrix} -8189 \\ 16381 \end{bmatrix}, \quad \text{which, divided by 16381, normalizes to} \quad \begin{bmatrix} -0.49908 \\ 1 \end{bmatrix} \approx \mathbf{v}_1.$$

□

The Power method has the disadvantage that it is unknown at the outset whether or not the matrix has a single dominant eigenvalue. Nor is it known how $\mathbf{x}^{(0)}$ should be chosen so as to ensure that its representation in terms of the eigenvectors of the matrix will contain a nonzero contribution from the eigenvector associated with the dominant eigenvalue, should it exist.

Algorithm 9.1 implements the Power method.

ALGORITHM 9.1

Power Method

To approximate the dominant eigenvalue and an associated eigenvector of the $n \times n$ matrix A given a nonzero vector $\mathbf{x}$:

INPUT dimension n; matrix A; vector $\mathbf{x}$; tolerance TOL; maximum number of iterations N.

OUTPUT approximate eigenvalue μ; approximate eigenvector $\mathbf{x}$ (with $||\mathbf{x}||_\infty = 1$) or a message that the maximum number of iterations was exceeded.

Step 1 Set $k = 1$.

Step 2 Find the smallest integer p with $1 \leq p \leq n$ and $|x_p| = ||\mathbf{x}||_\infty$.

Step 3 Set $\mathbf{x} = \mathbf{x}/x_p$.

Step 4 While ($k \leq N$) do Steps 5–11.

Step 5 Set $\mathbf{y} = A\mathbf{x}$.

Step 6 Set $\mu = y_p$.

Step 7 Find the smallest integer p with $1 \leq p \leq n$ and $|y_p| = \|\mathbf{y}\|_\infty$.

Step 8 If $y_p = 0$ then OUTPUT ('Eigenvector', $\mathbf{x}$);
OUTPUT ('A has the eigenvalue 0, select a new vector $\mathbf{x}$ and restart');
STOP.

Step 9 Set $ERR = ||\mathbf{x} - (\mathbf{y}/y_p)||_\infty$;
$\mathbf{x} = \mathbf{y}/y_p$.

Step 10 If $ERR < TOL$ then OUTPUT ($\mu, \mathbf{x}$);
(*The procedure was successful.*)
STOP.

Step 11 Set $k = k + 1$.

Step 12 OUTPUT ('The maximum number of iterations exceeded');
(*The procedure was unsuccessful.*)
STOP. ■

Accelerating Convergence

Choosing, in Step 7, the smallest integer p_m for which $|y_{p_m}^{(m)}| = \|\mathbf{y}^{(m)}\|_\infty$ will generally ensure that this index eventually becomes invariant. The rate at which $\{\mu^{(m)}\}_{m=1}^{\infty}$ converges to λ_1 is determined by the ratios $|\lambda_j/\lambda_1|^m$, for $j = 2, 3, \ldots, n$, and in particular by $|\lambda_2/\lambda_1|^m$. The rate of convergence is $O(|\lambda_2/\lambda_1|^m)$ (see [IK, p. 148]), so there is a constant k such that for large m,

$$|\mu^{(m)} - \lambda_1| \approx k \left|\frac{\lambda_2}{\lambda_1}\right|^m,$$

which implies that

$$\lim_{m\to\infty} \frac{|\mu^{(m+1)} - \lambda_1|}{|\mu^{(m)} - \lambda_1|} \approx \left|\frac{\lambda_2}{\lambda_1}\right| < 1.$$

The sequence $\{\mu^{(m)}\}$ converges linearly to λ_1, so Aitken's Δ^2 procedure discussed in Section 2.5 can be used to speed the convergence. Implementing the Δ^2 procedure in Algorithm 9.1 is accomplished by modifying the algorithm as follows:

Step 1 Set $k = 1$;
$\mu_0 = 0$;
$\mu_1 = 0$.

Step 6 Set $\mu = y_p$;

$$\hat{\mu} = \mu_0 - \frac{(\mu_1 - \mu_0)^2}{\mu - 2\mu_1 + \mu_0}.$$

Step 10 If $ERR < TOL$ and $k \geq 4$ then OUTPUT $(\hat{\mu}, \mathbf{x})$;
STOP.

Step 11 Set $k = k + 1$;
$\mu_0 = \mu_1$;
$\mu_1 = \mu$.

In actuality, it is not necessary for the matrix to have distinct eigenvalues for the Power method to converge. If the matrix has a unique dominant eigenvalue, λ_1, with multiplicity r greater than 1 and $\mathbf{v}^{(1)}, \mathbf{v}^{(2)}, \ldots, \mathbf{v}^{(r)}$ are linearly independent eigenvectors associated with λ_1, the procedure will still converge to λ_1. The sequence of vectors $\{\mathbf{x}^{(m)}\}_{m=0}^{\infty}$ will, in this case, converge to an eigenvector of λ_1 of l_∞ norm equal to one that depends on the choice of the initial vector $\mathbf{x}^{(0)}$ and is a linear combination of $\mathbf{v}^{(1)}, \mathbf{v}^{(2)}, \ldots, \mathbf{v}^{(r)}$. (See [Wil2], page 570.)

Example 1 Use the Power method to approximate the dominant eigenvalue of the matrix

$$A = \begin{bmatrix} -4 & 14 & 0 \\ -5 & 13 & 0 \\ -1 & 0 & 2 \end{bmatrix},$$

and then apply Aitken's Δ^2 method to the approximations to the eigenvalue of the matrix to accelerate the convergence.

Solution This matrix has eigenvalues $\lambda_1 = 6, \lambda_2 = 3$, and $\lambda_3 = 2$, so the Power method described in Algorithm 9.1 will converge. Let $\mathbf{x}^{(0)} = (1, 1, 1)^t$, then

$$\mathbf{y}^{(1)} = A\mathbf{x}^{(0)} = (10, 8, 1)^t,$$

so

$$||\mathbf{y}^{(1)}||_\infty = 10, \quad \mu^{(1)} = y_1^{(1)} = 10, \quad \text{and} \quad \mathbf{x}^{(1)} = \frac{\mathbf{y}^{(1)}}{10} = (1, 0.8, 0.1)^t.$$

Continuing in this manner leads to the values in Table 9.1, where $\hat{\mu}^{(m)}$ represents the sequence generated by the Aitken's Δ^2 procedure. An approximation to the dominant

Table 9.1

m	$(\mathbf{x}^{(m)})^t$	$\mu^{(m)}$	$\hat{\mu}^{(m)}$
0	$(1, 1, 1)$		
1	$(1, 0.8, 0.1)$	10	6.266667
2	$(1, 0.75, -0.111)$	7.2	6.062473
3	$(1, 0.730769, -0.188803)$	6.5	6.015054
4	$(1, 0.722200, -0.220850)$	6.230769	6.004202
5	$(1, 0.718182, -0.235915)$	6.111000	6.000855
6	$(1, 0.716216, -0.243095)$	6.054546	6.000240
7	$(1, 0.715247, -0.246588)$	6.027027	6.000058
8	$(1, 0.714765, -0.248306)$	6.013453	6.000017
9	$(1, 0.714525, -0.249157)$	6.006711	6.000003
10	$(1, 0.714405, -0.249579)$	6.003352	6.000000
11	$(1, 0.714346, -0.249790)$	6.001675	
12	$(1, 0.714316, -0.249895)$	6.000837	

eigenvalue, 6, at this stage is $\hat{\mu}^{(10)} = 6.000000$. The approximate l_∞-unit eigenvector for the eigenvalue 6 is $(\mathbf{x}^{(12)})^t = (1, 0.714316, -0.249895)^t$.

Although the approximation to the eigenvalue is correct to the places listed, the eigenvector approximation is considerably less accurate to the true eigenvector, $(1, 5/7, -1/4)^t \approx (1, 0.714286, -0.25)^t$. ■

Symmetric Matrices

When A is symmetric, a variation in the choice of the vectors $\mathbf{x}^{(m)}$ and $\mathbf{y}^{(m)}$ and the scalars $\mu^{(m)}$ can be made to significantly improve the rate of convergence of the sequence $\{\mu^{(m)}\}_{m=1}^{\infty}$ to the dominant eigenvalue λ_1. In fact, although the rate of convergence of the general Power method is $O(|\lambda_2/\lambda_1|^m)$, the rate of convergence of the modified procedure given in Algorithm 9.2 for symmetric matrices is $O(|\lambda_2/\lambda_1|^{2m})$. (See [IK, pp. 149 ff].) Because the sequence $\{\mu^{(m)}\}$ is still linearly convergent, Aitken's Δ^2 procedure can also be applied.

Symmetric Power Method

To approximate the dominant eigenvalue and an associated eigenvector of the $n \times n$ symmetric matrix A, given a nonzero vector $\mathbf{x}$:

INPUT dimension n; matrix A; vector $\mathbf{x}$; tolerance TOL; maximum number of iterations N.

OUTPUT approximate eigenvalue μ; approximate eigenvector $\mathbf{x}$ (with $\|\mathbf{x}\|_2 = 1$) or a message that the maximum number of iterations was exceeded.

Step 1 Set $k = 1$;
$\mathbf{x} = \mathbf{x}/\|\mathbf{x}\|_2$.

Step 2 While ($k \le N$) do Steps 3–8.

Step 3 Set $\mathbf{y} = A\mathbf{x}$.

Step 4 Set $\mu = \mathbf{x}^t\mathbf{y}$.

Step 5 If $\|\mathbf{y}\|_2 = 0$, then OUTPUT ('Eigenvector', $\mathbf{x}$);
OUTPUT ('A has eigenvalue 0, select new vector $\mathbf{x}$ and restart');
STOP.

Step 6 Set $ERR = \left\| \mathbf{x} - \dfrac{\mathbf{y}}{\|\mathbf{y}\|_2} \right\|_2$;
$\mathbf{x} = \mathbf{y}/\|\mathbf{y}\|_2$.

Step 7 If $ERR < TOL$ then OUTPUT ($\mu, \mathbf{x}$);
(*The procedure was successful.*)
STOP.

Step 8 Set $k = k + 1$.

Step 9 OUTPUT ('Maximum number of iterations exceeded');
(*The procedure was unsuccessful.*)
STOP. ■

Example 2 Apply both the Power method and the Symmetric Power method to the matrix

$$A = \begin{bmatrix} 4 & -1 & 1 \\ -1 & 3 & -2 \\ 1 & -2 & 3 \end{bmatrix},$$

using Aitken's Δ^2 method to accelerate the convergence.

Solution This matrix has eigenvalues $\lambda_1 = 6, \lambda_2 = 3$, and $\lambda_3 = 1$. An eigenvector for the eigenvalue 6 is $(1, -1, 1)^t$. Applying the Power method to this matrix with initial vector $(1, 0, 0)^t$ gives the values in Table 9.2. ▪

Table 9.2

m	$(\mathbf{y}^{(m)})^t$	$\mu^{(m)}$	$\hat{\mu}^{(m)}$	$(\mathbf{x}^{(m)})^t$ with $\\|\mathbf{x}^{(m)}\\|_\infty = 1$
0				$(1, 0, 0)$
1	$(4, -1, 1)$	4		$(1, -0.25, 0.25)$
2	$(4.5, -2.25, 2.25)$	4.5	7	$(1, -0.5, 0.5)$
3	$(5, -3.5, 3.5)$	5	6.2	$(1, -0.7, 0.7)$
4	$(5.4, -4.5, 4.5)$	5.4	6.047617	$(1, -0.833\bar{3}, 0.833\bar{3})$
5	$(5.66\bar{6}, -5.166\bar{6}, 5.166\bar{6})$	$5.66\bar{6}$	6.011767	$(1, -0.911765, 0.911765)$
6	$(5.823529, -5.558824, 5.558824)$	5.823529	6.002931	$(1, -0.954545, 0.954545)$
7	$(5.909091, -5.772727, 5.772727)$	5.909091	6.000733	$(1, -0.976923, 0.976923)$
8	$(5.953846, -5.884615, 5.884615)$	5.953846	6.000184	$(1, -0.988372, 0.988372)$
9	$(5.976744, -5.941861, 5.941861)$	5.976744		$(1, -0.994163, 0.994163)$
10	$(5.988327, -5.970817, 5.970817)$	5.988327		$(1, -0.997076, 0.997076)$

We will now apply the Symmetric Power method to this matrix with the same initial vector $(1, 0, 0)^t$. The first steps are

$$\mathbf{x}^{(0)} = (1, 0, 0)^t, \quad A\mathbf{x}^{(0)} = (4, -1, 1)^t, \mu^{(1)} = 4,$$

and

$$\mathbf{x}^{(1)} = \frac{1}{||A\mathbf{x}^{(0)}||_2} \cdot A\mathbf{x}^{(0)} = (0.942809, -0.235702, 0.235702)^t.$$

The remaining entries are shown in Table 9.3.

Table 9.3

m	$(\mathbf{y}^{(m)})^t$	$\mu^{(m)}$	$\hat{\mu}^{(m)}$	$(\mathbf{x}^{(m)})^t$ with $\\|\mathbf{x}^{(m)}\\|_2 = 1$
0	$(1, 0, 0)$			$(1, 0, 0)$
1	$(4, -1, 1)$	4	7	$(0.942809, -0.235702, 0.235702)$
2	$(4.242641, -2.121320, 2.121320$	5	6.047619	$(0.816497, -0.408248, 0.408248)$
3	$(4.082483, -2.857738, 2.857738)$	5.666667	6.002932	$(0.710669, -0.497468, 0.497468)$
4	$(3.837613, -3.198011, 3.198011)$	5.909091	6.000183	$(0.646997, -0.539164, 0.539164)$
5	$(3.666314, -3.342816, 3.342816)$	5.976744	6.000012	$(0.612836, -0.558763, 0.558763)$
6	$(3.568871, -3.406650, 3.406650)$	5.994152	6.000000	$(0.595247, -0.568190, 0.568190)$
7	$(3.517370, -3.436200, 3.436200)$	5.998536	6.000000	$(0.586336, -0.572805, 0.572805)$
8	$(3.490952, -3.450359, 3.450359)$	5.999634		$(0.581852, -0.575086, 0.575086)$
9	$(3.477580, -3.457283, 3.457283)$	5.999908		$(0.579603, -0.576220, 0.576220)$
10	$(3.470854, -3.460706, 3.460706)$	5.999977		$(0.578477, -0.576786, 0.576786)$

The Symmetric Power method gives considerably faster convergence for this matrix than the Power method. The eigenvector approximations in the Power method converge to $(1, -1, 1)^t$, a vector with unit l_∞-norm. In the Symmetric Power method, the convergence is to the parallel vector $(\sqrt{3}/3, -\sqrt{3}/3, \sqrt{3}/3)^t$, which has unit l_2-norm.

If λ is a real number that approximates an eigenvalue of a symmetric matrix A and $\mathbf{x}$ is an associated approximate eigenvector, then $A\mathbf{x} - \lambda\mathbf{x}$ is approximately the zero vector. The following theorem relates the norm of this vector to the accuracy of λ to the eigenvalue.

Theorem 9.19 Suppose that A is an $n \times n$ symmetric matrix with eigenvalues $\lambda_1, \lambda_2, \ldots, \lambda_n$. If we have $\|A\mathbf{x} - \lambda\mathbf{x}\|_2 < \varepsilon$ for some real number λ and vector $\mathbf{x}$ with $\|\mathbf{x}\|_2 = 1$, then

$$\min_{1 \le j \le n} |\lambda_j - \lambda| < \varepsilon. \qquad \blacksquare$$

Proof Suppose that $\mathbf{v}^{(1)}, \mathbf{v}^{(2)}, \ldots, \mathbf{v}^{(n)}$ form an orthonormal set of eigenvectors of A associated, respectively, with the eigenvalues $\lambda_1, \lambda_2, \ldots, \lambda_n$. By Theorems 9.5 and 9.3, $\mathbf{x}$ can be expressed, for some unique set of constants $\beta_1, \beta_2, \ldots, \beta_n$, as

$$\mathbf{x} = \sum_{j=1}^{n} \beta_j \mathbf{v}^{(j)}.$$

Thus

$$\|A\mathbf{x} - \lambda\mathbf{x}\|_2^2 = \left\| \sum_{j=1}^{n} \beta_j(\lambda_j - \lambda)\mathbf{v}^{(j)} \right\|_2^2 = \sum_{j=1}^{n} |\beta_j|^2 |\lambda_j - \lambda|^2 \ge \min_{1 \le j \le n} |\lambda_j - \lambda|^2 \sum_{j=1}^{n} |\beta_j|^2.$$

But

$$\sum_{j=1}^{n} |\beta_j|^2 = \|\mathbf{x}\|_2^2 = 1, \quad \text{so} \quad \varepsilon \ge \|A\mathbf{x} - \lambda\mathbf{x}\|_2 > \min_{1 \le j \le n} |\lambda_j - \lambda|. \qquad \blacksquare\ \blacksquare\ \blacksquare$$

Inverse Power Method

The **Inverse Power method** is a modification of the Power method that gives faster convergence. It is used to determine the eigenvalue of A that is closest to a specified number q.

Suppose the matrix A has eigenvalues $\lambda_1, \ldots, \lambda_n$ with linearly independent eigenvectors $\mathbf{v}^{(1)}, \ldots, \mathbf{v}^{(n)}$. The eigenvalues of $(A - qI)^{-1}$, where $q \ne \lambda_i$, for $i = 1, 2, \ldots, n$, are

$$\frac{1}{\lambda_1 - q}, \quad \frac{1}{\lambda_2 - q}, \quad \ldots, \quad \frac{1}{\lambda_n - q},$$

with these same eigenvectors $\mathbf{v}^{(1)}, \mathbf{v}^{(2)}, \ldots, \mathbf{v}^{(n)}$. (See Exercise 15 of Section 7.2.)

Applying the Power method to $(A - qI)^{-1}$ gives

$$\mathbf{y}^{(m)} = (A - qI)^{-1}\mathbf{x}^{(m-1)},$$

$$\mu^{(m)} = y_{p_{m-1}}^{(m)} = \frac{y_{p_{m-1}}^{(m)}}{x_{p_{m-1}}^{(m-1)}} = \frac{\sum_{j=1}^{n} \beta_j \dfrac{1}{(\lambda_j - q)^m} v_{p_{m-1}}^{(j)}}{\sum_{j=1}^{n} \beta_j \dfrac{1}{(\lambda_j - q)^{m-1}} v_{p_{m-1}}^{(j)}}, \tag{9.4}$$

and

$$\mathbf{x}^{(m)} = \frac{\mathbf{y}^{(m)}}{y_{p_m}^{(m)}},$$

where, at each step, p_m represents the smallest integer for which $|y_{p_m}^{(m)}| = ||\mathbf{y}^{(m)}||_\infty$. The sequence $\{\mu^{(m)}\}$ in Eq. (9.4) converges to $1/(\lambda_k - q)$, where

$$\frac{1}{|\lambda_k - q|} = \max_{1 \le i \le n} \frac{1}{|\lambda_i - q|},$$

and $\lambda_k \approx q + 1/\mu^{(m)}$ is the eigenvalue of A closest to q.

With k known, Eq. (9.4) can be written as

$$\mu^{(m)} = \frac{1}{\lambda_k - q} \left[\frac{\beta_k v_{p_{m-1}}^{(k)} + \sum_{\substack{j=1 \\ j \ne k}}^{n} \beta_j \left[\frac{\lambda_k - q}{\lambda_j - q} \right]^m v_{p_{m-1}}^{(j)}}{\beta_k v_{p_{m-1}}^{(k)} + \sum_{\substack{j=1 \\ j \ne k}}^{n} \beta_j \left[\frac{\lambda_k - q}{\lambda_j - q} \right]^{m-1} v_{p_{m-1}}^{(j)}} \right]. \tag{9.5}$$

Thus, the choice of q determines the convergence, provided that $1/(\lambda_k - q)$ is a unique dominant eigenvalue of $(A - qI)^{-1}$ (although it may be a multiple eigenvalue). The closer q is to an eigenvalue λ_k, the faster the convergence since the convergence is of order

$$O\left(\left| \frac{(\lambda - q)^{-1}}{(\lambda_k - q)^{-1}} \right|^m \right) = O\left(\left| \frac{(\lambda_k - q)}{(\lambda - q)} \right|^m \right),$$

where λ represents the eigenvalue of A that is second closest to q.

The vector $\mathbf{y}^{(m)}$ is obtained by solving the linear system

$$(A - qI)\mathbf{y}^{(m)} = \mathbf{x}^{(m-1)}.$$

In general, Gaussian elimination with pivoting is used, but as in the case of the *LU* factorization, the multipliers can be saved to reduce the computation. The selection of q can be based on the Geršgorin Circle Theorem or on another means of localizing an eigenvalue.

Algorithm 9.3 computes q from an initial approximation to the eigenvector $\mathbf{x}^{(0)}$ by

$$q = \frac{\mathbf{x}^{(0)t} A \mathbf{x}^{(0)}}{\mathbf{x}^{(0)t} \mathbf{x}^{(0)}}.$$

This choice of q results from the observation that if $\mathbf{x}$ is an eigenvector of A with respect to the eigenvalue λ, then $A\mathbf{x} = \lambda \mathbf{x}$. So $\mathbf{x}^t A \mathbf{x} = \lambda \mathbf{x}^t \mathbf{x}$ and

$$\lambda = \frac{\mathbf{x}^t A \mathbf{x}}{\mathbf{x}^t \mathbf{x}} = \frac{\mathbf{x}^t A \mathbf{x}}{\|\mathbf{x}\|_2^2}.$$

If q is close to an eigenvalue, the convergence will be quite rapid, but a pivoting technique should be used in Step 6 to avoid contamination by round-off error.

Algorithm 9.3 is often used to approximate an eigenvector when an approximate eigenvalue q is known.

Inverse Power Method

To approximate an eigenvalue and an associated eigenvector of the $n \times n$ matrix A given a nonzero vector $\mathbf{x}$:

INPUT dimension n; matrix A; vector $\mathbf{x}$; tolerance TOL; maximum number of iterations N.

OUTPUT approximate eigenvalue μ; approximate eigenvector $\mathbf{x}$ (with $\|\mathbf{x}\|_\infty = 1$) or a message that the maximum number of iterations was exceeded.

Step 1 Set $q = \dfrac{\mathbf{x}^t A \mathbf{x}}{\mathbf{x}^t \mathbf{x}}$.

Step 2 Set $k = 1$.

Step 3 Find the smallest integer p with $1 \le p \le n$ and $|x_p| = \|\mathbf{x}\|_\infty$.

Step 4 Set $\mathbf{x} = \mathbf{x}/x_p$.

Step 5 While ($k \le N$) do Steps 6–12.

Step 6 Solve the linear system $(A - qI)\mathbf{y} = \mathbf{x}$.

Step 7 If the system does not have a unique solution, then
OUTPUT ('q is an eigenvalue', q);
STOP.

Step 8 Set $\mu = y_p$.

Step 9 Find the smallest integer p with $1 \le p \le n$ and $|y_p| = \|\mathbf{y}\|_\infty$.

Step 10 Set $ERR = \left\|\mathbf{x} - (\mathbf{y}/y_p)\right\|_\infty$;
$\mathbf{x} = \mathbf{y}/y_p$.

Step 11 If $ERR < TOL$ then set $\mu = (1/\mu) + q$;
OUTPUT $(\mu, \mathbf{x})$;
(*The procedure was successful.*)
STOP.

Step 12 Set $k = k + 1$.

Step 13 OUTPUT ('Maximum number of iterations exceeded');
(*The procedure was unsuccessful.*)
STOP. ■

The convergence of the Inverse Power method is linear, so Aitken Δ^2 method can again be used to speed convergence. The following example illustrates the fast convergence of the Inverse Power method if q is close to an eigenvalue.

Example 3 Apply the Inverse Power method with $\mathbf{x}^{(0)} = (1, 1, 1)^t$ to the matrix

$$A = \begin{bmatrix} -4 & 14 & 0 \\ -5 & 13 & 0 \\ -1 & 0 & 2 \end{bmatrix} \quad \text{with} \quad q = \frac{\mathbf{x}^{(0)t} A \mathbf{x}^{(0)}}{\mathbf{x}^{(0)t}\mathbf{x}^{(0)}} = \frac{19}{3},$$

and use Aitken's Δ^2 method to accelerate the convergence.

Solution The Power method was applied to this matrix in Example 1 using the initial vector $\mathbf{x}^{(0)} = (1, 1, 1)^t$. It gave the approximate eigenvalue $\mu^{(12)} = 6.000837$ and eigenvector $(\mathbf{x}^{(12)})^t = (1, 0.714316, -0.249895)^t$.

For the Inverse Power method we consider

$$A - qI = \begin{bmatrix} -\frac{31}{3} & 14 & 0 \\ -5 & \frac{20}{3} & 0 \\ -1 & 0 & -\frac{13}{3} \end{bmatrix}$$

With $\mathbf{x}^{(0)} = (1, 1, 1)^t$, the method first finds $\mathbf{y}^{(1)}$ by solving $(A - qI)\mathbf{y}^{(1)} = \mathbf{x}^{(0)}$. This gives

$$\mathbf{y}^{(1)} = \left(-\frac{33}{5}, -\frac{24}{5}, \frac{84}{65}\right)^t = (-6.6, -4.8, 1.292\overline{307692})^t.$$

So

$$\|\mathbf{y}^{(1)}\|_\infty = 6.6, \quad \mathbf{x}^{(1)} = \frac{1}{-6.6}\mathbf{y}^{(1)} = (1, 0.7272727, -0.1958042)^t,$$

and

$$\mu^{(1)} = -\frac{1}{6.6} + \frac{19}{3} = 6.1818182.$$

Subsequent results are listed in Table 9.4, and the right column lists the results of Aitken's Δ^2 method applied to the $\mu^{(m)}$. These are clearly superior results to those obtained with the Power method. ■

Table 9.4

m	$\mathbf{x}^{(m)t}$	$\mu^{(m)}$	$\hat{\mu}^{(m)}$
0	(1, 1, 1)		
1	(1, 0.7272727, −0.1958042)	6.1818182	6.000098
2	(1, 0.7155172, −0.2450520)	6.0172414	6.000001
3	(1, 0.7144082, −0.2495224)	6.0017153	6.000000
4	(1, 0.7142980, −0.2499534)	6.0001714	6.000000
5	(1, 0.7142869, −0.2499954)	6.0000171	
6	(1, 0.7142858, −0.2499996)	6.0000017	

If A is symmetric, then for any real number q, the matrix $(A - qI)^{-1}$ is also symmetric, so the Symmetric Power method, Algorithm 9.2, can be applied to $(A - qI)^{-1}$ to speed the convergence to

$$O\left(\left|\frac{\lambda_k - q}{\lambda - q}\right|^{2m}\right).$$

Deflation Methods

Numerous techniques are available for obtaining approximations to the other eigenvalues of a matrix once an approximation to the dominant eigenvalue has been computed. We will restrict our presentation to **deflation techniques**.

Deflation techniques involve forming a new matrix B whose eigenvalues are the same as those of A, except that the dominant eigenvalue of A is replaced by the eigenvalue 0 in B. The following result justifies the procedure. The proof of this theorem can be found in [Wil2], p. 596.

Theorem 9.20 Suppose $\lambda_1, \lambda_2, \ldots, \lambda_n$ are eigenvalues of A with associated eigenvectors $\mathbf{v}^{(1)}, \mathbf{v}^{(2)}, \ldots, \mathbf{v}^{(n)}$ and that λ_1 has multiplicity 1. Let $\mathbf{x}$ be a vector with $\mathbf{x}^t\mathbf{v}^{(1)} = 1$. Then the matrix

$$B = A - \lambda_1 \mathbf{v}^{(1)}\mathbf{x}^t$$

has eigenvalues $0, \lambda_2, \lambda_3, \ldots, \lambda_n$ with associated eigenvectors $\mathbf{v}^{(1)}$, $\mathbf{w}^{(2)}$, $\mathbf{w}^{(3)}, \ldots, \mathbf{w}^{(n)}$, where $\mathbf{v}^{(i)}$ and $\mathbf{w}^{(i)}$ are related by the equation

$$\mathbf{v}^{(i)} = (\lambda_i - \lambda_1)\mathbf{w}^{(i)} + \lambda_1(\mathbf{x}^t\mathbf{w}^{(i)})\mathbf{v}^{(1)}, \tag{9.6}$$

for each $i = 2, 3, \ldots, n$. ■

There are many choices of the vector $\mathbf{x}$ that could be used in Theorem 9.20. **Wielandt deflation** proceeds from defining

$$\mathbf{x} = \frac{1}{\lambda_1 v_i^{(1)}}(a_{i1}, a_{i2}, \ldots, a_{in})^t, \tag{9.7}$$

where $v_i^{(1)}$ is a nonzero coordinate of the eigenvector $\mathbf{v}^{(1)}$, and the values $a_{i1}, a_{i2}, \ldots, a_{in}$ are the entries in the ith row of A.

Helmut Wielandt (1910–2001) originally worked in permutation groups, but during during World War II he was engaged in research on meteorology, cryptology, and aerodynamics. This involved vibration problems that required the estimation of eigenvalues associated with differential equations and matrices.

With this definition,

$$\mathbf{x}^t\mathbf{v}^{(1)} = \frac{1}{\lambda_1 v_i^{(1)}}[a_{i1}, a_{i2}, \ldots, a_{in}](v_1^{(1)}, v_2^{(1)}, \ldots, v_n^{(1)})^t = \frac{1}{\lambda_1 v_i^{(1)}}\sum_{j=1}^{n} a_{ij}v_j^{(1)},$$

where the sum is the ith coordinate of the product $A\mathbf{v}^{(1)}$. Since $A\mathbf{v}^{(1)} = \lambda_1\mathbf{v}^{(1)}$, we have

$$\sum_{j=1}^{n} a_{ij}v_j^{(1)} = \lambda_1 v_i^{(1)},$$

which implies that

$$\mathbf{x}^t\mathbf{v}^{(1)} = \frac{1}{\lambda_1 v_i^{(1)}}(\lambda_1 v_i^{(1)}) = 1.$$

So $\mathbf{x}$ satisfies the hypotheses of Theorem 9.20. Moreover (see Exercise 20), the ith row of $B = A - \lambda_1\mathbf{v}^{(1)}\mathbf{x}^t$ consists entirely of zero entries.

If $\lambda \neq 0$ is an eigenvalue with associated eigenvector $\mathbf{w}$, the relation $B\mathbf{w} = \lambda\mathbf{w}$ implies that the ith coordinate of $\mathbf{w}$ must also be zero. Consequently the ith column of the matrix B makes no contribution to the product $B\mathbf{w} = \lambda\mathbf{w}$. Thus, the matrix B can be replaced by an $(n-1) \times (n-1)$ matrix B' obtained by deleting the ith row and column from B. The matrix B' has eigenvalues $\lambda_2, \lambda_3, \ldots, \lambda_n$.

If $|\lambda_2| > |\lambda_3|$, the Power method is reapplied to the matrix B' to determine this new dominant eigenvalue and an eigenvector, $\mathbf{w}^{(2)'}$, associated with λ_2, with respect to the matrix B'. To find the associated eigenvector $\mathbf{w}^{(2)}$ for the matrix B, insert a zero coordinate between the coordinates $w_{i-1}^{(2)'}$ and $w_i^{(2)'}$ of the $(n-1)$-dimensional vector $\mathbf{w}^{(2)'}$ and then calculate $\mathbf{v}^{(2)}$ by the use of Eq. (9.6).

Example 4 The matrix

$$A = \begin{bmatrix} 4 & -1 & 1 \\ -1 & 3 & -2 \\ 1 & -2 & 3 \end{bmatrix}$$

has the dominant eigenvalue $\lambda_1 = 6$ with associated unit eigenvector $\mathbf{v}^{(1)} = (1, -1, 1)^t$. Assume that this dominant eigenvalue is known and apply deflation to approximate the other eigenvalues and eigenvectors.

Solution The procedure for obtaining a second eigenvalue λ_2 proceeds as follows:

$$\mathbf{x} = \frac{1}{6}\begin{bmatrix} 4 \\ -1 \\ 1 \end{bmatrix} = \left(\frac{2}{3}, -\frac{1}{6}, \frac{1}{6}\right)^t,$$

$$\mathbf{v}^{(1)}\mathbf{x}^t = \begin{bmatrix} 1 \\ -1 \\ 1 \end{bmatrix}\begin{bmatrix} \frac{2}{3}, & -\frac{1}{6}, & \frac{1}{6} \end{bmatrix} = \begin{bmatrix} \frac{2}{3} & -\frac{1}{6} & \frac{1}{6} \\ -\frac{2}{3} & \frac{1}{6} & -\frac{1}{6} \\ \frac{2}{3} & -\frac{1}{6} & \frac{1}{6} \end{bmatrix},$$

and

$$B = A - \lambda_1 \mathbf{v}^{(1)}\mathbf{x}^t = \begin{bmatrix} 4 & -1 & 1 \\ -1 & 3 & -2 \\ 1 & -2 & 3 \end{bmatrix} - 6\begin{bmatrix} \frac{2}{3} & -\frac{1}{6} & \frac{1}{6} \\ -\frac{2}{3} & \frac{1}{6} & -\frac{1}{6} \\ \frac{2}{3} & -\frac{1}{6} & \frac{1}{6} \end{bmatrix} = \begin{bmatrix} 0 & 0 & 0 \\ 3 & 2 & -1 \\ -3 & -1 & 2 \end{bmatrix}.$$

Deleting the first row and column gives

$$B' = \begin{bmatrix} 2 & -1 \\ -1 & 2 \end{bmatrix},$$

which has eigenvalues $\lambda_2 = 3$ and $\lambda_3 = 1$. For $\lambda_2 = 3$, the eigenvector $\mathbf{w}^{(2)'}$ can be obtained by solving the linear system

$$(B' - 3I)\mathbf{w}^{(2)'} = \mathbf{0}, \quad \text{resulting in} \quad \mathbf{w}^{(2)'} = (1, -1)^t.$$

Adding a zero for the first component gives $\mathbf{w}^{(2)} = (0, 1, -1)^t$ and, from Eq. (9.6), we have the eigenvector $\mathbf{v}^{(2)}$ of A corresponding to $x_2 = 3$:

$$\begin{aligned} \mathbf{v}^{(2)} &= (\lambda_2 - \lambda_1)\mathbf{w}^{(2)} + \lambda_1(\mathbf{x}^t\mathbf{w}^{(2)})\mathbf{v}^{(1)} \\ &= (3-6)(0, 1, -1)^t + 6\left[\left(\frac{2}{3}, -\frac{1}{6}, \frac{1}{6}\right)(0, 1, -1)^t\right](1, -1, 1)^t = (-2, -1, 1)^t. \quad \blacksquare \end{aligned}$$

Although this deflation process can be used to find approximations to all of the eigenvalues and eigenvectors of a matrix, the process is susceptible to round-off error. After deflation is used to approximate an eigenvalue of a matrix, the approximation should be used as a starting value for the Inverse Power method applied to the original matrix. This will ensure convergence to an eigenvalue of the original matrix, not to one of the reduced matrix, which likely contains errors. When all the eigenvalues of a matrix are required, techniques considered in Section 9.5, based on similarity transformations, should be used.

We close this section with Algorithm 9.4, which calculates the second most dominant eigenvalue and associated eigenvector for a matrix, once the dominant eigenvalue and associated eigenvector have been determined.

Wielandt Deflation

To approximate the second most dominant eigenvalue and an associated eigenvector of the $n \times n$ matrix A given an approximation λ to the dominant eigenvalue, an approximation $\mathbf{v}$ to a corresponding eigenvector, and a vector $\mathbf{x} \in \mathbb{R}^{n-1}$:

INPUT dimension n; matrix A; approximate eigenvalue λ with eigenvector $\mathbf{v} \in \mathbb{R}^n$; vector $\mathbf{x} \in \mathbb{R}^{n-1}$, tolerance TOL, maximum number of iterations N.

OUTPUT approximate eigenvalue μ; approximate eigenvector $\mathbf{u}$ or a message that the method fails.

Step 1 Let i be the smallest integer with $1 \le i \le n$ and $|v_i| = \max_{1\le j\le n} |v_j|$.

Step 2 If $i \neq 1$ then
for $k = 1, \ldots, i-1$
for $j = 1, \ldots, i-1$

$$\text{set } b_{kj} = a_{kj} - \frac{v_k}{v_i} a_{ij}.$$

Step 3 If $i \neq 1$ and $i \neq n$ then
for $k = i, \ldots, n-1$
for $j = 1, \ldots, i-1$

$$\text{set } b_{kj} = a_{k+1,j} - \frac{v_{k+1}}{v_i} a_{ij};$$

$$b_{jk} = a_{j,k+1} - \frac{v_j}{v_i} a_{i,k+1}.$$

Step 4 If $i \neq n$ then
for $k = i, \ldots, n-1$
for $j = i, \ldots, n-1$

$$\text{set } b_{kj} = a_{k+1,j+1} - \frac{v_{k+1}}{v_i} a_{i,j+1}.$$

Step 5 Perform the power method on the $(n-1) \times (n-1)$ matrix $B' = (b_{kj})$ with $\mathbf{x}$ as initial approximation.

Step 6 If the method fails, then OUTPUT ('Method fails');
STOP
else let μ be the approximate eigenvalue and
$\mathbf{w}' = (w'_1, \ldots, w'_{n-1})^t$ the approximate eigenvector.

Step 7 If $i \neq 1$ then for $k = 1, \ldots, i-1$ set $w_k = w'_k$.

Step 8 Set $w_i = 0$.

Step 9 If $i \neq n$ then for $k = i+1, \ldots, n$ set $w_k = w'_{k-1}$.

Step 10 For $k = 1, \ldots, n$

$$\text{set } u_k = (\mu - \lambda) w_k + \left(\sum_{j=1}^{n} a_{ij} w_j \right) \frac{v_k}{v_i}.$$

(*Compute the eigenvector using Eq.* (9.6).)

Step 11 OUTPUT $(\mu, \mathbf{u})$; (*The procedure was successful.*)
STOP. ■

EXERCISE SET 9.3

1. Find the first three iterations obtained by the Power method applied to the following matrices.

a. $\begin{bmatrix} 2 & 1 & 1 \\ 1 & 2 & 1 \\ 1 & 1 & 2 \end{bmatrix}$; Use $\mathbf{x}^{(0)} = (1, -1, 2)^t$.

b. $\begin{bmatrix} 1 & 1 & 1 \\ 1 & 1 & 0 \\ 1 & 0 & 1 \end{bmatrix}$; Use $\mathbf{x}^{(0)} = (-1, 0, 1)^t$.

c. $\begin{bmatrix} 1 & -1 & 0 \\ -2 & 4 & -2 \\ 0 & -1 & 2 \end{bmatrix}$; Use $\mathbf{x}^{(0)} = (-1, 2, 1)^t$.

d. $\begin{bmatrix} 4 & 1 & 1 & 1 \\ 1 & 3 & -1 & 1 \\ 1 & -1 & 2 & 0 \\ 1 & 1 & 0 & 2 \end{bmatrix}$; Use $\mathbf{x}^{(0)} = (1, -2, 0, 3)^t$.

2. Find the first three iterations obtained by the Power method applied to the following matrices.

a. $\begin{bmatrix} 4 & 2 & 1 \\ 0 & 3 & 2 \\ 1 & 1 & 4 \end{bmatrix}$; Use $\mathbf{x}^{(0)} = (1, 2, 1)^t$.

b. $\begin{bmatrix} 1 & 1 & 0 & 0 \\ 1 & 2 & 0 & 1 \\ 0 & 0 & 3 & 3 \\ 0 & 1 & 3 & 2 \end{bmatrix}$; Use $\mathbf{x}^{(0)} = (1, 1, 0, 1)^t$.

c. $\begin{bmatrix} 5 & -2 & -\frac{1}{2} & \frac{3}{2} \\ -2 & 5 & \frac{3}{2} & -\frac{1}{2} \\ -\frac{1}{2} & \frac{3}{2} & 5 & -2 \\ \frac{3}{2} & -\frac{1}{2} & -2 & 5 \end{bmatrix}$; Use $\mathbf{x}^{(0)} = (1, 1, 0, -3)^t$.

d. $\begin{bmatrix} -4 & 0 & \frac{1}{2} & \frac{1}{2} \\ \frac{1}{2} & -2 & 0 & \frac{1}{2} \\ \frac{1}{2} & \frac{1}{2} & 0 & 0 \\ 0 & 1 & 1 & 4 \end{bmatrix}$; Use $\mathbf{x}^{(0)} = (0, 0, 0, 1)^t$.

3. Repeat Exercise 1 using the Inverse Power method.

4. Repeat Exercise 2 using the Inverse Power method.

5. Find the first three iterations obtained by the Symmetric Power method applied to the following matrices.

a. $\begin{bmatrix} 2 & 1 & 1 \\ 1 & 2 & 1 \\ 1 & 1 & 2 \end{bmatrix}$; Use $\mathbf{x}^{(0)} = (1, -1, 2)^t$.

b. $\begin{bmatrix} 1 & 1 & 1 \\ 1 & 1 & 0 \\ 1 & 0 & 1 \end{bmatrix}$; Use $\mathbf{x}^{(0)} = (-1, 0, 1)^t$.

c. $\begin{bmatrix} 4.75 & 2.25 & -0.25 \\ 2.25 & 4.75 & 1.25 \\ -0.25 & 1.25 & 4.75 \end{bmatrix}$; Use $\mathbf{x}^{(0)} = (0, 1, 0)^t$.

d. $\begin{bmatrix} 4 & 1 & -1 & 0 \\ 1 & 3 & -1 & 0 \\ -1 & -1 & 5 & 2 \\ 0 & 0 & 2 & 4 \end{bmatrix}$; Use $\mathbf{x}^{(0)} = (0, 1, 0, 0)^t$.

6. Find the first three iterations obtained by the Symmetric Power method applied to the following matrices.

a. $\begin{bmatrix} -2 & 1 & 3 \\ 1 & 3 & -1 \\ 3 & -1 & 2 \end{bmatrix}$; Use $\mathbf{x}^{(0)} = (1, -1, 2)^t$.

b. $\begin{bmatrix} 4 & 2 & -1 \\ 2 & 0 & 2 \\ -1 & 2 & 0 \end{bmatrix}$; Use $\mathbf{x}^{(0)} = (-1, 0, 1)^t$.

c. $\begin{bmatrix} 4 & 1 & 1 & 1 \\ 1 & 3 & -1 & 1 \\ 1 & -1 & 2 & 0 \\ 1 & 1 & 0 & 2 \end{bmatrix}$; Use $\mathbf{x}^{(0)} = (1, 0, 0, 0)^t$.

d. $\begin{bmatrix} 5 & -2 & -\frac{1}{2} & \frac{3}{2} \\ -2 & 5 & \frac{3}{2} & -\frac{1}{2} \\ -\frac{1}{2} & \frac{3}{2} & 5 & -2 \\ \frac{3}{2} & -\frac{1}{2} & -2 & 5 \end{bmatrix}$; Use $\mathbf{x}^{(0)} = (1, 1, 0, -3)^t$.

7. Use the Power method to approximate the most dominant eigenvalue of the matrices in Exercise 1. Iterate until a tolerance of 10^{-4} is achieved or until the number of iterations exceeds 25.

8. Use the Power method to approximate the most dominant eigenvalue of the matrices in Exercise 2. Iterate until a tolerance of 10^{-4} is achieved or until the number of iterations exceeds 25.

9. Use the Inverse Power method to approximate the most dominant eigenvalue of the matrices in Exercise 1. Iterate until a tolerance of 10^{-4} is achieved or until the number of iterations exceeds 25.

10. Use the Inverse Power method to approximate the most dominant eigenvalue of the matrices in Exercise 2. Iterate until a tolerance of 10^{-4} is achieved or until the number of iterations exceeds 25.

11. Use the Symmetric Power method to approximate the most dominant eigenvalue of the matrices in Exercise 5. Iterate until a tolerance of 10^{-4} is achieved or until the number of iterations exceeds 25.

12. Use the Symmetric Power method to approximate the most dominant eigenvalue of the matrices in Exercise 6. Iterate until a tolerance of 10^{-4} is achieved or until the number of iterations exceeds 25.

13. Use Wielandt deflation and the results of Exercise 7 to approximate the second most dominant eigenvalue of the matrices in Exercise 1. Iterate until a tolerance of 10^{-4} is achieved or until the number of iterations exceeds 25.

14. Use Wielandt deflation and the results of Exercise 8 to approximate the second most dominant eigenvalue of the matrices in Exercise 2. Iterate until a tolerance of 10^{-4} is achieved or until the number of iterations exceeds 25.

15. Repeat Exercise 7 using Aitken's Δ^2 technique and the Power method for the most dominant eigenvalue.

16. Repeat Exercise 8 using Aitken's Δ^2 technique and the Power method for the most dominant eigenvalue.

17. **Hotelling Deflation** Assume that the largest eigenvalue λ_1 in magnitude and an associated eigenvector $\mathbf{v}^{(1)}$ have been obtained for the $n \times n$ symmetric matrix A. Show that the matrix

$$B = A - \frac{\lambda_1}{(\mathbf{v}^{(1)})^t \mathbf{v}^{(1)}} \mathbf{v}^{(1)} (\mathbf{v}^{(1)})^t$$

has the same eigenvalues $\lambda_2, \ldots, \lambda_n$ as A, except that B has eigenvalue 0 with eigenvector $\mathbf{v}^{(1)}$ instead of eigenvector λ_1. Use this deflation method to find λ_2 for each matrix in Exercise 5. Theoretically, this method can be continued to find more eigenvalues, but round-off error soon makes the effort worthless.

18. **Annihilation Technique** Suppose the $n \times n$ matrix A has eigenvalues $\lambda_1, \ldots, \lambda_n$ ordered by

$$|\lambda_1| > |\lambda_2| > |\lambda_3| \geq \cdots \geq |\lambda_n|,$$

with linearly independent eigenvectors $\mathbf{v}^{(1)}, \mathbf{v}^{(2)}, \ldots, \mathbf{v}^{(n)}$.

a. Show that if the Power method is applied with an initial vector $\mathbf{x}^{(0)}$ given by

$$\mathbf{x}^{(0)} = \beta_2 \mathbf{v}^{(2)} + \beta_3 \mathbf{v}^{(3)} + \cdots + \beta_n \mathbf{v}^{(n)},$$

then the sequence $\{\mu^{(m)}\}$ described in Algorithm 9.1 will converge to λ_2.

b. Show that for any vector $\mathbf{x} = \sum_{i=1}^{n} \beta_i \mathbf{v}^{(i)}$, the vector $\mathbf{x}^{(0)} = (A - \lambda_1 I)\mathbf{x}$ satisfies the property given in part (a).

c. Obtain an approximation to λ_2 for the matrices in Exercise 1.

d. Show that this method can be continued to find λ_3 using $\mathbf{x}^{(0)} = (A - \lambda_2 I)(A - \lambda_1 I)\mathbf{x}$.

19. Following along the line of Exercise 11 in Section 6.3 and Exercise 15 in Section 7.2, suppose that a species of beetle has a life span of 4 years, and that a female in the first year has a survival rate of $\frac{1}{2}$, in the second year a survival rate of $\frac{1}{4}$, and in the third year a survival rate of $\frac{1}{8}$. Suppose additionally that a female gives birth, on the average, to two new females in the third year and to four new females in the fourth year. The matrix describing a single female's contribution in 1 year to the female population in the succeeding year is

$$A = \begin{bmatrix} 0 & 0 & 2 & 4 \\ \frac{1}{2} & 0 & 0 & 0 \\ 0 & \frac{1}{4} & 0 & 0 \\ 0 & 0 & \frac{1}{8} & 0 \end{bmatrix},$$

where again the entry in the ith row and jth column denotes the probabilistic contribution that a female of age j makes on the next year's female population of age i.

a. Use the Geršgorin Circle Theorem to determine a region in the complex plane containing all the eigenvalues of A.

b. Use the Power method to determine the dominant eigenvalue of the matrix and its associated eigenvector.

c. Use Algorithm 9.4 to determine any remaining eigenvalues and eigenvectors of A.

d. Find the eigenvalues of A by using the characteristic polynomial of A and Newton's method.

e. What is your long-range prediction for the population of these beetles?

20. Show that the ith row of $B = A - \lambda_1 \mathbf{v}^{(1)}\mathbf{x}^t$ is zero, where λ_1 is the largest value of A in absolute value, $\mathbf{v}^{(1)}$ is the associated eigenvector of A for λ_1, and $\mathbf{x}$ is the vector defined in Eq. (9.7).

21. The $(m-1) \times (m-1)$ tridiagonal matrix

$$A = \begin{bmatrix} 1+2\alpha & -\alpha & 0 & \cdots & 0 \\ -\alpha & 1+2\alpha & -\alpha & \ddots & \vdots \\ 0 & \ddots & \ddots & \ddots & 0 \\ \vdots & \ddots & \ddots & \ddots & -\alpha \\ 0 & \cdots & 0 & -\alpha & 1+2\alpha \end{bmatrix}$$

is involved in the Backward Difference method to solve the heat equation. (See Section 12.2.) For the stability of the method we need $\rho(A^{-1}) < 1$.With $m = 11$, approximate $\rho(A^{-1})$ for each of the following.

a. $\alpha = \frac{1}{4}$ **b.** $\alpha = \frac{1}{2}$ **c.** $\alpha = \frac{3}{4}$

When is the method stable?

22. The eigenvalues of the matrix A in Exercise 21 are

$$\lambda_i = 1 + 4\alpha\left(\sin\frac{\pi i}{2m}\right)^2, \quad \text{for } i = 1, \ldots, m-1.$$

Compare the approximation in Exercise 21 to the actual value of $\rho(A^{-1})$. Again, when is the method stable?

23. The $(m-1) \times (m-1)$ matrices A and B given by

$$A = \begin{bmatrix} 1+\alpha & -\frac{\alpha}{2} & 0 & \cdots & 0 \\ -\frac{\alpha}{2} & 1+\alpha & -\frac{\alpha}{2} & \ddots & \vdots \\ 0 & \ddots & \ddots & \ddots & 0 \\ \vdots & \ddots & \ddots & \ddots & -\frac{\alpha}{2} \\ 0 & \cdots & 0 & -\frac{\alpha}{2} & 1+\alpha \end{bmatrix} \quad \text{and} \quad B = \begin{bmatrix} 1+\alpha & \frac{\alpha}{2} & 0 & \cdots & 0 \\ \frac{\alpha}{2} & 1+\alpha & \frac{\alpha}{2} & \ddots & \vdots \\ 0 & \ddots & \ddots & \ddots & 0 \\ \vdots & \ddots & \ddots & \ddots & \frac{\alpha}{2} \\ 0 & \cdots & 0 & \frac{\alpha}{2} & 1+\alpha \end{bmatrix}$$

are involved in the Crank-Nicolson method to solve the heat equation (see Section 12.2). With $m = 11$, approximate $\rho(A^{-1}B)$ for each of the following.

a. $\alpha = \frac{1}{4}$ **b.** $\alpha = \frac{1}{2}$ **c.** $\alpha = \frac{3}{4}$

24. A linear dynamical system can be represented by the equations

$$\frac{d\mathbf{x}}{dt} = A(t)\mathbf{x}(t) + B(t)\mathbf{u}(t), \quad \mathbf{y}(t) = C(t)\mathbf{x}(t) + D(t)\mathbf{u}(t),$$

where A is an $n \times n$ variable matrix, B is an $n \times r$ variable matrix, C is an $m \times n$ variable matrix, D is an $m \times r$ variable matrix, $\mathbf{x}$ is an n-dimensional vector variable, $\mathbf{y}$ is an m-dimensional vector variable, and $\mathbf{u}$ is an r-dimensional vector variable. For the system to be stable, the matrix A must have all its eigenvalues with nonpositive real part for all t. Is the system stable if

a.

$$A(t) = \begin{bmatrix} -1 & 2 & 0 \\ -2.5 & -7 & 4 \\ 0 & 0 & -5 \end{bmatrix}?$$

b.

$$A(t) = \begin{bmatrix} -1 & 1 & 0 & 0 \\ 0 & -2 & 1 & 0 \\ 0 & 0 & -5 & 1 \\ -1 & -1 & -2 & -3 \end{bmatrix}?$$

9.4 Householder's Method

In Section 9.5 we will use the QR method to reduce a symmetric tridiagonal matrix to a similar matrix that is nearly diagonal. The diagonal entries of the reduced matrix are approximations to the eigenvalues of the given matrix. In this section, we present a method devised by Alston Householder for reducing an arbitrary symmetric matrix to a similar tridiagonal matrix. Although there is a clear connection between the problems we are solving in these two sections, Householder's method has a such wide application in areas other than eigenvalue approximation, that it deserves special treatment.

Alston Householder (1904–1993) did research in mathematical biology before becoming the Director of the Oak Ridge National Laboratory in Tennessee in 1948. He began work on solving linear systems in the 1950s, which was when these methods were developed.

Householder's method is used to find a symmetric tridiagonal matrix B that is similar to a given symmetric matrix A. Theorem 9.16 implies that A is similar to a diagonal matrix D since an orthogonal matrix Q exists with the property that $D = Q^{-1}AQ = Q^tAQ$. Because the matrix Q (and consequently D) is generally difficult to compute, Householder's method offers a compromise. After Householder's method has been implemented, efficient methods such as the QR algorithm can be used for accurate approximation of the eigenvalues of the resulting symmetric tridiagonal matrix.

Householder Transformations

Definition 9.21 Let $\mathbf{w} \in \mathbb{R}^n$ with $\mathbf{w}^t\mathbf{w} = 1$. The $n \times n$ matrix

$$P = I - 2\mathbf{w}\mathbf{w}^t$$

is called a **Householder transformation**. ■

Householder transformations are used to selectively zero out blocks of entries in vectors or columns of matrices in a manner that is extremely stable with respect to round-off error. (See [Wil2], pp. 152–162, for further discussion.) Properties of Householder transformations are given in the following theorem.

Theorem 9.22 A Householder transformation, $P = I - 2\mathbf{w}\mathbf{w}^t$, is symmetric and orthogonal, so $P^{-1} = P$. ■

Proof It follows from

$$(\mathbf{w}\mathbf{w}^t)^t = (\mathbf{w}^t)^t\mathbf{w}^t = \mathbf{w}\mathbf{w}^t,$$

that

$$P^t = (I - 2\mathbf{w}\mathbf{w}^t)^t = I - 2\mathbf{w}\mathbf{w}^t = P.$$

Further, $\mathbf{w}^t\mathbf{w} = 1$, so

$$PP^t = (I - 2\mathbf{w}\mathbf{w}^t)(I - 2\mathbf{w}\mathbf{w}^t) = I - 2\mathbf{w}\mathbf{w}^t - 2\mathbf{w}\mathbf{w}^t + 4\mathbf{w}\mathbf{w}^t\mathbf{w}\mathbf{w}^t$$
$$= I - 4\mathbf{w}\mathbf{w}^t + 4\mathbf{w}\mathbf{w}^t = I,$$

and $P^{-1} = P^t = P$. ■ ■ ■

Householder's method begins by determining a transformation $P^{(1)}$ with the property that $A^{(2)} = P^{(1)}AP^{(1)}$ zero's out the entries in the first column of A beginning with the third row. That is, such that

$$a_{j1}^{(2)} = 0, \quad \text{for each } j = 3, 4, \ldots, n. \tag{9.8}$$

By symmetry, we also have $a_{1j}^{(2)} = 0$.

We now choose a vector $\mathbf{w} = (w_1, w_2, \ldots, w_n)^t$ so that $\mathbf{w}^t\mathbf{w} = 1$, Eq. (9.8) holds, and in the matrix

$$A^{(2)} = P^{(1)}AP^{(1)} = (I - 2\mathbf{w}\mathbf{w}^t)A(I - 2\mathbf{w}\mathbf{w}^t),$$

we have $a_{11}^{(2)} = a_{11}$ and $a_{j1}^{(2)} = 0$, for each $j = 3, 4, \ldots, n$. This choice imposes n conditions on the n unknowns $w_1, w_2, \ldots, w_n$.

Setting $w_1 = 0$ ensures that $a_{11}^{(2)} = a_{11}$. We want

$$P^{(1)} = I - 2\mathbf{w}\mathbf{w}^t$$

to satisfy

$$P^{(1)}(a_{11}, a_{21}, a_{31}, \ldots, a_{n1})^t = (a_{11}, \alpha, 0, \ldots, 0)^t, \tag{9.9}$$

where α will be chosen later. To simplify notation, let

$$\hat{\mathbf{w}} = (w_2, w_3, \ldots, w_n)^t \in \mathbb{R}^{n-1}, \quad \hat{\mathbf{y}} = (a_{21}, a_{31}, \ldots, a_{n1})^t \in \mathbb{R}^{n-1},$$

and $\hat{P}$ be the $(n-1) \times (n-1)$ Householder transformation

$$\hat{P} = I_{n-1} - 2\hat{\mathbf{w}}\hat{\mathbf{w}}^t.$$

Eq. (9.9) then becomes

$$P^{(1)}\begin{bmatrix} a_{11} \\ a_{21} \\ a_{31} \\ \vdots \\ a_{n1} \end{bmatrix} = \left[\begin{array}{c|ccc} 1 & 0 & \cdots & 0 \\ \hline 0 & & & \\ \vdots & & \hat{P} & \\ 0 & & & \end{array}\right] \cdot \left[\begin{array}{c} a_{11} \\ \hline \hat{\mathbf{y}} \end{array}\right] = \left[\begin{array}{c} a_{11} \\ \hline \hat{P}\hat{\mathbf{y}} \end{array}\right] = \left[\begin{array}{c} a_{11} \\ \hline \alpha \\ 0 \\ \vdots \\ 0 \end{array}\right]$$

with

$$\hat{P}\hat{\mathbf{y}} = (I_{n-1} - 2\hat{\mathbf{w}}\hat{\mathbf{w}}^t)\hat{\mathbf{y}} = \hat{\mathbf{y}} - 2(\hat{\mathbf{w}}^t\hat{\mathbf{y}})\hat{\mathbf{w}} = (\alpha, 0, \ldots, 0)^t. \tag{9.10}$$

Let $r = \hat{\mathbf{w}}^t\hat{\mathbf{y}}$. Then

$$(\alpha, 0, \ldots, 0)^t = (a_{21} - 2rw_2, a_{31} - 2rw_3, \ldots, a_{n1} - 2rw_n)^t,$$

and we can determine all of the w_i once we know α and r. Equating components gives

$$\alpha = a_{21} - 2rw_2$$

and

$$0 = a_{j1} - 2rw_j, \quad \text{for each } j = 3, \ldots, n.$$

Thus

$$2rw_2 = a_{21} - \alpha \tag{9.11}$$

and

$$2rw_j = a_{j1}, \quad \text{for each } j = 3, \ldots, n. \tag{9.12}$$

Squaring both sides of each of the equations and adding corresponding terms gives

$$4r^2 \sum_{j=2}^{n} w_j^2 = (a_{21} - \alpha)^2 + \sum_{j=3}^{n} a_{j1}^2.$$

Since $\mathbf{w}^t\mathbf{w} = 1$ and $w_1 = 0$, we have $\sum_{j=2}^{n} w_j^2 = 1$, and

$$4r^2 = \sum_{j=2}^{n} a_{j1}^2 - 2\alpha a_{21} + \alpha^2. \tag{9.13}$$

Equation (9.10) and the fact that P is orthogonal imply that

$$\alpha^2 = (\alpha, 0, \ldots, 0)(\alpha, 0, \ldots, 0)^t = (\hat{P}\hat{\mathbf{y}})^t\hat{P}\hat{\mathbf{y}} = \hat{\mathbf{y}}^t\hat{P}^t\hat{P}\hat{\mathbf{y}} = \hat{\mathbf{y}}^t\hat{\mathbf{y}}.$$

Thus

$$\alpha^2 = \sum_{j=2}^{n} a_{j1}^2,$$

which, when substituted into Eq. (9.13), gives

$$2r^2 = \sum_{j=2}^{n} a_{j1}^2 - \alpha a_{21}.$$

To ensure that $2r^2 = 0$ only if $a_{21} = a_{31} = \cdots = a_{n1} = 0$, we choose

$$\alpha = -\text{sgn}(a_{21}) \left(\sum_{j=2}^{n} a_{j1}^2 \right)^{1/2},$$

which implies that

$$2r^2 = \sum_{j=2}^{n} a_{j1}^2 + |a_{21}| \left(\sum_{j=2}^{n} a_{j1}^2 \right)^{1/2}.$$

With this choice of α and $2r^2$, we solve Eqs. (9.11) and (9.12) to obtain

$$w_2 = \frac{a_{21} - \alpha}{2r} \quad \text{and} \quad w_j = \frac{a_{j1}}{2r}, \quad \text{for each } j = 3, \ldots, n.$$

To summarize the choice of $P^{(1)}$, we have

$$\alpha = -\operatorname{sgn}(a_{21})\left(\sum_{j=2}^{n} a_{j1}^2\right)^{1/2},$$

$$r = \left(\frac{1}{2}\alpha^2 - \frac{1}{2}a_{21}\alpha\right)^{1/2},$$

$$w_1 = 0,$$

$$w_2 = \frac{a_{21} - \alpha}{2r},$$

and

$$w_j = \frac{a_{j1}}{2r}, \quad \text{for each } j = 3, \ldots, n.$$

With this choice,

$$A^{(2)} = P^{(1)}AP^{(1)} = \begin{bmatrix} a_{11}^{(2)} & a_{12}^{(2)} & 0 & \cdots & 0 \\ a_{21}^{(2)} & a_{22}^{(2)} & a_{23}^{(2)} & \cdots & a_{2n}^{(2)} \\ 0 & a_{32}^{(2)} & a_{33}^{(2)} & \cdots & a_{3n}^{(2)} \\ \vdots & \vdots & \vdots & & \vdots \\ 0 & a_{n2}^{(2)} & a_{n3}^{(2)} & \cdots & a_{nn}^{(2)} \end{bmatrix}.$$

Having found $P^{(1)}$ and computed $A^{(2)}$, the process is repeated for $k = 2, 3, \ldots, n-2$ as follows:

$$\alpha = -\operatorname{sgn}(a_{k+1,k}^{(k)})\left(\sum_{j=k+1}^{n} (a_{jk}^{(k)})^2\right)^{1/2},$$

$$r = \left(\frac{1}{2}\alpha^2 - \frac{1}{2}\alpha a_{k+1,k}^{(k)}\right)^{1/2},$$

$$w_1^{(k)} = w_2^{(k)} = \ldots = w_k^{(k)} = 0,$$

$$w_{k+1}^{(k)} = \frac{a_{k+1,k}^{(k)} - \alpha}{2r},$$

$$w_j^{(k)} = \frac{a_{jk}^{(k)}}{2r}, \quad \text{for each} \quad j = k+2, k+3, \ldots, n,$$

$$P^{(k)} = I - 2\mathbf{w}^{(k)} \cdot (\mathbf{w}^{(k)})^t,$$

and

$$A^{(k+1)} = P^{(k)}A^{(k)}P^{(k)},$$

where

$$A^{(k+1)} = \begin{bmatrix} a_{11}^{(k+1)} & a_{12}^{(k+1)} & 0 & \cdots & \cdots & 0 \\ a_{21}^{(k+1)} & \ddots & \ddots & \ddots & & \vdots \\ 0 & \ddots & \ddots & \ddots & 0 & 0 \\ \vdots & & a_{k+1,k}^{(k+1)} & a_{k+1,k+1}^{(k+1)} & a_{k+1,k+2}^{(k+1)} \cdots & a_{k+1,n}^{(k+1)} \\ \vdots & & 0 & \vdots & \ddots & \vdots \\ 0 & \cdots & 0 & a_{n,k+1}^{(k+1)} & \cdots & a_{nn}^{(k+1)} \end{bmatrix}.$$

Continuing in this manner, the tridiagonal and symmetric matrix $A^{(n-1)}$ is formed, where

$$A^{(n-1)} = P^{(n-2)}P^{(n-3)} \cdots P^{(1)}AP^{(1)} \cdots P^{(n-3)}P^{(n-2)}.$$

Example 1 Apply Householder transformations to the symmetric 4×4 matrix

$$A = \begin{bmatrix} 4 & 1 & -2 & 2 \\ 1 & 2 & 0 & 1 \\ -2 & 0 & 3 & -2 \\ 2 & 1 & -2 & -1 \end{bmatrix}$$

to produce a symmetric tridiagonal matrix that is similar to A.

Solution For the first application of a Householder transformation,

$$\alpha = -(1)\left(\sum_{j=2}^{4} a_{j1}^2\right)^{1/2} = -3, \; r = \left(\frac{1}{2}(-3)^2 - \frac{1}{2}(1)(-3)\right)^{1/2} = \sqrt{6},$$

$$\mathbf{w} = \left(0, \frac{\sqrt{6}}{3}, -\frac{\sqrt{6}}{6}, \frac{\sqrt{6}}{6}\right)^t,$$

$$P^{(1)} = \begin{bmatrix} 1 & 0 & 0 & 0 \\ 0 & 1 & 0 & 0 \\ 0 & 0 & 1 & 0 \\ 0 & 0 & 0 & 1 \end{bmatrix} - 2\left(\frac{\sqrt{6}}{6}\right)^2 \begin{bmatrix} 0 \\ 2 \\ -1 \\ 1 \end{bmatrix} \cdot (0, 2, -1, 1)$$

$$= \begin{bmatrix} 1 & 0 & 0 & 0 \\ 0 & -\frac{1}{3} & \frac{2}{3} & -\frac{2}{3} \\ 0 & \frac{2}{3} & \frac{2}{3} & \frac{1}{3} \\ 0 & -\frac{2}{3} & \frac{1}{3} & \frac{2}{3} \end{bmatrix},$$

and

$$A^{(2)} = \begin{bmatrix} 4 & -3 & 0 & 0 \\ -3 & \frac{10}{3} & 1 & \frac{4}{3} \\ 0 & 1 & \frac{5}{3} & -\frac{4}{3} \\ 0 & \frac{4}{3} & -\frac{4}{3} & -1 \end{bmatrix}.$$

Continuing to the second iteration,

$$\alpha = -\frac{5}{3}, \quad r = \frac{2\sqrt{5}}{3}, \quad \mathbf{w} = \left(0, 0, 2\sqrt{5}, \frac{\sqrt{5}}{5}\right)^t,$$

$$P^{(2)} = \begin{bmatrix} 1 & 0 & 0 & 0 \\ 0 & 1 & 0 & 0 \\ 0 & 0 & -\frac{3}{5} & -\frac{4}{5} \\ 0 & 0 & -\frac{4}{5} & \frac{3}{5} \end{bmatrix},$$

and the symmetric tridiagonal matrix is

$$A^{(3)} = \begin{bmatrix} 4 & -3 & 0 & 0 \\ -3 & \frac{10}{3} & -\frac{5}{3} & 0 \\ 0 & -\frac{5}{3} & -\frac{33}{25} & \frac{68}{75} \\ 0 & 0 & \frac{68}{75} & \frac{149}{75} \end{bmatrix}.$$

■

Algorithm 9.5 performs Householder's method as described here, although the actual matrix multiplications are circumvented.

ALGORITHM 9.5

Householder's

To obtain a symmetric tridiagonal matrix $A^{(n-1)}$ similar to the symmetric matrix $A = A^{(1)}$, construct the following matrices $A^{(2)}, A^{(3)}, \ldots, A^{(n-1)}$, where $A^{(k)} = (a_{ij}^{(k)})$ for each $k = 1, 2, \ldots, n-1$:

INPUT dimension n; matrix A.

OUTPUT $A^{(n-1)}$. *(At each step, A can be overwritten.)*

Step 1 For $k = 1, 2, \ldots, n-2$ do Steps 2–14.

Step 2 Set

$$q = \sum_{j=k+1}^{n} \left(a_{jk}^{(k)}\right)^2.$$

Step 3 If $a_{k+1,k}^{(k)} = 0$ then set $\alpha = -q^{1/2}$

$$\text{else set } \alpha = -\frac{q^{1/2} a_{k+1,k}^{(k)}}{|a_{k+1,k}^{(k)}|}.$$

Step 4 Set $RSQ = \alpha^2 - \alpha a_{k+1,k}^{(k)}$. *(Note:* $RSQ = 2r^2$*)*

Step 5 Set $v_k = 0$; *(Note:* $v_1 = \cdots = v_{k-1} = 0$*, but are not needed.)*
$v_{k+1} = a_{k+1,k}^{(k)} - \alpha$;
For $j = k+2, \ldots, n$ set $v_j = a_{jk}^{(k)}$.

$$\left(\textit{Note: } \mathbf{w} = \left(\frac{1}{\sqrt{2RSQ}}\right)\mathbf{v} = \frac{1}{2r}\mathbf{v}.\right)$$

Step 6 For $j = k, k+1, \ldots, n$ set $u_j = \left(\frac{1}{RSQ}\right) \sum_{i=k+1}^{n} a_{ji}^{(k)} v_i$.

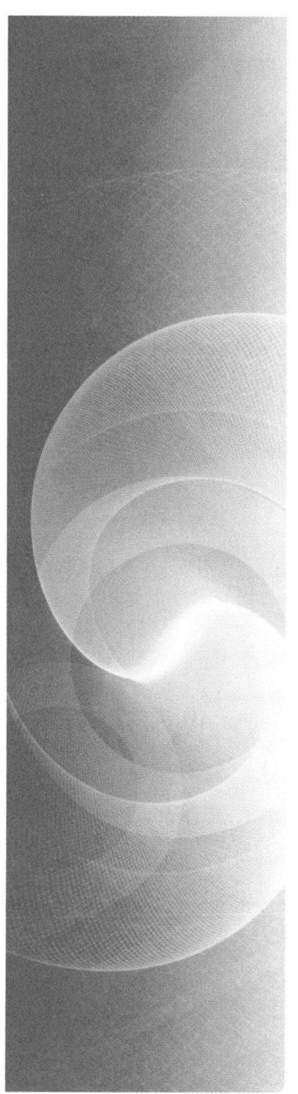

$$\left(\textit{Note: } \mathbf{u} = \left(\frac{1}{RSQ}\right) A^{(k)}\mathbf{v} = \frac{1}{2r^2} A^{(k)}\mathbf{v} = \frac{1}{r} A^{(k)}\mathbf{w}.\right)$$

Step 7 Set $PROD = \sum_{i=k+1}^{n} v_i u_i$.

$$\left(\textit{Note: } PROD = \mathbf{v}^t\mathbf{u} = \frac{1}{2r^2}\mathbf{v}^t A^{(k)}\mathbf{v}.\right)$$

Step 8 For $j = k, k+1, \ldots, n$ set $z_j = u_j - \left(\frac{PROD}{2RSQ}\right) v_j$.

$$\left(\textit{Note: } \mathbf{z} = \mathbf{u} - \frac{1}{2RSQ}\mathbf{v}^t\mathbf{u}\mathbf{v} = \mathbf{u} - \frac{1}{4r^2}\mathbf{v}^t\mathbf{u}\mathbf{v}\right.$$

$$\left. = \mathbf{u} - \mathbf{w}\mathbf{w}^t\mathbf{u} = \frac{1}{r}A^{(k)}\mathbf{w} - \mathbf{w}\mathbf{w}^t\frac{1}{r}A^{(k)}\mathbf{w}.\right)$$

Step 9 For $l = k+1, k+2, \ldots, n-1$ do Steps 10 and 11.
(*Note: Compute* $A^{(k+1)} = A^{(k)} - \mathbf{v}\mathbf{z}^t - \mathbf{z}\mathbf{v}^t = (I - 2\mathbf{w}\mathbf{w}^t)A^{(k)}(I - 2\mathbf{w}\mathbf{w}^t)$.)

Step 10 For $j = l+1, \ldots, n$ set
$a_{jl}^{(k+1)} = a_{jl}^{(k)} - v_l z_j - v_j z_l$;
$a_{lj}^{(k+1)} = a_{jl}^{(k+1)}$.

Step 11 Set $a_{ll}^{(k+1)} = a_{ll}^{(k)} - 2v_l z_l$.

Step 12 Set $a_{nn}^{(k+1)} = a_{nn}^{(k)} - 2v_n z_n$.

Step 13 For $j = k+2, \ldots, n$ set $a_{kj}^{(k+1)} = a_{jk}^{(k+1)} = 0$.

Step 14 Set $a_{k+1,k}^{(k+1)} = a_{k+1,k}^{(k)} - v_{k+1} z_k$;
$a_{k,k+1}^{(k+1)} = a_{k+1,k}^{(k+1)}$.
(*Note: The other elements of* $A^{(k+1)}$ *are the same as* $A^{(k)}$.)

Step 15 OUTPUT $(A^{(n-1)})$;
(*The process is complete.* $A^{(n-1)}$ *is symmetric, tridiagonal, and similar to* A.)
STOP. ■

Householder's method can be implemented in Maple with the *LinearAlgebra* package. For the matrix in Example 1 we would do the following.

with(LinearAlgebra): $A := Matrix([[4, 1, -2, 2], [1, 2, 0, 1], [-2, 0, 3, -2], [2, 1, -2, -1]])$

Then an orthogonal matrix Q and a tridiagonal matrix T with $A = QTQ^t$ are found using

$Q := TridiagonalForm(A, output = 'Q')$; $T := TridiagonalForm(A, output = 'T')$

The matrices produced by Maple are the 10-digit approximations to

$$Q = \begin{bmatrix} 1 & 0 & 0 & 0 \\ 0 & -0.\overline{3} & 0.1\overline{3} & -0.9\overline{3} \\ 0 & 0.\overline{6} & -0.\overline{6} & -0.\overline{3} \\ 0 & -0.\overline{6} & 0.7\overline{3} & 0.1\overline{3} \end{bmatrix} \quad \text{and} \quad T = \begin{bmatrix} 4 & -3 & 0 & 0 \\ -3 & 3.\overline{3} & -0.1\overline{6} & 0 \\ 0 & -0.1\overline{6} & -1.32 & 0.90\overline{6} \\ 0 & 0 & 0.90\overline{6} & 1.98\overline{6} \end{bmatrix}$$

In the next section, we will examine how the QR algorithm can be applied to determine the eigenvalues of $A^{(n-1)}$, which are the same as those of the original matrix A.

Householder's Algorithm can be applied to an arbitrary $n \times n$ matrix, but modifications must be made to account for a possible lack of symmetry. The resulting matrix $A^{(n-1)}$ will not be tridiagonal unless the original matrix A is symmetric, but all the entries below the lower subdiagonal will be 0. A matrix of this type is called *upper Hessenberg*. That is, $H = (h_{ij})$ is **upper Hessenberg** if $h_{ij} = 0$, for all $i \geq j + 2$.

The required modifications for arbitrary matrices are:

Step 6 For $j = 1, 2, \ldots, n$ set $u_j = \dfrac{1}{RSQ} \sum_{i=k+1}^{n} a_{ji}^{(k)} v_i;$

$$y_j = \frac{1}{RSQ} \sum_{i=k+1}^{n} a_{ij}^{(k)} v_i.$$

Step 8 For $j = 1, 2, \ldots, n$ set $z_j = u_j - \dfrac{PROD}{RSQ} v_j.$

Step 9 For $l = k + 1, k + 2, \ldots, n$ do Steps 10 and 11.

Step 10 For $j = 1, 2, \ldots, k$ set $a_{jl}^{(k+1)} = a_{jl}^{(k)} - z_j v_l;$

$$a_{lj}^{(k+1)} = a_{lj}^{(k)} - y_j v_l.$$

Step 11 For $j = k + 1, \ldots, n$ set $a_{jl}^{(k+1)} = a_{jl}^{(k)} - z_j v_l - y_l v_j.$

After these steps are modified, delete Steps 12 through 14 and output $A^{(n-1)}$.

EXERCISE SET 9.4

1. Use Householder's method to place the following matrices in tridiagonal form.

a. $\begin{bmatrix} 12 & 10 & 4 \\ 10 & 8 & -5 \\ 4 & -5 & 3 \end{bmatrix}$ **b.** $\begin{bmatrix} 2 & -1 & -1 \\ -1 & 2 & -1 \\ -1 & -1 & 2 \end{bmatrix}$

c. $\begin{bmatrix} 1 & 1 & 1 \\ 1 & 1 & 0 \\ 1 & 0 & 1 \end{bmatrix}$ **d.** $\begin{bmatrix} 4.75 & 2.25 & -0.25 \\ 2.25 & 4.75 & 1.25 \\ -0.25 & 1.25 & 4.75 \end{bmatrix}$

2. Use Householder's method to place the following matrices in tridiagonal form.

a. $\begin{bmatrix} 4 & -1 & -1 & 0 \\ -1 & 4 & 0 & -1 \\ -1 & 0 & 4 & -1 \\ 0 & -1 & -1 & 4 \end{bmatrix}$ **b.** $\begin{bmatrix} 5 & -2 & -0.5 & 1.5 \\ -2 & 5 & 1.5 & -0.5 \\ -0.5 & 1.5 & 5 & -2 \\ 1.5 & -0.5 & -2 & 5 \end{bmatrix}$

c. $\begin{bmatrix} 8 & 0.25 & 0.5 & 2 & -1 \\ 0.25 & -4 & 0 & 1 & 2 \\ 0.5 & 0 & 5 & 0.75 & -1 \\ 2 & 1 & 0.75 & 5 & -0.5 \\ -1 & 2 & -1 & -0.5 & 6 \end{bmatrix}$

d. $\begin{bmatrix} 2 & -1 & -1 & 0 & 0 \\ -1 & 3 & 0 & -2 & 0 \\ -1 & 0 & 4 & 2 & 1 \\ 0 & -2 & 2 & 8 & 3 \\ 0 & 0 & 1 & 3 & 9 \end{bmatrix}$

3. Modify Householder's Algorithm 9.5 to compute similar upper Hessenberg matrices for the following nonsymmetric matrices.

a. $\begin{bmatrix} 2 & -1 & 3 \\ 2 & 0 & 1 \\ -2 & 1 & 4 \end{bmatrix}$ **b.** $\begin{bmatrix} -1 & 2 & 3 \\ 2 & 3 & -2 \\ 3 & 1 & -1 \end{bmatrix}$

c. $\begin{bmatrix} 5 & -2 & -3 & 4 \\ 0 & 4 & 2 & -1 \\ 1 & 3 & -5 & 2 \\ -1 & 4 & 0 & 3 \end{bmatrix}$ **d.** $\begin{bmatrix} 4 & -1 & -1 & -1 \\ -1 & 4 & 0 & -1 \\ -1 & -1 & 4 & -1 \\ -1 & -1 & -1 & 4 \end{bmatrix}$

9.5 The QR Algorithm

The deflation methods discussed in Section 9.3 are not generally suitable for calculating all the eigenvalues of a matrix because of the growth of round-off error. In this section we consider the QR Algorithm, a matrix reduction technique used to simultaneously determine all the eigenvalues of a symmetric matrix.

To apply the QR method, we begin with a symmetric matrix in tridiagonal form; that is, the only nonzero entries in the matrix lie either on the diagonal or on the subdiagonals directly above or below the diagonal. If this is not the form of the symmetric matrix, the first step is to apply Householder's method to compute a symmetric, tridiagonal matrix similar to the given matrix.

In the remainder of this section it will be assumed that the symmetric matrix for which these eigenvalues are to be calculated is tridiagonal. If we let A denote a matrix of this type, we can simplify the notation somewhat by labeling the entries of A as follows:

$$A = \begin{bmatrix} a_1 & b_2 & 0 & \cdots & \cdots & 0 \\ b_2 & a_2 & b_3 & \ddots & & \vdots \\ 0 & b_3 & a_3 & \ddots & \ddots & 0 \\ \vdots & \ddots & \ddots & \ddots & \ddots & b_n \\ 0 & \cdots & \cdots & 0 & b_n & a_n \end{bmatrix}. \tag{9.14}$$

If $b_2 = 0$ or $b_n = 0$, then the 1×1 matrix $[a_1]$ or $[a_n]$ immediately produces an eigenvalue a_1 or a_n of A. The QR method takes advantage of this observation by successively decreasing the values of the entries below the main diagonal until $b_2 \approx 0$ or $b_n \approx 0$.

When $b_j = 0$ for some j, where $2 < j < n$, the problem can be reduced to considering, instead of A, the smaller matrices

$$\begin{bmatrix} a_1 & b_2 & 0 & \cdots & \cdots & 0 \\ b_2 & a_2 & b_3 & \ddots & & \vdots \\ 0 & b_3 & a_3 & \ddots & \ddots & 0 \\ \vdots & \ddots & \ddots & \ddots & \ddots & b_{j-1} \\ 0 & \cdots & \cdots & 0 & b_{j-1} & a_{j-1} \end{bmatrix} \quad \text{and} \quad \begin{bmatrix} a_j & b_{j+1} & 0 & \cdots & \cdots & 0 \\ b_{j+1} & a_{j+1} & b_{j+2} & \ddots & & \vdots \\ 0 & b_{j+2} & a_{j+2} & \ddots & \ddots & 0 \\ \vdots & \ddots & \ddots & \ddots & \ddots & b_n \\ 0 & \cdots & \cdots & 0 & b_n & a_n \end{bmatrix}. \tag{9.15}$$

If none of the b_j are zero, the QR method proceeds by forming a sequence of matrices $A = A^{(1)}, A^{(2)}, A^{(3)}, \ldots,$ as follows:

1. $A^{(1)} = A$ is factored as a product $A^{(1)} = Q^{(1)}R^{(1)}$, where $Q^{(1)}$ is orthogonal and $R^{(1)}$ is upper triangular.

2. $A^{(2)}$ is defined as $A^{(2)} = R^{(1)}Q^{(1)}$.

In general, $A^{(i)}$ is factored as a product $A^{(i)} = Q^{(i)}R^{(i)}$ of an orthogonal matrix $Q^{(i)}$ and an upper triangular matrix $R^{(i)}$. Then $A^{(i+1)}$ is defined by the product of $R^{(i)}$ and $Q^{(i)}$ in the reverse direction $A^{(i+1)} = R^{(i)}Q^{(i)}$. Since $Q^{(i)}$ is orthogonal, $R^{(i)} = Q^{(i)^t}A^{(i)}$ and

$$A^{(i+1)} = R^{(i)}Q^{(i)} = (Q^{(i)^t}A^{(i)})Q^{(i)} = Q^{(i)^t}A^{(i)}Q^{(i)}. \tag{9.16}$$

This ensures that $A^{(i+1)}$ is symmetric with the same eigenvalues as $A^{(i)}$. By the manner in which we define $R^{(i)}$ and $Q^{(i)}$, we also ensure that $A^{(i+1)}$ is tridagonal.

Continuing by induction, $A^{(i+1)}$ has the same eigenvalues as the original matrix A, and $A^{(i+1)}$ tends to a diagonal matrix with the eigenvalues of A along the diagonal.

Rotation Matrices

To describe the construction of the factoring matrices $Q^{(i)}$ and $R^{(i)}$, we need the notion of a *rotation matrix*.

Definition 9.23 A **rotation matrix** P differs from the identity matrix in at most four elements. These four elements are of the form

$$p_{ii} = p_{jj} = \cos\theta \quad \text{and} \quad p_{ij} = -p_{ji} = \sin\theta,$$

for some θ and some $i \neq j$. ■

If A is the 2×2 rotation matrix

$$A = \begin{bmatrix} \cos\theta & -\sin\theta \\ \sin\theta & \cos\theta \end{bmatrix},$$

then $A\mathbf{x}$ is $\mathbf{x}$ rotated counter-clockwise by the angle θ.

It is easy to show (see Exercise 8) that, for any rotation matrix P, the matrix AP differs from A only in the ith and jth columns and the matrix PA differs from A only in the ith and jth rows. For any $i \neq j$, the angle θ can be chosen so that the product PA has a zero entry for $(PA)_{ij}$. In addition, every rotation matrix P is orthogonal, because the definition implies that $PP^t = I$.

Example 1 Find a rotation matrix P with the property that PA has a zero entry in the second row and first column, where

$$A = \begin{bmatrix} 3 & 1 & 0 \\ 1 & 3 & 1 \\ 0 & 1 & 3 \end{bmatrix}.$$

These are often called Givens rotations because they were used by James Wallace Givens (1910–1993) in the 1950s when he was at Argonne National Laboratories.

Solution The form of P is

$$P = \begin{bmatrix} \cos\theta & \sin\theta & 0 \\ -\sin\theta & \cos\theta & 0 \\ 0 & 0 & 1 \end{bmatrix} \quad \text{so} \quad PA = \begin{bmatrix} 3\cos\theta + \sin\theta & \cos\theta + 3\sin\theta & \sin\theta \\ -3\sin\theta + \cos\theta & -\sin\theta + 3\cos\theta & \cos\theta \\ 0 & 1 & 3 \end{bmatrix}.$$

The angle θ is chosen so that $-3\sin\theta + \cos\theta = 0$, that is, so that $\tan\theta = \dfrac{1}{3}$. Hence

$$\cos\theta = \frac{3\sqrt{10}}{10}, \quad \sin\theta = \frac{\sqrt{10}}{10}$$

and

$$PA = \begin{bmatrix} \frac{3\sqrt{10}}{10} & \frac{\sqrt{10}}{10} & 0 \\ -\frac{\sqrt{10}}{10} & \frac{3\sqrt{10}}{10} & 0 \\ 0 & 0 & 1 \end{bmatrix} \begin{bmatrix} 3 & 1 & 0 \\ 1 & 3 & 1 \\ 0 & 1 & 3 \end{bmatrix} = \begin{bmatrix} \sqrt{10} & \frac{3}{5}\sqrt{10} & \frac{1}{10}\sqrt{10} \\ 0 & \frac{4}{5}\sqrt{10} & \frac{3}{10}\sqrt{10} \\ 0 & 1 & 3 \end{bmatrix}.$$

Note that the resulting matrix is neither symmetric nor tridiagonal. ■

The factorization of $A^{(1)}$ into $A^{(1)} = Q^{(1)}R^{(1)}$ uses a product of $n-1$ rotation matrices to construct

$$R^{(1)} = P_n P_{n-1} \cdots P_2 A^{(1)}.$$

We first choose the rotation matrix P_2 with

$$p_{11} = p_{22} = \cos\theta_2 \quad \text{and} \quad p_{12} = -p_{21} = \sin\theta_2,$$

where

$$\sin\theta_2 = \frac{b_2}{\sqrt{b_2^2 + a_1^2}} \quad \text{and} \quad \cos\theta_2 = \frac{a_1}{\sqrt{b_2^2 + a_1^2}}.$$

This choice gives

$$(-\sin\theta_2)a_1 + (\cos\theta_2)b_2 = \frac{-b_2 a_1}{\sqrt{b_2^2 + a_1^2}} + \frac{a_1 b_2}{\sqrt{b_2^2 + a_1^2}} = 0.$$

for the entry in the (2, 1) position, that is, in the second row and first column of the product $P_2A^{(1)}$. So the matrix

$$A_2^{(1)} = P_2 A^{(1)}$$

has a zero in the (2, 1) position.

The multiplication $P_2A^{(1)}$ affects both rows 1 and 2 of $A^{(1)}$, so the matrix $A_2^{(1)}$ does not necessarily retain zero entries in positions $(1,3), (1,4), \ldots,$ and $(1,n)$. However, $A^{(1)}$ is tridiagonal, so the $(1,4), \ldots, (1,n)$ entries of $A_2^{(1)}$ must also be 0. Only the $(1,3)$-entry, the one in the first row and third column, can become nonzero in $A_2^{(1)}$.

In general, the matrix P_k is chosen so that the $(k, k-1)$ entry in $A_k^{(1)} = P_k A_{k-1}^{(1)}$ is zero. This results in the $(k-1, k+1)$-entry becoming nonzero. The matrix $A_k^{(1)}$ has the form

$$A_k^{(1)} = \begin{bmatrix} z_1 & q_1 & r_1 & 0 & \cdots & \cdots & \cdots & \cdots & \cdots & 0 \\ 0 & \ddots & \ddots & \ddots & \ddots & & & & & \vdots \\ 0 & \ddots & \ddots & \ddots & \ddots & \ddots & & & & \vdots \\ \vdots & & \ddots & 0 & z_{k-1} & q_{k-1} & r_{k-1} & \ddots & & \vdots \\ \vdots & & & \ddots & 0 & x_k & y_k & 0 & \ddots & \vdots \\ \vdots & & & & \ddots & b_{k+1} & a_{k+1} & b_{k+2} & \ddots & 0 \\ \vdots & & & & & \ddots & \ddots & \ddots & \ddots & 0 \\ \vdots & & & & & & \ddots & \ddots & \ddots & b_n \\ 0 & \cdots & \cdots & \cdots & \cdots & \cdots & \cdots & 0 & b_n & a_n \end{bmatrix}$$

and P_{k+1} has the form

$$P_{k+1} = \left[\begin{array}{c|cc|c} I_{k-1} & O & & O \\ \hline & c_{k+1} & s_{k+1} & \\ O & & & O \\ & -s_{k+1} & c_{k+1} & \\ \hline O & O & & I_{n-k-1} \end{array}\right] \begin{array}{l} \\ \leftarrow \text{row } k \\ \\ \\ \\ \end{array} \qquad (9.17)$$

$\uparrow$ column k

where 0 denotes the appropriately dimensional matrix with all zero entries.

The constants $c_{k+1} = \cos\theta_{k+1}$ and $s_{k+1} = \sin\theta_{k+1}$ in P_{k+1} are chosen so that the $(k+1, k)$-entry in $A_{k+1}^{(1)}$ is zero; that is, $-s_{k+1}x_k + c_{k+1}b_{k+1} = 0$.

Since $c_{k+1}^2 + s_{k+1}^2 = 1$, the solution to this equation is

$$s_{k+1} = \frac{b_{k+1}}{\sqrt{b_{k+1}^2 + x_k^2}} \quad \text{and} \quad c_{k+1} = \frac{x_k}{\sqrt{b_{k+1}^2 + x_k^2}},$$

and $A_{k+1}^{(1)}$ has the form

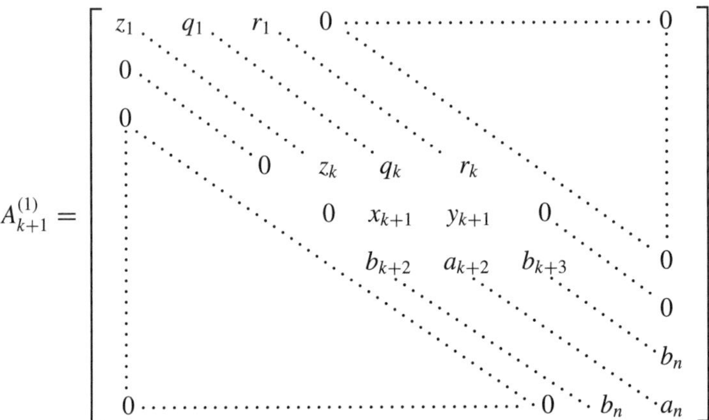

Proceeding with this construction in the sequence $P_2, \ldots, P_n$ produces the upper triangular matrix

$$R^{(1)} \equiv A_n^{(1)} = \begin{bmatrix} z_1 & q_1 & r_1 & 0 & \cdots & 0 \\ 0 & \ddots & \ddots & \ddots & \ddots & \vdots \\ \vdots & \ddots & \ddots & \ddots & \ddots & 0 \\ \vdots & & \ddots & \ddots & \ddots & r_{n-2} \\ \vdots & & & \ddots & z_{n-1} & q_{n-1} \\ 0 & \cdots & \cdots & \cdots & 0 & x_n \end{bmatrix}.$$

The other half of the QR factorization is the matrix

$$Q^{(1)} = P_2^t P_3^t \cdots P_n^t,$$

because the orthogonality of the rotation matrices implies that

$$Q^{(1)}R^{(1)} = (P_2^t P_3^t \cdots P_n^t) \cdot (P_n \cdots P_3 P_2)A^{(1)} = A^{(1)}.$$

The matrix $Q^{(1)}$ is orthogonal because

$$(Q^{(1)})^t Q^{(1)} = (P_2^t P_3^t \cdots P_n^t)^t (P_2^t P_3^t \cdots P_n^t) = (P_n \cdots P_3 P_2) \cdot (P_2^t P_3^t \cdots P_n^t) = I.$$

In addition, $Q^{(1)}$ is an upper-Hessenberg matrix. To see why this is true, you can follow the steps in Exercises 9 and 10.

As a consequence, $A^{(2)} = R^{(1)}Q^{(1)}$ is also an upper-Hessenberg matrix, because multiplying $Q^{(1)}$ on the left by the upper triangular matrix $R^{(1)}$ does not affect the entries in the lower triangle. We already know that it is symmetric, so $A^{(2)}$ is tridiagonal.

The entries off the diagonal of $A^{(2)}$ will generally be smaller in magnitude than the corresponding entries of $A^{(1)}$, so $A^{(2)}$ is closer to being a diagonal matrix than is $A^{(1)}$. The process is repeated to construct $A^{(3)}, A^{(4)}, \ldots$ until satisfactory convergence is obtained. (See [Wil2], pages 516–523.)

Example 2 Apply one iteration of the QR method to the matrix that was given in Example 1:

$$A = \begin{bmatrix} 3 & 1 & 0 \\ 1 & 3 & 1 \\ 0 & 1 & 3 \end{bmatrix}.$$

Solution Let $A^{(1)} = A$ be the given matrix and P_2 represent the rotation matrix determined in Example 1. We found, using the notation introduced in the QR method, that

$$A_2^{(1)} = P_2 A^{(1)} = \begin{bmatrix} \frac{3\sqrt{10}}{10} & \frac{\sqrt{10}}{10} & 0 \\ -\frac{\sqrt{10}}{10} & \frac{3\sqrt{10}}{10} & 0 \\ 0 & 0 & 1 \end{bmatrix} \begin{bmatrix} 3 & 1 & 0 \\ 1 & 3 & 1 \\ 0 & 1 & 3 \end{bmatrix} = \begin{bmatrix} \sqrt{10} & \frac{3}{5}\sqrt{10} & \frac{\sqrt{10}}{10} \\ 0 & \frac{4\sqrt{10}}{5} & \frac{3\sqrt{10}}{10} \\ 0 & 1 & 3 \end{bmatrix}$$

$$\equiv \begin{bmatrix} z_1 & q_1 & r_1 \\ 0 & x_2 & y_2 \\ 0 & b_3^{(1)} & a_3^{(1)} \end{bmatrix}.$$

Continuing, we have

$$s_3 = \frac{b_3^{(1)}}{\sqrt{x_2^2 + (b_3^{(1)})^2}} = 0.36761 \quad \text{and} \quad c_3 = \frac{x_2}{\sqrt{x_2^2 + (b_3^{(1)})^2}} = 0.92998.$$

so

$$R^{(1)} \equiv A_3^{(1)} = P_3 A_2^{(1)} = \begin{bmatrix} 1 & 0 & 0 \\ 0 & 0.92998 & 0.36761 \\ 0 & -0.36761 & 0.92998 \end{bmatrix} \begin{bmatrix} \sqrt{10} & \frac{3}{5}\sqrt{10} & \frac{\sqrt{10}}{10} \\ 0 & \frac{4\sqrt{10}}{5} & \frac{3\sqrt{10}}{10} \\ 0 & 1 & 3 \end{bmatrix}$$

$$= \begin{bmatrix} \sqrt{10} & \frac{3}{5}\sqrt{10} & \frac{\sqrt{10}}{10} \\ 0 & 2.7203 & 1.9851 \\ 0 & 0 & 2.4412 \end{bmatrix},$$

and

$$Q^{(1)} = P_2^t P_3^t = \begin{bmatrix} \frac{3\sqrt{10}}{10} & -\frac{\sqrt{10}}{10} & 0 \\ \frac{\sqrt{10}}{10} & \frac{3\sqrt{10}}{10} & 0 \\ 0 & 0 & 1 \end{bmatrix} \begin{bmatrix} 1 & 0 & 0 \\ 0 & 0.92998 & -0.36761 \\ 0 & 0.36761 & 0.92998 \end{bmatrix}$$

$$= \begin{bmatrix} 0.94868 & -0.29409 & 0.11625 \\ 0.31623 & 0.88226 & -0.34874 \\ 0 & 0.36761 & 0.92998 \end{bmatrix}.$$

As a consequence,

$$A^{(2)} = R^{(1)} Q^{(1)} = \begin{bmatrix} \sqrt{10} & \frac{3}{5}\sqrt{10} & \frac{\sqrt{10}}{10} \\ 0 & 2.7203 & 1.9851 \\ 0 & 0 & 2.4412 \end{bmatrix} \begin{bmatrix} 0.94868 & -0.29409 & 0.11625 \\ 0.31623 & 0.88226 & -0.34874 \\ 0 & 0.36761 & 0.92998 \end{bmatrix}$$

$$= \begin{bmatrix} 3.6 & 0.86024 & 0 \\ 0.86024 & 3.12973 & 0.89740 \\ 0 & 0.89740 & 2.27027 \end{bmatrix}.$$

The off-diagonal elements of $A^{(2)}$ are smaller than those of $A^{(1)}$ by about 14%, so we have a reduction but it is not substantial. To decrease to below 0.001 we would need to perform 13 iterations of the QR method. Doing this gives

$$A^{(13)} = \begin{bmatrix} 4.4139 & 0.01941 & 0 \\ 0.01941 & 3.0003 & 0.00095 \\ 0 & 0.00095 & 1.5858 \end{bmatrix}$$

This would give an approximate eigenvalue of 1.5858 and the remaining eigenvalues could be approximated by considering the reduced matrix

$$\begin{bmatrix} 4.4139 & 0.01941 \\ 0.01941 & 3.0003 \end{bmatrix}$$

■

Accelerating Convergence

If the eigenvalues of A have distinct moduli with $|\lambda_1| > |\lambda_2| > \cdots > |\lambda_n|$, then the rate of convergence of the entry $b_{j+1}^{(i+1)}$ to 0 in the matrix $A^{(i+1)}$ depends on the ratio $|\lambda_{j+1}/\lambda_j|$ (see [Fr]). The rate of convergence of $b_{j+1}^{(i+1)}$ to 0 determines the rate at which the entry $a_j^{(i+1)}$ converges to the jth eigenvalue λ_j. Thus, the rate of convergence can be slow if $|\lambda_{j+1}/\lambda_j|$ is not significantly less than 1.

To accelerate this convergence, a shifting technique is employed similar to that used with the Inverse Power method in Section 9.3. A constant σ is selected close to an eigenvalue of A. This modifies the factorization in Eq. (9.16) to choosing $Q^{(i)}$ and $R^{(i)}$ so that

$$A^{(i)} - \sigma I = Q^{(i)}R^{(i)}, \tag{9.18}$$

and, correspondingly, the matrix $A^{(i+1)}$ is defined to be

$$A^{(i+1)} = R^{(i)}Q^{(i)} + \sigma I. \tag{9.19}$$

With this modification, the rate of convergence of $b_{j+1}^{(i+1)}$ to 0 depends on the ratio $|(\lambda_{j+1} - \sigma)/(\lambda_j - \sigma)|$. This can result in a significant improvement over the original rate of convergence of $a_j^{(i+1)}$ to λ_j if σ is close to λ_{j+1} but not close to λ_j.

We change σ at each step so that when A has eigenvalues of distinct modulus, $b_n^{(i+1)}$ converges to 0 faster than $b_j^{(i+1)}$ for any integer j less than n. When $b_n^{(i+1)}$ is sufficiently small, we assume that $\lambda_n \approx a_n^{(i+1)}$, delete the nth row and column of the matrix, and proceed in the same manner to find an approximation to λ_{n-1}. The process is continued until an approximation has been determined for each eigenvalue.

The shifting technique chooses, at the ith step, the shifting constant σ_i, where σ_i is the eigenvalue of the matrix

$$E^{(i)} = \begin{bmatrix} a_{n-1}^{(i)} & b_n^{(i)} \\ b_n^{(i)} & a_n^{(i)} \end{bmatrix}$$

that is closest to $a_n^{(i)}$. This shift translates the eigenvalues of A by a factor σ_i. With this shifting technique, the convergence is usually cubic. (See [WR], p. 270.) The method accumulates these shifts until $b_n^{(i+1)} \approx 0$ and then adds the shifts to $a_n^{(i+1)}$ to approximate the eigenvalue λ_n.

If A has eigenvalues of the same modulus, $b_j^{(i+1)}$ may tend to 0 for some $j \neq n$ at a faster rate than $b_n^{(i+1)}$. In this case, the matrix-splitting technique described in (9.14) can be employed to reduce the problem to one involving a pair of matrices of reduced order.

Example 3 Incorporate shifting into the QR method for the matrix

$$A = \begin{bmatrix} 3 & 1 & 0 \\ 1 & 3 & 1 \\ 0 & 1 & 3 \end{bmatrix} = \begin{bmatrix} a_1^{(1)} & b_2^{(1)} & 0 \\ b_2^{(1)} & a_2^{(1)} & b_3^{(1)} \\ 0 & b_3^{(1)} & a_3^{(1)} \end{bmatrix}.$$

Solution To find the acceleration parameter for shifting requires finding the eigenvalues of

$$\begin{bmatrix} a_2^{(1)} & b_3^{(1)} \\ b_3^{(1)} & a_3^{(1)} \end{bmatrix} = \begin{bmatrix} 3 & 1 \\ 1 & 3 \end{bmatrix},$$

which are $\mu_1 = 4$ and $\mu_2 = 2$. The choice of eigenvalue closest to $a_3^{(1)} = 3$ is arbitrary, and we choose $\mu_2 = 2$ and shift by this amount. Then $\sigma_1 = 2$ and

$$\begin{bmatrix} d_1 & b_2^{(1)} & 0 \\ b_2^{(1)} & d_2 & b_3^{(1)} \\ 0 & b_3^{(1)} & d_3 \end{bmatrix} = \begin{bmatrix} 1 & 1 & 0 \\ 1 & 1 & 1 \\ 0 & 1 & 1 \end{bmatrix}.$$

Continuing the computation gives

$$x_1 = 1, \quad y_1 = 1, \quad z_1 = \sqrt{2}, \quad c_2 = \frac{\sqrt{2}}{2}, \quad s_2 = \frac{\sqrt{2}}{2},$$

$$q_1 = \sqrt{2}, \quad x_2 = 0, \quad r_1 = \frac{\sqrt{2}}{2}, \quad \text{and} \quad y_2 = \frac{\sqrt{2}}{2},$$

so

$$A_2^{(1)} = \begin{bmatrix} \sqrt{2} & \sqrt{2} & \frac{\sqrt{2}}{2} \\ 0 & 0 & \frac{\sqrt{2}}{2} \\ 0 & 1 & 1 \end{bmatrix}.$$

Further,

$$z_2 = 1, \quad c_3 = 0, \quad s_3 = 1, \quad q_2 = 1, \quad \text{and} \quad x_3 = -\frac{\sqrt{2}}{2},$$

so

$$R^{(1)} = A_3^{(1)} = \begin{bmatrix} \sqrt{2} & \sqrt{2} & \frac{\sqrt{2}}{2} \\ 0 & 1 & 1 \\ 0 & 0 & -\frac{\sqrt{2}}{2} \end{bmatrix}.$$

To compute $A^{(2)}$, we have

$$z_3 = -\frac{\sqrt{2}}{2}, \quad a_1^{(2)} = 2, \quad b_2^{(2)} = \frac{\sqrt{2}}{2}, \quad a_2^{(2)} = 1, \quad b_3^{(2)} = -\frac{\sqrt{2}}{2}, \quad \text{and} \quad a_3^{(2)} = 0,$$

so

$$A^{(2)} = R^{(1)}Q^{(1)} = \begin{bmatrix} 2 & \frac{\sqrt{2}}{2} & 0 \\ \frac{\sqrt{2}}{2} & 1 & -\frac{\sqrt{2}}{2} \\ 0 & -\frac{\sqrt{2}}{2} & 0 \end{bmatrix}.$$

One iteration of the QR method is complete. Neither $b_2^{(2)} = \sqrt{2}/2$ nor $b_3^{(2)} = -\sqrt{2}/2$ is small, so another iteration of the QR method is performed. For this iteration, we calculate the eigenvalues $\frac{1}{2} \pm \frac{1}{2}\sqrt{3}$ of the matrix

$$\begin{bmatrix} a_2^{(2)} & b_3^{(2)} \\ b_3^{(2)} & a_3^{(2)} \end{bmatrix} = \begin{bmatrix} 1 & -\frac{\sqrt{2}}{2} \\ -\frac{\sqrt{2}}{2} & 0 \end{bmatrix},$$

and choose $\sigma_2 = \frac{1}{2} - \frac{1}{2}\sqrt{3}$, the closest eigenvalue to $a_3^{(2)} = 0$. Completing the calculations gives

$$A^{(3)} = \begin{bmatrix} 2.6720277 & 0.37597448 & 0 \\ 0.37597448 & 1.4736080 & 0.030396964 \\ 0 & 0.030396964 & -0.047559530 \end{bmatrix}.$$

If $b_3^{(3)} = 0.030396964$ is sufficiently small, then the approximation to the eigenvalue λ_3 is 1.5864151, the sum of $a_3{}^{(3)}$ and the shifts $\sigma_1 + \sigma_2 = 2 + (1 - \sqrt{3})/2$. Deleting the third row and column gives

$$A^{(3)} = \begin{bmatrix} 2.6720277 & 0.37597448 \\ 0.37597448 & 1.4736080 \end{bmatrix},$$

which has eigenvalues $\mu_1 = 2.7802140$ and $\mu_2 = 1.3654218$. Adding the shifts gives the approximations

$$\lambda_1 \approx 4.4141886 \quad \text{and} \quad \lambda_2 \approx 2.9993964.$$

The actual eigenvalues of the matrix A are 4.41420, 3.00000, and 1.58579, so the QR method gave four significant digits of accuracy in only two iterations. ■

Algorithm 9.6 implements the QR method with shifting incorporated to accelerate convergence.

QR

To obtain the eigenvalues of the symmetric, tridiagonal $n \times n$ matrix

$$A \equiv A_1 = \begin{bmatrix} a_1^{(1)} & b_2^{(1)} & 0 & \cdots & 0 \\ b_2^{(1)} & a_2^{(1)} & \ddots & \ddots & \vdots \\ 0 & \ddots & \ddots & \ddots & 0 \\ \vdots & \ddots & \ddots & \ddots & b_n^{(1)} \\ 0 & \cdots & 0 & b_n^{(1)} & a_n^{(1)} \end{bmatrix}$$

INPUT $n; a_1^{(1)}, \ldots, a_n^{(1)}, b_2^{(1)}, \ldots, b_n^{(1)}$; tolerance TOL; maximum number of iterations M.

OUTPUT eigenvalues of A, or recommended splitting of A, or a message that the maximum number of iterations was exceeded.

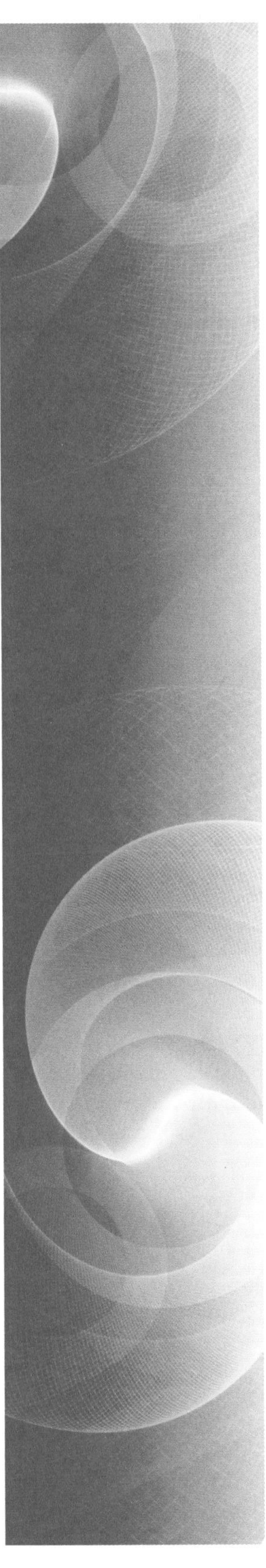

Step 1 Set $k = 1$;
$SHIFT = 0$. *(Accumulated shift.)*

Step 2 While $k \leq M$ do Steps 3–19.
(Steps 3–7 test for success.)

Step 3 If $|b_n^{(k)}| \leq TOL$ then set $\lambda = a_n^{(k)} + SHIFT$;
OUTPUT (λ);
set $n = n - 1$.

Step 4 If $|b_2^{(k)}| \leq TOL$ then set $\lambda = a_1^{(k)} + SHIFT$;
OUTPUT (λ);
set $n = n - 1$;
$a_1^{(k)} = a_2^{(k)}$;
for $j = 2, \ldots, n$
set $a_j^{(k)} = a_{j+1}^{(k)}$;
$b_j^{(k)} = b_{j+1}^{(k)}$.

Step 5 If $n = 0$ then
STOP.

Step 6 If $n = 1$ then
set $\lambda = a_1^{(k)} + SHIFT$;
OUTPUT (λ);
STOP.

Step 7 For $j = 3, \ldots, n - 1$
if $|b_j^{(k)}| \leq TOL$ then
OUTPUT ('split into', $a_1^{(k)}, \ldots, a_{j-1}^{(k)}, b_2^{(k)}, \ldots, b_{j-1}^{(k)}$,
'and',
$a_j^{(k)}, \ldots, a_n^{(k)}, b_{j+1}^{(k)}, \ldots, b_n^{(k)}, SHIFT$);
STOP.

Step 8 *(Compute shift.)*
Set $b = -(a_{n-1}^{(k)} + a_n^{(k)})$;
$c = a_n^{(k)} a_{n-1}^{(k)} - \left[b_n^{(k)}\right]^2$;
$d = (b^2 - 4c)^{1/2}$.

Step 9 If $b > 0$ then set $\mu_1 = -2c/(b + d)$;
$\mu_2 = -(b + d)/2$
else set $\mu_1 = (d - b)/2$;
$\mu_2 = 2c/(d - b)$.

Step 10 If $n = 2$ then set $\lambda_1 = \mu_1 + SHIFT$;
$\lambda_2 = \mu_2 + SHIFT$;
OUTPUT (λ_1, λ_2);
STOP.

Step 11 Choose σ so that $|\sigma - a_n^{(k)}| = \min\{|\mu_1 - a_n^{(k)}|, |\mu_2 - a_n^{(k)}|\}$.

Step 12 *(Accumulate the shift.)*
Set $SHIFT = SHIFT + \sigma$.

Step 13 *(Perform shift.)*
For $j = 1, \ldots, n$, set $d_j = a_j^{(k)} - \sigma$.

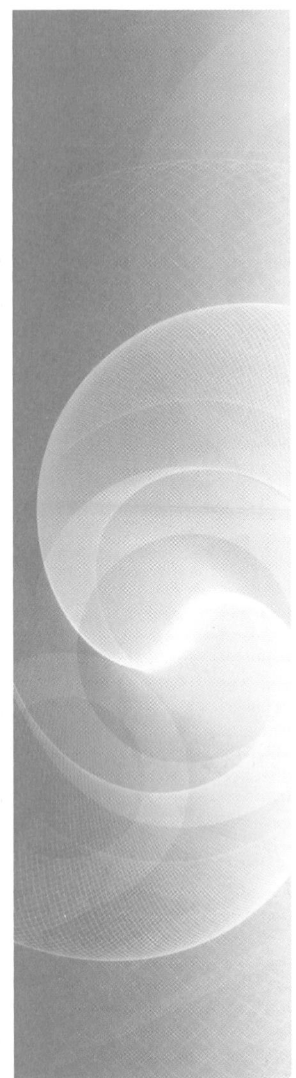

Step 14 (*Steps* 14 *and* 15 *compute* $R^{(k)}$.)
Set $x_1 = d_1$;
$y_1 = b_2$.

Step 15 For $j = 2, \ldots, n$

set $z_{j-1} = \left\{x_{j-1}^2 + \left[b_j^{(k)}\right]^2\right\}^{1/2}$;

$c_j = \dfrac{x_{j-1}}{z_{j-1}}$;

$\sigma_j = \dfrac{b_j^{(k)}}{z_{j-1}}$;

$q_{j-1} = c_j y_{j-1} + s_j d_j$;
$x_j = -\sigma_j y_{j-1} + c_j d_j$;
If $j \neq n$ then set $r_{j-1} = \sigma_j b_{j+1}^{(k)}$;
$y_j = c_j b_{j+1}^{(k)}$.
$\left(A_j^{(k)} = P_j A_{j-1}^{(k)}\right.$ *has just been computed and* $\left.R^{(k)} = A_n^{(k)}.\right)$

Step 16 (*Steps* 16–18 *compute* $A^{(k+1}$.)
Set $z_n = x_n$;

$a_1^{(k+1)} = \sigma_2 q_1 + c_2 z_1$;

$b_2^{(k+1)} = \sigma_2 z_2$.

Step 17 For $j = 2, 3, \ldots, n-1$
set $a_j^{(k+1)} = \sigma_{j+1} q_j + c_j c_{j+1} z_j$;

$b_{j+1}^{(k+1)} = \sigma_{j+1} z_{j+1}$.

Step 18 Set $a_n^{(k+1)} = c_n z_n$.

Step 19 Set $k = k + 1$.

Step 20 OUTPUT ('Maximum number of iterations exceeded');
(*The procedure was unsuccessful.*)
STOP. ▪

A similar procedure can be used to find approximations to the eigenvalues of a nonsymmetric $n \times n$ matrix. The matrix is first reduced to a similar upper-Hessenberg matrix H using the Householder Algorithm for nonsymmetric matrices described at the end of Section 9.4.

The QR factoring process assumes the following form. First

$$H \equiv H^{(1)} = Q^{(1)} R^{(1)}. \tag{9.20}$$

Then $H^{(2)}$ is defined by

$$H^{(2)} = R^{(1)} Q^{(1)} \tag{9.21}$$

and factored into

$$H^{(2)} = Q^{(2)} R^{(2)}. \tag{9.22}$$

The method of factoring proceeds with the same aim as the QR Algorithm for Symmetric Matrices. That is, the matrices are chosen to introduce zeros at appropriate entries of the matrix, and a shifting procedure is used similar to that in the QR method. However, the shifting is somewhat more complicated for nonsymmetric matrices since complex eigenvalues with the same modulus can occur. The shifting process modifies the calculations in Eqs. (9.20), (9.21), and (9.22) to obtain the double QR method

$$H^{(1)} - \sigma_1 I = Q^{(1)}R^{(1)}, \qquad H^{(2)} = R^{(1)}Q^{(1)} + \sigma_1 I,$$
$$H^{(2)} - \sigma_2 I = Q^{(2)}R^{(2)}, \qquad H^{(3)} = R^{(2)}Q^{(2)} + \sigma_2 I,$$

where σ_1 and σ_2 are complex conjugates and $H^{(1)}, H^{(2)}, \ldots$ are real upper-Hessenberg matrices.

James Hardy Wilkinson (1919–1986) is best known for his extensive work on numerical methods for solving systems of linear equations and eigenvalue problems. He also developed the numerical linear algebra technique of backward error analysis.

A complete description of the QR method can be found in works of Wilkinson [Wil2]. Detailed algorithms and programs for this method and most other commonly employed methods are given in [WR]. We refer the reader to these works if the method we have discussed does not give satisfactory results.

The QR method can be performed in a manner that will produce the eigenvectors of a matrix as well as its eigenvalues, but Algorithm 9.6 has not been designed to accomplish this. If the eigenvectors of a symmetric matrix are needed as well as the eigenvalues, we suggest either using the Inverse Power method after Algorithms 9.5 and 9.6 have been employed or using one of the more powerful techniques listed in [WR].

EXERCISE SET 9.5

1. Apply two iterations of the QR method without shifting to the following matrices.

a. $\begin{bmatrix} 2 & -1 & 0 \\ -1 & 2 & -1 \\ 0 & -1 & 2 \end{bmatrix}$
b. $\begin{bmatrix} 3 & 1 & 0 \\ 1 & 4 & 2 \\ 0 & 2 & 1 \end{bmatrix}$

c. $\begin{bmatrix} 4 & -1 & 0 \\ -1 & 3 & -1 \\ 0 & -1 & 2 \end{bmatrix}$
d. $\begin{bmatrix} 1 & 1 & 0 & 0 \\ 1 & 2 & -1 & 0 \\ 0 & -1 & 3 & 1 \\ 0 & 0 & 1 & 4 \end{bmatrix}$

e. $\begin{bmatrix} -2 & 1 & 0 & 0 \\ 1 & -3 & -1 & 0 \\ 0 & -1 & 1 & 1 \\ 0 & 0 & 1 & 3 \end{bmatrix}$
f. $\begin{bmatrix} 0.5 & 0.25 & 0 & 0 \\ 0.25 & 0.8 & 0.4 & 0 \\ 0 & 0.4 & 0.6 & 0.1 \\ 0 & 0 & 0.1 & 1 \end{bmatrix}$

2. Apply two iterations of the QR method without shifting to the following matrices.

a. $\begin{bmatrix} 2 & -1 & 0 \\ -1 & -1 & -2 \\ 0 & -2 & 3 \end{bmatrix}$
b. $\begin{bmatrix} 3 & 1 & 0 \\ 1 & 4 & 2 \\ 0 & 2 & 3 \end{bmatrix}$

c. $\begin{bmatrix} 4 & 2 & 0 & 0 & 0 \\ 2 & 4 & 2 & 0 & 0 \\ 0 & 2 & 4 & 2 & 0 \\ 0 & 0 & 2 & 4 & 2 \\ 0 & 0 & 0 & 2 & 4 \end{bmatrix}$
d. $\begin{bmatrix} 5 & -1 & 0 & 0 & 0 \\ -1 & 4.5 & 0.2 & 0 & 0 \\ 0 & 0.2 & 1 & -0.4 & 0 \\ 0 & 0 & -0.4 & 3 & 1 \\ 0 & 0 & 0 & 1 & 3 \end{bmatrix}$

3. Use the QR Algorithm to determine, to within 10^{-5}, all the eigenvalues for the matrices given in Exercise 1.

4. Use the QR Algorithm to determine, to within 10^{-5}, all the eigenvalues of the following matrices.

a. $\begin{bmatrix} 2 & -1 & 0 \\ -1 & -1 & -2 \\ 0 & -2 & 3 \end{bmatrix}$ b. $\begin{bmatrix} 3 & 1 & 0 \\ 1 & 4 & 2 \\ 0 & 2 & 3 \end{bmatrix}$

c. $\begin{bmatrix} 4 & 2 & 0 & 0 & 0 \\ 2 & 4 & 2 & 0 & 0 \\ 0 & 2 & 4 & 2 & 0 \\ 0 & 0 & 2 & 4 & 2 \\ 0 & 0 & 0 & 2 & 4 \end{bmatrix}$ d. $\begin{bmatrix} 5 & -1 & 0 & 0 & 0 \\ -1 & 4.5 & 0.2 & 0 & 0 \\ 0 & 0.2 & 1 & -0.4 & 0 \\ 0 & 0 & -0.4 & 3 & 1 \\ 0 & 0 & 0 & 1 & 3 \end{bmatrix}$

5. Use the Inverse Power method to determine, to within 10^{-5}, the eigenvectors of the matrices in Exercise 1.

6. Use the Inverse Power method to determine, to within 10^{-5}, the eigenvectors of the matrices in Exercise 2.

7. a. Show that the rotation matrix

$$\begin{bmatrix} \cos\theta & -\sin\theta \\ \sin\theta & \cos\theta \end{bmatrix}$$

applied to the vector $\mathbf{x} = (x_1, x_2)^t$ has the geometric effect of rotating $\mathbf{x}$ through the angle θ without changing its magnitude with respect to the l_2 norm.

b. Show that the magnitude of $\mathbf{x}$ with respect to the l_∞ norm can be changed by a rotation matrix.

8. Let P be the rotation matrix with $p_{ii} = p_{jj} = \cos\theta$ and $p_{ij} = -p_{ji} = \sin\theta$, for $j < i$. Show that for any $n \times n$ matrix A:

$$(AP)_{pq} = \begin{cases} a_{pq}, & \text{if } q \neq i, j, \\ (\cos\theta)a_{pj} + (\sin\theta)a_{pi}, & \text{if } q = j, \\ (\cos\theta)a_{pi} - (\sin\theta)a_{pj}, & \text{if } q = i. \end{cases}$$

$$(PA)_{pq} = \begin{cases} a_{pq}, & \text{if } p \neq i, j, \\ (\cos\theta)a_{jq} - (\sin\theta)a_{iq}, & \text{if } p = j, \\ (\sin\theta)a_{jq} + (\cos\theta)a_{iq}, & \text{if } p = i. \end{cases}$$

9. Show that the product of an upper triangular matrix (on the left) and an upper Hessenberg matrix produces an upper Hessenberg matrix.

10. Let P_k denote a rotation matrix of the form given in (9.17).

a. Show that $P_2^t P_3^t$ differs from an upper triangular matrix only in at most the (2,1) and (3,2) positions.

b. Assume that $P_2^t P_3^t \cdots P_k^t$ differs from an upper triangular matrix only in at most the $(2,1)$, $(3,2), \ldots, (k, k-1)$ positions. Show that $P_2^t P_3^t \cdots P_k^t P_{k+1}^t$ differs from an upper triangular matrix only in at most the $(2,1), (3,2), \ldots, (k, k-1), (k+1, k)$ positions.

c. Show that the matrix $P_2^t P_3^t \cdots P_n^t$ is upper Hessenberg.

11. **Jacobi's method** for a symmetric matrix A is described by

$$A_1 = A,$$

$$A_2 = P_1 A_1 P_1^t$$

and, in general,

$$A_{i+1} = P_i A_i P_i^t.$$

The matrix A_{i+1} tends to a diagonal matrix, where P_i is a rotation matrix chosen to eliminate a large off-diagonal element in A_i. Suppose $a_{j,k}$ and $a_{k,j}$ are to be set to 0, where $j \neq k$. If $a_{jj} \neq a_{kk}$, then

$$(P_i)_{jj} = (P_i)_{kk} = \sqrt{\frac{1}{2}\left(1 + \frac{b}{\sqrt{c^2 + b^2}}\right)}, \quad (P_i)_{kj} = \frac{c}{2(P_i)_{jj}\sqrt{c^2 + b^2}} = -(P_i)_{jk},$$

where

$$c = 2a_{jk}\text{sgn}(a_{jj} - a_{kk}) \quad \text{and} \quad b = |a_{jj} - a_{kk}|,$$

or if $a_{jj} = a_{kk}$,

$$(P_i)_{jj} = (P_i)_{kk} = \frac{\sqrt{2}}{2}$$

and

$$(P_i)_{kj} = -(P_i)_{jk} = \frac{\sqrt{2}}{2}.$$

Develop an algorithm to implement Jacobi's method by setting $a_{21} = 0$. Then set $a_{31}, a_{32}, a_{41}, a_{42}, a_{43}, \ldots, a_{n,1}, \ldots, a_{n,n-1}$ in turn to zero.This is repeated until a matrix A_k is computed with

$$\sum_{i=1}^{n} \sum_{\substack{j=1 \\ j \neq i}}^{n} |a_{ij}^{(k)}|$$

sufficiently small. The eigenvalues of A can then be approximated by the diagonal entries of A_k.

12. Repeat Exercise 3 using the Jacobi method.

13. In the lead example of this chapter, the linear system $A\mathbf{w} = -0.04(\rho/p)\lambda\mathbf{w}$ must be solved for $\mathbf{w}$ and λ in order to approximate the eigenvalues λ_k of the Strum-Liouville system.

    a. Find all four eigenvalues $\mu_1, \ldots, \mu_4$ of the matrix

$$A = \begin{bmatrix} 2 & -1 & 0 & 0 \\ -1 & 2 & -1 & 0 \\ 0 & -1 & 2 & -1 \\ 0 & 0 & -1 & 2 \end{bmatrix}$$

    to within 10^{-5}.

    b. Approximate the eigenvalues $\lambda_1, \ldots, \lambda_4$ of the system in terms of ρ and p.

14. The $(m-1) \times (m-1)$ tridiagonal matrix

$$A = \begin{bmatrix} 1-2\alpha & \alpha & 0 & \cdots & \cdots & 0 \\ \alpha & 1-2\alpha & \alpha & \ddots & & \vdots \\ 0 & \ddots & \ddots & \ddots & \ddots & 0 \\ \vdots & & \ddots & \ddots & \ddots & \alpha \\ 0 & \cdots & \cdots & 0 & \alpha & 1-2\alpha \end{bmatrix}$$

is involved in the Forward Difference method to solve the heat equation (see Section 12.2). For the stability of the method we need $\rho(A) < 1$. With $m = 11$, approximate the eigenvalues of A for each of the following.

    a. $\alpha = \frac{1}{4}$ **b.** $\alpha = \frac{1}{2}$ **c.** $\alpha = \frac{3}{4}$

    When is the method stable?

15. The eigenvalues of the matrix A in Exercise 14 are

$$\lambda_i = 1 - 4\alpha\left(\sin\frac{\pi i}{2m}\right)^2, \quad \text{for } i = 1, \ldots, m-1.$$

Compare the approximations in Exercise 14 to the actual eigenvalues. Again, when is the method stable?

9.6 Singular Value Decomposition

In this section we consider the factorization of a general $m \times n$ matrix A into what is called a Singular Value Decomposition. This factorization takes the form

$$A = U\,S\,V^t,$$

where U is an $m \times m$ orthogonal matrix, V is an $n \times n$ orthogonal matrix, and S is an $m \times n$ matrix whose only nonzero elements lie along the main diagonal. We will assume throughout this section that $m \geq n$, and in many important applications m is much larger than n.

Singular Value Decomposition has quite a long history, being first considered by mathematicians in the latter part of the 19th century. However, the important applications of the technique had to wait until computing power became available in the second half of the 20th century, when algorithms could be developed for its efficient implementation. These were primarily the work of Gene Golub (1932–2007) in a series of papers in the 1960s and 1970s (see, in particular, [GK] and [GR]). A quite complete history of the technique can be found in a paper by G. W. Stewart, which is available through the internet at the address given in [Stew3].

To factor A we consider the $n \times n$ matrix A^tA and the $m \times m$ matrix AA^t. The following definition is used to describe some essential properties of arbitrary matrices.

Definition 9.24 Let A be an $m \times n$ matrix.

(i) The **Rank** of A, denoted $\text{Rank}(A)$ is the number of linearly independent rows in A.

(ii) The **Nullity** of A, denoted $\text{Nullity}(A)$, is $n - \text{Rank}(A)$, and describes the largest set of linearly independent vectors $\mathbf{v}$ in $\mathbb{R}^n$ for which $A\mathbf{v} = \mathbf{0}$. ■

The Rank and Nullity of a matrix are important in characterizing the behavior of the matrix. When the matrix is square, for example, the matrix is invertible if and only if its Nullity is 0 and its Rank is the same as the size of the matrix.

The following is one of the basic theorems in linear algebra.

Theorem 9.25 The number of linearly independent rows of an $m \times n$ matrix A is the same as the number of linearly independent columns of A. ■

The next result gives some useful facts about the matrices AA^t and A^tA.

Theorem 9.26 Let A be $m \times n$ matrix.

(i) The matrices A^tA and AA^t are symmetric.

(ii) $\text{Nullity}(A) = \text{Nullity}(A^tA)$.

(iii) $\text{Rank}(A) = \text{Rank}(A^tA)$.

(iv) The eigenvalues of A^tA and AA^t are real and nonnegative.

(v) The nonzero eigenvalues of AA^t and A^tA are the same. ■

Proof **(i)** Because $\left(A^tA\right)^t = A^t\left(A^t\right)^t = A^tA$, this matrix is symmetric, and similarly, so is AA^t.

(ii) Let $\mathbf{v} \neq \mathbf{0}$ be a vector with $A\mathbf{v} = \mathbf{0}$. Then

$$(A^tA)\mathbf{v} = A^t(A\mathbf{v}) = A^t\mathbf{0} = \mathbf{0}, \quad \text{so} \quad \text{Nullity}(A) \leq \text{Nullity}(A^tA).$$

Now suppose that $\mathbf{v}$ is a vector with $A^tA\mathbf{v} = \mathbf{0}$. Then

$$0 = \mathbf{v}^tA^tA\mathbf{v} = (A\mathbf{v})^tA\mathbf{v} = ||A\mathbf{v}||_2^2, \quad \text{which implies that} \quad A\mathbf{v} = \mathbf{0}.$$

Hence $\text{Nullity}(A^tA) \leq \text{Nullity}(A)$. As a consequence, $\text{Nullity}(A^tA) = \text{Nullity}(A)$.

(iii) The matrices A and A^tA both have n columns and their Nullities agree, so

$$\text{Rank}(A) = n - \text{Nullity}(A) = n - \text{Nullity}(A^tA) = \text{Rank}(A^tA).$$

(iv) The matrix A^tA is symmetric so by Corollary 9.17 its eigenvalues are real numbers. Suppose that $\mathbf{v}$ is an eigenvector of A^tA with $||\mathbf{v}||_2 = 1$ corresponding to the eigenvalue λ. Then

$$0 \leq ||A\mathbf{v}||_2^2 = (A\mathbf{v})^t(A\mathbf{v}) = \mathbf{v}^tA^tA\mathbf{v} = \mathbf{v}^t\left(A^tA\mathbf{v}\right) = \mathbf{v}^t(\lambda\mathbf{v}) = \lambda\mathbf{v}^t\mathbf{v} = \lambda||\mathbf{v}||_2^2 = \lambda.$$

(v) Let $\mathbf{v}$ be an eigenvector corresponding to the nonzero eigenvalue λ of A^tA. Then

$$A^tA\mathbf{v} = \lambda\mathbf{v} \quad \text{implies that} \quad (AA^t)A\mathbf{v} = \lambda A\mathbf{v}.$$

If $A\mathbf{v} = \mathbf{0}$, then $A^tA\mathbf{v} = A^t\mathbf{0} = \mathbf{0}$, which contradicts the assumption that $\lambda \neq 0$. Hence $A\mathbf{v} \neq \mathbf{0}$ and $A\mathbf{v}$ is an eigenvector of AA^t associated with λ. The reverse conclusion also follows from this argument because if λ is a nonzero eigenvalue of $AA^t = \left(A^t\right)^t A^t$, then λ is also an eigenvalue of $A^t\left(A^t\right)^t = A^tA$. ■ ■ ■

In Section 5 of Chapter 6 we saw how effective factorization can be when solving linear systems of the form $A\mathbf{x} = \mathbf{b}$ when the matrix A is used repeatedly for varying $\mathbf{b}$. We now consider a technique for factoring a general $m \times n$ matrix. It has application in many areas, including least squares fitting of data, image compression, signal processing, and statistics.

Constructing a Singular Value Decomposition

A non-square matrix A, that is, a matrix with a different number of rows and columns, cannot have an eigenvalue because $A\mathbf{x}$ and $\mathbf{x}$ will be vectors of different sizes. However, there are numbers that play roles for non-square matrices that are similar to those played by eigenvalues for square matrices. One of the important features of the Singular Value Decomposition of a general matrix is that it permits a generalization of eigenvalues and eigenvectors in this situation.

Our objective is to determine a factorization of the $m \times n$ matrix A, where $m \geq n$, in the form

$$A = U\,S\,V^t,$$

where U is an $m \times m$ orthogonal matrix, V is $n \times n$ an orthogonal matrix, and S is an $m \times n$ diagonal matrix, that is, its only nonzero entries are $(S)_{ii} \equiv s_i \geq 0$, for $i = 1, \ldots, n$. (See Figure 9.2.)

Figure 9.2

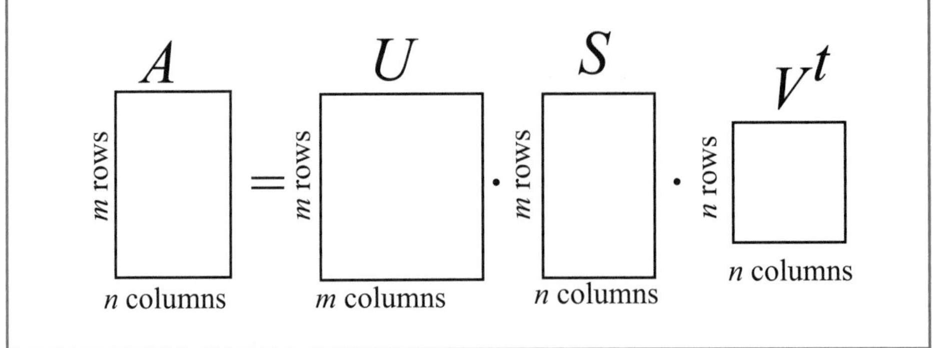

Constructing S in the factorization $A = U\,S\,V^t$.

We construct the matrix S by finding the eigenvalues of the $n \times n$ symmetric matrix A^tA. These eigenvalues are all non-negative real numbers, and we order them from largest to smallest and denote them by

$$s_1^2 \geq s_2^2 \geq \cdots \geq s_k^2 > s_{k+1} = \cdots = s_n = 0.$$

That is, we denote by s_k^2 the smallest nonzero eigenvalue of A^tA. The positive square roots of these eigenvalues of A^tA give the diagonal entries in S. They are called the *singular values* of A. Hence,

$$S = \begin{bmatrix} s_1 & 0 & \cdots & 0 \\ 0 & s_2 & \ddots & \vdots \\ \vdots & \ddots & \ddots & 0 \\ 0 & \cdots & 0 & s_n \\ 0 & \cdots & \cdots & 0 \\ \vdots & & & \vdots \\ 0 & \cdots & \cdots & 0 \end{bmatrix},$$

where $s_i = 0$ when $k < i \leq n$.

Definition 9.27 The **singular values** of an $m \times n$ matrix A are the positive square roots of the nonzero eigenvalues of the $n \times n$ symmetric matrix A^tA. ■

Example 1 Determine the singular values of the 5×3 matrix

$$A = \begin{bmatrix} 1 & 0 & 1 \\ 0 & 1 & 0 \\ 0 & 1 & 1 \\ 0 & 1 & 0 \\ 1 & 1 & 0 \end{bmatrix}.$$

Solution We have

$$A^t = \begin{bmatrix} 1 & 0 & 0 & 0 & 1 \\ 0 & 1 & 1 & 1 & 1 \\ 1 & 0 & 1 & 0 & 0 \end{bmatrix} \quad \text{so} \quad A^tA = \begin{bmatrix} 2 & 1 & 1 \\ 1 & 4 & 1 \\ 1 & 1 & 2 \end{bmatrix}.$$

The characteristic polynomial of A^tA is

$$p(A^tA) = \lambda^3 - 8\lambda^2 + 17\lambda - 10 = (\lambda - 5)(\lambda - 2)(\lambda - 1),$$

so the eigenvalues of A^tA are $\lambda_1 = s_1^2 = 5$, $\lambda_2 = s_2^2 = 2$, and $\lambda_3 = s_3^2 = 1$. As a consequence, the singular values of A are $s_1 = \sqrt{5}$, $s_2 = \sqrt{2}$, $s_3 = 1$, and in the singular value decomposition of A we have

$$S = \begin{bmatrix} \sqrt{5} & 0 & 0 \\ 0 & \sqrt{2} & 0 \\ 0 & 0 & 1 \\ 0 & 0 & 0 \\ 0 & 0 & 0 \end{bmatrix}.$$ ■

When A is a symmetric $n \times n$ matrix, all the s_i^2 are eigenvalues of $A^2 = A^tA$, and these are the squares of the eigenvalues of A. (See Exercise 15 of Section 7.2.) So in this case the singular values are the absolute values of the eigenvalues of A. In the special case when A is positive definite, or even nonnegative definite, the eigenvalues and singular values of A are the same.

Constructing V in the factorization $A = U\,S\,V^t$.

The $n \times n$ matrix A^tA is symmetric, so by Theorem 9.16 in Section 9.2 (see page 572), we have a factorization

$$A^tA = V\,D\,V^t,$$

where D is a diagonal matrix whose diagonal entries are the eigenvalues of A^tA, and V is an orthogonal matrix whose ith column is an eigenvector with l_2 norm 1 corresponding to the eigenvalue on the ith diagonal entry of D. The specific diagonal matrix depends on the order of the eigenvalues along the diagonal. We choose D so that these are written in decreasing order. The columns, denoted $\mathbf{v}_1^t, \mathbf{v}_2^t, \ldots, \mathbf{v}_n^t$, of the $n \times n$ orthogonal matrix V are orthonormal eigenvectors corresponding to these eigenvalues. Multiple eigenvalues of A^tA permit multiple choices of the corresponding eigenvectors, so although D is uniquely defined, the matrix V might not be. No problem, though, we can choose any such V. Because the eigenvalues of A^tA are all nonnegative we have $D = S^2$.

Constructing U in the factorization $A = U\,S\,V^t$.

To construct the $m \times m$ matrix U, we first consider the nonzero values $s_1 \geq s_2 \geq \cdots \geq s_k > 0$ and the corresponding columns in V given by $\mathbf{v}_1, \mathbf{v}_2, \ldots, \mathbf{v}_k$. We define

$$\mathbf{u}_i = \frac{1}{s_i}A\mathbf{v}_i, \quad \text{for } i = 1, 2, \ldots, k.$$

We use these as the first k of the m columns of U. Because A is $m \times n$ and each $\mathbf{v}_i$ is $n \times 1$, the vector $\mathbf{u}_i$ is $m \times 1$, as required. In addition, for each $1 \leq i \leq k$ and $1 \leq j \leq k$, the fact that the vectors $\mathbf{v}_1, \mathbf{v}_2, \ldots, \mathbf{v}_n$ are eigenvectors of A^tA that form an orthonormal set implies that

$$\mathbf{u}_i^t\mathbf{u}_j = \left(\frac{1}{s_i}A\mathbf{v}_i\right)^t \frac{1}{s_j}A\mathbf{v}_j = \frac{1}{s_is_j}\mathbf{v}_i^tA^tA\mathbf{v}_j = \frac{1}{s_is_j}\mathbf{v}_i^ts_j^2\mathbf{v}_j = \frac{s_j}{s_i}\mathbf{v}_i^t\mathbf{v}_j = \begin{cases} 0 & \text{if } i \neq j, \\ 1 & \text{if } i = j. \end{cases}$$

So the first k columns of U form an orthonormal set of vectors in $\mathbb{R}^m$. However, we need $m - k$ additional columns of U. For this we first need to find $m - k$ vectors which when added to the vectors from the first k columns will give us a linearly independent set. Then we can apply the Gram-Schmidt process to obtain appropriate additional columns.

The matrix U will not be unique unless $k = m$, and then only if all the eigenvalues of A^tA are unique. Non-uniqueness is of no concern;, we only need one such matrix U.

Verifying the factorization $A = USV^t$.

To verify that this process actually gives the factorization $A = USV^t$, first recall that the transpose of an orthogonal matrix is also the inverse of the matrix. (See part (i) of Theorem 9.10 on page 570.) Hence to show that $A = USV^t$ we can show the equivalent statement $AV = US$.

The vectors $\mathbf{v}_1, \mathbf{v}_2, \ldots, \mathbf{v}_n$ form a basis for $\mathbb{R}^n$, $A\mathbf{v}_i = s_i\mathbf{u}_i$, for $i = 1, \ldots, k$, and $A\mathbf{v}_i = \mathbf{0}$, for $i = k+1, \ldots, n$. Only the first k columns of U produce nonzero entries in the product US, so we have

$$\begin{aligned}
AV &= A\left[\mathbf{v}_1\ \mathbf{v}_2\ \cdots \mathbf{v}_k\ \mathbf{v}_{k+1}\ \cdots \mathbf{v}_n\right] \\
&= \left[A\mathbf{v}_1\ A\mathbf{v}_2\ \cdots A\mathbf{v}_k\ A\mathbf{v}_{k+1}\ \cdots A\mathbf{v}_n\right] \\
&= [s_1\mathbf{u}_1\ s_2\mathbf{u}_2\ \cdots s_k\mathbf{u}_k\ \mathbf{0}\ \cdots \mathbf{0}] \\
&= [\mathbf{u}_1\ \mathbf{u}_2\ \cdots \mathbf{u}_k\ \mathbf{0}\ \cdots \mathbf{0}]
\begin{bmatrix}
s_1 & 0 & \cdots & 0 & 0 & \cdots & 0 \\
0 & \ddots & \ddots & \vdots & \vdots & & \vdots \\
\vdots & \ddots & \ddots & \vdots & \vdots & & \vdots \\
0 & \cdots & 0 & s_k & 0 & \cdots & 0 \\
0 & \cdots & \cdots & 0 & 0 & \cdots & 0 \\
\vdots & & & \vdots & \vdots & & \vdots \\
0 & \cdots & \cdots & 0 & 0 & \cdots & 0
\end{bmatrix} = US,
\end{aligned}$$

This completes the construction of the Singular Value Decomposition of A.

Example 2 Determine the singular value decomposition of the 5×3 matrix

$$A = \begin{bmatrix} 1 & 0 & 1 \\ 0 & 1 & 0 \\ 0 & 1 & 1 \\ 0 & 1 & 0 \\ 1 & 1 & 0 \end{bmatrix}.$$

Solution We found in Example 1 that A has the singular values $s_1 = \sqrt{5}$, $s_2 = \sqrt{2}$, and $s_3 = 1$, so

$$S = \begin{bmatrix} \sqrt{5} & 0 & 0 \\ 0 & \sqrt{2} & 0 \\ 0 & 0 & 1 \\ 0 & 0 & 0 \\ 0 & 0 & 0 \end{bmatrix}.$$

Eigenvectors of A^tA corresponding to $s_1 = \sqrt{5}$, $s_2 = \sqrt{2}$ and $s_3 = 1$, are, respectively, $(1, 2, 1)^t$, $(1, -1, 1)^t$, and $(-1, 0, 1)^t$ (see Exercise 5). Normalizing these vectors and using the values for the columns of V gives

$$V = \begin{bmatrix} \frac{\sqrt{6}}{6} & \frac{\sqrt{3}}{3} & -\frac{\sqrt{2}}{2} \\ \frac{\sqrt{6}}{3} & -\frac{\sqrt{3}}{3} & 0 \\ \frac{\sqrt{6}}{6} & \frac{\sqrt{3}}{3} & \frac{\sqrt{2}}{2} \end{bmatrix} \quad \text{and} \quad V^t = \begin{bmatrix} \frac{\sqrt{6}}{6} & \frac{\sqrt{6}}{3} & \frac{\sqrt{6}}{6} \\ \frac{\sqrt{3}}{3} & -\frac{\sqrt{3}}{3} & \frac{\sqrt{3}}{3} \\ -\frac{\sqrt{2}}{2} & 0 & \frac{\sqrt{2}}{2} \end{bmatrix}$$

The first 3 columns of U are therefore

$$\mathbf{u}_1 = \frac{1}{\sqrt{5}} \cdot A \left(\frac{\sqrt{6}}{6}, \frac{\sqrt{6}}{3}, \frac{\sqrt{6}}{6} \right)^t = \left(\frac{\sqrt{30}}{15}, \frac{\sqrt{30}}{15}, \frac{\sqrt{30}}{10}, \frac{\sqrt{30}}{15}, \frac{\sqrt{30}}{10} \right)^t$$

$$\mathbf{u}_2 = \frac{1}{\sqrt{2}} \cdot A \left(\frac{\sqrt{3}}{3}, -\frac{\sqrt{3}}{3}, \frac{\sqrt{3}}{3} \right)^t = \left(\frac{\sqrt{6}}{3}, -\frac{\sqrt{6}}{6}, 0, -\frac{\sqrt{6}}{6}, 0 \right)^t$$

$$\mathbf{u}_3 = 1 \cdot A \left(-\frac{\sqrt{2}}{2}, 0, \frac{\sqrt{2}}{2} \right)^t = \left(0, 0, \frac{\sqrt{2}}{2}, 0, -\frac{\sqrt{2}}{2} \right)^t$$

To determine the two remaining columns of U we first need two vectors $\mathbf{x}_4$ and $\mathbf{x}_5$ so that $\{\mathbf{u}_1, \mathbf{u}_2, \mathbf{u}_3, \mathbf{x}_4, \mathbf{x}_5\}$ is a linearly independent set. Then we apply the Gram Schmidt process to obtain $\mathbf{u}_4$ and $\mathbf{u}_5$ so that $\{\mathbf{u}_1, \mathbf{u}_2, \mathbf{u}_3, \mathbf{u}_4, \mathbf{u}_5\}$ is an orthogonal set. Two vectors that satisfy are

$$(1, 1, -1, 1, -1)^t \quad \text{and} \quad (0, 1, 0, -1, 0)^t.$$

Normalizing the vectors $\mathbf{u}_i$, for $i = 1, 2, 3, 4$, and 5 produces the matrix U and the singular value decomposition as

$$A = U\,S\,V^t = \begin{bmatrix} \frac{\sqrt{30}}{15} & \frac{\sqrt{6}}{3} & 0 & \frac{\sqrt{5}}{5} & 0 \\ \frac{\sqrt{30}}{15} & -\frac{\sqrt{6}}{6} & 0 & \frac{\sqrt{5}}{5} & \frac{\sqrt{2}}{2} \\ \frac{\sqrt{30}}{10} & 0 & \frac{\sqrt{2}}{2} & -\frac{\sqrt{5}}{5} & 0 \\ \frac{\sqrt{30}}{15} & -\frac{\sqrt{6}}{3} & 0 & \frac{\sqrt{5}}{5} & -\frac{\sqrt{2}}{2} \\ \frac{\sqrt{30}}{10} & 0 & -\frac{\sqrt{2}}{2} & -\frac{\sqrt{5}}{5} & 0 \end{bmatrix} \begin{bmatrix} \sqrt{5} & 0 & 0 \\ 0 & \sqrt{2} & 0 \\ 0 & 0 & 1 \\ 0 & 0 & 0 \\ 0 & 0 & 0 \end{bmatrix}$$

$$\times \begin{bmatrix} \frac{\sqrt{6}}{6} & \frac{\sqrt{6}}{3} & \frac{\sqrt{6}}{6} \\ \frac{\sqrt{3}}{3} & -\frac{\sqrt{3}}{3} & \frac{\sqrt{3}}{3} \\ -\frac{\sqrt{2}}{2} & 0 & \frac{\sqrt{2}}{2} \end{bmatrix}.$$

■

A difficulty with the process in Example 2 is the need to determine the additional vectors $\mathbf{x}_4$ and $\mathbf{x}_5$ to give a linearly independent set on which we can apply the Gram Schmidt process. We will now consider a way to simplify the process in many instances.

An alternative method for finding U

Part (v) of Theorem 9.26 states that the nonzero eigenvalues of A^tA and those of AA^t are the same. In addition, the corresponding eigenvectors of the symmetric matrices A^tA and AA^t form complete orthonormal subsets of $\mathbb{R}^n$ and $\mathbb{R}^m$, respectively. So the orthonormal set of n eigenvectors for A^tA form the columns of V, as outlined above, and the orthonormal set of m eigenvectors for AA^t form the columns of U in the same way.

In summary, then, to determine the Singular Value Decomposition of the $m \times n$ matrix A we can:

- Find the eigenvalues $s_1^2 \geq s_2^2 \geq \cdots \geq s_k^2 \geq s_{k+1} = \cdots = s_n = 0$ for the symmetric matrix A^tA, and place the positive square root of s_i^2 in the entry $(S)_{ii}$ of the $n \times n$ diagonal matrix S.

- Find a set of orthonormal eigenvectors $\{\mathbf{v}_1, \mathbf{v}_2, \ldots, \mathbf{v}_n\}$ corresponding to the eigenvalues of A^tA and construct the $n \times n$ matrix V with these vectors as columns.

- Find a set of orthonormal eigenvectors $\{\mathbf{u}_1, \mathbf{u}_2, \ldots, \mathbf{u}_m\}$ corresponding to the eigenvalues of AA^t and construct the $m \times m$ matrix U with these vectors as columns.

  Then A has the Singular Value Decomposition $A = U\,S\,V^t$.

Example 3 Determine the singular value decomposition of the 5×3 matrix

$$A = \begin{bmatrix} 1 & 0 & 1 \\ 0 & 1 & 0 \\ 0 & 1 & 1 \\ 0 & 1 & 0 \\ 1 & 1 & 0 \end{bmatrix}$$

by determining U from the eigenvectors of AA^t.

Solution We have

$$AA^t = \begin{bmatrix} 1 & 0 & 1 \\ 0 & 1 & 0 \\ 0 & 1 & 1 \\ 0 & 1 & 0 \\ 1 & 1 & 0 \end{bmatrix} \begin{bmatrix} 1 & 0 & 0 & 0 & 1 \\ 0 & 1 & 1 & 1 & 1 \\ 1 & 0 & 1 & 0 & 0 \end{bmatrix} = \begin{bmatrix} 2 & 0 & 1 & 0 & 1 \\ 0 & 1 & 1 & 1 & 1 \\ 1 & 1 & 2 & 1 & 1 \\ 0 & 1 & 1 & 1 & 1 \\ 1 & 1 & 1 & 1 & 2 \end{bmatrix},$$

which has the same nonzero eigenvalues as A^tA, that is, $\lambda_1 = 5$, $\lambda_2 = 2$, and $\lambda_3 = 1$, and, additionally, $\lambda_4 = 0$, and $\lambda_5 = 0$. Eigenvectors corresponding to these eigenvalues are, respectively,

$$\mathbf{x}_1 = (2,2,3,2,3)^t, \quad \mathbf{x}_2 = (2,-1,0,-1,0)^t, \quad \mathbf{x}_3 = (0,0,1,0,-1)^t, \quad \mathbf{x}_4 = (1,2,-1,0,-1)^t,$$

and $\quad \mathbf{x}_5 = (0,1,0,-1,0)^t$.

Both the sets $\{\mathbf{x}_1, \mathbf{x}_2, \mathbf{x}_3, \mathbf{x}_4\}$ and $\{\mathbf{x}_1, \mathbf{x}_2, \mathbf{x}_3, \mathbf{x}_5\}$ are orthogonal because they are eigenvectors associated with distinct eigenvalues of the symmetric matrix AA^t. However, $\mathbf{x}_4$ is not orthogonal to $\mathbf{x}_5$. We will keep $\mathbf{x}_4$ as one of the eigenvectors used to form U and determine the fifth vector that will give an orthogonal set. For this we use the Gram Schmidt process as described in Theorem 9.8 on page 567. Using the notation in that theorem we have

$$\mathbf{v}_1 = \mathbf{x}_1, \mathbf{v}_2 = \mathbf{x}_2, \mathbf{v}_3 = \mathbf{x}_3, \mathbf{v}_4 = \mathbf{x}_4,$$

and, because $\mathbf{x}_5$ is orthogonal to all but $\mathbf{x}_4$,

$$\begin{aligned} \mathbf{v}_5 &= \mathbf{x}_5 - \frac{\mathbf{v}_4^t\mathbf{x}_5}{\mathbf{v}_4^t\mathbf{v}_4}\mathbf{x}_4 \\ &= (0,1,0,-1,0)^t - \frac{(1,2,-1,0,-1)\cdot(0,1,0,-1,0)^t}{\|(1,2,-1,0,-1)^t\|_2^2}(1,2,-1,0,-1) \\ &= (0,1,0,-1,0)^t - \frac{2}{7}(1,2,-1,0,-1)^t = -\frac{1}{7}(2,-3,-2,7,-2)^t. \end{aligned}$$

It is easily verified that $\mathbf{v}_5$ is orthogonal to $\mathbf{v}_4 = \mathbf{x}_4$. It is also orthogonal to the vectors in $\{\mathbf{v}_1, \mathbf{v}_2, \mathbf{v}_3\}$ because it is a linear combination of $\mathbf{x}_4$ and $\mathbf{x}_5$. Normalizing these vectors gives the columns of the matrix U in the factorization. Hence

$$U = [\mathbf{u}_1, \mathbf{u}_2, \mathbf{u}_3, \mathbf{u}_4, \mathbf{u}_5] = \begin{bmatrix} \frac{\sqrt{30}}{15} & \frac{\sqrt{6}}{3} & 0 & \frac{\sqrt{7}}{7} & \frac{\sqrt{70}}{35} \\ \frac{\sqrt{30}}{15} & -\frac{\sqrt{6}}{6} & 0 & \frac{2\sqrt{7}}{7} & -\frac{3\sqrt{70}}{70} \\ \frac{\sqrt{30}}{10} & 0 & \frac{\sqrt{2}}{2} & -\frac{\sqrt{7}}{7} & -\frac{\sqrt{70}}{35} \\ \frac{\sqrt{30}}{15} & -\frac{\sqrt{6}}{6} & 0 & 0 & \frac{\sqrt{70}}{10} \\ \frac{\sqrt{30}}{10} & 0 & -\frac{\sqrt{2}}{2} & -\frac{\sqrt{7}}{7} & -\frac{\sqrt{70}}{35} \end{bmatrix}.$$

This is a different U from the one found in Example 2, but it gives a valid factorization $A = U\,S\,V^t$ using the same S and V as in that example. ■

Maple has a *SingularValues* command in its *LinearAlgebra* package. It can be used to output the singular values of a matrix A as well as the orthogonal matrices U and V. For example, for the matrix A in Examples 2 and 3 the command

$U, S, Vt := SingularValues\,(A, output = ['U', 'S', 'Vt'])$

produces orthogonal matrices U and V and a column vector S containing the singular values of A. By default, Maple uses 18 digits of precision for the calculations.

Least Squares Approximation

The singular value decomposition has application in many areas, one of which is an alternative means for finding the least squares polynomials for fitting data. Let A be an $m \times n$ matrix, with $m > n$, and $\mathbf{b}$ is a vector in $\mathbb{R}^m$. The least squares objective is to find a vector $\mathbf{x}$ in $\mathbb{R}^n$ that will minimize $||A\mathbf{x} - \mathbf{b}||_2$.

Suppose that the singular value decomposition of A is known, that is

$$A = U\,S\,V^t,$$

where U is an $m \times m$ orthogonal matrix, V is an $n \times n$ orthogonal matrix, and S is an $m \times n$ matrix that contains the nonzero singular values in decreasing order along the main diagonal in the first $k \leq n$ rows, and zero entries elsewhere. Because U and V are both orthogonal we have $U^{-1} = U^t$, $V^{-1} = V^t$, and by part (iii) of Theorem 9.10 in Section 9.2 on page 570, U and V are both l_2-norm preserving. As a consequence,

$$||A\mathbf{x} - \mathbf{b}||_2 = ||U\,S\,V^t\mathbf{x} - U\,U^t\mathbf{b}||_2 = ||S\,V^t\mathbf{x} - U^t\mathbf{b}||_2.$$

Define $\mathbf{z} = V^t\mathbf{x}$ and $\mathbf{c} = U^t\mathbf{b}$. Then

$$\begin{aligned} ||A\mathbf{x} - \mathbf{b}||_2 &= ||(s_1z_1 - c_1, s_2z_2 - c_2, \ldots, s_kz_k - c_k, -c_{k+1}, \ldots, -c_m)^t||_2 \\ &= \left\{ \sum_{i=1}^{k} (s_iz_i - c_i)^2 + \sum_{i=k+1}^{m} (c_i)^2 \right\}^{1/2}. \end{aligned}$$

The norm is minimized when the vector $\mathbf{z}$ is chosen with

$$z_i = \begin{cases} \dfrac{c_i}{s_i}, & \text{when } i \leq k, \\ \text{arbitrarily}, & \text{when } k < i \leq n. \end{cases}$$

Because $\mathbf{c} = U^t\mathbf{b}$ and $\mathbf{x} = V\mathbf{z}$ are both easy to compute, the least squares solution is also easily found.

Example 4 Use the singular value decomposition technique to determine the least squares polynomial of degree two for the data given in Table 9.5.

Table 9.5

i	x_i	y_i
1	0	1.0000
2	0.25	1.2840
3	0.50	1.6487
4	0.75	2.1170
5	1.00	2.7183

Solution This problem was solved using normal equations as Example 2 in Section 8.1. Here we first need to determine the appropriate form for A, $\mathbf{x}$, and $\mathbf{b}$. In Example 2 in Section 8.1 the problem was described as finding a_0, a_1, and a_2 with

$$P_2(x) = a_0 + a_1 x + a_2 x^2.$$

In order to express this in matrix form, we let

$$\mathbf{x} = \begin{bmatrix} a_0 \\ a_1 \\ a_2 \end{bmatrix}, \quad \mathbf{b} = \begin{bmatrix} y_0 \\ y_1 \\ y_2 \\ y_3 \\ y_4 \end{bmatrix} = \begin{bmatrix} 1.0000 \\ 1.2840 \\ 1.6487 \\ 2.1170 \\ 2.7183 \end{bmatrix}, \quad \text{and}$$

$$A = \begin{bmatrix} 1 & x_0 & x_0^2 \\ 1 & x_1 & x_1^2 \\ 1 & x_2 & x_2^2 \\ 1 & x_3 & x_3^2 \\ 1 & x_4 & x_4^2 \end{bmatrix} = \begin{bmatrix} 1 & 0 & 0 \\ 1 & 0.25 & 0.0625 \\ 1 & 0.5 & 0.25 \\ 1 & 0.75 & 0.5625 \\ 1 & 1 & 1 \end{bmatrix}.$$

The singular value decomposition of A has the form $A = U\,S\,V^t$, where

$$U = \begin{bmatrix} -0.2945 & -0.6327 & 0.6314 & -0.0143 & -0.3378 \\ -0.3466 & -0.4550 & -0.2104 & 0.2555 & 0.7505 \\ -0.4159 & -0.1942 & -0.5244 & -0.6809 & -0.2250 \\ -0.5025 & 0.1497 & -0.3107 & 0.6524 & -0.4505 \\ -0.6063 & 0.5767 & 0.4308 & -0.2127 & 0.2628 \end{bmatrix},$$

$$S = \begin{bmatrix} 2.7117 & 0 & 0 \\ 0 & 0.9371 & 0 \\ 0 & 0 & 0.1627 \\ 0 & 0 & 0 \\ 0 & 0 & 0 \end{bmatrix}, \quad \text{and} \quad V^t = \begin{bmatrix} -0.7987 & -0.4712 & -0.3742 \\ -0.5929 & 0.5102 & 0.6231 \\ 0.1027 & -0.7195 & 0.6869 \end{bmatrix}.$$

So

$$\mathbf{c} = U^t \begin{bmatrix} y_0 \\ y_1 \\ y_2 \\ y_3 \\ y_4 \end{bmatrix} = \begin{bmatrix} -0.2945 & -0.6327 & 0.6314 & -0.0143 & -0.3378 \\ -0.3466 & -0.4550 & -0.2104 & 0.2555 & 0.7505 \\ -0.4159 & -0.1942 & -0.5244 & -0.6809 & -0.2250 \\ -0.5025 & 0.1497 & -0.3107 & 0.6524 & -0.4505 \\ -0.6063 & 0.5767 & 0.4308 & -0.2127 & 0.2628 \end{bmatrix}^t \begin{bmatrix} 1 \\ 1.284 \\ 1.6487 \\ 2.117 \\ 2.7183 \end{bmatrix}$$

$$= \begin{bmatrix} -4.1372 \\ 0.3473 \\ 0.0099 \\ -0.0059 \\ 0.0155 \end{bmatrix},$$

and the components of $\mathbf{z}$ are

$$z_1 = \frac{c_1}{s_1} = \frac{-4.1372}{2.7117} = -1.526, \quad z_2 = \frac{c_2}{s_2} = \frac{0.3473}{0.9371} = 0.3706, \quad \text{and}$$

$$z_3 = \frac{c_3}{s_3} = \frac{0.0099}{0.1627} = 0.0609.$$

This gives the least squares coefficients in $P_2(x)$ as

$$\begin{bmatrix} a_0 \\ a_1 \\ a_2 \end{bmatrix} = \mathbf{x} = V\,\mathbf{z} = \begin{bmatrix} -0.7987 & -0.5929 & 0.1027 \\ -0.4712 & 0.5102 & -0.7195 \\ -0.3742 & 0.6231 & 0.6869 \end{bmatrix} \begin{bmatrix} -1.526 \\ 0.3706 \\ 0.0609 \end{bmatrix} = \begin{bmatrix} 1.005 \\ 0.8642 \\ 0.8437 \end{bmatrix},$$

which agrees with the results in Example 2 of Section 8.1. The least squares error using these values uses the last two components of $\mathbf{c}$, and is

$$||A\mathbf{x} - \mathbf{b}||_2 = \sqrt{c_4^2 + c_5^2} = \sqrt{(-0.0059)^2 + (0.0155)^2} = 0.0165. \quad \blacksquare$$

Additional Applications

The reason for the importance of the singular value decomposition in many applications is that it permits us to obtain the most important features of an $m \times n$ matrix using a matrix that is often of significantly smaller size. Because the singular values are listed on the diagonal of S in decreasing order, retaining only the first k rows and columns of S produces the best possible approximation of this size to the matrix A. As an illustration, recall the figure, reproduced for reference as Figure 9.3, that indicates the singular value decomposition of the $m \times n$ matrix A.

Figure 9.3

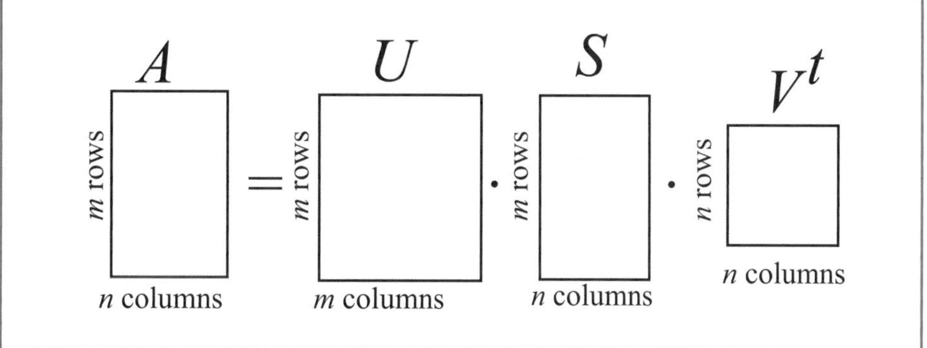

Replace the $n \times n$ matrix S with the $k \times k$ matrix S_k that contains the most significant singular values. These would certainly be only those that are nonzero, but we might also delete some singular values that are relatively small.

Determine corresponding $k \times n$ and $m \times k$ matrices U_k and V_k^t, respectively, in accordance with the singular value decomposition procedure. This is shown shaded in Figure 9.4.

Figure 9.4

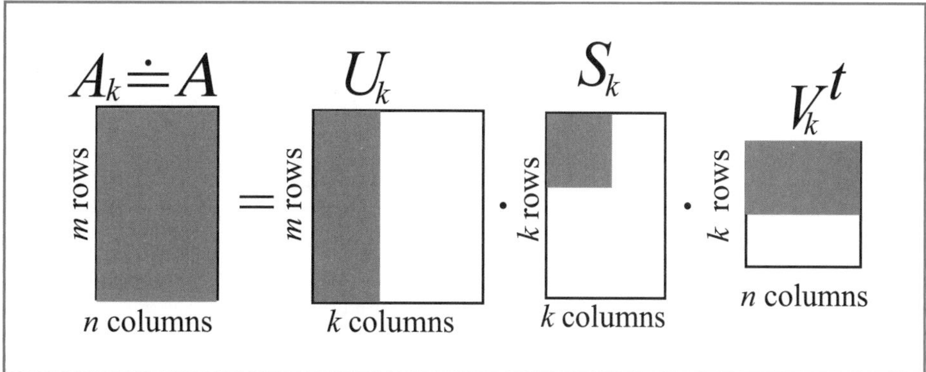

Then the new matrix $A_k = U_k\, S_k\, V_k^t$ is still of size $m \times n$ and would require $m \cdot n$ storage registers for its representation. However, in factored form, the storage requirement for the data is $m \cdot k$, for U_k, k for S_k, and $n \cdot k$ for V_k^t, for a total of $k(m+n+1)$.

Suppose, for example, that $m = 2n$, and $k = n/3$. Then the original matrix A contains $mn = 2n^2$ items of data. The factorization producing A_k however, contains only $mk = 2n^2/3$, for U_k, k for S_k, and $nk = n^2/3$ for V_k^t, items of data which occupy a total of $(n/3)(3n^2+1)$ storage registers. This is a reduction of approximately 50% from the amount required to store the entire matrix A, and results in what is called *data compression*.

Illustration In Example 2 we demonstrated that

$$A = U\,S\,V^t = \begin{bmatrix} \frac{\sqrt{30}}{15} & \frac{\sqrt{6}}{3} & 0 & \frac{\sqrt{5}}{5} & 0 \\ \frac{\sqrt{30}}{15} & -\frac{\sqrt{6}}{6} & 0 & \frac{\sqrt{5}}{5} & \frac{\sqrt{2}}{2} \\ \frac{\sqrt{30}}{10} & 0 & \frac{\sqrt{2}}{2} & -\frac{\sqrt{5}}{5} & 0 \\ \frac{\sqrt{30}}{15} & -\frac{\sqrt{6}}{3} & 0 & \frac{\sqrt{5}}{5} & -\frac{\sqrt{2}}{2} \\ \frac{\sqrt{30}}{10} & 0 & -\frac{\sqrt{2}}{2} & -\frac{\sqrt{5}}{5} & 0 \end{bmatrix} \begin{bmatrix} \sqrt{5} & 0 & 0 \\ 0 & \sqrt{2} & 0 \\ 0 & 0 & 1 \\ 0 & 0 & 0 \\ 0 & 0 & 0 \end{bmatrix} \times \begin{bmatrix} \frac{\sqrt{6}}{6} & \frac{\sqrt{6}}{3} & \frac{\sqrt{6}}{6} \\ \frac{\sqrt{3}}{3} & -\frac{\sqrt{3}}{3} & \frac{\sqrt{3}}{3} \\ -\frac{\sqrt{2}}{2} & 0 & \frac{\sqrt{2}}{2} \end{bmatrix}.$$

Consider the reduced matrices associated with this factorization

$$U_3 = \begin{bmatrix} \frac{\sqrt{30}}{15} & \frac{\sqrt{6}}{3} & 0 \\ \frac{\sqrt{30}}{15} & -\frac{\sqrt{6}}{6} & 0 \\ \frac{\sqrt{30}}{10} & 0 & \frac{\sqrt{2}}{2} \\ \frac{\sqrt{30}}{15} & -\frac{\sqrt{6}}{3} & 0 \\ \frac{\sqrt{30}}{10} & 0 & -\frac{\sqrt{2}}{2} \end{bmatrix}, \quad S_3 = \begin{bmatrix} \sqrt{5} & 0 & 0 \\ 0 & \sqrt{2} & 0 \\ 0 & 0 & 1 \end{bmatrix}, \quad \text{and} \quad V_3^t = \begin{bmatrix} \frac{\sqrt{6}}{6} & \frac{\sqrt{6}}{3} & \frac{\sqrt{6}}{6} \\ \frac{\sqrt{3}}{3} & -\frac{\sqrt{3}}{3} & \frac{\sqrt{3}}{3} \\ -\frac{\sqrt{2}}{2} & 0 & \frac{\sqrt{2}}{2} \end{bmatrix}.$$

Then

$$S_3 V_3^t = \begin{bmatrix} \frac{\sqrt{30}}{6} & \frac{\sqrt{30}}{3} & \frac{\sqrt{30}}{6} \\ \frac{\sqrt{6}}{3} & -\frac{\sqrt{6}}{3} & \frac{\sqrt{6}}{3} \\ -\frac{\sqrt{2}}{2} & 0 & \frac{\sqrt{2}}{2} \end{bmatrix} \quad \text{and} \quad A_3 = U_3 S_3 V_3^t = \begin{bmatrix} 1 & 0 & 1 \\ 0 & 1 & 0 \\ 0 & 1 & 1 \\ 0 & 1 & 0 \\ 1 & 1 & 0 \end{bmatrix}$$

□

Because the calculations in the Illustration were done using exact arithmetic, the matrix A_3 agreed precisely with the original matrix A. In general, finite-digit arithmetic would be used to perform the calculations, and absolute agreement would not be expected. The hope is that the data compression does not result in a matrix A_k that significantly differs from the original matrix A, and this depends on the relative magnitudes of the singular values of A. When the rank of the matrix A is k there will be no deterioration since there are only k rows of the original matrix A that are linearly independent and the matrix could, in theory, be reduced to a matrix which has all zeros in its last $m-k$ rows or $n-k$ columns. When k is less than the rank of A, then A_k will differ from A, but this is not always to its detriment.

Consider the situation when A is a matrix consisting of pixels in a gray-scale photograph, perhaps taken from a great distance, such as a satellite photo of a portion of the earth. The photograph likely includes *noise*, that is, data that doesn't truly represent the image, but rather represents the deterioration of the image by atmospheric particles, quality of the lens and reproduction process, etc. The noise data is incorporated in the data given in A, but hopefully this noise is much less significant than the true image. We expect the larger singular values to represent the true image and the smaller singular values, those closest to zero, to be contributions of the noise. By performing a singular value decomposition that retains only those singular values above a certain threshold we might be able to eliminate much of the noise, and actually obtain an image that is not only smaller is size but a truer representation than the original photograph. (See [AP] for further details; in particular, Figure 3.)

Additional important applications of the singular value decomposition include determining effective condition numbers for square matrices (see Exercise 15), determining the effective rank of a matrix, and removing signal noise. For more information on this important topic and a geometric interpretation of the factorization see the survey paper by Kalman [Ka] and the references in that paper. For a more complete and extensive study of the theory see Golub and Van Loan [GV].

EXERCISE SET 9.6

1. Determine the singular values of the following matrices.

   a. $A = \begin{bmatrix} 2 & 1 \\ 1 & 0 \end{bmatrix}$ **b.** $A = \begin{bmatrix} 2 & 1 \\ 1 & 0 \\ 0 & 1 \end{bmatrix}$

   c. $A = \begin{bmatrix} 2 & 1 \\ -1 & 1 \\ 1 & 1 \\ 2 & -1 \end{bmatrix}$ **d.** $A = \begin{bmatrix} 1 & 1 & 0 \\ -1 & 0 & 1 \\ 0 & 1 & -1 \\ 1 & 1 & -1 \end{bmatrix}$

2. Determine the singular values of the following matrices.

   a. $A = \begin{bmatrix} -1 & 1 \\ 1 & 1 \end{bmatrix}$ **b.** $A = \begin{bmatrix} 1 & 1 & 0 \\ 1 & 0 & 1 \\ 0 & 1 & 1 \end{bmatrix}$

   c. $A = \begin{bmatrix} 1 & -1 \\ 1 & 1 \\ 0 & 1 \\ 1 & 0 \\ -1 & 1 \end{bmatrix}$ **d.** $A = \begin{bmatrix} 0 & 1 & 1 \\ 0 & 1 & 0 \\ 1 & 1 & 0 \\ 0 & 1 & 0 \\ 1 & 0 & 1 \end{bmatrix}$

3. Determine a singular value decomposition for the matrices in Exercise 1.
4. Determine a singular value decomposition for the matrices in Exercise 2.
5. Let A be the matrix given in Example 2. Show that $(1, 2, 1)^t$, $(1, -1, 1)^t$, and $(-1, 0, 1)^t$ are eigenvectors of A^tA corresponding to, respectively, the eigenvalues $\lambda_1 = 5$, $\lambda_2 = 2$ and $\lambda_3 = 1$.
6. Suppose that A is an $m \times n$ matrix A. Show that Rank(A) is the same as the Rank(A^t).
7. Show that Nullity(A) = Nullity(A^t) if and only if A is a square matrix.
8. Suppose that A has the singular value decomposition $A = U\,S\,V^t$. Determine, with justification a singular value decomposition of A^t.
9. Suppose that A has the singular value decomposition $A = U\,S\,V^t$. Show that Rank(A) = Rank(S).
10. Suppose that the $m \times n$ matrix A has the singular value decomposition $A = U\,S\,V^t$. Express the Nullity(A) in terms of Rank(S).
11. Suppose that the $n \times n$ matrix A has the singular value decomposition $A = U\,S\,V^t$. Show that A^{-1} exists if and only if S^{-1} exists and find a singular value decomposition for A^{-1} when it exists.
12. Part (ii) of Theorem 9.26 states that Nullity(A) = Nullity(A^tA). Is it also true that Nullity(A) = Nullity(AA^t)?

13. Part (iii) of Theorem 9.26 states that $\text{Rank}(A) = \text{Rank}(A^tA)$. Is it also true that $\text{Rank}(A) = \text{Rank}(AA^t)$?

14. Show that if A is an $m \times n$ matrix and P is an $n \times n$ orthogonal matrix, then PA has the same singular values as A.

15. Show that if A is an $n \times n$ nonsingular matrix with singular values $s_1, s_2, \ldots, s_n$, then the l_2 condition number of A is $K_2(A) = (s_1/s_n)$.

16. Use the result in Exercise 15 to determine the condition numbers of the nonsingular square matrices in Exercises 1 and 2.

17. Given the data

x_i	1.0	2.0	3.0	4.0	5.0
y_i	1.3	3.5	4.2	5.0	7.0

,

a. Use the singular value decomposition technique to determine the least squares polynomial of degree 1.

b. Use the singular value decomposition technique to determine the least squares polynomial of degree 2.

18. Given the data

x_i	1.0	1.1	1.3	1.5	1.9	2.1
y_i	1.84	1.96	2.21	2.45	2.94	3.18

,

a. Use the singular value decomposition technique to determine the least squares polynomial of degree 2.

b. Use the singular value decomposition technique to determine the least squares polynomial of degree 3.

9.7 Survey of Methods and Software

The general theme of this chapter is the approximation of eigenvalues and eigenvectors. It concluded with a technique for factoring an arbitrary matrix that requires these approximation methods.

The Geršgorin circles give a crude approximation to the location of the eigenvalues of a matrix. The Power method can be used to find the dominant eigenvalue and an associated eigenvector for an arbitrary matrix A. If A is symmetric, the Symmetric Power method gives faster convergence to the dominant eigenvalue and an associated eigenvector. The Inverse Power method will find the eigenvalue closest to a given value and an associated eigenvector. This method is often used to refine an approximate eigenvalue and to compute an eigenvector once an eigenvalue has been found by some other technique.

Deflation methods, such as Wielandt deflation, obtain other eigenvalues once the dominant eigenvalue is known. These methods are used if only a few eigenvalues are required since they are susceptible to round-off error. The Inverse Power method should be used to improve the accuracy of approximate eigenvalues obtained from a deflation technique.

Methods based on similarity transformations, such as Householder's method, are used to convert a symmetric matrix into a similar matrix that is tridiagonal (or upper Hessenberg if the matrix is not symmetric). Techniques such as the QR method can then be applied to the tridiagonal (or upper-Hessenberg) matrix to obtain approximations to all the eigenvalues. The associated eigenvectors can be found by using an iterative method, such as the Inverse Power method, or by modifying the QR method to include the approximation of eigenvectors. We restricted our study to symmetric matrices and presented the QR method only to compute eigenvalues for the symmetric case.

The Singular Value Decomposition is discussed in Section 9.6. It is used to factor an $m \times n$ matrix into the form $U S V^t$, where U is an $m \times m$ orthogonal matrix, V is an $n \times n$ orthogonal matrix, and S is an $m \times n$ matrix whose only nonzero entries are located along the main diagonal. This factorization has important applications that include image processing, data compression, and solving over-determined linear systems that arise in least squares approximations. The singular value decomposition requires the computation of eigenvalues and eigenvectors so it is appropriate to have this technique conclude the chapter.

The subroutines in the IMSL and NAG libraries, as well as the routines in Netlib and the commands in MATLAB, Maple, and Mathematica are based on those contained in EISPACK and LAPACK, packages that were discussed in Section 1.4. In general, the subroutines transform a matrix into the appropriate form for the QR method or one of its modifications, such as the QL method. The subroutines approximate all the eigenvalues and can approximate an associated eigenvector for each eigenvalue. Nonsymmetric matrices are generally balanced so that the sums of the magnitudes of the entries in each row and in each column are about the same. Householder's method is then applied to determine a similar upper Hessenberg matrix. Eigenvalues can then be computed using the QR or QL method. It is also possible to compute the Schur form $S D S^t$, where S is orthogonal and the diagonal of D holds the eigenvalues of A. The corresponding eigenvectors can then be determined. For a symmetric matrix a similar tridiagonal matrix is computed. Eigenvalues and eigenvectors can then be computed using the QR or QL method.

There are special routines that find all the eigenvalues within an interval or region or that find only the largest or smallest eigenvalue. Subroutines are also available to determine the accuracy of the eigenvalue approximation and the sensitivity of the process to round-off error.

One MATLAB procedure that computes a selected number of eigenvalues and eigenvectors is based on the implicitly restarted Arnoldi method by Sorensen [So]. There is software package contained in Netlib to solve large sparse eigenvalue problems, that is also based on the implicitly restarted Arnoldi method. The implicitly restarted Arnoldi method is a Krylov subspace method that finds a sequence of Krylov subspaces that converge to a subspace containing the eigenvalues.

The books by Wilkinson [Wil2] and Wilkinson and Reinsch [WR] are classics in the study of eigenvalue problems. Stewart [Stew2] is also a good source of information on the general problem, and Parlett [Par] considers the symmetric problem. A study of the nonsymmetric problem can be found in Saad [Sa1].

CHAPTER

10 Numerical Solutions of Nonlinear Systems of Equations

Introduction

The amount of pressure required to sink a large heavy object into soft, homogeneous soil lying above a hard base soil can be predicted by the amount of pressure required to sink smaller objects in the same soil. Specifically, the amount of pressure p to sink a circular plate of radius r a distance d in the soft soil, where the hard base soil lies a distance $D > d$ below the surface, can be approximated by an equation of the form

$$p = k_1 e^{k_2 r} + k_3 r,$$

where k_1, k_2, and k_3 are constants depending on d and the consistency of the soil, but not on the radius of the plate.

There are three unknown constants in this equation, so three small plates with differing radii are sunk to the same distance. This will determine the minimal size plate required to sustain a large load. The loads required for this sinkage are recorded, as shown in the accompanying figure.

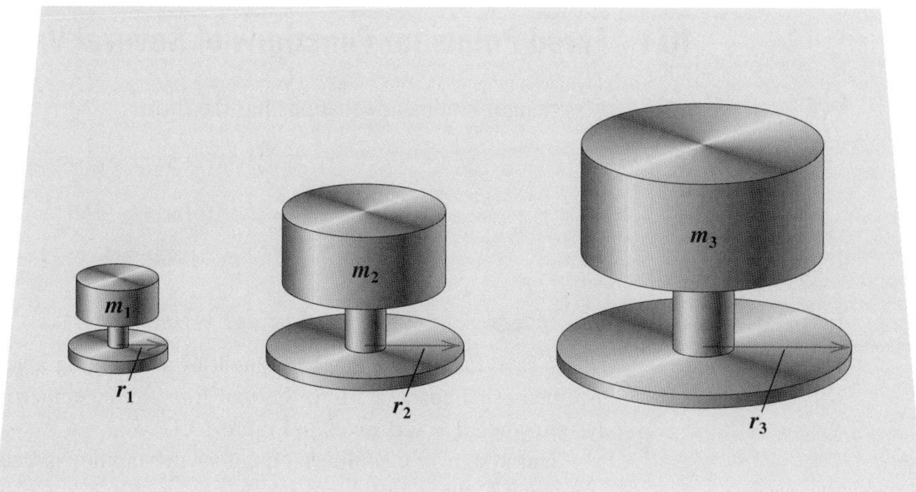

This produces the three nonlinear equations

$$m_1 = k_1 e^{k_2 r_1} + k_3 r_1,$$
$$m_2 = k_1 e^{k_2 r_2} + k_3 r_2,$$
$$m_3 = k_1 e^{k_2 r_3} + k_3 r_3,$$

in the three unknowns k_1, k_2, and k_3. Numerical approximation methods are usually needed for solving systems of equations when the equations are nonlinear. Exercise 12 of Section 10.2 concerns an application of the type described here.

Solving a system of nonlinear equations is a problem that is avoided when possible, customarily by approximating the nonlinear system by a system of linear equations. When this is unsatisfactory, the problem must be tackled directly. The most straightforward approach is to adapt the methods from Chapter 2, which approximate the solutions of a single nonlinear equation in one variable, to apply when the single-variable problem is replaced by a vector problem that incorporates all the variables.

The principal tool in Chapter 2 was Newton's method, a technique that is generally quadratically convergent. This is the first technique we modify to solve systems of nonlinear equations. Newton's method, as modified for systems of equations, is quite costly to apply, and in Section 10.3 we describe how a modified Secant method can be used to obtain approximations more easily, although with a loss of the extremely rapid convergence that Newton's method can produce.

Section 10.4 describes the method of Steepest Descent. It is only linearly convergent, but it does not require the accurate starting approximations needed for more rapidly converging techniques. It is often used to find a good initial approximation for Newton's method or one of its modifications.

In Section 10.5, we give an introduction to continuation methods, which use a parameter to move from a problem with an easily determined solution to the solution of the original nonlinear problem.

Many of the proofs of the theoretical results in this chapter are omitted because they involve methods that are usually studied in advanced calculus. A good general reference for this material is Ortega's book entitled *Numerical Analysis–A Second Course* [Or2]. A more complete reference is [OR].

10.1 Fixed Points for Functions of Several Variables

A system of nonlinear equations has the form

$$\begin{aligned} f_1(x_1, x_2, \ldots, x_n) &= 0, \\ f_2(x_1, x_2, \ldots, x_n) &= 0, \\ \vdots \qquad & \quad \vdots \\ f_n(x_1, x_2, \ldots, x_n) &= 0, \end{aligned} \tag{10.1}$$

where each function f_i can be thought of as mapping a vector $\mathbf{x} = (x_1, x_2, \ldots, x_n)^t$ of the n-dimensional space $\mathbb{R}^n$ into the real line $\mathbb{R}$. A geometric representation of a nonlinear system when $n = 2$ is given in Figure 10.1.

This system of n nonlinear equations in n unknowns can also be represented by defining a function $\mathbf{F}$ mapping $\mathbb{R}^n$ into $\mathbb{R}^n$ as

$$\mathbf{F}(x_1, x_2, \ldots, x_n) = (f_1(x_1, x_2, \ldots, x_n), f_2(x_1, x_2, \ldots, x_n), \ldots, f_n(x_1, x_2, \ldots, x_n))^t.$$

If vector notation is used to represent the variables $x_1, x_2, \ldots, x_n$, then system (10.1) assumes the form

$$\mathbf{F}(\mathbf{x}) = \mathbf{0}. \tag{10.2}$$

The functions $f_1, f_2, \ldots, f_n$ are called the **coordinate functions** of $\mathbf{F}$.

Figure 10.1

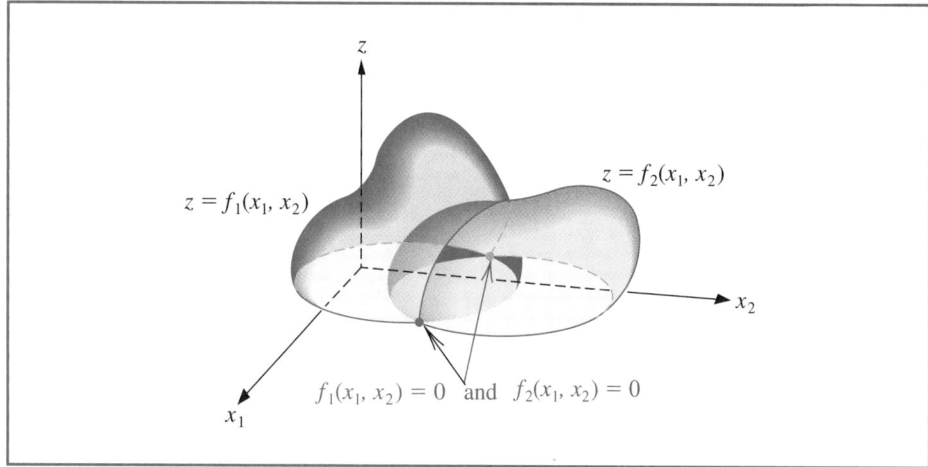

Example 1 Place the 3×3 nonlinear system

$$\begin{aligned} 3x_1 - \cos(x_2x_3) - \frac{1}{2} &= 0, \\ x_1^2 - 81(x_2 + 0.1)^2 + \sin x_3 + 1.06 &= 0, \\ e^{-x_1x_2} + 20x_3 + \frac{10\pi - 3}{3} &= 0 \end{aligned}$$

in the form (10.2).

Solution Define the three coordinate functions f_1, f_2, and f_3 from $\mathbb{R}^3$ to $\mathbb{R}$ as

$$\begin{aligned} f_1(x_1, x_2, x_3) &= 3x_1 - \cos(x_2x_3) - \frac{1}{2}, \\ f_2(x_1, x_2, x_3) &= x_1^2 - 81(x_2 + 0.1)^2 + \sin x_3 + 1.06, \\ f_3(x_1, x_2, x_3) &= e^{-x_1x_2} + 20x_3 + \frac{10\pi - 3}{3}, \end{aligned}$$

Then define $\mathbf{F}$ from $\mathbb{R}^3 \to \mathbb{R}^3$ by

$$\begin{aligned} \mathbf{F}(\mathbf{x}) &= \mathbf{F}(x_1, x_2, x_3) \\ &= (f_1(x_1, x_2, x_3), f_2(x_1, x_2, x_3), f_3(x_1, x_2, x_3))^t \\ &= \left(3x_1 - \cos(x_2x_3) - \frac{1}{2}, x_1^2 - 81(x_2 + 0.1)^2 \right. \\ &\qquad \left. + \sin x_3 + 1.06, e^{-x_1x_2} + 20x_3 + \frac{10\pi - 3}{3}\right)^t. \end{aligned}$$

■

Before discussing the solution of a system given in the form (10.1) or (10.2), we need some results concerning continuity and differentiability of functions from $\mathbb{R}^n$ into $\mathbb{R}^n$. Although this study could be presented directly (see Exercise 12), we use an alternative method that permits us to present the more theoretically difficult concepts of limits and continuity in terms of functions from $\mathbb{R}^n$ into $\mathbb{R}$.

Definition 10.1 Let f be a function defined on a set $D \subset \mathbb{R}^n$ and mapping into $\mathbb{R}$. The function f is said to have the **limit** L at $\mathbf{x}_0$, written

$$\lim_{\mathbf{x}\to\mathbf{x}_0} f(\mathbf{x}) = L,$$

if, given any number $\varepsilon > 0$, a number $\delta > 0$ exists with

$$|f(\mathbf{x}) - L| < \varepsilon,$$

whenever $\mathbf{x} \in D$ and

$$0 < ||\mathbf{x} - \mathbf{x}_0|| < \delta.$$

■

The existence of a limit is also independent of the particular vector norm being used, as discussed in Section 7.1. Any convenient norm can be used to satisfy the condition in this definition. The specific value of δ will depend on the norm chosen, but the existence of a δ is independent of the norm.

The notion of a limit permits us to define continuity for functions from $\mathbb{R}^n$ into $\mathbb{R}$. Although various norms can be used, continuity is independent of the particular choice.

Definition 10.2 Let f be a function from a set $D \subset \mathbb{R}^n$ into $\mathbb{R}$. The function f is **continuous** at $\mathbf{x}_0 \in D$ provided $\lim_{\mathbf{x}\to\mathbf{x}_0} f(\mathbf{x})$ exists and

$$\lim_{\mathbf{x}\to\mathbf{x}_0} f(\mathbf{x}) = f(\mathbf{x}_0).$$

Moreover, f is **continuous** on a set D if f is continuous at every point of D. This concept is expressed by writing $f \in C(D)$. ■

Continuous definitions for functions of n variables follow from those for a single variable by replacing, where necessary, absolute values by norms.

We can now define the limit and continuity concepts for functions from $\mathbb{R}^n$ into $\mathbb{R}^n$ by considering the coordinate functions from $\mathbb{R}^n$ into $\mathbb{R}$.

Definition 10.3 Let $\mathbf{F}$ be a function from $D \subset \mathbb{R}^n$ into $\mathbb{R}^n$ of the form

$$\mathbf{F}(\mathbf{x}) = (f_1(\mathbf{x}), f_2(\mathbf{x}), \ldots, f_n(\mathbf{x}))^t,$$

where f_i is a mapping from $\mathbb{R}^n$ into $\mathbb{R}$ for each i. We define

$$\lim_{\mathbf{x}\to\mathbf{x}_0} \mathbf{F}(\mathbf{x}) = \mathbf{L} = (L_1, L_2, \ldots, L_n)^t,$$

if and only if $\lim_{\mathbf{x}\to\mathbf{x}_0} f_i(\mathbf{x}) = L_i$, for each $i = 1, 2, \ldots, n$. ■

The function $\mathbf{F}$ is **continuous** at $\mathbf{x}_0 \in D$ provided $\lim_{\mathbf{x}\to\mathbf{x}_0} \mathbf{F}(\mathbf{x})$ exists and $\lim_{\mathbf{x}\to\mathbf{x}_0} \mathbf{F}(\mathbf{x}) = \mathbf{F}(\mathbf{x}_0)$. In addition, $\mathbf{F}$ is continuous on the set D if $\mathbf{F}$ is continuous at each $\mathbf{x}$ in D. This concept is expressed by writing $\mathbf{F} \in C(D)$.

For functions from $\mathbb{R}$ into $\mathbb{R}$, continuity can often be shown by demonstrating that the function is differentiable (see Theorem 1.6). Although this theorem generalizes to functions of several variables, the derivative (or total derivative) of a function of several variables is quite involved and will not be presented here. Instead we state the following theorem, which relates the continuity of a function of n variables at a point to the partial derivatives of the function at the point.

Theorem 10.4 Let f be a function from $D \subset \mathbb{R}^n$ into $\mathbb{R}$ and $\mathbf{x}_0 \in D$. Suppose that all the partial derivatives of f exist and constants $\delta > 0$ and $K > 0$ exist so that whenever $\|\mathbf{x} - \mathbf{x}_0\| < \delta$ and $\mathbf{x} \in D$, we have

$$\left| \frac{\partial f(\mathbf{x})}{\partial x_j} \right| \le K, \quad \text{for each } j = 1, 2, \ldots, n.$$

Then f is continuous at $\mathbf{x}_0$. ■

Fixed Points in $\mathbb{R}^n$

In Chapter 2, an iterative process for solving an equation $f(x) = 0$ was developed by first transforming the equation into the fixed-point form $x = g(x)$. A similar procedure will be investigated for functions from $\mathbb{R}^n$ into $\mathbb{R}^n$.

Definition 10.5 A function $\mathbf{G}$ from $D \subset \mathbb{R}^n$ into $\mathbb{R}^n$ has a **fixed point** at $\mathbf{p} \in D$ if $\mathbf{G}(\mathbf{p}) = \mathbf{p}$. ■

The following theorem extends the Fixed-Point Theorem 2.4 on page 62 to the n-dimensional case. This theorem is a special case of the Contraction Mapping Theorem, and its proof can be found in [Or2], p. 153.

Theorem 10.6 Let $D = \{ (x_1, x_2, \ldots, x_n)^t \mid a_i \le x_i \le b_i, \text{ for each } i = 1, 2, \ldots, n \}$ for some collection of constants $a_1, a_2, \ldots, a_n$ and $b_1, b_2, \ldots, b_n$. Suppose $\mathbf{G}$ is a continuous function from $D \subset \mathbb{R}^n$ into $\mathbb{R}^n$ with the property that $\mathbf{G}(\mathbf{x}) \in D$ whenever $\mathbf{x} \in D$. Then $\mathbf{G}$ has a fixed point in D.

Moreover, suppose that all the component functions of $\mathbf{G}$ have continuous partial derivatives and a constant $K < 1$ exists with

$$\left| \frac{\partial g_i(\mathbf{x})}{\partial x_j} \right| \le \frac{K}{n}, \quad \text{whenever } \mathbf{x} \in D,$$

for each $j = 1, 2, \ldots, n$ and each component function g_i. Then the sequence $\{\mathbf{x}^{(k)}\}_{k=0}^{\infty}$ defined by an arbitrarily selected $\mathbf{x}^{(0)}$ in D and generated by

$$\mathbf{x}^{(k)} = G(\mathbf{x}^{(k-1)}), \quad \text{for each } k \ge 1$$

converges to the unique fixed point $\mathbf{p} \in D$ and

$$\left\| \mathbf{x}^{(k)} - \mathbf{p} \right\|_\infty \le \frac{K^k}{1-K} \left\| \mathbf{x}^{(1)} - \mathbf{x}^{(0)} \right\|_\infty . \tag{10.3}$$

■

Example 2 Place the nonlinear system

$$\begin{aligned}
3x_1 - \cos(x_2 x_3) - \frac{1}{2} &= 0, \\
x_1^2 - 81(x_2 + 0.1)^2 + \sin x_3 + 1.06 &= 0, \\
e^{-x_1 x_2} + 20x_3 + \frac{10\pi - 3}{3} &= 0.
\end{aligned}$$

in a fixed-point form $\mathbf{x} = \mathbf{G}(\mathbf{x})$ by solving the ith equation for x_i, show that there is a unique solution on

$$D = \{ (x_1, x_2, x_3)^t \mid -1 \le x_i \le 1, \quad \text{for each } i = 1, 2, 3\}.$$

and iterate starting with $\mathbf{x}^{(0)} = (0.1, 0.1, -0.1)^t$ until accuracy within 10^{-5} in the l_∞ norm is obtained.

Solution Solving the ith equation for x_i gives the fixed-point problem

$$
\begin{aligned}
x_1 &= \frac{1}{3}\cos(x_2x_3) + \frac{1}{6},\\
x_2 &= \frac{1}{9}\sqrt{x_1^2 + \sin x_3 + 1.06} - 0.1,\\
x_3 &= -\frac{1}{20}e^{-x_1x_2} - \frac{10\pi - 3}{60}.
\end{aligned}
\tag{10.4}
$$

Let $\mathbf{G} : \mathbb{R}^3 \to \mathbb{R}^3$ be defined by $\mathbf{G}(\mathbf{x}) = (g_1(\mathbf{x}), g_2(\mathbf{x}), g_3(\mathbf{x}))^t$, where

$$
\begin{aligned}
g_1(x_1,x_2,x_3) &= \frac{1}{3}\cos(x_2x_3) + \frac{1}{6},\\
g_2(x_1,x_2,x_3) &= \frac{1}{9}\sqrt{x_1^2 + \sin x_3 + 1.06} - 0.1,\\
g_3(x_1,x_2,x_3) &= -\frac{1}{20}e^{-x_1x_2} - \frac{10\pi - 3}{60}.
\end{aligned}
$$

Theorems 10.4 and 10.6 will be used to show that $\mathbf{G}$ has a unique fixed point in

$$D = \{\, (x_1,x_2,x_3)^t \mid -1 \le x_i \le 1, \quad \text{for each } i = 1,2,3\}.$$

For $\mathbf{x} = (x_1,x_2,x_3)^t$ in D,

$$|g_1(x_1,x_2,x_3)| \le \frac{1}{3}|\cos(x_2x_3)| + \frac{1}{6} \le 0.50,$$

$$|g_2(x_1,x_2,x_3)| = \left|\frac{1}{9}\sqrt{x_1^2 + \sin x_3 + 1.06} - 0.1\right| \le \frac{1}{9}\sqrt{1 + \sin 1 + 1.06} - 0.1 < 0.09,$$

and

$$|g_3(x_1,x_2,x_3)| = \frac{1}{20}e^{-x_1x_2} + \frac{10\pi - 3}{60} \le \frac{1}{20}e + \frac{10\pi - 3}{60} < 0.61.$$

So we have, for each $i = 1,2,3$,

$$-1 \le g_i(x_1,x_2,x_3) \le 1.$$

Thus $\mathbf{G}(\mathbf{x}) \in D$ whenever $\mathbf{x} \in D$.

Finding bounds for the partial derivatives on D gives

$$\left|\frac{\partial g_1}{\partial x_1}\right| = 0, \quad \left|\frac{\partial g_2}{\partial x_2}\right| = 0, \quad \text{and} \quad \left|\frac{\partial g_3}{\partial x_3}\right| = 0,$$

as well as

$$\left|\frac{\partial g_1}{\partial x_2}\right| \le \frac{1}{3}|x_3| \cdot |\sin x_2x_3| \le \frac{1}{3}\sin 1 < 0.281, \quad \left|\frac{\partial g_1}{\partial x_3}\right| \le \frac{1}{3}|x_2| \cdot |\sin x_2x_3| \le \frac{1}{3}\sin 1 < 0.281,$$

$$\left|\frac{\partial g_2}{\partial x_1}\right| = \frac{|x_1|}{9\sqrt{x_1^2 + \sin x_3 + 1.06}} < \frac{1}{9\sqrt{0.218}} < 0.238,$$

$$\left|\frac{\partial g_2}{\partial x_3}\right| = \frac{|\cos x_3|}{18\sqrt{x_1^2 + \sin x_3 + 1.06}} < \frac{1}{18\sqrt{0.218}} < 0.119,$$

$$\left|\frac{\partial g_3}{\partial x_1}\right| = \frac{|x_2|}{20}e^{-x_1x_2} \le \frac{1}{20}e < 0.14, \quad \text{and} \quad \left|\frac{\partial g_3}{\partial x_2}\right| = \frac{|x_1|}{20}e^{-x_1x_2} \le \frac{1}{20}e < 0.14.$$

The partial derivatives of g_1, g_2, and g_3 are all bounded on D, so Theorem 10.4 implies that these functions are continuous on D. Consequently, $\mathbf{G}$ is continuous on D. Moreover, for every $\mathbf{x} \in D$,

$$\left|\frac{\partial g_i(\mathbf{x})}{\partial x_j}\right| \leq 0.281, \quad \text{for each } i = 1, 2, 3 \quad \text{and} \quad j = 1, 2, 3,$$

and the condition in the second part of Theorem 10.6 holds with $K = 3(0.281) = 0.843$.

In the same manner it can also be shown that $\partial g_i/\partial x_j$ is continuous on D for each $i = 1, 2, 3$ and $j = 1, 2, 3$. (This is considered in Exercise 3.) Consequently, $\mathbf{G}$ has a unique fixed point in D, and the nonlinear system has a solution in D.

Note that $\mathbf{G}$ having a unique fixed point in D does not imply that the solution to the original system is unique on this domain, because the solution for x_2 in (10.4) involved the choice of the principal square root. Exercise 7(d) examines the situation that occurs if the negative square root is instead chosen in this step.

To approximate the fixed point $\mathbf{p}$, we choose $\mathbf{x}^{(0)} = (0.1, 0.1, -0.1)^t$. The sequence of vectors generated by

$$x_1^{(k)} = \frac{1}{3}\cos x_2^{(k-1)} x_3^{(k-1)} + \frac{1}{6},$$

$$x_2^{(k)} = \frac{1}{9}\sqrt{\left(x_1^{(k-1)}\right)^2 + \sin x_3^{(k-1)} + 1.06} - 0.1,$$

$$x_3^{(k)} = -\frac{1}{20}e^{-x_1^{(k-1)} x_2^{(k-1)}} - \frac{10\pi - 3}{60}$$

converges to the unique solution of the system in (10.4). The results in Table 10.1 were generated until

$$\left\|\mathbf{x}^{(k)} - \mathbf{x}^{(k-1)}\right\|_\infty < 10^{-5}. \quad \blacksquare$$

Table 10.1

k	$x_1^{(k)}$	$x_2^{(k)}$	$x_3^{(k)}$	$\left\|\mathbf{x}^{(k)} - \mathbf{x}^{(k-1)}\right\|_\infty$
0	0.10000000	0.10000000	−0.10000000	
1	0.49998333	0.00944115	−0.52310127	0.423
2	0.49999593	0.00002557	−0.52336331	9.4×10^{-3}
3	0.50000000	0.00001234	−0.52359814	2.3×10^{-4}
4	0.50000000	0.00000003	−0.52359847	1.2×10^{-5}
5	0.50000000	0.00000002	−0.52359877	3.1×10^{-7}

We could use the error bound (10.3) with $K = 0.843$ in the previous example. This gives

$$\|\mathbf{x}^{(5)} - \mathbf{p}\|_\infty \leq \frac{(0.843)^5}{1 - 0.843}(0.423) < 1.15,$$

which does not indicate the true accuracy of $\mathbf{x}^{(5)}$. The actual solution is

$$\mathbf{p} = \left(0.5, 0, -\frac{\pi}{6}\right)^t \approx (0.5, 0, -0.5235987757)^t, \quad \text{so} \quad \|\mathbf{x}^{(5)} - \mathbf{p}\|_\infty \leq 2 \times 10^{-8}.$$

Accelerating Convergence

One way to accelerate convergence of the fixed-point iteration is to use the latest estimates $x_1^{(k)}, \ldots, x_{i-1}^{(k)}$ instead of $x_1^{(k-1)}, \ldots, x_{i-1}^{(k-1)}$ to compute $x_i^{(k)}$, as in the Gauss-Seidel method for linear systems. The component equations for the problem in the example then become

$$x_1^{(k)} = \frac{1}{3}\cos\left(x_2^{(k-1)}x_3^{(k-1)}\right) + \frac{1}{6},$$

$$x_2^{(k)} = \frac{1}{9}\sqrt{\left(x_1^{(k)}\right)^2 + \sin x_3^{(k-1)} + 1.06} - 0.1,$$

$$x_3^{(k)} = -\frac{1}{20}e^{-x_1^{(k)}x_2^{(k)}} - \frac{10\pi - 3}{60}.$$

With $\mathbf{x}^{(0)} = (0.1, 0.1, -0.1)^t$, the results of these calculations are listed in Table 10.2.

Table 10.2

k	$x_1^{(k)}$	$x_2^{(k)}$	$x_3^{(k)}$	$\|\mathbf{x}^{(k)} - \mathbf{x}^{(k-1)}\|_\infty$
0	0.10000000	0.10000000	−0.10000000	
1	0.49998333	0.02222979	−0.52304613	0.423
2	0.49997747	0.00002815	−0.52359807	2.2×10^{-2}
3	0.50000000	0.00000004	−0.52359877	2.8×10^{-5}
4	0.50000000	0.00000000	−0.52359877	3.8×10^{-8}

The iterate $\mathbf{x}^{(4)}$ is accurate to within 10^{-7} in the l_∞ norm; so the convergence was indeed accelerated for this problem by using the Gauss-Seidel method. However, this method does not *always* accelerate the convergence.

Maple provides the function *fsolve* to solve systems of equations. The fixed-point problem of Example 2 can be solved with the following commands:

$g1 := x1 = \frac{1}{3}\cos(x_2 x_3) + \frac{1}{6}:\quad g2 := x2 = \frac{1}{9}\sqrt{(x_1)^2 + \sin(x_3) + 1.06} - 0.1:$
$g3 := x3 = -\frac{1}{20}e^{-x_1 \cdot x_2} - \frac{10\pi - 3}{60}:$
fsolve$(\{g1, g2, g3\}, \{x1, x2, x3\},\ \{x1 = -1..1,\ x2 = -1..1,\ x3 = -1..1\});$

The first three commands define the system, and the last command invokes the procedure *fsolve*. Maple displays the answer as

$$\{x1 = 0.5000000000,\ x2 = -2.079196195\ 10^{-11},\ x3 = -0.5235987758\}$$

In general, *fsolve(eqns,vars,options)* solves the system of equations represented by the parameter *eqns* for the variables represented by the parameter *vars* under optional parameters represented by *options*. Under *options* we specify a region in which the routine is required to search for a solution. This specification is not mandatory, and Maple determines its own search space if the options are omitted.

EXERCISE SET 10.1

1. Show that the function $\mathbf{F}: \mathbb{R}^3 \to \mathbb{R}^3$ defined by

$$\mathbf{F}(x_1, x_2, x_3) = (x_1 + 2x_3, x_1\cos x_2, x_2^2 + x_3)^t$$

is a continuous at each point of $\mathbb{R}^3$.

2. Give an example of a function $\mathbf{F}: \mathbb{R}^2 \to \mathbb{R}^2$ that is continuous at each point of $\mathbb{R}^2$, except at $(1, 0)$.

3. Show that the first partial derivatives in Example 2 are continuous on D.
4. The nonlinear system

$$-x_1(x_1+1)+2x_2=18, \quad (x_1-1)^2+(x_2-6)^2=25$$

has two solutions.

a. Approximate the solutions graphically.

b. Use the approximations from part (a) as initial approximations for an appropriate function iteration, and determine the solutions to within 10^{-5} in the l_∞ norm.

5. The nonlinear system

$$x_1^2-10x_1+x_2^2+8=0, \quad x_1x_2^2+x_1-10x_2+8=0$$

can be transformed into the fixed-point problem

$$x_1=g_1(x_1,x_2)=\frac{x_1^2+x_2^2+8}{10}, \quad x_2=g_1(x_1,x_2)=\frac{x_1x_2^2+x_1+8}{10}.$$

a. Use Theorem 10.6 to show that $\mathbf{G}=(g_1,g_2)^t$ mapping $D\subset\mathbb{R}^2$ into $\mathbb{R}^2$ has a unique fixed point in

$$D=\{\,(x_1,x_2)^t \mid 0\le x_1,x_2\le 1.5\,\}.$$

b. Apply functional iteration to approximate the solution.

c. Does the Gauss-Seidel method accelerate convergence?

6. The nonlinear system

$$5x_1^2-x_2^2=0, \quad x_2-0.25(\sin x_1+\cos x_2)=0$$

has a solution near $\left(\frac{1}{4},\frac{1}{4}\right)^t$.

a. Find a function $\mathbf{G}$ and a set D in $\mathbb{R}^2$ such that $\mathbf{G}:D\to\mathbb{R}^2$ and $\mathbf{G}$ has a unique fixed point in D.

b. Apply functional iteration to approximate the solution to within 10^{-5} in the l_∞ norm.

c. Does the Gauss-Seidel method accelerate convergence?

7. Use Theorem 10.6 to show that $\mathbf{G}:D\subset\mathbb{R}^3\to\mathbb{R}^3$ has a unique fixed point in D. Apply functional iteration to approximate the solution to within 10^{-5}, using the l_∞ norm.

a. $$\mathbf{G}(x_1,x_2,x_3)=\left(\frac{\cos(x_2x_3)+0.5}{3},\frac{1}{25}\sqrt{x_1^2+0.3125}-0.03,-\frac{1}{20}e^{-x_1x_2}-\frac{10\pi-3}{60}\right)^t;$$

$$D=\{\,(x_1,x_2,x_3)^t \mid -1\le x_i\le 1, i=1,2,3\,\}$$

b. $$\mathbf{G}(x_1,x_2,x_3)=\left(\frac{13-x_2^2+4x_3}{15},\frac{11+x_3-x_1^2}{10},\frac{22+x_2^3}{25}\right);$$

$$D=\{\,(x_1,x_2,x_3)^t \mid 0\le x_1\le 1.5, i=1,2,3\,\}$$

c. $$\mathbf{G}(x_1,x_2,x_3)=(1-\cos(x_1x_2x_3),1-(1-x_1)^{1/4}-0.05x_3^2+0.15x_3,x_1^2 +0.1x_2^2-0.01x_2+1)^t;$$

$$D=\{\,(x_1,x_2,x_3)^t \mid -0.1\le x_1\le 0.1,-0.1\le x_2\le 0.3,0.5\le x_3\le 1.1\,\}$$

d. $$\mathbf{G}(x_1,x_2,x_3)=\left(\frac{1}{3}\cos(x_2x_3)+\frac{1}{6},-\frac{1}{9}\sqrt{x_1^2+\sin x_3+1.06}-0.1, -\frac{1}{20}e^{-x_1x_2}-\frac{10\pi-3}{60}\right)^t;$$

$$D=\{\,(x_1,x_2,x_3)^t \mid -1\le x_i\le 1, i=1,2,3\,\}$$

8. Use functional iteration to find solutions to the following nonlinear systems, accurate to within 10^{-5}, using the l_∞ norm.

a. $$x_2^2 + x_2^2 - x_1 = 0$$ $$x_1^2 - x_2^2 - x_2 = 0.$$

b. $$3x_1^2 - x_2^2 = 0,$$ $$3x_1x_2^2 - x_1^3 - 1 = 0.$$

c. $$x_1^2 + x_2 - 37 = 0,$$ $$x_1 - x_2^2 - 5 = 0,$$ $$x_1 + x_2 + x_3 - 3 = 0.$$

d. $$x_1^2 + 2x_2^2 - x_2 - 2x_3 = 0,$$ $$x_1^2 - 8x_2^2 + 10x_3 = 0,$$ $$\frac{x_1^2}{7x_2x_3} - 1 = 0.$$

9. Use the Gauss-Seidel method to approximate the fixed points in Exercise 7 to within 10^{-5}, using the l_∞ norm.

10. Repeat Exercise 8 using the Gauss-Seidel method.

11. In Exercise 10 of Section 5.9, we considered the problem of predicting the population of two species that compete for the same food supply. In the problem, we made the assumption that the populations could be predicted by solving the system of equations

$$\frac{dx_1(t)}{dt} = x_1(t)(4 - 0.0003x_1(t) - 0.0004x_2(t))$$

and

$$\frac{dx_2(t)}{dt} = x_2(t)(2 - 0.0002x_1(t) - 0.0001x_2(t)).$$

In this exercise, we would like to consider the problem of determining equilibrium populations of the two species. The mathematical criteria that must be satisfied in order for the populations to be at equilibrium is that, simultaneously,

$$\frac{dx_1(t)}{dt} = 0 \quad \text{and} \quad \frac{dx_2(t)}{dt} = 0.$$

This occurs when the first species is extinct and the second species has a population of 20,000 or when the second species is extinct and the first species has a population of 13,333. Can an equilibrium occur in any other situation?

12. Show that a function $\mathbf{F}$ mapping $D \subset \mathbb{R}^n$ into $\mathbb{R}^n$ is continuous at $\mathbf{x}_0 \in D$ precisely when, given any number $\varepsilon > 0$, a number $\delta > 0$ can be found with property that for any vector norm $\|\cdot\|$,

$$\|\mathbf{F}(\mathbf{x}) - \mathbf{F}(\mathbf{x}_0)\| < \varepsilon,$$

whenever $\mathbf{x} \in D$ and $\|\mathbf{x} - \mathbf{x}_0\| < \delta$.

13. Let A be an $n \times n$ matrix and $\mathbf{F}$ be the function from $\mathbb{R}^n$ to $\mathbb{R}^n$ defined by $\mathbf{F}(\mathbf{x}) = A\mathbf{x}$. Use the result in Exercise 12 to show that $\mathbf{F}$ is continuous on $\mathbb{R}^n$.

10.2 Newton's Method

The problem in Example 2 of Section 10.1 is transformed into a convergent fixed-point problem by algebraically solving the three equations for the three variables x_1, x_2, and x_3. It is, however, unusual to be able to find an explicit representation for all the variables. In this section, we consider an algorithmic procedure to perform the transformation in a more general situation.

To construct the algorithm that led to an appropriate fixed-point method in the one-dimensional case, we found a function ϕ with the property that

$$g(x) = x - \phi(x)f(x)$$

gives quadratic convergence to the fixed point p of the function g (see Section 2.4). From this condition Newton's method evolved by choosing $\phi(x) = 1/f'(x)$, assuming that $f'(x) \neq 0$.

A similar approach in the n-dimensional case involves a matrix

$$A(\mathbf{x}) = \begin{bmatrix} a_{11}(\mathbf{x}) & a_{12}(\mathbf{x}) & \cdots & a_{1n}(\mathbf{x}) \\ a_{21}(\mathbf{x}) & a_{22}(\mathbf{x}) & \cdots & a_{2n}(\mathbf{x}) \\ \vdots & \vdots & & \vdots \\ a_{n1}(\mathbf{x}) & a_{n2}(\mathbf{x}) & \cdots & a_{nn}(\mathbf{x}) \end{bmatrix}, \tag{10.5}$$

where each of the entries $a_{ij}(\mathbf{x})$ is a function from $\mathbb{R}^n$ into $\mathbb{R}$. This requires that $A(\mathbf{x})$ be found so that

$$\mathbf{G}(\mathbf{x}) = \mathbf{x} - A(\mathbf{x})^{-1}\mathbf{F}(\mathbf{x})$$

gives quadratic convergence to the solution of $\mathbf{F}(\mathbf{x}) = \mathbf{0}$, assuming that $A(\mathbf{x})$ is nonsingular at the fixed point $\mathbf{p}$ of $\mathbf{G}$.

The following theorem parallels Theorem 2.8 on page 80. Its proof requires being able to express $\mathbf{G}$ in terms of its Taylor series in n variables about $\mathbf{p}$.

Theorem 10.7 Let $\mathbf{p}$ be a solution of $\mathbf{G}(\mathbf{x}) = \mathbf{x}$. Suppose a number $\delta > 0$ exists with

(i) $\partial g_i/\partial x_j$ is continuous on $N_\delta = \{\mathbf{x} \mid \|\mathbf{x} - \mathbf{p}\| < \delta\}$, for each $i = 1, 2, \ldots, n$ and $j = 1, 2, \ldots, n$;

(ii) $\partial^2 g_i(\mathbf{x})/(\partial x_j \partial x_k)$ is continuous, and $|\partial^2 g_i(\mathbf{x})/(\partial x_j \partial x_k)| \leq M$ for some constant M, whenever $\mathbf{x} \in N_\delta$, for each $i = 1, 2, ..., n$, $j = 1, 2, \ldots, n$, and $k = 1, 2, \ldots, n$;

(iii) $\partial g_i(\mathbf{p})/\partial x_k = 0$, for each $i = 1, 2, \ldots, n$ and $k = 1, 2, \ldots, n$.

Then a number $\hat{\delta} \leq \delta$ exists such that the sequence generated by $\mathbf{x}^{(k)} = \mathbf{G}(\mathbf{x}^{(k-1)})$ converges quadratically to $\mathbf{p}$ for any choice of $\mathbf{x}^{(0)}$, provided that $\left\|\mathbf{x}^{(0)} - \mathbf{p}\right\| < \hat{\delta}$. Moreover,

$$\|\mathbf{x}^{(k)} - \mathbf{p}\|_\infty \leq \frac{n^2 M}{2} \|\mathbf{x}^{(k-1)} - \mathbf{p}\|_\infty^2, \quad \text{for each } k \geq 1. \quad \blacksquare$$

To apply Theorem 10.7, suppose that $A(\mathbf{x})$ is an $n \times n$ matrix of functions from $\mathbb{R}^n$ into $\mathbb{R}$ in the form of Eq. (10.5), where the specific entries will be chosen later. Assume, moreover, that $A(\mathbf{x})$ is nonsingular near a solution $\mathbf{p}$ of $\mathbf{F}(\mathbf{x}) = \mathbf{0}$, and let $b_{ij}(\mathbf{x})$ denote the entry of $A(\mathbf{x})^{-1}$ in the ith row and jth column.

For $\mathbf{G}(\mathbf{x}) = \mathbf{x} - A(\mathbf{x})^{-1}\mathbf{F}(\mathbf{x})$, we have $g_i(\mathbf{x}) = x_i - \sum_{j=1}^n b_{ij}(\mathbf{x}) f_j(\mathbf{x})$. So

$$\frac{\partial g_i}{\partial x_k}(\mathbf{x}) = \begin{cases} 1 - \displaystyle\sum_{j=1}^n \left(b_{ij}(\mathbf{x}) \frac{\partial f_j}{\partial x_k}(\mathbf{x}) + \frac{\partial b_{ij}}{\partial x_k}(\mathbf{x}) f_j(\mathbf{x}) \right), & \text{if } i = k, \\ -\displaystyle\sum_{j=1}^n \left(b_{ij}(\mathbf{x}) \frac{\partial f_j}{\partial x_k}(\mathbf{x}) + \frac{\partial b_{ij}}{\partial x_k}(\mathbf{x}) f_j(\mathbf{x}) \right), & \text{if } i \neq k. \end{cases}$$

Theorem 10.7 implies that we need $\partial g_i(\mathbf{p})/\partial x_k = 0$, for each $i = 1, 2, \ldots, n$ and $k = 1, 2, \ldots, n$. This means that for $i = k$,

$$0 = 1 - \sum_{j=1}^n b_{ij}(\mathbf{p}) \frac{\partial f_j}{\partial x_i}(\mathbf{p}),$$

that is,

$$\sum_{j=1}^{n} b_{ij}(\mathbf{p})\frac{\partial f_j}{\partial x_i}(\mathbf{p}) = 1. \tag{10.6}$$

When $k \neq i$,

$$0 = -\sum_{j=1}^{n} b_{ij}(\mathbf{p})\frac{\partial f_j}{\partial x_k}(\mathbf{p}),$$

so

$$\sum_{j=1}^{n} b_{ij}(\mathbf{p})\frac{\partial f_j}{\partial x_k}(\mathbf{p}) = 0. \tag{10.7}$$

The Jacobian Matrix

Define the matrix $J(\mathbf{x})$ by

$$J(\mathbf{x}) = \begin{bmatrix} \frac{\partial f_1}{\partial x_1}(\mathbf{x}) & \frac{\partial f_1}{\partial x_2}(\mathbf{x}) & \cdots & \frac{\partial f_1}{\partial x_n}(\mathbf{x}) \\ \frac{\partial f_2}{\partial x_1}(\mathbf{x}) & \frac{\partial f_2}{\partial x_2}(\mathbf{x}) & \cdots & \frac{\partial f_2}{\partial x_n}(\mathbf{x}) \\ \vdots & \vdots & & \vdots \\ \frac{\partial f_n}{\partial x_1}(\mathbf{x}) & \frac{\partial f_n}{\partial x_2}(\mathbf{x}) & \cdots & \frac{\partial f_n}{\partial x_n}(\mathbf{x}) \end{bmatrix}. \tag{10.8}$$

Then conditions (10.6) and (10.7) require that

$$A(\mathbf{p})^{-1}J(\mathbf{p}) = I, \text{the identity matrix,} \quad \text{so } A(\mathbf{p}) = J(\mathbf{p}).$$

An appropriate choice for $A(\mathbf{x})$ is, consequently, $A(\mathbf{x}) = J(\mathbf{x})$ since this satisfies condition (iii) in Theorem 10.7. The function $\mathbf{G}$ is defined by

$$\mathbf{G}(\mathbf{x}) = \mathbf{x} - J(\mathbf{x})^{-1}\mathbf{F}(\mathbf{x}),$$

and the functional iteration procedure evolves from selecting $\mathbf{x}^{(0)}$ and generating, for $k \geq 1$,

$$\mathbf{x}^{(k)} = \mathbf{G}\left(\mathbf{x}^{(k-1)}\right) = \mathbf{x}^{(k-1)} - J\left(\mathbf{x}^{(k-1)}\right)^{-1}\mathbf{F}\left(\mathbf{x}^{(k-1)}\right). \tag{10.9}$$

This is called **Newton's method for nonlinear systems**, and it is generally expected to give quadratic convergence, provided that a sufficiently accurate starting value is known and that $J(\mathbf{p})^{-1}$ exists. The matrix $J(\mathbf{x})$ is called the **Jacobian** matrix and has a number of applications in analysis. It might, in particular, be familiar to the reader due to its application in the multiple integration of a function of several variables over a region that requires a change of variables to be performed.

The Jacobian matrix first appeared in a 1815 paper by Cauchy, but Jacobi wrote *De determinantibus functionalibus* in 1841 and proved numerous results about this matrix.

A weakness in Newton's method arises from the need to compute and invert the matrix $J(\mathbf{x})$ at each step. In practice, explicit computation of $J(\mathbf{x})^{-1}$ is avoided by performing the operation in a two-step manner. First, a vector $\mathbf{y}$ is found that satisfies $J(\mathbf{x}^{(k-1)})\mathbf{y} = -\mathbf{F}(\mathbf{x}^{(k-1)})$. Then the new approximation, $\mathbf{x}^{(k)}$, is obtained by adding $\mathbf{y}$ to $\mathbf{x}^{(k-1)}$. Algorithm 10.1 uses this two-step procedure.

Newton's Method for Systems

To approximate the solution of the nonlinear system $\mathbf{F}(\mathbf{x}) = \mathbf{0}$ given an initial approximation $\mathbf{x}$:

INPUT number n of equations and unknowns; initial approximation $\mathbf{x} = (x_1, \ldots, x_n)^t$, tolerance TOL; maximum number of iterations N.

OUTPUT approximate solution $\mathbf{x} = (x_1, \ldots, x_n)^t$ or a message that the number of iterations was exceeded.

Step 1 Set $k = 1$.

Step 2 While ($k \leq N$) do Steps 3–7.

Step 3 Calculate $\mathbf{F}(\mathbf{x})$ and $J(\mathbf{x})$, where $J(\mathbf{x})_{i,j} = (\partial f_i(\mathbf{x})/\partial x_j)$ for $1 \leq i, j \leq n$.

Step 4 Solve the $n \times n$ linear system $J(\mathbf{x})\mathbf{y} = -\mathbf{F}(\mathbf{x})$.

Step 5 Set $\mathbf{x} = \mathbf{x} + \mathbf{y}$.

Step 6 If $||\mathbf{y}|| < TOL$ then OUTPUT ($\mathbf{x}$);
(*The procedure was successful.*)
STOP.

Step 7 Set $k = k + 1$.

Step 8 OUTPUT ('Maximum number of iterations exceeded');
(*The procedure was unsuccessful.*)
STOP. ■

Example 1 The nonlinear system

$$\begin{aligned} 3x_1 - \cos(x_2x_3) - \frac{1}{2} &= 0, \\ x_1^2 - 81(x_2 + 0.1)^2 + \sin x_3 + 1.06 &= 0, \\ e^{-x_1x_2} + 20x_3 + \frac{10\pi - 3}{3} &= 0 \end{aligned}$$

was shown in Example 2 of Section 10.1 to have the approximate solution $(0.5, 0, -0.52359877)^t$. Apply Newton's method to this problem with $\mathbf{x}^{(0)} = (0.1, 0.1, -0.1)^t$.

Solution Define

$$\mathbf{F}(x_1, x_2, x_3) = (f_1(x_1, x_2, x_3), f_2(x_1, x_2, x_3), f_3(x_1, x_2, x_3))^t,$$

where

$$f_1(x_1, x_2, x_3) = 3x_1 - \cos(x_2x_3) - \frac{1}{2},$$

$$f_2(x_1, x_2, x_3) = x_1^2 - 81(x_2 + 0.1)^2 + \sin x_3 + 1.06,$$

and

$$f_3(x_1, x_2, x_3) = e^{-x_1x_2} + 20x_3 + \frac{10\pi - 3}{3}.$$

The Jacobian matrix $J(\mathbf{x})$ for this system is

$$J(x_1,x_2,x_3) = \begin{bmatrix} 3 & x_3 \sin x_2x_3 & x_2 \sin x_2x_3 \\ 2x_1 & -162(x_2+0.1) & \cos x_3 \\ -x_2e^{-x_1x_2} & -x_1e^{-x_1x_2} & 20 \end{bmatrix}.$$

Let $\mathbf{x}^{(0)} = (0.1, 0.1, -0.1)^t$. Then $\mathbf{F}(\mathbf{x}^{(0)}) = (-0.199995, -2.269833417, 8.462025346)^t$ and

$$J(\mathbf{x}^{(0)}) = \begin{bmatrix} 3 & 9.999833334 \times 10^{-4} & 9.999833334 \times 10^{-4} \\ 0.2 & -32.4 & 0.9950041653 \\ -0.09900498337 & -0.09900498337 & 20 \end{bmatrix}.$$

Solving the linear system, $J(\mathbf{x}^{(0)})\mathbf{y}^{(0)} = -\mathbf{F}(\mathbf{x}^{(0)})$ gives

$$\mathbf{y}^{(0)} = \begin{bmatrix} 0.3998696728 \\ -0.08053315147 \\ -0.4215204718 \end{bmatrix} \quad \text{and} \quad \mathbf{x}^{(1)} = \mathbf{x}^{(0)} + \mathbf{y}^{(0)} = \begin{bmatrix} 0.4998696782 \\ 0.01946684853 \\ -0.5215204718 \end{bmatrix}.$$

Continuing for $k = 2, 3, \ldots$, we have

$$\begin{bmatrix} x_1^{(k)} \\ x_2^{(k)} \\ x_3^{(k)} \end{bmatrix} = \begin{bmatrix} x_1^{(k-1)} \\ x_2^{(k-1)} \\ x_3^{(k-1)} \end{bmatrix} + \begin{bmatrix} y_1^{(k-1)} \\ y_2^{(k-1)} \\ y_3^{(k-1)} \end{bmatrix},$$

where

$$\begin{bmatrix} y_1^{(k-1)} \\ y_2^{(k-1)} \\ y_3^{(k-1)} \end{bmatrix} = -\left(J\left(x_1^{(k-1)}, x_2^{(k-1)}, x_3^{(k-1)}\right)\right)^{-1} \mathbf{F}\left(x_1^{(k-1)}, x_2^{(k-1)}, x_3^{(k-1)}\right).$$

Thus, at the kth step, the linear system $J\left(\mathbf{x}^{(k-1)}\right)\mathbf{y}^{(k-1)} = -\mathbf{F}\left(\mathbf{x}^{(k-1)}\right)$ must be solved, where

$$J\left(\mathbf{x}^{(k-1)}\right) = \begin{bmatrix} 3 & x_3^{(k-1)} \sin x_2^{(k-1)}x_3^{(k-1)} & x_2^{(k-1)} \sin x_2^{(k-1)}x_3^{(k-1)} \\ 2x_1^{(k-1)} & -162\left(x_2^{(k-1)} + 0.1\right) & \cos x_3^{(k-1)} \\ -x_2^{(k-1)}e^{-x_1^{(k-1)}x_2^{(k-1)}} & -x_1^{(k-1)}e^{-x_1^{(k-1)}x_2^{(k-1)}} & 20 \end{bmatrix},$$

$$\mathbf{y}^{(k-1)} = \begin{bmatrix} y_1^{(k-1)} \\ y_2^{(k-1)} \\ y_3^{(k-1)} \end{bmatrix},$$

and

$$\mathbf{F}\left(\mathbf{x}^{(k-1)}\right) = \begin{bmatrix} 3x_1^{(k-1)} - \cos x_2^{(k-1)}x_3^{(k-1)} - \frac{1}{2} \\ \left(x_1^{(k-1)}\right)^2 - 81\left(x_2^{(k-1)} + 0.1\right)^2 + \sin x_3^{(k-1)} + 1.06 \\ e^{-x_1^{(k-1)}x_2^{(k-1)}} + 20x_3^{(k-1)} + \frac{10\pi-3}{3} \end{bmatrix}.$$

The results using this iterative procedure are shown in Table 10.3. ■

Table 10.3

k	$x_1^{(k)}$	$x_2^{(k)}$	$x_3^{(k)}$	$\|\mathbf{x}^{(k)} - \mathbf{x}^{(k-1)}\|_\infty$
0	0.1000000000	0.1000000000	−0.1000000000	
1	0.4998696728	0.0194668485	−0.5215204718	0.4215204718
2	0.5000142403	0.0015885914	−0.5235569638	1.788×10^{-2}
3	0.5000000113	0.0000124448	−0.5235984500	1.576×10^{-3}
4	0.5000000000	8.516×10^{-10}	−0.5235987755	1.244×10^{-5}
5	0.5000000000	-1.375×10^{-11}	−0.5235987756	8.654×10^{-10}

The previous example illustrates that Newton's method can converge very rapidly once a good approximation is obtained that is near the true solution. However, it is not always easy to determine good starting values, and the method is comparatively expensive to employ. In the next section, we consider a method for overcoming the latter weakness. Good starting values can usually be found using the Steepest Descent method, which will be discussed in Section 10.4.

Using Maple for Initial Approximations

The graphing facilities of Maple can assist in finding initial approximations to the solutions of 2×2 and often 3×3 nonlinear systems. For example, the nonlinear system

$$x_1^2 - x_2^2 + 2x_2 = 0, \quad 2x_1 + x_2^2 - 6 = 0$$

has two solutions, $(0.625204094, 2.179355825)^t$ and $(2.109511920, -1.334532188)^t$. To use Maple we first define the two equations

$eq1 := x1^2 - x2^2 + 2x2 = 0;\ eq2 := 2x1 + x2^2 - 6 = 0;$

To obtain a graph of the two equations for $-3 \le x_1, x_2 \le 3$, enter the commands

with(plots): *implicitplot*$(\{eq1, eq2\}, x1 = -6..6, x2 = -6..6)$;

From the graph shown in Figure 10.2, we are able to estimate that there are solutions near $(2.1, -1.3)^t$, $(0.64, 2.2)^t$, $(-1.9, 3.0)^t$, and $(-5.0, -4.0)^t$. This gives us good starting values for Newton's method.

Figure 10.2

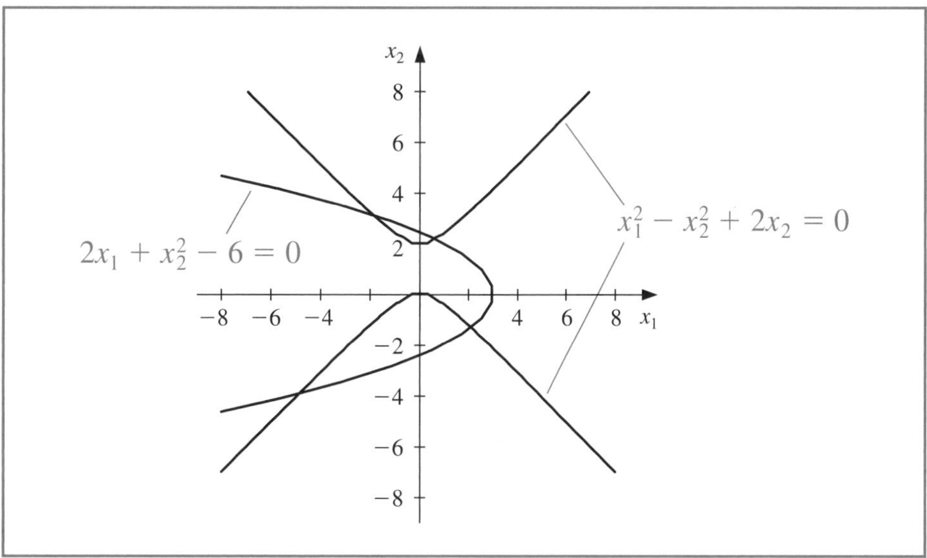

The problem is more difficult in three dimensions. Consider the nonlinear system

$$2x_1 - 3x_2 + x_3 - 4 = 0, \quad 2x_1 + x_2 - x_3 + 4 = 0, \quad x_1^2 + x_2^2 + x_3^2 - 4 = 0.$$

Define three equations using the Maple commands

$eq1 := 2x1{-}3x2{+}x3{-}4 = 0; eq2 := 2x1{+}x2{-}x3{+}4 = 0; eq3 := x1^2{+}x2^2{+}x3^2{-}4 = 0;$

The third equation describes a sphere of radius 2 and center $(0, 0, 0)$, so $x1$, $x2$, and $x3$ are in $[-2, 2]$. The Maple commands to obtain the graph in this case are

with(plots): *implicitplot3d*$(\{eq1, eq2, eq3\}, x1 = -2..2, x2 = -2..2, x3 = -2..2);$

Various three-dimensional plotting options are available in Maple for isolating a solution to the nonlinear system. For example, we can rotate the graph to better view the sections of the surfaces. Then we can zoom into regions where the intersections lie and alter the display form of the axes for a more accurate view of the intersection's coordinates. For this problem, a reasonable initial approximation is $(x_1, x_2, x_3)^t = (-0.5, -1.5, 1.5)^t$.

EXERCISE SET 10.2

1. Use Newton's method with $\mathbf{x}^{(0)} = \mathbf{0}$ to compute $\mathbf{x}^{(2)}$ for each of the following nonlinear systems.

   a. $4x_1^2 - 20x_1 + \frac{1}{4}x_2^2 + 8 = 0,$
   $\frac{1}{2}x_1x_2^2 + 2x_1 - 5x_2 + 8 = 0.$

   b. $\sin(4\pi x_1 x_2) - 2x_2 - x_1 = 0,$
   $\left(\frac{4\pi - 1}{4\pi}\right)(e^{2x_1} - e) + 4ex_2^2 - 2ex_1 = 0.$

   c. $x_1(1 - x_1) + 4x_2 = 12,$
   $(x_1 - 2)^2 + (2x_2 - 3)^2 = 25.$

   d. $5x_1^2 - x_2^2 = 0,$
   $x_2 - 0.25(\sin x_1 + \cos x_2) = 0.$

2. Use Newton's method with $\mathbf{x}^{(0)} = \mathbf{0}$ to compute $\mathbf{x}^{(2)}$ for each of the following nonlinear systems.

   a. $3x_1 - \cos(x_2x_3) - \frac{1}{2} = 0,$
   $4x_1^2 - 625x_2^2 + 2x_2 - 1 = 0,$
   $e^{-x_1x_2} + 20x_3 + \frac{10\pi - 3}{3} = 0.$

   b. $x_1^2 + x_2 - 37 = 0,$
   $x_1 - x_2^2 - 5 = 0,$
   $x_1 + x_2 + x_3 - 3 = 0.$

   c. $15x_1 + x_2^2 - 4x_3 = 13,$
   $x_1^2 + 10x_2 - x_3 = 11,$
   $x_2^3 - 25x_3 = -22.$

   d. $10x_1 - 2x_2^2 + x_2 - 2x_3 - 5 = 0,$
   $8x_2^2 + 4x_3^2 - 9 = 0,$
   $8x_2x_3 + 4 = 0.$

3. Use the graphing facilities of Maple to approximate solutions to the following nonlinear systems.

   a. $4x_1^2 - 20x_1 + \frac{1}{4}x_2^2 + 8 = 0,$
   $\frac{1}{2}x_1x_2^2 + 2x_1 - 5x_2 + 8 = 0.$

   b. $\sin(4\pi x_1 x_2) - 2x_2 - x_1 = 0,$
   $\left(\frac{4\pi - 1}{4\pi}\right)(e^{2x_1} - e) + 4ex_2^2 - 2ex_1 = 0.$

   c. $x_1(1 - x_1) + 4x_2 = 12,$
   $(x_1 - 2)^2 + (2x_2 - 3)^2 = 25.$

   d. $5x_1^2 - x_2^2 = 0,$
   $x_2 - 0.25(\sin x_1 + \cos x_2) = 0.$

4. Use the graphing facilities of Maple to approximate solutions to the following nonlinear systems within the given limits.

a. $$3x_1 - \cos(x_2x_3) - \frac{1}{2} = 0,$$
$$4x_1^2 - 625x_2^2 + 2x_2 - 1 = 0,$$
$$e^{-x_1x_2} + 20x_3 + \frac{10\pi - 3}{3} = 0.$$
$$-1 \le x_1 \le 1, -1 \le x_2 \le 1, -1 \le x_3 \le 1$$

b. $$x_1^2 + x_2 - 37 = 0,$$
$$x_1 - x_2^2 - 5 = 0,$$
$$x_1 + x_2 + x_3 - 3 = 0.$$
$$-4 \le x_1 \le 8, -2 \le x_2 \le 2, -6 \le x_3 \le 0$$

c. $$15x_1 + x_2^2 - 4x_3 = 13,$$
$$x_1^2 + 10x_2 - x_3 = 11,$$
$$x_2^3 - 25x_3 = -22.$$
$$0 \le x_1 \le 2, 0 \le x_2 \le 2, 0 \le x_3 \le 2$$
and $0 \le x_1 \le 2, 0 \le x_2 \le 2, -2 \le x_3 \le 0$

d. $$10x_1 - 2x_2^2 + x_2 - 2x_3 - 5 = 0,$$
$$8x_2^2 + 4x_3^2 - 9 = 0,$$
$$8x_2x_3 + 4 = 0.$$
$$0 \le x_1 \le 2, -2 \le x_2 \le 0, 0 \le x_3 \le 2$$

5. Use the answers obtained in Exercise 3 as initial approximations to Newton's method. Iterate until $\|\mathbf{x}^{(k)} - \mathbf{x}^{(k-1)}\|_\infty < 10^{-6}$.

6. Use the answers obtained in Exercise 4 as initial approximations to Newton's method. Iterate until $\|\mathbf{x}^{(k)} - \mathbf{x}^{(k-1)}\|_\infty < 10^{-6}$.

7. Use Newton's method to find a solution to the following nonlinear systems in the given domain. Iterate until $\|\mathbf{x}^{(k)} - \mathbf{x}^{(k-1)}\|_\infty < 10^{-6}$.

a. $$3x_1^2 - x_2^2 = 0,$$
$$3x_1x_2^2 - x_1^3 - 1 = 0.$$
Use $\mathbf{x}^{(0)} = (1, 1)^t$.

b. $$\ln\left(x_1^2 + x_2^2\right) - \sin(x_1x_2) = \ln 2 + \ln \pi,$$
$$e^{x_1 - x_2} + \cos(x_1x_2) = 0.$$
Use $\mathbf{x}^{(0)} = (2, 2)^t$.

c. $$x_1^3 + x_1^2x_2 - x_1x_3 + 6 = 0,$$
$$e^{x_1} + e^{x_2} - x_3 = 0,$$
$$x_2^2 - 2x_1x_3 = 4.$$
Use $\mathbf{x}^{(0)} = (-1, -2, 1)^t$.

d. $$6x_1 - 2\cos(x_2x_3) - 1 = 0,$$
$$9x_2 + \sqrt{x_1^2 + \sin x_3 + 1.06} + 0.9 = 0,$$
$$60x_3 + 3e^{-x_1x_2} + 10\pi - 3 = 0.$$
Use $\mathbf{x}^{(0)} = (0, 0, 0)^t$.

8. The nonlinear system

$$E_1 : 4x_1 - x_2 + x_3 = x_1x_4, \qquad E_2 : -x_1 + 3x_2 - 2x_3 = x_2x_4,$$
$$E_3 : x_1 - 2x_2 + 3x_3 = x_3x_4, \qquad E_4 : x_1^2 + x_2^2 + x_3^2 = 1$$

has six solutions.

a. Show that if $(x_1, x_2, x_3, x_4)^t$ is a solution then $(-x_1, -x_2, -x_3, x_4)^t$ is a solution.

b. Use Newton's method three times to approximate all solutions. Iterate until $\|\mathbf{x}^{(k)} - \mathbf{x}^{(k-1)}\|_\infty < 10^{-5}$.

9. The nonlinear system

$$3x_1 - \cos(x_2x_3) - \frac{1}{2} = 0,$$
$$x_1^2 - 625x_2^2 - \frac{1}{4} = 0,$$
$$e^{-x_1x_2} + 20x_3 + \frac{10\pi - 3}{3} = 0$$

has a singular Jacobian matrix at the solution. Apply Newton's method with $\mathbf{x}^{(0)} = (1, 1 - 1)^t$. Note that convergence may be slow or may not occur within a reasonable number of iterations.

10. What does Newton's method reduce to for the linear system $A\mathbf{x} = \mathbf{b}$ given by

$$a_{11}x_1 + a_{12}x_2 + \cdots + a_{1n}x_n = b_1,$$
$$a_{21}x_1 + a_{22}x_2 + \cdots + a_{2n}x_n = b_2,$$
$$\vdots$$
$$a_{n1}x_1 + a_{n2}x_2 + \cdots + a_{nn}x_n = b_n,$$

where A is a nonsingular matrix?

11. Show that when $n = 1$, Newton's method given by Eq. (10.9) reduces to the familiar Newton's method given by in Section 2.3.

12. The amount of pressure required to sink a large, heavy object in a soft homogeneous soil that lies above a hard base soil can be predicted by the amount of pressure required to sink smaller objects in the same soil. Specifically, the amount of pressure p required to sink a circular plate of radius r a distance d in the soft soil, where the hard base soil lies a distance $D > d$ below the surface, can be approximated by an equation of the form

$$p = k_1 e^{k_2 r} + k_3 r,$$

where k_1, k_2, and k_3 are constants, with $k_2 > 0$, depending on d and the consistency of the soil but not on the radius of the plate. (See [Bek], pp. 89–94.)

a. Find the values of k_1, k_2, and k_3 if we assume that a plate of radius 1 in. requires a pressure of 10 lb/in.^{2} to sink 1 ft in a muddy field, a plate of radius 2 in. requires a pressure of 12 lb/in.^{2} to sink 1 ft, and a plate of radius 3 in. requires a pressure of 15 lb/in.^{2} to sink this distance (assuming that the mud is more than 1 ft deep).

b. Use your calculations from part (a) to predict the minimal size of circular plate that would be required to sustain a load of 500 lb on this field with sinkage of less than 1 ft.

13. In calculating the shape of a gravity-flow discharge chute that will minimize transit time of discharged granular particles, C. Chiarella, W. Charlton, and A. W. Roberts [CCR] solve the following equations by Newton's method:

(i) $f_n(\theta_1, \ldots, \theta_N) = \dfrac{\sin \theta_{n+1}}{v_{n+1}}(1 - \mu w_{n+1}) - \dfrac{\sin \theta_n}{v_n}(1 - \mu w_n) = 0$, for each $n = 1, 2, \ldots, N - 1$.

(ii) $f_N(\theta_1, \ldots, \theta_N) = \Delta y \sum_{i=1}^{N} \tan \theta_i - X = 0$, where

a. $v_n^2 = v_0^2 + 2gn\Delta y - 2\mu \Delta y \sum_{j=1}^{n} \dfrac{1}{\cos \theta_j}$, for each $n = 1, 2, \ldots, N$, and

b. $w_n = -\Delta y v_n \sum_{i=1}^{N} \dfrac{1}{v_i^3 \cos \theta_i}$, for each $n = 1, 2, \ldots, N$.

The constant v_0 is the initial velocity of the granular material, X is the x-coordinate of the end of the chute, μ is the friction force, N is the number of chute segments, and $g = 32.17\text{ft/s}^2$ is the gravitational constant. The variable θ_i is the angle of the ith chute segment from the vertical, as shown in the following figure, and v_i is the particle velocity in the ith chute segment. Solve (i) and (ii) for $\theta = (\theta_1, \ldots, \theta_N)^t$ with $\mu = 0$, $X = 2$, $\Delta y = 0.2$, $N = 20$, and $v_0 = 0$, where the values for v_n and w_n can be obtained directly from (a) and (b). Iterate until $\|\theta^{(k)} - \theta^{(k-1)}\|_\infty < 10^{-2}$.

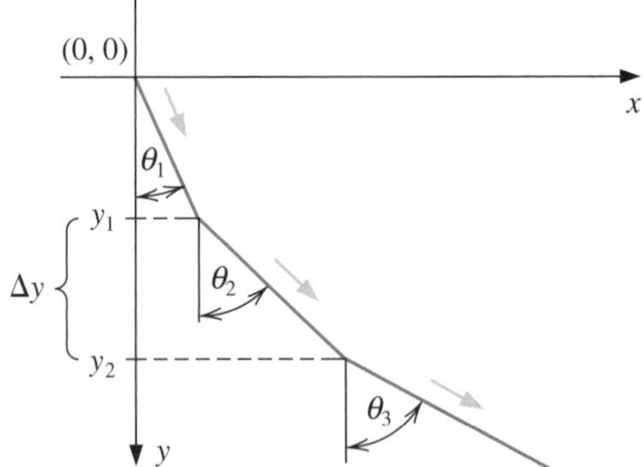

14. An interesting biological experiment (see [Schr2]) concerns the determination of the maximum water temperature, X_M, at which various species of hydra can survive without shortened life expectancy. One approach to the solution of this problem uses a weighted least squares fit of the form $f(x) = y = a/(x - b)^c$ to a collection of experimental data. The x-values of the data refer to water temperature. The constant b is the asymptote of the graph of f and as such is an approximation to X_M.

a. Show that choosing a, b, and c to minimize

$$\sum_{i=1}^{n}\left[w_i y_i - \frac{a}{(x_i-b)^c}\right]^2$$

reduces to solving the nonlinear system

$$a = \sum_{i=1}^{n}\frac{w_i y_i}{(x_i-b)^c} \Bigg/ \sum_{i=1}^{n}\frac{1}{(x_i-b)^{2c}},$$

$$0 = \sum_{i=1}^{n}\frac{w_i y_i}{(x_i-b)^c}\cdot\sum_{i=1}^{n}\frac{1}{(x_i-b)^{2c+1}} - \sum_{i=1}^{n}\frac{w_i y_i}{(x_i-b)^{c+1}}\cdot\sum_{i=1}^{n}\frac{1}{(x_i-b)^{2c}},$$

$$0 = \sum_{i=1}^{n}\frac{w_i y_i}{(x_i-b)^c}\cdot\sum_{i=1}^{n}\frac{\ln(x_i-b)}{(x_i-b)^{2c}} - \sum_{i=1}^{n}\frac{w_i y_i\ln(x_i-b)}{(x_i-b)^c}\cdot\sum_{i=1}^{n}\frac{1}{(x_i-b)^{2c}}.$$

b. Solve the nonlinear system for the species with the following data. Use the weights $w_i = \ln y_i$.

i	1	2	3	4
y_i	2.40	3.80	4.75	21.60
x_i	31.8	31.5	31.2	30.2

10.3 Quasi-Newton Methods

A significant weakness of Newton's method for solving systems of nonlinear equations is the need, at each iteration, to determine a Jacobian matrix and solve an $n \times n$ linear system that involves this matrix. Consider the amount of computation associated with one iteration of Newton's method. The Jacobian matrix associated with a system of n nonlinear equations written in the form $\mathbf{F}(\mathbf{x}) = \mathbf{0}$ requires that the n^2 partial derivatives of the n component functions of $\mathbf{F}$ be determined and evaluated. In most situations, the exact evaluation of the partial derivatives is inconvenient, although the problem has been made more tractable with the widespread use of symbolic computation systems, such as Maple.

When the exact evaluation is not practical, we can use finite difference approximations to the partial derivatives. For example,

$$\frac{\partial f_j}{\partial x_k}(\mathbf{x}^{(i)}) \approx \frac{f_j(\mathbf{x}^{(i)} + \mathbf{e}_k h) - f_j(\mathbf{x}^{(i)})}{h}, \tag{10.10}$$

where h is small in absolute value and $\mathbf{e}_k$ is the vector whose only nonzero entry is a 1 in the kth coordinate. This approximation, however, still requires that at least n^2 scalar functional evaluations be performed to approximate the Jacobian and does not decrease the amount of calculation, in general $O(n^3)$, required for solving the linear system involving this approximate Jacobian.

The total computational effort for just one iteration of Newton's method is consequently at least $n^2 + n$ scalar functional evaluations (n^2 for the evaluation of the Jacobian matrix and n for the evaluation of $\mathbf{F}$) together with $O(n^3)$ arithmetic operations to solve the linear system. This amount of computational effort is extensive, except for relatively small values of n and easily evaluated scalar functions.

In this section we consider a generalization of the Secant method to systems of nonlinear equations, a technique known as **Broyden's method** (see [Broy]). The method requires only n scalar functional evaluations per iteration and also reduces the number of arithmetic calculations to $O(n^2)$. It belongs to a class of methods known as *least-change secant updates* that produce algorithms called **quasi-Newton**. These methods replace the Jacobian matrix in Newton's method with an approximation matrix that is easily updated at each iteration.

The disadvantage of the quasi-Newton methods is that the quadratic convergence of Newton's method is lost, being replaced, in general, by a convergence called **superlinear**. This implies that

$$\lim_{i\to\infty} \frac{\left\|\mathbf{x}^{(i+1)} - \mathbf{p}\right\|}{\left\|\mathbf{x}^{(i)} - \mathbf{p}\right\|} = 0,$$

where $\mathbf{p}$ denotes the solution to $\mathbf{F}(\mathbf{x}) = \mathbf{0}$ and $\mathbf{x}^{(i)}$ and $\mathbf{x}^{(i+1)}$ are consecutive approximations to $\mathbf{p}$.

In most applications, the reduction to superlinear convergence is a more than acceptable trade-off for the decrease in the amount of computation. An additional disadvantage of quasi-Newton methods is that, unlike Newton's method, they are not self-correcting. Newton's method will generally correct for roundoff error with successive iterations, but unless special safeguards are incorporated, Broyden's method will not.

To describe Broyden's method, suppose that an initial approximation $\mathbf{x}^{(0)}$ is given to the solution $\mathbf{p}$ of $\mathbf{F}(\mathbf{x}) = \mathbf{0}$. We calculate the next approximation $\mathbf{x}^{(1)}$ in the same manner as Newton's method. If it is inconvenient to determine $J(\mathbf{x}^{(0)})$ exactly, we use the difference equations given by (10.10) to approximate the partial derivatives. To compute $\mathbf{x}^{(2)}$, however, we depart from Newton's method and examine the Secant method for a single nonlinear equation. The Secant method uses the approximation

$$f'(x_1) \approx \frac{f(x_1) - f(x_0)}{x_1 - x_0}$$

as a replacement for $f'(x_1)$ in the single-variable Newton's method.

For nonlinear systems, $\mathbf{x}^{(1)} - \mathbf{x}^{(0)}$ is a vector, so the corresponding quotient is undefined. However, the method proceeds similarly in that we replace the matrix $J\left(\mathbf{x}^{(1)}\right)$ in Newton's method for systems by a matrix A_1 with the property that

$$A_1\left(\mathbf{x}^{(1)} - \mathbf{x}^{(0)}\right) = \mathbf{F}\left(\mathbf{x}^{(1)}\right) - \mathbf{F}\left(\mathbf{x}^{(0)}\right). \tag{10.11}$$

Any nonzero vector in $\mathbb{R}^n$ can be written as the sum of a multiple of $\mathbf{x}^{(1)} - \mathbf{x}^{(0)}$ and a multiple of a vector in the orthogonal complement of $\mathbf{x}^{(1)} - \mathbf{x}^{(0)}$. So, to uniquely define the matrix A_1, we also need to specify how it acts on the orthogonal complement of $\mathbf{x}^{(1)} - \mathbf{x}^{(0)}$. No information is available about the change in $\mathbf{F}$ in a direction orthogonal to $\mathbf{x}^{(1)} - \mathbf{x}^{(0)}$, so we specify that no change be made in this direction, that is,

$$A_1\mathbf{z} = J\left(\mathbf{x}^{(0)}\right)\mathbf{z}, \quad \text{whenever} \quad \left(\mathbf{x}^{(1)} - \mathbf{x}^{(0)}\right)^t \mathbf{z} = 0. \tag{10.12}$$

Thus, any vector orthogonal to $\mathbf{x}^{(1)} - \mathbf{x}^{(0)}$ is unaffected by the update from $J\left(\mathbf{x}^{(0)}\right)$, which was used to compute $\mathbf{x}^{(1)}$, to A_1, which is used in the determination of $\mathbf{x}^{(2)}$.

Conditions (10.11) and (10.12) uniquely define A_1 (see [DM]) as

$$A_1 = J\left(\mathbf{x}^{(0)}\right) + \frac{\left[\mathbf{F}(\mathbf{x}^{(1)}) - \mathbf{F}(\mathbf{x}^{(0)}) - J(\mathbf{x}^{(0)})\left(\mathbf{x}^{(1)} - \mathbf{x}^{(0)}\right)\right]\left(\mathbf{x}^{(1)} - \mathbf{x}^{(0)}\right)^t}{\left\|\mathbf{x}^{(1)} - \mathbf{x}^{(0)}\right\|_2^2}.$$

It is this matrix that is used in place of $J\left(\mathbf{x}^{(1)}\right)$ to determine $\mathbf{x}^{(2)}$ as

$$\mathbf{x}^{(2)} = \mathbf{x}^{(1)} - A_1^{-1}\mathbf{F}\big(\mathbf{x}^{(1)}\big).$$

Once $\mathbf{x}^{(2)}$ has been determined, the method is repeated to determine $\mathbf{x}^{(3)}$, using A_1 in place of $A_0 \equiv J\left(\mathbf{x}^{(0)}\right)$, and with $\mathbf{x}^{(2)}$ and $\mathbf{x}^{(1)}$ in place of $\mathbf{x}^{(1)}$ and $\mathbf{x}^{(0)}$.

In general, once $\mathbf{x}^{(i)}$ has been determined, $\mathbf{x}^{(i+1)}$ is computed by

$$A_i = A_{i-1} + \frac{\mathbf{y}_i - A_{i-1}\mathbf{s}_i}{||\mathbf{s}_i||_2^2}\mathbf{s}_i^t \tag{10.13}$$

and

$$\mathbf{x}^{(i+1)} = \mathbf{x}^{(i)} - A_i^{-1}\mathbf{F}\left(\mathbf{x}^{(i)}\right), \tag{10.14}$$

where the notation $\mathbf{y}_i = \mathbf{F}\big(\mathbf{x}^{(i)}\big) - \mathbf{F}\big(\mathbf{x}^{(i-1)}\big)$ and $\mathbf{s}_i = \mathbf{x}^{(i)} - \mathbf{x}^{(i-1)}$ is introduced to simplify the equations.

If the method was performed as outlined in Eqs. (10.13) and (10.14), the number of scalar functional evaluations would be reduced from $n^2 + n$ to n (those required for evaluating $\mathbf{F}\big(\mathbf{x}^{(i)}\big)$), but $O\big(n^3\big)$ calculations would still required to solve the associated $n \times n$ linear system (see Step 4 in Algorithm 10.1)

$$A_i\mathbf{s}_{i+1} = -\mathbf{F}\big(\mathbf{x}^{(i)}\big). \tag{10.15}$$

Employing the method in this form would not be justified because of the reduction to superlinear convergence from the quadratic convergence of Newton's method.

Sherman-Morrison Formula

A considerable improvement can be incorporated, however, by employing a matrix inversion formula of Sherman and Morrison (see, for example, [DM], p. 55).

Theorem 10.8 **(Sherman-Morrison Formula)**

Suppose that A is a nonsingular matrix and that $\mathbf{x}$ and $\mathbf{y}$ are vectors with $\mathbf{y}^tA^{-1}\mathbf{x} \neq -1$. Then $A + \mathbf{x}\mathbf{y}^t$ is nonsingular and

$$\left(A + \mathbf{x}\mathbf{y}^t\right)^{-1} = A^{-1} - \frac{A^{-1}\mathbf{x}\mathbf{y}^tA^{-1}}{1 + \mathbf{y}^tA^{-1}\mathbf{x}}. \quad \blacksquare$$

The Sherman-Morrison formula permits A_i^{-1} to be computed directly from A_{i-1}^{-1}, eliminating the need for a matrix inversion with each iteration.

Letting $A = A_{i-1}$, $\mathbf{x} = (\mathbf{y}_i - A_{i-1}\mathbf{s}_i)/||\mathbf{s}_i||_2^2$, and $\mathbf{y} = \mathbf{s}_i$, in Eq. (10.13) gives

$$\begin{aligned}
A_i^{-1} &= \left(A_{i-1} + \frac{\mathbf{y}_i - A_{i-1}\mathbf{s}_i}{||\mathbf{s}_i||_2^2}\mathbf{s}_i^t\right)^{-1} \\
&= A_{i-1}^{-1} - \frac{A_{i-1}^{-1}\left(\frac{\mathbf{y}_i - A_{i-1}\mathbf{s}_i}{||\mathbf{s}_i||_2^2}\mathbf{s}_i^t\right)A_{i-1}^{-1}}{1 + \mathbf{s}_i^tA_{i-1}^{-1}\left(\dfrac{\mathbf{y}_i - A_{i-1}\mathbf{s}_i}{||\mathbf{s}_i||_2^2}\right)} \\
&= A_{i-1}^{-1} - \frac{\left(A_{i-1}^{-1}\mathbf{y}_i - \mathbf{s}_i\right)\mathbf{s}_i^tA_{i-1}^{-1}}{||\mathbf{s}_i||_2^2 + \mathbf{s}_i^tA_{i-1}^{-1}\mathbf{y}_i - ||\mathbf{s}_i||_2^2},
\end{aligned}$$

so

$$A_i^{-1} = A_{i-1}^{-1} + \frac{\left(\mathbf{s}_i - A_{i-1}^{-1}\mathbf{y}_i\right)\mathbf{s}_i^t A_{i-1}^{-1}}{\mathbf{s}_i^t A_{i-1}^{-1}\mathbf{y}_i}. \tag{10.16}$$

This computation involves only matrix-vector multiplications at each step and therefore requires only $O(n^2)$ arithmetic calculations. The calculation of A_i is bypassed, as is the necessity of solving the linear system (10.15).

Algorithm 10.2 follows directly from this construction, incorporating (10.16) into the iterative technique (10.14).

Broyden

To approximate the solution of the nonlinear system $\mathbf{F}(\mathbf{x}) = \mathbf{0}$ given an initial approximation $\mathbf{x}$:

INPUT number n of equations and unknowns; initial approximation $\mathbf{x} = (x_1, \ldots, x_n)^t$; tolerance TOL; maximum number of iterations N.

OUTPUT approximate solution $\mathbf{x} = (x_1, \ldots, x_n)^t$ or a message that the number of iterations was exceeded.

Step 1 Set $A_0 = J(\mathbf{x})$ where $J(\mathbf{x})_{i,j} = \frac{\partial f_i}{\partial x_j}(\mathbf{x})$ for $1 \le i,j \le n$;
$\mathbf{v} = \mathbf{F}(\mathbf{x})$. (*Note:* $\mathbf{v} = \mathbf{F}(\mathbf{x}^{(0)})$.)

Step 2 Set $A = A_0^{-1}$. (*Use Gaussian elimination.*)

Step 3 Set $\mathbf{s} = -A\mathbf{v}$; (*Note:* $\mathbf{s} = \mathbf{s}_1$.)
$\mathbf{x} = \mathbf{x} + \mathbf{s}$; (*Note:* $\mathbf{x} = \mathbf{x}^{(1)}$.)
$k = 2$.

Step 4 While ($k \le N$) do Steps 5–13.

Step 5 Set $\mathbf{w} = \mathbf{v}$; (*Save* $\mathbf{v}$.)
$\mathbf{v} = \mathbf{F}(\mathbf{x})$; (*Note:* $\mathbf{v} = \mathbf{F}(\mathbf{x}^{(k)})$.)
$\mathbf{y} = \mathbf{v} - \mathbf{w}$. (*Note:* $\mathbf{y} = \mathbf{y}_k$.)

Step 6 Set $\mathbf{z} = -A\mathbf{y}$. (*Note:* $\mathbf{z} = -A_{k-1}^{-1}\mathbf{y}_k$.)

Step 7 Set $p = -\mathbf{s}^t\mathbf{z}$. (*Note:* $p = \mathbf{s}_k^t A_{k-1}^{-1}\mathbf{y}_k$.)

Step 8 Set $\mathbf{u}^t = \mathbf{s}^t A$.

Step 9 Set $A = A + \frac{1}{p}(\mathbf{s} + \mathbf{z})\mathbf{u}^t$. (*Note:* $A = A_k^{-1}$.)

Step 10 Set $\mathbf{s} = -A\mathbf{v}$. (*Note:* $\mathbf{s} = -A_k^{-1}\mathbf{F}(\mathbf{x}^{(k)})$.)

Step 11 Set $\mathbf{x} = \mathbf{x} + \mathbf{s}$. (*Note:* $\mathbf{x} = \mathbf{x}^{(k+1)}$.)

Step 12 If $||\mathbf{s}|| < TOL$ then OUTPUT ($\mathbf{x}$);
(*The procedure was successful.*)
STOP.

Step 13 Set $k = k + 1$.

Step 14 OUTPUT ('Maximum number of iterations exceeded');
(*The procedure was unsuccessful.*)
STOP. ▪

Example 1 Use Broyden's method with $\mathbf{x}^{(0)} = (0.1, 0.1, -0.1)^t$ to approximate the solution to the nonlinear system

$$3x_1 - \cos(x_2x_3) - \frac{1}{2} = 0,$$

$$x_1^2 - 81(x_2 + 0.1)^2 + \sin x_3 + 1.06 = 0,$$

$$e^{-x_1x_2} + 20x_3 + \frac{10\pi - 3}{3} = 0.$$

Solution This system was solved by Newton's method in Example 1 of Section 10.2. The Jacobian matrix for this system is

$$J(x_1, x_2, x_3) = \begin{bmatrix} 3 & x_3 \sin x_2x_3 & x_2 \sin x_2x_3 \\ 2x_1 & -162(x_2 + 0.1) & \cos x_3 \\ -x_2e^{-x_1x_2} & -x_1e^{-x_1x_2} & 20 \end{bmatrix}.$$

Let $\mathbf{x}^{(0)} = (0.1, 0.1, -0.1)^t$ and

$$\mathbf{F}(x_1, x_2, x_3) = (f_1(x_1, x_2, x_3), f_2(x_1, x_2, x_3), f_3(x_1, x_2, x_3))^t,$$

where

$$f_1(x_1, x_2, x_3) = 3x_1 - \cos(x_2x_3) - \frac{1}{2},$$

$$f_2(x_1, x_2, x_3) = x_1^2 - 81(x_2 + 0.1)^2 + \sin x_3 + 1.06,$$

and

$$f_3(x_1, x_2, x_3) = e^{-x_1x_2} + 20x_3 + \frac{10\pi - 3}{3}.$$

Then

$$\mathbf{F}\left(\mathbf{x}^{(0)}\right) = \begin{bmatrix} -1.199950 \\ -2.269833 \\ 8.462025 \end{bmatrix}.$$

Because

$$A_0 = J\left(x_1^{(0)}, x_2^{(0)}, x_3^{(0)}\right)$$

$$= \begin{bmatrix} 3 & 9.999833 \times 10^{-4} & -9.999833 \times 10^{-4} \\ 0.2 & -32.4 & 0.9950042 \\ -9.900498 \times 10^{-2} & -9.900498 \times 10^{-2} & 20 \end{bmatrix},$$

we have

$$A_0^{-1} = J\left(x_1^{(0)}, x_2^{(0)}, x_3^{(0)}\right)^{-1}$$

$$= \begin{bmatrix} 0.3333332 & 1.023852 \times 10^{-5} & 1.615701 \times 10^{-5} \\ 2.108607 \times 10^{-3} & -3.086883 \times 10^{-2} & 1.535836 \times 10^{-3} \\ 1.660520 \times 10^{-3} & -1.527577 \times 10^{-4} & 5.000768 \times 10^{-2} \end{bmatrix}.$$

So

$$\mathbf{x}^{(1)} = \mathbf{x}^{(0)} - A_0^{-1}\mathbf{F}\left(\mathbf{x}^{(0)}\right) = \begin{bmatrix} 0.4998697 \\ 1.946685 \times 10^{-2} \\ -0.5215205 \end{bmatrix},$$

$$\mathbf{F}\left(\mathbf{x}^{(1)}\right) = \begin{bmatrix} -3.394465 \times 10^{-4} \\ -0.3443879 \\ 3.188238 \times 10^{-2} \end{bmatrix},$$

$$\mathbf{y}_1 = \mathbf{F}(\mathbf{x}^{(1)}) - \mathbf{F}(\mathbf{x}^{(0)}) = \begin{bmatrix} 1.199611 \\ 1.925445 \\ -8.430143 \end{bmatrix},$$

$$\mathbf{s}_1 = \begin{bmatrix} 0.3998697 \\ -8.053315 \times 10^{-2} \\ -0.4215204 \end{bmatrix},$$

$$\mathbf{s}_1^t A_0^{-1} \mathbf{y}_1 = 0.3424604,$$

$$A_1^{-1} = A_0^{-1} + (1/0.3424604)\left[\left(\mathbf{s}_1 - A_0^{-1}\mathbf{y}_1\right)\mathbf{s}_1^t A_0^{-1}\right]$$

$$= \begin{bmatrix} 0.3333781 & 1.11050 \times 10^{-5} & 8.967344 \times 10^{-6} \\ -2.021270 \times 10^{-3} & -3.094849 \times 10^{-2} & 2.196906 \times 10^{-3} \\ 1.022214 \times 10^{-3} & -1.650709 \times 10^{-4} & 5.010986 \times 10^{-2} \end{bmatrix},$$

and

$$\mathbf{x}^{(2)} = \mathbf{x}^{(1)} - A_1^{-1}\mathbf{F}(\mathbf{x}^{(1)}) = \begin{bmatrix} 0.4999863 \\ 8.737833 \times 10^{-3} \\ -0.5231746 \end{bmatrix}.$$

Additional iterations are listed in Table 10.4. The fifth iteration of Broyden's method is slightly less accurate than was the fourth iteration of Newton's method given in the example at the end of the preceding section. ■

Table 10.4

k	$x_1^{(k)}$	$x_2^{(k)}$	$x_3^{(k)}$	$\Vert\mathbf{x}^{(k)} - \mathbf{x}^{(k-1)}\Vert_2$
3	0.5000066	8.672157×10^{-4}	−0.5236918	7.88×10^{-3}
4	0.5000003	6.083352×10^{-5}	−0.5235954	8.12×10^{-4}
5	0.5000000	-1.448889×10^{-6}	−0.5235989	6.24×10^{-5}
6	0.5000000	6.059030×10^{-9}	−0.5235988	1.50×10^{-6}

Procedures are also available that maintain quadratic convergence but significantly reduce the number of required functional evaluations. Methods of this type were originally proposed by Brown [Brow,K]. A survey and comparison of some commonly used methods of this type can be found in [MC]. In general, however, these methods are much more difficult to implement efficiently than Broyden's method.

EXERCISE SET 10.3

1. Use Broyden's method with $\mathbf{x}^{(0)} = \mathbf{0}$ to compute $\mathbf{x}^{(2)}$ for each of the following nonlinear systems.

   a. $4x_1^2 - 20x_1 + \frac{1}{4}x_2^2 + 8 = 0,$

   $\frac{1}{2}x_1x_2^2 + 2x_1 - 5x_2 + 8 = 0.$

   b. $\sin(4\pi x_1 x_2) - 2x_2 - x_1 = 0,$

   $\left(\frac{4\pi - 1}{4\pi}\right)(e^{2x_1} - e) + 4ex_2^2 - 2ex_1 = 0.$

   c. $3x_1^2 - x_2^2 = 0,$

   $3x_1x_2^2 - x_1^3 - 1 = 0.$

   Use $\mathbf{x}^{(0)} = (1, 1)^t$.

   d. $\ln(x_1^2 + x_2^2) - \sin(x_1x_2) = \ln 2 + \ln \pi,$

   $e^{x_1 - x_2} + \cos(x_1x_2) = 0.$

   Use $\mathbf{x}^{(0)} = (2, 2)^t$.

2. Use Broyden's method with $\mathbf{x}^{(0)} = \mathbf{0}$ to compute $\mathbf{x}^{(2)}$ for each of the following nonlinear systems.

a.
$$3x_1 - \cos(x_2x_3) - \frac{1}{2} = 0,$$
$$4x_1^2 - 625x_2^2 + 2x_2 - 1 = 0,$$
$$e^{-x_1x_2} + 20x_3 + \frac{10\pi - 3}{3} = 0.$$

b.
$$x_1^2 + x_2 - 37 = 0,$$
$$x_1 - x_2^2 - 5 = 0,$$
$$x_1 + x_2 + x_3 - 3 = 0.$$

c.
$$x_1^3 + x_1^2x_2 - x_1x_3 + 6 = 0,$$
$$e^{x_1} + e^{x_2} - x_3 = 0,$$
$$x_2^2 - 2x_1x_3 = 4.$$

Use $\mathbf{x}^{(0)} = (-1, -2, 1)^t$.

d.
$$6x_1 - 2\cos(x_2x_3) - 1 = 0,$$
$$9x_2 + \sqrt{x_1^2 + \sin x_3 + 1.06} + 0.9 = 0,$$
$$60x_3 + 3e^{-x_1x_2} + 10\pi - 3 = 0.$$

Use $\mathbf{x}^{(0)} = (0, 0, 0)^t$.

3. Use Broyden's method to approximate solutions to the nonlinear systems in Exercise 1 using the following initial approximations $\mathbf{x}^{(0)}$.

a. $(0, 0)^t$ **b.** $(0, 0)^t$ **c.** $(1, 1)^t$ **d.** $(2, 2)^t$

4. Use Broyden's method to approximate solutions to the nonlinear systems in Exercise 2 using the following initial approximations $\mathbf{x}^{(0)}$.

a. $(1, 1, 1)^t$ **b.** $(2, 1, -1)^t$ **c.** $(-1, -2, 1)^t$ **d.** $(0, 0, 0)^t$

5. Use Broyden's method to approximate solutions to the following nonlinear systems. Iterate until $\left\|\mathbf{x}^{(k)} - \mathbf{x}^{(k-1)}\right\|_\infty < 10^{-6}$.

a.
$$x_1(1 - x_1) + 4x_2 = 12,$$
$$(x_1 - 2)^2 + (2x_2 - 3)^2 = 25.$$

b.
$$5x_1^2 - x_2^2 = 0,$$
$$x_2 - 0.25(\sin x_1 + \cos x_2) = 0.$$

c.
$$15x_1 + x_2^2 - 4x_3 = 13,$$
$$x_1^2 + 10x_2 - x_3 = 11,$$
$$x_2^3 - 25x_3 = -22.$$

d.
$$10x_1 - 2x_2^2 + x_2 - 2x_3 - 5 = 0,$$
$$8x_2^2 + 4x_3^2 - 9 = 0,$$
$$8x_2x_3 + 4 = 0.$$

6. The nonlinear system

$$4x_1 - x_2 + x_3 = x_1x_4,$$
$$-x_1 + 3x_2 - 2x_3 = x_2x_4,$$
$$x_1 - 2x_2 + 3x_3 = x_3x_4,$$
$$x_1^2 + x_2^2 + x_3^2 = 1$$

has six solutions.

a. Show that if $(x_1, x_2, x_3, x_4)^t$ is a solution then $(-x_1, -x_2, -x_3, x_4)^t$ is a solution.

b. Use Broyden's method three times to approximate each solution. Iterate until $\left\|\mathbf{x}^{(k)} - \mathbf{x}^{(k-1)}\right\|_\infty < 10^{-5}$.

7. The nonlinear system

$$3x_1 - \cos(x_2x_3) - \frac{1}{2} = 0, \quad x_1^2 - 625x_2^2 - \frac{1}{4} = 0, \quad e^{-x_1x_2} + 20x_3 + \frac{10\pi - 3}{3} = 0$$

has a singular Jacobian matrix at the solution. Apply Broyden's method with $\mathbf{x}^{(0)} = (1, 1 - 1)^t$. Note that convergence may be slow or may not occur within a reasonable number of iterations.

8. Show that if $\mathbf{0} \neq \mathbf{y} \in \mathbb{R}^n$ and $\mathbf{z} \in \mathbb{R}^n$, then $\mathbf{z} = \mathbf{z}_1 + \mathbf{z}_2$, where $\mathbf{z}_1 = (\mathbf{y}^t\mathbf{z}/\|\mathbf{y}\|_2^2)\mathbf{y}$ is parallel to $\mathbf{y}$ and $\mathbf{z}_2$ is orthogonal to $\mathbf{y}$.

9. Show that if $\mathbf{u}, \mathbf{v} \in \mathbb{R}^n$, then $\det(I + \mathbf{u}\mathbf{v}^t) = 1 + \mathbf{v}^t\mathbf{u}$.

10. **a.** Use the result in Exercise 9 to show that if A^{-1} exists and $\mathbf{x}, \mathbf{y} \in \mathbb{R}^n$, then $(A + \mathbf{x}\mathbf{y}^t)^{-1}$ exists if and only if $\mathbf{y}^tA^{-1}\mathbf{x} \neq -1$.

b. By multiplying on the right by $A + \mathbf{x}\mathbf{y}^t$, show that when $\mathbf{y}^tA^{-1}\mathbf{x} \neq -1$ we have

$$(A + \mathbf{x}\mathbf{y}^t)^{-1} = A^{-1} - \frac{A^{-1}\mathbf{x}\mathbf{y}^tA^{-1}}{1 + \mathbf{y}^tA^{-1}\mathbf{x}}.$$

11. Exercise 13 of Section 8.1 dealt with determining an exponential least squares relationship of the form $R = bw^a$ to approximate a collection of data relating the weight and respiration rule of *Modest sphinx* moths. In that exercise, the problem was converted to a log-log relationship, and in part (c), a quadratic term was introduced in an attempt to improve the approximation. Instead of converting the problem, determine the constants a and b that minimize $\sum_{i=1}^{n}(R_i - bw_i^a)^2$ for the data listed in Exercise 13 of 8.1. Compute the error associated with this approximation, and compare this to the error of the previous approximations for this problem.

10.4 Steepest Descent Techniques

The advantage of the Newton and quasi-Newton methods for solving systems of nonlinear equations is their speed of convergence once a sufficiently accurate approximation is known. A weakness of these methods is that an accurate initial approximation to the solution is needed to ensure convergence. The **Steepest Descent** method considered in this section converges only linearly to the solution, but it will usually converge even for poor initial approximations. As a consequence, this method is used to find sufficiently accurate starting approximations for the Newton-based techniques in the same way the Bisection method is used for a single equation.

The name for the Steepest Descent method follows from the three-dimensional application of pointing in the steepest downward direction.

The method of Steepest Descent determines a local minimum for a multivariable function of the form $g : \mathbb{R}^n \to \mathbb{R}$. The method is valuable quite apart from the application as a starting method for solving nonlinear systems. (Some other applications are considered in the exercises.)

The connection between the minimization of a function from $\mathbb{R}^n$ to $\mathbb{R}$ and the solution of a system of nonlinear equations is due to the fact that a system of the form

$$\begin{aligned} f_1(x_1, x_2, \ldots, x_n) &= 0, \\ f_2(x_1, x_2, \ldots, x_n) &= 0, \\ \vdots \qquad & \quad \vdots \\ f_n(x_1, x_2, \ldots, x_n) &= 0, \end{aligned}$$

has a solution at $\mathbf{x} = (x_1, x_2, \ldots, x_n)^t$ precisely when the function g defined by

$$g(x_1, x_2, \ldots, x_n) = \sum_{i=1}^{n} [f_i(x_1, x_2, \ldots, x_n)]^2$$

has the minimal value 0.

The method of Steepest Descent for finding a local minimum for an arbitrary function g from $\mathbb{R}^n$ into $\mathbb{R}$ can be intuitively described as follows:

1. Evaluate g at an initial approximation $\mathbf{x}^{(0)} = \left(x_1^{(0)}, x_2^{(0)}, \ldots, x_n^{(0)}\right)^t$.
2. Determine a direction from $\mathbf{x}^{(0)}$ that results in a decrease in the value of g.
3. Move an appropriate amount in this direction and call the new value $\mathbf{x}^{(1)}$.
4. Repeat steps 1 through 3 with $\mathbf{x}^{(0)}$ replaced by $\mathbf{x}^{(1)}$.

The Gradient of a Function

Before describing how to choose the correct direction and the appropriate distance to move in this direction, we need to review some results from calculus. The Extreme Value 1.9 Theorem states that a differentiable single-variable function can have a relative minimum only when the derivative is zero. To extend this result to multivariable functions, we need the following definition.

Definition 10.9 For $g : \mathbb{R}^n \to \mathbb{R}$, the **gradient** of g at $\mathbf{x} = (x_1, x_2, \ldots, x_n)^t$ is denoted $\nabla g(\mathbf{x})$ and defined by

$$\nabla g(\mathbf{x}) = \left(\frac{\partial g}{\partial x_1}(\mathbf{x}), \frac{\partial g}{\partial x_2}(\mathbf{x}), \ldots, \frac{\partial g}{\partial x_n}(\mathbf{x}) \right)^t. \quad \blacksquare$$

The root gradient comes from the Latin word *gradi*, meaning "to walk". In this sense, the gradient of a surface is the rate at which it "walks uphill".

The gradient for a multivariable function is analogous to the derivative of a single-variable function in the sense that a differentiable multivariable function can have a relative minimum at $\mathbf{x}$ only when the gradient at $\mathbf{x}$ is the zero vector. The gradient has another important property connected with the minimization of multivariable functions. Suppose that $\mathbf{v} = (v_1, v_2, \ldots, v_n)^t$ is a unit vector in $\mathbb{R}^n$; that is,

$$||\mathbf{v}||_2^2 = \sum_{i=1}^{n} v_i^2 = 1.$$

The **directional derivative** of g at $\mathbf{x}$ in the direction of $\mathbf{v}$ measures the change in the value of the function g relative to the change in the variable in the direction of $\mathbf{v}$. It is defined by

$$D_{\mathbf{v}} g(\mathbf{x}) = \lim_{h \to 0} \frac{1}{h} [g(\mathbf{x} + h\mathbf{v}) - g(\mathbf{x})] = \mathbf{v}^t \cdot \nabla g(\mathbf{x}).$$

When g is differentiable, the direction that produces the maximum value for the directional derivative occurs when $\mathbf{v}$ is chosen to be parallel to $\nabla g(\mathbf{x})$, provided that $\nabla g(\mathbf{x}) \neq \mathbf{0}$. As a consequence, the direction of greatest decrease in the value of g at $\mathbf{x}$ is the direction given by $-\nabla g(\mathbf{x})$. Figure 10.3 is an illustration when g is a function of two variables.

Figure 10.3

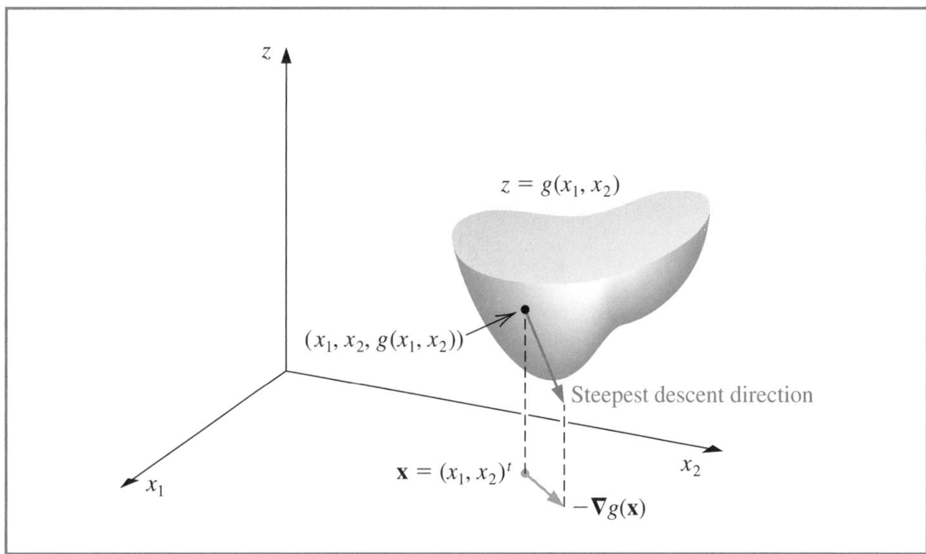

The object is to reduce $g(\mathbf{x})$ to its minimal value of zero, so an appropriate choice for $\mathbf{x}^{(1)}$ is to move away from $\mathbf{x}^{(0)}$ in the direction that gives the greatest decrease in the value of $g(\mathbf{x})$. Hence we let

$$\mathbf{x}^{(1)} = \mathbf{x}^{(0)} - \alpha \nabla g(\mathbf{x}^{(0)}), \quad \text{for some constant } \alpha > 0. \tag{10.17}$$

The problem now reduces to choosing an appropriate value of α so that $g(\mathbf{x}^{(1)})$ will be significantly less than $g(\mathbf{x}^{(0)})$.

To determine an appropriate choice for the value α, we consider the single-variable function

$$h(\alpha) = g(\mathbf{x}^{(0)} - \alpha \nabla g(\mathbf{x}^{(0)})). \tag{10.18}$$

The value of α that minimizes h is the value needed for Eq. (10.17).

Finding a minimal value for h directly would require differentiating h and then solving a root-finding problem to determine the critical points of h. This procedure is generally too costly. Instead, we choose three numbers $\alpha_1 < \alpha_2 < \alpha_3$ that, we hope, are close to where the minimum value of $h(\alpha)$ occurs. We then construct the quadratic polynomial $P(x)$ that interpolates h at α_1, α_2, and α_3. The minimum of a quadratic polynomial is easily found in a manner similar to that used in Müller's method in section 2.6.

We define $\hat{\alpha}$ in $[\alpha_1, \alpha_3]$ so that $P(\hat{\alpha})$ is a minimum in $[\alpha_1, \alpha_3]$ and use $P(\hat{\alpha})$ to approximate the minimal value of $h(\alpha)$. Then $\hat{\alpha}$ is used to determine the new iterate for approximating the minimal value of g:

$$\mathbf{x}^{(1)} = \mathbf{x}^{(0)} - \hat{\alpha} \nabla g(\mathbf{x}^{(0)}).$$

Because $g(\mathbf{x}^{(0)})$ is available, to minimize the computation we first choose $\alpha_1 = 0$. Next a number α_3 is found with $h(\alpha_3) < h(\alpha_1)$. (Since α_1 does not minimize h, such a number α_3 does exist.) Finally, α_2 is chosen to be $\alpha_3/2$.

The minimum value of P on $[\alpha_1, \alpha_3]$ occurs either at the only critical point of P or at the right endpoint α_3 because, by assumption, $P(\alpha_3) = h(\alpha_3) < h(\alpha_1) = P(\alpha_1)$. Because $P(x)$ is a quadratic polynomial, the critical point can be found by solving a linear equation.

Example 1 Use the Steepest Descent method with $\mathbf{x}^{(0)} = (0, 0, 0)^t$ to find a reasonable starting approximation to the solution of the nonlinear system

$$f_1(x_1, x_2, x_3) = 3x_1 - \cos(x_2 x_3) - \frac{1}{2} = 0,$$

$$f_2(x_1, x_2, x_3) = x_1^2 - 81(x_2 + 0.1)^2 + \sin x_3 + 1.06 = 0,$$

$$f_3(x_1, x_2, x_3) = e^{-x_1 x_2} + 20x_3 + \frac{10\pi - 3}{3} = 0.$$

Solution Let $g(x_1, x_2, x_3) = [f_1(x_1, x_2, x_3)]^2 + [f_2(x_1, x_2, x_3)]^2 + [f_3(x_1, x_2, x_3)]^2$. Then

$$\begin{aligned} \nabla g(x_1, x_2, x_3) \equiv \nabla g(\mathbf{x}) &= \Big(2f_1(\mathbf{x}) \frac{\partial f_1}{\partial x_1}(\mathbf{x}) + 2f_2(\mathbf{x}) \frac{\partial f_2}{\partial x_1}(\mathbf{x}) + 2f_3(\mathbf{x}) \frac{\partial f_3}{\partial x_1}(\mathbf{x}), \\ &\quad 2f_1(\mathbf{x}) \frac{\partial f_1}{\partial x_2}(\mathbf{x}) + 2f_2(\mathbf{x}) \frac{\partial f_2}{\partial x_2}(\mathbf{x}) + 2f_3(\mathbf{x}) \frac{\partial f_3}{\partial x_2}(\mathbf{x}), \\ &\quad 2f_1(\mathbf{x}) \frac{\partial f_1}{\partial x_3}(\mathbf{x}) + 2f_2(\mathbf{x}) \frac{\partial f_2}{\partial x_3}(\mathbf{x}) + 2f_3(\mathbf{x}) \frac{\partial f_3}{\partial x_3}(\mathbf{x}) \Big) \\ &= 2\mathbf{J}(\mathbf{x})^t \mathbf{F}(\mathbf{x}). \end{aligned}$$

For $\mathbf{x}^{(0)} = (0, 0, 0)^t$, we have

$$g(\mathbf{x}^{(0)}) = 111.975 \quad \text{and} \quad z_0 = ||\nabla g(\mathbf{x}^{(0)})||_2 = 419.554.$$

Let

$$\mathbf{z} = \frac{1}{z_0} \nabla g(\mathbf{x}^{(0)}) = (-0.0214514, -0.0193062, 0.999583)^t.$$

With $\alpha_1 = 0$, we have $g_1 = g(\mathbf{x}^{(0)} - \alpha_1 \mathbf{z}) = g(\mathbf{x}^{(0)}) = 111.975$. We arbitrarily let $\alpha_3 = 1$ so that

$$g_3 = g(\mathbf{x}^{(0)} - \alpha_3 \mathbf{z}) = 93.5649.$$

Because $g_3 < g_1$, we accept α_3 and set $\alpha_2 = \alpha_3/2 = 0.5$. Thus

$$g_2 = g(\mathbf{x}^{(0)} - \alpha_2 \mathbf{z}) = 2.53557.$$

We now find the quadratic polynomial that interpolates the data $(0, 111.975)$, $(1, 93.5649)$, and $(0.5, 2.53557)$. It is most convenient to use Newton's forward divided-difference interpolating polynomial for this purpose, which has the form

$$P(\alpha) = g_1 + h_1\alpha + h_3\alpha(\alpha - \alpha_2).$$

This interpolates

$$g(\mathbf{x}^{(0)} - \alpha \nabla g(\mathbf{x}^{(0)})) = g(\mathbf{x}^{(0)} - \alpha \mathbf{z})$$

at $\alpha_1 = 0$, $\alpha_2 = 0.5$, and $\alpha_3 = 1$ as follows:

$$\alpha_1 = 0, \quad g_1 = 111.975,$$

$$\alpha_2 = 0.5, \quad g_2 = 2.53557, \quad h_1 = \frac{g_2 - g_1}{\alpha_2 - \alpha_1} = -218.878,$$

$$\alpha_3 = 1, \quad g_3 = 93.5649, \quad h_2 = \frac{g_3 - g_2}{\alpha_3 - \alpha_2} = 182.059, \quad h_3 = \frac{h_2 - h_1}{\alpha_3 - \alpha_1} = 400.937.$$

Thus

$$P(\alpha) = 111.975 - 218.878\alpha + 400.937\alpha(\alpha - 0.5).$$

We have $P'(\alpha) = 0$ when $\alpha = \alpha_0 = 0.522959$. Since $g_0 = g(\mathbf{x}^{(0)} - \alpha_0 \mathbf{z}) = 2.32762$ is smaller than g_1 and g_3, we set

$$\mathbf{x}^{(1)} = \mathbf{x}^{(0)} - \alpha_0 \mathbf{z} = \mathbf{x}^{(0)} - 0.522959\mathbf{z} = (0.0112182, 0.0100964, -0.522741)^t$$

and

$$g(\mathbf{x}^{(1)}) = 2.32762.$$

Table 10.5 contains the remainder of the results. A true solution to the nonlinear system is $(0.5, 0, -0.5235988)^t$, so $\mathbf{x}^{(2)}$ would likely be adequate as an initial approximation for Newton's method or Broyden's method. One of these quicker converging techniques would be appropriate at this stage, since 70 iterations of the Steepest Descent method are required to find $\|\mathbf{x}^{(k)} - \mathbf{x}\|_\infty < 0.01$. ■

Table 10.5

k	$x_1^{(k)}$	$x_2^{(k)}$	$x_3^{(k)}$	$g(x_1^{(k)}, x_2^{(k)}, x_3^{(k)})$
2	0.137860	−0.205453	−0.522059	1.27406
3	0.266959	0.00551102	−0.558494	1.06813
4	0.272734	−0.00811751	−0.522006	0.468309
5	0.308689	−0.0204026	−0.533112	0.381087
6	0.314308	−0.0147046	−0.520923	0.318837
7	0.324267	−0.00852549	−0.528431	0.287024

Algorithm 10.3 applies the method of Steepest Descent to approximate the minimal value of $g(\mathbf{x})$. To begin an iteration, the value 0 is assigned to α_1 and the value 1 is assigned to α_3. If $h(\alpha_3) \geq h(\alpha_1)$, then successive divisions of α_3 by 2 are performed and the value of α_3 is reassigned until $h(\alpha_3) < h(\alpha_1)$ and $\alpha_3 = 2^{-k}$ for some value of k.

To employ the method to approximate the solution to the system

$$\begin{aligned} f_1(x_1, x_2, \ldots, x_n) &= 0, \\ f_2(x_1, x_2, \ldots, x_n) &= 0, \\ \vdots \qquad & \quad \vdots \\ f_n(x_1, x_2, \ldots, x_n) &= 0, \end{aligned}$$

we simply replace the function g with $\sum_{i=1}^{n} f_i^2$.

Steepest Descent

To approximate a solution $\mathbf{p}$ to the minimization problem

$$g(\mathbf{p}) = \min_{\mathbf{x} \in \mathbb{R}^n} g(\mathbf{x})$$

given an initial approximation $\mathbf{x}$:

INPUT number n of variables; initial approximation $\mathbf{x} = (x_1, \ldots, x_n)^t$; tolerance TOL; maximum number of iterations N.

OUTPUT approximate solution $\mathbf{x} = (x_1, \ldots, x_n)^t$ or a message of failure.

Step 1 Set $k = 1$.

Step 2 While ($k \leq N$) do Steps 3–15.

Step 3 Set $g_1 = g(x_1, \ldots, x_n)$; (*Note:* $g_1 = g\left(\mathbf{x}^{(k)}\right)$.)
$\mathbf{z} = \nabla g(x_1, \ldots, x_n)$; (*Note:* $\mathbf{z} = \nabla g\left(\mathbf{x}^{(k)}\right)$.)
$z_0 = ||\mathbf{z}||_2$.

Step 4 If $z_0 = 0$ then OUTPUT ('Zero gradient');
OUTPUT $(x_1, \ldots, x_n, g_1)$;
(*The procedure completed, might have a minimum.*)
STOP.

Step 5 Set $\mathbf{z} = \mathbf{z}/z_0$; (*Make* $\mathbf{z}$ *a unit vector.*)
$\alpha_1 = 0$;
$\alpha_3 = 1$;
$g_3 = g(\mathbf{x} - \alpha_3 \mathbf{z})$.

Step 6 While ($g_3 \geq g_1$) do Steps 7 and 8.

Step 7 Set $\alpha_3 = \alpha_3/2$;
$g_3 = g(\mathbf{x} - \alpha_3 \mathbf{z})$.

Step 8 If $\alpha_3 < TOL/2$ then
OUTPUT ('No likely improvement');
OUTPUT $(x_1, \ldots, x_n, g_1)$;
(*The procedure completed, might have a minimum.*)
STOP.

Step 9 Set $\alpha_2 = \alpha_3/2$;
$g_2 = g(\mathbf{x} - \alpha_2\mathbf{z})$.

Step 10 Set $h_1 = (g_2 - g_1)/\alpha_2$;
$h_2 = (g_3 - g_2)/(\alpha_3 - \alpha_2)$;
$h_3 = (h_2 - h_1)/\alpha_3$.
(Note: Newton's forward divided-difference formula is used to find the quadratic $P(\alpha) = g_1 + h_1\alpha + h_3\alpha(\alpha - \alpha_2)$ that interpolates $h(\alpha)$ at $\alpha = 0, \alpha = \alpha_2, \alpha = \alpha_3$.)

Step 11 Set $\alpha_0 = 0.5(\alpha_2 - h_1/h_3)$; *(The critical point of P occurs at α_0.)*
$g_0 = g(\mathbf{x} - \alpha_0\mathbf{z})$.

Step 12 Find α from $\{\alpha_0, \alpha_3\}$ so that $g = g(\mathbf{x} - \alpha\mathbf{z}) = \min\{g_0, g_3\}$.

Step 13 Set $\mathbf{x} = \mathbf{x} - \alpha\mathbf{z}$.

Step 14 If $|g - g_1| < TOL$ then
OUTPUT $(x_1, \ldots, x_n, g)$;
(The procedure was successful.)
STOP.

Step 15 Set $k = k + 1$.

Step 16 OUTPUT ('Maximum iterations exceeded');
(The procedure was unsuccessful.)
STOP. ■

There are many variations of the method of Steepest Descent, some of which involve more intricate methods for determining the value of α that will produce a minimum for the single-variable function h defined in Eq. (10.18). Other techniques use a multidimensional Taylor polynomial to replace the original multivariable function g and minimize the polynomial instead of g. Although there are advantages to some of these methods over the procedure discussed here, all the Steepest Descent methods are, in general, linearly convergent and converge independent of the starting approximation. In some instances, however, the methods may converge to something other than the absolute minimum of the function g.

A more complete discussion of Steepest Descent methods can be found in [OR] or [RR].

EXERCISE SET 10.4

1. Use the method of Steepest Descent with $TOL = 0.05$ to approximate the solutions of the following nonlinear systems.

a. $$4x_1^2 - 20x_1 + \frac{1}{4}x_2^2 + 8 = 0,$$
$$\frac{1}{2}x_1x_2^2 + 2x_1 - 5x_2 + 8 = 0.$$

b. $$3x_1^2 - x_2^2 = 0,$$
$$3x_1x_2^2 - x_1^3 - 1 = 0.$$

c. $$\ln(x_1^2 + x_2^2) - \sin(x_1x_2) = \ln 2 + \ln \pi,$$
$$e^{x_1 - x_2} + \cos(x_1x_2) = 0.$$

d. $$\sin(4\pi x_1x_2) - 2x_2 - x_1 = 0,$$
$$\left(\frac{4\pi - 1}{4\pi}\right)(e^{2x_1} - e) + 4ex_2^2 - 2ex_1 = 0.$$

2. Use the method of Steepest Descent with $TOL = 0.05$ to approximate the solutions of the following nonlinear systems.

   a. $15x_1 + x_2^2 - 4x_3 = 13,$
   $x_1^2 + 10x_2 - x_3 = 11,$
   $x_2^3 - 25x_3 = -22.$

   b. $10x_1 - 2x_2^2 + x_2 - 2x_3 - 5 = 0,$
   $8x_2^2 + 4x_3^2 - 9 = 0,$
   $8x_2x_3 + 4 = 0.$

   c. $x_1^3 + x_1^2x_2 - x_1x_3 + 6 = 0,$
   $e^{x_1} + e^{x_2} - x_3 = 0,$
   $x_2^2 - 2x_1x_3 = 4.$

   d. $x_1 + \cos(x_1x_2x_3) - 1 = 0,$
   $(1 - x_1)^{1/4} + x_2 + 0.05x_3^2 - 0.15x_3 - 1 = 0,$
   $-x_1^2 - 0.1x_2^2 + 0.01x_2 + x_3 - 1 = 0.$

3. Use the results in Exercise 1 and Newton's method to approximate the solutions of the nonlinear systems in Exercise 1 to within 10^{-6}.

4. Use the results of Exercise 2 and Newton's method to approximate the solutions of the nonlinear systems in Exercise 2 to within 10^{-6}.

5. Use the method of Steepest Descent to approximate minima to within 0.005 for the following functions.

   a. $g(x_1, x_2) = \cos(x_1 + x_2) + \sin x_1 + \cos x_2$
   b. $g(x_1, x_2) = 100(x_1^2 - x_2)^2 + (1 - x_1)^2$
   c. $g(x_1, x_2, x_3) = x_1^2 + 2x_2^2 + x_3^2 - 2x_1x_2 + 2x_1 - 2.5x_2 - x_3 + 2$
   d. $g(x_1, x_2, x_3) = x_1^4 + 2x_2^4 + 3x_3^4 + 1.01$

6. a. Show that the quadratic polynomial

$$P(\alpha) = g_1 + h_1\alpha + h_3\alpha(\alpha - \alpha_2)$$

   interpolates the function h defined in (10.18):

$$h(\alpha) = g\big(\mathbf{x}^{(0)} - \alpha\nabla g(\mathbf{x}^{(0)})\big)$$

   at $\alpha = 0, \alpha_2$, and α_3.

   b. Show that a critical point of P occurs at

$$\alpha_0 = \frac{1}{2}\left(\alpha_2 - \frac{h_1}{h_3}\right).$$

10.5 Homotopy and Continuation Methods

Homotopy, or *continuation*, methods for nonlinear systems embed the problem to be solved within a collection of problems. Specifically, to solve a problem of the form

$$\mathbf{F}(\mathbf{x}) = \mathbf{0},$$

which has the unknown solution $\mathbf{x}^*$, we consider a family of problems described using a parameter λ that assumes values in $[0, 1]$. A problem with a known solution $\mathbf{x}(0)$ corresponds to the situation when $\lambda = 0$, and the problem with the unknown solution $\mathbf{x}(1) \equiv \mathbf{x}^*$ corresponds to $\lambda = 1$.

A homotopy is a continuous deformation; a function that takes a real interval continuously into a set of functions.

For example, suppose $\mathbf{x}(0)$ is an initial approximation to the solution of $\mathbf{F}(\mathbf{x}^*) = \mathbf{0}$. Define

$$\mathbf{G} : [0, 1] \times \mathbb{R}^n \to \mathbb{R}^n$$

by

$$\mathbf{G}(\lambda, \mathbf{x}) = \lambda\mathbf{F}(\mathbf{x}) + (1 - \lambda)\left[\mathbf{F}(\mathbf{x}) - \mathbf{F}(\mathbf{x}(0))\right] = \mathbf{F}(\mathbf{x}) + (\lambda - 1)\mathbf{F}(\mathbf{x}(0)). \tag{10.19}$$

We will determine, for various values of λ, a solution to

$$\mathbf{G}(\lambda, \mathbf{x}) = \mathbf{0}.$$

When $\lambda = 0$, this equation assumes the form

$$\mathbf{0} = \mathbf{G}(0, \mathbf{x}) = \mathbf{F}(\mathbf{x}) - \mathbf{F}(\mathbf{x}(0)),$$

and $\mathbf{x}(0)$ is a solution. When $\lambda = 1$, the equation assumes the form

$$\mathbf{0} = \mathbf{G}(1, \mathbf{x}) = \mathbf{F}(\mathbf{x}),$$

and $\mathbf{x}(1) = \mathbf{x}^*$ is a solution.

The function $\mathbf{G}$, with the parameter λ, provides us with a family of functions that can lead from the known value $\mathbf{x}(0)$ to the solution $\mathbf{x}(1) = \mathbf{x}^*$. The function $\mathbf{G}$ is called a **homotopy** between the function $\mathbf{G}(0, \mathbf{x}) = \mathbf{F}(\mathbf{x}) - \mathbf{F}(\mathbf{x}(0))$ and the function $\mathbf{G}(1, \mathbf{x}) = \mathbf{F}(\mathbf{x})$.

Continuation

The **continuation** problem is to:

- Determine a way to proceed from the known solution $\mathbf{x}(0)$ of $\mathbf{G}(0, \mathbf{x}) = \mathbf{0}$ to the unknown solution $\mathbf{x}(1) = \mathbf{x}^*$ of $\mathbf{G}(1, \mathbf{x}) = \mathbf{0}$, that is, the solution to $\mathbf{F}(\mathbf{x}) = \mathbf{0}$.

We first assume that $\mathbf{x}(\lambda)$ is the unique solution to the equation

$$\mathbf{G}(\lambda, \mathbf{x}) = \mathbf{0}, \tag{10.20}$$

for each $\lambda \in [0, 1]$. The set $\{\, \mathbf{x}(\lambda) \mid 0 \le \lambda \le 1 \,\}$ can be viewed as a curve in $\mathbb{R}^n$ from $\mathbf{x}(0)$ to $\mathbf{x}(1) = \mathbf{x}^*$ parameterized by λ. A continuation method finds a sequence of steps along this curve corresponding to $\{\mathbf{x}(\lambda_k)\}_{k=0}^{m}$, where $\lambda_0 = 0 < \lambda_1 < \cdots < \lambda_m = 1$.

If the functions $\lambda \to \mathbf{x}(\lambda)$ and $\mathbf{G}$ are differentiable, then differentiating Eq. (10.20) with respect to λ gives

$$\mathbf{0} = \frac{\partial \mathbf{G}(\lambda, \mathbf{x}(\lambda))}{\partial \lambda} + \frac{\partial \mathbf{G}(\lambda, \mathbf{x}(\lambda))}{\partial \mathbf{x}} \mathbf{x}'(\lambda),$$

and solving for $\mathbf{x}'(\lambda)$ gives

$$\mathbf{x}'(\lambda) = -\left[\frac{\partial \mathbf{G}(\lambda, \mathbf{x}(\lambda))}{\partial \mathbf{x}} \right]^{-1} \frac{\partial \mathbf{G}(\lambda, \mathbf{x}(\lambda))}{\partial \lambda}.$$

This is a a system of differential equations with the initial condition $\mathbf{x}(0)$.

Since

$$\mathbf{G}(\lambda, \mathbf{x}(\lambda)) = \mathbf{F}(\mathbf{x}(\lambda)) + (\lambda - 1)\mathbf{F}(\mathbf{x}(0)),$$

we can determine both

$$\frac{\partial \mathbf{G}}{\partial \mathbf{x}}(\lambda, \mathbf{x}(\lambda)) = \begin{bmatrix} \dfrac{\partial f_1}{\partial x_1}(\mathbf{x}(\lambda)) & \dfrac{\partial f_1}{\partial x_2}(\mathbf{x}(\lambda)) & \cdots & \dfrac{\partial f_1}{\partial x_n}(\mathbf{x}(\lambda)) \\ \dfrac{\partial f_2}{\partial x_1}(\mathbf{x}(\lambda)) & \dfrac{\partial f_2}{\partial x_2}(\mathbf{x}(\lambda)) & \cdots & \dfrac{\partial f_2}{\partial x_n}(\mathbf{x}(\lambda)) \\ \vdots & \vdots & & \vdots \\ \dfrac{\partial f_n}{\partial x_1}(\mathbf{x}(\lambda)) & \dfrac{\partial f_n}{\partial x_2}(\mathbf{x}(\lambda)) & \cdots & \dfrac{\partial f_n}{\partial x_n}(\mathbf{x}(\lambda)) \end{bmatrix} = J(\mathbf{x}(\lambda)),$$

the Jacobian matrix, and

$$\frac{\partial \mathbf{G}(\lambda, \mathbf{x}(\lambda))}{\partial \lambda} = \mathbf{F}(\mathbf{x}(0)).$$

Therefore, the system of differential equations becomes

$$\mathbf{x}'(\lambda) = -[J(\mathbf{x}(\lambda))]^{-1}\mathbf{F}(\mathbf{x}(0)), \quad \text{for} \quad 0 \le \lambda \le 1, \tag{10.21}$$

with the initial condition $\mathbf{x}(0)$. The following theorem (see [OR], pp. 230–231) gives conditions under which the continuation method is feasible.

Theorem 10.10 Let $\mathbf{F}(\mathbf{x})$ be continuously differentiable for $\mathbf{x} \in \mathbb{R}^n$. Suppose that the Jacobian matrix $J(\mathbf{x})$ is nonsingular for all $\mathbf{x} \in \mathbb{R}^n$ and that a constant M exists with $\|J(\mathbf{x})^{-1}\| \le M$, for all $\mathbf{x} \in \mathbb{R}^n$. Then, for any $\mathbf{x}(0)$ in $\mathbb{R}^n$, there exists a unique function $\mathbf{x}(\lambda)$, such that

$$\mathbf{G}(\lambda, \mathbf{x}(\lambda)) = \mathbf{0},$$

for all λ in $[0, 1]$. Moreover, $\mathbf{x}(\lambda)$ is continuously differentiable and

$$\mathbf{x}'(\lambda) = -J(\mathbf{x}(\lambda))^{-1}\mathbf{F}(\mathbf{x}(0)), \quad \text{for each } \lambda \in [0, 1]. \quad ■$$

The following shows the form of the system of differential equations associated with a nonlinear system of equations.

Illustration Consider the nonlinear system

$$f_1(x_1, x_2, x_3) = 3x_1 - \cos(x_2x_3) - 0.5 = 0,$$

$$f_2(x_1, x_2, x_3) = x_1^2 - 81(x_2 + 0.1)^2 + \sin x_3 + 1.06 = 0,$$

$$f_3(x_1, x_2, x_3) = e^{-x_1x_2} + 20x_3 + \frac{10\pi - 3}{3} = 0.$$

The Jacobian matrix is

$$J(\mathbf{x}) = \begin{bmatrix} 3 & x_3 \sin x_2x_3 & x_2 \sin x_2x_3 \\ 2x_1 & -162(x_2 + 0.1) & \cos x_3 \\ -x_2e^{-x_1x_2} & -x_1e^{-x_1x_2} & 20 \end{bmatrix}.$$

Let $\mathbf{x}(0) = (0, 0, 0)^t$, so that

$$\mathbf{F}(\mathbf{x}(0)) = \begin{bmatrix} -1.5 \\ 0.25 \\ 10\pi/3 \end{bmatrix}.$$

The system of differential equations is

$$\begin{bmatrix} x_1'(\lambda) \\ x_2'(\lambda) \\ x_3'(\lambda) \end{bmatrix} = -\begin{bmatrix} 3 & x_3 \sin x_2x_3 & x_2 \sin x_2x_3 \\ 2x_1 & -162(x_2 + 0.1) & \cos x_3 \\ -x_2e^{-x_1x_2} & -x_1e^{-x_1x_2} & 20 \end{bmatrix}^{-1} \begin{bmatrix} -1.5 \\ 0.25 \\ 10\pi/3 \end{bmatrix}. \quad □$$

In general, the system of differential equations that we need to solve for our continuation problem has the form

$$\frac{dx_1}{d\lambda} = \phi_1(\lambda, x_1, x_2, \ldots, x_n),$$
$$\frac{dx_2}{d\lambda} = \phi_2(\lambda, x_1, x_2, \ldots, x_n),$$
$$\vdots$$
$$\frac{dx_n}{d\lambda} = \phi_n(\lambda, x_1, x_2, \ldots, x_n),$$

where

$$\begin{bmatrix} \phi_1(\lambda, x_1, \ldots, x_n) \\ \phi_2(\lambda, x_1, \ldots, x_n) \\ \vdots \\ \phi_n(\lambda, x_1, \ldots, x_n) \end{bmatrix} = -J(x_1, \ldots, x_n)^{-1} \begin{bmatrix} f_1(\mathbf{x}(0)) \\ f_2(\mathbf{x}(0)) \\ \vdots \\ f_n(\mathbf{x}(0)) \end{bmatrix}. \tag{10.22}$$

To use the Runge-Kutta method of order four to solve this system, we first choose an integer $N > 0$ and let $h = (1-0)/N$. Partition the interval $[0, 1]$ into N subintervals with the mesh points

$$\lambda_j = jh, \quad \text{for each} \quad j = 0, 1, \ldots, N.$$

We use the notation w_{ij}, for each $j = 0, 1, \ldots, N$ and $i = 1, \ldots, n$, to denote an approximation to $x_i(\lambda_j)$. For the initial conditions, set

$$w_{1,0} = x_1(0), \quad w_{2,0} = x_2(0), \quad \ldots, \quad w_{n,0} = x_n(0).$$

Suppose $w_{1,j}$, $w_{2,j}$, $\ldots$, $w_{n,j}$ have been computed. We obtain $w_{1,j+1}$, $w_{2,j+1}$, $\ldots$, $w_{n,j+1}$ using the equations

$$k_{1,i} = h\phi_i(\lambda_j, w_{1,j}, w_{2,j}, \ldots, w_{n,j}), \quad \text{for each} \quad i = 1, 2, \ldots, n;$$

$$k_{2,i} = h\phi_i\left(\lambda_j + \frac{h}{2}, w_{1,j} + \frac{1}{2}k_{1,1}, \ldots, w_{n,j} + \frac{1}{2}k_{1,n}\right), \quad \text{for each } i = 1, 2, \ldots, n;$$

$$k_{3,i} = h\phi_i\left(\lambda_j + \frac{h}{2}, w_{1,j} + \frac{1}{2}k_{2,1}, \ldots, w_{n,j} + \frac{1}{2}k_{2,n}\right), \quad \text{for each } i = 1, 2, \ldots, n;$$

$$k_{4,i} = h\phi_i(\lambda_j + h, w_{1,j} + k_{3,1}, w_{2,j} + k_{3,2}, \ldots, w_{n,j} + k_{3,n}), \quad \text{for each } i = 1, 2, \ldots, n;$$

and, finally

$$w_{i,j+1} = w_{i,j} + \frac{1}{6}\left(k_{1,i} + 2k_{2,i} + 2k_{3,i} + k_{4,i}\right), \quad \text{for each } i = 1, 2, \ldots, n.$$

The vector notation

$$\mathbf{k}_1 = \begin{bmatrix} k_{1,1} \\ k_{1,2} \\ \vdots \\ k_{1,n} \end{bmatrix}, \mathbf{k}_2 = \begin{bmatrix} k_{2,1} \\ k_{2,2} \\ \vdots \\ k_{2,n} \end{bmatrix}, \mathbf{k}_3 = \begin{bmatrix} k_{3,1} \\ k_{3,2} \\ \vdots \\ k_{3,n} \end{bmatrix}, \mathbf{k}_4 = \begin{bmatrix} k_{4,1} \\ k_{4,2} \\ \vdots \\ k_{4,n} \end{bmatrix}, \quad \text{and} \quad \mathbf{w}_j = \begin{bmatrix} w_{1,j} \\ w_{2,j} \\ \vdots \\ w_{n,j} \end{bmatrix}$$

Simplifies the presentation. Then Eq. (10.22) gives us $\mathbf{x}(0) = \mathbf{x}(\lambda_0) = \mathbf{w}_0$, and for each $j = 0, 1, \ldots, N$,

$$\mathbf{k}_1 = h \begin{bmatrix} \phi_1(\lambda_j, w_{1,j}, \ldots, w_{n,j}) \\ \phi_2(\lambda_j, w_{1,j}, \ldots, w_{n,j}) \\ \vdots \\ \phi_n(\lambda_j, w_{1,j}, \ldots, w_{n,j}) \end{bmatrix} = h\left[-J(w_{1,j}, \ldots, w_{n,j})\right]^{-1} \mathbf{F}(\mathbf{x}(0))$$

$$= h\left[-J(\mathbf{w}_j)\right]^{-1} \mathbf{F}(\mathbf{x}(0));$$

$$\mathbf{k}_2 = h\left[-J\left(\mathbf{w}_j + \frac{1}{2}\mathbf{k}_1\right)\right]^{-1} \mathbf{F}(\mathbf{x}(0));$$

$$\mathbf{k}_3 = h\left[-J\left(\mathbf{w}_j + \frac{1}{2}\mathbf{k}_2\right)\right]^{-1} \mathbf{F}(\mathbf{x}(0));$$

$$\mathbf{k}_4 = h\left[-J\left(\mathbf{w}_j + \mathbf{k}_3\right)\right]^{-1} \mathbf{F}(\mathbf{x}(0));$$

and

$$\mathbf{x}(\lambda_{j+1}) = \mathbf{x}(\lambda_j) + \frac{1}{6}(\mathbf{k}_1 + 2\mathbf{k}_2 + 2\mathbf{k}_3 + \mathbf{k}_4) = \mathbf{w}_j + \frac{1}{6}(\mathbf{k}_1 + 2\mathbf{k}_2 + 2\mathbf{k}_3 + \mathbf{k}_4).$$

Finally, $\mathbf{x}(\lambda_n) = \mathbf{x}(1)$ is our approximation to $\mathbf{x}^*$.

Example 1 Use the Continuation method with $\mathbf{x}(0) = (0, 0, 0)^t$ to approximate the solution to

$$f_1(x_1, x_2, x_3) = 3x_1 - \cos(x_2 x_3) - 0.5 = 0,$$

$$f_2(x_1, x_2, x_3) = x_1^2 - 81(x_2 + 0.1)^2 + \sin x_3 + 1.06 = 0,$$

$$f_3(x_1, x_2, x_3) = e^{-x_1 x_2} + 20x_3 + \frac{10\pi - 3}{3} = 0.$$

Solution The Jacobian matrix is

$$J(\mathbf{x}) = \begin{bmatrix} 3 & x_3 \sin x_2 x_3 & x_2 \sin x_2 x_3 \\ 2x_1 & -162(x_2 + 0.1) & \cos x_3 \\ -x_2 e^{-x_1 x_2} & -x_1 e^{-x_1 x_2} & 20 \end{bmatrix}$$

and

$$F(\mathbf{x}(0)) = (-1.5, 0.25, 10\pi/3)^t.$$

With $N = 4$ and $h = 0.25$, we have

$$\mathbf{k}_1 = h[-J(\mathbf{x}^{(0)})]^{-1} F(\mathbf{x}(0)) = 0.25 \begin{bmatrix} 3 & 0 & 0 \\ 0 & -16.2 & 1 \\ 0 & 0 & 20 \end{bmatrix}^{-1} \begin{bmatrix} -1.5 \\ 0.25 \\ 10\pi/3 \end{bmatrix}$$

$$= (0.125, -0.004222203325, -0.1308996939)^t;$$

$$\mathbf{k}_2 = h[-J(0.0625, -0.002111101663, -0.06544984695)]^{-1}(-1.5, 0.25, 10\pi/3)^t$$

$$= 0.25\begin{bmatrix} 3 & -0.9043289149 \times 10^{-5} & -0.2916936196 \times 10^{-6} \\ 0.125 & -15.85800153 & 0.9978589232 \\ 0.002111380229 & -0.06250824706 & 20 \end{bmatrix}^{-1}\begin{bmatrix} -1.5 \\ 0.25 \\ 10\pi/3 \end{bmatrix}$$

$$= (0.1249999773, -0.003311761993, -0.1309232406)^t;$$

$$\mathbf{k}_3 = h[-J(0.06249998865, -0.001655880997, -0.0654616203)]^{-1}(-1.5, 0.25, 10\pi/3)^t$$

$$= (0.1249999844, -0.003296244825, -0.130920346)^t;$$

$$\mathbf{k}_4 = h[-J(0.1249999844, -0.003296244825, -0.130920346)]^{-1}(-1.5, 0.25, 10\pi/3)^t$$

$$= (0.1249998945, -0.00230206762, -0.1309346977)^t;$$

and

$$\mathbf{x}(\lambda_1) = \mathbf{w}_1 = \mathbf{w}_0 + \frac{1}{6}(\mathbf{k}_1 + 2\mathbf{k}_2 + 2\mathbf{k}_3 + \mathbf{k}_4)$$
$$= (0.1249999697, -0.00329004743, -0.1309202608)^t.$$

Continuing, we have

$$\mathbf{x}(\lambda_2) = \mathbf{w}_2 = (0.2499997679, -0.004507400128, -0.2618557619)^t,$$
$$\mathbf{x}(\lambda_3) = \mathbf{w}_3 = (0.3749996956, -0.003430352103, -0.3927634423)^t,$$

and

$$\mathbf{x}(\lambda_4) = \mathbf{x}(1) = \mathbf{w}_4 = (0.4999999954, 0.126782 \times 10^{-7}, -0.5235987758)^t.$$

These results are very accurate because the actual solution is $(0.5, 0, -0.52359877)^t$. ■

Note that in the Runge-Kutta methods, the steps similar to

$$\mathbf{k}_i = h[-J(\mathbf{x}(\lambda_i) + \alpha_{i-1}\mathbf{k}_{i-1})]^{-1}\mathbf{F}(\mathbf{x}(0))$$

can be written as solving for $\mathbf{k}_i$ in the linear system

$$J\left(\mathbf{x}(\lambda_i) + \alpha_{i-1}\mathbf{k}_{i-1}\right)\mathbf{k}_i = -h\mathbf{F}(\mathbf{x}(0)).$$

So in the Runge-Kutta method of order four, the calculation of each $\mathbf{w}_j$ requires four linear systems to be solved, one each when computing $\mathbf{k}_1$, $\mathbf{k}_2$, $\mathbf{k}_3$, and $\mathbf{k}_4$. Thus using N steps requires solving $4N$ linear systems. By comparison, Newton's method requires solving one linear system per iteration. Therefore, the work involved for the Runge-Kutta method is roughly equivalent to $4N$ iterations of Newton's method.

An alternative is to use a Runge-Kutta method of order two, such as the modified Euler method or even Euler's method, to decrease the number of linear systems that need to be solved. Another possibility is to use smaller values of N. The following illustrates these ideas.

Illustration Table 10.6 summarizes a comparison of Euler's method, the Midpoint method, and the Runge-Kutta method of order four applied to the problem in the example, with initial approximation $\mathbf{x}(0) = (0, 0, 0)^t$. The right-hand column in the table lists the number of linear systems that are required for the solution. □

Table 10.6

Method	N	$\mathbf{x}(1)$	Systems
Euler	1	$(0.5, -0.0168888133, -0.5235987755)^t$	1
Euler	4	$(0.499999379, -0.004309160698, -0.523679652)^t$	4
Midpoint	1	$(0.4999966628, -0.00040240435, -0.523815371)^t$	2
Midpoint	4	$(0.500000066, -0.00001760089, -0.5236127761)^t$	8
Runge-Kutta	1	$(0.4999989843, -0.1676151 \times 10^{-5}, -0.5235989561)^t$	4
Runge-Kutta	4	$(0.4999999954, 0.126782 \times 10^{-7}, -0.5235987758)^t$	16

The continuation method can be used as a stand-alone method, and does not require a particularly good choice of $\mathbf{x}(0)$. However, the method can also be used to give an initial approximation for Newton's or Broyden's method. For example, the result obtained in Example 2 using Euler's method and $N = 2$ might easily be sufficient to start the more efficient Newton's or Broyden's methods and be better for this purpose than the continuation methods, which require more calculation.

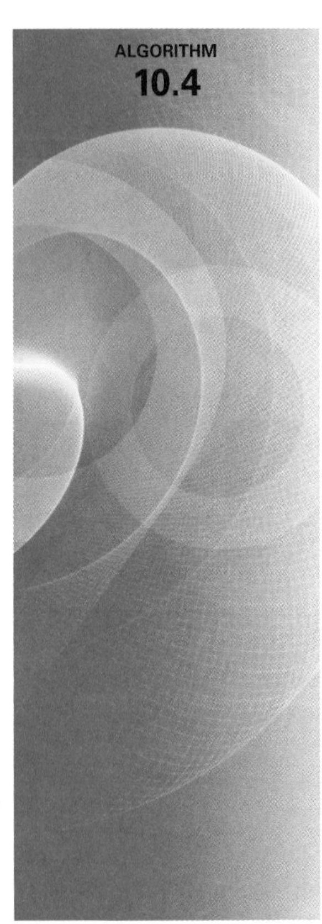

Continuation Algorithm

To approximate the solution of the nonlinear system $\mathbf{F}(\mathbf{x}) = \mathbf{0}$ given an initial approximation $\mathbf{x}$:

INPUT number n of equations and unknowns; integer $N > 0$; initial approximation $\mathbf{x} = (x_1, x_2, \ldots, x_n)^t$.

OUTPUT approximate solution $\mathbf{x} = (x_1, x_2, \ldots, x_n)^t$.

Step 1 Set $h = 1/N$;
$\mathbf{b} = -h\mathbf{F}(\mathbf{x})$.

Step 2 For $i = 1, 2, \ldots, N$ do Steps 3–7.

Step 3 Set $A = J(\mathbf{x})$;
Solve the linear system $A\mathbf{k}_1 = \mathbf{b}$.

Step 4 Set $A = J(\mathbf{x} + \frac{1}{2}\mathbf{k}_1)$;
Solve the linear system $A\mathbf{k}_2 = \mathbf{b}$.

Step 5 Set $A = J(\mathbf{x} + \frac{1}{2}\mathbf{k}_2)$;
Solve the linear system $A\mathbf{k}_3 = \mathbf{b}$.

Step 6 Set $A = J(\mathbf{x} + \mathbf{k}_3)$;
Solve the linear system $A\mathbf{k}_3 = \mathbf{b}$.

Step 7 Set $\mathbf{x} = \mathbf{x} + (\mathbf{k}_1 + 2\mathbf{k}_2 + 2\mathbf{k}_3 + \mathbf{k}_4)/6$.

Step 8 OUTPUT $(x_1, x_2, \ldots, x_n)$;
STOP. ■

EXERCISE SET 10.5

1. The nonlinear system

$$f_1(x_1, x_2) = x_1^2 - x_2^2 + 2x_2 = 0, \quad f_2(x_1, x_2) = 2x_1 + x_2^2 - 6 = 0$$

has two solutions, $(0.625204094, 2.179355825)^t$ and $(2.109511920, -1.334532188)^t$. Use the continuation method and Euler's method with $N = 2$ to approximate the solutions where

a. $\mathbf{x}(0) = (0, 0)^t$ **b.** $\mathbf{x}(0) = (1, 1)^t$ **c.** $\mathbf{x}(0) = (3, -2)^t$

2. Repeat Exercise 1 using the Runge-Kutta method of order four with $N = 1$.

3. Use the continuation method and Euler's method with $N = 2$ on the following nonlinear systems.

a. $4x_1^2 - 20x_1 + \frac{1}{4}x_2^2 + 8 = 0,$

$\frac{1}{2}x_1x_2^2 + 2x_1 - 5x_2 + 8 = 0.$

b. $\sin(4\pi x_1 x_2) - 2x_2 - x_1 = 0,$

$\left(\frac{4\pi - 1}{4\pi}\right)(e^{2x_1} - e) + 4ex_2^2 - 2ex_1 = 0.$

c. $3x_1 - \cos(x_2x_3) - \frac{1}{2} = 0,$

$4x_1^2 - 625x_2^2 + 2x_2 - 1 = 0,$

$e^{-x_1x_2} + 20x_3 + \frac{10\pi - 3}{3} = 0.$

d. $x_1^2 + x_2 - 37 = 0,$

$x_1 - x_2^2 - 5 = 0,$

$x_1 + x_2 + x_3 - 3 = 0.$

4. Use the continuation method and the Runge-Kutta method of order four with $N = 1$ on the following nonlinear systems using $\mathbf{x}(0) = \mathbf{0}$. Are the answers here comparable to Newton's method or are they suitable initial approximations for Newton's method?

a. $x_1(1 - x_1) + 4x_2 = 12,$

$(x_1 - 2)^2 + (2x_2 - 3)^2 = 25.$

Compare to 10.2(5c).

b. $5x_1^2 - x_2^2 = 0,$

$x_2 - 0.25(\sin x_1 + \cos x_2) = 0.$

Compare to 10.2(5d).

c. $15x_1 + x_2^2 - 4x_3 = 13,$

$x_1^2 + 10x_2 - x_3 = 11.$

$x_2^3 - 25x_3 = -22$

Compare to 10.2(6c).

d. $10x_1 - 2x_2^2 + x_2 - 2x_3 - 5 = 0,$

$8x_2^2 + 4x_3^2 - 9 = 0.$

$8x_2x_3 + 4 = 0$

Compare to 10.2(6d).

5. Repeat Exercise 4 using the initial approximations obtained as follows.

a. From 10.2(3c) **b.** From 10.2(3d) **c.** From 10.2(4c) **d.** From 10.2(4d)

6. Use the continuation method and the Runge-Kutta method of order four with $N = 1$ on Exercise 7 of Section 10.2. Are the results as good as those obtained there?

7. Repeat Exercise 5 using $N = 2$.

8. Repeat Exercise 8 of Section 10.2 using the continuation method and the Runge-Kutta method of order four with $N = 1$.

9. Repeat Exercise 9 of Section 10.2 using the continuation method and the Runge-Kutta method of order four with $N = 2$.

10. Show that the continuation method and Euler's method with $N = 1$ gives the same result as Newton's method for the first iteration; that is, with $\mathbf{x}(0) = \mathbf{x}^{(0)}$ we always obtain $\mathbf{x}(1) = \mathbf{x}^{(1)}$.

11. Show that the homotopy

$$G(\lambda, \mathbf{x}) = F(\mathbf{x}) - e^{-\lambda}F(\mathbf{x}(0))$$

used in the continuation method with Euler's method and $h = 1$ also duplicates Newton's method for any $\mathbf{x}^{(0)}$; that is, with $\mathbf{x}(0) = \mathbf{x}^{(0)}$, we have $\mathbf{x}(1) = \mathbf{x}^{(1)}$.

12. Let the continuation method with the Runge-Kutta method of order four be abbreviated CMRK4. After completing Exercises 4, 5, 6, 7, 8, and 9, answer the following questions.

a. Is CMRK4 with $N = 1$ comparable to Newton's method? Support your answer with the results of earlier exercises.

b. Should CMRK4 with $N = 1$ be used as a means to obtain an initial approximation for Newton's method? Support your answer with the results of earlier exercises.

c. Repeat part (a) for CMRK4 with $N = 2$.

d. Repeat part (b) for CMRK4 with $N = 2$.

10.6 Survey of Methods and Software

In this chapter we considered methods to approximate solutions to nonlinear systems

$$\begin{aligned} f_1(x_1, x_2, \ldots, x_n) &= 0, \\ f_2(x_1, x_2, \ldots, x_n) &= 0, \\ &\vdots \\ f_n(x_1, x_2, \ldots, x_n) &= 0. \end{aligned}$$

Newton's method for systems requires a good initial approximation $\left(x_1^{(0)}, x_2^{(0)}, \ldots, x_n^{(0)}\right)^t$ and generates a sequence

$$\mathbf{x}^{(k)} = \mathbf{x}^{(k-1)} - J\left(\mathbf{x}^{(k-1)}\right)^{-1}\mathbf{F}\left(\mathbf{x}^{(k-1)}\right),$$

that converges rapidly to a solution $\mathbf{x}$ if $\mathbf{x}^{(0)}$ is sufficiently close to $\mathbf{p}$. However, Newton's method requires evaluating, or approximating, n^2 partial derivatives and solving an n by n linear system at each step. Solving the linear system requires $O\left(n^3\right)$ computations.

Broyden's method reduces the amount of computation at each step without significantly degrading the speed of convergence. This technique replaces the Jacobian matrix J with a matrix A_{k-1} whose inverse is directly determined at each step. This reduces the arithmetic computations from $O\left(n^3\right)$ to $O\left(n^2\right)$. Moreover, the only scalar function evaluations required are in evaluating the f_i, saving n^2 scalar function evaluations per step. Broyden's method also requires a good initial approximation.

The Steepest Descent method was presented as a way to obtain good initial approximations for Newton's and Broyden's methods. Although Steepest Descent does not give a rapidly convergent sequence, it does not require a good initial approximation. The Steepest Descent method approximates a minimum of a multivariable function g. For our application we choose

$$g(x_1, x_2, \ldots, x_n) = \sum_{i=1}^{n} [f_i(x_1, x_2, \ldots, x_n)]^2.$$

The minimum value of g is 0, which occurs when the functions f_i are simultaneously 0.

Homotopy and continuation methods are also used for nonlinear systems and are the subject of current research (see [AG]). In these methods, a given problem

$$\mathbf{F}(\mathbf{x}) = \mathbf{0}$$

is embedded in a one-parameter family of problems using a parameter λ that assumes values in $[0, 1]$. The original problem corresponds to $\lambda = 1$, and a problem with a known solution corresponds to $\lambda = 0$. For example, the set of problems

$$G(\lambda, \mathbf{x}) = \lambda\mathbf{F}(\mathbf{x}) + (1-\lambda)(\mathbf{F}(\mathbf{x}) - \mathbf{F}(\mathbf{x_0})) = \mathbf{0}, \quad \text{for } 0 \le \lambda \le 1,$$

with fixed $\mathbf{x}_0 \in \mathbb{R}^n$ forms a homotopy. When $\lambda = 0$, the solution is $\mathbf{x}(\lambda = 0) = \mathbf{x}_0$. The solution to the original problem corresponds to $\mathbf{x}(\lambda = 1)$. A continuation method attempts to determine $\mathbf{x}(\lambda = 1)$ by solving the sequence of problems corresponding to $\lambda_0 = 0 < \lambda_1 < \lambda_2 < \cdots < \lambda_m = 1$. The initial approximation to the solution of

$$\lambda_i\mathbf{F}(\mathbf{x}) + (1-\lambda_i)(\mathbf{F}(\mathbf{x}) - \mathbf{F}(\mathbf{x_0})) = \mathbf{0}$$

would be the solution, $\mathbf{x}(\lambda = \lambda_{i-1})$, to the problem

$$\lambda_{i-1}\mathbf{F}(\mathbf{x}) + (1-\lambda_{i-1})(\mathbf{F}(\mathbf{x}) - \mathbf{F}(\mathbf{x}_0)) = \mathbf{0}.$$

The package Hompack in netlib solves a system of nonlinear equations by using various homotopy methods.

The nonlinear systems methods in the IMSL and NAG libraries use the Levenberg-Marquardt method, which is a weighted average of Newton's method and the Steepest Descent method. The weight is biased toward the Steepest Descent method until convergence is detected, at which time the weight is shifted toward the more rapidly convergent Newton's method. In either routine a finite difference approximation to the Jacobian can be used or a user-supplied subroutine entered to compute the Jacobian.

A comprehensive treatment of methods for solving nonlinear systems of equations can be found in Ortega and Rheinbolt [OR] and in Dennis and Schnabel [DenS]. Recent developments on iterative methods can be found in Argyros and Szidarovszky [AS], and information on the use of continuation methods is available in Allgower and Georg [AG].

CHAPTER

11 Boundary-Value Problems for Ordinary Differential Equations

Introduction

A common problem in civil engineering concerns the deflection of a beam of rectangular cross section subject to uniform loading while the ends of the beam are supported so that they undergo no deflection.

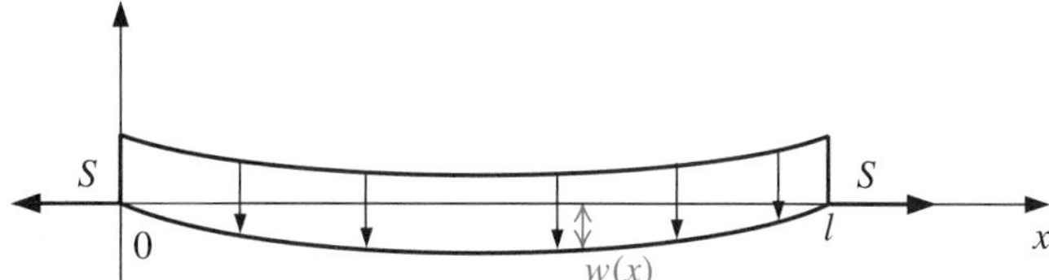

Suppose that l, q, E, S, and I represent, respectively, the length of the beam, the intensity of the uniform load, the modulus of elasticity, the stress at the endpoints, and the central moment of inertia. The differential equation approximating the physical situation is of the form

$$\frac{d^2w}{dx^2}(x) = \frac{S}{EI}w(x) + \frac{qx}{2EI}(x - l),$$

where $w(x)$ is the deflection a distance x from the left end of the beam. Since no deflection occurs at the ends of the beam, there are two boundary conditions

$$w(0) = 0 \quad \text{and} \quad w(l) = 0.$$

When the beam is of uniform thickness, the product EI is constant. In this case the exact solution is easily obtained. When the thickness is not uniform, the moment of inertia I is a function of x, and approximation techniques are required. Problems of this type are considered in Exercises 7 of Section 11.3 and 6 of Section 11.4.

The differential equations in Chapter 5 are of first order and have one initial condition to satisfy. Later in the chapter we saw that the techniques could be extended to systems of equations and then to higher-order equations, but all the specified conditions are on the same endpoint. These are initial-value problems. In this chapter we show how to approximate the solution to **boundary-value** problems, differential equations with conditions imposed at different points. For first-order differential equations, only one condition is specified, so there is no distinction between initial-value and boundary-value problems. We will be considering second-order equations with two boundary values.

Physical problems that are position-dependent rather than time-dependent are often described in terms of differential equations with conditions imposed at more than one point.

The two-point boundary-value problems in this chapter involve a second-order differential equation of the form

$$y'' = f(x, y, y'), \quad \text{for } a \le x \le b, \tag{11.1}$$

together with the boundary conditions

$$y(a) = \alpha \quad \text{and} \quad y(b) = \beta. \tag{11.2}$$

11.1 The Linear Shooting Method

The following theorem gives general conditions that ensure the solution to a second-order boundary value problem exists and is unique. The proof of this theorem can be found in [Keller, H].

Theorem 11.1 Suppose the function f in the boundary-value problem

$$y'' = f(x, y, y'), \quad \text{for } a \le x \le b, \text{ with } y(a) = \alpha \text{ and } y(b) = \beta,$$

is continuous on the set

$$D = \{\, (x, y, y') \mid \text{for } a \le x \le b, \text{ with } -\infty < y < \infty \text{ and } -\infty < y' < \infty \,\},$$

and that the partial derivatives f_y and $f_{y'}$ are also continuous on D. If

(i) $f_y(x, y, y') > 0$, for all $(x, y, y') \in D$, and

(ii) a constant M exists, with

$$\left| f_{y'}(x, y, y') \right| \le M, \quad \text{for all } (x, y, y') \in D,$$

then the boundary-value problem has a unique solution. ∎

Example 1 Use Theorem 11.1 to show that the boundary-value problem

$$y'' + e^{-xy} + \sin y' = 0, \quad \text{for } 1 \le x \le 2, \text{ with } y(1) = y(2) = 0,$$

has a unique solution.

Solution We have

$$f(x, y, y') = -e^{-xy} - \sin y'.$$

and for all x in $[1, 2]$,

$$f_y(x, y, y') = xe^{-xy} > 0 \quad \text{and} \quad \left| f_{y'}(x, y, y') \right| = |-\cos y'| \le 1.$$

So the problem has a unique solution. ∎

Linear Boundary-Value Problems

The differential equation

$$y'' = f(x, y, y')$$

is linear when functions $p(x)$, $q(x)$, and $r(x)$ exist with

A linear equation involves only linear powers of y and its derivatives.

$$f(x, y, y') = p(x)y' + q(x)y + r(x).$$

Problems of this type frequently occur, and in this situation, Theorem 11.1 can be simplified.

Corollary 11.2 Suppose the linear boundary-value problem

$$y'' = p(x)y' + q(x)y + r(x), \quad \text{for } a \le x \le b, \text{ with } y(a) = \alpha \text{ and } y(b) = \beta,$$

satisfies

(i) $p(x)$, $q(x)$, and $r(x)$ are continuous on $[a, b]$,

(ii) $q(x) > 0$ on $[a, b]$.

Then the boundary-value problem has a unique solution. ■

To approximate the unique solution to this linear problem, we first consider the initial-value problems

$$y'' = p(x)y' + q(x)y + r(x), \text{with} \quad a \le x \le b, \quad y(a) = \alpha, \text{and} \quad y'(a) = 0, \tag{11.3}$$

and

$$y'' = p(x)y' + q(x)y, \text{with} \quad a \le x \le b, \quad y(a) = 0, \text{and} \quad y'(a) = 1. \tag{11.4}$$

Theorem 5.17 in Section 5.9 (see page 329) ensures that under the hypotheses in Corollary 11.2, both problems have a unique solution.

Let $y_1(x)$ denote the solution to (11.3), and let $y_2(x)$ denote the solution to (11.4). Assume that $y_2(b) \neq 0$. (That $y_2(b) = 0$ is in conflict with the hypotheses of Corollary 11.2 is considered in Exercise 8.) Define

$$y(x) = y_1(x) + \frac{\beta - y_1(b)}{y_2(b)} y_2(x). \tag{11.5}$$

Then $y(x)$ is the solution to the linear boundary problem (11.3). To see this, first note that

$$y'(x) = y_1'(x) + \frac{\beta - y_1(b)}{y_2(b)} y_2'(x)$$

and

$$y''(x) = y_1''(x) + \frac{\beta - y_1(b)}{y_2(b)} y_2''(x).$$

Substituting for $y_1''(x)$ and $y_2''(x)$ in this equation gives

$$\begin{aligned}
y'' &= p(x)y_1' + q(x)y_1 + r(x) + \frac{\beta - y_1(b)}{y_2(b)} \left(p(x)y_2' + q(x)y_2\right) \\
&= p(x)\left(y_1' + \frac{\beta - y_1(b)}{y_2(b)} y_2'\right) + q(x)\left(y_1 + \frac{\beta - y_1(b)}{y_2(b)} y_2\right) + r(x) \\
&= p(x)y'(x) + q(x)y(x) + r(x).
\end{aligned}$$

Moreover,

$$y(a) = y_1(a) + \frac{\beta - y_1(b)}{y_2(b)} y_2(a) = \alpha + \frac{\beta - y_1(b)}{y_2(b)} \cdot 0 = \alpha$$

and

$$y(b) = y_1(b) + \frac{\beta - y_1(b)}{y_2(b)} y_2(b) = y_1(b) + \beta - y_1(b) = \beta.$$

Linear Shooting

This "shooting" hits the target after one trial shot. In the next section we see that nonlinear problems require multiple shots.

The Shooting method for linear equations is based on the replacement of the linear boundary-value problem by the two initial-value problems (11.3) and (11.4). Numerous methods are available from Chapter 5 for approximating the solutions $y_1(x)$ and $y_2(x)$, and once these approximations are available, the solution to the boundary-value problem is approximated using Eq. (11.5). Graphically, the method has the appearance shown in Figure 11.1.

Figure 11.1

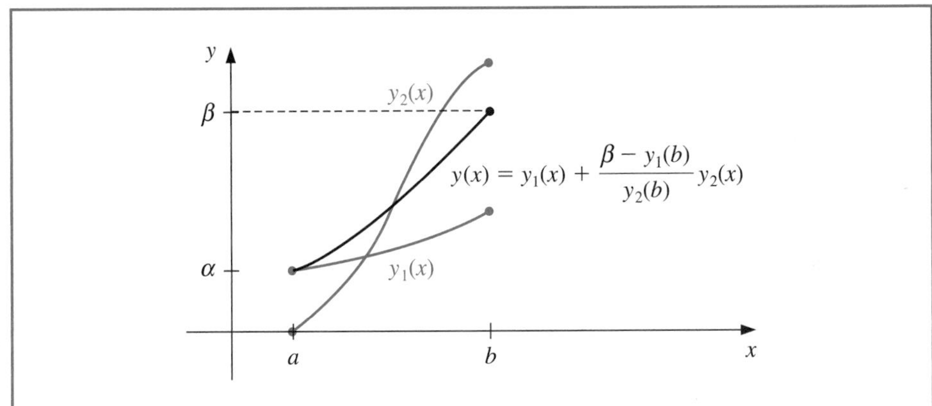

Algorithm 11.1 uses the fourth-order Runge-Kutta technique to find the approximations to $y_1(x)$ and $y_2(x)$, but other techniques for approximating the solutions to initial-value problems can be substituted into Step 4.

The algorithm has the additional feature of obtaining approximations for the derivative of the solution to the boundary-value problem as well as to the solution of the problem itself. The use of the algorithm is not restricted to those problems for which the hypotheses of Corollary 11.2 can be verified; it will work for many problems that do not satisfy these hypotheses. One such example can be found in Exercise 4.

Linear Shooting

To approximate the solution of the boundary-value problem

$$-y'' + p(x)y' + q(x)y + r(x) = 0, \quad \text{for } a \le x \le b, \text{ with } y(a) = \alpha \text{ and } y(b) = \beta,$$

(Note: Equations (11.3) *and* (11.4) *are written as first-order systems and solved.)*

INPUT endpoints a, b; boundary conditions α, β; number of subintervals N.

OUTPUT approximations $w_{1,i}$ to $y(x_i)$; $w_{2,i}$ to $y'(x_i)$ for each $i = 0, 1, \ldots, N$.

Step 1 Set $h = (b - a)/N$;
$u_{1,0} = \alpha$;
$u_{2,0} = 0$;
$v_{1,0} = 0$;
$v_{2,0} = 1$.

Step 2 For $i = 0, \dots, N - 1$ do Steps 3 and 4.
(*The Runge-Kutta method for systems is used in Steps* 3 *and* 4.)

Step 3 Set $x = a + ih$.

Step 4 Set $k_{1,1} = hu_{2,i}$;

$$
\begin{aligned}
k_{1,2} &= h\left[p(x)u_{2,i} + q(x)u_{1,i} + r(x)\right];\\
k_{2,1} &= h\left[u_{2,i} + \tfrac{1}{2}k_{1,2}\right];\\
k_{2,2} &= h\left[p(x + h/2)\left(u_{2,i} + \tfrac{1}{2}k_{1,2}\right)\right.\\
&\qquad \left. + q(x + h/2)\left(u_{1,i} + \tfrac{1}{2}k_{1,1}\right) + r(x + h/2)\right];\\
k_{3,1} &= h\left[u_{2,i} + \tfrac{1}{2}k_{2,2}\right];\\
k_{3,2} &= h\left[p(x + h/2)\left(u_{2,i} + \tfrac{1}{2}k_{2,2}\right)\right.\\
&\qquad \left. + q(x + h/2)(u_{1,i} + \tfrac{1}{2}k_{2,1}) + r(x + h/2)\right];\\
k_{4,1} &= h\left[u_{2,i} + k_{3,2}\right];\\
k_{4,2} &= h\left[p(x + h)(u_{2,i} + k_{3,2}) + q(x + h)(u_{1,i} + k_{3,1}) + r(x + h)\right];\\
u_{1,i+1} &= u_{1,i} + \tfrac{1}{6}\left[k_{1,1} + 2k_{2,1} + 2k_{3,1} + k_{4,1}\right];\\
u_{2,i+1} &= u_{2,i} + \tfrac{1}{6}\left[k_{1,2} + 2k_{2,2} + 2k_{3,2} + k_{4,2}\right];\\
k'_{1,1} &= hv_{2,i};\\
k'_{1,2} &= h\left[p(x)v_{2,i} + q(x)v_{1,i}\right];\\
k'_{2,1} &= h\left[v_{2,i} + \tfrac{1}{2}k'_{1,2}\right];\\
k'_{2,2} &= h\left[p(x + h/2)\left(v_{2,i} + \tfrac{1}{2}k'_{1,2}\right) + q(x + h/2)\left(v_{1,i} + \tfrac{1}{2}k'_{1,1}\right)\right];\\
k'_{3,1} &= h\left[v_{2,i} + \tfrac{1}{2}k'_{2,2}\right];\\
k'_{3,2} &= h\left[p(x + h/2)\left(v_{2,i} + \tfrac{1}{2}k'_{2,2}\right) + q(x + h/2)\left(v_{1,i} + \tfrac{1}{2}k'_{2,1}\right)\right];\\
k'_{4,1} &= h\left[v_{2,i} + k'_{3,2}\right];\\
k'_{4,2} &= h\left[p(x + h)(v_{2,i} + k'_{3,2}) + q(x + h)(v_{1,i} + k'_{3,1})\right];\\
v_{1,i+1} &= v_{1,i} + \tfrac{1}{6}\left[k'_{1,1} + 2k'_{2,1} + 2k'_{3,1} + k'_{4,1}\right];\\
v_{2,i+1} &= v_{2,i} + \tfrac{1}{6}\left[k'_{1,2} + 2k'_{2,2} + 2k'_{3,2} + k'_{4,2}\right].
\end{aligned}
$$

Step 5 Set $w_{1,0} = \alpha$;

$$w_{2,0} = \frac{\beta - u_{1,N}}{v_{1,N}};$$

OUTPUT $(a, w_{1,0}, w_{2,0})$.

Step 6 For $i = 1, \dots, N$
set $W1 = u_{1,i} + w_{2,0}v_{1,i}$;
$W2 = u_{2,i} + w_{2,0}v_{2,i}$;
$x = a + ih$;
OUTPUT $(x, W1, W2)$. (*Output is* $x_i, w_{1,i}, w_{2,i}$.)

Step 7 STOP. (*The process is complete.*) ■

Example 2 Apply the Linear Shooting technique with $N = 10$ to the boundary-value problem

$$y'' = -\frac{2}{x}y' + \frac{2}{x^2}y + \frac{\sin(\ln x)}{x^2}, \quad \text{for } 1 \le x \le 2, \text{ with } y(1) = 1 \text{ and } y(2) = 2,$$

and compare the results to those of the exact solution

$$y = c_1 x + \frac{c_2}{x^2} - \frac{3}{10}\sin(\ln x) - \frac{1}{10}\cos(\ln x),$$

where

$$c_2 = \frac{1}{70}[8 - 12\sin(\ln 2) - 4\cos(\ln 2)] \approx -0.03920701320$$

and

$$c_1 = \frac{11}{10} - c_2 \approx 1.1392070132.$$

Solution Applying Algorithm 11.1 to this problem requires approximating the solutions to the initial-value problems

$$y_1'' = -\frac{2}{x}y_1' + \frac{2}{x^2}y_1 + \frac{\sin(\ln x)}{x^2}, \quad \text{for } 1 \le x \le 2, \text{ with } y_1(1) = 1 \text{ and } y_1'(1) = 0,$$

and

$$y_2'' = -\frac{2}{x}y_2' + \frac{2}{x^2}y_2, \quad \text{for } 1 \le x \le 2, \text{ with } y_2(1) = 0 \text{ and } y_2'(1) = 1.$$

The results of the calculations, using Algorithm 11.1 with $N = 10$ and $h = 0.1$, are given in Table 11.1. The value listed as $u_{1,i}$ approximates $y_1(x_i)$, the value $v_{1,i}$ approximates $y_2(x_i)$, and w_i approximates

$$y(x_i) = y_1(x_i) + \frac{2 - y_1(2)}{y_2(2)} y_2(x_i).$$

■

Table 11.1

x_i	$u_{1,i} \approx y_1(x_i)$	$v_{1,i} \approx y_2(x_i)$	$w_i \approx y(x_i)$	$y(x_i)$	$\lvert y(x_i) - w_i \rvert$
1.0	1.00000000	0.00000000	1.00000000	1.00000000	
1.1	1.00896058	0.09117986	1.09262917	1.09262930	1.43×10^{-7}
1.2	1.03245472	0.16851175	1.18708471	1.18708484	1.34×10^{-7}
1.3	1.06674375	0.23608704	1.28338227	1.28338236	9.78×10^{-8}
1.4	1.10928795	0.29659067	1.38144589	1.38144595	6.02×10^{-8}
1.5	1.15830000	0.35184379	1.48115939	1.48115942	3.06×10^{-8}
1.6	1.21248372	0.40311695	1.58239245	1.58239246	1.08×10^{-8}
1.7	1.27087454	0.45131840	1.68501396	1.68501396	5.43×10^{-10}
1.8	1.33273851	0.49711137	1.78889854	1.78889853	5.05×10^{-9}
1.9	1.39750618	0.54098928	1.89392951	1.89392951	4.41×10^{-9}
2.0	1.46472815	0.58332538	2.00000000	2.00000000	

The accurate results in this example are due to the fact that the fourth-order Runge-Kutta method gives $O(h^4)$ approximations to the solutions of the initial-value problems. Unfortunately, because of roundoff errors, there can be problems hidden in this technique.

Reducing Round-Off Error

Round-off problems can occur if $y_1(x)$ rapidly increases as x goes from a to b. In this case $u_{1,N} \approx y_1(b)$ will be large and if β is small in magnitude compared to $u_{1,N}$, the term $w_{2,0} = (\beta - u_{1,N})/v_{1,N}$ will be approximately $-u_{1,N}/v_{1,N}$. The computations in Step 6 then become

$$W1 = u_{1,i} + w_{2,0}v_{1,i} \approx u_{1,i} - \left(\frac{u_{1,N}}{v_{1,N}}\right) v_{1,i},$$

$$W2 = u_{2,i} + w_{2,0}v_{2,i} \approx u_{2,i} - \left(\frac{u_{1,N}}{v_{1,N}}\right) v_{2,i},$$

which allows a possibility of a loss of significant digits due to cancelation. However, because $u_{1,i}$ is an approximation to $y_1(x_i)$, the behavior of y_1 can easily be monitored, and if $u_{1,i}$ increases rapidly from a to b, the shooting technique can be employed backward from $x_0 = b$ to $x_N = a$. This changes the initial-value problems that need to be solved to

$$y'' = p(x)y' + q(x)y + r(x), \quad \text{for } a \le x \le b, \text{ with } y(b) = \alpha \text{ and } y'(b) = 0,$$

and

$$y'' = p(x)y' + q(x)y, \quad \text{for } a \le x \le b, \text{ with } y(b) = 0 \text{ and } y'(b) = 1.$$

If this reverse shooting technique still gives cancellation of significant digits and if increased precision does not yield greater accuracy, other techniques must be used. Some of these are presented later in this chapter. In general, however, if $u_{1,i}$ and $v_{1,i}$ are $O(h^n)$ approximations to $y_1(x_i)$ and $y_2(x_i)$, respectively, for each $i = 0, 1, \ldots, N$, then $w_{1,i}$ will be an $O(h^n)$ approximation to $y(x_i)$. In particular,

$$|w_{1,i} - y(x_i)| \le Kh^n \left|1 + \frac{v_{1,i}}{v_{1,N}}\right|,$$

for some constant K (see [IK], p. 426).

EXERCISE SET 11.1

1. The boundary-value problem

$$y'' = 4(y - x), \quad 0 \le x \le 1, \quad y(0) = 0, \quad y(1) = 2,$$

has the solution $y(x) = e^2(e^4 - 1)^{-1}(e^{2x} - e^{-2x}) + x$. Use the Linear Shooting method to approximate the solution, and compare the results to the actual solution.

   a. With $h = \frac{1}{2}$; **b.** With $h = \frac{1}{4}$.

2. The boundary-value problem

$$y'' = y' + 2y + \cos x, \quad 0 \le x \le \tfrac{\pi}{2}, \quad y(0) = -0.3, \quad y\left(\tfrac{\pi}{2}\right) = -0.1$$

has the solution $y(x) = -\frac{1}{10}(\sin x + 3\cos x)$. Use the Linear Shooting method to approximate the solution, and compare the results to the actual solution.

   a. With $h = \frac{\pi}{4}$; **b.** With $h = \frac{\pi}{8}$.

3. Use the Linear Shooting method to approximate the solution to the following boundary-value problems.

   a. $y'' = -3y' + 2y + 2x + 3, \quad 0 \le x \le 1, \ y(0) = 2, \ y(1) = 1;$ use $h = 0.1$.

   b. $y'' = -4x^{-1}y' - 2x^{-2}y + 2x^{-2}\ln x, \quad 1 \le x \le 2, \ y(1) = -\frac{1}{2}, \ y(2) = \ln 2;$ use $h = 0.05$.

c. $y'' = -(x+1)y' + 2y + (1-x^2)e^{-x}, \quad 0 \le x \le 1,\ y(0) = -1,\ y(1) = 0;$ use $h = 0.1$.

d. $y'' = x{-}1y' + 3x^{-2}y + x^{-1}\ln x - 1, \quad 1 \le x \le 2,\ y(1) = y(2) = 0;$ use $h = 0.1$.

4. Although $q(x) < 0$ in the following boundary-value problems, unique solutions exist and are given. Use the Linear Shooting Algorithm to approximate the solutions to the following problems, and compare the results to the actual solutions.

   a. $y'' + y = 0, \quad 0 \le x \le \frac{\pi}{4},\ y(0) = 1,\ y(\frac{\pi}{4}) = 1;$ use $h = \frac{\pi}{20}$; actual solution $y(x) = \cos x + (\sqrt{2} - 1)\sin x$.

   b. $y'' + 4y = \cos x, \quad 0 \le x \le \frac{\pi}{4},\ y(0) = 0,\ y(\frac{\pi}{4}) = 0;$ use $h = \frac{\pi}{20}$; actual solution $y(x) = -\frac{1}{3}\cos 2x - \frac{\sqrt{2}}{6}\sin 2x + \frac{1}{3}\cos x$.

   c. $y'' = -4x^{-1}y' - 2x^{-2}y + 2x^{-2}\ln x, \quad 1 \le x \le 2,\ y(1) = \frac{1}{2},\ y(2) = \ln 2;$ use $h = 0.05$; actual solution $y(x) = 4x^{-1} - 2x^{-2} + \ln x - 3/2$.

   d. $y'' = 2y' - y + xe^x - x, \quad 0 \le x \le 2,\ y(0) = 0,\ y(2) = -4;$ use $h = 0.2$; actual solution $y(x) = \frac{1}{6}x^3e^x - \frac{5}{3}xe^x + 2e^x - x - 2$.

5. Use the Linear Shooting Algorithm to approximate the solution $y = e^{-10x}$ to the boundary-value problem

$$y'' = 100y, \quad 0 \le x \le 1, \quad y(0) = 1, \quad y(1) = e^{-10}.$$

   Use $h = 0.1$ and 0.05.

6. Write the second-order initial-value problems (11.3) and (11.4) as first-order systems, and derive the equations necessary to solve the systems using the fourth-order Runge-Kutta method for systems.

7. Let u represent the electrostatic potential between two concentric metal spheres of radii R_1 and R_2 ($R_1 < R_2$). The potential of the inner sphere is kept constant at V_1 volts, and the potential of the outer sphere is 0 volts. The potential in the region between the two spheres is governed by Laplace's equation, which, in this particular application, reduces to

$$\frac{d^2u}{dr^2} + \frac{2}{r}\frac{du}{dr} = 0, \quad R_1 \le r \le R_2, \quad u(R_1) = V_1, \quad u(R_2) = 0.$$

   Suppose $R_1 = 2$ in., $R_2 = 4$ in., and $V_1 = 110$ volts.

   a. Approximate $u(3)$ using the Linear Shooting Algorithm.

   b. Compare the results of part (a) with the actual potential $u(3)$, where

$$u(r) = \frac{V_1R_1}{r}\left(\frac{R_2 - r}{R_2 - R_1}\right).$$

8. Show that, under the hypothesis of Corollary 11.2, if y_2 is the solution to $y'' = p(x)y' + q(x)y$ and $y_2(a) = y_2(b) = 0$, then $y_2 \equiv 0$.

9. Consider the boundary-value problem

$$y'' + y = 0, \quad 0 \le x \le b, \quad y(0) = 0, \quad y(b) = B.$$

   Find choices for b and B so that the boundary-value problem has

   a. No solution b. Exactly one solution c. Infinitely many solutions.

10. Attempt to apply Exercise 9 to the boundary-value problem

$$y'' - y = 0, \quad 0 \le x \le b, \quad y(0) = 0, \quad y(b) = B.$$

   What happens? How do both problems relate to Corollary 11.2?

11.2 The Shooting Method for Nonlinear Problems

The shooting technique for the nonlinear second-order boundary-value problem

$$y'' = f(x, y, y'), \quad \text{for } a \le x \le b, \text{ with } y(a) = \alpha \text{ and } y(b) = \beta, \tag{11.6}$$

is similar to the linear technique, except that the solution to a nonlinear problem cannot be expressed as a linear combination of the solutions to two initial-value problems. Instead, we approximate the solution to the boundary-value problem by using the solutions to a *sequence* of initial-value problems involving a parameter t. These problems have the form

$$y'' = f(x, y, y'), \quad \text{for } a \leq x \leq b, \text{ with } y(a) = \alpha \text{ and } y'(a) = t. \tag{11.7}$$

We do this by choosing the parameters $t = t_k$ in a manner to ensure that

$$\lim_{k \to \infty} y(b, t_k) = y(b) = \beta,$$

where $y(x, t_k)$ denotes the solution to the initial-value problem (11.7) with $t = t_k$, and $y(x)$ denotes the solution to the boundary-value problem (11.6).

Shooting methods for nonlinear problems require iterations to approach the "target".

This technique is called a "shooting" method by analogy to the procedure of firing objects at a stationary target. (See Figure 11.2.) We start with a parameter t_0 that determines the initial elevation at which the object is fired from the point (a, α) and along the curve described by the solution to the initial-value problem:

$$y'' = f(x, y, y'), \quad \text{for } a \leq x \leq b, \text{ with } y(a) = \alpha \text{ and } y'(a) = t_0.$$

Figure 11.2

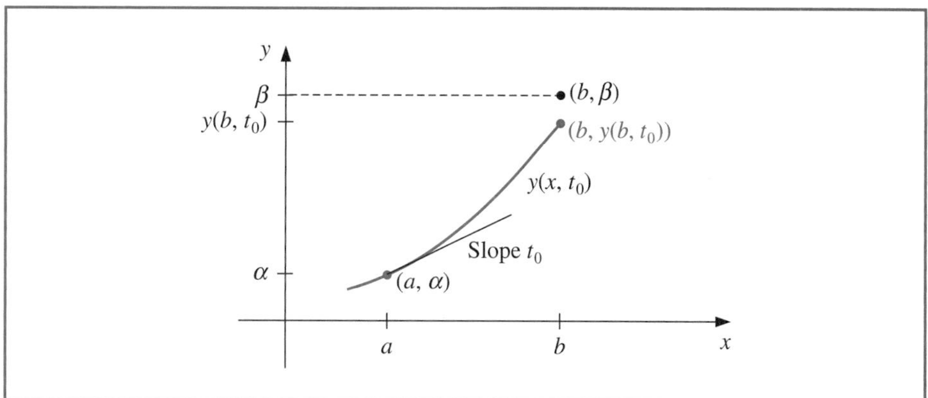

If $y(b, t_0)$ is not sufficiently close to β, we correct our approximation by choosing elevations t_1, t_2, and so on, until $y(b, t_k)$ is sufficiently close to "hitting" β. (See Figure 11.3.)

To determine the parameters t_k, suppose a boundary-value problem of the form (11.6) satisfies the hypotheses of Theorem 11.1. If $y(x, t)$ denotes the solution to the initial-value problem (11.7), we next determine t with

$$y(b, t) - \beta = 0. \tag{11.8}$$

This is a nonlinear equation in the variable t. Problems of this type were considered in Chapter 2, and a number of methods are available.

To use the Secant method to solve the problem, we need to choose initial approximations t_0 and t_1, and then generate the remaining terms of the sequence by

$$t_k = t_{k-1} - \frac{(y(b, t_{k-1}) - \beta)(t_{k-1} - t_{k-2})}{y(b, t_{k-1}) - y(b, t_{k-2})}, \quad k = 2, 3, \ldots.$$

Figure 11.3

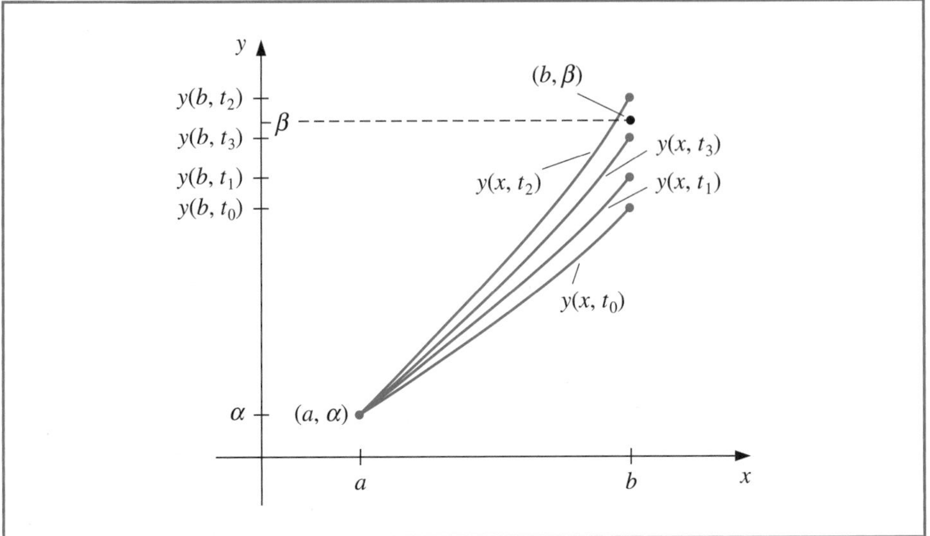

Newton Iteration

To use the more powerful Newton's method to generate the sequence $\{t_k\}$, only one initial approximation, t_0, is needed. However, the iteration has the form

$$t_k = t_{k-1} - \frac{y(b, t_{k-1}) - \beta}{\frac{dy}{dt}(b, t_{k-1})}, \tag{11.9}$$

and it requires the knowledge of $(dy/dt)(b, t_{k-1})$. This presents a difficulty because an explicit representation for $y(b, t)$ is not known; we know only the values $y(b, t_0)$, $y(b, t_1)$, $\ldots, y(b, t_{k-1})$.

Suppose we rewrite the initial-value problem (11.7), emphasizing that the solution depends on both x and the parameter t:

$$y''(x, t) = f(x, y(x, t), y'(x, t)), \quad \text{for } a \le x \le b, \text{ with } y(a, t) = \alpha \text{ and } y(a, t) = t. \tag{11.10}$$

We have retained the prime notation to indicate differentiation with respect to x. We need to determine $(dy/dt)(b, t)$ when $t = t_{k-1}$, so we first take the partial derivative of (11.10) with respect to t. This implies that

$$\begin{aligned}\frac{\partial y''}{\partial t}(x, t) &= \frac{\partial f}{\partial t}(x, y(x, t), y'(x, t)) \\ &= \frac{\partial f}{\partial x}(x, y(x, t), y'(x, t))\frac{\partial x}{\partial t} + \frac{\partial f}{\partial y}(x, y(x, t), y'(x, t))\frac{\partial y}{\partial t}(x, t) \\ &\quad + \frac{\partial f}{\partial y'}(x, y(x, t), y'(x, t))\frac{\partial y'}{\partial t}(x, t).\end{aligned}$$

Since x and t are independent,we have $\partial x/\partial t = 0$ and the equation simplifies to

$$\frac{\partial y''}{\partial t}(x, t) = \frac{\partial f}{\partial y}(x, y(x, t), y'(x, t))\frac{\partial y}{\partial t}(x, t) + \frac{\partial f}{\partial y'}(x, y(x, t), y'(x, t))\frac{\partial y'}{\partial t}(x, t), \tag{11.11}$$

for $a \le x \le b$. The initial conditions give

$$\frac{\partial y}{\partial t}(a, t) = 0 \quad \text{and} \quad \frac{\partial y'}{\partial t}(a, t) = 1.$$

If we simplify the notation by using $z(x,t)$ to denote $(\partial y/\partial t)(x,t)$ and assume that the order of differentiation of x and t can be reversed, (11.11) with the initial conditions becomes the initial-value problem

$$z''(x,t) = \frac{\partial f}{\partial y}(x,y,y')z(x,t) + \frac{\partial f}{\partial y'}(x,y,y')z'(x,t), \quad \text{for } a \le x \le b, \tag{11.12}$$

with $z(a,t) = 0$ and $z(a,t) = 1$.

Newton's method therefore requires that two initial-value problems, (11.10) and (11.12), be solved for each iteration. Then from Eq. (11.9), we have

$$t_k = t_{k-1} - \frac{y(b,t_{k-1}) - \beta}{z(b,t_{k-1})}. \tag{11.13}$$

Of course, none of these initial-value problems is solved exactly; the solutions are approximated by one of the methods discussed in Chapter 5. Algorithm 11.2 uses the Runge-Kutta method of order four to approximate both solutions required by Newton's method. A similar procedure for the Secant method is considered in Exercise 5.

ALGORITHM 11.2

Nonlinear Shooting with Newton's Method

To approximate the solution of the nonlinear boundary-value problem

$$y'' = f(x,y,y'), \quad \text{for } a \le x \le b, \text{ with } y(a) = \alpha \text{ and } y(b) = \beta:$$

(*Note: Equations* (11.10) *and* (11.12) *are written as first-order systems and solved.*)

INPUT endpoints a, b; boundary conditions α, β; number of subintervals $N \ge 2$; tolerance TOL; maximum number of iterations M.

OUTPUT approximations $w_{1,i}$ to $y(x_i)$; $w_{2,i}$ to $y'(x_i)$ for each $i = 0, 1, \ldots, N$ or a message that the maximum number of iterations was exceeded.

Step 1 Set $h = (b-a)/N$;
$k = 1$;
$TK = (\beta - \alpha)/(b-a)$. (*Note: TK could also be input.*)

Step 2 While ($k \le M$) do Steps 3–10.

Step 3 Set $w_{1,0} = \alpha$;
$w_{2,0} = TK$;
$u_1 = 0$;
$u_2 = 1$.

Step 4 For $i = 1, \ldots, N$ do Steps 5 and 6.
(*The Runge-Kutta method for systems is used in Steps* 5 *and* 6.)

Step 5 Set $x = a + (i-1)h$.

Step 6 Set $k_{1,1} = hw_{2,i-1}$;
$k_{1,2} = hf(x, w_{1,i-1} w_{2,i-1})$;
$k_{2,1} = h\left(w_{2,i-1} + \frac{1}{2}k_{1,2}\right)$;
$k_{2,2} = hf\left(x + h/2, w_{1,i-1} + \frac{1}{2}k_{1,1}, w_{2,i-1} + \frac{1}{2}k_{1,2}\right)$;
$k_{3,1} = h\left(w_{2,i-1} + \frac{1}{2}k_{2,2}\right)$;

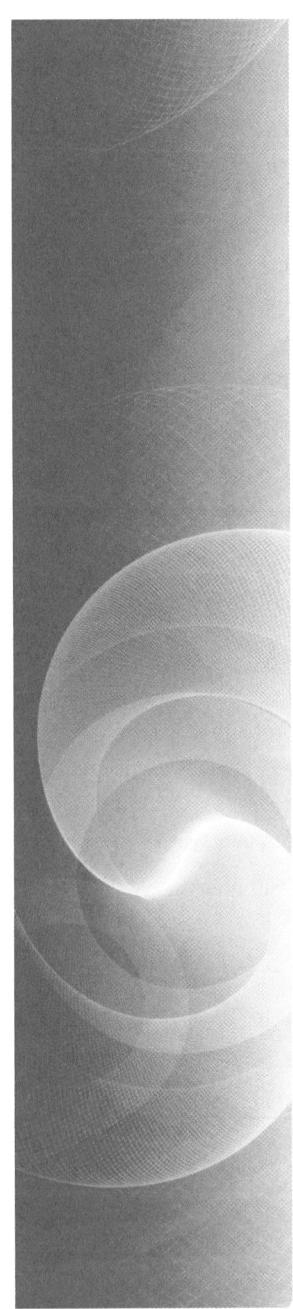

$$k_{3,2} = hf\left(x + h/2, w_{1,i-1} + \tfrac{1}{2}k_{2,1}, w_{2,i-1} + \tfrac{1}{2}k_{2,2}\right);$$
$$k_{4,1} = h(w_{2,i-1} + k_{3,2});$$
$$k_{4,2} = hf(x + h, w_{1,i-1} + k_{3,1}, w_{2,i-1} + k_{3,2});$$
$$w_{1,i} = w_{1,i-1} + (k_{1,1} + 2k_{2,1} + 2k_{3,1} + k_{4,1})/6;$$
$$w_{2,i} = w_{2,i-1} + (k_{1,2} + 2k_{2,2} + 2k_{3,2} + k_{4,2})/6;$$
$$k'_{1,1} = hu_2;$$
$$k'_{1,2} = h[f_y(x, w_{1,i-1}, w_{2,i-1})u_1 + f_{y'}(x, w_{1,i-1}, w_{2,i-1})u_2];$$
$$k'_{2,1} = h\left[u_2 + \tfrac{1}{2}k'_{1,2}\right];$$
$$k'_{2,2} = h\left[f_y(x + h/2, w_{1,i-1}, w_{2,i-1})\left(u_1 + \tfrac{1}{2}k'_{1,1}\right) + f_{y'}(x + h/2, w_{1,i-1}, w_{2,i-1})\left(u_2 + \tfrac{1}{2}k'_{1,2}\right)\right];$$
$$k'_{3,1} = h\left(u_2 + \tfrac{1}{2}k'_{2,2}\right);$$
$$k'_{3,2} = h\left[f_y(x + h/2, w_{1,i-1}, w_{2,i-1})\left(u_1 + \tfrac{1}{2}k'_{2,1}\right) + f_{y'}(x + h/2, w_{1,i-1}, w_{2,i-1})\left(u_2 + \tfrac{1}{2}k'_{2,2}\right)\right];$$
$$k'_{4,1} = h(u_2 + k'_{3,2});$$
$$k'_{4,2} = h\left[f_y(x + h, w_{1,i-1}, w_{2,i-1})\left(u_1 + k'_{3,1}\right) + f_{y'}(x + h, w_{1,i-1}, w_{2,i-1})\left(u_2 + k'_{3,2}\right)\right];$$
$$u_1 = u_1 + \tfrac{1}{6}[k'_{1,1} + 2k'_{2,1} + 2k'_{3,1} + k'_{4,1}];$$
$$u_2 = u_2 + \tfrac{1}{6}[k'_{1,2} + 2k'_{2,2} + 2k'_{3,2} + k'_{4,2}].$$

Step 7 If $|w_{1,N} - \beta| \leq TOL$ then do Steps 8 and 9.

Step 8 For $i = 0, 1, \ldots, N$
set $x = a + ih$;
OUTPUT $(x, w_{1,i}, w_{2,i})$.

Step 9 (*The procedure is complete.*)
STOP.

Step 10 Set $TK = TK - \dfrac{w_{1,N} - \beta}{u_1}$;
(*Newton's method is used to compute* TK.)
$k = k + 1$.

Step 11 OUTPUT ('Maximum number of iterations exceeded');
(*The procedure was unsuccessful.*)
STOP. ■

The value $t_0 = TK$ selected in Step 1 is the slope of the straight line through (a, α) and (b, β). If the problem satisfies the hypotheses of Theorem 11.1, any choice of t_0 will give convergence, but a good choice of t_0 will improve convergence, and the procedure will even work for many problems that do not satisfy these hypotheses. One such example can be found in Exercise 3(d).

Example 1 Apply the Shooting method with Newton's Method to the boundary-value problem

$$y'' = \frac{1}{8}(32 + 2x^3 - yy'), \quad \text{for } 1 \leq x \leq 3, \text{ with } y(1) = 17 \text{ and } y(3) = \frac{43}{3}.$$

Use $N = 20$, $M = 10$, and $TOL = 10^{-5}$, and compare the results with the exact solution $y(x) = x^2 + 16/x$.

Solution We need approximate solutions to the initial-value problems

$$y'' = \frac{1}{8}(32 + 2x^3 - yy'), \quad \text{for } 1 \le x \le 3, \text{ with } y(1) = 17 \text{ and } y'(1) = t_k,$$

and

$$z'' = \frac{\partial f}{\partial y} z + \frac{\partial f}{\partial y'} z' = -\frac{1}{8}(y'z + yz'), \quad \text{for } 1 \le x \le 3, \text{ with } z(1) = 0 \text{ and } z'(1) = 1,$$

at each step in the iteration. If the stopping technique in Algorithm 11.2 requires

$$|w_{1,N}(t_k) - y(3)| \le 10^{-5},$$

then we need four iterations and $t_4 = -14.000203$. The results obtained for this value of t are shown in Table 11.2. ■

Table 11.2

x_i	$w_{1,i}$	$y(x_i)$	$\lvert w_{1,i} - y(x_i)\rvert$
1.0	17.000000	17.000000	
1.1	15.755495	15.755455	4.06×10^{-5}
1.2	14.773389	14.773333	5.60×10^{-5}
1.3	13.997752	13.997692	5.94×10^{-5}
1.4	13.388629	13.388571	5.71×10^{-5}
1.5	12.916719	12.916667	5.23×10^{-5}
1.6	12.560046	12.560000	4.64×10^{-5}
1.7	12.301805	12.301765	4.02×10^{-5}
1.8	12.128923	12.128889	3.14×10^{-5}
1.9	12.031081	12.031053	2.84×10^{-5}
2.0	12.000023	12.000000	2.32×10^{-5}
2.1	12.029066	12.029048	1.84×10^{-5}
2.2	12.112741	12.112727	1.40×10^{-5}
2.3	12.246532	12.246522	1.01×10^{-5}
2.4	12.426673	12.426667	6.68×10^{-6}
2.5	12.650004	12.650000	3.61×10^{-6}
2.6	12.913847	12.913845	9.17×10^{-7}
2.7	13.215924	13.215926	1.43×10^{-6}
2.8	13.554282	13.554286	3.46×10^{-6}
2.9	13.927236	13.927241	5.21×10^{-6}
3.0	14.333327	14.333333	6.69×10^{-6}

Although Newton's method used with the shooting technique requires the solution of an additional initial-value problem, it will generally give faster convergence than the Secant method. However both methods are only locally convergent because they require good initial approximations.

For a general discussion of the convergence of the shooting techniques for nonlinear problems, the reader is referred to the excellent book by Keller [Keller, H]. In that reference, more general boundary conditions are discussed. It is also noted that the shooting technique for nonlinear problems is sensitive to roundoff errors, especially if the solution $y(x)$ and $z(x, t)$ are rapidly increasing functions of x on $[a, b]$.

EXERCISE SET 11.2

1. Use the Nonlinear Shooting Algorithm with $h = 0.5$ to approximate the solution to the boundary-value problem

$$y'' = -(y')^2 - y + \ln x, \quad 1 \le x \le 2, \quad y(1) = 0, \quad y(2) = \ln 2.$$

Compare your results to the actual solution $y = \ln x$.

2. Use the Nonlinear Shooting Algorithm with $h = 0.25$ to approximate the solution to the boundary-value problem

$$y'' = 2y^3, \quad -1 \le x \le 0, \quad y(-1) = \frac{1}{2}, \quad y(0) = \frac{1}{3}.$$

Compare your results to the actual solution $y(x) = 1/(x + 3)$.

3. Use the Nonlinear Shooting method with $TOL = 10^{-4}$ to approximate the solution to the following boundary-value problems. The actual solution is given for comparison to your results.
   a. $y'' = -e^{-2y}$, $\quad 1 \le x \le 2$, $y(1) = 0$, $y(2) = \ln 2$; use $N = 10$; actual solution $y(x) = \ln x$.
   b. $y'' = y' \cos x - y \ln y$, $\quad 0 \le x \le \frac{\pi}{2}$, $y(0) = 1$, $y\left(\frac{\pi}{2}\right) = e$; use $N = 10$; actual solution $y(x) = e^{\sin x}$.
   c. $y'' = -\left(2(y')^3 + y^2 y'\right) \sec x$, $\quad \frac{\pi}{4} \le x \le \frac{\pi}{3}$, $y\left(\frac{\pi}{4}\right) = 2^{-1/4}$, $y\left(\frac{\pi}{3}\right) = \frac{1}{2}\sqrt[4]{12}$; use $N = 5$; actual solution $y(x) = \sqrt{\sin x}$.
   d. $y'' = \frac{1}{2}\left(1 - (y')^2 - y \sin x\right)$, $\quad 0 \le x \le \pi$, $y(0) = 2$, $y(\pi) = 2$; use $N = 20$; actual solution $y(x) = 2 + \sin x$.

4. Use the Nonlinear Shooting method with $TOL = 10^{-4}$ to approximate the solution to the following boundary-value problems. The actual solution is given for comparison to your results.
   a. $y'' = y^3 - yy'$, $\quad 1 \le x \le 2$, $y(1) = \frac{1}{2}$, $y(2) = \frac{1}{3}$; use $h = 0.1$; actual solution $y(x) = (x + 1)^{-1}$.
   b. $y'' = 2y^3 - 6y - 2x^3$, $\quad 1 \le x \le 2$, $y(1) = 2$, $y(2) = \frac{5}{2}$; use $h = 0.1$; actual solution $y(x) = x + x^{-1}$.
   c. $y'' = y' + 2(y - \ln x)^3 - x^{-1}$, $\quad 2 \le x \le 3$, $y(2) = \frac{1}{2} + \ln 2$, $y(3) = \frac{1}{3} + \ln 3$; use $h = 0.1$; actual solution $y(x) = x^{-1} + \ln x$.
   d. $y'' = 2(y')^2 x^{-3} - 9y^2 x^{-5} + 4x$, $\quad 1 \le x \le 2$, $y(1) = 0$, $y(2) = \ln 256$; use $h = 0.05$; actual solution $y(x) = x^3 \ln x$.

5. **a.** Change Algorithm 11.2 to incorporate the Secant method instead of Newton's method. Use $t_0 = (\beta - \alpha)/(b - a)$ and $t_1 = t_0 + (\beta - y(b, t_0))/(b - a)$.
   b. Repeat Exercise 4(a) and 4(c) using the Secant algorithm derived in part (a), and compare the number of iterations required for the two methods.

6. The Van der Pol equation,

$$y'' - \mu(y^2 - 1)y' + y = 0, \quad \mu > 0,$$

governs the flow of current in a vacuum tube with three internal elements. Let $\mu = \frac{1}{2}$, $y(0) = 0$, and $y(2) = 1$. Approximate the solution $y(t)$ for $t = 0.2i$, where $1 \le i \le 9$.

11.3 Finite-Difference Methods for Linear Problems

The linear and nonlinear Shooting methods for boundary-value problems can present problems of instability. The methods in this section have better stability characteristics, but they generally require more computation to obtain a specified accuracy.

Methods involving finite differences for solving boundary-value problems replace each of the derivatives in the differential equation with an appropriate difference-quotient approximation of the type considered in Section 4.1. The particular difference quotient and step size h are chosen to maintain a specified order of truncation error. However, h cannot be chosen too small because of the general instability of the derivative approximations.

Discrete Approximation

The finite difference method for the linear second-order boundary-value problem,

$$y'' = p(x)y' + q(x)y + r(x), \quad \text{for } a \le x \le b, \text{ with } y(a) = \alpha \text{ and } y(b) = \beta, \tag{11.14}$$

requires that difference-quotient approximations be used to approximate both y' and y''. First, we select an integer $N > 0$ and divide the interval $[a, b]$ into $(N+1)$ equal subintervals whose endpoints are the mesh points $x_i = a+ih$, for $i = 0, 1, \ldots, N+1$, where $h = (b-a)/(N+1)$. Choosing the step size h in this manner facilitates the application of a matrix algorithm from Chapter 6, which solves a linear system involving an $N \times N$ matrix.

At the interior mesh points, x_i, for $i = 1, 2, \ldots, N$, the differential equation to be approximated is

$$y''(x_i) = p(x_i)y'(x_i) + q(x_i)y(x_i) + r(x_i). \tag{11.15}$$

Expanding y in a third Taylor polynomial about x_i evaluated at x_{i+1} and x_{i-1}, we have, assuming that $y \in C^4[x_{i-1}, x_{i+1}]$,

$$y(x_{i+1}) = y(x_i + h) = y(x_i) + hy'(x_i) + \frac{h^2}{2}y''(x_i) + \frac{h^3}{6}y'''(x_i) + \frac{h^4}{24}y^{(4)}(\xi_i^+),$$

for some ξ_i^+ in (x_i, x_{i+1}), and

$$y(x_{i-1}) = y(x_i - h) = y(x_i) - hy'(x_i) + \frac{h^2}{2}y''(x_i) - \frac{h^3}{6}y'''(x_i) + \frac{h^4}{24}y^{(4)}(\xi_i^-),$$

for some ξ_i^- in (x_{i-1}, x_i). If these equations are added, we have

$$y(x_{i+1}) + y(x_{i-1}) = 2y(x_i) + h^2y''(x_i) + \frac{h^4}{24}[y^{(4)}(\xi_i^+) + y^{(4)}(\xi_i^-)],$$

and solving for $y''(x_i)$ gives

$$y''(x_i) = \frac{1}{h^2}[y(x_{i+1}) - 2y(x_i) + y(x_{i-1})] - \frac{h^2}{24}[y^{(4)}(\xi_i^+) + y^{(4)}(\xi_i^-)].$$

The Intermediate Value Theorem 1.11 can be used to simplify the error term to give

$$y''(x_i) = \frac{1}{h^2}[y(x_{i+1}) - 2y(x_i) + y(x_{i-1})] - \frac{h^2}{12}y^{(4)}(\xi_i), \tag{11.16}$$

for some ξ_i in (x_{i-1}, x_{i+1}). This is called the **centered-difference formula** for $y''(x_i)$.

A centered-difference formula for $y'(x_i)$ is obtained in a similar manner (the details were considered in Section 4.1), resulting in

$$y'(x_i) = \frac{1}{2h}[y(x_{i+1}) - y(x_{i-1})] - \frac{h^2}{6}y'''(\eta_i), \tag{11.17}$$

for some η_i in (x_{i-1}, x_{i+1}).

The use of these centered-difference formulas in Eq. (11.15) results in the equation

$$\frac{y(x_{i+1}) - 2y(x_i) + y(x_{i-1})}{h^2} = p(x_i)\left[\frac{y(x_{i+1}) - y(x_{i-1})}{2h}\right] + q(x_i)y(x_i)$$
$$+ r(x_i) - \frac{h^2}{12}\left[2p(x_i)y'''(\eta_i) - y^{(4)}(\xi_i)\right].$$

A Finite-Difference method with truncation error of order $O(h^2)$ results by using this equation together with the boundary conditions $y(a) = \alpha$ and $y(b) = \beta$ to define the system of linear equations

$$w_0 = \alpha, \qquad w_{N+1} = \beta$$

and

$$\left(\frac{-w_{i+1} + 2w_i - w_{i-1}}{h^2}\right) + p(x_i)\left(\frac{w_{i+1} - w_{i-1}}{2h}\right) + q(x_i)w_i = -r(x_i), \tag{11.18}$$

for each $i = 1, 2, \ldots, N$.

In the form we will consider, Eq. (11.18) is rewritten as

$$-\left(1 + \frac{h}{2}p(x_i)\right)w_{i-1} + \left(2 + h^2 q(x_i)\right)w_i - \left(1 - \frac{h}{2}p(x_i)\right)w_{i+1} = -h^2 r(x_i),$$

and the resulting system of equations is expressed in the tridiagonal $N \times N$ matrix form

$$A\mathbf{w} = \mathbf{b}, \quad \text{where} \tag{11.19}$$

$$A = \begin{bmatrix} 2 + h^2 q(x_1) & -1 + \frac{h}{2}p(x_1) & 0 & \cdots & \cdots & 0 \\ -1 - \frac{h}{2}p(x_2) & 2 + h^2 q(x_2) & -1 + \frac{h}{2}p(x_2) & \ddots & & \vdots \\ 0 & \ddots & \ddots & \ddots & \ddots & 0 \\ \vdots & & \ddots & \ddots & \ddots & -1 + \frac{h}{2}p(x_{N-1}) \\ 0 & \cdots & 0 & & -1 - \frac{h}{2}p(x_N) & 2 + h^2 q(x_N) \end{bmatrix},$$

$$\mathbf{w} = \begin{bmatrix} w_1 \\ w_2 \\ \vdots \\ w_{N-1} \\ w_N \end{bmatrix}, \quad \text{and} \quad \mathbf{b} = \begin{bmatrix} -h^2 r(x_1) + \left(1 + \frac{h}{2}p(x_1)\right)w_0 \\ -h^2 r(x_2) \\ \vdots \\ -h^2 r(x_{N-1}) \\ -h^2 r(x_N) + \left(1 - \frac{h}{2}p(x_N)\right)w_{N+1} \end{bmatrix}.$$

The following theorem gives conditions under which the tridiagonal linear system (11.19) has a unique solution. Its proof is a consequence of Theorem 6.31 on page 424 and is considered in Exercise 9.

Theorem 11.3 Suppose that p, q, and r are continuous on $[a, b]$. If $q(x) \geq 0$ on $[a, b]$, then the tridiagonal linear system (11.19) has a unique solution provided that $h < 2/L$, where $L = \max_{a \leq x \leq b} |p(x)|$. ■

It should be noted that the hypotheses of Theorem 11.3 guarantee a unique solution to the boundary-value problem (11.14), but they do not guarantee that $y \in C^4[a, b]$. We need to establish that $y^{(4)}$ is continuous on $[a, b]$ to ensure that the truncation error has order $O(h^2)$.

Algorithm 11.3 implements the Linear Finite-Difference method.

Linear Finite-Difference

To approximate the solution of the boundary-value problem

$$y'' = p(x)y' + q(x)y + r(x), \quad \text{for } a \le x \le b, \text{ with } y(a) = \alpha \text{ and } y(b) = \beta :$$

INPUT endpoints a, b; boundary conditions α, β; integer $N \ge 2$.

OUTPUT approximations w_i to $y(x_i)$ for each $i = 0, 1, \ldots, N + 1$.

Step 1 Set $h = (b - a)/(N + 1)$;
$x = a + h$;
$a_1 = 2 + h^2 q(x)$;
$b_1 = -1 + (h/2)p(x)$;
$d_1 = -h^2 r(x) + (1 + (h/2)p(x))\alpha$.

Step 2 For $i = 2, \ldots, N - 1$
set $x = a + ih$;
$a_i = 2 + h^2 q(x)$;
$b_i = -1 + (h/2)p(x)$;
$c_i = -1 - (h/2)p(x)$;
$d_i = -h^2 r(x)$.

Step 3 Set $x = b - h$;
$a_N = 2 + h^2 q(x)$;
$c_N = -1 - (h/2)p(x)$;
$d_N = -h^2 r(x) + (1 - (h/2)p(x))\beta$.

Step 4 Set $l_1 = a_1$; (*Steps 4–8 solve a tridiagonal linear system using Algorithm* 6.7.)
$u_1 = b_1/a_1$;
$z_1 = d_1/l_1$.

Step 5 For $i = 2, \ldots, N - 1$ set $l_i = a_i - c_i u_{i-1}$;
$u_i = b_i/l_i$;
$z_i = (d_i - c_i z_{i-1})/l_i$.

Step 6 Set $l_N = a_N - c_N u_{N-1}$;
$z_N = (d_N - c_N z_{N-1})/l_N$.

Step 7 Set $w_0 = \alpha$;
$w_{N+1} = \beta$.
$w_N = z_N$.

Step 8 For $i = N - 1, \ldots, 1$ set $w_i = z_i - u_i w_{i+1}$.

Step 9 For $i = 0, \ldots, N + 1$ set $x = a + ih$;
OUTPUT (x, w_i).

Step 10 STOP. (*The procedure is complete.*) ■

Example 1 Use Algorithm 11.3 with $N = 9$ to approximate the solution to the linear boundary-value problem

$$y'' = -\frac{2}{x}y' + \frac{2}{x^2}y + \frac{\sin(\ln x)}{x^2}, \quad \text{for } 1 \le x \le 2, \text{ with } y(1) = 1 \text{ and } y(2) = 2,$$

and compare the results to those obtained using the Shooting method in Example 2 of Section 11.1.

Solution For this example, we will use $N = 9$, so $h = 0.1$, and we have the same spacing as in Example 2 of Section 11.1. The complete results are listed in Table 11.3.

Table 11.3

x_i	w_i	$y(x_i)$	$\lvert w_i - y(x_i)\rvert$
1.0	1.00000000	1.00000000	
1.1	1.09260052	1.09262930	2.88×10^{-5}
1.2	1.18704313	1.18708484	4.17×10^{-5}
1.3	1.28333687	1.28338236	4.55×10^{-5}
1.4	1.38140205	1.38144595	4.39×10^{-5}
1.5	1.48112026	1.48115942	3.92×10^{-5}
1.6	1.58235990	1.58239246	3.26×10^{-5}
1.7	1.68498902	1.68501396	2.49×10^{-5}
1.8	1.78888175	1.78889853	1.68×10^{-5}
1.9	1.89392110	1.89392951	8.41×10^{-6}
2.0	2.00000000	2.00000000	

These results are considerably less accurate than those obtained in Example 2 of Section 11.1. This is because the method used in that example involved a Runge-Kutta technique with local truncation error of order $O(h^4)$, whereas the difference method used here has local truncation error of order $O(h^2)$. ■

To obtain a difference method with greater accuracy, we can proceed in a number of ways. Using fifth-order Taylor series for approximating $y''(x_i)$ and $y'(x_i)$ results in a truncation error term involving h^4. However, this process requires using multiples not only of $y(x_{i+1})$ and $y(x_{i-1})$, but also of $y(x_{i+2})$ and $y(x_{i-2})$ in the approximation formulas for $y''(x_i)$ and $y'(x_i)$. This leads to difficulty at $i = 0$, because we do not know w_{-1}, and at $i = N$, because we do not know w_{N+2}. Moreover, the resulting system of equations analogous to (11.19) is not in tridiagonal form, and the solution to the system requires many more calculations.

Employing Richardson's Extrapolation

Instead of attempting to obtain a difference method with a higher-order truncation error in this manner, it is generally more satisfactory to consider a reduction in step size. In addition, Richardson's extrapolation technique can be used effectively for this method because the error term is expressed in even powers of h with coefficients independent of h, provided y is sufficiently differentiable (see, for example, [Keller, H], p. 81).

Example 2 Apply Richardson's extrapolation to approximate the solution to the boundary-value problem

$$y'' = -\frac{2}{x}y' + \frac{2}{x^2}y + \frac{\sin(\ln x)}{x^2}, \quad \text{for } 1 \le x \le 2, \text{ with } y(1) = 1 \text{ and } y(2) = 2,$$

using $h = 0.1, 0.05$, and 0.025.

Solution The results are listed in Table 11.4. The first extrapolation is

$$\text{Ext}_{1i} = \frac{4w_i(h = 0.05) - w_i(h = 0.1)}{3};$$

the second extrapolation is

$$\text{Ext}_{2i} = \frac{4w_i(h = 0.025) - w_i(h = 0.05)}{3};$$

and the final extrapolation is

$$\text{Ext}_{3i} = \frac{16\text{Ext}_{2i} - \text{Ext}_{1i}}{15}.$$

Table 11.4

x_i	$w_i(h = 0.05)$	$w_i(h = 0.025)$	Ext_{1i}	Ext_{2i}	Ext_{3i}
1.0	1.00000000	1.00000000	1.00000000	1.00000000	1.00000000
1.1	1.09262207	1.09262749	1.09262925	1.09262930	1.09262930
1.2	1.18707436	1.18708222	1.18708477	1.18708484	1.18708484
1.3	1.28337094	1.28337950	1.28338230	1.28338236	1.28338236
1.4	1.38143493	1.38144319	1.38144589	1.38144595	1.38144595
1.5	1.48114959	1.48115696	1.48115937	1.48115941	1.48115942
1.6	1.58238429	1.58239042	1.58239242	1.58239246	1.58239246
1.7	1.68500770	1.68501240	1.68501393	1.68501396	1.68501396
1.8	1.78889432	1.78889748	1.78889852	1.78889853	1.78889853
1.9	1.89392740	1.89392898	1.89392950	1.89392951	1.89392951
2.0	2.00000000	2.00000000	2.00000000	2.00000000	2.00000000

The values of $w_i(h = 0.1)$ are omitted from the table to save space, but they are listed in Table 11.3. The results for $w_i(h = 0.025)$ are accurate to approximately 3×10^{-6}. However, the results of Ext_{3i} are correct to the decimal places listed. In fact, if sufficient digits had been used, this approximation would agree with the exact solution with maximum error of 6.3×10^{-11} at the mesh points, an impressive improvement. ■

EXERCISE SET 11.3

1. The boundary-value problem

$$y'' = 4(y - x), \quad 0 \le x \le 1, \quad y(0) = 0, \quad y(1) = 2$$

has the solution $y(x) = e^2(e^4 - 1)^{-1}(e^{2x} - e^{-2x}) + x$. Use the Linear Finite-Difference method to approximate the solution, and compare the results to the actual solution.

a. With $h = \frac{1}{2}$; **b.** With $h = \frac{1}{4}$.

c. Use extrapolation to approximate $y(1/2)$.

2. The boundary-value problem

$$y'' = y' + 2y + \cos x, \quad 0 \le x \le \tfrac{\pi}{2}, \quad y(0) = -0.3, \quad y\left(\tfrac{\pi}{2}\right) = -0.1$$

has the solution $y(x) = -\frac{1}{10}(\sin x + 3\cos x)$. Use the Linear Finite-Difference method to approximate the solution, and compare the results to the actual solution.

a. With $h = \frac{\pi}{4}$; b. With $h = \frac{\pi}{8}$.

c. Use extrapolation to approximate $y(\pi/4)$.

3. Use the Linear Finite-Difference Algorithm to approximate the solution to the following boundary-value problems.

a. $y'' = -3y' + 2y + 2x + 3, \quad 0 \le x \le 1, y(0) = 2, y(1) = 1$; use $h = 0.1$.

b. $y'' = -4x^{-1}y' + 2x^{-2}y - 2x^{-2}\ln x, \quad 1 \le x \le 2, y(1) = -\frac{1}{2}, y(2) = \ln 2$; use $h = 0.05$.

c. $y'' = -(x+1)y' + 2y + (1-x^2)e^{-x}, \quad 0 \le x \le 1, y(0) = -1, y(1) = 0$; use $h = 0.1$.

d. $y'' = x^{-1}y' + 3x^{-2}y + x^{-1}\ln x - 1, \quad 1 \le x \le 2, y(1) = y(2) = 0$; use $h = 0.1$.

4. Although $q(x) < 0$ in the following boundary-value problems, unique solutions exist and are given. Use the Linear Finite-Difference Algorithm to approximate the solutions, and compare the results to the actual solutions.

a. $y'' + y = 0, \quad 0 \le x \le \frac{\pi}{4}, y(0) = 1, y(\frac{\pi}{4}) = 1$; use $h = \frac{\pi}{20}$; actual solution $y(x) = \cos x + (\sqrt{2} - 1)\sin x$.

b. $y'' + 4y = \cos x, \quad 0 \le x \le \frac{\pi}{4}, y(0) = 0, y(\frac{\pi}{4}) = 0$; use $h = \frac{\pi}{20}$; actual solution $y(x) = -\frac{1}{3}\cos 2x - \frac{\sqrt{2}}{6}\sin 2x + \frac{1}{3}\cos x$.

c. $y'' = -4x^{-1}y' + 2x^{-2}y - 2x^{-2}\ln x, y(1) = \frac{1}{2}, y(2) = \ln 2$; use $h = 0.05$; actual solution $y(x) = 4x^{-1} - 2x^{-2} + \ln x - 3/2$.

d. $y'' = 2y' - y + xe^x - x, \quad 0 \le x \le 2, y(0) = 0, y(2) = -4$; use $h = 0.2$; actual solution $y(x) = \frac{1}{6}x^3e^x - \frac{5}{3}xe^x + 2e^x - x - 2$.

5. Use the Linear Finite-Difference Algorithm to approximate the solution $y = e^{-10x}$ to the boundary-value problem

$$y'' = 100y, \quad 0 \le x \le 1, \quad y(0) = 1, \quad y(1) = e^{-10}.$$

Use $h = 0.1$ and 0.05. Can you explain the consequences?

6. Repeat Exercise 3(a) and (b) using the extrapolation discussed in Example 2.

7. The lead example of this chapter concerned the deflection of a beam with supported ends subject to uniform loading. The boundary-value problem governing this physical situation is

$$\frac{d^2w}{dx^2} = \frac{S}{EI}w + \frac{qx}{2EI}(x - l), \quad 0 < x < l,$$

with boundary conditions $w(0) = 0$ and $w(l) = 0$.

Suppose the beam is a W10-type steel I-beam with the following characteristics: length $l = 120$ in., intensity of uniform load $q = 100$ lb/ft, modulus of elasticity $E = 3.0 \times 10^7$ lb/in.^{2}, stress at ends $S = 1000$ lb, and central moment of inertia $I = 625$ in.^{4}.

a. Approximate the deflection $w(x)$ of the beam every 6 in.

b. The actual relationship is given by

$$w(x) = c_1e^{ax} + c_2e^{-ax} + b(x - l)x + c,$$

where $c_1 = 7.7042537 \times 10^4$, $c_2 = 7.9207462 \times 10^4$, $a = 2.3094010 \times 10^{-4}$, $b = -4.1666666 \times 10^{-3}$, and $c = -1.5625 \times 10^5$. Is the maximum error on the interval within 0.2 in.?

c. State law requires that $\max_{0<x<l} w(x) < 1/300$. Does this beam meet state code?

8. The deflection of a uniformly loaded, long rectangular plate under an axial tension force is governed by a second-order differential equation. Let S represent the axial force and q the intensity of the uniform load. The deflection W along the elemental length is given by

$$W''(x) - \frac{S}{D}W(x) = \frac{-ql}{2D}x + \frac{q}{2D}x^2, \quad 0 \le x \le l, \ W(0) = W(l) = 0,$$

where l is the length of the plate and D is the flexual rigidity of the plate. Let $q = 200$ lb/in.^{2}, $S = 100$ lb/in., $D = 8.8 \times 10^7$ lb/in., and $l = 50$ in. Approximate the deflection at 1-in. intervals.

9. Prove Theorem 11.3. [*Hint:* To use Theorem 6.31, first show that $\left|\frac{h}{2}p(x_i)\right| < 1$ implies that $\left|-1 - \frac{h}{2}p(x_i)\right| + \left|-1 + \frac{h}{2}p(x_i)\right| = 2$.]

10. Show that if $y \in C^6[a, b]$ and if $w_0, w_1, \ldots, w_{N+1}$ satisfy Eq. (11.18), then

$$w_i - y(x_i) = Ah^2 + O(h^4),$$

where A is independent of h, provided $q(x) \geq w > 0$ on $[a, b]$ for some w.

11.4 Finite-Difference Methods for Nonlinear Problems

For the general nonlinear boundary-value problem

$$y'' = f(x, y, y'), \quad \text{for } a \leq x \leq b, \text{ with } y(a) = \alpha \text{ and } y(b) = \beta,$$

the difference method is similar to the method applied to linear problems in Section 11.3. Here, however, the system of equations will not be linear, so an iterative process is required to solve it.

For the development of the procedure, we assume throughout that f satisfies the following conditions:

- f and the partial derivatives f_y and $f_{y'}$ are all continuous on

$$D = \{\, (x, y, y') \mid a \leq x \leq b, \text{ with } -\infty < y < \infty \text{ and } -\infty < y' < \infty \,\};$$

- $f_y(x, y, y') \geq \delta$ on D, for some $\delta > 0$;
- Constants k and L exist, with

$$k = \max_{(x,y,y') \in D} |f_y(x, y, y')| \quad \text{and} \quad L = \max_{(x,y,y') \in D} |f_{y'}(x, y, y')|.$$

This ensures, by Theorem 11.1, that a unique solution exists.

As in the linear case, we divide $[a, b]$ into $(N + 1)$ equal subintervals whose endpoints are at $x_i = a + ih$, for $i = 0, 1, \ldots, N + 1$. Assuming that the exact solution has a bounded fourth derivative allows us to replace $y''(x_i)$ and $y'(x_i)$ in each of the equations

$$y''(x_i) = f(x_i, y(x_i), y'(x_i))$$

by the appropriate centered-difference formula given in Eqs. (11.16) and (11.17) o page 685. This gives, for each $i = 1, 2, \ldots, N$,

$$\frac{y(x_{i+1}) - 2y(x_i) + y(x_{i-1})}{h^2} = f\left(x_i, y(x_i), \frac{y(x_{i+1}) - y(x_{i-1})}{2h} - \frac{h^2}{6}y'''(\eta_i)\right) + \frac{h^2}{12}y^{(4)}(\xi_i),$$

for some ξ_i and η_i in the interval (x_{i-1}, x_{i+1}).

As in the linear case, the difference method results from deleting the error terms and employing the boundary conditions:

$$w_0 = \alpha, \quad w_{N+1} = \beta,$$

and

$$-\frac{w_{i+1} - 2w_i + w_{i-1}}{h^2} + f\left(x_i, w_i, \frac{w_{i+1} - w_{i-1}}{2h}\right) = 0,$$

for each $i = 1, 2, \ldots, N$.

The $N \times N$ nonlinear system obtained from this method,

$$
\begin{aligned}
2w_1 - w_2 + h^2 f\left(x_1, w_1, \frac{w_2 - \alpha}{2h}\right) - \alpha &= 0,\\
-w_1 + 2w_2 - w_3 + h^2 f\left(x_2, w_2, \frac{w_3 - w_1}{2h}\right) &= 0,\\
&\ \vdots\\
-w_{N-2} + 2w_{N-1} - w_N + h^2 f\left(x_{N-1}, w_{N-1}, \frac{w_N - w_{N-2}}{2h}\right) &= 0,\\
-w_{N-1} + 2w_N + h^2 f\left(x_N, w_N, \frac{\beta - w_{N-1}}{2h}\right) - \beta &= 0
\end{aligned}
\tag{11.20}
$$

has a unique solution provided that $h < 2/L$, as shown in [Keller, H], p. 86.

Newton's Method for Iterations

We use Newton's method for nonlinear systems, discussed in Section 10.2, to approximate the solution to this system. A sequence of iterates $\{(w_1^{(k)}, w_2^{(k)}, \ldots, w_N^{(k)})^t\}$ is generated that converges to the solution of system (11.20), provided that the initial approximation $(w_1^{(0)}, w_2^{(0)}, \ldots, w_N^{(0)})^t$ is sufficiently close to the solution $(w_1, w_2, \ldots, w_N)^t$, and that the Jacobian matrix for the system is nonsingular. For system (11.20), the Jacobian matrix $J(w_1, \ldots, w_N)$ is tridiagonal with ij-th entry

$$
J(w_1, \ldots, w_N)_{ij} = \begin{cases}
-1 + \dfrac{h}{2} f_{y'}\left(x_i, w_i, \dfrac{w_{i+1} - w_{i-1}}{2h}\right), & \text{for } i = j - 1 \text{ and } j = 2, \ldots, N,\\[2ex]
2 + h^2 f_y\left(x_i, w_i, \dfrac{w_{i+1} - w_{i-1}}{2h}\right), & \text{for } i = j \text{ and } j = 1, \ldots, N,\\[2ex]
-1 - \dfrac{h}{2} f_{y'}\left(x_i, w_i, \dfrac{w_{i+1} - w_{i-1}}{2h}\right), & \text{for } i = j + 1 \text{ and } j = 1, \ldots, N - 1,
\end{cases}
$$

where $w_0 = \alpha$ and $w_{N+1} = \beta$.

Newton's method for nonlinear systems requires that at each iteration the $N \times N$ linear system

$$
\begin{aligned}
&J(w_1, \ldots, w_N)(v_1, \ldots, v_n)^t\\
&\quad = -\left(2w_1 - w_2 - \alpha + h^2 f\left(x_1, w_1, \frac{w_2 - \alpha}{2h}\right),\right.\\
&\qquad -w_1 + 2w_2 - w_3 + h^2 f\left(x_2, w_2, \frac{w_3 - w_1}{2h}\right), \ldots,\\
&\qquad -w_{N-2} + 2w_{N-1} - w_N + h^2 f\left(x_{N-1}, w_{N-1}, \frac{w_N - w_{N-2}}{2h}\right),\\
&\qquad \left. -w_{N-1} + 2w_N + h^2 f\left(x_N, w_N, \frac{\beta - w_{N-1}}{2h}\right) - \beta\right)^t
\end{aligned}
$$

be solved for $v_1, v_2, \ldots, v_N$, since

$$
w_i^{(k)} = w_i^{(k-1)} + v_i, \quad \text{for each } i = 1, 2, \ldots, N.
$$

Because J is tridiagonal this is not as formidable a problem as it might at first appear. In particular the Crout Factorization Algorithm 6.7 on page 424 can be applied. The process is detailed in Algorithm 11.4.

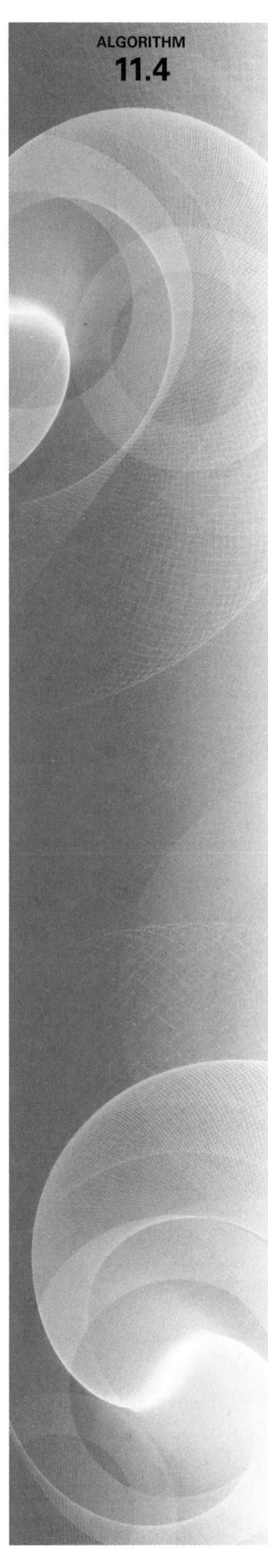

Nonlinear Finite-Difference

To approximate the solution to the nonlinear boundary-value problem

$$y'' = f(x, y, y'), \quad \text{for } a \le x \le b, \text{ with } y(a) = \alpha \text{ and } y(b) = \beta :$$

INPUT endpoints a, b; boundary conditions α, β; integer $N \ge 2$; tolerance TOL; maximum number of iterations M.

OUTPUT approximations w_i to $y(x_i)$ for each $i = 0, 1, \ldots, N+1$ or a message that the maximum number of iterations was exceeded.

Step 1 Set $h = (b-a)/(N+1)$;
$w_0 = \alpha$;
$w_{N+1} = \beta$.

Step 2 For $i = 1, \ldots, N$ set $w_i = \alpha + i\left(\dfrac{\beta-\alpha}{b-a}\right)h$.

Step 3 Set $k = 1$.

Step 4 While $k \le M$ do Steps 5–16.

Step 5 Set $x = a + h$;
$t = (w_2 - \alpha)/(2h)$;
$a_1 = 2 + h^2 f_y(x, w_1, t)$;
$b_1 = -1 + (h/2) f_{y'}(x, w_1, t)$;
$d_1 = -\big(2w_1 - w_2 - \alpha + h^2 f(x, w_1, t)\big)$.

Step 6 For $i = 2, \ldots, N-1$
set $x = a + ih$;
$t = (w_{i+1} - w_{i-1})/(2h)$;
$a_i = 2 + h^2 f_y(x, w_i, t)$;
$b_i = -1 + (h/2) f_{y'}(x, w_i, t)$;
$c_i = -1 - (h/2) f_{y'}(x, w_i, t)$;
$d_i = -\big(2w_i - w_{i+1} - w_{i-1} + h^2 f(x, w_i, t)\big)$.

Step 7 Set $x = b - h$;
$t = (\beta - w_{N-1})/(2h)$;
$a_N = 2 + h^2 f_y(x, w_N, t)$;
$c_N = -1 - (h/2) f_{y'}(x, w_N, t)$;
$d_N = -\big(2w_N - w_{N-1} - \beta + h^2 f(x, w_N, t)\big)$.

Step 8 Set $l_1 = a_1$; (*Steps* 8–12 *solve a tridiagonal linear system using Algorithm* 6.7.)
$u_1 = b_1/a_1$;
$z_1 = d_1/l_1$.

Step 9 For $i = 2, \ldots, N-1$ set $l_i = a_i - c_i u_{i-1}$;
$u_i = b_i/l_i$;
$z_i = (d_i - c_i z_{i-1})/l_i$.

Step 10 Set $l_N = a_N - c_N u_{N-1}$;
$z_N = (d_N - c_N z_{N-1})/l_N$.

Step 11 Set $v_N = z_N$;
$w_N = w_N + v_N$.

Step 12 For $i = N-1, \ldots, 1$ set $v_i = z_i - u_i v_{i+1}$;
$w_i = w_i + v_i$.

Step 13 If $\|\mathbf{v}\| \leq TOL$ then do Steps 14 and 15.

Step 14 For $i = 0, \ldots, N+1$ set $x = a + ih$;
OUTPUT (x, w_i).

Step 15 STOP. (*The procedure was successful.*)

Step 16 Set $k = k + 1$.

Step 17 OUTPUT ('Maximum number of iterations exceeded');
(*The procedure was unsuccessful.*)
STOP. ▪

It can be shown (see [IK], p. 433) that this Nonlinear Finite-Difference method is of order $O(h^2)$.

A good initial approximation is required when the satisfaction of the conditions given at the beginning of this presentation cannot be verified, so an upper bound for the number of iterations should be specified and, if exceeded, a new initial approximation or a reduction in step size considered. Unless contradictory information is available it is reasonable to begin the procedure by assuming that the solution is linear. So the initial approximations $w_i^{(0)}$ to w_i, for each $i = 1, 2, \ldots, N$, are obtained in Step 2 by passing a straight line through the known endpoints (a, α) and (b, β) and evaluating at x_i.

Example 1 Apply Algorithm 11.4, with $h = 0.1$, to the nonlinear boundary-value problem

$$y'' = \frac{1}{8}(32 + 2x^3 - yy'), \quad \text{for } 1 \leq x \leq 3, \text{ with } y(1) = 17 \text{ and } y(3) = \frac{43}{3},$$

and compare the results to those obtained in Example 1 of Section 11.2.

Solution The stopping procedure used in Algorithm 11.4 was to iterate until values of successive iterates differed by less than 10^{-8}. This was accomplished with four iterations. This gives the results in Table 11.5. They are less accurate than those obtained using the nonlinear shooting method, which gave results in the middle of the table accurate on the order of 10^{-5}. ▪

Employing Richardson's Extrapolation

Richardson's extrapolation procedure can also be used for the Nonlinear Finite-Difference method. Table 11.6 lists the results when this method is applied to our example using $h = 0.1$, 0.05, and 0.025, with four iterations in each case. The values of $w_i(h = 0.1)$ are omitted from the table to save space, but they are listed in Table 11.5. The values of $w_i(h = 0.25)$ are accurate to within about 1.5×10^{-4}. However, the values of Ext_{3i} are all accurate to the places listed, with an actual maximum error of 3.68×10^{-10}.

Table 11.5

x_i	w_i	$y(x_i)$	$\|w_i - y(x_i)\|$
1.0	17.000000	17.000000	
1.1	15.754503	15.755455	9.520×10^{-4}
1.2	14.771740	14.773333	1.594×10^{-3}
1.3	13.995677	13.997692	2.015×10^{-3}
1.4	13.386297	13.388571	2.275×10^{-3}
1.5	12.914252	12.916667	2.414×10^{-3}
1.6	12.557538	12.560000	2.462×10^{-3}
1.7	12.299326	12.301765	2.438×10^{-3}
1.8	12.126529	12.128889	2.360×10^{-3}
1.9	12.028814	12.031053	2.239×10^{-3}
2.0	11.997915	12.000000	2.085×10^{-3}
2.1	12.027142	12.029048	1.905×10^{-3}
2.2	12.111020	12.112727	1.707×10^{-3}
2.3	12.245025	12.246522	1.497×10^{-3}
2.4	12.425388	12.426667	1.278×10^{-3}
2.5	12.648944	12.650000	1.056×10^{-3}
2.6	12.913013	12.913846	8.335×10^{-4}
2.7	13.215312	13.215926	6.142×10^{-4}
2.8	13.553885	13.554286	4.006×10^{-4}
2.9	13.927046	13.927241	1.953×10^{-4}
3.0	14.333333	14.333333	

Table 11.6

x_i	$w_i(h = 0.05)$	$w_i(h = 0.025)$	Ext_{1i}	Ext_{2i}	Ext_{3i}
1.0	17.00000000	17.00000000	17.00000000	17.00000000	17.00000000
1.1	15.75521721	15.75539525	15.75545543	15.75545460	15.75545455
1.2	14.77293601	14.77323407	14.77333479	14.77333342	14.77333333
1.3	13.99718996	13.99756690	13.99769413	13.99769242	13.99769231
1.4	13.38800424	13.38842973	13.38857346	13.38857156	13.38857143
1.5	12.91606471	12.91651628	12.91666881	12.91666680	12.91666667
1.6	12.55938618	12.55984665	12.56000217	12.56000014	12.56000000
1.7	12.30115670	12.30161280	12.30176684	12.30176484	12.30176471
1.8	12.12830042	12.12874287	12.12899094	12.12888902	12.12888889
1.9	12.03049438	12.03091316	12.03105457	12.03105275	12.03105263
2.0	11.99948020	11.99987013	12.00000179	12.00000011	12.00000000
2.1	12.02857252	12.02892892	12.02902924	12.02904772	12.02904762
2.2	12.11230149	12.11262089	12.11272872	12.11272736	12.11272727
2.3	12.24614846	12.24642848	12.24652299	12.24652182	12.24652174
2.4	12.42634789	12.42658702	12.42666773	12.42666673	12.42666667
2.5	12.64973666	12.64993420	12.65000086	12.65000005	12.65000000
2.6	12.91362828	12.91379422	12.91384683	12.91384620	12.91384615
2.7	13.21577275	13.21588765	13.21592641	13.21592596	13.21592593
2.8	13.55418579	13.55426075	13.55428603	13.55428573	13.55428571
2.9	13.92719268	13.92722921	13.92724153	13.92724139	13.92724138
3.0	14.33333333	14.33333333	14.33333333	14.33333333	14.33333333

EXERCISE SET 11.4

1. Use the Nonlinear Finite-Difference method with $h = 0.5$ to approximate the solution to the boundary-value problem

$$y'' = -(y')^2 - y + \ln x, \quad 1 \le x \le 2, \quad y(1) = 0, \quad y(2) = \ln 2.$$

Compare your results to the actual solution $y = \ln x$.

2. Use the Nonlinear Finite-Difference method with $h = 0.25$ to approximate the solution to the boundary-value problem

$$y'' = 2y^3, \quad -1 \le x \le 0, \quad y(-1) = \frac{1}{2}, \quad y(0) = \frac{1}{3}$$

Compare your results to the actual solution $y(x) = 1/(x+3)$.

3. Use the Nonlinear Finite-Difference Algorithm with $TOL = 10^{-4}$ to approximate the solution to the following boundary-value problems. The actual solution is given for comparison to your results.
   a. $y'' = -e^{-2y}$, $1 \le x \le 2$, $y(1) = 0$, $y(2) = \ln 2$; use $N = 9$; actual solution $y(x) = \ln x$.
   b. $y'' = y' \cos x - y \ln y$, $0 \le x \le \frac{\pi}{2}$, $y(0) = 1$, $y\left(\frac{\pi}{2}\right) = e$; use $N = 9$; actual solution $y(x) = e^{\sin x}$.
   c. $y'' = -\left(2(y')^3 + y^2 y'\right) \sec x$, $\frac{\pi}{4} \le x \le \frac{\pi}{3}$, $y\left(\frac{\pi}{4}\right) = 2^{-1/4}$, $y\left(\frac{\pi}{3}\right) = \frac{1}{2}\sqrt[4]{12}$; use $N = 4$; actual solution $y(x) = \sqrt{\sin x}$.
   d. $y'' = \frac{1}{2}\left(1 - (y')^2 - y \sin x\right)$, $0 \le x \le \pi$, $y(0) = 2$, $y(\pi) = 2$; use $N = 19$; actual solution $y(x) = 2 + \sin x$.

4. Use the Nonlinear Finite-Difference Algorithm with $TOL = 10^{-4}$ to approximate the solution to the following boundary-value problems. The actual solution is given for comparison to your results.
   a. $y'' = y^3 - yy'$, $1 \le x \le 2$, $y(1) = \frac{1}{2}$, $y(2) = \frac{1}{3}$; use $h = 0.1$; actual solution $y(x) = (x+1)^{-1}$.
   b. $y'' = 2y^3 - 6y - 2x^3$, $1 \le x \le 2$, $y(1) = 2$, $y(2) = \frac{5}{2}$; use $h = 0.1$; actual solution $y(x) = x + x^{-1}$.
   c. $y'' = y' + 2(y - \ln x)^3 - x^{-1}$, $2 \le x \le 3$, $y(2) = \frac{1}{2} + \ln 2$, $y(3) = \frac{1}{3} + \ln 3$; use $h = 0.1$; actual solution $y(x) = x^{-1} + \ln x$.
   d. $y'' = (y')^2 x^{-3} - 9y^2 x^{-5} + 4x$, $1 \le x \le 2$, $y(1) = 0$, $y(2) = \ln 256$; use $h = 0.05$; actual solution $y(x) = x^3 \ln x$.

5. Repeat Exercise 4(a) and 4(b) using extrapolation.

6. In Exercise 7 of Section 11.3, the deflection of a beam with supported ends subject to uniform loading was approximated. Using a more appropriate representation of curvature gives the differential equation

$$[1 + (w'(x))^2]^{-3/2} w''(x) = \frac{S}{EI} w(x) + \frac{qx}{2EI}(x - l), \quad \text{for } 0 < x < l.$$

Approximate the deflection $w(x)$ of the beam every 6 in., and compare the results to those of Exercise 7 of Section 11.3.

7. Show that the hypotheses listed at the beginning of the section ensure the nonsingularity of the Jacobian matrix J for $h < 2/L$.

11.5 The Rayleigh-Ritz Method

John William Strutt Lord Rayleigh (1842–1919), a mathematical physicist who was particularly interested in wave propagation, received a Nobel Prize in physics in 1904.

The Shooting method for approximating the solution to a boundary-value problem replaced the boundary-value problem with pair of initial-value problems. The finite-difference approach replaces the continuous operation of differentiation with the discrete operation of finite differences. The Rayleigh-Ritz method is a variational technique that attacks the problem from a third approach. The boundary-value problem is first reformulated as a problem of choosing, from the set of all sufficiently differentiable functions satisfying the boundary

Walter Ritz (1878–1909), a theoretical physicist at Göttigen University, published a paper on a variational problem in 1909 [Ri]. He died of tuberculosis at the age of 31.

conditions, the function to minimize a certain integral. Then the set of feasible functions is reduced in size, and an approximation is found from this set to minimize the integral. This gives our approximation to the solution of the boundary-value problem.

To describe the Rayleigh-Ritz method, we consider approximating the solution to a linear two-point boundary-value problem from beam-stress analysis. This boundary-value problem is described by the differential equation

$$-\frac{d}{dx}\left(p(x)\frac{dy}{dx}\right) + q(x)y = f(x), \quad \text{for } 0 \le x \le 1, \tag{11.21}$$

with the boundary conditions

$$y(0) = y(1) = 0. \tag{11.22}$$

This differential equation describes the deflection $y(x)$ of a beam of length 1 with variable cross section represented by $q(x)$. The deflection is due to the added stresses $p(x)$ and $f(x)$. More general boundary conditions are considered in Exercises 6 and 9.

In the discussion that follows, we assume that $p \in C^1[0, 1]$ and $q, f \in C[0, 1]$. Further, we assume that there exists a constant $\delta > 0$ such that

$$p(x) \ge \delta, \quad \text{and that} \quad q(x) \ge 0, \quad \text{for each } x \text{ in } [0, 1].$$

These assumptions are sufficient to guarantee that the boundary-value problem given in (11.21) and (11.22) has a unique solution (see [BSW]).

Variational Problems

As is the case in many boundary-value problems that describe physical phenomena, the solution to the beam equation satisfies an integral minimization **variational** property. The variational principle for the beam equation is fundamental to the development of the Rayleigh-Ritz method and characterizes the solution to the beam equation as the function that minimizes an integral over all functions in $C_0^2[0, 1]$, the set of those functions u in $C^2[0, 1]$ with the property that $u(0) = u(1) = 0$. The following theorem gives the characterization.

Theorem 11.4 Let $p \in C^1[0, 1]$, $q, f \in C[0, 1]$, and

$$p(x) \ge \delta > 0, \quad q(x) \ge 0, \quad \text{for } 0 \le x \le 1.$$

The function $y \in C_0^2[0, 1]$ is the unique solution to the differential equation

$$-\frac{d}{dx}\left(p(x)\frac{dy}{dx}\right) + q(x)y = f(x), \quad \text{for } 0 \le x \le 1, \tag{11.23}$$

if and only if y is the unique function in $C_0^2[0, 1]$ that minimizes the integral

$$I[u] = \int_0^1 \{p(x)[u'(x)]^2 + q(x)[u(x)]^2 - 2f(x)u(x)\}\, dx. \tag{11.24}$$

■

Details of the proof of this theorem can be found in [Shul], pp. 88-89. It proceeds in three steps. First it is shown that any solution y to (11.23) also satisfies the equation

- $$\int_0^1 f(x)u(x)dx = \int_0^1 p(x)\frac{dy}{dx}(x)\frac{du}{dx}(x) + q(x)y(x)u(x)dx, \tag{11.25}$$

for all $u \in C_0^2[0.1]$.

- The second step shows that $y \in C_0^2[0,1]$ is a solution to (11.24) if and only if (11.25) holds for all $u \in C_0^2[0,1]$.

- The final step shows that (11.25) has a unique solution. This unique solution will also be a solution to (11.24) and to (11.23), so the solutions to (11.23) and (11.24) are identical.

The Rayleigh-Ritz method approximates the solution y by minimizing the integral, not over all the functions in $C_0^2[0,1]$, but over a smaller set of functions consisting of linear combinations of certain basis functions $\phi_1, \phi_2, \ldots, \phi_n$. The basis functions are linearly independent and satisfy

$$\phi_i(0) = \phi_i(1) = 0, \qquad \text{for each } i = 1, 2, \ldots, n.$$

An approximation $\phi(x) = \sum_{i=1}^n c_i\phi_i(x)$ to the solution $y(x)$ of Eq. (11.23) is then obtained by finding constants $c_1, c_2, \ldots, c_n$ to minimize the integral $I\left[\sum_{i=1}^n c_i\phi_i\right]$.

From Eq. (11.24),

$$I[\phi] = I\left[\sum_{i=1}^n c_i\phi_i\right] \tag{11.26}$$
$$= \int_0^1 \left\{p(x)\left[\sum_{i=1}^n c_i\phi_i'(x)\right]^2 + q(x)\left[\sum_{i=1}^n c_i\phi_i(x)\right]^2 - 2f(x)\sum_{i=1}^n c_i\phi_i(x)\right\} dx,$$

and, for a minimum to occur, it is necessary, when considering I as a function of $c_1, c_2, \ldots, c_n$, to have

$$\frac{\partial I}{\partial c_j} = 0, \qquad \text{for each } j = 1, 2, \ldots, n. \tag{11.27}$$

Differentiating (11.26) gives

$$\frac{\partial I}{\partial c_j} = \int_0^1 \left\{2p(x)\sum_{i=1}^n c_i\phi_i'(x)\phi_j'(x) + 2q(x)\sum_{i=1}^n c_i\phi_i(x)\phi_j(x) - 2f(x)\phi_j(x)\right\} dx,$$

and substituting into Eq. (11.27) yields

$$0 = \sum_{i=1}^n \left[\int_0^1 \{p(x)\phi_i'(x)\phi_j'(x) + q(x)\phi_i(x)\phi_j(x)\}\, dx\right] c_i - \int_0^1 f(x)\phi_j(x)\, dx, \tag{11.28}$$

for each $j = 1, 2, \ldots, n$.

The **normal equations** described in Eq. (11.28) produce an $n \times n$ linear system $A\mathbf{c} = \mathbf{b}$ in the variables $c_1, c_2, \ldots, c_n$, where the symmetric matrix A has

$$a_{ij} = \int_0^1 [p(x)\phi_i'(x)\phi_j'(x) + q(x)\phi_i(x)\phi_j(x)]\, dx,$$

and $\mathbf{b}$ is defined by

$$b_i = \int_0^1 f(x)\phi_i(x)\, dx.$$

Piecewise-Linear Basis

The simplest choice of basis functions involves piecewise-linear polynomials. The first step is to form a partition of $[0, 1]$ by choosing points $x_0, x_1, \ldots, x_{n+1}$ with

$$0 = x_0 < x_1 < \cdots < x_n < x_{n+1} = 1.$$

Letting $h_i = x_{i+1} - x_i$, for each $i = 0, 1, \ldots, n$, we define the basis functions $\phi_1(x)$, $\phi_2(x), \ldots, \phi_n(x)$ by

$$\phi_i(x) = \begin{cases} 0, & \text{if } 0 \le x \le x_{i-1}, \\ \dfrac{1}{h_{i-1}}(x - x_{i-1}), & \text{if } x_{i-1} < x \le x_i, \\ \dfrac{1}{h_i}(x_{i+1} - x), & \text{if } x_i < x \le x_{i+1}, \\ 0, & \text{if } x_{i+1} < x \le 1, \end{cases} \tag{11.29}$$

for each $i = 1, 2, \ldots, n$. (See Figure 11.4.)

Figure 11.4

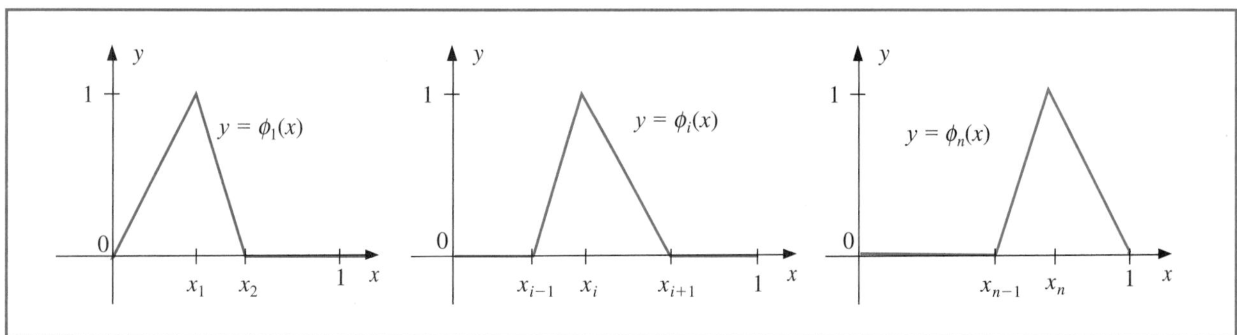

The functions ϕ_i are piecewise-linear, so the derivatives ϕ_i', while not continuous, are constant on (x_j, x_{j+1}), for each $j = 0, 1, \ldots, n$, and

$$\phi_i'(x) = \begin{cases} 0, & \text{if } 0 < x < x_{i-1}, \\ \dfrac{1}{h_{i-1}}, & \text{if } x_{i-1} < x < x_i, \\ -\dfrac{1}{h_i}, & \text{if } x_i < x < x_{i+1}, \\ 0, & \text{if } x_{i+1} < x < 1, \end{cases} \tag{11.30}$$

for each $i = 1, 2, \ldots, n$.

Because ϕ_i and ϕ_i' are nonzero only on (x_{i-1}, x_{i+1}),

$$\phi_i(x)\phi_j(x) \equiv 0 \qquad \text{and} \qquad \phi_i'(x)\phi_j'(x) \equiv 0,$$

except when j is $i-1$, i, or $i+1$. As a consequence, the linear system given by (11.28) reduces to an $n \times n$ tridiagonal linear system. The nonzero entries in A are

$$\begin{aligned} a_{ii} &= \int_0^1 \left\{p(x)[\phi_i'(x)]^2 + q(x)[\phi_i(x)]^2\right\} dx \\ &= \left(\frac{1}{h_{i-1}}\right)^2 \int_{x_{i-1}}^{x_i} p(x)\,dx + \left(\frac{-1}{h_i}\right)^2 \int_{x_i}^{x_{i+1}} p(x)\,dx \\ &\quad + \left(\frac{1}{h_{i-1}}\right)^2 \int_{x_{i-1}}^{x_i} (x-x_{i-1})^2 q(x)\,dx + \left(\frac{1}{h_i}\right)^2 \int_{x_i}^{x_{i+1}} (x_{i+1}-x)^2 q(x)\,dx, \end{aligned}$$

for each $i = 1, 2, \ldots, n$;

$$\begin{aligned} a_{i,i+1} &= \int_0^1 \{p(x)\phi_i'(x)\phi_{i+1}'(x) + q(x)\phi_i(x)\phi_{i+1}(x)\}\,dx \\ &= -\left(\frac{1}{h_i}\right)^2 \int_{x_i}^{x_{i+1}} p(x)\,dx + \left(\frac{1}{h_i}\right)^2 \int_{x_i}^{x_{i+1}} (x_{i+1}-x)(x-x_i)q(x)\,dx, \end{aligned}$$

for each $i = 1, 2, \ldots, n-1$; and

$$\begin{aligned} a_{i,i-1} &= \int_0^1 \{p(x)\phi_i'(x)\phi_{i-1}'(x) + q(x)\phi_i(x)\phi_{i-1}(x)\}\,dx \\ &= -\left(\frac{1}{h_{i-1}}\right)^2 \int_{x_{i-1}}^{x_i} p(x)\,dx + \left(\frac{1}{h_{i-1}}\right)^2 \int_{x_{i-1}}^{x_i} (x_i-x)(x-x_{i-1})q(x)\,dx, \end{aligned}$$

for each $i = 2, \ldots, n$. The entries in $\mathbf{b}$ are

$$b_i = \int_0^1 f(x)\phi_i(x)\,dx = \frac{1}{h_{i-1}} \int_{x_{i-1}}^{x_i} (x-x_{i-1})f(x)\,dx + \frac{1}{h_i} \int_{x_i}^{x_{i+1}} (x_{i+1}-x)f(x)\,dx,$$

for each $i = 1, 2, \ldots, n$.

There are six types of integrals to be evaluated:

$$Q_{1,i} = \left(\frac{1}{h_i}\right)^2 \int_{x_i}^{x_{i+1}} (x_{i+1}-x)(x-x_i)q(x)\,dx, \quad \text{for each } i = 1, 2, \ldots, n-1,$$

$$Q_{2,i} = \left(\frac{1}{h_{i-1}}\right)^2 \int_{x_{i-1}}^{x_i} (x-x_{i-1})^2 q(x)\,dx, \quad \text{for each } i = 1, 2, \ldots, n,$$

$$Q_{3,i} = \left(\frac{1}{h_i}\right)^2 \int_{x_i}^{x_{i+1}} (x_{i+1}-x)^2 q(x)\,dx, \quad \text{for each } i = 1, 2, \ldots, n,$$

$$Q_{4,i} = \left(\frac{1}{h_{i-1}}\right)^2 \int_{x_{i-1}}^{x_i} p(x)\,dx, \quad \text{for each } i = 1, 2, \ldots, n+1,$$

$$Q_{5,i} = \frac{1}{h_{i-1}} \int_{x_{i-1}}^{x_i} (x-x_{i-1})f(x)\,dx, \quad \text{for each } i = 1, 2, \ldots, n,$$

and

$$Q_{6,i} = \frac{1}{h_i} \int_{x_i}^{x_{i+1}} (x_{i+1}-x)f(x)\,dx, \quad \text{for each } i = 1, 2, \ldots, n.$$

The matrix A and the vector $\mathbf{b}$ in the linear system $A\mathbf{c} = \mathbf{b}$ have the entries

$$a_{i,i} = Q_{4,i} + Q_{4,i+1} + Q_{2,i} + Q_{3,i}, \quad \text{for each } i = 1, 2, \ldots, n,$$

$$a_{i,i+1} = -Q_{4,i+1} + Q_{1,i}, \quad \text{for each } i = 1, 2, \ldots, n-1,$$

$$a_{i,i-1} = -Q_{4,i} + Q_{1,i-1}, \quad \text{for each } i = 2, 3, \ldots, n,$$

and

$$b_i = Q_{5,i} + Q_{6,i}, \quad \text{for each } i = 1, 2, \ldots, n.$$

The entries in $\mathbf{c}$ are the unknown coefficients $c_1, c_2, \ldots, c_n$, from which the Rayleigh-Ritz approximation ϕ, given by $\phi(x) = \sum_{i=1}^{n} c_i\phi_i(x)$, is constructed.

To employ this method requires evaluating $6n$ integrals, which can be evaluated either directly or by a quadrature formula such as Composite Simpson's rule.

An alternative approach for the integral evaluation is to approximate each of the functions p, q, and f with its piecewise-linear interpolating polynomial and then integrate the approximation. Consider, for example, the integral $Q_{1,i}$. The piecewise-linear interpolation of q is

$$P_q(x) = \sum_{i=0}^{n+1} q(x_i)\phi_i(x),$$

where $\phi_1, \ldots, \phi_n$ are defined in (11.30) and

$$\phi_0(x) = \begin{cases} \dfrac{x_1 - x}{x_1}, & \text{if } \ 0 \le x \le x_1 \\ 0, & \text{elsewhere} \end{cases} \quad \text{and} \quad \phi_{n+1}(x) = \begin{cases} \dfrac{x - x_n}{1 - x_n}, & \text{if } \ x_n \le x \le 1 \\ 0, & \text{elsewhere.} \end{cases}$$

The interval of integration is $[x_i, x_{i+1}]$, so the piecewise polynomial $P_q(x)$ reduces to

$$P_q(x) = q(x_i)\phi_i(x) + q(x_{i+1})\phi_{i+1}(x).$$

This is the first-degree interpolating polynomial studied in Section 3.1. By Theorem 3.3 on page 112,

$$|q(x) - P_q(x)| = O(h_i^2), \quad \text{for } x_i \le x \le x_{i+1},$$

if $q \in C^2[x_i, x_{i+1}]$. For $i = 1, 2, \ldots, n-1$, the approximation to $Q_{1,i}$ is obtained by integrating the approximation to the integrand

$$\begin{aligned} Q_{1,i} &= \left(\frac{1}{h_i}\right)^2 \int_{x_i}^{x_{i+1}} (x_{i+1} - x)(x - x_i)q(x)\, dx \\ &\approx \left(\frac{1}{h_i}\right)^2 \int_{x_i}^{x_{i+1}} (x_{i+1} - x)(x - x_i)\left[\frac{q(x_i)(x_{i+1} - x)}{h_i} + \frac{q(x_{i+1})(x - x_i)}{h_i}\right] dx \\ &= \frac{h_i}{12}[q(x_i) + q(x_{i+1})]. \end{aligned}$$

Further, if $q \in C^2[x_i, x_{i+1}]$, then

$$\left| Q_{1,i} - \frac{h_i}{12}[q(x_i) + q(x_{i+1})] \right| = O(h_i^3).$$

Approximations to the other integrals are derived in a similar manner and are given by

$$Q_{2,i} \approx \frac{h_{i-1}}{12}[3q(x_i) + q(x_{i-1})], \qquad Q_{3,i} \approx \frac{h_i}{12}[3q(x_i) + q(x_{i+1})],$$

$$Q_{4,i} \approx \frac{h_{i-1}}{2}[p(x_i) + p(x_{i-1})], \qquad Q_{5,i} \approx \frac{h_{i-1}}{6}[2f(x_i) + f(x_{i-1})],$$

and

$$Q_{6,i} \approx \frac{h_i}{6}[2f(x_i) + f(x_{i+1})].$$

Algorithm 11.5 sets up the tridiagonal linear system and incorporates the Crout Factorization Algorithm 6.7 to solve the system. The integrals $Q_{1,i}, \ldots, Q_{6,i}$ can be computed by one of the methods mentioned previously.

Piecewise Linear Rayleigh-Ritz

To approximate the solution to the boundary-value problem

$$-\frac{d}{dx}\left(p(x)\frac{dy}{dx}\right) + q(x)y = f(x), \quad \text{for } 0 \le x \le 1, \text{ with } y(0) = 0 \text{ and } y(1) = 0$$

with the piecewise linear function

$$\phi(x) = \sum_{i=1}^{n} c_i\phi_i(x):$$

INPUT integer $n \ge 1$; points $x_0 = 0 < x_1 < \cdots < x_n < x_{n+1} = 1$.

OUTPUT coefficients $c_1, \ldots, c_n$.

Step 1 For $i = 0, \ldots, n$ set $h_i = x_{i+1} - x_i$.

Step 2 For $i = 1, \ldots, n$ define the piecewise linear basis ϕ_i by

$$\phi_i(x) = \begin{cases} 0, & 0 \le x \le x_{i-1}, \\ \dfrac{x - x_{i-1}}{h_{i-1}}, & x_{i-1} < x \le x_i, \\ \dfrac{x_{i+1} - x}{h_i}, & x_i < x \le x_{i+1}, \\ 0, & x_{i+1} < x \le 1. \end{cases}$$

Step 3 For each $i = 1, 2, \ldots, n-1$ compute $Q_{1,i}, Q_{2,i}, Q_{3,i}, Q_{4,i}, Q_{5,i}, Q_{6,i}$;
Compute $Q_{2,n}, Q_{3,n}, Q_{4,n}, Q_{4,n+1}, Q_{5,n}, Q_{6,n}$.

Step 4 For each $i = 1, 2, \ldots, n-1$, set $\alpha_i = Q_{4,i} + Q_{4,i+1} + Q_{2,i} + Q_{3,i}$;
$\beta_i = Q_{1,i} - Q_{4,i+1}$;
$b_i = Q_{5,i} + Q_{6,i}$.

Step 5 Set $\alpha_n = Q_{4,n} + Q_{4,n+1} + Q_{2,n} + Q_{3,n}$;
$b_n = Q_{5,n} + Q_{6,n}$.

Step 6 Set $a_1 = \alpha_1$; (*Steps* 6–10 *solve a symmetric tridiagonal linear system using Algorithm* 6.7.)

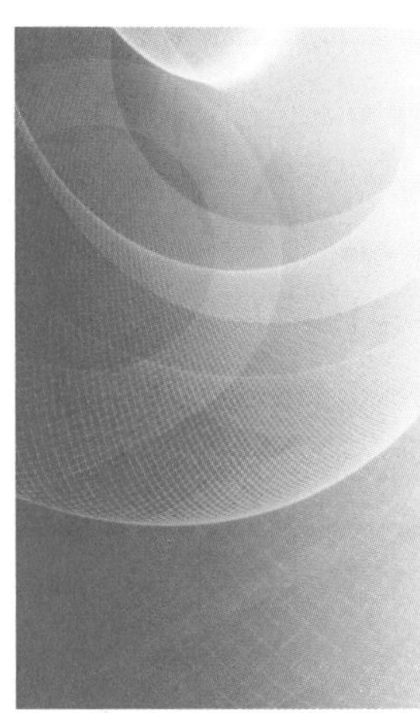

$\zeta_1 = \beta_1/\alpha_1;$
$z_1 = b_1/a_1.$

Step 7 For $i = 2, \ldots, n-1$ set $a_i = \alpha_i - \beta_{i-1}\zeta_{i-1};$
$\zeta_i = \beta_i/a_i;$
$z_i = (b_i - \beta_{i-1}z_{i-1})/a_i.$

Step 8 Set $a_n = \alpha_n - \beta_{n-1}\zeta_{n-1};$
$z_n = (b_n - \beta_{n-1}z_{n-1})/a_n.$

Step 9 Set $c_n = z_n;$
OUTPUT (c_n).

Step 10 For $i = n-1, \ldots, 1$ set $c_i = z_i - \zeta_i c_{i+1};$
OUTPUT (c_i).

Step 11 STOP. (*The procedure is complete.*) ■

The following uses Algorithm 11.5. Because of the elementary nature of this example, the integrals in Steps 3, 4, and 5 were found directly.

Illustration Consider the boundary-value problem

$$-y'' + \pi^2 y = 2\pi^2 \sin(\pi x), \quad \text{for } 0 \le x \le 1, \text{ with } y(0) = y(1) = 0.$$

Let $h_i = h = 0.1$, so that $x_i = 0.1i$, for each $i = 0, 1, \ldots, 9$. The integrals are

$$Q_{1,i} = 100 \int_{0.1i}^{0.1i+0.1} (0.1i + 0.1 - x)(x - 0.1i)\pi^2\, dx = \frac{\pi^2}{60},$$

$$Q_{2,i} = 100 \int_{0.1i-0.1}^{0.1i} (x - 0.1i + 0.1)^2\pi^2\, dx = \frac{\pi^2}{30},$$

$$Q_{3,i} = 100 \int_{0.1i}^{0.1i+0.1} (0.1i + 0.1 - x)^2\pi^2\, dx = \frac{\pi^2}{30},$$

$$Q_{4,i} = 100 \int_{0.1i-0.1}^{0.1i} dx = 10,$$

$$Q_{5,i} = 10 \int_{0.1i-0.1}^{0.1i} (x - 0.1i + 0.1)2\pi^2 \sin \pi x\, dx$$
$$= -2\pi \cos 0.1\pi i + 20[\sin(0.1\pi i) - \sin((0.1i - 0.1)\pi)],$$

and

$$Q_{6,i} = 10 \int_{0.1i}^{0.1i+0.1} (0.1i + 0.1 - x)2\pi^2 \sin \pi x\, dx$$
$$= 2\pi \cos 0.1\pi i - 20[\sin((0.1i + 0.1)\pi) - \sin(0.1\pi i)].$$

The linear system $A\mathbf{c} = \mathbf{b}$ has

$$a_{i,i} = 20 + \frac{\pi^2}{15}, \qquad \text{for each } i = 1, 2, \ldots, 9,$$

$$a_{i,i+1} = -10 + \frac{\pi^2}{60}, \qquad \text{for each } i = 1, 2, \ldots, 8,$$

$$a_{i,i-1} = -10 + \frac{\pi^2}{60}, \qquad \text{for each } i = 2, 3, \ldots, 9,$$

and

$$b_i = 40 \sin(0.1\pi i)[1 - \cos 0.1\pi], \qquad \text{for each } i = 1, 2, \ldots, 9.$$

The solution to the tridiagonal linear system is

$$c_9 = 0.3102866742, \; c_8 = 0.5902003271, \; c_7 = 0.8123410598,$$
$$c_6 = 0.9549641893, \; c_5 = 1.004108771, \; c_4 = 0.9549641893,$$
$$c_3 = 0.8123410598, \; c_2 = 0.5902003271, \; c_1 = 0.3102866742.$$

The piecewise-linear approximation is

$$\phi(x) = \sum_{i=1}^{9} c_i \phi_i(x),$$

and the actual solution to the boundary-value problem is $y(x) = \sin \pi x$. Table 11.7 lists the error in the approximation at x_i, for each $i = 1, \ldots, 9$. □

Table 11.7

i	x_i	$\phi(x_i)$	$y(x_i)$	$\lvert\phi(x_i) - y(x_i)\rvert$
1	0.1	0.3102866742	0.3090169943	0.00127
2	0.2	0.5902003271	0.5877852522	0.00241
3	0.3	0.8123410598	0.8090169943	0.00332
4	0.4	0.9549641896	0.9510565162	0.00390
5	0.5	1.0041087710	1.0000000000	0.00411
6	0.6	0.9549641893	0.9510565162	0.00390
7	0.7	0.8123410598	0.8090169943	0.00332
8	0.8	0.5902003271	0.5877852522	0.00241
9	0.9	0.3102866742	0.3090169943	0.00127

It can be shown that the tridiagonal matrix A given by the piecewise-linear basis functions is positive definite (see Exercise 12), so, by Theorem 6.26 on page 417, the linear system is stable with respect to roundoff error. Under the hypotheses presented at the beginning of this section, we have

$$|\phi(x) - y(x)| = O(h^2), \qquad \text{for each } x \text{ in } [0, 1].$$

A proof of this result can be found in [Schul], pp. 103–104.

B-Spline Basis

The use of piecewise-linear basis functions results in an approximate solution to Eqs. (11.22) and (11.23) that is continuous but not differentiable on [0, 1]. A more sophisticated set of basis functions is required to construct an approximation that belongs to $C_0^2[0, 1]$. These basis functions are similar to the cubic interpolatory splines discussed in Section 3.5.

Recall that the cubic *interpolatory* spline S on the five nodes x_0, x_1, x_2, x_3, and x_4 for a function f is defined by:

(a) $S(x)$ is a cubic polynomial, denoted $S_j(x)$, on the subinterval $[x_j, x_{j+1}]$ for each $j = 0, 1, 2, 3$;

(b) $S_j(x_j) = f(x_j)$ and $S_j(x_{j+1}) = f(x_{j+1})$ for each $j = 0, 1, 2, 3$;

(c) $S_{j+1}(x_{j+1}) = S_j(x_{j+1})$ for each $j = 0, 1, 2$; (*Implied by* **(b)**.)

(d) $S'_{j+1}(x_{j+1}) = S'_j(x_{j+1})$ for each $j = 0, 1, 2$;

(e) $S''_{j+1}(x_{j+1}) = S''_j(x_{j+1})$ for each $j = 0, 1, 2$;

(f) One of the following sets of boundary conditions is satisfied:

(i) $S''(x_0) = S''(x_n) = 0$ *(natural (or free) boundary)*;

(ii) $S'(x_0) = f'(x_0)$ and $S'(x_n) = f'(x_n)$ *(clamped boundary)*.

Since uniqueness of solution requires the number of constants in **(a)**, 16, to equal the number of conditions in **(b)** through **(f)**, only one of the boundary conditions in **(f)** can be specified for the interpolatory cubic splines.

B- (for "Basis") splines were introduced in 1946 by I. J. Schoenberg [Scho], but for more than a decade were difficult to compute. In 1972, Carl de Boor (1937–) [Deb1] described recursion formulae for evaluation which improved their stability and utility.

The cubic spline functions we will use for our basis functions are called **B-splines**, or *bell-shaped splines*. These differ from interpolatory splines in that both sets of boundary conditions in **(f)** are satisfied. This requires the relaxation of two of the conditions in **(b)** through **(e)**. Since the spline must have two continuous derivatives on $[x_0, x_4]$, we delete two of the interpolation conditions from the description of the interpolatory splines. In particular, we modify condition **(b)** to

b. $S(x_j) = f(x_j)$ for $j = 0, 2, 4$.

For example, the basic B-spline S defined next and shown in Figure 11.5 uses the equally spaced nodes $x_0 = -2$, $x_1 = -1$, $x_2 = 0$, $x_3 = 1$, and $x_4 = 2$. It satisfies the interpolatory conditions

b. $S(x_0) = 0, \quad S(x_2) = 1, \quad S(x_4) = 0$;

as well as both sets of conditions

(i) $S''(x_0) = S''(x_4) = 0$ and **(ii)** $S'(x_0) = S'(x_4) = 0$.

Figure 11.5

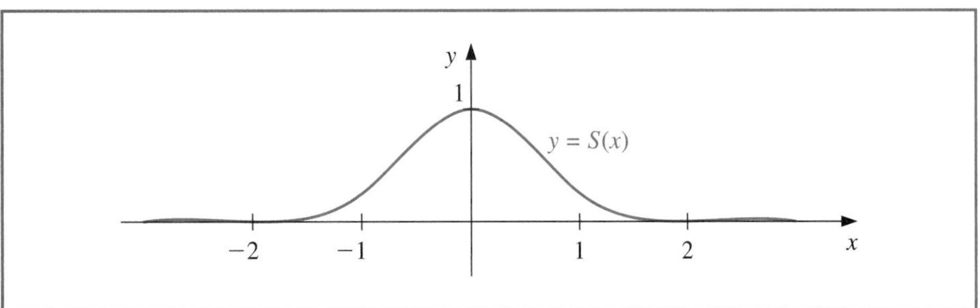

As a consequence, $S \in C_0^2(-\infty, \infty)$, and is given specifically as

$$S(x) = \begin{cases} 0, & \text{if } x \le -2, \\ \frac{1}{4}(2+x)^3, & \text{if } -2 \le x \le -1, \\ \frac{1}{4}\left[(2+x)^3 - 4(1+x)^3\right], & \text{if } -1 < x \le 0, \\ \frac{1}{4}\left[(2-x)^3 - 4(1-x)^3\right], & \text{if } 0 < x \le 1, \\ \frac{1}{4}(2-x)^3, & \text{if } 1 < x \le 2, \\ 0, & \text{if } 2 < x. \end{cases} \tag{11.31}$$

We will now use this basic B-spline to construct the basis functions ϕ_i in $C_0^2[0, 1]$. We first partition $[0, 1]$ by choosing a positive integer n and defining $h = 1/(n + 1)$. This produces the equally-spaced nodes $x_i = ih$, for each $i = 0, 1, \ldots, n + 1$. We then define the basis functions $\{\phi_i\}_{i=0}^{n+1}$ as

$$\phi_i(x) = \begin{cases} S\left(\dfrac{x}{h}\right) - 4S\left(\dfrac{x+h}{h}\right), & \text{if } i = 0, \\[2ex] S\left(\dfrac{x-h}{h}\right) - S\left(\dfrac{x+h}{h}\right), & \text{if } i = 1, \\[2ex] S\left(\dfrac{x-ih}{h}\right), & \text{if } 2 \le i \le n-1, \\[2ex] S\left(\dfrac{x-nh}{h}\right) - S\left(\dfrac{x-(n+2)h}{h}\right), & \text{if } i = n, \\[2ex] S\left(\dfrac{x-(n+1)h}{h}\right) - 4S\left(\dfrac{x-(n+2)h}{h}\right), & \text{if } i = n+1. \end{cases}$$

It is not difficult to show that $\{\phi_i\}_{i=0}^{n+1}$ is a linearly independent set of cubic splines satisfying $\phi_i(0) = \phi_i(1) = 0$, for each $i = 0, 1, \ldots, n, n + 1$ (see Exercise 11). The graphs of ϕ_i, for $2 \le i \le n - 1$, are shown in Figure 11.6, and the graphs of ϕ_0, ϕ_1, ϕ_n, and ϕ_{n+1} are in Figure 11.7.

Figure 11.6

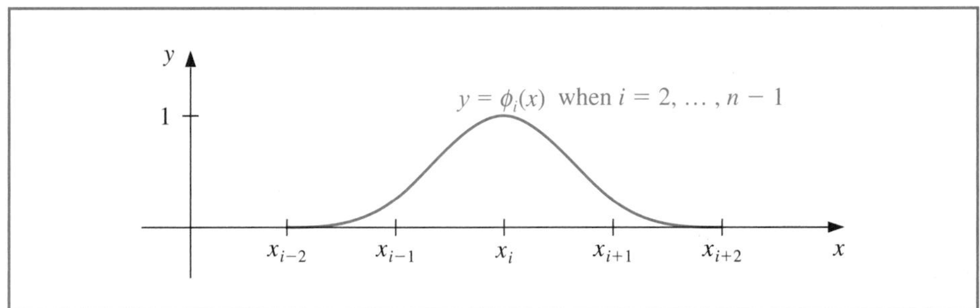

Since $\phi_i(x)$ and $\phi_i'(x)$ are nonzero only for $x \in [x_{i-2}, x_{i+2}]$, the matrix in the Rayleigh-Ritz approximation is a band matrix with bandwidth at most seven:

$$A = \begin{bmatrix} a_{00} & a_{01} & a_{02} & a_{03} & 0 & \cdots & \cdots & \cdots & 0 \\ a_{10} & a_{11} & a_{12} & a_{13} & a_{14} & \ddots & & & \vdots \\ a_{20} & a_{21} & a_{22} & a_{23} & a_{24} & a_{25} & \ddots & & \vdots \\ a_{30} & a_{31} & a_{32} & a_{33} & a_{34} & a_{35} & a_{36} & \ddots & \vdots \\ 0 & \ddots & \ddots & \ddots & \ddots & \ddots & \ddots & \ddots & 0 \\ \vdots & \ddots & \ddots & \ddots & \ddots & \ddots & \ddots & \ddots & a_{n-2,n+1} \\ \vdots & & \ddots & \ddots & \ddots & \ddots & \ddots & \ddots & a_{n-1,n+1} \\ \vdots & & & \ddots & \ddots & \ddots & \ddots & \ddots & a_{n,n+1} \\ 0 & \cdots & \cdots & \cdots & 0 & a_{n+1,n-2} & a_{n+1,n-1} & a_{n+1,n} & a_{n+1,n+1} \end{bmatrix}, \tag{11.32}$$

Figure 11.7

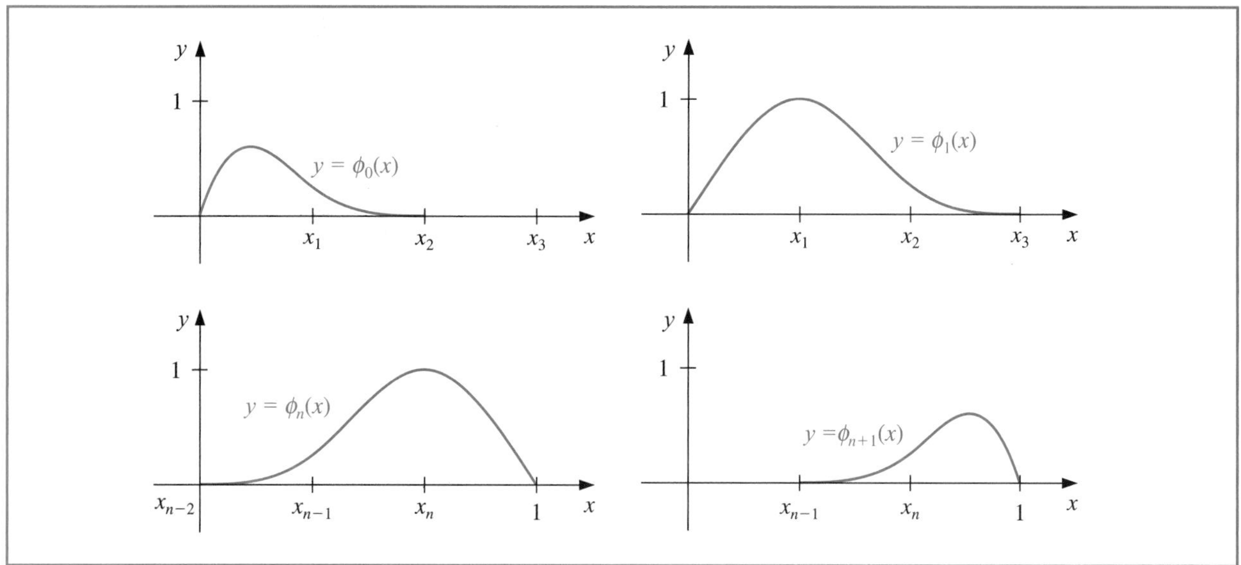

where

$$a_{ij} = \int_0^1 \{p(x)\phi_i'(x)\phi_j'(x) + q(x)\phi_i(x)\phi_j(x)\}\, dx,$$

for each $i, j = 0, 1, \ldots, n+1$. The vector $\mathbf{b}$ has the entries

$$b_i = \int_0^1 f(x)\phi_i(x)dx.$$

The matrix A is positive definite (see Exercise 13), so the linear system $A\mathbf{c} = \mathbf{b}$ can be solved by Cholesky's Algorithm 6.6 or by Gaussian elimination. Algorithm 11.6 details the construction of the cubic spline approximation $\phi(x)$ by the Rayleigh-Ritz method for the boundary-value problem (11.21) and (11.22) given at the beginning of this section.

Cubic Spline Rayleigh-Ritz

To approximate the solution to the boundary-value problem

$$-\frac{d}{dx}\left(p(x)\frac{dy}{dx}\right) + q(x)y = f(x), \quad \text{for } 0 \le x \le 1, \text{ with } y(0) = 0 \text{ and } y(1) = 0$$

with the sum of cubic splines

$$\phi(x) = \sum_{i=0}^{n+1} c_i\phi_i(x):$$

INPUT integer $n \ge 1$.

OUTPUT coefficients $c_0, \ldots, c_{n+1}$.

Step 1 Set $h = 1/(n+1)$.

Step 2 For $i = 0, \ldots, n+1$ set $x_i = ih$.
Set $x_{-2} = x_{-1} = 0$; $x_{n+2} = x_{n+3} = 1$.

Step 3 Define the function S by

$$S(x) = \begin{cases} 0, & x \le -2, \\ \frac{1}{4}(2+x)^3, & -2 < x \le -1, \\ \frac{1}{4}\left[(2+x)^3 - 4(1+x)^3\right], & -1 < x \le 0, \\ \frac{1}{4}\left[(2-x)^3 - 4(1-x)^3\right], & 0 < x \le 1, \\ \frac{1}{4}(2-x)^3, & 1 < x \le 2, \\ 0, & 2 < x \end{cases}$$

Step 4 Define the cubic spline basis $\{\phi_i\}_{i=0}^{n+1}$ by

$$\phi_0(x) = S\left(\frac{x}{h}\right) - 4S\left(\frac{x+h}{h}\right),$$

$$\phi_1(x) = S\left(\frac{x-x_1}{h}\right) - S\left(\frac{x+h}{h}\right),$$

$$\phi_i(x) = S\left(\frac{x-x_i}{h}\right), \text{ for } i = 2, \ldots, n-1,$$

$$\phi_n(x) = S\left(\frac{x-x_n}{h}\right) - S\left(\frac{x-(n+2)h}{h}\right),$$

$$\phi_{n+1}(x) = S\left(\frac{x-x_{n+1}}{h}\right) - 4S\left(\frac{x-(n+2)h}{h}\right).$$

Step 5 For $i = 0, \ldots, n+1$ do Steps 6–9.
(*Note: The integrals in Steps* 6 *and* 9 *can be evaluated using a numerical integration procedure.*)

Step 6 For $j = i, i+1, \ldots, \min\{i+3, n+1\}$
set $L = \max\{x_{j-2}, 0\}$;
$U = \min\{x_{i+2}, 1\}$;
$a_{ij} = \int_L^U \left[p(x)\phi_i'(x)\phi_j'(x) + q(x)\phi_i(x)\phi_j(x)\right] dx$;
if $i \ne j$, then set $a_{ji} = a_{ij}$. (*Since A is symmetric.*)

Step 7 If $i \ge 4$ then for $j = 0, \ldots, i-4$ set $a_{ij} = 0$.

Step 8 If $i \le n-3$ then for $j = i+4, \ldots, n+1$ set $a_{ij} = 0$.

Step 9 Set $L = \max\{x_{i-2}, 0\}$;
$U = \min\{x_{i+2}, 1\}$;
$b_i = \int_L^U f(x)\phi_i(x)\, dx$.

Step 10 Solve the linear system $A\mathbf{c} = \mathbf{b}$, where $A = (a_{ij})$, $\mathbf{b} = (b_0, \ldots, b_{n+1})^t$ and $\mathbf{c} = (c_0, \ldots, c_{n+1})^t$.

Step 11 For $i = 0, \ldots, n+1$
OUTPUT (c_i).

Step 12 STOP. (*The procedure is complete.*) ■

Illustration Consider the boundary-value problem

$$-y'' + \pi^2 y = 2\pi^2 \sin(\pi x), \quad \text{for } 0 \le x \le 1, \text{ with } y(0) = y(1) = 0.$$

In the Illustration following Algorithm 11.5 we let $h = 0.1$ and generated approximations using piecewise-linear basis functions. Table 11.8 lists the results obtained by applying the B-splines as detailed in Algorithm 11.6 with this same choice of nodes. □

Table 11.8

i	c_i	x_i	$\phi(x_i)$	$y(x_i)$	$\lvert y(x_i) - \phi(x_i)\rvert$
0	$0.50964361 \times 10^{-5}$	0	0.00000000	0.00000000	0.00000000
1	0.20942608	0.1	0.30901644	0.30901699	0.00000055
2	0.39835678	0.2	0.58778549	0.58778525	0.00000024
3	0.54828946	0.3	0.80901687	0.80901699	0.00000012
4	0.64455358	0.4	0.95105667	0.95105652	0.00000015
5	0.67772340	0.5	1.00000002	1.00000000	0.00000020
6	0.64455370	0.6	0.95105713	0.95105652	0.00000061
7	0.54828951	0.7	0.80901773	0.80901699	0.00000074
8	0.39835730	0.8	0.58778690	0.58778525	0.00000165
9	0.20942593	0.9	0.30901810	0.30901699	0.00000111
10	$0.74931285 \times 10^{-5}$	1.0	0.00000000	0.00000000	0.00000000

We recommend that the integrations in Steps 6 and 9 be performed in two steps. First, construct cubic spline interpolatory polynomials for p, q, and f using the methods presented in Section 3.5. Then approximate the integrands by products of cubic splines or derivatives of cubic splines. The integrands are now piecewise polynomials and can be integrated exactly on each subinterval, and then summed. This leads to accurate approximations of the integrals.

The hypotheses assumed at the beginning of this section are sufficient to guarantee that

$$\left\{ \int_0^1 |y(x) - \phi(x)|^2 \, dx \right\}^{1/2} = O(h^4), \quad \text{if} \quad 0 \le x \le 1.$$

For a proof of this result, see [Schul], pp. 107–108.

B-splines can also be defined for unequally-spaced nodes, but the details are more complicated. A presentation of the technique can be found in [Schul], p. 73. Another commonly used basis is the piecewise cubic Hermite polynomials. For an excellent presentation of this method, again see [Schul], pp. 24ff.

Boris Grigorievich Galerkin (1871–1945) did fundamental work applying approximation techniques to solve boundary-value problems associated with civil engineering problems. His initial paper on finite-element analysis was published in 1915, and his fundamental manuscript on thin elastic plates in 1937.

Other methods that receive considerable attention are Galerkin, or "weak form," methods. For the boundary-value problem we have been considering,

$$-\frac{d}{dx}\left(p(x)\frac{dy}{dx}\right) + q(x)y = f(x), \quad \text{for } 0 \le x \le 1, \text{ with } y(0) = 0 \text{ and } y(1) = 0,$$

under the assumptions listed at the beginning of this section, the Galerkin and Rayleigh-Ritz methods are both determined by Eq. (11.29). However, this is not the case for an arbitrary boundary-value problem. A treatment of the similarities and differences in the two methods and a discussion of the wide application of the Galerkin method can be found in [Schul] and in [SF].

The word collocation has its root in the Latin "co-" and "locus" indicating together with and place. It is equivalent to what we call interpolation.

Another popular technique for solving boundary-value problems is the **method of collocation**.

This procedure begins by selecting a set of basis functions $\{\phi_1, \ldots, \phi_N\}$, a set of numbers $\{x_i, \ldots, x_n\}$ in $[0, 1]$, and requiring that an approximation

$$\sum_{i=1}^{N} c_i \phi_i(x)$$

satisfy the differential equation at each of the numbers x_j, for $1 \le j \le n$. If, in addition, it is required that $\phi_i(0) = \phi_i(1) = 0$, for $1 \le i \le N$, then the boundary conditions are automatically satisfied. Much attention in the literature has been given to the choice of the numbers $\{x_j\}$ and the basis functions $\{\phi_i\}$. One popular choice is to let the ϕ_i be the basis functions for spline functions relative to a partition of $[0, 1]$, and to let the nodes $\{x_j\}$ be the Gaussian points or roots of certain orthogonal polynomials, transformed to the proper subinterval.

A comparison of various collocation methods and finite difference methods is contained in [Ru]. The conclusion is that the collocation methods using higher-degree splines are competitive with finite-difference techniques using extrapolation. Other references for collocation methods are [DebS] and [LR].

EXERCISE SET 11.5

1. Use the Piecewise Linear Algorithm to approximate the solution to the boundary-value problem

$$y'' + \frac{\pi^2}{4}y = \frac{\pi^2}{16}\cos\frac{\pi}{4}x, \quad 0 \le x \le 1, \quad y(0) = y(1) = 0$$

using $x_0 = 0, x_1 = 0.3, x_2 = 0.7, x_3 = 1$. Compare your results to the actual solution $y(x) = -\frac{1}{3}\cos\frac{\pi}{2}x - \frac{\sqrt{2}}{6}\sin\frac{\pi}{2}x + \frac{1}{3}\cos\frac{\pi}{4}x$.

2. Use the Piecewise Linear Algorithm to approximate the solution to the boundary-value problem

$$-\frac{d}{dx}(xy') + 4y = 4x^2 - 8x + 1, \quad 0 \le x \le 1, \quad y(0) = y(1) = 0$$

using $x_0 = 0, x_1 = 0.4, x_2 = 0.8, x_3 = 1$. Compare your results to the actual solution $y(x) = x^2 - x$.

3. Use the Piecewise Linear Algorithm to approximate the solutions to the following boundary-value problems, and compare the results to the actual solution:
   a. $-x^2y'' - 2xy' + 2y = -4x^2$, $0 \le x \le 1$, $y(0) = y(1) = 0$; use $h = 0.1$; actual solution $y(x) = x^2 - x$.
   b. $-\frac{d}{dx}(e^x y') + e^x y = x + (2 - x)e^x$, $0 \le x \le 1$, $y(0) = y(1) = 0$; use $h = 0.1$; actual solution $y(x) = (x - 1)(e^{-x} - 1)$.
   c. $-\frac{d}{dx}(e^{-x}y') + e^{-x}y = (x - 1) - (x + 1)e^{-(x-1)}$, $0 \le x \le 1$, $y(0) = y(1) = 0$; use $h = 0.05$; actual solution $y(x) = x(e^x - e)$.
   d. $-(x + 1)y'' - y' + (x + 2)y = [2 - (x + 1)^2]e\ln 2 - 2e^x$, $0 \le x \le 1$, $y(0) = y(1) = 0$; use $h = 0.05$; actual solution $y(x) = e^x\ln(x + 1) - (e\ln 2)x$.
4. Use the Cubic Spline Algorithm with $n = 3$ to approximate the solution to each of the following boundary-value problems, and compare the results to the actual solutions given in Exercises 1 and 2:
   a. $y'' + \frac{\pi^2}{4}y = \frac{\pi^2}{16}\cos\frac{\pi}{4}x$, $0 \le x \le 1$, $y(0) = 0$, $y(1) = 0$
   b. $-\frac{d}{dx}(xy') + 4y = 4x^2 - 8x + 1$, $0 \le x \le 1$, $y(0) = 0$, $y(1) = 0$
5. Repeat Exercise 3 using the Cubic Spline Algorithm.

6. Show that the boundary-value problem

$$-\frac{d}{dx}(p(x)y') + q(x)y = f(x), \quad 0 \le x \le 1, \quad y(0) = \alpha, \quad y(1) = \beta,$$

can be transformed by the change of variable

$$z = y - \beta x - (1 - x)\alpha$$

into the form

$$-\frac{d}{dx}(p(x)z') + q(x)z = F(x), \quad 0 \le x \le 1, \quad z(0) = 0, \quad z(1) = 0.$$

7. Use Exercise 6 and the Piecewise Linear Algorithm with $n = 9$ to approximate the solution to the boundary-value problem

$$-y'' + y = x, \quad 0 \le x \le 1, \quad y(0) = 1, \quad y(1) = 1 + e^{-1}.$$

8. Repeat Exercise 7 using the Cubic Spline Algorithm.

9. Show that the boundary-value problem

$$-\frac{d}{dx}(p(x)y') + q(x)y = f(x), \quad a \le x \le b, \quad y(a) = \alpha, \quad y(b) = \beta,$$

can be transformed into the form

$$-\frac{d}{dw}(p(w)z') + q(w)z = F(w), \quad 0 \le w \le 1, \quad z(0) = 0, \quad z(1) = 0,$$

by a method similar to that given in Exercise 6.

10. Show that the piecewise-linear basis functions $\{\phi_i\}_{i=1}^{n}$ are linearly independent.

11. Show that the cubic spline basis functions $\{\phi_i\}_{i=0}^{n+1}$ are linearly independent.

12. Show that the matrix given by the piecewise linear basis functions is positive definite.

13. Show that the matrix given by the cubic spline basis functions is positive definite.

11.6 Survey of Methods and Software

In this chapter we discussed methods for approximating solutions to boundary-value problems. For the linear boundary-value problem

$$y'' = p(x)y' + q(x)y + r(x), \quad a \le x \le b, \quad y(a) = \alpha, \quad y(b) = \beta,$$

we considered both a linear shooting method and a finite-difference method to approximate the solution. The shooting method uses an initial-value technique to solve the problems

$$y'' = p(x)y' + q(x)y + r(x), \quad \text{for } a \le x \le b, \text{ with } y(a) = \alpha \text{ and } y'(a) = 0,$$

and

$$y'' = p(x)y' + q(x)y, \quad \text{for } a \le x \le b, \text{ with } y(a) = 0 \text{ and } y'(a) = 1.$$

A weighted average of these solutions produces a solution to the linear boundary-value problem, although in certain situations there are problems with round-off error.

In the finite-difference method, we replaced y'' and y' with difference approximations and solved a linear system. Although the approximations may not be as accurate as the shooting method, there is less sensitivity to roundoff error. Higher-order difference methods are available, or extrapolation can be used to improve accuracy.

For the nonlinear boundary problem

$$y'' = f(x, y, y'), \quad \text{for } a \le x \le b, \text{ with } y(a) = \alpha \text{ and } y(b) = \beta,$$

we also considered two methods. The nonlinear shooting method requires the solution of the initial-value problem

$$y'' = f(x, y, y'), \quad \text{for } a \le x \le b, \text{ with } y(a) = \alpha \text{ and } y'(a) = t,$$

for an initial choice of t. We improved the choice of t by using Newton's method to approximate the solution to $y(b, t) = \beta$. This method required solving two initial-value problems at each iteration. The accuracy is dependent on the choice of method for solving the initial-value problems.

The finite-difference method for the nonlinear equation requires the replacement of y'' and y' by difference quotients, which results in a nonlinear system. This system is solved using Newton's method. Higher-order differences or extrapolation can be used to improve accuracy. Finite-difference methods tend to be less sensitive to roundoff error than shooting methods.

The Rayleigh-Ritz-Galerkin method was illustrated by approximating the solution to the boundary-value problem

$$-\frac{d}{dx}\left(p(x)\frac{dy}{dx}\right) + q(x)y = f(x), \quad 0 \le x \le 1, \quad y(0) = y(1) = 0.$$

A piecewise-linear approximation or a cubic spline approximation can be obtained.

Most of the material concerning second-order boundary-value problems can be extended to problems with boundary conditions of the form

$$\alpha_1 y(a) + \beta_1 y'(a) = \alpha \quad \text{and} \quad \alpha_2 y(b) + \beta_2 y'(b) = \beta,$$

where $|\alpha_1| + |\beta_1| \ne 0$ and $|\alpha_2| + |\beta_2| \ne 0$, but some of the techniques become quite complicated. The reader who is interested in problems of this type is advised to consider a book specializing in boundary-value problems, such as [Keller, H].

The IMSL library has many subroutines for boundary-value problems. There are both shooting and finite difference methods. The shooting methods use the Runge-Kutta-Verner technique for solving the associated initial-value problems.

The NAG Library also has a multitude of subroutines for solving boundary-value problems. Some of these are a shooting method using the Runge-Kutta-Merson initial-value method in conjunction with Newton's method, a finite-difference method with Newton's method to solve the nonlinear system, and a linear finite-difference method based on collocation.

There are subroutines in the ODE package contained in the Netlib library for solving both linear and nonlinear two-point boundary-value problems, respectively. These routines are based on multiple shooting methods.

Further information on the general problems involved with the numerical solution to two-point boundary-value problems can be found in Keller [Keller, H] and Bailey, Shampine and Waltman [BSW]. Roberts and Shipman [RS] focuses on the shooting methods for the two-point boundary-value problem, and Pryce [Pr] restricts attention to Sturm-Liouville problems. The book by Ascher, Mattheij, and Russell [AMR] has a comprehensive presentation of multiple shooting and parallel shooting methods.

CHAPTER

12 Numerical Solutions to Partial Differential Equations

Introduction

A body is *isotropic* if the thermal conductivity at each point in the body is independent of the direction of heat flow through the point. Suppose that k, c, and ρ are functions of (x, y, z) and represent, respectively, the thermal conductivity, specific heat, and density of an isotropic the body at the point (x, y, z). Then the temperature, $u \equiv u(x, y, z, t)$, in a body can be found by solving the partial differential equation

$$\frac{\partial}{\partial x}\left(k\frac{\partial u}{\partial x}\right) + \frac{\partial}{\partial y}\left(k\frac{\partial u}{\partial y}\right) + \frac{\partial}{\partial z}\left(k\frac{\partial u}{\partial z}\right) = c\rho\frac{\partial u}{\partial t},$$

When k, c, and ρ are constants, this equation is known as the simple three-dimensional heat equation and is expressed as

$$\frac{\partial^2 u}{\partial x^2} + \frac{\partial^2 u}{\partial y^2} + \frac{\partial^2 u}{\partial z^2} = \frac{c\rho}{k}\frac{\partial u}{\partial t}.$$

If the boundary of the body is relatively simple, the solution to this equation can be found using Fourier series.

In most situations where k, c, and ρ are not constant or when the boundary is irregular, the solution to the partial differential equation must be obtained by approximation techniques. An introduction to techniques of this type is presented in this chapter.

Elliptic Equations

Common partial differential equations are categorized in a manner similar to the conic sections. The partial differential equation we will consider in Section 12.1 involves $u_{xx}(x, y) + u_{yy}(x, y)$ and is an **elliptic** equation. The particular elliptic equation we will consider is known as the **Poisson equation**:

$$\frac{\partial^2 u}{\partial x^2}(x, y) + \frac{\partial^2 u}{\partial y^2}(x, y) = f(x, y).$$

In this equation we assume that f describes the input to the problem on a plane region R with boundary S. Equations of this type arise in the study of various time-independent physical problems such as the steady-state distribution of heat in a plane region, the potential energy of a point in a plane acted on by gravitational forces in the plane, and two-dimensional steady-state problems involving incompressible fluids.

Siméon-Denis Poisson (1781–1840) was a student of Laplace and Legendre during the Napoleonic years in France. Later he assumed Fourier's professorship at the École Polytechnique where he worked on ordinary and partial differential equations, and later in life on probability theory.

Additional constraints must be imposed to obtain a unique solution to the Poisson equation. For example, the study of the steady-state distribution of heat in a plane region requires that $f(x, y) \equiv 0$, resulting in a simplification to **Laplace's equation**

$$\frac{\partial^2 u}{\partial x^2}(x, y) + \frac{\partial^2 u}{\partial y^2}(x, y) = 0.$$

If the temperature within the region is determined by the temperature distribution on the boundary of the region, the constraints are called the **Dirichlet boundary conditions**, given by

$$u(x, y) = g(x, y),$$

for all (x, y) on S, the boundary of the region R. (See Figure 12.1.)

Figure 12.1

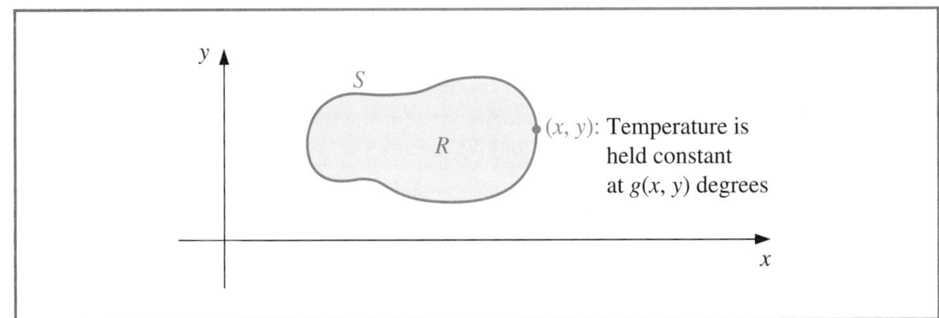

Pierre-Simon Laplace (1749–1827) worked in many mathematical areas, producing seminal papers in probability and mathematical physics. He published his major work on the theory of heat during the period 1817–1820.

Johann Peter Gustav Lejeune Dirichlet (1805–1859) made major contributions to the areas of number theory and the convergence of series. In fact, he could be considered the founder of Fourier series, since according to Riemann he was the first to write a profound paper on this subject.

Parabolic Equations

In Section 12.2 we consider the numerical solution to a problem involving a **parabolic** partial differential equation of the form

$$\frac{\partial u}{\partial t}(x, t) - \alpha^2 \frac{\partial^2 u}{\partial x^2}(x, t) = 0.$$

The physical problem considered here concerns the flow of heat along a rod of length l (see Figure 12.2) which has a uniform temperature within each cross-sectional element. This requires the rod to be perfectly insulated on its lateral surface. The constant α is assumed to be independent of the position in the rod. It is determined by the heat-conductive properties of the material of which the rod is composed.

Figure 12.2

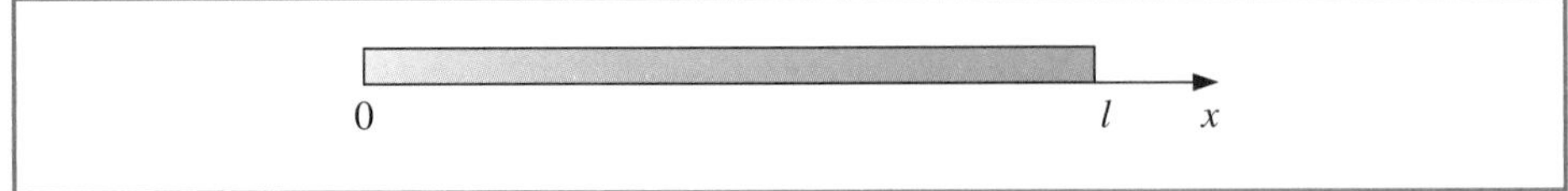

One of the typical sets of constraints for a heat-flow problem of this type is to specify the initial heat distribution in the rod,

$$u(x, 0) = f(x),$$

and to describe the behavior at the ends of the rod. For example, if the ends are held at constant temperatures U_1 and U_2, the boundary conditions have the form

$$u(0, t) = U_1 \quad \text{and} \quad u(l, t) = U_2,$$

and the heat distribution approaches the limiting temperature distribution

$$\lim_{t\to\infty} u(x, t) = U_1 + \frac{U_2 - U_1}{l}x.$$

If, instead, the rod is insulated so that no heat flows through the ends, the boundary conditions are

$$\frac{\partial u}{\partial x}(0, t) = 0 \quad \text{and} \quad \frac{\partial u}{\partial x}(l, t) = 0.$$

Then no heat escapes from the rod and in the limiting case the temperature on the rod is constant. The parabolic partial differential equation is also of importance in the study of gas diffusion; in fact, it is known in some circles as the **diffusion equation**.

Hyperbolic Equations

The problem studied in Section 12.3 is the one-dimensional **wave equation** and is an example of a **hyperbolic** partial differential equation. Suppose an elastic string of length l is stretched between two supports at the same horizontal level (see Figure 12.3).

Figure 12.3

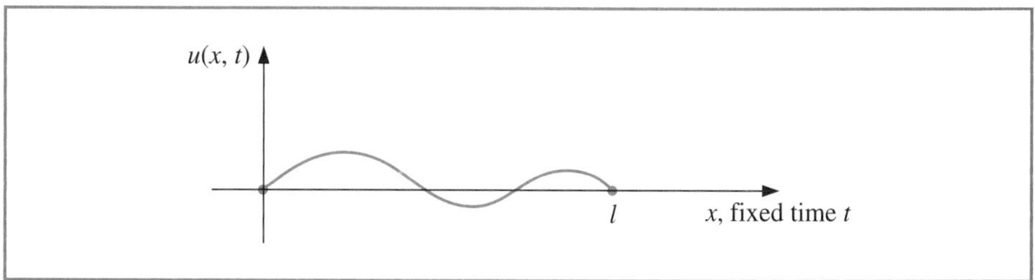

If the string is set to vibrate in a vertical plane, the vertical displacement $u(x, t)$ of a point x at time t satisfies the partial differential equation

$$\alpha^2\frac{\partial^2 u}{\partial x^2}(x, t) - \frac{\partial^2 u}{\partial t^2}(x, t) = 0, \quad \text{for } 0 < x < l \quad \text{and} \quad 0 < t,$$

provided that damping effects are neglected and the amplitude is not too large. To impose constraints on this problem, assume that the initial position and velocity of the string are given by

$$u(x, 0) = f(x) \quad \text{and} \quad \frac{\partial u}{\partial t}(x, 0) = g(x), \quad \text{for } 0 \le x \le l.$$

If the endpoints are fixed, we also have $u(0, t) = 0$ and $u(l, t) = 0$.

Other physical problems involving the hyperbolic partial differential equation occur in the study of vibrating beams with one or both ends clamped and in the transmission of electricity on a long line where there is some leakage of current to the ground.

12.1 Elliptic Partial Differential Equations

The *elliptic* partial differential equation we consider is the Poisson equation,

$$\nabla^2 u(x,y) \equiv \frac{\partial^2 u}{\partial x^2}(x,y) + \frac{\partial^2 u}{\partial y^2}(x,y) = f(x,y) \tag{12.1}$$

on $R = \{(x,y) \mid a < x < b, c < y < d\}$, with $u(x,y) = g(x,y)$ for $(x,y) \in S$, where S denotes the boundary of R. If f and g are continuous on their domains, then there is a unique solution to this equation.

Selecting a Grid

The method used is a two-dimensional adaptation of the Finite-Difference method for linear boundary-value problems, which was discussed in Section 11.3. The first step is to choose integers n and m to define step sizes $h = (b - a)/n$ and $k = (d - c)/m$. Partition the interval $[a, b]$ into n equal parts of width h and the interval $[c, d]$ into m equal parts of width k (see Figure 12.4).

Figure 12.4

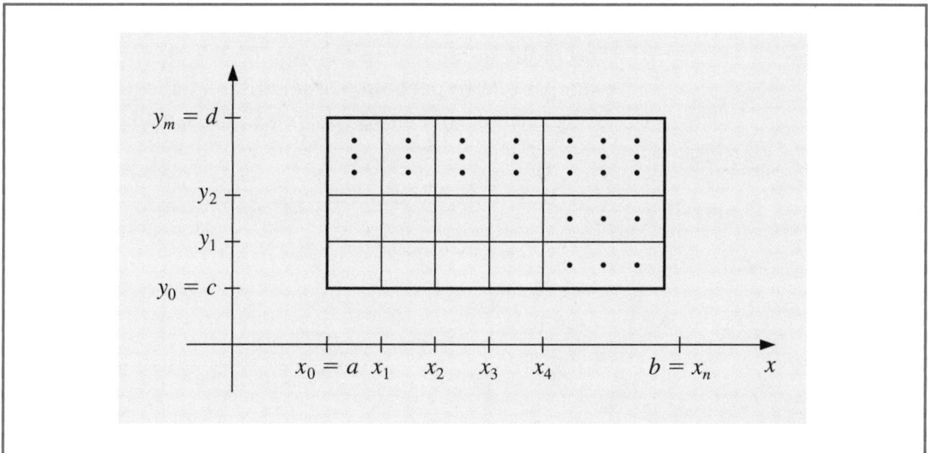

Place a grid on the rectangle R by drawing vertical and horizontal lines through the points with coordinates (x_i, y_j), where

$$x_i = a + ih, \quad \text{for each } i = 0, 1, \ldots, n, \quad \text{and} \quad y_j = c + jk, \quad \text{for each } j = 0, 1, \ldots, m.$$

The lines $x = x_i$ and $y = y_j$ are **grid lines**, and their intersections are the **mesh points** of the grid. For each mesh point in the interior of the grid, (x_i, y_j), for $i = 1, 2, \ldots, n-1$ and $j = 1, 2, \ldots, m-1$, we can use the Taylor series in the variable x about x_i to generate the centered-difference formula

$$\frac{\partial^2 u}{\partial x^2}(x_i, y_j) = \frac{u(x_{i+1}, y_j) - 2u(x_i, y_j) + u(x_{i-1}, y_j)}{h^2} - \frac{h^2}{12}\frac{\partial^4 u}{\partial x^4}(\xi_i, y_j), \tag{12.2}$$

where $\xi_i \in (x_{i-1}, x_{i+1})$. We can also use the Taylor series in the variable y about y_j to generate the centered-difference formula

$$\frac{\partial^2 u}{\partial y^2}(x_i, y_j) = \frac{u(x_i, y_{j+1}) - 2u(x_i, y_j) + u(x_i, y_{j-1})}{k^2} - \frac{k^2}{12}\frac{\partial^4 u}{\partial y^4}(x_i, \eta_j), \tag{12.3}$$

where $\eta_j \in (y_{j-1}, y_{j+1})$.

Using these formulas in Eq. (12.1) allows us to express the Poisson equation at the points (x_i, y_j) as

$$\frac{u(x_{i+1}, y_j) - 2u(x_i, y_j) + u(x_{i-1}, y_j)}{h^2} + \frac{u(x_i, y_{j+1}) - 2u(x_i, y_j) + u(x_i, y_{j-1})}{k^2}$$
$$= f(x_i, y_j) + \frac{h^2}{12}\frac{\partial^4 u}{\partial x^4}(\xi_i, y_j) + \frac{k^2}{12}\frac{\partial^4 u}{\partial y^4}(x_i, \eta_j),$$

for each $i = 1, 2, \ldots, n-1$ and $j = 1, 2, \ldots, m-1$. The boundary conditions are

$$u(x_0, y_j) = g(x_0, y_j) \quad \text{and} \quad u(x_n, y_j) = g(x_n, y_j), \quad \text{for each } j = 0, 1, \ldots, m;$$
$$u(x_i, y_0) = g(x_i, y_0) \quad \text{and} \quad u(x_i, y_m) = g(x_i, y_m), \quad \text{for each } i = 1, 2, \ldots, n-1.$$

Finite-Difference Method

In difference-equation form, this results in the **Finite-Difference method**:

$$2\left[\left(\frac{h}{k}\right)^2 + 1\right]w_{ij} - (w_{i+1,j} + w_{i-1,j}) - \left(\frac{h}{k}\right)^2 (w_{i,j+1} + w_{i,j-1}) = -h^2 f(x_i, y_j), \quad (12.4)$$

for each $i = 1, 2, \ldots, n-1$ and $j = 1, 2, \ldots, m-1$, and

$$w_{0j} = g(x_0, y_j) \quad \text{and} \quad w_{nj} = g(x_n, y_j), \quad \text{for each } j = 0, 1, \ldots, m; \quad (12.5)$$
$$w_{i0} = g(x_i, y_0) \quad \text{and} \quad w_{im} = g(x_i, y_m), \quad \text{for each } i = 1, 2, \ldots, n-1;$$

where w_{ij} approximates $u(x_i, y_j)$. This method has local truncation error of order $O(h^2 + k^2)$

The typical equation in (12.4) involves approximations to $u(x, y)$ at the points

$$(x_{i-1}, y_j), \quad (x_i, y_j), \quad (x_{i+1}, y_j), \quad (x_i, y_{j-1}), \quad \text{and} \quad (x_i, y_{j+1}).$$

Reproducing the portion of the grid where these points are located (see Figure 12.5) shows that each equation involves approximations in a star-shaped region about the blue X at (x_i, y_j).

Figure 12.5

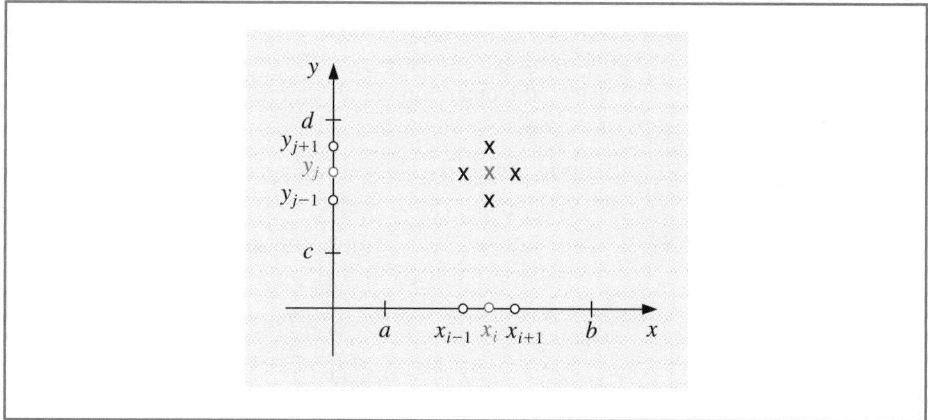

We use the information from the boundary conditions (12.5) whenever appropriate in the system given by (12.4); that is, at all points (x_i, y_j) adjacent to a boundary mesh point.

This produces an $(n-1)(m-1) \times (n-1)(m-1)$ linear system with the unknowns being the approximations $w_{i,j}$ to $u(x_i, y_j)$ at the interior mesh points.

The linear system involving these unknowns is expressed for matrix calculations more efficiently if a relabeling of the interior mesh points is introduced. A recommended labeling of these points (see [Var1], p. 210) is to let

$$P_l = (x_i, y_j) \quad \text{and} \quad w_l = w_{i,j},$$

where $l = i + (m-1-j)(n-1)$, for each $i = 1, 2, \ldots, n-1$ and $j = 1, 2, \ldots, m-1$. This labels the mesh points consecutively from left to right and top to bottom. Labeling the points in this manner ensures that the system needed to determine the $w_{i,j}$ is a banded matrix with band width at most $2n-1$.

For example, with $n = 4$ and $m = 5$, the relabeling results in a grid whose points are shown in Figure 12.6.

Figure 12.6

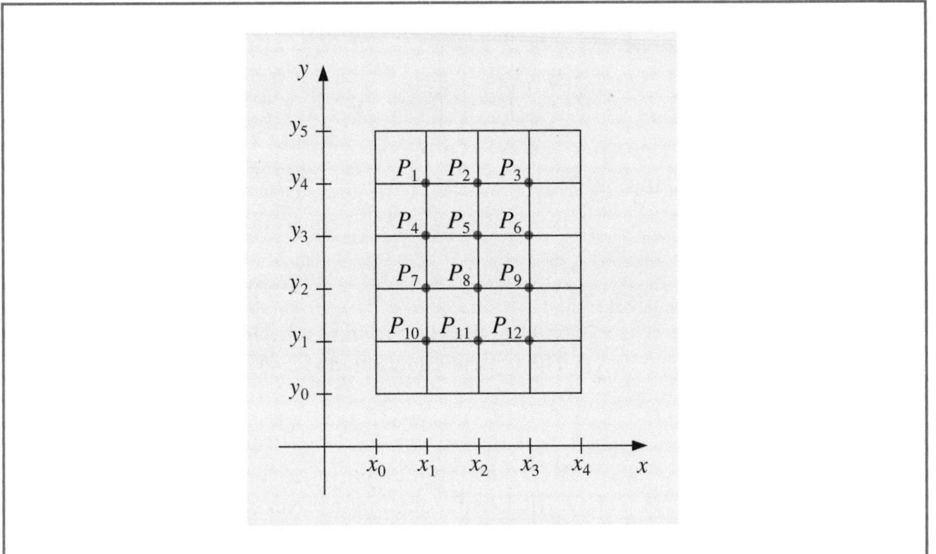

Example 1 Determine the steady-state heat distribution in a thin square metal plate with dimensions 0.5 m by 0.5 m using $n = m = 4$. Two adjacent boundaries are held at 0°C, and the heat on the other boundaries increases linearly from 0°C at one corner to 100°C where the sides meet.

Solution Place the sides with the zero boundary conditions along the x- and y-axes. Then the problem is expressed as

$$\frac{\partial^2 u}{\partial x^2}(x, y) + \frac{\partial^2 u}{\partial y^2}(x, y) = 0,$$

for (x, y) in the set $R = \{(x, y) \mid 0 < x < 0.5,\ 0 < y < 0.5\}$. The boundary conditions are

$$u(0, y) = 0, \quad u(x, 0) = 0, \quad u(x, 0.5) = 200x, \quad \text{and } u(0.5, y) = 200y.$$

If $n = m = 4$, the problem has the grid given in Figure 12.7, and the difference equation (12.4) is

$$4w_{i,j} - w_{i+1,j} - w_{i-1,j} - w_{i,j-1} - w_{i,j+1} = 0,$$

for each $i = 1, 2, 3$ and $j = 1, 2, 3$.

Figure 12.7

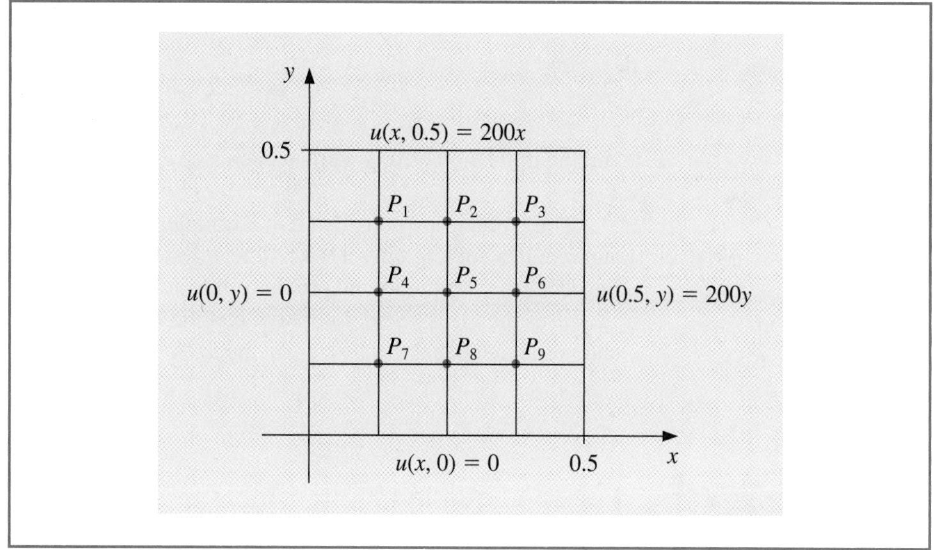

Expressing this in terms of the relabeled interior grid points $w_i = u(P_i)$ implies that the equations at the points P_i are:

$$
\begin{aligned}
P_1: &\quad 4w_1 - w_2 - w_4 = w_{0,3} + w_{1,4},\\
P_2: &\quad 4w_2 - w_3 - w_1 - w_5 = w_{2,4},\\
P_3: &\quad 4w_3 - w_2 - w_6 = w_{4,3} + w_{3,4},\\
P_4: &\quad 4w_4 - w_5 - w_1 - w_7 = w_{0,2},\\
P_5: &\quad 4w_5 - w_6 - w_4 - w_2 - w_8 = 0,\\
P_6: &\quad 4w_6 - w_5 - w_3 - w_9 = w_{4,2},\\
P_7: &\quad 4w_7 - w_8 - w_4 = w_{0,1} + w_{1,0},\\
P_8: &\quad 4w_8 - w_9 - w_7 - w_5 = w_{2,0},\\
P_9: &\quad 4w_9 - w_8 - w_6 = w_{3,0} + w_{4,1},
\end{aligned}
$$

where the right sides of the equations are obtained from the boundary conditions.

In fact, the boundary conditions imply that

$$w_{1,0} = w_{2,0} = w_{3,0} = w_{0,1} = w_{0,2} = w_{0,3} = 0,$$

$$w_{1,4} = w_{4,1} = 25, \quad w_{2,4} = w_{4,2} = 50, \quad \text{and} \quad w_{3,4} = w_{4,3} = 75.$$

Table 12.1

i	w_i
1	18.75
2	37.50
3	56.25
4	12.50
5	25.00
6	37.50
7	6.25
8	12.50
9	18.75

So the linear system associated with this problem has the form

$$
\begin{bmatrix}
4 & -1 & 0 & -1 & 0 & 0 & 0 & 0 & 0\\
-1 & 4 & -1 & 0 & -1 & 0 & 0 & 0 & 0\\
0 & -1 & 4 & 0 & 0 & -1 & 0 & 0 & 0\\
-1 & 0 & 0 & 4 & -1 & 0 & -1 & 0 & 0\\
0 & -1 & 0 & -1 & 4 & -1 & 0 & -1 & 0\\
0 & 0 & -1 & 0 & -1 & 4 & 0 & 0 & -1\\
0 & 0 & 0 & -1 & 0 & 0 & 4 & -1 & 0\\
0 & 0 & 0 & 0 & -1 & 0 & -1 & 4 & -1\\
0 & 0 & 0 & 0 & 0 & -1 & 0 & -1 & 4
\end{bmatrix}
\begin{bmatrix} w_1\\ w_2\\ w_3\\ w_4\\ w_5\\ w_6\\ w_7\\ w_8\\ w_9 \end{bmatrix}
=
\begin{bmatrix} 25\\ 50\\ 150\\ 0\\ 0\\ 50\\ 0\\ 0\\ 25 \end{bmatrix}.
$$

The values of $w_1, w_2, \ldots, w_9$, found by applying the Gauss-Seidel method to this matrix, are given in Table 12.1.

These answers are exact, because the true solution, $u(x, y) = 400xy$, has

$$\frac{\partial^4 u}{\partial x^4} = \frac{\partial^4 u}{\partial y^4} \equiv 0,$$

and the truncation error is zero at each step. ▪

The problem we considered in Example 1 has the same mesh size, 0.125, on each axis and requires solving only a 9×9 linear system. This simplifies the situation and does not introduce the computational problems that are present when the system is larger. Algorithm 12.1 uses the Gauss-Seidel iterative method for solving the linear system that is produced and permits unequal mesh sizes on the axes.

Poisson Equation Finite-Difference

To approximate the solution to the Poisson equation

$$\frac{\partial^2 u}{\partial x^2}(x, y) + \frac{\partial^2 u}{\partial y^2}(x, y) = f(x, y), \quad a \leq x \leq b, \quad c \leq y \leq d,$$

subject to the boundary conditions

$$u(x, y) = g(x, y) \quad \text{if } x = a \text{ or } x = b \quad \text{and} \quad c \leq y \leq d$$

and

$$u(x, y) = g(x, y) \quad \text{if } y = c \text{ or } y = d \quad \text{and} \quad a \leq x \leq b:$$

INPUT endpoints a, b, c, d; integers $m \geq 3$, $n \geq 3$; tolerance TOL; maximum number of iterations N.

OUTPUT approximations $w_{i,j}$ to $u(x_i, y_j)$ for each $i=1,\ldots,n-1$ and for each $j = 1, \ldots, m-1$ or a message that the maximum number of iterations was exceeded.

Step 1 Set $h = (b - a)/n$;
$k = (d - c)/m$.

Step 2 For $i = 1, \ldots, n-1$ set $x_i = a + ih$. (*Steps 2 and 3 construct mesh points.*)

Step 3 For $j = 1, \ldots, m-1$ set $y_j = c + jk$.

Step 4 For $i = 1, \ldots, n-1$
for $j = 1, \ldots, m-1$ set $w_{i,j} = 0$.

Step 5 Set $\lambda = h^2/k^2$;
$\mu = 2(1 + \lambda)$;
$l = 1$.

Step 6 While $l \leq N$ do Steps 7–20. (*Steps 7–20 perform Gauss-Seidel iterations.*)

Step 7 Set $z = \left(-h^2 f(x_1, y_{m-1}) + g(a, y_{m-1}) + \lambda g(x_1, d) + \lambda w_{1,m-2} + w_{2,m-1}\right)/\mu$;
$NORM = |z - w_{1,m-1}|$;
$w_{1,m-1} = z$.

Step 8 For $i = 2, \ldots, n-2$

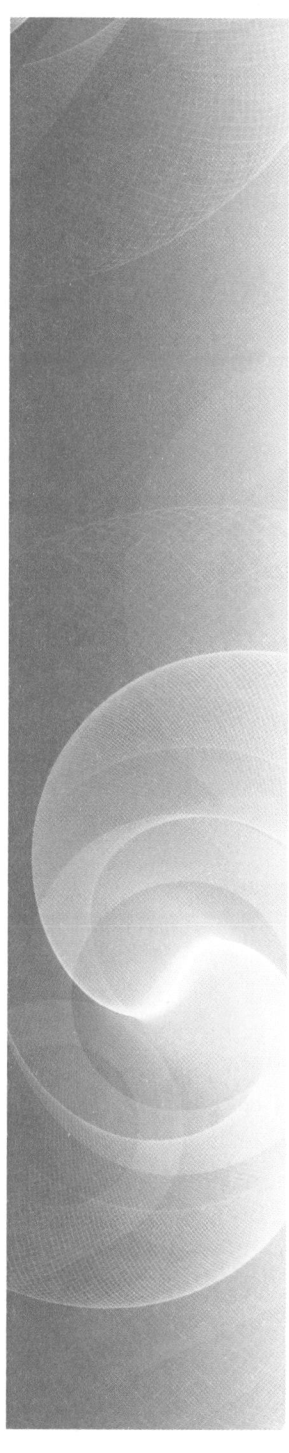

set $z = \big(- h^2 f(x_i, y_{m-1}) + \lambda g(x_i, d) + w_{i-1,m-1}$
$+ w_{i+1,m-1} + \lambda w_{i,m-2}\big)/\mu;$
if $|w_{i,m-1} - z| > NORM$ then set $NORM = |w_{i,m-1} - z|$;
set $w_{i,m-1} = z$.

Step 9 Set $z = \big(- h^2 f(x_{n-1}, y_{m-1}) + g(b, y_{m-1}) + \lambda g(x_{n-1}, d)$
$+ w_{n-2,m-1} + \lambda w_{n-1,m-2}\big)/\mu;$
if $|w_{n-1,m-1} - z| > NORM$ then set $NORM = |w_{n-1,m-1} - z|$;
set $w_{n-1,m-1} = z$.

Step 10 For $j = m - 2, \ldots, 2$ do Steps 11, 12, and 13.

Step 11 Set $z = \big(-h^2 f(x_1, y_j) + g(a, y_j) + \lambda w_{1,j+1} + \lambda w_{1,j-1} + w_{2,j}\big)/\mu;$
if $|w_{1,j} - z| > NORM$ then set $NORM = |w_{1,j} - z|$;
set $w_{1,j} = z$.

Step 12 For $i = 2, \ldots, n - 2$
set $z = \big(-h^2 f(x_i, y_j) + w_{i-1,j} + \lambda w_{i,j+1} + w_{i+1,j} + \lambda w_{i,j-1}\big)/\mu;$
if $|w_{i,j} - z| > NORM$ then set $NORM = |w_{i,j} - z|$;
set $w_{i,j} = z$.

Step 13 Set $z = \big(- h^2 f(x_{n-1}, y_j) + g(b, y_j) + w_{n-2,j}$
$+ \lambda w_{n-1,j+1} + \lambda w_{n-1,j-1}\big)/\mu;$
if $|w_{n-1,j} - z| > NORM$ then set $NORM = |w_{n-1,j} - z|$;
set $w_{n-1,j} = z$.

Step 14 Set $z = \big(-h^2 f(x_1, y_1) + g(a, y_1) + \lambda g(x_1, c) + \lambda w_{1,2} + w_{2,1}\big)/\mu;$
if $|w_{1,1} - z| > NORM$ then set $NORM = |w_{1,1} - z|$;
set $w_{1,1} = z$.

Step 15 For $i = 2, \ldots, n - 2$
set $z = \big(-h^2 f(x_i, y_1) + \lambda g(x_i, c) + w_{i-1,1} + \lambda w_{i,2} + w_{i+1,1}\big)/\mu;$
if $|w_{i,1} - z| > NORM$ then set $NORM = |w_{i,1} - z|$;
set $w_{i,1} = z$.

Step 16 Set $z = \big(-h^2 f(x_{n-1}, y_1) + g(b, y_1) + \lambda g(x_{n-1}, c) + w_{n-2,1} + \lambda w_{n-1,2}\big)/\mu;$
if $|w_{n-1,1} - z| > NORM$ then set $NORM = |w_{n-1,1} - z|$;
set $w_{n-1,1} = z$.

Step 17 If $NORM \leq TOL$ then do Steps 18 and 19.

Step 18 For $i = 1, \ldots, n - 1$
for $j = 1, \ldots, m - 1$ OUTPUT $(x_i, y_j, w_{i,j})$.

Step 19 STOP. (*The procedure was successful.*)

Step 20 Set $l = l + 1$.

Step 21 OUTPUT ('Maximum number of iterations exceeded');
(*The procedure was unsuccessful.*)
STOP. ■

Although the Gauss-Seidel iterative procedure is incorporated into Algorithm 12.1 for simplicity, it is advisable to use a direct technique such as Gaussian elimination when the system is small, on the order of 100 or less, because the positive definiteness ensures stability with respect to round-off errors. In particular, a generalization of the Crout Factorization

Algorithm 6.7 (see [Var1], p. 221), is efficient for solving this system because the matrix is in the symmetric-block tridiagonal form

$$\begin{bmatrix} A_1 & C_1 & 0 & \cdots & \cdots & 0 \\ C_1 & A_2 & C_2 & \ddots & & \vdots \\ 0 & C_2 & \ddots & \ddots & \ddots & \vdots \\ \vdots & \ddots & \ddots & \ddots & \ddots & 0 \\ \vdots & & \ddots & \ddots & \ddots & C_{m-1} \\ 0 & \cdots & \cdots & 0 & C_{m-1} & A_{m-1} \end{bmatrix},$$

with square blocks of size $(n-1) \times (n-1)$.

Choice of Iterative Method

For large systems, an iterative method should be used—specifically, the SOR method discussed in Algorithm 7.3. The choice of ω that is optimal in this situation comes from the fact that when A is decomposed into its diagonal D and upper- and lower-triangular parts U and L,

$$A = D - L - U,$$

and B is the matrix for the Jacobi method,

$$B = D^{-1}(L + U),$$

then the spectral radius of B is (see [Var1])

$$\rho(B) = \frac{1}{2}\left[\cos\left(\frac{\pi}{m}\right) + \cos\left(\frac{\pi}{n}\right)\right].$$

The value of ω to be used is, consequently,

$$\omega = \frac{2}{1 + \sqrt{1 - [\rho(B)]^2}} = \frac{4}{2 + \sqrt{4 - \left[\cos\left(\frac{\pi}{m}\right) + \cos\left(\frac{\pi}{n}\right)\right]^2}}.$$

A block technique can be incorporated into the algorithm for faster convergence of the SOR procedure. For a presentation of this technique, see [Var1], pp. 219–223.

Example 2 Use the Poisson finite-difference method with $n = 6$, $m = 5$, and a tolerance of 10^{-10} to approximate the solution to

$$\frac{\partial^2 u}{\partial x^2}(x, y) + \frac{\partial^2 u}{\partial y^2}(x, y) = xe^y, \quad 0 < x < 2, \quad 0 < y < 1,$$

with the boundary conditions

$$u(0, y) = 0, \quad u(2, y) = 2e^y, \quad 0 \le y \le 1,$$
$$u(x, 0) = x, \quad u(x, 1) = ex, \quad 0 \le x \le 2,$$

and compare the results with the exact solution $u(x, y) = xe^y$.

Solution Using Algorithm 12.1 with a maximum number of iterations set at $N = 100$ gives the results in Table 12.2. The stopping criterion for the Gauss-Seidel method in Step 17 requires that

$$\left|w_{ij}^{(l)} - w_{ij}^{(l-1)}\right| \le 10^{-10},$$

for each $i = 1, \ldots, 5$ and $j = 1, \ldots, 4$. The solution to the difference equation was accurately obtained, and the procedure stopped at $l = 61$. The results, along with the correct values, are presented in Table 12.2. ■

Table 12.2

i	j	x_i	y_j	$w_{i,j}^{(61)}$	$u(x_i, y_j)$	$\left\|u(x_i, y_j) - w_{i,j}^{(61)}\right\|$
1	1	0.3333	0.2000	0.40726	0.40713	1.30×10^{-4}
1	2	0.3333	0.4000	0.49748	0.49727	2.08×10^{-4}
1	3	0.3333	0.6000	0.60760	0.60737	2.23×10^{-4}
1	4	0.3333	0.8000	0.74201	0.74185	1.60×10^{-4}
2	1	0.6667	0.2000	0.81452	0.81427	2.55×10^{-4}
2	2	0.6667	0.4000	0.99496	0.99455	4.08×10^{-4}
2	3	0.6667	0.6000	1.2152	1.2147	4.37×10^{-4}
2	4	0.6667	0.8000	1.4840	1.4837	3.15×10^{-4}
3	1	1.0000	0.2000	1.2218	1.2214	3.64×10^{-4}
3	2	1.0000	0.4000	1.4924	1.4918	5.80×10^{-4}
3	3	1.0000	0.6000	1.8227	1.8221	6.24×10^{-4}
3	4	1.0000	0.8000	2.2260	2.2255	4.51×10^{-4}
4	1	1.3333	0.2000	1.6290	1.6285	4.27×10^{-4}
4	2	1.3333	0.4000	1.9898	1.9891	6.79×10^{-4}
4	3	1.3333	0.6000	2.4302	2.4295	7.35×10^{-4}
4	4	1.3333	0.8000	2.9679	2.9674	5.40×10^{-4}
5	1	1.6667	0.2000	2.0360	2.0357	3.71×10^{-4}
5	2	1.6667	0.4000	2.4870	2.4864	5.84×10^{-4}
5	3	1.6667	0.6000	3.0375	3.0369	6.41×10^{-4}
5	4	1.6667	0.8000	3.7097	3.7092	4.89×10^{-4}

EXERCISE SET 12.1

1. Use Algorithm 12.1 to approximate the solution to the elliptic partial differential equation

$$\frac{\partial^2 u}{\partial x^2} + \frac{\partial^2 u}{\partial y^2} = 4, \quad 0 < x < 1, \quad 0 < y < 2;$$

$$u(x, 0) = x^2, \quad u(x, 2) = (x - 2)^2, \quad 0 \le x \le 1;$$

$$u(0, y) = y^2, \quad u(1, y) = (y - 1)^2, \quad 0 \le y \le 2.$$

Use $h = k = \frac{1}{2}$, and compare the results to the actual solution $u(x, y) = (x - y)^2$.

2. Use Algorithm 12.1 to approximate the solution to the elliptic partial differential equation

$$\frac{\partial^2 u}{\partial x^2} + \frac{\partial^2 u}{\partial y^2} = 0, \quad 1 < x < 2, \quad 0 < y < 1;$$

$$u(x, 0) = 2 \ln x, \qquad u(x, 1) = \ln(x^2 + 1), \quad 1 \le x \le 2;$$

$$u(1, y) = \ln(y^2 + 1), \quad u(2, y) = \ln(y^2 + 4), \quad 0 \le y \le 1.$$

Use $h = k = \frac{1}{3}$, and compare the results to the actual solution $u(x, y) = \ln(x^2 + y^2)$.

3. Approximate the solutions to the following elliptic partial differential equations, using Algorithm 12.1:

a. $\dfrac{\partial^2 u}{\partial x^2} + \dfrac{\partial^2 u}{\partial y^2} = 0, \quad 0 < x < 1, \quad 0 < y < 1;$

$$u(x,0) = 0, \quad u(x,1) = x, \quad 0 \le x \le 1;$$
$$u(0,y) = 0, \quad u(1,y) = y, \quad 0 \le y \le 1.$$

Use $h = k = 0.2$, and compare the results to the actual solution $u(x,y) = xy$.

b. $\dfrac{\partial^2 u}{\partial x^2} + \dfrac{\partial^2 u}{\partial y^2} = -(\cos(x+y) + \cos(x-y)), \quad 0 < x < \pi, \quad 0 < y < \dfrac{\pi}{2};$

$$u(0,y) = \cos y, \quad u(\pi,y) = -\cos y, \quad 0 \le y \le \frac{\pi}{2},$$
$$u(x,0) = \cos x, \quad u\left(x, \frac{\pi}{2}\right) = 0, \quad 0 \le x \le \pi.$$

Use $h = \pi/5$ and $k = \pi/10$, and compare the results to the actual solution $u(x,y) = \cos x \cos y$.

c. $\dfrac{\partial^2 u}{\partial x^2} + \dfrac{\partial^2 u}{\partial y^2} = (x^2 + y^2)e^{xy}, \quad 0 < x < 2,\ 0 < y < 1;$

$$u(0,y) = 1, \quad u(2,y) = e^{2y}, \quad 0 \le y \le 1;$$
$$u(x,0) = 1, \quad u(x,1) = e^x, \quad 0 \le x \le 2.$$

Use $h = 0.2$ and $k = 0.1$, and compare the results to the actual solution $u(x,y) = e^{xy}$.

d. $\dfrac{\partial^2 u}{\partial x^2} + \dfrac{\partial^2 u}{\partial y^2} = \dfrac{x}{y} + \dfrac{y}{x}, \quad 1 < x < 2, \quad 1 < y < 2;$

$$u(x,1) = x \ln x, \quad u(x,2) = x \ln\left(4x^2\right), \quad 1 \le x \le 2;$$
$$u(1,y) = y \ln y, \quad u(2,y) = 2y \ln(2y), \quad 1 \le y \le 2.$$

Use $h = k = 0.1$, and compare the results to the actual solution $u(x,y) = xy \ln xy$.

4. Repeat Exercise 3(a) using extrapolation with $h_0 = 0.2$, $h_1 = h_0/2$, and $h_2 = h_0/4$.

5. Construct an algorithm similar to Algorithm 12.1, except use the SOR method with optimal ω instead of the Gauss-Seidel method for solving the linear system.

6. Repeat Exercise 3 using the algorithm constructed in Exercise 5.

7. A coaxial cable is made of a 0.1-in.-square inner conductor and a 0.5-in.-square outer conductor. The potential at a point in the cross section of the cable is described by Laplace's equation. Suppose the inner conductor is kept at 0 volts and the outer conductor is kept at 110 volts. Find the potential between the two conductors by placing a grid with horizontal mesh spacing $h = 0.1$ in. and vertical mesh spacing $k = 0.1$ in. on the region

$$D = \{\, (x,y) \mid 0 \le x, y \le 0.5 \,\}.$$

Approximate the solution to Laplace's equation at each grid point, and use the two sets of boundary conditions to derive a linear system to be solved by the Gauss-Seidel method.

8. A 6-cm by 5-cm rectangular silver plate has heat being uniformly generated at each point at the rate $q = 1.5$ cal/cm^3·s. Let x represent the distance along the edge of the plate of length 6 cm and y be the distance along the edge of the plate of length 5 cm. Suppose the temperature u along the edges is kept at the following temperatures:

$$u(x,0) = x(6-x),\ u(x,5) = 0, \quad 0 \le x \le 6,$$
$$u(0,y) = y(5-y),\ u(6,y) = 0, \quad 0 \le y \le 5,$$

where the origin lies at a corner of the plate with coordinates $(0, 0)$ and the edges lie along the positive x- and y-axes. The steady-state temperature $u = u(x, y)$ satisfies Poisson's equation:

$$\frac{\partial^2 u}{\partial x^2}(x, y) + \frac{\partial^2 u}{\partial y^2}(x, y) = -\frac{q}{K}, \quad 0 < x < 6,\ 0 < y < 5,$$

where K, the thermal conductivity, is 1.04 cal/cm·deg·s. Approximate the temperature $u(x, y)$ using Algorithm 12.1 with $h = 0.4$ and $k = \frac{1}{3}$.

12.2 Parabolic Partial Differential Equations

The *parabolic* partial differential equation we consider is the heat, or diffusion, equation

$$\frac{\partial u}{\partial t}(x, t) = \alpha^2 \frac{\partial^2 u}{\partial x^2}(x, t), \quad 0 < x < l, \quad t > 0, \tag{12.6}$$

subject to the conditions

$$u(0, t) = u(l, t) = 0, \quad t > 0, \quad \text{and} \quad u(x, 0) = f(x), \quad 0 \le x \le l.$$

The approach we use to approximate the solution to this problem involves finite differences and is similar to the method used in Section 12.1.

First select an integer $m > 0$ and define the x-axis step size $h = l/m$. Then select a time-step size k. The grid points for this situation are (x_i, t_j), where $x_i = ih$, for $i = 0, 1, \ldots, m$, and $t_j = jk$, for $j = 0, 1, \ldots$.

Forward Difference Method

We obtain the difference method using the Taylor series in t to form the difference quotient

$$\frac{\partial u}{\partial t}(x_i, t_j) = \frac{u(x_i, t_j + k) - u(x_i, t_j)}{k} - \frac{k}{2}\frac{\partial^2 u}{\partial t^2}(x_i, \mu_j), \tag{12.7}$$

for some $\mu_j \in (t_j, t_{j+1})$, and the Taylor series in x to form the difference quotient

$$\frac{\partial^2 u}{\partial x^2}(x_i, t_j) = \frac{u(x_i + h, t_j) - 2u(x_i, t_j) + u(x_i - h, t_j)}{h^2} - \frac{h^2}{12}\frac{\partial^4 u}{\partial x^4}(\xi_i, t_j), \tag{12.8}$$

where $\xi_i \in (x_{i-1}, x_{i+1})$.

The parabolic partial differential equation (12.6) implies that at interior gridpoints (x_i, t_j), for each $i = 1, 2, \ldots, m-1$ and $j = 1, 2, \ldots$, we have

$$\frac{\partial u}{\partial t}(x_i, t_j) - \alpha^2 \frac{\partial^2 u}{\partial x^2}(x_i, t_j) = 0,$$

so the difference method using the difference quotients (12.7) and (12.8) is

$$\frac{w_{i,j+1} - w_{ij}}{k} - \alpha^2 \frac{w_{i+1,j} - 2w_{ij} + w_{i-1,j}}{h^2} = 0, \tag{12.9}$$

where w_{ij} approximates $u(x_i, t_j)$.

The local truncation error for this difference equation is

$$\tau_{ij} = \frac{k}{2}\frac{\partial^2 u}{\partial t^2}(x_i, \mu_j) - \alpha^2 \frac{h^2}{12}\frac{\partial^4 u}{\partial x^4}(\xi_i, t_j). \tag{12.10}$$

Solving Eq. (12.9) for $w_{i,j+1}$ gives

$$w_{i,j+1} = \left(1 - \frac{2\alpha^2 k}{h^2}\right) w_{ij} + \alpha^2 \frac{k}{h^2}(w_{i+1,j} + w_{i-1,j}), \tag{12.11}$$

for each $i = 1, 2, \ldots, m-1$ and $j = 1, 2, \ldots$.

So we have

$$w_{0,0} = f(x_0), \quad w_{1,0} = f(x_1), \quad \ldots w_{m,0} = f(x_m).$$

Then we generate the next t-row by

$$\begin{aligned}
w_{0,1} &= u(0, t_1) = 0;\\
w_{1,1} &= \left(1 - \frac{2\alpha^2 k}{h^2}\right) w_{1,0} + \alpha^2 \frac{k}{h^2}(w_{2,0} + w_{0,0});\\
w_{2,1} &= \left(1 - \frac{2\alpha^2 k}{h^2}\right) w_{2,0} + \alpha^2 \frac{k}{h^2}(w_{3,0} + w_{1,0});\\
&\vdots\\
w_{m-1,1} &= \left(1 - \frac{2\alpha^2 k}{h^2}\right) w_{m-1,0} + \alpha^2 \frac{k}{h^2}(w_{m,0} + w_{m-2,0});\\
w_{m,1} &= u(m, t_1) = 0.
\end{aligned}$$

Now we can use the $w_{i,1}$ values to generate all the $w_{i,2}$ values and so on.

The explicit nature of the difference method implies that the $(m-1) \times (m-1)$ matrix associated with this system can be written in the tridiagonal form

$$A = \begin{bmatrix} (1-2\lambda) & \lambda & 0 & \cdots & 0 \\ \lambda & (1-2\lambda) & \lambda & \ddots & \vdots \\ 0 & \ddots & \ddots & \ddots & 0 \\ \vdots & & \ddots & \ddots & \lambda \\ 0 & \cdots & 0 & \lambda & (1-2\lambda) \end{bmatrix},$$

where $\lambda = \alpha^2(k/h^2)$. If we let

$$\mathbf{w}^{(0)} = (f(x_1), f(x_2), \ldots, f(x_{m-1}))^t$$

and

$$\mathbf{w}^{(j)} = (w_{1j}, w_{2j}, \ldots, w_{m-1,j})^t, \quad \text{for each } j = 1, 2, \ldots,$$

then the approximate solution is given by

$$\mathbf{w}^{(j)} = A\mathbf{w}^{(j-1)}, \quad \text{for each } j = 1, 2, \ldots,$$

so $\mathbf{w}^{(j)}$ is obtained from $\mathbf{w}^{(j-1)}$ by a simple matrix multiplication. This is known as the **Forward-Difference method**, and the approximation at the cyan point shown in Figure 12.8 uses information from the other points marked on that figure. If the solution to the partial differential equation has four continuous partial derivatives in x and two in t, then Eq. (12.10) implies that the method is of order $O(k + h^2)$.

Figure 12.8

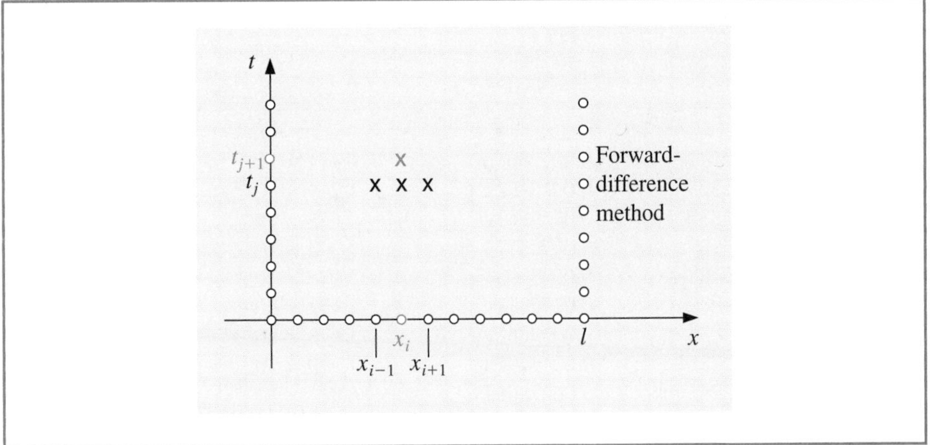

Example 1 Use steps sizes **(a)** $h = 0.1$ and $k = 0.0005$ and **(b)** $h = 0.1$ and $k = 0.01$ to approximate the solution to the heat equation

$$\frac{\partial u}{\partial t}(x,t) - \frac{\partial^2 u}{\partial x^2}(x,t) = 0, \quad 0 < x < 1, \quad 0 \le t,$$

with boundary conditions

$$u(0,t) = u(1,t) = 0, \quad 0 < t,$$

and initial conditions

$$u(x,0) = \sin(\pi x), \quad 0 \le x \le 1.$$

Compare the results at $t = 0.5$ to the exact solution

$$u(x,t) = e^{-\pi^2 t}\sin(\pi x).$$

Solution **(a)** Forward-Difference method with $h = 0.1$, $k = 0.0005$ and $\lambda = (1)^2(0.0005/(0.1)^2) = 0.05$ gives the results in the third column of Table 12.3. As can be seem from the fourth column, these results are quite accurate.

(b) Forward-Difference method with $h = 0.1$, $k = 0.01$ and $\lambda = (1)^2(0.01/(0.1)^2) = 1$ gives the results in the fifth column of Table 12.3. As can be seem from the sixth column, these results are worthless. ■

Stability Considerations

A truncation error of order $O(k + h^2)$ is expected in Example 1. Although this is obtained with $h = 0.1$ and $k = 0.0005$, it certainly is not obtained when $h = 0.1$ and $k = 0.01$. To explain the difficulty, we need to look at the stability of the Forward-Difference method.

Suppose that an error $\mathbf{e}^{(0)} = \left(e_1^{(0)}, e_2^{(0)}, \ldots, e_{m-1}^{(0)}\right)^t$ is made in representing the initial data

$$\mathbf{w}^{(0)} = \left(f(x_1), f(x_2), \ldots, f(x_{m-1})\right)^t$$

(or in any particular step, the choice of the initial step is simply for convenience). An error of $A\mathbf{e}^{(0)}$ propagates in $\mathbf{w}^{(1)}$, because

Table 12.3

x_i	$u(x_i, 0.5)$	$w_{i,1000}$ $k = 0.0005$	$\lvert u(x_i, 0.5) - w_{i,1000}\rvert$	$w_{i,50}$ $k = 0.01$	$\lvert u(x_i, 0.5) - w_{i,50}\rvert$
0.0	0	0		0	
0.1	0.00222241	0.00228652	6.411×10^{-5}	8.19876×10^{7}	8.199×10^{7}
0.2	0.00422728	0.00434922	1.219×10^{-4}	-1.55719×10^{8}	1.557×10^{8}
0.3	0.00581836	0.00598619	1.678×10^{-4}	2.13833×10^{8}	2.138×10^{8}
0.4	0.00683989	0.00703719	1.973×10^{-4}	-2.50642×10^{8}	2.506×10^{8}
0.5	0.00719188	0.00739934	2.075×10^{-4}	2.62685×10^{8}	2.627×10^{8}
0.6	0.00683989	0.00703719	1.973×10^{-4}	-2.49015×10^{8}	2.490×10^{8}
0.7	0.00581836	0.00598619	1.678×10^{-4}	2.11200×10^{8}	2.112×10^{8}
0.8	0.00422728	0.00434922	1.219×10^{-4}	-1.53086×10^{8}	1.531×10^{8}
0.9	0.00222241	0.00228652	6.511×10^{-5}	8.03604×10^{7}	8.036×10^{7}
1.0	0	0		0	

$$\mathbf{w}^{(1)} = A\left(\mathbf{w}^{(0)} + \mathbf{e}^{(0)}\right) = A\mathbf{w}^{(0)} + A\mathbf{e}^{(0)}.$$

This process continues. At the nth time step, the error in $\mathbf{w}^{(n)}$ due to $\mathbf{e}^{(0)}$ is $A^n\mathbf{e}^{(0)}$. The method is consequently stable precisely when these errors do not grow as n increases. But this is true if and only if for any initial error $\mathbf{e}^{(0)}$, we have $\|A^n\mathbf{e}^{(0)}\| \le \|\mathbf{e}^{(0)}\|$ for all n. Hence, we must have $||A^n|| \le 1$, a condition that, by Theorem 7.15 on page 446, requires that $\rho(A^n) = (\rho(A))^n \le 1$. The Forward-Difference method is therefore stable only if $\rho(A) \le 1$.

The eigenvalues of A can be shown (see Exercise 13) to be

$$\mu_i = 1 - 4\lambda\left(\sin\left(\frac{i\pi}{2m}\right)\right)^2, \quad \text{for each } i = 1, 2, \ldots, m-1.$$

So the condition for stability consequently reduces to determining whether

$$\rho(A) = \max_{1\le i\le m-1}\left|1 - 4\lambda\left(\sin\left(\frac{i\pi}{2m}\right)\right)^2\right| \le 1,$$

and this simplifies to

$$0 \le \lambda\left(\sin\left(\frac{i\pi}{2m}\right)\right)^2 \le \frac{1}{2}, \quad \text{for each } i = 1, 2, \ldots, m-1.$$

Stability requires that this inequality condition hold as $h \to 0$, or, equivalently, as $m \to \infty$. The fact that

$$\lim_{m\to\infty}\left[\sin\left(\frac{(m-1)\pi}{2m}\right)\right]^2 = 1$$

means that stability will occur only if $0 \le \lambda \le \frac{1}{2}$.

By definition $\lambda = \alpha^2(k/h^2)$, so this inequality requires that h and k be chosen so that

$$\alpha^2\frac{k}{h^2} \le \frac{1}{2}.$$

In Example 1 we have $\alpha^2 = 1$, so this condition is satisfied when $h = 0.1$ and $k = 0.0005$. But when k was increased to 0.01 with no corresponding increase in h, the ratio was

$$\frac{0.01}{(0.1)^2} = 1 > \frac{1}{2},$$

and stability problems became immediately apparent and dramatic.

Consistent with the terminology of Chapter 5, we call the Forward-Difference method **conditionally stable**. The method converges to the solution of Eq. (12.6) with rate of convergence $O(k + h^2)$, provided

$$\alpha^2 \frac{k}{h^2} \leq \frac{1}{2}$$

and the required continuity conditions on the solution are met. (For a detailed proof of this fact, see [IK, pp. 502–505].)

Backward-Difference Method

To obtain a method that is **unconditionally stable**, we consider an implicit-difference method that results from using the backward-difference quotient for $(\partial u/\partial t)(x_i, t_j)$ in the form

$$\frac{\partial u}{\partial t}(x_i, t_j) = \frac{u(x_i, t_j) - u(x_i, t_{j-1})}{k} + \frac{k}{2}\frac{\partial^2 u}{\partial t^2}(x_i, \mu_j),$$

where μ_j is in (t_{j-1}, t_j). Substituting this equation, together with Eq. (12.8) for $\partial^2 u/\partial x^2$, into the partial differential equation gives

$$\frac{u(x_i, t_j) - u(x_i, t_{j-1})}{k} - \alpha^2 \frac{u(x_{i+1}, t_j) - 2u(x_i, t_j) + u(x_{i-1}, t_j)}{h^2}$$
$$= -\frac{k}{2}\frac{\partial^2 u}{\partial t^2}(x_i, \mu_j) - \alpha^2 \frac{h^2}{12}\frac{\partial^4 u}{\partial x^4}(\xi_i, t_j),$$

for some $\xi_i \in (x_{i-1}, x_{i+1})$. The **Backward-Difference method** that results is

$$\frac{w_{ij} - w_{i,j-1}}{k} - \alpha^2 \frac{w_{i+1,j} - 2w_{ij} + w_{i-1,j}}{h^2} = 0, \tag{12.12}$$

for each $i = 1, 2, \ldots, m-1$ and $j = 1, 2, \ldots$.

The Backward-Difference method involves the mesh points (x_i, t_{j-1}), (x_{i-1}, t_j), and (x_{i+1}, t_j) to approximate the value at (x_i, t_j), as illustrated in Figure 12.9.

Figure 12.9

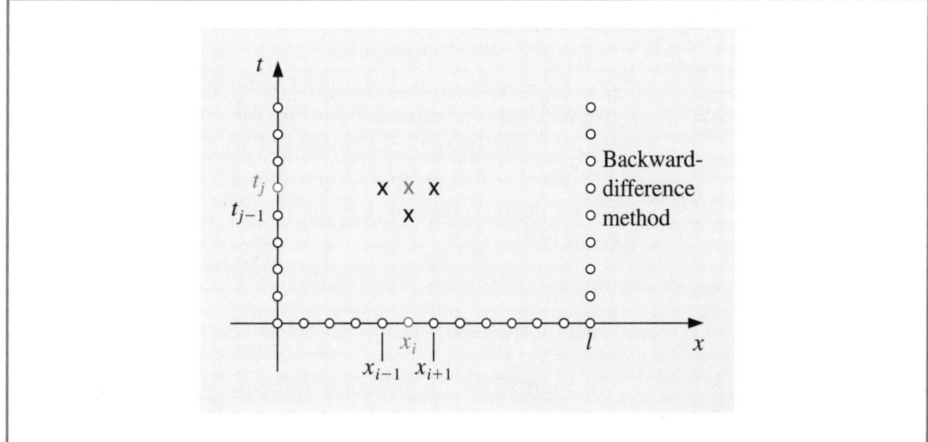

Since the boundary and initial conditions associated with the problem give information at the circled mesh points, the figure shows that no explicit procedures can be used to solve Eq. (12.12). Recall that in the Forward-Difference method (see Figure 12.10), approximations at (x_{i-1}, t_{j-1}), (x_i, t_{j-1}), and (x_{i+1}, t_{j-1}) were used to find the approximation at (x_i, t_j). So an explicit method could be used to find the approximations, based on the information from the initial and boundary conditions.

Figure 12.10

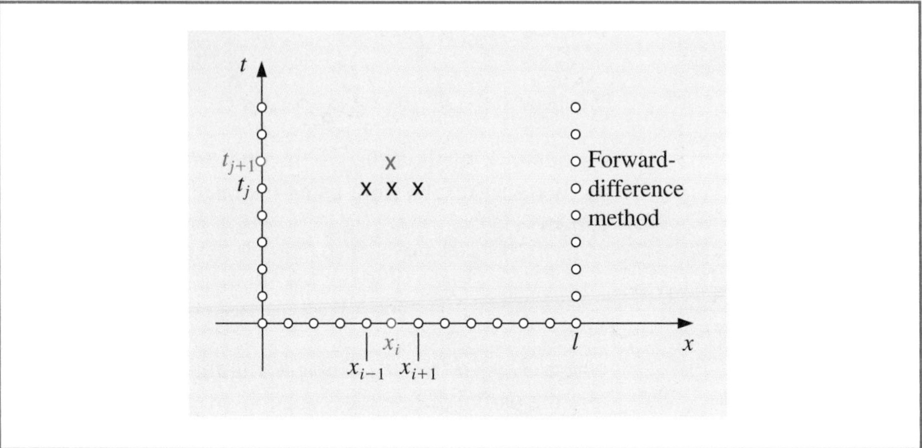

If we again let λ denote the quantity $\alpha^2(k/h^2)$, the Backward-Difference method becomes

$$(1+2\lambda)w_{ij} - \lambda w_{i+1,j} - \lambda w_{i-1,j} = w_{i,j-1},$$

for each $i = 1, 2, \ldots, m-1$ and $j = 1, 2, \ldots$. Using the knowledge that $w_{i,0} = f(x_i)$, for each $i = 1, 2, \ldots, m-1$ and $w_{m,j} = w_{0,j} = 0$, for each $j = 1, 2, \ldots$, this difference method has the matrix representation:

$$\begin{bmatrix} (1+2\lambda) & -\lambda & 0 & \cdots & 0 \\ -\lambda & \ddots & \ddots & \ddots & \vdots \\ 0 & \ddots & \ddots & \ddots & 0 \\ \vdots & \ddots & \ddots & \ddots & -\lambda \\ 0 & \cdots & 0 & -\lambda & (1+2\lambda) \end{bmatrix} \begin{bmatrix} w_{1,j} \\ w_{2,j} \\ \vdots \\ w_{m-1,j} \end{bmatrix} = \begin{bmatrix} w_{1,j-1} \\ w_{2,j-1} \\ \vdots \\ w_{m-1,j-1} \end{bmatrix}, \tag{12.13}$$

or $A\mathbf{w}^{(j)} = \mathbf{w}^{(j-1)}$, for each $i = 1, 2, \ldots$.

Hence, we must now solve a linear system to obtain $\mathbf{w}^{(j)}$ from $\mathbf{w}^{(j-1)}$. However $\lambda > 0$, so the matrix A is positive definite and strictly diagonally dominant, as well as being tridiagonal. We can consequently use either the Crout Factorization Algorithm 6.7 or the SOR Algorithm 7.3 to solve this system. Algorithm 12.2 solves (12.13) using Crout factorization, which is acceptable unless m is large. In this algorithm we assume, for stopping purposes, that a bound is given for t.

ALGORITHM **12.2**

Heat Equation Backward-Difference

To approximate the solution to the parabolic partial differential equation

$$\frac{\partial u}{\partial t}(x,t) - \alpha^2 \frac{\partial^2 u}{\partial x^2}(x,t) = 0, \quad 0 < x < l, \quad 0 < t < T,$$

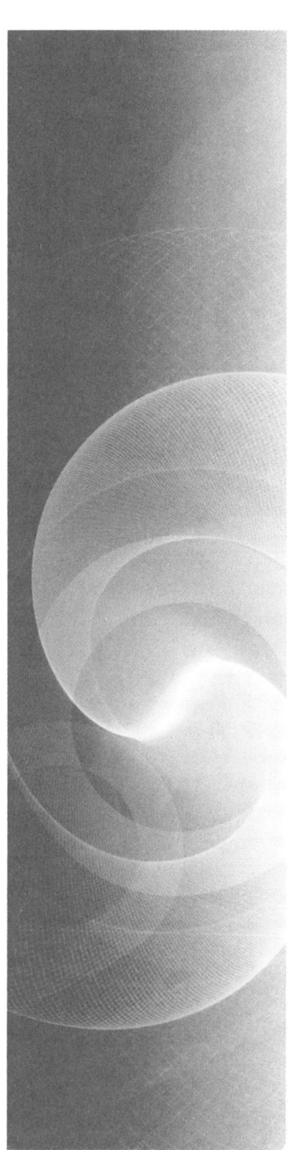

subject to the boundary conditions

$$u(0,t) = u(l,t) = 0, \quad 0 < t < T,$$

and the initial conditions

$$u(x,0) = f(x), \quad 0 \le x \le l:$$

INPUT endpoint l; maximum time T; constant α; integers $m \ge 3$, $N \ge 1$.

OUTPUT approximations $w_{i,j}$ to $u(x_i, t_j)$ for each $i = 1, \ldots, m-1$ and $j = 1, \ldots, N$.

Step 1 Set $h = l/m$;
$k = T/N$;
$\lambda = \alpha^2 k/h^2$.

Step 2 For $i = 1, \ldots, m-1$ set $w_i = f(ih)$. (*Initial values.*)
(*Steps* 3–11 *solve a tridiagonal linear system using Algorithm* 6.7.)

Step 3 Set $l_1 = 1 + 2\lambda$;
$u_1 = -\lambda/l_1$.

Step 4 For $i = 2, \ldots, m-2$ set $l_i = 1 + 2\lambda + \lambda u_{i-1}$;
$u_i = -\lambda/l_i$.

Step 5 Set $l_{m-1} = 1 + 2\lambda + \lambda u_{m-2}$.

Step 6 For $j = 1, \ldots, N$ do Steps 7–11.

Step 7 Set $t = jk$; (*Current* t_j.)
$z_1 = w_1/l_1$.

Step 8 For $i = 2, \ldots, m-1$ set $z_i = (w_i + \lambda z_{i-1})/l_i$.

Step 9 Set $w_{m-1} = z_{m-1}$.

Step 10 For $i = m-2, \ldots, 1$ set $w_i = z_i - u_i w_{i+1}$.

Step 11 OUTPUT (t); (*Note:* $t = t_j$.)
For $i = 1, \ldots, m-1$ set $x = ih$;
OUTPUT (x, w_i). (*Note*: $w_i = w_{i,j}$.)

Step 12 STOP. (*The procedure is complete.*) ■

Example 2 Use the Backward-Difference method (Algorithm 12.2) with $h = 0.1$ and $k = 0.01$ to approximate the solution to the heat equation

$$\frac{\partial u}{\partial t}(x,t) - \frac{\partial^2 u}{\partial x^2}(x,t) = 0, \quad 0 < x < 1, \quad 0 < t,$$

subject to the constraints

$$u(0,t) = u(1,t) = 0, \quad 0 < t, \quad u(x,0) = \sin \pi x, \quad 0 \le x \le 1.$$

Solution This problem was considered in Example 1 where we found that choosing $h = 0.1$ and $k = 0.0005$ gave quite accurate results. However,with the values in this example, $h = 0.1$ and $k = 0.01$, the results were exceptionally poor. To demonstrate the unconditional stability of the Backward-Difference method, we will use $h = 0.1$ and $k = 0.01$ and again compare $w_{i,50}$ to $u(x_i, 0.5)$, where $i = 0, 1, \ldots, 10$.

The results listed in Table 12.4 have the same values of h and k as those in the fifth and sixth columns of Table 12.3, which illustrates the stability of this method. ■

Table 12.4

x_i	$w_{i,50}$	$u(x_i, 0.5)$	$\lvert w_{i,50} - u(x_i, 0.5)\rvert$
0.0	0	0	
0.1	0.00289802	0.00222241	6.756×10^{-4}
0.2	0.00551236	0.00422728	1.285×10^{-3}
0.3	0.00758711	0.00581836	1.769×10^{-3}
0.4	0.00891918	0.00683989	2.079×10^{-3}
0.5	0.00937818	0.00719188	2.186×10^{-3}
0.6	0.00891918	0.00683989	2.079×10^{-3}
0.7	0.00758711	0.00581836	1.769×10^{-3}
0.8	0.00551236	0.00422728	1.285×10^{-3}
0.9	0.00289802	0.00222241	6.756×10^{-4}
1.0	0	0	

The reason that the Backward-Difference method does not have the stability problems of the Forward-Difference method can be seen by analyzing the eigenvalues of the matrix A. For the Backward-Difference method (see Exercise 14), the eigenvalues are

$$\mu_i = 1 + 4\lambda\left[\sin\left(\frac{i\pi}{2m}\right)\right]^2, \quad \text{for each } i = 1, 2, \ldots, m-1.$$

Since $\lambda > 0$, so we have $\mu_i > 1$ for all $i = 1, 2, \ldots, m-1$. Since the eigenvalues of A^{-1} are the reciprocals of those of A, the spectral radius of A^{-1}, $\rho(A^{-1}) < 1$. This implies that A^{-1} is a convergent matrix.

An error $\mathbf{e}^{(0)}$ in the initial data produces an error $(A^{-1})^n\mathbf{e}^{(0)}$ at the nth step of the Backward-Difference method. Since A^{-1} is convergent,

$$\lim_{n\to\infty} (A^{-1})^n\mathbf{e}^{(0)} = \mathbf{0}.$$

So the method is stable, independent of the choice of $\lambda = \alpha^2(k/h^2)$. In the terminology of Chapter 5, we call the Backward-Difference method an **unconditionally stable** method. The local truncation error for the method is of order $O(k + h^2)$, provided the solution of the differential equation satisfies the usual differentiability conditions. In this case, the method converges to the solution of the partial differential equation with this same rate of convergence (see [IK], p. 508).

The weakness of the Backward-Difference method results from the fact that the local truncation error has one of order $O(h^2)$, and another of order $O(k)$. This requires that time intervals be made much smaller than the x-axis intervals. It would clearly be desirable to have a procedure with local truncation error of order $O(k^2 + h^2)$. The first step in this direction is to use a difference equation that has $O(k^2)$ error for $u_t(x, t)$ instead of those we have used previously, whose error was $O(k)$. This can be done by using the Taylor series in t for the function $u(x, t)$ at the point (x_i, t_j) and evaluating at (x_i, t_{j+1}) and (x_i, t_{j-1}) to obtain the Centered-Difference formula

L. E. Richardson, who we saw associated with extrapolation, did substantial work in the approximation of partial-differential equations.

$$\frac{\partial u}{\partial t}(x_i, t_j) = \frac{u(x_i, t_{j+1}) - u(x_i, t_{j-1})}{2k} + \frac{k^2}{6}\frac{\partial^3 u}{\partial t^3}(x_i, \mu_j),$$

where $\mu_j \in (t_{j-1}, t_{j+1})$. The difference method that results from substituting this and the usual difference quotient for $(\partial^2 u/\partial x^2)$, Eq. (12.8), into the differential equation is called **Richardson's method** and is given by

$$\frac{w_{i,j+1} - w_{i,j-1}}{2k} - \alpha^2\frac{w_{i+1,j} - 2w_{ij} + w_{i-1,j}}{h^2} = 0. \tag{12.14}$$

This method has local truncation error of order $O(k^2 + h^2)$, but unfortunately, like the Forward-Difference method, it has serious stability problems (see Exercises 11 and 12).

Crank-Nicolson Method

A more rewarding method is derived by averaging the Forward-Difference method at the jth step in t,

$$\frac{w_{i,j+1} - w_{i,j}}{k} - \alpha^2 \frac{w_{i+1,j} - 2w_{i,j} + w_{i-1,j}}{h^2} = 0,$$

which has local truncation error

$$\tau_F = \frac{k}{2}\frac{\partial^2 u}{\partial t^2}(x_i, \mu_j) + O(h^2),$$

and the Backward-Difference method at the $(j+1)$st step in t,

$$\frac{w_{i,j+1} - w_{i,j}}{k} - \alpha^2 \frac{w_{i+1,j+1} - 2w_{i,j+1} + w_{i-1,j+1}}{h^2} = 0,$$

which has local truncation error

$$\tau_B = -\frac{k}{2}\frac{\partial^2 u}{\partial t^2}(x_i, \hat{u}_j) + O(h^2).$$

If we assume that

$$\frac{\partial^2 u}{\partial t^2}(x_i, \hat{\mu}_j) \approx \frac{\partial^2 u}{\partial t^2}(x_i, \mu_j),$$

then the averaged-difference method,

$$\frac{w_{i,j+1} - w_{ij}}{k} - \frac{\alpha^2}{2}\left[\frac{w_{i+1,j} - 2w_{i,j} + w_{i-1,j}}{h^2} + \frac{w_{i+1,j+1} - 2w_{i,j+1} + w_{i-1,j+1}}{h^2}\right] = 0,$$

Following work as a mathematical physicist during World War II, John Crank (1916–2006) did research in the numerical solution of partial differential equations; in particular, heat-conduction problems. The Crank-Nicolson method is based on work done with Phyllis Nicolson (1917–1968), a physicist at Leeds University. Their original paper on the method appeared in 1947 [CN].

has local truncation error of order $O(k^2 + h^2)$, provided, of course, that the usual differentiability conditions are satisfied.

This is known as the **Crank-Nicolson method** and is represented in the matrix form

$$A\mathbf{w}^{(j+1)} = B\mathbf{w}^{(j)}, \quad \text{for each } j = 0, 1, 2, \ldots, \tag{12.15}$$

where

$$\lambda = \alpha^2 \frac{k}{h^2}, \quad \mathbf{w}^{(j)} = (w_{1,j}, w_{2,j}, \ldots, w_{m-1,j})^t,$$

and the matrices A and B are given by:

$$A = \begin{bmatrix} (1+\lambda) & -\frac{\lambda}{2} & 0 & \cdots & 0 \\ -\frac{\lambda}{2} & \ddots & \ddots & \ddots & \vdots \\ 0 & \ddots & \ddots & \ddots & 0 \\ \vdots & \ddots & \ddots & \ddots & -\frac{\lambda}{2} \\ 0 & \cdots & 0 & -\frac{\lambda}{2} & (1+\lambda) \end{bmatrix}$$

and

$$B = \begin{bmatrix} (1-\lambda) & \frac{\lambda}{2} & 0 & \cdots & 0 \\ \frac{\lambda}{2} & \ddots & \ddots & \ddots & \vdots \\ 0 & \ddots & \ddots & \ddots & 0 \\ \vdots & \ddots & \ddots & \ddots & \frac{\lambda}{2} \\ 0 & \cdots & 0 & \frac{\lambda}{2} & (1-\lambda) \end{bmatrix}.$$

The nonsingular matrix A is positive definite, strictly diagonally dominant, and tridiagonal matrix. Either the Crout Factorization 6.7 or the SOR Algorithm 7.3 can be used to obtain $\mathbf{w}^{(j)}$ from $\mathbf{w}^{(j-1)}$, for each $j = 0, 1, 2, \ldots$. Algorithm 12.3 incorporates Crout factorization into the Crank-Nicolson technique. As in Algorithm 12.2, a finite length for the time interval must be specified to determine a stopping procedure. The verification that the Crank-Nicolson method is unconditionally stable and has order of convergence $O(k^2 + h^2)$ can be found in [IK], pp. 508–512. A diagram showing the interaction of the nodes for determining an approximation at (x_i, t_j) is shown in Figure 12.11.

Figure 12.11

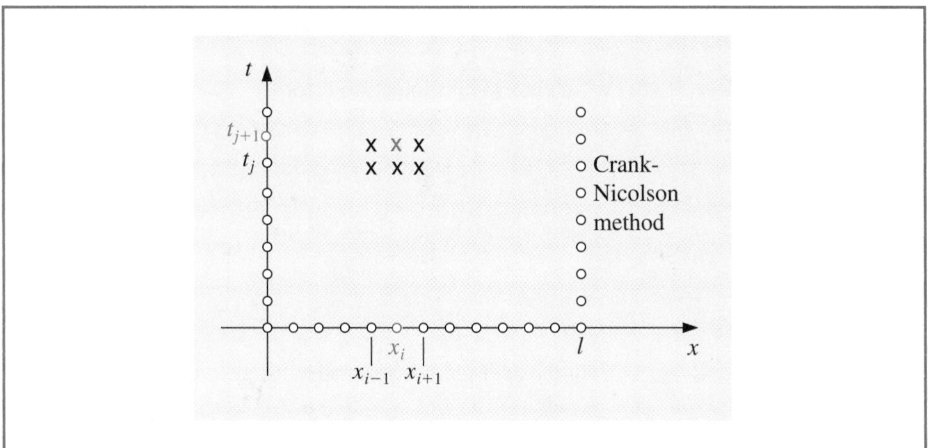

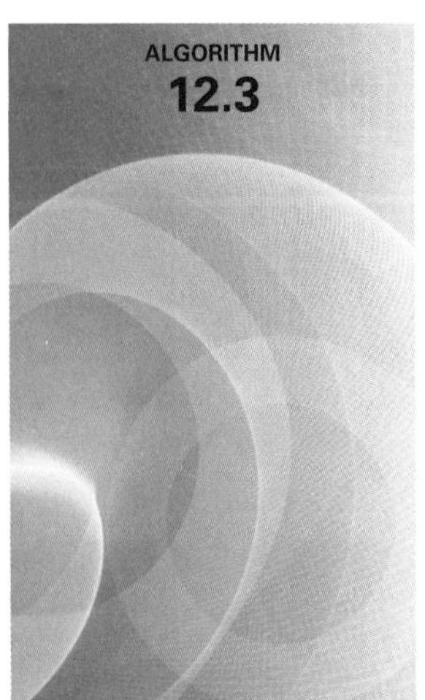

Crank-Nicolson

To approximate the solution to the parabolic partial differential equation

$$\frac{\partial u}{\partial t}(x,t) - \alpha^2 \frac{\partial^2 u}{\partial x^2}(x,t) = 0, \quad 0 < x < l, \quad 0 < t < T,$$

subject to the boundary conditions

$$u(0,t) = u(l,t) = 0, \quad 0 < t < T,$$

and the initial conditions

$$u(x,0) = f(x), \quad 0 \le x \le l:$$

INPUT endpoint l; maximum time T; constant α; integers $m \ge 3$, $N \ge 1$.

OUTPUT approximations $w_{i,j}$ to $u(x_i, t_j)$ for each $i = 1, \ldots, m-1$ and $j = 1, \ldots, N$.

Step 1 Set $h = l/m$;
$k = T/N$;
$\lambda = \alpha^2 k/h^2$;
$w_m = 0$.

Step 2 For $i = 1, \ldots, m-1$ set $w_i = f(ih)$. (*Initial values.*)
(*Steps* 3–11 *solve a tridiagonal linear system using Algorithm* 6.7.)

Step 3 Set $l_1 = 1 + \lambda$;
$u_1 = -\lambda/(2l_1)$.

Step 4 For $i = 2, \ldots, m-2$ set $l_i = 1 + \lambda + \lambda u_{i-1}/2$;
$u_i = -\lambda/(2l_i)$.

Step 5 Set $l_{m-1} = 1 + \lambda + \lambda u_{m-2}/2$.

Step 6 For $j = 1, \ldots, N$ do Steps 7–11.

Step 7 Set $t = jk$; (*Current* t_j.)

$$z_1 = \left[(1-\lambda)w_1 + \frac{\lambda}{2}w_2\right] \Big/ l_1.$$

Step 8 For $i = 2, \ldots, m-1$ set

$$z_i = \left[(1-\lambda)w_i + \frac{\lambda}{2}(w_{i+1} + w_{i-1} + z_{i-1})\right] \Big/ l_i.$$

Step 9 Set $w_{m-1} = z_{m-1}$.

Step 10 For $i = m-2, \ldots, 1$ set $w_i = z_i - u_i w_{i+1}$.

Step 11 OUTPUT (t); (*Note:* $t = t_j$.)
For $i = 1, \ldots, m-1$ set $x = ih$;
OUTPUT (x, w_i). (*Note:* $w_i = w_{i,j}$.)

Step 12 STOP. (*The procedure is complete.*) ■

Example 3 Use the Crank-Nicolson method with $h = 0.1$ and $k = 0.01$ to approximate the solution to the problem

$$\frac{\partial u}{\partial t}(x,t) - \frac{\partial^2 u}{\partial x^2}(x,t) = 0, \quad 0 < x < 1 \quad 0 < t,$$

subject to the conditions

$$u(0,t) = u(1,t) = 0, \quad 0 < t,$$

and

$$u(x,0) = \sin(\pi x), \quad 0 \le x \le 1.$$

Solution Choosing $h = 0.1$ and $k = 0.01$ gives $m = 10$, $N = 50$, and $\lambda = 1$ in Algorithm 12.3. Recall that the Forward-Difference method gave dramatically poor results for this choice of h and k, but the Backward-Difference method gave results that were accurate to about 2×10^{-3} for entries in the middle of the table. The results in Table 12.5 indicate the increase in accuracy of the Crank-Nicolson method over the Backward-Difference method, the best of the two previously discussed techniques. ■

Table 12.5

x_i	$w_{i,50}$	$u(x_i, 0.5)$	$\lvert w_{i,50} - u(x_i, 0.5)\rvert$
0.0	0	0	
0.1	0.00230512	0.00222241	8.271×10^{-5}
0.2	0.00438461	0.00422728	1.573×10^{-4}
0.3	0.00603489	0.00581836	2.165×10^{-4}
0.4	0.00709444	0.00683989	2.546×10^{-4}
0.5	0.00745954	0.00719188	2.677×10^{-4}
0.6	0.00709444	0.00683989	2.546×10^{-4}
0.7	0.00603489	0.00581836	2.165×10^{-4}
0.8	0.00438461	0.00422728	1.573×10^{-4}
0.9	0.00230512	0.00222241	8.271×10^{-5}
1.0	0	0	

EXERCISE SET 12.2

1. Approximate the solution to the following partial differential equation using the Backward-Difference method.

$$\frac{\partial u}{\partial t} - \frac{\partial^2 u}{\partial x^2} = 0, \quad 0 < x < 2,\ 0 < t;$$

$$u(0,t) = u(2,t) = 0, \quad 0 < t, \quad u(x,0) = \sin\frac{\pi}{2}x, \quad 0 \le x \le 2.$$

Use $m = 4$, $T = 0.1$, and $N = 2$, and compare your results to the actual solution $u(x,t) = e^{-(\pi^2/4)t}\sin\frac{\pi}{2}x$.

2. Approximate the solution to the following partial differential equation using the Backward-Difference method.

$$\frac{\partial u}{\partial t} - \frac{1}{16}\frac{\partial^2 u}{\partial x^2} = 0, \quad 0 < x < 1,\ 0 < t;$$

$$u(0,t) = u(1,t) = 0, \quad 0 < t, \quad u(x,0) = 2\sin 2\pi x, \quad 0 \le x \le 1.$$

Use $m = 3$, $T = 0.1$, and $N = 2$, and compare your results to the actual solution $u(x,t) = 2e^{-(\pi^2/4)t}\sin 2\pi x$.

3. Repeat Exercise 1 using the Crank-Nicolson Algorithm.
4. Repeat Exercise 2 using the Crank-Nicolson Algorithm.
5. Use the Forward-Difference method to approximate the solution to the following parabolic partial differential equations.

a. $$\frac{\partial u}{\partial t} - \frac{\partial^2 u}{\partial x^2} = 0, \quad 0 < x < 2,\ 0 < t;$$

$$u(0,t) = u(2,t) = 0, \quad 0 < t,$$

$$u(x,0) = \sin 2\pi x, \quad 0 \le x \le 2.$$

Use $h = 0.4$ and $k = 0.1$, and compare your results at $t = 0.5$ to the actual solution $u(x,t) = e^{-4\pi^2 t}\sin 2\pi x$. Then use $h = 0.4$ and $k = 0.05$, and compare the answers.

b. $$\frac{\partial u}{\partial t} - \frac{\partial^2 u}{\partial x^2} = 0, \quad 0 < x < \pi,\ 0 < t;$$

$$u(0,t) = u(\pi,t) = 0, \quad 0 < t,$$

$$u(x,0) = \sin x, \quad 0 \le x \le \pi.$$

Use $h = \pi/10$ and $k = 0.05$, and compare your results at $t = 0.5$ to the actual solution $u(x,t) = e^{-t}\sin x$.

6. Use the Forward-Difference method to approximate the solution to the following parabolic partial differential equations.

a. $$\frac{\partial u}{\partial t} - \frac{4}{\pi^2}\frac{\partial^2 u}{\partial x^2} = 0, \quad 0 < x < 4, \; 0 < t;$$

$$u(0,t) = u(4,t) = 0, \quad 0 < t,$$

$$u(x,0) = \sin\tfrac{\pi}{4}x\left(1 + 2\cos\tfrac{\pi}{4}x\right), \quad 0 \le x \le 4.$$

Use $h = 0.2$ and $k = 0.04$, and compare your results at $t = 0.4$ to the actual solution $u(x,t) = e^{-t}\sin\frac{\pi}{2}x + e^{-t/4}\sin\frac{\pi}{4}x$.

b. $$\frac{\partial u}{\partial t} - \frac{1}{\pi^2}\frac{\partial^2 u}{\partial x^2} = 0, \quad 0 < x < 1, \; 0 < t;$$

$$u(0,t) = u(1,t) = 0, \quad 0 < t,$$

$$u(x,0) = \cos\pi\left(x - \tfrac{1}{2}\right), \quad 0 \le x \le 1.$$

Use $h = 0.1$ and $k = 0.04$, and compare your results at $t = 0.4$ to the actual solution $u(x,t) = e^{-t}\cos\pi(x - \frac{1}{2})$.

7. Repeat Exercise 5 using the Backward-Difference Algorithm.

8. Repeat Exercise 6 using the Backward-Difference Algorithm.

9. Repeat Exercise 5 using the Crank-Nicolson Algorithm.

10. Repeat Exercise 6 using the Crank-Nicolson Algorithm.

11. Repeat Exercise 5 using Richardson's method.

12. Repeat Exercise 6 using Richardson's method.

13. Show that the eigenvalues for the $(m-1)$ by $(m-1)$ tridiagonal method matrix A given by

$$a_{ij} = \begin{cases} \lambda, & j = i-1 \text{ or } j = i+1, \\ 1 - 2\lambda, & j = i, \\ 0, & \text{otherwise} \end{cases}$$

are

$$\mu_i = 1 - 4\lambda\left(\sin\frac{i\pi}{2m}\right)^2, \quad \text{for each } i = 1, 2, \ldots, m-1,$$

with corresponding eigenvectors $\mathbf{v}^{(i)}$, where $v_j^{(i)} = \sin(ij\pi/m)$.

14. Show that the $(m-1)$ by $(m-1)$ tridiagonal method matrix A given by

$$a_{ij} = \begin{cases} -\lambda, & j = i-1 \text{ or } j = i+1, \\ 1 + 2\lambda, & j = i, \\ 0, & \text{otherwise,} \end{cases}$$

where $\lambda > 0$, is positive definite and diagonally dominant and has eigenvalues

$$\mu_i = 1 + 4\lambda\left(\sin\frac{i\pi}{2m}\right)^2, \quad \text{for each } i = 1, 2, \ldots, m-1,$$

with corresponding eigenvectors $\mathbf{v}^{(i)}$, where $v_j^{(i)} = \sin(ij\pi/m)$.

15. Modify Algorithms 12.2 and 12.3 to include the parabolic partial differential equation

$$\frac{\partial u}{\partial t} - \frac{\partial^2 u}{\partial x^2} = F(x), \quad 0 < x < l, \; 0 < t;$$

$$u(0,t) = u(l,t) = 0, \quad 0 < t;$$

$$u(x,0) = f(x), \quad 0 \le x \le l.$$

16. Use the results of Exercise 15 to approximate the solution to

$$\frac{\partial u}{\partial t} - \frac{\partial^2 u}{\partial x^2} = 2, \quad 0 < x < 1,\ 0 < t;$$
$$u(0,t) = u(1,t) = 0, \quad 0 < t;$$
$$u(x,0) = \sin \pi x + x(1-x),$$

with $h = 0.1$ and $k = 0.01$. Compare your answer at $t = 0.25$ to the actual solution $u(x,t) = e^{-\pi^2 t}\sin \pi x + x(1-x)$.

17. Change Algorithms 12.2 and 12.3 to accommodate the partial differential equation

$$\frac{\partial u}{\partial t} - \alpha^2 \frac{\partial^2 u}{\partial x^2} = 0, \quad 0 < x < l,\ 0 < t;$$
$$u(0,t) = \phi(t),\ u(l,t) = \Psi(t), \quad 0 < t;$$
$$u(x,0) = f(x), \quad 0 \le x \le l,$$

where $f(0) = \phi(0)$ and $f(l) = \Psi(0)$.

18. The temperature $u(x,t)$ of a long, thin rod of constant cross section and homogeneous conducting material is governed by the one-dimensional heat equation. If heat is generated in the material, for example, by resistance to current or nuclear reaction, the heat equation becomes

$$\frac{\partial^2 u}{\partial x^2} + \frac{Kr}{\rho C} = K\frac{\partial u}{\partial t}, \quad 0 < x < l, \quad 0 < t,$$

where l is the length, ρ is the density, C is the specific heat, and K is the thermal diffusivity of the rod. The function $r = r(x,t,u)$ represents the heat generated per unit volume. Suppose that

$$l = 1.5 \text{ cm}, \quad K = 1.04 \text{ cal/cm} \cdot \text{deg} \cdot \text{s}, \quad \rho = 10.6 \text{ g/cm}^3, \quad C = 0.056 \text{ cal/g} \cdot \text{deg},$$

and

$$r(x,t,u) = 5.0 \text{ cal/cm}^3 \cdot \text{s}.$$

If the ends of the rod are kept at 0°C, then

$$u(0,t) = u(l,t) = 0, \quad t > 0.$$

Suppose the initial temperature distribution is given by

$$u(x,0) = \sin \frac{\pi x}{l}, \quad 0 \le x \le l.$$

Use the results of Exercise 15 to approximate the temperature distribution with $h = 0.15$ and $k = 0.0225$.

19. Sagar and Payne [SP] analyze the stress-strain relationships and material properties of a cylinder alternately subjected to heating and cooling and consider the equation

$$\frac{\partial^2 T}{\partial r^2} + \frac{1}{r}\frac{\partial T}{\partial r} = \frac{1}{4K}\frac{\partial T}{\partial t}, \quad \frac{1}{2} < r < 1,\ 0 < T,$$

where $T = T(r,t)$ is the temperature, r is the radial distance from the center of the cylinder, t is time, and K is a diffusivity coefficient.

a. Find approximations to $T(r, 10)$ for a cylinder with outside radius 1, given the initial and boundary conditions:

$$T(1,t) = 100 + 40t, \quad T\left(\frac{1}{2}, t\right) = t, \quad 0 \le t \le 10;$$
$$T(r,0) = 200(r - 0.5), \quad 0.5 \le r \le 1.$$

Use a modification of the Backward-Difference method with $K = 0.1$, $k=0.5$, and $h=\Delta r=0.1$.

b. Use the temperature distribution of part (a) to calculate the strain I by approximating the integral

$$I = \int_{0.5}^{1} \alpha T(r,t) r \, dr,$$

where $\alpha = 10.7$ and $t = 10$. Use the Composite Trapezoidal method with $n = 5$.

12.3 Hyperbolic Partial Differential Equations

In this section, we consider the numerical solution to the **wave equation**, an example of a *hyperbolic* partial differential equation. The wave equation is given by the differential equation

$$\frac{\partial^2 u}{\partial t^2}(x,t) - \alpha^2 \frac{\partial^2 u}{\partial x^2}(x,t) = 0, \quad 0 < x < l, \quad t > 0, \tag{12.16}$$

subject to the conditions

$$u(0,t) = u(l,t) = 0, \quad \text{for} \quad t > 0,$$

$$u(x,0) = f(x), \quad \text{and} \quad \frac{\partial u}{\partial t}(x,0) = g(x), \quad \text{for} \quad 0 \le x \le l,$$

where α is a constant dependent on the physical conditions of the problem.

Select an integer $m > 0$ to define the x-axis grid points using $h = l/m$. In addition, select a time-step size $k > 0$. The mesh points (x_i, t_j) are defined by

$$x_i = ih \quad \text{and} \quad t_j = jk,$$

for each $i = 0, 1, \dots, m$ and $j = 0, 1, \dots$.

At any interior mesh point (x_i, t_j), the wave equation becomes

$$\frac{\partial^2 u}{\partial t^2}(x_i,t_j) - \alpha^2 \frac{\partial^2 u}{\partial x^2}(x_i,t_j) = 0. \tag{12.17}$$

The difference method is obtained using the centered-difference quotient for the second partial derivatives given by

$$\frac{\partial^2 u}{\partial t^2}(x_i,t_j) = \frac{u(x_i,t_{j+1}) - 2u(x_i,t_j) + u(x_i,t_{j-1})}{k^2} - \frac{k^2}{12}\frac{\partial^4 u}{\partial t^4}(x_i,\mu_j),$$

where $\mu_j \in (t_{j-1}, t_{j+1})$, and

$$\frac{\partial^2 u}{\partial x^2}(x_i,t_j) = \frac{u(x_{i+1},t_j) - 2u(x_i,t_j) + u(x_{i-1},t_j)}{h^2} - \frac{h^2}{12}\frac{\partial^4 u}{\partial x^4}(\xi_i,t_j),$$

where $\xi_i \in (x_{i-1}, x_{i+1})$. Substituting these into Eq. (12.17) gives

$$\frac{u(x_i,t_{j+1}) - 2u(x_i,t_j) + u(x_i,t_{j-1})}{k^2} - \alpha^2 \frac{u(x_{i+1},t_j) - 2u(x_i,t_j) + u(x_{i-1},t_j)}{h^2}$$
$$= \frac{1}{12}\left[k^2 \frac{\partial^4 u}{\partial t^4}(x_i,\mu_j) - \alpha^2 h^2 \frac{\partial^4 u}{\partial x^4}(\xi_i,t_j)\right].$$

Neglecting the error term

$$\tau_{i,j} = \frac{1}{12}\left[k^2 \frac{\partial^4 u}{\partial t^4}(x_i,\mu_j) - \alpha^2 h^2 \frac{\partial^4 u}{\partial x^4}(\xi_i,t_j)\right], \tag{12.18}$$

leads to the difference equation

$$\frac{w_{i,j+1} - 2w_{i,j} + w_{i,j-1}}{k^2} - \alpha^2 \frac{w_{i+1,j} - 2w_{i,j} + w_{i-1,j}}{h^2} = 0.$$

Define $\lambda = \alpha k/h$. Then we can write the difference equation as

$$w_{i,j+1} - 2w_{i,j} + w_{i,j-1} - \lambda^2 w_{i+1,j} + 2\lambda^2 w_{i,j} - \lambda^2 w_{i-1,j} = 0$$

and solve for $w_{i,j+1}$, the most advanced time-step approximation, to obtain

$$w_{i,j+1} = 2(1 - \lambda^2)w_{i,j} + \lambda^2(w_{i+1,j} + w_{i-1,j}) - w_{i,j-1}. \tag{12.19}$$

This equation holds for each $i = 1, 2, \ldots, m-1$ and $j = 1, 2, \ldots$. The boundary conditions give

$$w_{0,j} = w_{m,j} = 0, \quad \text{for each } j = 1, 2, 3, \ldots, \tag{12.20}$$

and the initial condition implies that

$$w_{i,0} = f(x_i), \quad \text{for each } i = 1, 2, \ldots, m-1. \tag{12.21}$$

Writing this set of equations in matrix form gives

$$\begin{bmatrix} w_{1,j+1} \\ w_{2,j+1} \\ \vdots \\ w_{m-1,j+1} \end{bmatrix} = \begin{bmatrix} 2(1-\lambda^2) & \lambda^2 & 0 & \cdots & 0 \\ \lambda^2 & 2(1-\lambda^2) & \lambda^2 & \ddots & \vdots \\ 0 & \ddots & \ddots & \ddots & 0 \\ \vdots & \ddots & \ddots & \ddots & \lambda^2 \\ 0 & \cdots & 0 & \lambda^2 & 2(1-\lambda^2) \end{bmatrix} \begin{bmatrix} w_{1,j} \\ w_{2,j} \\ \vdots \\ w_{m-1,j} \end{bmatrix} - \begin{bmatrix} w_{1,j-1} \\ w_{2,j-1} \\ \vdots \\ w_{m-1,j-1} \end{bmatrix}. \tag{12.22}$$

Equations (12.18) and (12.19) imply that the $(j+1)$st time step requires values from the jth and $(j-1)$st time steps. (See Figure 12.12.) This produces a minor starting problem because values for $j = 0$ are given by Eq. (12.20), but values for $j = 1$, which are needed in Eq. (12.18) to compute $w_{i,2}$, must be obtained from the initial-velocity condition

$$\frac{\partial u}{\partial t}(x, 0) = g(x), \quad 0 \leq x \leq l.$$

Figure 12.12

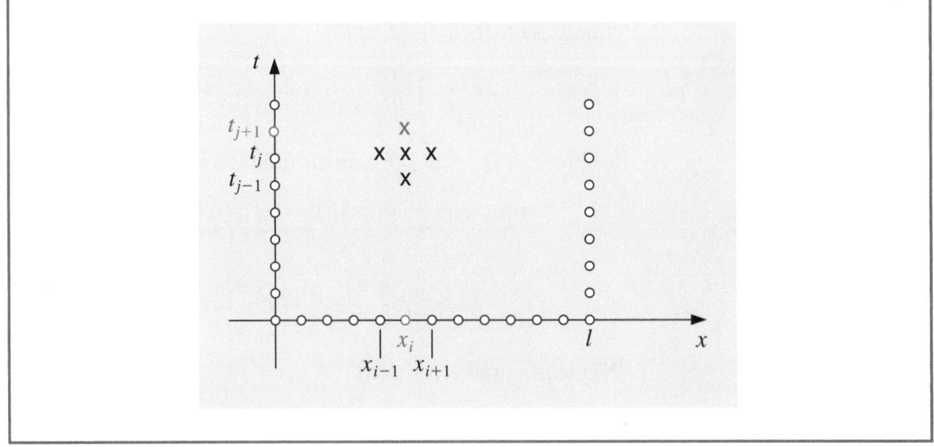

One approach is to replace $\partial u/\partial t$ by a forward-difference approximation,

$$\frac{\partial u}{\partial t}(x_i, 0) = \frac{u(x_i, t_1) - u(x_i, 0)}{k} - \frac{k}{2}\frac{\partial^2 u}{\partial t^2}(x_i, \tilde{\mu}_i), \tag{12.23}$$

for some $\tilde{\mu}_i$ in $(0, t_1)$. Solving for $u(x_i, t_1)$ in the equation gives

$$\begin{aligned} u(x_i, t_1) &= u(x_i, 0) + k\frac{\partial u}{\partial t}(x_i, 0) + \frac{k^2}{2}\frac{\partial^2 u}{\partial t^2}(x_i, \tilde{\mu}_i) \\ &= u(x_i, 0) + kg(x_i) + \frac{k^2}{2}\frac{\partial^2 u}{\partial t^2}(x_i, \tilde{\mu}_i). \end{aligned}$$

Deleting the truncation term gives the approximation,

$$w_{i,1} = w_{i,0} + kg(x_i), \quad \text{for each } i = 1, \ldots, m-1. \tag{12.24}$$

However, this approximation has truncation error of only $O(k)$ whereas the truncation error in Eq. (12.19) is $O(k^2)$.

Improving the Initial Approximation

To obtain a better approximation to $u(x_i, 0)$, expand $u(x_i, t_1)$ in a second Maclaurin polynomial in t. Then

$$u(x_i, t_1) = u(x_i, 0) + k\frac{\partial u}{\partial t}(x_i, 0) + \frac{k^2}{2}\frac{\partial^2 u}{\partial t^2}(x_i, 0) + \frac{k^3}{6}\frac{\partial^3 u}{\partial t^3}(x_i, \hat{\mu}_i),$$

for some $\hat{\mu}_i$ in $(0, t_1)$. If f'' exists, then

$$\frac{\partial^2 u}{\partial t^2}(x_i, 0) = \alpha^2\frac{\partial^2 u}{\partial x^2}(x_i, 0) = \alpha^2\frac{d^2 f}{dx^2}(x_i) = \alpha^2 f''(x_i)$$

and

$$u(x_i, t_1) = u(x_i, 0) + kg(x_i) + \frac{\alpha^2 k^2}{2}f''(x_i) + \frac{k^3}{6}\frac{\partial^3 u}{\partial t^3}(x_i, \hat{\mu}_i).$$

This produces an approximation with error $O(k^3)$:

$$w_{i1} = w_{i0} + kg(x_i) + \frac{\alpha^2 k^2}{2}f''(x_i).$$

If $f \in C^4[0, 1]$ but $f''(x_i)$ is not readily available, we can use the difference equation in Eq. (4.9) to write

$$f''(x_i) = \frac{f(x_{i+1}) - 2f(x_i) + f(x_{i-1})}{h^2} - \frac{h^2}{12}f^{(4)}(\tilde{\xi}_i),$$

for some $\tilde{\xi}_i$ in (x_{i-1}, x_{i+1}). This implies that

$$u(x_i, t_1) = u(x_i, 0) + kg(x_i) + \frac{k^2\alpha^2}{2h^2}[f(x_{i+1}) - 2f(x_i) + f(x_{i-1})] + O(k^3 + h^2k^2).$$

Because $\lambda = k\alpha/h$, we can write this as

$$\begin{aligned} u(x_i, t_1) &= u(x_i, 0) + kg(x_i) + \frac{\lambda^2}{2}[f(x_{i+1}) - 2f(x_i) + f(x_{i-1})] + O(k^3 + h^2k^2) \\ &= (1 - \lambda^2)f(x_i) + \frac{\lambda^2}{2}f(x_{i+1}) + \frac{\lambda^2}{2}f(x_{i-1}) + kg(x_i) + O(k^3 + h^2k^2). \end{aligned}$$

Thus, the difference equation,

$$w_{i,1} = (1-\lambda^2)f(x_i) + \frac{\lambda^2}{2}f(x_{i+1}) + \frac{\lambda^2}{2}f(x_{i-1}) + kg(x_i), \tag{12.25}$$

can be used to find $w_{i,1}$, for each $i = 1, 2, \ldots, m-1$. To determine subsequent approximates we use the system in (12.22).

Algorithm 12.4 uses Eq. (12.25) to approximate $w_{i,1}$, although Eq. (12.24) could also be used. It is assumed that there is an upper bound for the value of t to be used in the stopping technique, and that $k = T/N$, where N is also given.

Wave Equation Finite-Difference

To approximate the solution to the wave equation

$$\frac{\partial^2 u}{\partial t^2}(x,t) - \alpha^2\frac{\partial^2 u}{\partial x^2}(x,t) = 0, \quad 0 < x < l, \quad 0 < t < T,$$

subject to the boundary conditions

$$u(0,t) = u(l,t) = 0, \quad 0 < t < T,$$

and the initial conditions

$$u(x,0) = f(x), \quad \text{and} \quad \frac{\partial u}{\partial t}(x,0) = g(x), \quad \text{for} \quad 0 \le x \le l,$$

INPUT endpoint l; maximum time T; constant α; integers $m \ge 2, N \ge 2$.

OUTPUT approximations $w_{i,j}$ to $u(x_i, t_j)$ for each $i = 0, \ldots, m$ and $j = 0, \ldots, N$.

Step 1 Set $h = l/m$;
$k = T/N$;
$\lambda = k\alpha/h$.

Step 2 For $j = 1, \ldots, N$ set $w_{0,j} = 0$;
$w_{m,j} = 0$;

Step 3 Set $w_{0,0} = f(0)$;
$w_{m,0} = f(l)$.

Step 4 For $i = 1, \ldots, m-1$ (*Initialize for $t = 0$ and $t = k$.*)
set $w_{i,0} = f(ih)$;
$w_{i,1} = (1-\lambda^2)f(ih) + \frac{\lambda^2}{2}[f((i+1)h) + f((i-1)h)] + kg(ih)$.

Step 5 For $j = 1, \ldots, N-1$ (*Perform matrix multiplication.*)
for $i = 1, \ldots, m-1$
set $w_{i,j+1} = 2(1-\lambda^2)w_{i,j} + \lambda^2(w_{i+1,j} + w_{i-1,j}) - w_{i,j-1}$.

Step 6 For $j = 0, \ldots, N$
set $t = jk$;
for $i = 0, \ldots, m$
set $x = ih$;
OUTPUT $(x, t, w_{i,j})$.

Step 7 STOP. (*The procedure is complete.*) ■

Example 1 Approximate the solution to the hyperbolic problem

$$\frac{\partial^2 u}{\partial t^2}(x,t) - 4\frac{\partial^2 u}{\partial x^2}(x,t) = 0, \quad 0 < x < 1, \quad 0 < t,$$

with boundary conditions

$$u(0,t) = u(1,t) = 0, \quad \text{for} \quad 0 < t,$$

and initial conditions

$$u(x,0) = \sin(\pi x), \quad 0 \le x \le 1, \quad \text{and} \quad \frac{\partial u}{\partial t}(x,0) = 0, \quad 0 \le x \le 1,$$

using $h = 0.1$ and $k = 0.05$. Compare the results with the exact solution

$$u(x,t) = \sin \pi x \cos 2\pi t.$$

Solution Choosing $h = 0.1$ and $k = 0.05$ gives $\lambda = 1$, $m = 10$, and $N = 20$. We will choose a maximum time $T = 1$ and apply the Finite-Difference Algorithm 12.4. This produces the approximations $w_{i,N}$ to $u(0.1i, 1)$ for $i = 0, 1, \ldots, 10$. These results are shown in Table 12.6 and are correct to the places given. ■

Table 12.6

x_i	$w_{i,20}$
0.0	0.0000000000
0.1	0.3090169944
0.2	0.5877852523
0.3	0.8090169944
0.4	0.9510565163
0.5	1.0000000000
0.6	0.9510565163
0.7	0.8090169944
0.8	0.5877852523
0.9	0.3090169944
1.0	0.0000000000

The results of the example were very accurate, more so than the truncation error $O(k^2 + h^2)$ would lead us to believe. This is because the true solution to the equation is infinitely differentiable. When this is the case, Taylor series gives

$$\frac{u(x_{i+1},t_j) - 2u(x_i,t_j) + u(x_{i-1},t_j)}{h^2} = \frac{\partial^2 u}{\partial x^2}(x_i,t_j) + 2\left[\frac{h^2}{4!}\frac{\partial^4 u}{\partial x^4}(x_i,t_j) + \frac{h^4}{6!}\frac{\partial^6 u}{\partial x^6}(x_i,t_j) + \cdots\right]$$

and

$$\frac{u(x_i,t_{j+1}) - 2u(x_i,t_j) + u(x_i,t_{j-1})}{k^2} = \frac{\partial^2 u}{\partial t^2}(x_i,t_j) + 2\left[\frac{k^2}{4!}\frac{\partial^4 u}{\partial t^4}(x_i,t_j) + \frac{h^4}{6!}\frac{\partial^6 u}{\partial t^6}(x_i,t_j) + \cdots\right].$$

Since $u(x,t)$ satisfies the partial differential equation,

$$\frac{u(x_i,t_{j+1}) - 2u(x_i,t_j) + u(x_i,t_{j-1})}{k^2} - \alpha^2\frac{u(x_{i+1},t_j) - 2u(x_i,t_j) + u(x_{i-1},t_j)}{h^2}$$
$$= 2\left[\frac{1}{4!}\left(k^2\frac{\partial^4 u}{\partial t^4}(x_i,t_j) - \alpha^2 h^2\frac{\partial^4 u}{\partial x^4}(x_i,t_j)\right)\right.$$
$$\left. + \frac{1}{6!}\left(k^4\frac{\partial^6 u}{\partial t^6}(x_i,t_j) - \alpha^2 h^4\frac{\partial^6 u}{\partial x^6}(x_i,t_j)\right) + \cdots\right]. \tag{12.26}$$

However, differentiating the wave equation gives

$$k^2\frac{\partial^4 u}{\partial t^4}(x_i,t_j) = k^2\frac{\partial^2}{\partial t^2}\left[\alpha^2\frac{\partial^2 u}{\partial x^2}(x_i,t_j)\right] = \alpha^2 k^2\frac{\partial^2}{\partial x^2}\left[\frac{\partial^2 u}{\partial t^2}(x_i,t_j)\right]$$
$$= \alpha^2 k^2\frac{\partial^2}{\partial x^2}\left[\alpha^2\frac{\partial^2 u}{\partial x^2}(x_i,t_j)\right] = \alpha^4 k^2\frac{\partial^4 u}{\partial x^4}(x_i,t_j),$$

and we see that since $\lambda^2 = (\alpha^2 k^2/h^2) = 1$, we have

$$\frac{1}{4!}\left[k^2\frac{\partial^4 u}{\partial t^4}(x_i, t_j) - \alpha^2 h^2 \frac{\partial^4 u}{\partial x^4}(x_i, t_j)\right] = \frac{\alpha^2}{4!}[\alpha^2 k^2 - h^2]\frac{\partial^4 u}{\partial x^4}(x_i, t_j) = 0.$$

Continuing in this manner, all the terms on the right-hand side of (12.26) are 0, implying that the local truncation error is 0. The only errors in Example 1 are those due to the approximation of $w_{i,1}$ and to round-off.

As in the case of the Forward-Difference method for the heat equation, the Explicit Finite-Difference method for the wave equation has stability problems. In fact, it is necessary that $\lambda = \alpha k/h \leq 1$ for the method to be stable. (See [IK], p. 489.) The explicit method given in Algorithm 12.4, with $\lambda \leq 1$, is $O(h^2 + k^2)$ convergent if f and g are sufficiently differentiable. For verification of this, see [IK], p. 491.

Although we will not discuss them, there are implicit methods that are unconditionally stable. A discussion of these methods can be found in [Am], p. 199, [Mi], or [Sm,G].

EXERCISE SET 12.3

1. Approximate the solution to the wave equation

$$\frac{\partial^2 u}{\partial t^2} - \frac{\partial^2 u}{\partial x^2} = 0, \quad 0 < x < 1, \quad 0 < t;$$
$$u(0, t) = u(1, t) = 0, \quad 0 < t,$$
$$u(x, 0) = \sin \pi x, \quad 0 \leq x \leq 1,$$
$$\frac{\partial u}{\partial t}(x, 0) = 0, \quad 0 \leq x \leq 1,$$

using the Finite-Difference Algorithm 12.4 with $m = 4$, $N = 4$, and $T = 1.0$. Compare your results at $t = 1.0$ to the actual solution $u(x, t) = \cos \pi t \sin \pi x$.

2. Approximate the solution to the wave equation

$$\frac{\partial^2 u}{\partial t^2} - \frac{1}{16\pi^2}\frac{\partial^2 u}{\partial x^2} = 0, \quad 0 < x < 0.5, \ 0 < t;$$
$$u(0, t) = u(0.5, t) = 0, \quad 0 < t,$$
$$u(x, 0) = 0, \quad 0 \leq x \leq 0.5,$$
$$\frac{\partial u}{\partial t}(x, 0) = \sin 4\pi x, \quad 0 \leq x \leq 0.5,$$

using the Finite-Difference Algorithm 12.4 with $m = 4$, $N = 4$ and $T = 0.5$. Compare your results at $t = 0.5$ to the actual solution $u(x, t) = \sin t \sin 4\pi x$.

3. Approximate the solution to the wave equation

$$\frac{\partial^2 u}{\partial t^2} - \frac{\partial^2 u}{\partial x^2} = 0, \quad 0 < x < \pi, \ 0 < t;$$
$$u(0, t) = u(\pi, t) = 0, \quad 0 < t,$$
$$u(x, 0) = \sin x, \quad 0 \leq x \leq \pi,$$
$$\frac{\partial u}{\partial t}(x, 0) = 0, \quad 0 \leq x \leq \pi,$$

using the Finite-Difference Algorithm with $h = \pi/10$ and $k = 0.05$, with $h = \pi/20$ and $k = 0.1$, and then with $h = \pi/20$ and $k = 0.05$. Compare your results at $t = 0.5$ to the actual solution $u(x, t) = \cos t \sin x$.

4. Repeat Exercise 3, using in Step 4 of Algorithm 12.4 the approximation

$$w_{i,1} = w_{i,0} + kg(x_i), \quad \text{for each } i = 1, \ldots, m-1.$$

5. Approximate the solution to the wave equation

$$\frac{\partial^2 u}{\partial t^2} - \frac{\partial^2 u}{\partial x^2} = 0, \quad 0 < x < 1,\ 0 < t;$$

$$u(0,t) = u(1,t) = 0, \quad 0 < t,$$

$$u(x,0) = \sin 2\pi x, \quad 0 \le x \le 1,$$

$$\frac{\partial u}{\partial t}(x,0) = 2\pi \sin 2\pi x, \quad 0 \le x \le 1,$$

using Algorithm 12.4 with $h = 0.1$ and $k = 0.1$. Compare your results at $t = 0.3$ to the actual solution $u(x,t) = \sin 2\pi x(\cos 2\pi t + \sin 2\pi t)$.

6. Approximate the solution to the wave equation

$$\frac{\partial^2 u}{\partial t^2} - \frac{\partial^2 u}{\partial x^2} = 0, \quad 0 < x < 1,\ 0 < t;$$

$$u(0,t) = u(1,t) = 0, \quad 0 < t,$$

$$u(x,0) = \begin{cases} 1, & 0 \le x \le \frac{1}{2}, \\ -1, & \frac{1}{2} < x \le 1, \end{cases}$$

$$\frac{\partial u}{\partial t}(x,0) = 0, \quad 0 \le x \le 1.$$

using Algorithm 12.4 with $h = 0.1$ and $k = 0.1$.

7. The air pressure $p(x,t)$ in an organ pipe is governed by the wave equation

$$\frac{\partial^2 p}{\partial x^2} = \frac{1}{c^2}\frac{\partial^2 p}{\partial t^2}, \quad 0 < x < l,\ 0 < t,$$

where l is the length of the pipe, and c is a physical constant. If the pipe is open, the boundary conditions are given by

$$p(0,t) = p_0 \quad \text{and} \quad p(l,t) = p_0.$$

If the pipe is closed at the end where $x = l$, the boundary conditions are

$$p(0,t) = p_0 \quad \text{and} \quad \frac{\partial p}{\partial x}(l,t) = 0.$$

Assume that $c = 1$, $l = 1$, and the initial conditions are

$$p(x,0) = p_0 \cos 2\pi x, \quad \text{and} \quad \frac{\partial p}{\partial t}(x,0) = 0, \quad 0 \le x \le 1.$$

a. Approximate the pressure for an open pipe with $p_0 = 0.9$ at $x = \frac{1}{2}$ for $t = 0.5$ and $t = 1$, using Algorithm 12.4 with $h = k = 0.1$.

b. Modify Algorithm 12.4 for the closed-pipe problem with $p_0 = 0.9$, and approximate $p(0.5, 0.5)$ and $p(0.5, 1)$ using $h = k = 0.1$.

8. In an electric transmission line of length l that carries alternating current of high frequency (called a "lossless" line), the voltage V and current i are described by

$$\frac{\partial^2 V}{\partial x^2} = LC\frac{\partial^2 V}{\partial t^2}, \quad 0 < x < l,\ 0 < t;$$

$$\frac{\partial^2 i}{\partial x^2} = LC\frac{\partial^2 i}{\partial t^2}, \quad 0 < x < l,\ 0 < t;$$

where L is the inductance per unit length, and C is the capacitance per unit length. Suppose the line is 200 ft long and the constants C and L are given by

$$C = 0.1 \text{ farads/ft} \quad \text{and} \quad L = 0.3 \text{ henries/ft}.$$

Suppose the voltage and current also satisfy

$$V(0,t) = V(200,t) = 0, \quad 0 < t;$$
$$V(x,0) = 110\sin\frac{\pi x}{200}, \quad 0 \le x \le 200;$$
$$\frac{\partial V}{\partial t}(x,0) = 0, \quad 0 \le x \le 200;$$
$$i(0,t) = i(200,t) = 0, \quad 0 < t;$$
$$i(x,0) = 5.5\cos\frac{\pi x}{200}, \quad 0 \le x \le 200;$$

and

$$\frac{\partial i}{\partial t}(x,0) = 0, \quad 0 \le x \le 200.$$

Approximate the voltage and current at $t = 0.2$ and $t = 0.5$ using Algorithm 12.4 with $h = 10$ and $k = 0.1$.

12.4 An Introduction to the Finite-Element Method

Finite elements began in the 1950s in the aircraft industry. Use of the techniques followed a paper by Turner, Clough, Martin, and Topp [TCMT] that was published in 1956. Wide spread application of the methods required large computer recourses that were not available until the early 1970s.

The **Finite-Element method** is similar to the Rayleigh-Ritz method for approximating the solution to two-point boundary-value problems that was introduced in Section 11.5. It was originally developed for use in civil engineering, but it is now used for approximating the solutions to partial differential equations that arise in all areas of applied mathematics.

One advantage the Finite-Element method has over finite-difference methods is the relative ease with which the boundary conditions of the problem are handled. Many physical problems have boundary conditions involving derivatives and irregularly shaped boundaries. Boundary conditions of this type are difficult to handle using finite-difference techniques because each boundary condition involving a derivative must be approximated by a difference quotient at the grid points, and irregular shaping of the boundary makes placing the grid points difficult. The Finite-Element method includes the boundary conditions as integrals in a functional that is being minimized, so the construction procedure is independent of the particular boundary conditions of the problem.

In our discussion, we consider the partial differential equation

$$\frac{\partial}{\partial x}\left(p(x,y)\frac{\partial u}{\partial x}\right) + \frac{\partial}{\partial y}\left(q(x,y)\frac{\partial u}{\partial y}\right) + r(x,y)u(x,y) = f(x,y), \tag{12.27}$$

with $(x, y) \in \mathcal{D}$, where $\mathcal{D}$ is a plane region with boundary $\mathcal{S}$.

Boundary conditions of the form

$$u(x,y) = g(x,y) \tag{12.28}$$

are imposed on a portion, $\mathcal{S}_1$, of the boundary. On the remainder of the boundary, $\mathcal{S}_2$, the solution $u(x, y)$ is required to satisfy

$$p(x,y)\frac{\partial u}{\partial x}(x,y)\cos\theta_1 + q(x,y)\frac{\partial u}{\partial y}(x,y)\cos\theta_2 + g_1(x,y)u(x,y) = g_2(x,y), \qquad (12.29)$$

where θ_1 and θ_2 are the direction angles of the outward normal to the boundary at the point (x, y). (See Figure 12.13.)

Figure 12.13

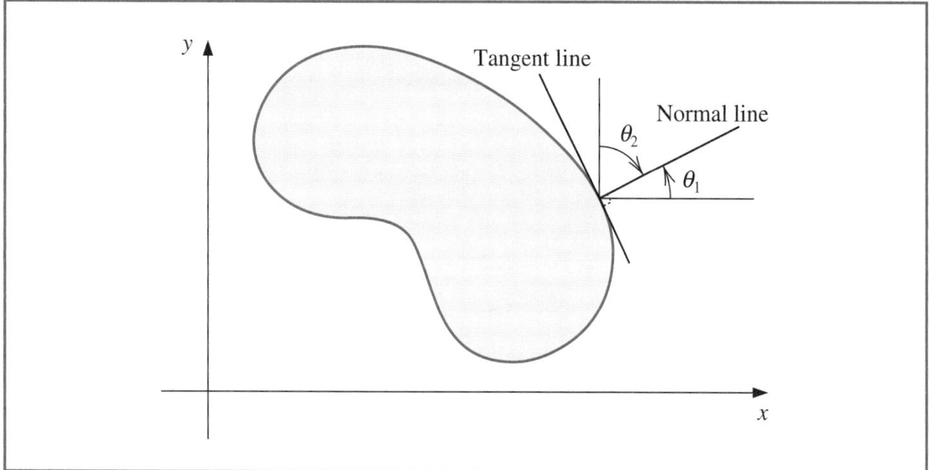

Physical problems in the areas of solid mechanics and elasticity have associated partial differential equations similar to Eq. (12.26). The solution to a problem of this type typically minimizes a certain functional, involving integrals, over a class of functions determined by the problem.

Suppose p, q, r, and f are all continuous on $\mathcal{D} \cup \mathcal{S}$, p and q have continuous first partial derivatives, and g_1 and g_2 are continuous on $\mathcal{S}_2$. Suppose, in addition, that $p(x, y) > 0$, $q(x, y) > 0$, $r(x, y) \le 0$, and $g_1(x, y) > 0$. Then a solution to Eq. (12.27) uniquely minimizes the functional

$$I[w] = \iint_{\mathcal{D}} \left\{ \frac{1}{2}\left[p(x,y)\left(\frac{\partial w}{\partial x}\right)^2 + q(x,y)\left(\frac{\partial w}{\partial y}\right)^2 - r(x,y)w^2 \right] + f(x,y)w \right\} dx\, dy$$
$$+ \int_{\mathcal{S}_2} \left\{ -g_2(x,y)w + \frac{1}{2}g_1(x,y)w^2 \right\} dS \qquad (12.30)$$

over all twice continuously-differentiable functions w satisfying Eq. (12.28) on $\mathcal{S}_1$. The Finite-Element method approximates this solution by minimizing the functional I over a smaller class of functions, just as the Rayleigh-Ritz method did for the boundary-value problem considered in Section 11.5.

Defining the Elements

The first step is to divide the region into a finite number of sections, or elements, of a regular shape, either rectangles or triangles. (See Figure 12.14.)

The set of functions used for approximation is generally a set of piecewise polynomials of fixed degree in x and y, and the approximation requires that the polynomials be pieced together in such a manner that the resulting function is continuous with an integrable or continuous first or second derivative on the entire region. Polynomials of linear type in x and y,

Figure 12.14

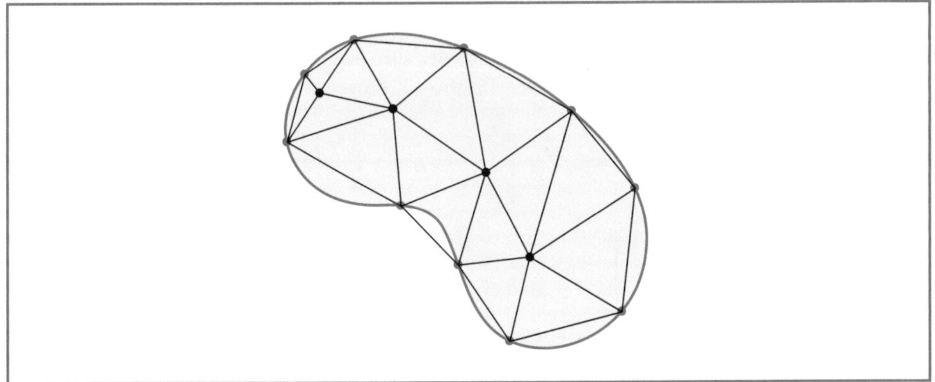

$$\phi(x, y) = a + bx + cy,$$

are commonly used with triangular elements, whereas polynomials of bilinear type in x and y,

$$\phi(x, y) = a + bx + cy + dxy,$$

are used with rectangular elements.

Suppose that the region $\mathcal{D}$ has been subdivided into triangular elements. The collection of triangles is denoted D, and the vertices of these triangles are called **nodes**. The method seeks an approximation of the form

$$\phi(x, y) = \sum_{i=1}^{m} \gamma_i \phi_i(x, y), \tag{12.31}$$

where $\phi_1, \phi_2, \ldots, \phi_m$ are linearly independent piecewise-linear polynomials, and $\gamma_1, \gamma_2, \ldots, \gamma_m$ are constants. Some of these constants, for example, $\gamma_{n+1}, \gamma_{n+2}, \ldots, \gamma_m$, are used to ensure that the boundary condition,

$$\phi(x, y) = g(x, y),$$

is satisfied on $\mathcal{S}_1$, and the remaining constants, $\gamma_1, \gamma_2, \ldots, \gamma_n$, are used to minimize the functional $I\left[\sum_{i=1}^{m} \gamma_i \phi_i\right]$.

Inserting the form of $\phi(x, y)$ given in Eq. (12.31) for w in Eq. (12.30) produces

$$\begin{aligned}
I[\phi] &= I\left[\sum_{i=1}^{m} \gamma_i \phi_i\right] \\
&= \iint_{\mathcal{D}} \left(\frac{1}{2} \left\{ p(x, y) \left[\sum_{i=1}^{m} \gamma_i \frac{\partial \phi_i}{\partial x}(x, y) \right]^2 + q(x, y) \left[\sum_{i=1}^{m} \gamma_i \frac{\partial \phi_i}{\partial y}(x, y) \right]^2 \right.\right. \\
&\quad \left.\left. - r(x, y) \left[\sum_{i=1}^{m} \gamma_i \phi_i(x, y) \right]^2 \right\} + f(x, y) \sum_{i=1}^{m} \gamma_i \phi_i(x, y) \right) dy\, dx \\
&\quad + \int_{\mathcal{S}_2} \left\{ - g_2(x, y) \sum_{i=1}^{m} \gamma_i \phi_i(x, y) + \frac{1}{2} g_1(x, y) \left[\sum_{i=1}^{m} \gamma_i \phi_i(x, y) \right]^2 \right\} dS.
\end{aligned} \tag{12.32}$$

Consider I as a function of $\gamma_1, \gamma_2, \ldots, \gamma_n$. For a minimum to occur we must have

$$\frac{\partial I}{\partial \gamma_j} = 0, \quad \text{for each } j = 1, 2, \ldots, n.$$

Differentiating (12.32) gives

$$\begin{aligned}
\frac{\partial I}{\partial \gamma_j} = \iint_{\mathcal{D}} \Bigg\{ & p(x,y) \sum_{i=1}^{m} \gamma_i \frac{\partial \phi_i}{\partial x}(x,y) \frac{\partial \phi_j}{\partial x}(x,y) \\
& + q(x,y) \sum_{i=1}^{m} \gamma_i \frac{\partial \phi_i}{\partial y}(x,y) \frac{\partial \phi_j}{\partial y}(x,y) \\
& - r(x,y) \sum_{i=1}^{m} \gamma_i \phi_i(x,y) \phi_j(x,y) + f(x,y)\phi_j(x,y) \Bigg\} \, dx \, dy \\
& + \int_{\mathcal{S}_2} \Bigg\{ - g_2(x,y)\phi_j(x,y) + g_1(x,y) \sum_{i=1}^{m} \gamma_i \phi_i(x,y) \phi_j(x,y) \Bigg\} \, dS,
\end{aligned}$$

so

$$\begin{aligned}
0 = \sum_{i=1}^{m} \Bigg[\iint_{\mathcal{D}} \Bigg\{ & p(x,y) \frac{\partial \phi_i}{\partial x}(x,y) \frac{\partial \phi_j}{\partial x}(x,y) + q(x,y) \frac{\partial \phi_i}{\partial y}(x,y) \frac{\partial \phi_j}{\partial y}(x,y) \\
& - r(x,y)\phi_i(x,y)\phi_j(x,y) \Bigg\} \, dx \, dy \\
& + \int_{\mathcal{S}_2} g_1(x,y)\phi_i(x,y)\phi_j(x,y) \, dS \Bigg] \gamma_i \\
& + \iint_{\mathcal{D}} f(x,y)\phi_j(x,y) \, dx \, dy - \int_{\mathcal{S}_2} g_2(x,y)\phi_j(x,y) \, dS,
\end{aligned}$$

for each $j = 1, 2, \ldots, n$. This set of equations can be written as a linear system:

$$A\mathbf{c} = \mathbf{b},$$

where $\mathbf{c} = (\gamma_1, \ldots, \gamma_n)^t$, and where $A = (\alpha_{ij})$ and $\mathbf{b} = (\beta_1, \ldots, \beta_n)^t$ are defined by

$$\begin{aligned}
\alpha_{ij} = \iint_{\mathcal{D}} \Bigg[& p(x,y) \frac{\partial \phi_i}{\partial x}(x,y) \frac{\partial \phi_j}{\partial x}(x,y) + q(x,y) \frac{\partial \phi_i}{\partial y}(x,y) \frac{\partial \phi_j}{\partial y}(x,y) \\
& - r(x,y)\phi_i(x,y)\phi_j(x,y) \Bigg] \, dx \, dy + \int_{\mathcal{S}_2} g_1(x,y)\phi_i(x,y)\phi_j(x,y) \, dS, \qquad (12.33)
\end{aligned}$$

for each $i = 1, 2, \ldots, n$ and $j = 1, 2, \ldots, m$, and

$$\beta_i = -\iint_{\mathcal{D}} f(x,y)\phi_i(x,y) \, dx \, dy + \int_{\mathcal{S}_2} g_2(x,y)\phi_i(x,y) \, dS - \sum_{k=n+1}^{m} \alpha_{ik}\gamma_k, \qquad (12.34)$$

for each $i = 1, \ldots, n$.

The particular choice of basis functions is important because the appropriate choice can often make the matrix A positive definite and banded. For the second-order problem (12.27), we assume that $\mathcal{D}$ is polygonal, so that $\mathcal{D} = D$, and that $\mathcal{S}$ is a contiguous set of straight lines.

Triangulating the Region

To begin the procedure, we divide the region D into a collection of triangles $T_1, T_2, \ldots, T_M$, with the ith triangle having three vertices, or nodes, denoted

$$V_j^{(i)} = \left(x_j^{(i)}, y_j^{(i)}\right), \quad \text{for } j = 1, 2, 3.$$

To simplify the notation, we write $V_j^{(i)}$ simply as $V_j = (x_j, y_j)$ when working with the fixed triangle T_i. With each vertex V_j we associate a linear polynomial

$$N_j^{(i)}(x, y) \equiv N_j(x, y) = a_j + b_j x + c_j y, \quad \text{where} \quad N_j^{(i)}(x_k, y_k) = \begin{cases} 1, & \text{if } j = k, \\ 0, & \text{if } j \neq k. \end{cases}$$

This produces linear systems of the form

$$\begin{bmatrix} 1 & x_1 & y_1 \\ 1 & x_2 & y_2 \\ 1 & x_3 & y_3 \end{bmatrix} \begin{bmatrix} a_j \\ b_j \\ c_j \end{bmatrix} = \begin{bmatrix} 0 \\ 1 \\ 0 \end{bmatrix},$$

with the element 1 occurring in the jth row in the vector on the right (here $j = 2$).

Let $E_1, \ldots, E_n$ be a labeling of the nodes lying in $D \cup \mathcal{S}$. With each node E_k, we associate a function ϕ_k that is linear on each triangle, has the value 1 at E_k, and is 0 at each of the other nodes. This choice makes ϕ_k identical to $N_j^{(i)}$ on triangle T_i when the node E_k is the vertex denoted $V_j^{(i)}$.

Illustration Suppose that a finite-element problem contains the triangles T_1 and T_2 shown in Figure 12.15.

Figure 12.15

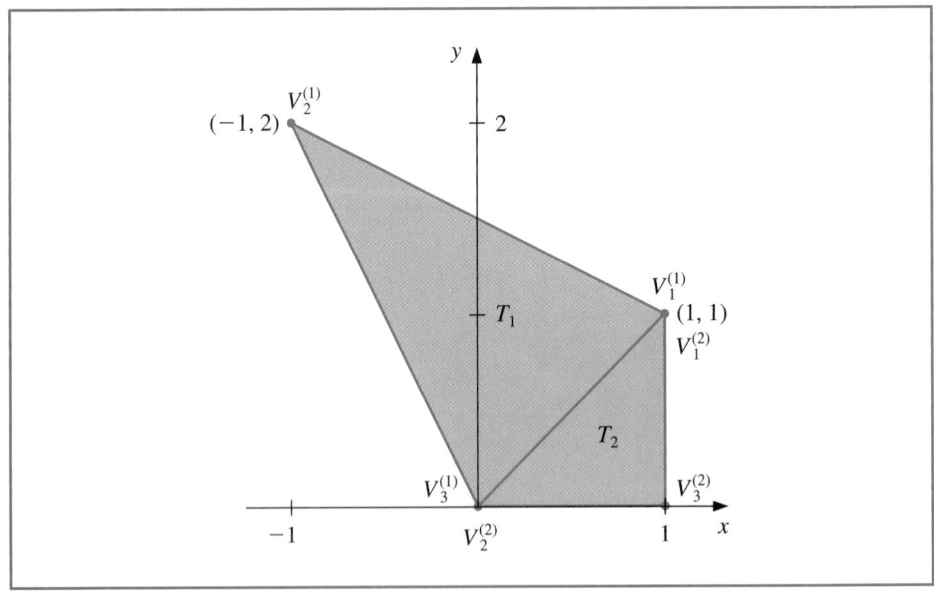

The linear function $N_1^{(1)}(x, y)$ that assumes the value 1 at $(1, 1)$ and the value 0 at both $(0, 0)$ and $(-1, 2)$ satisfies

$$a_1^{(1)} + b_1^{(1)}(1) + c_1^{(1)}(1) = 1,$$

$$a_1^{(1)} + b_1^{(1)}(-1) + c_1^{(1)}(2) = 0,$$

and

$$a_1^{(1)} + b_1^{(1)}(0) + c_1^{(1)}(0) = 0.$$

The solution to this system is $a_1^{(1)} = 0$, $b_1^{(1)} = \frac{2}{3}$,and $c_1^{(1)} = \frac{1}{3}$, so

$$N_1^{(1)}(x, y) = \frac{2}{3}x + \frac{1}{3}y.$$

In a similar manner, the linear function $N_1^{(2)}(x, y)$ that assumes the value 1 at $(1, 1)$ and the value 0 at both $(0, 0)$ and $(1, 0)$ satisfies

$$a_1^{(2)} + b_1^{(2)}(1) + c_1^{(2)}(1) = 1,$$

$$a_1^{(2)} + b_1^{(2)}(0) + c_1^{(2)}(0) = 0,$$

and

$$a_1^{(2)} + b_1^{(2)}(1) + c_1^{(2)}(0) = 0.$$

This implies that $a_1^{(2)} = 0, b_1^{(2)} = 0$, and $c_1^{(2)} = 1$. As a consequence, $N_1^{(2)}(x, y) = y$. Note that $N_1^{(1)}(x, y) = N_1^{(2)}(x, y)$ on the common boundary of T_1 and T_2, because $y = x$. □

Consider Figure 12.16, the upper left portion of the region shown in Figure 12.12. We will generate the entries in the matrix A that correspond to the nodes shown in this figure.

Figure 12.16

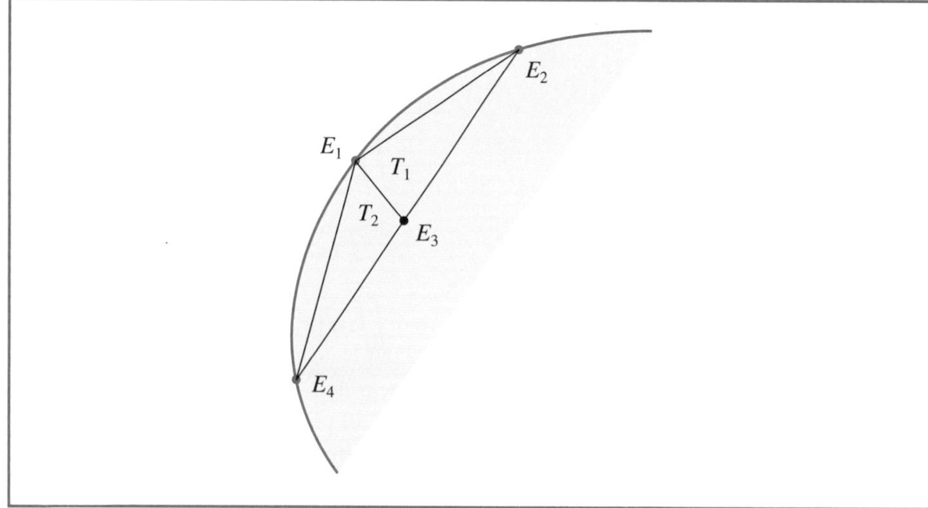

For simplicity, we assume that E_1 is one of the nodes on $\mathcal{S}_1$, where the boundary condition $u(x, y) = g(x, y)$ is imposed. The relationship between the nodes and the vertices of the triangles for this portion is

$$E_1 = V_3^{(1)} = V_1^{(2)},\ E_4 = V_2^{(2)},\ E_3 = V_2^{(1)} = V_3^{(2)}, \quad \text{and} \quad E_2 = V_1^{(1)}.$$

Since ϕ_1 and ϕ_3 are both nonzero on T_1 and T_2, the entries $\alpha_{1,3} = \alpha_{3,1}$ are computed by

$$\begin{aligned}\alpha_{1,3} &= \iint_D \left[p\frac{\partial \phi_1}{\partial x}\frac{\partial \phi_3}{\partial x} + q\frac{\partial \phi_1}{\partial y}\frac{\partial \phi_3}{\partial y} - r\phi_1\phi_3 \right] dx\, dy \\ &= \iint_{T_1} \left[p\frac{\partial \phi_1}{\partial x}\frac{\partial \phi_3}{\partial x} + q\frac{\partial \phi_1}{\partial y}\frac{\partial \phi_3}{\partial y} - r\phi_1\phi_3 \right] dx\, dy \\ &\quad + \iint_{T_2} \left[p\frac{\partial \phi_1}{\partial x}\frac{\partial \phi_3}{\partial x} + q\frac{\partial \phi_1}{\partial y}\frac{\partial \phi_3}{\partial y} - r\phi_1\phi_3 \right] dx\, dy.\end{aligned}$$

On triangle T_1,

$$\phi_1(x, y) = N_3^{(1)}(x, y) = a_3^{(1)} + b_3^{(1)}x + c_3^{(1)}y$$

and

$$\phi_3(x, y) = N_2^{(1)}(x, y) = a_2^{(1)} + b_2^{(1)}x + c_2^{(1)}y,$$

so for all (x, y),

$$\frac{\partial \phi_1}{\partial x} = b_3^{(1)}, \quad \frac{\partial \phi_1}{\partial y} = c_3^{(1)}, \quad \frac{\partial \phi_3}{\partial x} = b_2^{(1)}, \quad \text{and} \quad \frac{\partial \phi_3}{\partial y} = c_2^{(1)}.$$

Similarly, on T_2,

$$\phi_1(x, y) = N_1^{(2)}(x, y) = a_1^{(2)} + b_1^{(2)}x + c_1^{(2)}y$$

and

$$\phi_3(x, y) = N_3^{(2)}(x, y) = a_3^{(2)} + b_3^{(2)}x + c_3^{(2)}y,$$

so for all (x, y),

$$\frac{\partial \phi_1}{\partial x} = b_1^{(2)}, \quad \frac{\partial \phi_1}{\partial y} = c_1^{(2)}, \quad \frac{\partial \phi_3}{\partial x} = b_3^{(2)}, \quad \text{and} \quad \frac{\partial \phi_3}{\partial y} = c_3^{(2)}.$$

Thus,

$$\begin{aligned}\alpha_{1,3} &= b_3^{(1)}b_2^{(1)} \iint_{T_1} p\, dx\, dy + c_3^{(1)}c_2^{(1)} \iint_{T_1} q\, dx\, dy \\ &\quad - \iint_{T_1} r\left(a_3^{(1)} + b_3^{(1)}x + c_3^{(1)}y\right)\left(a_2^{(1)} + b_2^{(1)}x + c_2^{(1)}y\right) dx\, dy \\ &\quad + b_1^{(2)}b_3^{(2)} \iint_{T_2} p\, dx\, dy + c_1^{(2)}c_3^{(2)} \iint_{T_2} q\, dx\, dy \\ &\quad - \iint_{T_2} r\left(a_1^{(2)} + b_1^{(2)}x + c_1^{(2)}y\right)\left(a_3^{(2)} + b_3^{(2)}x + c_3^{(2)}y\right) dx\, dy.\end{aligned}$$

All the double integrals over D reduce to double integrals over triangles. The usual procedure is to compute all possible integrals over the triangles and accumulate them into the correct entry α_{ij} in A. Similarly, the double integrals of the form

$$\iint_D f(x, y)\phi_i(x, y)\, dx\, dy$$

are computed over triangles and then accumulated into the correct entry β_i of the vector **b**. For example, to determine β_1, we need

$$-\iint_D f(x,y)\phi_1(x,y)\,dx\,dy = -\iint_{T_1} f(x,y)\left[a_3^{(1)} + b_3^{(1)}x + c_3^{(1)}y\right] dx\,dy$$
$$-\iint_{T_2} f(x,y)\left[a_1^{(2)} + b_1^{(2)}x + c_1^{(2)}y\right] dx\,dy.$$

Because E_1 is a vertex of both T_1 and T_2, part of β_1 is contributed by ϕ_1 restricted to T_1 and the remainder by ϕ_1 restricted to T_2. In addition, nodes that lie on $\mathcal{S}_2$ have line integrals added to their entries in A and **b**.

Algorithm 12.5 performs the Finite-Element method on a second-order elliptic differential equation. The algorithm sets all values of the matrix A and vector **b** initially to 0 and, after all the integrations have been performed on all the triangles, adds these values to the appropriate entries in A and **b**.

Finite-Element

To approximate the solution to the partial differential equation

$$\frac{\partial}{\partial x}\left(p(x,y)\frac{\partial u}{\partial x}\right) + \frac{\partial}{\partial y}\left(q(x,y)\frac{\partial u}{\partial y}\right) + r(x,y)u = f(x,y), \quad (x,y) \in D$$

subject to the boundary conditions

$$u(x,y) = g(x,y), \quad (x,y) \in \mathcal{S}_1$$

and

$$p(x,y)\frac{\partial u}{\partial x}(x,y)\cos\theta_1 + q(x,y)\frac{\partial u}{\partial y}(x,y)\cos\theta_2 + g_1(x,y)u(x,y) = g_2(x,y),$$
$$(x,y) \in \mathcal{S}_2,$$

where $\mathcal{S}_1 \cup \mathcal{S}_2$ is the boundary of D, and θ_1 and θ_2 are the direction angles of the normal to the boundary:

Step 0 Divide the region D into triangles $T_1, \ldots, T_M$ such that:
$T_1, \ldots, T_K$ are the triangles with no edges on $\mathcal{S}_1$ or $\mathcal{S}_2$;
(*Note:* $K = 0$ *implies that no triangle is interior to* D.)
$T_{K+1}, \ldots, T_N$ are the triangles with at least one edge on $\mathcal{S}_2$;
$T_{N+1}, \ldots, T_M$ are the remaining triangles.
(*Note:* $M = N$ *implies that all triangles have edges on* $\mathcal{S}_2$.)
Label the three vertices of the triangle T_i by
$\left(x_1^{(i)}, y_1^{(i)}\right), \left(x_2^{(i)}, y_2^{(i)}\right)$, and $\left(x_3^{(i)}, y_3^{(i)}\right)$.
Label the nodes (vertices) $E_1, \ldots, E_m$ where
$E_1, \ldots, E_n$ are in $D \cup \mathcal{S}_2$ and $E_{n+1}, \ldots, E_m$ are on $\mathcal{S}_1$.
(*Note:* $n = m$ *implies that* $\mathcal{S}_1$ *contains no nodes.*)

INPUT integers K, N, M, n, m; vertices $\left(x_1^{(i)}, y_1^{(i)}\right), \left(x_2^{(i)}, y_2^{(i)}\right), \left(x_3^{(i)}, y_3^{(i)}\right)$ for each $i = 1, \ldots, M$; nodes E_j for each $j = 1, \ldots, m$.

(*Note: All that is needed is a means of corresponding a vertex* $\left(x_k^{(i)}, y_k^{(i)}\right)$ *to a node* $E_j = (x_j, y_j)$.)

OUTPUT constants $\gamma_1, \ldots, \gamma_m$; $a_j^{(i)}, b_j^{(i)}, c_j^{(i)}$ for each $j = 1, 2, 3$ and $i = 1, \ldots, M$.

Step 1 For $l = n+1, \ldots, m$ set $\gamma_l = g(x_l, y_l)$. (*Note:* $E_l = (x_l, y_l)$.)

Step 2 For $i = 1, \ldots, n$
set $\beta_i = 0$;
for $j = 1, \ldots, n$ set $\alpha_{i,j} = 0$.

Step 3 For $i = 1, \ldots, M$

$$\text{set } \Delta_i = \det \begin{vmatrix} 1 & x_1^{(i)} & y_1^{(i)} \\ 1 & x_2^{(i)} & y_2^{(i)} \\ 1 & x_3^{(i)} & y_3^{(i)} \end{vmatrix};$$

$$a_1^{(i)} = \frac{x_2^{(i)}y_3^{(i)} - y_2^{(i)}x_3^{(i)}}{\Delta_i}; \quad b_1^{(i)} = \frac{y_2^{(i)} - y_3^{(i)}}{\Delta_i}; \quad c_1^{(i)} = \frac{x_3^{(i)} - x_2^{(i)}}{\Delta_i};$$

$$a_2^{(i)} = \frac{x_3^{(i)}y_1^{(i)} - y_3^{(i)}x_1^{(i)}}{\Delta_i}; \quad b_2^{(i)} = \frac{y_3^{(i)} - y_1^{(i)}}{\Delta_i}; \quad c_2^{(i)} = \frac{x_1^{(i)} - x_3^{(i)}}{\Delta_i};$$

$$a_3^{(i)} = \frac{x_1^{(i)}y_2^{(i)} - y_1^{(i)}x_2^{(i)}}{\Delta_i}; \quad b_3^{(i)} = \frac{y_1^{(i)} - y_2^{(i)}}{\Delta_i}; \quad c_3^{(i)} = \frac{x_2^{(i)} - x_1^{(i)}}{\Delta_i};$$

for $j = 1, 2, 3$
define $N_j^{(i)}(x, y) = a_j^{(i)} + b_j^{(i)}x + c_j^{(i)}y$.

Step 4 For $i = 1, \ldots, M$ (*The integrals in Steps 4 and 5 can be evaluated using numerical integration.*)
for $j = 1, 2, 3$
for $k = 1, \ldots, j$ (*Compute all double integrals over the triangles.*)

$$\text{set } z_{j,k}^{(i)} = b_j^{(i)}b_k^{(i)} \iint_{T_i} p(x, y)\,dx\,dy + c_j^{(i)}c_k^{(i)} \iint_{T_i} q(x, y)\,dx\,dy - \iint_{T_i} r(x, y)N_j^{(i)}(x, y)N_k^{(i)}(x, y)\,dx\,dy;$$

$$\text{set } H_j^{(i)} = -\iint_{T_i} f(x, y)N_j^{(i)}(x, y)\,dx\,dy.$$

Step 5 For $i = K+1, \ldots, N$ (*Compute all line integrals.*)
for $j = 1, 2, 3$
for $k = 1, \ldots, j$

$$\text{set } J_{j,k}^{(i)} = \int_{\mathcal{S}_2} g_1(x, y)N_j^{(i)}(x, y)N_k^{(i)}(x, y)\,dS;$$

$$\text{set } I_j^{(i)} = \int_{\mathcal{S}_2} g_2(x, y)N_j^{(i)}(x, y)\,dS.$$

Step 6 For $i = 1, \ldots, M$ do Steps 7–12. (*Assembling the integrals over each triangle into the linear system.*)

Step 7 For $k = 1, 2, 3$ do Steps 8–12.

Step 8 Find l so that $E_l = \left(x_k^{(i)}, y_k^{(i)}\right)$.

Step 9 If $k > 1$ then for $j = 1, \ldots, k-1$ do Steps 10, 11.

Step 10 Find t so that $E_t = \left(x_j^{(i)}, y_j^{(i)}\right)$.

Step 11 If $l \leq n$ then

if $t \leq n$ then set $\alpha_{lt} = \alpha_{lt} + z_{k,j}^{(i)}$;

$\alpha_{tl} = \alpha_{tl} + z_{k,j}^{(i)}$

else set $\beta_l = \beta_l - \gamma_t z_{k,j}^{(i)}$

else

if $t \leq n$ then set $\beta_t = \beta_t - \gamma_l z_{k,j}^{(i)}$.

Step 12 If $l \leq n$ then set $a_{ll} = \alpha_{ll} + z_{k,k}^{(i)}$;

$\beta_l = \beta_l + H_k^{(i)}$.

Step 13 For $i = K+1, \ldots, N$ do Steps 14–19. *(Assembling the line integrals into the linear system.)*

Step 14 For $k = 1, 2, 3$ do Steps 15–19.

Step 15 Find l so that $E_l = \left(x_k^{(i)}, y_k^{(i)}\right)$.

Step 16 If $k > 1$ then for $j = 1, \ldots, k-1$ do Steps 17, 18.

Step 17 Find t so that $E_t = \left(x_j^{(i)}, y_j^{(i)}\right)$.

Step 18 If $l \leq n$ then

if $t \leq n$ then set $\alpha_{lt} = \alpha_{lt} + J_{k,j}^{(i)}$;

$\alpha_{tl} = \alpha_{tl} + J_{k,j}^{(i)}$

else set $\beta_l = \beta_l - \gamma_t J_{k,j}^{(i)}$

else

if $t \leq n$ then set $\beta_t = \beta_t - \gamma_l J_{k,j}^{(i)}$.

Step 19 If $l \leq n$ then set $\alpha_{ll} = \alpha_{ll} + J_{k,k}^{(i)}$;

$\beta_l = \beta_l + I_k^{(i)}$.

Step 20 Solve the linear system $A\mathbf{c} = \mathbf{b}$ where $A = (\alpha_{l,t}), \mathbf{b} = (\beta_l)$ and $\mathbf{c} = (\gamma_t)$ for $1 \leq l \leq n$ and $1 \leq t \leq n$.

Step 21 OUTPUT $(\gamma_1, \ldots, \gamma_m)$.
*(For each $k = 1, \ldots, m$ let $\phi_k = N_j^{(i)}$ on T_i if $E_k = \left(x_j^{(i)}, y_j^{(i)}\right)$.
Then $\phi(x, y) = \sum_{k=1}^{m} \gamma_k \phi_k(x, y)$ approximates $u(x, y)$ on $D \cup \mathcal{S}_1 \cup \mathcal{S}_2$.)*

Step 22 For $i = 1, \ldots, M$

for $j = 1, 2, 3$ OUTPUT $\left(a_j^{(i)}, b_j^{(i)}, c_j^{(i)}\right)$.

Step 23 STOP. *(The procedure is complete.)* ■

Illustration The temperature, $u(x, y)$, in a two-dimensional region D satisfies Laplace's equation

$$\frac{\partial^2 u}{\partial x^2}(x, y) + \frac{\partial^2 u}{\partial y^2}(x, y) = 0 \quad \text{on } D.$$

Consider the region D shown in Figure 12.17 with boundary conditions given by

$$
\begin{aligned}
u(x,y) &= 4, && \text{for } (x,y) \in L_6 \text{ and } (x,y) \in L_7; \\
\frac{\partial u}{\partial \mathbf{n}}(x,y) &= x, && \text{for } (x,y) \in L_2 \text{ and } (x,y) \in L_4; \\
\frac{\partial u}{\partial \mathbf{n}}(x,y) &= y, && \text{for } (x,y) \in L_5; \\
\frac{\partial u}{\partial \mathbf{n}}(x,y) &= \frac{x+y}{\sqrt{2}}, && \text{for } (x,y) \in L_1 \text{ and } (x,y) \in L_3,
\end{aligned}
$$

where $\partial u/\partial \mathbf{n}$ denotes the directional derivative in the direction of the normal $\mathbf{n}$ to the boundary of the region D at the point (x, y).

Figure 12.17

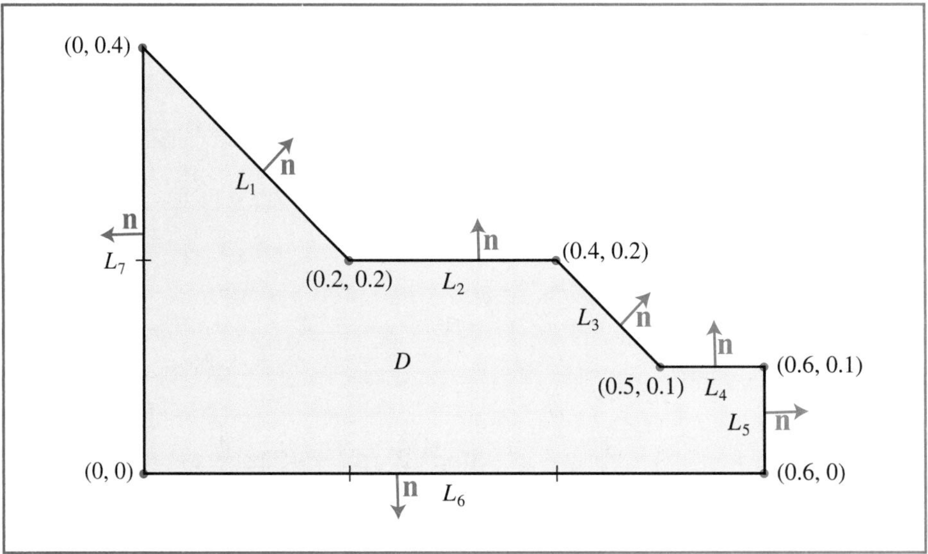

We first subdivide D into triangles with the labeling suggested in Step 0 of the algorithm. For this example, $\mathcal{S}_1 = L_6 \cup L_7$ and $\mathcal{S}_2 = L_1 \cup L_2 \cup L_3 \cup L_4 \cup L_5$. The labeling of triangles is shown in Figure 12.18.

The boundary condition $u(x, y) = 4$ on L_6 and L_7 implies that $\gamma_t = 4$ when $t = 6, 7, \ldots, 11$, that is, at the nodes $E_6, E_7, \ldots, E_{11}$. To determine the values of γ_l for $l = 1, 2, \ldots, 5$, apply the remaining steps of the algorithm and generate the matrix

$$
A = \begin{bmatrix}
2.5 & 0 & -1 & 0 & 0 \\
0 & 1.5 & -1 & -0.5 & 0 \\
-1 & -1 & 4 & 0 & 0 \\
0 & -0.5 & 0 & 2.5 & -0.5 \\
0 & 0 & 0 & -0.5 & 1
\end{bmatrix}
$$

Figure 12.18

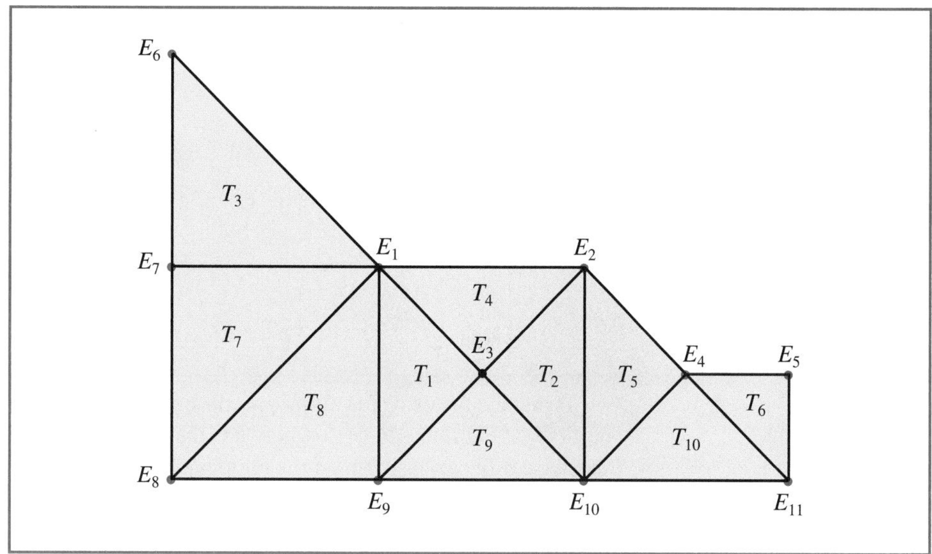

and the vector

$$\mathbf{b} = \begin{bmatrix} 6.066\bar{6} \\ 0.063\bar{3} \\ 8.0000 \\ 6.056\bar{6} \\ 2.031\bar{6} \end{bmatrix}.$$

The solution to the equation $A\mathbf{c} = \mathbf{b}$ is

$$\mathbf{c} = \begin{bmatrix} \gamma_1 \\ \gamma_2 \\ \gamma_3 \\ \gamma_4 \\ \gamma_5 \end{bmatrix} = \begin{bmatrix} 4.0383 \\ 4.0782 \\ 4.0291 \\ 4.0496 \\ 4.0565 \end{bmatrix}.$$

Solving this system gives the following approximation to the solution of Laplace's equation and the boundary conditions on the respective triangles:

$$T_1: \quad \phi(x,y) = 4.0383(1 - 5x + 5y) + 4.0291(-2 + 10x) + 4(2 - 5x - 5y),$$

$$T_2: \quad \phi(x,y) = 4.0782(-2 + 5x + 5y) + 4.0291(4 - 10x) + 4(-1 + 5x - 5y),$$

$$T_3: \quad \phi(x,y) = 4(-1 + 5y) + 4(2 - 5x - 5y) + 4.0383(5x),$$

$$T_4: \quad \phi(x,y) = 4.0383(1 - 5x + 5y) + 4.0782(-2 + 5x + 5y) + 4.0291(2 - 10y),$$

$$T_5: \quad \phi(x,y) = 4.0782(2 - 5x + 5y) + 4.0496(-4 + 10x) + 4(3 - 5x - 5y),$$

$$T_6: \quad \phi(x,y) = 4.0496(6 - 10x) + 4.0565(-6 + 10x + 10y) + 4(1 - 10y),$$

$$T_7: \quad \phi(x,y) = 4(-5x + 5y) + 4.0383(5x) + 4(1 - 5y),$$

$$T_8: \quad \phi(x,y) = 4.0383(5y) + 4(1 - 5x) + 4(5x - 5y),$$

$$T_9: \quad \phi(x,y) = 4.0291(10y) + 4(2 - 5x - 5y) + 4(-1 + 5x - 5y),$$

$$T_{10}: \quad \phi(x,y) = 4.0496(10y) + 4(3 - 5x - 5y) + 4(-2 + 5x - 5y).$$

The actual solution to the boundary-value problem is $u(x, y) = xy+4$. Table 12.7 compares the value of u to the value of ϕ at E_i, for each $i = 1, \ldots, 5$. □

Table 12.7

x	y	$\phi(x, y)$	$u(x, y)$	$\lvert\phi(x, y) - u(x, y)\rvert$
0.2	0.2	4.0383	4.04	0.0017
0.4	0.2	4.0782	4.08	0.0018
0.3	0.1	4.0291	4.03	0.0009
0.5	0.1	4.0496	4.05	0.0004
0.6	0.1	4.0565	4.06	0.0035

Typically, the error for elliptic second-order problems of the type (12.27) with smooth coefficient functions is $O(h^2)$, where h is the maximum diameter of the triangular elements. Piecewise bilinear basis functions on rectangular elements are also expected to give $O(h^2)$ results, where h is the maximum diagonal length of the rectangular elements. Other classes of basis functions can be used to give $O(h^4)$ results, but the construction is more complex. Efficient error theorems for finite-element methods are difficult to state and apply because the accuracy of the approximation depends on the regularity of the boundary as well as on the continuity properties of the solution.

The Finite-Element method can also be applied to parabolic and hyperbolic partial differential equations, but the minimization procedure is more difficult. A good survey on the advantages and techniques of the Finite-Element method applied to various physical problems can be found in a paper by [Fi]. For a more extensive discussion, refer to [SF], [ZM], or [AB].

EXERCISE SET 12.4

1. Use Algorithm 12.5 to approximate the solution to the following partial differential equation (see the figure):

$$\frac{\partial}{\partial x}\left(y^2\frac{\partial u}{\partial x}(x, y)\right) + \frac{\partial}{\partial y}\left(y^2\frac{\partial u}{\partial y}(x, y)\right) - yu(x, y) = -x, \quad (x, y) \in D,$$

$$u(x, 0.5) = 2x, \quad 0 \le x \le 0.5, \quad u(0, y) = 0, \quad 0.5 \le y \le 1,$$

$$y^2\frac{\partial u}{\partial x}(x, y)\cos\theta_1 + y^2\frac{\partial u}{\partial y}(x, y)\cos\theta_2 = \frac{\sqrt{2}}{2}(y - x) \quad \text{for } (x, y) \in \mathcal{S}_2.$$

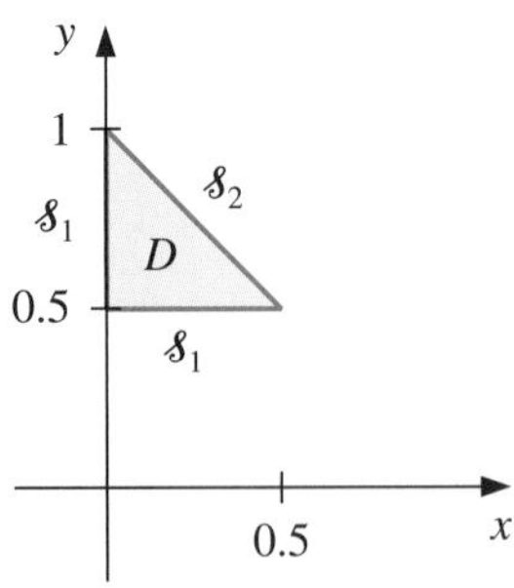

Let $M = 2$; T_1 have vertices $(0, 0.5)$, $(0.25, 0.75)$, $(0, 1)$; and T_2 have vertices $(0, 0.5)$, $(0.5, 0.5)$, and $(0.25, 0.75)$.

2. Repeat Exercise 1, using instead the triangles

$$T_1: \quad (0, 0.75), (0, 1), (0.25, 0.75);$$
$$T_2: \quad (0.25, 0.5), (0.25, 0.75), (0.5, 0.5);$$
$$T_3: \quad (0, 0.5), (0, 0.75), (0.25, 0.75);$$
$$T_4: \quad (0, 0.5), (0.25, 0.5), (0.25, 0.75).$$

3. Approximate the solution to the partial differential equation

$$\frac{\partial^2 u}{\partial x^2}(x, y) + \frac{\partial^2 u}{\partial y^2}(x, y) - 12.5\pi^2 u(x, y) = -25\pi^2 \sin\frac{5\pi}{2}x \sin\frac{5\pi}{2}y, \quad 0 < x,\ y < 0.4,$$

subject to the Dirichlet boundary condition

$$u(x, y) = 0,$$

using the Finite-Element Algorithm 12.5 with the elements given in the accompanying figure. Compare the approximate solution to the actual solution,

$$u(x, y) = \sin\frac{5\pi}{2}x \sin\frac{5\pi}{2}y,$$

at the interior vertices and at the points (0.125, 0.125), (0.125, 0.25), (0.25, 0.125), and (0.25, 0.25).

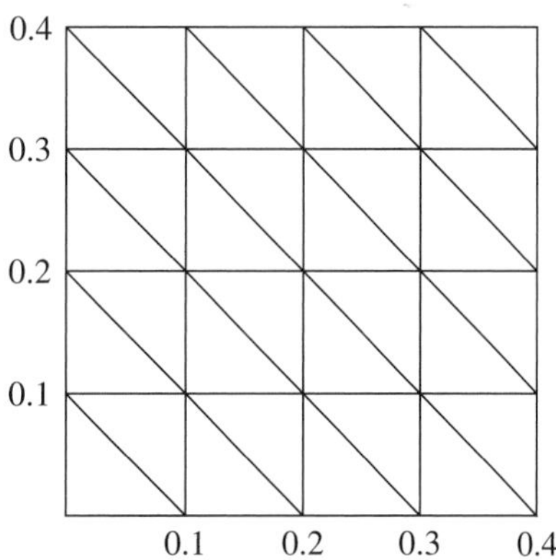

4. Repeat Exercise 3 with $f(x, y) = -25\pi^2 \cos\frac{5\pi}{2}x \cos\frac{5\pi}{2}y$, using the Neumann boundary condition

$$\frac{\partial u}{\partial n}(x, y) = 0.$$

The actual solution for this problem is

$$u(x, y) = \cos\frac{5\pi}{2}x \cos\frac{5\pi}{2}y.$$

5. A silver plate in the shape of a trapezoid (see the accompanying figure) has heat being uniformly generated at each point at the rate $q = 1.5\ \text{cal/cm}^3 \cdot \text{s}$. The steady-state temperature $u(x, y)$ of the plate satisfies the Poisson equation

$$\frac{\partial^2 u}{\partial x^2}(x, y) + \frac{\partial^2 u}{\partial y^2}(x, y) = \frac{-q}{k},$$

where k, the thermal conductivity, is 1.04 cal/cm·deg·s. Assume that the temperature is held at 15°C on L_2, that heat is lost on the slanted edges L_1 and L_3 according to the boundary condition $\partial u/\partial n = 4$, and that no heat is lost on L_4; that is, $\partial u/\partial n = 0$. Approximate the temperature of the plate at $(1, 0)$, $(4, 0)$, and $\left(\frac{5}{2}, \sqrt{3}/2\right)$ by using Algorithm 12.5.

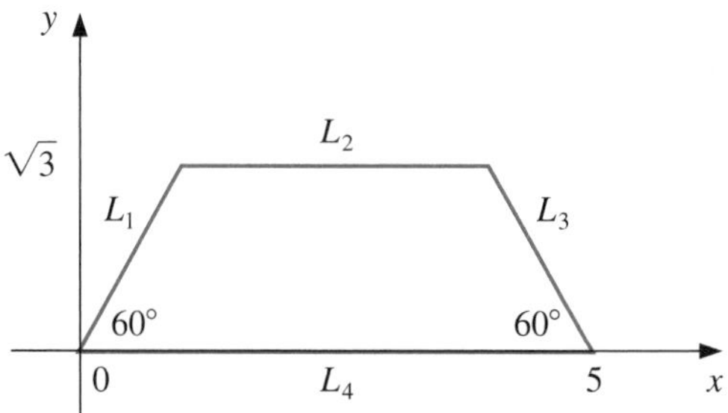

12.5 Survey of Methods and Software

In this chapter, methods to approximate solutions to partial differential equations were considered. We restricted our attention to Poisson's equation as an example of an elliptic partial differential equation, the heat or diffusion equation as an example of a parabolic partial differential equation, and the wave equation as an example of a hyperbolic partial differential equation. Finite-difference approximations were discussed for these three examples.

Poisson's equation on a rectangle required the solution of a large sparse linear system, for which iterative techniques, such as the SOR method, are recommended. Four finite-difference methods were presented for the heat equation. The Forward-Difference and Richardson's methods had stability problems, so the Backward-Difference method and the Crank-Nicolson methods were introduced. Although a tridiagonal linear system must be solved at each time step with these implicit methods, they are more stable than the explicit Forward-Difference and Richardson's methods. The Finite-Difference method for the wave equation is explicit and can also have stability problems for certain choice of time and space discretizations.

In the last section of the chapter, we presented an introduction to the Finite-Element method for a self-adjoint elliptic partial differential equation on a polygonal domain. Although our methods will work adequately for the problems and examples in the textbook, more powerful generalizations and modifications of these techniques are required for commercial applications.

One of the subroutines from the IMSL Library is used to solve the partial differential equation

$$\frac{\partial u}{\partial t} = F\left(x, t, u, \frac{\partial u}{\partial x}, \frac{\partial^2 u}{\partial x^2}\right),$$

with boundary conditions

$$\alpha(x, t)u(x, t) + \beta(x, t)\frac{\partial u}{\partial x}(x, t) = \gamma(x, t).$$

The routine is based on collocation at Gaussian points on the x-axis for each value of t and uses cubic Hermite splines as basis functions. Another subroutine from IMSL is used to solve Poisson's equation on a rectangle. The method of solution is based on a choice of second- or fourth-order finite differences on a uniform mesh.

The NAG Library has a number of subroutines for partial differential equations. One subroutine is used for Laplace's equation on an arbitrary domain in the xy-plane, and another is used to solve a single parabolic partial differential equation by the method of lines.

There are specialized packages, such as NASTRAN, consisting of codes for the Finite-Element method. These packages are popular in engineering applications. The package FISHPACK in the netlib library is used to solve separable elliptic partial differential equations. General codes for partial differential equations are difficult to write because of the problem of specifying domains other than common geometrical figures. Research in the area of solution of partial differential equations is currently very active.

We have only presented a small sample of the many techniques used for approximating the solutions to the problems involving partial differential equations. Further information on the general topic can be found in Lapidus and Pinder [LP], Twizell [Tw], and the recent book by Morton and Mayers [MM]. Software information can be found in Rice and Boisvert [RB] and in Bank [Ban].

Books that focus on finite-difference methods include Strikwerda [Stri], Thomas [Th], and Shashkov and Steinberg [ShS]. Strange and Fix [SF] and Zienkiewicz and Morgan [ZM] are good sources for information on the finite-element method. Time-dependent equations are treated in Schiesser [Schi] and in Gustafsson, Kreiss, and Oliger [GKO]. Birkhoff and Lynch [BL] and Roache [Ro] discuss the solution to elliptic problems.

Multigrid methods use coarse grid approximations and iterative techniques to provide approximations on finer grids. References on these techniques include Briggs [Brigg], Mc Cormick [Mc], and Bramble [Bram].

Bibliography

Text pages referring to the items are given in italics at the end of each reference.

[AHU] Aho, A. V., J. E. Hopcroft, and J. D. Ullman, *The design and analysis of computer algorithms*, Addison-Wesley, Reading, MA, 1974, 470 pp. QA76.6.A36 *557*

[Ai] Aitken, A. C. *On interpolation by iteration of proportional parts, without the use of differences.* Proc. Edinburgh Math. Soc. 3(2): 56-76 (1932) QA1.E23 *87*

[AG] Allgower, E. and K. Georg, *Numerical continuation methods: an introduction*, Springer-Verlag, New York, 1990, 388 pp. QA377.A56 *668, 669*

[Am] Ames, W. F., *Numerical methods for partial differential equations*, (Third edition), Academic Press, New York, 1992, 451 pp. QA374.A46 *744*

[AP] Andrews, H. C. and C. L. Patterson, *Outer product expansions and their uses in digital image processing*, American Mathematical Monthly **82**, No. 1 (1975), 1–13, QA1.A515 *625*

[AS] Argyros, I. K. and F. Szidarovszky, *The theory and applications of iteration methods*, CRC Press, Boca Raton, FL, 1993, 355 pp. QA297.8.A74 *669*

[AMR] Ascher, U. M., R. M. M. Mattheij, and R. D. Russell, *Numerical solution of boundary value problems for ordinary differential equations*, Prentice-Hall, Englewood Cliffs, NJ, 1988, 595 pp. QA379.A83 *712*

[Ax] Axelsson, O., *Iterative solution methods*, Cambridge University Press, New York, 1994, 654 pp. QA297.8.A94 *495*

[AB] Axelsson, O. and V. A. Barker, *Finite element solution of boundary value problems: theory and computation*, Academic Press, Orlando, FL, 1984, 432 pp. QA379.A9 *758*

[Ba1] Bailey, N. T. J., *The mathematical approach to biology and medicine*, John Wiley & Sons, New York, 1967, 269 pp. QH324.B28 *301*

[Ba2] Bailey, N. T. J., *The mathematical theory of epidemics*, Hafner, New York, 1957, 194 pp. RA625.B3 *301, 302*

[BSW] Bailey, P. B., L. F. Shampine, and P. E. Waltman, *Nonlinear two-point boundary-value problems*, Academic Press, New York, 1968, 171 pp. QA372.B27 *697, 712*

[Ban] Bank, R. E., *PLTMG, A software package for solving elliptic partial differential equations: Users' Guide 7.0*, SIAM Publications, Philadelphia, PA, 1994, 128 pp. QA377.B26 *761*

[Barr] Barrett, R., et al., *Templates for the solution of linear systems: building blocks for iterative methods*, SIAM Publications, Philadelphia, PA, 1994, 112 pp. QA297.8.T45 *495*

[Bart] Bartle, R. G., *The elements of real analysis*, (Second edition), John Wiley & Sons, New York, 1976, 480 pp. QA300.B29 *106, 117*

[Bek] Bekker, M. G., *Introduction to terrain vehicle systems*, University of Michigan Press, Ann Arbor, MI, 1969, 846 pp. TL243.B39 *78, 646*

[Ber] Bernadelli, H., *Population waves*, Journal of the Burma Research Society **31** (1941), 1–18, DS527.B85 *394*

[BD] Birkhoff, G. and C. De Boor, *Error bounds for spline interpolation*, Journal of Mathematics and Mechanics **13** (1964), 827–836, QA1.J975 *160*

[BL] Birkhoff, G. and R. E. Lynch, *Numerical solution of elliptic problems*, SIAM Publications, Philadelphia, PA, 1984, 319 pp. QA377.B57 *761*

[BiR] Birkhoff, G. and G. Rota, *Ordinary differential equations*, (Fourth edition), John Wiley & Sons, New York, 1989, 399 pp. QA372.B57 *262, 263, 329*

[BP] Botha, J. F. and G. F. Pinder, *Fundamental concepts in the numerical solution of differential equations*, Wiley-Interscience, New York, 1983, 202 pp. QA374.B74 *356*

[Brac] Bracewell, R., *The Fourier transform and its application*, (Third edition), McGraw-Hill, New York, 2000, 616 pp. QA403.5.B7 *557*

[Bram] Bramble, J. H., *Multigrid methods*, John Wiley & Sons, New York, 1993, 161 pp. QA377.B73 *761*

[Bre] Brent, R., *Algorithms for minimization without derivatives*, Prentice-Hall, Englewood Cliffs, NJ, 1973, 195 pp. QA402.5.B74 *102, 103*

[Brigg] Briggs, W. L., *A multigrid tutorial*, SIAM Publications, Philadelphia, PA, 1987, 88 pp. QA377.B75 *761*

[BH] Briggs, W. L. and V. E. Henson, *The DFT: an owner's manual for the discrete Fourier transform*, SIAM Publications, Philadelphia, PA, 1995, 434 pp. QA403.5.B75 *559*

[Brigh] Brigham, E. O., *The fast Fourier transform*, Prentice-Hall, Englewood Cliffs, NJ, 1974, 252 pp. QA403.B74 *548*

[Brow,K] Brown, K. M., *A quadratically convergent Newton-like method based upon Gaussian elimination*, SIAM Journal on Numerical Analysis **6**, No. 4 (1969), 560–569, QA297.A1S2 *652*

[Brow, W] Brown, W. S., *A simple but realistic model of floating point computation*, ACM transactions of Mathematical Software **7** (1981), 445–480, QA76.A8 *42, 44*

[Broy] Broyden, C. G., *A class of methods for solving nonlinear simultaneous equations*, Mathematics of Computation **19** (1965), 577–593, QA1.M4144 *648*

[BS1] Bulirsch R. and J. Stoer, *Numerical treatment of ordinary differential equations by extrapolation methods*, Numerische Mathematik **8** (1966), 1–13, QA241.N9 *327*

[BS2] Bulirsch, R. and J. Stoer, *Fehlerabschätzungen und extrapolation mit rationalen Funktionen bei Verfahren von Richardson-typus*, Numerische Mathematik **6** (1964), 413–427, QA241.N9 *327*

[BS3] Bulirsch, R. and J. Stoer, *Asymptotic upper and lower bounds for results of extrapolation methods*, Numerische Mathematik **8** (1966), 93–104, QA241.N9 *327*

[BuR] Bunch, J. R. and D. J. Rose (eds.), *Sparse matrix computations* (Proceedings of a conference held at Argonne National Laboratories, September 9–11, 1975), Academic Press, New York, 1976, 453 pp. QA188.S9 *429*

[BFR] Burden, R. L., J. D. Faires, and A. C. Reynolds, *Numerical Analysis*, (Second edition), Prindle, Weber & Schmidt, Boston, MA, 1981, 598 pp. QA297.B84 *96*

[Bur] Burrage, K., 1995, *Parallel and sequential methods for ordinary differential equations*, Oxford University Press, New York, 446 pp. QA372.B883 *356*

[But] Butcher, J. C., *The non-existence of ten-stage eighth-order explicit Runge-Kutta methods*, BIT **25** (1985), 521–542, QA76.N62 *290*

[CF] Chaitin-Chatelin, F. and Fraysse, V., *Lectures on finite precision computations*, SIAM Publications, Philadelphia, PA, 1996, 235 pp. QA297.C417 *46*

[CGGG] Char, B. W., K. O. Geddes, W. M. Gentlemen, G. H. Gonnet, *The design of Maple: A compact, portable, and powerful computer algebra system*, Computer Algebra. Lecture Notes in Computer

Science No. 162, (J. A. Van Hulzen, ed.), Springer-Verlag, Berlin, 1983, 101–115 pp. QA155.7 E4 E85 *6, 46*

[CCR] Chiarella, C., W. Charlton, and A. W. Roberts, *Optimum chute profiles in gravity flow of granular materials: a discrete segment solution method*, Transactions of the ASME, Journal of Engineering for Industry Series B **97** (1975), 10–13, TJ1.A712 *646*

[Ch] Cheney, E. W., *Introduction to approximation theory*, McGraw-Hill, New York, 1966, 259 pp. QA221.C47 *559*

[CW] Cody, W. J. and W. Waite, *Software manual for the elementary functions*, Prentice-Hall, Englewood Cliffs, NJ, 1980, 269 pp. QA331.C635 *46*

[CV] Coleman, T. F. and C. Van Loan, *Handbook for matrix computations*, SIAM Publications, Philadelphia, PA, 1988, 264 pp. QA188.C65 *44, 430*

[CT] Cooley, J. W. and J. W. Tukey, *An algorithm for the machine calculation of complex Fourier series*, Mathematics of Computation **19**, No. 90 (1965), 297–301, QA1.M4144 *548*

[CLRS] Cormen, T. H., C. E. Leiserson, R. I. Rivest, C. Stein, *Introduction to algorithms*, (Second Edition) The MIT Press, Cambridge MA, 2001, 1180 pp. QA76.66.I5858 *38*

[Co] Cowell, W. (ed.), *Sources and development of mathematical software*, Prentice-Hall, Englewood Cliffs, NJ, 1984, 404 pp. QA76.95.S68 *44*

[CN] Crank, J. and P Nicolson. *A practical method for numerical evaluation of solutions of partial differential equations of the heat-conduction type*, Proc. Cambridge Philos. Soc. 43 (1947), 0–67, Q41.C17 *733*

[DaB] Dahlquist, G. and Å. Björck (Translated by N. Anderson), *Numerical methods*, Prentice-Hall, Englewood Cliffs, NJ, 1974, 573 pp. QA297.D3313 *86*

[Da] Davis, P. J., *Interpolation and approximation*, Dover, New York, 1975, 393 pp. QA221.D33 *172, 559*

[DR] Davis, P. J. and P. Rabinowitz, *Methods of numerical integration*, (Second edition), Academic Press, New York, 1984, 612 pp. QA299.3.D28 *257*

[Deb1] De Boor, C., *On calculating with B-splines*, Journal of Approximation Theory, **6**, (1972), 50–62, QA221.J63 *705*

[Deb2] De Boor, C., *A practical guide to splines*, Springer-Verlag, New York, 1978, 392 pp. QA1.A647 vol. 27 *161, 172*

[DebS] De Boor, C. and B. Swartz, *Collocation at Gaussian points*, SIAM Journal on Numerical Analysis **10**, No. 4 (1973), 582–606, QA297.A1S2 *710*

[DG] DeFranza, J. and D. Gagliardi, *Introduction to linear algebra*, McGraw-Hill, New York, 2009, 488 pp. QA184.2.D44 *563*

[DM] Dennis, J. E., Jr. and J. J. Moré, *Quasi-Newton methods, motivation and theory*, SIAM Review **19**, No. 1 (1977), 46–89, QA1.S2 *648, 649*

[DenS] Dennis, J. E., Jr. and R. B. Schnabel, *Numerical methods for unconstrained optimization and nonlinear equations*, Prentice-Hall, Englewood Cliffs, NJ, 1983, 378 pp. QA402.5.D44 *669*

[Di] Dierckx, P., *Curve and surface fitting with splines*, Oxford University Press, New York, 1993, 285 pp. QA297.6.D54 *172*

[DBMS] Dongarra, J. J., J. R. Bunch, C. B. Moler, and G. W. Stewart, *LINPACK users guide*, SIAM Publications, Philadephia, PA, 1979, 367 pp. QA214.L56 *44*

[DRW] Dongarra, J. J., T. Rowan, and R. Wade, *Software distributions using Xnetlib*, ACM Transactions on Mathematical Software **21**, No. 1 (1995), 79–88 QA76.6.A8 *45*

[DW] Dongarra, J. and D. W. Walker, *Software libraries for linear algebra computation on high performance computers*, SIAM Review **37**, No. 2 (1995), 151–180 QA1.S2 *46*

[Do] Dormand, J. R., *Numerical methods for differential equations: a computational approach*, CRC Press, Boca Raton, FL, 1996, 368 pp. QA372.D67 *356*

[DoB] Dorn, G. L. and A. B. Burdick, *On the recombinational structure of complementation relationships in the m-dy complex of the* Drosophila melanogaster, Genetics **47** (1962), 503–518, QH431.G43 *428*

[E] Engels, H., *Numerical quadrature and cubature*, Academic Press, New York, 1980, 441 pp. QA299.3.E5 *257*

[Fe] Fehlberg, E., *Klassische Runge-Kutta Formeln vierter und niedrigerer Ordnung mit Schrittweiten-Kontrolle und ihre Anwendung auf Wärmeleitungsprobleme*, Computing **6** (1970), 61–71, QA76.C777 *296*

[Fi] Fix, G., *A survey of numerical methods for selected problems in continuum mechanics*, Proceedings of a Conference on Numerical Methods of Ocean Circulation, National Academy of Sciences (1975), 268–283, Q11.N26 *758*

[FVFH] Foley, J., A. van Dam, S. Feiner, J. Hughes *Computer graphics: principles and practice*, (Second Edition), Addison-Wesley, Reading, MA, 1996, 1175 pp. T385 .C5735 *165, 170*

[FM] Forsythe, G. E. and C. B. Moler, *Computer solution of linear algebraic systems*, Prentice-Hall, Englewood Cliffs, NJ, 1967, 148 pp. QA297.F57 *430, 471*

[Fr] Francis, J. G. F., *The QR transformation*, Computer Journal **4** (1961–2), Part I, 265–271; Part II, 332–345, QA76.C57 *606*

[Fu] Fulks, W., *Advanced calculus*, (Third edition), John Wiley & Sons, New York, 1978, 731 pp. QA303.F954 *10, 283*

[Gar] Garbow, B. S., et al., *Matrix eigensystem routines: EISPACK guide extension*, Springer-Verlag, New York, 1977, 343 pp. QA193.M38 *44*

[Ge1] Gear, C. W., *Numerical initial-value problems in ordinary differential equations*, Prentice-Hall, Englewood Cliffs, NJ, 1971, 253 pp. QA372.G4 *356*

[Ge2] Gear, C. W., *Numerical solution of ordinary differential equations: Is there anything left to do?*, SIAM Review **23**, No. 1 (1981), 10–24, QA1.S2 *353*

[Ger] Geršgorin, S. A. *Über die Abgrenzung der Eigenwerte einer Matrix*. Dokl. Akad. Nauk.(A), Otd. Fiz-Mat. Nauk. (1931), 749–754, QA1.A3493 *562*

[GL] George, A. and J. W. Liu, *Computer solution of large sparse positive definite systems*, Prentice-Hall, Englewood Cliffs, NJ, 1981, 324 pp. QA188.G46 *430*

[Go] Goldberg, D., *What every scientist should know about floating-point arithmetic*, ACM Computing Surveys **23**, No. 1 (1991), 5–48, QA76.5.A1 *46*

[Golds] Goldstine, H. H. *A History of Numerical Analysis from the 16th through the 19th Centuries*. Springer-Verlag, 348 pp. QA297.G64 *xiii, 229*

[GK] Golub, G.H. and W. Kahan, *Calculating the singular values and pseudo-inverse of a matrix*, SIAM J. Numer. Anal. 2,Ser. B (1965) 205–224, QA297.A1S2 *614*

[GO] Golub, G. H. and J. M. Ortega, *Scientific computing: an introduction with parallel computing*, Academic Press, Boston, MA, 1993, 442 pp. QA76.58.G64 *46, 356*

[GR] Golub, G. H. and C. Reinsch, *Singular value decomposition and least squares solutions*, Numerische Mathematik 14 (1970) 403–420, QA241.N9 *614*

[GV] Golub, G. H. and C. F. Van Loan, *Matrix computations*, (Third edition), Johns Hopkins University Press, Baltimore, MD, 1996, 694 pp. QA188.G65 *414, 430, 625*

[Gr] Gragg, W. B., *On extrapolation algorithms for ordinary initial-value problems*, SIAM Journal on Numerical Analysis **2** (1965), 384–403, QA297.A1S2 *321, 327*

[GKO] Gustafsson, B., H. Kreiss, and J. Oliger, *Time dependent problems and difference methods*, John Wiley & Sons, New York, 1995, 642 pp. QA374.G974 *761*

[Hac] Hackbusch, W., *Iterative solution of large sparse systems of equations*, Springer-Verlag, New York, 1994, 429 pp. QA1.A647 vol. 95 *495*

[HY] Hageman, L. A. and D. M. Young, *Applied iterative methods*, Academic Press, New York, 1981, 386 pp. QA297.8.H34 *495*

[HNW1] Hairer, E., S. P. Nörsett, and G. Wanner, *Solving ordinary differential equations. Vol. 1: Nonstiff equations*, (Second revised edition), Springer-Verlag, Berlin, 1993, 519 pp. QA372.H16 *336, 356*

[HNW2] Hairer, E., S. P. Nörsett, and G. Wanner, *Solving ordinary differential equations. Vol. 2: Stiff and differential-algebraic problems*, (Second revised edition), Springer, Berlin, 1996, 614 pp. QA372.H16 *356*

[Ham] Hamming, R. W., *Numerical methods for scientists and engineers*, (Second edition), McGraw-Hill, New York, 1973, 721 pp. QA297.H28 *557*

[He1] Henrici, P., *Discrete variable methods in ordinary differential equations*, John Wiley & Sons, New York, 1962, 407 pp. QA372.H48 *356*

[He2] Henrici, P., *Elements of numerical analysis*, John Wiley & Sons, New York, 1964, 328 pp. QA297.H54 *89, 345*

[HS] Hestenes, M. R. and E. Steifel, *Conjugate gradient methods in optimization*, Journal of Research of the National Bureau of Standards **49**, (1952), 409–436, Q1.N34 *479*

[Heu] Heun, K., *Neue methode zur approximativen integration der differntialqleichungen einer unabhängigen veränderlichen*, Zeitschrift für Mathematik und Physik, **45** , (1900), 23–38, QA1.Z48 *287*

[Hild] Hildebrand, F. B., *Introduction to numerical analysis*, (Second edition), McGraw-Hill, New York, 1974, 669 pp. QA297.H54 *133*

[Ho] Householder, A. S., *The numerical treatment of a single nonlinear equation*, McGraw-Hill, New York, 1970, 216 pp. QA218.H68 *102, 103*

[IK] Issacson, E. and H. B. Keller, *Analysis of numerical methods*, John Wiley & Sons, New York, 1966, 541 pp. QA297.I8 *89, 198, 200, 343, 345, 449, 677, 694, 732, 734, 744*

[JT] Jenkins, M. A. and J. F. Traub, *A three-stage algorithm for real polynomials using quadratic iteration*, SIAM Journal on Numerical Analysis **7**, No. 4 (1970), 545–566, QA297.A1S2 *102*

[Joh] Johnston, R. L., *Numerical methods: a software approach*, John Wiley & Sons, New York, 1982, 276 pp. QA297.J64 *218*

[Joy] Joyce, D. C., *Survey of extrapolation processes in numerical analysis*, SIAM Review **13**, No. 4 (1971), 435–490, QA1.S2 *185*

[Ka] Kalman, D., *A singularly valuable decomposition: The SVD of a matrix*, The College Mathematics Journal, vol. 27 (1996), 2–23, QA11.A1 T9. *625*

[Keller,H] Keller, H. B., *Numerical methods for two-point boundary-value problems*, Blaisdell, Waltham, MA, 1968, 184 pp. QA372.K42 *683, 688*

[Keller,J] Keller, J. B., *Probability of a shutout in racquetball*, SIAM Review **26**, No. 2 (1984), 267–268, QA1.S2 *78*

[Kelley] Kelley, C. T., *Iterative methods for linear and nonlinear equations*, SIAM Publications, Philadelphia, PA, 1995, 165 pp. QA297.8.K45 *492, 495*

[Ko] Köckler, N., *Numerical methods and scientific computing: using software libraries for problem solving*, Oxford University Press, New York, 1994, 328 pp. TA345.K653 *46*

[Lam] Lambert, J. D., *The initial value problem for ordinary differential equations. The state of art in numerical analysis* (D. Jacobs, ed.), Academic Press, New York, 1977, 451–501 pp. QA297.C646 *353*

[LP] Lapidus, L. and G. F. Pinder, *Numerical solution of partial differential equations in science and engineering*, John Wiley & Sons, New York, 1982, 677 pp. Q172.L36 *761*

[Lar] Larson, H. J., *Introduction to probability theory and statistical inference*, (Third edition), John Wiley & Sons, New York, 1982, 637 pp. QA273.L352 *499*

[Lau] Laufer, H. B., *Discrete mathematics and applied modern algebra*, PWS-Kent Publishing, Boston, MA, 1984, 538 pp. QA162.L38 *557*

[LH] Lawson, C. L. and R. J. Hanson, *Solving least squares problems*, SIAM Publications, Philadelphia, PA, 1995, 337 pp. QA275.L38 *559*

[Lo1] Lotka, A.J., *Relation between birth rates and death rates*, Science, **26**(1907), 121–130 Q1 *338*

[Lo2] Lotka, A.J., *Natural selection as a physical principle*, Science, Proc. Natl. Acad. Sci., **8** (1922), 151–154 Q11.N26 *338*

[LR] Lucas, T. R. and G. W. Reddien, Jr., *Some collocation methods for nonlinear boundary value problems*, SIAM Journal on Numerical Analysis **9**, No. 2 (1972), 341–356, QA297.A1S2 *710*

[Lu] Luenberger, D. G., *Linear and nonlinear programming*, (Second edition), Addison-Wesley, Reading, MA, 1984, 245 pp. T57.7L8 *485*

[Mc] McCormick, S. F., *Multigrid methods*, SIAM Publications, Philadelphia, PA, 1987, 282 pp. QA374.M84 *761*

[Mi] Mitchell, A. R., *Computation methods in partial differential equations*, John Wiley & Sons, New York, 1969, 255 pp. QA374.M68 *744*

[Mo] Moler, C. B., *Demonstration of a matrix laboratory. Lecture notes in mathematics* (J. P. Hennart, ed.), Springer-Verlag, Berlin, 1982, 84–98 *45*

[MC] Moré J. J. and M. Y. Cosnard, *Numerical solution of nonlinear equations*, ACM Transactions on Mathematical Software **5**, No. 1 (1979), 64–85, QA76.6.A8 *652*

[MM] Morton, K. W. and D. F. Mayers, *Numerical solution of partial differential equations: an introduction*, Cambridge University Press, New York, 1994, 227 pp. QA377.M69 *761*

[Mu] Müller, D. E., *A method for solving algebraic equations using an automatic computer*, Mathematical Tables and Other Aids to Computation **10** (1956), 208–215, QA47.M29 *96*

[N] Neville, E. H. *Iterative Interpolation*, J. Indian Math Soc. 20: 87-120 (1934) *120*

[ND] Noble, B. and J. W. Daniel, *Applied linear algebra*, (Third edition), Prentice-Hall, Englewood Cliffs, NJ, 1988, 521 pp. QA184.N6 *563*

[Or1] Ortega, J. M., *Introduction to parallel and vector solution of linear systems*, Plenum Press, New York, 1988, 305 pp. QA218.O78 *46*

[Or2] Ortega, J. M., *Numerical analysis; a second course*, Academic Press, New York, 1972, 201 pp. QA297.O78 *438, 446, 447, 458, 465, 476, 563, 630, 633*

[OP] Ortega, J. M. and W. G. Poole, Jr., *An introduction to numerical methods for differential equations*, Pitman Publishing, Marshfield, MA, 1981, 329 pp. QA371.O65 *356*

[OR] Ortega, J. M. and W. C. Rheinboldt, *Iterative solution of nonlinear equations in several variables*, Academic Press, New York, 1970, 572 pp. QA297.8.O77 *630, 659, 662, 669*

[Os] Ostrowski, A. M., *Solution of equations and systems of equations*, (Second edition), Academic Press, New York, 1966, 338 pp. QA3.P8 vol. 9 *103*

[Par] Parlett, B. N., *The symmetric eigenvalue problem*, Prentice-Hall, Englewood Cliffs, NJ, 1980, 348 pp. QA188.P37 *627*

[Pat] Patterson, T. N. L., *The optimum addition of points to quadrature formulae*, Mathematics of Computation **22**, No. 104 (1968), 847–856, QA1.M4144 *256*

[PF] Phillips, C. and T. L. Freeman, *Parallel numerical algorithms*, Prentice-Hall, New York, 1992, 315 pp. QA76.9.A43 F74 *46*

[Ph] Phillips, J., *The NAG Library: a beginner's guide*, Clarendon Press, Oxford, 1986, 245 pp. QA297.P35 *45*

[PDUK] Piessens, R., E. de Doncker-Kapenga, C. W. Überhuber, and D. K. Kahaner, *QUADPACK: a subroutine package for automatic integration*, Springer-Verlag, New York, 1983, 301 pp. QA299.3.Q36 *256*

[Pi] Pissanetzky, S., *Sparse matrix technology*, Academic Press, New York, 1984, 321 pp. QA188.P57 *430*

[Poo] Poole, *Linear algebra: A modern introduction*, (Second Edition), Thomson Brooks/Cole,Belmont CA, 2006, 712 pp. QA184.2.P66 *563*

[Pow] Powell, M. J. D., *Approximation theory and methods*, Cambridge University Press, Cambridge, 1981, 339 pp. QA221.P65 *139, 141, 172, 537, 559*

[Pr] Pryce, J. D., *Numerical solution of Sturm-Liouville problems*, Oxford University Press, New York, 1993, 322 pp. QA379.P79 *712*

[RR] Ralston, A. and P. Rabinowitz, *A first course in numerical analysis*, (Second edition), McGraw-Hill, New York, 1978, 556 pp. QA297.R3 *213, 533, 537, 659*

[Ra] Rashevsky, N., *Looking at history through mathematics*, Massachusetts Institute of Technology Press, Cambridge, MA, 1968, 199 pp. D16.25.R3 *276*

[RB] Rice, J. R. and R. F. Boisvert, *Solving elliptic problems using ELLPACK*, Springer-Verlag, New York, 1985, 497 pp. QA377.R53 *761*

[RG] Richardson, L. F. and J. A. Gaunt, *The deferred approach to the limit*, Philosophical Transactions of the Royal Society of London **226A** (1927), 299–361, Q41.L82 *185*

[Ri] Ritz, W., *Über eine neue methode zur lösung gewisser variationsprobleme der mathematischen physik*, Journal für die reine und angewandte Mathematik, **135**(1909), pp. 1–61, QA1.J95 *697*

[Ro] Roache, P. J., *Elliptic marching methods and domain decomposition*, CRC Press, Boca Raton, FL, 1995, 190 pp. QA377.R63 *761*

[RS] Roberts, S. and J. Shipman, *Two-point boundary value problems: shooting methods*, Elsevier, New York, 1972, 269 pp. QA372.R76 *712*

[RW] Rose, D. J. and R. A. Willoughby (eds.), *Sparse matrices and their applications* (Proceedings of a conference held at IBM Research, New York, September 9–10, 1971. 215 pp.), Plenum Press, New York, 1972, QA263.S94 *429*

[Ru] Russell, R. D., *A comparison of collocation and finite differences for two-point boundary value problems*, SIAM Journal on Numerical Analysis **14**, No. 1 (1977), 19–39, QA297.A1S2 *710*

[Sa1] Saad, Y., *Numerical methods for large eigenvalue problems*, Halsted Press, New York, 1992, 346 pp. QA188.S18 *627*

[Sa2] Saad, Y., *Iterative methods for sparse linear systems*, (Second Edition), SIAM, Philadelphia, PA 2003, 528 pp. QA188.S17 *495*

[SaS] Saff, E. B. and A. D. Snider, *Fundamentals of complex analysis for mathematics, science, and engineering*, (Third edition), Prentice-Hall, Upper Saddle River, NJ, 2003, 511 pp. QA300.S18 *91*

[SP] Sagar, V. and D. J. Payne, *Incremental collapse of thick-walled circular cylinders under steady axial tension and torsion loads and cyclic transient heating*, Journal of the Mechanics and Physics of Solids **21**, No. 1 (1975), 39–54, TA350.J68 *738*

[SD] Sale, P. F. and R. Dybdahl, *Determinants of community structure for coral-reef fishes in experimental habitat*, Ecology **56** (1975), 1343–1355, QH540.E3 *508*

[Sche] Schendel, U., *Introduction to numerical methods for parallel computers*, (Translated by B.W. Conolly), Halsted Press, New York, 1984, 151 pp. QA297.S3813 *46*

[Schi] Schiesser, W. E., *Computational mathematics in engineering and applied science: ODE's, DAE's, and PDE's*, CRC Press, Boca Raton, FL, 1994, 587 pp. TA347.D45 S34 *761*

[Scho] Schoenberg, I. J., *Contributions to the problem of approximation of equidistant data by analytic functions*, Quarterly of Applied Mathematics **4**, (1946), Part A, 45–99; Part B, 112–141, QA1.A26 *172, 705*

[Schr1] Schroeder, L. A., *Energy budget of the larvae of the moth* Pachysphinx modesta, Oikos **24** (1973), 278–281, QH540.O35 *509*

[Schr2] Schroeder, L. A., *Thermal tolerances and acclimation of two species of hydras*, Limnology and Oceanography **26**, No. 4 (1981), 690–696, GC1.L5 *646*

[Schul] Schultz, M. H., *Spline analysis*, Prentice-Hall, Englewood Cliffs, NJ, 1973, 156 pp. QA211.S33 *160, 172, 704, 709*

[Schum] Schumaker, L. L., *Spline functions: basic theory*, Wiley-Interscience, New York, 1981, 553 pp. QA224.S33 *172*

[Schw] Schwartzman, S., *The words of mathematics*, The Mathematical Association of America, Washington, 1994, 261 pp. QA5.S375 *xiii*

[Se] Searle, S. R., *Matrix algebra for the biological sciences*, John Wiley & Sons, New York, 1966, 296 pp. QH324.S439 *394*

[SH] Secrist, D. A. and R. W. Hornbeck, *An analysis of heat transfer and fade in disk brakes*, Transactions of the ASME, Journal of Engineering for Industry Series B **98** No. 2 (1976), 385–390, TJ1.A712 *212*

[Sh] Shampine, L. F., *Numerical solution of ordinary differential equations*, Chapman & Hall, New York, 1994, 484 pp. QA372.S417 *356*

[SGe] Shampine, L. F. and C. W. Gear, *A user's view of solving stiff ordinary differential equations*, SIAM Review **21**, No. 1 (1979), 1–17, QA1.S2

[ShS] Shashkov, M. and S. Steinberg, *Conservative finite-difference methods on general grids*, CRC Press, Boca Raton, FL, 1996, 359 pp. QA431.S484 *761*

[Si] Singh, V. P., *Investigations of attentuation and internal friction of rocks by ultrasonics*, International Journal of Rock Mechanics and Mining Sciences (1976), 69–72, TA706.I45 *509*

[SJ] Sloan, I. H. and S. Joe, *Lattice methods for multiple integration*, Oxford University Press, New York, 1994, 239 pp. QA311.S56 *257*

[Sm,B] Smith, B. T., et al., *Matrix eigensystem routines: EISPACK guide*, (Second edition), Springer-Verlag, New York, 1976, 551 pp. QA193.M37 *44*

[Sm,G] Smith, G. D., *Numerical solution of partial differential equations*, Oxford University Press, New York, 1965, 179 pp. QA377.S59 *744*

[So] Sorenson, D. C., *Implicitly restarted Arnoldi/Lanczos methods for large scale eigenvalue calculations*, *Parallel numerical algorithms* (David E. Keyes, Ahmed Sameh and V. Vankatakrishan, eds.), Kluwer Academic Publishers, Dordrecht, 1997, 119-166 QA76.9.A43 P35 *627*

[Stee] Steele, J. Michael, *The Cauchy-Schwarz Master Class*. Cambridge University Press, 2004, 306 pp. QA295.S78 *434*

[Stet] Stetter, H. J., *Analysis of discretization methods for ordinary differential equations. From tracts in natural philosophy*, Springer-Verlag, New York, 1973, 388 pp. QA372.S84 *327*

[Stew1] Stewart, G. W., *Afternotes on numerical analysis*, SIAM Publications, Philadelphia, PA, 1996, 200 pp. QA297.S785 *430*

[Stew2] Stewart, G. W., *Introduction to matrix computations*, Academic Press, New York, 1973, 441 pp. QA188.S7 *416, 627*

[Stew3] Stewart, G. W., *On the early history of the singular value decomposition.* http://www.lib.umd.edu/drum/bitstream/1903/566/4/CS-TR-2855.pdf *614*

[SF] Strang, W. G. and G. J. Fix, *An analysis of the finite element method*, Prentice-Hall, Englewood Cliffs, NJ, 1973, 306 pp. TA335.S77 *709, 758, 761*

[Stri] Strikwerda, J. C., *Finite difference schemes and partial differential equations*, (second Edition), SIAM Publications, Philadelphia, PA, 2004, 435 pp. QA374.S88 *761*

[Stro] Stroud, A. H., *Approximate calculation of multiple integrals*, Prentice-Hall, Englewood Cliffs, NJ, 1971, 431 pp. QA311.S85 *257*

[StS] Stroud, A. H. and D. Secrest, *Gaussian quadrature formulas*, Prentice-Hall, Englewood Cliffs, NJ, 1966, 374 pp. QA299.4.G4 S7 *257*

[Sz] Szüsz, P., *Math bite*, Mathematics Magazine **68**, No. 2, 1995, 97, QA1.N28 *442*

[Th] Thomas, J. W., *Numerical partial differential equations*, Springer-Verlag, New York, 1998, 445 pp. QA377.T495 *761*

[TCMT] Turner, M. J., R. W. Clough, H. C. Martin, L. J. Topp, *Stiffness and deflection of complex structures*, Journal of the Aeronautical Sciences, **23**, (1956), 805–824, TL501.I522 *746*

[Tr] Traub, J. F., *Iterative methods for the solution of equations*, Prentice-Hall, Englewood Cliffs, NJ, 1964, 310 pp. QA297.T7 *103*

[Tw] Twizell, E. H., *Computational methods for partial differential equations*, Ellis Horwood Ltd., Chichester, West Sussex, England, 1984, 276 pp. QA377.T95 *761*

[Van] Van Loan, C. F., *Computational frameworks for the fast Fourier transform*, SIAM Publications, Philadelphia, PA, 1992, 273 pp. QA403.5.V35 *559*

[Var1] Varga, R. S., *Matrix iterative analysis*, (Second edition), Springer, New York, 2000, 358 pp. QA263.V3 *495, 718, 722*

[Var2] Varga, R. S., *Geršgorin and his circles*, Springer, New York, 2004, 226 pp. QA184.V37 *562, 563*

[Ve] Verner, J. H., *Explicit Runge-Kutta methods with estimates of the local trucation error*, SIAM Journal on Numerical Analysis **15**, No. 4 (1978), 772–790, QA297.A1S2 *301*

[Vo] Volterra, V., *Variazioni e fluttuazioni del numero d'individui in specie animali conviventi*, Mem. Acad. Lineci Roma, **2**, (1926), 31–113, QA297.A1S2 *338*

[We] Wendroff, B., *Theoretical numerical analysis*, Academic Press, New York, 1966, 239 pp. QA297.W43 *413, 416*

[Wil1] Wilkinson, J. H., *Rounding errors in algebraic processes*, Prentice-Hall, Englewood Cliffs, NJ, 1963, 161 pp. QA76.5.W53 *476*

[Wil2] Wilkinson, J. H., *The algebraic eigenvalue problem*, Clarendon Press, Oxford, 1965, 662 pp. QA218.W5 *476, 580, 586, 593, 604, 611, 627*

[WR] Wilkinson, J. H. and C. Reinsch (eds.), *Handbook for automatic computation. Vol. 2: Linear algebra*, Springer-Verlag, New York, 1971, 439 pp. QA251.W67 *44, 606, 611, 627*

[Win] Winograd, S., *On computing the discrete Fourier transform*, Mathematics of Computation **32** (1978), 175–199, QA1.M4144 *557*

[Y] Young, D. M., *Iterative solution of large linear systems*, Academic Press, New York, 1971, 570 pp. QA195.Y68 *459, 495*

[YG] Young, D. M. and R. T. Gregory, *A survey of numerical mathematics. Vol. 1*, Addison-Wesley, Reading, MA, 1972, 533 pp. QA297.Y63 *102*

[ZM] Zienkiewicz, O. C. and K. Morgan, *Finite elements and approximation*, John Wiley & Sons, New York, 1983, 328 pp. QA297.5.Z53 *758, 761*

Answers for Selected Exercises

Exercise Set 1.1 (Page 14)

1. For each part, $f \in C[a,b]$ on the given interval. Since $f(a)$ and $f(b)$ are of opposite sign, the Intermediate Value Theorem implies that a number c exists with $f(c) = 0$.

3. For each part, $f \in C[a,b]$, f' exists on (a,b) and $f(a) = f(b) = 0$. Rolle's Theorem implies that a number c exists in (a,b) with $f'(c) = 0$. For part (d), we can use $[a,b] = [-1,0]$ or $[a,b] = [0,2]$.

5. For $x < 0$, $f(x) < 2x + k < 0$, provided that $x < -\frac{1}{2}k$. Similarly, for $x > 0$, $f(x) > 2x + k > 0$, provided that $x > -\frac{1}{2}k$. By Theorem 1.11, there exists a number c with $f(c) = 0$. If $f(c) = 0$ and $f(c') = 0$ for some $c' \neq c$, then by Theorem 1.7, there exists a number p between c and c' with $f'(p) = 0$. However, $f'(x) = 3x^2 + 2 > 0$ for all x.

7. **a.** $P_2(x) = 0$ **b.** $R_2(0.5) = 0.125$; actual error $= 0.125$

c. $P_2(x) = 1 + 3(x-1) + 3(x-1)^2$ **d.** $R_2(0.5) = -0.125$; actual error $= -0.125$

9. Since

$$P_2(x) = 1 + x \quad \text{and} \quad R_2(x) = \frac{-2e^{\xi}(\sin\xi + \cos\xi)}{6}x^3$$

for some ξ between x and 0, we have the following:

a. $P_2(0.5) = 1.5$ and $|f(0.5) - P_2(0.5)| \leq 0.0532$; **b.** $|f(x) - P_2(x)| \leq 1.252$;

c. $\int_0^1 f(x)\,dx \approx 1.5$;

d. $|\int_0^1 f(x)\,dx - \int_0^1 P_2(x)\,dx| \leq \int_0^1 |R_2(x)|\,dx \leq 0.313$, and the actual error is 0.122.

11. **a.** $P_3(x) = -4 + 6x - x^2 - 4x^3$; $P_3(0.4) = -2.016$

b. $|R_3(0.4)| \leq 0.05849$; $|f(0.4) - P_3(0.4)| = 0.013365367$

c. $P_4(x) = -4 + 6x - x^2 - 4x^3$; $P_4(0.4) = -2.016$

d. $|R_4(0.4)| \leq 0.01366$; $|f(0.4) - P_4(0.4)| = 0.013365367$

13. $P_4(x) = x + x^3$

a. $|f(x) - P_4(x)| \leq 0.012405$ **b.** $\int_0^{0.4} P_4(x)\,dx = 0.0864$, $\int_0^{0.4} xe^{x^2}\,dx = 0.086755$

c. 8.27×10^{-4}

d. $P_4'(0.2) = 1.12$, $f'(0.2) = 1.124076$. The actual error is 4.076×10^{-3}.

15. Since $42^\circ = 7\pi/30$ radians, use $x_0 = \pi/4$. Then

$$\left|R_n\left(\frac{7\pi}{30}\right)\right| \leq \frac{\left(\frac{\pi}{4} - \frac{7\pi}{30}\right)^{n+1}}{(n+1)!} < \frac{(0.053)^{n+1}}{(n+1)!}.$$

For $|R_n(\frac{7\pi}{30})| < 10^{-6}$, it suffices to take $n = 3$. To 7 digits, $\cos 42^\circ = 0.7431448$ and $P_3(42^\circ) = P_3(\frac{7\pi}{30}) = 0.7431446$, so the actual error is 2×10^{-7}.

17. **a.** $P_3(x) = \ln(3) + \frac{2}{3}(x-1) + \frac{1}{9}(x-1)^2 - \frac{10}{81}(x-1)^3$ **b.** $\max_{0\leq x\leq 1} |f(x) - P_3(x)| = |f(0) - P_3(0)| = 0.02663366$

c. $\tilde{P}_3(x) = \ln(2) + \frac{1}{2}x^2$ **d.** $\max_{0\leq x\leq 1} |f(x) - \tilde{P}_3(x)| = |f(1) - \tilde{P}_3(1)| = 0.09453489$

e. $P_3(0)$ approximates $f(0)$ better than $\tilde{P}_3(1)$ approximates $f(1)$.

19. $P_n(x) = \sum_{k=0}^{n} \frac{1}{k!}x^k$, $n \geq 7$

21. A bound for the maximum error is 0.0026.

23. a. The assumption is that $f(x_i) = 0$ for each $i = 0, 1, \ldots, n$. Applying Rolle's Theorem on each on the intervals $[x_i, x_{i+1}]$ implies that for each $i = 0, 1, \ldots, n-1$ there exists a number z_i with $f'(z_i) = 0$. In addition, we have $a \le x_0 < z_0 < x_1 < z_1 < \cdots < z_{n-1} < x_n \le b$.

b. Apply the logic in part (a) to the function $g(x) = f'(x)$ with the number of zeros of g in $[a, b]$ reduced by 1. This implies that numbers w_i, for $i = 0, 1, \ldots, n-2$ exist with $g'(w_i) = f''(w_i) = 0$, and $z_0 < w_0 < z_1 < w_1 < \cdots < z_{n-2} < w_{n-2} < z_{n-1}$.

c. Continuing by induction following the logic in parts (a) and (b) provides $n - j + 1$ distinct zeros of $f^{(j)}$ in $[a, b]$.

d. The conclusion of the theorem follows from part (c) when $j = n$, for in this case there will be (at least) $n - (n-1) = 1$ zero of $f^{(n)}$ in $[a, b]$.

25. Since $R_2(1) = \frac{1}{6}e^{\xi}$, for some ξ in $(0, 1)$, we have $|E - R_2(1)| = \frac{1}{6}|1 - e^{\xi}| \le \frac{1}{6}(e - 1)$.

27. a. Let x_0 be any number in $[a, b]$. Given $\varepsilon > 0$, let $\delta = \varepsilon/L$. If $|x - x_0| < \delta$ and $a \le x \le b$, then $|f(x) - f(x_0)| \le L|x - x_0| < \varepsilon$.

b. Using the Mean Value Theorem, we have

$$|f(x_2) - f(x_1)| = |f'(\xi)||x_2 - x_1|,$$

for some ξ between x_1 and x_2, so

$$|f(x_2) - f(x_1)| \le L|x_2 - x_1|.$$

c. One example is $f(x) = x^{1/3}$ on $[0, 1]$.

29. a. Since f is continuous at p and $f(p) \ne 0$, there exists a $\delta > 0$ with

$$|f(x) - f(p)| < \frac{|f(p)|}{2},$$

for $|x - p| < \delta$ and $a < x < b$. We restrict δ so that $[p - \delta, p + \delta]$ is a subset of $[a, b]$. Thus, for $x \in [p - \delta, p + \delta]$, we have $x \in [a, b]$. So

$$-\frac{|f(p)|}{2} < f(x) - f(p) < \frac{|f(p)|}{2}$$

and

$$f(p) - \frac{|f(p)|}{2} < f(x) < f(p) + \frac{|f(p)|}{2}.$$

If $f(p) > 0$, then

$$f(p) - \frac{|f(p)|}{2} = \frac{f(p)}{2} > 0, \quad \text{so} \quad f(x) > f(p) - \frac{|f(p)|}{2} > 0.$$

If $f(p) < 0$, then $|f(p)| = -f(p)$, and

$$f(x) < f(p) + \frac{|f(p)|}{2} = f(p) - \frac{f(p)}{2} = \frac{f(p)}{2} < 0.$$

In either case, $f(x) \ne 0$, for $x \in [p - \delta, p + \delta]$.

b. Since f is continuous at p and $f(p) = 0$, there exists a $\delta > 0$ with

$$|f(x) - f(p)| < k, \quad \text{for} \quad |x - p| < \delta \quad \text{and} \quad a < x < b.$$

We restrict δ so that $[p - \delta, p + \delta]$ is a subset of $[a, b]$. Thus, for $x \in [p - \delta, p + \delta]$, we have

$$|f(x)| = |f(x) - f(p)| < k.$$

Exercise Set 1.2 (Page 28)

1.	Absolute Error	Relative Error
a.	7.346×10^{-6}	2.338×10^{-6}
b.	0.001264	4.025×10^{-4}
c.	2.136×10^{-4}	1.510×10^{-4}
d.	2.818×10^{-4}	1.037×10^{-4}
e.	1.454×10^{1}	1.050×10^{-2}
f.	2.647×10^{1}	1.202×10^{-3}
g.	420	1.042×10^{-2}
h.	3.343×10^{3}	9.213×10^{-3}

3. The largest intervals are

a. (149.85, 150.15) **b.** (899.1, 900.9) **c.** (1498.5, 1501.5) **d.** (89.91, 90.09)

5.	Approximation	Absolute Error	Relative Error
a.	134	0.079	5.90×10^{-4}
b.	133	0.499	3.77×10^{-3}
c.	2.00	0.327	0.195
d.	1.67	0.003	1.79×10^{-3}
e.	1.80	0.154	0.0786
f.	-15.1	0.0546	3.60×10^{-3}
g.	0.286	2.86×10^{-4}	10^{-3}
h.	0.00	0.0215	1.00

7.	Approximation	Absolute Error	Relative Error
a.	133	0.921	6.88×10^{-3}
b.	132	0.501	3.78×10^{-3}
c.	1.00	0.673	0.402
d.	1.67	0.003	1.79×10^{-3}
e.	3.55	1.60	0.817
f.	-15.2	0.0454	0.00299
g.	0.284	0.00171	0.00600
h.	0	0.02150	1

9.	Approximation	Absolute Error	Relative Error
a.	3.14557613	3.983×10^{-3}	1.268×10^{-3}
b.	3.14162103	2.838×10^{-5}	9.032×10^{-6}

11. a. $\lim_{x\to 0} \frac{x\cos x - \sin x}{x - \sin x} = \lim_{x\to 0} \frac{-x\sin x}{1 - \cos x} = \lim_{x\to 0} \frac{-\sin x - x\cos x}{\sin x} = \lim_{x\to 0} \frac{-2\cos x + x\sin x}{\cos x} = -2$

b. -1.941

c. $\frac{x(1 - \frac{1}{2}x^2) - (x - \frac{1}{6}x^3)}{x - (x - \frac{1}{6}x^3)} = -2$

d. The relative error in part (b) is 0.029. The relative error in part (c) is 0.00050.

13.

	x_1	Absolute Error	Relative Error	x_2	Absolute Error	Relative Error
a.	92.26	0.01542	1.672×10^{-4}	0.005419	6.273×10^{-7}	1.157×10^{-4}
b.	0.005421	1.264×10^{-6}	2.333×10^{-4}	−92.26	4.580×10^{-3}	4.965×10^{-5}
c.	10.98	6.875×10^{-3}	6.257×10^{-4}	0.001149	7.566×10^{-8}	6.584×10^{-5}
d.	−0.001149	7.566×10^{-8}	6.584×10^{-5}	−10.98	6.875×10^{-3}	6.257×10^{-4}

15. The machine numbers are equivalent to

a. 3224 **b.** −3224 **c.** 1.32421875

d. 1.3242187500000002220446049250313080847263336181640625

17. b. The first formula gives −0.00658, and the second formula gives −0.0100. The true three-digit value is −0.0116.

19. The approximate solutions to the systems are

a. $x = 2.451$, $y = -1.635$ **b.** $x = 507.7$, $y = 82.00$

21. a. In nested form, we have $f(x) = (((1.01e^x - 4.62)e^x - 3.11)e^x + 12.2)e^x - 1.99$.

b. −6.79

c. −7.07

23. a. $n = 77$ **b.** $n = 35$

25. a. $m = 17$

b. $$\binom{m}{k} = \frac{m!}{k!(m-k)!} = \frac{m(m-1)\cdots(m-k-1)(m-k)!}{k!(m-k)!}$$
$$= \left(\frac{m}{k}\right)\left(\frac{m-1}{k-1}\right)\cdots\left(\frac{m-k-1}{1}\right)$$

c. $m = 181707$

d. 2,597,000; actual error 1960; relative error 7.541×10^{-4}

27. a. 124.03 **b.** 124.03 **c.** −124.03 **d.** −124.03

e. 0.0065 **f.** 0.0065 **g.** −0.0065 **h.** −0.0065

Exercise Set 1.3 (Page 39)

1. a. The approximate sums are 1.53 and 1.54, respectively. The actual value is 1.549. Significant roundoff error occurs earlier with the first method.

3. a. 2000 terms **b.** 20,000,000,000 terms

5. 3 terms

7. The rates of convergence are:

a. $O(h^2)$ **b.** $O(h)$ **c.** $O(h)$ **d.** $O(h^2)$

13. a. If $|\alpha_n - \alpha|/(1/n^p) \le K$, then $|\alpha_n - \alpha| \le K(1/n^p) \le K(1/n^q)$ since $0 < q < p$. Thus, $|\alpha_n - \alpha|/(1/n^p) \le K$ and $\{\alpha_n\}_{n=1}^{\infty} \to \alpha$ with rate of convergence $O(1/n^p)$.

b.

n	$1/n$	$1/n^2$	$1/n^3$	$1/n^4$
5	0.2	0.04	0.008	0.0016
10	0.1	0.01	0.001	0.0001
50	0.02	0.0004	8×10^{-6}	1.6×10^{-7}
100	0.01	10^{-4}	10^{-6}	10^{-8}

$O(1/n^4)$ is the most rapid convergence rate.

15. Suppose that for sufficiently small $|x|$ we have positive constants k_1 and k_2 independent of x, for which

$$|F_1(x) - L_1| \le K_1|x|^\alpha \quad \text{and} \quad |F_2(x) - L_2| \le K_2|x|^\beta.$$

Let $c = \max(|c_1|, |c_2|, 1)$, $K = \max(K_1, K_2)$, and $\delta = \max(\alpha, \beta)$.

a. We have

$$\begin{aligned} |F(x) - c_1L_1 - c_2L_2| &= |c_1(F_1(x) - L_1) + c_2(F_2(x) - L_2)| \\ &\le |c_1|K_1|x|^\alpha + |c_2|K_2|x|^\beta \\ &\le cK[|x|^\alpha + |x|^\beta] \\ &\le cK|x|^\gamma[1 + |x|^{\delta-\gamma}] \\ &\le \tilde{K}|x|^\gamma, \end{aligned}$$

for sufficiently small $|x|$ and some constant $\tilde{K}$. Thus, $F(x) = c_1L_1 + c_2L_2 + O(x^\gamma)$.

b. We have

$$\begin{aligned} |G(x) - L_1 - L_2| &= |F_1(c_1x) + F_2(c_2x) - L_1 - L_2| \\ &\le K_1|c_1x|^\alpha + K_2|c_2x|^\beta \\ &\le Kc^\delta[|x|^\alpha + |x|^\beta] \\ &\le Kc^\delta|x|^\gamma[1 + |x|^{\delta-\gamma}] \\ &\le \tilde{K}|x|^\gamma, \end{aligned}$$

for sufficiently small $|x|$ and some constant $\tilde{K}$. Thus, $G(x) = L_1 + L_2 + O(x^\gamma)$.

17. a. 354224848179261915075 **b.** $0.3542248538 \times 10^{21}$

c. The result in part (a) is computed using exact integer arithmetic, and the result in part (b) is computed using 10-digit rounding arithmetic.

d. The result in part (a) required traversing a loop 98 times.

e. The result is the same as the result in part (a).

Exercise Set 2.1 (Page 54)

1. $p_3 = 0.625$

3. The Bisection method gives:

a. $p_7 = -1.414$ **b.** $p_8 = 1.414$ **c.** $p_7 = 2.727$ **d.** $p_7 = -0.7265$

5. The Bisection method gives:

a. $p_{17} = 1.51213837$ **b.** $p_{17} = 0.97676849$

c. For the interval $[1, 2]$, we have $p_{17} = 1.41239166$, and for the interval $[2, 4]$, we have $p_{18} = 3.05710602$.

d. For the interval $[0, 0.5]$, we have $p_{16} = 0.20603180$, and for the interval $[0.5, 1]$, we have $p_{16} = 0.68196869$.

7. a.

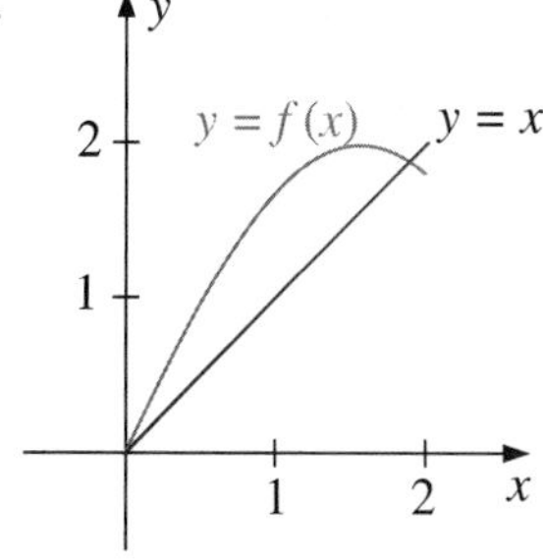

b. Using $[1.5, 2]$ from part (a) gives $p_{16} = 1.89550018$.

9. a.

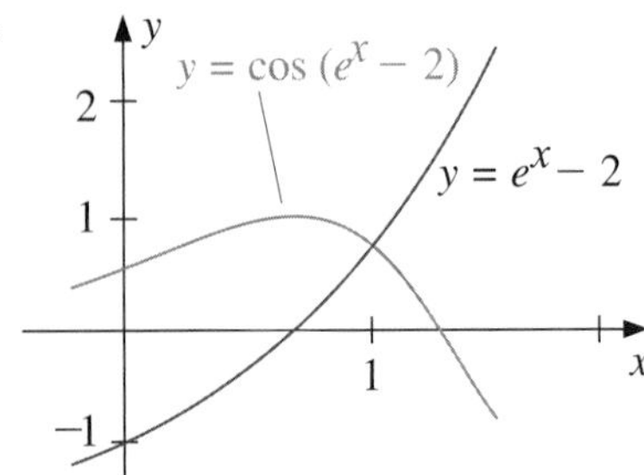

b. $p_{17} = 1.00762177$

11. a. 2 **b.** -2 **c.** -1 **d.** 1

13. The third root of 25 is approximately $p_{14} = 2.92401$, using $[2, 3]$.

15. A bound is $n \geq 14$, and $p_{14} = 1.32477$.

17. Since $\lim_{n\to\infty}(p_n - p_{n-1}) = \lim_{n\to\infty} 1/n = 0$, the difference in the terms goes to zero. However, p_n is the nth term of the divergent harmonic series, so $\lim_{n\to\infty} p_n = \infty$.

19. The depth of the water is 0.838 ft.

Exercise Set 2.2 (Page 64)

1. For the value of x under consideration we have

a. $x = (3 + x - 2x^2)^{1/4} \Leftrightarrow x^4 = 3 + x - 2x^2 \Leftrightarrow f(x) = 0$

b. $x = \left(\dfrac{x + 3 - x^4}{2}\right)^{1/2} \Leftrightarrow 2x^2 = x + 3 - x^4 \Leftrightarrow f(x) = 0$

c. $x = \left(\dfrac{x + 3}{x^2 + 2}\right)^{1/2} \Leftrightarrow x^2(x^2 + 2) = x + 3 \Leftrightarrow f(x) = 0$

d. $x = \dfrac{3x^4 + 2x^2 + 3}{4x^3 + 4x - 1} \Leftrightarrow 4x^4 + 4x^2 - x = 3x^4 + 2x^2 + 3 \Leftrightarrow f(x) = 0$

3. The sequence in (c) converges faster than in (d). The sequences in (a) and (b) diverge.

5. With $g(x) = \sqrt{1 + \frac{1}{x}}$ and $p_0 = 1$, we have $p_4 = 1.324$.

7. Since $g'(x) = \frac{1}{4}\cos\frac{x}{2}$, g is continuous and g' exists on $[0, 2\pi]$. Further, $g'(x) = 0$ only when $x = \pi$, so that $g(0) = g(2\pi) = \pi \leq g(x) =\leq g(\pi) = \pi + \frac{1}{2}$ and $|g'(x)| \leq \frac{1}{4}$, for $0 \leq x \leq 2\pi$. Theorem 2.3 implies that a unique fixed point p exists in $[0, 2\pi]$. With $k = \frac{1}{4}$ and $p_0 = \pi$, we have $p_1 = \pi + \frac{1}{2}$. Corollary 2.5 implies that

$$|p_n - p| \leq \frac{k^n}{1-k}|p_1 - p_0| = \frac{2}{3}\left(\frac{1}{4}\right)^n.$$

For the bound to be less than 0.1, we need $n \geq 4$. However, $p_3 = 3.626996$ is accurate to within 0.01.

9. For $p_0 = 1.0$ and $g(x) = 0.5(x + \frac{3}{x})$, we have $\sqrt{3} \approx p_4 = 1.73205$.

11. a. With $[0, 1]$ and $p_0 = 0$, we have $p_9 = 0.257531$.
b. With $[2.5, 3.0]$ and $p_0 = 2.5$, we have $p_{17} = 2.690650$.
c. With $[0.25, 1]$ and $p_0 = 0.25$, we have $p_{14} = 0.909999$.
d. With $[0.3, 0.7]$ and $p_0 = 0.3$, we have $p_{39} = 0.469625$.
e. With $[0.3, 0.6]$ and $p_0 = 0.3$, we have $p_{48} = 0.448059$.
f. With $[0, 1]$ and $p_0 = 0$, we have $p_6 = 0.704812$.

13. For $g(x) = (2x^2 - 10\cos x)/(3x)$, we have the following:

$$p_0 = 3 \Rightarrow p_8 = 3.16193; \quad p_0 = -3 \Rightarrow p_8 = -3.16193.$$

For $g(x) = \arccos(-0.1x^2)$, we have the following:

$$p_0 = 1 \Rightarrow p_{11} = 1.96882; \quad p_0 = -1 \Rightarrow p_{11} = -1.96882.$$

15. With $g(x) = \frac{1}{\pi}\arcsin\left(-\frac{x}{2}\right) + 2$, we have $p_5 = 1.683855$.

17. One of many examples is $g(x) = \sqrt{2x - 1}$ on $\left[\frac{1}{2}, 1\right]$.

21. Replace the second sentence in the proof with: "Since g satisfies a Lipschitz condition on $[a, b]$ with a Lipschitz constant $L < 1$, we have, for each n,

$$|p_n - p| = |g(p_{n-1}) - g(p)| \leq L|p_{n-1} - p|."$$

The rest of the proof is the same, with k replaced by L.

23. With $g(t) = 501.0625 - 201.0625e^{-0.4t}$ and $p_0 = 5.0$, $p_3 = 6.0028$ is within 0.01 s of the actual time.

Exercise Set 2.3 (Page 75)

1. $p_2 = 2.60714$

3. a. -1.25208 **b.** -0.841355

5. a. For $p_0 = 0$, we have $p_3 = 0.96434$. **b.** For $p_0 = -3$, we have $p_3 = -2.87939$.
c. For $p_0 = 0$, we have $p_4 = 0.73909$. **d.** For $p_0 = 2$, we have $p_5 = 2.69065$.

7. Using the endpoints of the intervals as p_0 and p_1, we have:
a. $p_5 = 0.96433$ **b.** $p_7 = -2.87939$ **c.** $p_6 = 0.73909$ **d.** $p_{11} = 2.69065$

9. Using the endpoints of the intervals as p_0 and p_1, we have:
a. $p_6 = 0.96433$ **b.** $p_6 = -2.87938$ **c.** $p_7 = 0.73908$ **d.** $p_{16} = 2.69060$

11. a. Newton's method with $p_0 = 1.5$ gives $p_3 = 1.51213455$.
The Secant method with $p_0 = 1$ and $p_1 = 2$ gives $p_{10} = 1.51213455$.
The Method of False Position with $p_0 = 1$ and $p_1 = 2$ gives $p_{17} = 1.51212954$.
b. Newton's method with $p_0 = 0.5$ gives $p_5 = 0.976773017$.
The Secant method with $p_0 = 0$ and $p_1 = 1$ gives $p_5 = 10.976773017$.
The Method of False Position with $p_0 = 0$ and $p_1 = 1$ gives $p_5 = 0.976772976$.

13. For $p_0 = 1$, we have $p_5 = 0.589755$. The point has the coordinates $(0.589755, 0.347811)$.

15. The equation of the tangent line is

$$y - f(p_{n-1}) = f'(p_{n-1})(x - p_{n-1}).$$

To complete this problem, set $y = 0$ and solve for $x = p_n$.

17. a. For $p_0 = -1$ and $p_1 = 0$, we have $p_{17} = -0.04065850$, and for $p_0 = 0$ and $p_1 = 1$, we have $p_9 = 0.9623984$.
b. For $p_0 = -1$ and $p_1 = 0$, we have $p_5 = -0.04065929$, and for $p_0 = 0$ and $p_1 = 1$, we have $p_{12} = -0.04065929$.
c. For $p_0 = -0.5$, we have $p_5 = -0.04065929$, and for $p_0 = 0.5$, we have $p_{21} = 0.9623989$.

19. This formula involves the subtraction of nearly equal numbers in both the numerator and denominator if p_{n-1} and p_{n-2} are nearly equal.

21. a. $p_0 = -10, p_{11} = -4.30624527$ **b.** $p_0 = -5, p_5 = -4.30624527$
c. $p_0 = -3, p_5 = 0.824498585$ **d.** $p_0 = -1, p_4 = -0.824498585$
e. $p_0 = 0$, and you cannot compute p_1, since $f'(0) = 0$ **f.** $p_0 = 1, p_4 = 0.824498585$
g. $p_0 = 3, p_5 = -0.824498585$ **h.** $p_0 = 5, p_5 = 4.30624527$
i. $p_0 = 10, p_{11} = 4.30624527$

23. For $f(x) = \ln(x^2 + 1) - e^{0.4x} \cos \pi x$, we have the following roots.
a. For $p_0 = -0.5$, we have $p_3 = -0.4341431$.
b. For $p_0 = 0.5$, we have $p_3 = 0.4506567$.
For $p_0 = 1.5$, we have $p_3 = 1.7447381$.
For $p_0 = 2.5$, we have $p_5 = 2.2383198$.
For $p_0 = 3.5$, we have $p_4 = 3.7090412$.
c. The initial approximation $n - 0.5$ is quite reasonable.
d. For $p_0 = 24.5$, we have $p_2 = 24.4998870$.

25. The two numbers are approximately 6.512849 and 13.487151.

27. The borrower can afford to pay at most 8.10%.

29. a. *solve*$(3^{(3x+1)} - 7 \cdot 5^{(2x)}, x)$ and *fsolve*$(3^{(3x+1)} - 7 \cdot 5^{(2x)}, x)$ both fail.
b. *plot*$(3^{(3x+1)} - 7 \cdot 5^{(2x)}, x = a\, ..\, b)$ generally yields no useful information. However, with $a = 10.5$ and $b = 11.5$ in the plot command shows that $f(x)$ has a root near $x = 11$.
c. With $p_0 = 11$, $p_5 = 11.0094386442681716$ is accurate to 10^{-16}.
d. $p = \dfrac{\ln(3/7)}{\ln(25/27)}$

31. We have $P_L = 265816$, $c = -0.75658125$, and $k = 0.045017502$. The 1980 population is $P(30) = 222{,}248{,}320$, and the 2010 population is $P(60) = 252{,}967{,}030$.

33. Using $p_0 = 0.5$ and $p_1 = 0.9$, the Secant method gives $p_5 = 0.842$.

Exercise Set 2.4 (Page 85)

1. a. For $p_0 = 0.5$, we have $p_{13} = 0.567135$.
b. For $p_0 = 0.5$, we have $p_{22} = 0.641166$.
c. For $p_0 = -1.5$, we have $p_{23} = -1.414325$.
d. For $p_0 = -0.5$, we have $p_{23} = -0.183274$.

3. Modified Newton's method in Equation (2.11) gives the following:
a. For $p_0 = 0.5$, we have $p_3 = 0.567143$.
b. For $p_0 = 0.5$, we have $p_3 = 0.641274$.
c. For $p_0 = -1.5$, we have $p_2 = -1.414158$.
d. For $p_0 = -0.5$, we have $p_5 = -0.183319$.

5. Newton's method with $p_0 = -0.5$ gives $p_{13} = -0.169607$. Modified Newton's method in Eq. (2.11) with $p_0 = -0.5$ gives $p_{11} = -0.169607$.

7. a. For $k > 0$,

$$\lim_{n\to\infty} \frac{|p_{n+1} - 0|}{|p_n - 0|} = \lim_{n\to\infty} \frac{\frac{1}{(n+1)^k}}{\frac{1}{n^k}} = \lim_{n\to\infty} \left(\frac{n}{n+1}\right)^k = 1,$$

so the convergence is linear.

b. We need to have $N > 10^{m/k}$.

9. Typical examples are
a. $p_n = 10^{-3^n}$
b. $p_n = 10^{-\alpha^n}$

11. This follows from the fact that $\lim_{n\to\infty} \frac{\left|\frac{b-a}{2^{n+1}}\right|}{\left|\frac{b-a}{2^n}\right|} = \frac{1}{2}$.

13. If $\frac{|p_{n+1}-p|}{|p_n-p|^3} = 0.75$ and $|p_0 - p| = 0.5$, then

$$|p_n - p| = (0.75)^{(3^n-1)/2} |p_0 - p|^{3^n}.$$

To have $|p_n - p| \le 10^{-8}$ requires that $n \ge 3$.

Exercise Set 2.5 (Page 90)

1. The results are listed in the following table.

	a.	**b.**	**c.**	**d.**
$\hat{p}_0$	0.731385	0.548101	0.907859	0.258684
$\hat{p}_1$	0.736087	0.547915	0.909568	0.257613
$\hat{p}_2$	0.737653	0.547847	0.909917	0.257536
$\hat{p}_3$	0.738469	0.547823	0.909989	0.257531
$\hat{p}_4$	0.738798	0.547814	0.910004	0.257530
$\hat{p}_5$	0.738958	0.547810	0.910007	0.257530

3. $p_0^{(1)} = 0.826427$

5. $p_1^{(0)} = 1.5$

7. For $g(x) = \sqrt{1 + \frac{1}{x}}$ and $p_0 = 1$, we have $p_3 = 1.32472$.

9. For $g(x) = 0.5(x + \frac{3}{x})$ and $p_0 = 0.5$, we have $p_4 = 1.73205$.

11. a. For $g(x) = \left(2 - e^x + x^2\right)/3$ and $p_0 = 0$, we have $p_3 = 0.257530$.
b. For $g(x) = 0.5(\sin x + \cos x)$ and $p_0 = 0$, we have $p_4 = 0.704812$.
c. With $p_0 = 0.25$, $p_4 = 0.910007572$.
d. With $p_0 = 0.3$, $p_4 = 0.469621923$.

13. Aitken's Δ^2 method gives:
a. $\hat{p}_{10} = 0.0\overline{45}$
b. $\hat{p}_2 = 0.0363$

15. We have

$$\frac{|p_{n+1}-p_n|}{|p_n-p|} = \frac{|p_{n+1}-p+p-p_n|}{|p_n-p|} = \left|\frac{p_{n+1}-p}{p_n-p} - 1\right|,$$

so

$$\lim_{n\to\infty}\frac{|p_{n+1}-p_n|}{|p_n-p|} = \lim_{n\to\infty}\left|\frac{p_{n+1}-p}{p_n-p} - 1\right| = 1.$$

17. a. *Hint*: First show that $p_n - p = -\frac{1}{(n+1)!}e^{\xi}x^{n+1}$, where ξ is between 0 and 1.

b.

n	p_n	$\hat{p}_n$
0	1	3
1	2	2.75
2	2.5	$2.7\overline{2}$
3	$2.\overline{6}$	2.71875
4	$2.708\overline{3}$	$2.718\overline{3}$
5	$2.71\overline{6}$	2.7182870
6	$2.7180\overline{5}$	2.7182823
7	2.7182539	2.7182818
8	2.7182787	2.7182818
9	2.7182815	
10	2.7182818	

Exercise Set 2.6 (Page 100)

1. a. For $p_0 = 1$, we have $p_5 = 1.32472$.
b. For $p_0 = 1$, we have $p_{22} = 2.69065$.
c. For $p_0 = 1$, we have $p_5 = 0.53209$; for $p_0 = -1$, we have $p_3 = -0.65270$; and for $p_0 = -3$, we have $p_3 = -2.87939$.
d. For $p_0 = 1$, we have $p_4 = 1.12412$; and for $p_0 = 0$, we have $p_8 = -0.87605$.
e. For $p_0 = 0$, we have $p_6 = -0.47006$; for $p_0 = -1$, we have $p_4 = -0.88533$; and for $p_0 = -3$, we have $p_4 = -2.64561$.
f. For $p_0 = 0$, we have $p_{10} = 1.49819$.

3. The following table lists the initial approximation and the roots.

	p_0	p_1	p_2	Approximate roots	Complex conjugate roots
a.	0	1	2	$p_5 = 1.32472$	
	−2	−1	0	$p_7 = -0.66236 - 0.56228i$	$-0.66236 + 0.56228i$
b.	−1	0	1	$p_7 = -0.34532 - 1.31873i$	$-0.34532 + 1.31873i$
	0	1	2	$p_6 = 2.69065$	
c.	0	1	2	$p_6 = 0.53209$	
	1	2	3	$p_9 = -0.65270$	
	−2	−3	−2.5	$p_4 = -2.87939$	
d.	0	1	2	$p_5 = 1.12412$	
	2	3	4	$p_{12} = -0.12403 + 1.74096i$	$-0.12403 - 1.74096i$
	−2	0	−1	$p_5 = -0.87605$	
e.	0	1	2	$p_{10} = -0.88533$	
	1	0	−0.5	$p_5 = -0.47006$	
	−1	−2	−3	$p_5 = -2.64561$	
f.	0	1	2	$p_6 = 1.49819$	
	−1	−2	−3	$p_{10} = -0.51363 - 1.09156i$	$-0.51363 + 1.09156i$
	1	0	−1	$p_8 = 0.26454 - 1.32837i$	$0.26454 + 1.32837i$

5. **a.** The roots are 1.244, 8.847, and −1.091, and the critical points are 0 and 6.

b. The roots are 0.5798, 1.521, 2.332, and −2.432, and the critical points are 1, 2.001, and −1.5.

7. The only real zero is $\dfrac{(54+6\sqrt{129})^{2/3}-12}{3(54+6\sqrt{129})^{1/3}}$.

9. The methods all find the solution 0.23235.

11. The minimal material is approximately 573.64895 cm^2.

Exercise Set 3.1 (Page 114)

1. **a.** $P_1(x) = 0.467251x + 1$; $P_2(x) = -0.0780026x^2 + 0.490652x + 1$; $P_1(0.45) = 1.210263$; $|f(0.45) - P_1(0.45)| = 0.006104$; $P_2(0.45) = 1.204998$; $|f(0.45) - P_2(0.45)| = 0.000839$

b. $P_1(x) = -0.148878x + 1$; $P_2(x) = -0.452592x^2 - 0.0131009x + 1$; $P_1(0.45) = 0.933005$; $|f(0.45) - P_1(0.45)| = 0.032558$; $P_2(0.45) = 0.902455$; $|f(0.45) - P_2(0.45)| = 0.002008$

c. $P_1(x) = 0.874548x$; $P_2(x) = -0.268961x^2 + 0.955236x$; $P_1(0.45) = 0.393546$; $|f(0.45) - P_1(0.45)| = 0.0212983$; $P_2(0.45) = 0.375392$; $|f(0.45) - P_2(0.45)| = 0.003828$

d. $P_1(x) = 1.031121x$; $P_2(x) = 0.615092x^2 + 0.846593x$; $P_1(0.45) = 0.464004$; $|f(0.45) - P_1(0.45)| = 0.019051$; $P_2(0.45) = 0.505523$; $|f(0.45) - P_2(0.45)| = 0.022468$

3. **a.** $\left|\frac{f''(\xi)}{2}(0.45-0)(0.45-0.6)\right| \le 0.03375$; $\left|\frac{f'''(\xi)}{6}(0.45-0)(0.45-0.6)(0.45-0.9)\right| \le 0.001898$

b. $\left|\frac{f''(\xi)}{2}(0.45-0)(0.45-0.6)\right| \le 0.135$; $\left|\frac{f'''(\xi)}{6}(0.45-0)(0.45-0.6)(0.45-0.9)\right| \le 0.00397$

c. $\left|\frac{f''(\xi)}{2}(0.45-0)(0.45-0.6)\right| \le 0.135$; $\left|\frac{f'''(\xi)}{6}(0.45-0)(0.45-0.6)(0.45-0.9)\right| \le 0.010125$

d. $\left|\frac{f''(\xi)}{2}(0.45-0)(0.45-0.6)\right| \le 0.06779$; $\left|\frac{f'''(\xi)}{6}(0.45-0)(0.45-0.6)(0.45-0.9)\right| \le 0.151$

5. **a.**

n	$x_0, x_1, \ldots, x_n$	$P_n(8.4)$
1	8.3, 8.6	17.87833
2	8.3, 8.6, 8.7	17.87716
3	8.3, 8.6, 8.7, 8.1	17.87714

b.

n	$x_0, x_1, \ldots, x_n$	$P_n(0.25)$
1	0.2, 0.3	−0.13869287
2	0.2, 0.3, 0.4	−0.13259734
3	0.2, 0.3, 0.4, 0.1	−0.13277477

c.

n	$x_0, x_1, \ldots, x_n$	$P_n(-1/3)$
1	−0.5, −0.25	0.21504167
2	−0.5, −0.25, 0.0	0.16988889
3	−0.5, −0.25, 0.0, −0.75	0.17451852

d.

n	$x_0, x_1, \ldots, x_n$	$P_n(0.9)$
1	0.8, 1.0	0.44086280
2	0.8, 1.0, 0.7	0.43841352
3	0.8, 1.0, 0.7, 0.6	0.44198500

7. **a.**

n	Actual Error	Error Bound
1	1.180×10^{-3}	1.200×10^{-3}
2	1.367×10^{-5}	1.452×10^{-5}

b.

n	Actual Error	Error Bound
1	5.921×10^{-3}	6.097×10^{-3}
2	1.746×10^{-4}	1.813×10^{-4}

c.

n	Actual Error	Error Bound
1	4.052×10^{-2}	4.515×10^{-2}
2	4.630×10^{-3}	4.630×10^{-3}

d.

n	Actual Error	Error Bound
1	2.730×10^{-3}	1.408×10^{-2}
2	5.179×10^{-3}	9.222×10^{-3}

9. $y = 1.25$

11. We have $f(1.09) \approx 0.2826$. The actual error is 4.3×10^{-5}, and an error bound is 7.4×10^{-6}. The discrepancy is due to the fact that the data are given to only four decimal places, and only four-digit arithmetic is used.

13. **a.** $P_2(x) = -11.22388889x^2 + 3.810500000x + 1$, and an error bound is 0.11371294.

b. $P_2(x) = -0.1306344167x^2 + 0.8969979335x - 0.63249693$, and an error bound is 9.45762×10^{-4}.

c. $P_3(x) = 0.1970056667x^3 - 1.06259055x^2 + 2.532453189x - 1.666868305$, and an error bound is 10^{-4}.

d. $P_3(x) = -0.07932x^3 - 0.545506x^2 + 1.0065992x + 1$, and an error bound is 1.591376×10^{-3}.

15. Using 10 digits gives

$$P_3(x) = 1.302637066x^3 - 3.511333118x^2 + 4.071141936x - 1.670043560, P_3(1.09) = 0.282639050,$$

and $|f(1.09) - P_3(1.09)| = 3.8646 \times 10^{-6}$.

17. The largest possible step size is 0.004291932, so 0.004 would be a reasonable choice.

19. **a.** Sample 1: $P_6(x) = 6.67 - 42.6434x + 16.1427x^2 - 2.09464x^3 + 0.126902x^4 - 0.00367168x^5 + 0.0000409458x^6$;
Sample 2: $P_6(x) = 6.67 - 5.67821x + 2.91281x^2 - 0.413799x^3 + 0.0258413x^4 - 0.000752546x^5 + 0.00000836160x^6$

b. Sample 1: 42.71 mg; Sample 2: 19.42 mg

21. Since $g(x) = g(x_0) = 0$, there exists a number ξ_1 between x and x_0, for which $g'(\xi_1) = 0$. Also, $g'(x_0) = 0$, so there exists a number ξ_2 between x_0 and ξ_1, for which $g''(\xi_2) = 0$. The process is continued by induction to show that a number ξ_{n+1} between x_0 and ξ_n exists with $g^{(n+1)}(\xi_{n+1}) = 0$. The error formula for Taylor polynomials follows.

23. **a.** (i) $B_3(x) = x$ (ii) $B_3(x) = 1$ **b.** $n \geq 250{,}000$

Exercise Set 3.2 (Page 123)

1. The approximations are the same as in Exercise 5 of Section 3.1.

3. **a.** We have $\sqrt{3} \approx P_4(1/2) = 1.708\overline{3}$. **b.** We have $\sqrt{3} \approx P_4(3) = 1.690607$.

c. Absolute error in part (a) is approximately 0.0237, and the absolute error in part (b) is 0.0414, so part (a) is more accurate.

5. $P_2 = f(0.5) = 4$

7. $P_{0,1,2,3}(2.5) = 2.875$

9. The incorrect approximation is $-f(2)/6 + 2f(1)/3 + 4/3 + 2f(-1)/3 - f(-2)/6$. and the correct approximation is $-f(2)/6 + 2f(1)/3 + 2f(-1)/3 - f(-2)/6$, so the incorrect approximation is 4/3 too large.

11. The first ten terms of the sequence are 0.038462, 0.333671, 0.116605, −0.371760, −0.0548919, 0.605935, 0.190249, −0.513353, −0.0668173, and 0.448335. Since $f(1+\sqrt{10}) = 0.0545716$, the sequence does not appear to converge.

13. Change Algorithm 3.1 as follows:

INPUT numbers $y_0, y_1, \ldots, y_n$; values $x_0, x_1, \ldots, x_n$ as the first column $Q_{0,0}, Q_{1,0}, \ldots, Q_{n,0}$ of Q.

OUTPUT the table Q with $Q_{n,n}$ approximating $f^{-1}(0)$.

Step 1 For $i = 1, 2, \ldots, n$
for $j = 1, 2, \ldots, i$
set $Q_{i,j} = \dfrac{y_i Q_{i-1,j-1} - y_{i-j} Q_{i,j-1}}{y_i - y_{i-j}}$.

Exercise Set 3.3 (Page 133)

1. **a.** $P_1(x) = 16.9441 + 3.1041(x - 8.1); P_1(8.4) = 17.87533$ $P_2(x) = P_1(x) + 0.06(x - 8.1)(x - 8.3); P_2(8.4) = 17.87713$
$P_3(x) = P_2(x) + -0.00208333(x - 8.1)(x - 8.3)(x - 8.6); P_3(8.4) = 17.87714$

b. $P_1(x) = -0.1769446 + 1.9069687(x - 0.6); P_1(0.9) = 0.395146$
$P_2(x) = P_1(x) + 0.959224(x - 0.6)(x - 0.7); P_2(0.9) = 0.4526995$
$P_3(x) = P_2(x) - 1.785741(x - 0.6)(x - 0.7)(x - 0.8); P_3(0.9) = 0.4419850$

3. In the following equations, we have $s = \dfrac{1}{h}(x - x_0)$.

a. $P_1(s) = -0.62049958 + 0.3365129s$; $P_1(0.25) = -0.1157302$
$P_2(s) = P_1(s) - 0.04592527s(s-1)/2$; $P_2(0.25) = -0.1329522$
$P_3(s) = P_2(s) - 0.00283891s(s-1)(s-2)/6$; $P_3(0.25) = -0.1327748$

b. $P_1(s) = -0.718125 - 0.0470625s$; $P_1\left(-\frac{1}{3}\right) = -0.006625$
$P_2(s) = P_1(s) + 0.312625s(s-1)/2$; $P_2\left(-\frac{1}{3}\right) = 0.1803056$
$P_3(s) = P_2(s) + 0.09375s(s-1)(s-2)/6$; $P_3\left(-\frac{1}{3}\right) = 0.1745185$

5. In the following equations, we have $s = \dfrac{1}{h}(x - x_n)$.

a. $P_1(s) = 0.2484244 + 0.2418235s$; $f(0.25) \approx P_1(-1.5) = -0.1143108$
$P_2(s) = P_1(s) - 0.04876419s(s+1)/2$; $f(0.25) \approx P_2(-1.5) = -0.1325973$
$P_3(s) = P_2(s) - 0.00283891s(s+1)(s+2)/6$; $f(0.25) \approx P_3(-1.5) = -0.1327748$

b. $P_1(s) = 1.101 + 0.7660625s;\quad f(-\frac{1}{3}) \approx P_1(-\frac{4}{3}) = 0.07958333$
$P_2(s) = P_1(s) + 0.406375s(s+1)/2;\quad f(-\frac{1}{3}) \approx P_2(-\frac{4}{3}) = 0.1698889$
$P_3(s) = P_2(s) + 0.09375s(s+1)(s+2)/6;\quad f(-\frac{1}{3}) \approx P_3(-\frac{4}{3}) = 0.1745185$

7. a. $P_3(x) = 5.3 - 33(x+0.1) + 129.8\overline{3}(x+0.1)x - 556.\overline{6}(x+0.1)x(x-0.2)$

b. $P_4(x) = P_3(x) + 2730.243387(x+0.1)x(x-0.2)(x-0.3)$

9. a. $f(0.05) \approx 1.05126$ **b.** $f(0.65) \approx 1.91555$ **c.** $f(0.43) \approx 1.53725$

11. a. $P(-2) = Q(-2) = -1,\ P(-1) = Q(-1) = 3,\ P(0) = Q(0) = 1,\ P(1) = Q(1) = -1,\ P(2) = Q(2) = 3$

b. The format of the polynomial is not unique. If $P(x)$ and $Q(x)$ are expanded, they are identical. There is only one interpolating polynomial if the degree is less than or equal to four for the given data. However, it can be expressed in various ways depending on the application.

13. The coefficient of x^2 is 3.5.

15. The approximation to $f(0.3)$ should be increased by 5.9375.

17. $f[x_0] = f(x_0) = 1,\ f[x_1] = f(x_1) = 3,\ f[x_0, x_1] = 5$

19. Since $f[x_2] = f[x_0] + f[x_0, x_1](x_2 - x_0) + a_2(x_2 - x_0)(x_2 - x_1)$,

$$a_2 = \frac{f[x_2] - f[x_0]}{(x_2 - x_0)(x_2 - x_1)} - \frac{f[x_0, x_1]}{(x_2 - x_1)}.$$

This simplifies to $f[x_0, x_1, x_2]$.

21. Let $\tilde{P}(x) = f[x_{i_0}] + \sum_{k=1}^{n} f[x_{i_0}, \ldots, x_{i_k}](x - x_{i_0}) \cdots (x - x_{i_k})$ and $\hat{P}(x) = f[x_0] + \sum_{k=1}^{n} f[x_0, \ldots, x_k](x - x_0) \cdots (x - x_k)$. The polynomial $\tilde{P}(x)$ interpolates $f(x)$ at the nodes $x_{i_0}, \ldots, x_{i_n}$, and the polynomial $\hat{P}(x)$ interpolates $f(x)$ at the nodes $x_0, \ldots, x_n$. Since both sets of nodes are the same and the interpolating polynomial is unique, we have $\tilde{P}(x) = \hat{P}(x)$. The coefficient of x^n in $\tilde{P}(x)$ is $f[x_{i_0}, \ldots, x_{i_n}]$, and the coefficient of x^n in $\hat{P}(x)$ is $f[x_0, \ldots, x_n]$. Thus, $f[x_{i_0}, \ldots, x_{i_n}] = f[x_0, \ldots, x_n]$.

Exercise Set 3.4 (Page 142)

1. The coefficients for the polynomials in divided-difference form are given in the following tables. For example, the polynomial in part (a) is

$$H_3(x) = 17.56492 + 3.116256(x - 8.3) + 0.05948(x - 8.3)^2 - 0.00202222(x - 8.3)^2(x - 8.6).$$

a.	b.	c.	d.
17.56492	0.22363362	−0.02475	−0.62049958
3.116256	2.1691753	0.751	3.5850208
0.05948	0.01558225	2.751	−2.1989182
−0.00202222	−3.2177925	1	−0.490447
		0	0.037205
		0	0.040475
			−0.0025277777
			0.0029629628

3.

	x	Approximation to $f(x)$	Actual $f(x)$	Error
a.	8.4	17.877144	17.877146	2.33×10^{-6}
b.	0.9	0.44392477	0.44359244	3.3323×10^{-4}
c.	$-\frac{1}{3}$	0.1745185	0.17451852	1.85×10^{-8}
d.	0.25	−0.1327719	−0.13277189	5.42×10^{-9}

5. a. We have $\sin 0.34 \approx H_5(0.34) = 0.33349$.

b. The formula gives an error bound of 3.05×10^{-14}, but the actual error is 2.91×10^{-6}. The discrepancy is due to the fact that the data are given to only five decimal places.

c. We have $\sin 0.34 \approx H_7(0.34) = 0.33350$. Although the error bound is now 5.4×10^{-20}, the accuracy of the given data dominates the calculations. This result is actually less accurate than the approximation in part (b), since $\sin 0.34 = 0.333487$.

7. For 3(a), we have an error bound of 5.9×10^{-8}. The error bound for 3(c) is 0 since $f^{(n)}(x) \equiv 0$, for $n > 3$.

9. $H_3(1.25) = 1.169080403$ with an error bound of 4.81×10^{-5}, and $H_5(1.25) = 1.169016064$ with an error bound of 4.43×10^{-4}.

Exercise Set 3.5 (Page 161)

1. $S(x) = x$ on $[0, 2]$.

3. The equations of the respective free cubic splines are

$$S(x) = S_i(x) = a_i + b_i(x - x_i) + c_i(x - x_i)^2 + d_i(x - x_i)^3,$$

for x in $[x_i, x_{i+1}]$, where the coefficients are given in the following tables.

a.

i	a_i	b_i	c_i	d_i
0	0.22363362	2.17229175	0.00000000	0.00000000

b.

i	a_i	b_i	c_i	d_i
0	17.564920	3.13410000	0.00000000	0.00000000

c.

i	a_i	b_i	c_i	d_i
0	−0.02475000	1.03237500	0.00000000	6.50200000
1	0.33493750	2.25150000	4.87650000	−6.50200000

d.

i	a_i	b_i	c_i	d_i
0	−0.62049958	3.45508693	0.00000000	−8.9957933
1	−0.28398668	3.18521313	−2.69873800	−0.94630333
2	0.00660095	2.61707643	−2.98262900	9.9420966

5.

	x	Approximation to $f(x)$	Actual $f(x)$	Error
a.	0.9	0.4408628	0.44359244	2.7296×10^{-3}
b.	8.4	17.87833	17.877146	1.1840×10^{-3}
c.	$-\frac{1}{3}$	0.1774144	0.17451852	2.8959×10^{-3}
d.	0.25	−0.1315912	−0.13277189	1.1807×10^{-3}

	x	Approximation to $f'(x)$	Actual $f'(x)$	Error
a.	0.9	2.172292	2.204367	0.0320747
b.	8.4	3.134100	3.128232	5.86829×10^{-3}
c.	$-\frac{1}{3}$	1.574208	1.668000	0.093792
d.	0.25	2.908242	2.907061	1.18057×10^{-3}

7. The equations of the respective clamped cubic splines are

$$s(x) = s_i(x) = a_i + b_i(x - x_i) + c_i(x - x_i)^2 + d_i(x - x_i)^3,$$

for x in $[x_i, x_{i+1}]$, where the coefficients are given in the following tables.

a.

i	a_i	b_i	c_i	d_i
0	0.22363362	2.1691753	0.65914075	−3.2177925

b.

i	a_i	b_i	c_i	d_i
0	17.564920	3.1162560	0.060087	−0.002022

c.

i	a_i	b_i	c_i	d_i
0	−0.02475000	0.75100000	2.5010000	1.0000000
1	0.33493750	2.18900000	3.2510000	1.0000000

d.

i	a_i	b_i	c_i	d_i
0	−0.62049958	3.5850208	−2.1498407	−0.49077413
1	−0.28398668	3.1403294	−2.2970730	−0.47458360
2	0.006600950	2.6666773	−2.4394481	−0.44980146

9.

	x	Approximation to $f(x)$	Actual $f(x)$	Error
a.	0.9	0.4439248	0.44359244	3.323×10^{-4}
b.	8.4	17.877144	17.877146	0.188×10^{-6}
c.	$-\frac{1}{3}$	0.17451852	0.17451852	0
d.	0.25	−0.13277221	−0.13277189	3.19×10^{-7}

	x	Approximation to $f'(x)$	Actual $f'(x)$	Error
a.	0.9	2.204470	2.204367	1.0296×10^{-4}
b.	8.4	3.128213	3.128232	1.90×10^{-5}
c.	$-\frac{1}{3}$	1.668000	1.668000	0
d.	0.25	2.908242	2.907061	1.18057×10^{-3}

11. $b = -1,\ c = -3,\ d = 1$

13. $B = \frac{1}{4},\ D = \frac{1}{4},\ b = -\frac{1}{2},\ d = \frac{1}{4}$

15. The equation of the spline is

$$S(x) = S_i(x) = a_i + b_i(x - x_i) + c_i(x - x_i)^2 + d_i(x - x_i)^3,$$

for x in $[x_i, x_{i+1}]$, where the coefficients are given in the following table.

x_i	a_i	b_i	c_i	d_i
0	1.0	−0.7573593	0.0	−6.627417
0.25	0.7071068	−2.0	−4.970563	6.627417
0.5	0.0	−3.242641	0.0	6.627417
0.75	−0.7071068	−2.0	4.970563	−6.627417

$\int_0^1 S(x)dx = 0.000000$, $S'(0.5) = -3.24264$, and $S''(0.5) = 0.0$

17. The equation of the spline is

$$s(x) = s_i(x) = a_i + b_i(x - x_i) + c_i(x - x_i)^2 + d_i(x - x_i)^3,$$

for x in $[x_i, x_{i+1}]$, where the coefficients are given in the following table.

x_i	a_i	b_i	c_i	d_i
0	1.0	0.0	−5.193321	2.028118
0.25	0.7071068	−2.216388	−3.672233	4.896310
0.5	0.0	−3.134447	0.0	4.896310
0.75	−0.7071068	−2.216388	3.672233	2.028118

$\int_0^1 s(x)\,dx = 0.000000$, $s'(0.5) = -3.13445$, and $s''(0.5) = 0.0$

19. Let $f(x) = a + bx + cx^2 + dx^3$. Clearly, f satisfies properties (a), (c), (d), and (e) of Definition 3.10, and f interpolates itself for any choice of $x_0, \ldots, x_n$. Since (ii) of property (f) in Definition 3.10 holds, f must be its own clamped cubic spline. However, $f''(x) = 2c + 6dx$ can be zero only at $x = -c/3d$. Thus, part (i) of property (f) in Definition 3.10 cannot hold at two values x_0 and x_n. Thus, f cannot be a natural cubic spline.

21. The piecewise linear approximation to f is given by

$$F(x) = \begin{cases} 20(e^{0.1} - 1)x + 1, & \text{for } x \text{ in } [0, 0.05] \\ 20(e^{0.2} - e^{0.1})x + 2e^{0.1} - e^{0.2}, & \text{for } x \text{ in } (0.05, 1]. \end{cases}$$

We have

$$\int_0^{0.1} F(x)\,dx = 0.1107936 \quad \text{and} \quad \int_0^{0.1} f(x)\,dx = 0.1107014.$$

25. a. On $[0, 0.05]$, we have $s(x) = 1.000000 + 1.999999x + 1.998302x^2 + 1.401310x^3$, and on $(0.05, 0.1]$, we have $s(x) = 1.105170 + 2.210340(x - 0.05) + 2.208498(x - 0.05)^2 + 1.548758(x - 0.05)^3$.

b. $\int_0^{0.1} s(x)\,dx = 0.110701$

c. 1.6×10^{-7}

d. On $[0, 0.05]$, we have $S(x) = 1 + 2.04811x + 22.12184x^3$, and on $(0.05, 0.1]$, we have $S(x) = 1.105171 + 2.214028(x - 0.05) + 3.318277(x - 0.05)^2 - 22.12184(x - 0.05)^3$. $S(0.02) = 1.041139$ and $S(0.02) = 1.040811$.

27. $S(x) = \begin{cases} 2x - x^2, & 0 \le x \le 1 \\ 1 + (x-1)^2, & 1 \le x \le 2 \end{cases}$

29. The spline has the equation

$$s(x) = s_i(x) = a_i + b_i(x - x_i) + c_i(x - x_i)^2 + d_i(x - x_i)^3,$$

for x in $[x_i, x_{i+1}]$, where the coefficients are given in the following table.

x_i	a_i	b_i	c_i	d_i
0	0	75	−0.659292	0.219764
3	225	76.9779	1.31858	−0.153761
5	383	80.4071	0.396018	−0.177237
8	623	77.9978	−1.19912	0.0799115

The spline predicts a position of $s(10) = 774.84$ ft and a speed of $s'(10) = 74.16$ ft/s. To maximize the speed, we find the single critical point of $s'(x)$, and compare the values of $s(x)$ at this point and the endpoints. We find that max $s'(x) = s'(5.7448) = 80.7$ ft/s $= 55.02$ mi/h. The speed 55 mi/h was first exceeded at approximately 5.5 s.

31. The equation of the spline is

$$S(x) = S_i(x) = a_i + b_i(x - x_i) + c_i(x - x_i)^2 + d_i(x - x_i)^3,$$

for x in $[x_i, x_{i+1}]$, where the coefficients are given in the following table.

	Sample 1				Sample 2			
x_i	a_i	b_i	c_i	d_i	a_i	b_i	c_i	d_i
0	6.67	−0.44687	0	0.06176	6.67	1.6629	0	−0.00249
6	17.33	6.2237	1.1118	−0.27099	16.11	1.3943	−0.04477	−0.03251
10	42.67	2.1104	−2.1401	0.28109	18.89	−0.52442	−0.43490	0.05916
13	37.33	−3.1406	0.38974	−0.01411	15.00	−1.5365	0.09756	0.00226
17	30.10	−0.70021	0.22036	−0.02491	10.56	−0.64732	0.12473	−0.01113
20	29.31	−0.05069	−0.00386	0.00016	9.44	−0.19955	0.02453	−0.00102

33. The three natural splines have equations of the form

$$S_i(x) = a_i + b_i(x - x_i) + c_i(x - x_i)^2 + d_i(x - x_i)^3,$$

for x in $[x_i, x_{i+1}]$, where the values of the coefficients are given in the following tables.

Spline 1

i	x_i	$a_i = f(x_i)$	b_i	c_i	d_i
0	1	3.0	0.786	0.0	−0.086
1	2	3.7	0.529	−0.257	0.034
2	5	3.9	−0.086	0.052	0.334
3	6	4.2	1.019	1.053	−0.572
4	7	5.7	1.408	−0.664	0.156
5	8	6.6	0.547	−0.197	0.024
6	10	7.1	0.049	−0.052	−0.003
7	13	6.7	−0.342	−0.078	0.007
8	17	4.5			

Spline 2

i	x_i	$a_i = f(x_i)$	b_i	c_i	d_i
0	17	4.5	1.106	0.0	−0.030
1	20	7.0	0.289	−0.272	0.025
2	23	6.1	−0.660	−0.044	0.204
3	24	5.6	−0.137	0.567	−0.230
4	25	5.8	0.306	−0.124	−0.089
5	27	5.2	−1.263	−0.660	0.314
6	27.7	4.1			

Spline 3

i	x_i	$a_i = f(x_i)$	b_i	c_i	d_i
0	27.7	4.1	0.749	0.0	−0.910
1	28	4.3	0.503	−0.819	0.116
2	29	4.1	−0.787	−0.470	0.157
3	30	3.0			

Exercise Set 3.6 (Page 170)

1. a. $x(t) = -10t^3 + 14t^2 + t, \quad y(t) = -2t^3 + 3t^2 + t$
b. $x(t) = -10t^3 + 14.5t^2 + 0.5t, \quad y(t) = -3t^3 + 4.5t^2 + 0.5t$
c. $x(t) = -10t^3 + 14t^2 + t, \quad y(t) = -4t^3 + 5t^2 + t$
d. $x(t) = -10t^3 + 13t^2 + 2t, \quad y(t) = 2t$

3. a. $x(t) = -11.5t^3 + 15t^2 + 1.5t + 1, \quad y(t) = -4.25t^3 + 4.5t^2 + 0.75t + 1$
b. $x(t) = -6.25t^3 + 10.5t^2 + 0.75t + 1, \quad y(t) = -3.5t^3 + 3t^2 + 1.5t + 1$

c. For t between $(0,0)$ and $(4,6)$, we have

$$x(t) = -5t^3 + 7.5t^2 + 1.5t, \quad y(t) = -13.5t^3 + 18t^2 + 1.5t,$$

and for t between $(4,6)$ and $(6,1)$, we have

$$x(t) = -5.5t^3 + 6t^2 + 1.5t + 4, \quad y(t) = 4t^3 - 6t^2 - 3t + 6.$$

d. For t between $(0,0)$ and $(2,1)$, we have

$$x(t) = -5.5t^3 + 6t^2 + 1.5t, \quad y(t) = -0.5t^3 + 1.5t,$$

for t between $(2,1)$ and $(4,0)$, we have

$$x(t) = -4t^3 + 3t^2 + 3t + 2, \quad y(t) = -t^3 + 1,$$

and for t between $(4,0)$ and $(6,-1)$, we have

$$x(t) = -8.5t^3 + 13.5t^2 - 3t + 4, \quad y(t) = -3.25t^3 + 5.25t^2 - 3t.$$

5. a. Using the forward divided difference gives the following table.

0	u_0			
0	u_0	$3(u_1 - u_0)$		
1	u_3	$u_3 - u_0$	$u_3 - 3u_1 + 2u_0$	
1	u_3	$3(u_3 - u_2)$	$2u_3 - 3u_2 + u_0$	$u_3 - 3u_2 + 3u_1 - u_0$

Therefore

$$\begin{aligned} u(t) &= u_0 + 3(u_1 - u_0)t + (u_3 - 3u_1 + 2u_0)t^2 + (u_3 - 3u_2 + 3u_1 - u_0)t^2(t-1) \\ &= u_0 + 3(u_1 - u_0)t + (-6u_1 + 3u_0 + 3u_2)t^2 + (u_3 - 3u_2 + 3u_1 - u_0)t^3. \end{aligned}$$

Similarly, $v(t) = v_0 + 3(v_1 - v_0)t + (3v_2 - 6v_1 + 3v_0)t^2 + (v_3 - 3v_2 + 3v_1 - v_0)t^3$.

b. Using the formula for Bernstein polynomials gives

$$\begin{aligned} u(t) &= u_0(1-t)^3 + 3u_1t(1-t)^2 + 3u_2t^2(1-t) + u_3t^3 \\ &= u_0 + 3(u_1 - u_0)t + (3u_2 - 6u_1 + 3u_0)t^2 + (u_3 - 3u_2 + 3u_1 - u_0)t^3. \end{aligned}$$

Similarly,

$$\begin{aligned} v(t) &= \sum_{k=0}^{3} \binom{3}{k} v_k t^k (1-t)^{3-k} \\ &= v_0 + 3(v_1 - v_0)t + (3v_2 - 6v_1 + 3v_0)t^2 + (v_3 - 3v_2 + 3v_1 - v_0)t^3. \end{aligned}$$

Exercise Set 4.1 (Page 182)

1. From the forward-backward difference formula (4.1), we have the following approximations:

a. $f'(0.0) \approx 3.7070$, $f'(0.2) \approx 3.1520$, $f'(0.4) \approx 3.1520$

b. $f'(0.5) \approx 0.8520$, $f'(0.6) \approx 0.8520$, $f'(0.7) \approx 0.7960$

3. a.

x	Actual Error	Error Bound
0.0	0.2930	0.3000
0.2	0.2694	0.2779
0.4	0.2602	0.2779

b.

x	Actual Error	Error Bound
0.5	0.0255	0.0282
0.6	0.0267	0.0282
0.7	0.0312	0.0322

5. For the endpoints of the tables, we use Formula (4.4). The other approximations come from Formula (4.5).

a. $f'(8.1) \approx 3.092050$, $f'(8.3) \approx 3.116150$, $f'(8.5) \approx 3.139975$, $f'(8.7) \approx 3.163525$

b. $f'(1.1) \approx 17.769705$, $f'(1.2) \approx 22.193635$, $f'(1.3) \approx 27.107350$, $f'(1.4) \approx 32.150850$

c. $f'(2.0) \approx 0.13533150$, $f'(2.1) \approx -0.09989550$, $f'(2.2) \approx -0.3298960$, $f'(2.3) \approx -0.5546700$

d. $f'(2.9) \approx 5.101375$, $f'(3.0) \approx 6.654785$, $f'(3.1) \approx 8.216330$, $f'(3.2) \approx 9.786010$

7. a.

x	Actual Error	Error Bound
8.1	0.00018594	0.000020322
8.3	0.00010551	0.000010161
8.5	9.116×10^{-5}	0.000009677
8.7	0.00020197	0.000019355

b.

x	Actual Error	Error Bound
1.1	0.280322	0.359033
1.2	0.147282	0.179517
1.3	0.179874	0.219262
1.4	0.378444	0.438524

c.

x	Actual Error	Error Bound
2.0	0.00252235	0.00410304
2.1	0.00142882	0.00205152
2.2	0.00204851	0.00260034
2.3	0.00437954	0.00520068

d.

x	Actual Error	Error Bound
2.9	0.011956	0.0180988
3.0	0.0049251	0.00904938
3.1	0.0004765	0.00493920
3.2	0.0013745	0.00987840

9. The approximations and the formulas used are:

a. $f'(2.1) \approx 3.899344$ from (4.7) $f'(2.2) \approx 2.876876$ from (4.7) $f'(2.3) \approx 2.249704$ from (4.6) $f'(2.4) \approx 1.837756$ from (4.6) $f'(2.5) \approx 1.544210$ from (4.7) $f'(2.6) \approx 1.355496$ from (4.7)

b. $f'(-3.0) \approx -5.877358$ from (4.7) $f'(-2.8) \approx -5.468933$ from (4.7) $f'(-2.6) \approx -5.059884$ from (4.6) $f'(-2.4) \approx -4.650223$ from (4.6) $f'(-2.2) \approx -4.239911$ from (4.7) $f'(-2.0) \approx -3.828853$ from (4.7)

11. a.

x	Actual Error	Error Bound
2.1	0.0242312	0.109271
2.2	0.0105138	0.0386885
2.3	0.0029352	0.0182120
2.4	0.0013262	0.00644808
2.5	0.0138323	0.109271
2.6	0.0064225	0.0386885

b.

x	Actual Error	Error Bound
−3.0	1.55×10^{-5}	6.33×10^{-7}
−2.8	1.32×10^{-5}	6.76×10^{-7}
−2.6	7.95×10^{-7}	1.05×10^{-7}
−2.4	6.79×10^{-7}	1.13×10^{-7}
−2.2	1.28×10^{-5}	6.76×10^{-7}
−2.0	7.96×10^{-6}	6.76×10^{-7}

13. $f'(3) \approx \frac{1}{12}[f(1) - 8f(2) + 8f(4) - f(5)] = 0.21062$, with an error bound given by

$$\max_{1\le x\le 5} \frac{|f^{(5)}(x)|h^4}{30} \le \frac{23}{30} = 0.7\overline{6}.$$

15. From the forward-backward difference formula (4.1), we have the following approximations:

a. $f'(0.0) \approx 3.707$, $f'(0.2) \approx 3.153$, $f'(0.4) \approx 3.153$

b. $f'(0.5) \approx 0.852$, $f'(0.6) \approx 0.852$, $f'(0.7) \approx 0.7960$

17. For the endpoints of the tables, we use Formula (4.7). The other approximations come from Formula (4.6).

a. $f'(2.1) \approx 3.884$ $f'(2.2) \approx 2.896$ $f'(2.3) \approx 2.249$ $f'(2.4) \approx 1.836$ $f'(2.5) \approx 1.550$ $f'(2.6) \approx 1.348$

b. $f'(-3.0) \approx -5.883$ $f'(-2.8) \approx -5.467$ $f'(-2.6) \approx -5.059$ $f'(-2.4) \approx -4.650$ $f'(-2.2) \approx -4.208$ $f'(-2.0) \approx -3.875$

19. The approximation is -4.8×10^{-9}. $f''(0.5) = 0$. The error bound is 0.35874. The method is very accurate since the function is symmetric about $x = 0.5$.

21. a. $f'(0.2) \approx -0.1951027$ **b.** $f'(1.0) \approx -1.541415$ **c.** $f'(0.6) \approx -0.6824175$

23. $f'(0.4) \approx -0.4249840$ and $f'(0.8) \approx -1.032772$.

25. The three-point formulas give the results in the following table.

Time	0	3	5	8	10	13
Speed	79	82.4	74.2	76.8	69.4	71.2

27. The approximations eventually become zero because the numerator becomes zero.

29. Since $e'(h) = -\varepsilon/h^2 + hM/3$, we have $e'(h) = 0$ if and only if $h = \sqrt[3]{3\varepsilon/M}$. Also, $e'(h) < 0$ if $h < \sqrt[3]{3\varepsilon/M}$ and $e'(h) > 0$ if $h > \sqrt[3]{3\varepsilon/M}$, so an absolute minimum for $e(h)$ occurs at $h = \sqrt[3]{3\varepsilon/M}$.

Exercise Set 4.2 (Page 191)

1. **a.** $f'(1) \approx 1.0000109$ **b.** $f'(0) \approx 2.0000000$ **c.** $f'(2.3) \approx -19.646799$ **d.** $f'(1.05) \approx 2.2751459$

3. **a.** $f'(1) \approx 1.001$ **b.** $f'(0) \approx 1.999$ **c.** $f'(2.3) \approx -19.61$ **d.** $f'(1.05) \approx 2.283$

5. $\int_0^{\pi} \sin x \, dx \approx 1.999999$

9. Let

$$N_2(h) = N\left(\frac{h}{3}\right) + \left(\frac{N\left(\frac{h}{3}\right) - N(h)}{2}\right) \text{ and } N_3(h) = N_2\left(\frac{h}{3}\right) + \left(\frac{N_2\left(\frac{h}{3}\right) - N_2(h)}{8}\right).$$

Then $N_3(h)$ is an $O(h^3)$ approximation to M.

11. Let $N(h) = (1+h)^{1/h}$, $N_2(h) = 2N\left(\frac{h}{2}\right) - N(h)$, $N_3(h) = N_2\left(\frac{h}{2}\right) + \frac{1}{3}\left(N_2\left(\frac{h}{2}\right) - N_2(h)\right)$.

a. $N(0.04) = 2.665836331$, $N(0.02) = 2.691588029$, $N(0.01) = 2.704813829$

b. $N_2(0.04) = 2.717339727$, $N_2(0.02) = 2.718039629$. The $O(h^3)$ approximation is $N_3(0.04) = 2.718272931$.

c. Yes, since the errors seem proportioned to h for $N(h)$, to h^2 for $N_2(h)$, and to h^3 for $N_3(h)$.

15. **c.**

k	4	8	16	32	64	128	256	512
p_k	$2\sqrt{2}$	3.0614675	3.1214452	3.1365485	3.1403312	3.1412723	3.1415138	3.1415729
P_k	4	3.3137085	3.1825979	3.1517249	3.144184	3.1422236	3.1417504	3.1416321

d. Values of p_k and P_k are given in the following tables, together with the extrapolation results:

For p_k:

2.8284271				
3.0614675	3.1391476			
3.1214452	3.1414377	3.1415904		
3.1365485	3.1415829	3.1415926	3.1415927	
3.1403312	3.1415921	3.1415927	3.1415927	3.1415927

For P_k:

4				
3.3137085	3.0849447			
3.1825979	3.1388943	3.1424910		
3.1517249	3.1414339	3.1416032	3.1415891	
3.1441184	3.1415829	3.1415928	3.1415926	3.1415927

Exercise Set 4.3 (Page 202)

1. The Trapezoidal rule gives the following approximations.

a. 0.265625 **b.** −0.2678571 **c.** 0.1839397 **d.** −0.17776434

e. −0.1777643 **f.** −0.8666667 **g.** 0.2180895 **h.** 4.1432597

3.

	Actual Error	Error Bound
a.	0.071875	0.125
b.	7.943×10^{-4}	9.718×10^{-4}
c.	0.0233369	0.1666667
d.	0.0358147	0.0396972
e.	9.443×10^{-4}	1.0707×10^{-3}
f.	0.1326975	0.5617284
g.	0.0663431	0.0807455
h.	1.554631	2.298827

5. Simpson's rule gives the following approximations.

a. 0.1940104 **b.** −0.2670635 **c.** 0.16240168 **d.** 0.1922453
e. −0.1768216 **f.** −0.7391053 **g.** 0.1513826 **h.** 2.5836964

7.

	Actual Error	Error Bound
a.	2.604×10^{-4}	2.6042×10^{-4}
b.	7.14×10^{-7}	9.92×10^{-7}
c.	1.7989×10^{-3}	4.1667×10^{-4}
d.	1.406×10^{-5}	2.170×10^{-5}
e.	1.549×10^{-6}	2.095×10^{-6}
f.	5.1361×10^{-3}	0.063280
g.	3.6381×10^{-4}	4.1507×10^{-4}
h.	4.9322×10^{-3}	0.1302826

9. The Midpoint rule gives the following approximations.

a. 0.1582031 **b.** −0.2666667 **c.** 0.1516327 **d.** 0.1743309
e. −0.1768200 **f.** −0.6753247 **g.** 0.1180292 **h.** 1.8039148

11.

	Actual Error	Error Bound
a.	0.0355469	0.0625
b.	3.961×10^{-4}	4.859×10^{-4}
c.	8.9701×10^{-3}	0.0833333
d.	0.0179285	0.0198486
e.	4.698×10^{-4}	5.353×10^{-4}
f.	0.0564448	0.2808642
g.	0.0337172	0.0403728
h.	0.7847138	1.1494136

13. $f(1) = \frac{1}{2}$

15. The degree of precision is 3.

17. $c_0 = \frac{1}{3}$, $c_1 = \frac{4}{3}$, $c_2 = \frac{1}{3}$

19. $c_0 = c_1 = \frac{1}{2}$ gives the highest degree of precision, 1.

21. The following approximations are obtained from Formula (4.23) through Formula (4.30), respectively.

a. 0.1024404, 0.1024598, 0.1024598, 0.1024598, 0.1024695, 0.1024663, 0.1024598, and 0.1024598

b. 0.7853982, 0.7853982, 0.7853982, 0.7853982, 0.7853982, 0.7853982, 0.7853982, and 0.7853982

c. 1.497171, 1.477536, 1.477529, 1.477523, 1.467719, 1.470981, 1.477512, and 1.477515

d. 4.950000, 2.740909, 2.563393, 2.385700, 1.636364, 1.767857, 2.074893, and 2.116379

e. 3.293182, 2.407901, 2.359772, 2.314751, 1.965260, 2.048634, 2.233251, and 2.249001

f. 0.5000000, 0.6958004, 0.7126032, 0.7306341, 0.7937005, 0.7834709, 0.7611137, and 0.7593572

23. The errors in Exercise 22 are 1.6×10^{-6}, 5.3×10^{-8}, -6.7×10^{-7}, -7.2×10^{-7}, and -1.3×10^{-6}, respectively.

25. If $E(x^k) = 0$, for all $k = 0, 1, \ldots, n$ and $E(x^{n+1}) \neq 0$, then with $p_{n+1}(x) = x^{n+1}$, we have a polynomial of degree $n+1$ for which $E(p_{n+1}(x)) \neq 0$. Let $p(x) = a_n x^n + \cdots + a_1 x + a_0$ be any polynomial of degree less than or equal to n. Then $E(p(x)) = a_n E(x^n) + \cdots + a_1 E(x) + a_0 E(1) = 0$. Conversely, if $E(p(x)) = 0$, for all polynomials of degree less than or equal to n, it follows that $E(x^k) = 0$, for all $k = 0, 1, \ldots, n$. Let $p_{n+1}(x) = a_{n+1}x^{n+1} + \cdots + a_0$ be a polynomial of degree $n+1$ for which $E(p_{n+1}(x)) \neq 0$. Since $a_{n+1} \neq 0$, we have

$$x^{n+1} = \frac{1}{a_{n+1}} p_{n+1}(x) - \frac{a_n}{a_{n+1}} x^n - \cdots - \frac{a_0}{a_{n+1}}.$$

Then

$$E(x^{n+1}) = \frac{1}{a_{n+1}}E(p_{n+1}(x)) - \frac{a_n}{a_{n+1}}E(x^n) - \cdots - \frac{a_0}{a_{n+1}}E(1) = \frac{1}{a_{n+1}}E(p_{n+1}(x)) \neq 0.$$

Thus, the quadrature formula has degree of precision n.

Exercise Set 4.4 (Page 210)

1. The Composite Trapezoidal rule approximations are:

a. 31.3653 **b.** 0.639900 **c.** 0.476977 **d.** −6.42872

e. −13.5760 **f.** 0.784241 **g.** 0.605498 **h.** 0.970926

3. a. 22.47713 **b.** 0.6363098 **c.** 0.4777547 **d.** −6.274868

e. −14.18334 **f.** 0.783980 **g.** 0.6043941 **h.** 0.9610554

5. The Composite Midpoint rule approximations are:

a. 11.1568 **b.** 0.633096 **c.** 0.478751 **d.** −6.11274

e. −14.9985 **f.** 0.786700 **g.** 0.602961 **h.** 0.947868

7. a. 3.15947567 **b.** 3.10933713 **c.** 3.00906003

9. $\alpha = 0.75$

11. a. The Composite Trapezoidal rule requires $h < 0.000922295$ and $n \geq 2168$.

b. The Composite Simpson's rule requires $h < 0.037658$ and $n \geq 54$.

c. The Composite Midpoint rule requires $h < 0.00065216$ and $n \geq 3066$.

13. a. The Composite Trapezoidal rule requires $h < 0.04382$ and $n \geq 46$. The approximation is 0.405471.

b. The Composite Simpson's rule requires $h < 0.44267$ and $n \geq 6$. The approximation is 0.405466.

c. The Composite Midpoint rule requires $h < 0.03098$ and $n \geq 64$. The approximation is 0.405460.

15. a. Because the right and left limits at 0.1 and 0.2 for f, f', and f'' are the same, the functions are continuous on $[0, 0.3]$. However,

$$f'''(x) = \begin{cases} 6, & 0 \leq x \leq 0.1 \\ 12, & 0.1 < x \leq 0.2 \\ 12, & 0.2 < x \leq 0.3 \end{cases}$$

is discontinuous at $x = 0.1$.

b. We have 0.302506 with an error bound of 1.9×10^{-4}.

c. We have 0.302425, and the value of the actual integral is the same.

17. a. For the Composite Trapezoidal rule, we have

$$E(f) = -\frac{h^3}{12}\sum_{j=1}^{n} f''(\xi_j) = -\frac{h^2}{12}\sum_{j=1}^{n} f''(\xi_j)h = -\frac{h^2}{12}\sum_{j=1}^{n} f''(\xi_j)\Delta x_j,$$

where $\Delta x_j = x_{j+1} - x_j = h$ for each j. Since $\sum_{j=1}^{n} f''(\xi_j)\Delta x_j$ is a Riemann sum for $\int_a^b f''(x)\,dx = f'(b) - f'(a)$, we have

$$E(f) \approx -\frac{h^2}{12}[f'(b) - f'(a)].$$

b. For the Composite Midpoint rule, we have

$$E(f) = \frac{h^3}{3}\sum_{j=1}^{n/2} f''(\xi_j) = \frac{h^2}{6}\sum_{j=1}^{n/2} f''(\xi_j)(2h).$$

But $\sum_{j=1}^{n/2} f''(\xi_j)(2h)$ is a Riemann sum for $\int_a^b f''(x)\,dx = f'(b) - f'(a)$, so

$$E(f) \approx \frac{h^2}{6}[f'(b) - f'(a)].$$

19. a. The estimate using the Composite Trapezoidal rule is $-\frac{1}{2}h^2 \ln 2 = -6.296 \times 10^{-6}$.

b. The estimate using the Composite Simpson's rule is $-\frac{1}{240}h^2 = -3.75 \times 10^{-6}$.

c. The estimate using the Composite Midpoint rule is $\frac{1}{6}h^2 \ln 2 = 6.932 \times 10^{-6}$.

21. The length is approximately 15.8655.

23. Composite Simpson's rule with $h = 0.25$ gives 2.61972 s.

25. The length is approximately 58.47082, using $n = 100$ in the Composite Simpson's rule.

Exercise Set 4.5 (Page 218)

1. Romberg integration gives $R_{3,3}$ as follows:

a. 0.1606105 **b.** 0.1922593 **c.** −0.7341567 **d.** 0.08875677
e. 2.5879685 **f.** −0.1768200 **g.** 0.6362135 **h.** 0.6426970

3. Romberg integration gives $R_{4,4}$ as follows:

a. 0.1606028 **b.** 0.1922594 **c.** −0.7339728 **d.** 0.08875528
e. 2.5886272 **f.** −0.1768200 **g.** 0.6362134 **h.** 0.6426991

5. Romberg integration gives:

a. 0.19225936 with $n = 4$ **b.** 0.16060279 with $n = 5$ **c.** −0.17682002 with $n = 4$ **d.** 0.088755284 with $n = 5$
e. 2.5886286 with $n = 6$ **f.** −0.73396918 with $n = 6$ **g.** 0.63621335 with $n = 4$ **h.** 0.64269908 with $n = 5$

7. $R_{33} = 11.5246$

9. $f(2.5) \approx 0.43459$

11. $R_{31} = 5$

13. We have

$$R_{k,2} = \frac{4R_{k,1} - R_{k-1,1}}{3} = \frac{1}{3}\left[R_{k-1,1} + 2h_{k-1}\sum_{i=1}^{2^{k-2}} f(a + (i - 1/2))h_{k-1})\right], \quad \text{from (4.34)},$$

$$= \frac{1}{3}\left[\frac{h_{k-1}}{2}(f(a) + f(b)) + h_{k-1}\sum_{i=1}^{2^{k-2}-1} f(a + ih_{k-1})\right.$$

$$\left. + 2h_{k-1}\sum_{i=1}^{2^{k-2}} f(a + (i - 1/2)h_{k-1})\right], \quad \text{from (4.34) with } k-1 \text{ instead of } k,$$

$$= \frac{1}{3}\left[h_k(f(a) + f(b)) + 2h_k\sum_{i=1}^{2^{k-2}-1} f(a + 2ih_k) + 4h_k\sum_{i=1}^{2^{k-2}} f(a + (2i - 1)h)\right]$$

$$= \frac{h}{3}\left[f(a) + f(b) + 2\sum_{i=1}^{M-1} f(a + 2ih) + 4\sum_{i=1}^{M} f(a + (2i - 1)h)\right], \quad \text{where } h = h_k \text{ and } M = 2^{k-2}.$$

15. Equation (4.34) follows from

$$R_{k,1} = \frac{h_k}{2}\left[f(a) + f(b) + 2\sum_{i=1}^{2^{k-1}-1} f(a + ih_k)\right]$$

$$= \frac{h_k}{2}\left[f(a) + f(b) + 2\sum_{i=1}^{2^{k-1}-1} f(a + \frac{i}{2}h_{k-1})\right]$$

$$= \frac{h_k}{2}\left[f(a) + f(b) + 2\sum_{i=1}^{2^{k-1}-1} f(a + ih_{k-1}) + 2\sum_{i=1}^{2^{k-2}} f(a + (i - 1/2)h_{k-1})\right]$$

$$= \frac{1}{2}\left\{\frac{h_{k-1}}{2}\left[f(a)+f(b)+2\sum_{i=1}^{2^{k-2}-1} f(a+ih_{k-1})\right]+h_{k-1}\sum_{i=1}^{2^{k-2}} f(a+(i-1/2)h_{k-1})\right\}$$

$$= \frac{1}{2}\left[R_{k-1,1}+h_{k-1}\sum_{i=1}^{2^{k-2}} f(a+(i-1/2)h_{k-1})\right].$$

Exercise Set 4.6 (Page 227)

1. Simpson's rule gives

a. $S(0,1) = 0.16240168$, $S(0,0.5) = 0.028861071$, $S(0.5,1) = 0.13186140$, and the actual value is 0.16060279.

b. $S(1,1.5) = 0.19224530$, $S(1,1.25) = 0.039372434$, $S(1.25,1.5) = 0.15288602$, and the actual value is 0.19225935.

c. $S(1,1.6) = -0.73910533$, $S(1,1.3) = -0.26141244$, $S(1.3,1.6) = -0.47305351$, and the actual value is -0.73396917.

d. $S(0,\frac{\pi}{4}) = 0.087995669$, $S(0,\frac{\pi}{8}) = 0.0058315797$, $S(\frac{\pi}{8},\frac{\pi}{4}) = 0.082877624$, and the actual value is 0.088755285.

e. $S(0,\frac{\pi}{4}) = 2.5836964$, $S(0,\frac{\pi}{8}) = 0.33088926$, $S(\frac{\pi}{8},\frac{\pi}{4}) = 2.2568121$, and the actual value is 2.5886286.

f. $S(0,0.35) = -0.17682156$, $S(0,0.175) = -0.087724382$, $S(0.175,0.35) = -0.089095736$, and the actual value is -0.17682002.

g. $S(3,3.5) = 0.63623873$, $S(3,3.25) = 0.32567095$, $S(3.25,3.5) = 0.31054412$, and the actual value is 0.63621334.

h. $S(0,\frac{\pi}{4}) = 0.64326905$, $S(0,\frac{\pi}{8}) = 0.37315002$, $S(\frac{\pi}{8},\frac{\pi}{4}) = 0.26958270$, and the actual value is 0.64269908.

3. Adaptive quadrature gives:

a. 2.00000103 **b.** 1.37296499 **c.** 0.23222233 **d.** 5.11383291

5.

	Simpson's Rule	Number Evaluation	Error	Adaptive Quadrature	Number Evaluation	Error
a.	−0.21515695	57	6.3×10^{-6}	−0.21515062	229	1.0×10^{-8}
b.	0.95135226	83	9.6×10^{-6}	0.95134257	217	1.1×10^{-7}
c.	−6.2831813	41	4.0×10^{-6}	−6.2831852	109	1.1×10^{-7}
d.	5.8696024	27	2.6×10^{-6}	5.8696044	109	4.0×10^{-9}

7. $\int_0^{2\pi} u(t)\,dt \approx 0.00001$

9. We have, for $h = b - a$,

$$\left|T(a,b) - T\left(a,\frac{a+b}{2}\right) - T\left(\frac{a+b}{2},b\right)\right| \approx \frac{h^3}{16}\left|f''(\mu)\right|$$

and

$$\left|\int_a^b f(x)\,dx - T\left(a,\frac{a+b}{2}\right) - T\left(\frac{a+b}{2},b\right)\right| \approx \frac{h^3}{48}\left|f''(\mu)\right|.$$

So

$$\left|\int_a^b f(x)\,dx - T\left(a,\frac{a+b}{2}\right) - T\left(\frac{a+b}{2},b\right)\right| \approx \frac{1}{3}\left|T(a,b) - T\left(a,\frac{a+b}{2}\right) - T\left(\frac{a+b}{2},b\right)\right|.$$

Exercise Set 4.7 (Page 234)

1. Gaussian quadrature gives:

a. 0.1594104 **b.** 0.1922687 **c.** −0.7307230 **d.** 0.08926302

e. 2.5913247 **f.** −0.1768190 **g.** 0.6361966 **h.** 0.6423172

3. Gaussian quadrature gives:

a. 0.1606028 **b.** 0.1922594 **c.** −0.7339604 **d.** 0.08875529

e. 2.5886327 **f.** −0.1768200 **g.** 0.6362133 **h.** 0.6426991

5. $a = 1$, $b = 1$, $c = \frac{1}{3}$, $d = -\frac{1}{3}$

9. The exact value to 10 digits is 0.878884623. Part (a) gives 0.878884623, with absolute error 4×10^{-10}. Part (b) gives 0.878884546, with absolute error 7.66×10^{-8}. Part (c) gives 0.878387796, with absolute error 4.97×10^{-4}. All the approximations require 8 function evaluations, and Gaussian quadrature for a given n chooses the interpolation nodes optimally. The composite methods in (b) and (c) do not use these nodes so they should not be expected to give as accurate results.

Exercise Set 4.8 (Page 248)

1. Algorithm 4.4 with $n = m = 4$ gives:
a. 0.2552526 **b.** 0.3115733 **c.** 16.50864 **d.** 1.476684

3. Algorithm 4.4 with $n = 4$ and $m = 8$, $n = 8$ and $m = 4$, and $n = m = 6$ gives:
a. 1.718857, 1.718220, 1.718385
b. 0.5119875, 0.5118533, 0.5118722
c. 2.004596, 2.000879, 2.000980
d. 0.7838542, 0.7833659, 0.7834362
e. −1.985611, −1.999182, −1.997353
f. 1.001953, 1.000122, 1.000386
g. 0.3084277, 0.3084562, 0.3084323
h. −22.61612, −19.85408, −20.14117

5. Algorithm 4.5 with $n = m = 2$ gives:
a. 0.2552446 **b.** 0.3115733 **c.** 16.50863 **d.** 1.488875

7. Algorithm 4.5 with $n = m = 3$, $n = 3$ and $m = 4$, $n = 4$ and $m = 3$, and $n = m = 4$ gives:
a. 1.718163, 1.718302, 1.718139, 1.718277, 1.2×10^{-4}, 2.0×10^{-5}, 1.4×10^{-4}, 4.8×10^{-6}
b. 0.5118655, 0.5118445, 0.5118655, 0.5118445, 2.1×10^{-5}, 1.3×10^{-7}, 2.1×10^{-5}, 1.3×10^{-7}
c. 2.001494, 2.000080, 2.001388, 1.999984, 1.5×10^{-3}, 8×10^{-5}, 1.4×10^{-3}, 1.6×10^{-5}
d. 0.7833333, 0.7833333, 0.7833333, 0.7833333, 0, 0, 0, 0
e. −1.991878, −2.000124, −1.991878, −2.000124, 8.1×10^{-3}, 1.2×10^{-4}, 8.1×10^{-3}, 1.2×10^{-4}
f. 1.000000, 1.000000, 1.0000000, 1.000000, 0, 0, 0, 0
g. 0.3084151, 0.3084145, 0.3084246, 0.3084245, 10^{-5}, 5.5×10^{-7}, 1.1×10^{-5}, 6.4×10^{-7}
h. −12.74790, −21.21539, −11.83624, −20.30373, 7.0, 1.5, 7.9, 0.564

9. Algorithm 4.4 with $n = m = 14$ gives 0.1479103, and Algorithm 4.5 with $n = m = 4$ gives 0.1506823.

11. The approximation to the center of mass is $(\overline{x}, \overline{y})$, where $\overline{x} = 0.3806333$ and $\overline{y} = 0.3822558$.

13. The area is approximately 1.0402528.

15. Algorithm 4.6 with $n = m = p = 2$ gives the first listed value. The second is the exact result.
a. 5.204036, $e(e^{0.5} - 1)(e - 1)^2$
b. 0.08429784, $\frac{1}{12}$
c. 0.08641975, $\frac{1}{14}$
d. 0.09722222, $\frac{1}{12}$
e. 7.103932, $2 + \frac{1}{2}\pi^2$
f. 1.428074, $\frac{1}{2}(e^2 + 1) - e$

17. Algorithm 4.6 with $n = m = p = 4$ gives the first listed value. The second is from Algorithm 4.6 with $n = m = p = 5$.
a. 5.206447, 5.206447
b. 0.08333333,0.08333333
c. 0.07142857,0.07142857
d. 0.08333333,0.08333333
e. 6.934912,6.934801
f. 1.476207, 1.476246

19. The approximation 20.41887 requires 125 functional evaluations.

Exercise Set 4.9 (Page 254)

1. The Composite Simpson's rule gives:
a. 0.5284163 **b.** 4.266654 **c.** 0.4329748 **d.** 0.8802210

3. The Composite Simpson's rule gives:
a. 0.4112649 **b.** 0.2440679 **c.** 0.05501681 **d.** 0.2903746

5. The escape velocity is approximately 6.9450 mi/s.

7. a. $\int_0^\infty e^{-x} f(x)\,dx \approx 0.8535534\ f(0.5857864) + 0.1464466\ f(3.4142136)$
b. $\int_0^\infty e^{-x} f(x)\,dx \approx 0.7110930\ f(0.4157746) + 0.2785177\ f(2.2942804) + 0.0103893\ f(6.2899451)$

9. $n = 2$: 2.9865139 $n = 3$: 2.9958198

Exercise Set 5.1 (Page 264)

1. a. Since $f(t, y) = y \cos t$, we have $\frac{\partial f}{\partial y}(t, y) = \cos t$, and f satisfies a Lipschitz condition in y with $L = 1$ on

$$D = \{(t, y) | 0 \le t \le 1, -\infty < y < \infty\}.$$

Also, f is continuous on D, so there exists a unique solution, which is $y(t) = e^{\sin t}$.

b. Since $f(t, y) = -\frac{2}{t}y + t^2 e^t$, we have $\frac{\partial f}{\partial y} = -\frac{2}{t}$, and f satisfies a Lipschitz condition in y with $L = 2$ on

$$D = \{(t, y) | 1 \le t \le 2, -\infty < y < \infty\}.$$

Also, f is continuous on D, so there exists a unique solution, which is

$$y(t) = (t^4 e^t - 4t^3 e^t + 12t^2 e^t - 24te^t + 24e^t + (\sqrt{2} - 9)e)/t^2.$$

c. Since $f(t, y) = \frac{2}{t}y + t^2 e^t$, we have $\frac{\partial f}{\partial y} = \frac{2}{t}$, and f satisfies a Lipschitz condition in y with $L = 2$ on

$$D = \{(t, y) | 1 \le t \le 2, -\infty < y < \infty\}.$$

Also, f is continuous on D, so there exists a unique solution, which is $y(t) = t^2(e^t - e)$.

d. Since $f(t, y) = \frac{4t^3 y}{1 + t^4}$, we have $\frac{\partial f}{\partial y} = \frac{4t^3}{1 + t^4}$, and f satisfies a Lipschitz condition in y with $L = 2$ on

$$D = \{(t, y) | 0 \le t \le 1, -\infty < y < \infty\}.$$

Also, f is continuous on D, so there exists a unique solution, which is $y(t) = 1 + t^4$.

3. a. Lipschitz constant $L = 1$; it is a well-posed problem.

b. Lipschitz constant $L = 1$; it is a well-posed problem.

c. Lipschitz constant $L = 1$; it is a well-posed problem.

d. The function f does not a satisfy Lipschitz condition, so Theorem 5.6 cannot be used.

5. a. Differentiating $y^3 t + yt = 2$ gives $3y^2 y' t + y^3 + y' t + y = 0$. Solving for y' gives the original differential equation, and setting $t = 1$ and $y = 1$ verifies the initial condition. To approximate $y(2)$, use Newton's method to solve the equation $y^3 + y - 1 = 0$. This gives $y(2) \approx 0.6823278$.

b. Differentiating $y \sin t + t^2 e^y + 2y - 1 = 0$ gives $y' \sin t + y \cos t + 2te^y + t^2 e^y y' + 2y' = 0$. Solving for y' gives the original differential equation, and setting $t = 1$ and $y = 0$ verifies the initial condition. To approximate $y(2)$, use Newton's method to solve the equation $(2 + \sin 2)y + 4e^y - 1 = 0$. This gives $y(2) \approx -0.4946599$.

7. Let (t_1, y_1) and (t_2, y_2) be in D, with $a \le t_1 \le b$, $a \le t_2 \le b$, $-\infty < y_1 < \infty$, and $-\infty < y_2 < \infty$. For $0 \le \lambda \le 1$, we have $(1 - \lambda)a \le (1 - \lambda)t_1 \le (1 - \lambda)b$ and $\lambda a \le \lambda t_2 \le \lambda b$. Hence, $a = (1 - \lambda)a + \lambda a \le (1 - \lambda)t_1 + \lambda t_2 \le (1 - \lambda)b + \lambda b = b$. Also, $-\infty < (1 - \lambda)y_1 + \lambda y_2 < \infty$, so D is convex.

9. a. Since $y' = f(t, y(t))$, we have

$$\int_a^t y'(z)\, dz = \int_a^t f(z, y(z))\, dz.$$

So $y(t) - y(a) = \int_a^t f(z, y(z))\, dz$ and $y(t) = \alpha + \int_a^t f(z, y(z))\, dz$.
The iterative method follows from this equation.

b. We have $y_0(t) = 1$, $y_1(t) = 1 + \frac{1}{2}t^2$, $y_2(t) = 1 + \frac{1}{2}t^2 - \frac{1}{6}t^3$, and $y_3(t) = 1 + \frac{1}{2}t^2 - \frac{1}{6}t^3 + \frac{1}{24}t^4$.

c. We have $y(t) = 1 + \frac{1}{2}t^2 - \frac{1}{6}t^3 + \frac{1}{24}t^4 - \frac{1}{120}t^5 + \cdots$.

Exercise Set 5.2 (Page 273)

1. Euler's method gives the approximations in the following table.

a.

i	t_i	w_i	$y(t_i)$
1	0.500	0.0000000	0.2836165
2	1.000	1.1204223	3.2190993

b.

i	t_i	w_i	$y(t_i)$
1	1.250	2.7500000	2.7789294
2	1.500	3.5500000	3.6081977
3	1.750	4.3916667	4.4793276
4	2.000	5.2690476	5.3862944

c.

i	t_i	w_i	$y(t_i)$
1	2.500	2.0000000	1.8333333
2	3.000	2.6250000	2.5000000

d.

i	t_i	w_i	$y(t_i)$
1	0.250	1.2500000	1.3291498
2	0.500	1.6398053	1.7304898
3	0.750	2.0242547	2.0414720
4	1.000	2.2364573	2.1179795

3. a.

t	Actual Error	Error bound
0.5	0.2836165	11.3938
1.0	2.0986771	42.3654

b.

t	Actual Error	Error bound
1.25	0.0289294	0.0355032
1.50	0.0581977	0.0810902
1.75	0.0876610	0.139625
2.00	0.117247	0.214785

c.

t	Actual Error	Error bound
2.5	0.166667	0.429570
3.0	0.125000	1.59726

d.

t	Actual Error
0.25	0.0791498
0.50	0.0906844
0.75	0.0172174
1.00	0.118478

For Part (d), error bound formula (5.10) cannot be applied since $L = 0$.

5. Euler's method gives the approximations in the following tables.

a.

i	t_i	w_i	$y(t_i)$
2	1.200	1.0082645	1.0149523
4	1.400	1.0385147	1.0475339
6	1.600	1.0784611	1.0884327
8	1.800	1.1232621	1.1336536
10	2.000	1.1706516	1.1812322

b.

i	t_i	w_i	$y(t_i)$
2	1.400	0.4388889	0.4896817
4	1.800	1.0520380	1.1994386
6	2.200	1.8842608	2.2135018
8	2.600	3.0028372	3.6784753
10	3.000	4.5142774	5.8741000

c.

i	t_i	w_i	$y(t_i)$
2	0.400	−1.6080000	−1.6200510
4	0.800	−1.3017370	−1.3359632
6	1.200	−1.1274909	−1.1663454
8	1.600	−1.0491191	−1.0783314
10	2.000	−1.0181518	−1.0359724

d.

i	t_i	w_i	$y(t_i)$
2	0.2	0.1083333	0.1626265
4	0.4	0.1620833	0.2051118
6	0.6	0.3455208	0.3765957
8	0.8	0.6213802	0.6461052
10	1.0	0.9803451	1.0022460

7. The actual errors for the approximations in Exercise 3 are in the following tables.

a.

t	Actual Error
1.2	0.0066879
1.5	0.0095942
1.7	0.0102229
2.0	0.0105806

b.

t	Actual Error
1.4	0.0507928
2.0	0.2240306
2.4	0.4742818
3.0	1.3598226

c.

t	Actual Error
0.4	0.0120510
1.0	0.0391546
1.4	0.0349030
2.0	0.0178206

d.

t	Actual Error
0.2	0.0542931
0.5	0.0363200
0.7	0.0273054
1.0	0.0219009

9. Euler's method gives the approximations in the following table.

a.

i	t_i	w_i	$y(t_i)$
1	1.1	0.271828	0.345920
5	1.5	3.18744	3.96767
6	1.6	4.62080	5.70296
9	1.9	11.7480	14.3231
10	2.0	15.3982	18.6831

b. Linear interpolation gives the approximations in the following table.

t	Approximation	$y(t)$	Error
1.04	0.108731	0.119986	0.01126
1.55	3.90412	4.78864	0.8845
1.97	14.3031	17.2793	2.976

c. $h < 0.00064$

11. a. Euler's method produces the following approximation to $y(5) = 5.00674$.

	$h = 0.2$	$h = 0.1$	$h = 0.05$
w_N	5.00377	5.00515	5.00592

b. $h = \sqrt{2 \times 10^{-6}} \approx 0.0014142$.

13. a. $1.021957 = y(1.25) \approx 1.014978$, $1.164390 = y(1.93) \approx 1.153902$

b. $1.924962 = y(2.1) \approx 1.660756$, $4.394170 = y(2.75) \approx 3.526160$

c. $-1.138277 = y(1.3) \approx -1.103618$, $-1.041267 = y(1.93) \approx -1.022283$

d. $0.3140018 = y(0.54) \approx 0.2828333$, $0.8866318 = y(0.94) \approx 0.8665521$

15. a. $h = 10^{-n/2}$

b. The minimal error is $10^{-n/2}(e - 1) + 5e10^{-n-1}$.

c.

t	$w(h = 0.1)$	$w(h = 0.01)$	$y(t)$	Error ($n = 8$)
0.5	0.40951	0.39499	0.39347	1.5×10^{-4}
1.0	0.65132	0.63397	0.63212	3.1×10^{-4}

17. b. $w_{50} = 0.10430 \approx p(50)$

c. Since $p(t) = 1 - 0.99e^{-0.002t}$, $p(50) = 0.10421$.

Exercise Set 5.3 (Page 281)

1. a.

t_i	w_i	$y(t_i)$
0.50	0.12500000	0.28361652
1.00	2.02323897	3.21909932

b.

t_i	w_i	$y(t_i)$
1.25	2.78125000	2.77892944
1.50	3.61250000	3.60819766
1.75	4.48541667	4.47932763
2.00	5.39404762	5.38629436

c.

t_i	w_i	$y(t_i)$
2.50	1.75000000	1.83333333
3.00	2.42578125	2.50000000

d.

t_i	w_i	$y(t_i)$
0.25	1.34375000	1.32914981
0.50	1.77218707	1.73048976
0.75	2.11067606	2.04147203
1.00	2.20164395	2.11797955

3. a.

t_i	w_i	$y(t_i)$
0.50	0.25781250	0.28361652
1.00	3.05529474	3.21909932

b.

t_i	w_i	$y(t_i)$
1.25	2.77897135	2.77892944
1.50	3.60826562	3.60819766
1.75	4.47941561	4.47932763
2.00	5.38639966	5.38629436

c.

t_i	w_i	$y(t_i)$
2.50	1.81250000	1.83333333
3.00	2.48591644	2.50000000

d.

t_i	w_i	$y(t_i)$
0.25	1.32893880	1.32914981
0.50	1.72966730	1.73048976
0.75	2.03993417	2.04147203
1.00	2.11598847	2.11797955

5. a.

		Order 2	
i	t_i	w_i	$y(t_i)$
1	0.5	0.5000000	0.5158868
2	1.0	1.076858	1.091818

b.

		Order 2	
i	t_i	w_i	$y(t_i)$
1	1.1	1.214999	1.215886
2	1.2	1.465250	1.467570

c.

		Order 2	
i	t_i	w_i	$y(t_i)$
1	1.5	−2.000000	−1.500000
2	2.0	−1.777776	−1.333333
3	2.5	−1.585732	−1.250000
4	3.0	−1.458882	−1.200000

d.

		Order 2	
i	t_i	w_i	$y(t_i)$
1	0.25	1.093750	1.087088
2	0.50	1.312319	1.289805
3	0.75	1.538468	1.513490
4	1.0	1.720480	1.701870

7. a.

		Order 4	
i	t_i	w_i	$y(t_i)$
1	0.5	0.5156250	0.5158868
2	1.0	1.091267	1.091818

b.

		Order 4	
i	t_i	w_i	$y(t_i)$
1	1.1	1.215883	1.215886
2	1.2	1.467561	1.467570

c.

		Order 4	
i	t_i	w_i	$y(t_i)$
1	1.5	−2.000000	−1.500000
2	2.0	−1.679012	−1.333333
3	2.5	−1.484493	−1.250000
4	3.0	−1.374440	−1.200000

d.

		Order 4	
i	t_i	w_i	$y(t_i)$
1	0.25	1.086426	1.087088
2	0.50	1.288245	1.289805
3	0.75	1.512576	1.513490
4	1.0	1.701494	1.701870

9. a. Taylor's method of order two gives the results in the following table.

i	t_i	w_i	$y(t_i)$
1	1.1	0.3397852	0.3459199
5	1.5	3.910985	3.967666
6	1.6	5.643081	5.720962
9	1.9	14.15268	14.32308
10	2.0	18.46999	18.68310

b. Linear interpolation gives $y(1.04) \approx 0.1359139$, $y(1.55) \approx 4.777033$, and $y(1.97) \approx 17.17480$. Actual values are $y(1.04) = 0.1199875$, $y(1.55) = 4.788635$, and $y(1.97) = 17.27930$.

c. Taylor's method of order four gives the results in the following table.

i	t_i	w_i
1	1.1	0.3459127
5	1.5	3.967603
6	1.6	5.720875
9	1.9	14.32290
10	2.0	18.68287

d. Cubic Hermite interpolation gives $y(1.04) \approx 0.1199704$, $y(1.55) \approx 4.788527$, and $y(1.97) \approx 17.27904$.

11. a.

i	t_i	Order 2	Order 4
2	0.2	5.86595	5.86433
5	0.5	2.82145	2.81789
7	0.7	0.84926	0.84455
10	1.0	−2.08606	−2.09015

b. 0.8 s

Exercise Set 5.4 (Page 291)

1. a.

t	Modified Euler	$y(t)$
0.5	0.5602111	0.2836165
1.0	5.3014898	3.2190993

b.

t	Modified Euler	$y(t)$
1.25	2.7750000	2.7789294
1.50	3.6008333	3.6081977
1.75	4.4688294	4.4793276
2.00	5.3728586	5.3862944

c.

t	Modified Euler	$y(t)$
2.5	1.8125000	1.8333333
3.0	2.4815531	2.5000000

d.

t	Modified Euler	$y(t)$
0.25	1.3199027	1.3291498
0.50	1.7070300	1.7304898
0.75	2.0053560	2.0414720
1.00	2.0770789	2.1179795

3. a.

	Modified Euler	
t_i	w_i	$y(t_i)$
1.4	0.4850495	0.4896817
2.0	1.6384229	1.6612818
2.4	2.8250651	2.8765514
3.0	5.7075699	5.8741000

b.

	Modified Euler	
t_i	w_i	$y(t_i)$
1.2	1.0147137	1.0149523
1.5	1.0669093	1.0672624
1.7	1.1102751	1.1106551
2.0	1.1808345	1.1812322

c.

	Modified Euler	
t_i	w_i	$y(t_i)$
0.4	−1.6229206	−1.6200510
1.0	−1.2442903	−1.2384058
1.4	−1.1200763	−1.1146484
2.0	−1.0391938	−1.0359724

d.

	Modified Euler	
t_i	w_i	$y(t_i)$
0.2	0.1742708	0.1626265
0.5	0.2878200	0.2773617
0.7	0.5088359	0.5000658
1.0	1.0096377	1.0022460

5. a.

t	Midpoint	$y(t)$
0.5	0.2646250	0.2836165
1.0	3.1300023	3.2190993

b.

t	Midpoint	$y(t)$
1.25	2.7777778	2.7789294
1.50	3.6060606	3.6081977
1.75	4.4763015	4.4793276
2.00	5.3824398	5.3862944

c.

t	Midpoint	$y(t)$
2.5	1.7812500	1.8333333
3.0	2.4550638	2.5000000

d.

t	Midpoint	$y(t)$
0.25	1.3337962	1.3291498
0.50	1.7422854	1.7304898
0.75	2.0596374	2.0414720
1.00	2.1385560	2.1179795

7. a.

	Midpoint	
t_i	w_i	$y(t_i)$
1.4	0.4861770	0.4896817
2.0	1.6438889	1.6612818
2.4	2.8364357	2.8765514
3.0	5.7386475	5.8741000

b.

	Midpoint	
t_i	w_i	$y(t_i)$
1.2	1.0153257	1.0149523
1.5	1.0677427	1.0672624
1.7	1.1111478	1.1106551
2.0	1.1817275	1.1812322

c.

	Midpoint	
t_i	w_i	$y(t_i)$
0.4	−1.6192966	−1.6200510
1.0	−1.2402470	−1.2384058
1.4	−1.1175165	−1.1146484
2.0	−1.0382227	−1.0359724

d.

	Midpoint	
t_i	w_i	$y(t_i)$
0.2	0.1722396	0.1626265
0.5	0.2848046	0.2773617
0.7	0.5056268	0.5000658
1.0	1.0063347	1.0022460

9. a.

	Heun	
t_i	w_i	$y(t_i)$
0.50	0.2710885	0.2836165
1.00	3.1327255	3.2190993

b.

	Heun	
t_i	w_i	$y(t_i)$
1.25	2.7788462	2.7789294
1.50	3.6080529	3.6081977
1.75	4.4791319	4.4793276
2.00	5.3860533	5.3862944

c.

	Heun	
t_i	w_i	$y(t_i)$
2.50	1.8464828	1.8333333
3.00	2.5094123	2.5000000

d.

	Heun	
t_i	w_i	$y(t_i)$
0.25	1.3295717	1.3291498
0.50	1.7310350	1.7304898
0.75	2.0417476	2.0414720
1.00	2.1176975	2.1179795

11. a.

	Heun	
t_i	w_i	$y(t_i)$
1.4	0.4895074	0.4896817
2.0	1.6602954	1.6612818
2.4	2.8741491	2.8765514
3.0	5.8652189	5.8741000

b.

	Heun	
t_i	w_i	$y(t_i)$
1.2	1.0149305	1.0149523
1.5	1.0672363	1.0672624
1.7	1.1106289	1.1106551
2.0	1.1812064	1.1812322

c.

	Heun	
t_i	w_i	$y(t_i)$
0.4	−1.6201023	−1.6200510
1.0	−1.2383500	−1.2384058
1.4	−1.1144745	−1.1146484
2.0	−1.0357989	−1.0359724

d.

	Heun	
t_i	w_i	$y(t_i)$
0.2	0.1614497	0.1626265
0.5	0.2765100	0.2773617
0.7	0.4994538	0.5000658
1.0	1.0018114	1.0022460

13. a.

	Runge-Kutta	
t_i	w_i	$y(t_i)$
0.5	0.2969975	0.2836165
1.0	3.3143118	3.2190993

b.

	Runge-Kutta	
t_i	w_i	$y(t_i)$
1.25	2.7789095	2.7789294
1.50	3.6081647	3.6081977
1.75	4.4792846	4.4793276
2.00	5.3862426	5.3862944

c.

	Runge-Kutta	
t_i	w_i	$y(t_i)$
2.5	1.8333234	1.8333333
3.0	2.4999712	2.5000000

d.

	Runge-Kutta	
t_i	w_i	$y(t_i)$
0.25	1.3291650	1.3291498
0.50	1.7305336	1.7304898
0.75	2.0415436	2.0414720
1.00	2.1180636	2.1179795

15. a.

	Runge-Kutta	
t_i	w_i	$y(t_i)$
1.4	0.4896842	0.4896817
2.0	1.6612651	1.6612818
2.4	2.8764941	2.8765514
3.0	5.8738386	5.8741000

b.

	Runge-Kutta	
t_i	w_i	$y(t_i)$
1.2	1.0149520	1.0149523
1.5	1.0672620	1.0672624
1.7	1.1106547	1.1106551
2.0	1.1812319	1.1812322

c.

	Runge-Kutta	
t_i	w_i	$y(t_i)$
0.4	−1.6200576	−1.6200510
1.0	−1.2384307	−1.2384058
1.4	−1.1146769	−1.1146484
2.0	−1.0359922	−1.0359724

d.

	Runge-Kutta	
t_i	w_i	$y(t_i)$
0.2	0.1627655	0.1626265
0.5	0.2774767	0.2773617
0.7	0.5001579	0.5000658
1.0	1.0023207	1.0022460

17. a. $1.9086500 \approx y(2.1) = 1.9249616$, $4.3105913 \approx y(2.75) = 4.3941697$
b. $1.0221167 \approx y(1.25) = 1.0219569$, $1.1640347 \approx y(1.93) = 1.1643901$
c. $-1.1461434 \approx y(1.3) = -1.1382768$, $-1.0454854 \approx y(1.93) = -1.0412665$
d. $0.3271470 \approx y(0.54) = 0.3140018$, $0.8967073 \approx y(0.94) = 0.8866318$

19. a. $1.9153749 \approx y(2.1) = 1.9249616$, $4.3312939 \approx y(2.75) = 4.3941697$
b. $1.0227863 \approx y(1.25) = 1.0219569$, $1.1649247 \approx y(1.93) = 1.1643901$
c. $-1.1432070 \approx y(1.3) = -1.1382768$, $-1.0443743 \approx y(1.93) = -1.0412665$
d. $0.3240839 \approx y(0.54) = 0.3140018$, $0.8934152 \approx y(0.94) = 0.8866318$

21. a. $1.88084805 \approx y(2.1) = 1.9249616$, $4.40842612 \approx y(2.75) = 4.3941697$
b. $1.02235985 \approx y(1.25) = 1.0219569$, $1.16440371 \approx y(1.93) = 1.1643901$
c. $-1.14034696 \approx y(1.3) = -1.1382768$, $-1.04182026 \approx y(1.93) = -1.0412665$
d. $0.31625699 \approx y(0.54) = 0.3140018$, $0.88866134 \approx y(0.94) = 0.8866318$

23. a. $1.9373672 \approx y(2.1) = 1.9249616,\ \ 4.4134745 \approx y(2.75) = 4.3941697$

b. $1.0223826 \approx y(1.25) = 1.0219569,\ \ 1.1644292 \approx y(1.93) = 1.1643901$

c. $-1.1405252 \approx y(1.3) = -1.1382768,\ \ -1.0420211 \approx y(1.93) = -1.0412665$

d. $0.31716526 \approx y(0.54) = 0.3140018,\ \ 0.88919730 \approx y(0.94) = 0.8866318$

25. a. $1.9249617 = y(2.10) \approx 1.9249217, 4.3941697 = y(2.75) \approx 4.3939943$

b. $1.0219569 = y(1.25) \approx 1.0219550, 1.1643902 = y(1.93) \approx 1.1643898$

c. $-1.138268 = y(1.3) \approx -1.1383036, -1.0412666 = y(1.93) \approx -1.0412862$

d. $0.31400184 = y(0.54) \approx 0.31410579, 0.88663176 = y(0.94) \approx 0.88670653$

27. With $f(t, y) = -y + t + 1$, we have both

$$w_i + hf\left(t_i + \frac{h}{2}, w_i + \frac{h}{2} f(t_i, w_i)\right) = w_i\left(1 - h + \frac{h^2}{2}\right) + t_i\left(h - \frac{h^2}{2}\right) + h$$

and

$$w_i + \frac{h}{2}\left[f(t_i, w_i) + f(t_{i+1}, w_i + hf(t_i, w_i))\right] = w_i\left(1 - h + \frac{h^2}{2}\right) + t_i\left(h - \frac{h^2}{2}\right) + h,$$

because $f(t, y)$ is linear in both variables.

29. In 0.2 s we have approximately 2099 units of KOH.

Exercise Set 5.5 (Page 300)

1. The Runge-Kutta-Fehlberg Algorithm gives the results in the following tables.

a.

i	t_i	w_i	h_i	y_i
1	0.2093900	0.0298184	0.2093900	0.0298337
3	0.5610469	0.4016438	0.1777496	0.4016860
5	0.8387744	1.5894061	0.1280905	1.5894600
7	1.0000000	3.2190497	0.0486737	3.2190993

b.

i	t_i	w_i	h_i	y_i
1	1.2500000	2.7789299	0.2500000	2.7789294
2	1.5000000	3.6081985	0.2500000	3.6081977
3	1.7500000	4.4793288	0.2500000	4.4793276
4	2.0000000	5.3862958	0.2500000	5.3862944

c.

i	t_i	w_i	h_i	y_i
1	2.2500000	1.4499988	0.2500000	1.4500000
2	2.5000000	1.8333332	0.2500000	1.8333333
3	2.7500000	2.1785718	0.2500000	2.1785714
4	3.0000000	2.5000005	0.2500000	2.5000000

d.

i	t_i	w_i	h_i	y_i
1	0.2500000	1.3291478	0.2500000	1.3291498
2	0.5000000	1.7304857	0.2500000	1.7304898
3	0.7500000	2.0414669	0.2500000	2.0414720
4	1.0000000	2.1179750	0.2500000	2.1179795

3. The Runge-Kutta-Fehlberg Algorithm gives the results in the following tables.

a.

i	t_i	w_i	h_i	y_i
4	1.5482238	0.7234123	0.1256486	0.7234119
7	1.8847226	1.3851234	0.1073571	1.3851226
10	2.1846024	2.1673514	0.0965027	2.1673499
16	2.6972462	4.1297939	0.0778628	4.1297904
21	3.0000000	5.8741059	0.0195070	5.8741000

b.

i	t_i	w_i	h_i	y_i
1	1.1101946	1.0051237	0.1101946	1.0051237
5	1.7470584	1.1213948	0.2180472	1.1213947
7	2.3994350	1.2795396	0.3707934	1.2795395
11	4.0000000	1.6762393	0.1014853	1.6762391

c.

i	t_i	w_i	h_i	y_i
1	0.1633541	−1.8380836	0.1633541	−1.8380836
5	0.7585763	−1.3597623	0.1266248	−1.3597624
9	1.1930325	−1.1684827	0.1048224	−1.1684830
13	1.6229351	−1.0749509	0.1107510	−1.0749511
17	2.1074733	−1.0291158	0.1288897	−1.0291161
23	3.0000000	−1.0049450	0.1264618	−1.0049452

d.

i	t_i	w_i	h_i	y_i
1	0.3986051	0.3108201	0.3986051	0.3108199
3	0.9703970	0.2221189	0.2866710	0.2221186
5	1.5672905	0.1133085	0.3042087	0.1133082
8	2.0000000	0.0543454	0.0902302	0.0543455

5. a. The number of infectives is $y(30) \approx 80295.7$.

b. The limiting value for the number of infectives for this model is $\lim_{t\to\infty} y(t) = 100{,}000$.

Exercise Set 5.6 (Page 314)

1. The Adams-Bashforth methods give the results in the following tables.

a.

t	2-step	3-step	4-step	5-step	$y(t)$
0.2	0.0268128	0.0268128	0.0268128	0.0268128	0.0268128
0.4	0.1200522	0.1507778	0.1507778	0.1507778	0.1507778
0.6	0.4153551	0.4613866	0.4960196	0.4960196	0.4960196
0.8	1.1462844	1.2512447	1.2961260	1.3308570	1.3308570
1.0	2.8241683	3.0360680	3.1461400	3.1854002	3.2190993

b.

t	2-step	3-step	4-step	5-step	$y(t)$
1.2	2.6187859	2.6187859	2.6187859	2.6187859	2.6187859
1.4	3.2734823	3.2710611	3.2710611	3.2710611	3.2710611
1.6	3.9567107	3.9514231	3.9520058	3.9520058	3.9520058
1.8	4.6647738	4.6569191	4.6582078	4.6580160	4.6580160
2.0	5.3949416	5.3848058	5.3866452	5.3862177	5.3862944

c.

t	2-step	3-step	4-step	5-step	$y(t)$
2.2	1.3666667	1.3666667	1.3666667	1.3666667	1.3666667
2.4	1.6750000	1.6857143	1.6857143	1.6857143	1.6857143
2.6	1.9632431	1.9794407	1.9750000	1.9750000	1.9750000
2.8	2.2323184	2.2488759	2.2423065	2.2444444	2.2444444
3.0	2.4884512	2.5051340	2.4980306	2.5011406	2.5000000

d.

t	2-step	3-step	4-step	5-step	$y(t)$
0.2	1.2529306	1.2529306	1.2529306	1.2529306	1.2529306
0.4	1.5986417	1.5712255	1.5712255	1.5712255	1.5712255
0.6	1.9386951	1.8827238	1.8750869	1.8750869	1.8750869
0.8	2.1766821	2.0844122	2.0698063	2.0789180	2.0789180
1.0	2.2369407	2.1115540	2.0998117	2.1180642	2.1179795

3. The Adams-Bashforth methods give the results in the following tables.

a.

t	2-step	3-step	4-step	5-step	$y(t)$
1.4	0.4867550	0.4896842	0.4896842	0.4896842	0.4896817
1.8	1.1856931	1.1982110	1.1990422	1.1994320	1.1994386
2.2	2.1753785	2.2079987	2.2117448	2.2134792	2.2135018
2.6	3.5849181	3.6617484	3.6733266	3.6777236	3.6784753
3.0	5.6491203	5.8268008	5.8589944	5.8706101	5.8741000

b.

t	2-step	3-step	4-step	5-step	$y(t)$
1.2	1.0161982	1.0149520	1.0149520	1.0149520	1.0149523
1.4	1.0497665	1.0468730	1.0477278	1.0475336	1.0475339
1.6	1.0910204	1.0875837	1.0887567	1.0883045	1.0884327
1.8	1.1363845	1.1327465	1.1340093	1.1334967	1.1336536
2.0	1.1840272	1.1803057	1.1815967	1.1810689	1.1812322

c.

t	2-step	3-step	4-step	5-step	$y(t)$
0.5	−1.5357010	−1.5381988	−1.5379372	−1.5378676	−1.5378828
1.0	−1.2374093	−1.2389605	−1.2383734	−1.2383693	−1.2384058
1.5	−1.0952910	−1.0950952	−1.0947925	−1.0948481	−1.0948517
2.0	−1.0366643	−1.0359996	−1.0359497	−1.0359760	−1.0359724

d.

t	2-step	3-step	4-step	5-step	$y(t)$
0.2	0.1739041	0.1627655	0.1627655	0.1627655	0.1626265
0.4	0.2144877	0.2026399	0.2066057	0.2052405	0.2051118
0.6	0.3822803	0.3747011	0.3787680	0.3765206	0.3765957
0.8	0.6491272	0.6452640	0.6487176	0.6471458	0.6461052
1.0	1.0037415	1.0020894	1.0064121	1.0073348	1.0022460

5. a.

t_i	w_i	$y(t_i)$
0.2	0.0269059	0.0268128
0.4	0.1510468	0.1507778
0.6	0.4966479	0.4960196
0.8	1.3408657	1.3308570
1.0	3.2450881	3.2190993

b.

t_i	w_i	$y(t_i)$
1.2	2.6187787	2.6187859
1.4	3.2710491	3.2710611
1.6	3.9519900	3.9520058
1.8	4.6579968	4.6580160
2.0	5.3862715	5.3862944

c.

t_i	w_i	$y(t_i)$
2.2	1.3666610	1.3666667
2.4	1.6857079	1.6857143
2.6	1.9749941	1.9750000
2.8	2.2446995	2.2444444
3.0	2.5003083	2.5000000

d.

t_i	w_i	$y(t_i)$
0.2	1.2529350	1.2529306
0.4	1.5712383	1.5712255
0.6	1.8751097	1.8750869
0.8	2.0796618	2.0789180
1.0	2.1192575	2.1179795

7. The Adams Fourth-order Predictor-Corrector Algorithm gives the results in the following tables.

a.

t	w	$y(t)$
1.4	0.4896842	0.4896817
1.8	1.1994245	1.1994386
2.2	2.2134701	2.2135018
2.6	3.6784144	3.6784753
3.0	5.8739518	5.8741000

b.

t	w	$y(t)$
1.2	1.0149520	1.0149523
1.4	1.0475227	1.0475339
1.6	1.0884141	1.0884327
1.8	1.1336331	1.1336536
2.0	1.1812112	1.1812322

c.

t	w	$y(t)$
0.5	−1.5378788	−1.5378828
1.0	−1.2384134	−1.2384058
1.5	−1.0948609	−1.0948517
2.0	−1.0359757	−1.0359724

d.

t	w	$y(t)$
0.2	0.1627655	0.1626265
0.4	0.2048557	0.2051118
0.6	0.3762804	0.3765957
0.8	0.6458949	0.6461052
1.0	1.0021372	1.0022460

9. a. With $h = 0.01$, the three-step Adams-Moulton method gives the values in the following table.

i	t_i	w_i
10	0.1	1.317218
20	0.2	1.784511

b. Newton's method will reduce the number of iterations per step from three to two, using the stopping criterion

$$|w_i^{(k)} - w_i^{(k-1)}| \leq 10^{-6}.$$

15. To derive Milne's method, integrate $y'(t) = f(t, y(t))$ on the interval $[t_{i-3}, t_{i+1}]$ to obtain

$$y(t_{i+1}) - y(t_{i-3}) = \int_{t_{i-3}}^{t_{i+1}} f(t, y(t))\, dt.$$

Using the open Newton-Cotes formula (4.31) on page 201, we have

$$y(t_{i+1}) - y(t_{i-3}) = \frac{4h[2f(t_i, y(t_i)) - f(t_{i-1}, y(t_{i-1})) + 2f(t_{i-2}, y(t_{i-2}))]}{3} + \frac{14h^5 f^{(4)}(\xi, y(\xi))}{45}.$$

The difference equation becomes

$$w_{i+1} = w_{i-3} + \frac{h[8f(t_i, w_i) - 4f(t_{i-1}, w_{i-1}) + 8f(t_{i-2}, w_{i-2})]}{3},$$

with local truncation error

$$\tau_{i+1}(h) = \frac{14h^4 y^{(5)}(\xi)}{45}.$$

Exercise Set 5.7 (Page 320)

1. The Adams Variable Step-Size Predictor-Corrector Algorithm gives the results in the following tables.

a.

i	t_i	w_i	h_i	y_i
1	0.04275596	0.00096891	0.04275596	0.00096887
5	0.22491460	0.03529441	0.05389076	0.03529359
12	0.60214994	0.50174348	0.05389076	0.50171761
17	0.81943926	1.45544317	0.04345786	1.45541453
22	0.99830392	3.19605697	0.03577293	3.19602842
26	1.00000000	3.21912776	0.00042395	3.21909932

b.

i	t_i	w_i	h_i	y_i
1	1.06250000	2.18941363	0.06250000	2.18941366
4	1.25000000	2.77892931	0.06250000	2.77892944
8	1.85102559	4.84179835	0.15025640	4.84180141
12	2.00000000	5.38629105	0.03724360	5.38629436

c.

i	t_i	w_i	h_i	y_i
1	2.06250000	1.12132350	0.06250000	1.12132353
5	2.31250000	1.55059834	0.06250000	1.55059524
9	2.62471924	2.00923157	0.09360962	2.00922829
13	2.99915773	2.49895243	0.09360962	2.49894707
17	3.00000000	2.50000535	0.00021057	2.50000000

d.

i	t_i	w_i	h_i	y_i
1	0.06250000	1.06817960	0.06250000	1.06817960
5	0.31250000	1.42861668	0.06250000	1.42861361
10	0.62500000	1.90768386	0.06250000	1.90767015
13	0.81250000	2.08668486	0.06250000	2.08666541
16	1.00000000	2.11800208	0.06250000	2.11797955

3. The following tables list representative results from the Adams Variable Step-Size Predictor-Corrector Algorithm.

a.

i	t_i	w_i	h_i	y_i
5	1.18519603	0.20333499	0.03703921	0.20333497
15	1.55558810	0.73586642	0.03703921	0.73586631
25	1.92598016	1.48072467	0.03703921	1.48072442
35	2.29637222	2.51764797	0.03703921	2.51764743
45	2.65452689	3.92602442	0.03092051	3.92602332
55	2.94341188	5.50206466	0.02584049	5.50206279
61	3.00000000	5.87410206	0.00122679	5.87409998

b.

i	t_i	w_i	h_i	y_i
5	1.10431651	1.00463041	0.02086330	1.00463045
15	1.31294952	1.03196889	0.02086330	1.03196898
25	1.59408142	1.08714711	0.03122028	1.08714722
35	2.00846205	1.18327922	0.04824992	1.18327937
45	2.66272188	1.34525123	0.07278716	1.34525143
52	3.40193112	1.52940900	0.11107035	1.52940924
57	4.00000000	1.67623887	0.12174963	1.67623914

c.

i	t_i	w_i	h_i	y_i
5	0.16854008	−1.83303780	0.03370802	−1.83303783
17	0.64833341	−1.42945306	0.05253230	−1.42945304
27	1.06742915	−1.21150951	0.04190957	−1.21150932
41	1.75380240	−1.05819340	0.06681937	−1.05819325
51	2.50124702	−1.01335240	0.07474446	−1.01335258
61	3.00000000	−1.00494507	0.01257155	−1.00494525

d.

i	t_i	w_i	h_i	y_i
5	0.28548652	0.32153668	0.05709730	0.32153674
15	0.85645955	0.24281066	0.05709730	0.24281095
20	1.35101725	0.15096743	0.09891154	0.15096772
25	1.66282314	0.09815109	0.06236118	0.09815137
29	1.91226786	0.06418555	0.06236118	0.06418579
33	2.00000000	0.05434530	0.02193303	0.05434551

5. The current after 2 seconds is approximately $i(2) = 8.693$ amperes.

Exercise Set 5.8 (Page 327)

1. The Extrapolation Algorithm gives the results in the following tables.

a.

i	t_i	w_i	h	k	y_i
1	0.25	0.04543132	0.25	3	0.04543123
2	0.50	0.28361684	0.25	3	0.28361652
3	0.75	1.05257634	0.25	4	1.05257615
4	1.00	3.21909944	0.25	4	3.21909932

b.

i	t_i	w_i	h	k	y_i
1	1.25	2.77892942	0.25	3	2.77892944
2	1.50	3.60819763	0.25	3	3.60819766
3	1.75	4.47932759	0.25	3	4.47932763
4	2.00	5.38629431	0.25	3	5.38629436

c.

i	t_i	w_i	h	k	y_i
1	2.25	1.44999987	0.25	3	1.45000000
2	2.50	1.83333321	0.25	3	1.83333333
3	2.75	2.17857133	0.25	3	2.17857143
4	3.00	2.49999993	0.25	3	2.50000000

d.

i	t_i	w_i	h	k	y_i
1	0.25	1.32914981	0.25	3	1.32914981
2	0.50	1.73048976	0.25	3	1.73048976
3	0.75	2.04147203	0.25	3	2.04147203
4	1.00	2.11797954	0.25	3	2.11797955

3. The Extrapolation Algorithm gives the results in the following tables.

a.

i	t_i	w_i	h	k	y_i
1	1.50	0.64387537	0.50	4	0.64387533
2	2.00	1.66128182	0.50	5	1.66128176
3	2.50	3.25801550	0.50	5	3.25801536
4	3.00	5.87410027	0.50	5	5.87409998

b.

i	t_i	w_i	h	k	y_i
1	1.50	1.06726237	0.50	4	1.06726235
2	2.00	1.18123223	0.50	3	1.18123222
3	2.50	1.30460372	0.50	3	1.30460371
4	3.00	1.42951608	0.50	3	1.42951607
5	3.50	1.55364771	0.50	3	1.55364770
6	4.00	1.67623915	0.50	3	1.67623914

c.

i	t_i	w_i	h	k	y_i
1	0.50	−1.53788284	0.50	4	−1.53788284
2	1.00	−1.23840584	0.50	5	−1.23840584
3	1.50	−1.09485175	0.50	5	−1.09485175
4	2.00	−1.03597242	0.50	5	−1.03597242
5	2.50	−1.01338570	0.50	5	−1.01338570
6	3.00	−1.00494526	0.50	4	−1.00494525

d.

i	t_i	w_i	h	k	y_i
1	0.50	0.29875177	0.50	4	0.29875178
2	1.00	0.21662642	0.50	4	0.21662642
3	1.50	0.12458565	0.50	4	0.12458565
4	2.00	0.05434552	0.50	4	0.05434551

Exercise Set 5.9 (Page 337)

1. The Runge-Kutta for Systems Algorithm gives the results in the following tables.

a.

t_i	w_{1i}	u_{1i}	w_{2i}	u_{2i}
0.500	0.95671390	0.95672798	−1.08381950	−1.08383310
1.000	1.30654440	1.30655930	−0.83295364	−0.83296776
1.500	1.34416716	1.34418117	−0.56980329	−0.56981634
2.000	1.14332436	1.14333672	−0.36936318	−0.36937457

b.

t_i	w_{1i}	u_{1i}	w_{2i}	u_{2i}
0.200	2.12036583	2.12500839	1.50699185	1.51158743
0.400	4.44122776	4.46511961	3.24224021	3.26598528
0.600	9.73913329	9.83235869	8.16341700	8.25629549
0.800	22.67655977	23.00263945	21.34352778	21.66887674
1.000	55.66118088	56.73748265	56.03050296	57.10536209

c.

t_i	w_{1i}	u_{1i}	w_{2i}	u_{2i}	w_{3i}	u_{3i}
0.5	0.70787076	0.70828683	−1.24988663	−1.25056425	0.39884862	0.39815702
1.0	−0.33691753	−0.33650854	−3.01764179	−3.01945051	−0.29932294	−0.30116868
1.5	−2.41332734	−2.41345688	−5.40523279	−5.40844686	−0.92346873	−0.92675778
2.0	−5.89479008	−5.89590551	−8.70970537	−8.71450036	−1.32051165	−1.32544426

d.

t_i	w_{1i}	u_{1i}	w_{2i}	u_{2i}	w_{3i}	u_{3i}
0.2	1.38165297	1.38165325	1.00800000	1.00800000	−0.61833075	−0.61833075
0.5	1.90753116	1.90753184	1.12500000	1.12500000	−0.09090565	−0.09090566
0.7	2.25503524	2.25503620	1.34300000	1.34000000	0.26343971	0.26343970
1.0	2.83211921	2.83212056	2.00000000	2.00000000	0.88212058	0.88212056

3. The Runge-Kutta for Systems Algorithm gives the results in the following tables.

a.

t_i	w_{1i}	y_i
0.200	0.00015352	0.00015350
0.500	0.00742968	0.00743027
0.700	0.03299617	0.03299805
1.000	0.17132224	0.17132880

b.

t_i	w_{1i}	y_i
1.200	0.96152437	0.96152583
1.500	0.77796897	0.77797237
1.700	0.59373369	0.59373830
2.000	0.27258237	0.27258872

c.

t_i	w_{1i}	y_i
1.000	3.73162695	3.73170445
2.000	11.31424573	11.31452924
3.000	34.04395688	34.04517155

d.

t_i	w_{1i}	w_{2i}
1.200	0.27273759	0.27273791
1.500	1.08849079	1.08849259
1.700	2.04353207	2.04353642
2.000	4.36156675	4.36157780

5. To approximate the solution of the mth–order system of first–order initial–value problems

$$u_j' = f_j(t, u_1, u_2, \dots, u_m), \ \ j = 1, 2, \dots, m, \quad \text{for } \ a \le t \le b, \ u_j(a) = \alpha_j, \ j = 1, 2, \dots, m$$

at $(n+1)$ equally spaced numbers in the interval $[a, b]$;

INPUT endpoints a, b; number of equations m; integer N; initial conditions $\alpha_1, \dots, \alpha_m$.

OUTPUT approximations $w_{i,j}$ to $u_j(t_i)$.

Step 1 Set $h = (b - a)/N$;

Step 2 For $j = 1, 2, \dots, m$ set $w_{0,j} = \alpha_j$.

Step 3 OUTPUT $(t_0, w_{0,1}, w_{0,2}, \dots, w_{0,m})$.

Step 4 For $i = 1, 2, 3$ do Steps 5–11.

Step 5 For $j = 1, 2, \dots, m$ set

$$k_{1,j} = hf_j(t_{i-1}, w_{i-1,1}, \dots, w_{i-1,m}).$$

Step 6 For $j = 1, 2, \dots, m$ set

$$k_{2,j} = hf_j\left(t_{i-1} + \tfrac{h}{2}, w_{i-1,1} + \tfrac{1}{2}k_{1,1}, w_{i-1,2} + \tfrac{1}{2}k_{1,2}, \dots, w_{i-1,m} + \tfrac{1}{2}k_{1,m}\right).$$

Step 7 For $j = 1, 2, \ldots, m$ set

$$k_{3,j} = hf_j\left(t_{i-1} + \tfrac{h}{2}, w_{i-1,1} + \tfrac{1}{2}k_{2,1}, w_{i-1,2} + \tfrac{1}{2}k_{2,2}, \ldots, w_{i-1,m} + \tfrac{1}{2}k_{2,m}\right).$$

Step 8 For $j = 1, 2, \ldots, m$ set

$$k_{4,j} = hf_j(t_{i-1} + h, w_{i-1,1} + k_{3,1}, w_{i-1,2} + k_{3,2}, \ldots, w_{i-1,m} + k_{3,m}).$$

Step 9 For $j = 1, 2, \ldots, m$ set

$$w_{i,j} = w_{i-1,j} + (k_{1,j} + 2k_{2,j} + 2k_{3,j} + k_{4,j})/6.$$

Step 10 Set $t_i = a + ih$.

Step 11 OUTPUT $(t_i, w_{i,1}, w_{i,2}, \ldots, w_{i,m})$.

Step 12 For $i = 4, \ldots, N$ do Steps 13–16.

Step 13 Set $t_i = a + ih$.

Step 14 For $j = 1, 2, \ldots, m$ set

$$\begin{aligned} w_{i,j}^{(0)} =& w_{i-1,j} + h\big[55f_j(t_{i-1}, w_{i-1,1}, \ldots, w_{i-1,m}) - 59f_j(t_{i-2}, w_{i-2,1}, \ldots, w_{i-2,m}) \\ &+ 37f_j(t_{i-3}, w_{i-3,1}, \ldots, w_{i-3,m}) - 9f_j(t_{i-4}, w_{i-4,1}, \ldots, w_{i-4,m})\big]/24. \end{aligned}$$

Step 15 For $j = 1, 2, \ldots, m$ set

$$\begin{aligned} w_{i,j} =& w_{i-1,j} + h\Big[9f_j\left(t_i, w_{i,1}^{(0)}, \ldots, w_{i,m}^{(0)}\right) + 19f_j(t_{i-1}, w_{i-1,1}, \ldots, w_{i-1,m}) \\ &- 5f_j(t_{i-2}, w_{i-2,1}, \ldots, w_{i-2,m}) + f_j(t_{i-3}, w_{i-3,1}, \ldots, w_{i-3,m})\Big]/24. \end{aligned}$$

Step 16 OUTPUT $(t_i, w_{i,1}, w_{i,2}, \ldots, w_{i,m})$.

Step 17 STOP

7. The Adams fourth-order predictor-corrector method for systems applied to the problems in Exercise 1 gives the results in the following tables.

a.

t_i	w_{1i}	u_{1i}	w_{2i}	u_{2i}
0.500	0.95675505	0.95672798	−1.08385916	−1.08383310
1.000	1.30659995	1.30655930	−0.83300571	−0.83296776
1.500	1.34420613	1.34418117	−0.56983853	−0.56981634
2.000	1.14334795	1.14333672	−0.36938396	−0.36937457

b.

t_i	w_{1i}	u_{1i}	w_{2i}	u_{2i}
0.200	2.12036583	2.12500839	1.50699185	1.51158743
0.400	4.44122776	4.46511961	3.24224021	3.26598528
0.600	9.73913329	9.83235869	8.16341700	8.25629549
0.800	22.52673210	23.00263945	21.20273983	21.66887674
1.000	54.81242211	56.73748265	55.20490157	57.10536209

c.

t_i	w_{1i}	u_{1i}	w_{2i}	u_{2i}	w_{3i}	u_{3i}
0.5	0.70787076	0.70828683	−1.24988663	−1.25056425	0.39884862	0.39815702
1.0	−0.33691753	−0.33650854	−3.01764179	−3.01945051	−0.29932294	−0.30116868
1.5	−2.41332734	−2.41345688	−5.40523279	−5.40844686	−0.92346873	−0.92675778
2.0	−5.88968402	−5.89590551	−8.72213325	−8.71450036	−1.32972524	−1.32544426

d.

t_i	w_{1i}	u_{1i}	w_{2i}	u_{2i}	w_{3i}	u_{3i}
0.2	1.38165297	1.38165325	1.00800000	1.00800000	−0.61833075	−0.61833075
0.5	1.90752882	1.90753184	1.12500000	1.12500000	−0.09090527	−0.09090566
0.7	2.25503040	2.25503620	1.34300000	1.34300000	0.26344040	0.26343970
1.0	2.83211032	2.83212056	2.00000000	2.00000000	0.88212163	0.88212056

9. The predicted number of prey, x_{1i}, and predators, x_{2i}, are given in the following table.

i	t_i	x_{1i}	x_{2i}
10	1.0	4393	1512
20	2.0	288	3175
30	3.0	32	2042
40	4.0	25	1258

Exercise Set 5.10 (Page 347)

1. Let L be the Lipschitz constant for ϕ. Then

$$u_{i+1} - v_{i+1} = u_i - v_i + h[\phi(t_i, u_i, h) - \phi(t_i, v_i, h)],$$

so

$$|u_{i+1} - v_{i+1}| \le (1 + hL)|u_i - v_i| \le (1 + hL)^{i+1}|u_0 - v_0|.$$

3. By Exercise 32 in Section 5.4, we have

$$\begin{aligned}\phi(t, w, h) = {} & \frac{1}{6}f(t, w) + \frac{1}{3}f\left(t + \frac{1}{2}h, w + \frac{1}{2}hf(t, w)\right) \\ & + \frac{1}{3}f\left(t + \frac{1}{2}h, w + \frac{1}{2}hf\left(t + \frac{1}{2}h, w + \frac{1}{2}hf(t, w)\right)\right) \\ & + \frac{1}{6}f\left(t + h, w + hf\left(t + \frac{1}{2}h, w + \frac{1}{2}hf\left(t + \frac{1}{2}h, w + \frac{1}{2}hf(t, w)\right)\right)\right),\end{aligned}$$

so

$$\phi(t, w, 0) = \frac{1}{6}f(t, w) + \frac{1}{3}f(t, w) + \frac{1}{3}f(t, w) + \frac{1}{6}f(t, w) = f(t, w).$$

5. a. The local truncation error is $\tau_{i+1} = \frac{1}{4}h^3 y^{(4)}(\xi_i)$, for some ξ, where $t_{i-2} < \xi_i < t_{i+1}$.

b. The method is consistent but unstable and not convergent.

7. The method is unstable.

Exercise Set 5.11 (Page 354)

1. Euler's method gives the results in the following tables.

a.

t_i	w_i	y_i
0.200	0.027182818	0.449328964
0.500	0.000027183	0.030197383
0.700	0.000000272	0.004991594
1.000	0.000000000	0.000335463

b.

t_i	w_i	y_i
0.500	16.47925	0.479470939
1.000	256.7930	0.841470987
1.500	4096.142	0.997494987
2.000	65523.12	0.909297427

c.

t_i	w_i	y_i
0.200	0.373333333	0.046105213
0.500	−0.093333333	0.250015133
0.700	0.146666667	0.490000277
1.000	1.333333333	1.000000001

d.

t_i	w_i	y_i
0.200	6.128259	1.000000001
0.500	−378.2574	1.000000000
0.700	−6052.063	1.000000000
1.000	387332.0	1.000000000

3. The Runge-Kutta fourth order method gives the results in the following tables.

a.

t_i	w_i	y_i
0.200	0.45881186	0.44932896
0.500	0.03181595	0.03019738
0.700	0.00537013	0.00499159
1.000	0.00037239	0.00033546

b.

t_i	w_i	y_i
0.500	188.3082	0.47947094
1.000	35296.68	0.84147099
1.500	6632737	0.99749499
2.000	1246413200	0.90929743

c.

t_i	w_i	y_i
0.200	0.07925926	0.04610521
0.500	0.25386145	0.25001513
0.700	0.49265127	0.49000028
1.000	1.00250560	1.00000000

d.

t_i	w_i	y_i
0.200	−215.7459	1.00000000
0.500	−555750.0	1.00000000
0.700	−104435653	1.00000000
1.000	−269031268010	1.00000000

5. The Adams Fourth-Order Predictor-Corrector Algorithm gives the results in the following tables.

a.

t_i	w_i	y_i
0.200	0.4588119	0.4493290
0.500	−0.0112813	0.0301974
0.700	0.0013734	0.0049916
1.000	0.0023604	0.0003355

b.

t_i	w_i	y_i
.500	188.3082	0.4794709
1.000	38932.03	0.8414710
1.500	9073607	0.9974950
2.000	2115741299	0.9092974

c.

t_i	w_i	y_i
0.200	0.0792593	0.0461052
0.500	0.1554027	0.2500151
0.700	0.5507445	0.4900003
1.000	0.7278557	1.0000000

d.

t_i	w_i	y_i
0.200	−215.7459	1.000000001
0.500	−682637.0	1.000000000
0.700	−159172736	1.000000000
1.000	−566751172258	1.000000000

7. The Trapezoidal Algorithm gives the results in the following tables.

a.

t_i	w_i	k	y_i
0.200	0.39109643	2	0.44932896
0.500	0.02134361	2	0.03019738
0.700	0.00307084	2	0.00499159
1.000	0.00016759	2	0.00033546

b.

t_i	w_i	k	y_i
0.500	0.66291133	2	0.47947094
1.000	0.87506346	2	0.84147099
1.500	1.00366141	2	0.99749499
2.000	0.91053267	2	0.90929743

c.

t_i	w_i	k	y_i
0.200	0.04000000	2	0.04610521
0.500	0.25000000	2	0.25001513
0.700	0.49000000	2	0.49000028
1.000	1.00000000	2	1.00000000

d.

t_i	w_i	k	y_i
0.200	−1.07568307	4	1.00000000
0.500	−0.97868360	4	1.00000000
0.700	−0.99046408	3	1.00000000
1.000	−1.00284456	3	1.00000000

9. a.

t_i	w_{1i}	u_{1i}	w_{2i}	u_{2i}
0.100	−96.33011	0.66987648	193.6651	−0.33491554
0.200	−28226.32	0.67915383	56453.66	−0.33957692
0.300	−8214056	0.69387881	16428113	−0.34693941
0.400	−2390290586	0.71354670	4780581173	−0.35677335
0.500	−695574560790	0.73768711	1391149121600	−0.36884355

b.

t_i	w_{1i}	u_{1i}	w_{2i}	u_{2i}
0.100	0.61095960	0.66987648	−0.21708179	−0.33491554
0.200	0.66873489	0.67915383	−0.31873903	−0.33957692
0.300	0.69203679	0.69387881	−0.34325535	−0.34693941
0.400	0.71322103	0.71354670	−0.35612202	−0.35677335
0.500	0.73762953	0.73768711	−0.36872840	−0.36884355

11. Using (4.25) on page 199 gives $\tau_{i+1} = -\frac{1}{12}y'''(\xi_i)h^2$, for some $t_i < \xi_i < t_{i+1}$, and by Definition 5.18, the Trapezoidal method is consistent. Once again using (4.25) gives

$$y(t_{i+1}) = y(t_i) + \frac{h}{2}\left[f(t_i, y(t_i)) + f(t_{i+1}, y(t_{i+1}))\right] - \frac{y'''(\xi_i)}{12}h^3.$$

Subtracting the difference equation and using the Lipschitz constant L for f gives

$$|y(t_{i+1}) - w_{i+1}| \leq |y(t_i) - w_i| + \frac{hL}{2}|y(t_i) - w_i| + \frac{hL}{2}|y(t_{i+1}) - w_{i+1}| + \frac{h^3}{12}\left|y'''(\xi_i)\right|.$$

Let $M = \max_{a \leq x \leq b} |y'''(x)|$. Then, assuming $hL \neq 2$,

$$|y(t_{i+1}) - w_{i+1}| \leq \frac{2 + hL}{2 - hL}|y(t_i) - w_i| + \frac{h^3}{6(2 - hL)}M.$$

Using Lemma 5.8 on page 270 gives

$$|y(t_{i+1}) - w_{i+1}| \leq e^{2(b-a)L/(2-hL)}\left[\frac{Mh^2}{12L} + |\alpha - w_0|\right] - \frac{Mh^2}{12L}.$$

Thus, if $hL \neq 2$, the Trapezoidal method is convergent, and consequently stable.

13. b. The following tables list the results of the Backward Euler method applied to the problems in Exercise 1.

a.

i	t_i	w_i	k	y_i
2	0.20	0.75298666	2	0.44932896
5	0.50	0.10978082	2	0.03019738
7	0.70	0.03041020	2	0.00499159
10	1.00	0.00443362	2	0.00033546

b.

i	t_i	w_i	k	y_i
2	0.50	0.50495522	2	0.47947094
4	1.00	0.83751817	2	0.84147099
6	1.50	0.99145076	2	0.99749499
8	2.00	0.90337560	2	0.90929743

c.

i	t_i	w_i	k	y_i
2	0.20	0.08148148	2	0.04610521
5	0.50	0.25635117	2	0.25001513
7	0.70	0.49515013	2	0.49000028
10	1.00	1.00500556	2	1.00000000

d.

i	t_i	w_i	k	y_i
2	0.20	1.00348713	3	1.00000000
5	0.50	1.00000262	2	1.00000000
7	0.70	1.00000002	1	1.00000000
10	1.00	1.00000000	1	1.00000000

15. a. The Trapezoidal method applied to the test equation gives

$$w_{j+1} = \frac{1+\frac{h\lambda}{2}}{1-\frac{h\lambda}{2}} w_j, \quad \text{so} \quad Q(h\lambda) = \frac{2+h\lambda}{2-h\lambda}.$$

Thus, $|Q(h\lambda)| < 1$, whenever $\text{Re}(h\lambda) < 0$.

b. The Backward Euler method applied to the test equation gives

$$w_{j+1} = \frac{w_j}{1-h\lambda}, \quad \text{so} \quad Q(h\lambda) = \frac{1}{1-h\lambda}.$$

Thus, $|Q(h\lambda)| < 1$, whenever $\text{Re}(h\lambda) < 0$.

Exercise Set 6.1 (Page 368)

1. a. One line, so there is an infinite number of solutions with $x_2 = \frac{3}{2} - \frac{1}{2}x_1$.

b. Intersecting lines with solution $x_1 = x_2 = 1$.

c. One line, so there is an infinite number of solutions with $x_2 = -\frac{1}{2}x_1$.

d. Intersecting lines with solution $x_1 = \frac{2}{7}$ and $x_2 = -\frac{11}{7}$.

3. a. $x_1 = 1.1$, $x_2 = -1.1$, $x_3 = 2.9$ **b.** $x_1 = 1.0$, $x_2 = -0.98$, $x_3 = 2.9$

5. Gaussian elimination gives the following solutions.

a. $x_1 = 1.1875, x_2 = 1.8125, x_3 = 0.875$ with one row interchange required

b. $x_1 = -1, x_2 = 0, x_3 = 1$ with no interchange required

c. $x_1 = 1.5, x_2 = 2, x_3 = -1.2, x_4 = 3$ with no interchange required

d. No unique solution

7. Gaussian elimination with single precision arithmetic gives the following solutions:

a. $x_1 = -227.0769$, $x_2 = 476.9231$, $x_3 = -177.6923$;

b. $x_1 = 1.001291$, $x_2 = 1$, $x_3 = 1.00155$;

c. $x_1 = -0.03174600$, $x_2 = 0.5952377$, $x_3 = -2.380951$, $x_4 = 2.777777$;

d. $x_1 = 1.918129$, $x_2 = 1.964912$, $x_3 = -0.9883041$, $x_4 = -3.192982$, $x_5 = -1.134503$.

9. a. When $\alpha = -1/3$, there is no solution.

b. When $\alpha = 1/3$, there is an infinite number of solutions with $x_1 = x_2 + 1.5$, and x_2 is arbitrary.

c. If $\alpha \neq \pm 1/3$, then the unique solution is

$$x_1 = \frac{3}{2(1+3\alpha)} \quad \text{and} \quad x_2 = \frac{-3}{2(1+3\alpha)}.$$

13. The Gauss-Jordan method gives the following results.

a. $x_1 = 1.1, x_2 = -1.0, x_3 = 2.9$ **b.** $x_1 = 0.98, x_2 = -0.98, x_3 = 2.9$

15. b. The results for this exercise are listed in the following table. (The abbreviations M/D and A/S are used for multiplications/divisions and additions/subtractions, respectively.)

	Gaussian Elimination		Gauss-Jordan	
n	M/D	A/S	M/D	A/S
3	17	11	21	12
10	430	375	595	495
50	44150	42875	64975	62475
100	343300	338250	509950	499950

17. The Gaussian-Elimination–Gauss-Jordan hybrid method gives the following results.

a. $x_1 = 1.0, x_2 = -1.0, x_3 = 2.9$ **b.** $x_1 = 1.0, x_2 = -0.98, x_3 = 2.9$

19. a. There is sufficient food to satisfy the average daily consumption.

b. We could add 200 of species 1, or 150 of species 2, or 100 of species 3, or 100 of species 4.

c. Assuming none of the increases indicated in part (b) was selected, species 2 could be increased by 650, or species 3 could be increased by 150, or species 4 could be increased by 150.

d. Assuming none of the increases indicated in parts (b) or (c) were selected, species 3 could be increased by 150, or species 4 could be increased by 150.

Exercise Set 6.2 (Page 379)

1. a. none **b.** none

c. Interchange rows 2 and 3. **d.** Interchange rows 1 and 2.

3. a. Interchange rows 1 and 2. **b.** Interchange rows 1 and 2, then interchange rows 2 and 3.

c. Interchange rows 1 and 3. **d.** Interchange rows 1 and 2.

5. a. Interchange rows 1 and 3, then interchange rows 2 and 3. **b.** Interchange rows 2 and 3.

c. Interchange rows 2 and 3. **d.** Interchange rows 1 and 3, then interchange rows 2 and 3.

7. a. Interchange rows 1 and 2, and columns 1 and 3, then interchange rows 2 and 3, and columns 2 and 3.

b. Interchange rows 1 and 2, and columns 1 and 3, then interchange rows 2 and 3.

c. Interchange rows 1 and 2, and columns 1 and 3, then interchange rows 2 and 3.

d. Interchange rows 1 and 2, and columns 1 and 2, then interchange rows 2 and 3; and columns 2 and 3.

9. Gaussian elimination with three-digit chopping arithmetic gives the following results.

a. $x_1 = 30.0, x_2 = 0.990$ **b.** $x_1 = 0.00, x_2 = 10.0, x_3 = 0.142$

c. $x_1 = 0.206, x_2 = 0.0154, x_3 = -0.0156, x_4 = -0.716$ **d.** $x_1 = 0.828, x_2 = -3.32, x_3 = 0.153, x_4 = 4.91$

11. Gaussian elimination with three-digit rounding arithmetic gives the following results.

a. $x_1 = -10.0,\ x_2 = 1.01$ **b.** $x_1 = 0.00,\ x_2 = 10.0,\ x_3 = 0.143$

c. $x_1 = 0.185,\ x_2 = 0.0103,\ x_3 = -0.0200,\ x_4 = -1.12$ **d.** $x_1 = 0.799,\ x_2 = -3.12,\ x_3 = 0.151,\ x_4 = 4.56$

13. Gaussian elimination with partial pivoting and three-digit chopping arithmetic gives the following results.

a. $x_1 = 10.0,\ x_2 = 1.00$ **b.** $x_1 = -0.163,\ x_2 = 9.98, x_3 = 0.142$

c. $x_1 = 0.177,\ x_2 = -0.0072,\ x_3 = -0.0208,\ x_4 = -1.18$ **d.** $x_1 = 0.777,\ x_2 = -3.10,\ x_3 = 0.161,\ x_4 = 4.50$

15. Gaussian elimination with partial pivoting and three-digit rounding arithmetic gives the following results.

a. $x_1 = 10.0,\ x_2 = 1.00$ **b.** $x_1 = 0.00,\ x_2 = 10.0,\ x_3 = 0.143$

c. $x_1 = 0.178,\ x_2 = 0.0127,\ x_3 = -0.0204,\ x_4 = -1.16$ **d.** $x_1 = 0.845,\ x_2 = -3.37,\ x_3 = 0.182,\ x_4 = 5.07$

17. Gaussian elimination with scaled partial pivoting and three-digit chopping arithmetic gives the following results.

a. $x_1 = 10.0,\ x_2 = 1.00$ **b.** $x_1 = -0.163,\ x_2 = 9.98,\ x_3 = 0.142$

c. $x_1 = 0.171,\ x_2 = 0.0102,\ x_3 = -0.0217,\ x_4 = -1.27$ **d.** $x_1 = 0.687,\ x_2 = -2.66,\ x_3 = 0.117,\ x_4 = 3.59$

19. Gaussian elimination with scaled partial pivoting and three-digit rounding arithmetic gives the following results.

a. $x_1 = 10.0,\ x_2 = 1.00$ **b.** $x_1 = 0.00,\ x_2 = 10.0,\ x_3 = 0.143$

c. $x_1 = 0.180,\ x_2 = 0.0128,\ x_3 = -0.0200,\ x_4 = -1.13$ **d.** $x_1 = 0.783,\ x_2 = -3.12,\ x_3 = 0.147,\ x_4 = 4.53$

21. Using Algorithm 6.1 in Maple with *Digits:=10* gives

a. $x_1 = 10.00000000, x_2 = 1.000000000$

b. $x_1 = 0.000000033, x_2 = 10.00000001, x_3 = 0.1428571429$

c. $x_1 = 0.1768252958, x_2 = 0.0126926913, x_3 = -0.0206540503, x_4 = -1.182608714$

d. $x_1 = 0.7883937842, x_2 = -3.125413672, x_3 = 0.1675965951, x_4 = 4.557002521$

23. Using Algorithm 6.2 in Maple with *Digits:=10* gives

a. $x_1 = 10.00000000, x_2 = 1.000000000$ **b.** $x_1 = 0.000000000, x_2 = 10.00000000, x_3 = 0.142857142$

c. $x_1 = 0.1768252975, x_2 = 0.0126926909, x_3 = -0.0206540502, x_4 = -1.182608696$

d. $x_1 = 0.7883937863, x_2 = -3.125413680, x_3 = 0.1675965980, x_4 = 4.557002510$

25. Using Algorithm 6.3 in Maple with *Digits:=10* gives

a. $x_1 = 10.00000000, x_2 = 1.000000000$

b. $x_1 = 0.000000000, x_2 = 10.00000000, x_3 = 0.1428571429$

c. $x_1 = 0.1768252977, x_2 = 0.0126926909, x_3 = -0.0206540501, x_4 = -1.182608693$

d. $x_1 = 0.7883937842, x_2 = -3.125413672, x_3 = 0.1675965952, x_4 = 4.55700252$

27. a. $x_1 = 9.98, x_2 = 1.00$ **b.** $x_1 = 0.0724, x_2 = 10.0, x_3 = 0.0952$

c. $x_1 = 0.161, x_2 = 0.0125, x_3 = -0.0232, x_4 = -1.42$ **d.** $x_1 = 0.719, x_2 = -2.86, x_3 = 0.146, x_4 = 4.00$

29. a. $x_1 = 10.0, x_2 = 1.00$ **b.** $x_1 = 0.00, x_2 = 10.0, x_3 = 0.143$

c. $x_1 = 0.179, x_2 = 0.0127, x_3 = -0.0203, x_4 = -1.15$ **d.** $x_1 = 0.874, x_2 = -3.49, x_3 = 0.192, x_4 = 5.33$

31. Only for (a), where $\alpha = 6$.

33. Using the Complete Pivoting Algorithm in Maple with *Digits:=10* gives

a. $x_1 = 10.00000000, x_2 = 1.000000000$

b. $x_1 = 0.000000000, x_2 = 10.00000000, x_3 = 0.1428571429$

c. $x_1 = 0.1768252974, x_2 = 0.01269269087, x_3 = -0.02065405015, x_4 = -1.182608697$

d. $x_1 = 0.17883937840, x_2 = -3.125413669, x_3 = 0.1675965971, x_4 = 4.557002516$

Exercise Set 6.3 (Page 390)

1. a. $\begin{bmatrix} 3 \\ 0 \end{bmatrix}$ **b.** $\begin{bmatrix} 1 \\ 2 \end{bmatrix}$ **c.** $\begin{bmatrix} 9 \\ -1 \\ 14 \end{bmatrix}$ **d.** $\begin{bmatrix} 10 & -9 & -4 \end{bmatrix}$

3. a. $\begin{bmatrix} 11 & 4 & -8 \\ 6 & 13 & -12 \end{bmatrix}$ **b.** $\begin{bmatrix} -4 & 10 \\ 1 & 15 \end{bmatrix}$ **c.** $\begin{bmatrix} -1 & 5 & -3 \\ 3 & 4 & -11 \\ -6 & -7 & -4 \end{bmatrix}$ **d.** $\begin{bmatrix} -2 & 1 \\ -14 & 7 \\ 6 & 1 \end{bmatrix}$

5. a. The matrix is singular. **b.** $\begin{bmatrix} -\frac{1}{4} & \frac{1}{4} & \frac{1}{4} \\ \frac{5}{8} & -\frac{1}{8} & -\frac{1}{8} \\ \frac{1}{8} & -\frac{5}{8} & \frac{3}{8} \end{bmatrix}$ **c.** The matrix is singular. **d.** $\begin{bmatrix} \frac{1}{4} & 0 & 0 & 0 \\ -\frac{3}{14} & \frac{1}{7} & 0 & 0 \\ \frac{3}{28} & -\frac{11}{7} & 1 & 0 \\ -\frac{1}{2} & 1 & -1 & 1 \end{bmatrix}$

7. The solutions to the linear systems obtained in parts (a) and (b) are, from left to right,

$$3, -6, -2, -1 \quad \text{and} \quad 1, 1, 1, 1.$$

9. a. Suppose $\tilde{A}$ and $\hat{A}$ are both inverses of A. Then $A\tilde{A} = \tilde{A}A = I$ and $A\hat{A} = \hat{A}A = I$. Thus,

$$\tilde{A} = \tilde{A}I = \tilde{A}(A\hat{A}) = (\tilde{A}A)\hat{A} = I\hat{A} = \hat{A}.$$

b. $(AB)(B^{-1}A^{-1}) = A(BB^{-1})A^{-1} = AIA^{-1} = AA^{-1} = I$ and $(B^{-1}A^{-1})(AB) = B^{-1}(A^{-1}A)B = B^{-1}IB = B^{-1}B = I$, so $(AB)^{-1} = B^{-1}A^{-1}$ since there is only one inverse.

c. Since $A^{-1}A = AA^{-1} = I$, it follows that A^{-1} is nonsingular. Since the inverse is unique, we have $(A^{-1})^{-1} = A$.

11. a. If $C = AB$, where A and B are lower triangular, then $a_{ik} = 0$ if $k > i$ and $b_{kj} = 0$ if $k < j$. Thus,

$$c_{ij} = \sum_{k=1}^{n} a_{ik}b_{kj} = \sum_{k=j}^{i} a_{ik}b_{kj},$$

which will have the sum zero unless $j \le i$. Hence C is lower triangular.

b. We have $a_{ik} = 0$ if $k < i$ and $b_{kj} = 0$ if $k > j$. The steps are similar to those in part (a).

c. Let L be a nonsingular lower triangular matrix. To obtain the ith column of L^{-1}, solve n linear systems of the form

$$\begin{bmatrix} l_{11} & 0 & \cdots & & \cdots & 0 \\ l_{21} & l_{22} & \ddots & & & \vdots \\ \vdots & \vdots & & \ddots & & \vdots \\ l_{i1} & l_{i2} & \cdots & l_{ii} & \ddots & \vdots \\ \vdots & \vdots & & & \ddots & 0 \\ l_{n1} & l_{n2} & \cdots & & \cdots & l_{nn} \end{bmatrix} \begin{bmatrix} x_1 \\ x_2 \\ \vdots \\ x_i \\ \vdots \\ x_n \end{bmatrix} = \begin{bmatrix} 0 \\ 0 \\ \vdots \\ 0 \\ 1 \\ 0 \\ \vdots \\ 0 \end{bmatrix},$$

where the 1 appears in the ith position to obtain the ith column of L^{-1}.

13. The answers are the same as those in Exercise 5.

15. a. $A^2 = \begin{bmatrix} 0 & 2 & 0 \\ 0 & 0 & 3 \\ \frac{1}{6} & 0 & 0 \end{bmatrix}$, $\quad A^3 = \begin{bmatrix} 1 & 0 & 0 \\ 0 & 1 & 0 \\ 0 & 0 & 1 \end{bmatrix}$, $\quad A^4 = A, \quad A^5 = A^2, \quad A^6 = I, \ldots$

b.

	Year 1	Year 2	Year 3	Year 4
Age 1	6000	36000	12000	6000
Age 2	6000	3000	18000	6000
Age 3	6000	2000	1000	6000

c. $A^{-1} = \begin{bmatrix} 0 & 2 & 0 \\ 0 & 0 & 3 \\ \frac{1}{6} & 0 & 0 \end{bmatrix}$. The i,j-entry is the number of beetles of age i necessary to produce one beetle of age j.

17. a. We have

$$\begin{bmatrix} 7 & 4 & 4 & 0 \\ -6 & -3 & -6 & 0 \\ 0 & 0 & 3 & 0 \\ 0 & 0 & 0 & 1 \end{bmatrix} \begin{bmatrix} 2(x_0 - x_1) + \alpha_0 + \alpha_1 \\ 3(x_1 - x_0) - \alpha_1 - 2\alpha_0 \\ \alpha_0 \\ x_0 \end{bmatrix} = \begin{bmatrix} 2(x_0 - x_1) + 3\alpha_0 + 3\alpha_1 \\ 3(x_1 - x_0) - 3\alpha_1 - 6\alpha_0 \\ 3\alpha_0 \\ x_0 \end{bmatrix}$$

b. $B = A^{-1} = \begin{bmatrix} -1 & -\frac{4}{3} & -\frac{4}{3} & 0 \\ 2 & \frac{7}{3} & 2 & 0 \\ 0 & 0 & \frac{1}{3} & 0 \\ 0 & 0 & 0 & 1 \end{bmatrix}$

Exercise Set 6.4 (Page 399)

1. The determinants of the matrices are:

a. 8 **b.** −8 **c.** 0 **d.** 0

3. The answers are the same as in Exercise 1.

5. $\alpha = -\frac{3}{2}$ and $\alpha = 2$

7. $\alpha = -5$

9. When $n = 2$, $\det A = a_{11}a_{22} - a_{12}a_{21}$ requires 2 Multiplications and 1 Subtraction. Since

$$2!\sum_{k=1}^{1}\frac{1}{k!} = 2 \quad \text{and} \quad 2! - 1 = 1,$$

the formula holds for $n = 2$. Assume the formula is true for $n = 2, \ldots, m$, and let A be an $(m+1) \times (m+1)$ matrix. Then

$$\det A = \sum_{j=1}^{m+1} a_{ij}A_{ij},$$

for any i, where $1 \le i \le m+1$. To compute each A_{ij} requires

$$m!\sum_{k=1}^{m-1}\frac{1}{k!} \quad \text{Multiplications} \quad \text{and} \quad m! - 1 \quad \text{Additions/Subtractions.}$$

Thus, the number of Multiplications for $\det A$ is

$$(m+1)\left[m!\sum_{k=1}^{m-1}\frac{1}{k!}\right] + (m+1) = (m+1)!\left[\sum_{k=1}^{m-1}\frac{1}{k!} + \frac{1}{m!}\right] = (m+1)!\sum_{k=1}^{m}\frac{1}{k!},$$

and the number of Additions/Subtractions is

$$(m+1)\,[m! - 1] + m = (m+1)! - 1.$$

By the principle of mathematical induction, the formula is valid for any $n \ge 2$.

11. The result follows from $\det AB = \det A \cdot \det B$ and Theorem 6.17.

13. a. If D_i is the determinant of the matrix formed by replacing the ith column of A with $\mathbf{b}$ and if $D = \det A$, then

$$x_i = D_i/D, \quad \text{for } i = 1, \ldots, n.$$

b. $(n+1)!\left(\sum_{k=1}^{n-1}\frac{1}{k!}\right) + n$ Multiplications/Divisions
$(n+1)! - n - 1$ Additions/Subtractions.

Exercise Set 6.5 (Page 409)

1. a. $x_1 = 11/20$, $x_2 = 3/10$, $x_3 = 2/5$ **b.** $x_1 = 176$, $x_2 = -50$, $x_3 = 24$

3. a. $P = \begin{bmatrix} 0 & 1 & 0 \\ 1 & 0 & 0 \\ 0 & 0 & 1 \end{bmatrix}$ **b.** $P = \begin{bmatrix} 1 & 0 & 0 \\ 0 & 0 & 1 \\ 0 & 1 & 0 \end{bmatrix}$ **c.** $P = \begin{bmatrix} 1 & 0 & 0 & 0 \\ 0 & 0 & 0 & 1 \\ 0 & 0 & 1 & 0 \\ 0 & 1 & 0 & 0 \end{bmatrix}$ **d.** $P = \begin{bmatrix} 1 & 0 & 0 & 0 \\ 0 & 0 & 1 & 0 \\ 0 & 1 & 0 & 0 \\ 0 & 0 & 0 & 1 \end{bmatrix}$

5. a. $L = \begin{bmatrix} 1 & 0 & 0 \\ 1.5 & 1 & 0 \\ 1.5 & 1 & 1 \end{bmatrix}$ and $U = \begin{bmatrix} 2 & -1 & 1 \\ 0 & 4.5 & 7.5 \\ 0 & 0 & -4 \end{bmatrix}$

b. $L = \begin{bmatrix} 1 & 0 & 0 \\ -2.106719 & 1 & 0 \\ 3.067193 & 1.197756 & 1 \end{bmatrix}$ and $U = \begin{bmatrix} 1.012 & -2.132 & 3.104 \\ 0 & -0.3955257 & -0.4737443 \\ 0 & 0 & -8.939141 \end{bmatrix}$

c. $L = \begin{bmatrix} 1 & 0 & 0 & 0 \\ 0.5 & 1 & 0 & 0 \\ 0 & -2 & 1 & 0 \\ 1 & -1.33333 & 2 & 1 \end{bmatrix}$ and $U = \begin{bmatrix} 2 & 0 & 0 & 0 \\ 0 & 1.5 & 0 & 0 \\ 0 & 0 & 0.5 & 0 \\ 0 & 0 & 0 & 1 \end{bmatrix}$

d. $L = \begin{bmatrix} 1 & 0 & 0 & 0 \\ -1.849190 & 1 & 0 & 0 \\ -0.4596433 & -0.2501219 & 1 & 0 \\ 2.768661 & -0.3079435 & -5.352283 & 1 \end{bmatrix}$ and $U = \begin{bmatrix} 2.175600 & 4.023099 & -2.173199 & 5.196700 \\ 0 & 13.43947 & -4.018660 & 10.80698 \\ 0 & 0 & -0.8929510 & 5.091692 \\ 0 & 0 & 0 & 12.03614 \end{bmatrix}$

7. a. $x_1 = 1, x_2 = 2, x_3 = -1$ **b.** $x_1 = 1, x_2 = 1, x_3 = 1$

c. $x_1 = 1.5, x_2 = 2, x_3 = -1.199998, x_4 = 3$

d. $x_1 = 2.939851, x_2 = 0.07067770, x_3 = 5.677735, x_4 = 4.379812$

9. a. $P^tLU = \begin{bmatrix} 0 & 1 & 0 \\ 1 & 0 & 0 \\ 0 & 0 & 1 \end{bmatrix} \begin{bmatrix} 1 & 0 & 0 \\ 0 & 1 & 0 \\ 0 & -\frac{1}{2} & 1 \end{bmatrix} \begin{bmatrix} 1 & 1 & -1 \\ 0 & 2 & 3 \\ 0 & 0 & \frac{5}{2} \end{bmatrix}$ **b.** $P^tLU = \begin{bmatrix} 1 & 0 & 0 \\ 0 & 0 & 1 \\ 0 & 1 & 0 \end{bmatrix} \begin{bmatrix} 1 & 0 & 0 \\ 2 & 1 & 0 \\ 1 & 0 & 1 \end{bmatrix} \begin{bmatrix} 1 & 2 & -1 \\ 0 & -5 & 6 \\ 0 & 0 & 4 \end{bmatrix}$

c. $P^tLU = \begin{bmatrix} 1 & 0 & 0 & 0 \\ 0 & 0 & 0 & 1 \\ 0 & 1 & 0 & 0 \\ 0 & 0 & 1 & 0 \end{bmatrix} \begin{bmatrix} 1 & 0 & 0 & 0 \\ 2 & 1 & 0 & 0 \\ 1 & 0 & 1 & 0 \\ 3 & 0 & 0 & 1 \end{bmatrix} \begin{bmatrix} 1 & -2 & 3 & 0 \\ 0 & 5 & -2 & 1 \\ 0 & 0 & -1 & -2 \\ 0 & 0 & 0 & 3 \end{bmatrix}$

d. $P^tLU = \begin{bmatrix} 1 & 0 & 0 & 0 \\ 0 & 0 & 0 & 1 \\ 0 & 0 & 1 & 0 \\ 0 & 1 & 0 & 0 \end{bmatrix} \begin{bmatrix} 1 & 0 & 0 & 0 \\ 2 & 1 & 0 & 0 \\ 1 & 0 & 1 & 0 \\ 1 & 0 & 0 & 1 \end{bmatrix} \begin{bmatrix} 1 & -2 & 3 & 0 \\ 0 & 5 & -3 & -1 \\ 0 & 0 & -1 & -2 \\ 0 & 0 & 0 & 1 \end{bmatrix}$

11. c.

	Multiplications/Divisions	Additions/Subtractions
Factoring into LU	$\frac{1}{3}n^3 - \frac{1}{3}n$	$\frac{1}{3}n^3 - \frac{1}{2}n^2 + \frac{1}{6}n$
Solving $Ly = b$	$\frac{1}{2}n^2 - \frac{1}{2}n$	$\frac{1}{2}n^2 - \frac{1}{2}n$
Solving $Ux = y$	$\frac{1}{2}n^2 + \frac{1}{2}n$	$\frac{1}{2}n^2 - \frac{1}{2}n$
Total	$\frac{1}{3}n^3 + n^2 - \frac{1}{3}n$	$\frac{1}{3}n^3 + \frac{1}{2}n^2 - \frac{5}{6}n$

d.

	Multiplications/Divisions	Additions/Subtractions
Factoring into LU	$\frac{1}{3}n^3 - \frac{1}{3}n$	$\frac{1}{3}n^3 - \frac{1}{2}n^2 + \frac{1}{6}n$
Solving $Ly^{(k)} = b^{(k)}$	$(\frac{1}{2}n^2 - \frac{1}{2}n)m$	$(\frac{1}{2}n^2 - \frac{1}{2}n)m$
Solving $Ux^{(k)} = y^{(k)}$	$(\frac{1}{2}n^2 + \frac{1}{2}n)m$	$(\frac{1}{2}n^2 - \frac{1}{2}n)m$
Total	$\frac{1}{3}n^3 + mn^2 - \frac{1}{3}n$	$\frac{1}{3}n^3 + (m - \frac{1}{2})n^2 - (m - \frac{1}{6})n$

Exercise Set 6.6 (Page 425)

1. i. Matrices (a) and (c) are symmetric.

ii. Matrices (a), (b), and (c) are nonsingular.

iii. Matrices (a) and (b) are strictly diagonally dominant.

iv. Matrices (b) and (c) are positive definite.

3. The LDL^t factorization of the matrices A have the following forms.

a. $L = \begin{bmatrix} 1 & 0 & 0 \\ -0.25 & 1 & 0 \\ 0.25 & 0.09090909 & 1 \end{bmatrix}, \quad D = \begin{bmatrix} 4 & 0 & 0 \\ 0 & 2.75 & 0 \\ 0 & 0 & 1.72727273 \end{bmatrix}$

b. $L = \begin{bmatrix} 1 & 0 & 0 \\ 0.5 & 1 & 0 \\ 0.5 & 0.2 & 1 \end{bmatrix}, \quad D = \begin{bmatrix} 4 & 0 & 0 \\ 0 & 5 & 0 \\ 0 & 0 & 3.6 \end{bmatrix}$

c. $L = \begin{bmatrix} 1 & 0 & 0 & 0 \\ 0 & 1 & 0 & 0 \\ 0.5 & -0.3333333 & 1 & 0 \\ 0.25 & 0.3333333 & 0.60714286 & 1 \end{bmatrix}, \quad D = \begin{bmatrix} 4 & 0 & 0 & 0 \\ 0 & 3 & 0 & 0 \\ 0 & 0 & 4.666667 & 0 \\ 0 & 0 & 0 & 5.696429 \end{bmatrix}$

d. $L = \begin{bmatrix} 1 & 0 & 0 & 0 \\ 0.25 & 1 & 0 & 0 \\ 0.25 & -0.9090909 & 1 & 0 \\ 0.25 & -0.4545455 & 0.3684211 & 1 \end{bmatrix}$, $D = \begin{bmatrix} 4 & 0 & 0 & 0 \\ 0 & 2.75 & 0 & 0 \\ 0 & 0 & 1.727273 & 0 \\ 0 & 0 & 0 & 2.947368 \end{bmatrix}$

5. Cholesky's Algorithm gives the following results.

a. $L = \begin{bmatrix} 2 & 0 & 0 \\ -1/2 & \sqrt{11}/2 & 0 \\ 1/2 & \sqrt{11}/22 & \sqrt{209}/11 \end{bmatrix}$ **b.** $L = \begin{bmatrix} 2 & 0 & 0 \\ 1 & \sqrt{5} & 0 \\ 1 & \sqrt{5}/5 & \sqrt{95}/5 \end{bmatrix}$

c. $L = \begin{bmatrix} 2 & 0 & 0 & 0 \\ 0 & \sqrt{3} & 0 & 0 \\ 1 & -\sqrt{3}/3 & \sqrt{42}/3 & 0 \\ 1/20 & \sqrt{3}/3 & 17\sqrt{42}/84 & \sqrt{4466}/28 \end{bmatrix}$ **d.** $L = \begin{bmatrix} 2 & 0 & 0 & 0 \\ 1/2 & \sqrt{11}/2 & 0 & 0 \\ 1/2 & -\sqrt{11}/22 & \sqrt{209}/11 & 0 \\ 1/2 & -5\sqrt{11}/22 & 7\sqrt{209}/209 & 2\sqrt{266}/19 \end{bmatrix}$

7. The modified factorization algorithm gives the following results.

a. $x_1 = 1,\ x_2 = -1,\ x_3 = 0$ **b.** $x_1 = 0.2,\ x_2 = -0.2,\ x_3 = -0.2,\ x_4 = 0.25$

c. $x_1 = 1,\ x_2 = 2,\ x_3 = -1,\ x_4 = 2$ **d.** $x_1 = -0.8586387,\ x_2 = 2.418848,\ x_3 = -0.9581152,\ x_4 = -1.272251$

9. The modified Cholesky's algorithm gives the following results.

a. $x_1 = 1, x_2 = -1, x_3 = 0$ **b.** $x_1 = 0.2, x_2 = -0.2, x_3 = -0.2, x_4 = 0.25$

c. $x_1 = 1, x_2 = 2, x_3 = -1, x_4 = 2$ **d.** $x_1 = -0.85863874,\ x_2 = 2.4188482,\ x_3 = -0.95811518,\ x_4 = -1.2722513$

11. The Crout Factorization Algorithm gives the following results.

a. $x_1 = 0.5,\ x_2 = 0.5,\ x_3 = 1$ **b.** $x_1 = -0.9999995,\ x_2 = 1.999999,\ x_3 = 1$

c. $x_1 = 1,\ x_2 = -1,\ x_3 = 0$ **d.** $x_1 = -0.09357798,\ x_2 = 1.587156,\ x_3 = -1.167431,\ x_4 = 0.5412844$

13. We have $x_i = 1$, for each $i = 1, \ldots, 10$.

15. Only the matrix in (d) is positive definite.

17. $-2 < \alpha < \frac{3}{2}$

19. $0 < \beta < 1$ and $3 < \alpha < 5 - \beta$

21. **a.** No, for example, consider $\begin{bmatrix} 1 & 0 \\ 0 & 1 \end{bmatrix}$.

b. Yes, since $A = A^t$.

c. Yes, since $\mathbf{x}^t(A + B)\mathbf{x} = \mathbf{x}^t A\mathbf{x} + \mathbf{x}^t B\mathbf{x}$.

d. Yes, since $\mathbf{x}^t A^2 \mathbf{x} = \mathbf{x}^t A^t A\mathbf{x} = (A\mathbf{x})^t(A\mathbf{x}) \geq 0$, and because A is nonsingular, equality holds only if $\mathbf{x} = \mathbf{0}$.

e. No, for example, consider $A = \begin{bmatrix} 1 & 0 \\ 0 & 1 \end{bmatrix}$ and $B = \begin{bmatrix} 10 & 0 \\ 0 & 10 \end{bmatrix}$.

23. **a.** Since $\det A = 3\alpha - 2\beta$, A is singular if and only if $\alpha = 2\beta/3$. **b.** $|\alpha| > 1, |\beta| < 1$

c. $\beta = 1$ **d.** $\alpha > \frac{2}{3}, \beta = 1$

25. One example is $A = \begin{bmatrix} 1.0 & 0.2 \\ 0.1 & 1.0 \end{bmatrix}$.

27. The Crout Factorization Algorithm can be rewritten as follows:

Step 1 Set $l_1 = a_1;\ u_1 = c_1/l_1$.
Step 2 For $i = 2, \ldots, n-1$ set $l_i = a_i - b_i u_{i-1};\ u_i = c_i/l_i$.
Step 3 Set $l_n = a_n - b_n u_{n-1}$.
Step 4 Set $z_1 = d_1/l_1$.
Step 5 For $i = 2, \ldots, n$ set $z_i = (d_i - b_i z_{i-1})/l_i$.
Step 6 Set $x_n = z_n$.
Step 7 For $i = n-1, \ldots, 1$ set $x_i = z_i - u_i x_{i+1}$.
Step 8 OUTPUT $(x_1, \ldots, x_n)$;
STOP.

29. $i_1 = 0.6785047,\quad i_2 = 0.4214953,\quad i_3 = 0.2570093,\quad i_4 = 0.1542056,\quad i_5 = 0.1028037$

31. **a.** Mating male i with female j produces offspring with the same wing characteristics as mating male j with female i.

b. No. Consider, for example, $\mathbf{x} = (1, 0, -1)^t$.

Exercise Set 7.1 (Page 441)

1. a. We have $||\mathbf{x}||_\infty = 4$ and $||\mathbf{x}||_2 = 5.477226$.

b. We have $||\mathbf{x}||_\infty = 4$ and $||\mathbf{x}||_2 = 5.220153$.

c. We have $||\mathbf{x}||_\infty = 2^k$ and $||\mathbf{x}||_2 = (1+4^k)^{1/2}$.

d. We have $||\mathbf{x}||_\infty = 4/(k+1)$ and $||\mathbf{x}||_2 = (16/(k+1)^2 + 4/k^4 + k^4e^{-2k})^{1/2}$.

3. a. We have $\lim_{k\to\infty} \mathbf{x}^{(k)} = (0,0,0)^t$.

b. We have $\lim_{k\to\infty} \mathbf{x}^{(k)} = (0,1,3)^t$.

c. We have $\lim_{k\to\infty} \mathbf{x}^{(k)} = (0,0,\frac{1}{2})^t$.

d. We have $\lim_{k\to\infty} \mathbf{x}^{(k)} = (1,-1,1)^t$.

5. a. We have $||\mathbf{x}-\hat{\mathbf{x}}||_\infty = 8.57\times 10^{-4}$ and $||A\hat{\mathbf{x}}-\mathbf{b}||_\infty = 2.06\times 10^{-4}$.

b. We have $||\mathbf{x}-\hat{\mathbf{x}}||_\infty = 0.5$ and $||A\hat{\mathbf{x}}-\mathbf{b}||_\infty = 0.3$.

c. We have $||\mathbf{x}-\hat{\mathbf{x}}||_\infty = 0.90$ and $||A\hat{\mathbf{x}}-\mathbf{b}||_\infty = 0.27$.

d. We have $||\mathbf{x}-\hat{\mathbf{x}}||_\infty = 6.55\times 10^{-2}$, and $||A\hat{\mathbf{x}}-\mathbf{b}||_\infty = 0.32$.

7. Let $A = \begin{bmatrix} 1 & 1 \\ 0 & 1 \end{bmatrix}$ and $B = \begin{bmatrix} 1 & 0 \\ 1 & 1 \end{bmatrix}$. Then $\|AB\|_\otimes = 2$, but $\|A\|_\otimes \cdot \|B\|_\otimes = 1$.

9. b. We have

4a. $\|A\|_F = \sqrt{326}$

4b. $\|A\|_F = \sqrt{326}$

4c. $\|A\|_F = 4$

4d. $\|A\|_F = \sqrt{148}$.

15. First note that the right-hand side of the inequality is unchanged if $\mathbf{x}$ is replaced by any vector $\hat{\mathbf{x}}$ with $|x_i| = |\hat{x}_i|$ for each $i = 1, 2, \ldots n$. Then choose the new vector $\hat{\mathbf{x}}$ so that $\hat{x}_i y_i \geq 0$ for each i, and apply the inequality to $\hat{\mathbf{x}}$ and $\mathbf{y}$.

Exercise Set 7.2 (Page 449)

1. a. The eigenvalue $\lambda_1 = 0$ has the eigenvector $\mathbf{x}_1 = (1,-1)^t$, and the eigenvalue $\lambda_2 = -1$ has the eigenvector $\mathbf{x}_2 = (1,-2)^t$.

b. The eigenvalue $\lambda_1 = -1/3$ has the eigenvector $\mathbf{x}_1 = (-3/2, 1)^t$, and the eigenvalue $\lambda_2 = -1/2$ has the eigenvector $\mathbf{x}_2 = (-2,1)^t$.

c. The eigenvalue $\lambda_1 = -1$ has the eigenvector $\mathbf{x}_1 = (1,-1)^t$, and the eigenvalue $\lambda_2 = 4$ has the eigenvector $\mathbf{x}_2 = (4,1)^t$.

d. The eigenvalue $\lambda_1 = 3$ has the eigenvector $\mathbf{x}_1 = (-1,1,2)^t$, the eigenvalue $\lambda_2 = 4$ has the eigenvector $\mathbf{x}_2 = (0,1,2)^t$, and the eigenvalue $\lambda_3 = -2$ has the eigenvector $\mathbf{x} = (-3,8,1)^t$.

e. The eigenvalue $\lambda_1 = \lambda_2 = 1/2$ has the eigenvector $\mathbf{x}_1 = (0,5,12)^t$, and the eigenvalue $\lambda_3 = -1/3$ has the eigenvector $\mathbf{x}_3 = (0,0,1)^t$.

f. The eigenvalue $\lambda_1 = 2+2i$ has the eigenvector $\mathbf{x}_1 = (0,-2i,1)^t$, the eigenvalue $\lambda_2 = 2-2i$ has the eigenvector $\mathbf{x}_2 = (0,2i,1)^t$, and the eigenvalue $\lambda_3 = 2$ has the eigenvector $\mathbf{x}_3 = (1,0,0)^t$.

3. a. The eigenvalues $\lambda_1 = 2+\sqrt{2}\,i$ and $\lambda_2 = 2-\sqrt{2}\,i$ have eigenvectors $\mathbf{x}_1 = (-\sqrt{2}\,i, 1)^t$ and $\mathbf{x}_2 = (\sqrt{2}\,i, 1)^t$.

b. The eigenvalues $\lambda_1 = (3+\sqrt{7}\,i)/2$ and $\lambda_2 = (3-\sqrt{7}\,i)/2$ have eigenvectors $\mathbf{x}_1 = ((1-\sqrt{7}\,i)/2, 1)^t$ and $\mathbf{x}_2 = ((1+\sqrt{7}\,i)/2, 1)^t$.

5. The spectral radii for the matrices in Exercise 1 are:

a. 1 **b.** $\frac{1}{2}$ **c.** 4 **d.** 4 **e.** $\frac{1}{2}$ **f.** $2\sqrt{2}$

7. Only the matrices in 1(b) and 1(e) are convergent.

9. The l_2 norms for the matrices in Exercise 1 are:

a. 3.162278 **b.** 1.458020 **c.** 5.036796 **d.** 5.601152 **e.** 2.896954 **f.** 4.701562

11. Since

$$A_1^k = \begin{bmatrix} 1 & 0 \\ \frac{2^k-1}{2^{k+1}} & 2^{-k} \end{bmatrix}, \text{ we have } \lim_{k\to\infty} A_1^k = \begin{bmatrix} 1 & 0 \\ \frac{1}{2} & 0 \end{bmatrix}.$$

Also,

$$A_2^k = \begin{bmatrix} 2^{-k} & 0 \\ \frac{16k}{2^{k-1}} & 2^{-k} \end{bmatrix}, \text{ so } \lim_{k\to\infty} A_2^k = \begin{bmatrix} 0 & 0 \\ 0 & 0 \end{bmatrix}.$$

13. Let A be an $n \times n$ matrix. Expanding across the first row gives the characteristic polynomial

$$p(\lambda) = \det(A - \lambda I) = (a_{11} - \lambda)M_{11} + \sum_{j=2}^{n}(-1)^{j+1}a_{1j}M_{1j}.$$

The determinants M_{1j} are of the form

$$M_{1j} = \det\begin{bmatrix} a_{21} & a_{22}-\lambda & \cdots & a_{2,j-1} & a_{2,j+1} & \cdots & a_{2n} \\ a_{31} & a_{32} & \cdots & a_{3,j-1} & a_{3,j+1} & \cdots & a_{3n} \\ \vdots & \vdots & \ddots & \vdots & \vdots & \ddots & \vdots \\ a_{j-1,1} & a_{j-1,2} & \cdots & a_{j-1,j-1}-\lambda & a_{j-1,j+1} & \cdots & a_{j-1,n} \\ a_{j,1} & a_{j,2} & \cdots & a_{j,j-1} & a_{j,j+1} & \cdots & a_{j,n} \\ a_{j+1,1} & a_{j+1,2} & \cdots & a_{j+1,j-1} & a_{j+1,j+1}-\lambda & \cdots & a_{j+1,n} \\ \vdots & \vdots & \ddots & \vdots & \vdots & \ddots & \vdots \\ a_{n1} & a_{n2} & \cdots & a_{n,j-1} & a_{n,j+1} & \cdots & a_{nn}-\lambda \end{bmatrix},$$

for $j = 2, \ldots, n$. Note that each M_{1j} has $n-2$ entries of the form $a_{ii} - \lambda$. Thus,

$$p(\lambda) = \det(A - \lambda I) = (a_{11} - \lambda)M_{11} + \{\text{terms of degree } n-2 \text{ or less}\}.$$

Since

$$M_{11} = \det\begin{bmatrix} a_{22}-\lambda & a_{23} & \cdots & \cdots & a_{2n} \\ a_{32} & a_{33}-\lambda & \ddots & & \vdots \\ \vdots & \ddots & \ddots & \ddots & \vdots \\ \vdots & & \ddots & \ddots & a_{n-1,n} \\ a_{n2} & \cdots & \cdots & a_{n,n-1} & a_{nn}-\lambda \end{bmatrix}$$

is of the same form as $\det(A - \lambda I)$, the same argument can be repeatedly applied to determine

$$p(\lambda) = (a_{11} - \lambda)(a_{22} - \lambda)\cdots(a_{nn} - \lambda) + \{\text{terms of degree } n-2 \text{ or less in } \lambda\}.$$

Thus, $p(\lambda)$ is a polynomial of degree n.

15. **a.** $\det(A - \lambda I) = \det((A - \lambda I)^t) = \det(A^t - \lambda I)$

b. If $A\mathbf{x} = \lambda\mathbf{x}$, then $A^2\mathbf{x} = \lambda A\mathbf{x} = \lambda^2\mathbf{x}$, and by induction, $A^k\mathbf{x} = \lambda^k\mathbf{x}$.

c. If $A\mathbf{x} = \lambda\mathbf{x}$ and A^{-1} exists, then $\mathbf{x} = \lambda A^{-1}\mathbf{x}$. By Exercise 8 (b), $\lambda \neq 0$, so $\frac{1}{\lambda}\mathbf{x} = A^{-1}\mathbf{x}$.

d. Since $A^{-1}\mathbf{x} = \frac{1}{\lambda}\mathbf{x}$, we have $(A^{-1})^2\mathbf{x} = \frac{1}{\lambda}A^{-1}\mathbf{x} = \frac{1}{\lambda^2}\mathbf{x}$. Mathematical induction gives

$$(A^{-1})^k\mathbf{x} = \frac{1}{\lambda^k}\mathbf{x}.$$

e. If $A\mathbf{x} = \lambda\mathbf{x}$, then

$$q(A)\mathbf{x} = q_0\mathbf{x} + q_1A\mathbf{x} + \ldots + q_kA^k\mathbf{x} = q_0\mathbf{x} + q_1\lambda\mathbf{x} + \ldots + q_k\lambda^k\mathbf{x} = q(\lambda)\mathbf{x}.$$

f. Let $A - \alpha I$ be nonsingular. Since $A\mathbf{x} = \lambda\mathbf{x}$,

$$(A - \alpha I)\mathbf{x} = A\mathbf{x} - \alpha I\mathbf{x} = \lambda\mathbf{x} - \alpha\mathbf{x} = (\lambda - \alpha)\mathbf{x}.$$

Thus

$$\frac{1}{\lambda - \alpha}\mathbf{x} = (A - \alpha I)^{-1}\mathbf{x}.$$

17. **a.** We have the real eigenvalue $\lambda = 1$ with the eigenvector $\mathbf{x} = (6, 3, 1)^t$.

b. Choose any multiple of the vector $(6, 3, 1)^t$.

19. Let $A\mathbf{x} = \lambda\mathbf{x}$. Then $|\lambda|\ \|\mathbf{x}\| = \|A\mathbf{x}\| \leq \|A\|\ \|\mathbf{x}\|$, which implies $|\lambda| \leq \|A\|$. Also, $(1/\lambda)\mathbf{x} = A^{-1}\mathbf{x}$ so $1/|\lambda| \leq \|A^{-1}\|$ and $\|A^{-1}\|^{-1} \leq |\lambda|$.

Exercise Set 7.3 (Page 459)

1. Two iterations of Jacobi's method gives the following results.

a. $\mathbf{x}^{(2)} = (1.2500000, -1.3333333, 0.2000000)^t$

b. $\mathbf{x}^{(2)} = (-1.0000000, 1.0000000, -1.3333333)^t$

c. $\mathbf{x}^{(2)} = (-0.5208333, -0.04166667, -0.2166667, 0.4166667)^t$

d. $\mathbf{x}^{(2)} = (0.6875, 1.125, 0.6875, 1.375, 0.5625, 1.375)^t$

3. Two iterations of the Gauss-Seidel method give the following results.

a. $\mathbf{x}^{(2)} = (1.250000000, -0.9166666667, 0.06666666666)^t$

b. $\mathbf{x}^{(2)} = (-1.666666667, 1.333333334, -0.8888888894)^t$

c. $\mathbf{x}^{(2)} = (-0.625, 0, -0.225, 0.6166667)^t$

d. $\mathbf{x}^{(2)} = (0.6875, 1.546875, 0.7929688, 1.71875, 0.7226563, 1.878906)^t$

5. Jacobi's Algorithm gives the following results.

a. $\mathbf{x}^{(10)} = (1.447642384, -0.8355647882, -0.0450226618)^t$

b. $\mathbf{x}^{(21)} = (-1.45485795, 1.45485795, -0.72704396)^t$

c. $\mathbf{x}^{(12)} = (-0.75205599, 0.04027028, -0.28025957, 0.69008536)^t$

d. $\mathbf{x}^{(9)} = (0.35705566, 1.42852883, 0.35705566, 1.57141113, 0.28552246, 1.57141113)^t$

7. The Gauss-Seidel Algorithm gives the following results.

a. $\mathbf{x}^{(6)} = (1.447816350, -0.8358173037, -0.0447996186)^t$

b. $\mathbf{x}^{(8)} = (-1.45480420, 1.45441316, -0.72720658)^t$

c. $\mathbf{x}^{(8)} = (-0.7531763, 0.04101049, -0.2807047, 0.6916305)^t$

d. $\mathbf{x}^{(6)} = (0.35713196, 1.42856598, 0.35714149, 1.57140350, 0.28570175, 1.57142544)^t$

9. a.

$$T_j = \begin{bmatrix} 0 & \frac{1}{2} & -\frac{1}{2} \\ -1 & 0 & -1 \\ \frac{1}{2} & \frac{1}{2} & 0 \end{bmatrix} \quad \text{and} \quad \det(\lambda I - T_j) = \lambda^3 + \frac{5}{4}\lambda.$$

Thus, the eigenvalues of T_j are 0 and $\pm\frac{\sqrt{5}}{2}i$, so $\rho(T_j) = \frac{\sqrt{5}}{2} > 1$.

b. $\mathbf{x}^{(25)} = (-20.827873, 2.0000000, -22.827873)^t$

c.

$$T_g = \begin{bmatrix} 0 & \frac{1}{2} & -\frac{1}{2} \\ 0 & -\frac{1}{2} & -\frac{1}{2} \\ 0 & 0 & -\frac{1}{2} \end{bmatrix} \quad \text{and} \quad \det(\lambda I - T_g) = \lambda\left(\lambda + \frac{1}{2}\right)^2.$$

Thus, the eigenvalues of T_g are 0, $-1/2$, and $-1/2$; and $\rho(T_g) = 1/2$.

d. $\mathbf{x}^{(23)} = (1.0000023, 1.9999975, -1.0000001)^t$ is within 10^{-5} in the l_∞ norm.

11. a. A is not strictly diagonally dominant.

b.

$$T_j = \begin{bmatrix} 0 & 0 & 1 \\ 0.5 & 0 & 0.25 \\ -1 & 0.5 & 0 \end{bmatrix} \quad \text{and} \quad \rho(T_j) = 0.97210521.$$

Since T_j is convergent, the Jacobi method will converge.

c. With $\mathbf{x}^{(0)} = (0, 0, 0)^t$, $\mathbf{x}^{(187)} = (0.90222655, -0.79595242, 0.69281316)^t$.

d. $\rho(T_j) = 1.39331779371$. Since T_j is not convergent, the Jacobi method will not converge.

13. a. Subtract $\mathbf{x} = T\mathbf{x} + \mathbf{c}$ from $\mathbf{x}^{(k)} = T\mathbf{x}^{(k-1)} + \mathbf{c}$ to obtain $\mathbf{x}^{(k)} - \mathbf{x} = T(\mathbf{x}^{(k-1)} - \mathbf{x})$. Thus,

$$\|\mathbf{x}^{(k)} - \mathbf{x}\| \leq \|T\| \ \|\mathbf{x}^{(k-1)} - \mathbf{x}\|.$$

Inductively, we have

$$\|\mathbf{x}^{(k)} - \mathbf{x}\| \leq \|T\|^k \|\mathbf{x}^{(0)} - \mathbf{x}\|.$$

The remainder of the proof is similar to the proof of Corollary 2.5.

b. The last column has no entry when $||T||_\infty = 1$.

	$\|\mathbf{x}^{(2)} - \mathbf{x}\|_\infty$	$\|T\|_\infty$	$\|T\|_\infty^2 \|\mathbf{x}^{(0)} - \mathbf{x}\|_\infty$	$\frac{\|T\|_\infty^2}{1-\|T\|_\infty} \|\mathbf{x}^{(1)} - \mathbf{x}^{(0)}\|_\infty$
1 (a)	0.22932	0.857143	0.48335	2.9388
1 (b)	0.051579	0.3	0.089621	0.11571
1 (c)	1.1453	0.9	2.2642	20.25
1 (d)	0.27511	1	0.75342	
1 (e)	0.59743	1	1.9897	
1 (f)	0.875	0.75	1.125	3.375

15. The results for this exercise are listed on page 827 in Exercise 11, where additional results are given for a method presented in Section 7.4.

Exercise Set 7.4 (Page 467)

1. Two iterations of the SOR method with $\omega = 1.1$ give the following results.

a. $\mathbf{x}^{(2)} = (1.512775000, -0.8298491667, -0.0843373667)^t$

b. $\mathbf{x}^{(2)} = (-1.58523750, 1.37885688, -0.7039212812)^t$

c. $\mathbf{x}^{(2)} = (-0.6604902, 0.03700749, -0.2493513, 0.6561139)^t$

d. $\mathbf{x}^{(2)} = (0.3781250000, 1.445468750, 0.3596914062, 1.458531250, 0.3071921875, 1.572124727)^t$

3. Two iterations of the SOR method with $\omega = 1.3$ give the following results.

a. $\mathbf{x}^{(2)} = (1.455783334, -0.7721494442, -0.0805396228)^t$

b. $\mathbf{x}^{(2)} = (-1.42073750, 1.595758125, -0.8597927812)^t$

c. $\mathbf{x}^{(2)} = (-0.7268893, 0.1251483, -0.2923371, 0.7037018)^t$

d. $\mathbf{x}^{(2)} = (0.5281250000, 1.480781250, 0.322816406, 1.359718750, 0.4288171875, 1.505949961)^t$

5. The SOR Algorithm with $\omega = 1.2$ gives the following results.

a. $\mathbf{x}^{(8)} = (1.447503814, -0.8359297624, -0.0445516532)^t$

b. $\mathbf{x}^{(6)} = (-1.454582850, 1.454498863, -0.7273302714)^t$

c. $\mathbf{x}^{(6)} = (-0.75308971, 0.04117281, -0.28074817, 0.69163506)^t$

d. $\mathbf{x}^{(7)} = (0.3571284945, 1.428582240, 0.3571489731, 1.571440116, 0.2857000650, 1.571445036)^t$

7. The tridiagonal matrix is in part (d).
(1d): For $\omega = 1.033370453$ we have

$$\mathbf{x}^{(5)} = (0.3571407017, 1.428570817, 0.357142771, 1.571421010, 0.2857118407, 1.571428256)^t.$$

9. Let $\lambda_1, \ldots, \lambda_n$ be the eigenvalues of T_ω. Then

$$\prod_{i=1}^{n} \lambda_i = \det T_\omega = \det\left((D - \omega L)^{-1}[(1-\omega)D + \omega U]\right)$$

$$= \det(D - \omega L)^{-1} \det((1-\omega)D + \omega U) = \det\left(D^{-1}\right) \det((1-\omega)D)$$

$$= \left(\frac{1}{(a_{11}a_{22}\ldots a_{nn})}\right)\left((1-\omega)^n a_{11}a_{22}\ldots a_{nn}\right) = (1-\omega)^n.$$

Thus

$$\rho(T_\omega) = \max_{1\leq i\leq n} |\lambda_i| \geq |\omega - 1|,$$

and $|\omega - 1| < 1$ if and only if $0 < \omega < 2$.

11.

	Jacobi 33 iterations	Gauss-Seidel 8 iterations	SOR ($\omega = 1.2$) 13 iterations
x_1	1.53873501	1.53873270	1.53873549
x_2	0.73142167	0.73141966	0.73142226
x_3	0.10797136	0.10796931	0.10797063
x_4	0.17328530	0.17328340	0.17328480
x_5	0.04055865	0.04055595	0.04055737
x_6	0.08525019	0.08524787	0.08524925
x_7	0.16645040	0.16644711	0.16644868
x_8	0.12198156	0.12197878	0.12198026
x_9	0.10125265	0.10124911	0.10125043
x_{10}	0.09045966	0.09045662	0.09045793
x_{11}	0.07203172	0.07202785	0.07202912
x_{12}	0.07026597	0.07026266	0.07026392
x_{13}	0.06875835	0.06875421	0.06875546
x_{14}	0.06324659	0.06324307	0.06324429
x_{15}	0.05971510	0.05971083	0.05971200
x_{16}	0.05571199	0.05570834	0.05570949
x_{17}	0.05187851	0.05187416	0.05187529
x_{18}	0.04924911	0.04924537	0.04924648
x_{19}	0.04678213	0.04677776	0.04677885
x_{20}	0.04448679	0.04448303	0.04448409
x_{21}	0.04246924	0.04246493	0.04246597
x_{22}	0.04053818	0.04053444	0.04053546
x_{23}	0.03877273	0.03876852	0.03876952
x_{24}	0.03718190	0.03717822	0.03717920
x_{25}	0.03570858	0.03570451	0.03570548
x_{26}	0.03435107	0.03434748	0.03434844
x_{27}	0.03309542	0.03309152	0.03309246
x_{28}	0.03192212	0.03191866	0.03191958
x_{29}	0.03083007	0.03082637	0.03082727
x_{30}	0.02980997	0.02980666	0.02980755
x_{31}	0.02885510	0.02885160	0.02885248
x_{32}	0.02795937	0.02795621	0.02795707
x_{33}	0.02711787	0.02711458	0.02711543
x_{34}	0.02632478	0.02632179	0.02632262
x_{35}	0.02557705	0.02557397	0.02557479
x_{36}	0.02487017	0.02486733	0.02486814
x_{37}	0.02420147	0.02419858	0.02419938
x_{38}	0.02356750	0.02356482	0.02356560
x_{39}	0.02296603	0.02296333	0.02296410
x_{40}	0.02239424	0.02239171	0.02239247
x_{41}	0.02185033	0.02184781	0.02184855
x_{42}	0.02133203	0.02132965	0.02133038
x_{43}	0.02083782	0.02083545	0.02083615
x_{44}	0.02036585	0.02036360	0.02036429
x_{45}	0.01991483	0.01991261	0.01991324

	Jacobi 33 iterations	Gauss-Seidel 8 iterations	SOR ($\omega = 1.2$) 13 iterations
x_{46}	0.01948325	0.01948113	0.01948175
x_{47}	0.01907002	0.01906793	0.01906846
x_{48}	0.01867387	0.01867187	0.01867239
x_{49}	0.01829386	0.01829190	0.01829233
x_{50}	0.71792896	0.01792707	0.01792749
x_{51}	0.01757833	0.01757648	0.01757683
x_{52}	0.01724113	0.01723933	0.01723968
x_{53}	0.01691660	0.01691487	0.01691517
x_{54}	0.01660406	0.01660237	0.01660267
x_{55}	0.01630279	0.01630127	0.01630146
x_{56}	0.01601230	0.01601082	0.01601101
x_{57}	0.01573198	0.01573087	0.01573077
x_{58}	0.01546129	0.01546020	0.01546010
x_{59}	0.01519990	0.01519909	0.01519878
x_{60}	0.01494704	0.01494626	0.01494595
x_{61}	0.01470181	0.01470085	0.01470077
x_{62}	0.01446510	0.01446417	0.01446409
x_{63}	0.01423556	0.01423437	0.01423461
x_{64}	0.01401350	0.01401233	0.01401256
x_{65}	0.01380328	0.01380234	0.01380242
x_{66}	0.01359448	0.01359356	0.01359363
x_{67}	0.01338495	0.01338434	0.01338418
x_{68}	0.01318840	0.01318780	0.01318765
x_{69}	0.01297174	0.01297109	0.01297107
x_{70}	0.01278663	0.01278598	0.01278597
x_{71}	0.01270328	0.01270263	0.01270271
x_{72}	0.01252719	0.01252656	0.01252663
x_{73}	0.01237700	0.01237656	0.01237654
x_{74}	0.01221009	0.01220965	0.01220963
x_{75}	0.01129043	0.01129009	0.01129008
x_{76}	0.01114138	0.01114104	0.01114102
x_{77}	0.01217337	0.01217312	0.01217313
x_{78}	0.01201771	0.01201746	0.01201746
x_{79}	0.01542910	0.01542896	0.01542896
x_{80}	0.01523810	0.01523796	0.01523796

Exercise Set 7.5 (Page 476)

1. The $\|\cdot\|_\infty$ condition numbers are:

a. 50 **b.** 600,002 **c.** 241.37 (d) 339,866

3.

	$\|\mathbf{x} - \hat{\mathbf{x}}\|_\infty$	$K_\infty(A)\|\mathbf{b} - A\hat{\mathbf{x}}\|_\infty / \|A\|_\infty$
a.	8.571429×10^{-4}	1.238095×10^{-2}
b.	0.04	0.8
c.	0.1	3.832060
d.	20	1.152440×10^5

5. Gaussian elimination and iterative refinement give the following results.

a. (i) $(-10.0, 1.01)^t$, (ii) $(10.0, 1.00)^t$

b. (i) $(12.0, 0.499, -1.98)^t$, (ii) $(1.00, 0.500, -1.00)^t$

c. (i) $(0.185, 0.0103, -0.0200, -1.12)^t$, (ii) $(0.177, 0.0127, -0.0207, -1.18)^t$

d. (i) $(0.799, -3.12, 0.151, 4.56)^t$, (ii) $(0.758, -3.00, 0.159, 4.30)^t$

7. The matrix is ill-conditioned since $K_\infty = 60002$. We have $\tilde{\mathbf{x}} = (-1.0000, 2.0000)^t$.

9. For any vector $\mathbf{x}$, we have

$$\|\mathbf{x}\| = \left\|A^{-1}A\mathbf{x}\right\| \le \left\|A^{-1}\right\| \; \|A\mathbf{x}\|, \quad \text{so} \quad \|A\mathbf{x}\| \ge \frac{\|\mathbf{x}\|}{\left\|A^{-1}\right\|}.$$

Let $\mathbf{x} \neq \mathbf{0}$ be such that $\|\mathbf{x}\| = 1$ and $B\mathbf{x} = \mathbf{0}$. Then

$$\|(A - B)\mathbf{x}\| = \|A\mathbf{x}\| \ge \frac{\|\mathbf{x}\|}{\left\|A^{-1}\right\|}$$

and

$$\frac{\|(A - B)\mathbf{x}\|}{\|A\|} \ge \frac{1}{\left\|A^{-1}\right\| \; \|A\|} = \frac{1}{K(A)}.$$

Since $\|\mathbf{x}\| = 1$,

$$\|(A - B)\mathbf{x}\| \le \|A - B\| \; \|\mathbf{x}\| = \|A - B\| \quad \text{and} \quad \frac{\|A - B\|}{\|A\|} \ge \frac{1}{K(A)}.$$

11. a. $K_\infty\left(H^{(4)}\right) = 28,375$

b. $K_\infty\left(H^{(5)}\right) = 943,656$

c. actual solution $\mathbf{x} = (-124, 1560, -3960, 2660)^t$;
approximate solution $\tilde{\mathbf{x}} = (-124.2, 1563.8, -3971.8, 2668.8)^t$; $\|\mathbf{x} - \tilde{\mathbf{x}}\|_\infty = 11.8$; $\frac{\|\mathbf{x}-\tilde{\mathbf{x}}\|_\infty}{\|\mathbf{x}\|_\infty} = 0.02980$;

$$\frac{K_\infty(A)}{1 - K_\infty(A)\left(\frac{\|\delta A\|_\infty}{\|A\|_\infty}\right)}\left[\frac{\|\delta b\|_\infty}{\|b\|_\infty} + \frac{\|\delta A\|_\infty}{\|A\|_\infty}\right] = \frac{28375}{1 - 28375\left(\frac{6.\bar{6}\times 10^{-6}}{2.08\bar{3}}\right)}\left[0 + \frac{6.\bar{6} \times 10^{-6}}{2.08\bar{3}}\right]$$

$$= 0.09987.$$

Exercise Set 7.6 (Page 492)

1. a. $(0.18, 0.13)^t$

b. $(0.19, 0.10)^t$

c. Gaussian elimination gives the best answer since $\mathbf{v}^{(2)} = (0, 0)^t$ in the conjugate gradient method.

d. $(0.13, 0.21)^t$. There is no improvement, although $\mathbf{v}^{(2)} \neq \mathbf{0}$.

3. a. $(1.00, -1.00, 1.00)^t$

b. $(0.827, 0.0453, -0.0357)^t$

c. Partial pivoting and scaled partial pivoting also give $(1.00, -1.00, 1.00)^t$.

d. $(0.776, 0.238, -0.185)^t$;
The residual from (3b) is $(-0.0004, -0.0038, 0.0037)^t$, and the residual from part (3d) is $(0.0022, -0.0038, 0.0024)^t$. There does not appear to be much improvement, if any. Rounding error is more prevalent because of the increase in the number of matrix multiplications.

5. a. $\mathbf{x}^{(2)} = (0.1535933456, -0.1697932117, 0.5901172091)^t$, $\|\mathbf{r}^{(2)}\|_\infty = 0.221$.

b. $\mathbf{x}^{(2)} = (0.9993129510, 0.9642734456, 0.7784266575)^t$, $\|\mathbf{r}^{(2)}\|_\infty = 0.144$.

c. $\mathbf{x}^{(2)} = (-0.7290954114, 2.515782452, -0.6788904058, -2.331943982)^t$, $\|\mathbf{r}^{(2)}\|_\infty = 2.2$.

d. $\mathbf{x}^{(2)} = (-0.7071108901, -0.0954748881, -0.3441074093, 0.5256091497)^t$, $\|\mathbf{r}^{(2)}\|_\infty = 0.39$.

e. $\mathbf{x}^{(2)} = (0.5335968381, 0.9367588935, 1.339920949, 1.743083004, 1.743083004)^t$, $\|\mathbf{r}^{(2)}\|_\infty = 1.3$.

f. $\mathbf{x}^{(2)} = (1.022375671, 1.686451893, 1.022375671, 2.060919568, 0.8310997764, 2.060919568)^t$, $\|\mathbf{r}^{(2)}\|_\infty = 1.13$.

7. a. $\mathbf{x}^{(3)} = (0.06185567013, -0.1958762887, 0.6185567010)^t$, $\|\mathbf{r}^{(3)}\|_\infty = 0.4 \times 10^{-9}$.

b. $\mathbf{x}^{(3)} = (0.9957894738, 0.9578947369, 0.7915789474)^t$, $\|\mathbf{r}^{(3)}\|_\infty = 0.1 \times 10^{-9}$.

c. $\mathbf{x}^{(4)} = (-0.7976470579, 2.795294120, -0.2588235305, -2.251764706)^t$, $\|\mathbf{r}^{(4)}\|_\infty = 0.39 \times 10^{-7}$.

d. $\mathbf{x}^{(4)} = (-0.7534246575, 0.04109589039, -0.2808219179, 0.6917808219)^t$, $\|\mathbf{r}^{(4)}\|_\infty = 0.11 \times 10^{-9}$.

e. $\mathbf{x}^{(5)} = (0.4516129032, 0.7096774197, 1.677419355, 1.741935483, 1.806451613)^t$, $\|\mathbf{r}^{(5)}\|_\infty = 0.2 \times 10^{-9}$.

f. $\mathbf{x}^{(4)} = (1.000000000, 2.000000000, 1.000000000, 2.000000000, 0.9999999997, 2.000000000)^t$, $\|\mathbf{r}^{(4)}\|_\infty = 0.44 \times 10^{-9}$.

9. a.

	Jacobi 49 iterations	Gauss-Seidel 28 iterations	SOR ($\omega = 1.3$) 13 iterations	Conjugate Gradient 9 iterations
x_1	0.93406183	0.93406917	0.93407584	0.93407713
x_2	0.97473885	0.97475285	0.97476180	0.97476363
x_3	1.10688692	1.10690302	1.10691093	1.10691243
x_4	1.42346150	1.42347226	1.42347591	1.42347699
x_5	0.85931331	0.85932730	0.85933633	0.85933790
x_6	0.80688119	0.80690725	0.80691961	0.80692197
x_7	0.85367746	0.85370564	0.85371536	0.85372011
x_8	1.10688692	1.10690579	1.10691075	1.10691250
x_9	0.87672774	0.87674384	0.87675177	0.87675250
x_{10}	0.80424512	0.80427330	0.80428301	0.80428524
x_{11}	0.80688119	0.80691173	0.80691989	0.80692252
x_{12}	0.97473885	0.97475850	0.97476265	0.97476392
x_{13}	0.93003466	0.93004542	0.93004899	0.93004987
x_{14}	0.87672774	0.87674661	0.87675155	0.87675298
x_{15}	0.85931331	0.85933296	0.85933709	0.85933979
x_{16}	0.93406183	0.93407462	0.93407672	0.93407768

b.

	Jacobi 60 iterations	Gauss-Seidel 35 iterations	SOR ($\omega = 1.2$) 23 iterations	Conjugate Gradient 11 iterations
x_1	0.39668038	0.39668651	0.39668915	0.39669775
x_2	0.07175540	0.07176830	0.07177348	0.07178516
x_3	−0.23080396	−0.23078609	−0.23077981	−0.23076923
x_4	0.24549277	0.24550989	0.24551535	0.24552253
x_5	0.83405412	0.83406516	0.83406823	0.83407148
x_6	0.51497606	0.51498897	0.51499414	0.51500583
x_7	0.12116003	0.12118683	0.12119625	0.12121212
x_8	−0.24044414	−0.24040991	−0.24039898	−0.24038462
x_9	0.37873579	0.37876891	0.37877812	0.37878788
x_{10}	1.09073364	1.09075392	1.09075899	1.09076341
x_{11}	0.54207872	0.54209658	0.54210286	0.54211344
x_{12}	0.13838259	0.13841682	0.13842774	0.13844211
x_{13}	−0.23083868	−0.23079452	−0.23078224	−0.23076923
x_{14}	0.41919067	0.41923122	0.41924136	0.41925019
x_{15}	1.15015953	1.15018477	1.15019025	1.15019425
x_{16}	0.51497606	0.51499318	0.51499864	0.51500583
x_{17}	0.12116003	0.12119315	0.12120236	0.12121212
x_{18}	−0.24044414	−0.24040359	−0.24039345	−0.24038462
x_{19}	0.37873579	0.37877365	0.37878188	0.37878788
x_{20}	1.09073364	1.09075629	1.09076069	1.09076341
x_{21}	0.39668038	0.39669142	0.39669449	0.39669775
x_{22}	0.07175540	0.07177567	0.07178074	0.07178516
x_{23}	−0.23080396	−0.23077872	−0.23077323	−0.23076923
x_{24}	0.24549277	0.24551542	0.24551982	0.24552253
x_{25}	0.83405412	0.83406793	0.83407025	0.83407148

c.

	Jacobi 15 iterations	Gauss-Seidel 9 iterations	SOR ($\omega = 1.1$) 8 iterations	Conjugate Gradient 8 iterations
x_1	−3.07611424	−3.07611739	−3.07611796	−3.07611794
x_2	−1.65223176	−1.65223563	−1.65223579	−1.65223582
x_3	−0.53282391	−0.53282528	−0.53282531	−0.53282528
x_4	−0.04471548	−0.04471608	−0.04471609	−0.04471604
x_5	0.17509673	0.17509661	0.17509661	0.17509661
x_6	0.29568226	0.29568223	0.29568223	0.29568218
x_7	0.37309012	0.37309011	0.37309011	0.37309011
x_8	0.42757934	0.42757934	0.42757934	0.42757927
x_9	0.46817927	0.46817927	0.46817927	0.46817927
x_{10}	0.49964748	0.49964748	0.49964748	0.49964748
x_{11}	0.52477026	0.52477026	0.52477026	0.52477027
x_{12}	0.54529835	0.54529835	0.54529835	0.54529836
x_{13}	0.56239007	0.56239007	0.56239007	0.56239009
x_{14}	0.57684345	0.57684345	0.57684345	0.57684347
x_{15}	0.58922662	0.58922662	0.58922662	0.58922664
x_{16}	0.59995522	0.59995522	0.59995522	0.59995523
x_{17}	0.60934045	0.60934045	0.60934045	0.60934045
x_{18}	0.61761997	0.61761997	0.61761997	0.61761998
x_{19}	0.62497846	0.62497846	0.62497846	0.62497847
x_{20}	0.63156161	0.63156161	0.63156161	0.63156161
x_{21}	0.63748588	0.63748588	0.63748588	0.63748588
x_{22}	0.64284553	0.64284553	0.64284553	0.64284553
x_{23}	0.64771764	0.64771764	0.64771764	0.64771764
x_{24}	0.65216585	0.65216585	0.65216585	0.65216585
x_{25}	0.65624320	0.65624320	0.65624320	0.65624320
x_{26}	0.65999423	0.65999423	0.65999423	0.65999422
x_{27}	0.66345660	0.66345660	0.66345660	0.66345660
x_{28}	0.66666242	0.66666242	0.66666242	0.66666242
x_{29}	0.66963919	0.66963919	0.66963919	0.66963919
x_{30}	0.67241061	0.67241061	0.67241061	0.67241060
x_{31}	0.67499722	0.67499722	0.67499722	0.67499721
x_{32}	0.67741692	0.67741692	0.67741691	0.67741691
x_{33}	0.67968535	0.67968535	0.67968535	0.67968535
x_{34}	0.68181628	0.68181628	0.68181628	0.68181628
x_{35}	0.68382184	0.68382184	0.68382184	0.68382184
x_{36}	0.68571278	0.68571278	0.68571278	0.68571278
x_{37}	0.68749864	0.68749864	0.68749864	0.68749864
x_{38}	0.68918652	0.68918652	0.68918652	0.68918652
x_{39}	0.69067718	0.69067718	0.69067718	0.69067717
x_{40}	0.68363346	0.68363346	0.68363346	0.68363349

11. a.

Solution	Residual
2.55613420	0.00668246
4.09171393	−0.00533953
4.60840390	−0.01739814
3.64309950	−0.03171624
5.13950533	0.01308093
7.19697808	−0.02081095
7.68140405	−0.04593118
5.93227784	0.01692180
5.81798997	0.04414047
5.85447806	0.03319707
5.94202521	−0.00099947
4.42152959	−0.00072826
3.32211695	0.02363822
4.49411604	0.00982052
4.80968966	0.00846967
3.81108707	−0.01312902

This converges in 6 iterations with tolerance 5.00×10^{-2} in the l_∞ norm and $\|\mathbf{r}^{(6)}\|_\infty = 0.046$.

b.

Solution	Residual
2.55613420	0.00668246
4.09171393	−0.00533953
4.60840390	−0.01739814
3.64309950	−0.03171624
5.13950533	0.01308093
7.19697808	−0.02081095
7.68140405	−0.04593118
5.93227784	0.01692180
5.81798996	0.04414047
5.85447805	0.03319706
5.94202521	−0.00099947
4.42152959	−0.00072826
3.32211694	0.02363822
4.49411603	0.00982052
4.80968966	0.00846967
3.81108707	−0.01312902

This converges in 6 iterations with tolerance 5.00×10^{-2} in the l_∞ norm and $\|\mathbf{r}^{(6)}\|_\infty = 0.046$.

c. All tolerances lead to the same convergence specifications.

13. a. Let $\{\mathbf{v}^{(1)}, \ldots \mathbf{v}^{(n)}\}$ be a set of nonzero A-orthogonal vectors for the symmetric positive definite matrix A. Then $\langle \mathbf{v}^{(i)}, A\mathbf{v}^{(j)} \rangle = 0$, if $i \neq j$. Suppose

$$c_1\mathbf{v}^{(1)} + c_2\mathbf{v}^{(2)} + \cdots + c_n\mathbf{v}^{(n)} = \mathbf{0},$$

where not all c_i are zero. Suppose k is the smallest integer for which $c_k \neq 0$. Then

$$c_k\mathbf{v}^{(k)} + c_{k+1}\mathbf{v}^{(k+1)} + \cdots + c_n\mathbf{v}^{(n)} = \mathbf{0}.$$

We solve for $\mathbf{v}^{(k)}$ to obtain

$$\mathbf{v}^{(k)} = -\frac{c_{k+1}}{c_k}\mathbf{v}^{(k+1)} - \cdots - \frac{c_n}{c_k}\mathbf{v}^{(n)}.$$

Multiplying by A gives

$$A\mathbf{v}^{(k)} = -\frac{c_{k+1}}{c_k}A\mathbf{v}^{(k+1)} - \cdots - \frac{c_n}{c_k}A\mathbf{v}^{(n)},$$

so

$$\begin{aligned}(\mathbf{v}^{(k)})^t A\mathbf{v}^{(k)} &= -\frac{c_{k+1}}{c_k}(\mathbf{v}^{(k)})^t A\mathbf{v}^{(k+1)} - \cdots - \frac{c_n}{c_k}(\mathbf{v}^{(k)t})A\mathbf{v}^{(n)}\\ &= -\frac{c_{k+1}}{c_k}\langle \mathbf{v}^{(k)}, A\mathbf{v}^{(k+1)}\rangle - \cdots - \frac{c_n}{c_k}\langle \mathbf{v}^{(k)}, A\mathbf{v}^{(n)}\rangle\\ &= -\frac{c_{k+1}}{c_k}\cdot 0 - \cdots - \frac{c_n}{c_k}\cdot 0.\end{aligned}$$

Since A is positive definite, $\mathbf{v}^{(k)} = \mathbf{0}$, which is a contradiction. Thus, all c_i must be zero, and $\{\mathbf{v}^{(1)}, \ldots, \mathbf{v}^{(n)}\}$ is linearly independent.

b. Let $\{\mathbf{v}^{(1)}, \ldots, \mathbf{v}^{(n)}\}$ be a set of nonzero A-orthogonal vectors for the symmetric positive definite matrix A, and let $\mathbf{z}$ be orthogonal to $\mathbf{v}^{(i)}$, for each $i = 1, \ldots, n$. From part (a), the set $\{\mathbf{v}^{(1)}, \ldots \mathbf{v}^{(n)}\}$ is linearly independent, so there is a collection of constants $\beta_1, \ldots, \beta_n$ with

$$\mathbf{z} = \sum_{i=1}^{n} \beta_i \mathbf{v}^{(i)}.$$

Hence,

$$\langle \mathbf{z}, \mathbf{z}\rangle = \mathbf{z}^t\mathbf{z} = \sum_{i=1}^{n} \beta_i \mathbf{z}^t \mathbf{v}^{(i)} = \sum_{i=1}^{n} \beta_i \cdot 0 = 0,$$

and Theorem 7.30, part (v), implies that $\mathbf{z} = \mathbf{0}$.

15. If A is a positive definite matrix whose eigenvalues are $0 < \lambda_1 \le \cdots \le \lambda_n$, then $||A||_2 = \lambda_n$ and $||A^{-1}||_2 = 1/\lambda_1$, so $K_2(A) = \lambda_n/\lambda_1$.
For the matrix A in Example 3 we have

$$K_2(A) = \frac{\lambda_5}{\lambda_1} = \frac{700.031}{0.0570737} = 12265.2,$$

and the matrix AH has

$$K_2(AH) = \frac{\lambda_5}{\lambda_1} = \frac{1.88052}{0.156370} = 12.0261.$$

Maple gives $ConditionNumber(A, 2) = 12265.15914$ and $ConditionNumber(AH, 2) = 12.02598124$.

Exercise Set 8.1 (Page 506)

1. The linear least-squares polynomial is $1.70784x + 0.89968$.

3. The least-squares polynomials with their errors are, respectively,
$P_1(x) = 0.9295140 + 0.5281021x$, with 2.457×10^{-2};
$P_2(x) = 1.011341 - 0.3256988x + 1.147330x^2$, with 9.453×10^{-4}; and
$P_3(x) = 1.000440 - 0.001540986x - 0.011505675x^2 + 1.021023x^3$ with 1.112×10^{-4}.

5. a. The linear least-squares polynomial is $P_1(x) = 1.665540x - 0.5124568$, with error 0.33559.

b. The least-squares polynomial of degree two is $P_2(x) = 1.129424x^2 - 0.3114035x + 0.08514393$, with error 2.4199×10^{-3}.

c. The least-squares polynomial of degree three is $P_3(x) = 0.2662081x^3 + 0.4029322x^2 + 0.2483857x - 0.01840140$, with error 5.0747×10^{-6}.

d. The least-squares approximation of the form be^{ax} is $f(x) = 0.04570748e^{2.707295x}$, with error 1.0750.

e. The least-squares approximation of the form bx^a is $f(x) = 0.9501565x^{1.872009}$, with error 0.054477.

7. a. $k = 0.8996$, $E(k) = 0.295$

b. $k = 0.9052$, $E(k) = 0.128$ Part (b) fits the total experimental data best.

9. The least squares line for the point average is 0.101 (ACT score) + 0.487.

11. The linear least-squares polynomial gives $y \approx 0.17952x + 8.2084$.

13. a. $\ln R = \ln 1.304 + 0.5756 \ln W$

b. $E = 25.25$

c. $\ln R = \ln 1.051 + 0.7006 \ln W + 0.06695(\ln W)^2$

d. $E = \sum_{i=1}^{37}\left(R_i - bW_i^a e^{c(\ln W_i)^2}\right)^2 = 20.30$

Exercise Set 8.2 (Page 518)

1. The linear least-squares approximations are:

a. $P_1(x) = 1.833333 + 4x$ **b.** $P_1(x) = -1.600003 + 3.600003x$ **c.** $P_1(x) = 0.1945267 + 3.000001x$

d. $P_1(x) = 1.140981 - 0.2958375x$ **e.** $P_1(x) = 0.6109245 + 0.09167105x$ **f.** $P_1(x) = -1.861455 + 1.666667x$

3. The least squares approximations of degree two are:

a. $P_2(x) = 2.000002 + 2.999991x + 1.000009x^2$ **b.** $P_2(x) = 0.4000163 - 2.400054x + 3.000028x^2$

c. $P_2(x) = 1.167179 + 0.08204442x + 1.458979x^2$ **d.** $P_2(x) = 1.723551 - 0.9313682x + 0.1588827x^2$

e. $P_2(x) = 0.4880058 + 0.8291830x - 0.7375119x^2$ **f.** $P_2(x) = -0.9089523 + 0.6275723x + 0.2597736x^2$

5. **a.** 0.3427×10^{-9} **b.** 0.0457142 **c.** 0.0106445

d. 0.000358354 **e.** 0.0000134621 **f.** 0.0000967795

7. The Gram-Schmidt process produces the following collections of polynomials:

a. $\phi_0(x) = 1, \phi_1(x) = x - 0.5,\quad \phi_2(x) = x^2 - x + \frac{1}{6},$ and $\phi_3(x) = x^3 - 1.5x^2 + 0.6x - 0.05$

b. $\phi_0(x) = 1, \phi_1(x) = x - 1,\quad \phi_2(x) = x^2 - 2x + \frac{2}{3},$ and $\phi_3(x) = x^3 - 3x^2 + \frac{12}{5}x - \frac{2}{5}$

c. $\phi_0(x) = 1, \phi_1(x) = x - 2,\quad \phi_2(x) = x^2 - 4x + \frac{11}{3},$ and $\phi_3(x) = x^3 - 6x^2 + 11.4x - 6.8$

9. The least-squares polynomials of degree two are:

a. $P_2(x) = 3.833333\phi_0(x) + 4\phi_1(x) + 0.9999998\phi_2(x)$

b. $P_2(x) = 2\phi_0(x) + 3.6\phi_1(x) + 3\phi_2(x)$

c. $P_2(x) = 3.194528\phi_0(x) + 3\phi_1(x) + 1.458960\phi_2(x)$

d. $P_2(x) = 0.5493061\phi_0(x) - 0.2958369\phi_1(x) + 0.1588785\phi_2(x)$

e. $P_2(x) = 0.6567600\phi_0(x) + 0.09167105\phi_1(x) - 0.73751218\phi_2(x)$

f. $P_2(x) = 1.471878\phi_0(x) + 1.666667\phi_1(x) + 0.2597705\phi_2(x)$

11. The Laguerre polynomials are $L_1(x) = x - 1$, $L_2(x) = x^2 - 4x + 2$ and $L_3(x) = x^3 - 9x^2 + 18x - 6$.

Exercise Set 8.3 (Page 527)

1. The interpolating polynomials of degree two are:

a. $P_2(x) = 2.377443 + 1.590534(x - 0.8660254) + 0.5320418(x - 0.8660254)x$

b. $P_2(x) = 0.7617600 + 0.8796047(x - 0.8660254)$

c. $P_2(x) = 1.052926 + 0.4154370(x - 0.8660254) - 0.1384262x(x - 0.8660254)$

d. $P_2(x) = 0.5625 + 0.649519(x - 0.8660254) + 0.75x(x - 0.8660254)$

3. Bounds for the maximum errors of polynomials in Exercise 1 are:

a. 0.1132617 **b.** 0.04166667 **c.** 0.08333333 **d.** 1.000000

5. The zeros of $\tilde{T}_3$ produce the following interpolating polynomials of degree two.

a. $P_2(x) = 0.3489153 - 0.1744576(x - 2.866025) + 0.1538462(x - 2.866025)(x - 2)$

b. $P_2(x) = 0.1547375 - 0.2461152(x - 1.866025) + 0.1957273(x - 1.866025)(x - 1)$

c. $P_2(x) = 0.6166200 - 0.2370869(x - 0.9330127) - 0.7427732(x - 0.9330127)(x - 0.5)$

d. $P_2(x) = 3.0177125 + 1.883800(x - 2.866025) + 0.2584625(x - 2.866025)(x - 2)$

7. The cubic polynomial $\frac{383}{384}x - \frac{5}{32}x^3$ approximates $\sin x$ with error at most 7.19×10^{-4}.

9. The change of variable $x = \cos\theta$ produces

$$\int_{-1}^{1} \frac{T_n^2(x)}{\sqrt{1-x^2}}\,dx = \int_{-1}^{1} \frac{[\cos(n\arccos x)]^2}{\sqrt{1-x^2}}\,dx = \int_0^{\pi} (\cos(n\theta))^2\,dx = \frac{\pi}{2}.$$

11. It was shown in text (see Eq. (8.13)) that the zeros of $T_n'(x)$ occur at $x_k' = \cos(k\pi/n)$ for $k = 1, \ldots, n-1$. Because $x_0' = \cos(0) = 1$, $x_n' = \cos(\pi) = -1$, and all values of the cosine lie in the interval $[-1, 1]$ it remains only to show that the zeros are distinct. This follows from the fact that for each $k = 1, \ldots, n-1$, we have x_k' in the interval $(0, \pi)$ and on this interval $D_x \cos(x) = -\sin x < 0$. As a consequence, $T_n'(x)$ is one-to-one on $(0, \pi)$, and these $n-1$ zeros of $T_n'(x)$ are distinct.

Exercise Set 8.4 (Page 537)

1. The Padé approximations of degree two for $f(x) = e^{2x}$ are:

$$n = 2, m = 0 : r_{2,0}(x) = 1 + 2x + 2x^2$$

$$n = 1, m = 1 : r_{1,1}(x) = (1 + x)/(1 - x)$$

$$n = 0, m = 2 : r_{0,2}(x) = (1 - 2x + 2x^2)^{-1}$$

i	x_i	$f(x_i)$	$r_{2,0}(x_i)$	$r_{1,1}(x_i)$	$r_{0,2}(x_i)$
1	0.2	1.4918	1.4800	1.5000	1.4706
2	0.4	2.2255	2.1200	2.3333	1.9231
3	0.6	3.3201	2.9200	4.0000	1.9231
4	0.8	4.9530	3.8800	9.0000	1.4706
5	1.0	7.3891	5.0000	undefined	1.0000

3. $r_{2,3}(x) = (1 + \frac{2}{5}x + \frac{1}{20}x^2)/(1 - \frac{3}{5}x + \frac{3}{20}x^2 - \frac{1}{60}x^3)$

i	x_i	$f(x_i)$	$r_{2,3}(x_i)$
1	0.2	1.22140276	1.22140277
2	0.4	1.49182470	1.49182561
3	0.6	1.82211880	1.82213210
4	0.8	2.22554093	2.22563652
5	1.0	2.71828183	2.71875000

5. $r_{3,3}(x) = (x - \frac{7}{60}x^3)/(1 + \frac{1}{20}x^2)$

i	x_i	$f(x_i)$	MacLaurin polynomial of degree 6	$r_{3,3}(x_i)$
0	0.0	0.00000000	0.00000000	0.00000000
1	0.1	0.09983342	0.09966675	0.09938640
2	0.2	0.19866933	0.19733600	0.19709571
3	0.3	0.29552021	0.29102025	0.29246305
4	0.4	0.38941834	0.37875200	0.38483660
5	0.5	0.47942554	0.45859375	0.47357724

7. The Padé approximations of degree five are:

a. $r_{0,5}(x) = (1 + x + \frac{1}{2}x^2 + \frac{1}{6}x^3 + \frac{1}{24}x^4 + \frac{1}{120}x^5)^{-1}$

b. $r_{1,4}(x) = (1 - \frac{1}{5}x)/(1 + \frac{4}{5}x + \frac{3}{10}x^2 + \frac{1}{15}x^3 + \frac{1}{120}x^4)$

c. $r_{3,2}(x) = (1 - \frac{3}{5}x + \frac{3}{20}x^2 - \frac{1}{60}x^3)/(1 + \frac{2}{5}x + \frac{1}{20}x^2)$

d. $r_{4,1}(x) = (1 - \frac{4}{5}x + \frac{3}{10}x^2 - \frac{1}{15}x^3 + \frac{1}{120}x^4)/(1 + \frac{1}{5}x)$

i	x_i	$f(x_i)$	$r_{0,5}(x_i)$	$r_{1,4}(x_i)$	$r_{2,3}(x_i)$	$r_{4,1}(x_i)$
1	0.2	0.81873075	0.81873081	0.81873074	0.81873075	0.81873077
2	0.4	0.67032005	0.67032276	0.67031942	0.67031963	0.67032099
3	0.6	0.54881164	0.54883296	0.54880635	0.54880763	0.54882143
4	0.8	0.44932896	0.44941181	0.44930678	0.44930966	0.44937931
5	1.0	0.36787944	0.36809816	0.36781609	0.36781609	0.36805556

9. $r_{T_{2,0}}(x) = (1.266066T_0(x) - 1.130318T_1(x) + 0.2714953T_2(x))/T_0(x)$

$r_{T_{1,1}}(x) = (0.9945705T_0(x) - 0.4569046T_1(x))/(T_0(x) + 0.48038745T_1(x))$

$r_{T_{0,2}}(x) = 0.7940220T_0(x)/(T_0(x) + 0.8778575T_1(x) + 0.1774266T_2(x))$

i	x_i	$f(x_i)$	$r_{T_{2,0}}(x_i)$	$r_{T_{1,1}}(x_i)$	$r_{T_{0,2}}(x_i)$
1	0.25	0.77880078	0.74592811	0.78595377	0.74610974
2	0.50	0.60653066	0.56515935	0.61774075	0.58807059
3	1.00	0.36787944	0.40724330	0.36319269	0.38633199

11. $r_{T_{2,2}}(x) = \dfrac{0.91747T_1(x)}{T_0(x) + 0.088914T_2(x)}$

i	x_i	$f(x_i)$	$r_{T_{2,2}}(x_i)$
0	0.00	0.00000000	0.00000000
1	0.10	0.09983342	0.09093843
2	0.20	0.19866933	0.18028797
3	0.30	0.29552021	0.26808992
4	0.40	0.38941834	0.35438412

13. a. $e^x = e^{M \ln \sqrt{10} + s} = e^{M \ln \sqrt{10}} e^s = e^{\ln 10^{\frac{M}{2}}} e^s = 10^{\frac{M}{2}} e^s$

b. $e^s \approx \left(1 + \frac{1}{2}s + \frac{1}{10}s^2 + \frac{1}{120}s^3\right) / \left(1 - \frac{1}{2}s + \frac{1}{10}s^2 - \frac{1}{120}s^3\right)$, with $|\text{error}| \le 3.75 \times 10^{-7}$.

c. Set $M = \text{round}(0.8685889638x)$, $s = x - M/(0.8685889638)$, and $\hat{f} = \left(1 + \frac{1}{2}s + \frac{1}{10}s^2 + \frac{1}{120}s^3\right) / \left(1 - \frac{1}{2}s + \frac{1}{10}s^2 - \frac{1}{120}s^3\right)$. Then $f = (3.16227766)^M \hat{f}$.

Exercise Set 8.5 (Page 546)

1. $S_3(x) = 2\sin x - \sin 2x$

3. $S_3(x) = 3.676078 - 3.676078\cos x + 1.470431\cos 2x - 0.7352156\cos 3x + 3.676078\sin x - 2.940862\sin 2x$

5. $S_n(x) = \displaystyle\sum_{k=1}^{n-1} \frac{2}{k\pi}\left(1 - (-1)^k\right)\sin kx$

7. The trigonometric least-squares polynomials are:

a. $S_2(x) = \cos 2x$

b. $S_2(x) = 0$

c. $S_3(x) = -2.046326 + 3.883872\cos x - 2.320482\cos 2x + 0.7310818\cos 3x$

d. $S_3(x) = 1.566453 + 0.5886815\cos x - 0.2700642\cos 2x + 0.2175679\cos 3x + 0.8341640\sin x - 0.3097866\sin 2x$

9. The trigonometric least-squares polynomial is $S_3(x) = -0.4968929 + 0.2391965\cos x + 1.515393\cos 2x + 0.2391965\cos 3x - 1.150649\sin x$, with error $E(S_3) = 7.271197$.

11. The trigonometric least-squares polynomials and their errors are

a. $S_3(x) = -0.08676065 - 1.446416\cos\pi(x-3) - 1.617554\cos 2\pi(x-3) + 3.980729\cos 3\pi(x-3) - 2.154320\sin\pi(x-3) + 3.907451\sin 2\pi(x-3)$ with $E(S_3) = 210.90453$

b. $S_3(x) = -0.0867607 - 1.446416\cos\pi(x-3) - 1.617554\cos 2\pi(x-3) + 3.980729\cos 3\pi(x-3) - 2.354088\cos 4\pi(x-3) - 2.154320\sin\pi(x-3) + 3.907451\sin 2\pi(x-3) - 1.166181\sin 3\pi(x-3)$ with $E(S_4) = 169.4943$

13. Let $f(-x) = -f(x)$. The integral $\int_{-a}^{0} f(x)\,dx$ under the change of variable $t = -x$ transforms to

$$-\int_a^0 f(-t)\,dt = \int_0^a f(-t)\,dt = -\int_0^a f(t)\,dt = -\int_0^a f(x)\,dx.$$

Thus,

$$\int_{-a}^{a} f(x)\,dx = \int_{-a}^{0} f(x)\,dx + \int_0^a f(x)\,dx = -\int_0^a f(x)\,dx + \int_0^a f(x)\,dx = 0.$$

17. The steps are nearly identical to those for determining the constants b_k except for the additional constant term a_0 in the cosine series. In this case

$$0 = \frac{\partial E}{\partial a_0} = 2\sum_{j=0}^{2m-1}[y_j - S_n(x_j)](-1/2) = \sum_{j=0}^{2m-1} y_j - \sum_{j=0}^{2m-1}\left(\frac{a_0}{2} + a_n\cos nx_j + \sum_{k=1}^{n-1}(a_k\cos kx_j + b_k\sin kx_j)\right),$$

The orthogonality implies that only the constant term remains in the second sum, and we have

$$0 = \sum_{j=0}^{2m-1} y_j - \frac{a_0}{2}(2m) \quad \text{which implies that} \quad a_0 = \frac{1}{m}\sum_{j=0}^{2m-1} y_j.$$

Exercise Set 8.6 (Page 557)

1. The trigonometric interpolating polynomials are:

a. $S_2(x) = -6.168503 + 9.869604 \cos x - 3.701102 \cos 2x + 4.934802 \sin x$

b. $S_2(x) = -12.33701 + 4.934802 \cos x - 2.467401 \cos 2x + 4.934802 \sin x$

c. $S_2(x) = 1.570796 - 1.570796 \cos x$

d. $S_2(x) = -0.5 - 0.5 \cos 2x + \sin x$

3. The Fast Fourier Transform Algorithm gives the following trigonometric interpolating polynomials.

a. $S_4(x) = -11.10331 + 2.467401 \cos x - 2.467401 \cos 2x + 2.467401 \cos 3x - 1.233701 \cos 4x + 5.956833 \sin x - 2.467401 \sin 2x + 1.022030 \sin 3x$

b. $S_4(x) = 1.570796 - 1.340759 \cos x - 0.2300378 \cos 3x$

c. $S_4(x) = -0.1264264 + 0.2602724 \cos x - 0.3011140 \cos 2x + 1.121372 \cos 3x + 0.04589648 \cos 4x - 0.1022190 \sin x + 0.2754062 \sin 2x - 2.052955 \sin 3x$

d. $S_4(x) = -0.1526819 + 0.04754278 \cos x + 0.6862114 \cos 2x - 1.216913 \cos 3x + 1.176143 \cos 4x - 0.8179387 \sin x + 0.1802450 \sin 2x + 0.2753402 \sin 3x$

5.

	Approximation	Actual
a.	−69.76415	−62.01255
b.	9.869602	9.869604
c.	−0.7943605	−0.2739383
d.	−0.9593287	−0.9557781

7. The b_j terms are all zero. The a_j terms are as follows:

$a_0 = -4.0008033$	$a_1 = 3.7906715$	$a_2 = -2.2230259$	$a_3 = 0.6258042$
$a_4 = -0.3030271$	$a_5 = 0.1813613$	$a_6 = -0.1216231$	$a_7 = 0.0876136$
$a_8 = -0.0663172$	$a_9 = 0.0520612$	$a_{10} = -0.0420333$	$a_{11} = 0.0347040$
$a_{12} = -0.0291807$	$a_{13} = 0.0249129$	$a_{14} = -0.0215458$	$a_{15} = 0.0188421$
$a_{16} = -0.0166380$	$a_{17} = 0.0148174$	$a_{18} = -0.0132962$	$a_{19} = 0.0120123$
$a_{20} = -0.0109189$	$a_{21} = 0.0099801$	$a_{22} = -0.0091683$	$a_{23} = 0.0084617$
$a_{24} = -0.0078430$	$a_{25} = 0.0072984$	$a_{26} = -0.0068167$	$a_{27} = 0.0063887$
$a_{28} = -0.0060069$	$a_{29} = 0.0056650$	$a_{30} = -0.0053578$	$a_{31} = 0.0050810$
$a_{32} = -0.0048308$	$a_{33} = 0.0046040$	$a_{34} = -0.0043981$	$a_{35} = 0.0042107$
$a_{36} = -0.0040398$	$a_{37} = 0.0038837$	$a_{38} = -0.0037409$	$a_{39} = 0.0036102$
$a_{40} = -0.0034903$	$a_{41} = 0.0033803$	$a_{42} = -0.0032793$	$a_{43} = 0.0031866$
$a_{44} = -0.0031015$	$a_{45} = 0.0030233$	$a_{46} = -0.0029516$	$a_{47} = 0.0028858$
$a_{48} = -0.0028256$	$a_{49} = 0.0027705$	$a_{50} = -0.0027203$	$a_{51} = 0.0026747$
$a_{52} = -0.0026333$	$a_{53} = 0.0025960$	$a_{54} = -0.0025626$	$a_{55} = 0.0025328$
$a_{56} = -0.0025066$	$a_{57} = 0.0024837$	$a_{58} = -0.0024642$	$a_{59} = 0.0024478$
$a_{60} = -0.0024345$	$a_{61} = 0.0024242$	$a_{62} = -0.0024169$	$a_{63} = 0.0024125$

Exercise Set 9.1 (Page 568)

1. a. The eigenvalues and associated eigenvectors are $\lambda_1 = 2, \mathbf{v}^{(1)} = (1, 0, 0)^t$; $\lambda_2 = 1, \mathbf{v}^{(2)} = (0, 2, 1)^t$; and $\lambda_3 = -1, \mathbf{v}^{(3)} = (-1, 1, 1)^t$. The set is linearly independent.

b. The eigenvalues and associated eigenvectors are $\lambda_1 = 2, \mathbf{v}^{(1)} = (0, 1, 0)^t$; $\lambda_2 = 3, \mathbf{v}^{(2)} = (1, 0, 1)^t$; and $\lambda_3 = 1, \mathbf{v}^{(3)} = (1, 0, -1)^t$. The set is linearly independent.

c. The eigenvalues and associated eigenvectors are $\lambda_1 = 1, \mathbf{v}^{(1)} = (0, -1, 1)^t$; $\lambda_2 = 1 + \sqrt{2}, \mathbf{v}^{(2)} = (\sqrt{2}, 1, 1)^t$; and $\lambda_3 = 1 - \sqrt{2}, \mathbf{v}^{(3)} = (-\sqrt{2}, 1, 1)^t$; The set is linearly independent.

d. The eigenvalues and associated eigenvectors are $\lambda_1 = \lambda_2 = 2$, $\mathbf{v}^{(1)} = \mathbf{v}^{(2)} = (1, 0, 0)^t$; $\lambda_3 = 3$ with $\mathbf{v}^{(3)} = (0, 1, 1)^t$. There are only 2 linearly independent eigenvectors.

3. a. The three eigenvalues are within $\{\lambda \mid |\lambda| \le 2\} \cup \{\lambda \mid |\lambda - 2| \le 2\}$ so $\rho(A) \le 4$.

b. The three eigenvalues are within $\{\lambda \mid |\lambda - 4| \le 2\}$ so $\rho(A) \le 6$.

c. The three real eigenvalues satisfy $0 \le \lambda \le 6$ so $\rho(A) \le 6$.

d. The three real eigenvalues satisfy $1.25 \le \lambda \le 8.25$ so $1.25 \le \rho(A) \le 8.25$.

5. All the matrices except (d) have 3 linearly independent eigenvectors. The matrix in part (d) has only 2 linearly independent eigenvectors. One choice for P is each case is

a. $\begin{bmatrix} -1 & 0 & 1 \\ 1 & 2 & 0 \\ 1 & 1 & 0 \end{bmatrix}$, **b.** $\begin{bmatrix} 0 & -1 & 1 \\ 1 & 0 & 0 \\ 0 & 1 & 1 \end{bmatrix}$, **c.** $\begin{bmatrix} 0 & \sqrt{2} & -\sqrt{2} \\ -1 & 1 & 1 \\ 1 & 1 & 1 \end{bmatrix}$,

7. The vectors are linearly dependent since $-2\mathbf{v}_1 + 7\mathbf{v}_2 - 3\mathbf{v}_3 = \mathbf{0}$.

9. If $c_1\mathbf{v}_1 + \cdots + c_k\mathbf{v}_k = \mathbf{0}$, then for any j, with $1 \le j \le k$, we have $c_1\mathbf{v}_j^t\mathbf{v}_1 + \cdots + c_k\mathbf{v}_j^t\mathbf{v}_k = 0$. But orthogonality gives $c_i\mathbf{v}_j^t\mathbf{v}_i = 0$, for $i \ne j$, so $c_j\mathbf{v}_j^t\mathbf{v}_j = 0$ and since $\mathbf{v}_j^t\mathbf{v}_j \ne 0$, we have $c_j = 0$.

11. Since $\{\mathbf{v}_i\}_{i=1}^n$ is linearly independent in $\mathbb{R}^n$, there exist numbers $c_1, \ldots, c_n$ with

$$\mathbf{x} = c_1\mathbf{v}_1 + \cdots + c_n\mathbf{v}_n.$$

Hence, for any k, with $1 \le k \le n$,

$$\mathbf{v}_k^t\mathbf{x} = c_1\mathbf{v}_k^t\mathbf{v}_1 + \cdots + c_n\mathbf{v}_k^t\mathbf{v}_n = c_k\mathbf{v}_k^t\mathbf{v}_k = c_k.$$

13. a. i. $\mathbf{0} = c_1(1, 1)^t + c_2(-2, 1)^t$ implies that $\begin{bmatrix} 1 & -2 \\ 1 & 1 \end{bmatrix}\begin{bmatrix} c_1 \\ c_2 \end{bmatrix} = \begin{bmatrix} 0 \\ 0 \end{bmatrix}$. But $\det\begin{bmatrix} 1 & -2 \\ 1 & 1 \end{bmatrix} = 3 \ne 0$ so by Theorem 6.7 we have $c_1 = c_2 = 0$.

ii. $\{(1, 1)^t, (-3/2, 3/2)^t\}$.

iii. $\{(\sqrt{2}/2, \sqrt{2}/2)^t, (-\sqrt{2}/2, \sqrt{2}/2)^t\}$.

b. i. The determinant of the matrix with these vectors as columns is $-2 \ne 0$, so $\{(1, 1, 0)^t, (1, 0, 1)^t, (0, 1, 1)^t\}$ is a linearly independent set.

ii. $\{(1, 1, 0)^t, (1/2, -1/2, 1)^t, (-2/3, 2/3, 2/3)^t\}$

iii. $\{(\sqrt{2}/2, \sqrt{2}/2, 0)^t, (\sqrt{6}/6, -\sqrt{6}/6, \sqrt{6}/3)^t, (-\sqrt{3}/3, \sqrt{3}/3, \sqrt{3}/3)^t\}$

c. i. If $\mathbf{0} = c_1(1, 1, 1, 1)^t + c_2(0, 2, 2, 2)^t + c_3(1, 0, 0, 1)^t$, then we have

$$(E_1): c_1 + c_3 = 0, \quad (E_2): c_1 + 2c_2 = 0, \quad (E_3): c_1 + 2c_2 = 0, \quad (E_4): c_1 + 2c_2 + c_3 = 0.$$

Subtracting (E_3) from (E_4) implies that $c_3 = 0$. Hence, from (E_1) we have $c_1 = 0$, and from (E_2) we have $c_2 = 0$. The vectors are linearly independent.

ii. $\{(1, 1, 1, 1)^t, (-3/2, 1/2, 1/2, 1/2)^t, (0, -1/3, -1/3, 2/3)^t\}$

iii. $\{(1/2, 1/2, 1/2, 1/2)^t, (-\sqrt{3}/2, \sqrt{3}/6, \sqrt{3}/6, \sqrt{3}/6)^t, (0, -\sqrt{6}/6, -\sqrt{6}/6, \sqrt{6}/3)^t\}$

d. i. If A is the matrix whose columns are the vectors $\mathbf{v}_1, \mathbf{v}_2, \mathbf{v}_3, \mathbf{v}_4, \mathbf{v}_5$, then $\det A = 60 \ne 0$, so the vectors are linearly independent.

ii. $\{(2, 2, 3, 2, 3)^t, (2, -1, 0, -1, 0)^t, (0, 0, 1, 0, -1)^t, (1, 2, -1, 0, -1)^t, (-2/7, 3/7, 2/7, -1, 2/7)^t\}$

iii. $\{(\sqrt{30}/15, \sqrt{30}/15, \sqrt{30}/10, \sqrt{30}/15, \sqrt{30}/10)^t, (\sqrt{6}/3, -\sqrt{6}/6, 0, -\sqrt{6}/6, 0)^t,$
$(0, 0, \sqrt{2}/2, 0, -\sqrt{2}/2)^t, (\sqrt{7}/7, 2\sqrt{7}/7, -\sqrt{7}/7, 0, -\sqrt{7}/7)^t, (-\sqrt{70}/35, 3\sqrt{70}/70, \sqrt{70}/35, -\sqrt{70}/10, \sqrt{70}/35)^t\}$

15. A strictly diagonally dominant matrix has all its diagonal elements larger in magnitude than the sum of the magnitudes of all the other elements in its row. As a consequence, the magnitude of the center of each Geršgorin circle exceeds in magnitude the radius of the circle. No circle can therefore include the origin. Hence 0 cannot be an eigenvalue of the matrix, and the matrix is nonsingular.

Exercise Set 9.2 (Page 573)

1. In each instance we will compare the characteristic polynomial of A, denoted $C(A)$, to that of B, denoted $C(B)$. They must agree if the matrices are to be similar.

a. $C(A) = x^2 - 4x + 3 \ne x^2 - 2x - 3 = C(B)$.

b. $C(A) = x^2 - 5x + 6 \ne x^2 - 6x + 6 = C(B)$.

c. $C(A) = x^3 - 4x^2 + 5x - 2 \neq x^3 - 4x^2 + 5x - 6 = C(B)$.

d. $C(A) = x^3 - 5x^2 + 12x - 11 \neq x^3 - 4x^2 + 4x + 11 = C(B)$.

3. In each case we have $A^3 = (PDP^{(-1)})(PDP^{(-1)})(PDP^{(-1)}) = PD^3P^{(-1)}$.

a. $\begin{bmatrix} \frac{26}{5} & -\frac{14}{5} \\ -\frac{21}{5} & \frac{19}{5} \end{bmatrix}$

b. $\begin{bmatrix} 1 & 9 \\ 0 & -8 \end{bmatrix}$

c. $\begin{bmatrix} \frac{9}{5} & -\frac{8}{5} & \frac{7}{5} \\ \frac{4}{5} & -\frac{3}{5} & \frac{2}{5} \\ -\frac{2}{5} & \frac{4}{5} & -\frac{6}{5} \end{bmatrix}$

d. $\begin{bmatrix} 8 & 0 & 0 \\ 0 & 8 & 0 \\ 0 & 0 & 8 \end{bmatrix}$

5. They are all diagonalizable with P and D as follows.

a. $P = \begin{bmatrix} -1 & \frac{1}{4} \\ 1 & 1 \end{bmatrix}$ and $D = \begin{bmatrix} 5 & 0 \\ 0 & 0 \end{bmatrix}$

b. $P = \begin{bmatrix} 1 & -1 \\ 1 & 1 \end{bmatrix}$ and $D = \begin{bmatrix} 1 & 0 \\ 0 & 3 \end{bmatrix}$

c. $P = \begin{bmatrix} 1 & -1 & 0 \\ 0 & 0 & 1 \\ 1 & 1 & 0 \end{bmatrix}$ and $D = \begin{bmatrix} 3 & 0 & 0 \\ 0 & 1 & 0 \\ 0 & 0 & 1 \end{bmatrix}$

d. $P = \begin{bmatrix} \sqrt{2} & -\sqrt{2} & 0 \\ 1 & 1 & -1 \\ 1 & 1 & 1 \end{bmatrix}$ and $D = \begin{bmatrix} 1+\sqrt{2} & 0 & 0 \\ 0 & 1-\sqrt{2} & 0 \\ 0 & 0 & 1 \end{bmatrix}$

7. Only the matrices in parts (a) and (c) are positive definite.

a. $Q = \begin{bmatrix} -\frac{\sqrt{2}}{2} & \frac{\sqrt{2}}{2} \\ \frac{\sqrt{2}}{2} & \frac{\sqrt{2}}{2} \end{bmatrix}$ and $D = \begin{bmatrix} 1 & 0 \\ 0 & 3 \end{bmatrix}$

c. $Q = \begin{bmatrix} \frac{\sqrt{2}}{2} & 0 & -\frac{\sqrt{2}}{2} \\ 0 & 1 & 0 \\ \frac{\sqrt{2}}{2} & 0 & \frac{\sqrt{2}}{2} \end{bmatrix}$ and $D = \begin{bmatrix} 3 & 0 & 0 \\ 0 & 2 & 0 \\ 0 & 0 & 1 \end{bmatrix}$

9. In each case the matrix fails to have 3 linearly independent eigenvectors.

a. $\det(A) = 12$, so A is nonsingular.

b. $\det(A) = -1$, so A is nonsingular.

c. $\det(A) = 12$, so A is nonsingular.

d. $\det(A) = 1$, so A is nonsingular.

11. **a.** The eigenvalues and associated eigenvectors are
$\lambda_1 = 5.307857563$, $(0.59020967, 0.51643129, 0.62044441)^t$;
$\lambda_2 = -0.4213112993$, $(0.77264234, -0.13876278, -0.61949069)^t$;
$\lambda_3 = -0.1365462647$, $(0.23382978, -0.84501102, 0.48091581)^t$.

b. A is not positive definite because $\lambda_2 < 0$ and $\lambda_3 < 0$.

13. Because A is similar to B and B is similar to C, there exist invertible matrices S and T with $A = S^{-1}BS$ and $B = T^{-1}CT$. Hence A is similar to C because

$$A = S^{-1}BS = S^{-1}(T^{-1}CT)S = (S^{-1}T^{-1})C(TS) = (TS)^{-1}C(TS).$$

15. The matrix A has an eigenvalue of multiplicity 1 at $\lambda_1 = 3$ with eigenvector $\mathbf{s}_1 = (0, 1, 1)^t$, and an eigenvalue of multiplicity 2 at $\lambda_2 = 2$ with linearly independent eigenvectors $\mathbf{s}_2 = (1, 1, 0)^t$ and $\mathbf{s}_3 = (-2, 0, 1)^t$. Let $S_1 = \{\mathbf{s}_1, \mathbf{s}_2, \mathbf{s}_3\}$, $S_2 = \{\mathbf{s}_2, \mathbf{s}_1, \mathbf{s}_3\}$, and $S_3 = \{\mathbf{s}_2, \mathbf{s}_3, \mathbf{s}_1\}$. Then $A = S_1^{-1}D_1S_1 = S_2^{-1}D_2S_2 = S_3^{-1}D_3S_3$, so A is similar to D_1, D_2, and D_3.

17. The matrix A has an eigenvalue of multiplicity 1 at $\lambda_1 = 3$, and an eigenvalue of multiplicity 2 at $\lambda_2 = 2$. However, $\lambda_2 = 2$ has only one linearly independent eigenvector, so by Theorem 9.13, A is not similar to a diagonal matrix.

19. The proof of Theorem 9.13 follows by considering the form the diagonal matrix must assume. The matrix A is similar to a diagonal matrix D if and only if an invertible matrix S exists with $D = S^{-1}AS$, which is equivalent to $AS = SD$, with S invertible. Suppose that we have $AS = SD$ with the columns of S denoted $\mathbf{s}_1, \mathbf{s}_2, \ldots, \mathbf{s}_n$ and the diagonal elements of D denoted $d_1, d_2, \ldots, d_n$. Then $A\mathbf{s}_i = d_i\mathbf{s}_i$ for each $i = 1, 2, \ldots, n$. Hence each d_i is an eigenvalue of A with corresponding eigenvector $\mathbf{s}_i$. The matrix S is invertible, and consequently A is similar to D, if and only if there are n linearly independent eigenvectors that can be placed in the columns of S.

Exercise Set 9.3 (Page 590)

1. The approximate eigenvalues and approximate eigenvectors are:

a. $\mu^{(3)} = 3.666667$, $\mathbf{x}^{(3)} = (0.9772727, 0.9318182, 1)^t$

b. $\mu^{(3)} = 2.000000$, $\mathbf{x}^{(3)} = (1, 1, 0.5)^t$

c. $\mu^{(3)} = 5.000000$, $\mathbf{x}^{(3)} = (-0.2578947, 1, -0.2842105)^t$

d. $\mu^{(3)} = 5.038462$, $\mathbf{x}^{(3)} = (1, 0.2213741, 0.3893130, 0.4045802)^t$

3. The approximate eigenvalues and approximate eigenvectors are:

a. $\mu^{(3)} = 1.027730$, $\mathbf{x}^{(3)} = (-0.1889082, 1, -0.7833622)^t$

b. $\mu^{(3)} = -0.4166667$, $\mathbf{x}^{(3)} = (1, -0.75, -0.6666667)^t$

c. $\mu^{(3)} = 17.64493$, $\mathbf{x}^{(3)} = (-0.3805794, -0.09079132, 1)^t$

d. $\mu^{(3)} = 1.378684$, $\mathbf{x}^{(3)} = (-0.3690277, -0.2522880, 0.2077438, 1)^t$

5. The approximate eigenvalues and approximate eigenvectors are:

a. $\mu^{(3)} = 3.959538$, $\mathbf{x}^{(3)} = (0.5816124, 0.5545606, 0.5951383)^t$

b. $\mu^{(3)} = 2.0000000$, $\mathbf{x}^{(3)} = (-0.6666667, -0.6666667, -0.3333333)^t$

c. $\mu^{(3)} = 7.189567$, $\mathbf{x}^{(3)} = (0.5995308, 0.7367472, 0.3126762)^t$

d. $\mu^{(3)} = 6.037037$, $\mathbf{x}^{(3)} = (0.5073714, 0.4878571, -0.6634857, -0.2536857)^t$

7. The approximate eigenvalues and approximate eigenvectors are:

a. $\lambda_1 \approx \mu^{(9)} = 3.999908$, $\mathbf{x}^{(9)} = (0.9999943, 0.9999828, 1)^t$

b. $\lambda_1 \approx \mu^{(13)} = 2.414214$, $\mathbf{x}^{(13)} = (1, 0.7071429, 0.7070707)^t$

c. $\lambda_1 \approx \mu^{(9)} = 5.124749$, $\mathbf{x}^{(9)} = (-0.2424476, 1, -0.3199733)^t$

d. $\lambda_1 \approx \mu^{(24)} = 5.235861$, $\mathbf{x}^{(24)} = (1, 0.6178361, 0.1181667, 0.4999220)^t$

9. a. $\mu^{(9)} = 1.00001523$ with $\mathbf{x}^{(9)} = (-0.19999391, 1, -0.79999087)^t$

b. $\mu^{(12)} = -0.41421356$ with $\mathbf{x}^{(12)} = (1, -0.70709184, -0.707121720^t$

c. The method did not converge in 25 iterations. However, convergence occurred with $\mu^{(42)} = 1.63663642$ with $\mathbf{x}^{(42)} = (-0.57068151, 0.3633658, 1)^t$

d. $\mu^{(9)} = 1.38195929$ with $\mathbf{x}^{(9)} = (-0.38194003, -0.23610068, 0.23601909, 1)^t$

11. The approximate eigenvalues and approximate eigenvectors are:

a. $\mu^{(8)} = 4.0000000$, $\mathbf{x}^{(8)} = (0.5773547, 0.5773282, 0.5773679)^t$

b. $\mu^{(13)} = 2.414214$, $\mathbf{x}^{(13)} = (-0.7071068, -0.5000255, -0.4999745)^t$

c. $\mu^{(16)} = 7.223663$, $\mathbf{x}^{(16)} = (0.6247845, 0.7204271, 0.3010466)^t$

d. $\mu^{(20)} = 7.086130$, $\mathbf{x}^{(20)} = (0.3325999, 0.2671862, -0.7590108, -0.4918246)^t$

13. The approximate eigenvalues and approximate eigenvectors are:

a. $\lambda_2 \approx \mu^{(1)} = 1.000000$, $\mathbf{x}^{(1)} = (-2.999908, 2.999908, 0)^t$

b. $\lambda_2 \approx \mu^{(1)} = 1.000000$, $\mathbf{x}^{(1)} = (0, -1.414214, 1.414214)^t$

c. $\lambda_2 \approx \mu^{(6)} = 1.636734$, $\mathbf{x}^{(6)} = (1.783218, -1.135350, -3.124733)^t$

d. $\lambda_2 \approx \mu^{(10)} = 3.618177$, $\mathbf{x}^{(10)} = (0.7236390, -1.170573, 1.170675, -0.2763374)^t$

15. The approximate eigenvalues and approximate eigenvectors are:

a. $\mu^{(8)} = 4.000001$, $\mathbf{x}^{(8)} = (0.9999773, 0.99993134, 1)^t$

b. The method fails because of division by zero.

c. $\mu^{(7)} = 5.124890$, $\mathbf{x}^{(7)} = (-0.2425938, 1, -0.3196351)^t$

d. $\mu^{(15)} = 5.236112$, $\mathbf{x}^{(15)} = (1, 0.6125369, 0.1217216, 0.4978318)^t$

17. The approximate eigenvalues and approximate eigenvectors are:

a. $\mu^{(2)} = 1.000000$, $\mathbf{x}^{(2)} = (0.1542373, -0.7715828, 0.6171474)^t$

b. $\mu^{(13)} = 1.000000$, $\mathbf{x}^{(13)} = (0.00007432, -0.7070723, 0.7071413)^t$

c. $\mu^{(14)} = 4.961699$, $\mathbf{x}^{(14)} = (-0.4814472, 0.05180473, 0.8749428)^t$

d. $\mu^{(17)} = 4.428007$, $\mathbf{x}^{(17)} = (0.7194230, 0.4231908, 0.1153589, 0.5385466)^t$

19. **a.** We have $|\lambda| \leq 6$ for all eigenvalues λ.

b. The approximate eigenvalue and approximate eigenvector are $\mu^{(133)} = 0.69766854$, $\mathbf{x}^{(133)} = (1, 0.7166727, 0.2568099, 0.04601217)^t$.

c. The characteristic polynomial is $P(\lambda) = \lambda^4 - \frac{1}{4}\lambda - \frac{1}{16}$, and the eigenvalues are $\lambda_1 = 0.6976684972$, $\lambda_2 = -0.2301775942 + 0.56965884i$, $\lambda_3 = -0.2301775942 - 0.56965884i$, and $\lambda_4 = -0.237313308$.

d. The beetle population should approach zero since A is convergent.

21. Using the Inverse Power method with $\mathbf{x}^{(0)} = (1, 0, 0, 1, 0, 0, 1, 0, 0, 1)^t$ and $q = 0$ gives the following results:

a. $\mu^{(49)} = 1.0201926$, so $\rho(A^{-1}) \approx 1/\mu^{(49)} = 0.9802071$;

b. $\mu^{(30)} = 1.0404568$, so $\rho(A^{-1}) \approx 1/\mu^{(30)} = 0.9611163$;

c. $\mu^{(22)} = 1.0606974$, so $\rho(A^{-1}) \approx 1/\mu^{(22)} = 0.9427760$. The method appears to be stable for all α in $[\frac{1}{4}, \frac{3}{4}]$.

23. Forming $A^{-1}B$ and using the Power method with $\mathbf{x}^{(0)} = (1, 0, 0, 1, 0, 0, 1, 0, 0, 1)^t$ gives the following results:

a. The spectral radius is approximately $\mu^{(46)} = 0.9800021$.

b. The spectral radius is approximately $\mu^{(25)} = 0.9603543$.

c. The spectral radius is approximately $\mu^{(18)} = 0.9410754$.

Exercise Set 9.4 (Page 600)

1. Householder's method produces the following tridiagonal matrices.

a. $\begin{bmatrix} 12.00000 & -10.77033 & 0.0 \\ -10.77033 & 3.862069 & 5.344828 \\ 0.0 & 5.344828 & 7.137931 \end{bmatrix}$ **b.** $\begin{bmatrix} 2.0000000 & 1.414214 & 0.0 \\ 1.414214 & 1.000000 & 0.0 \\ 0.0 & 0.0 & 3.0 \end{bmatrix}$

c. $\begin{bmatrix} 1.0000000 & -1.414214 & 0.0 \\ -1.414214 & 1.000000 & 0.0 \\ 0.0 & 0.0 & 1.000000 \end{bmatrix}$ **d.** $\begin{bmatrix} 4.750000 & -2.263846 & 0.0 \\ -2.263846 & 4.475610 & -1.219512 \\ 0.0 & -1.219512 & 5.024390 \end{bmatrix}$

3. Householder's method produces the following tridiagonal matrices.

a. $\begin{bmatrix} 2.0000000 & 2.8284271 & 1.4142136 \\ -2.8284271 & 1.0000000 & 2.0000000 \\ 0.0000000 & 2.0000000 & 3.0000000 \end{bmatrix}$

b. $\begin{bmatrix} -1.0000000 & -3.0655513 & 0.0000000 \\ -3.6055513 & -0.23076923 & 3.1538462 \\ 0.0000000 & 0.15384615 & 2.2307692 \end{bmatrix}$

c. $\begin{bmatrix} 5.0000000 & 4.9497475 & -1.4320780 & -1.5649769 \\ -1.4142136 & -2.0000000 & -2.4855515 & 1.8226448 \\ 0.0000000 & -5.4313902 & -1.4237288 & -2.6486542 \\ 0.0000000 & 0.0000000 & 1.5939865 & 5.4237288 \end{bmatrix}$

d. $\begin{bmatrix} 4.0000000 & 1.7320508 & 0.0000000 & 0.0000000 \\ 1.7320508 & 2.3333333 & 0.23570226 & 0.40824829 \\ 0.0000000 & -0.47140452 & 4.6666667 & -0.57735027 \\ 0.0000000 & 0.0000000 & 0.0000000 & 5.0000000 \end{bmatrix}$

Exercise Set 9.5 (Page 611)

1. Two iterations of the QR Algorithm produce the following matrices.

a. $A^{(3)} = \begin{bmatrix} 3.142857 & -0.559397 & 0.0 \\ -0.559397 & 2.248447 & -0.187848 \\ 0.0 & -0.187848 & 0.608696 \end{bmatrix}$

b. $A^{(3)} = \begin{bmatrix} 4.549020 & 1.206958 & 0.0 \\ 1.206958 & 3.519688 & 0.000725 \\ 0.0 & 0.000725 & -0.068708 \end{bmatrix}$

c. $A^{(3)} = \begin{bmatrix} 4.592920 & -0.472934 & 0.0 \\ -0.472934 & 3.108760 & -0.232083 \\ 0.0 & -0.232083 & 1.298319 \end{bmatrix}$

d. $A^{(3)} = \begin{bmatrix} 3.071429 & 0.855352 & 0.0 & 0.0 \\ 0.855352 & 3.314192 & -1.161046 & 0.0 \\ 0.0 & -1.161046 & 3.331770 & 0.268898 \\ 0.0 & 0.0 & 0.268898 & 0.282609 \end{bmatrix}$

e. $A^{(3)} = \begin{bmatrix} -3.607843 & 0.612882 & 0.0 & 0.0 \\ 0.612882 & -1.395227 & -1.111027 & 0.0 \\ 0.0 & 0.346353 & 3.133919 & 0.346353 \\ 0.0 & 0.0 & 0.346353 & 0.869151 \end{bmatrix}$

f. $A^{(3)} = \begin{bmatrix} 1.013260 & 0.279065 & 0.0 & 0.0 \\ 0.279065 & 0.696255 & 0.107448 & 0.0 \\ 0.0 & 0.107448 & 0.843061 & 0.310832 \\ 0.0 & 0.0 & 0.310832 & 0.347424 \end{bmatrix}$

3. The matrices in Exercise 1 have the following eigenvalues, accurate to within 10^{-5}.

a. 3.414214, 2.000000, 0.58578644

b. $-0.06870782, 5.346462, 2.722246$

c. 1.267949, 4.732051, 3.000000

d. 4.745281, 3.177283, 1.822717, 0.2547188

e. $3.438803, 0.8275517, -1.488068, -3.778287$

f. 0.9948440, 1.189091, 0.5238224, 0.1922421

5. The matrices in Exercise 1 have the following eigenvectors, accurate to within 10^{-5}.

a. $(-0.7071067, 1, -0.7071067)^t$, $(1, 0, -1)^t$, $(0.7071068, 1, 0.7071068)^t$

b. $(0.1741299, -0.5343539, 1)^t$, $(0.4261735, 1, 0.4601443)^t$, $(1, -0.2777544, -0.3225491)^t$

c. $(0.2679492, 0.7320508, 1)^t$, $(1, -0.7320508, 0.2679492)^t$, $(1, 1, -1)^t$

d. $(-0.08029447, -0.3007254, 0.7452812, 1)^t$, $(0.4592880, 1, -0.7179949, 0.8727118)^t$, $(0.8727118, 0.7179949, 1, -0.4592880)^t$ $(1, -0.7452812, -0.3007254, 0.08029447)^t$

e. $(-0.01289861, -0.07015299, 0.4388026, 1)^t$, $(-0.1018060, -0.2878618, 1, -0.4603102)^t$, $(1, 0.5119322, 0.2259932, -0.05035423)^t$ $(-0.5623391, 1, 0.2159474, -0.03185871)^t$

f. $(-0.1520150, -0.3008950, -0.05155956, 1)^t$, $(0.3627966, 1, 0.7459807, 0.3945081)^t$, $(1, 0.09528962, -0.6907921, 0.1450703)^t$, $(0.8029403, -0.9884448, 1, -0.1237995)^t$

7. a. Let

$$P = \begin{bmatrix} \cos\theta & -\sin\theta \\ \sin\theta & \cos\theta \end{bmatrix}$$

and $\mathbf{y} = P\mathbf{x}$. Show that $\|\mathbf{x}\|_2 = \|\mathbf{y}\|_2$. Use the relationship $x_1 + ix_2 = re^{i\alpha}$, where $r = \|\mathbf{x}\|_2$ and $\alpha = \tan^{-1}(x_2/x_1)$, and $y_1 + iy_2 = re^{i(\alpha+\theta)}$.

b. Let $\mathbf{x} = (1, 0)^t$ and $\theta = \pi/4$.

9. Let $C = RQ$, where R is upper triangular and Q is upper Hessenberg. Then $c_{ij} = \sum_{k=1}^{n} r_{ik}q_{kj}$. Since R is an upper triangular matrix, $r_{ik} = 0$ if $k < i$. Thus $c_{ij} = \sum_{k=i}^{n} r_{ik}q_{kj}$. Since Q is an upper Hessenberg matrix, $q_{kj} = 0$ if $k > j + 1$. Thus, $c_{ij} = \sum_{k=i}^{j+1} r_{ik}q_{kj}$. The sum will be zero if $i > j + 1$. Hence, $c_{ij} = 0$ if $i \geq j + 2$. This means that C is an upper Hessenberg matrix.

11. INPUT: dimension n, matrix $A = (a_{ij})$, tolerance TOL, maximum number of iterations N.

OUTPUT: eigenvalues $\lambda_1, \ldots, \lambda_n$ of A or a message that the number of iterations was exceeded.

Step 1 Set $FLAG = 1$; $k1 = 1$.

Step 2 While ($FLAG = 1$) do Steps 3 – 10

Step 3 For $i = 2, \ldots, n$ do Steps 4 – 8.

Step 4 For $j = 1, \ldots, i-1$ do Steps 5 – 8.

Step 5 If $a_{ii} = a_{jj}$ then set

$CO = 0.5\sqrt{2}$;
$SI = CO$

else set

$b = |a_{ii} - a_{jj}|$;
$c = 2a_{ij}\,\text{sign}(a_{ii} - a_{jj})$;
$CO = 0.5\left(1 + b/\left(c^2 + b^2\right)^{\frac{1}{2}}\right)^{\frac{1}{2}}$;
$SI = 0.5c/\left(CO\left(c^2 + b^2\right)^{\frac{1}{2}}\right)$.

Step 6 For $k = 1, \ldots, n$
if ($k \neq i$) and ($k \neq j$) then
set $x = a_{k,j}$;
$y = a_{k,i}$;
$a_{k,j} = CO \cdot x + SI \cdot y$;
$a_{k,i} = CO \cdot y + SI \cdot x$;
$x = a_{j,k}$;
$y = a_{i,k}$;
$a_{j,k} = CO \cdot x + SI \cdot y$;
$a_{i,k} = CO \cdot y - SI \cdot x$.

Step 7 Set $x = a_{j,j}$;
$y = a_{i,i}$;
$a_{j,j} = CO \cdot CO \cdot x + 2 \cdot SI \cdot CO \cdot a_{j,i} + SI \cdot SI \cdot y$;
$a_{i,i} = SI \cdot SI \cdot x - 2 \cdot SI \cdot CO \cdot a_{i,j} + CO \cdot CO \cdot y$.

Step 8 Set $a_{i,j} = 0$; $a_{j,i} = 0$.

Step 9 Set

$$s = \sum_{i=1}^{n} \sum_{\substack{j=1 \\ j\neq i}}^{n} |a_{ij}|.$$

Step 10 If $s < TOL$ then for $i = 1, \ldots, n$ set

$\lambda_i = a_{ii}$;
OUTPUT $(\lambda_1, \ldots, \lambda_n)$;
set $FLAG = 0$.

else set $k1 = k1 + 1$;
if $k1 > N$ then set $FLAG = 0$.

Step 11 If $k1 > N$ then
OUTPUT ('Maximum number of iterations exceeded');
STOP.

13. a. To within 10^{-5}, the eigenvalues are 2.618034, 3.618034, 1.381966, and 0.3819660.

b. In terms of p and ρ the eigenvalues are $-65.45085p/\rho$, $-90.45085p/\rho$, $-34.54915p/\rho$, and $-9.549150p/\rho$.

15. The actual eigenvalues are as follows:

a. When $\alpha = 1/4$ we have 0.97974649, 0.92062677, 0.82743037, 0.70770751, 0.57115742, 0.42884258, 0.29229249, 0.17256963, 0.07937323, and 0.02025351.

b. When $\alpha = 1/2$ we have 0.95949297, 0.84125353, 0.65486073, 0.41541501, 0.14231484, −0.14231484, −0.41541501, −0.65486073, −0.84125353, and −0.95949297.

c. When $\alpha = 3/4$ we have 0.93923946, 0.76188030, 0.48229110, 0.12312252, −0.28652774, −0.71347226, −1.12312252, −1.48229110, −1.76188030, and −1.93923946. The method appears to be stable for $\alpha \leq \frac{1}{2}$.

Exercise Set 9.6 (Page 625)

1. a. $s_1 = 1 + \sqrt{2}$, $s_2 = -1 + \sqrt{2}$ **b.** $s_1 = \sqrt{6}$, $s_2 = 1$

c. $s_1 = \sqrt{11}$, $s_2 = \sqrt{6}$ **d.** $s_1 = \sqrt{7}$, $s_2 = 1$, $s_3 = 1$

3. a.

$$U = \begin{bmatrix} -0.923880 & -0.382683 \\ -0.3826831 & 0.923880 \end{bmatrix}, \quad S = \begin{bmatrix} 2.414214 & 0 \\ 0 & 0.414214 \end{bmatrix}, \quad V^t = \begin{bmatrix} -0.923880 & -0.382683 \\ 0.382683 & -0.923880 \end{bmatrix}$$

b.

$$U = \begin{bmatrix} -0.912871 & 0 & -0.408248 \\ -0.365148 & -0.447214 & 0.816497 \\ -0.182574 & 0.894427 & 0.408248 \end{bmatrix}, \quad S = \begin{bmatrix} 2.449490 & 0 \\ 0 & 1 \\ 0 & 0 \end{bmatrix}, \quad V^t = \begin{bmatrix} -0.894427 & -0.447214 \\ -0.447214 & 0.894427 \end{bmatrix}$$

c.

$$U = \begin{bmatrix} -0.632456 & -0.5 & -0.522293 & -0.277867 \\ 0.316228 & -0.5 & -0.301969 & 0.747539 \\ -0.316228 & -0.5 & 0.797047 & 0.121309 \\ -0.632456 & 0.5 & -0.027215 & 0.590982 \end{bmatrix}, \quad S = \begin{bmatrix} 3.162278 & 0 \\ 0 & 2.0 \\ 0 & 0 \\ 0 & 0 \end{bmatrix},$$

$$V^t = \begin{bmatrix} -1.0 & 0.0 \\ 0.0 & -1.0 \end{bmatrix}$$

d.

$$U = \begin{bmatrix} -0.436436 & 0.707107 & 0.408248 & -0.377964 \\ 0.436436 & 0.707107 & -0.408248 & 0.377964 \\ -0.436436 & 0 & -0.816497 & -0.377964 \\ -0.654654 & 0 & 0 & 0.755929 \end{bmatrix}, \quad S = \begin{bmatrix} 2.645751 & 0 & 0 \\ 0 & 1 & 0 \\ 0 & 0 & 1 \\ 0 & 0 & 0 \end{bmatrix},$$

$$V^t = \begin{bmatrix} -0.577350 & -0.577350 & 0.577350 \\ 0 & 0.707107 & 0.707107 \\ 0.816497 & -0.408248 & 0.408248 \end{bmatrix}$$

5. For the matrix A in Example 2 we have

$$A^tA = \begin{bmatrix} 1 & 0 & 0 & 0 & 1 \\ 0 & 1 & 1 & 1 & 1 \\ 1 & 0 & 1 & 0 & 0 \end{bmatrix} \begin{bmatrix} 1 & 0 & 1 \\ 0 & 1 & 0 \\ 0 & 1 & 1 \\ 0 & 1 & 0 \\ 1 & 1 & 0 \end{bmatrix} = \begin{bmatrix} 2 & 1 & 1 \\ 1 & 4 & 1 \\ 1 & 1 & 2 \end{bmatrix}$$

So $A^tA(1,2,1)^t = (5,10,5)^t = 5(1,2,1)^t$, $A^tA(1,-1,1)^t = (2,-2,2)^t = 2(1,-1,1)^t$, and $A^tA(-1,0,1)^t = (-1,0,1)^t$.

7. Let A be an $m \times n$ matrix. Theorem 9.25 implies that $\text{Rank}(A) = \text{Rank}(A^t)$, so $\text{Nullity}(A) = n - \text{Rank}(A)$ and $\text{Nullity}(A^t) = m - \text{Rank}(A^t) = m - \text{Rank}(A)$. Hence $\text{Nullity}(A) = \text{Nullity}(A^t)$ if and only if $n = m$.

9. $\text{Rank}(S)$ is the number of nonzero entries on the diagonal of S. This corresponds to the number of nonzero eigenvalues (counting multiplicities) of A^tA. So $\text{Rank}(S) = \text{Rank}(A^tA)$, and by part (ii) of Theorem 9.26 this is the same as $\text{Rank}(A)$.

11. Because both $U^{-1} = U^t$ and $V^{-1} = V^t$ exist, $A = USV^t$ implies that $A^{-1} = (USV^t)^{-1} = VS^{-1}U^t$ if and only if S^{-1} exists.

13. Yes. By Theorem 9.25 we have $\text{Rank}(A^tA) = \text{Rank}((A^tA)^t) = \text{Rank}(AA^t)$. Applying part (iii) of Theorem 9.26 gives $\text{Rank}(AA^t) = \text{Rank}(A^tA) = \text{Rank}(A)$.

15. If the $n \times n$ matrix A has the singular values $s_1 \geq s_2 \geq \cdots \geq s_n > 0$, then $||A||_2 = \sqrt{\rho(A^tA)} = s_1$. In addition, the singular values of A^{-1} are $\frac{1}{s_n} \geq \cdots \geq \frac{1}{s_2} \geq \frac{1}{s_1} > 0$, so $||A^{-1}||_2 = \sqrt{\frac{1}{s_n}} = \frac{1}{s_n}$. Hence $K_2(A) = ||A||_2 \cdot ||A^{-1}||_2 = s_1/s_n$.

17. a.

$$A = \begin{bmatrix} 1 & 1 \\ 1 & 2 \\ 1 & 3 \\ 1 & 4 \\ 1 & 5 \end{bmatrix}, \quad S = \begin{bmatrix} 7.691213 & 0 \\ 0 & 0.919370 \\ 0 & 0 \\ 0 & 0 \\ 0 & 0 \end{bmatrix}, \quad V^t = \begin{bmatrix} 0.266934 & 0.963715 \\ 0.963715 & -0.266934 \end{bmatrix}$$

and

$$U = \begin{bmatrix} 0.160007 & 0.757890 & -0.414912 & -0.362646 & -0.310381 \\ 0.285308 & 0.467546 & 0.067225 & 0.399603 & 0.731982 \\ 0.410609 & 0.177202 & 0.837705 & -0.201287 & -0.240279 \\ 0.535909 & -0.113142 & -0.217438 & 0.654348 & -0.473867 \\ 0.661210 & -0.403486 & -0.272580 & -0.490018 & 0.292544 \end{bmatrix}$$

This produces $P(x) = 0.33 + 1.29x$.

b.

$$A = \begin{bmatrix} 1 & 1 & 1 \\ 1 & 2 & 4 \\ 1 & 3 & 9 \\ 1 & 4 & 16 \\ 1 & 5 & 25 \end{bmatrix}, \quad S = \begin{bmatrix} 32.15633 & 0 & 0 \\ 0 & 2.197733 & 0 \\ 0 & 0 & 0.374376 \\ 0 & 0 & 0 \\ 0 & 0 & 0 \end{bmatrix},$$

$$V^t = \begin{bmatrix} -0.055273 & -0.224442 & -0.972919 \\ -0.602286 & -0.769677 & 0.211773 \\ 0.796364 & -0.597681 & 0.092637 \end{bmatrix}$$

and

$$U = \begin{bmatrix} -0.038954 & -0.527903 & 0.778148 & -0.008907 & -0.337944 \\ -0.136702 & -0.589038 & -0.075997 & 0.243571 & 0.754483 \\ -0.294961 & -0.457453 & -0.435258 & -0.677268 & -0.235783 \\ -0.513732 & -0.133148 & -0.299632 & 0.659453 & -0.440105 \\ -0.793015 & 0.383877 & 0.330878 & -0.216849 & 0.259350 \end{bmatrix}$$

This produces $P(x) = 0.18 + 1.418571x - 0.0214286x^2$.

Exercise Set 10.1 (Page 636)

1. Use Theorem 10.5.

3. Use Theorem 10.5 for each of the partial derivatives.

5. b. With $\mathbf{x}^{(0)} = (0, 0)^t$ and tolerance 10^{-5}, we have $\mathbf{x}^{(13)} = (0.9999973, 0.9999973)^t$.

c. With $\mathbf{x}^{(0)} = (0, 0)^t$ and tolerance 10^{-5}, we have $\mathbf{x}^{(11)} = (0.9999984, 0.9999991)^t$.

7. a. With $\mathbf{x}^{(0)} = (1, 1, 1)^t$, we have $\mathbf{x}^{(5)} = (5.0000000, 0.0000000, -0.5235988)^t$.

b. With $\mathbf{x}^{(0)} = (1, 1, 1)^t$, we have $\mathbf{x}^{(9)} = (1.0364011, 1.0857072, 0.93119113)^t$.

c. With $\mathbf{x}^{(0)} = (0, 0, 0.5)^t$, we have $\mathbf{x}^{(5)} = (0.00000000, 0.09999999, 1.0000000)^t$.

d. With $\mathbf{x}^{(0)} = (0, 0, 0)^t$, we have $\mathbf{x}^{(5)} = (0.49814471, -0.19960600, -0.52882595)^t$.

9. a. With $\mathbf{x}^{(0)} = (1, 1, 1)^t$, we have $\mathbf{x}^{(3)} = (0.5000000, 0, -0.5235988)^t$.

b. With $\mathbf{x}^{(0)} = (1, 1, 1)^t$, we have $\mathbf{x}^{(4)} = (1.036400, 1.085707, 0.9311914)^t$.

c. With $\mathbf{x}^{(0)} = (0, 0, 0)^t$, we have $\mathbf{x}^{(3)} = (0, 0.1000000, 1.0000000)^t$.

d. With $\mathbf{x}^{(0)} = (0, 0, 0)^t$, we have $\mathbf{x}^{(4)} = (0.4981447, -0.1996059, -0.5288260)^t$.

11. A stable solution occurs when $x_1 = 8000$ and $x_2 = 4000$.

13. In this situation we have, for any matrix norm,

$$||\mathbf{F}(\mathbf{x}) - \mathbf{F}(\mathbf{x}_0)|| = ||A\mathbf{x} - A\mathbf{x}_0|| = ||A(\mathbf{x} - \mathbf{x}_0)|| \leq ||A|| \cdot ||\mathbf{x} - \mathbf{x}_0||.$$

The result follows by selecting $\delta = \varepsilon/||A||$, provided that $||A|| \neq 0$. When $||A|| = 0$, δ can be arbitrarily chosen, because A is the zero matrix.

Exercise Set 10.2 (Page 644)

1. a. $\mathbf{x}^{(2)} = (0.4958936, 1.983423)^t$ **b.** $\mathbf{x}^{(2)} = (-0.5131616, -0.01837622)^t$

c. $\mathbf{x}^{(2)} = (-23.942626, 7.6086797)^t$ **d.** $\mathbf{x}^{(1)}$ cannot be computed since $J(0)$ is singular.

3. a. $(0.5, 0.2)^t$ and $(1.1, 6.1)^t$ **b.** $(-0.35, 0.05)^t$, $(0.2, -0.45)^t$, $(0.4, -0.5)^t$ and $(1, -0.3)^t$

c. $(-1, 3.5)^t$, $(2.5, 4)^t$ **d.** $(0.11, 0.27)^t$

5. a. With $\mathbf{x}^{(0)} = (0.5, 2)^t, \mathbf{x}^{(3)} = (0.5, 2)^t$ With $\mathbf{x}^{(0)} = (1.1, 6.1), \mathbf{x}^{(3)} = (1.0967197, 6.0409329)^t$

b. With $\mathbf{x}^{(0)} = (-0.35, 0.05)^t, \mathbf{x}^{(3)} = (-0.37369822, 0.056266490^t$.
With $\mathbf{x}^{(0)} = (0.2, -0.45)^t, \mathbf{x}^{(4)} = (0.14783924, -0.43617762)^t$.
With $\mathbf{x}^{(0)} = (0.4, -0.5)^t, \mathbf{x}^{(3)} = (0.40809566, -0.49262939)^t$.
With $\mathbf{x}^{(0)} = (1, -0.3)^t, \mathbf{x}^{(4)} = (1.0330715, -0.27996184)^t$

c. With $\mathbf{x}^{(0)} = (-1, 3.5)^t$, $\mathbf{x}^{(1)} = (-1, 3.5)^t$ and $\mathbf{x}^{(0)} = (2.5, 4)^t$, $\mathbf{x}^{(3)} = (2.546947, 3.984998)^t$.

d. With $\mathbf{x}^{(0)} = (0.11, 0.27)^t$, $\mathbf{x}^{(6)} = (0.1212419, 0.2711051)^t$.

7. a. $\mathbf{x}^{(5)} = (0.5000000, 0.8660254)^t$ **b.** $\mathbf{x}^{(6)} = (1.772454, 1.772454)^t$

c. $\mathbf{x}^{(5)} = (-1.456043, -1.664230, 0.4224934)^t$ **d.** $\mathbf{x}^{(4)} = (0.4981447, -0.1996059, -0.5288260)^t$

9. With $\mathbf{x}^{(0)} = (1, 1 - 1)^t$ and $TOL = 10^{-6}$, we have $\mathbf{x}^{(20)} = (0.5, 9.5 \times 10^{-7}, -0.5235988)^t$.

11. When the dimension n is 1, $\mathbf{F}(\mathbf{x})$ is a one-component function $f(\mathbf{x}) = f_1(\mathbf{x})$, and the vector $\mathbf{x}$ has only one component $x_1 = x$. In this case, the Jacobian matrix $J(\mathbf{x})$ reduces to the 1×1 matrix $\left[\frac{\partial f_1}{\partial x_1}(\mathbf{x})\right] = f'(\mathbf{x}) = f'(x)$. Thus the vector equation

$$\mathbf{x}^{(k)} = \mathbf{x}^{(k-1)} - J(\mathbf{x}^{(k-1)})^{-1}\mathbf{F}(\mathbf{x}^{(k-1)})$$

becomes the scalar equation

$$x_k = x_{k-1} - f(x_{k-1})^{-1} f(x_{k-1}) = x_{k-1} - \frac{f(x_{k-1})}{f'(x_{k-1})}.$$

13. With $\theta_i^{(0)} = 1$, for each $i = 1, 2, \ldots, 20$, the following results are obtained.

i	1	2	3	4	5	6
$\theta_i^{(5)}$	0.14062	0.19954	0.24522	0.28413	0.31878	0.35045

i	7	8	9	10	11	12	13
$\theta_i^{(5)}$	0.37990	0.40763	0.43398	0.45920	0.48348	0.50697	0.52980

i	14	15	16	17	18	19	20
$\theta_i^{(5)}$	0.55205	0.57382	0.59516	0.61615	0.63683	0.65726	0.67746

Exercise Set 10.3 (Page 652)

1. a. $\mathbf{x}^{(2)} = (0.4777920, 1.927557)^t$ **b.** $\mathbf{x}^{(2)} = (-0.3250070, -0.1386967)^t$

c. $\mathbf{x}^{(2)} = (0.52293721, 0.82434906)^t$ **d.** $\mathbf{x}^{(2)} = (1.77949990, 1.74339606)^t$

3. a. $\mathbf{x}^{(8)} = (0.5, 2)^t$. **b.** $\mathbf{x}^{(9)} = (-0.3736982, 0.05626649)^t$.

c. $\mathbf{x}^{(9)} = (0.5, 0.8660254)^t$ **d.** $\mathbf{x}^{(8)} = (1.772454, 1.772454)^t$

5. **a.** With $\mathbf{x}^{(0)} = (2.5, 4)^t$, we have $\mathbf{x}^{(3)} = (2.546947, 3.984998)^t$.

b. With $\mathbf{x}^{(0)} = (0.11, 0.27)^t$, we have $\mathbf{x}^{(4)} = (0.1212419, 0.2711052)^t$.

c. With $\mathbf{x}^{(0)} = (1, 1, 1)^t$, we have $\mathbf{x}^{(3)} = (1.036401, 1.085707, 0.9311914)^t$.

d. With $\mathbf{x}^{(0)} = (1, -1, 1)^t$, we have $\mathbf{x}^{(8)} = (0.9, -1, 0.5)^t$; and with $\mathbf{x}^{(0)} = (1, 1, -1)^t$, we have $\mathbf{x}^{(8)} = (0.5, 1, -0.5)^t$.

7. With $\mathbf{x}^{(0)} = (1, 1 - 1)^t$, we have $\mathbf{x}^{(56)} = (0.5000591, 0.01057235, -0.5224818)^t$.

9. Let λ be an eigenvalue of $M = \left(I + \mathbf{u}v^t\right)$ with eigenvector $\mathbf{x} \neq \mathbf{0}$. Then $\lambda\mathbf{x} = M\mathbf{x} = \left(I + \mathbf{u}v^t\right)\mathbf{x} = \mathbf{x} + \left(v^t\mathbf{x}\right)\mathbf{u}$. Thus, $(\lambda - 1)\mathbf{x} = \left(v^t\mathbf{x}\right)\mathbf{u}$. If $\lambda = 1$, then $v^t\mathbf{x} = 0$. So $\lambda = 1$ is an eigenvalue of M with multiplicity $n - 1$ and eigenvectors $\mathbf{x}^{(1)}, \ldots, \mathbf{x}^{(n-1)}$ where $v^t\mathbf{x}^{(j)} = 0$, for $j = 1, \ldots, n - 1$. Assuming $\lambda \neq 1$ implies $\mathbf{x}$ and $\mathbf{u}$ are parallel. Suppose $\mathbf{x} = \alpha\mathbf{u}$. Then $(\lambda - 1)\alpha\mathbf{u} = \left(v^t(\alpha\mathbf{u})\right)\mathbf{u}$. Thus, $\alpha(\lambda - 1)\mathbf{u} = \alpha\left(v^t\mathbf{u}\right)\mathbf{u}$, which implies that $\lambda - 1 = v^t\mathbf{u}$ or $\lambda = 1 + v^t\mathbf{u}$. Hence, M has eigenvalues λ_i, $1 \leq i \leq n$ where $\lambda_i = 1$, for $i = 1, \ldots, n - 1$ and $\lambda_n = 1 + v^t\mathbf{u}$. Since $\det M = \prod_{i=1}^{n} \lambda_i$, we have $\det M = 1 + v^t\mathbf{u}$.

11. With $\mathbf{x}^{(0)} = (0.75, 1.25)^t$, we have $\mathbf{x}^{(4)} = (0.7501948, 1.184712)^t$. Thus, $a = 0.7501948, b = 1.184712$, and the error is 19.796.

Exercise Set 10.4 (Page 659)

1. **a.** With $\mathbf{x}^{(0)} = (0, 0)^t$, we have $\mathbf{x}^{(11)} = (0.4943541, 1.948040)^t$.

b. With $\mathbf{x}^{(0)} = (1, 1)^t$, we have $\mathbf{x}^{(2)} = (0.4970073, 0.8644143)^t$.

c. With $\mathbf{x}^{(0)} = (2, 2)^t$, we have $\mathbf{x}^{(1)} = (1.736083, 1.804428)^t$.

d. With $\mathbf{x}^{(0)} = (0, 0)^t$, we have $\mathbf{x}^{(2)} = (-0.3610092, 0.05788368)^t$.

3. **a.** $\mathbf{x}^{(3)} = (0.5, 2)^t$

b. $\mathbf{x}^{(3)} = (0.5, 0.8660254)^t$

c. $\mathbf{x}^{(4)} = (1.772454, 1.772454)^t$

d. $\mathbf{x}^{(3)} = (-0.3736982, 0.05626649)^t$

5. **a.** $\mathbf{x}^{(3)} = (1.036400, 1.085707, 0.9311914)^t$

b. $\mathbf{x}^{(3)} = (0.5, 1, -0.5)^t$

c. $\mathbf{x}^{(5)} = (-1.456043, -1.664230, 0.4224934)^t$

d. $\mathbf{x}^{(6)} = (0.0000000, 0.10000001, 1.0000000)^t$

Exercise Set 10.5 (Page 666)

1. **a.** $(3, -2.25)^t$

b. $(0.42105263, 2.6184211)^t$

c. $(2.173110, -1.3627731)^t$

3. Using $\mathbf{x}(0) = \mathbf{0}$ in all parts gives:

a. $(0.44006047, 1.8279835)^t$

b. $(-0.41342613, 0.096669468)^t$

c. $(0.49858909, 0.24999091, -0.52067978)^t$

d. $(6.1935484, 18.532258, -21.725806)^t$

5. **a.** With $x^(0) = (-1, 3.5)^t$ the result is $(-1, 3.5)^t$. With $\mathbf{x}(0) = (2.5, 4)^t$ the result is $(-1, 3.5)^t$.

b. With $\mathbf{x}(0) = (0.11, 0.27)^t$ the result is $(0.12124195, 0.27110516)^t$.

c. With $\mathbf{x}(0) = (1, 1, 1)^t$ the result is $(1.03640047, 1.08570655, 0.93119144)^t$.

d. With $\mathbf{x}(0) = (1, -1, 1)^t$ the result is $(0.90016074, -1.00238008, 0.496610937)^t$. With $\mathbf{x}(0) = (1, 1, -1)^t$ the result is $(0.50104035, 1.00238008, -0.49661093)^t$.

7. **a.** With $\mathbf{x}(0) = (-1, 3.5)^t$ the result is $(-1, 3.5)^t$. With $\mathbf{x}(0) = (2.5, 4)^t$ the result is $(2.5469465, 3.9849975)^t$.

b. With $\mathbf{x}(0) = (0.11, 0.27)^t$ the result is $(0.12124191, 0.27110516)^t$.

c. With $\mathbf{x}(0) = (1, 1, 1)^t$ the result is $(1.03640047, 1.08570655, 0.93119144)^t$.

d. With $\mathbf{x}(0) = (1, -1, 1)^t$ the result is $(0.90015964, -1.00021826, 0.49968944)^t$.
With $\mathbf{x}(0) = (1, 1, -1)^t$ the result is $(0.5009653, 1.00021826, -0.49968944)^t$.

9. $(0.50024553, 0.078230039, -0.52156996)^t$

11. For each λ, we have

$$0 = G(\lambda, \mathbf{x}(\lambda)) = F(\mathbf{x}(\lambda)) - e^{-\lambda}F(\mathbf{x}(0)),$$

so

$$0 = \frac{\partial F(\mathbf{x}(\lambda))}{\partial \mathbf{x}} \frac{d\mathbf{x}}{d\lambda} + e^{-\lambda}F(\mathbf{x}(0)) = J(\mathbf{x}(\lambda))\mathbf{x}'(\lambda) + e^{-\lambda}F(\mathbf{x}(0))$$

and

$$J(\mathbf{x}(\lambda))\mathbf{x}'(\lambda) = -e^{-\lambda}F(\mathbf{x}(0)) = -F(\mathbf{x}(0)).$$

Thus

$$\mathbf{x}'(\lambda) = -J(\mathbf{x}(\lambda))^{-1}F(\mathbf{x}(0)).$$

With $N = 1$, we have $h = 1$ so that

$$\mathbf{x}(1) = \mathbf{x}(0) - J(\mathbf{x}(0))^{-1}F(\mathbf{x}(0)).$$

However, Newton's method gives

$$\mathbf{x}^{(1)} = \mathbf{x}^{(0)} - J(\mathbf{x}^{(0)})^{-1}F(\mathbf{x}^{(0)}).$$

Since $\mathbf{x}(0) = \mathbf{x}^{(0)}$, we have $\mathbf{x}(1) = \mathbf{x}^{(1)}$.

Exercise Set 11.1 (Page 677)

1. The Linear Shooting Algorithm gives the results in the following tables.

a.

i	x_i	w_{1i}	$y(x_i)$
1	0.5	0.82432432	0.82402714

b.

i	x_i	w_{1i}	$y(x_i)$
1	0.25	0.3937095	0.3936767
2	0.50	0.8240948	0.8240271
3	0.75	1.337160	1.337086

3. The Linear Shooting Algorithm gives the results in the following tables.

a.

i	x_i	w_{1i}	$y(x_i)$
3	0.3	0.7833204	0.7831923
6	0.6	0.6023521	0.6022801
9	0.9	0.8568906	0.8568760

b.

i	x_i	w_{1i}	$y(x_i)$
5	1.25	0.1676179	0.1676243
10	1.50	0.4581901	0.4581935
15	1.75	0.6077718	0.6077740

c.

i	x_i	w_{1i}	$y(x_i)$
3	0.3	−0.5185754	−0.5185728
6	0.6	−0.2195271	−0.2195247
9	0.9	−0.0406577	−0.0406570

d.

i	x_i	w_{1i}	$y(x_i)$
3	1.3	0.0655336	0.06553420
6	1.6	0.0774590	0.07745947
9	1.9	0.0305619	0.03056208

5. The Linear Shooting Algorithm with $h = 0.05$ gives the following results.

i	x_i	w_{1i}
6	0.3	0.04990547
10	0.5	0.00673795
16	0.8	0.00033755

The Linear Shooting Algorithm with $h = 0.1$ gives the following results.

i	x_i	w_{1i}
3	0.3	0.05273437
5	0.5	0.00741571
8	0.8	0.00038976

7. a. The approximate potential is $u(3) \approx 36.66702$ using $h = 0.1$.

b. The actual potential is $u(3) = 36.66667$.

9. a. There are no solutions if b is an integer multiple of π and $B \neq 0$.

b. A unique solution exists whenever b is not an integer multiple of π.

c. There is an infinite number of solutions if b is an multiple integer of π and $B = 0$.

Exercise Set 11.2 (Page 684)

1. The Nonlinear Shooting Algorithm gives $w_1 = 0.405505 \approx \ln 1.5 = 0.405465$.

3. The Nonlinear Shooting Algorithm gives the results in the following tables.

a.

i	x_i	w_{1i}	$y(x_i)$	w_{2i}
2	1.20000000	0.18232094	0.18232156	0.83333370
4	1.40000000	0.33647129	0.33647224	0.71428547
6	1.60000000	0.47000243	0.47000363	0.62499939
8	1.80000000	0.58778522	0.58778666	0.55555468

Convergence in 4 iterations $t = 1.0000017$.

b.

i	x_i	w_{1i}	$y(x_i)$	w_{2i}
2	0.31415927	1.36209813	1.36208552	1.29545926
4	0.62831853	1.80002060	1.79999746	1.45626846
6	0.94247780	2.24572329	2.24569937	1.32001776
8	1.25663706	2.58845757	2.58844295	0.79988757

Convergence in 4 iterations $t = 1.0000301$.

c.

i	x_i	w_{1i}	$y(x_i)$	w_{2i}
1	0.83775804	0.86205941	0.86205848	0.38811718
2	0.89011792	0.88156057	0.88155882	0.35695076
3	0.94247780	0.89945618	0.89945372	0.32675844
4	0.99483767	0.91579268	0.91578959	0.29737141

Convergence in 3 iterations $t = 0.42046725$.

d.

i	x_i	w_{1i}	$y(x_i)$	w_{2i}
4	0.62831853	2.58784539	2.58778525	0.80908243
8	1.25663706	2.95114591	2.95105652	0.30904693
12	1.88495559	2.95115520	2.95105652	−0.30901625
16	2.51327412	2.58787536	2.58778525	−0.80904433

Convergence in 6 iterations $t = 1.0001253$.

5. a. Modify Algorithm 11.2 as follows:

Step 1 Set $h = (b - a)/N$;
$k = 2$;
$TK1 = (\beta - \alpha)/(b - a)$.

Step 2 Set $w_{1,0} = \alpha$;
$w_{2,0} = TK1$.

Step 3 For $i = 1, \ldots, N$ do Steps 4 and 5.

Step 4 Set $x = a + (i - 1)h$.

Step 5 Set

$k_{1,1} = hw_{2,i-1}$;
$k_{1,2} = hf(x, w_{1,i-1}, w_{2,i-1})$;
$k_{2,1} = h(w_{2,i-1} + k_{1,2}/2)$;
$k_{2,2} = hf(x + h/2, w_{1,i-1} + k_{1,1}/2, w_{2,i-1} + k_{1,2}/2)$;
$k_{3,1} = h(w_{2,i-1} + k_{2,2}/2)$;
$k_{3,2} = hf(x + h/2, w_{1,i-1} + k_{2,1}/2, w_{2,i-1} + k_{2,2}/2)$;
$k_{4,1} = h(w_{2,i-1} + k_{3,2}/2)$;
$k_{4,2} = hf(x + h/2, w_{1,i-1} + k_{3,1}, w_{2,i-1} + k_{3,2})$;
$w_{1,i} = w_{1,i-1} + (k_{1,1} + 2k_{2,1} + 2k_{3,1} + k_{4,1})/6$;
$w_{2,i} = w_{2,i-1} + (k_{1,2} + 2k_{2,2} + 2k_{3,2} + k_{4,2})/6$.

Step 6 Set $TK2 = TK1 + (\beta - w_{1,N})/(b - a)$.

Step 7 While ($k \le M$) do Steps 8–15.

Step 8 Set $w_{2,0} = TK2$;
$HOLD = w_{1,N}$.

Step 9 For $i = 1, \ldots, N$ do Steps 10 and 11.

Step 10 (Same as Step 4)

Step 11 (Same as Step 5)

Step 12 If $|w_{1,N} - \beta| \le TOL$ then do Steps 13 and 14.

Step 13 For $i = 0, \ldots, N$ set $x = a + ih$;
OUTPUT($x, w_{1,i}, w_{2,i}$).

Step 14 STOP.

Step 15 Set
$TK = TK2 - (w_{1,N} - \beta)(TK2 - TK1)/(w_{1,N} - HOLD)$;
$TK1 = TK2$;
$TK2 = TK$;
$k = k + 1$.

Step 16 OUTPUT('Maximum number of iterations exceeded.');
STOP.

b. (3a) 3 iterations:

i	x_i	w_i	$y(x_i)$
1	1.2	0.45453896	0.45454545
2	1.4	0.41665348	0.41666667
3	1.6	0.38459538	0.38461538
4	1.8	0.35711592	0.35714286

(3c) 3 iterations:

i	x_i	w_i	$y(x_i)$
1	2.2	1.24299575	1.24300281
2	2.4	1.29211897	1.29213540
3	2.6	1.34009800	1.34012683
4	2.8	1.38671706	1.38676227

Exercise Set 11.3 (Page 689)

1. The Linear Finite-Difference Algorithm gives following results.

a.

i	x_i	w_{1i}	$y(x_i)$
1	0.5	0.83333333	0.82402714

b.

i	x_i	w_{1i}	$y(x_i)$
1	0.25	0.39512472	0.39367669
2	0.5	0.82653061	0.82402714
3	0.75	1.33956916	1.33708613

c. $\dfrac{4(0.82653061) - 0.83333333}{3} = 0.82426304$

3. The Linear Finite-Difference Algorithm gives the results in the following tables.

a.

i	x_i	w_i	$y(x_i)$
2	0.2	1.018096	1.0221404
5	0.5	0.5942743	0.59713617
7	0.7	0.6514520	0.65290384

b.

i	x_i	w_i	$y(x_i)$
5	1.25	0.16797186	0.16762427
10	1.50	0.45842388	0.45819349
15	1.75	0.60787334	0.60777401

c.

i	x_i	w_{1i}	$y(x_i)$
3	0.3	−0.5183084	−0.5185728
6	0.6	−0.2192657	−0.2195247
9	0.9	−0.0405748	−0.04065697

d.

i	x_i	w_{1i}	$y(x_i)$
3	1.3	0.0654387	0.0655342
6	1.6	0.0773936	0.0774595
9	1.9	0.0305465	0.0305621

5. The Linear Finite-Difference Algorithm gives the results in the following tables.

i	x_i	$w_i(h = 0.1)$
3	0.3	0.05572807
6	0.6	0.00310518
9	0.9	0.00016516

i	x_i	$w_i(h = 0.05)$
6	0.3	0.05132396
12	0.6	0.00263406
18	0.9	0.00013340

7. **a.** The approximate deflections are shown in the following table.

i	x_i	w_{1i}
5	30	0.0102808
10	60	0.0144277
15	90	0.0102808

b. Yes.

c. Yes. Maximum deflection occurs at $x = 60$. The exact solution is within tolerance, but the approximation is not.

Exercise Set 11.4 (Page 696)

1. The Nonlinear Finite-Difference Algorithm gives the following results.

i	x_i	w_i	$y(x_i)$
1	1.5	0.4067967	0.4054651

3. The Nonlinear Finite-Difference Algorithm gives the results in the following tables.

a.

i	x_i	w_i	$y(x_i)$
2	1.20000000	0.18220299	0.18232156
4	1.40000000	0.33632929	0.33647224
6	1.60000000	0.46988413	0.47000363
8	1.80000000	0.58771808	0.58778666

Convergence in 3 iterations

b.

i	x_i	w_i	$y(x_i)$
2	0.31415927	1.36244080	1.36208552
4	0.62831853	1.80138559	1.79999746
6	0.94247780	2.24819259	2.24569937
8	1.25663706	2.59083695	2.58844295

Convergence in 3 iterations

c.

i	x_i	w_i	$y(x_i)$
1	0.83775804	0.86205907	0.86205848
2	0.89011792	0.88155964	0.88155882
3	0.94247780	0.89945447	0.89945372
4	0.99483767	0.91579005	0.91578959

Convergence in 2 iterations

d.

i	x_i	w_i	$y(x_i)$
4	0.62831853	2.58932301	2.58778525
8	1.25663706	2.95378037	2.95105652
12	1.88495559	2.95378037	2.95105652
16	2.51327412	2.58932301	2.58778525

Convergence in 4 iterations

5. b. For (4a)

x_i	$w_i(h=0.2)$	$w_i(h=0.1)$	$w_i(h=0.05)$	$EXT_{1,i}$	$EXT_{2,i}$	$EXT_{3,i}$
1.2	0.45458862	0.45455753	0.45454935	0.45454717	0.45454662	0.45454659
1.4	0.41672067	0.41668202	0.41667179	0.41666914	0.41666838	0.41666833
1.6	0.38466137	0.38462855	0.38461984	0.38461761	0.38461694	0.38461689
1.8	0.35716943	0.35715045	0.35714542	0.35714412	0.35714374	0.35714372

For (4c)

x_i	$w_i(h=0.2)$	$w_i(h=0.1)$	$w_i(h=0.05)$	$EXT_{1,i}$	$EXT_{2,i}$	$EXT_{3,i}$
1.2	2.0340273	2.0335158	2.0333796	2.0333453	2.0333342	2.0333334
1.4	2.1148732	2.1144386	2.1143243	2.1142937	2.1142863	2.1142858
1.6	2.2253630	2.2250937	2.2250236	2.2250039	2.2250003	2.2250000
1.8	2.3557284	2.3556001	2.3555668	2.3555573	2.3555556	2.3355556

7. The Jacobian matrix $J = (a_{i,j})$ is tridiagonal with entries given in (11.21). So

$$\begin{aligned}
a_{1,1} &= 2 + h^2 f_y\left(x_1, w_1, \frac{1}{2h}(w_2 - \alpha)\right),\\
a_{1,2} &= -1 + \frac{h}{2} f_{y'}\left(x_1, w_1, \frac{1}{2h}(w_2 - \alpha)\right),\\
a_{i,i-1} &= -1 - \frac{h}{2} f_{y'}\left(x_i, w_i, \frac{1}{2h}(w_{i+1} - w_{i-1})\right), \quad \text{for } 2 \le i \le N-1\\
a_{i,i} &= 2 + h^2 f_y\left(x_i, w_i, \frac{1}{2h}(w_{i+1} - w_{i-1})\right), \quad \text{for } 2 \le i \le N-1\\
a_{i,i+1} &= -1 + \frac{h}{2} f_{y'}\left(x_i, w_i, \frac{1}{2h}(w_{i+1} - w_{i-1})\right), \quad \text{for } 2 \le i \le N-1\\
a_{N,N-1} &= -1 - \frac{h}{2} f_{y'}\left(x_N, w_N, \frac{1}{2h}(\beta - w_{N-1})\right),\\
a_{N,N} &= 2 + h^2 f_y\left(x_N, w_N, \frac{1}{2h}(\beta - w_{N-1})\right).
\end{aligned}$$

Thus, $|a_{i,i}| \ge 2 + h^2\delta$, for $i = 1, \ldots, N$. Since $|f_{y'}(x, y, y')| \le L$ and $h < 2/L$,

$$\left|\frac{h}{2} f_{y'}(x, y, y')\right| \le \frac{hL}{2} < 1.$$

So

$$|a_{1,2}| = \left| -1 + \frac{h}{2} f_{y'}\left(x_1, w_1, \frac{1}{2h}(w_2 - \alpha)\right)\right| < 2 < |a_{1,1}|,$$

$$\begin{aligned}
|a_{i,i-1}| + |a_{i,i+1}| &= -a_{i,i-1} - a_{i,i+1}\\
&= 1 + \frac{h}{2} f_{y'}\left(x_i, w_i, \frac{1}{2h}(w_{i+1} - w_{i-1})\right) + 1 - \frac{h}{2} f_{y'}\left(x_i, w_i, \frac{1}{2h}(w_{i+1} - w_{i-1})\right)\\
&= 2 \le |a_{i,i}|,
\end{aligned}$$

and

$$|a_{N,N-1}| = -a_{N,N-1} = 1 + \frac{h}{2} f_{y'}\left(x_N, w_N, \frac{1}{2h}(\beta - w_{N-1})\right) < 2 < |a_{N,N}|.$$

By Theorem 6.31, the matrix J is nonsingular.

Exercise Set 11.5 (Page 710)

1. The Piecewise Linear Algorithm gives $\phi(x) = -0.07713274\phi_1(x) - 0.07442678\phi_2(x)$. The actual values are $y(x_1) = -0.07988545$ and $y(x_2) = -0.07712903$.

3. The Piecewise Linear Algorithm gives the results in the following tables.

a.

i	x_i	$\phi(x_i)$	$y(x_i)$
3	0.3	−0.212333	−0.21
6	0.6	−0.241333	−0.24
9	0.9	−0.090333	−0.09

b.

i	x_i	$\phi(x_i)$	$y(x_i)$
3	0.3	0.1815138	0.1814273
6	0.6	0.1805502	0.1804753
9	0.9	0.05936468	0.05934303

c.

i	x_i	$\phi(x_i)$	$y(x_i)$
5	0.25	−0.3585989	−0.3585641
10	0.50	−0.5348383	−0.5347803
15	0.75	−0.4510165	−0.4509614

d.

i	x_i	$\phi(x_i)$	$y(x_i)$
5	0.25	−0.1846134	−0.1845204
10	0.50	−0.2737099	−0.2735857
15	0.75	−0.2285169	−0.2284204

5. The Cubic Spline Algorithm gives the results in the following tables.

a.

i	x_i	$\phi(x_i)$	$y(x_i)$
3	0.3	−0.2100000	−0.21
6	0.6	−0.2400000	−0.24
9	0.9	−0.0900000	−0.09

b.

i	x_i	$\phi(x_i)$	$y(x_i)$
3	0.3	0.1814269	0.1814273
6	0.6	0.1804753	0.1804754
9	0.9	0.05934321	0.05934303

c.

i	x_i	$\phi(x_i)$	$y(x_i)$
5	0.25	−0.3585639	−0.3585641
10	0.50	−0.5347779	−0.5347803
15	0.75	−0.4509109	−0.4509614

d.

i	x_i	$\phi(x_i)$	$y(x_i)$
5	0.25	−0.1845191	−0.1845204
10	0.50	−0.2735833	−0.2735857
15	0.75	−0.2284186	−0.2284204

7.

i	x_i	$\phi(x_i)$	$y(x_i)$
3	0.3	1.0408182	1.0408182
6	0.6	1.1065307	1.1065306
9	0.9	1.3065697	1.3065697

9. A change in variable $w = (x - a)/(b - a)$ gives the boundary value problem

$$-\frac{d}{dw}(p((b-a)w + a)y') + (b-a)^2 q((b-a)w + a)y = (b-a)^2 f((b-a)w + a),$$

where $0 < w < 1$, $y(0) = \alpha$, and $y(1) = \beta$. Then Exercise 6 can be used.

13. For $\mathbf{c} = (c_0, c_1, \ldots, c_{n+1})^t$ and $\phi(x) = \sum_{i=0}^{n+1} c_i\phi_i(x)$, we have

$$\mathbf{c}^t A\mathbf{c} = \int_0^1 p(x)[\phi'(x)]^2 + q(x)[\phi(x)]^2\, dx.$$

But $p(x) > 0$ and $q(x)[\phi(x)]^2 \geq 0$, so $\mathbf{c}^t A\mathbf{c} \geq 0$, and it can be 0, for $\mathbf{x} \neq 0$, only if $\phi'(x) \equiv 0$ on $[0, 1]$. However, $\{\phi_0', \phi_1', \ldots, \phi_{n+1}'\}$ is linearly independent, so $\phi'(x) \neq 0$ on $[0, 1]$ and $\mathbf{c}^t A\mathbf{c} = 0$ if and only if $\mathbf{c} = \mathbf{0}$.

Exercise Set 12.1 (Page 723)

1. The Poisson Equation Finite-Difference Algorithm gives the following results.

i	j	x_i	y_j	$w_{i,j}$	$u(x_i, y_j)$
1	1	0.5	0.5	0.0	0
1	2	0.5	1.0	0.25	0.25
1	3	0.5	1.5	1.0	1

3. The Poisson Equation Finite-Difference Algorithm gives the following results.

a. 30 iterations required:

i	j	x_i	y_j	$w_{i,j}$	$u(x_i, y_j)$
2	2	0.4	0.4	0.1599988	0.16
2	4	0.4	0.8	0.3199988	0.32
4	2	0.8	0.4	0.3199995	0.32
4	4	0.8	0.8	0.6399996	0.64

b. 29 iterations required:

i	j	x_i	y_j	$w_{i,j}$	$u(x_i, y_j)$
2	1	1.256637	0.3141593	0.2951855	0.2938926
2	3	1.256637	0.9424778	0.1830822	0.1816356
4	1	2.513274	0.3141593	−0.7721948	−0.7694209
4	3	2.513274	0.9424778	−0.4785169	−0.4755283

c. 126 iterations required:

i	j	x_i	y_j	$w_{i,j}$	$u(x_i, y_j)$
4	3	0.8	0.3	1.2714468	1.2712492
4	7	0.8	0.7	1.7509414	1.7506725
8	3	1.6	0.3	1.6167917	1.6160744
8	7	1.6	0.7	3.0659184	3.0648542

d. 127 iterations required:

i	j	x_i	y_j	$w_{i,j}$	$u(x_i, y_j)$
2	2	1.2	1.2	0.5251533	0.5250861
4	4	1.4	1.4	1.3190830	1.3189712
6	6	1.6	1.6	2.4065150	2.4064186
8	8	1.8	1.8	3.8088995	3.8088576

7. The approximate potential at some typical points are as follows.

i	j	x_i	y_j	$w_{i,j}$
1	4	0.1	0.4	88
2	1	0.2	0.1	66
4	2	0.4	0.2	66

Exercise Set 12.2 (Page 736)

1. The Heat Equation Backward-Difference Algorithm gives the following results.

a.

i	j	x_i	t_j	w_{ij}	$u(x_i, t_j)$
1	1	0.5	0.05	0.632952	0.652037
2	1	1.0	0.05	0.895129	0.883937
3	1	1.5	0.05	0.632952	0.625037
1	2	0.5	0.1	0.566574	0.552493
2	2	1.0	0.1	0.801256	0.781344
3	2	1.5	0.1	0.566574	0.552493

3. The Crank-Nicolson Algorithm gives the following results.

i	j	x_i	t_j	w_{ij}	$u(x_i, t_j)$
1	1	0.5	0.05	0.628848	0.652037
2	1	1.0	0.05	0.889326	0.883937
3	1	1.5	0.05	0.628848	0.625037
1	2	0.5	0.1	0.559251	0.552493
2	2	1.0	0.1	0.790901	0.781344
3	2	1.5	0.1	0.559252	0.552493

5. The Forward-Difference Algorithm gives the following results.

a. For $h = 0.4$ and $k = 0.1$:

i	j	x_i	t_j	w_{ij}	$u(x_i, t_j)$
2	5	0.8	0.5	3.035630	0
3	5	1.2	0.5	−3.035630	0
4	5	1.6	0.5	1.876122	0

For $h = 0.4$ and $k = 0.05$:

i	j	x_i	t_j	w_{ij}	$u(x_i, t_j)$
2	10	0.8	0.5	0	0
3	10	1.2	0.5	0	0
4	10	1.6	0.5	0	0

b. For $h = \frac{\pi}{10}$ and $k = 0.05$:

i	j	x_i	t_j	w_{ij}	$u(x_i, t_j)$
3	10	0.94247780	0.5	0.4864832	0.4906936
6	10	1.88495559	0.5	0.5718943	0.5768449
9	10	2.82743339	0.5	0.1858197	0.1874283

7. a. For $h = 0.4$ and $k = 0.1$:

i	j	x_i	t_j	$w_{i,j}$	$u(x_i, t_j)$
2	5	0.8	0.5	−0.00258	0
3	5	1.2	0.5	0.00258	0
4	5	1.6	0.5	−0.00159	0

For $h = 0.4$ and $k = 0.05$:

i	j	x_i	t_j	$w_{i,j}$	$u(x_i, t_j)$
2	10	0.8	0.5	-4.93×10^{-4}	0
3	10	1.2	0.5	4.93×10^{-4}	0
4	10	1.6	0.5	-3.05×10^{-4}	0

b. For $h = \frac{\pi}{10}$ and $k = 0.05$:

i	j	x_i	t_j	$w_{i,j}$	$u(x_i, t_j)$
3	10	0.94247780	0.5	0.4986092	0.4906936
6	10	1.88495559	0.5	0.5861503	0.5768449
9	10	2.82743339	0.5	0.1904518	0.1874283

9. The Crank-Nicolson Algorithm gives the following results.

a. For $h = 0.4$ and $k = 0.1$:

i	j	x_i	t_j	w_{ij}	$u(x_i, t_j)$
2	5	0.8	0.5	8.2×10^{-7}	0
3	5	1.2	0.5	-8.2×10^{-7}	0
4	5	1.6	0.5	5.1×10^{-7}	0

For $h = 0.4$ and $k = 0.05$:

i	j	x_i	t_j	w_{ij}	$u(x_i, t_j)$
2	10	0.8	0.5	-2.6×10^{-6}	0
3	10	1.2	0.5	2.6×10^{-6}	0
4	10	1.6	0.5	-1.6×10^{-6}	0

b. For $h = \frac{\pi}{10}$ and $k = 0.05$:

i	j	x_i	t_j	w_{ij}	$u(x_i, t_j)$
3	10	0.94247780	0.5	0.4926589	0.4906936
6	10	1.88495559	0.5	0.5791553	0.5768449
9	10	2.82743339	0.5	0.1881790	0.1874283

11. a. Using $h = 0.4$ and $k = 0.1$ leads to meaningless results. Using $h = 0.4$ and $k = 0.05$ again gives meaningless answers. Letting $h = 0.4$ and $k = 0.005$ produces the following:

i	j	x_i	t_j	w_{ij}
1	100	0.4	0.5	−165.405
2	100	0.8	0.5	267.613
3	100	1.2	0.5	−267.613
4	100	1.6	0.5	165.405

b.

i	j	x_i	t_j	$w(x_{ij})$
3	10	0.94247780	0.5	0.46783396
6	10	1.8849556	0.5	0.54995267
9	10	2.8274334	0.5	0.17871220

13. We have

$$a_{11}v_1^{(i)} + a_{12}v_2^{(i)} = (1-2\lambda)\sin\frac{i\pi}{m} + \lambda\sin\frac{2\pi i}{m}$$

and

$$\begin{aligned}\mu_i v_1^{(i)} &= \left[1-4\lambda\left(\sin\frac{i\pi}{2m}\right)^2\right]\sin\frac{i\pi}{m} = \left[1-4\lambda\left(\sin\frac{i\pi}{2m}\right)^2\right]\left(2\sin\frac{i\pi}{2m}\cos\frac{i\pi}{2m}\right)\\ &= 2\sin\frac{i\pi}{2m}\cos\frac{i\pi}{2m} - 8\lambda\left(\sin\frac{i\pi}{2m}\right)^3\cos\frac{i\pi}{2m}.\end{aligned}$$

However,

$$\begin{aligned}(1-2\lambda)\sin\frac{i\pi}{m} + \lambda\sin\frac{2\pi i}{m} =& 2(1-2\lambda)\sin\frac{i\pi}{2m}\cos\frac{i\pi}{2m} + 2\lambda\sin\frac{i\pi}{m}\cos\frac{i\pi}{m}\\ =& 2(1-2\lambda)\sin\frac{i\pi}{2m}\cos\frac{i\pi}{2m}\\ &+ 2\lambda\left[2\sin\frac{i\pi}{2m}\cos\frac{i\pi}{2m}\right]\left[1-2\left(\sin\frac{i\pi}{2m}\right)^2\right]\\ =& 2\sin\frac{i\pi}{2m}\cos\frac{i\pi}{2m} - 8\lambda\cos\frac{i\pi}{2m}\left[\sin\frac{i\pi}{2m}\right]^3.\end{aligned}$$

Thus

$$a_{11}v_1^{(i)} + a_{12}v_2^{(i)} = \mu_i v_1^{(i)}.$$

Further

$$\begin{aligned}a_{j,j-1}v_{j-1}^{(i)} + a_{j,j}v_j^{(i)} + a_{j,j+1}v_{j+1}^{(i)} =& \lambda\sin\frac{i(j-1)\pi}{m} + (1-2\lambda)\sin\frac{ij\pi}{m} + \lambda\sin\frac{i(j+1)\pi}{m}\\ =& \lambda\left(\sin\frac{ij\pi}{m}\cos\frac{i\pi}{m} - \sin\frac{i\pi}{m}\cos\frac{ij\pi}{m}\right) + (1-2\lambda)\sin\frac{ij\pi}{m}\\ &+ \lambda\left(\sin\frac{ij\pi}{m}\cos\frac{i\pi}{m} + \sin\frac{i\pi}{m}\cos\frac{ij\pi}{m}\right)\\ =& \sin\frac{ij\pi}{m} - 2\lambda\sin\frac{ij\pi}{m} + 2\lambda\sin\frac{ij\pi}{m}\cos\frac{i\pi}{m}\\ =& \sin\frac{ij\pi}{m} + 2\lambda\sin\frac{ij\pi}{m}\left(\cos\frac{i\pi}{m} - 1\right)\end{aligned}$$

and

$$\begin{aligned}\mu_i v_j^{(i)} &= \left[1-4\lambda\left(\sin\frac{i\pi}{2m}\right)^2\right]\sin\frac{ij\pi}{m} = \left[1-4\lambda\left(\frac{1}{2}-\frac{1}{2}\cos\frac{i\pi}{m}\right)\right]\sin\frac{ij\pi}{m}\\ &= \left[1+2\lambda\left(\cos\frac{i\pi}{m}-1\right)\right]\sin\frac{ij\pi}{m},\end{aligned}$$

so

$$a_{j,j-1}v_{j-1}^{(i)} + a_{j,j}v_j^{(i)} + a_{j,j+1}v_j^{(i)} = \mu_i v_j^{(i)}.$$

Similarly,

$$a_{m-2,m-1}v_{m-2}^{(i)} + a_{m-1,m-1}v_{m-1}^{(i)} = \mu_i v_{m-1}^{(i)},$$

so $A\mathbf{v}^{(i)} = \mu_i \mathbf{v}^{(i)}$.

15. To modify Algorithm 12.2, change the following:

Step 7 Set

$$t = jk;$$
$$z_1 = (w_1 + kF(h))/l_1.$$

Step 8 For $i = 2, \ldots, m-1$ set

$$z_i = (w_i + kF(ih) + \lambda z_{i-1})/l_i.$$

To modify Algorithm 12.3, change the following:
Step 7 Set

$$t = jk;$$
$$z_1 = \left[(1-\lambda)w_1 + \frac{\lambda}{2}w_2 + kF(h)\right]\Big/ l_1.$$

Step 8 For $i = 2, \ldots, m-1$ set

$$z_i = \left[(1-\lambda)w_i + \frac{\lambda}{2}(w_{i+1} + w_{i-1} + z_{i-1}) + kF(ih)\right]\Big/ l_i.$$

17. To modify Algorithm 12.2, change the following:

Step 7 Set
$t = jk;$
$w_0 = \phi(t);$
$z_1 = (w_1 + \lambda w_0)/l_1.$
$w_m = \psi(t).$

Step 8 For $i = 2, \ldots, m-2$ set

$$z_i = (w_i + \lambda z_{i-1})/l_i;$$

Set
$z_{m-1} = (w_{m-1} + \lambda w_m + \lambda z_{m-2})/l_{m-1}.$

Step 11 OUTPUT (t);
For $i = 0, \ldots, m$ set $x = ih$;
OUTPUT (x, w_i).

To modify Algorithm 12.3, change the following:

Step 1 Set
$h = l/m;$
$k = T/N;$
$\lambda = \alpha^2 k/h^2;$
$w_m = \psi(0);$
$w_0 = \phi(0).$

Step 7 Set
$t = jk;$
$z_1 = \left[(1-\lambda)w_1 + \frac{\lambda}{2}w_2 + \frac{\lambda}{2}{}_0 + \frac{\lambda}{2}\phi(t)\right]/l_1;$
$w_0 = \phi(t).$

Step 8 For $i = 2, \ldots, m-2$ set

$$z_i = \left[(1-\lambda)w_i + \tfrac{\lambda}{2}(w_{i+1} + w_{i-1} + z_{i-1})\right]/l_i;$$

Set
$z_{m-1} = \left[(1-\lambda)w_{m-1} + \frac{\lambda}{2}(w_m + w_{m-2} + z_{m-2} + \psi(t))\right]/l_{m-1};$
$w_m = \psi(t).$

Step 11 OUTPUT (t);
For $i = 0, \ldots, m$ set $x = ih$;
OUTPUT (x, w_i).

19. a. The approximate temperature at some typical points is given in the table.

i	j	r_i	t_j	$w_{i,j}$
1	20	0.6	10	137.6753
2	20	0.7	10	245.9678
3	20	0.8	10	340.2862
4	20	0.9	10	424.1537

The strain is approximately $I = 1242.537$.

Exercise Set 12.3 (Page 744)

1. The Wave Equation Finite-Difference Algorithm gives the following results.

i	j	x_i	t_j	w_{ij}	$u(x_i, t_j)$
2	4	0.25	1.0	−0.7071068	−0.7071068
3	4	0.50	1.0	−1.0000000	−1.0000000
4	4	0.75	1.0	−0.7071068	−0.7071068

3. The Wave Equation Finite-Difference Algorithm with $h = \frac{\pi}{10}$ and $k = 0.05$ gives the following results.

i	j	x_i	t_j	w_{ij}	$u(x_i, t_j)$
2	10	$\frac{\pi}{5}$	0.5	0.5163933	0.5158301
5	10	$\frac{\pi}{2}$	0.5	0.8785407	0.8775826
8	10	$\frac{4\pi}{5}$	0.5	0.5163933	0.5158301

The Wave Equation Finite-Difference Algorithm with $h = \frac{\pi}{20}$ and $k = 0.1$ gives the following results.

i	j	x_i	t_j	w_{ij}
4	5	$\frac{\pi}{5}$	0.5	0.5159163
10	5	$\frac{\pi}{2}$	0.5	0.8777292
16	5	$\frac{4\pi}{5}$	0.5	0.5159163

The Wave Equation Finite-Difference Algorithm with $h = \frac{\pi}{20}$ and $k = 0.05$ gives the following results.

i	j	x_i	t_j	w_{ij}
4	10	$\frac{\pi}{5}$	0.5	0.5159602
10	10	$\frac{\pi}{2}$	0.5	0.8778039
16	10	$\frac{4\pi}{5}$	0.5	0.5159602

5. The Wave Equation Finite-Difference Algorithm gives the following results.

i	j	x_i	t_j	w_{ij}	$u(x_i, t_j)$
2	3	0.2	0.3	0.6729902	0.61061587
5	3	0.5	0.3	0	0
8	3	0.8	0.3	−0.6729902	−0.61061587

7. a. The air pressure for the open pipe is $p(0.5, 0.5) \approx 0.9$ and $p(0.5, 1.0) \approx 2.7$.

b. The air pressure for the closed pipe is $p(0.5, 0.5) \approx 0.9$ and $p(0.5, 1.0) \approx 0.9187927$.

Exercise Set 12.4 (Page 758)

1. With $E_1 = (0.25, 0.75)$, $E_2 = (0, 1)$, $E_3 = (0.5, 0.5)$, and $E_4 = (0, 0.5)$, the basis functions are

$$\phi_1(x,y) = \begin{cases} 4x & \text{on } T_1 \\ -2+4y & \text{on } T_2, \end{cases}$$

$$\phi_2(x,y) = \begin{cases} -1-2x+2y & \text{on } T_1 \\ 0 & \text{on } T_2, \end{cases}$$

$$\phi_3(x,y) = \begin{cases} 0 & \text{on } T_1 \\ 1+2x-2y & \text{on } T_2, \end{cases}$$

$$\phi_4(x,y) = \begin{cases} 2-2x-2y & \text{on } T_1 \\ 2-2x-2y & \text{on } T_2, \end{cases}$$

and $\gamma_1 = 0.323825$, $\gamma_2 = 0$, $\gamma_3 = 1.0000$, and $\gamma_4 = 0$.

3. The Finite-Element Algorithm with $K = 8, N = 8, M = 32, n = 9, m = 25$, and $NL = 0$ gives the following results, where the labeling is as shown in the diagram.

```
10———11———12———13———14
|\ 10 |\ 11 |\ 12 |\ 13 |
| 9 \ | 23 \ | 24 \ | 25 \|
15——— 1 ——— 2 ——— 3 ———16
|\ 26 |\  2 |\  4 |\ 15 |
| 14 \|  1 \|  3 \| 27 \|
17——— 4 ——— 5 ——— 6 ———18
|\ 28 |\  6 |\  8 |\ 17 |
| 16 \|  5 \|  7 \| 29 \|
19——— 7 ——— 8 ——— 9 ———20
|\ 30 |\ 31 |\ 32 |\ 22 |
| 18 \| 19 \| 20 \| 21 \|
21———22———23———24———25
```

$$\gamma_1 = 0.511023$$
$$\gamma_2 = 0.720476$$
$$\gamma_3 = 0.507899$$
$$\gamma_4 = 0.720476$$
$$\gamma_5 = 1.01885$$
$$\gamma_6 = 0.720476$$
$$\gamma_7 = 0.507896$$
$$\gamma_8 = 0.720476$$
$$\gamma_9 = 0.511023$$
$$\gamma_i = 0 \quad 10 \le i \le 25$$
$$u(0.125, 0.125) \approx 0.614187$$
$$u(0.125, 0.25) \approx 0.690343$$
$$u(0.25, 0.125) \approx 0.690343$$
$$u(0.25, 0.25) \approx 0.720476$$

5. The Finite-Element Algorithm with $K = 0, N = 12, M = 32, n = 20, m = 27$, and $NL = 14$ gives the following results, where the labeling is as shown in the diagram.

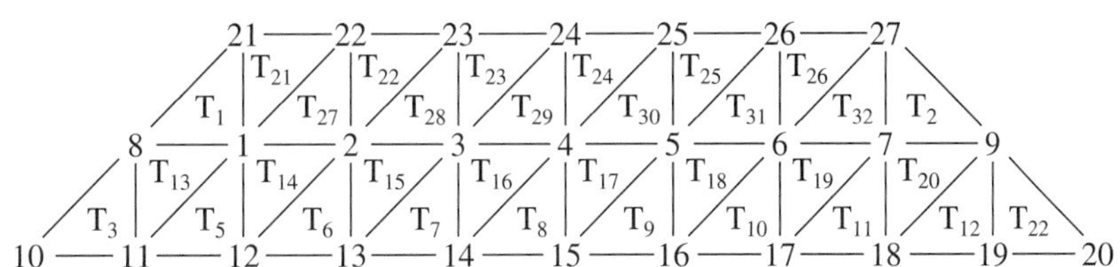

$\gamma_1 = 21.40335$ $\quad\gamma_8 = 24.19855$ $\quad\gamma_{15} = 20.23334$ $\quad\gamma_{22} = 15$

$\gamma_2 = 19.87372$ $\quad\gamma_9 = 24.16799$ $\quad\gamma_{16} = 20.50056$ $\quad\gamma_{23} = 15$

$\gamma_3 = 19.10019$ $\quad\gamma_{10} = 27.55237$ $\quad\gamma_{17} = 21.35070$ $\quad\gamma_{24} = 15$

$\gamma_4 = 18.85895$ $\quad\gamma_{11} = 25.11508$ $\quad\gamma_{18} = 22.84663$ $\quad\gamma_{25} = 15$

$\gamma_5 = 19.08533$ $\quad\gamma_{12} = 22.92824$ $\quad\gamma_{19} = 24.98178$ $\quad\gamma_{26} = 15$

$\gamma_6 = 19.84115$ $\quad\gamma_{13} = 21.39741$ $\quad\gamma_{20} = 27.41907$ $\quad\gamma_{27} = 15$

$\gamma_7 = 21.34694$ $\quad\gamma_{14} = 20.52179$ $\quad\gamma_{21} = 15$

$$u(1, 0) \approx 22.92824$$

$$u(4, 0) \approx 22.84663$$

$$u\left(\frac{5}{2}, \frac{\sqrt{3}}{2}\right) \approx 18.85895$$

Index

Trigonometry

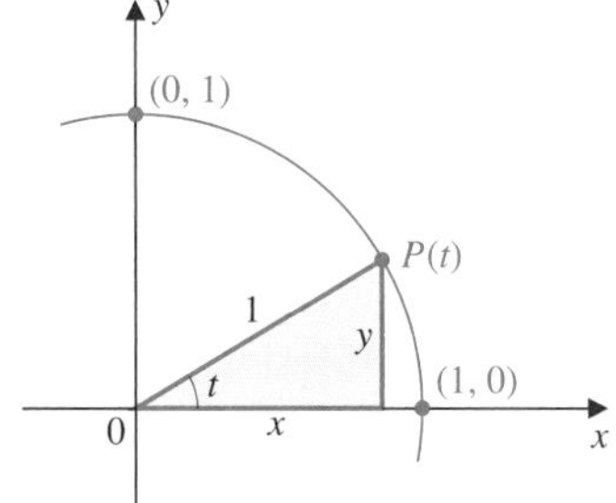

$$\sin t = y \qquad \cos t = x$$

$$\tan t = \frac{\sin t}{\cos t} \qquad \cot t = \frac{\cos t}{\sin t}$$

$$\sec t = \frac{1}{\cos t} \qquad \csc t = \frac{1}{\sin t}$$

$$(\sin t)^2 + (\cos t)^2 = 1$$

$$\sin(t_1 \pm t_2) = \sin t_1 \cos t_2 \pm \cos t_1 \sin t_2$$

$$\cos(t_1 \pm t_2) = \cos t_1 \cos t_2 \mp \sin t_1 \sin t_2$$

$$\sin t_1 \sin t_2 = \frac{1}{2}[\cos(t_1 - t_2) - \cos(t_1 + t_2)]$$

$$\cos t_1 \cos t_2 = \frac{1}{2}[\cos(t_1 - t_2) + \cos(t_1 + t_2)]$$

$$\sin t_1 \cos t_2 = \frac{1}{2}[\sin(t_1 - t_2) + \sin(t_1 + t_2)]$$

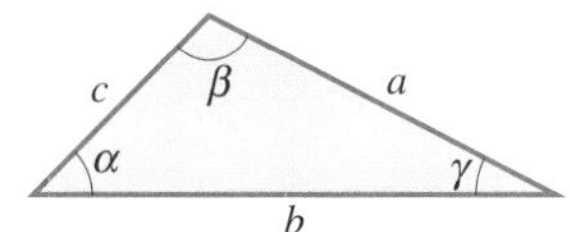

Law of Sines: $\dfrac{\sin \alpha}{\alpha} = \dfrac{\sin \beta}{\beta} = \dfrac{\sin \gamma}{\gamma}$

Law of Cosines: $c^2 = a^2 + b^2 - 2ab \cos \gamma$

Common Series

$$\sin t = \sum_{n=0}^{\infty} \frac{(-1)^n t^{2n+1}}{(2n+1)!} = t - \frac{t^3}{3!} + \frac{t^5}{5!} - \cdots$$

$$e^t = \sum_{n=0}^{\infty} \frac{t^n}{n!} = 1 + t + \frac{t^2}{2!} + \frac{t^3}{3!} + \cdots$$

$$\cos t = \sum_{n=0}^{\infty} \frac{(-1)^n t^{2n}}{(2n)!} = 1 - \frac{t^2}{2!} + \frac{t^4}{4!} - \cdots$$

$$\frac{1}{1-t} = \sum_{n=0}^{\infty} t^n = 1 + t + t^2 + \cdots, \qquad |t| < 1$$

The Greek Alphabet

Alpha	A	α	Eta	H	η	Nu	N	ν	Tau	T	τ
Beta	B	β	Theta	Θ	θ	Xi	Ξ	ξ	Upsilon	Υ	υ
Gamma	Γ	γ	Iota	I	ι	Omicron	O	o	Phi	Φ	ϕ
Delta	Δ	δ	Kappa	K	κ	Pi	Π	π	Chi	X	χ
Epsilon	E	ϵ	Lambda	Λ	λ	Rho	P	ρ	Psi	Ψ	ψ
Zeta	Z	ζ	Mu	M	μ	Sigma	Σ	σ	Omega	Ω	ω

Common Graphs

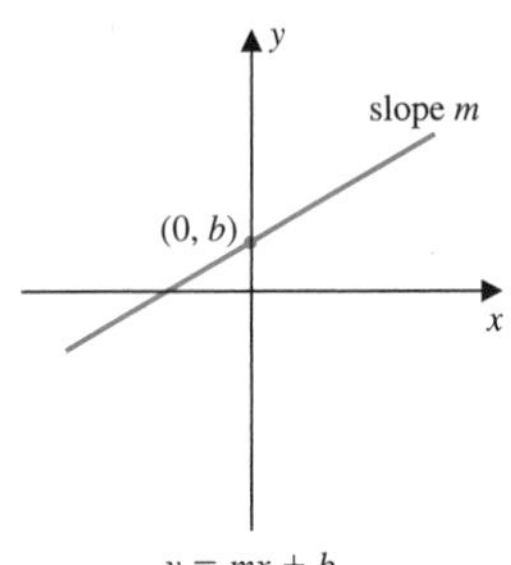

$y = mx + b$

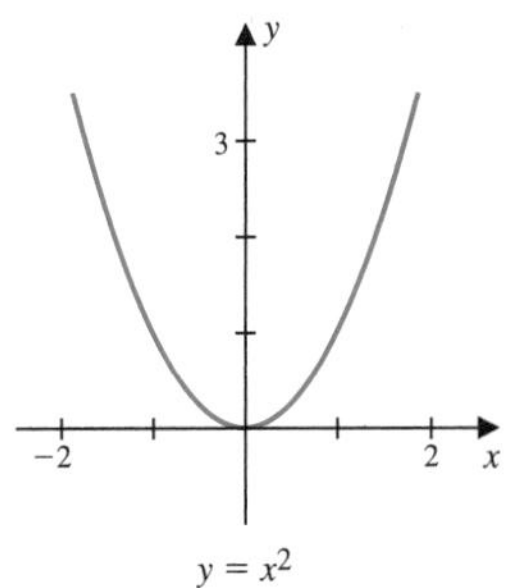

$y = x^2$

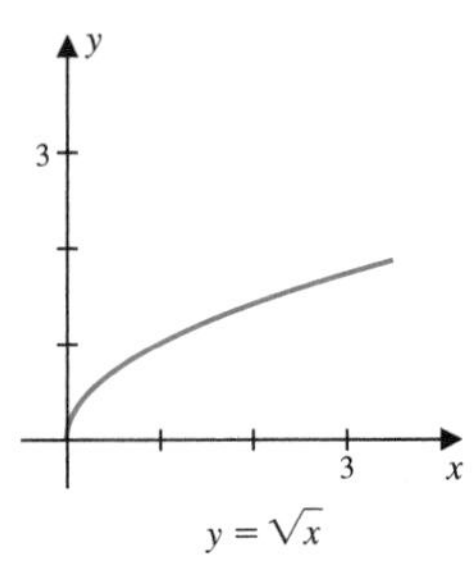

$y = \sqrt{x}$

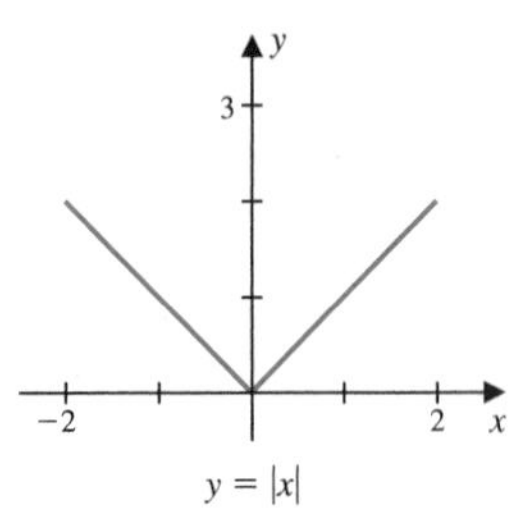

$y = |x|$

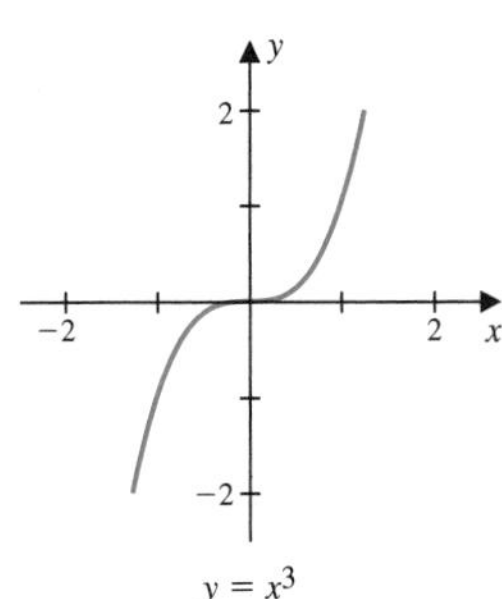

$y = x^3$

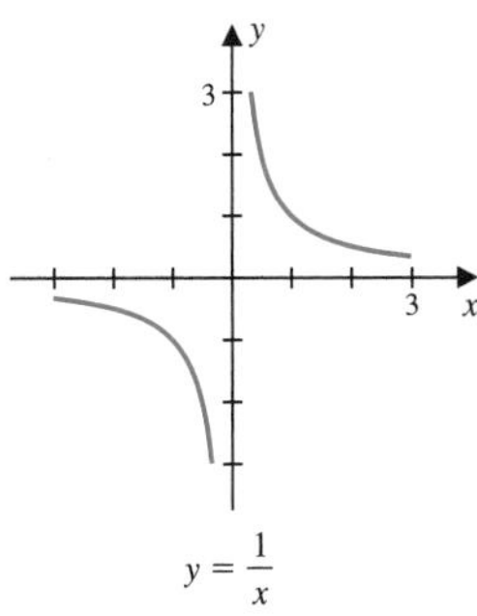

$y = \dfrac{1}{x}$

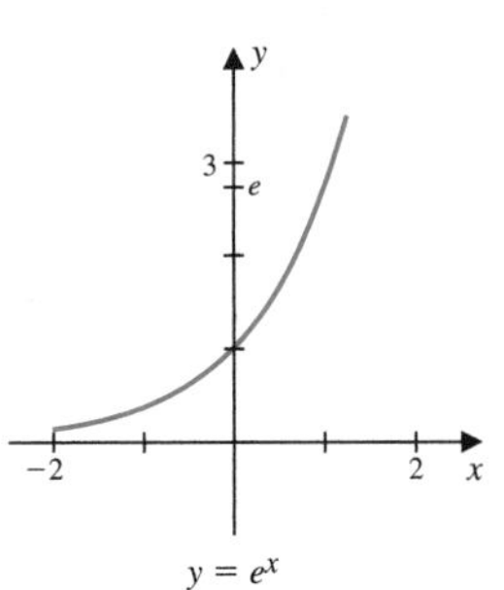

$y = e^x$

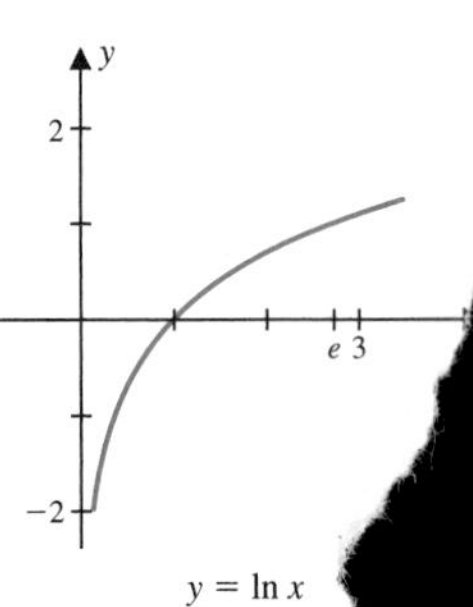

$y = \ln x$

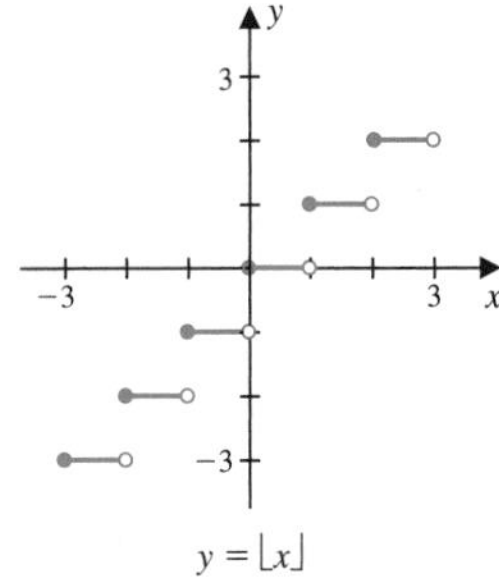

$y = \lfloor x \rfloor$

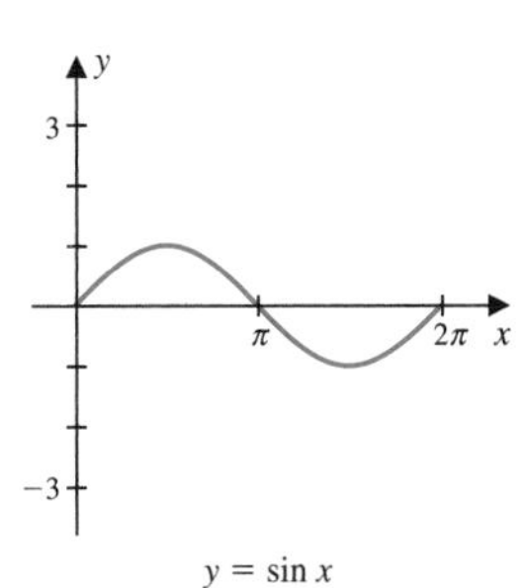

$y = \sin x$

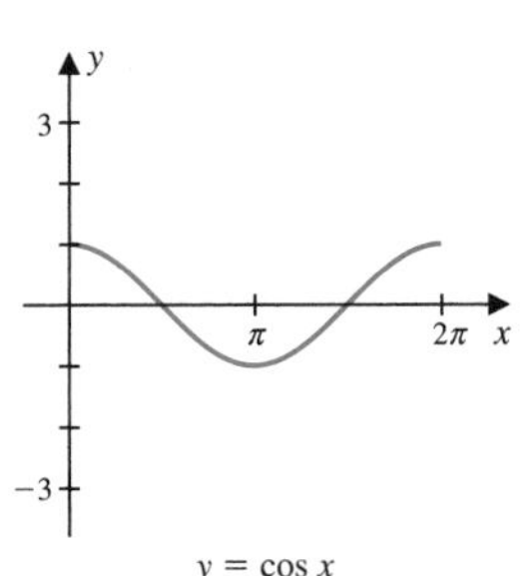

$y = \cos x$

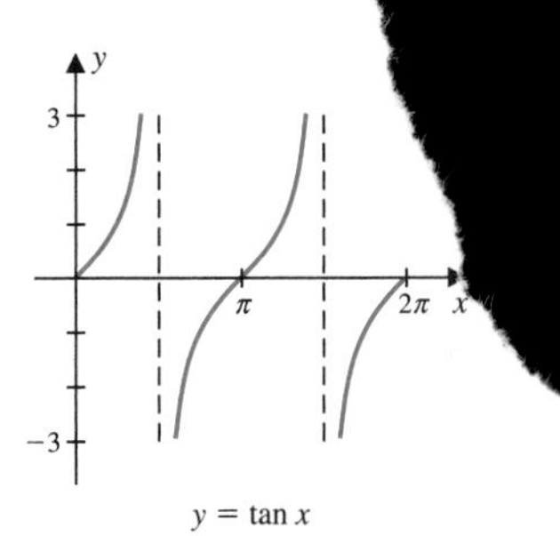

$y = \tan x$

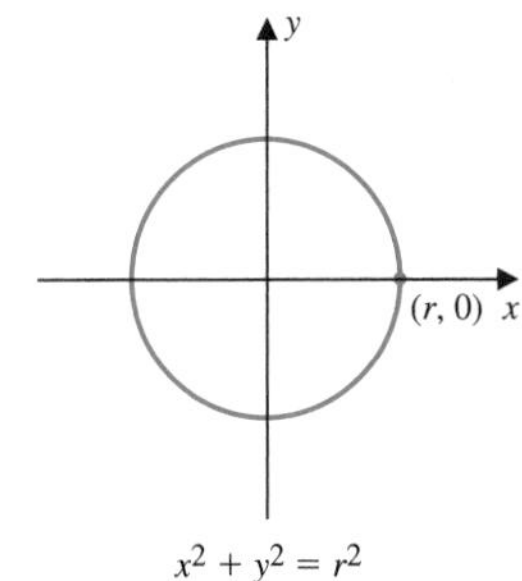

$x^2 + y^2 = r^2$

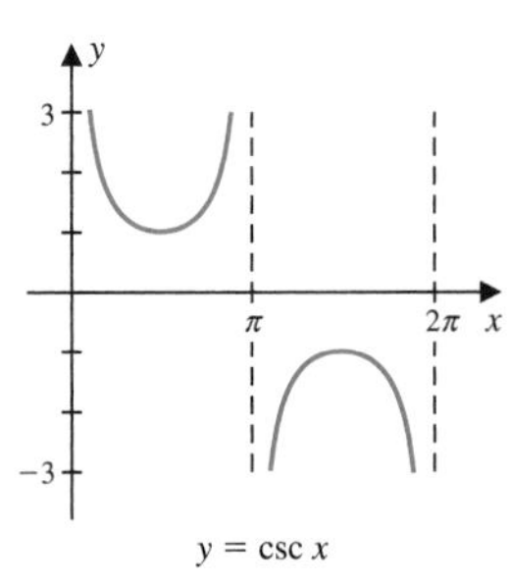

$y = \csc x$

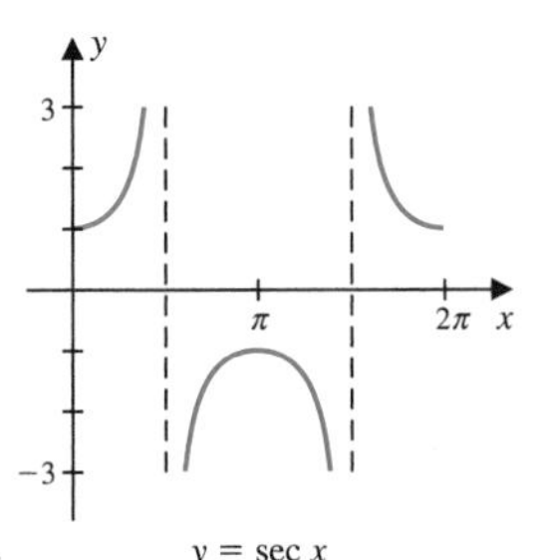

$y = \sec x$

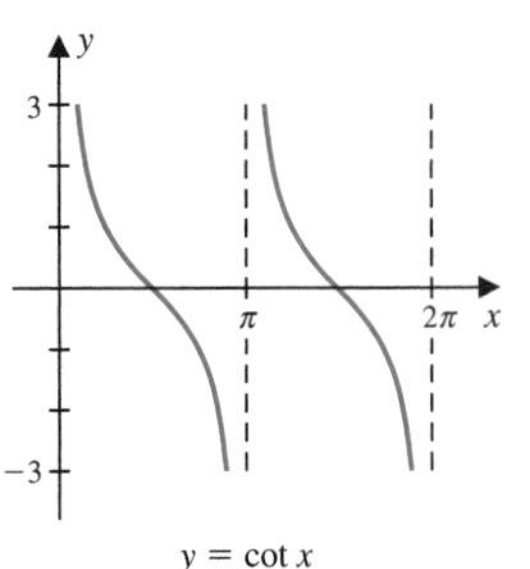

$y = \cot x$